Multivariable Analysis

G. Baley Price

Multivariable Analysis

With 93 Illustrations

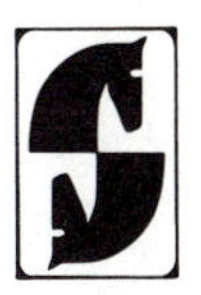

Springer-Verlag
New York Berlin Heidelberg Tokyo

G. Baley Price
Department of Mathematics
University of Kansas
Lawrence, Kansas 66045
U.S.A.

AMS Classification: 26-01

Library of Congress Cataloging in Publication Data
Price, G. Baley (Griffith Baley)
Multivariable analysis
Bibliography: p.
Includes index.
1. Calculus. 2. Functions of several real variables.
I. Title.
QA303.P917 1984 515 83-20328

Typeset by Asco Trade Typesetting Ltd., Hong Kong
Printed and bound by R. R. Donnelley & Sons, Harrisonburg, Virginia
Printed in the United States of America

9 8 7 6 5 4 3 2 1

ISBN 0-387-90934-6 Springer-Verlag New York Berlin Heidelberg Tokyo
ISBN 3-540-90934-6 Springer-Verlag Berlin Heidelberg New York Tokyo

To

C. L. B. P.

"Nullum quod tetigit non ornavit."

A mathematician, like a painter or a poet, is a maker of patterns. . . .
The mathematician's patterns, like the painter's or the poet's, must be *beautiful*; the ideas, like the colours or the words, must fit together in a harmonious way.

—G. H. Hardy, *A Mathematician's Apology*

It is undeniable that some of the best inspirations in mathematics—in those parts of it which are as pure mathematics as one can imagine—have come from the natural sciences. We will mention the two most monumental facts.

The first example is, as it should be, geometry. . . .

The second example is calculus—or rather all of analysis, which sprang from it. The calculus was the first achievement of modern mathematics, and it is difficult to overestimate its importance. I think it defines more unequivocally than anything else the inception of modern mathematics, and the system of mathematical analysis, which is its logical development, still constitutes the greatest technicaladvance in exact thinking.

—John von Neumann, "The Mathematician"

A movement for the reform of the teaching of mathematics, which some decades ago made quite a stir in Germany under the leadership of the great mathematician Felix Klein, adopted the slogan "functional thinking." The important thing with the average educated man should have learned in his mathematics classes, so the reformers claimed, is thinking in terms of *variables and functions*. A function describes how one variable y depends on another x; or more generally, it maps one variety, the range of a variable element x, upon another (or the same) variety. This idea of function or mapping is certainly one of the most fundamental concepts, with accompanies mathematics at every step in theory and application.

. . . But I should have completely failed if you had not realized at least this much, that mathematics, in spite of its age, is not doomed to progressive sclerosis by its growing complexity, but is still intensely alive, drawing nourishment from its deep roots in mind and nature.

—Hermann Weyl, "The Mathematical Way of Thinking"

Preface

This book contains an introduction to the theory of functions, with emphasis on functions of several variables. The central topics are the differentiation and integration of such functions. Although many of the topics are familiar, the treatment is new; the book developed from a new approach to the theory of differentiation. If f is a function of two real variables x and y, its derivatives at a point p_0 can be approximated and found as follows. Let p_1, p_2 be two points near p_0 such that p_0, p_1, p_2 are not on a straight line. The linear function of x and y whose values at p_0, p_1, p_2 are equal to those of f at these points approximates f near p_0; determinants can be used to find an explicit representation of this linear function (think of the equation of the plane through three points in three-dimensional space). The (partial) derivatives of this linear function are approximations to the derivatives of f at p_0; each of these (partial) derivatives of the linear function is the ratio of two determinants. The derivatives of f at p_0 are defined to be the limits of these ratios as p_1 and p_2 approach p_0 (subject to an important regularity condition). This simple example is only the beginning, but it hints at a theory of differentiation for functions which map sets in $\mathbb{R}^n$ into $\mathbb{R}^m$ which is both general and powerful, and which reduces to the standard theory of differentiation in the one-dimensional case.

This book develops general theories in which both the methods and the results for functions of several variables are similar to those for functions of a single variable. Although general methods are always employed rather than *ad hoc* methods, the results (theorems) are similar to the standard one-dimensional theorems and are sometimes better than the traditional theorems for functions of several variables. The approach and the general methods employed succeed in unifying many aspects of the theory. The book is elementary in the sense that it does not employ Lebesgue measure or Lebesgue integration.

The treatment is geometric in nature, and the principal geometric tool is the simplex (in the example above, p_0, p_1, p_2 are the vertices of a simplex). Often the simplex occurs as an element in an oriented Euclidean complex. Chapter 3 is an introduction to the geometry of n-dimensional Euclidean space; it treats convex sets, simplexes, the orientation of simplexes, complexes and chains, boundaries of simplexes and chains, the volumes of simplexes, and simplicial subdivisions of cubes and simplexes. Because of the geometric nature of the treatment, numerous figures have been included to make the reading of the text easier. The principal analytic tools are the determinant and the Stolz condition. With each n-simplex in $\mathbb{R}^n$ there is associated an $(n + 1)$ by $(n + 1)$ matrix whose determinant is proportional to the volume of the simplex. Appendix 1 contains a complete treatment of determinants, including proofs of all of the theorems used in this book. Some of these theorems are not at all well known, but they find natural and important applications in the theories developed in this book. The Stolz condition, introduced by Otto Stolz in 1893, states that, in the approximation of the increment of a function by a linear function, the remainder term has a certain specified form. This book introduces a Stolz condition for the increment of a function which maps a set in $\mathbb{R}^n$ into $\mathbb{R}^m$.

The subject matter of this book includes the theory of differentiation and (Riemann) integration, and a number of related topics in analysis. Chapter 4 treats Sperner's lemma by novel methods which fit easily and naturally into this book's general methods based on oriented simplicial complexes and determinants. Sperner's lemma is used to prove a very general form of the intermediate-value theorem; it is applied in Chapter 5 to prove a very general inverse-function theorem. The most important theorem in the theory of integration is the fundamental theorem of the integral calculus. By defining both derivatives and integrals by means of simplexes, it becomes easy to establish a connection between differentiation and integration. The fundamental theorem results from properties of determinants and from properties of the boundary of a chain in an oriented simplicial complex. As is well known, Stokes' theorem is a corollary of the fundamental theorem. This book shows that the evaluation of integrals by iterated integrals and Cauchy's integral theorem are also corollaries of the fundamental theorem of the integral calculus (see Chapters 8 and 10). Chapter 9 contains a treatment of Kronecker's integral; the Kronecker integral formula is closely related to the fundamental theorem. Chapter 10 is an introduction to the differentiation and integration of functions of a single complex variable and of several complex variables by the methods developed in earlier chapters of the book. As stated above, the fundamental theorem of the integral calculus becomes Cauchy's integral theorem for functions of one complex variable and also of several complex variables.

The prerequisites for the study of this book are two: a first course in calculus and the ability to read and understand mathematical definitions, theorems, and proofs. For the student who knows elementary calculus, the book contains everything needed to read and understand the book. Appendix 1 presents a treatment of determinants (including several relatively unknown theorems) and several topics in linear algebra; Appendix 2 contains the basic theorems on numbers, sets, and functions. Although the book treats only elementary mathematics, it is not always easy. As a result, some readers may desire a more extensive background and more maturity than they have acquired from an elementary course in calculus.

The book consists of a Table of Contents, ten chapters, two appendices, References and Notes, an Index of Symbols, and an Index. The ten chapters and two appendices are divided into 97 sections, numbered in order from 1 for the first section in Chapter 1 to 97 for the last section in Appendix 2. There is a set of exercises at the end of each section in the ten chapters; these exercises are designed to illustrate and to supplement the material in the text. For easy reference throughout the book, the important definitions, theorems, corollaries, lemmas, and examples in each section are numbered with boldface numbers containing a decimal point; the digits before the decimal point indicate the number of the section, and the digits after the decimal point are the number of the item in the section. For example, Theorem 20.5 is the fifth numbered item in Section 20. In each section, the equations, formulas, and other special items to which reference is made mostly within the section are numbered (1) to (n) on the right margin.

The relationship of the chapters in this book is indicated by the following diagram.

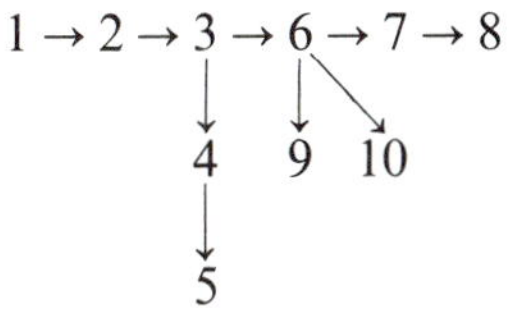

```
1 → 2 → 3 → 6 → 7 → 8
        ↓   ↓ ↘
        4   9  10
        ↓
        5
```

Theorem 62.7 at the end of Chapter 9 requires Chapter 4, but except for this one theorem, Chapter 9 is independent of Chapter 4 as indicated by the diagram. The diagram shows that a minimum course on the differentiation and integration of functions of several real variables can be taught from Chapters 1, 2, 3, and 6. More extensive courses, corresponding to the needs of students and the interests of the instructor, can be obtained by adding chapters to this minimum course in accordance with the diagram above. As the diagram suggests, many different courses are possible.

I am pleased to take this opportunity to acknowledge with appreciation and thanks the assistance that I have received in the preparation and publication of this book. This assistance includes a Guggenheim Fellowship in 1946–1947 and a sabbatical leave from The University of Kansas in 1972–1973. Also, I am indebted to the editorial staff of Springer-Verlag for suggestions that have led to many improvements and for their help in preparing the manuscript and publishing the book. Finally, I gratefully acknowledge the assistance of my wife, Cora Lee Beers Price, without whose help and support this book would not have been written. To all of those who have assisted in the writing and publication of this book, I extend my hearty thanks.

Lawrence, Kansas
February 28, 1983

G. BALEY PRICE

Contents

CHAPTER 1

Differentiable Functions and Their Derivatives

1. Introduction

One of the important problems in mathematics and in its applications in science and engineering is the following: if $y = f(x)$, find the rate of increase of y with respect to x. For example, if $y = 2x + 5$, then $y_0 = 2x_0 + 5$, $y_1 = 2x_1 + 5$, and $y_1 - y_0 = 2(x_1 - x_0)$. Thus

$$\frac{y_1 - y_0}{x_1 - x_0} = 2, \tag{1}$$

and the rate of increase of y with respect to x is 2. This example shows that the problem has a simple solution in all cases in which f is a linear function. Thus, if $y = ax + b$, then

$$\frac{y_1 - y_0}{x_1 - x_0} = a. \tag{2}$$

Observe that, for this linear function f, the rate of increase of y with respect to x is the same, namely a, for every x_0 and x_1.

If f is not a linear function, then the rate of increase of y with respect to x is not the same for every x_0 and x_1, but it is possible to find an average rate of increase for each particular interval. If $f(x) = x^2$, then $f(x_0) = x_0^2$, $f(x_1) = x_1^2$, and the average rate of increase for this interval is

$$\frac{f(x_1) - f(x_0)}{x_1 - x_0} = \frac{x_1^2 - x_0^2}{x_1 - x_0} = x_1 + x_0. \tag{3}$$

The rate of increase is a variable which depends both on x_0 and x_1. The rate of increase of y with respect to x at the point x_0 is defined to be the following limit.

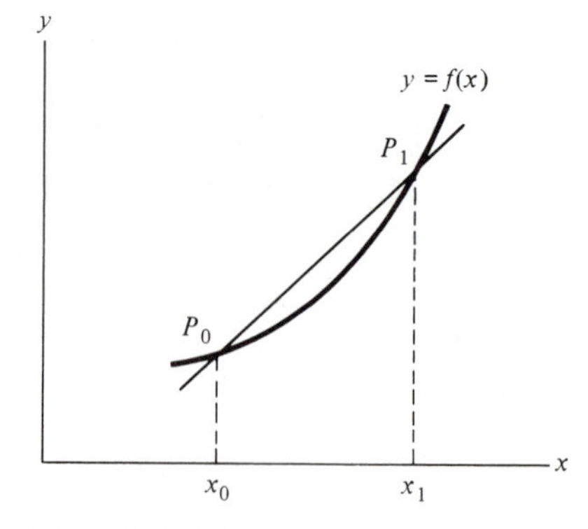

Figure 1.1. The line through P_0 and P_1.

$$\lim_{x_1 \to x_0} \frac{f(x_1) - f(x_0)}{x_1 - x_0} = \lim_{x_1 \to x_0} (x_1 + x_0) = 2x_0. \tag{4}$$

There is a slightly different way of looking at the results in (3) and (4) which is informative. In (1) and (2) we found the rate of increase of a linear function. In (3) and (4) we approximated the given function f by a linear function which has the same value as f at x_0 and x_1. To find this linear function, find the equation of the straight line which passes through the points $P_0 : [x_0, f(x_0)]$ and $P_1 : [x_1, f(x_0)]$. The equation

$$\begin{vmatrix} x & y & 1 \\ x_1 & f(x_1) & 1 \\ x_0 & f(x_0) & 1 \end{vmatrix} = 0 \tag{5}$$

is a linear equation in x and y; since it is satisfied by the coordinates of P_0 and P_1, it is the equation of the line through these points [see Figure 1.1]. In (5) subtract the third row from the first row [see Theorem 77.11], and then expand the determinant by elements in the first row of the matrix [see Theorem 79.1]. Simple transformations convert the equation of the line into the following form:

$$y - f(x_0) = \frac{\begin{vmatrix} f(x_1) & 1 \\ f(x_0) & 1 \end{vmatrix}}{\begin{vmatrix} x_1 & 1 \\ x_0 & 1 \end{vmatrix}} (x - x_0). \tag{6}$$

Then this equation and (2) show that the average rate of increase of f on the interval from x_0 to x_1 is the coefficient of x, which is

$$\frac{\begin{vmatrix} f(x_1) & 1 \\ f(x_0) & 1 \end{vmatrix}}{\begin{vmatrix} x_1 & 1 \\ x_0 & 1 \end{vmatrix}}, \qquad x_1 \neq x_0. \tag{7}$$

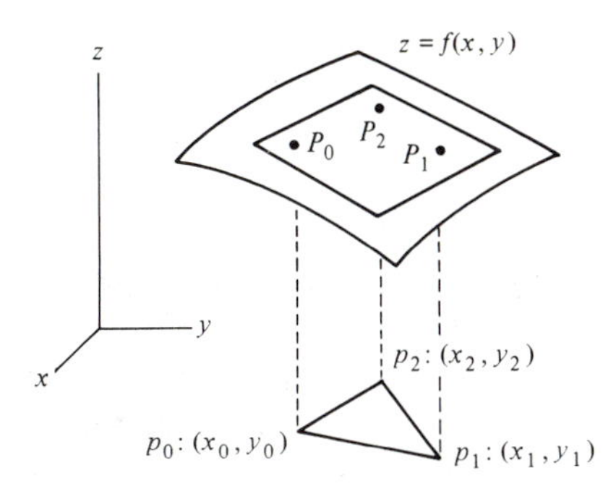

Figure 1.2. The plane through $P_i : [x_i, y_i, f(p_i)]$, $i = 0, 1, 2$, on $z = f(x, y)$.

By definition, the rate of increase of f [or y if $y = f(x)$] at x_0 is

$$\lim_{x_1 \to x_0} \frac{\begin{vmatrix} f(x_1) & 1 \\ f(x_0) & 1 \end{vmatrix}}{\begin{vmatrix} x_1 & 1 \\ x_0 & 1 \end{vmatrix}}, \qquad x_1 \neq x_0, \tag{8}$$

if this limit exists; the rate of increase of f at x_0 is not defined if the limit in (8) does not exist. Observe that (8) is

$$\lim_{x_1 \to x_0} \frac{f(x_1) - f(x_0)}{x_1 - x_0}, \qquad x_1 \neq x_0. \tag{9}$$

Written in the form (9), the limit in (8) is easily recognized as the derivative of f at x_0; it is denoted by $f'(x_0)$. The limits in (8) and (9) are interpreted also as the slope of the line which is tangent to the graph of $y = f(x)$ at x_0; thus the equation of this tangent line is

$$y - f(x_0) = f'(x_0)(x - x_0) \tag{10}$$

if $f'(x_0)$ exists; if the derivative does not exist, the tangent line is not defined.

In the past, the denominator in (7), (8), (9) has frequently been denoted by Δx and called the increment in x; also, the numerator has been denoted by Δf and called the increment in f corresponding to the increment Δx in x. Thus

$$\Delta x = \begin{vmatrix} x_1 & 1 \\ x_0 & 1 \end{vmatrix} = x_1 - x_0, \quad \Delta f = \begin{vmatrix} f(x_1) & 1 \\ f(x_0) & 1 \end{vmatrix} = f(x_1) - f(x_0),$$
$$f'(x_0) = \lim_{\Delta x \to 0} \frac{\Delta f}{\Delta x}. \tag{11}$$

But there are other problems to be solved. Let $f: N(p_0, r) \to \mathbb{R}$ be a function of two independent variables x and y which is defined in a neighborhood $N(p_0, r)$ of $p_0 : (x_0, y_0)$; the graph of $z = f(x, y)$ is the surface shown in Figure 1.2. In this case there are two rates of increase to be investigated, namely, the rate of increase of z with respect to x and the rate of increase of z with respect to y. If $z = ax + by + c$, then the values of z at (x_0, y_0) and (x_1, y_0) are

$$z_0 = ax_0 + by_0 + c, \qquad z_1 = ax_1 + by_0 + c,$$

and $z_1 - z_0 \doteq a(x_1 - x_0)$. Then the rate of increase of z with respect to x is a, and a similar calculation shows that the rate of increase of z from (x_0, y_0) to (x_0, y_1) is b. Since the function f is linear, these rates of increase are the same for every pair of points (x_0, y_0), (x_1, y_0) and (x_0, y_0), (x_0, y_1), respectively. If the function f is not linear, then it is possible to proceed as follows: apply the procedure described in equations (7), (8), (9) to the two functions $f(\cdot, y_0)$ and $f(x_0, \cdot)$. Then the average rates of increase of f for the pairs of points (x_0, y_0), (x_1, y_0) and (x_0, y_0), (x_0, y_1) are

$$\frac{\begin{vmatrix} f(x_1, y_0) & 1 \\ f(x_0, y_0) & 1 \end{vmatrix}}{\begin{vmatrix} x_1 & 1 \\ x_0 & 1 \end{vmatrix}}, \qquad \frac{\begin{vmatrix} f(x_0, y_1) & 1 \\ f(x_0, y_0) & 1 \end{vmatrix}}{\begin{vmatrix} y_1 & 1 \\ y_0 & 1 \end{vmatrix}} \tag{12}$$

respectively, and the rates of change of f with respect to x and y at (x_0, y_0) are

$$\lim_{x_1 \to x_0} \frac{\begin{vmatrix} f(x_1, y_0) & 1 \\ f(x_0, y_0) & 1 \end{vmatrix}}{\begin{vmatrix} x_1 & 1 \\ x_0 & 1 \end{vmatrix}}, \qquad \lim_{y_1 \to y_0} \frac{\begin{vmatrix} f(x_0, y_1) & 1 \\ f(x_0, y_0) & 1 \end{vmatrix}}{\begin{vmatrix} y_1 & 1 \\ y_0 & 1 \end{vmatrix}}. \tag{13}$$

These limits are the partial derivatives of f with respect to x and y, respectively, at (x_0, y_0); they are denoted by

$$f_x(x_0, y_0), \qquad f_y(x_0, y_0) \quad \text{or} \quad \frac{\partial f(x_0, y_0)}{\partial x}, \qquad \frac{\partial f(x_0, y_0)}{\partial y}. \tag{14}$$

Under certain circumstances, these partial derivatives define the equation of the tangent plane to the surface $z = f(x, y)$ as follows:

$$z - f(x_0, y_0) = f_x(x_0, y_0)(x - x_0) + f_y(x_0, y_0)(y - y_0). \tag{15}$$

The partial derivatives (14) have great importance in the study of mathematics and its applications, but, as defined in (13), they have serious deficiencies as indicated by the following example.

1.1 Example. Let $f\colon \mathbb{R}^2 \to \mathbb{R}$ be defined as follows:

$$\begin{aligned} f(x, y) &= \frac{2xy}{x^2 + y^2}, \qquad (x, y) \neq (0, 0), \\ f(0, 0) &= 0. \end{aligned} \tag{16}$$

Now this function has the value zero on the x-axis and on the y-axis; then (13) shows that the partial derivatives of f exist at (0, 0) and that $f_x(0, 0) = 0$, $f_y(0, 0) = 0$. But (15) now reduces to $z = 0$—the equation of the xy-plane—and this plane cannot qualify as the tangent plane to the surface $z = f(x, y)$ at (0, 0). The function f is continuous everywhere in the xy-plane except at the origin, but there it has a serious discontinuity. On every ray which issues from the origin, the function f is constant. Perhaps the easiest way to prove this statement is to represent the function in polar coordinates. Since $x = r \cos \theta$, $y = r \sin \theta$, then

$$f(r\cos\theta, r\sin\theta) = \frac{r^2\, 2\sin\theta\cos\theta}{r^2(\sin^2\theta + \cos^2\theta)} = \sin 2\theta, \quad r \neq 0, \quad 0 \leqq \theta \leqq 2\pi,$$
$$f(0, 0) = 0. \tag{17}$$

Then in every neighborhood of (0, 0), the function f takes on every value from $+1$ to -1, and f is discontinuous [see Definition 96.2]. The difficulty is clear: the partial derivatives $f_x(0, 0)$, $f_y(0, 0)$ depend only on the values of f on the x-axis and y-axis, and these values are not very representative of the totality of values of f in the neighborhood of the origin. We need a definition of derivatives of f which depends on all values of the function in the neighborhood of the point where the derivatives are being calculated.

A review of the one-dimensional case suggests a different approach to the problem of finding the rates of increase of a function $f\colon N(p_0, r) \to \mathbb{R}$, $N(p_0, r) \subset \mathbb{R}^2$. The problem is equivalent to the problem of finding the tangent plane to a surface $z = f(x, y)$. To find the tangent line to a curve $y = f(x)$ at P_0 [see Figure 1.1], we began by considering all the lines through P_0 and a nearby point on the curve. For the function $f\colon N(p_0, r) \to \mathbb{R}$, the corresponding procedure is to consider all planes through $P_0\colon [x_0, y_0, f(x_0, y_0)]$ and two nearby points $P_i\colon [x_i, y_i, f(x_i, y_i)]$, $i = 1, 2$. The rates of increase of linear functions are easy to calculate, and in considering the planes through P_0, P_1, P_2 we are essentially approximating f by a linear function and using the rates of increase of the linear function as approximations to the rates of increase of the given function f. We seek a good approximation; the plane through P_0, P_1, P_2 should lie near the surface $z = f(x, y)$ at least near the three points [see Figure 1.2]. In order to obtain a good approximation to the surface near $P_0\colon [x_0, y_0, f(x_0, y_0)]$, the points $p_i\colon (x_i, y_i)$, $i = 1, 2$, must be chosen with some care. If p_0, p_1, p_2 lie on a straight line, then the points $P_i\colon [x_i, y_i, f(x_i, y_i)]$, $i = 0, 1, 2$, either do not determine a plane or they determine a plane which is parallel to the z-axis. Thus, initially the approximating planes admitted to consideration are all those planes, but only those planes, through P_0 and two nearby points which satisfy the following restriction:

$$\text{the points } p_i,\ i = 0, 1, 2, \text{ are not on a line.} \tag{18}$$

The equation of the plane through the three points $P_i\colon [x_i, y_i, f(x_i, y_i)]$,

$i = 0, 1, 2$, is

$$\begin{vmatrix} x & y & z & 1 \\ x_1 & y_1 & f(x_1, y_1) & 1 \\ x_2 & y_2 & f(x_2, y_2) & 1 \\ x_0 & y_0 & f(x_0, y_0) & 1 \end{vmatrix} = 0 \tag{19}$$

since this equation is linear in x, y, z and is satisfied by the coordinates of P_0, P_1, P_2 [see also (29) in Section 89]. In (19), subtract the fourth row from the first row, and then expand the determinant by elements in the first row of the matrix [see Section 77, Appendix 1, for elementary properties of determinants]. Simple transformations convert the equation of the plane to the following form.

$$z - f(p_0) = \frac{\begin{vmatrix} f(p_1) & y_1 & 1 \\ f(p_2) & y_2 & 1 \\ f(p_0) & y_0 & 1 \end{vmatrix}}{\begin{vmatrix} x_1 & y_1 & 1 \\ x_2 & y_2 & 1 \\ x_0 & y_0 & 1 \end{vmatrix}}(x - x_0) + \frac{\begin{vmatrix} x_1 & f(p_1) & 1 \\ x_2 & f(p_2) & 1 \\ x_0 & f(p_0) & 1 \end{vmatrix}}{\begin{vmatrix} x_1 & y_1 & 1 \\ x_2 & y_2 & 1 \\ x_0 & y_0 & 1 \end{vmatrix}}(y - y_0). \tag{20}$$

The determinant in the denominators in (20) is, by (25) in Section 89, twice the signed area of the triangle with vertices $p_i : (x_i, y_i)$, $i = 0, 1, 2$. Since the points p_i satisfy the restriction in (18), the area of this triangle is not zero and the denominators in (20) are not zero. Furthermore, the determinants in the numerators in (20) are twice the signed areas of certain triangles whose vertices depend on the function f.

For the approximating plane (20), the rates of increase of z with respect to x and y are the coefficients of x and y, respectively. Then, by definition, the rates of increase of f with respect to x and y at $p_0 : (x_0, y_0)$ are, respectively,

$$\lim_{\substack{p_1 \to p_0 \\ p_2 \to p_0}} \frac{\begin{vmatrix} f(p_1) & y_1 & 1 \\ f(p_2) & y_2 & 1 \\ f(p_0) & y_0 & 1 \end{vmatrix}}{\begin{vmatrix} x_1 & y_1 & 1 \\ x_2 & y_2 & 1 \\ x_0 & y_0 & 1 \end{vmatrix}}, \qquad \lim_{\substack{p_1 \to p_0 \\ p_2 \to p_0}} \frac{\begin{vmatrix} x_1 & f(p_1) & 1 \\ x_2 & f(p_2) & 1 \\ x_0 & f(p_0) & 1 \end{vmatrix}}{\begin{vmatrix} x_1 & y_1 & 1 \\ x_2 & y_2 & 1 \\ x_0 & y_0 & 1 \end{vmatrix}}, \tag{21}$$

if these limits exist. For the present, p_1 and p_2 are allowed to approach p_0 in any manner as long as they satisfy the restriction in (18); this restriction on p_1, p_2 is equivalent to the requirement that the denominators in (21) be different from zero. Examples will show very soon that other restrictions, in addition to (18), must be imposed on the approach of p_1, p_2 to p_0 if a significant theory of differentiation is to be obtained. If the limits in (21) exist and are denoted by A and B for the present, then the equation of the tangent plane to the surface $z = f(x, y)$ at $P_0 : [x_0, y_0, f(x_0, y_0)]$ is $z - f(x_0, y_0) = A(x - x_0) + B(y - y_0)$.

1.2 Example. Evaluate the limits in (21) for the function f such that $f(x, y) = ax + by + c$. By elementary properties of determinants [see Section 77],

$$\begin{vmatrix} f(p_1) & y_1 & 1 \\ f(p_2) & y_2 & 1 \\ f(p_0) & y_0 & 1 \end{vmatrix} = \begin{vmatrix} ax_1 + by_1 + c & y_1 & 1 \\ ax_2 + by_2 + c & y_2 & 1 \\ ax_0 + by_0 + c & y_0 & 1 \end{vmatrix} = \begin{vmatrix} ax_1 + by_1 & y_1 & 1 \\ ax_2 + by_2 & y_2 & 1 \\ ax_0 + by_0 & y_0 & 1 \end{vmatrix}$$

$$= \begin{vmatrix} ax_1 & y_1 & 1 \\ ax_2 & y_2 & 1 \\ ax_0 & y_0 & 1 \end{vmatrix} = a \begin{vmatrix} x_1 & y_1 & 1 \\ x_2 & y_2 & 1 \\ x_0 & y_0 & 1 \end{vmatrix}.$$

Then the first ratio in (21) has the value a, and the limit exists and has the value a. In the same way, the second limit in (21) is b.

1.3 Example. Show that, for the function defined in (16) in Example 1.1, the limits in (21) do not exist. First, find the limits for the points

$$p_0 : (0, 0), \quad p_1 : (n^{-1}, 0), \quad p_2 : (n^{-1}, n^{-1}), \qquad n = 1, 2, \cdots. \tag{22}$$

The numerators in (21) are 0 and n^{-1}; the denominators are n^{-2}. Then the limits in (21), for the points in (22), as $n \to \infty$ are 0 and $+\infty$. Next, find the limits in (21) for the points

$$p_0 : (0, 0), \quad p_1 : (n^{-1}, n^{-1}), \quad p_2 : (0, n^{-1}), \qquad n = 1, 2, \cdots. \tag{23}$$

For these points, the numerators in (21) are n^{-1} and 0, and the denominators are n^{-2}. Then the limits in (21), for the points in (23), as $n \to \infty$ are $+\infty$ and 0. These facts are more than sufficient to prove that the limits in (21) do not exist for the function defined in (16). The limits in (21) fail to exist in this case because f is badly discontinuous at $p_0 : (0, 0)$, and f cannot be closely approximated in the neighborhood of p_0 by any linear function.

1.4 Example. Let f be the function $f: \mathbb{R}^2 \to \mathbb{R}$ such that $f(x, y) = x^2 - y^2$. For this function, evaluate the first limit in (21) with respect to each of the following four sets of points.

(a) $p_0 : (0, 0), p_1 : (n^{-1}, n^{-1}), p_2 : (n^{-1}, -n^{-1}), \quad n = 1, 2, \cdots.$
(b) $p_0 : (0, 0), p_1 : (n^{-2}, n^{-1}), p_2 : (n^{-2}, -n^{-1}), \quad n = 1, 2, \cdots.$
(c) $p_0 : (0, 0), p_1 : (n^{-4}, n^{-1}), p_2 : (n^{-4}, -n^{-1}), \quad n = 1, 2, \cdots.$
(d) $p_0 : (0, 0), p_1 : (n^{-4}, n^{-1}), p_2 : (n^{-4}, 2n^{-1}), \quad n = 1, 2, \cdots.$

In each of the four cases (a), $\cdots$, (d), the points p_1 and p_2 approach p_0 as n tends to infinity, and the denominator in (21), evaluated for the points

p_1, p_2, p_0, is different from zero for every value of n. Thus in each of the four cases (a), $\cdots$, (d), the points meet the minimum condition (18) for the determination of the limits in (21). For the points in (a), the first limit in (21), as $n \to \infty$, is 0; for (b), it is -1; for (c), it is $-\infty$; and for (d) it is $+\infty$. The limit in (a) is the value at $p_0 : (0, 0)$ of the partial derivative of f with respect to x; in (b), the limit exists, but its value is not that of a derivative of f; and in (c) and (d) the limit does not exist. These results show that, if p_1 and p_2 are allowed to approach p_0 in any way as long as they satisfy (18), the first limit in (21) does not exist. In Example 1.3 the limits in (21) fail to exist because the function in (16) is discontinuous at $p_0 : (0, 0)$ and cannot be closely approximated by a linear function. In the present example, however, the function f such that $f(x, y) = x^2 - y^2$ is continuous and has continuous partial derivatives of all orders. Thus the fault would seem to lie, not with f, but rather with the manner in which the limits in (21) are evaluated—with the manner in which p_1 and p_2 approach p_0. This example suggests that, if the limits in (21) are to serve as definitions of derivatives, then some further restriction must be imposed on the manner in which p_1 and p_2 approach p_0. Furthermore, the example suggests the restriction which may be necessary (it still must be shown to be sufficient). Equation (9) in Section 84 contains the standard formula for the cosine of the angle between two vectors. In (a) the vectors from p_0 to p_1 and p_2 are obviously orthogonal for $n = 1, 2, \cdots$; in (b) and (c) the cosine formula shows that the angle between the two vectors approaches π; and in (d) the formula shows that this angle approaches 0. This example suggests that p_1 and p_2 must be restricted to approach p_0 in such a way that the angle between the vectors $p_0 p_1$ and $p_0 p_2$ is bounded away from 0 and from π. The problems raised by this example will be resolved in the next section.

Equation (8) contains, in unfamiliar notation, the classical definition of the derivative of a function of a single variable, and equation (21) contains a generalization, from one to two dimensions, of this ancient formula. But there is a second generalization of (8) which is quite different. In (8), the function f maps, through the relation $y = f(x)$, a set E in the one-dimensional space $\mathbb{R}$ into a set in the one-dimensional space $\mathbb{R}$ [see Figure 1.3(a)]. In the generalization, the function (f, g) maps, through the relation

$$\begin{aligned} u &= f(x, y), \\ v &= g(x, y), \end{aligned} \tag{24}$$

a set E in the two-dimensional space $\mathbb{R}^2$ into a set in the two-dimensional space $\mathbb{R}^2$ [see Figure 1.3(b)]. In (8), f maps two points p_0, p_1 into two points $f(p_0), f(p_1)$, and the determinants are the signed lengths of the segments bounded by these points. Similarly, (f, g) in (24) maps three points $p_i : (x_i, y_i)$, $i = 0, 1, 2$, into three points $q_i : [f(p_i), g(p_i)]$, and the limit which corresponds to (8) is

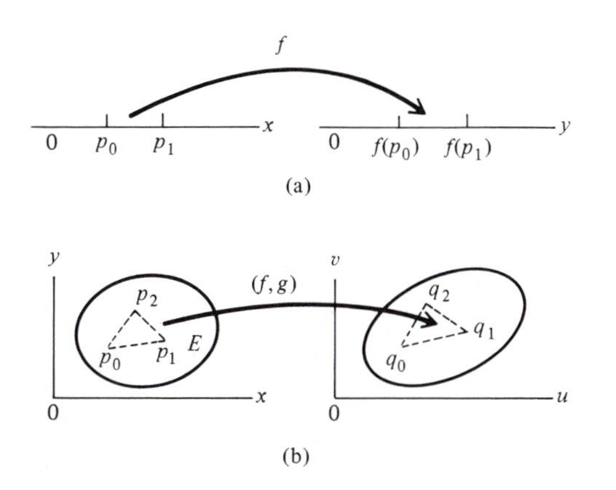

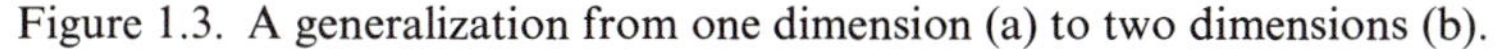

Figure 1.3. A generalization from one dimension (a) to two dimensions (b).

$$
\lim_{\substack{p_1 \to p_0 \\ p_2 \to p_0}} \frac{\begin{vmatrix} f(p_1) & g(p_1) & 1 \\ f(p_2) & g(p_2) & 1 \\ f(p_0) & g(p_0) & 1 \end{vmatrix}}{\begin{vmatrix} x_1 & y_1 & 1 \\ x_2 & y_2 & 1 \\ x_0 & y_0 & 1 \end{vmatrix}}. \tag{25}
$$

As shown in (25) in Section 89, the denominator in (25) is twice the signed area of the triangle (p_1, p_2, p_0), and the numerator is twice the signed area of the triangle (q_1, q_2, q_0). In order for (25) to have meaning, the limit must be taken in such a way that the denominator is never zero; thus, as a minimum requirement, p_1 and p_2 must be restricted so that the area of the triangle (p_1, p_2, p_0) is always different from zero. The denominator in (25) is the same as the denominators in (21), and the limits in (21) seem to be special cases of (25). The limit in (25) is not ordinarily used as the definition of any derivative, but examples show that it is the Jacobian of f and g at least in some cases.

1.5 Example. Let $(f, g): \mathbb{R}^2 \to \mathbb{R}^2$ be the function such that $f(x, y) = ax + by + c$, $g(x, y) = dx + ey + k$. Then the numerator in (25) is

$$
\begin{vmatrix} ax_1 + by_1 + c & dx_1 + ey_1 + k & 1 \\ ax_2 + by_2 + c & dx_2 + ey_2 + k & 1 \\ ax_0 + by_0 + c & dx_0 + ey_0 + k & 1 \end{vmatrix} = \begin{vmatrix} ax_1 + by_1 & dx_1 + ey_1 & 1 \\ ax_2 + by_2 & dx_2 + ey_2 & 1 \\ ax_0 + by_0 & dx_0 + ey_0 & 1 \end{vmatrix}. \tag{26}
$$

To obtain the determinant on the right from the one on the left, multiply the third column by c (by k) and subtract it from the first column (the second column). Next, recall the Binet–Cauchy multiplication theorem for determinants in (2) in Theorem 80.1 in Appendix 1. By that theorem,

$$\begin{vmatrix} ax_1 + by_1 & dx_1 + ey_1 & 1 \\ ax_2 + by_2 & dx_2 + ey_2 & 1 \\ ax_0 + by_0 & dx_0 + ey_0 & 1 \end{vmatrix} = \begin{vmatrix} x_1 & y_1 & 1 \\ x_2 & y_2 & 1 \\ x_0 & y_0 & 1 \end{vmatrix} \begin{vmatrix} a & d & 0 \\ b & e & 0 \\ 0 & 0 & 1 \end{vmatrix}. \tag{27}$$

Then the limit in (25) in this case is

$$\begin{vmatrix} a & d \\ b & e \end{vmatrix}. \tag{28}$$

The determinant of the partial derivatives of f and g is called the Jacobian of f and g.

This book develops a theory of differentiation based on definitions of derivatives by limits such as those in (8), (21), and (25). The next section begins the systematic development of this theory of differentiation.

Exercises

1.1. Let $f: E \to \{c\}$ be a constant function defined on an open set E in $\mathbb{R}^2$. Show that the limits in (21) exist and have the value zero if p_1, p_2 satisfy the restriction in (18).

1.2. For special choices of the points p_1, p_2, the limits at p_0 in (21) reduce to partial derivatives.

(a) Let $f: E \to \mathbb{R}$ be defined in an open set E in $\mathbb{R}^2$, and let $p_0 : (x_0, y_0)$ be a fixed point in E. Also, let $p_1 : (x_1, y_1)$, $p_2 : (x_2, y_2)$ be restricted to be points in E of the special form $p_1 : (x_0 + \Delta x, y_0)$, $p_2 : (x_0, y_0 + \Delta y)$ with $\Delta x \neq 0$ and $\Delta y \neq 0$. Show that

$$\begin{vmatrix} x_1 & y_1 & 1 \\ x_2 & y_2 & 1 \\ x_0 & y_0 & 1 \end{vmatrix} = \Delta x \Delta y \neq 0,$$

and hence that p_0, p_1, p_2 satisfy the condition in (18).

(b) Show that, for the special points p_0, p_1, p_2 in (a),

$$\frac{\begin{vmatrix} f(x_0 + \Delta x, y_0) & y_0 & 1 \\ f(x_0, y_0 + \Delta y) & y_0 + \Delta y & 1 \\ f(x_0, y_0) & y_0 & 1 \end{vmatrix}}{\begin{vmatrix} x_0 + \Delta x & y_0 & 1 \\ x_0 & y_0 + \Delta y & 1 \\ x_0 & y_0 & 1 \end{vmatrix}} = \frac{f(x_0 + \Delta x, y_0) - f(x_0, y_0)}{\Delta x},$$

$$\frac{\begin{vmatrix} x_0 + \Delta x & f(x_0 + \Delta x, y_0) & 1 \\ x_0 & f(x_0, y_0 + \Delta y) & 1 \\ x_0 & f(x_0, y_0) & 1 \end{vmatrix}}{\begin{vmatrix} x_0 + \Delta x & y_0 & 1 \\ x_0 & y_0 + \Delta y & 1 \\ x_0 & y_0 & 1 \end{vmatrix}} = \frac{f(x_0, y_0 + \Delta y) - f(x_0, y_0)}{\Delta y}.$$

(c) If the limits in (21) exist, show that f has partial derivatives $f_x(x_0, y_0)$, $f_y(x_0, y_0)$, and that the limits in (21) are equal to these partial derivatives.

(d) If f has partial derivatives $f_x(x_0, y_0)$, $f_y(x_0, y_0)$, do the limits in (21) always exist? [Hint. Examples 1.1 and 1.3.]

1.3. (a) The graph of $z = x^2 + y^2$ is a smooth surface which has a tangent plane at each point. Consider the following sets of points; in each case $n = 1, 2, \cdots$.

(i) $p_0 : (1, 1), p_1 : (1 + n^{-1}, 1 - n^{-1}), p_2 : (1 + n^{-1}, 1 + n^{-1})$;
(ii) $p_0 : (1, 1), p_1 : (1 + n^{-1}, 1 + n^{-1}), p_2 : (1 - n^{-1}, 1 + n^{-1})$;
(iii) $p_0 : (1, 1), p_1 : (1 - n^{-1}, 1 + n^{-1}), p_2 : (1 - n^{-1}, 1 - n^{-1})$;
(iv) $p_0 : (1, 1), p_1 : (1 - n^{-1}, 1 - n^{-1}), p_2 : (1 + n^{-1}, 1 - n^{-1})$.

(b) Find the equation of the plane through the three points in each of the sets (i), $\cdots$, (iv) in (a).

(c) Show that each limit in (21) exists and has the value 2 for each of the sets of points (i), $\cdots$, (iv) in (a) as n tends to infinity. The existence of the limits is to be expected since the geometry of the situation suggests that the limit of the secant planes through p_0, p_1, p_2 exists and is the tangent plane to $z = x^2 + y^2$ at p_0.

(d) Show that the vectors p_0p_1 and p_0p_2 are orthogonal in each of the sets (i), $\cdots$, (iv) in (a).

1.4. Let $f: E \to \mathbb{R}$ and $g: E \to \mathbb{R}$ be functions which have derivatives of the type defined by the limits in (21), and let a and b be constants. Show that the function $(af + bg): E \to \mathbb{R}$ also has derivatives of the type (21), and find the values of these derivatives in terms of those of f and g. [Hint. Theorems 77.1 and 77.3 in Appendix 1.]

1.5. Let u and v be vectors with components (a_1, a_2) and (b_1, b_2) respectively, and let θ be the angle between u and v. Then by (7) in Theorem 84.2,

$$\cos \theta = \frac{\sum_{j=1}^{2} a_j b_j}{\left[\sum_{j=1}^{2} a_j^2\right]^{1/2} \left[\sum_{j=1}^{2} b_j^2\right]^{1/2}}.$$

In the inner product notation defined in Section 84, $\cos \theta = (u, v)/[(u, u)^{1/2}(v, v)^{1/2}]$.

(a) Let θ denote the angle between the vectors p_0p_1 and p_0p_2 in Example 1.4. Prove the following: in (a), $\theta = \pi/2$ for every n; in (b) and (c), θ approaches π as $n \to \infty$; and in (d), θ approaches 0 as $n \to \infty$.

1.6. Let $f: \mathbb{R}^2 \to \mathbb{R}$ be the function such that $f(x, y) = x^2 - y^2$. Find the value of the second limit in (21) for each of the four sets of points (a), $\cdots$, (d) in Example 1.4.

1.7. Let $f: \mathbb{R}^2 \to \mathbb{R}$ be the function such that $f(x, y) = x^2 + y^2$. Consider the limits in (21) with respect to the following three sets of points.

(i) $p_0 : (1, 0), p_1 : (1 + n^{-1}, n^{-1}), p_2 : (1 + n^{-1}, -n^{-1}), n = 1, 2, \cdots$.
(ii) $p_0 : (1, 0), p_1 : (1 + n^{-2}, n^{-1}), p_2 : (1 + n^{-2}, -n^{-1}), n = 1, 2, \cdots$.
(iii) $p_0 : (1, 0), p_1 : (1 + n^{-4}, n^{-1}), p_2 : (1 + n^{-4}, -n^{-1}), n = 1, 2, \cdots$.

(a) For each class in (i), (ii), (iii), show that the denominators in (21) are not zero for $n = 1, 2, \cdots$; show also that p_1 and p_2 approach p_0 as $n \to \infty$.

(b) For the class (i) show that the first limit in (21) exists and has the value 2. Show also that $f_x(1, 0) = 2$. Find the value of the second limit in (21).

(c) For the class (ii) show that the first limit in (21) exists and has the value 3. Find the value of the second limit.

(d) For the class (iii) show that the first limit in (21) is $+\infty$. Find the value of the second limit.

(e) Use the formula for $\cos\theta$ in Exercise 1.5 to find the cosine of the angle between the vectors p_0p_1 and p_0p_2 in each of the classes (i), (ii), (iii). For each of the classes find the limit of the angle as $n \to \infty$.

1.8. (a) Let f_1, f_2 be two functions defined by the following statements:

$$f_1(x, y) = \frac{a_1x + b_1y + c_1}{a_3x + b_3y + c_3}, \qquad f_2(x, y) = \frac{a_2x + b_2y + c_2}{a_3x + b_3y + c_3},$$

$$(x, y) \in \mathbb{R}^2, \qquad a_3x + b_3y + c_3 \neq 0.$$

Let D be the determinant of the 3 by 3 matrix whose i-th row is (a_i, b_i, c_i), $i = 1, 2, 3$. If (f_1, f_2) are defined at (x_0, y_0), show that (f_1, f_2) has the derivative (25) at this point [the single restriction in computing the limit is that the denominator is not zero], and show that the value of this derivative [the limit in (25)] is

$$\frac{D}{(a_3x_0 + b_3y_0 + c_3)^3}.$$

(b) Find the Jacobian of (f_1, f_2) at (x_0, y_0) and compare it with the value of the limit (25) found in (a).

2. Definitions and Notation

This section begins a systematic development of the theory of differentiation based on the ideas introduced in Section 1; it introduces some necessary notation and terminology, describes the functions which are the object of study, and defines their derivatives. The two examples which begin the section provide background which helps the reader to understand the developments which follow.

2.1 Example. Let $f(x, y) = x^2 - y^2$ as in Example 1.4. The ratios of two determinants in (21) in Section 1 exist for each three points $p_i: (x_i, y_i)$, $i = 0, 1, 2$, such that

$$\begin{vmatrix} x_1 & y_1 & 1 \\ x_2 & y_2 & 1 \\ x_0 & y_0 & 1 \end{vmatrix} \neq 0. \tag{1}$$

Example 1.4 shows, however, that the restriction (1) is not strong enough to guarantee the existence of the limits in (21) in Section 1. That example suggests that difficulties result if the angle θ between the vectors p_0p_1 and p_0p_2 is not bounded away from zero and π. Let the determinant in (1) be denoted by $\Delta(p_1, p_2, p_0)$ [see (11) in Section 1], and let $|p_1 - p_0|$ and $|p_2 - p_0|$ denote the lengths of the vectors p_0p_1 and p_0p_2 [see Section 89 for definitions of the norm and distance in $\mathbb{R}^3$]. By a familiar formula in trigonometry, the area of a triangle equals one-half the product of the lengths of two sides of the triangle multiplied by the sine of the included angle. Since $(1/2!)|\Delta(p_1, p_2, p_0)|$ is the area of the triangle p_1, p_2, p_0 [see (25) in Section 89],

$$|\Delta(p_1, p_2, p_0)| = |p_1 - p_0||p_2 - p_0| \sin\theta. \tag{2}$$

Let ρ be a constant such that $0 < \rho \leqq 1$; if

$$|\Delta(p_1, p_2, p_0)| \geqq \rho|p_1 - p_0||p_2 - p_0| > 0, \tag{3}$$

then $\sin\theta \geqq \rho > 0$ and the angle between the vectors p_0p_1 and p_0p_2 is bounded away from 0 and π.

2.2 Example. Let $f(x, y) = x^2 + y^2$ and show that the limits in (21) in Section 1 exist if $p_i : (x_i, y_i)$, $i = 1, 2$, approach $p_0 : (x_0, y_0)$ in such a way that $\Delta(p_1, p_2, p_0)$ always satisfies the restriction in (3). Now

$$\begin{aligned} f(x_1, y_1) &= x_1^2 + y_1^2 \\ &= [x_0 + (x_1 - x_0)]^2 + [y_0 + (y_1 - y_0)]^2 \\ &= x_0^2 + y_0^2 + 2x_0(x_1 - x_0) + 2y_0(y_1 - y_0) \\ &\quad + (x_1 - x_0)^2 + (y_1 - y_0)^2 \\ &= f(x_0, y_0) + 2x_0(x_1 - x_0) + 2y_0(y_1 - y_0) + |p_1 - p_0|^2. \\ f(x_2, y_2) &= f(x_0, y_0) + 2x_0(x_2 - x_0) + 2y_0(y_2 - y_0) + |p_2 - p_0|^2. \end{aligned} \tag{4}$$

Then by elementary properties of determinants [see Section 77],

$$\begin{aligned} &\begin{vmatrix} f(p_1) & y_1 & 1 \\ f(p_2) & y_2 & 1 \\ f(p_0) & y_0 & 1 \end{vmatrix} \\ &= \begin{vmatrix} 2x_0(x_1 - x_0) + 2y_0(y_1 - y_0) + |p_1 - p_0|^2 & y_1 - y_0 & 0 \\ 2x_0(x_2 - x_0) + 2y_0(y_2 - y_0) + |p_2 - p_0|^2 & y_2 - y_0 & 0 \\ f(x_0, y_0) & y_0 & 1 \end{vmatrix} \\ &= \begin{vmatrix} 2x_0(x_1 - x_0) + 2y_0(y_1 - y_0) + |p_1 - p_0|^2 & y_1 - y_0 \\ 2x_0(x_2 - x_0) + 2y_0(y_2 - y_0) + |p_2 - p_0|^2 & y_2 - y_0 \end{vmatrix} \\ &= \begin{vmatrix} 2x_0(x_1 - x_0) + |p_1 - p_0|^2 & y_1 - y_0 \\ 2x_0(x_2 - x_0) + |p_2 - p_0|^2 & y_2 - y_0 \end{vmatrix} \\ &= 2x_0\Delta(p_1, p_2, p_0) + \begin{vmatrix} |p_1 - p_0|^2 & y_1 - y_0 \\ |p_2 - p_0|^2 & y_2 - y_0 \end{vmatrix}. \end{aligned} \tag{5}$$

Now divide both sides of this equation by $\Delta(p_1, p_2, p_0)$, which is (1). The result on the right is

$$2x_0 + \frac{\begin{vmatrix} |p_1 - p_0|^2 & y_1 - y_0 \\ |p_2 - p_0|^2 & y_2 - y_0 \end{vmatrix}}{\Delta(p_1, p_2, p_0)}. \tag{6}$$

Then the first limit in (21) in Section 1 exists and equals $2x_0$ if it can be shown that the limit of the fraction in (6) is zero. Now since $\Delta(p_1, p_2, p_0)$ satisfies (3), the absolute value of this fraction is equal to or less than

$$\frac{\text{abs. val.} \begin{vmatrix} |p_1 - p_0|^2 & y_1 - y_0 \\ |p_2 - p_0|^2 & y_2 - y_0 \end{vmatrix}}{\rho |p_1 - p_0| |p_2 - p_0|}. \tag{7}$$

Divide the numerator and the denominator of this fraction by $|p_1 - p_0|$ $|p_2 - p_0|$; divide the numerator by this product by dividing the first row of the matrix by $|p_1 - p_0|$ and the second row by $|p_2 - p_0|$. Thus the absolute value of the fraction in (6) is equal to or less than

$$(1/\rho) \text{ abs. val.} \begin{vmatrix} |p_1 - p_0| & (y_1 - y_0)/|p_1 - p_0| \\ |p_2 - p_0| & (y_2 - y_0)/|p_2 - p_0| \end{vmatrix}. \tag{8}$$

Since

$$|p_i - p_0| = [(x_i - x_0)^2 + (y_i - y_0)^2]^{1/2}, \qquad i = 1, 2,$$

the absolute value of each element in the second column of the matrix is equal to or less than 1. Furthermore, the elements in the first column approach zero as $p_1 \to p_0$ and $p_2 \to p_0$. Then the limit of the expression in (8), and thus of the fraction in (6), is zero as p_1, p_2 tend to p_0 subject to the restriction in (3). More formally, Hadamard's determinant theorem [see (19) in Corollary 87.2] shows that (8) is equal to or less than

$$(\sqrt{2}/\rho)[|p_1 - p_0|^2 + |p_2 - p_0|^2]^{1/2}. \tag{9}$$

Thus the first limit in (21) in Section 1 exists, for the function $f(x, y) = x^2 + y^2$, at (x_0, y_0) and has the value $2x_0$. In the same way, it can be shown that the second limit exists and has the value $2y_0$. These limits equal the classical partial derivatives of f. This example shows that, at least for the function f such that $f(x, y) = x^2 + y^2$, the condition (3) is sufficient to guarantee the existence of the limits in (21) in Section 1 [compare Example 1.4 and Exercises 1.3 and 1.7].

The next step is to introduce definitions, notation, and terminology so that problems similar to the ones in Examples 2.1 and 2.2 can be treated for functions defined on a set E in Euclidean n-dimensional space $\mathbb{R}^n$. Section 90 in Appendix 2 contains an introduction to the basic features of this space. A point x in $\mathbb{R}^n$ has coordinates $(x^1, \cdots, x^n)$, and the norm of x, denoted by $|x|$, is the square root of the sum of the squares of the coordinates of x.

An n-dimensional multivector $\mathbf{x}$, or n-vector $\mathbf{x}$, at a point x_0 in $\mathbb{R}^n$ is an ordered set $(x_1 - x_0, \cdots, x_n - x_0)$ of vectors $x_i - x_0$ whose terminal points are $x_1, \cdots, x_n$ and whose initial points are x_0. A convenient representation for this n-vector $\mathbf{x}$ is the $(n+1) \times n$ matrix whose rows are the coordinates of $x_1, \cdots, x_n, x_0$. Thus

$$\mathbf{x} = \begin{bmatrix} x_1^1 & x_1^2 & \cdots & x_1^n \\ \cdots & \cdots & \cdots & \cdots \\ x_n^1 & x_n^2 & \cdots & x_n^n \\ x_0^1 & x_0^2 & \cdots & x_0^n \end{bmatrix}. \tag{10}$$

The matrix (10) is sometimes represented by the briefer notation $[x_i^j]$, $i = 1, \cdots, n, 0, j = 1, \cdots, n$. Let $[x_i^j 1]$ be the $(n+1) \times (n+1)$ matrix obtained by bordering the matrix $[x_i^j]$ in (10) with a column of 1's on the right. Then the increment of x corresponding to the n-vector $\mathbf{x}$ is denoted by $\Delta(\mathbf{x})$ and defined thus:

$$\Delta(\mathbf{x}) = \det[x_i^j 1] = \begin{vmatrix} x_1^1 & x_1^2 & \cdots & x_1^n & 1 \\ \cdots & \cdots & \cdots & \cdots & \cdots \\ x_n^1 & x_n^2 & \cdots & x_n^n & 1 \\ x_0^1 & x_0^2 & \cdots & x_0^n & 1 \end{vmatrix}. \tag{11}$$

As for the notation, recall (11) in Section 1 and the fact that $\Delta(p_1, p_2, p_0)$ denoted the determinant in (1) in Example 2.1. As for geometric interpretation, equation (26) in Section 89 in Appendix 2 shows that $(1/3!)|\Delta(\mathbf{x})|$ is the volume of the tetrahedron with vertices $x_1, \cdots, x_3, x_0$ if $n = 3$, and Section 20 contains a proof that $(1/n!)|\Delta(\mathbf{x})|$ is the volume of the simplex with vertices $x_1, \cdots, x_n, x_0$ in the general case. Elementary properties of determinants show that

$$\Delta(\mathbf{x}) = \det[x_i^j - x_0^j], \qquad i, j = 1, \cdots, n. \tag{12}$$

2.3 Examples. If $n = 2$, then $\mathbf{x}$ is the 2-vector $(x_1 - x_0, x_2 - x_0)$ whose initial point is $x_0 : (x_0^1, x_0^2)$ and whose terminal points are $x_1 : (x_1^1, x_1^2)$ and $x_2 : (x_2^1, x_2^2)$. Then (11) in this case is

$$\Delta(\mathbf{x}) = \begin{vmatrix} x_1^1 & x_1^2 & 1 \\ x_2^1 & x_2^2 & 1 \\ x_0^1 & x_0^2 & 1 \end{vmatrix}.$$

Subtract the third row from each of the first two rows and then expand the determinant. Thus, as indicated in (12),

$$\Delta(\mathbf{x}) = \det[x_i^j - x_0^j] = \begin{vmatrix} x_1^1 - x_0^1 & x_1^2 - x_0^2 \\ x_2^1 - x_0^1 & x_2^2 - x_0^2 \end{vmatrix}, \qquad i, j = 1, 2.$$

As shown in (25) in Section 89, $|\Delta(\mathbf{x})|$ is 2! times the area of the triangle whose edges at x_0 are the vectors $x_1 - x_0$ and $x_2 - x_0$. For given lengths of the edges, the area is maximum when the vectors are orthogonal; thus

$$
\begin{aligned}
(1/2!)\,|\Delta(\mathbf{x})| &\leqq (1/2)|x_1 - x_0|\,|x_2 - x_0|, \\
|\Delta(\mathbf{x})| &\leqq |x_1 - x_0|\,|x_2 - x_0|.
\end{aligned}
\tag{13}
$$

Similar relations hold in all dimensions. Thus, if $n = 3$, then

$$
\Delta(\mathbf{x}) = \begin{vmatrix} x_1^1 & x_1^2 & x_1^3 & 1 \\ \cdots & \cdots & \cdots & \cdots \\ x_3^1 & x_3^2 & x_3^3 & 1 \\ x_0^1 & x_0^2 & x_0^3 & 1 \end{vmatrix},
$$

$$
\Delta(\mathbf{x}) = \det[x_i^j - x_0^j] = \begin{vmatrix} x_1^1 - x_0^1 & \cdots & x_1^3 - x_0^3 \\ \cdots & \cdots & \cdots \\ x_3^1 - x_0^1 & \cdots & x_3^3 - x_0^3 \end{vmatrix}.
$$

Also, $(1/3!)|\Delta(\mathbf{x})|$ is the volume of the tetrahedron whose edges at x_0 are the vectors $x_i - x_0$, $i = 1, 2, 3$ [see (26) in Section 89]. Then

$$
|\Delta(\mathbf{x})| \leqq |x_1 - x_0|\,|x_2 - x_0|\,|x_3 - x_0|, \tag{14}
$$

and the equality holds if and only if the edges are mutually orthogonal. As a continuation of the long-standing notation and terminology in the one-dimensional case [see (11) in Section 1], this book denotes the determinant in (11) by $\Delta(\mathbf{x})$ and calls it an *increment in x at x_0*. The inequalities in (13) and (14) are special cases of Hadamard's determinant theorem [see Corollary 87.2].

Examples 1.4, 2.1, and 2.2 have shown already that $\mathbf{x}$ must satisfy some stronger condition than merely $\Delta(\mathbf{x}) \neq 0$ if interesting and useful results are to be obtained; the following definition contains this condition [compare (3)] and defines a class of n-vectors at x_0 which is important throughout the remainder of this book. As already indicated by the examples [see especially (2) and (3) in Example 2.1], the condition bounds the angle between each two vectors in $\mathbf{x}$ away from 0 and from π.

2.4 Definition. Let ρ be a number (constant) such that $0 < \rho \leqq 1$. Let

$$
X(x_0, \rho) = \left\{ \mathbf{x} : |\Delta(\mathbf{x})| \geqq \rho \prod_{i=1}^{n} |x_i - x_0| > 0 \right\}. \tag{15}
$$

Then $X(x_0, \rho)$ is called the *ρ-class of n-vectors at x_0*. The inequality in the definition of $X(x_0, \rho)$ is called the *regularity condition* satisfied by $\mathbf{x}$ at x_0, and ρ is the *constant of regularity*.

By (12) and Hadamard's determinant theorem [see (18) in Corollary 87.2],

$$
|\Delta(\mathbf{x})| \leqq \prod_{i=1}^{n} |x_i - x_0|; \tag{16}
$$

thus, if $\rho > 1$, the set $X(x_0, \rho)$ in (15) is empty. For each ρ such that $0 < \rho \leqq 1$, the class $X(x_0, \rho)$ is not empty. If the vectors $x_i - x_0$, $i = 1, \cdots, n$, are mutually orthogonal, then $|\Delta(\mathbf{x})| = \prod_1^n |x_i - x_0|$, and $\mathbf{x}$ satisfies the regularity condition with $\rho = 1$ [see Corollary 87.2]. Thus $X(x_0, 1)$ is not empty. If $\rho_2 < \rho_1$, then $X(x_0, \rho_1) \subset X(x_0, \rho_2)$, and $X(x_0, 1) \subset X(x_0, \rho)$ for every ρ such that $0 < \rho \leqq 1$. Also, if $\rho_2 < \rho_1$, then $X(x_0, \rho_1)$ is a proper subset of $X(x_0, \rho_2)$; to prove this statement, recall first that a determinant is a polynomial in the elements of its matrix and therefore continuous as a function of these elements considered as independent variables. Thus if $|\Delta(\mathbf{x})| \geqq \rho_1 \prod_1^n |x_i - x_0|$, then one of the vectors $x_i - x_0$ in $\mathbf{x}$ can be changed so that the new n-vector is in $X(x_0, \rho_2)$ but no longer in $X(x_0, \rho_1)$. Hence, if $\rho_2 < \rho_1$, there are n-vectors which are in $X(x_0, \rho_2)$ but not in $X(x_0, \rho_1)$, and $X(x_0, \rho_1)$ is a proper subset of $X(x_0, \rho_2)$.

Let m and n be integers such that $m \leqq n$. There will be a frequent need for indexes to describe, and also to order, the $C(n, m)$ sets of m objects which can be selected from n objects. Let $(j_1, \cdots, j_m)$ be an ordered set of m integers selected from $\{1, \cdots, n\}$ such that $1 \leqq j_1 < \cdots < j_m \leqq n$. The sets $(j_1, \cdots, j_m)$ will serve as indexes to identify the $C(n, m)$ subsets of m objects selected from n objects. There is a simple way to order the sets $(j_1, \cdots, j_m)$ as follows: if $(j_1, \cdots, j_m)$ and $(k_1, \cdots, k_m)$ are two sets, and if $(j_1, \cdots, j_m)$ has the smaller integer in the first position in which the two sets differ, then $(j_1, \cdots, j_m)$ precedes $(k_1, \cdots, k_m)$. These ordered sets of integers are denoted by (m/n), and this method of ordering sets of integers is called *lexicographical ordering* because the same principle is used to order the words in a dictionary (lexicon). For example,

$$(4/5) = ((1, 2, 3, 4), (1, 2, 3, 5), (1, 2, 4, 5), (1, 3, 4, 5), (2, 3, 4, 5)).$$

Other examples of lexicographically ordered sets can be found in Section 78.

Let $f: E \to \mathbb{R}^m$, $E \subset \mathbb{R}^n$, denote a function whose domain is E and whose range is in $\mathbb{R}^m$ [see Section 96 for an introduction to the study of these functions]. Unless there is a statement to the contrary, E is assumed to be an open set. Then if $x: (x^1, \cdots, x^n)$ is a point in E, the value $f(x)$ of f at this point is a point $y: (y^1, \cdots, y^m)$ in $\mathbb{R}^m$, and we write $y = f(x)$. Since the range of f is in $\mathbb{R}^m$, the function $f: E \to \mathbb{R}^m$ has m components $(f^1, \cdots, f^m)$, and

$$\begin{aligned} y^1 &= f^1(x^1, \cdots, x^n), \\ &\cdots\cdots\cdots\cdots\cdots \\ y^m &= f^m(x^1, \cdots, x^n). \end{aligned} \tag{17}$$

The n-vector $\mathbf{x}: (x_1, \cdots, x_n, x_0)$ is said to be in E if and only if the vertices $(x_1, \cdots, x_n, x_0)$ are in E. Let x_0 be a point in E; then since E is open, there are n-vectors $\mathbf{x}: (x_1, \cdots, x_n, x_0)$ in E in every neighborhood $N(x_0, \varepsilon)$ of x_0. If $\mathbf{x}$ is in E, the function f has increments at x_0 which correspond to the increment $\Delta(\mathbf{x})$ of x at x_0. The following examples illustrate the general definition which follows.

2.5 Examples. Let $\mathbf{x}: (x_1, \cdots, x_n, x_0)$ be an n-vector in E in $\mathbb{R}^n$. If $m = 1$ and $n = 3$, then f has a single component (denoted by f); there are three increments of f at x_0 which correspond to the increment $\Delta(\mathbf{x})$ of x at x_0, and which are denoted by $\Delta_1 f(\mathbf{x})$, $\Delta_2 f(\mathbf{x})$, $\Delta_3 f(\mathbf{x})$ and defined as follows [compare (21) in Section 1].

$$\Delta_1 f(\mathbf{x}) = \begin{vmatrix} f(x_1) & x_1^2 & x_1^3 & 1 \\ \cdots & \cdots & \cdots & \cdots \\ f(x_3) & x_3^2 & x_3^3 & 1 \\ f(x_0) & x_0^2 & x_0^3 & 1 \end{vmatrix}, \qquad \Delta_2 f(\mathbf{x}) = \begin{vmatrix} x_1^1 & f(x_1) & x_1^3 & 1 \\ \cdots & \cdots & \cdots & \cdots \\ x_3^1 & f(x_3) & x_3^3 & 1 \\ x_0^1 & f(x_0) & x_0^3 & 1 \end{vmatrix},$$

$$\Delta_3 f(\mathbf{x}) = \begin{vmatrix} x_1^1 & x_1^2 & f(x_1) & 1 \\ \cdots & \cdots & \cdots & \cdots \\ x_3^1 & x_3^2 & f(x_3) & 1 \\ x_0^1 & x_0^2 & f(x_0) & 1 \end{vmatrix}. \tag{18}$$

If $m = 2$ and $n = 3$, then f has two components (f^1, f^2), and there are three increments $\Delta_{(j_1, j_2)}(f^1, f^2)(\mathbf{x})$, $(j_1, j_2) \in (2/3)$, which are defined as follows.

$$\Delta_{(1,2)}(f^1, f^2)(\mathbf{x}) = \begin{vmatrix} f^1(x_1) & f^2(x_1) & x_1^3 & 1 \\ \cdots & \cdots & \cdots & \cdots \\ f^1(x_3) & f^2(x_3) & x_3^3 & 1 \\ f^1(x_0) & f^2(x_0) & x_0^3 & 1 \end{vmatrix},$$

$$\Delta_{(1,3)}(f^1, f^2)(\mathbf{x}) = \begin{vmatrix} f^1(x_1) & x_1^2 & f^2(x_1) & 1 \\ \cdots & \cdots & \cdots & \cdots \\ f^1(x_3) & x_3^2 & f^2(x_3) & 1 \\ f^1(x_0) & x_0^2 & f^2(x_0) & 1 \end{vmatrix}, \tag{19}$$

$$\Delta_{(2,3)}(f^1, f^2)(\mathbf{x}) = \begin{vmatrix} x_1^1 & f^1(x_1) & f^2(x_1) & 1 \\ \cdots & \cdots & \cdots & \cdots \\ x_3^1 & f^1(x_3) & f^2(x_3) & 1 \\ x_0^1 & f^1(x_0) & f^2(x_0) & 1 \end{vmatrix}.$$

Finally, if $m = 3$ and $n = 3$, then f has three components (f^1, f^2, f^3) and there is a single increment $\Delta_{(1,2,3)}(f^1, f^2, f^3)(\mathbf{x})$ which is defined as follows [compare (25) in Section 1].

$$\Delta_{(1,2,3)}(f^1, f^2, f^3)(\mathbf{x}) = \begin{vmatrix} f^1(x_1) & f^2(x_1) & f^3(x_1) & 1 \\ \cdots & \cdots & \cdots & \cdots \\ f^1(x_3) & f^2(x_3) & f^3(x_3) & 1 \\ f^1(x_0) & f^2(x_0) & f^3(x_0) & 1 \end{vmatrix}. \tag{20}$$

Observe that the absolute value of each of the determinants in (18), (19), and (20) is 3! times the volume of a certain tetrahedron whose vertices depend on the functions f, f^1, f^2, f^3. Furthermore, the notation $\Delta_1 f(\mathbf{x}), \cdots, \Delta_3 f(\mathbf{x}), \Delta_{(1,2)}(f^1, f^2)(\mathbf{x}), \cdots, \Delta_{(2,3)}(f^1, f^2)(\mathbf{x}), \Delta_{(1,2,3)}(f^1, f^2, f^3)(\mathbf{x})$ is an extension of the notation in (11) in Section 1 which has long been in use in the one-dimensional case.

2.6 Definition. Let $f: E \to \mathbb{R}^m$, $E \subset \mathbb{R}^n$, be a function with domain E and components $(f^1, \cdots, f^m)$; assume that $m \leqq n$. Let $\mathbf{x}: (x_1, \cdots, x_n, x_0)$ be the n-vector at x_0 shown in (10); assume that $\mathbf{x}$ is in E. If $(j_1, \cdots, j_m)$ is in (m/n), replace columns $j_1, \cdots, j_m$ in the matrix $\mathbf{x}$ in (10) in order by the following columns.

$$\begin{bmatrix} f^1(x_1) \\ \cdot\cdot\cdot\cdot\cdot \\ f^1(x_n) \\ f^1(x_0) \end{bmatrix}, \quad \cdots, \quad \begin{bmatrix} f^m(x_1) \\ \cdot\cdot\cdot\cdot\cdot \\ f^m(x_n) \\ f^m(x_0) \end{bmatrix}. \tag{21}$$

The determinant of the resulting matrix is called the $(j_1, \cdots, j_m)$*-increment of f at x_0* corresponding to the increment $\Delta(\mathbf{x})$ of x at x_0; it is denoted by

$$\Delta_{(j_1, \cdots, j_m)} f(\mathbf{x}) \quad \text{or} \quad \Delta_{(j_1, \cdots, j_m)}(f^1, \cdots, f^m)(\mathbf{x}), \qquad (j_1, \cdots, j_m) \in (m/n). \tag{22}$$

In all cases there are $C(n, m)$ increments of f at x_0 in (22). If $m = n$, there is a single increment as illustrated in (20). If $m < n$, there are $C(n, m)$ increments as illustrated in (18) and (19). If $m = 1$, there are n increments of f at x_0 in (22) as illustrated in (18).

2.7 Definition. Let $f: E \to \mathbb{R}^m$, $E \subset \mathbb{R}^n$, be a function with components $(f^1, \cdots, f^m)$ whose domain E is an open set E in $\mathbb{R}^n$; let x_0 be a point in E; and let $X(x_0, \rho)$ be the ρ-class of n-vectors at x_0. Then

$$\lim_{\mathbf{x} \to x_0} \frac{\Delta_{(j_1, \cdots, j_m)}(f^1, \cdots, f^m)(\mathbf{x})}{\Delta(\mathbf{x})}, \qquad \mathbf{x} \in X(x_0, \rho), \qquad \mathbf{x} \text{ is in } E, \tag{23}$$

exists and has the value L if and only if for each $\varepsilon > 0$ there exists a $\delta(\varepsilon, x_0)$ such that

$$\left| \frac{\Delta_{(j_1, \cdots, j_m)}(f^1, \cdots, f^m)(\mathbf{x})}{\Delta(\mathbf{x})} - L \right| < \varepsilon \tag{24}$$

for every $\mathbf{x}$ in $X(x_0, \rho)$ and in E such that $|x_i - x_0| < \delta(\varepsilon, x_0)$, $i = 1, \cdots, n$.

Example 2.2 contains an example of the evaluation of a limit of the type defined in Definition 2.7.

2.8 Definition. Let f be the function described in Definition 2.7. Then f *has a* $(j_1, \cdots, j_m)$*-derivative at* x_0 in E if and only if the limit in (23) exists. If this limit exists, its value is called *the* $(j_1, \cdots, j_m)$*-derivative of f at* x_0 and denoted by

$$D_{(j_1, \cdots, j_m)} f(x_0) \quad \text{or} \quad D_{(j_1, \cdots, j_m)}(f^1, \cdots, f^m)(x_0). \tag{25}$$

Thus

$$D_{(j_1, \cdots, j_m)} f(x_0) = \lim_{\mathbf{x} \to x_0} \frac{\Delta_{(j_1, \cdots, j_m)} f(\mathbf{x})}{\Delta(\mathbf{x})}, \qquad \mathbf{x} \in X(x_0, \rho), \qquad \mathbf{x} \text{ is in } E. \tag{26}$$

Also, f *is differentiable at* x_0 if and only if $D_{(j_1,\cdots,j_m)}f(x_0)$ exists for each $(j_1, \cdots, j_m)$ in (m/n). If f is differentiable at x_0, its derivative at x_0 is the vector in $\mathbb{R}^{C(n,m)}$ whose components are $D_{(j_1,\cdots,j_m)}f(x_0)$. Finally, f *is differentiable on* E if and only if it is differentiable at each point x_0 in E.

This definition requires a word of explanation. Example 1.4 and Exercise 1.7 have shown that the limits (21) in Section 1 (their existence as well as their values) depend on special properties of the class of increments with respect to which the limits are evaluated. Thus, strictly speaking, Definition 2.8 defines "$(j_1, \cdots, j_m)$-derivative of f at x_0 with respect to $X(x_0, \rho)$," "differentiable at x_0 with respect to $X(x_0, \rho)$," and so on. However, Corollary 3.7 in the next section shows that the qualification "with respect to $X(x_0, \rho)$" is unnecessary for functions $f: E \to \mathbb{R}$, $E \subset \mathbb{R}^n$; if $f: E \to \mathbb{R}$ is differentiable with respect to at least one class $X(x_0, \rho)$, then it is differentiable with respect to every class $X(x_0, \rho)$, $0 < \rho \leqq 1$, and the values of the derivatives $D_j f(x_0)$, $j = 1, \cdots, n$, are the same for every ρ. As will be explained in Section 6, the class of functions $(f^1, \cdots, f^m): E \to \mathbb{R}^m$ studied in this book is restricted to those whose components $f^1, \cdots, f^m$ are differentiable. With this restriction on the class of differentiable functions, the qualification is unnecessary. For this reason, the cumbersome qualification has been omitted in the statement of Definition 2.8.

If $m = 1$, then $(m/n) = ((1), (2), \cdots, (n))$, and the derivative of $f: E \to \mathbb{R}$ at x_0 is denoted by the simpler and more customary notation $(D_1 f(x_0), \cdots, D_n f(x_0))$, and occasionally by $(D_{x^1} f(x_0), \cdots, D_{x^n} f(x_0))$. In this case, the derivative of f at x_0 is a vector in $\mathbb{R}^n$; later it will be shown that this vector is the one known as the gradient of f at x_0 and denoted by $\nabla f(x_0)$. If $m = n$, then (m/n) contains the single set $(1, 2, \cdots, n)$, and the derivative $D_{(1,\cdots,n)}f(x_0)$ is a point in $\mathbb{R}$.

2.9 Example. If the components of f are linear functions with constant coefficients, then the derivatives of f exist with respect to the class of n-vectors $\{\mathbf{x} : \Delta(\mathbf{x}) \neq 0\}$, and its derivatives are constants. For example, if

$$\begin{aligned} f^1(x) &= a_{11}x^1 + a_{12}x^2, \\ f^2(x) &= a_{21}x^1 + a_{22}x^2, \end{aligned} \qquad A = \begin{bmatrix} a_{11} & a_{12} \\ a_{21} & a_{22} \end{bmatrix}, \tag{27}$$

then the multiplication theorem for determinants [see Theorem 80.1] can be used to show that

$$\begin{aligned} \Delta_{(1,2)}(f^1, f^2)(\mathbf{x}) &= \Delta(\mathbf{x}) \det A, \\ D_{(1,2)}(f^1, f^2)(x_0) &= \det A, \qquad x_0 \in \mathbb{R}^2. \end{aligned}$$

The function f in (27) transforms the triangle (x_1, x_2, x_0) into the triangle $(f(x_1), f(x_2), f(x_0))$ and multiplies its area by $|\det A|$. If $f: \mathbb{R}^3 \to \mathbb{R}^2$ is the function such that

$$\begin{aligned} f^1(x) &= a_{11}x^1 + a_{12}x^2 + a_{13}x^3, \\ f^2(x) &= a_{21}x^1 + a_{22}x^2 + a_{23}x^3, \end{aligned} \tag{28}$$

then it can be shown in the same way that

$$D_{(1,2)}f(x_0) = \begin{vmatrix} a_{11} & a_{12} \\ a_{21} & a_{22} \end{vmatrix}, \qquad D_{(1,3)}f(x_0) = \begin{vmatrix} a_{11} & a_{13} \\ a_{21} & a_{23} \end{vmatrix},$$

$$D_{(2,3)}f(x_0) = \begin{vmatrix} a_{12} & a_{13} \\ a_{22} & a_{23} \end{vmatrix}.$$

2.10 Example. Let E be an open set in $\mathbb{R}^2$, and let $f: E \to \mathbb{R}$ be a function which is differentiable at x_0 in E. Then Section 1 has shown that the equation of the tangent plane to the surface $y = f(x)$, x in E, at $(x_0^1, x_0^2, f(x_0))$ is

$$y - f(x_0) = D_1 f(x_0)(x^1 - x_0^1) + D_2 f(x_0)(x^2 - x_0^2). \tag{29}$$

If E is a set in $\mathbb{R}^n$ and $f: E \to \mathbb{R}$ is a function which is differentiable at x_0 in E, then the methods employed in Section 1 can be used to show that the equation of the hyperplane which is tangent to the surface $y = f(x)$, x in E, at $(x_0^1, \cdots, x_0^n, f(x_0))$ is

$$y - f(x_0) = \sum_{j=1}^{n} D_j f(x_0)(x^j - x_0^j). \tag{30}$$

2.11 Example. Let E be an open set in $\mathbb{R}^n$, and let $f: E \to \mathbb{R}^m$ be a function whose components $(f^1, \cdots, f^m)$ are differentiable at x_0 in E. Then, by the result in (30),

$$\begin{aligned} y^1 &= f^1(x_0) + \sum_{j=1}^{n} D_j f^1(x_0)(x^j - x_0^j), \\ &\cdots\cdots\cdots\cdots\cdots\cdots \\ y^m &= f^m(x_0) + \sum_{j=1}^{n} D_j f^m(x_0)(x^j - x_0^j), \end{aligned} \tag{31}$$

are the equations of the affine space which is tangent at $(x_0^1, \cdots, x_0^n)$ to the graph of the surface whose equations are

$$\begin{aligned} y^1 &= f^1(x), \\ &\cdots\cdots \\ y^m &= f^m(x). \end{aligned} \tag{32}$$

2.12 Examples. The derivatives defined in this section can be used for all the purposes for which derivatives have always been used. Furthermore, they permit the development of general methods in $\mathbb{R}^n$ which are the same as those used traditionally for functions of a single variable, and in most cases the theorems are as good (the same hypotheses and the same conclusions) in $\mathbb{R}^n$ as they are in $\mathbb{R}$. Thus the derivatives defined in this chapter are the first step in the development of methods which are very general and which lead to general results. For example, the same general method leads to integrals for the length of a curve, the area of a surface in $\mathbb{R}^3$ (or $\mathbb{R}^n$), and the area of a hypersurface in $\mathbb{R}^n$ [the details are found in Section 45]. The following proof of the fundamental theorem of the integral calculus for functions of a single variable has an exact generalization in $\mathbb{R}^n$ which is made possible by the derivatives introduced in this chapter.

$$\begin{aligned} f(b) - f(a) &= \lim_{n\to\infty} \sum_{i=1}^{n} [f(x_i) - f(x_{i-1})] = \lim_{n\to\infty} \sum_{i=1}^{n} f'(x_i^*)(x_i - x_{i-1}) \\ &= \int_a^b f'(x)\,dx. \end{aligned} \tag{33}$$

EXERCISES

2.1. Let $f(x, y) = x^2 + y^2$ for (x, y) in the open set E in $\mathbb{R}^2$. Example 2.2 has proved that $D_x f(x_0, y_0) = 2x_0$. Use the methods of that example to prove that $D_y f(x_0, y_0) = 2y_0$.

2.2. Each of the functions described below is defined in an open set E in $\mathbb{R}^2$ which contains the point (x_0, y_0). Use the method of Example 2.2 to find $D_x f(x_0, y_0)$ and $D_y f(x_0, y_0)$ for each of these functions.
(a) $f(x, y) = xy$.
(b) $f(x, y) = x^2 + 2xy + y^2$.
(c) $f(x, y) = x^2 - y^2$.

2.3. Let $X(x_0, \rho)$ be the ρ-class of 2-vectors $\mathbf{x}$ at x_0 in $\mathbb{R}^2$. If $\mathbf{x} = [x_i^j]$, $i = 1, 2, 0$ and $j = 1, 2$, and if θ, $0 \leqq \theta < \pi$ is the angle formed by the vectors $x_1 - x_0$ and $x_2 - x_0$, then $|\Delta(\mathbf{x})| = |x_1 - x_0||x_2 - x_0| \sin\theta$ [see Example 2.1].
(a) Describe the class $X(x_0, \rho)$ of 2-vectors if $\rho = \sqrt{3}/2$; also, if $\rho = \sqrt{2}/2$ and $\rho = 1/2$.
(b) Prove that $X(x_0, \sqrt{3}/2) \subset X(x_0, \sqrt{2}/2)$. Prove also that $X(x_0, \sqrt{3}/2)$ is a proper subset of $X(x_0, \sqrt{2}/2)$ by finding a 2-vector $\mathbf{x}$ which is in $X(x_0, \sqrt{2}/2)$ but not in $X(x_0, \sqrt{3}/2)$.

2.4. If the vectors $x_i - x_0$, $i = 1, \cdots, n$, are mutually orthogonal in $\mathbb{R}^n$, prove that $|\Delta(\mathbf{x})| = \Pi_1^n |x_i - x_0|$.

2.5. Prove the following: if $\rho_2 < \rho_1$, then $X(x_0, \rho_1) \subset X(x_0, \rho_2)$, and $X(x_0, 1) \subset X(x_0, \rho)$ for every ρ such that $0 < \rho \leqq 1$.

2.6. (a) Let ρ_1, ρ_2 be two numbers such that $0 < \rho_2 < \rho_1 \leqq 1$. Prove that, if

$$\lim_{\mathbf{x}\to x_0} \frac{\Delta_{(j_1,\cdots,j_m)} f(\mathbf{x})}{\Delta(\mathbf{x})} = L, \qquad \mathbf{x} \in X(x_0, \rho_2), \tag{i}$$

then

$$\lim_{\mathbf{x}\to x_0} \frac{\Delta_{(j_1,\cdots,j_m)} f(\mathbf{x})}{\Delta(\mathbf{x})} = L, \qquad \mathbf{x} \in X(x_0, \rho_1). \tag{ii}$$

(b) It is easy to see that the existence of the limit in (i) with respect to $X(x_0, \rho_2)$ implies the existence of the limit in (ii) with respect to the subset $X(x_0, \rho_1)$ of $X(x_0, \rho_2)$. Explain why (ii) does not imply (i). [See Corollary 3.7 for the important results that can be established for functions $f: E \to \mathbb{R}$, $E \subset \mathbb{R}^n$, that is, for functions with a single component.]

(c) For certain functions f, limits exist with respect to some subsets of $X(x_0, \rho)$ although they do not exist with respect to $X(x_0, \rho)$ itself. For example, there are functions which have partial derivatives but are not differentiable in the sense of Definition 2.8. Describe the subset of $X(x_0, 1)$ which defines partial derivatives [see Examples 1.1 and 1.3 and Exercise 1.2].

2.7. (a) Let A be an $n \times n$ matrix, and let $f: \mathbb{R}^n \to \mathbb{R}^n$ be the linear transformation whose matrix is A. Show that f is differentiable at each point in $\mathbb{R}^n$, and show that $D_{(1,\cdots,n)}f(x_0) = \det A$, x_0 in $\mathbb{R}^n$. [Hint. Theorem 80.1 contains the Cauchy–Binet multiplication theorem for the product of two determinants.]

(b) Let B be an $m \times n$ matrix, $m \leqq n$, and let $f: \mathbb{R}^n \to \mathbb{R}^m$ be the linear transformation whose matrix is B. Show that f is differentiable at each point in $\mathbb{R}^n$, and that $D_{(j_1,\cdots,j_m)}f(x_0)$ equals the determinant of the minor of B in columns $j_1, \cdots, j_m$.

2.8. Prove that there is no ρ, $0 < \rho \leqq 1$, for which the 2-vectors in (ii) and (iii) in Exercise 1.7 satisfy, for $n = 1, 2, \cdots$, the ρ-regularity condition at $(1, 0)$. Show that the 2-vectors in (i) belong to a class $X(x_0, \rho)$, $x_0 = (1, 0)$, and find the largest ρ for which this statement is true. Is there a smallest ρ?

2.9. Show that $\Delta(\mathbf{x}) = \det[x_i^j - x_0^j]$, $i, j = 1, \cdots, n$ [see (12)].

2.10. (a) Which of the words "institute" and "instantly" is listed first in a dictionary? Explain how you decided.

(b) In the lexicographical ordering of the sets of six integers selected from $\{1, 2, \cdots, 12\}$, which set in the following pairs precedes the other?

(i) (1, 2, 3, 7, 10, 12), (1, 2, 3, 5, 11, 12).
(ii) (3, 4, 6, 9, 11, 12), (2, 8, 9, 10, 11, 12).
(iii) (5, 6, 7, 8, 10, 11), (5, 6, 7, 8, 10, 12).

(c) Arrange the following triples in lexicographical order.

(i) (3, 4, 6, 7, 8), (1, 3, 4, 5, 6), (1, 3, 4, 5, 8).
(ii) (2, 3, 4, 5, 8), (2, 3, 4, 5, 6), (2, 3, 4, 5, 7).
(iii) (2, 4, 6, 7, 8), (1, 2, 3, 4, 5), (3, 5, 6, 7, 8).

3. Elementary Properties of Differentiable Functions

This section treats the analytic and geometric properties of differentiable functions $f: E \to \mathbb{R}$ and $f: E \to \mathbb{R}^m$, $E \subset \mathbb{R}^n$. It begins with the definition of the Stolz condition for functions $f: E \to \mathbb{R}$. This condition was introduced by Otto Stolz in 1893 in connection with his study of partial differentiation [see 4 in References and Notes at the end of this book]; the Stolz condition is a necessary and sufficient condition for differentiability for functions $f: E \to \mathbb{R}$.

3.1 Definition (The Stolz Condition). The function $f: E \to \mathbb{R}$ satisfies the *Stolz condition at* x_0 in the open set E in $\mathbb{R}^n$ if and only if there exist constants

$A_1, \cdots, A_n$ and a real-valued function of x, denoted by $r(f; x_0, x)$ and defined in a neighborhood of x_0 in E, such that

$$f(x) - f(x_0) = \sum_{j=1}^{n} A_j(x^j - x_0^j) + r(f; x_0, x)|x - x_0|,$$
$$\lim_{x \to x_0} r(f; x_0, x) = 0, \qquad r(f; x_0, x_0) = 0. \tag{1}$$

3.2 Example. The function $f\colon \mathbb{R}^2 \to \mathbb{R}$, $f(x, y) = x^2 + y^2$, in Example 2.2 satisfies the Stolz condition at every point (x_0, y_0) in $\mathbb{R}^2$. Equation (4) in Section 2 shows that

$$f(x, y) - f(x_0, y_0) = 2x_0(x - x_0) + 2y_0(y - y_0) + |p - p_0||p - p_0|. \tag{2}$$

Thus f satisfies the Stolz condition (1) with $A_1 = 2x_0$, $A_2 = 2y_0$, and $r(f; p_0, p) = |p - p_0|$.

3.3 Theorem. *Let $f\colon E \to \mathbb{R}$, $E \subset \mathbb{R}^n$, be a function which satisfies the Stolz condition with constants $A_1, \cdots, A_n$ at x_0 in E [see Definition 3.1]. Then the derivatives $D_j f(x_0)$ exist with respect to each class $X(x_0, \rho)$, $0 < \rho \leqq 1$, and $D_j f(x_0) = A_j, j = 1, \cdots, n$.*

PROOF. Let ρ be a number such that $0 < \rho \leqq 1$. To prove the theorem, it is necessary to show that

$$\lim_{\mathbf{x} \to x_0} \frac{\Delta_j f(\mathbf{x})}{\Delta(\mathbf{x})} = A_j, \qquad \mathbf{x} \in X(x_0, \rho), \qquad j = 1, \cdots, n. \tag{3}$$

The numerator in (3) is

$$\Delta_j f(\mathbf{x}) = \begin{vmatrix} x_1^1 & \cdots & f(x_1) & \cdots & x_1^n & 1 \\ \cdots & \cdots & \cdots & \cdots & \cdots & \cdots \\ x_n^1 & \cdots & f(x_n) & \cdots & x_n^n & 1 \\ x_0^1 & \cdots & f(x_0) & \cdots & x_0^n & 1 \end{vmatrix}; \tag{4}$$

in the determinant in (4), the elements $f(x_i)$ are in the j-th column of the matrix. Subtract the last row in the matrix of $\Delta_j f(\mathbf{x})$ from the preceding rows and expand the determinant by elements in the last column. The element in the i-th row and j-th column is now $f(x_i) - f(x_0)$, whose value by the Stolz condition (1) is

$$\sum_{k=1}^{n} A_k(x_i^k - x_0^k) + r(f; x_0, x_i)|x_i - x_0|. \tag{5}$$

Multiply the k-th column, $k \neq j$, by A_k and subtract it from the j-th column; after all simplifications of this type [see Section 77 in Appendix 1 for the properties of determinants used in this proof], the (i, j)-th element is $A_j(x_i^j - x_0^j) + r(f; x_0, x_i)|x_i - x_0|$. Write the determinant as the sum of two determinants; factor out A_j from the first one; then divide by $\Delta(\mathbf{x})$ to obtain

$$\frac{\Delta_j f(\mathbf{x})}{\Delta(\mathbf{x})} = A_j + \frac{\det D}{\Delta(\mathbf{x})}, \tag{6}$$

$$D = \begin{bmatrix} x_1^1 - x_0^1 & \cdots & r(f; x_0, x_1)|x_1 - x_0| & \cdots & x_1^n - x_0^n \\ \cdots & \cdots & \cdots & \cdots & \cdots \\ x_n^1 - x_0^1 & \cdots & r(f; x_0, x_n)|x_n - x_0| & \cdots & x_n^n - x_0^n \end{bmatrix}. \tag{7}$$

Since $\mathbf{x}$ is in $X(x_0, \rho)$, then

$$\left|\frac{\det D}{\Delta(\mathbf{x})}\right| \leqq \frac{|\det D|}{\rho \prod_1^n |x_i - x_0|}. \tag{8}$$

Divide the i-th row of D by $|x_i - x_0|$, and divide the denominator on the right in (8) by $\prod_1^n |x_i - x_0|$. Then the elements in the j-th column are $r(f; x_0, x_i)$, $i = 1, \cdots, n$, and the absolute value of each element not in the j-th column is equal to or less than 1 since $|x_i^k - x_0^k| \leqq |x_i - x_0|$. Then (6) and (8) show that

$$\lim_{\mathbf{x} \to x_0} \frac{\det D}{\Delta(\mathbf{x})} = 0, \tag{9}$$

$$\lim_{\mathbf{x} \to x_0} \frac{\Delta_j f(\mathbf{x})}{\Delta(\mathbf{x})} = A_j + \lim_{\mathbf{x} \to x_0} \frac{\det D}{\Delta(\mathbf{x})} = A_j, \qquad j = 1, \cdots, n. \tag{10}$$

A more formal proof of (9) can be obtained by applying Hadamard's determinant theorem [see (19) in Corollary 87.2]; this inequality for determinants shows that, after the modifications described, the expression on the right in (8) is equal to or less than

$$(1/\rho) n^{(n-1)/2} \left\{ \sum_{i=1}^n [r(f; x_0, x_i)]^2 \right\}^{1/2}. \tag{11}$$

Since $x_i \to x_0$ as $\mathbf{x} \to x_0$, the limit of (11) is zero as $\mathbf{x} \to x_0$ [see (1)] and (9) and (10) follow. Thus (3) is true for each ρ such that $0 < \rho \leqq 1$, and the proof of Theorem 3.3 is complete. □

3.4 Example. The function $f: \mathbb{R}^2 \to \mathbb{R}$ in Example 3.2 satisfies the Stolz condition at (x_0, y_0) with $A_1 = 2x_0$, $A_2 = 2y_0$ as shown in (2). Example 2.2 has shown already that $D_x f(x_0, y_0) = 2x_0$ and Exercise 2.1 asserts that $D_y f(x_0, y_0) = 2y_0$.

Theorem 3.3 shows that the Stolz condition [see Definition 3.1] is sufficient for the differentiability of $f: E \to \mathbb{R}$, $E \subset \mathbb{R}^n$, and the next theorem proves that this condition is necessary. The proof depends on an obscure theorem for determinants known in the literature as Sylvester's theorem of 1839 and 1851, but to be called Sylvester's interchange theorem hereafter in this book. This theorem (with its original name) is proved in Theorem 81.1 in Appendix 1 [see also 2 and 5 in References and Notes]; the proof depends on a matrix identity which is not difficult to establish, but which unfortunately discloses little about the deeper reasons for the truth of the theorem.

Before Sylvester's interchange theorem can be stated, it is necessary to describe some special notation. Let A and B be $n \times n$ matrices with elements a_{rs} and b_{rs}. Let m be an integer such that $m \leqq n$. Let $(k_1, \cdots, k_m)$ be an arbitrary, but fixed, index set in (m/n), and let $(j_1, \cdots, j_m)$ be a variable index set in (m/n). Let $A(b_{(k_1,\cdots,k_m)}/a_{(j_1,\cdots,j_m)})$ denote the matrix obtained by replacing the $(j_1, \cdots, j_m)$ columns of A in order by the $(k_1, \cdots, k_m)$ columns of B, and let $B(a_{(j_1,\cdots,j_m)}/b_{(k_1,\cdots,k_m)})$ denote the matrix obtained by replacing the $(k_1, \cdots, k_m)$ columns of B by the $(j_1, \cdots, j_m)$ columns of A. This notation is similar to that used in proving Theorem 81.1.

3.5 Theorem (Sylvester's Interchange Theorem). *The following identity holds for the matrices just described. For each* $(k_1, \cdots, k_m)$ *in* (m/n),

$$\det A \det B = \sum_{(j_1,\cdots,j_m)} \det A(b_{(k_1,\cdots,k_m)}/a_{(j_1,\cdots,j_m)}) \det B(a_{(j_1,\cdots,j_m)}/b_{(k_1,\cdots,k_m)}). \tag{12}$$

3.6 Theorem. *Let* $f: E \to \mathbb{R}$, $E \subset \mathbb{R}^n$, *be differentiable at* x_0 *with respect to at least one class* $X(x_0, \rho)$, $0 < \rho \leqq 1$. *Then* f *satisfies the Stolz condition at* x_0; *if* A_j *denotes the constants in the Stolz condition in Definition 3.1, then* $A_j = D_j f(x_0)$, $j = 1, \cdots, n$.

Proof. Let $X(x_0, \rho)$ be a class with respect to which f is differentiable; by hypothesis there is at least one such class. Let x_1 be a point in E near x_0. To prove the theorem, it is necessary to show that $f(x_1) - f(x_0)$ can be represented in the manner indicated in (1). Now x_1 is a vertex of an n-vector $\mathbf{x}$ in $X(x_0, \rho)$; the orthogonal n-vector which contains $x_1 - x_0$ is one such $\mathbf{x}$. Then let $x_1, \cdots, x_n, x_0$ be the vertices of an n-vector $\mathbf{x}$ in $X(x_0, \rho)$ which is in a neighborhood $N(x_0, r) \subset E$, and let A be the matrix $[x_i^j 1]$ in (11) in Section 2. Let B be the following $(n + 1) \times (n + 1)$ matrix.

$$B = \begin{bmatrix} f(x_1) & 0 & \cdots & 0 & 1 \\ f(x_2) & 1 & \cdots & 0 & 1 \\ \vdots & \vdots & \ddots & \vdots & \vdots \\ f(x_n) & 0 & \cdots & 1 & 1 \\ f(x_0) & 0 & \cdots & 0 & 1 \end{bmatrix}. \tag{13}$$

Let $m = 1$, and choose $(k_1, \cdots, k_m)$ to be the index set in $(1/n)$ which contains the single integer 1; with m and $(k_1, \cdots, k_m)$ chosen thus, apply Theorem 3.5 (Sylvester's interchange theorem) to the matrices A and B. By expanding $\det B$ and the other determinants based on B [see Theorems 79.1 and 79.3], the result can be simplified to the following equation.

$$\Delta(\mathbf{x})[f(x_1) - f(x_0)] = \sum_{j=1}^{n} \Delta_j f(\mathbf{x})(x_1^j - x_0^j). \tag{14}$$

Then

$$f(x_1)-f(x_0)=\sum_{j=1}^{n} D_jf(x_0)(x_1^j-x_0^j)+\sum_{j=1}^{n}\left\{\frac{\Delta_j f(\mathbf{x})}{\Delta(\mathbf{x})}-D_jf(x_0)\right\}(x_1^j-x_0^j). \tag{15}$$

Set

$$\begin{aligned} r(f;x_0,x_1)&=\frac{1}{|x_1-x_0|}\sum_{j=1}^{n}\left\{\frac{\Delta_j f(\mathbf{x})}{\Delta(\mathbf{x})}-D_jf(x_0)\right\}(x_1^j-x_0^j),\\ r(f;x_0,x_0)&=0. \end{aligned} \tag{16}$$

Then

$$f(x_1)-f(x_0)=\sum_{j=1}^{n} D_jf(x_0)(x_1^j-x_0^j)+r(f;x_0,x_1)|x_1-x_0|.$$

The individual terms in the sum in (16) are functions of $\mathbf{x}$, but (15) shows that the value of the sum depends only on x_0 and x_1. The situation is simple but slightly subtle; it should be understood because it will occur again later. Since $|x_1^j-x_0^j|\leqq|x_1-x_0|$,

$$|r(f;x_0,x_1)|\leqq\sum_{j=1}^{n}\left|\frac{\Delta_j f(\mathbf{x})}{\Delta(\mathbf{x})}-D_jf(x_0)\right|, \tag{17}$$

and (16) and Schwarz's inequality [see (7) and Corollary 86.2 in Section 86 in Appendix 1] show that

$$|r(f;x_0,x_1)|\leqq\left\{\sum_{j=1}^{n}\left|\frac{\Delta_j f(\mathbf{x})}{\Delta(\mathbf{x})}-D_jf(x_0)\right|^2\right\}^{1/2}. \tag{18}$$

Since f is differentiable at x_0 by hypothesis, each of the inequalities in (17) and (18) shows that

$$\lim_{\mathbf{x}\to x_0} r(f;x_0,x_1)=\lim_{x_1\to x_0} r(f;x_0,x_1)=0,\qquad x\in X(x_0,\rho). \tag{19}$$

Then (15), (16), and (19) show that f satisfies Stolz's condition in Definition 3.1 with $A_j=D_jf(x_0)$, $j=1,\cdots,n$, and the proof of Theorem 3.6 is complete. □

3.7 Corollary. *If $f: E\to\mathbb{R}$, $E\subset\mathbb{R}^n$, is differentiable with respect to at least one class $X(x_0,\rho)$, then f is differentiable with respect to every class $X(x_0,\rho)$, and the values of the derivatives $D_jf(x_0)$ are the same for every ρ.*

Proof. If f is differentiable with respect to one class $X(x_0,\rho^*)$, then, by Theorem 3.6, f satisfies the Stolz condition (1) at x_0, and the Stolz constants A_j are $A_j=D_j^*f(x_0)$, $j=1,\cdots,n$. Then by Theorem 3.3, f is differentiable with respect to every class $X(x_0,\rho)$ and $D_jf(x_0)=A_j=D_j^*f(x_0)$, $j=1,\cdots,$ n. □

3.8 Corollary. *The function $f\colon E \to \mathbb{R}$, $E \subset \mathbb{R}^n$, is differentiable at x_0 in E if and only if it satisfies the Stolz condition at x_0.*

PROOF. If f satisfies the Stolz condition in Definition 3.1 at x_0, then f is differentiable at x_0 by Theorem 3.3. If f is differentiable at x_0, then f satisfies the Stolz condition at x_0 by Theorem 3.6. □

3.9 Corollary. *If $f\colon E \to \mathbb{R}$, $E \subset \mathbb{R}^n$, is differentiable at x_0 in E, then*

$$|f(x_1) - f(x_0)| \leqq \left\{\left[\sum_{j=1}^{n} |D_j f(x_0)|^2\right]^{1/2} + |r(f; x_0, x_1)|\right\}|x_1 - x_0|, \tag{20}$$

and f is continuous at x_0.

PROOF. If f is differentiable at x_0 in E, then as shown in (15) and (16) in the proof of Theorem 3.6,

$$f(x_1) - f(x_0) = \sum_{j=1}^{n} D_j f(x_0)(x_1^j - x_0^j) + r(f; x_0, x_1)|x_1 - x_0|. \tag{21}$$

Then

$$|f(x_1) - f(x_0)| \leqq \left|\sum_{j=1}^{n} D_j f(x_0)(x_1^j - x_0^j)\right| + |r(f; x_0, x_1)||x_1 - x_0|. \tag{22}$$

By Schwarz's inequality [see Corollary 86.2],

$$\left|\sum_{j=1}^{n} D_j f(x_0)(x_1^j - x_0^j)\right| \leqq \left[\sum_{j=1}^{n} |D_j f(x_0)|^2\right]^{1/2}\left[\sum_{j=1}^{n} |x_1^j - x_0^j|^2\right]^{1/2}. \tag{23}$$

The second square bracket on the right in (23) is $|x_1 - x_0|$. Substitute from (23) in (22); the result can be simplified easily to (20).

To show that f is continuous at x_0, begin by choosing η so that the neighborhood $N(x_0, \eta)$ of x_0 is in the open set E and so that $|r(f; x_0, x_1)| < 1$ for all x_1 in $N(x_0, \eta)$; this choice is possible by (19). Let $\varepsilon > 0$ be given; choose $\delta(\varepsilon, x_0)$ so that $0 < \delta(\varepsilon, x_0) < \eta$ and so that

$$\delta(\varepsilon, x_0) < \varepsilon \Big/ \left\{\left[\sum_{j=1}^{n} |D_j f(x_0)|^2\right]^{1/2} + 1\right\}. \tag{24}$$

Then by (20)

$$\begin{aligned} |f(x_1) - f(x_0)| &< \left\{\left[\sum_{j=1}^{n} |D_j f(x_0)|^2\right]^{1/2} + 1\right\}|x_1 - x_0|, && x_1 \in N(x_0, \eta), \\ &< \varepsilon, && |x_1 - x_0| < \delta(\varepsilon, x_0) < \eta. \end{aligned} \tag{25}$$

Then f is continuous at x_0 by Definition 96.2 in Appendix 2, and the proof of Corollary 3.9 is complete. □

If $f: E \to \mathbb{R}^m$, $E \subset \mathbb{R}^n$, is a function whose components $(f^1, \cdots, f^m)$ are differentiable at x_0 in E, one would hope and conjecture that f itself is differentiable at x_0. This conjecture becomes Theorem 3.12; its proof depends on another obscure identity in the theory of determinants. This identity is known as the Bazin–Reiss–Picquet theorem; Section 83 in Appendix 1 contains a proof of this theorem and also numerous examples. With the proper background (all of which is provided in Appendix 1), the proof is an algebraic exercise in matrix theory which is not difficult; unfortunately, this proof discloses nothing about the deeper reasons why the theorem is true. The following special notation is needed for a statement of the Bazin–Reiss–Picquet theorem. Let A and B be $n \times n$ matrices, let $i, j = 1, \cdots, n$, and let $A(b_i/a_j)$ denote the matrix obtained by replacing column j in A by column i in B. Then $[A(b_i/a_j)]$ is a new $n \times n$ matrix whose elements are $n \times n$ matrices.

3.10 Theorem (The Bazin–Reiss–Picquet Theorem). *The following identity holds for the matrices just described.*

$$[\det A]^{n-1}[\det B] = \det[\det A(b_i/a_j)], \qquad i, j = 1, \cdots, n. \tag{26}$$

The identity in (26) is only a special case of the general Bazin–Reiss–Picquet theorem which is proved in Section 83, but it is the only case of the theorem which is applied in this book. The notation used in (26) is a simplification of the notation used in the general theorem in Section 83. Exercise 3.12 contains a simple proof of the special case in Theorem 3.10; this proof throws some light on the theorem in this special case.

3.11 Example. If

$$A = \begin{bmatrix} x_1^1 & x_1^2 & 1 \\ x_2^1 & x_2^2 & 1 \\ x_0^1 & x_0^2 & 1 \end{bmatrix}, \qquad B = \begin{bmatrix} f^1(x_1) & f^2(x_1) & 1 \\ f^1(x_2) & f^2(x_2) & 1 \\ f^1(x_0) & f^2(x_0) & 1 \end{bmatrix}, \tag{27}$$

then $n = 3$ and the identity in (26) is

$$[\det A]^2[\det B] = \begin{vmatrix} \det A(b_1/a_1) & \det A(b_1/a_2) & \det A(b_1/a_3) \\ \det A(b_2/a_1) & \det A(b_2/a_2) & \det A(b_2/a_3) \\ \det A(b_3/a_1) & \det A(b_3/a_2) & \det A(b_3/a_3) \end{vmatrix}. \tag{28}$$

Now $\det A(b_3/a_1) = 0$ and $\det A(b_3/a_2) = 0$ because each of these determinants has two columns which are the same [two columns of 1's], and $\det A(b_3/a_3) = \det A$. Then the identity in (28) can be simplified to the following.

$$[\det A]^2[\det B] = \begin{vmatrix} \det A(b_1/a_1) & \det A(b_1/a_2) \\ \det A(b_2/a_1) & \det A(b_2/a_2) \end{vmatrix} [\det A]. \tag{29}$$

If A is the 2-vector $\mathbf{x}$ in $X(x_0, \rho)$, then (29) can be written in more familiar notation and simplified as follows.

$$\Delta(\mathbf{x})\Delta_{(1,2)}(f^1, f^2)(\mathbf{x}) = \begin{vmatrix} \Delta_1 f^1(\mathbf{x}) & \Delta_2 f^1(\mathbf{x}) \\ \Delta_1 f^2(\mathbf{x}) & \Delta_2 f^2(\mathbf{x}) \end{vmatrix}. \tag{30}$$

Divide each side of this equation by $[\Delta(\mathbf{x})]^2$; divide the determinant on the right by $[\Delta(\mathbf{x})]^2$ by dividing each row of its matrix by $\Delta(\mathbf{x})$. Then (30) becomes

$$\frac{\Delta_{(1,2)}(f^1, f^2)(\mathbf{x})}{\Delta(\mathbf{x})} = \begin{vmatrix} \dfrac{\Delta_1 f^1(\mathbf{x})}{\Delta(\mathbf{x})} & \dfrac{\Delta_2 f^1(\mathbf{x})}{\Delta(\mathbf{x})} \\ \dfrac{\Delta_1 f^2(\mathbf{x})}{\Delta(\mathbf{x})} & \dfrac{\Delta_2 f^2(\mathbf{x})}{\Delta(\mathbf{x})} \end{vmatrix}. \tag{31}$$

3.12 Theorem. *Let $f: E \to \mathbb{R}^n$, $E \subset \mathbb{R}^n$, be a function whose components $(f^1, \cdots, f^n)$ are differentiable at x_0 in the open set E. Then f is differentiable at x_0 with respect to every class $X(x_0, \rho)$, $0 < \rho \leqq 1$, and*

$$D_{(1,\cdots,n)}f(x_0) = \det[D_j f^i(x_0)], \qquad i, j = 1, \cdots, n. \tag{32}$$

Proof. Let $X(x_0, \rho)$ be a class of n-vectors at x_0, and let $x_1, \cdots, x_n, x_0$ be the vertices of an n-vector $\mathbf{x}$ in this class. Apply the Bazin–Reiss–Picquet theorem [Theorem 3.10] to the following $(n+1) \times (n+1)$ matrices: A is the matrix $[x_i^j 1]$, and B is the matrix whose i-th row is $[f^1(x_i) \cdots f^n(x_i)\ 1]$, $i = 1, \cdots, n, 0$. The result can be simplified to

$$\frac{\Delta_{(1,\cdots,n)}f(\mathbf{x})}{\Delta(\mathbf{x})} = \det\left[\frac{\Delta_j f^i(\mathbf{x})}{\Delta(\mathbf{x})}\right], \qquad i, j = 1, \cdots, n. \tag{33}$$

Example 3.11 describes the details of this simplification in the special case $n = 2$. Each of the functions f^i is differentiable at x_0 with respect to every class $X(x_0, \rho)$ [see Corollary 3.7]; hence, each element in the matrix on the right in (33) has a limit as $\mathbf{x} \to x_0$. Furthermore, a determinant is a continuous function of the elements of its matrix; therefore,

$$\begin{aligned} \lim_{\mathbf{x}\to x_0} \frac{\Delta_{(1,\cdots,n)}f(\mathbf{x})}{\Delta(\mathbf{x})} &= \det\left[\lim_{\mathbf{x}\to x_0} \frac{\Delta_j f^i(\mathbf{x})}{\Delta(\mathbf{x})}\right] \\ &= \det[D_j f^i(x_0)], \qquad i, j = 1, \cdots, n. \end{aligned} \tag{34}$$

□

3.13 Corollary. *If $f: E \to \mathbb{R}^m$, $E \subset \mathbb{R}^n$, $1 \leqq m \leqq n$, is a function whose components $(f^1, \cdots, f^m)$ are differentiable at x_0 in E, then f is differentiable at x_0 with respect to every class $X(x_0, \rho)$, and*

$$D_{(j_1,\cdots,j_m)}(f^1, \cdots, f^m)(x_0) = \begin{vmatrix} D_{j_1} f^1(x_0) & \cdots & D_{j_m} f^1(x_0) \\ \cdots & \cdots & \cdots \\ D_{j_1} f^m(x_0) & \cdots & D_{j_m} f^m(x_0) \end{vmatrix}, \qquad (j_1, \cdots, j_m) \in (m/n). \tag{35}$$

PROOF. If $m = 1$, there is nothing to prove; if $m = n$, this corollary is Theorem 3.12; and if $1 < m < n$, Theorem 3.12 can be used to prove the corollary. The functions $x^j : \mathbb{R}^n \to \mathbb{R}$, $j = 1, \cdots, n$, such that $x^j(x) = x^j$ are differentiable at every point in $\mathbb{R}^n$ and

$$D_k x^j(x_0) = \begin{cases} 1 & \text{if } k = j, \quad j = 1, \cdots, n. \\ 0 & \text{if } k \neq j, \end{cases} \tag{36}$$

To prove these statements, observe that $\Delta_j x^j(\mathbf{x}) = \Delta(\mathbf{x})$, and that $\Delta_k x^j(\mathbf{x}) = 0$, $k \neq j$, since $\Delta_k x^j(\mathbf{x})$ is the determinant of a matrix which has two identical columns. Prove (35) first for $(j_1, \cdots, j_m) = (1, \cdots, m)$. Define a function $g : E \to \mathbb{R}^n$ as follows.

$$\begin{aligned} g^j(x) &= f^j(x), \qquad j = 1, \cdots, m, \\ g^j(x) &= x^j(x), \qquad j = m + 1, \cdots, n, \qquad x \in E. \end{aligned} \tag{37}$$

Then g is a function whose components $(g^1, \cdots, g^n)$ are differentiable at x_0, and g is differentiable at x_0 by Theorem 3.12. Also

$$D_{(1, \cdots, n)} g(x_0) = \det[D_j g^i(x_0)], \qquad i, j = 1, \cdots, n. \tag{38}$$

Now the definition of g in (37) and Definition 2.8 show that $D_{(1, \cdots, n)} g(x_0) = D_{(1, \cdots, m)}(f^1, \cdots, f^m)(x_0)$. Also, (37) and (36) show that

$$\begin{aligned} \det[D_j g^i(x_0)] &= \begin{vmatrix} D_1 f^1(x_0) & \cdots & D_m f^1(x_0) & 0 & \cdots & 0 \\ \vdots & \cdots & \vdots & \vdots & \cdots & \vdots \\ D_1 f^m(x_0) & \cdots & D_m f^m(x_0) & 0 & \cdots & 0 \\ 0 & \cdots & 0 & 1 & \cdots & 0 \\ \vdots & \cdots & \vdots & \vdots & \ddots & \vdots \\ 0 & \cdots & 0 & 0 & \cdots & 1 \end{vmatrix} \\ i, j = 1, \cdots, n, \quad & \\ &= \det[D_j f^i(x_0)], \\ &\quad i, j = 1, \cdots, m. \end{aligned}$$

This equation contains (35) in the special case $(j_1, \cdots, j_m) = (1, \cdots, m)$, and the proof of (35) for the other index sets $(j_1, \cdots, j_m)$ in (m/n) is similar. □

3.14 Theorem. *If $f : E \to \mathbb{R}$, $E \subset \mathbb{R}^n$, is differentiable at x_0, then f has partial derivatives $f_{x^j}(x_0)$, and*

$$f_{x^j}(x_0) = D_j f(x_0), \qquad j = 1, \cdots, n. \tag{39}$$

PROOF. Let $\mathbf{x}$ be the n-vector $(x_1, \cdots, x_n, x_0)$ in E such that $x_j = (x_0^1, \cdots, x_0^j + h^j, \cdots, x_0^n)$, $h^j \neq 0$, $j = 1, \cdots, n$. Then $x_j - x_0$ is parallel to the x^j-axis, $\mathbf{x}$ is an orthogonal n-vector, and $|\Delta(\mathbf{x})| = \Pi_1^n |h^j|$. Let $X_A(x_0, 1)$ denote the class of these n-vectors at x_0. Then $X_A(x_0, 1) \subset X(x_0, \rho)$ for every ρ such that $0 < \rho \leqq 1$. Since f is differentiable at x_0 by hypothesis, its derivatives can be evaluated with respect to $X_A(x_0, 1)$. The definitions of $D_j f(x_0)$ and $f_{x^j}(x_0)$ and elementary simplifications of determinants show that

$$\begin{aligned} \Delta(\mathbf{x}) &= h^1 \cdots h^j \cdots h^n, \\ \Delta_j f(\mathbf{x}) &= h^1 \cdots [f(x_j) - f(x_0)] \cdots h^n. \end{aligned} \qquad \mathbf{x} \in X_A(x_0, 1), \tag{40}$$

Then

$$\begin{aligned} D_j f(x_0) &= \lim_{\mathbf{x} \to x_0} \frac{\Delta_j f(\mathbf{x})}{\Delta(\mathbf{x})}, \qquad \mathbf{x} \in X(x_0, \rho), \\ &= \lim_{\mathbf{x} \to x_0} \frac{\Delta_j f(\mathbf{x})}{\Delta(\mathbf{x})}, \qquad \mathbf{x} \in X_A(x_0, 1), \\ &= \lim_{h^j \to 0} \frac{f(x_j) - f(x_0)}{h^j}, \qquad h^j \neq 0, \\ &= f_{x^j}(x_0). \end{aligned} \tag{41}$$

□

Let $f^1, \cdots, f^n$ be functions which have partial derivatives at x_0. The following equation defines the symbol on the left, which is called the Jacobian of $f^1, \cdots, f^n$ at x_0.

$$\frac{\partial(f^1, \cdots, f^n)}{\partial(x^1, \cdots, x^n)}(x_0) = \det[f^i_{x^j}(x_0)], \qquad i, j = 1, \cdots, n. \tag{42}$$

3.15 Corollary. *If $f: E \to \mathbb{R}^n$, $E \subset \mathbb{R}^n$, is a function whose components $(f^1, \cdots, f^n)$ are differentiable at x_0, then both the derivative and the Jacobian of these functions exist at x_0, and*

$$D_{(1, \cdots, n)}(f^1, \cdots, f^n)(x_0) = \frac{\partial(f^1, \cdots, f^n)}{\partial(x^1, \cdots, x^n)}(x_0). \tag{43}$$

Proof. Since $f^1, \cdots, f^n$ are differentiable at x_0 by hypothesis, the derivative $D_{(1, \cdots, n)}(f^1, \cdots, f^n)(x_0)$ exists by Theorem 3.12. Also, the partial derivatives $f^i_{x^j}(x_0)$ exist by Theorem 3.14, and equation (43) follows from equations (32), (42), and (39). Thus the derivative of $(f^1, \cdots, f^n)$ at x_0 equals the Jacobian of these functions at x_0. □

3.16 Example. Let A be the square $[-a, a] \times [-a, a]$ in the (x, y)-plane. Define the function $f: A \to \mathbb{R}$ as follows:

$$\begin{aligned} f(x, y) &= 1, \quad (x, y) \in A, \quad x > 0 \text{ and } y > 0; \\ &= 0, \quad (x, y) \in A, \quad x \leqq 0 \text{ or } y \leqq 0. \end{aligned} \tag{44}$$

The graph of this function is shown in Figure 3.1. Since f equals zero on the x-axis and on the y-axis, the partial derivatives of f exist at $(0, 0)$. The

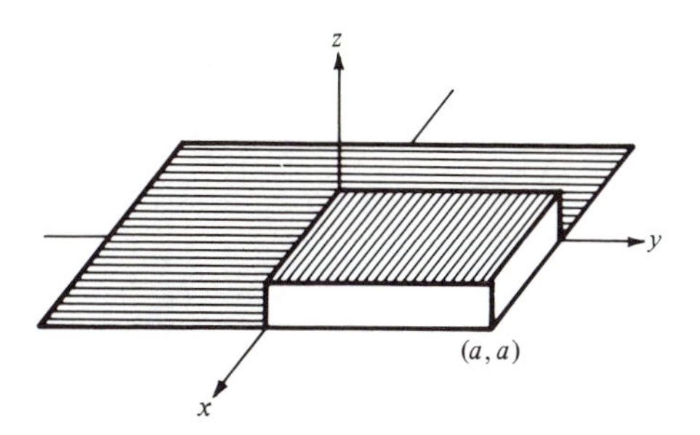

Figure 3.1. The graph of f in (44).

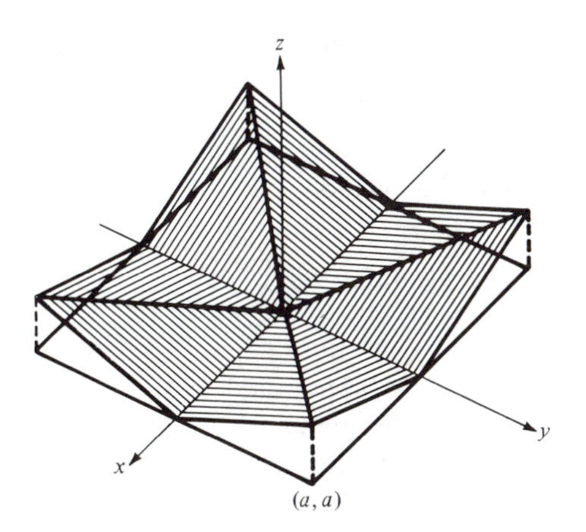

Figure 3.2. Graph of the function in Example 3.17.

derivatives $D_x f(0, 0)$ and $D_y f(0, 0)$ do not exist, however, since f is discontinuous at $(0, 0)$ [see Corollary 3.9]. Thus partial derivatives may exist at a point at which the function is discontinuous, but derivatives cannot exist at such a point.

3.17 Example. Let A be the square $[-a, a] \times [-a, a]$ in $\mathbb{R}^2$, and let $f: A \to \mathbb{R}$ be a function whose graph consists of eight triangles fitted together to make a continuous surface as shown in Figure 3.2. As the graph shows, f equals zero on the x-axis and on the y-axis; therefore the partial derivatives of f exist at the origin and $f_x(0, 0) = 0$, $f_y(0, 0) = 0$. But it is geometrically obvious from the figure that f is not differentiable at the origin although f is continuous and has partial derivatives there.

3.18 Example. If the components of f are differentiable, then f is differentiable [see Theorem 3.12], but an example shows that the converse is not true. Let $f: E \to \mathbb{R}^3$, $E \subset \mathbb{R}^3$, be a function which maps its domain E into a plane in $\mathbb{R}^3$ as indicated in Figure 3.3. The components (f^1, f^2, f^3) of f are not assumed to be differentiable nor even continuous; nevertheless,

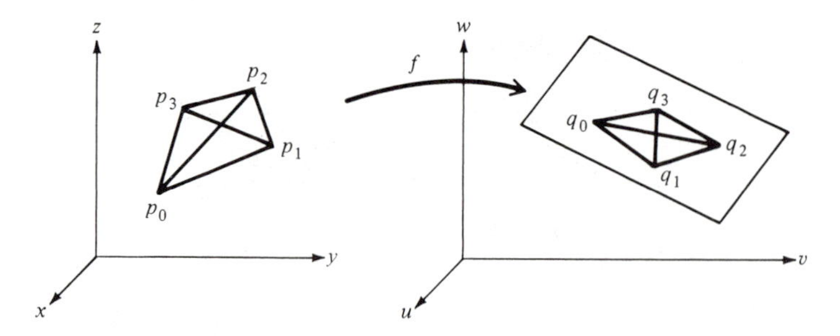

Figure 3.3. The function f maps E in $\mathbb{R}^3$ into a plane in $\mathbb{R}^3$.

$D_{(1,2,3)}f(x_0)$ exists and equals zero at each point x_0 in E. To prove this statement, let $\mathbf{x}$ be a 3-vector in $X(x_0, \rho)$ with vertices (p_1, p_2, p_3, p_0). Set $q_i = f(p_i)$, $i = 1, 2, 3, 0$. Then $(1/3!)|\Delta_{(1,2,3)}(f^1, f^2, f^3)(\mathbf{x})|$ is the volume of the tetrahedron with vertices q_1, q_2, q_3, q_0, and this volume is zero since the four vertices lie in a plane by hypothesis. Therefore,

$$D_{(1,2,3)}f(x_0) = \lim_{\mathbf{x} \to x_0} \frac{\Delta_{(1,2,3)}f(\mathbf{x})}{\Delta(\mathbf{x})} = 0. \tag{45}$$

Thus the derivative of f exists although its components may be highly pathological and certainly not differentiable.

The examples have shown that many functions $f: E \to \mathbb{R}$, $E \subset \mathbb{R}^n$, are not differentiable, but the next theorem shows that there is a very large class of such functions which are differentiable.

3.19 Theorem. *If $f: E \to \mathbb{R}$, $E \subset \mathbb{R}^n$, is a function which has partial derivatives f_{x^j}, $j = 1, \cdots, n$, in a neighborhood $N(x_0, r)$ in E, and if these partial derivatives are continuous at x_0, then f satisfies the Stolz condition at x_0, with $A_j = f_{x^j}(x_0)$ [see Definition 3.1], and f is therefore differentiable at x_0 with respect to each class $X(x_0, \rho)$, $0 < \rho \leqq 1$.*

PROOF. Let $(x_0^1, \cdots, x_0^n)$ be the coordinates of x_0, and let $(x^1, \cdots, x^n)$ be an arbitrary point x in $N(x_0, r)$; for convenience in notation designate x also by x_n. Define additional points $x_1, \cdots, x_{n-1}$ as follows: $x_j = (x^1, \cdots, x^j, x_0^{j+1}, \cdots, x_0^n)$, $j = 1, \cdots, n-1$. Then each of the points $x_0, x_1, \cdots, x_n$ is in $N(x_0, r)$ and

$$f(x_n) - f(x_0) = \sum_{j=1}^{n} [f(x_j) - f(x_{j-1})]. \tag{46}$$

Since x_{j-1} and x_j differ only in the j-th coordinate, the elementary mean-value theorem shows that

$$f(x_j) - f(x_{j-1}) = f_{x^j}(x_j^*)(x^j - x_0^j), \qquad j = 1, \cdots, n. \tag{47}$$

Here x_j^* is a point on the segment whose end-points are x_{j-1} and x_j. More precisely, $x_j^* = (x^1, \cdots, x^{j-1}, \xi^j, x_0^{j+1}, \cdots, x_0^n$, and ξ^j is a number between x_0^j and x^j if $x^j \neq x_0^j$. If $x^j = x_0^j$, set $\xi^j = x_0^j$; then (47) is a valid equation in all cases for $j = 1, \cdots, n$. Equations (46) and (47) show that

$$f(x_n) - f(x_0) = \sum_{j=1}^{n} f_{x^j}(x_0)(x^j - x_0^j) + \sum_{j=1}^{n} [f_{x^j}(x_j^*) - f_{x^j}(x_0)](x^j - x_0^j). \tag{48}$$

Define $r(f; x_0, x_n)$ as follows:

$$\begin{aligned} r(f; x_0, x_n) &= \frac{1}{|x_n - x_0|} \sum_{j=1}^{n} [f_{x^j}(x_j^*) - f_{x^j}(x_0)](x^j - x_0^j), \\ r(f; x_0, x_0) &= 0. \end{aligned} \tag{49}$$

Equations (48) and (49) show that

$$f(x_n) - f(x_0) = \sum_{j=1}^{n} f_{x^j}(x_0)(x^j - x_0^j) + r(f; x_0, x_n)|x_n - x_0|. \tag{50}$$

To complete the proof of the theorem, it is necessary and sufficient to prove that [see Definition 3.1]

$$\lim_{x_n \to x_0} r(f; x_0, x_n) = 0. \tag{51}$$

Since $|x^j - x_0^j| \leqq |x_n - x_0|, j = 1, \cdots, n$, equation (49) shows that

$$|r(f; x_0, x_n)| \leqq \sum_{j=1}^{n} |f_{x^j}(x_j^*) - f_{x^j}(x_0)|; \tag{52}$$

also, Schwarz's inequality [see (7) and Corollary 86.2 in Section 86] shows that

$$|r(f; x_0, x_n)| \leqq \left\{ \sum_{j=1}^{n} |f_{x^j}(x_j^*) - f_{x^j}(x_0)|^2 \right\}^{1/2}. \tag{53}$$

When $x_n \to x_0$, each $x^j \to x_0^j$, and the definition of x_j^* shows that $x_j^* \to x_0$. But the partial derivatives $f_{x^j}, j = 1, \cdots, n$, are continuous at x_0 by hypothesis. Then

$$\lim_{x_n \to x_0} |f_{x^j}(x_j^*) - f_{x^j}(x_0)| = 0, \qquad j = 1, \cdots, n, \tag{54}$$

and each of the inequalities (52) and (53) shows that (51) is true. The proof of Theorem 3.19 is complete. □

3.20 Corollary. *If $f: E \to \mathbb{R}^m$, $E \subset \mathbb{R}^n$, $1 \leqq m \leqq n$, is a function whose components $(f^1, \cdots, f^m)$ have partial derivatives in a neighborhood of x_0 in E which are continuous at x_0, then f is differentiable at x_0 with respect to every class $X(x_0, \rho)$, $0 < \rho \leqq 1$, and*

$$D_{(j_1, \cdots, j_m)}(f^1, \cdots, f^m)(x_0) = \begin{vmatrix} D_{j_1} f^1(x_0) & \cdots & D_{j_m} f^1(x_0) \\ \cdots & \cdots & \cdots \\ D_{j_1} f^m(x_0) & \cdots & D_{j_m} f^m(x_0) \end{vmatrix}, \tag{55}$$

$$(j_1, \cdots, j_m) \in (m/n).$$

Proof. Each of the components $(f^1, \cdots, f^m)$ is differentiable at x_0 by Theorem 3.19, and f is differentiable at x_0 with respect to every class $X(x_0, \rho)$, $0 < \rho \leqq 1$, by Corollary 3.13. Formula (55) is (35) in Corollary 3.13. □

3.21 Theorem. *Let $f: E \to \mathbb{R}$, $g: E \to \mathbb{R}$, $E \subset \mathbb{R}^n$, be two functions which are differentiable at x_0 in E, and let c be a constant. Then:*

A constant function is differentiable, and all of its derivatives are zero. (56)

$f + g$ is differentiable at x_0, and

$$D_j(f+g)(x_0) = D_j f(x_0) + D_j g(x_0), \qquad j = 1, \cdots, n. \tag{57}$$

cf is differentiable at x_0, and

$$D_j(cf)(x_0) = cD_j f(x_0), \qquad j = 1, \cdots, n. \tag{58}$$

fg is differentiable at x_0, and

$$D_j(fg)(x_0) = f(x_0)D_j g(x_0) + g(x_0)D_j f(x_0), \qquad j = 1, \cdots, n. \tag{59}$$

f/g is differentiable at x_0 if $g(x_0) \neq 0$, and

$$D_j(f/g)(x_0) = \frac{g(x_0)D_j f(x_0) - f(x_0)D_j g(x_0)}{[g(x_0)]^2}, \qquad j = 1, \cdots, n. \tag{60}$$

Proof. If f is constant and equal to c, then $\Delta_j f(\mathbf{x})$ has a column of c's; multiply the column of 1's by c and subtract it from the column of c's. Then $\Delta_j f(\mathbf{x})$ has a column of 0's and its value is zero. Hence, $D_j f(x) = 0$, $x \in E$, and the statement in (56) is true.

In (57), the j-th column of the matrix of the determinant $\Delta_j(f+g)(\mathbf{x})$ is $f(x_i) + g(x_i)$, $i = 1, \cdots, n, 0$. Then by Theorem 77.3 and 77.17, $\Delta_j(f+g)(\mathbf{x}) = \Delta_j f(\mathbf{x}) + \Delta_j g(\mathbf{x})$ and

$$D_j(f+g)(x_0) = \lim_{\mathbf{x} \to x_0} \frac{\Delta_j(f+g)(\mathbf{x})}{\Delta(\mathbf{x})}$$

$$= \lim_{\mathbf{x} \to x_0} \frac{\Delta_j f(\mathbf{x})}{\Delta(\mathbf{x})} + \lim_{\mathbf{x} \to x_0} \frac{\Delta_j g(\mathbf{x})}{\Delta(\mathbf{x})} = D_j f(x_0) + D_j g(x_0)$$

for $j = 1, \cdots, n$. Hence, (57) is true.

In (58), $\Delta_j(cf)(\mathbf{x}) = c\Delta_j f(\mathbf{x})$ by Theorems 77.1 and 77.17 for determinants. Then

$$D_j(cf)(x_0) = \lim_{\mathbf{x} \to x_0} \frac{\Delta_j(cf)(\mathbf{x})}{\Delta(\mathbf{x})} = c \lim_{\mathbf{x} \to x_0} \frac{\Delta_j f(\mathbf{x})}{\Delta(\mathbf{x})} = cD_j f(x_0), \qquad j = 1, \cdots, n.$$

In (59), subtract the last row in the matrix of $\Delta_j(fg)(\mathbf{x})$ from each of the other rows and expand. Then the element in the i-th row and j-th column is $f(x_i)g(x_i) - f(x_0)g(x_0)$, $i = 1, \cdots, n$. Use the following obvious identity to evaluate these elements in the j-th column.

$$f(x_i)g(x_i) - f(x_0)g(x_0) = f(x_0)[g(x_i) - g(x_0)] + g(x_0)[f(x_i) - f(x_0)] + [f(x_i) - f(x_0)][g(x_i) - g(x_0)]. \tag{61}$$

Then by elementary properties of determinants in Section 77,

$$\frac{\Delta_j(fg)(\mathbf{x})}{\Delta(\mathbf{x})} = f(x_0)\frac{\Delta_j g(\mathbf{x})}{\Delta(\mathbf{x})} + g(x_0)\frac{\Delta_j f(\mathbf{x})}{\Delta(\mathbf{x})} + \frac{\det M_j}{\Delta(\mathbf{x})}. \tag{62}$$

Here the j-th column of the matrix M_j contains the terms $[f(x_i) - f(x_0)] \times [g(x_i) - g(x_0)]$. Since $\mathbf{x}$ is in $X(x_0, \rho)$, then

$$\left|\frac{\det M_j}{\Delta(\mathbf{x})}\right| \leq \frac{|\det M_j|}{\rho \Pi_1^n |x_i - x_0|}. \tag{63}$$

In the expression on the right, divide the denominator by $\Pi_1^n |x_i - x_0|$ and divide the numerator by the same product by dividing the i-th row of M_j by $|x_i - x_0|$. All elements not in the j-th column are now equal to or less than 1 in absolute value, and the j-th column of the numerator contains the following term in the i-th row:

$$\frac{[f(x_i) - f(x_0)][g(x_i) - g(x_0)]}{|x_i - x_0|}, \qquad i = 1, \cdots, n. \tag{64}$$

Since f and g are differentiable and hence continuous at x_0, then

$$\lim_{x_i \to x_0} \frac{|f(x_i) - f(x_0)|}{|x_i - x_0|} \leq \left[\sum_{j=1}^{n} |D_j f(x_0)|^2\right]^{1/2}, \qquad \lim_{x_i \to x_0} |g(x_i) - g(x_0)| = 0, \tag{65}$$

by (20) in Corollary 3.9. The limits in (65) show that the limit of (64), as $x_i \to x_0$, is zero for $i = 1, \cdots, n$. These facts show that the limit, as $\mathbf{x} \to x_0$, of the expression on the right in (63) is zero; hence,

$$\lim_{\mathbf{x} \to x_0} \frac{\det M_j}{\Delta(\mathbf{x})} = 0, \qquad j = 1, \cdots, n. \tag{66}$$

Take the limit in (62) as $\mathbf{x} \to x_0$. Since f and g are differentiable at x_0, then (66) shows that $D_j(fg)(x_0)$ exists, and that the value of this derivative is given by the formula in (59).

Finally, consider the proof of (60). The first step is to prove the following special case of (60): the function $1/g$ is differentiable at x_0 and

$$D_j(1/g)(x_0) = -\frac{D_j g(x_0)}{[g(x_0)]^2}, \qquad j = 1, \cdots, n, \qquad g(x_0) \neq 0. \tag{67}$$

Then since $f/g = f(1/g)$, the function f/g is differentiable by (59), and formula (67) can be used with the formula in (59) for the derivative of a product to find the values of its derivatives. Thus

$$\begin{aligned} D_j(f/g)(x_0) &= f(x_0)\left\{-\frac{D_j g(x_0)}{[g(x_0)]^2}\right\} + \frac{1}{g(x_0)} D_j f(x_0) \\ &= \frac{g(x_0) D_j f(x_0) - f(x_0) D_j g(x_0)}{[g(x_0)]^2}. \end{aligned} \tag{68}$$

Thus the proof of (60) can be completed by proving (67).

Since $D_j g(x_0)$ exists, g is continuous at x_0. Then since $g(x_0) \neq 0$, there is a neighborhood $N(x_0, r)$ of x_0 in E in which $|g(x)| \geqq |g(x_0)|/2$ [see the proof of Theorem 96.9]. Let $\mathbf{x}$ be an n-vector in $X(x_0, \rho)$ and in $N(x_0, r)$. Subtract the last row in the matrix of $\Delta_j(1/g)(\mathbf{x})$ from each of the other rows and expand. Then the element in the i-th row and j-th column is $[1/g(x_i)] - [1/g(x_0)]$, $i = 1, \cdots, n$. Use the following easily verified identity to evaluate these elements in the j-th column.

$$\frac{1}{g(x_i)} - \frac{1}{g(x_0)} = -\frac{[g(x_i) - g(x_0)]}{[g(x_0)]^2} + \frac{[g(x_i) - g(x_0)]^2}{[g(x_0)]^2 g(x_i)}. \tag{69}$$

Then by elementary properties of determinants in Section 77,

$$\frac{\Delta_j(1/g)(\mathbf{x})}{\Delta(\mathbf{x})} = -\frac{1}{[g(x_0)]^2}\frac{\Delta_j g(\mathbf{x})}{\Delta(\mathbf{x})} + \frac{1}{[g(x_0)]^2}\frac{\det N_j}{\Delta(\mathbf{x})}. \tag{70}$$

Here the j-th column of N_j contains the terms $[g(x_i) - g(x_0)]^2/g(x_i)$, $i = 1, \cdots, n$. Then by methods similar to those used in proving (59), and by using the facts that

$$\lim_{x_i \to x_0} \frac{[g(x_i) - g(x_0)]^2}{|x_i - x_0|} = 0, \qquad \lim_{x_i \to x_0} g(x_i) = g(x_0) \neq 0, \tag{71}$$

it is easy to prove that

$$\lim_{\mathbf{x} \to x_0} \frac{\det \mathrm{N}_j}{\Delta(\mathbf{x})} = 0. \tag{72}$$

Take the limit in (70) as $\mathbf{x} \to x_0$. Since g is differentiable at x_0, then (72) shows that $D_j(1/g)(x_0)$ exists, and that the value of this derivative is given by the formula in (67). The proof of (67) completes the proof of (60) and of all parts of Theorem 3.21. □

Exercises

3.1. Define a function $f: \mathbb{R}^2 \to \mathbb{R}$ as follows:

$$f(x, y) = \frac{2xy}{x^2 + y^2}, \qquad x^2 + y^2 \neq 0,$$

$$f(0, 0) = 0.$$

Show that f has partial derivatives at every point (x, y) in $\mathbb{R}^2$, including $(0, 0)$. Show also that f is not differentiable at $(0, 0)$, but that it is differentiable at every other point in $\mathbb{R}^2$. [Hints. Example 1.1, Corollary 3.9, Exercises 2.1 and 2.2, and Theorem 3.21.]

3.2. Let $f: \mathbb{R}^3 \to \mathbb{R}$ be the function such that

$$f(x, y, z) = \frac{xy + xz + yz}{x^2 + y^2 + z^2}, \qquad x^2 + y^2 + z^2 \neq 0,$$

$$f(0, 0, 0) = 0.$$

(a) Show that f is discontinuous at $(0, 0, 0)$, and that it is not differentiable at $(0, 0, 0)$. Prove that f is differentiable at every point in $\mathbb{R}^3$ except $(0, 0, 0)$. [Hints. Corollary 3.9, Theorems 3.19 and 3.21.]
(b) Show that f has partial derivatives at $(0, 0, 0)$ and at all other points in $\mathbb{R}^3$. [Hint. Theorem 3.14.]

3.3. Let $f: \mathbb{R}^2 \to \mathbb{R}$ be the function such that $f(x, y) = |xy|^{1/2}$. Show that f is continuous at $(0, 0)$, and that the two partial derivatives of f exist and have the value 0 at $(0, 0)$. Finally, show that f is not differentiable at $(0, 0)$ since $D_1 f(0, 0)$ and $D_2 f(0, 0)$ do not exist. [Hints. Prove that $|xy|^{1/2} \leqq |x^2 + y^2|^{1/2}$. Definition 96.2.]

3.4. Let $f: \mathbb{R}^2 \to \mathbb{R}$ be the function such that $f(x, y) = |x| + |y|$. Show that f is continuous everywhere, but that the partial derivatives of f do not exist at $(0, 0)$. Find all the points at which this function has partial derivatives; at which it is differentiable.

3.5. Let $r = [(x^1)^2 + (x^2)^2]^{1/2}$ and define $f: \mathbb{R}^2 \to \mathbb{R}$ to be the function such that

$$\begin{aligned} f(x^1, x^2) &= r^n \sin(1/r), && r \neq 0, \\ &= 0, && r = 0. \end{aligned}$$

In each case below, sketch f as a function of r, for r greater than zero (the graph of f as a function of (x^1, x^2) is just this graph rotated around the vertical axis).
(a) If $n = 0$, show that f is discontinuous at $x_0 = (0, 0)$, that f does not have partial derivatives at x_0, and that f is not differentiable at x_0.
(b) If $n = 1$, show that f is continuous at x_0, but that the partial derivatives of f do not exist at x_0. Show that f is not differentiable at x_0.
(c) If $n = 2$, show that f has partial derivatives at every point (x^1, x^2) in $\mathbb{R}^2$, and that these partial derivatives are discontinuous at $x_0 = (0, 0)$ but continuous elsewhere. Nevertheless, show that f is differentiable at x_0, and thus that the sufficient condition for differentiability stated in Theorem 3.19 is not necessary.
(d) If $n = 3$, show that f has partial derivatives at every point in $\mathbb{R}^2$, that the partial derivatives of f are continuous at every point in $\mathbb{R}^2$, and that f is differentiable at every point in $\mathbb{R}^2$.

3.6. Define a function $(f^1, f^2): \mathbb{R}^2 \to \mathbb{R}^2$ as follows. Let n be a non-negative integer, let a_{ij}, $i, j = 1, 2$, be constants, and let $r = [(x^1)^2 + (x^2)^2]^{1/2}$. Set

$$\begin{aligned} f^1(x^1, x^2) &= a_{11}x^1 + a_{12}x^2 + r^n \sin(1/r), \\ f^2(x^1, x^2) &= a_{21}x^1 + a_{22}x^2 + r^n \sin(1/r), \qquad r \neq 0, \end{aligned}$$

and $f^1(0, 0) = f^2(0, 0) = 0$. If $x_0 = (0, 0)$, determine the values of n for which $D_{(1,2)}(f^1, f^2)(x_0)$ exists, and find the value of this derivative. [Hints. Definition 3.1, Theorem 3.12, Exercise 3.5.]

3.7. Let $f: E \to \mathbb{R}^n$, $E \subset \mathbb{R}^n$, have components $f^1, \cdots, f^n$, and let a_j, $j = 0, 1, \cdots, n$, be constants such that $\Sigma_1^n a_j^2 \neq 0$ and

$$\sum_{j=1}^{n} a_j f^j(x) = a_0, \qquad x \in E.$$

(a) If $\mathbf{x}$ is in $X(x_0, \rho)$ and in E, prove that $\Delta_{(1,\cdots,n)}f(\mathbf{x}) = 0$, and thus that $D_{(1,\cdots,n)}f(x_0)$ exists and equals zero at each x_0 in E. Give an algebraic proof but also a geometric explanation of this result. [Hint. Example 3.18.]

(b) Assume that $f^1, \cdots, f^n$ are differentiable in E and give a second proof of the result in (a). [Hint. Differentiate the two sides of the equation $\Sigma_1^n a_j f^j(x) = a_0$ with respect to $x^1, \cdots, x^n$ and consider the resulting system of equations. Apply Theorem 3.12.]

3.8. Let $f: E \to \mathbb{R}$, $E \subset \mathbb{R}^n$, be a function which is differentiable at $x_0 = (x_0^1, \cdots, x_0^m, \cdots, x_0^n)$ in E, and let g be a function of $x^1, \cdots, x^m$ which is defined as follows:

$$g(x^1, \cdots, x^m) = f(x^1, \cdots, x^m, x_0^{m+1}, \cdots, x_0^n),$$

$$(x^1, \cdots, x^m, x_0^{m+1}, \cdots, x_0^n) \in E.$$

Show that g is differentiable at $(x_0^1, \cdots, x_0^m)$; express its derivatives in terms of derivatives of f. In the case $m = 1$, relate your results to Theorem 3.14.

3.9. Let $f: E \to \mathbb{R}$, $E \subset \mathbb{R}^n$, be a function which is differentiable at $x_0 = (x_0^1, \cdots, x_0^n)$ in E, and let g be a function of $(x^1, \cdots, x^n, \cdots, x^m)$ in $\mathbb{R}^m$ defined as follows.

$$g(x^1, \cdots, x^n, \cdots, x^m) = f(x^1, \cdots, x^n), \qquad (x^1, \cdots, x^n) \in E.$$

Show that g is differentiable at every point x in $\mathbb{R}^m$ such that $x = (x_0^1, \cdots, x_0^n, x^{n+1}, \cdots, x^m)$, and that

$$\begin{aligned} D_j g(x) &= D_j f(x_0), &\quad j &= 1, \cdots, n, \\ &= 0, & j &= n+1, \cdots, m. \end{aligned}$$

3.10. Let $f: E \to \mathbb{R}^2$, $E \subset \mathbb{R}^2$, be a function whose components (f^1, f^2) are differentiable at x_0 in E, and assume that f is differentiable at x_0.

(a) Without using the Bazin–Reiss–Picquet theorem, prove that

$$D_{(1,2)}(f_1, f_2)(x_0) = \det[D_j f^i(x_0)], \qquad i, j = 1, 2.$$

(b) Explain what the Bazin–Reiss–Picquet theorem contributes to the proof of Theorem 3.12.

[Outline of the proof in (a). Let $X_A(x_0, 1)$ be the special class of 2-vectors described in the proof of Theorem 3.14. Since $D_{(1,2)}(f^1, f^2)(x_0)$ exists by hypothesis, then

$$D_{(1,2)}(f^1, f^2)(x_0) = \lim_{\mathbf{x} \to x_0} \frac{\Delta_{(1,2)}f(\mathbf{x})}{\Delta(\mathbf{x})}, \qquad \mathbf{x} \in X_A(x_0, 1) \subset X(x_0, \rho).$$

Let $\mathbf{x}$ be the 2-vector (x_1, x_2, x_0) such that $x_1 = (x_0^1 + h^1, x_0^2)$, $x_2 = (x_0^1, x_0^2 + h^2)$ $h^1 \neq 0$, $h^2 \neq 0$. For this $\mathbf{x}$ in $X_A(x_0, 1)$, show that

$$\frac{\Delta_{(1,2)}(f^1, f^2)(\mathbf{x})}{\Delta(\mathbf{x})}$$

$$= \begin{vmatrix} \dfrac{f^1(x_0^1 + h^1, x_0^2) - f^1(x_0^1, x_0^2)}{h^1} & \dfrac{f^2(x_0^1 + h^1, x_0^2) - f^2(x_0^1, x_0^2)}{h^1} \\ \dfrac{f^1(x_0^1, x_0^2 + h^2) - f^1(x_0^1, x_0^2)}{h^2} & \dfrac{f^2(x_0^1, x_0^2 + h^2) - f^2(x_0^1, x_0^2)}{h^2} \end{vmatrix}.$$

Use Theorem 3.14 and Theorem 77.14.]

3.11. Let $f: E \to \mathbb{R}^2$, $E \subset \mathbb{R}^2$, be a function whose components (f^1, f^2) are differentiable at x_0 in E. Without using the Bazin–Reiss–Picquet theorem, prove that f is differentiable at x_0 and that

$$D_{(1,2)}(f^1, f^2)(x_0) = \det[D_j f^i(x_0)], \qquad i, j = 1, 2.$$

[Outline of proof. Since f^1, f^2 are differentiable at x_0, then f^1, f^2 satisfy the Stolz condition at x_0 by Theorem 3.6 and

$$f^1(x) - f^1(x_0) = \sum_{j=1}^{2} D_j f^1(x_0)(x^j - x_0^j) + r(f^1; x_0, x)|x - x_0|,$$

$$f^2(x) - f^2(x_0) = \sum_{j=1}^{2} D_j f^2(x_0)(x^j - x_0^j) + r(f^2; x_0, x)|x - x_0|.$$

Let $\mathbf{x}$ be a 2-vector in $X(x_0, \rho)$. Then

$$\Delta_{(1,2)}(f^1, f^2)(\mathbf{x})$$

$$= \begin{vmatrix} f^1(x_1) - f^1(x_0) & f^2(x_1) - f^2(x_0) \\ f^1(x_2) - f^1(x_0) & f^2(x_2) - f^2(x_0) \end{vmatrix}$$

$$= \begin{vmatrix} \Sigma_1^2 D_j f^1(x_0)(x_1^j - x_0^j) & \Sigma_1^2 D_j f^2(x_0)(x_1^j - x_0^j) \\ \Sigma_1^2 D_j f^1(x_0)(x_2^j - x_0^j) & \Sigma_1^2 D_j f^2(x_0)(x_2^j - x_0^j) \end{vmatrix} \qquad [= \det A]$$

$$+ \begin{vmatrix} \Sigma_1^2 D_j f^1(x_0)(x_1^j - x_0^j) & r(f^2; x_0, x_1)|x_1 - x_0| \\ \Sigma_1^2 D_j f^1(x_0)(x_2^j - x_0^j) & r(f^2; x_0, x_2)|x_2 - x_0| \end{vmatrix} \qquad [= \det B]$$

$$+ \begin{vmatrix} r(f^1; x_0, x_1)|x_1 - x_0| & f^2(x_1) - f^2(x_0) \\ r(f^1; x_0, x_2)|x_2 - x_0| & f^2(x_2) - f^2(x_0) \end{vmatrix}. \qquad [= \det C]$$

Thus

$$\frac{\Delta_{(1,2)}(f^1, f^2)(\mathbf{x})}{\Delta(\mathbf{x})} = \frac{\det A}{\Delta(\mathbf{x})} + \frac{\det B}{\Delta(\mathbf{x})} + \frac{\det C}{\Delta(\mathbf{x})}.$$

Use the multiplication theorem for determinants in Theorem 80.1 to prove that

$$\det A = \det[D_j f^i(x_0)]\Delta(\mathbf{x}), \qquad i, j = 1, 2.$$

Use the fact that $\mathbf{x}$ satisfies the regularity condition at x_0 to prove that

$$\lim_{\mathbf{x} \to x_0} \frac{\det B}{\Delta(\mathbf{x})} = 0, \qquad \lim_{\mathbf{x} \to x_0} \frac{\det C}{\Delta(\mathbf{x})} = 0.$$

These proofs are similar to the proof of Theorem 3.3. Then

$$\lim_{\mathbf{x} \to x_0} \frac{\Delta_{(1,2)}(f^1, f^2)(\mathbf{x})}{\Delta(\mathbf{x})} = \det[D_j f^i(x_0)], \qquad i, j = 1, 2.]$$

3.12. Let A and B be $n \times n$ matrices, and assume that $\det B \neq 0$.

(a) Show that there is an $n \times n$ matrix C such that $A = BC$. This equation expresses the columns of A as linear combinations of the columns of B.

(b) Use the Binet–Cauchy multiplication theorem for determinants [Theorem 80.1] to show that $\det A = \det B \det C$.

(c) Find the elements of C by Cramer's rule and substitute their values in the equation $\det A = \det B \det C$. Show that the result is the Bazin–Reiss–Picquet theorem in Theorem 3.10 and a special case of the general theorem in Theorem 83.1.

4. Derivatives of Composite Functions

This section investigates the differentiability of the composition of two functions; precise statements of the results are contained in two theorems and several formulas (the chain rule).

Let G be a neighborhood of a point x_0 in $\mathbb{R}^n$, and let $x:(x^1, \cdots, x^n)$ denote points in G. Let H be a neighborhood of y_0 in $\mathbb{R}^m$, and let $y:(y^1, \cdots, y^m)$ denote points in H. Let $g: G \to H$ be a function with components $(g^1, \cdots, g^m)$, and assume that $y_0 = g(x_0)$. Then the composition of $h: H \to \mathbb{R}$ and $g: G \to H$ is the function $h \circ g: G \to \mathbb{R}$ whose value at x is $h(g^1(x), \cdots, g^m(x))$.

4.1 Theorem. *If $g^1, \cdots, g^m$ are differentiable at x_0 and h is differentiable at y_0, then $h \circ g$ is differentiable at x_0 and (the chain rule)*

$$D_j h \circ g(x_0) = \sum_{i=1}^{m} D_i h(y_0) D_j g^i(x_0), \qquad j = 1, \cdots, n. \tag{1}$$

Proof. Let $X(x_0, \rho)$ be a class of n-vectors $\mathbf{x}$ at x_0. Then $D_j h \circ g(x_0)$ exists if and only if

$$\lim_{\mathbf{x} \to x_0} \frac{\Delta_j h \circ g(\mathbf{x})}{\Delta(\mathbf{x})}, \qquad \mathbf{x} \in X(x_0, \rho), \tag{2}$$

exists; if this limit exists, its value is $D_j h \circ g(x_0)$. In the matrix of the determinant in the numerator, subtract the last row from each of the preceding rows and then expand by elements in the last column. The element in the r-th row and j-th column of the resulting matrix is $h \circ g(x_r) - h \circ g(x_0)$, and

$$\Delta_j h \circ g(\mathbf{x}) = \begin{vmatrix} x_1^1 - x_0^1 & \cdots & h \circ g(x_1) - h \circ g(x_0) & \cdots & x_1^n - x_0^n \\ \cdots & \cdots & \cdots & \cdots & \cdots \\ x_r^1 - x_0^1 & \cdots & h \circ g(x_r) - h \circ g(x_0) & \cdots & x_r^n - x_0^n \\ \cdots & \cdots & \cdots & \cdots & \cdots \\ x_n^1 - x_0^1 & \cdots & h \circ g(x_n) - h \circ g(x_0) & \cdots & x_n^n - x_0^n \end{vmatrix} \tag{3}$$

Since h is differentiable at y_0, it satisfies the Stolz condition at y_0 and

$$\begin{aligned} &h \circ g(x_r) - h \circ g(x_0) \\ &\quad = \sum_{i=1}^{m} D_i h(y_0)[g^i(x_r) - g^i(x_0)] + r[h; g(x_0), g(x_r)]|g(x_r) - g(x_0)|. \end{aligned} \tag{4}$$

By using (3) and this relation, the expression in (2) can be transformed into this:

$$\lim_{\mathbf{x}\to x_0} \sum_{i=1}^{m} D_i h(y_0) \frac{\Delta_j g^i(\mathbf{x})}{\Delta(\mathbf{x})} + \lim_{\mathbf{x}\to x_0} \frac{\det R}{\Delta(\mathbf{x})}, \qquad j = 1, \cdots, n. \tag{5}$$

Since the components of g are differentiable at x_0 by hypothesis, the first limit is the sum on the right in (1), and the proof can be completed by showing that the value of the second limit is zero. The element in the r-th row and j-th column of R is $r[h; g(x_0), g(x_r)]|g(x_r) - g(x_0)|$. In the denominator, apply the regularity condition satisfied by $\mathbf{x}$. For $r = 1, \cdots, n$, divide the denominator and the r-th row of R by $|x_r - x_0|$. Since $g^1, \cdots, g^m$ are differentiable at x_0 by hypothesis, they are continuous at x_0 by Corollary 3.9. Then g is continuous at x_0, and $g(x_r) \to g(x_0)$ as $x_r \to x_0$. Finally, h is differentiable at $g(x_0)$ by hypothesis, and $r[h; g(x_0), g(x_r)] \to 0$ as $\mathbf{x} \to x_0$. Since

$$|g(x_r) - g(x_0)| = \left\{ \sum_{i=1}^{m} [g^i(x_r) - g^i(x_0)]^2 \right\}^{1/2}, \tag{6}$$

and since $g^1, \cdots, g^m$ are differentiable at x_0, then Corollary 3.9 shows that

$$\frac{|g(x_r) - g(x_0)|}{|x_r - x_0|} \tag{7}$$

is bounded as $x_r \to x_0$. These facts prove that

$$\lim_{x_n \to x_0} \frac{r[h; g(x_0), g(x_r)]|g(x_r) - g(x_0)|}{|x_r - x_0|} = 0. \tag{8}$$

The absolute value of the elements in columns k, $k \neq j$, is equal to or less than 1. These facts and Hadamard's determinant theorem [see (19) in Corollary 87.2] prove that the second limit in (5) exists and has the value zero. This statement completes the proof of (1). □

4.2 Example. Let G be a neighborhood of x_0 in $\mathbb{R}^2$ and let H be a neighborhood of y_0 in $\mathbb{R}^2$. Let $g: G \to H$ be a function whose components (g^1, g^2) are differentiable at x_0, and assume that $g(x_0) = y_0$. Let $h: H \to \mathbb{R}^2$ be a function whose components (h^1, h^2) are differentiable at y_0. Then the composition of h and g is a function $h \circ g$ which maps G into $\mathbb{R}^2$, and $h \circ g$ has components $h^1 \circ g: G \to \mathbb{R}$ and $h^2 \circ g: G \to \mathbb{R}$ which are differentiable at x_0 by Theorem 4.1. Then $h \circ g: G \to \mathbb{R}^2$ is differentiable at x_0 by Theorem 3.12 and

$$D_{(1,2)} h \circ g(x_0) = \begin{vmatrix} D_1 h^1 \circ g(x_0) & D_2 h^1 \circ g(x_0) \\ D_1 h^2 \circ g(x_0) & D_2 h^2 \circ g(x_0) \end{vmatrix} \tag{9}$$

by (32) in Theorem 3.12. Then by (1) in Theorem 4.1,

$$D_{(1,2)} h \circ g(x_0) = \begin{vmatrix} \sum_{i=1}^{2} D_i h^1(y_0) D_1 g^i(x_0) & \sum_{i=1}^{2} D_i h^1(y_0) D_2 g^i(x_0) \\ \sum_{i=1}^{2} D_i h^2(y_0) D_1 g^i(x_0) & \sum_{i=1}^{2} D_i h^2(y_0) D_2 g^i(x_0) \end{vmatrix}. \tag{10}$$

Then the Binet–Cauchy multiplication theorem [see Theorem 80.1] shows that

$$D_{(1,2)}h\circ g(x_0)=\begin{vmatrix} D_1h^1(y_0) & D_2h^1(y_0)\\ D_1h^2(y_0) & D_2h^2(y_0)\end{vmatrix}\begin{vmatrix} D_1g^1(x_0) & D_2g^1(x_0)\\ D_1g^2(x_0) & D_2g^2(x_0)\end{vmatrix}. \tag{11}$$

Finally, (32) in Theorem 3.12 shows that

$$D_{(1,2)}h\circ g(x_0)=D_{(1,2)}h(y_0)D_{(1,2)}g(x_0). \tag{12}$$

4.3 Example. Let G be a neighborhood of x_0 in $\mathbb{R}^2$, and let H be a neighborhood of y_0 in $\mathbb{R}^3$. Let $g: G\to H$ be a function whose components (g^1, g^2, g^3) are differentiable at x_0, and assume that $g(x_0)=y_0$. Let $h: H\to\mathbb{R}^2$ be a function whose components (h^1, h^2) are differentiable at y_0. Then the composition of h and g is a function $h\circ g$ which maps G into $\mathbb{R}^2$, and $h\circ g$ has components $h^1\circ g$ and $h^2\circ g$ which are differentiable at x_0 by Theorem 4.1. Then $h\circ g$ is differentiable at x_0 by Theorem 3.12, and the formula in (9) is correct as before. There is a change in (10), however, since h is now defined on a set H in $\mathbb{R}^3$. Equation (1) in Theorem 4.1 shows that the correct formula is

$$D_{(1,2)}h\circ g(x_0)=\begin{vmatrix} \sum_{i=1}^{3} D_ih^1(y_0)D_1g^i(x_0) & \sum_{i=1}^{3} D_ih^1(y_0)D_2g^i(x_0)\\ \sum_{i=1}^{3} D_ih^2(y_0)D_1g^i(x_0) & \sum_{i=1}^{3} D_ih^2(y_0)D_2g^i(x_0)\end{vmatrix}. \tag{13}$$

The definition of the product of two matrices shows that $D_{(1,2)}h\circ g(x_0)$ is the determinant of the following matrix product:

$$\begin{bmatrix} D_1h^1(y_0) & D_2h^1(y_0) & D_3h^1(y_0)\\ D_1h^2(y_0) & D_2h^2(y_0) & D_3h^2(y_0)\end{bmatrix}\begin{bmatrix} D_1g^1(x_0) & D_2g^1(x_0)\\ D_1g^2(x_0) & D_2g^2(x_0)\\ D_1g^3(x_0) & D_2g^3(x_0)\end{bmatrix}. \tag{14}$$

Then the Binet–Cauchy multiplication theorem [see Theorem 80.1] shows that

$$\begin{aligned} D_{(1,2)}h\circ g(x_0)&=\begin{vmatrix} D_1h^1(y_0) & D_2h^1(y_0)\\ D_1h^2(y_0) & D_2h^2(y_0)\end{vmatrix}\begin{vmatrix} D_1g^1(x_0) & D_2g^1(x_0)\\ D_1g^2(x_0) & D_2g^2(x_0)\end{vmatrix}\\ &\quad+\begin{vmatrix} D_1h^1(y_0) & D_3h^1(y_0)\\ D_1h^2(y_0) & D_3h^2(y_0)\end{vmatrix}\begin{vmatrix} D_1g^1(x_0) & D_2g^1(x_0)\\ D_1g^3(x_0) & D_2g^3(x_0)\end{vmatrix}\\ &\quad+\begin{vmatrix} D_2h^1(y_0) & D_3h^1(y_0)\\ D_2h^2(y_0) & D_3h^2(y_0)\end{vmatrix}\begin{vmatrix} D_1g^2(x_0) & D_2g^2(x_0)\\ D_1g^3(x_0) & D_2g^3(x_0)\end{vmatrix}. \end{aligned} \tag{15}$$

By (35) in Corollary 3.13 this statement of the chain rule can be written in briefer notation as follows:

$$D_{(1,2)}h\circ g(x_0) = D_{(1,2)}(h^1,h^2)(y_0)D_{(1,2)}(g^1,g^2)(x_0) + D_{(1,3)}(h^1,h^2)(y_0)D_{(1,2)}(g^1,g^3)(x_0) + D_{(2,3)}(h^1,h^2)(y_0)D_{(1,2)}(g^2,g^3)(x_0), \tag{16}$$

$$D_{(1,2)}h\circ g(x_0) = \sum_{(j_1,j_2)} D_{(j_1,j_2)}(h^1,h^2)(y_0)D_{(1,2)}(g^{j_1},g^{j_2})(x_0), \quad (j_1,j_2)\in(2/3). \tag{17}$$

4.4 Example. This example illustrates a third situation which arises in the differentiation of composite functions. Let G be a neighborhood of x_0 in $\mathbb{R}^2$, and let H be a neighborhood of y_0 in $\mathbb{R}$. Let $g: G\to H$ be a function whose single component g is differentiable at x_0, and assume that $g(x_0)=y_0$. Let $h: H\to\mathbb{R}^2$ be a function whose components (h^1,h^2) are differentiable at y_0. Then the composition of h and g is a function $h\circ g$ which maps G into $\mathbb{R}^2$, and $h\circ g$ has components $h^1\circ g$ and $h^2\circ g$ which are differentiable at x_0 by Theorem 4.1. Then $h\circ g$ is differentiable at x_0 by Theorem 3.12, and the formula in (9) is correct as before. Once more there is a change in (10), however, since h is now defined on a set in $\mathbb{R}$. Equation (1) in Theorem 4.1 shows that the correct formula now is

$$D_{(1,2)}h\circ g(x_0) = \begin{vmatrix} D_1h^1(y_0)D_1g(x_0) & D_1h^1(y_0)D_2g(x_0) \\ D_1h^2(y_0)D_1g(x_0) & D_1h^2(y_0)D_2g(x_0)\end{vmatrix} = 0. \tag{18}$$

Algebraically, it is clear that the determinant on the right is zero, and the geometric explanation for this result is not hard to find. Let $\mathbf{x}$ be a 2-vector in some class $X(x_0,\rho)$. Then

$$D_{(1,2)}h\circ g(x_0) = \lim_{\mathbf{x}\to x_0}\frac{\Delta_{(1,2)}h\circ g(\mathbf{x})}{\Delta(\mathbf{x})}. \tag{19}$$

The function g maps the three vertices of $\mathbf{x}$ into H in $\mathbb{R}$, and h maps H into a curve in $\mathbb{R}^2$. Thus $h\circ g$ maps the vertices of $\mathbf{x}$ into three points on a curve in $\mathbb{R}^2$, and these three points lie almost on a straight line when $\mathbf{x}$ is small. Since $|\Delta_{(1,2)}h\circ g(\mathbf{x})|$ is twice the area of the triangle formed by the three points on the curve, it is not surprising that the limit in (19) is zero. Thus it is geometrically clear why $D_{(1,2)}h\circ g(x_0)=0$.

These three examples serve as an introduction to the proof of the general case of the chain rule. Let G be a neighborhood of x_0 in $\mathbb{R}^n$, and let H be a neighborhood of y_0 in $\mathbb{R}^m$. Let k be an integer such that $1\leq k\leq n$, and let $(i_1,\cdots,i_k)\in(k/m)$, $(j_1,\cdots,j_k)\in(k/n)$.

4.5 Theorem. *Let $g: G\to H$ be a function whose components $(g^1,\cdots,g^m)$ are differentiable at x_0, and assume that $g(x_0)=y_0$. Let $h: H\to\mathbb{R}^k$ be a function whose components $(h^1,\cdots,h^k)$ are differentiable at y_0. Then the components of the composite function $h\circ g: G\to\mathbb{R}^k$ are differentiable at x_0; furthermore, $h\circ g$ is differentiable at x_0, and its derivatives are given by the following chain rule:*

$$D_{(j_1,\cdots,j_k)}h\circ g(x_0)$$
$$= \sum_{(i_1,\cdots,i_k)} D_{(i_1,\cdots,i_k)}(h^1,\cdots,h^k)(y_0)D_{(j_1,\cdots,j_k)}(g^{i_1},\cdots,g^{i_k})(x_0), \tag{20}$$
$$(i_1,\cdots,i_k)\in(k/m), \quad (j_1,\cdots,j_k)\in(k/n).$$

If $k > m$, the class (k/m) is empty, and the sum on the right in (20) is correctly interpreted as having the value zero. If $k = m$, then (k/m) contains the single index set $(1,\cdots,k) = (1,\cdots,m)$. If $k < m$, the sum in (20) contains $C(m,k)$ terms. If $k = 1$, then (20) reduces to (1). Thus (20) contains all statements of the chain rule.

Proof of Theorem 4.5. Theorem 4.1 proves that the components $h^1\circ g,\cdots,h^k\circ g$ of $h\circ g$ are differentiable, and (1) is a formula for their derivatives. Then Corollary 3.13 states that $h\circ g$ is differentiable at x_0 and (35) in that corollary contains formulas for its derivatives. Let A and B denote the following matrices.

$$A = \begin{bmatrix} D_1h^1(y_0) & \cdots & D_mh^1(y_0) \\ \cdots & \cdots & \cdots \\ D_1h^k(y_0) & \cdots & D_mh^k(y_0)\end{bmatrix}, \qquad B = \begin{bmatrix} D_{j_1}g^1(x_0) & \cdots & D_{j_k}g^1(x_0) \\ \cdots & \cdots & \cdots \\ D_{j_1}g^m(x_0) & \cdots & D_{j_k}g^m(x_0)\end{bmatrix}. \tag{21}$$

Then Theorem 4.1 and Corollary 3.13 show that

$$D_{(j_1,\cdots,j_k)}h\circ g(x_0) = \det[AB]. \tag{22}$$

The determinant of the minor of A in columns $(i_1,\cdots,i_k)$ is $D_{(i_1,\cdots,i_k)}(h^1,\cdots,h^k)(y_0)$, and the determinant of the minor of B in rows $(i_1,\cdots,i_k)$ is $D_{(j_1,\cdots,j_k)}(g^{j_1},\cdots,g^{j_k})(x_0)$. Observe that the determinant of the minor of B in rows $(i_1,\cdots,i_k)$ equals the determinant of the minor of B^t in columns $(i_1,\cdots,i_k)$ [see Theorem 77.14]. Then by the Binet–Cauchy multiplication theorem in Theorem 80.1, for $(j_1,\cdots,j_k)$ in (k/n) and $(i_1,\cdots,i_k)$ in (k/m),

$$\begin{aligned} D_{(j_1,\cdots,j_k)}&h\circ g(x_0) \\ &= 0, && m < k; \\ &= D_{(1,\cdots,k)}(h^1,\cdots,h^k)(y_0)D_{(j_1,\cdots,j_k)}(g^1,\cdots,g^k)(x_0), && m = k; \\ &= \sum_{(i_1,\cdots,i_k)} D_{(i_1,\cdots,i_k)}(h^1,\cdots,h^k)(y_0)D_{(j_1,\cdots,j_k)}(g^{i_1},\cdots,g^{i_k})(x_0), && m > k. \end{aligned} \tag{23}$$

With the interpretation given for the summation with respect to $(i_1,\cdots,i_k)$ in (k/m), these equations are summarized in the single statement (20) of the chain rule. □

Exercises

4.1. Give a second proof of Theorem 4.1. [Hint. The functions $g^1,\cdots,g^m$ are differentiable at x_0 and hence satisfy the Stolz condition at x_0; also h satisfies the Stolz condition at y_0. Give a direct proof that $h \circ g$ satisfies the Stolz condition at x_0, and find the constants in this condition.]

4.2. Let $g: G \to \mathbb{R}^2$, $G \subset \mathbb{R}^2$, be the function such that

$$g^1(x) = (x^1)^2 + (x^2)^2, \qquad g^2(x) = (x^1)^2 - (x^2)^2,$$

and let $h: H \to \mathbb{R}$, $H \subset \mathbb{R}^2$, be the function such that $h(y) = y^1 y^2$. Assume that $g(G) \subset H$.

(a) Use the chain rule in (1) to find the derivatives of $h \circ g: G \to \mathbb{R}$.

(b) Find the function $h \circ g: G \to \mathbb{R}$ and then find $D_1 h \circ g(x)$ and $D_2 h \circ g(x)$. Compare the results with those found in (a).

4.3. Let $g: G \to \mathbb{R}^3$, $G \subset \mathbb{R}^2$, be the function such that

$$g^1(x) = (x^1)^2 + (x^2)^2, \quad g^2(x) = 2x^1 x^2, \quad g^3(x) = (x^1)^2 - (x^2)^2,$$

and let $h: H \to \mathbb{R}^2$, $H \subset \mathbb{R}^3$, be the function such that

$$h^1(y) = 2y^1 - 3y^2 + 4y^3, \qquad h^2(y) = 5y^1 + 2y^2 - 3y^3.$$

Assume that $g(G) \subset H$.

(a) Use the chain rule in (20) to find $D_{(1,2)} h \circ g(x)$. [Hint. Show that $n = 2$, $m = 3$, $k = 2$; use the third equation in (23).]

(b) Find the function $h \circ g: G \to \mathbb{R}^2$ and then use (32) in Theorem 3.12 to find $D_{(1,2)} h \circ g(x)$. Compare your result with the one found in (a).

4.4. Write out the proof of Theorem 4.5 in each of the following special cases:

(a) $n = 3$, $m = 1$, $k = 2$;
(b) $n = 3$, $m = 2$, $k = 2$;
(c) $n = 3$, $m = 3$, $k = 2$;
(d) $n = 3$, $m = 3$, $k = 3$.

Observe that there are three statements to be proved in each of (a), (b), and (c). How many are there in (d)?

4.5. The conclusion $D_{(j_1, \cdots, j_k)} h \circ g(x_0) = 0$, $(j_1, \cdots, j_k) \in (k/n)$, if $k > m$ in Theorem 4.5 comes as a surprise. Give both an algebraic and a geometric explanation of why this result is to be expected.

4.6. Let G be a neighborhood of x_0 in $\mathbb{R}^n$, and let H be a neighborhood of y_0 in $\mathbb{R}^n$. Make the following assumptions about the functions $h: H \to \mathbb{R}$ and $g^i: G \to \mathbb{R}$, $i = 1, \cdots, n$.

(i) h is differentiable at y_0.
(ii) $\Sigma_{j=1}^n [D_j h(y_0)]^2 \neq 0$.
(iii) $g(x_0) = y_0$ and g maps G into H.
(iv) g^i, $i = 1, \cdots, n$, is differentiable at x_0.
(v) $h \circ g(x) = 0$ for all x in G.

(a) Show that g maps G into the surface $h(y) = 0$ in H.

(b) Prove that $D_{(1, \cdots, n)} g(x_0) = 0$. [Hint. Use (i), (iii), and (iv) to differentiate the two sides of the equation $h \circ g(x) = 0$ with respect to each of the variables $x^1, \cdots, x^n$. Then use (ii).]

(c) Explain geometrically why the result $D_{(1, \cdots, n)} g(x_0) = 0$ is to be expected; consider the special cases $n = 2$ and $n = 3$. [Hint. Exercise 3.7.]

4.7. Consider Theorem 4.5 again. If $k = m = n$ then g maps an n-vector $\mathbf{x} = (x_1, \cdots, x_n, x_0)$ into an n-vector $\mathbf{y} = g(\mathbf{x}) = (g(x_1), \cdots, g(x_n), g(x_0))$ in H. Let $X(x_0, \rho)$, $0 < \rho \leqq 1$, and $Y(y_0, \sigma)$, $0 < \sigma \leqq 1$, be classes of n-vectors at x_0 in G and y_0 in H which satisfy regularity conditions as indicated. There is a simple proof in this case if the following hypothesis is added: g maps every n-vector $\mathbf{x}$ in $X(x_0, \rho)$ into an n-vector $\mathbf{y} = g(\mathbf{x})$ in $Y(y_0, \sigma)$. With this added hypothesis, prove the following identity and Theorem 4.5.

$$\frac{\Delta_{(1,\cdots,n)}h \circ g(\mathbf{x})}{\Delta(\mathbf{x})} = \frac{\Delta_{(1,\cdots,n)}h(\mathbf{y})}{\Delta_{(1,\cdots,n)}g(\mathbf{x})} \frac{\Delta_{(1,\cdots,n)}g(\mathbf{x})}{\Delta(\mathbf{x})}, \qquad \mathbf{x} \in X(x_0, \rho), \qquad \mathbf{y} = g(\mathbf{x}).$$

In this proof it is necessary to assume that h is differentiable at y_0, that g is differentiable at x_0, and that $g^1, \cdots, g^n$ are continuous at x_0, but it is not necessary to assume that $h^1, \cdots, h^n$ are differentiable at y_0 or that $g^1, \cdots, g^n$ are differentiable at x_0.

4.8. Use the method and notation of Exercise 4.7 to prove Theorem 4.5 in the case $k < m = n$. [Hint. Let $(j_1, \cdots, j_k)$ be an arbitrary, but fixed, index set in (k/m), and let $(i_1, \cdots, i_k)$ be a variable index set in (k/n). Prove formula (20) by first using Sylvester's interchange theorem [Theorem 3.5] to establish the following identities.

$$\Delta(\mathbf{y})\Delta_{(j_1,\cdots,j_k)}h \circ g(\mathbf{x}) = \sum_{(i_1,\cdots,i_k)} \Delta_{(i_1,\cdots,i_k)}h(\mathbf{y})\Delta_{(j_1,\cdots,j_k)}(g^{i_1}, \cdots, g^{i_k})(\mathbf{x}),$$

$$(j_1, \cdots, j_k) \in (k/n);$$

$$\begin{aligned}\frac{\Delta_{(j_1,\cdots,j_k)}h \circ g(\mathbf{x})}{\Delta(\mathbf{x})} &= \sum_{(i_1,\cdots,i_k)} D_{(i_1,\cdots,i_k)}h(y_0)\frac{\Delta_{(j_1,\cdots,j_k)}(g^{i_1}, \cdots, g^{i_k})(\mathbf{x})}{\Delta(\mathbf{x})} \\ &\quad + \sum_{(i_1,\cdots,i_k)} \left\{\frac{\Delta_{(i_1,\cdots,i_k)}h(\mathbf{y})}{\Delta(\mathbf{y})} - D_{(i_1,\cdots,i_k)}h(y_0)\right\} \\ &\quad \times \frac{\Delta_{(j_1,\cdots,j_k)}(g^{i_1}, \cdots, g^{i_k})(\mathbf{x})}{\Delta(\mathbf{x})}.\end{aligned}$$

State the minimum hypotheses needed for the proof of the theorem in this case.

4.9. Let $u : E \to \mathbb{R}$, $(x, y) \in E \subset \mathbb{R}^2$, be a function which satisfies Laplace's equation

$$\frac{\partial^2 u}{\partial x^2} + \frac{\partial^2 u}{\partial y^2} = 0;$$

then u is called a harmonic function. If polar coordinates are introduced by the equations $x = r \cos\theta$, $y = r \sin\theta$, then $u(x, y)$ becomes $w(r, \theta)$. Show that w satisfies Laplace's equation in polar coordinates, which is

$$\frac{\partial^2 w}{\partial r^2} + \frac{1}{r^2}\frac{\partial^2 w}{\partial \theta^2} + \frac{1}{r}\frac{\partial w}{\partial r} = 0.$$

5. Compositions with Linear Functions

Section 4 has developed formulas for the derivatives of composite functions. The purpose of the present section is to derive the special forms of the chain rule which arise when one of the functions is a linear function. Composite functions of this special form arise in connection with a change of basis in the domain or range of a function and in other situations.

Let G be a neighborhood of a point x_0 in $\mathbb{R}^n$, and let $x:(x^1, \cdots, x^n)$ denote points in G. Let H be a neighborhood of y_0 in $\mathbb{R}^n$, and let $y:(y^1, \cdots, y^n)$ denote points in H. Let $g: G \to H$ be a linear function with components $(g^1, \cdots, g^n)$, and assume that $y_0 = g(x_0)$. Then there is a constant matrix $[a_j^i]$, $i, j = 1, \cdots, n$, such that

$$\begin{aligned} y^1 &= g^1(x) = a_1^1 x^1 + \cdots + a_n^1 x^n, \\ &\cdots\cdots\cdots\cdots\cdots\cdots \\ y^n &= g^n(x) = a_1^n x^1 + \cdots + a_n^n x^n. \end{aligned} \tag{1}$$

Then in matrix notation,

$$\begin{bmatrix} g^1(x) \\ \vdots \\ g^n(x) \end{bmatrix} = \begin{bmatrix} a_1^1 & \cdots & a_n^1 \\ \cdots & \cdots & \cdots \\ a_1^n & \cdots & a_n^n \end{bmatrix} \begin{bmatrix} x^1 \\ \vdots \\ x^n \end{bmatrix}. \tag{2}$$

If A denotes the matrix $[a_j^i]$, then (2) is the matrix equation $g(x) = Ax$. Also, the composition of $h: H \to \mathbb{R}$ and $g: G \to H$ is the function $h \circ g: G \to \mathbb{R}$ whose value at x is $h(g^1(x), \cdots, g^n(x))$. If h is differentiable at y_0, then Theorem 4.1 shows that $h \circ g$ is differentiable at x_0. Let $[D_j h \circ g(x_0)]$ and $[D_i h(y_0)]$ denote the derivative vectors, written as column vectors, of $h \circ g$ and h at x_0 and y_0 respectively, and let A^t denote the transpose of A.

5.1 Theorem. *If $h: H \to \mathbb{R}$ is differentiable at y_0, then $h \circ g$ is differentiable at x_0, and*

$$D_j h \circ g(x_0) = \sum_{i=1}^{n} a_j^i D_i h(y_0), \qquad j = 1, \cdots, n, \tag{3}$$

$$[D_j h \circ g(x_0)] = A^t [D_i h(y_0)]. \tag{4}$$

PROOF. The formula in (3) follows from (1) above and the formula in (1) in Theorem 4.1. Then

$$\begin{bmatrix} D_1 h \circ g(x_0) \\ \vdots \\ D_n h \circ g(x_0) \end{bmatrix} = \begin{bmatrix} a_1^1 & \cdots & a_1^n \\ \cdots & \cdots & \cdots \\ a_n^1 & \cdots & a_n^n \end{bmatrix} \begin{bmatrix} D_1 h(y_0) \\ \vdots \\ D_n h(y_0) \end{bmatrix} \tag{5}$$

and this equation is the same as (4). □

Before stating the next theorem, it is necessary to define the k-th compound of the matrix A [see Definition 78.1 in Appendix 1]. Let k be an integer such that $1 \leqq k \leqq n$, and let $(i_1, \cdots, i_k)$ and $(j_1, \cdots, j_k)$ denote index sets in (k/n). Then the k-th compound matrix $A^{(k)}$ of A is a $C(n, k)$ by $C(n, k)$ matrix defined as follows. Let $A_{(j_1, \cdots, j_k)}^{(i_1, \cdots, i_k)}$ denote the minor of A in rows

$(i_1, \cdots, i_k)$ and columns $(j_1, \cdots, j_k)$. Then the element in $A^{(k)}$ in the row which is indexed $(i_1, \cdots, i_k)$ and column which is indexed $(j_1, \cdots, j_k)$ is $\det A^{(i_1, \cdots, i_k)}_{(j_1, \cdots, j_k)}$. The first compound $A^{(1)}$ of A is A itself, and the n-th compound $A^{(n)}$ is the matrix whose single element is $\det A$.

The setting for the next theorem is the same as that of Theorem 5.1 except that $h : H \to \mathbb{R}$ is replaced by the function $h : H \to \mathbb{R}^k$.

5.2 Theorem. *If $h : H \to \mathbb{R}^k$ is a function whose components $(h^1, \cdots, h^k)$ are differentiable at y_0, then $h \circ g$ has components $(h^1 \circ g, \cdots, h^k \circ g)$ which are differentiable at x_0, and*

$$D_{(j_1, \cdots, j_k)} h \circ g(x_0) = \sum_{(i_1, \cdots, i_k)} \det A^{(i_1, \cdots, i_k)}_{(j_1, \cdots, j_k)} D_{(i_1, \cdots, i_k)} h(y_0), \qquad (j_1, \cdots, j_k) \in (k/n), \tag{6}$$

$$[D_{(j_1, \cdots, j_k)} h \circ g(x_0)] = [A^{(k)}]^t [D_{(i_1, \cdots, i_k)} h(y_0)], \qquad (i_1, \cdots, i_k) \in (k/n). \tag{7}$$

PROOF. Theorem 4.5 can be applied to the composite function $h \circ g$ to obtain the following equation:

$$\begin{aligned} D_{(j_1, \cdots, j_k)} h \circ g(x_0) &= \sum_{(i_1, \cdots, i_k)} D_{(i_1, \cdots, i_k)} h(y_0) D_{(j_1, \cdots, j_k)}(g^{i_1}, \cdots, g^{i_k})(x_0), \qquad (8) \\ &= \sum_{(i_1, \cdots, i_k)} \det A^{(i_1, \cdots, i_k)}_{(j_1, \cdots, j_k)} D_{(i_1, \cdots, i_k)} h(y_0). \end{aligned}$$

In this equation, $(i_1, \cdots, i_k)$ and $(j_1, \cdots, j_k)$ are in (k/n). Equation (8) contains (6), and (7) follows from (6). □

If $k = 1$, then Theorem 5.2 reduces to Theorem 5.1. The following theorem is stated for reference; a proof of it is given in Section 82 in Appendix 1.

5.3 Theorem (Sylvester–Franke Theorem). *Let A be an $n \times n$ matrix, and let $A^{(k)}$ be its k-th compound matrix. Then*

$$\det A^{(k)} = [\det A]^{C(n-1, k-1)}. \tag{9}$$

In Theorems 5.1 and 5.2, the function g in the composite function $h \circ g$ is assumed to be a linear function. In the next theorem, h is assumed to be the linear function. Let G and H be neighborhoods of x_0 in $\mathbb{R}^n$ and y_0 in $\mathbb{R}^m$ respectively, and let $g : G \to H$ be a function whose components are differentiable at x_0. Assume that $m \leq n$. Let B be an $m \times m$ constant matrix $[b^i_j]$, and let $h : H \to \mathbb{R}^m$ be the linear function with components $(h^1, \cdots, h^m)$ such that

$$\begin{aligned} z^1 &= h^1(y) = b^1_1 y^1 + \cdots + b^1_m y^m, \\ &\cdots\cdots\cdots\cdots\cdots\cdots \\ z^m &= h^m(y) = b^m_1 y^1 + \cdots + b^m_m y^m. \end{aligned} \qquad y : (y^1, \cdots, y^m) \in H, \tag{10}$$

Then $h \circ g$ is the function with components $(h^1 \circ g, \cdots, h^m \circ g)$ such that

$$\begin{aligned} h^1 \circ g(x) &= b^1_1 g^1(x) + \cdots + b^1_m g^m(x), \\ &\cdots\cdots\cdots\cdots \\ h^m \circ g(x) &= b^m_1 g^1(x) + \cdots + b^m_m g^m(x). \end{aligned} \qquad x \in G, \tag{11}$$

5.4 Theorem. *Let $g : G \to H$ and $h : H \to \mathbb{R}^m$ be the functions just described. If the components $(g^1, \cdots, g^m)$ of g are differentiable at x_0, then the components $(h^1 \circ g, \cdots, h^m \circ g)$ of $h \circ g$, and $h \circ g$ itself, are differentiable at x_0, and*

$$D_{(j_1, \cdots, j_m)} h \circ g(x_0) = \det B\, D_{(j_1, \cdots, j_m)} g(x_0), \qquad (j_1, \cdots, j_m) \in (m/n). \tag{12}$$

PROOF. The components $(h^1, \cdots, h^m)$ of $h : H \to \mathbb{R}^m$ are differentiable functions; then the components $(h^1 \circ g, \cdots, h^m \circ g)$ of $h \circ g : G \to \mathbb{R}^m$ are differentiable by Theorem 4.1. In the present case it is obvious that $h^1 \circ g, \cdots, h^m \circ g$ are differentiable at x_0 since equations (11) show that these functions are linear combinations of functions $g^1, \cdots, g^m$ which are differentiable at x_0. Since h maps H in $\mathbb{R}^m$ into $\mathbb{R}^m$, then (in Theorem 4.5) $k = m$, and the sum in (20) contains a single term. Thus

$$\begin{aligned} &D_{(j_1, \cdots, j_m)} h \circ g(x_0) \\ &\quad = D_{(1, \cdots, m)}(h^1, \cdots, h^m)(y_0) D_{(j_1, \cdots, j_m)}(g^1, \cdots, g^m)(x_0), \\ &\quad (j_1, \cdots, j_m) \in (m/n). \end{aligned} \tag{13}$$

But

$$D_{(1, \cdots, m)}(h^1, \cdots, h^m)(y_0) = \begin{vmatrix} b^1_1 & \cdots & b^1_m \\ \cdots & \cdots & \cdots \\ b^m_1 & \cdots & b^m_m \end{vmatrix} = \det B. \tag{14}$$

Thus (13) simplifies to (12), and the proof is complete. □

EXERCISES

5.1. Let A be an $n \times n$ matrix with elements a^i_j. Define the co-factor A^i_j of a^i_j as follows: delete row i and column j of A; then A^i_j equals the determinant, multiplied by $(-1)^{i+j}$, of the minor of A which remains. The matrix of co-factors is $[A^i_j]$, $i, j = 1, \cdots, n$. The adjoint matrix of A, abbreviated adj A, is defined thus: adj $A = [A^i_j]^t$.
 (a) Prove that $A \operatorname{adj} A = (\operatorname{adj} A)A = (\det A)I$.
 (b) If $\det A \neq 0$, prove that $A^{-1} = (\det A)^{-1} \operatorname{adj} A$.

5.2. Let A and B be $n \times n$ matrices such that $BA = I$.
 (a) Use the Binet–Cauchy multiplication theorem to prove that $\det A \neq 0$ and $\det B \neq 0$.
 (b) Prove that $B = A^{-1}$. [Hint. Use Exercise 5.1 to show that A^{-1} exists.]

5.3. Let $a_j : (a^1_j, \cdots, a^n_j)$, $j = 1, \cdots, n$, be the j-th column of a matrix A, and assume that the vectors $a_1, \cdots, a_n$ are an orthonormal set; that is, assume that

$$|a_j| = (a_j, a_j)^{1/2} = 1, \qquad (a_j, a_k) = 0 \quad \text{if} \quad j \neq k.$$

A matrix A which has these properties is called an orthogonal matrix.

(a) Prove that $A^tA = I$ and that $A^t = A^{-1}$.

(b) Prove that $AA^t = I$, and thus prove that A^t is also an orthogonal matrix.

(c) Prove that $\det A = \det A^{-1} = \pm 1$. [Hint. Use the Binet–Cauchy multiplication theorem and the fact that $\det A = \det A^t = \det A^{-1}$.]

(d) If x and y are column vectors such that $y = Ax$, prove that $|y| = |x|$.

(e) Let $\mathbf{x}$ and $\mathbf{y}$ denote n-vectors as follows.

$$\mathbf{x} = \begin{bmatrix} x_1^1 & x_1^2 & \cdots & x_1^n \\ \cdots & \cdots & \cdots & \cdots \\ x_n^1 & x_n^2 & \cdots & x_n^n \\ x_0^1 & x_0^2 & \cdots & x_0^n \end{bmatrix}, \qquad \mathbf{y} = \begin{bmatrix} y_1^1 & y_1^2 & \cdots & y_1^n \\ \cdots & \cdots & \cdots & \cdots \\ y_n^1 & y_n^2 & \cdots & y_n^n \\ y_0^1 & y_0^2 & \cdots & y_0^n \end{bmatrix}.$$

Assume that the linear transformation $y = Ax$ transforms $\mathbf{x}$ into $\mathbf{y}$. Use the matrix equation

$$\begin{bmatrix} y_1^1 & \cdots & y_n^1 & y_0^1 \\ \cdots & \cdots & \cdots & \cdots \\ y_1^n & \cdots & y_n^n & y_0^n \\ 1 & \cdots & 1 & 1 \end{bmatrix} = \begin{bmatrix} a_1^1 & \cdots & a_n^1 & 0 \\ \cdots & \cdots & \cdots & \cdots \\ a_1^n & \cdots & a_n^n & 0 \\ 0 & \cdots & 0 & 1 \end{bmatrix} \begin{bmatrix} x_1^1 & \cdots & x_n^1 & x_0^1 \\ \cdots & \cdots & \cdots & \cdots \\ x_1^n & \cdots & x_n^n & x_0^n \\ 1 & \cdots & 1 & 1 \end{bmatrix}$$

to prove that $\Delta(\mathbf{y}) = \pm\Delta(\mathbf{x})$.

(f) Let $X(x_0, \rho)$ be a class of n-vectors $\mathbf{x}$ at x_0, and let $\{\mathbf{y} : \mathbf{y}^t = A\mathbf{x}^t, \mathbf{x} \in X(x_0, \rho)\}$ be the class of n-vectors at y_0 into which the linear transformation $y = Ax$ transforms $X(x_0, \rho)$. Show that

$$|\Delta(\mathbf{y})| \geqq \rho \prod_{i=1}^{n} |y_i - y_0|;$$

show also that the transformation $y = Ax$ maps $X(x_0, \rho)$ onto $Y(y_0, \rho)$.

5.4. Let A be the matrix described in Exercise 5.3. In Theorem 5.2 assume that the two functions are $h : H \to \mathbb{R}^n$, $H \subset \mathbb{R}^n$, and $g : G \to H$, $G \subset \mathbb{R}^n$. The function $y = g(x) = Ax$ is a change of variables in the set H. The derivatives $D_{(1,\cdots,n)}h(y_0)$ and $D_{(1,\cdots,n)}h \circ g(x_0)$ are the Jacobians of h and $h \circ g$ with respect to the original variables y and the new variables x, respectively. Prove that $D_{(1,\cdots,n)}h \circ g(x_0) = \pm D_{(1,\cdots,n)}h(y_0)$.

5.5. In Theorem 5.4 assume that the columns of B are an orthonormal set of vectors. Prove that $\det B$ in equation (12) is either $+1$ or -1.

5.6. Let $h : \mathbb{R}^2 \to \mathbb{R}$ be the function such that

$$h(y^1, y^2) = \frac{2y^1y^2}{(y^1)^2 + (y^2)^2}, \qquad (y^1)^2 + (y^2)^2 \neq 0,$$

$$h(0, 0) = 0.$$

Show that this function has partial derivatives at $(0, 0)$ and that $h_{y^1}(0, 0) = 0$, $h_{y^2}(0, 0) = 0$.

(a) Make the change of variables $y = g(x)$ described by the following equations:

$$y^1 = g^1(x^1, x^2) = x^1 + x^2,$$

$$y^2 = g^2(x^1, x^2) = x^1 - x^2.$$

Show that the new function is

$$h\circ g(x^1, x^2) = \frac{(x^1)^2 - (x^2)^2}{(x^1)^2 + (x^2)^2}, \qquad (x^1)^2 + (x^2)^2 \neq 0,$$

$$h\circ g(0, 0) = 0.$$

(b) Prove that $h\circ g$ does not have partial derivatives at the origin.
(c) Explain the significance of Theorem 5.1 with respect to a linear change of variables in the domain of a differentiable function.

5.7. Let $A = [a^i_j]$, $i, j = 1, \cdots, n$, be a matrix such that $\det A \neq 0$.
(a) Let $\mathbf{x}$ be the n-vector $(x_1, \cdots, x_n, x_0)$ in $\mathbb{R}^n$, and let $\mathbf{y} = (y_1, \cdots, y_n, y_0)$ be the n-vector into which $\mathbf{x}$ is transformed by the linear transformation $y = Ax$. Then $\mathbf{x}$ and $\mathbf{y}$ are conveniently represented by matrices as in Exercise 5.3(e). By the methods used in that exercise, show that $\Delta(\mathbf{y}) = \det A\, \Delta(\mathbf{x})$.
(b) Let $X(x_0, \rho)$ be a class of n-vectors $\mathbf{x}$ at x_0, and let $\{\mathbf{y} : \mathbf{y}^t = A\mathbf{x}^t, \mathbf{x} \in X(x_0, \rho)\}$ be the class of n-vectors $\mathbf{y}$ at y_0 into which $y = Ax$ transforms $X(x_0, \rho)$. Show that $\{\mathbf{y} : \mathbf{y}^t = A\mathbf{x}^t, \mathbf{x} \in X(x_0, \rho)\}$ is contained in a class $Y(y_0, \sigma)$; that is, show that there is a constant σ, $0 < \sigma \leqq 1$, such that

$$|\Delta(\mathbf{y})| \geqq \sigma \prod_{i=1}^{n} |y_i - y_0|, \qquad \mathbf{y}^t = A\mathbf{x}^t, \qquad \mathbf{x} \in X(x_0, \rho).$$

Find a value for σ. [Outline of solution. Since $\mathbf{y}^t = A\mathbf{x}^t$, the Binet–Cauchy multiplication theorem shows that $\Delta(\mathbf{y}) = \det A\, \Delta(\mathbf{x})$. Then since $\mathbf{x}$ is in $X(x_0, \rho)$,

$$|\Delta(\mathbf{y})| = |\det A|\,|\Delta(\mathbf{x})| \geqq \rho|\det A| \prod_{i=1}^{n} |x_i - x_0|.$$

Since $y = Ax$ is a linear transformation, Schwarz's inequality can be used to show that, if $y = Ax$ and $y_0 = Ax_0$ for x and x_0 in $\mathbb{R}^n$, then there is a constant $M > 0$ such that $|y - y_0| \leqq M|x - x_0|$. (There are other ways to establish the existence of M, but Schwarz's inequality [see (7) and Corollary 86.2 in Section 86] probably provides the easiest and most accessible proof for the beginning student.) Then $|x_i - x_0| \geqq M^{-1}|y_i - y_0|$ and

$$|\Delta(\mathbf{y})| \geqq \rho|\det A|M^{-n} \prod_{i=1}^{n} |y_i - y_0|.$$

If $\sigma = \rho|\det A|M^{-n}$, then $\mathbf{y} \in Y(y_0, \sigma)$.]

5.8. If A in Theorem 5.1 is an orthogonal matrix, prove that

$$\left\{\sum_{j=1}^{n} [D_j h\circ g(x_0)]^2\right\}^{1/2} = \left\{\sum_{j=1}^{n} [D_j h(y_0)]^2\right\}^{1/2}.$$

[Hint. If A is an orthogonal matrix, then A^t is an orthogonal matrix by Exercise 5.3(b). Then by (4) in Theorem 5.1, the vector $(D_1 h(y_0), \cdots, D_n h(y_0))$ is transformed into the vector $(D_1 h\circ g(x_0), \cdots, D_n h\circ g(x_0))$ by an orthogonal matrix. But transformation by an orthogonal matrix preserves the lengths of vectors by Exercise 5.3(d).]

5.9. In Theorem 5.2 the derivative vector $D_{(i_1,\cdots,i_k)}h(y_0)$, $(i_1, \cdots, i_k) \in (k/n)$, is transformed into the derivative vector $D_{(j_1,\cdots,j_k)}h \circ g(x_0)$, $(j_1, \cdots, j_k) \in (k/n)$, by a linear transformation whose matrix is $[A^{(k)}]^t$. If $\det A \neq 0$, prove that this transformation is non-singular; that is, prove that $\det[A^{(k)}]^t \neq 0$. [Hint. Theorem 5.3.]

6. Classes of Differentiable Functions

There are several possible definitions of differentiability for functions $f: E \to \mathbb{R}^m$, $E \subset \mathbb{R}^n$. This section investigates and compares these definitions. The section proves that the class of functions $f: E \to \mathbb{R}^m$ for which each component $f^1, \cdots, f^m$ of f is differentiable in the sense of Definition 2.8 is the same as the classes of functions which are differentiable according to other current definitions.

Let $f: E \to \mathbb{R}^m$, $E \subset \mathbb{R}^n$, be a function with components $(f^1, \cdots, f^m)$ whose domain E is an open set in $\mathbb{R}^n$. Then according to Definition 2.8, f is differentiable at x_0 in E if and only if the derivatives

$$D_{(j_1,\cdots,j_m)}(f^1, \cdots, f^m)(x_0), \qquad (j_1, \cdots, j_m) \in (m/n),$$

exist; that is, f is differentiable at x_0 if and only if the following limits exist:

$$\lim_{\mathbf{x} \to x_0} \frac{\Delta_{(j_1,\cdots,j_m)}(f^1, \cdots, f^m)(\mathbf{x})}{\Delta(\mathbf{x})}, \quad \mathbf{x} \in X(x_0, \rho), \quad (j_1, \cdots, j_m) \in (m/n).$$

Theorem 3.12 and Corollary 3.13 prove that $f: E \to \mathbb{R}^m$ is differentiable at x_0 in E if each of its components $(f^1, \cdots, f^m)$ is differentiable at x_0. This result raises the following question.

6.1 Question. Are there functions $f: E \to \mathbb{R}^m$, $E \subset \mathbb{R}^n$, such that

(i) f is differentiable at x_0 in E;
(ii) at least one of the components $(f^1, \cdots, f^m)$ is not differentiable at x_0?

Some examples will help to answer this question.

6.2 Example. The function $f: E \to \mathbb{R}^3$, $E \subset \mathbb{R}^3$, in Example 3.18 maps an open set E in $\mathbb{R}^3$ into a plane in $\mathbb{R}^3$. Then $D_{(1,2,3)}(f^1, f^2, f^3)(x) = 0$ for every x in E although the components (f^1, f^2, f^3) are not differentiable and need not even be continuous. Exercise 3.7 contains the generalization of this example for functions $f: E \to \mathbb{R}^n$, $E \subset \mathbb{R}^n$.

6.3 Example. Let L be a straight line in $\mathbb{R}^3$, and let f be an arbitrary, non-constant, function which maps E in $\mathbb{R}^3$ into L. If $\mathbf{x} = (x_1, \cdots, x_3, x_0)$, then $(1/3!)|\Delta_{(1,2,3)}f(\mathbf{x})|$ is the volume of a tetrahedron whose four vertices lie on the line L. Thus $\Delta_{(1,2,3)}f(\mathbf{x}) = 0$, and $D_{(1,2,3)}f(x_0)$ exists and equals zero at each x_0 in E even if the components (f^1, f^2, f^3) are not differentiable.

Observe, however, that the components are differentiable if f maps E into a single point in $\mathbb{R}^3$, because then each of the components is a constant function.

6.4 Example. Let $f\colon \mathbb{R}^2 \to \mathbb{R}^2$ be a function whose components (f^1, f^2) are defined as follows [see Exercise 3.5]:

$$f^1(x^1, x^2) = r^2 \sin(1/r), \qquad f^2(x^1, x^2) = r\cos(1/r),$$
$$r = [(x^1)^2 + (x^2)^2]^{1/2} \neq 0, \qquad f^1(0, 0) = 0, \qquad f^2(0, 0) = 0.$$

Then f^1 satisfies the Stolz condition at $0 : (0, 0)$ with $A_1 = A_2 = 0$ [see Definition 3.1] and is therefore differentiable at $(0, 0)$ by Theorem 3.3. Furthermore, f^2 does not satisfy the Stolz condition at $(0, 0)$ and is therefore not differentiable at $(0, 0)$ by Theorem 3.6. Another proof that f^2 is not differentiable at $(0, 0)$ can be obtained from Theorem 3.14: if f^2 were differentiable at $(0, 0)$, then f^2 would have partial derivatives at $(0, 0)$; but it is easy to show directly that f^2 does not have partial derivatives at the origin. Although f^2 is not differentiable at $0 : (0, 0)$, the function (f^1, f^2) : $\mathbb{R}^2 \to \mathbb{R}^2$ is differentiable at $0 : (0, 0)$ and its derivative is zero. To prove this statement, let $\mathbf{x} = (x_1, x_2, 0)$ be a 2-vector in $X(0, \rho)$; then

$$\Delta_{(1,2)}(f^1, f^2)(\mathbf{x}) = \begin{vmatrix} r_1^2 \sin(1/r_1) & r_1 \cos(1/r_1) & 1 \\ r_2^2 \sin(1/r_2) & r_2 \cos(1/r_2) & 1 \\ 0 & 0 & 1 \end{vmatrix},$$

$$\left| \frac{\Delta_{(1,2)}(f^1, f^2)(\mathbf{x})}{\Delta(\mathbf{x})} \right| \leqq (1/\rho)\text{abs. val.} \begin{vmatrix} r_1 \sin(1/r_1) & \cos(1/r_1) \\ r_2 \sin(1/r_2) & \cos(1/r_2) \end{vmatrix}.$$

As $\mathbf{x} \to 0$, the elements in the first column of the matrix on the right approach zero, and the elements in the second column are equal to or less than 1 in absolute value; hence,

$$D_{(1,2)}(f^1, f^2)(0) = \lim_{\mathbf{x} \to 0} \frac{\Delta_{(1,2)}(f^1, f^2)(\mathbf{x})}{\Delta(\mathbf{x})} = 0.$$

6.5 Example. A function $h : E \to \mathbb{R}$, $E \subset \mathbb{R}^n$, is said to satisfy a Lipschitz condition on E if and only if there exists a constant M such that $|h(x_2) - h(x_1)| \leqq M|x_2 - x_1|$ for every x_1, x_2 in E. Let E and G be two sets in $\mathbb{R}^2$. Let $f\colon G \to \mathbb{R}$ be a function which satisfies the following hypotheses.

(i) f is differentiable at each y in G.
(ii) $D_1 f(y) \neq 0$ for every y in G.
(iii) The set $\{y : y \in G, f(y) = 0\}$ is not empty. Let $g : E \to G$ be a function with components (g^1, g^2) such that
(iv) g^1 and g^2 satisfy Lipschitz conditions on E; that is, there are constants M_1, M_2 such that $|g^i(x_2) - g^i(x_1)| \leqq M_i|x_2 - x_1|$ for every x_1, x_2 in E and $i = 1, 2$.
(v) $f \circ g(x) = 0$ for every x in E.

Geometrically stated, g maps E into the set in G on which f vanishes. A proof will now be given that $D_{(1,2)}(g^1, g^2)(x)$ exists and equals zero for each x in E although neither g^1 nor g^2 is assumed to be differentiable. This example is similar to the one in Exercise 4.6, but it has much weaker hypotheses.

Let (x_1, x_2, x_0) be a 2-vector $\mathbf{x}$ in $X(x_0, \rho)$. Then

$$\Delta_{(1,2)}(g^1, g^2)(\mathbf{x}) = \begin{vmatrix} g^1(x_1) & g^2(x_1) & 1 \\ g^1(x_2) & g^2(x_2) & 1 \\ g^1(x_0) & g^2(x_0) & 1 \end{vmatrix} = \begin{vmatrix} g^1(x_1) - g^1(x_0) & g^2(x_1) - g^2(x_0) \\ g^1(x_2) - g^1(x_0) & g^2(x_2) - g^2(x_0) \end{vmatrix}. \tag{1}$$

Since f is differentiable at each point y_0 in G by (i), then f satisfies the Stolz condition and

$$f(y) - f(y_0) = \sum_{i=1}^{2} D_i f(y_0)(y^i - y_0^i) + r(f; y_0, y)|y - y_0|. \tag{2}$$

In this equation replace y_0 and y by $g(x_0)$ and $g(x_1)$, respectively, and then by $g(x_0)$ and $g(x_2)$. Since $f \circ g(x) = 0$ for every x in E by (v), the resulting equations are

$$0 = \sum_{i=1}^{2} D_i f(y_0)[g^i(x_1) - g^i(x_0)] + r[f; g(x_0), g(x_1)]|g(x_1) - g(x_0)|, \tag{3}$$

$$0 = \sum_{i=1}^{2} D_i f(y_0)[g^i(x_2) - g^i(x_0)] + r[f; g(x_0), g(x_1)]|g(x_2) - g(x_0)|. \tag{4}$$

By (ii), $D_1 f(y_0) \neq 0$; multiply the first column of the matrix on the right in (1) by $D_1 f(y_0)$ and then divide the determinant by $D_1 f(y_0)$. Next, multiply the second column of the matrix on the right in (1) by $D_2 f(y_0)$ and add it to the first column. As a result of these transformations and equations (3) and (4),

$$\Delta_{(1,2)}(g^1, g^2)(\mathbf{x}) = [D_1 f(y_0)]^{-1} \begin{vmatrix} \sum_{i=1}^{2} D_i f(y_0)[g^i(x_1) - g^i(x_0)] & g^2(x_1) - g^2(x_0) \\ \sum_{i=1}^{2} D_i f(y_0)[g^i(x_2) - g^i(x_0)] & g^2(x_2) - g^2(x_0) \end{vmatrix} \tag{5}$$

$$= -[D_1 f(y_0)]^{-1} \begin{vmatrix} r[f; g(x_0), g(x_1)]|g(x_1) - g(x_0)| & g^2(x_1) - g^2(x_0) \\ r[f; g(x_0), g(x_2)]|g(x_2) - g(x_0)| & g^2(x_2) - g^2(x_0) \end{vmatrix}. \tag{6}$$

In

$$\lim_{\mathbf{x} \to x_0} \frac{\Delta_{(1,2)}(g^1, g^2)(\mathbf{x})}{\Delta(\mathbf{x})}, \tag{7}$$

apply the regularity condition to $\mathbf{x}$, which is in $X(x_0, \rho)$, and then divide the numerator and the denominator by $|x_1 - x_0||x_2 - x_0|$ in the usual way. Since f is differentiable by (i), and since g^1 and g^2 satisfy Lipschitz conditions by (iv), then g is continuous on E and

$$\lim_{\mathbf{x} \to x_0} r[f; g(x_0), g(x_k)] = 0, \qquad k = 1, 2.$$

Also, since g^1, g^2 satisfy Lipschitz conditions,

$$\frac{|g(x_k) - g(x_0)|}{|x_k - x_0|} \quad \text{and} \quad \frac{|g^2(x_k) - g^2(x_0)|}{|x_k - x_0|}, \qquad k = 1, 2,$$

are bounded as $\mathbf{x} \to x_0$. These facts are sufficient to prove that the limit in (7) exists and has the value 0 [recall (6) and the transformation described for the quotient in (7)]. This proof shows that $D_{(1,2)}(g^1, g^2)(\mathbf{x})$ exists and has the value zero for each x in E. Furthermore, a similar proof can be given if (ii) is replaced by the weaker hypothesis that $[D_1 f(y)]^2 + [D_2 f(y)]^2 \neq 0$ for every y in G.

Examples 6.2, $\cdots$, 6.5 are sufficient to show that the answer to Question 6.1 is "yes." However, an examination of these examples shows that in each case the function has a somewhat pathological nature. The functions is these examples are differentiable only because they map E into a small set, and the derivative in each case is zero. The only examples of differentiable functions $f: E \to \mathbb{R}^m$ whose components $(f^1, \cdots, f^m)$ are not differentiable which have been found are of this pathological type. Furthermore, the theorems proved already [see Theorem 4.5, for example] as well as those established later in this book indicate that a complete and satisfactory theory can be constructed only for functions $f: E \to \mathbb{R}^m$ whose components $(f^1, \cdots, f^m)$ are differentiable. For this reason it is useful to define a more restricted class of differentiable functions as stated in the next definition.

6.6 Definition. Let $f: E \to \mathbb{R}^m$ be a function which is defined on an open set E in $\mathbb{R}^n$. Then f is *differentiable in the restricted sense at* x_0 in E if and only if each of its components $(f^1, \cdots, f^m)$ is differentiable in the sense of Definition 2.8 at x_0. Also, f is *differentiable in the restricted sense on* E if and only if it is differentiable in the restricted sense at each point x_0 in E.

Henceforth, the statement "$f: E \to \mathbb{R}^m$, $E \subset \mathbb{R}^n$, is differentiable at x_0 (or on E)" will mean "$f: E \to \mathbb{R}^m$, $E \subset \mathbb{R}^n$, is differentiable in the restricted sense at x_0 (or on E)." The phrase "in the restricted sense" will usually be omitted in order to simplify the writing.

To summarize the properties of functions $f: E \to \mathbb{R}^m$ which are differentiable in the restricted sense at x_0, observe first [see Theorem 3.6 and Corollaries 3.7 and 3.8] that each component $f^1, \cdots, f^m$ satisfies the Stolz condition at x_0 and hence that

$$\begin{aligned} f^1(x) &= f^1(x_0) + D_1 f^1(x_0)(x^1 - x_0^1) + \cdots + D_n f^1(x_0)(x^n - x_0^n) + r(f^1; x_0, x)|x - x_0|, \\ &\cdots\cdots\cdots\cdots\cdots\cdots\cdots\cdots\cdots\cdots\cdots\cdots \\ f^m(x) &= f^m(x_0) + D_1 f^m(x_0)(x^1 - x_0^1) + \cdots + D_n f^m(x_0)(x^n - x_0^n) + r(f^m; x_0, x)|x - x_0|. \end{aligned} \tag{8}$$

These equations can be written in vector form as follows:

$$f(x) = f(x_0) + D(x - x_0) + r(f; x_0, x)|x - x_0|. \tag{9}$$

Thus, if $f: E \to \mathbb{R}^m$ is differentiable in the restricted sense at x_0, there exists a linear transformation $D: \mathbb{R}^n \to \mathbb{R}^m$ which approximates f as indicated in equations (8) and (9). The matrix of this linear transformation is the following matrix of derivatives:

$$\begin{bmatrix} D_1 f^1(x_0) & \cdots & D_n f^1(x_0) \\ \cdots & \cdots & \cdots \\ D_1 f^m(x_0) & \cdots & D_n f^m(x_0) \end{bmatrix}. \tag{10}$$

This matrix is sometimes called the Jacobian matrix of f at x_0. If $f: E \to \mathbb{R}^m$ is differentiable in the restricted sense at x_0, then [see Theorem 3.12 and Corollary 3.13] the derivatives $D_{(j_1, \cdots, j_m)}(f^1, \cdots, f^m)(x_0)$, $(j_1, \cdots, j_m) \in (m/n)$, exist and

$$D_{(j_1, \cdots, j_m)}(f^1, \cdots, f^m)(x_0) = \lim_{\mathbf{x} \to x_0} \frac{\Delta_{(j_1, \cdots, j_m)}(f^1, \cdots, f^m)(\mathbf{x})}{\Delta(\mathbf{x})}, \tag{11}$$

$$D_{(j_1, \cdots, j_m)}(f^1, \cdots, f^m)(x_0) = \det \begin{bmatrix} D_{j_1} f^1(x_0) & \cdots & D_{j_m} f^1(x_0) \\ \cdots & \cdots & \cdots \\ D_{j_1} f^m(x_0) & \cdots & D_{j_m} f^m(x_0) \end{bmatrix}. \tag{12}$$

Thus $D_{(j_1, \cdots, j_m)}(f^1, \cdots, f^m)(x_0)$ is the determinant of the minor in columns $(j_1, \cdots, j_m)$ of the matrix in (10).

There is another way to describe the manner in which the linear transformation $D: \mathbb{R}^n \to \mathbb{R}^m$ approximates $f: E \to \mathbb{R}^m$ at x_0. Equations (8) show that

$$\begin{aligned} &\left\{ \sum_{i=1}^m \left[f^i(x) - f^i(x_0) - \sum_{j=1}^n D_j f^i(x_0)(x^j - x_0^j) \right]^2 \right\}^{1/2} \\ &\qquad = \left\{ \sum_{i=1}^m [r(f^i; x_0, x)]^2 \right\}^{1/2} |x - x_0|. \end{aligned} \tag{13}$$

Define $|r(f; x_0, x)|$ as follows:

$$r(f; x_0, x) = \left\{ \sum_{i=1}^m [r(f^i; x_0, x)]^2 \right\}^{1/2}. \tag{14}$$

Then

$$\lim_{x \to x_0} |r(f; x_0, x)| = 0, \qquad |r(f; x_0, x_0| = 0, \tag{15}$$

and (13) can be written in vector form thus:

$$|f(x) - f(x_0) - D(x - x_0)| = |r(f; x_0, x)|\,|x - x_0|. \tag{16}$$

Divide the two sides of this equation by $|x - x_0|$ and take the limit as $x \to x_0$. Then (15) shows that

$$\lim_{x \to x_0} \frac{|f(x) - f(x_0) - D(x - x_0)|}{|x - x_0|} = \lim_{x \to x_0} |r(f; x_0, x)| = 0. \tag{17}$$

6.7 Theorem. *The class of functions $f: E \to \mathbb{R}^m$, $E \subset \mathbb{R}^n$, which are differentiable in the restricted sense at x_0 is exactly the same as the class of functions $f: E \to \mathbb{R}^m$, $E \subset \mathbb{R}^n$, which are defined to be differentiable in current modern treatments of differentiation.*

PROOF. The function $f: E \to \mathbb{R}^m$, $E \subset \mathbb{R}^n$, is differentiable in the restricted sense at x_0 if and only if its components $(f^1, \cdots, f^m)$ satisfy the equations in (8). Apostol [see page 258 in 9 in References and Notes] defines $f: E \to \mathbb{R}^m$ to be differentiable if and only if there exists a matrix $[d_j^i]$, $i = 1, \cdots, m$, $j = 1, \cdots, n$, and functions $r(f^i; x_0, \cdot)$, $i = 1, \cdots, m$, such that

$$\begin{aligned} &f^1(x) = f^1(x_0) + \sum_{j=1}^{n} d_j^1(x^j - x_0^j) + r(f^1; x_0, x)|x - x_0|, \\ &\cdots\cdots\cdots\cdots\cdots\cdots\cdots\cdots\cdots\cdots\cdots\cdots \\ &f^m(x) = f^m(x_0) + \sum_{j=1}^{n} d_j^m(x^j - x_0^j) + r(f^m; x_0, x)|x - x_0|, \\ &\lim_{x \to x_0} r(f^i; x_0, x) = 0, \; r(f^i; x_0, x_0) = 0, \qquad i = 1, \cdots, m. \end{aligned} \tag{18}$$

Then differentiability in the restricted sense implies differentiability in the sense of Apostol. Conversely, if $f: E \to \mathbb{R}^m$ is differentiable in the sense of Apostol, then (18) shows that each of the functions $f^1, \cdots, f^m$ satisfies the Stolz condition at x_0 and is therefore differentiable in the sense of Definition 2.8. Then $d_j^i = D_j f^i(x_0)$ by Theorem 3.3; the equations in (8) hold; and f is differentiable in the restricted sense. These statements show that the linear transformation with matrix $[d_j^i]$ which satisfies (18) is unique.

Next, Spivak [see page 16 in 10 in References and Notes] defines $f: E \to \mathbb{R}^m$ to be differentiable if and only if there exists a linear transformation $D: \mathbb{R}^n \to \mathbb{R}^m$ with matrix $[d_j^i]$ such that

$$\lim_{x \to x_0} \frac{|f(x) - f(x_0) - D(x - x_0)|}{|x - x_0|} = 0. \tag{19}$$

If $f: E \to \mathbb{R}^m$ is differentiable in the restricted sense at x_0, then (17) shows that f is differentiable in the sense of Spivak with $[d_j^i] = [D_j f^i(x_0)]$. It remains only to show that differentiability in the sense of Spivak implies differentiability in the restricted sense at x_0. Assume that there is a linear transformation $D: \mathbb{R}^n \to \mathbb{R}^m$ with matrix $[d_j^i]$ which satisfies (19). Written out in detail, this equation is

$$\lim_{x \to x_0} \frac{\left\{ \sum_{i=1}^{m} \left[f^i(x) - f^i(x_0) - \sum_{j=1}^{n} d_j^i(x^j - x_0^j) \right]^2 \right\}^{1/2}}{|x - x_0|} = 0. \tag{20}$$

Define $r(f^i; x_0, x)$ by the following equation:

$$f^i(x) - f^i(x_0) - \sum_{j=1}^{n} d_j^i(x^j - x_0^j) = r(f^i; x_0, x)|x - x_0|, \qquad r(f^i; x_0, x_0) = 0, \tag{21}$$

Then the absolute value of the expression on the left in (21) is equal to or less than the numerator in (20); from this statement it follows that

$$|r(f^i; x_0, x)| \leq \frac{\left\{\sum_{i=1}^{m}\left[f^i(x) - f^i(x_0) - \sum_{j=1}^{n} d_j^i(x^j - x_0^j)\right]^2\right\}^{1/2}}{|x - x_0|}, \tag{22}$$

and (20) shows that

$$\lim_{x \to x_0} r(f^i; x_0, x) = 0, \qquad i = 1, \cdots, m. \tag{23}$$

This equation and (21) show that each f^i satisfies the Stolz condition at x_0; the equations (8) hold as before; and f is differentiable in the restricted sense at x_0. It follows from these statements that the linear transformation $D: \mathbb{R}^n \to \mathbb{R}^m$ which satisfies the Spivak condition (19) is unique, and that its matrix is $[D_j f^i(x_0)]$, $i = 1, \cdots, m$, $j = 1, \cdots, n$.

Thus, differentiability in the restricted sense, differentiability in the sense of Apostol, and differentiability in the sense of Spivak are equivalent, and the proof of Theorem 6.7 is complete. □

Exercises

6.1. Let $f: E \to \mathbb{R}^m$, $E \subset \mathbb{R}^n$, be a linear transformation with components $(f^1, \cdots, f^m)$ and matrix $[a_j^i]$, $i = 1, \cdots, m$, $j = 1, \cdots, n$.

(a) Does each of the functions f^i, $i = 1, \cdots, m$, satisfy the Stolz condition at x_0 in E? If so, what are the constants in the Stolz condition?

(b) Is $f: E \to \mathbb{R}^m$ differentiable in the sense of Definition 2.8? Is f differentiable in the restricted sense? Explain.

(c) Is f differentiable in the sense of Apostol? If so, find the matrix $[d_j^i]$ and functions $r(f^i; x_0, \cdot)$ in (18). Are the matrix $[d_j^i]$ and functions $r(f^i; x_0, \cdot)$ unique?

(d) Is f differentiable in the sense of Spivak? If so, find the linear transformation D with matrix $[d_j^i]$ in (19). Is the linear transformation D in (19) unique?

(e) Compare the linear transformation in (8) (differentiability in the restricted sense) with the linear transformations in (18) (differentiability in the sense of Apostol) and in (19) (differentiability in the sense of Spivak).

6.2. If $f: E \to \mathbb{R}^m$, $E \subset \mathbb{R}^n$, is differentiable in the sense of Definition 2.8, is it differentiable in the restricted sense? If f is differentiable in the restricted sense, is it differentiable in the sense of Definition 2.8? Explain.

6.3. Let $f: E \to \mathbb{R}^m$, $E \subset \mathbb{R}^n$, be a function whose components $(f^1, \cdots, f^m)$ have continuous partial derivatives in the open set E. Describe all types of differentiability possessed by f and by its components $(f^1, \cdots, f^m)$.

6.4. A function $f: E \to \mathbb{R}^m$, $E \subset \mathbb{R}^n$, may have one or more of four types of differentiability: (i) differentiability in the sense of Definition 2.8; (ii) differentiability in the restricted sense; (iii) differentiability in the sense of Apostol; (iv) differentiability in the sense of Spivak. Which, if any, of these types of differentiability at x_0 imply that f is continuous at x_0? Explain.

6.5. Let $f: E \to \mathbb{R}^m$ and $g: E \to \mathbb{R}^m$, $E \subset \mathbb{R}^n$, be two functions which are differentiable in the restricted sense at x_0 in E. If a and b are constants, then $af + bg$ is defined to be the function with components $(af^1 + bg^1, \cdots, af^m + bg^m)$. Prove that $af + bg$ is differentiable in the restricted sense at x_0. Can you prove this result if f and g are assumed to be differentiable at x_0 only in the sense of Definition 2.8? Explain.

6.6. Let $\mathscr{D}$ denote the class of all functions $f: E \to \mathbb{R}^m$, $E \subset \mathbb{R}^n$, which are differentiable in the restricted sense on E. Define two operations in $\mathscr{D}$ as follows:

(i) addition $\oplus$: if $f: (f^1, \cdots, f^m)$ and $g: (g^1, \cdots, g^m)$ are in $\mathscr{D}$, then $f \oplus g = (f^1 + g^1, \cdots, f^m + g^m)$;
(ii) scalar multiplication $\odot$: if a is a constant and $f: (f^1, \cdots, f^m)$ is in $\mathscr{D}$, then $a \odot f = (af^1, \cdots, af^m)$.

Prove that the system $(\mathscr{D}, \oplus, \odot)$ thus defined is a vector space. [Hint. Vector spaces are defined in Section 89 in Appendix 2; see also (3) in Section 96.]

6.7. Let $\mathscr{D}_0$ be the class of all functions $f: E \to \mathbb{R}^m$, $E \subset \mathbb{R}^n$, which are differentiable in the sense of Definition 2.8, and let $\oplus$ and $\odot$ be defined as in Exercise 6.6. Can you prove that $(\mathscr{D}_0, \oplus, \odot)$ is a vector space? If not, explain why not.

6.8. Let $\mathscr{D}_C$ denote the class of all functions $f: E \to \mathbb{R}^m$, $E \subset \mathbb{R}^n$, whose components $(f^1, \cdots, f^m)$ have continuous derivatives $D_j f^i$, $i = 1, \cdots, m$, $j = 1, \cdots, n$. Prove that the system $(\mathscr{D}_C, \oplus, \odot)$ is a vector space. [Hint. Continuous functions are defined in Definition 96.2; Section 96 contains examples of vector spaces whose elements are functions.]

7. The Derivative as an Operator

Let m and n be integers such that $1 \leqq m \leqq n$, and let $\mathscr{D}^m(E^n)$ denote the class of functions $f: E \to \mathbb{R}^m$, $E \subset \mathbb{R}^n$, which are differentiable in the restricted sense. Then f has components $(f^1, \cdots, f^m)$, and each of these components is differentiable. If f is in $\mathscr{D}^m(E^n)$, the derivative of f at x_0 in E is a point with coordinates $(D_{(j_1, \cdots, j_m)}(f^1, \cdots, f^m)(x_0): (j_1, \cdots, j_m) \in (m/n))$ in $\mathbb{R}^{C(n,m)}$. If the derivative of f is evaluated at every point x in E, the result is a function $h: E \to \mathbb{R}^{C(n,m)}$ with components $(h^{(j_1, \cdots, j_m)}: (j_1, \cdots, j_m) \in (m/n))$ such that $h^{(j_1, \cdots, j_m)}(x) = D_{(j_1, \cdots, j_m)}(f^1, \cdots, f^m)(x)$, $x \in E$. Let $\mathscr{F}^{C(n,m)}(E^n)$ denote the class of all functions with domain E in $\mathbb{R}^n$ and range in $\mathbb{R}^{C(n,m)}$. Thus differentiation can be considered as a function D which transforms the function $f: E \to \mathbb{R}^m$ in $\mathscr{D}^m(E^n)$ into the function $h: E \to \mathbb{R}^{C(n,m)}$ in $\mathscr{F}^{C(n,m)}(E^n)$, and we write $h = D(f)$. A function whose domain and range are classes of functions is frequently called an *operator*. In particular, the function D such that $h = D(f)$ is called the *differentiation operator*. This section establishes some of the elementary properties of the differentiation operator.

7.1 Example. Let $f: (f^1, f^2)$ be the function in $\mathscr{D}^2(E^2)$ such that

$$f^1(x) = 2x^1 - 3x^2, \qquad f^2(x) = 4x^1x^2, \qquad x \in E.$$

If $h: E \to \mathbb{R}$ is the function into which the differentiation operator $D: \mathscr{D}^2(E^2) \to \mathscr{F}(E^2)$ transforms f, then

$$h(x) = D_{(1,2)}f(x) = \begin{vmatrix} 2 & -3 \\ 4x^2 & 4x^1 \end{vmatrix} = 8x^1 + 12x^2, \qquad x \in E.$$

7.2 Example. Let $f: (f^1, f^2)$ be the function in $\mathscr{D}^2(E^3)$ such that

$$f^1(x) = a_1^1x^1 + a_2^1x^2 + a_3^1x^3,$$
$$f^2(x) = a_1^2x^1 + a_2^2x^2 + a_3^2x^3, \qquad x \in E.$$

If $h: E \to \mathbb{R}^3$ is the function with components $(h^{(1,2)}, h^{(1,3)}, h^{(2,3)})$ into which the differentiation operator $D: \mathscr{D}^2(E^3) \to \mathscr{F}^3(E^3)$ transforms f, then

$$h^{(1,2)}(x) = D_{(1,2)}(f^1, f^2)(x) = \begin{vmatrix} a_1^1 & a_2^1 \\ a_1^2 & a_2^2 \end{vmatrix} = a_1^1a_2^2 - a_1^2a_2^1;$$

$$h^{(1,3)}(x) = D_{(1,3)}(f^1, f^2)(x) = \begin{vmatrix} a_1^1 & a_3^1 \\ a_1^2 & a_3^2 \end{vmatrix} = a_1^1a_3^2 - a_1^2a_3^1;$$

$$h^{(2,3)}(x) = D_{(2,3)}(f^1, f^2)(x) = \begin{vmatrix} a_2^1 & a_3^1 \\ a_2^2 & a_3^2 \end{vmatrix} = a_2^1a_3^2 - a_2^2a_3^1.$$

In this example, h is a constant function in $\mathscr{F}^3(E^3)$.

7.3 Definition. Let V be a vector space with elements v [see Section 89 in Appendix 2]; let $V^k = V \times \cdots \times V$ (k factors); let a_1^i, a_2^i, $i = 1, \cdots, k$, belong to $\mathbb{R}$; and let W be a vector space. Then $T: V^k \to W$ is a *multilinear operator* (*transformation*) if and only if

$$\begin{aligned} &T(v^1, \cdots, a_1^iv_1^i + a_2^iv_2^i, \cdots, v^k) \\ &\qquad = a_1^iT(v^1, \cdots, v_1^i, \cdots, v^k) + a_2^iT(v^1, \cdots, v_2^i, \cdots, v^k) \end{aligned} \tag{1}$$

for $i = 1, \cdots, k$. If $W = \mathbb{R}$, then $T: V^k \to \mathbb{R}$ is called a *multilinear function*. If $k = 1$, then $T: V \to W$ is a *linear operator* or *function*.

7.4 Example. The inner product (u, v) of two vectors $u: (a_1, \cdots, a_n)$ and $v: (b_1, \cdots, b_n)$ in $\mathbb{R}^n$ is defined as follows in Definition 84.1:

$$(u, v) = \sum_{j=1}^{n} a_jb_j. \tag{2}$$

Then (u, v) is linear in u and also in v, and the inner product is a bilinear function defined on $\mathbb{R}^n \times \mathbb{R}^n$ [see Theorem 84.2].

7.5 Example. The elements $(a_{i1}, \cdots, a_{in})$ in the i-th row of an $n \times n$ matrix $A : [a_{ij}]$ form a vector $A_i : (a_{i1}, \cdots, a_{in})$ in $\mathbb{R}^n$. The definition of the determinant in Section 76 shows that $\det A$ is a function of the vectors $A_1, \cdots, A_n$ in its rows. Thus $\det A = \det(A_1, \cdots, A_n)$. Theorems 77.1 and 77.2 in Appendix 1 prove that

$$\det(A_1, \cdots, tA_i, \cdots, A_n) = t \det(A_1, \cdots, A_i, \cdots, A_n),$$
$$\det(A_1, \cdots, A_i + B_i, \cdots, A_n) = \det(A_1, \cdots, A_i, \cdots, A_n) + \det(A_1, \cdots, B_i, \cdots, A_n), \tag{3}$$

for $i = 1, \cdots, n$. These two properties show that the determinant of an $n \times n$ matrix is a multilinear function defined on $(\mathbb{R}^n)^n$.

7.6 Theorem. *Let $\mathscr{D}^m(E^n)$ be the class of differentiable functions $f: E \to \mathbb{R}^m$, $E \subset \mathbb{R}^n$, and let $\mathscr{F}^{C(n,m)}(E^n)$ be the class of functions whose domain is E and whose range is $\mathbb{R}^{C(n,m)}$, $1 \leqq m \leqq n$. The differentiation operator*

$$D : \mathscr{D}^m(E^n) \to \mathscr{F}^{C(n,m)}(E^n) \tag{4}$$

has the following properties.

(a) *If $m = 1$, then D is a linear operator.*
(b) *If $1 < m \leqq n$, then D is a multilinear operator.*
(c) *There exist bounded functions f in $\mathscr{D}^m(E^n)$ such that $D(f)$ is a function which has values arbitrarily far from the origin in $\mathbb{R}^{C(n,m)}$.*
(d) *D is a many-to-one operator.*
(e) *Let $f_1, \cdots, f_k, \cdots$ be a sequence of functions in $\mathscr{D}^m(E^n)$ which converges uniformly on E to a function f_0 in $\mathscr{D}^m(E^n)$ [see Definition 96.19]. In some cases the sequence $D(f_k)$, $k = 1, 2, \cdots$, does not converge uniformly on E to $D(f_0)$.*

PROOF. If $m = 1$, equations (57) and (58) in Theorem 3.21 show that

$$D_j(c_1 f_1 + c_2 f_2)(x) = c_1 D_j f_1(x) + c_2 D_j f_2(x), \qquad x \in E, \qquad j = 1, \cdots, n. \tag{5}$$

Then $D(c_1 f_1 + c_2 f_2) = c_1 D(f_1) + c_2 D(f_2)$ if $m = 1$, and D is a linear operator as stated in (a). If $m > 1$, then the components of $D(f)$ are functions with the following values [see (12) in Section 6]:

$$D_{(j_1, \cdots, j_m)}(f^1, \cdots, f^m)(x) = \det \begin{bmatrix} D_{j_1} f^1(x) & \cdots & D_{j_m} f^1(x) \\ \cdots & \cdots & \cdots \\ D_{j_1} f^m(x) & \cdots & D_{j_m} f^m(x) \end{bmatrix}, \qquad (j_1, \cdots, j_m) \in (m/n). \tag{6}$$

Now part (a) of the theorem shows that the differentiation operator is linear in the case $m = 1$. Also, Example 7.5 shows that the determinant is linear in each row of its matrix. These facts and (6) prove that

$$\begin{aligned} D(f^1, \cdots, c_1^i f_1^i + c_2^i f_2^i, \cdots, f^m) \\ = c_1^i D(f^1, \cdots, f_1^i, \cdots, f^m) + c_2^i D(f^1, \cdots, f_2^i, \cdots, f^m), \\ i = 1, \cdots, m, \end{aligned} \tag{7}$$

and D is a multilinear operator by Definition 7.3.

Examples can be used to prove (c) in Theorem 7.6. Consider the functions $f_k : \mathbb{R} \to \mathbb{R}$ such that

$$f_k(x) = \sin kx, \qquad x \in \mathbb{R}, \qquad k = 1, 2, \cdots. \tag{8}$$

Then

$$D_x f_k(x) = k \cos kx, \qquad x \in \mathbb{R}, \qquad k = 1, 2, \cdots. \tag{9}$$

Although $|f_k(x)| \leqq 1$ for x in $\mathbb{R}$ and $k = 1, 2, \cdots,$

$$\lim_{k \to \infty} \max\{|D_x f_k(x)| : x \in \mathbb{R}\} = +\infty, \tag{10}$$

and there are functions in the bounded set $\{f_k : k = 1, 2, \cdots\}$ whose derivatives have arbitrarily large values. Similar examples for other values of m and n can be constructed without difficulty. These examples prove (c).

To prove (d), let $(f^1, \cdots, f^m) : E \to \mathbb{R}^m$, $E \subset \mathbb{R}^n$, be a function in $\mathscr{D}^m(E^n)$, and let $c^1, \cdots, c^m$ be constants. Then by Theorem 3.21, each of the functions $(f^i + c^i) : E \to \mathbb{R}$, $i = 1, \cdots, m$, is differentiable and

$$D_j(f^i + c^i)(x) = D_j f^i(x), \qquad x \in E, \qquad j = 1, \cdots, n, \qquad i = 1, \cdots, m. \tag{11}$$

Then for each set $(c^1, \cdots, c^m)$ in $\mathbb{R}^m$, the function $(f^1 + c^1, \cdots, f^m + c^m)$: $E \to \mathbb{R}^m$ is in $\mathscr{D}^m(E^n)$, and (12) in Section 6 and the definition of the differentiation operator D show that

$$D(f^1 + c^1, \cdots, f^m + c^m) = D(f^1, \cdots, f^m). \tag{12}$$

This example shows that, if f is any function in $\mathscr{D}^m(E^n)$, then there is an infinite set of functions in $\mathscr{D}^m(E^n)$ which are mapped by D into $D(f)$. Example 7.7 and Exercise 7.5 describe even larger classes of functions which D maps into $D(f)$.

To prove Theorem 7.6(e), consider the sequence of functions $f_k : \mathbb{R} \to \mathbb{R}$, $k = 1, 2, \cdots$, such that

$$f_k(x) = \frac{\sin kx}{k}. \tag{13}$$

Then

$$\lim_{k \to \infty} f_k(x) = 0, \qquad \text{uniformly on } \mathbb{R}. \tag{14}$$

But $D_x f_k(x) = \cos kx$, $x \in \mathbb{R}$, and this sequence of functions does not converge to the zero function. Thus, if limits are interpreted as uniform limits, then

$$D\left(\lim_{k\to\infty} f_k\right) = D(0) = 0, \qquad [0 \text{ is the zero function on } \mathbb{R}]$$
$$\lim_{k\to\infty} D(f_k) \text{ does not exist.} \tag{15}$$

This example suggests how examples can be constructed for other values of m. The proof of Theorem 7.6 is complete. □

7.7 Example. Let $f: E \to \mathbb{R}^2$, $E \subset \mathbb{R}^3$, be a function whose components (f^1, f^2) are differentiable on E. Then $D(f)$, or $D(f^1, f^2)$ is a function in $\mathscr{F}^3(E^3)$, and the components of $D(f^1, f^2)$ are $D_{(j_1, j_2)}(f^1, f^2)$, $(j_1, j_2) \in (2/3)$. Let A be a matrix $[a^i_j]$, $i, j = 1, 2$, such that $\det A = 1$. Then the Binet–Cauchy multiplication theorem [see Theorem 80.1] can be used to prove that

$$\begin{aligned} &D_{(j_1, j_2)}(a^1_1 f^1 + a^1_2 f^2, a^2_1 f^1 + a^2_2 f^2)(x) \\ &\quad = \det \begin{bmatrix} a^1_1 D_{j_1} f^1(x) + a^1_2 D_{j_1} f^2(x) & a^1_1 D_{j_2} f^1(x) + a^1_2 D_{j_2} f^2(x) \\ a^2_1 D_{j_1} f^1(x) + a^2_2 D_{j_1} f^2(x) & a^2_1 D_{j_2} f^1(x) + a^2_2 D_{j_2} f^2(x) \end{bmatrix} \\ &\quad = \det \begin{bmatrix} a^1_1 & a^1_2 \\ a^2_1 & a^2_2 \end{bmatrix} \det \begin{bmatrix} D_{j_1} f^1(x) & D_{j_2} f^1(x) \\ D_{j_1} f^2(x) & D_{j_2} f^2(x) \end{bmatrix} \end{aligned} \tag{16}$$

for $x \in E$ and $(j_1, j_2) \in (2/3)$. Because $\det A = 1$, this equation shows that

$$D(a^1_1 f^1 + a^1_2 f^2, a^2_1 f^1 + a^2_2 f^2) = D(f^1, f^2). \tag{17}$$

Since there is a large class of matrices A such that $\det A = 1$, this example shows that there is a large class of functions in $\mathscr{D}^2(E^3)$ which are transformed by D into the function $D(f^1, f^2)$ in $\mathscr{F}^3(E^3)$.

Exercises

7.1. If $f: E \to \mathbb{R}$; $E \subset \mathbb{R}^3$, is a function such that $f(x) = (x^1)^2 + (x^2)^2 + (x^3)^2$, find $D(f)$. How many components does $D(f)$ have?

7.2. Let $f: E \to \mathbb{R}^2$, $E \subset \mathbb{R}^3$, be the function whose components (f^1, f^2) are defined as follows.

$$f^1(x) = (x^1)^2 + (x^2)^2 + (x^3)^2,$$
$$f^2(x) = x^1 x^2 + x^1 x^3 + x^2 x^3.$$

Find $D(f)$. How many components does $D(f)$ have?

7.3. Let $f: E \to \mathbb{R}^3$, $E \subset \mathbb{R}^3$, be the function whose components (f^1, f^2, f^3) are defined as follows.

$$f^1(x) = 3x^1 - 4x^2 + 6x^3,$$
$$f^2(x) = 3x^1x^2 + 4x^2x^3,$$
$$f^3(x) = 2x^1x^2x^3.$$

Find $D(f)$. How many components does $D(f)$ have?

7.4. (a) Let $f: E \to \mathbb{R}^m$ be a function in $\mathscr{D}^m(E^n)$. How many components does f have?
(b) How many components does $D(f)$ have? Give a formula for the components of $D(f)$. What is the domain of D? of $D(f)$?

7.5. Let $f: E \to \mathbb{R}^m$, $E \subset \mathbb{R}^n$, be a function whose components $(f^1, \cdots, f^m)$ are differentiable on E. Let A be an $m \times m$ matrix $[a_j^i]$ such that $\det A = 1$, and let B be an $m \times 1$ matrix $[b^i]$. Define a function $g: E \to \mathbb{R}^m$ with components $(g^1, \cdots, g^m)$ as follows:

$$\begin{bmatrix} g^1 \\ \cdots \\ g^m \end{bmatrix} = \begin{bmatrix} a_1^1 & \cdots & a_m^1 \\ \cdots & \cdots & \cdots \\ a_1^m & \cdots & a_m^m \end{bmatrix} \begin{bmatrix} f^1 \\ \cdots \\ f^m \end{bmatrix} + \begin{bmatrix} b^1 \\ \cdots \\ b^m \end{bmatrix}.$$

Prove that $D(g^1, \cdots, g^m) = D(f^1, \cdots, f^m)$. Compare this result with Theorem 7.6 (d) and Example 7.7.

7.6. (Rolle's Theorem for Polynomial Functions). Give an algebraic proof of the following theorem. If r_1 and r_2 are two real roots of the polynomial equation $p(x) = 0$, then $D_1 p(x) = 0$ has at least one real root between r_1 and r_2. [Hint. Show that $p(x) = (x - r_1)^r(x - r_2)^s q(x)$ and that $q(x)$ is not zero on $[r_1, r_2]$ if $p(x) = 0$ has no root between r_1 and r_2; then examine the sign of the derivative of the polynomial p at r_1 and at r_2.]

7.7. (Rolle's Theorem). Prove the following theorem. Hypotheses: (i) f is continuous on $[a, b]$; (ii) f has a derivative on (a, b); and (iii) $f(a) = f(b)$. Conclusion: there exists an x_0 in (a, b) such that $D_1 f(x_0) = 0$. [Hint. Theorem 96.14.]

7.8. Prove the following theorem. If $f: [a, b] \to \mathbb{R}$ is a function which has a derivative on $[a, b]$, and if $D_1 f(a) D_1 f(b) < 0$, then there exists an x_0 in (a, b) such that $D_1 f(x_0) = 0$.

7.9. (Intermediate-Value Theorem). Prove the following theorem. If $f: [a, b] \to \mathbb{R}$ is continuous on $[a, b]$, and if c is a number between $f(a)$ and $f(b)$, then there exists an x_0 in (a, b) such that $f(x_0) = c$.

7.10. (Darboux's Theorem). Prove the following theorem. If $f: [a, b] \to \mathbb{R}$ is a function which has a derivative $D_1 f$ on $[a, b]$, and if c is a number between $D_1 f(a)$ and $D_1 f(b)$, then there exists an x_0 in (a, b) such that $D_1 f(x_0) = c$. Compare this theorem with the intermediate-value theorem in Exercise 7.9; observe that the derivative of f is not assumed to be continuous. [Hint. Apply the theorem in Exercise 7.8 to the function $g(x) = f(x) - c(x - a)$.]

7.11. Prove the following theorem. If $f: [a, b] \to \mathbb{R}$ is a function which has a derivative on $[a, b]$, and if x_0 is a point in (a, b) at which $D_1 f$ has a limit on the right and a limit on the left, then these limits are each equal to $D_1 f(x_0)$, and the derivative $D_1 f$ is continuous at x_0. Prove also that $D_1 f$ is continuous if it is monotonic.

7.12. The derivative of each of the following functions is discontinuous at $x = 0$. For each function sketch the graph of the derivative $D_1 f$ in a small neighborhood of $x = 0$.

(a) $f(x) = x^2 \sin(1/x), \quad x \neq 0, \quad$ and $f(0) = 0$.
(b) $f(x) = 2x + 1 + x^2 \sin(1/x), \quad x \neq 0, \quad$ and $f(0) = 1$.
(c) $f(x) = x^2 \cos(1/x), \quad x > 0; f(x) = x^2, \quad x \leqq 0$.
(d) $f(x) = x^3 + x^2 \sin(1/x), \quad x \neq 0, \quad$ and $f(0) = 0$.

7.13. Construct other examples of functions which have discontinuous derivatives. Sketch a graph of the derivative of each function and use it to give a geometric explanation of why Darboux's theorem is true.

CHAPTER 2

Uniform Differentiability and Approximations; Mappings

8. Introduction

This chapter treats certain approximations to derivatives and increments of functions. The proofs of most of the results in the chapter depend in the last analysis on the mean-value theorem for functions $f: [a, b] \to \mathbb{R}$. The following notation was introduced in the first chapter.

$$\mathbf{x} = \begin{bmatrix} x_1 \\ x_0 \end{bmatrix}, \quad \Delta(\mathbf{x}) = \begin{vmatrix} x_1 & 1 \\ x_0 & 1 \end{vmatrix}, \quad \Delta_1 f(\mathbf{x}) = \begin{vmatrix} f(x_1) & 1 \\ f(x_0) & 1 \end{vmatrix}, \quad D_1 f(x) = f'(x). \tag{1}$$

8.1 Theorem (Mean-Value Theorem). *If f is continuous on $[x_0, x_1]$ and differentiable on (x_0, x_1), then*

$$\Delta_1 f(\mathbf{x}) = D_1 f(x^*)\Delta(\mathbf{x}), \qquad x^* \in (x_0, x_1). \tag{2}$$

The definition of the derivative shows that $\Delta_1 f(\mathbf{x})/\Delta(\mathbf{x})$ is an approximation to $D_1 f(x_0)$, and that $D_1 f(x_0)\Delta(\mathbf{x})$ is an approximation to $\Delta_1 f(\mathbf{x})$. Also

$$\frac{\Delta_1 f(\mathbf{x})}{\Delta(\mathbf{x})} = D_1 f(x_0) + \left\{ \frac{\Delta_1 f(\mathbf{x})}{\Delta(\mathbf{x})} - D_1 f(x_0) \right\}, \tag{3}$$

$$\Delta_1 f(\mathbf{x}) = D_1 f(x_0)\Delta(\mathbf{x}) + \left\{ \frac{\Delta_1 f(\mathbf{x})}{\Delta(\mathbf{x})} - D_1 f(x_0) \right\} \Delta(\mathbf{x}). \tag{4}$$

The approximations are good if and only if

$$\frac{\Delta_1 f(\mathbf{x})}{\Delta(\mathbf{x})} - D_1 f(x_0) \tag{5}$$

is small. If f is differentiable on (x_0, x_1), then (2) shows that

$$\frac{\Delta_1 f(\mathbf{x})}{\Delta(\mathbf{x})} - D_1 f(x_0) = D_1 f(x^*) - D_1 f(x_0), \qquad x^* \in (x_0, x_1). \tag{6}$$

If (5) is uniformly small, relative to $|x_1 - x_0|$, on $[a, b]$, then f is said to be *uniformly differentiable* and Section 10 will show that $D_1 f$ is continuous. If $D_1 f$ is continuous on $[a, b]$, then Theorem 96.18 shows that $D_1 f$ is uniformly continuous on $[a, b]$ and (6) shows that (5) is uniformly small. Thus if $f: [a, b] \to \mathbb{R}$ is differentiable on $[a, b]$, it is uniformly differentiable if and only if $D_1 f$ is continuous on $[a, b]$, and the approximations in (3) and (4) are uniformly close, relative to $|x_1 - x_0|$, on $[a, b]$ if and only if $D_1 f$ is continuous on $[a, b]$.

These remarks suggest the nature of the problems and results in Sections 9, 10, 11. These sections generalize, for functions $f: [a_1, b_2] \times \cdots \times [a_n, b_n] \to \mathbb{R}^m$, the results outlined in this introduction. Section 12 contains some applications; they are theorems on mappings which are important in later chapters.

Exercises

8.1. Prove the mean-value theorem in Theorem 8.1. [Hint. Exercise 7.7.]

8.2. Prove the following theorem. If $f: [a, b] \to \mathbb{R}$ is differentiable on $[a, b]$, and if $f'(x) = 0$ for all $\mathbf{x}$ in $[a, b]$, then f is a constant function: $f(x) = c$, x in $[a, b]$.

8.3. Prove the following theorem. If the function $f: [a, b] \to \mathbb{R}$ is continuous on $[a, b]$, and if $F(x) = \int_a^x f(t)\,dt$, $x \in [a, b]$, then F is differentiable on $[a, b]$ and $D_1 F(x_0) = f(x_0)$, $x_0 \in [a, b]$.

8.4. Prove the following theorem, which is known as the fundamental theorem of the integral calculus. If the function $f: [a, b] \to \mathbb{R}$ has a continuous derivative f' on $[a, b]$, then the integral $\int_a^b f'(x)\,dx$ exists, and

$$\int_a^b f'(x)\,dx = f(b) - f(a).$$

[Hint. Define a function $g: [a, b] \to \mathbb{R}$ as follows:

$$g(x) = f(x) - \int_a^x f'(t)\,dt, \qquad x \in [a, b].$$

Find $g'(x)$ and then use the theorem in Exercise 8.2.]

8.5. Prove the following theorem, which is known as the mean-value theorem for integrals. If $f: [a, b] \to \mathbb{R}$ is continuous on $[a, b]$, then there exists an x^* such that

$$\int_a^b f(x)\,dx = f(x^*)(b - a), \qquad x^* \in (a, b).$$

[Hint. Prove that

$$\min\{f(x) : a \leqq x \leqq b\} \leqq \frac{1}{b - a} \int_a^b f(x)\,dx \leqq \max\{f(x) : a \leqq x \leqq b\};$$

then use the intermediate-value theorem in Exercise 7.9.]

8.6. Use the fundamental theorem of the integral calculus in Exercise 8.4 and the mean-value theorem for integrals in Exercise 8.5 to prove the following theorem. If the function $f: [a, b] \to \mathbb{R}$ has a continuous derivative on $[a, b]$, and if $[x_0, x_1]$ is in $[a, b]$, then there exists an x^* such that $f(x_1) - f(x_0) = f'(x^*)(x_1 - x_0)$, $x_0 < x^* < x_1$. Compare this theorem with the mean-value theorem in Theorem 8.1. [Hint. $f(x_1) - f(x_0) = \int_{x_0}^{x_1} f'(x)\,dx = f'(x^*)(x_1 - x_0)$, $x_0 < x^* < x_1$.]

8.7. The mean-value theorem for derivatives [Theorem 8.1] implies the fundamental theorem of the integral calculus. Prove this statement by using Theorem 8.1 to prove the following theorem. If $f: [a, b] \to \mathbb{R}$ has a continuous derivative f' on $[a, b]$, then $f(b) - f(a) = \int_a^b f'(x)\,dx$. [Hint. $f(b) - f(a) = \Sigma_{i=1}^n [f(x_i) - f(x_{i-1})] = \Sigma_{i=1}^n f'(x_i^*)(x_i - x_{i-1}) \to \int_a^b f'(x)\,dx$.]

8.8. The mean-value theorem for derivatives [Theorem 8.1] implies the mean-value theorem for integrals [Exercise 8.5]. Prove this statement by using Theorem 8.1 and the theorem in Exercise 8.3 to prove the following theorem. If the function $f: [a, b] \to \mathbb{R}$ is continuous on $[a, b]$, then there exists an x^* such that $\int_a^b f(x)\,dx = f(x^*)(b - a)$, $a < x^* < b$. [Hint. Apply Theorem 8.1 to the function $g: [a, b] \to \mathbb{R}$ such that $g(x) = \int_a^x f(t)\,dt$.]

8.9. Let $f: (a, b) \to \mathbb{R}$ be a function which has a derivative on (a, b). Prove the following theorems:

(a) If $D_1 f(x) > 0$ for every x in (a, b), then f is a strictly increasing function.

(b) If $D_1 f(x) < 0$ for every x in (a, b), then f is a strictly decreasing function.

(c) If $D_1 f(x) \neq 0$ for x in (a, b), then f is either strictly monotonically increasing or strictly monotonically decreasing. [Hint. Darboux's theorem in Exercise 7.10.]

(d) If $D_1 f(x_0) \neq 0$ and $D_1 f$ is continuous at x_0, then the mapping $f: (a, b) \to \mathbb{R}$ is one-to-one in a neighborhood of x_0. [Hint. Theorem 96.9 in Appendix 2.]

8.10. Let $f: (a, b) \to \mathbb{R}$ be a function which has a continuous derivative $D_1 f$ on (a, b), and let I be a closed interval in (a, b). Prove the following theorem. For every $\varepsilon > 0$ there exists a $\delta(\varepsilon)$ such that, if $\mathbf{x}: (x_1, x_0)$ is in I and $0 < |x_1 - x_0| < \delta(\varepsilon)$, then

$$\left| \frac{\Delta_1 f(\mathbf{x})}{\Delta(\mathbf{x})} - D_1 f(x_0) \right| < \varepsilon.$$

[Hint. Equation (6), and Theorem 96.18 in Appendix 2.]

9. The Mean-Value Theorem: A Generalization

This section generalizes Theorem 8.1 by proving the mean-value theorem for functions $f: E \to \mathbb{R}$, $E \subset \mathbb{R}^n$. The new theorem is used to study the approximation of f by linear functions.

9.1 Theorem. *If $f: E \to \mathbb{R}$, $E \subset \mathbb{R}^n$, is a function which has a derivative on E, and if the set $\{x : x = x_0 + t(x_1 - x_0), 0 \leqq t \leqq 1\}$ joining $x_0 : (x_0^1, \cdots, x_0^n)$ to $x_1 : (x_1^1, \cdots, x_1^n)$ is in E, then there exists a point x^* such that*

$$f(x_1) - f(x_0) = \sum_{j=1}^{n} D_j f(x^*)(x_1^j - x_0^j), \quad x^* = x_0 + t^*(x_1 - x_0), \quad 0 < t^* < 1. \tag{1}$$

PROOF. Define the function $g : [0, 1] \to \mathbb{R}^n$ with components $(g^1, \cdots, g^n)$ as follows:

$$g^j(t) = x_0^j + t(x_1^j - x_0^j), \qquad 0 \leqq t \leqq 1, \qquad j = 1, \cdots, n. \tag{2}$$

Then h, the composite function $f \circ g$, is defined for t in $[0, 1]$, and

$$h(0) = f(x_0), \qquad h(1) = f(x_1). \tag{3}$$

Since f is differentiable on E and g is differentiable on $[0, 1]$, then h is differentiable on $[0, 1]$ by the chain rule in Theorem 4.1, and

$$h'(t) = \sum_{j=1}^{n} D_j f(x)(x_1^j - x_0^j), \quad x = x_0 + t(x_1 - x_0), \quad 0 \leqq t \leqq 1. \tag{4}$$

Thus h is differentiable, and therefore continuous, on $[0, 1]$. Then by the mean-value theorem in Theorem 8.1, there exists a t^* such that

$$h(1) - h(0) = h'(t^*)(1 - 0), \qquad 0 < t^* < 1. \tag{5}$$

Equations (3) and (4) show that this equation is equivalent to (1). □

9.2 Example. A set is *convex* if and only if, for each two points in the set, the segment joining the points is in the set. The triangle inequality [see Section 91 in Appendix 2] shows, as follows, that every neighborhood $N(y_0, r)$ in $\mathbb{R}^n$ is convex. If x_0 and x_1 are in $N(y_0, r)$, then $|x_0 - y_0| < r$ and $|x_1 - y_0| < r$. The segment joining x_0 and x_1 consists of the points $x_0 + t(x_1 - x_0)$, or $tx_1 + (1 - t)x_0$, $0 \leqq t \leqq 1$. Write y_0 as $ty_0 + (1 - t)y_0$. Then $|[tx_1 + (1 - t)x_0] - [ty_0 + (1 - t)y_0]| < tr + (1 - t)r = r$ for $0 \leqq t \leqq 1$, and the segment joining x_0 and x_1 is in $N(y_0, r)$.

9.3 Example. Let $f: E \to \mathbb{R}$, be a function which is differentiable on the open set E in $\mathbb{R}^n$. If y_0 is a point in E, then y_0 has a neighborhood $N(y_0, r)$ in E. If x_0 and x_1 are two points in $N(y_0, r)$, then formula (1) in Theorem 9.1 can be applied to obtain a representation of $f(x_1) - f(x_0)$. The following statement is also true: if x_0 is in E, then x_0 has a neighborhood $N(x_0, r)$ which is in E, and (1) is valid for every point x_1 in $N(x_0, r)$.

9.4 Example. Let $f: E \to \mathbb{R}$, $E \subset \mathbb{R}^2$, be a function such that $f(x^1, x^2) = \sin x^1 + \sin x^2$. Let $x_0 : (x_0^1, x_0^2)$ be a point in E, and let $N(x_0, r)$ be a neighborhood of x_0 which is in E. Let $x_1 : (x_1^1, x_1^2)$ be a point in $N(x_0, r)$. Then by Theorem 9.1 there is a point $x^* : (x^{1*}, x^{2*})$ such that

$$f(x_1) - f(x_0) = [\cos x^{1*}](x_1^1 - x_0^1) + [\cos x^{2*}](x_1^2 - x_0^2),$$
$$x^* = x_0 + t^*(x_1 - x_0), \qquad 0 < t^* < 1.$$

9.5 Theorem. *Let $f: E \to \mathbb{R}$, $E \subset \mathbb{R}^n$, be a function which is differentiable on the open set E in $\mathbb{R}^n$. If x_0 and x_1 are the end points of a segment in E, then*

$$r(f; x_0, x_1) = \frac{1}{|x_1 - x_0|} \sum_{j=1}^{n} [D_j f(x^*) - D_j f(x_0)](x_1^j - x_0^j), \tag{6}$$

$$|r(f; x_0, x_1)| \leq \left\{ \sum_{j=1}^{n} [D_j f(x^*) - D_j f(x_0)]^2 \right\}^{1/2}. \tag{7}$$

PROOF. Sylvester's interchange theorem has been used in the proof of Theorem 3.6 to show that

$$f(x_1) - f(x_0) = \sum_{j=1}^{n} D_j f(x_0)(x_1^j - x_0^j) + r(f; x_0, x_1)|x_1 - x_0|, \tag{8}$$

$$r(f; x_0, x_1) = \frac{1}{|x_1 - x_0|} \sum_{j=1}^{n} \left\{ \frac{\Delta_j f(\mathbf{x})}{\Delta(\mathbf{x})} - D_j f(x_0) \right\} (x_1^j - x_0^j), \quad r(f; x_0, x_0) = 0. \tag{9}$$

To prove (8) and (9), it is sufficient to assume that f is differentiable at the single point x_0. If f is differentiable in E, then Theorem 9.1 can be used to give a different, and very important, representation for $r(f; x_0, x_1)$. From (1) in Theorem 9.1 it follows that

$$f(x_1) - f(x_0) = \sum_{j=1}^{n} D_j f(x_0)(x_1^j - x_0^j) + \sum_{j=1}^{n} [D_j f(x^*) - D_j f(x_0)](x_1^j - x_0^j). \tag{10}$$

Comparing (10) with (8), we see that (6) is true. Since

$$|x_1 - x_0| = \left\{ \sum_{j=1}^{n} (x_1^j - x_0^j)^2 \right\}^{1/2},$$

equation (6) above and Schwarz's inequality in Section 86 show that (7) is true. The proof of Theorem 9.5 is complete. □

9.6 Remarks. Equation (9) shows that

$$\lim_{x_1 \to x_0} r(f; x_0, x_1) = 0 \tag{11}$$

because f is differentiable at x_0 [see the proof of Theorem 3.6]. Then (11) and (6) show that

$$\lim_{x_1 \to x_0} \frac{1}{|x_1 - x_0|} \sum_{j=1}^{n} [D_j f(x^*) - D_j f(x_0)](x_1^j - x_0^j) = 0 \tag{12}$$

although no assumption has been made about the continuity of the derivatives $D_j f$, $j = 1, \cdots, n$. If these derivatives are continuous, then (11) follows from (7). Formula (6) has its greatest usefulness if the derivatives $D_j f$ are assumed to be continuous.

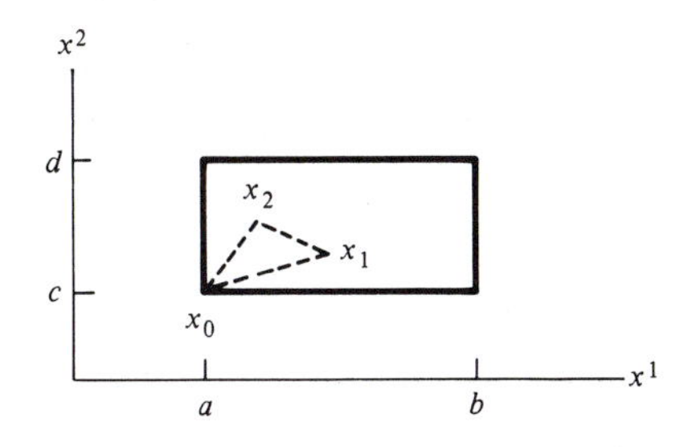

Figure 9.1. Figure for Example 9.8.

9.7 Remark. Some explanation is required about how derivatives are defined on the boundaries of sets of the form $[a_1, b_1] \times \cdots \times [a_n, b_n]$ in $\mathbb{R}^n$. The simplest case concerns the definition of derivatives of a function $f: [a, b] \to \mathbb{R}$ defined on a closed interval in $\mathbb{R}$. If x_0 is an interior point of $[a, b]$, then

$$D_1 f(x_0) = \lim_{\mathbf{x} \to x_0} \frac{\Delta_1 f(\mathbf{x})}{\Delta(\mathbf{x})}, \qquad \mathbf{x} = \begin{bmatrix} x_1 \\ x_0 \end{bmatrix}, \qquad \Delta(\mathbf{x}) \neq 0, \tag{13}$$

and x_1 may be either less than x_0 or greater than x_0. The only requirement is that $\mathbf{x}$ be in $[a, b]$. If $x_0 = a$, however, then x_1 must be greater than a in order for $\mathbf{x}$ to be in $[a, b]$; similarly, if $x_0 = b$, then x_1 must be less than b in order for $\mathbf{x}$ to be in $[a, b]$.

Let $A = [a, b] \times [c, d]$; then in a similar manner it is possible to define the derivatives of a function $f: A \to \mathbb{R}$ at points x_0 in A. There is no essential change in the definition of the derivative given in Definition 2.8, but some care must be exercised in choosing the class of n-vectors $X(x_0, \rho)$ if x_0 is on the boundary of A. Choose ρ, $0 < \rho \leqq 1$, small enough so that every point x_1, $x_1 \neq x_0$, in A is a vertex of a 2-vector $\mathbf{x}: (x_1, x_2, x_0)$ in $X(x_0, \rho)$. This restriction is needed in order to prove that f satisfies the Stolz condition at x_0 [see the proof of Theorem 3.6]. Then the formula for the derivative of composite functions holds as before [see Theorem 4.1], and no change is required in the proofs of Theorems 9.1 and 9.5 if the open set E in these theorems is replaced by the closed set A. Observe that A is a convex set.

Finally, if A is the closed set $[a_1, b_1] \times \cdots \times [a_n, b_n]$ in $\mathbb{R}^n$, then the discussion in the last paragraph indicates how the derivatives of a function $f: A \to \mathbb{R}$ are defined on the boundary of A. Since A is a convex set, the formulas in (1), (6), and (7) hold for every pair of points x_0, x_1 in A.

9.8 Example. Let $A = [a, b] \times [c, d]$ in $\mathbb{R}^2$. If x_0 is a point in A, then the class $X(x_0, \rho)$, $\rho = \frac{1}{2}$, of 2-vectors can be used to define the derivatives at x_0 of a function $f: A \to \mathbb{R}$. Example 2.1 shows that, if $\rho = \frac{1}{2}$, the angle between the vectors x_0x_1 and x_0x_2 in the 2-vector $\mathbf{x}: (x_1, x_2, x_0)$ is equal to or greater than $\pi/6$. Then, even if x_0 is a vertex of A, every point in A is a terminal point of a 2-vector $\mathbf{x}: (x_1, x_2, x_0)$ in $X(x_0, \frac{1}{2})$ and in A [see Figure 9.1]. The class $X(x_0, 1)$ contains only 2-vectors $\mathbf{x}: (x_1, x_2, x_0)$ for which the vectors x_0x_1 and x_0x_2 are orthogonal. Then if x_0 is a vertex of A, only one 2-vector $\mathbf{x}$ in $X(x_0, 1)$ is in A, and x_1 and x_2 for this 2-vector are on sides of A.

9.9 Theorem. *Let A be the closed, convex, bounded set $[a_1, b_1] \times \cdots \times [a_n, b_n]$ in $\mathbb{R}^n$, and let $f: A \to \mathbb{R}$ be a function which has continuous derivatives D_jf, $j = 1, \cdots, n$, on A. Then $r(f; x_0, x_1)$ is uniformly small, relative to $|x_1 - x_0|$, on A. More precisely, to each $\varepsilon > 0$ there corresponds a $\delta(\varepsilon) > 0$ such that*

$$|r(f; x_0, x_1)| < \varepsilon \tag{14}$$

for every x_0, x_1 in A for which $|x_1 - x_0| < \delta(\varepsilon)$.

PROOF. The remarks above have shown that Theorem 9.5 is true with the open set E replaced by A. Then, by (7),

$$|r(f; x_0, x_1)| \leq \left\{ \sum_{j=1}^{n} [D_jf(x^*) - D_jf(x_0)]^2 \right\}^{1/2}, \tag{15}$$

$$x^* = x_0 + t^*(x_1 - x_0), \qquad 0 < t^* < 1,$$

for every pair of points x_0, x_1 in A. Now A is closed and bounded; it is therefore compact by Definition 92.8 in Appendix 2. Also, by hypothesis, each derivative D_jf is continuous on the compact set A; it is therefore uniformly continuous by Theorem 96.18 in Appendix 2. Then to each $\varepsilon > 0$ there corresponds a $\delta_j(\varepsilon) > 0$ such that

$$|D_jf(x_1) - D_jf(x_0)| < \varepsilon/\sqrt{n} \tag{16}$$

for every x_0, x_1 in A for which $|x_1 - x_0| < \delta_j(\varepsilon)$. Since $x^* - x_0 = t^*(x_1 - x_0)$, then $|x^* - x_0| = t^*|x_1 - x_0| < |x_1 - x_0|$, and (16) shows that

$$|D_jf(x^*) - D_jf(x_0)| < \varepsilon/\sqrt{n}, \qquad |x_1 - x_0| < \delta_j(\varepsilon). \tag{17}$$

Set

$$\delta(\varepsilon) = \min\{\delta_j(\varepsilon) : j = 1, \cdots, n\}. \tag{18}$$

Then

$$|D_jf(x^*) - D_jf(x_0)| < \varepsilon/\sqrt{n}, \qquad j = 1, \cdots, n, \tag{19}$$

for every pair of points x_0, x_1 in A such that $|x_1 - x_0| < \delta(\varepsilon)$. Finally, (15) and (19) show that (14) holds for every x_0, x_1 in A for which $|x_1 - x_0| < \delta(\varepsilon)$. □

9.10 Example. Let $A = [a_1, b_1] \times \cdots \times [a_n, b_n]$ in $\mathbb{R}^n$, and let $f: A \to \mathbb{R}$ be the function such that $f(x) = \Sigma_1^n (x^j)^2$. Then $D_jf(x) = 2x^j, j = 1, \cdots, n$, and

$$f(x_1) - f(x_0) = \sum_{j=1}^{n} 2x_0^j(x_1^j - x_0^j) + \sum_{j=1}^{n} [2x^{j*} - 2x_0^j](x_1^j - x_0^j).$$

By (7) in Theorem 9.5,

$$|r(f; x_0, x_1)| \leqq \left\{ \sum_{j=1}^{n} [2(x^{j*} - x_0^j)]^2 \right\}^{1/2} = 2|x^* - x_0|.$$

Since $|x^* - x_0| < |x_1 - x_0|$, then $|r(f; x_0, x_1)| < 2|x_1 - x_0|$. Finally, if x_0, x_1 are any two points in A such that $|x_1 - x_0| < \varepsilon/2$, then $|r(f; x_0, x_1)| < \varepsilon$.

Exercises

9.1. Let $f: \mathbb{R}^n \to \mathbb{R}$ be the linear function such that $f(x) = \Sigma_1^n a_j x^j$. Then $D_j f(x) = a_j$, x in $\mathbb{R}^n$ and $j = 1, \cdots, n$, and

$$f(x_1) - f(x_0) = \sum_{j=1}^{n} D_j f(x_0)(x_1^j - x_0^j) + r(f; x_0, x_1)|x_1 - x_0|.$$

Show that $r(f; x_0, x_1) = 0$ for every x_0, x_1 in $\mathbb{R}^n$.

9.2. A set E in $\mathbb{R}^n$ is said to be *connected* if and only if each two points x_0, x_1 in E can be connected by a polygonal curve (a curve composed of line segments) which lies in E.

(a) Prove that the following sets are connected sets: a neighborhood $N(x_0, r)$ in $\mathbb{R}^n$; an interval $[a_1, b_1] \times \cdots \times [a_n, b_n]$ in $\mathbb{R}^n$; a convex set in $\mathbb{R}^n$.

(b) Prove the following theorem. If E_1 and E_2 are connected sets in $\mathbb{R}^n$, and if $E_1 \cap E_2$ is not empty, then $E_1 \cup E_2$ is a connected set.

9.3. Prove the following theorem. If E is an open connected set in $\mathbb{R}^n$, and if $f: E \to \mathbb{R}$ is a differentiable function such that $D_j f(x) = 0$, $j = 1, \cdots, n$, for every x in E, then f is a constant function. [Hint. Exercise 8.2; Theorem 9.1.]

9.4. Let A be the set $[a_1, b_1] \times [a_2, b_2]$ in $\mathbb{R}^2$, and let x_0 be a point on the boundary of A.

(a) Prove the following theorem. If $f: A \to \mathbb{R}$ is differentiable at x_0, then f satisfies the Stolz condition at x_0. [Hint. Theorem 3.6.]

(b) Assume that $f: A \to \mathbb{R}$ is differentiable on A and that $D_j f$, $j = 1, \cdots, n$, are continuous at x_0. Use equation (10) to prove that f satisfies the Stolz condition at x_0. Compare your proof with that of Theorem 3.19.

9.5. Let E be an open set in $\mathbb{R}^n$, let $f: E \to \mathbb{R}$ be a function which is differentiable on E, and assume that $\{x : x = x_0 + t(x_1 - x_0), 0 \leqq t \leqq 1\}$ is in E. Show that the following equations are valid.

(a) $$\begin{vmatrix} f(x_1) & 1 \\ f(x_0) & 1 \end{vmatrix} = \sum_{j=1}^{n} D_j f(x^*) \begin{vmatrix} x_1^j & 1 \\ x_0^j & 1 \end{vmatrix}, \qquad x^* = x_0 + t^*(x_1 - x_0), \qquad 0 < t^* < 1.$$

(b) $$\begin{vmatrix} f(x_1) & 1 \\ f(x_0) & 1 \end{vmatrix} = \sum_{j=1}^{n} D_j f(x_0) \begin{vmatrix} x_1^j & 1 \\ x_0^j & 1 \end{vmatrix} + r(f; x_0, x_1)|x_1 - x_0|.$$

(c) $$r(f; x_0, x_1) = \frac{1}{|x_1 - x_0|} \sum_{j=1}^{n} [D_j f(x^*) - D_j f(x_0)] \begin{vmatrix} x_1^j & 1 \\ x_0^j & 1 \end{vmatrix}.$$

(d) $\lim_{x_1 \to x_0} r(f; x_0, x_1) = 0$.

Section 11 uses these formulas to establish generalizations of (b), (c), and (d).

9.6. Use Theorem 9.5 to find the expression for $r(f; x_0, x_1)$ for each of the following functions.

(a) $f(x) = \sum_{i=1}^{n} a^i x^i$, $x : (x^1, \cdots, x^n)$ in $\mathbb{R}^n$.

(b) $f(x) = \sum_{i=1}^{n} a^i (x^i)^2$, $x : (x^1, \cdots, x^n)$ in $\mathbb{R}^n$.

(c) $f(x) = \sum_{i=1}^{n} a^i (x^i)^k$, $x : (x^1, \cdots, x^n)$ in $\mathbb{R}^n$, $k = 1, 2, \cdots$.

In each case verify that $\lim_{x_1 \to x_0} r(f; x_0, x_1) = 0$. Use Theorem 9.9 to show that to each $\varepsilon > 0$ there corresponds a $\delta(\varepsilon) > 0$ such that $|r(f; x_0, x_1)| < \varepsilon$ if x_0, x_1 are in a compact set $[a_1, b_1] \times \cdots \times [a_n, b_n]$ and $|x_1 - x_0| < \delta(\varepsilon)$.

10. Uniform Differentiability

This section defines uniform differentiability and establishes the relation between uniform differentiability and the continuity of derivatives. The first step is to give a definition of uniform differentiability.

Let E be a set in $\mathbb{R}^n$; for the present E may be an open set or a set of the form $[a_1, b_1] \times \cdots \times [a_n, b_n]$. Choose a fixed ρ small enough so that each x_0 in E has a class $X(x_0, \rho)$ of n-vectors $\mathbf{x} : (x_1, \cdots, x_n, x_0)$ in E [see Remarks 9.7 and Example 9.8]. Let $f: E \to \mathbb{R}^m$ be a function whose components are differentiable on E. Then for each $j : (j_1, \cdots, j_m)$ in (m/n), each x_0 in E, and each $\varepsilon > 0$, there is a $\delta(\varepsilon, x_0, j) > 0$ such that

$$\left| \frac{\Delta_{(j_1, \cdots, j_m)} f(\mathbf{x})}{\Delta(\mathbf{x})} - D_{(j_1, \cdots, j_m)} f(x_0) \right| < \varepsilon \tag{1}$$

for every $\mathbf{x}$ in $X(x_0, \rho)$ and in E for which $|x_i - x_0| < \delta(\varepsilon, x_0, j)$, $i = 1, \cdots, n$. Set $\delta(\varepsilon, x_0) = \min\{\delta(\varepsilon, x_0, j) : j \in (m/n)\}$. Then for each x_0 in E and each $\varepsilon > 0$, the inequality (1) holds for every $\mathbf{x}$ in $X(x_0, \rho)$ and in E for which $|x_i - x_0| < \delta(\varepsilon, x_0)$, $i = 1, \cdots, n$, and every $(j_1, \cdots, j_m)$ in (m/n).

10.1 Definition. Let $f: E \to \mathbb{R}^m$, with differentiable components $(f^1, \cdots, f^m)$, be the function just described. Then f is *uniformly differentiable on E* if and only if to each $\varepsilon > 0$ there corresponds a $\delta(\varepsilon) > 0$ [which does not depend on x_0] such that inequality (1) is satisfied for every x_0 in E, every $\mathbf{x} : (x_1, \cdots, x_n, x_0)$ in $X(x_0, \rho)$ and in E for which $|x_i - x_0| < \delta(\varepsilon)$, $i = 1, \cdots, n$, and for every $(j_1, \cdots, j_m)$ in (m/n).

10.2 Example. Let E be an open set in $\mathbb{R}^2$, and let $f: E \to \mathbb{R}$ be the function such that $f(x^1, x^2) = a_1 x^1 + a_2 x^2 + a_3$. Let ρ be a fixed number such that $0 < \rho \leqq 1$. Then $D_1 f(x_0) = a_1$, $D_2 f(x_0) = a_2$ for every x_0 in E. Also, elementary properties of determinants [see Section 77 in Appendix 1] show that

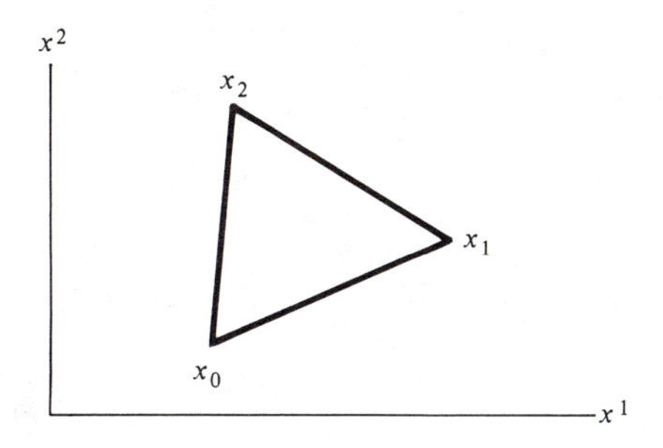

Figure 10.1. Figure for the proof of Theorem 10.3.

$$\frac{\Delta_1 f(\mathbf{x})}{\Delta(\mathbf{x})} = a_1, \qquad \frac{\Delta_2 f(\mathbf{x})}{\Delta(\mathbf{x})} = a_2, \qquad \mathbf{x} \in X(x_0, \rho).$$

Then

$$\left| \frac{\Delta_1 f(\mathbf{x})}{\Delta(\mathbf{x})} - D_1 f(x_0) \right| = 0, \quad \left| \frac{\Delta_2 f(\mathbf{x})}{\Delta(\mathbf{x})} - D_2 f(x_0) \right| = 0, \quad \mathbf{x} \in X(x_0, \rho).$$

Thus for each $\varepsilon > 0$, the inequality (1) is satisfied for an arbitrary choice of $\delta(\varepsilon)$; for convenience, choose $\delta(\varepsilon) = \varepsilon$. Then f is uniformly differentiable on E by Definition 10.1.

10.3 Theorem. *If the function $f: \mathbb{R}^2 \to \mathbb{R}^2$ is uniformly differentiable on $\mathbb{R}^2$, then $D_{(1,2)} f$ is uniformly continuous on $\mathbb{R}^2$.*

PROOF. In this proof, it is convenient to choose $\rho = \frac{1}{2}$; a smaller value for ρ would serve equally well. Let $\varepsilon > 0$ be given. Then since, by hypothesis, f is uniformly differentiable on $\mathbb{R}^2$, there exists a $\delta(\varepsilon)$ such that, for every p_0 in $\mathbb{R}^2$,

$$\left| \frac{\Delta_{(1,2)} f(\mathbf{x})}{\Delta(\mathbf{x})} - D_{(1,2)} f(p_0) \right| < \frac{\varepsilon}{2}, \qquad \mathbf{x} = (p_1, p_2, p_0), \tag{2}$$

for every $\mathbf{x}: (p_1, p_2, p_0)$ in $X(p_0, \rho)$ for which $|p_i - p_0| < \delta(\varepsilon)$, $i = 1, 2$. The proof will be completed by showing that

$$|D_{(1,2)} f(x_1) - D_{(1,2)} f(x_0)| < \varepsilon \tag{3}$$

for every pair of points x_0, x_1 in $\mathbb{R}^2$ for which

$$|x_1 - x_0| < \delta(\varepsilon). \tag{4}$$

Let x_0, x_1 be two points in $\mathbb{R}^2$ which satisfy (4). Choose a third point x_2 so that

$$|x_1 - x_0| = |x_2 - x_1| = |x_0 - x_2|. \tag{5}$$

Then x_0, x_1, x_2 are the vertices of an equilateral triangle [see Figure 10.1], and

$$\mathbf{x}_0 : (x_1, x_2, x_0) \in X(x_0, \rho), \qquad \mathbf{x}_1 : (x_2, x_0, x_1) \in X(x_1, \rho),$$
$$\mathbf{x}_2 : (x_0, x_1, x_2) \in X(x_2, \rho). \tag{6}$$

Then since $|x_i - x_0| < \delta(\varepsilon)$ for $i = 1, 2$, equation (2) shows that

$$\left| \frac{\Delta_{(1,2)} f(\mathbf{x}_0)}{\Delta(\mathbf{x}_0)} - D_{(1,2)} f(x_0) \right| < \frac{\varepsilon}{2}. \tag{7}$$

Also, since $|x_i - x_1| < \delta(\varepsilon)$ for $i = 0, 2$, equation (2) shows that

$$\left| D_{(1,2)} f(x_1) - \frac{\Delta_{(1,2)} f(\mathbf{x}_1)}{\Delta(\mathbf{x}_1)} \right| < \frac{\varepsilon}{2}. \tag{8}$$

Now $\Delta_{(1,2)} f(\mathbf{x}_1) = \Delta_{(1,2)} f(\mathbf{x}_0)$ because the first of these determinants can be obtained from the second by an even number of interchanges of two rows of its matrix [see Theorem 77.7 in Appendix 1]; for the same reason, $\Delta(\mathbf{x}_1) = \Delta(\mathbf{x}_0)$. Thus

$$\frac{\Delta_{(1,2)} f(\mathbf{x}_1)}{\Delta(\mathbf{x}_1)} = \frac{\Delta_{(1,2)} f(\mathbf{x}_0)}{\Delta(\mathbf{x}_0)}. \tag{9}$$

Then (3) follows from (7), (8), (9), and the triangle inequality and the proof of Theorem 10.3 is complete. □

10.4 Example. Let $f: [a, b] \to \mathbb{R}$ be a function which has a continuous derivative $D_1 f$ on $[a, b]$. If $\mathbf{x} = (x_1, x_0)$, then as shown in (6) in Section 8, there is an x^* between x_0 and x_1 such that

$$\frac{\Delta_1 f(\mathbf{x})}{\Delta(\mathbf{x})} - D_1 f(x_0) = D_1 f(x^*) - D_1 f(x_0). \tag{10}$$

The interval $[a, b]$ is a closed and bounded set, and it is therefore a compact set by Definition 92.8 in Appendix 2. Since $D_1 f$ is continuous on $[a, b]$, it is uniformly continuous on $[a, b]$ by Theorem 96.18. Then to each $\varepsilon > 0$ there corresponds a $\delta(\varepsilon) > 0$ such that $|D_1 f(x_1) - D_1 f(x_0)| < \varepsilon$ for every $\mathbf{x} : (x_1, x_0)$ in $[a, b]$ for which $|x_1 - x_0| < \delta(\varepsilon)$. Since $|x^* - x_0| < |x_1 - x_0|$, equation (10) shows that

$$\left| \frac{\Delta_1 f(\mathbf{x})}{\Delta(\mathbf{x})} - D_1 f(x_0) \right| < \varepsilon \tag{11}$$

for every $\mathbf{x} : (x_1, x_0)$ in $[a, b]$ for which $|x_1 - x_0| < \delta(\varepsilon)$. Therefore, the function $f: [a, b] \to \mathbb{R}$ is uniformly differentiable on $[a, b]$ by Definition 10.1.

10.5 Example. Let $f(x) = 1/x$ for $0 < x \leqq 1$. Then f has a continuous derivative on $(0, 1]$, but this interval is not compact because it is not closed. Let $\mathbf{x} = (x_1, x_0)$ with $0 < x_1 < x_0 \leqq 1$. Then a straightforward calculation shows that

$$\frac{\Delta_1 f(\mathbf{x})}{\Delta(\mathbf{x})} - D_1 f(x_0) = \frac{1}{x_0^2} - \frac{1}{x_0 x_1}. \tag{12}$$

Let δ be an arbitrary positive number, and choose x_0, x_1 so that $0 < x_1 < x_0 < \delta$. Then $|x_1 - x_0| < \delta$, but the absolute value of (12) is arbitrarily large if x_1 is sufficiently close to 0. Thus the function f for which $f(x) = 1/x$ is not uniformly differentiable on $(0, 1]$.

Theorem 10.3 and Examples 10.4 and 10.5 have examined the relation between uniform differentiability of f and uniform continuity of the derivatives of f. The remainder of this section is devoted to a proof of the following theorem: if $f: A \to \mathbb{R}^m$ is a function whose components $(f^1, \cdots, f^m)$ have continuous derivatives on the compact set $A: [a_1, b_1] \times \cdots \times [a_n, b_n]$, then f is uniformly differentiable on A. Theorem 10.6 states this result in the special case $m = 1$, and Theorem 10.9 is the general case.

10.6 Theorem. *Let $A = [a_1, b_1] \times \cdots \times [a_n, b_n]$, and let $f: A \to \mathbb{R}$ be a function which has continuous derivatives $D_j f$, $j = 1, \cdots, n$, on A. Then f is uniformly differentiable on A. More precisely, to each $\varepsilon > 0$ there corresponds a $\delta(\varepsilon) > 0$ such that*

$$\left| \frac{\Delta_j f(\mathbf{x})}{\Delta(\mathbf{x})} - D_j f(x_0) \right| < (1/\rho) n^{n/2} \varepsilon \tag{13}$$

for every x_0 in A, for every $\mathbf{x}: (x_1, \cdots, x_n, x_0)$ in $X(x_0, \rho)$ and in A for which $|x_i - x_0| < \delta(\varepsilon)$, $i = 1, \cdots, n$, and for $j = 1, \cdots, n$.

PROOF. Prove (13) first for $j = 1$. Now

$$\Delta_1 f(\mathbf{x}) = \begin{vmatrix} f(x_1) & x_1^2 & \cdots & x_1^n & 1 \\ \cdots & \cdots & \cdots & \cdots & \cdots \\ f(x_n) & x_n^2 & \cdots & x_n^n & 1 \\ f(x_0) & x_0^2 & \cdots & x_0^n & 1 \end{vmatrix}. \tag{14}$$

Subtract the last row of the matrix from each of the other rows and then expand the determinant by minors of elements in the last column. Then

$$\Delta_1 f(\mathbf{x}) = \begin{vmatrix} f(x_1) - f(x_0) & x_1^2 - x_0^2 & \cdots & x_1^n - x_0^n \\ \cdots & \cdots & \cdots & \cdots \\ f(x_n) - f(x_0) & x_n^2 - x_0^2 & \cdots & x_n^n - x_0^n \end{vmatrix}. \tag{15}$$

By equation (10) in Section 9 and Theorem 9.9,

$$f(x_i) - f(x_0) = \sum_{j=1}^{n} D_j f(x_0)(x_i^j - x_0^j) + r(f; x_0, x_i)|x_i - x_0|, \qquad i = 1, \cdots, n. \tag{16}$$

Use (16) to evaluate the terms in the first column of the matrix in (15). Then multiply the j-th column, $j = 2, \cdots, n$, by $D_j f(x_0)$ and subtract it from the first column. Theorems 77.1, 77.3, and 77.17 in Appendix 1 show that the determinant is unchanged by these transformations. The i-th row of the first column of the matrix now contains the element $D_1 f(x_0)(x_i^1 - x_0^1) + r(f; x_0, x_i)|x_i - x_0|$, $i = 1, \cdots, n$. Write the determinant as the sum of two determinants and factor out $D_1 f(x_0)$ from the first column of the matrix of the first determinant. Then

$$\Delta_1 f(\mathbf{x}) = D_1 f(x_0)\Delta(\mathbf{x}) + \det R_1, \tag{17}$$

$$\det R_1 = \begin{vmatrix} r(f; x_0, x_1)\,|x_1 - x_0| & x_1^2 - x_0^2 & \cdots & x_1^n - x_0^n \\ \cdots & \cdots & \cdots & \cdots \\ r(f; x_0, x_n)\,|x_n - x_0| & x_n^2 - x_0^2 & \cdots & x_n^n - x_0^n \end{vmatrix}. \tag{18}$$

Equation (17) shows that

$$\left|\frac{\Delta_1 f(\mathbf{x})}{\Delta(\mathbf{x})} - D_1 f(x_0)\right| = \left|\frac{\det R_1}{\Delta(\mathbf{x})}\right| \tag{19}$$

The proof can be completed by finding a bound for

$$\left|\frac{\det R_1}{\Delta(\mathbf{x})}\right|. \tag{20}$$

Now $\mathbf{x} \in X(x_0, \rho)$; hence $|\Delta(\mathbf{x})| \geqq \rho|x_1 - x_0| \cdots |x_n - x_0| > 0$. Divide the denominator of (20) by $\Pi_1^n |x_i - x_0|$; divide the numerator by the same quantity by dividing the i-th row of R_1 by $|x_i - x_0|$, $i = 1, \cdots, n$. Then (20) is equal to or less than the absolute value of

$$(1/\rho)\begin{vmatrix} r(f; x_0, x_1) & \dfrac{x_1^2 - x_0^2}{|x_1 - x_0|} & \cdots & \dfrac{x_1^n - x_0^n}{|x_1 - x_0|} \\ \cdots & \cdots & \cdots & \cdots \\ r(f; x_0, x_n) & \dfrac{x_n^2 - x_0^2}{|x_n - x_0|} & \cdots & \dfrac{x_n^n - x_0^n}{|x_n - x_0|} \end{vmatrix}. \tag{21}$$

Each element in columns $2, \cdots, n$ in (21) is equal to or less than 1 in absolute value. Then by Hadamard's determinant theorem [see (19) in Corollary 87.2 in Appendix 1], the absolute value of the determinant in (21) is equal to or less than the product of the lengths of the vectors in the columns of its matrix. The length of the vector in the first column is $\{\Sigma_1^n [r(f; x_0, x_i)]^2\}^{1/2}$, and the length of each vector in the remaining $n - 1$ columns is equal to or less than $n^{1/2}$. Hence,

$$\left|\frac{\det R_1}{\Delta(\mathbf{x})}\right| \leqq (1/\rho)n^{(n-1)/2}\left\{\sum_{i=1}^{n} [r(f; x_0, x_i)]^2\right\}^{1/2}. \tag{22}$$

Now since f has continuous derivatives $D_j f$, $j = 1, \cdots, n$, by hypothesis, Theorem 9.9 shows that to each $\varepsilon > 0$ there corresponds a $\delta(\varepsilon) > 0$ such that $|r(f; x_0, x_i)| < \varepsilon$ for every x_0, x_i in A for which $|x_i - x_0| < \delta(\varepsilon)$. Then, by (22),

$$\left|\frac{\det R_1}{\Delta(\mathbf{x})}\right| < (1/\rho)n^{n/2}\varepsilon \tag{23}$$

for every x_0 in A and every $\mathbf{x} : (x_1, \cdots, x_n, x_0)$ in $X(x_0, \rho)$ and in A such that $|x_i - x_0| < \delta(\varepsilon)$, $i = 1, \cdots, n$.

The proof of (13) is complete for $j = 1$; in exactly the same way, (13) can be proved for $j = 2, \cdots, n$ to complete the proof of Theorem 10.6. □

10.7 Example. Let $f: A \to \mathbb{R}$ be the function such that $f(x) = \Sigma_1^n (x^j)^2$ as in Example 9.10. Then $D_j f(x) = 2x^j$, $j = 1, \cdots, n$, and these derivatives are continuous on A. Let $\varepsilon > 0$ be given; set $\delta(\varepsilon) = \varepsilon/2$. Then Example 9.10 shows that $|r(f; x_0, x_1)| < \varepsilon$ for every pair of points x_0, x_1 in A such that $|x_1 - x_0| < \varepsilon/2$. Then f is uniformly differentiable on A, and (13) holds with $\delta(\varepsilon) = \varepsilon/2$.

10.8 Lemma. *Let $A = [a_1, b_1] \times \cdots \times [a_n, b_n]$, and let m be an integer such that $1 \leqq m \leqq n$. If $f: A \to \mathbb{R}^m$ is a function whose components $(f^1, \cdots, f^m)$ have continuous derivatives $D_j f^i, j = 1, \cdots, n, i = 1, \cdots, m$, on A, then there exists a constant $D > 0$ such that*

$$\frac{\left|\sum_{j=1}^{n} D_j f^i(x_0)(x_1^j - x_0^j)\right|}{|x_1 - x_0|} \leqq D, \qquad i = 1, \cdots, m, \tag{24}$$

$$\frac{|f^i(x_1) - f^i(x_0)|}{|x_1 - x_0|} \leqq D, \qquad i = 1, \cdots, m, \tag{25}$$

for every x_0, x_1 in A such that $|x_1 - x_0| > 0$.

Proof. Since the derivatives $D_j f^i, j = 1, \cdots, n$, are continuous by hypothesis, then

$$\left\{\sum_{j=1}^{n} [D_j f^i]^2\right\}^{1/2} \tag{26}$$

is a continuous function on the compact set A. Then this function has a maximum value by Theorem 96.14 in Appendix 2; call this maximum D_i. Let D be the maximum of $D_1, \cdots, D_m$. Then by Schwarz's inequality [see Corollary 86.2 in Appendix 1],

$$\frac{\left|\sum_{j=1}^{n} D_j f^i(x_0)(x_1^j - x_0^j)\right|}{|x_1 - x_0|} \leqq \left\{\sum_{j=1}^{n} [D_j f^i(x_0)]^2\right\}^{1/2} \leqq D_i \leqq D \tag{27}$$

for each pair of points x_0, x_1 in A and for $i = 1, \cdots, m$. Since

$$\frac{|f^i(x_1) - f^i(x_0)|}{|x_1 - x_0|} = \frac{\left|\sum_{j=1}^{n} D_j f(x^*)(x_1^j - x_0^j)\right|}{|x_1 - x_0|}, \qquad x^* = x_0 + t^*(x_1 - x_0), \quad 0 < t^* < 1, \tag{28}$$

by Theorem 9.1, the same proof establishes (25) and completes the proof of the lemma. □

10.9 Theorem. *Let m and n be integers such that $1 \leqq m \leqq n$; let $A = [a_1, b_1] \times \cdots \times [a_n, b_n]$; and let $f: A \to \mathbb{R}^m$ be a function whose components $(f^1, \cdots, f^m)$ have continuous derivatives $D_j f^i, j = 1, \cdots, n, i = 1, \cdots, m$, on A. Then f is uniformly differentiable on A. More precisely, to each $\varepsilon > 0$ there corre-*

sponds a $\delta(\varepsilon) > 0$ *such that*

$$\left| \frac{\Delta_{(j_1,\cdots,j_m)} f(\mathbf{x})}{\Delta(\mathbf{x})} - D_{(j_1,\cdots,j_m)} f(x_0) \right| < (1/\rho) m n^{n/2} D^{m-1} \varepsilon \tag{29}$$

for every x_0 *in* A, *for every* $\mathbf{x} : (x_1, \cdots, x_n, x_0)$ *in* $X(x_0, \rho)$ *and in* A *for which* $|x_i - x_0| < \delta(\varepsilon)$, $i = 1, \cdots, n$, *and for each* $(j_1, \cdots, j_m)$ *in* (m/n).

PROOF. The inequality (29) will be proved first for the case in which $(j_1, \cdots, j_m) = (1, \cdots, m)$. By a familiar transformation,

$$\Delta_{(1,\cdots,m)} f(\mathbf{x}) = \begin{vmatrix} f^1(x_1) - f^1(x_0) & \cdots & f^m(x_1) - f^m(x_0) & \cdots & x_1^n - x_0^n \\ \cdots & \cdots & \cdots & \cdots & \cdots \\ f^1(x_n) - f^1(x_0) & \cdots & f^m(x_n) - f^m(x_0) & \cdots & x_n^n - x_0^n \end{vmatrix}. \tag{30}$$

By equation (10) in Section 9 and Theorem 9.9,

$$f^k(x_i) - f^k(x_0) = \sum_{j=1}^{n} D_j f^k(x_0)(x_i^j - x_0^j) + r(f^k; x_0, x_i)|x_i - x_0| \tag{31}$$

for each two points x_0, x_i in A and for $k = 1, \cdots, m$. Use (31) to evaluate the terms in the first column of the matrix in (30). Then in an obvious manner this column is the sum of two columns; write the determinant as the sum of two determinants [see Section 77 in Appendix 1]. The matrix of the first of the two determinants contains terms of the form

$$\sum_{j=1}^{n} D_j f^1(x_0)(x_i^j - x_0^j), \qquad i = 1, \cdots, n, \tag{32}$$

and the matrix of the second determinant contains the terms

$$r(f^1; x_0, x_i)|x_i - x_0|, \qquad i = 1, \cdots, n. \tag{33}$$

In the matrix which contains the terms (32), use (31) to evaluate the terms in the second column; write the determinant of the resulting matrix as the sum of two determinants. Continue this process until (31) has been used to evaluate the terms in the first m columns of the matrix in (30). These evaluations have constructed $m + 1$ matrices, and the determinant in (30) equals the sum of the determinants of these matrices as follows.

$$\Delta_{(1,\cdots,m)} f(\mathbf{x}) = \begin{vmatrix} \sum_{j=1}^{n} D_j f^1(x_0)(x_1^j - x_0^j) & \cdots & \sum_{j=1}^{n} D_j f^m(x_0)(x_1^j - x_0^j) & \cdots & x_1^n - x_0^n \\ \cdots & \cdots & \cdots & \cdots & \cdots \\ \sum_{j=1}^{n} D_j f^1(x_0)(x_n^j - x_0^j) & \cdots & \sum_{j=1}^{n} D_j f^m(x_0)(x_n^j - x_0^j) & \cdots & x_n^n - x_0^n \end{vmatrix}$$

$$+ \sum_{k=1}^{m} \begin{vmatrix} \sum_{j=1}^{n} D_j f^1(x_0)(x_1^j - x_0^j) & \cdots & r(f^k; x_0, x_1)|x_1 - x_0| & \cdots & f^m(x_1) - f^m(x_0) & \cdots & x_1^n - x_0^n \\ \cdots & \cdots & \cdots & \cdots & \cdots & \cdots & \cdots \\ \sum_{j=1}^{n} D_j f^1(x_0)(x_n^j - x_0^j) & \cdots & r(f^k; x_0, x_n)|x_n - x_0| & \cdots & f^m(x_n) - f^m(x_0) & \cdots & x_n^n - x_0^n \end{vmatrix} \tag{34}$$

Now the Binet–Cauchy multiplication theorem [see Theorem 80.1] shows that the first determinant on the right in (34) equals

$$\begin{vmatrix} x_1^1 - x_0^1 & \cdots & x_1^m - x_0^m & \cdots & x_1^n - x_0^n \\ \cdots & \cdots & \cdots & \cdots & \cdots \\ x_m^1 - x_0^1 & \cdots & x_m^m - x_0^m & \cdots & x_m^n - x_0^n \\ x_{m+1}^1 - x_0^1 & \cdots & x_{m+1}^m - x_0^m & \cdots & x_{m+1}^n - x_0^n \\ \cdots & \cdots & \cdots & \cdots & \cdots \\ x_n^1 - x_0^1 & \cdots & x_n^m - x_0^m & \cdots & x_n^n - x_0^n \end{vmatrix} \begin{vmatrix} D_1 f^1(x_0) & \cdots & D_1 f^m(x_0) & 0 & \cdots & 0 \\ \cdots & \cdots & \cdots & \cdots & \cdots & \cdots \\ D_m f^1(x_0) & \cdots & D_m f^m(x_0) & 0 & \cdots & 0 \\ D_{m+1} f^1(x_0) & \cdots & D_{m+1} f^m(x_0) & 1 & \cdots & 0 \\ \cdots & \cdots & \cdots & \cdots & \cdots & \cdots \\ D_n f^1(x_0) & \cdots & D_n f^m(x_0) & 0 & \cdots & 1 \end{vmatrix}. \tag{35}$$

The first determinant in (35) is $\Delta(\mathbf{x})$. To evaluate the second determinant, expand by the columns of the matrix which contain the 0's. Transpose the matrix of the final determinant obtained in this expansion. Thus the second determinant in (35) equals

$$\begin{vmatrix} D_1 f^1(x_0) & \cdots & D_m f^1(x_0) \\ \cdots & \cdots & \cdots \\ D_1 f^m(x_0) & \cdots & D_m f^m(x_0) \end{vmatrix}, \tag{36}$$

and this determinant is $D_{(1,\cdots,m)}f(x_0)$ by Corollary 3.13. Thus the first determinant on the right in (34) equals $\Delta(\mathbf{x})D_{(1,\cdots,m)}f(x_0)$; substitute this value in (34) and rewrite the equation as follows:

$$\frac{\Delta_{(1,\cdots,m)}f(\mathbf{x})}{\Delta(\mathbf{x})} - D_{(1,\cdots,m)}f(x_0)$$
$$= \sum_{k=1}^{m} \frac{1}{\Delta(\mathbf{x})} \begin{vmatrix} \sum_{j=1}^{n} D_j f^1(x_0)(x_1^j - x_0^j) & \cdots & r(f^k; x_0, x_1)|x_1 - x_0| & \cdots & f^m(x_1) - f^m(x_0) & \cdots & x_1^n - x_0^n \\ \cdots & \cdots & \cdots & \cdots & \cdots & \cdots & \cdots \\ \sum_{j=1}^{n} D_j f^1(x_0)(x_n^j - x_0^j) & \cdots & r(f^k; x_0, x_n)|x_n - x_0| & \cdots & f^m(x_n) - f^m(x_0) & \cdots & x_n^n - x_0^n \end{vmatrix}. \tag{37}$$

The proof will be completed by finding a bound for the absolute value of the sum on the right in this equation. Begin by taking the sum of the absolute values of the terms. Since $\mathbf{x} \in X(x_0, \rho)$, then $|\Delta(\mathbf{x})| \geqq \rho \prod_1^n |x_i - x_0| > 0$. Divide the denominator of the k-th term, $k = 1, \cdots, m$, by $\prod_1^n |x_i - x_0|$, and divide the numerator by this product by dividing the i-th row by $|x_i - x_0|$ for $i = 1, \cdots, n$. Then Hadamard's determinant theorem can be used to find a bound for the result; in absolute value, each determinant, $k = 1, \cdots, m$, is equal to or less than the product of the lengths of the vectors in the columns of its matrix. By (24) in Lemma 10.8, each element on the left of the k-th column is equal to or less than D in absolute value; similarly, by (25), each element (on the right of the k-th column) of the form $[f^j(x_i) - f^j(x_0)]/|x_i - x_0|$ is equal to or less than D in absolute value. Thus the length of $(m-1)$ vectors is equal to or less than $n^{1/2}D$, the length of $n - m$ vectors is equal to or less than $n^{1/2}$, and the length of the vector in the k-th column [see (37)] is $\{\Sigma_1^n [r(f^k; x_0, x_i)]^2\}^{1/2}$. Collecting results, these statements show that the absolute value of the k-th term in the sum in (37) is equal to or less than

$$(1/\rho)[n^{1/2}D]^{m-1}[n^{1/2}]^{n-m}\left\{\sum_{i=1}^{n}[r(f^k;x_0,x_i)]^2\right\}^{1/2}$$
$$= (1/\rho)n^{(n-1)/2}D^{m-1}\left\{\sum_{i=1}^{n}[r(f^k;x_0,x_i)]^2\right\}^{1/2}. \tag{38}$$

Let $\varepsilon > 0$ be given. Then by Theorem 9.9 there exists a $\delta_k(\varepsilon) > 0$ such that

$$|r(f^k;x_0,x_i)| < \varepsilon \tag{39}$$

for every x_0, x_i in A for which $|x_i - x_0| < \delta_k(\varepsilon)$. Set $\delta(\varepsilon) = \min\{\delta_k(\varepsilon): k = 1, \cdots, m\}$. Thus for each k, $k = 1, \cdots, m$, the expression in (38) is less than

$$(1/\rho)n^{(n-1)/2}D^{m-1}n^{1/2}\varepsilon \tag{40}$$

for every x_0 and every $\mathbf{x}: (x_1, \cdots, x_n, x_0)$ in $X(x_0, \rho)$ and in A such that $|x_i - x_0| < \delta(\varepsilon)$, $i = 1, \cdots, n$. Since there are m terms in the sum in (37), then (37) and (40) show that

$$\left|\frac{\Delta_{(1,\cdots,m)}f(\mathbf{x})}{\Delta(\mathbf{x})} - D_{(1,\cdots,m)}f(x_0)\right| < (1/\rho)mn^{n/2}D^{m-1}\varepsilon \tag{41}$$

for each x_0 in A and each $\mathbf{x}: (x_1, \cdots, x_n, x_0)$ in $X(x_0, \rho)$ and in A such that $|x_i - x_0| < \delta(\varepsilon)$, $i = 1, \cdots, n$. The proof of (29) is complete for $(j_1, \cdots, j_m) = (1, \cdots, m)$, and the proof for the other index sets $(j_1, \cdots, j_m)$ in (m/n) is similar to that just given. □

10.10 Corollary. *If $m = n$ in Theorem 10.9, then to each $\varepsilon > 0$ there corresponds a $\delta(\varepsilon) > 0$ such that*

$$\left|\frac{\Delta_{(1,\cdots,n)}(f^1, \cdots, f^n)(\mathbf{x})}{\Delta(\mathbf{x})} - D_{(1,\cdots,n)}(f^1, \cdots, f^n)(x_0)\right| < (1/\rho)n^{(n+2)/2}D^{n-1}\varepsilon \tag{42}$$

for every x_0 in A and for every $\mathbf{x}: (x_1, \cdots, x_n, x_0)$ in $X(x_0, \rho)$ and in A for which $|x_i - x_0| < \delta(\varepsilon)$, $i = 1, \cdots, n$.

Proof. If $m = n$, there is a single index set in (m/n); it is $(1, \cdots, n)$. Inequality (42) is obtained by setting $m = n$ in (29). □

Exercises

10.1. Let $f: \mathbb{R} \to \mathbb{R}$ be a function such that $f(x) = \sin x$, $x \in \mathbb{R}$. Prove that f is uniformly differentiable on $\mathbb{R}$.

10.2. Prove the following theorem. If the function $f: [a, b] \to \mathbb{R}$ is uniformly differentiable on $[a, b]$, then $D_1 f$ is uniformly continuous on $[a, b]$.

10.3. Prove the following theorem. If the function $f: \mathbb{R}^3 \to \mathbb{R}$ is uniformly differentiable on $\mathbb{R}^3$, then the derivatives $D_j f$, $j = 1, 2, 3$, are uniformly continuous on $\mathbb{R}^3$.

10.4. Prove the following theorem. If $A = [a_1, b_1] \times [a_2, b_2]$, and if each of the functions $f^j : A \to \mathbb{R}$, $j = 1, 2$, is uniformly differentiable on A, then the function $(f^1, f^2) : A \to \mathbb{R}^2$ is uniformly differentiable on A.

10.5. Show that Theorem 10.9 reduces to Theorem 10.6 in the special case $m = 1$.

11. Approximation of Increments of Functions

Section 9 has established linear approximations for the increments of functions $f: E \to \mathbb{R}$, $E \subset \mathbb{R}^n$. Exercise 9.5 states that these approximations can be written in the following form:

$$\begin{vmatrix} f(x_1) & 1 \\ f(x_0) & 1 \end{vmatrix} = \sum_{j=1}^{n} D_j f(x_0) \begin{vmatrix} x_1^j & 1 \\ x_0^j & 1 \end{vmatrix} + r(f; x_0, x_1)|x_1 - x_0|. \tag{1}$$

The value of the determinant on the left in (1) is $f(x_1) - f(x_0)$, and accordingly this determinant is called the increment of f corresponding to the change in x from x_0 to x_1 [see Definition 2.6]. The form of the statement in (1) suggests that it is a special case of a more general formula. Let $f: E \to \mathbb{R}^2$, $E \subset \mathbb{R}^n$, be a function with components (f^1, f^2). If (x_1, x_2, x_0) is a 2-vector in E, then f maps each point x_i into a point $f(x_i)$ with coordinates $(f^1(x_i), f^2(x_i))$. Thus f maps the three points x_1, x_2, x_0 into three points $f(x_1), f(x_2), f(x_0)$, which are the vertices of a triangle in $\mathbb{R}^2$. Then by (25) in Section 89 in Appendix 2, the signed area of the triangle with vertices $f(x_i)$, $i = 1, 2, 0$, is

$$\frac{1}{2!}\begin{vmatrix} f^1(x_1) & f^2(x_1) & 1 \\ f^1(x_2) & f^2(x_2) & 1 \\ f^1(x_0) & f^2(x_0) & 1 \end{vmatrix}. \tag{2}$$

The determinant in (2) is called the increment of f corresponding to the 2-vector (x_1, x_2, x_0). The determinant in (2) is denoted by $\Delta f(x_1, x_2, x_0)$ or $\Delta f(\mathbf{x})$ for $\mathbf{x} : (x_1, x_2, x_0)$ in $\mathbb{R}^n$. Thus

$$\Delta f(x_1, x_2, x_0) = \Delta f(\mathbf{x}) = \begin{vmatrix} f^1(x_1) & f^2(x_1) & 1 \\ f^1(x_2) & f^2(x_2) & 1 \\ f^1(x_0) & f^2(x_0) & 1 \end{vmatrix}, \quad \mathbf{x} : (x_1, x_2, x_0) \text{ in } \mathbb{R}^n. \tag{3}$$

In the general case, $1 \leqq m \leqq n$, the function $f: E \to \mathbb{R}^m$, $E \subset \mathbb{R}^n$, maps the points x_i in the m-vector $\mathbf{x} : (x_1, \cdots, x_m, x_0)$ in $\mathbb{R}^n$ into the points $f(x_i)$ with coordinates $(f^1(x_i), \cdots, f^m(x_i))$, $i = 1, \cdots, m, 0$. The increment of f which corresponds to the m-vector $\mathbf{x}$ is denoted by $\Delta f(x_1, \cdots, x_m, x_0)$ or $\Delta f(\mathbf{x})$ and defined as follows:

$$\Delta f(x_1, \cdots, x_m, x_0) = \Delta f(\mathbf{x}) = \begin{vmatrix} f^1(x_1) & \cdots & f^m(x_1) & 1 \\ \cdots & \cdots & \cdots & \cdots \\ f^1(x_m) & \cdots & f^m(x_m) & 1 \\ f^1(x_0) & \cdots & f^m(x_0) & 1 \end{vmatrix}, \tag{4}$$

$\mathbf{x} : (x_1, \cdots, x_m, x_0)$ in $\mathbb{R}^n$.

By (26) in Section 89, $(1/3!)\Delta f(x_1, \cdots, x_3, x_0)$ is the signed volume of the tetrahedron with vertices $f(x_1), \cdots, f(x_3), f(x_0)$ in $\mathbb{R}^3$, and it will be shown in Section 20 that $(1/m!)|\Delta f(x_1, \cdots, x_m, x_0)|$ is the (hyper-) volume or measure of the figure (simplex) with vertices $f(x_1), \cdots, f(x_m), f(x_0)$ in $\mathbb{R}^m$. The purpose of this section is to find approximations for $\Delta f(x_1, \cdots, x_m, x_0)$ which correspond to the formula in (1).

11.1 Example. The remainder term in (1) has a very special form which is important; a simple example will illustrate why. Let $f: [a, b] \to \mathbb{R}$ be a function which has a continuous derivative $D_1 f$ on $[a, b]$. Let $a = x_0 < x_1 < \cdots < x_{i-1} < x_i < \cdots < x_k = b$ be a partition of $[a, b]$ into k equal subintervals. Then by (1) in the special case $n = 1$,

$$\sum_{i=1}^{k} \begin{vmatrix} f(x_i) & 1 \\ f(x_{i-1}) & 1 \end{vmatrix} = \sum_{i=1}^{k} D_1 f(x_{i-1}) \begin{vmatrix} x_i & 1 \\ x_{i-1} & 1 \end{vmatrix} + \sum_{i=1}^{k} r(f; x_{i-1}, x_i)|x_i - x_{i-1}|. \quad (5)$$

Let $\varepsilon > 0$ be given. Since f has a continuous derivative by hypothesis, then Theorem 9.9 shows that $|r(f; x_{i-1}, x_i)| < \varepsilon$, $i = 1, \cdots, k$, for all sufficiently large k. Thus, for each $\varepsilon > 0$ there is a $k(\varepsilon)$ such that

$$\begin{aligned} \left|\sum_{i=1}^{k} r(f; x_{i-1}, x_i)|x_i - x_{i-1}|\right| &\leqq \sum_{i=1}^{k} |r(f; x_{i-1}, x_i)|\,|x_i - x_{i-1}|, \\ &< \varepsilon \sum_{i=1}^{k} |x_i - x_{i-1}|, \\ &< \varepsilon |b - a|, \end{aligned} \quad (6)$$

for $k \geqq k(\varepsilon)$. This statement proves that

$$\lim_{k\to\infty} \sum_{i=1}^{k} r(f; x_{i-1}, x_i)|x_i - x_{i-1}| = 0. \quad (7)$$

Take the limit of the two sides in (5). Because of (7),

$$\lim_{k\to\infty} \sum_{i=1}^{k} \begin{vmatrix} f(x_i) & 1 \\ f(x_{i-1}) & 1 \end{vmatrix} = \lim_{k\to\infty} \sum_{i=1}^{k} D_1 f(x_{i-1}) \begin{vmatrix} x_i & 1 \\ x_{i-1} & 1 \end{vmatrix} = \int_a^b D_1 f(x)\,dx. \quad (8)$$

But

$$\begin{aligned} \lim_{k\to\infty} \sum_{i=1}^{k} \begin{vmatrix} f(x_i) & 1 \\ f(x_{i-1}) & 1 \end{vmatrix} &= \lim_{k\to\infty} \sum_{i=1}^{k} [f(x_i) - f(x_{i-1})] \\ &= \lim_{k\to\infty} [f(b) - f(a)] = f(b) - f(a). \end{aligned} \quad (9)$$

Then (8) and (9) show that

$$\int_a^b D_1 f(x)\,dx = f(b) - f(a). \quad (10)$$

The reader will recognize (10) as a statement of the fundamental theorem of the integral calculus. The fact that the remainder term in (1) is the product of $|x_1 - x_0|$ and a term which approaches zero uniformly with $|x_1 - x_0|$ is the key to the proof; see the inequalities in (6). We seek an approximation for $\Delta f(x_1, \cdots, x_n, x_0)$ in (4) in which the remainder term has a form similar to $r(f; x_0, x_1)|x_1 - x_0|$ in (1).

Some additional notation will be introduced before the next theorem is stated. It will be described first in $\mathbb{R}^3$ and later in $\mathbb{R}^n$. The points $x_i : (x_i^1, x_i^2, x_i^3)$, $i = 1, 2, 0$, are the terminal points and initial point of a 2-vector $\mathbf{x} : (x_1, x_2, x_0)$. Then $\mathbf{x}$ is conveniently represented by a 3×3 matrix as follows:

$$\mathbf{x} = (x_1, x_2, x_0) = \begin{bmatrix} x_1^1 & x_1^2 & x_1^3 \\ x_2^1 & x_2^2 & x_2^3 \\ x_0^1 & x_0^2 & x_0^3 \end{bmatrix}. \tag{11}$$

The matrix $\mathbf{x}$ has three 3×2 minors. Let the minor in columns (j_1, j_2) be denoted by $\mathbf{x}^{(j_1, j_2)}$ or $(x_1, x_2, x_0)^{(j_1, j_2)}$. Thus

$$\mathbf{x}^{(j_1, j_2)} = (x_1, x_2, x_0)^{(j_1, j_2)} = \begin{bmatrix} x_1^{j_1} & x_1^{j_2} \\ x_2^{j_1} & x_2^{j_2} \\ x_0^{j_1} & x_0^{j_2} \end{bmatrix}, \qquad (j_1, j_2) \in (2/3). \tag{12}$$

The matrix $\mathbf{x}^{(j_1, j_2)}$ has a geometric interpretation as follows: row i, for $i = 1, 2, 0$, contains the coordinates $(x_i^{j_1}, x_i^{j_2})$ of the projection of the point (x_i^1, x_i^2, x_i^3) into the (j_1, j_2) coordinate plane. The points x_1, x_2, x_0 project into three points which are the vertices of a triangle in the (j_1, j_2) coordinate plane. Define the symbols $\Delta[\mathbf{x}^{(j_1, j_2)}]$ and $\Delta[(x_1, x_2, x_0)^{(j_1, j_2)}]$ as follows:

$$\Delta[\mathbf{x}^{(j_1, j_2)}] = \Delta[(x_1, x_2, x_0)^{(j_1, j_2)}] = \det \begin{bmatrix} x_1^{j_1} & x_1^{j_2} & 1 \\ x_2^{j_1} & x_2^{j_2} & 1 \\ x_0^{j_1} & x_0^{j_2} & 1 \end{bmatrix}, \quad (j_1, j_2) \in (2/3). \tag{13}$$

Then by (25) in Section 89, the signed area of the projection of (x_1, x_2, x_0) into the (j_1, j_2) coordinate plane is $(1/2!)\Delta[\mathbf{x}^{(j_1, j_2)}]$. Call $\Delta[\mathbf{x}^{(j_1, j_2)}]$, or $\Delta[(x_1, x_2, x_0)^{(j_1, j_2)}]$, the *$(j_1, j_2)$-component of the 2-vector* (x_1, x_2, x_0). Then $\mathbf{x}$, or (x_1, x_2, x_0), has three components; they are $\Delta[\mathbf{x}^{(1,2)}]$, $\Delta[\mathbf{x}^{(1,3)}]$, $\Delta[\mathbf{x}^{(2,3)}]$. Define the symbols $\Delta(\mathbf{x})$ and $\Delta(x_1, x_2, x_0)$ as follows:

$$\begin{aligned} &\Delta(\mathbf{x}) = (\Delta[\mathbf{x}^{(1,2)}], \Delta[\mathbf{x}^{(1,3)}], \Delta[\mathbf{x}^{(2,3)}]), \\ &\Delta(x_1, x_2, x_0) \\ &\quad = (\Delta[(x_1, x_2, x_0)^{(1,2)}], \Delta[(x_1, x_2, x_0)^{(1,3)}], \Delta[(x_1, x_2, x_0)^{(2,3)}]). \end{aligned} \tag{14}$$

These equations show that $\Delta(\mathbf{x})$, or $\Delta(x_1, x_2, x_0)$, can be interpreted as a vector in $\mathbb{R}^3$. This vector has a length or norm which is defined in the usual way [see (18) in Section 89 in Appendix 2]; the length or norm is denoted by vertical bars placed around the symbol for the vector. Thus

$$
\begin{aligned}
|\Delta(\mathbf{x})| &= \left\{ \sum_{(j_1,j_2)} [\Delta[\mathbf{x}^{(j_1,j_2)}]]^2 \right\}^{1/2}, \qquad (j_1, j_2) \text{ in } (2/3);\\
|\Delta(x_1, x_2, x_0)| &= \left\{ \sum_{(j_1,j_2)} [\Delta[(x_1, x_2, x_0)^{(j_1,j_2)}]]^2 \right\}^{1/2}, \qquad (j_1, j_2) \text{ in } (2/3).
\end{aligned} \tag{15}
$$

The formula in (15), in a form less obscured by unfamiliar notation, is the following:

$$
|\Delta(\mathbf{x})| = |\Delta(x_1, x_2, x_0)| = \left[\sum_{(j_1,j_2)} \begin{vmatrix} x_1^{j_1} & x_1^{j_2} & 1 \\ x_2^{j_1} & x_2^{j_2} & 1 \\ x_0^{j_1} & x_0^{j_2} & 1 \end{vmatrix}^2 \right]^{1/2}. \tag{16}
$$

This formula has a geometric interpretation; it is shown in elementary analytic geometry [see Exercises 11.7 and 11.8, and also Section 20] that $(1/2!)|\Delta(\mathbf{x})|$ is the area of the triangle with vertices x_1, x_2, x_0 in $\mathbb{R}^3$. Finally, define the class of 2-vectors at x_0 in $\mathbb{R}^3$ which satisfy the regularity condition. Let ρ be a number such that $0 < \rho \leqq 1$; define $X_2(x_0, \rho)$ as follows:

$$
X_2(x_0, \rho) = \{(x_1, x_2, x_0) : |\Delta(x_1, x_2, x_0)| \geqq \rho |x_1 - x_0||x_2 - x_0| > 0\}. \tag{17}
$$

The next theorem is concerned with the following formula; in it, (j_1, j_2) are index sets in (2/3):

$$
\begin{aligned}
\begin{vmatrix} f^1(x_1) & f^2(x_1) & 1 \\ f^1(x_2) & f^2(x_2) & 1 \\ f^1(x_0) & f^2(x_0) & 1 \end{vmatrix} &= \sum_{(j_1,j_2)} D_{(j_1,j_2)}(f_1^1 f^2)(x_0) \begin{vmatrix} x_1^{j_1} & x_1^{j_2} & 1 \\ x_2^{j_1} & x_2^{j_2} & 1 \\ x_0^{j_1} & x_0^{j_2} & 1 \end{vmatrix} \\
&\quad + r(f; \mathbf{x}) \left[\sum_{(j_1,j_2)} \begin{vmatrix} x_1^{j_1} & x_1^{j_2} & 1 \\ x_2^{j_1} & x_2^{j_2} & 1 \\ x_0^{j_1} & x_0^{j_2} & 1 \end{vmatrix}^2 \right]^{1/2}.
\end{aligned} \tag{18}
$$

This formula, in the briefer notation introduced above, is the following:

$$
\Delta f(\mathbf{x}) = \sum_{(j_1,j_2)} D_{(j_1,j_2)} f(x_0) \Delta[\mathbf{x}^{(j_1,j_2)}] + r(f; \mathbf{x})|\Delta(\mathbf{x})|. \tag{19}
$$

11.2 Theorem. *Let $A = [a_1, b_1] \times \cdots \times [a_3, b_3]$, and let $f: A \to \mathbb{R}^2$ be a function whose components (f^1, f^2) have continuous derivatives on A. Then for each x_0 in A and each $\mathbf{x} : (x_1, x_2, x_0)$ in $X_2(x_0, \rho)$ there is a number $r(f; \mathbf{x})$ such that the formula in* (18) *and in* (19) *holds. Furthermore, to each $\varepsilon > 0$ there corresponds a $\delta(\varepsilon) > 0$ such that, if D is the constant in Lemma 10.8, then*

$$
|r(f; \mathbf{x})| < (1/\rho)4D\varepsilon \tag{20}
$$

for every x_0 in A and every $\mathbf{x} : (x_1, x_2, x_0)$ in $X_2(x_0, \rho)$ and in A for which $|x_i - x_0| < \delta(\varepsilon)$ for $i = 1, 2$.

Proof. Let $\mathbf{x} : (x_1, x_2, x_0)$ be a 2-vector in $X_2(x_0, \rho)$ and in A; then $|\Delta(\mathbf{x})| > 0$. To prove the theorem, it is necessary to find an approximation for $\Delta f(x_1,$

$x_2, x_0)$ in (3). Subtract the third row of the matrix of this determinant from each of the other rows and expand by elements in the third column. Then

$$\Delta f(x_1, x_2, x_0) = \begin{vmatrix} f^1(x_1) - f^1(x_0) & f^2(x_1) - f^2(x_0) \\ f^1(x_2) - f^1(x_0) & f^2(x_2) - f^2(x_0) \end{vmatrix}. \tag{21}$$

Use the formula in (1) to evaluate successively the terms in the columns of the matrix of this determinant. By the methods used in proving Theorem 10.9, the determinant in (21) can be written as the sum of the determinants of three matrices. Thus

$$\Delta f(x_1, x_2, x_0) = \begin{vmatrix} \sum_{j=1}^{3} D_j f^1(x_0)(x_1^j - x_0^j) & \sum_{j=1}^{3} D_j f^2(x_0)(x_1^j - x_0^j) \\ \sum_{j=1}^{3} D_j f^1(x_0)(x_2^j - x_0^j) & \sum_{j=1}^{3} D_j f^2(x_0)(x_2^j - x_0^j) \end{vmatrix} \tag{22, 23}$$

$$+ \begin{vmatrix} r(f^1; x_0, x_1)|x_1 - x_0| & f^2(x_1) - f^2(x_0) \\ r(f^1; x_0, x_2)|x_2 - x_0| & f^2(x_2) - f^2(x_0) \end{vmatrix} \tag{24}$$

$$+ \begin{vmatrix} \sum_{j=1}^{3} D_j f^1(x_0)(x_1^j - x_0^j) & r(f^2; x_0, x_1)|x_1 - x_0| \\ \sum_{j=1}^{3} D_j f^1(x_0)(x_2^j - x_0^j) & r(f^2; x_0, x_2)|x_2 - x_0| \end{vmatrix}. \tag{25}$$

The definition of matrix multiplication shows that (23) is the determinant of the following matrix product; observe that the second matrix is transposed.

$$\begin{bmatrix} x_1^1 - x_0^1 & x_1^2 - x_0^2 & x_1^3 - x_0^3 \\ x_2^1 - x_0^1 & x_2^2 - x_0^2 & x_2^3 - x_0^3 \end{bmatrix} \begin{bmatrix} D_1 f^1(x_0) & D_2 f^1(x_0) & D_3 f^1(x_0) \\ D_1 f^2(x_0) & D_2 f^2(x_0) & D_3 f^2(x_0) \end{bmatrix}^t.$$

Then by the Binet–Cauchy multiplication theorem [see Theorem 80.1 in Appendix 1] and Corollary 3.13, the determinant in (23) equals

$$\sum_{(j_1, j_2)} D_{(j_1, j_2)}(f^1, f^2)(x_0) \begin{vmatrix} x_1^{j_1} - x_0^{j_1} & x_1^{j_2} - x_0^{j_2} \\ x_2^{j_1} - x_0^{j_1} & x_2^{j_2} - x_0^{j_2} \end{vmatrix}. \tag{26}$$

Thus (23) equals the first sum on the right in (18) and (19). Define $r(f; \mathbf{x})$ to be the sum of the determinants in (24) and (25) divided by $|\Delta(\mathbf{x})|$; recall that $|\Delta(\mathbf{x})| > 0$ since $\mathbf{x}$ is in $X_2(x_0, \rho)$. Thus $r(f; \mathbf{x})$ is defined so that $r(f; \mathbf{x})|\Delta(\mathbf{x})|$ equals the sum of the determinants in (24) and (25). With this definition of $r(f; \mathbf{x})$, the statements in (22), $\cdots$, (26) show that the formula in equations (18) and (19) is valid. The proof of the theorem can be completed by showing that $r(f; \mathbf{x})$ has the properties described in the theorem.

Factor out $|x_1 - x_0|$ and $|x_2 - x_0|$ from the two rows of the matrices in (24) and (25). Then

$$r(f;\mathbf{x}) = \left[\begin{vmatrix} r(f^1; x_0, x_1) & \dfrac{f^2(x_1) - f^2(x_0)}{|x_1 - x_0|} \\ r(f^1; x_0, x_2) & \dfrac{f^2(x_2) - f^2(x_0)}{|x_2 - x_0|} \end{vmatrix} + \begin{vmatrix} \dfrac{\sum_1^3 D_j f^1(x_0)(x_1^j - x_0^j)}{|x_1 - x_0|} & r(f^2; x_0, x_1) \\ \dfrac{\sum_1^3 D_j f^1(x_0)(x_2^j - x_0^j)}{|x_2 - x_0|} & r(f^2; x_0, x_2) \end{vmatrix} \right] \frac{|x_1 - x_0||x_2 - x_0|}{|\Delta(\mathbf{x})|}. \tag{27}$$

Now f^1 and f^2 have continuous derivatives on the compact set A by hypothesis. Then by Lemma 10.8 there is a constant $D > 0$ such that

$$\begin{aligned} &\frac{|f^2(x_1) - f^2(x_0)|}{|x_1 - x_0|} \leqq D, & &\frac{|f^2(x_2) - f^2(x_0)|}{|x_2 - x_0|} \leqq D, \\ &\frac{|\sum_1^3 D_j f^1(x_0)(x_1^j - x_0^j)|}{|x_1 - x_0|} \leqq D, & &\frac{|\sum_1^3 D_j f^1(x_0)(x_2^j - x_0^j)|}{|x_2 - x_0|} \leqq D, \end{aligned} \tag{28}$$

since f^1, f^2 have continuous derivatives on A and $\mathbf{x}: (x_1, x_2, x_0)$ is in $X_2(x_0, \rho)$. Also since $\mathbf{x}$ is in $X_2(x_0, \rho)$, then

$$\frac{|x_1 - x_0||x_2 - x_0|}{|\Delta(\mathbf{x})|} \leqq \frac{1}{\rho}.$$

Next, Hadamard's determinant theorem [see (19) in Corollary 87.2], (27), and (28) show that, if $\mathbf{x}$ is in $X_2(x_0, \rho)$, then

$$\begin{aligned} |r(f;\mathbf{x})| \leqq (1/\rho)2^{1/2} D &\left\{ \sum_{i=1}^{2} [r(f^1; x_0, x_i)]^2 \right\}^{1/2} \\ + (1/\rho)2^{1/2} D &\left\{ \sum_{i=1}^{2} [r(f^2; x_0, x_i)]^2 \right\}^{1/2}. \end{aligned} \tag{29}$$

Let $\varepsilon > 0$ be given; then by Theorem 9.9 [see also equation (39) in the proof of Theorem 10.9] there is a $\delta(\varepsilon) > 0$ such that

$$|r(f^1; x_0, x_i)| < \varepsilon, \qquad |r(f^2; x_0, x_i)| < \varepsilon, \qquad i = 1, 2, \tag{30}$$

for every x_0, x_i in A for which $|x_i - x_0| < \delta(\varepsilon)$. Then (29) and (30) show that

$$|r(f;\mathbf{x})| < (1/\rho)4D\varepsilon \tag{31}$$

for every x_0 in A and every $\mathbf{x}: (x_1, x_2, x_0)$ in $X_2(x_0, \rho)$ and in A such that $|x_i - x_0| < \delta(\varepsilon)$, $i = 1, 2$. The proof of Theorem 11.2 is complete. □

Thus far this section has dealt with the special case of 2-vectors in $\mathbb{R}^3$; the general case will be described now. The points $x_i: (x_i^1, \cdots, x_i^n)$, $i = 1, \cdots, m$, 0, are the terminal points and initial point of an m-vector $\mathbf{x}: (x_1, \cdots, x_m, x_0)$

in $\mathbb{R}^n$, $1 \leqq m \leqq n$. Then $\mathbf{x}$ is conveniently represented by an $(m+1) \times n$ matrix as follows.

$$\mathbf{x} = (x_1, \cdots, x_m, x_0) = \begin{bmatrix} x_1^1 & x_1^2 & \cdots & x_1^n \\ \cdots & \cdots & \cdots & \cdots \\ x_m^1 & x_m^2 & \cdots & x_m^n \\ x_0^1 & x_0^2 & \cdots & x_0^n \end{bmatrix}. \tag{32}$$

The matrix $\mathbf{x}$ has $C(n, m)$ minors of dimension $(m+1) \times m$. Let the $(m+1) \times m$ minor of $\mathbf{x}$ in columns $(j_1, \cdots, j_m)$ be denoted by $\mathbf{x}^{(j_1, \cdots, j_m)}$ or $(x_1, \cdots, x_m, x_0)^{(j_1, \cdots, j_m)}$. Thus

$$\mathbf{x}^{(j_1, \cdots, j_m)} = (x_1, \cdots, x_m, x_0)^{(j_1, \cdots, j_m)}$$
$$= \begin{bmatrix} x_1^{j_1} & x_1^{j_2} & \cdots & x_1^{j_m} \\ \cdots & \cdots & \cdots & \cdots \\ x_m^{j_1} & x_m^{j_2} & \cdots & x_m^{j_m} \\ x_0^{j_1} & x_0^{j_2} & \cdots & x_0^{j_m} \end{bmatrix}, \qquad (j_1, \cdots, j_m) \in (m/n). \tag{33}$$

The matrix $\mathbf{x}^{(j_1, \cdots, j_m)}$ has a geometric interpretation as follows: row i, for $i = 1, \cdots, m, 0$, contains the coordinates $(x_i^{j_1}, \cdots, x_i^{j_m})$ of the projection of the point $(x_i^1, \cdots, x_i^n)$ into the $(j_1, \cdots, j_m)$ coordinate plane. The points $x_1, \cdots, x_m, x_0$ project into $m+1$ points which are the terminal points and initial point of an m-vector in the m-dimensional $(j_1, \cdots, j_m)$ coordinate plane. Define the symbols $\Delta[\mathbf{x}^{(j_1, \cdots, j_m)}]$ and $\Delta[(x_1, \cdots, x_m, x_0)^{(j_1, \cdots, j_m)}]$ as follows:

$$\Delta[\mathbf{x}^{(j_1, \cdots, j_m)}] = \Delta[(x_1, \cdots, x_m, x_0)^{(j_1, \cdots, j_m)}]$$
$$= \det \begin{bmatrix} x_1^{j_1} & x_1^{j_2} & \cdots & x_1^{j_m} & 1 \\ \cdots & \cdots & \cdots & \cdots & \cdots \\ x_m^{j_1} & x_m^{j_2} & \cdots & x_m^{j_m} & 1 \\ x_0^{j_1} & x_0^{j_2} & \cdots & x_0^{j_m} & 1 \end{bmatrix}. \tag{34}$$

Then the signed (hyper-) volume or measure of the projection of $(x_1, \cdots, x_m, x_0)$ into the $(j_1, \cdots, j_m)$ coordinate plane is $(1/m!)\Delta[\mathbf{x}^{(j_1, \cdots, j_m)}]$. Call $\Delta[\mathbf{x}^{(j_1, \cdots, j_m)}]$, or $\Delta[(x_1, \cdots, x_m, x_0)^{(j_1, \cdots, j_m)}]$, the $(j_1, \cdots, j_m)$-component of the m-vector $(x_1, \cdots, x_m, x_0)$. Then $\mathbf{x}$, or $(x_1, \cdots, x_m, x_0)$, has $C(n, m)$ components; they are $(\Delta[\mathbf{x}^{(j_1, \cdots, j_m)}] : (j_1, \cdots, j_m) \in (m/n))$. Define the symbols $\Delta(\mathbf{x})$ and $\Delta(x_1, \cdots, x_m, x_0)$ as follows:

$$\Delta(\mathbf{x}) = (\Delta[\mathbf{x}^{(j_1, \cdots, j_m)}] : (j_1, \cdots, j_m) \in (m/n)),$$
$$\Delta(x_1, \cdots, x_m, x_0) = (\Delta[(x_1, \cdots, x_m, x_0)^{(j_1, \cdots, j_m)}] : (j_1, \cdots, j_m) \in (m/n)). \tag{35}$$

These equations show that $\Delta(\mathbf{x})$, or $\Delta(x_1, \cdots, x_m, x_0)$, can be interpreted as a vector in $\mathbb{R}^{C(n,m)}$. This vector has a length or norm which is defined in the usual way [see (1) in Section 91 in Appendix 2]; the length or norm is denoted by vertical bars placed around the symbol for the vector. Thus

$$|\Delta(\mathbf{x})| = \left\{ \sum_{(j_1,\cdots,j_m)} [\Delta[\mathbf{x}^{(j_1,\cdots,j_m)}]]^2 \right\}^{1/2}, \quad (j_1, \cdots, j_m) \in (m/n),$$

$$|\Delta(x_1, \cdots, x_m, x_0)| = \left\{ \sum_{(j_1,\cdots,j_m)} [\Delta[(x_1, \cdots, x_m, x_0)^{(j_1,\cdots,j_m)}]]^2 \right\}^{1/2}, \tag{36}$$

$$(j_1, \cdots, j_m) \in (m/n).$$

The formula in (36), in a form less obscured by unfamiliar notation, is the following:

$$|\Delta(\mathbf{x})| = |\Delta(x_1, \cdots, x_m, x_0)| = \left[\sum_{(j_1,\cdots,j_m)} \begin{vmatrix} x_1^{j_1} & x_1^{j_2} & \cdots & x_1^{j_m} & 1 \\ \cdots & \cdots & \cdots & \cdots & \cdots \\ x_m^{j_1} & x_m^{j_2} & \cdots & x_m^{j_m} & 1 \\ x_0^{j_1} & x_0^{j_2} & \cdots & x_0^{j_m} & 1 \end{vmatrix}^2 \right]^{1/2}. \tag{37}$$

This formula has a geometric interpretation; Section 20 will show that $(1/m!)|\Delta(\mathbf{x})|$ is the (hyper-) volume or measure of the figure (simplex) with vertices $x_1, \cdots, x_m, x_0$, in $\mathbb{R}^n$. Finally, define the class of m-vectors at x_0 in $\mathbb{R}^n$ which satisfy the regularity condition. Let ρ be a number such that $0 < \rho \leqq 1$; define $X_m(x_0, \rho)$ as follows:

$$X_m(x_0, \rho) = \left\{(x_1, \cdots, x_m, x_0) : |\Delta(x_1, \cdots, x_m, x_0)| \geqq \rho \prod_{i=1}^{m} |x_i - x_0| > 0\right\}. \tag{38}$$

Observe that $X_n(x_0, \rho)$ is the class of n-vectors which has heretofore been denoted by $X(x_0, \rho)$.

The next theorem is a generalization of Theorem 11.2; the formula in (39) is a generalization of the formula in (1) and of the Stolz condition [see Definition 3.1 and Theorem 3.3].

11.3 Theorem. *Let $A = [a_1, b_1] \times \cdots \times [a_n, b_n]$, and let $f: A \to \mathbb{R}^m$ be a function whose components $(f^1, \cdots, f^m)$ have continuous derivatives on A. Then for each x_0 in A and each $\mathbf{x}: (x_1, \cdots, x_m, x_0)$ in $X_m(x_0, \rho)$ there is a number $r(f; \mathbf{x})$ such that*

$$\Delta f(x_1, \cdots, x_m, x_0) = \sum_{(j_1,\cdots,j_m)} D_{(j_1,\cdots,j_m)} f(x_0) \Delta[\mathbf{x}^{(j_1,\cdots,j_m)}] + r(f; \mathbf{x})|\Delta(\mathbf{x})|, \tag{39}$$

$$(j_1, \cdots, j_m) \in (m/n).$$

Furthermore, to each $\varepsilon > 0$ there corresponds a $\delta(\varepsilon) > 0$ such that, if D is the constant in Lemma 10.8, then

$$|r(f; \mathbf{x})| < (1/\rho) m^{(m+2)/2} D^{m-1} \varepsilon \tag{40}$$

for every x_0 in A and every $\mathbf{x}: (x_1, \cdots, x_m, x_0)$ in $X_m(x_0, \rho)$ and in A for which $|x_i - x_0| < \delta(\varepsilon)$ for $i = 1, \cdots, m$.

PROOF. The proof of Theorem 11.3 is similar to the proof of Theorem 11.2; the methods and the tools are the same. The reader should be able to write out the proof of Theorem 11.3 by following step-by-step the proof of Theorem 11.2. □

The next theorem is the special case of Theorem 11.3 in which $m = n$.

11.4 Theorem. *Let $A = [a_1, b_1] \times \cdots \times [a_n, b_n]$, and let $f: A \to \mathbb{R}^n$ be a function whose components $(f^1, \cdots, f^n)$ have continuous derivatives on A. Then for each x_0 in A and each $\mathbf{x}: (x_1, \cdots, x_n, x_0)$ in $X_n(x_0, \rho)$ there is a number $r(f; \mathbf{x})$ such that*

$$\Delta f(x_1, \cdots, x_n, x_0) = D_{(1,\cdots,n)} f(x_0) \Delta(x_1, \cdots, x_n, x_0) + r(f; \mathbf{x}) |\Delta(x_1, \cdots, x_n, x_0)|. \tag{41}$$

Furthermore, to each $\varepsilon > 0$ there corresponds a $\delta(\varepsilon) > 0$ such that, if D is the constant in Lemma 10.8, then

$$|r(f; \mathbf{x})| < (1/\rho) n^{(n+2)/2} D^{n-1} \varepsilon \tag{42}$$

for every x_0 in A and every $\mathbf{x}: (x_1, \cdots, x_m, x_0)$ in $X_n(x_0, \rho)$ and in A for which $|x_i - x_0| < \delta(\varepsilon)$ for $i = 1, \cdots, n$.

PROOF. The simplest proof of this theorem is obtained by setting $m = n$ in Theorem 11.3, but there is a direct proof which is more informative. Let $\mathbf{x}$ be an n-vector in $X_n(x_0, \rho)$. Then

$$\Delta(\mathbf{x}) = \det \begin{bmatrix} x_1^1 & x_1^2 & \cdots & x_1^n & 1 \\ \cdots & \cdots & \cdots & \cdots & \cdots \\ x_n^1 & x_n^2 & \cdots & x_n^n & 1 \\ x_0^1 & x_0^2 & \cdots & x_0^n & 1 \end{bmatrix}, \tag{43}$$

and the following is an obvious identity.

$$\Delta f(\mathbf{x}) = D_{(1,\cdots,n)} f(x_0) \Delta(\mathbf{x}) + \left[\frac{\Delta f(\mathbf{x})}{\Delta(\mathbf{x})} - D_{(1,\cdots,n)} f(x_0) \right] \Delta(\mathbf{x}). \tag{44}$$

Define $r(f; \mathbf{x})$ by the following equation:

$$r(f; \mathbf{x}) |\Delta(\mathbf{x})| = \left[\frac{\Delta f(\mathbf{x})}{\Delta(\mathbf{x})} - D_{(1,\cdots,n)} f(x_0) \right] \Delta(\mathbf{x}). \tag{45}$$

Then (44) and (45) show that (41) holds; also,

$$|r(f; \mathbf{x})| = \left| \frac{\Delta f(\mathbf{x})}{\Delta(\mathbf{x})} - D_{(1,\cdots,n)} f(x_0) \right|. \tag{46}$$

Finally, (46) and Corollary 10.10, equation (42), show that (42) holds for every x_0 in A and for every $\mathbf{x}:(x_1, \cdots, x_n, x_0)$ in $X_n(x_0, \rho)$ and in A for which $|x_i - x_0| < \delta(\varepsilon)$, $i = 1, \cdots, n$. The proof of Theorem 11.4 is complete. □

11.5 Remark. A slightly different form of (41) can be obtained. Define $s(f; \mathbf{x})$ as follows:

$$s(f; \mathbf{x}) = \left[\frac{\Delta f(\mathbf{x})}{\Delta(\mathbf{x})} - D_{(1,\cdots,n)} f(x_0)\right]. \tag{47}$$

Then the identity in (44) is

$$\Delta f(\mathbf{x}) = D_{(1,\cdots,n)} f(x_0)\Delta(\mathbf{x}) + s(f; \mathbf{x})\Delta(\mathbf{x}), \tag{48}$$

and (46) and (47) show that $|s(f; \mathbf{x})| = |r(f; \mathbf{x})|$.

Let $\mathbf{x}:(x_1, \cdots, x_m, x_0)$ be an m-vector in $X_m(x_0, \rho)$ of the following special form.

$$\mathbf{x} = \begin{bmatrix} x_1^1 & \cdots & x_1^m & x_0^{m+1} & \cdots & x_0^n \\ \cdots & \cdots & \cdots & \cdots & \cdots & \cdots \\ x_m^1 & \cdots & x_m^m & x_0^{m+1} & \cdots & x_0^n \\ x_0^1 & \cdots & x_0^m & x_0^{m+1} & \cdots & x_0^n \end{bmatrix}. \tag{49}$$

This m-vector is parallel to the $(x^1, \cdots, x^m)$-coordinate plane. Then $\Delta[\mathbf{x}^{(1,\cdots,m)}] \neq 0$, and $\Delta[\mathbf{x}^{(j_1,\cdots,j_m)}] = 0$ for $(j_1, \cdots, j_m) \neq (1, \cdots, m)$, $(j_1, \cdots, j_m)$ in (m/n). Hence, for this special m-vector, the formula in (39) becomes

$$\Delta f(x_1, \cdots, x_m, x_0) = D_{(1,\cdots,m)} f(x_0)\Delta[\mathbf{x}^{(1,\cdots,m)}] + r(f; \mathbf{x})|\Delta[\mathbf{x}^{(1,\cdots,m)}]|. \tag{50}$$

Equation (40) shows that

$$\lim_{\mathbf{x}\to x_0} r(f; \mathbf{x}) = 0. \tag{51}$$

Thus, if $\mathbf{x}$ is in $X_m(x_0, \rho)$ and is an m-vector of the form (49), then

$$\lim_{\mathbf{x}\to x_0} \frac{\Delta f(x_1, \cdots, x_m, x_0)}{\Delta[\mathbf{x}^{(1,\cdots,m)}]} = D_{(1,\cdots,m)} f(x_0). \tag{52}$$

The general case of this result is stated in the following theorem.

11.6 Theorem. *Let $f: A \to \mathbb{R}^m$ be the function described in Theorem 11.3. If $\mathbf{x}:(x_1, \cdots, x_m, x_0)$ is in $X_m(x_0, \rho)$ and parallel to the $(x^{j_1}, \cdots, x^{j_m})$-coordinate plane, then*

$$\begin{aligned} &\Delta f(x_1, \cdots, x_m, x_0) \\ &\quad = D_{(j_1,\cdots,j_m)} f(x_0)\Delta[\mathbf{x}^{(j_1,\cdots,j_m)}] + r(f; \mathbf{x})|\Delta[\mathbf{x}^{(j_1,\cdots,j_m)}]|. \end{aligned} \tag{53}$$

$$\lim_{\mathbf{x}\to x_0} \frac{\Delta f(x_1, \cdots, x_m, x_0)}{\Delta[\mathbf{x}^{(j_1,\cdots,j_m)}]} = D_{(j_1,\cdots,j_m)} f(x_0), \qquad (j_1, \cdots, j_m) \in (m/n). \tag{54}$$

Proof. The proof follows from the formula in (39) as in equations (49), $\cdots$, (52) above. The reader should compare (54) with the definition of $D_{(j_1,\cdots,j_m)}f(x_0)$ in equation (26) in Definition 2.8, and with the special case of partial derivatives in Theorem 3.14. □

Exercises

11.1. The points $x_0:(1, 1, 1)$, $x_1:(2, 3, 2)$, $x_2:(-2, 4, 3)$ are the vertices of a triangle in $\mathbb{R}^3$. Find the area of this triangle.

11.2. The initial point of a 2-vector $\mathbf{x}:(x_1, x_2, x_0)$ in $\mathbb{R}^3$ is $x_0:(1, 2, 1)$, and the terminal points are $x_1:(2, 3, 2)$ and $x_2:(3, -3, 4)$.

(a) Find $\mathbf{x}^{(1,2)}$, $\mathbf{x}^{(1,3)}$, $\mathbf{x}^{(2,3)}$.
(b) Find $\Delta[\mathbf{x}^{(1,2)}]$, $\Delta[\mathbf{x}^{(1,3)}]$, $\Delta[\mathbf{x}^{(2,3)}]$. What is $\Delta(\mathbf{x})$?
(c) Find $|\Delta(\mathbf{x})|$.
(d) Show that $|\Delta(\mathbf{x})| = |x_1 - x_0||x_2 - x_0|$.
(e) Show that $\mathbf{x}$ belongs to the class $X_2(x_0, \rho)$ of 2-vectors at x_0 for which $\rho = 1$.
(f) Show that the vectors $x_1 - x_0$ and $x_2 - x_0$ are orthogonal [see (7) in Theorem 84.2 in Appendix 1].

11.3. Let $f: \mathbb{R}^3 \to \mathbb{R}^2$ be the function with components (f^1, f^2) which are defined as follows:

$$f^1(x) = 2x^1 - 3x^2 + 5x^3,$$
$$f^2(x) = 5x^1 + 4x^2 - 3x^3, \qquad x:(x^1, x^2, x^3) \quad \text{in} \quad \mathbb{R}^3.$$

For this function, find the value of $r(f; \mathbf{x})$ in (27); then since $r(f; \mathbf{x}) = 0$ in (18) and (19), show that

$$\Delta f(x_1, x_2, x_0) = \sum_{(j_1,j_2)} D_{(j_1,j_2)}f(x_0)\Delta[\mathbf{x}^{(j_1,j_2)}], \qquad (j_1, j_2) \in (2/3),$$

for every $\mathbf{x}(x_1, x_2, x_0)$ in $X_2(x_0, \rho)$.

11.4. Let $f: \mathbb{R}^3 \to \mathbb{R}^2$ be the function with components (f^1, f^2) which are defined as follows.

$$f^1(x) = (x^1)^2 + (x^2)^2 + (x^3)^2,$$
$$f^2(x) = x^1 + x^2 + x^3, \qquad x:(x^1, x^2, x^3) \quad \text{in} \quad \mathbb{R}^3.$$

Let $\mathbf{x}:(x_1, x_2, x_0)$ be the 2-vector such that $x_1 = (2, 2, 2)$, $x_2 = (1, 3, 2)$ and $x_0 = (1, 2, 1)$.

(a) For this function f and 2-vector $\mathbf{x}$, verify the following:

$$|\Delta(\mathbf{x})| = \sqrt{3};$$
$$\Delta f(x_1, x_2, x_0) = -4;$$
$$\sum_{(j_1,j_2)} D_{(j_1,j_2)}f(x_0)\Delta[\mathbf{x}^{(j_1,j_2)}] = -4, \qquad (j_1, j_2) \text{ in } (2/3).$$

(b) Show that $r(f; \mathbf{x}) = 0$. [Hint. Recall that $r(f; \mathbf{x})$ is the sum of the determinants in (24) and (25) divided by $|\Delta(\mathbf{x})|$. Show that $r(f^2; x_0, x_1) = r(f^2; x_0, x_2) = 0$ in (25) because f^2 is a linear function. Find the exact value of each of the elements in the matrix in (24), and use these values to show that the determinant of this matrix is zero.]

(c) Show that the values found in (a) and (b) satisfy the formula in (19). How do you account for the fact that $r(f; \mathbf{x}) = 0$ in this exercise? Does $r(f; \mathbf{x})$ always equal zero? Explain.

11.5. (a) Let $A = [a_1, b_1] \times \cdots \times [a_n, b_n]$, and let $f: A \to \mathbb{R}^m$ be a function whose components $(f^1, \cdots, f^m)$ have continuous derivatives on A. Let x_0 be a point in A, and let $\mathbf{x}: (x_1, \cdots, x_m, x_0)$ be an m-vector in $X_m(x_0, \rho)$. Prove that

$$\lim_{\mathbf{x} \to x_0} \frac{\left|\Delta f(x_1, \cdots, x_m, x_0) - \sum_{(j_1, \cdots, j_m)} D_{(j_1, \cdots, j_m)} f(x_0) \Delta[\mathbf{x}^{(j_1, \cdots, j_m)}]\right|}{|\Delta(\mathbf{x})|} = 0,$$

$$\mathbf{x} \in X_m(x_0, \rho).$$

Compare this limit with the one in equation (17) in Section 6.

(b) Prove the following. For each $\varepsilon > 0$ there exists a $\delta(\varepsilon) > 0$ such that

$$\frac{\left|\Delta f(x_1, \cdots, x_m, x_0) - \sum_{(j_1, \cdots, j_m)} D_{(j_1, \cdots, j_m)} f(x_0) \Delta[\mathbf{x}^{(j_1, \cdots, j_m)}]\right|}{|\Delta(\mathbf{x})|} < \varepsilon$$

for each x_0 in A and each $\mathbf{x}: (x_1, \cdots, x_m, x_0)$ in $X_m(x_0, \rho)$ and in A for which $|x_i - x_0| < \delta(\varepsilon)$, $i = 1, \cdots, m$.

11.6. Let $x_i: (x_i^1, \cdots, x_i^n)$, $i = 0, 1, \cdots, m$, be $m + 1$ points in $\mathbb{R}^n$. Choose one of the points as the initial point and form the m vectors from this point to each of the other points in the set $\{x_0, x_1, \cdots, x_m\}$. The $m + 1$ points thus determine $(m + 1)$ m-vectors constructed as described. Each of these m-vectors has a (hyper-) volume or measure $(1/m!)|\Delta(\mathbf{x})|$ [see equations (36) and (37)]. Prove that the measures of the $(m + 1)$ m-vectors determined by $x_0, x_1, \cdots, x_m$ are all equal.

11.7. Prove the following theorem. Let $x_i: (x_i^1, \cdots, x_i^n)$, $i = 1, 2, 0$, be three points in $\mathbb{R}^n$, and let $v_i = x_i - x_0$, $i = 1, 2$. Then the area of the triangle whose vertices are x_1, x_2, x_0 is

$$\frac{1}{2!}\left\{\det\begin{bmatrix}(v_1, v_1) & (v_1, v_2)\\(v_2, v_1) & (v_2, v_2)\end{bmatrix}\right\}^{1/2}.$$

[Outline of the proof. The area is defined, as usual, as one-half the base times the altitude. See Section 84 in Appendix 1 for the definition and properties of the inner product of two vectors. Let b denote the point $x_0 + tv_1$, and observe that $x_2 = x_0 + v_2$. Show that the vector $x_2 - b$ is orthogonal to v_1 if and only if $t = (v_1, v_2)/(v_1, v_1)$. Show that the length of the altitude from x_2 to the base of the triangle at b is $\{\det[(v_i, v_j)]/(v_1, v_1)\}^{1/2}$, $i, j = 1, 2$. The length of the base is $(v_1, v_1)^{1/2}$. Then one-half the base times the altitude is given by the formula as stated in the theorem.]

11.8. Use the Binet–Cauchy multiplication theorem [see Theorem 80.1 in Appendix 1] to prove that the area of the triangle in Exercise 11.7 equals

$$(1/2!)\left\{\sum_{(j_1, j_2)} [\Delta[(x_1, x_2, x_0)^{(j_1, j_2)}]]^2\right\}^{1/2}, \qquad (j_1, j_2) \in (2/n).$$

Compare this formula with those in equations (16) and (37).

12. Applications: Theorems on Mappings

This section applies results in preceding sections of this chapter to establish four theorems on mappings from $\mathbb{R}^n$ into $\mathbb{R}^n$. These theorems are analogues of elementary theorems on mappings from $\mathbb{R}$ into $\mathbb{R}$. In each case, an example states and proves the elementary theorem [the case $n = 1$], and the theorem which follows contains the general theorem [the case $n > 1$].

12.1 Example. Let E be an open interval in $\mathbb{R}$, and let $f: E \to \mathbb{R}$ be a function which has a derivative $D_1 f$ on E. If $D_1 f(x) \neq 0$ for every x in E, then the mean-value theorem in Theorem 8.1 can be used to prove, as follows, that the mapping $f: E \to \mathbb{R}$ is one-to-one. If x_0, x_1 are distinct points in E, then $f(x_1) - f(x_0) = D_1 f(x^*)(x_1 - x_0) \neq 0$. Hence $f(x_1) \neq f(x_0)$, and the mapping is one-to-one. Since $D_1 f(x) \neq 0$ for x in E, Darboux's theorem [see Exercise 7.10] shows that either $D_1 f(x) > 0$ for all x in E, or $D_1 f(x) < 0$ for all x in E. If $D_1 f(x) > 0$, x in E, the mean-value theorem shows that $f(x_1) > f(x_0)$ and f is a strictly monotonically increasing function on E. Similarly, if $D_1 f(x) < 0$, x in E, then f is a strictly monotonically decreasing function. In both cases, the mapping $f: E \to \mathbb{R}$ is obviously one-to-one.

12.2 Theorem. *Let E be an open convex set in $\mathbb{R}^n$, and let $f: E \to \mathbb{R}^n$ be a function whose components $(f^1, \cdots, f^n)$ are differentiable on E. If $\det[D_j f^i(x_i^*)] \neq 0$, $i, j = 1, \cdots, n$, for every set of points $x_1^*, \cdots, x_n^*$ on an open segment in E, then the mapping $f: E \to \mathbb{R}^n$ is one-to-one on E.*

Proof. Assume the theorem false. Then there are two points x_0 and x_1 such that $x_1 \neq x_0$ and $f(x_1) = f(x_0)$, or $f^i(x_1) = f^i(x_0)$, $i = 1, \cdots, n$. Then by equation (1) in Theorem 9.1, there exist points $x_1^*, \cdots, x_n^*$ on the open segment with end points x_0 and x_1 such that

$$\sum_{j=1}^{n} D_j f^i(x_i^*)(x_1^j - x_0^j) = 0, \qquad i = 1, \cdots, n. \tag{1}$$

The determinant of this system of n equations in the unknowns $(x_1^j - x_0^j)$, $j = 1, \cdots, n$, is not zero by hypothesis. Then $x_1^j - x_0^j = 0$ and $x_1 = x_0$. This contradiction establishes the theorem. □

12.3 Example. Let E be an open interval in $\mathbb{R}$, and let $f: E \to \mathbb{R}$ be a function which has a derivative $D_1 f$ on E. If $D_1 f$ is continuous at the point x_0 in E, and if $D_1 f(x_0) \neq 0$, then there is a neighborhood $N(x_0, \varepsilon) \subset E$ on which the mapping $f: E \to \mathbb{R}$ is one-to-one; the proof follows. Since $D_1 f(x_0) \neq 0$ by hypothesis, then by Theorem 96.9 in Appendix 2 there is a neighborhood $N(x_0, \varepsilon)$, in E, of x_0 on which $D_1 f(x) \neq 0$. Then by Example 12.1, the mapping $f: E \to \mathbb{R}$ is one-to-one on $N(x_0, \varepsilon)$.

12.4 Theorem. *Let E be an open set in $\mathbb{R}^n$, and let $f: E \to \mathbb{R}^n$ be a function whose components $(f^1, \cdots, f^n)$ are differentiable on E. If the derivatives*

$D_j f^i$, $i, j = 1, \cdots, n$, are continuous at x_0 in E, and if $D_{(1,\cdots,n)}f(x_0) \neq 0$, then there is a neighborhood $N(x_0, \varepsilon)$, in E, of x_0 on which the mapping $f: E \to \mathbb{R}^n$ is one-to-one.

PROOF. A determinant is a polynomial in the elements of its matrix [see Section 76 in Appendix 1], and therefore a determinant is a continuous function of the elements of its matrix. Now the derivatives $D_j f^i$ are continuous functions of x at x_0, and $D_{(1,\cdots,n)}f = \det[D_j f^i]$, $i, j = 1, \cdots, n$, by Theorem 3.12. Also, $\det[D_j f^i(x_0)] \neq 0$ by hypothesis. Thus if $N(x_0, \varepsilon)$ is a sufficiently small neighborhood of x_0 and in E, and if $x_1^*, \cdots, x_n^*$ are points in $N(x_0, \varepsilon)$, then $\det[D_j f^i(x_i^*)] \neq 0$. The neighborhood $N(x_0, \varepsilon)$ is a convex set [see Example 9.2]. Thus, if $x_1^*, \cdots, x_n^*$ are points on an open segment whose end-points are in $N(x_0, \varepsilon)$, then these points are in $N(x_0, \varepsilon)$ and in E, and $\det[D_j f^i(x_i^*)] \neq 0$. Therefore, by Theorem 12.2, the mapping $f: E \to \mathbb{R}^n$ is one-to-one on $N(x_0, \varepsilon)$. □

12.5 Example. Let E be an open interval in $\mathbb{R}$, and let $f: E \to \mathbb{R}$ be a function which is differentiable at x_0. If $D_1 f(x_0)$ is positive (negative), then there exists a $\delta > 0$ such that $\Delta_1 f(\mathbf{x})\Delta(\mathbf{x})$ is positive (negative) for every $\mathbf{x}: (x_1, x_0)$ for which $|x_1 - x_0| < \delta$. To prove this statement, observe that to each $\varepsilon > 0$ there corresponds a $\delta(\varepsilon)$ such that

$$\left|\frac{\Delta_1 f(\mathbf{x})}{\Delta(\mathbf{x})} - D_1 f(x_0)\right| < \varepsilon, \qquad |x_1 - x_0| < \delta(\varepsilon). \tag{2}$$

Then

$$D_1 f(x_0) - \varepsilon < \frac{\Delta_1 f(\mathbf{x})}{\Delta(\mathbf{x})} < D_1 f(x_0) + \varepsilon, \qquad |x_1 - x_0| < \delta(\varepsilon). \tag{3}$$

Choose $\varepsilon = (1/2)|D_1 f(x_0)|$, and let δ be the corresponding $\delta(\varepsilon)$. Then (3) shows that $\Delta_1 f(\mathbf{x})\Delta(\mathbf{x})$ has the same sign as $D_1 f(x_0)$ for every $\mathbf{x}: (x_1, x_0)$ such that $|x_1 - x_0| < \delta$. In more geometric terms, $f(x_1) - f(x_0)$ has the same sign as $x_1 - x_0$ if $D_1 f(x_0) > 0$, and $f(x_1) - f(x_0)$ and $x_1 - x_0$ have opposite signs if $D_1 f(x_0) < 0$.

12.6 Theorem. *Let E be an open set in $\mathbb{R}^n$, and let $f: E \to \mathbb{R}^n$ be a function whose components $(f^1, \cdots, f^n)$ are differentiable at x_0 in E. If $D_{(1,\cdots,n)}f(x_0)$ is positive (negative), then there exists a $\delta > 0$ such that $\Delta_{(1,\cdots,n)}f(\mathbf{x})\Delta(\mathbf{x})$ is positive (negative) for every $\mathbf{x}: (x_1, \cdots, x_n, x_0)$ in $X(x_0, \rho)$ such that $|x_i - x_0| < \delta$, $i = 1, \cdots, n$.*

PROOF. The proof is similar to that in Example 12.5. If $\varepsilon = (1/2)|D_{(1,\cdots,n)}f(x_0)| > 0$, then there is a $\delta > 0$ such that

$$D_{(1,\cdots,n)}f(x_0) - \varepsilon < \frac{\Delta_{(1,\cdots,n)}f(\mathbf{x})}{\Delta(\mathbf{x})} < D_{(1,\cdots,n)}f(x_0) + \varepsilon \tag{4}$$

for all $\mathbf{x}: (x_1, \cdots, x_n, x_0)$ in $X(x_0, \rho)$ for which

$$|x_i - x_0| < \delta, \qquad i = 1, \cdots, n. \tag{5}$$

Thus if $D_{(1,\cdots,n)}f(x_0)$ is positive, then $\Delta_{(1,\cdots,n)}f(\mathbf{x})$ has the same sign as $\Delta(\mathbf{x})$ for all $\mathbf{x}$ which satisfy (5); if $D_{(1,\cdots,n)}f(x_0)$ is negative, the signs of $\Delta_{(1,\cdots,n)}f(\mathbf{x})$ and $\Delta(\mathbf{x})$ are opposite for all $\mathbf{x}$ which satisfy (5). □

12.7 Example. Let E be an open interval in $\mathbb{R}$, and let $f: E \to \mathbb{R}$ be a function which is differentiable on E. If $D_1 f(x)$ is positive on E, then f is monotonically increasing on E by Example 12.1, and $[f(x_1) - f(x_0)][x_1 - x_0]$ is positive for every $\mathbf{x}: (x_1, x_0)$ in E. Similarly, if $D_1 f(x)$ is negative on E, then f is monotonically decreasing on E, and $[f(x_1) - f(x_0)][x_1 - x_0]$ is negative for every $\mathbf{x}: (x_1, x_0)$ in E. The analogous statements for functions $f: E \to \mathbb{R}^2$, $E \subset \mathbb{R}^2$, are not true, as the following example shows.

If $f: \mathbb{R}^2 \to \mathbb{R}^2$ is the function such that

$$\begin{aligned} f^1(x^1, x^2) &= (x^1)^2 - (x^2)^2, \\ f^2(x^1, x^2) &= 2x^1x^2, \qquad (x^1, x^2) \text{ in } \mathbb{R}^2, \end{aligned} \tag{6}$$

then $D_{(1,2)}(f^1, f^2)(x) = 4[(x^1)^2 + (x^2)^2]$. Thus if the origin (0, 0) does not belong to E, then $D_{(1,2)}(f^1, f^2)(x) > 0$ for all x in E. The following statements describe the mapping of two 2-vectors by f.

$$\begin{gathered} x_1 = (2, -1), \qquad x_2 = (2, 1), \qquad x_0 = (1, 0); \qquad \Delta(x_1, x_2, x_0) = 2; \\ f(x_1) = (3, -4), \qquad f(x_2) = (3, 4), \qquad f(x_0) = (1, 0); \\ \Delta_{(1,2)}f(x_1, x_2, x_0) = 16. \end{gathered} \tag{7}$$

$$\begin{gathered} x_1 = (2, -3), \qquad x_2 = (2, 3), \qquad x_0 = (1, 0); \qquad \Delta(x_1, x_2, x_0) = 6; \\ f(x_1) = (-5, -12), \qquad f(x_2) = (-5, 12), \qquad f(x_0) = (1, 0); \\ \Delta_{(1,2)}f(x_1, x_2, x_0) = -144. \end{gathered} \tag{8}$$

Thus the function in (6) has $D_{(1,2)}f(x) > 0$ except at the origin (0, 0), but $\Delta_{(1,2)}f(\mathbf{x})$ is positive in (7) and negative in (8). Thus the situation for $n = 2$ is different from that for $n = 1$. The following is a theorem for $n > 1$.

12.8 Theorem. *Let $A = [a_1, b_1] \times \cdots \times [a_n, b_n]$, and let $f: A \to \mathbb{R}^n$ be a function whose components $(f^1, \cdots, f^n)$ have continuous derivatives on A. If $D_{(1,\cdots,n)}f(x)$ is positive (negative) on A, then there exists a $\delta > 0$ such that $\Delta_{(1,\cdots,n)}f(\mathbf{x})\Delta(\mathbf{x})$ is positive (negative) for every x_0 in A and every $\mathbf{x}: (x_1, \cdots, x_n, x_0)$ in $X(x_0, \rho)$ and in A for which $|x_i - x_0| < \delta$, $i = 1, \cdots, n$.*

Proof. Since the components $(f^1, \cdots, f^n)$ have continuous derivatives on A, then $D_{(1,\cdots,n)}f$ and $|D_{(1,\cdots,n)}f|$ are continuous on A. Since A is closed and bounded and therefore compact, then $|D_{(1,\cdots,n)}f|$ has a minimum value on A [see Theorem 96.14 in Appendix 2]. Since $D_{(1,\cdots,n)}f$ is positive (negative) on A, then $\min\{|D_{(1,\cdots,n)}f(x)| : x \in A\} > 0$. Define ε as follows:

$$\varepsilon = (1/2)\min\{|D_{(1,\cdots,n)}f(x)| : x \in A\} > 0. \tag{9}$$

Since the components of f have continuous derivatives on A, then corresponding to this ε there exists a $\delta > 0$ such that

$$\left|\frac{\Delta_{(1,\cdots,n)}f(\mathbf{x})}{\Delta(\mathbf{x})} - D_{(1,\cdots,n)}f(x_0)\right| < \varepsilon \tag{10}$$

for every x_0 in A and every $\mathbf{x} : (x_1, \cdots, x_n, x_0)$ in $X(x_0, \rho)$ and in A for which $|x_i - x_0| < \delta$, $i = 1, \cdots, n$ [see Corollary 10.10 and Theorem 11.4]. Thus for these same $\mathbf{x}$,

$$D_{(1,\cdots,n)}f(x_0) - \varepsilon < \frac{\Delta_{(1,\cdots,n)}f(\mathbf{x})}{\Delta(\mathbf{x})} < D_{(1,\cdots,n)}f(x_0) + \varepsilon; \tag{11}$$

and since $D_{(1,\cdots,n)}f$ is positive (negative) on A and ε is defined by (9), then $\Delta_{(1,\cdots,n)}f(\mathbf{x})\Delta(\mathbf{x})$ is positive (negative) for every x_0 in A and every $\mathbf{x} : (x_1, \cdots, x_n, x_0)$ in $X(x_0, \rho)$ and in A such that $|x_i - x_0| < \delta$, $i = 1, \cdots, n$. □

Exercises

12.1. Let $A = [-1, 1] \times [-1, 1]$, and let $f: A \to \mathbb{R}^2$ be the function whose components (f^1, f^2) are defined as follows.

$$f^1(x^1, x^2) = (x^1)^2 + (x^2)^2,$$
$$f^2(x^1, x^2) = x^1 + x^2, \qquad (x^1, x^2) \in A.$$

(a) Verify that f is not one-to-one on A by showing that $f(x^1, x^2) = f(x^2, x^1)$ for every (x^1, x^2) in A.

(b) Use Theorem 12.2 to prove that f is one-to-one on the set $[0, 1] \times [-1, 0]$.

12.2. Let $f : (f^1, \cdots, f^n)$ be a function with components $f^i : \mathbb{R}^n \to \mathbb{R}$ which are linear functions such that

$$f^i(x) = \sum_{j=1}^{n} a_i^j x^j, \qquad x : (x^1, \cdots, x^n) \text{ in } \mathbb{R}^n, \qquad i = 1, \cdots, n,$$

$$\det[a_i^j] \neq 0, \qquad i, j = 1, \cdots, n.$$

(a) Show that the function $f: \mathbb{R}^n \to \mathbb{R}^n$ satisfies the condition $\det[D_j f^i(x_i^*)] \neq 0$ in Theorem 12.2 and is therefore one-to-one.

(b) If $\det[a_i^j]$ is positive (negative), use Theorem 12.6 to prove that $\Delta_{(1,\cdots,n)} f(\mathbf{x})\Delta(\mathbf{x})$ is positive (negative) for every n-vector $\mathbf{x}$ in $\mathbb{R}^n$.

(c) Use the definition of matrix multiplication to prove that the determinant of the following matrix product is $\Delta_{(1,\cdots,n)}f(\mathbf{x})$.

$$\begin{bmatrix} x_1^1 & x_1^2 & \cdots & x_1^n & 1 \\ \cdots & \cdots & \cdots & \cdots & \cdots \\ x_n^1 & x_n^2 & \cdots & x_n^n & 1 \\ x_0^1 & x_0^2 & \cdots & x_0^n & 1 \end{bmatrix} \begin{bmatrix} a_1^1 & a_2^1 & \cdots & a_n^1 & 0 \\ \cdots & \cdots & \cdots & \cdots & \cdots \\ a_1^n & a_2^n & \cdots & a_n^n & 0 \\ 0 & 0 & \cdots & 0 & 1 \end{bmatrix}$$

Then use the Binet–Cauchy multiplication theorem for determinants [see Theorem 80.1] to show that $\Delta_{(1,\cdots,n)}f(\mathbf{x}) = \det[a_i^j]\Delta(\mathbf{x})$. Finally, use this result to prove that, if $\det[a_i^j]$ is positive (negative), then $\Delta_{(1,\cdots,n)}f(\mathbf{x})\Delta(\mathbf{x})$ is positive (negative) for every n-vector $\mathbf{x}$ in $\mathbb{R}^n$.

12.3. (a) Let $f\colon(f^1, f^2, f^3)$ be a function whose components are defined and differentiable on an open convex set E in $\mathbb{R}^2$. Assume that

$$\sum_{(i_1,i_2)}\begin{vmatrix} D_1 f^{i_1}(x_{i_1}^*) & D_2 f^{i_1}(x_{i_1}^*) \\ D_1 f^{i_2}(x_{i_2}^*) & D_2 f^{i_2}(x_{i_2}^*)\end{vmatrix}^2 \neq 0, \qquad (i_1, i_2)\in(2/3),$$

for each set of points on a line segment in E. Prove that the mapping $f\colon E\to\mathbb{R}^3$ is one-to-one on E.

(b) For functions $f\colon E\to\mathbb{R}^m$, $E\subset\mathbb{R}^n$, $m>n$, state and prove the generalization of the theorem in (a).

12.4. Let $f\colon(f^1, \cdots, f^m)$ be a function whose components are defined and differentiable on an open set E in $\mathbb{R}^n$, $m >$ n. Assume that the derivatives $D_j f^i$, $i=1, \cdots, m$, $j=1, \cdots, n$, are continuous at x_0 in E and that

$$\sum_{(i_1,\cdots,i_n)}[D_{(1,\cdots,n)}(f^{i_1}, \cdots, f^{i_n})(x_0)]^2 \neq 0, \qquad (i_1, \cdots, i_n)\in(n/m).$$

Prove that in a sufficiently small neighborhood $N(x_0, \varepsilon)$ of x_0 in E the mapping $f\colon E\to\mathbb{R}^m$ is one-to-one.

12.5. Let $f\colon\mathbb{R}^n\to\mathbb{R}^m$, $m>n$, be a function with components $f^i\colon\mathbb{R}^n\to\mathbb{R}$ which are linear functions such that

$$f^i(x) = \sum_{j=1}^n a_i^j x^j, \qquad i=1, \cdots, m, \qquad x\colon(x^1, \cdots, x^n)\in\mathbb{R}^n.$$

If

$$\sum_{(i_1,\cdots,i_n)}\begin{vmatrix} a_{i_1}^1 & a_{i_1}^2 & \cdots & a_{i_1}^n \\ a_{i_2}^1 & a_{i_2}^2 & \cdots & a_{i_2}^n \\ \cdots & \cdots & \cdots & \cdots \\ a_{i_n}^1 & a_{i_n}^2 & \cdots & a_{i_n}^n \end{vmatrix}^2 \neq 0, \qquad (i_1, \cdots, i_n)\in(n/m),$$

prove that the mapping $f\colon\mathbb{R}^n\to\mathbb{R}^m$ is one-to-one.

12.6. Let $f\colon\mathbb{R}^2\to\mathbb{R}^3$ be the linear function with components $f^i\colon\mathbb{R}^2\to\mathbb{R}$ such that

$$\begin{aligned} f^1(x) &= 2x^1 + 3x^2, \\ f^2(x) &= 4x^1 + 6x^2, \qquad x\colon(x^1, x^2)\in\mathbb{R}^2. \\ f^3(x) &= 8x^1 + 12x^2. \end{aligned}$$

(a) Show that the mapping $f\colon\mathbb{R}^2\to\mathbb{R}^3$ is not one-to-one.
(b) Show that f maps $\mathbb{R}^2$ onto a line in $\mathbb{R}^3$, and find the equations of this line.
(c) Exercise 12.5 contains a sufficient condition that the linear transformation in that exercise be one-to-one. Show that this sufficient condition is not satisfied by the linear function in the present exercise.

CHAPTER 3

Simplexes, Orientations, Boundaries, and Simplicial Subdivisions

13. Introduction

This chapter contains an introduction to a number of topics in the geometry of n-dimensional Euclidean space. It treats barycentric coordinates, convex sets, simplexes, the orientation of simplexes, chains of simplexes, boundaries of simplexes and chains, three identities and an inequality for determinants, affine and barycentric transformations, and simplicial subdivisions of cubes and simplexes in $\mathbb{R}^n$. The results in this chapter are developed in preparation for their use in the proof of the fundamental theorem of the integral calculus in $\mathbb{R}^n$ and of other theorems. This introduction begins the treatment of these subjects by reviewing some of the basic facts about $(n-1)$-dimensional planes in $\mathbb{R}^n$, convex sets, and linear independence and dependence of vectors in $\mathbb{R}^n$.

Let c be a constant, let $a:(a^1, \cdots, a^n)$ be a vector in $\mathbb{R}^n$ such that $|a|>0$, and let $x:(x^1, \cdots, x^n)$ be a point in $\mathbb{R}^n$. Then $\{x: \Sigma_1^n a^i x^i + c = 0\}$ is called an $(n-1)$-dimensional plane in $\mathbb{R}^n$, and $\Sigma_1^n a^i x^i + c = 0$ is called an equation of this plane. In the inner product notation of Section 84 in Appendix 1, $\Sigma_1^n a^i x^i$ is denoted by (a, x), and the equation of the plane can be written as $(a, x) + c = 0$.

Example 9.2 defines *convex set* as follows: a set E is convex if and only if, for each two points x_0, x_1 in E, the segment $\{x: x_0 + t(x_1 - x_0), 0 \leqq t \leqq 1\}$ joining the points is in E. Triangles, squares, (solid) circles, and (solid) ellipses are examples of convex sets in $\mathbb{R}^2$; the set $\{x: x \in \mathbb{R}^n, |x| = 1\}$ is not convex.

Section 85 in Appendix 1 contains a treatment of linearly independent and dependent vectors in $\mathbb{R}^n$. Let $x_i:(x_i^1, \cdots, x_i^n)$, $i = 0, 1, \cdots, m$, be points in $\mathbb{R}^n$. If $v_i = x_i - x_0$, then $v_i : (x_i^1 - x_0^1, \cdots, x_i^n - x_0^n)$ is a vector in $\mathbb{R}^n$. The vectors $v_1, \cdots, v_m$ are linearly dependent [see Definition 85.1 in Appendix 1] if and only if there exist constants $c_1, \cdots, c_m$, not all zero, such that

$$c_1 v_1 + \cdots + c_m v_m = 0; \tag{1}$$

they are linearly independent if and only if (1) implies that $c_1 = \cdots = c_m = 0$. Equation (1) is a vector equation. Write the vectors $v_1, \cdots, v_m$ as column vectors; then (1) in full detail is this.

$$c_1 \overset{v_1}{\begin{bmatrix} x_1^1 - x_0^1 \\ \vdots \\ x_1^n - x_0^n \end{bmatrix}} + \cdots + c_m \overset{v_m}{\begin{bmatrix} x_m^1 - x_0^1 \\ \vdots \\ x_m^n - x_0^n \end{bmatrix}} = \overset{0}{\begin{bmatrix} 0 \\ \vdots \\ 0 \end{bmatrix}}. \tag{2}$$

This vector equation is equivalent to the following system of linear equations:

$$\begin{gathered} c_1(x_1^1 - x_0^1) + \cdots + c_m(x_m^1 - x_0^1) = 0, \\ \cdots\cdots\cdots\cdots\cdots\cdots\cdots\cdots\cdots \\ c_1(x_1^n - x_0^n) + \cdots + c_m(x_m^n - x_0^n) = 0. \end{gathered} \tag{3}$$

If the only solution of this system of equations is the trivial solution $[c_1 = \cdots = c_m = 0]$, the vectors $v_1, \cdots, v_m$ are linearly independent; if (3) has a non-trivial solution $[c_1^2 + \cdots + c_m^2 \neq 0]$, the vectors are linearly dependent. In many cases, a simple inspection of the system (3) is sufficient to determine whether $v_1, \cdots, v_m$ are linearly independent or dependent, but the following theorem provides the answer in all cases.

13.1 Theorem. *Let $v_1, \cdots, v_m$ be the vectors in $\mathbb{R}^n$ just described. If $m > n$, these vectors are always linearly dependent. If $m \leqq n$, then $v_1, \cdots, v_m$ form an m-vector at x_0; moreover, $v_1, \cdots, v_m$ are linearly dependent if and only if $|\Delta(x_0, x_1, \cdots, x_m)| = 0$ and linearly independent if and only if $|\Delta(x_0, x_1, \cdots, x_m)| > 0$.*

PROOF. Let M be the $m \times n$ matrix whose rows are the vectors $v_1, \cdots, v_m$; thus

$$M = \begin{bmatrix} v_1 \\ \cdots \\ v_m \end{bmatrix} = \begin{bmatrix} x_1^1 - x_0^1 & \cdots & x_1^n - x_0^n \\ \cdots & \cdots & \cdots \\ x_m^1 - x_0^1 & \cdots & x_m^n - x_0^n \end{bmatrix}. \tag{4}$$

Then $MM^t = [(v_i, v_j)]$, $i, j = 1, \cdots, m$, and Section 85 establishes the following results: $v_1, \cdots, v_m$ are linearly dependent if and only if $\det[(v_i, v_j)] = 0$; they are linearly independent if and only if $\det[(v_i, v_j)] > 0$. If $m > n$, the Binet–Cauchy multiplication theorem [see Theorem 80.1 (1) in Appendix 1] shows that $\det[(v_i, v_j)] = 0$; therefore, if $m > n$, the m vectors $v_1, \cdots, v_m$ in $\mathbb{R}^n$ are always linearly dependent. If $m \leqq n$, then the Binet–Cauchy multiplication theorem shows that

$$\det[(v_i, v_j)] = \sum_{(j_1,\cdots,j_m)} \begin{vmatrix} x_1^{j_1} - x_0^{j_1} & \cdots & x_1^{j_m} - x_0^{j_m} \\ \cdots & \cdots & \cdots \\ x_m^{j_1} - x_0^{j_1} & \cdots & x_m^{j_m} - x_0^{j_m} \end{vmatrix}^2, \quad (j_1, \cdots, j_m) \in (m/n). \tag{5}$$

A comparison of this equation with equation (37) in Section 11 shows that

$$|\Delta(x_0, x_1, \cdots, x_m)| = \{\det[(v_i, v_j)]\}^{1/2}, \qquad i, j, = 1, \cdots, m. \tag{6}$$

Thus $\det[(v_i, v_j)]$ is zero (positive) if and only if $|\Delta(x_0, x_1, \cdots, x_m)|$ is zero (positive), and $v_1, \cdots, v_m$ are linearly dependent (linearly independent) if and only if $|\Delta(x_0, x_1, \cdots, x_m)| = 0$ ($|\Delta(x_0, x_1, \cdots, x_m)| > 0$). The proof of Theorem 13.1 is complete. □

Exercises

13.1. Let $(a, x) + c = 0$ be an equation of an $(n-1)$-dimensional plane in $\mathbb{R}^n$. Prove that each of the following sets is a convex set:

$$\{x : (a, x) + c \geqq 0\}, \qquad \{x : (a, x) + c > 0\}, \qquad \{x : (a, x) + c = 0\},$$

$$\{x : (a, x) + c \leqq 0\}, \qquad \{x : (a, x) + c < 0\}.$$

13.2. Let x_0 be a point in $\{x : (a, x) + c < 0\}$, and let x_1 be a point in $\{x : (a, x) + c > 0\}$. Use the intermediate-value theorem in Exercise 7.9 to prove that the segment $\{x : x = x_0 + t(x_1 - x_0), 0 \leqq t \leqq 1\}$ contains a point in the plane $\{x : (a, x) + c = 0\}$; find this point.

13.3. Let x_0 and x denote a fixed point and a variable point, respectively, in the plane $(a, x) + c = 0$. Then $(a, x_0) + c = 0$, $(a, x) + c = 0$, and $(a, x) - (a, x_0) = 0$. Show that $(a, x - x_0) = 0$ [see Section 84 in Appendix 1]. Show also that the plane is the set of points x such that the vector $x - x_0$ is orthogonal to the vector $a : (a^1, \cdots, a^n)$.

13.4. Exercise 13.2 shows that the plane $(a, x) + c = 0$ divides the space $\mathbb{R}^n$ into two half-spaces; in one of these half-spaces $(a, x) + c > 0$ and in the other, $(a, x) + c < 0$. Prove that the vector $a : (a^1, \cdots, a^n)$ points in the direction of the half-space in which $(a, x) + c > 0$. [Hint. Let x_0 be a point in the plane $\{x : (a, x) + c = 0\}$. Show that

$$\{x : x_0 + at, t > 0\} \subset \{x : (a, x) + c > 0\},$$

$$\{x : x_0 + at, t < 0\} \subset \{x : (a, x) + c < 0\}.]$$

13.5. The closed ball $B(x_0, r)$ with center x_0 in $\mathbb{R}^n$ and radius r is the set $\{x : x \in \mathbb{R}^n, |x - x_0| \leqq r\}$. Prove that $B(x_0, r)$ is a convex set.

13.6. Prove that the intersection of a collection (finite or infinite) of convex sets is a convex set.

13.7. Use the equations in (3) to show that the vectors $v_1 : (1, 2, 4)$, $v_2 : (2, -3, 1)$, $v_3 : (3, 4, 2)$ are linearly independent.

13.8. Use the equations in (3) to show that the vectors $v_1 : (2, -4, 5)$, $v_2 : (2, 2, 1)$, $v_3 : (4, -2, 6)$ are linearly dependent. Find constants c_1, c_2, c_3, not all zero, such that $c_1 v_1 + c_2 v_2 + c_3 v_3 = 0$.

13.9. It is known that a system of n homogeneous linear equations in m unknowns always has a non-trivial solution if $m > n$. Use this fact and the system of equations in (3) to prove that m vectors in $\mathbb{R}^n$ are linearly dependent if $m > n$. Compare Corollary 85.5 in Appendix 1.

13.10. Prove that the vectors $x_i - x_0$, $i = 1, \cdots, n$, in an n-vector in $X(x_0, \rho)$ in $\mathbb{R}^n$ are linearly independent.

13.11. Let $v_1 : (2, 3)$, $v_2 : (3, -4)$, $v_3 : (-5, 1)$ be three vectors in $\mathbb{R}^2$. Show that v_1, v_2, v_3 are linearly dependent by finding constants c_1, c_2, c_3, not all zero, such that $c_1 v_1 + c_2 v_2 + c_3 v_3 = 0$.

13.12. Let $x_0 : (1, 2, 4, 3)$, $x_1 : (5, -2, 4, 3)$, $x_2 : (7, 2, 6, -5)$, $x_3 : (6, -4, 3, 5)$ be four points in $\mathbb{R}^4$. These points determine three vectors $v_i = x_i - x_0$, $i = 1, 2, 3$.
(a) Use the formula in equation (6) to show that $|\Delta(x_0, x_1, \cdots, x_3)| = 32$ and thus to prove that the vectors v_1, v_2, v_3 are linearly independent.
(b) Use the formula in equation (36) in Section 11 to show that $|\Delta(x_0, x_1, \cdots, x_3)| = 32$ and thus to prove that the vectors v_1, v_2, v_3 are linearly independent.

13.13. Let $x_0 : (2, 4, 3, 5)$, $x_1 : (3, 6, 1, 8)$, $x_2 : (5, 5, 7, 7)$, $x_3 : (2, 9, -7, 12)$ be four points in $\mathbb{R}^4$. These points determine three vectors $v_i = x_i - x_0$, $i = 1, 2, 3$.
(a) Use the formula in equation (6) to show that $|\Delta(x_0, x_1, \cdots, x_3)| = 0$ and thus to prove that the vectors v_1, v_2, v_3 are linearly dependent.
(b) Use the formula in equation (36) in Section 11 to show that $|\Delta(x_0, x_1, \cdots, x_3)| = 0$ and thus to prove that the vectors v_1, v_2, v_3 are linearly dependent.
(c) Find the system of equations (3) for the vectors v_1, v_2, v_3. Then find constants c_1, c_2, c_3, not all zero, which satisfy these equations. Thus prove that v_1, v_2, v_3 are linearly dependent by using only the definition of linear dependence in Definition 85.1.

13.14. Prove the following theorem. Let $v_1, \cdots, v_m$ be a set of vectors in $\mathbb{R}^n$, and let $v_{i_1}, \cdots, v_{i_k}$ be a subset of these vectors. If $v_1, \cdots, v_m$ are linearly independent, then $v_{i_1}, \cdots, v_{i_k}$ are linearly independent; if $v_{i_1}, \cdots, v_{i_k}$ are linearly dependent, then $v_1, \cdots, v_m$ are linearly dependent.

14. Barycentric Coordinates, Convex Sets, and Simplexes

The purpose of this section is to define Euclidean simplexes and abstract simplexes and to derive their principal descriptions and properties. The section contains also a treatment of planes, barycentric coordinates, and convex sets since an understanding of these topics is needed for a study of simplexes.

Let $a_i : (a_i^1, \cdots, a_i^n)$, $i = 0, 1, \cdots, m$, be $m + 1$ points in $\mathbb{R}^n$. These points are said to be linearly independent if and only if the vectors $a_i - a_0$, $i = 1, \cdots, m$, are linearly independent; they are linearly dependent if they are not linearly independent. Then by Theorem 13.1, the points $a_0, a_1, \cdots, a_m$ are linearly dependent if $m > n$, and they are linearly independent if and only if $1 \leqq m \leqq n$ and $|\Delta(a_0, a_1, \cdots, a_m)| > 0$ [see (37) in Section 11 for the definition of $|\Delta(a_0, a_1, \cdots, a_m)|$]. It is clear that linear independence or dependence is a property of the set $\{a_0, a_1, \cdots, a_m\}$, and that it does not depend on the particular way in which an m-vector is constructed from these points [see Exercise 11.11].

Let a_i, $i = 0, 1, \cdots, m$, be $m + 1$ linearly independent points in $\mathbb{R}^n$. These points determine the m-dimensional plane $P(a_0, a_1, \cdots, a_m)$ which is defined as follows:

$$P(a_0, a_1, \cdots, a_m) = \{x : x \in \mathbb{R}^n, x = a_0 + \sum_{i=1}^{m} u^i(a_i - a_0), \quad (u^1, \cdots, u^m) \in \mathbb{R}^m\}. \tag{1}$$

Since the vectors $a_i - a_0$, $i = 1, \cdots, m$, are linearly independent, there is a unique set of parameters $(u^1, \cdots, u^m)$ for each x in the plane, because two representations for x would imply that the vectors $a_i - a_0$ are linearly dependent. Equation (1) shows that x is in $P(a_0, a_1, \cdots, a_m)$ if and only if

$$x = (1 - \sum_{i=1}^{m} u^i)a_0 + \sum_{i=1}^{m} u^i a_i. \tag{2}$$

Then x is in the plane if and only if there is a set of numbers $(t^0, t^1, \cdots, t^m)$, called the *barycentric coordinates of x with respect to $a_0, a_1, \cdots, a_m$*, such that

$$x = \sum_{i=0}^{m} t^i a_i, \qquad \sum_{i=0}^{m} t^i = 1. \tag{3}$$

14.1 Example. The points $a_0 : (4, 2, 3)$, $a_1 : (1, 5, 6)$, and $a_2 : (1, 3, 8)$ determine a 2-dimensional plane in $\mathbb{R}^3$ if and only if they are linearly independent, and, by the definition given above, these points are linearly independent if and only if $|\Delta(a_0, a_1, a_2)| > 0$. Section 11 [see especially equations (15) and (16)] defines $|\Delta(a_0, a_1, a_2)|$ and describes its geometric significance. By (16) in Section 11,

$$|\Delta(a_0, a_1, a_2)| = \left\{ \begin{vmatrix} 4 & 2 & 1 \\ 1 & 5 & 1 \\ 1 & 3 & 1 \end{vmatrix}^2 + \begin{vmatrix} 4 & 3 & 1 \\ 1 & 6 & 1 \\ 1 & 8 & 1 \end{vmatrix}^2 + \begin{vmatrix} 2 & 3 & 1 \\ 5 & 6 & 1 \\ 3 & 8 & 1 \end{vmatrix}^2 \right\}^{1/2} = 6\sqrt{6}. \tag{4}$$

Thus, the points a_0, a_1, a_2 are linearly independent, and they determine the 2-dimensional plane $P(a_0, a_1, a_2)$ in $\mathbb{R}^3$ defined as follows:

$$P(a_0, a_1, a_2) = \{x : x = a_0 + \sum_{i=1}^{2} u^i(a_i - a_0), (u^1, u^2) \in \mathbb{R}^2\}. \tag{5}$$

This description of the plane is a vector equation; in coordinate form, $P(a_0, a_1, a_2)$ is the set of points $x:(x^1, x^2, x^3)$ such that

$$\begin{aligned} x^1 &= 4 - 3u^1 - 3u^2, \\ x^2 &= 2 + 3u^1 + u^2, \qquad (u^1, u^2) \in \mathbb{R}^2. \\ x^3 &= 3 + 3u^1 + 5u^2. \end{aligned} \tag{6}$$

Equation (3) shows that the equations of $P(a_0, a_1, a_2)$ in barycentric coordinates are

$$\begin{aligned} x^1 &= 4t^0 + t^1 + t^2, \qquad (t^0, t^1, t^2) \in \mathbb{R}^3, \\ x^2 &= 2t^0 + 5t^1 + 3t^2, \qquad \sum_{i=0}^{2} t^i = 1. \\ x^3 &= 3t^0 + 6t^1 + 8t^2. \end{aligned} \tag{7}$$

There is still a third description of this plane; it is a single equation of the form $(a, x) + c = 0$ [see Section 13]. Equation (5) states that x is in $P(a_0, a_1, a_2)$ if and only if

$$(x - a_0) - \sum_{i=1}^{2} u^i(a_i - a_0) = 0. \tag{8}$$

Thus x is in $P(a_0, a_1, a_2)$ if and only if the three vectors $(x - a_0)$, $(a_1 - a_0)$, $(a_2 - a_0)$ are linearly dependent. Theorem 13.1 states that these vectors are linearly dependent if and only if $|\Delta(x, a_0, a_1, a_2)| = 0$, or

$$\Delta(x, a_0, a_1, a_2) = \begin{vmatrix} x^1 & x^2 & x^3 & 1 \\ 4 & 2 & 3 & 1 \\ 1 & 5 & 6 & 1 \\ 1 & 3 & 8 & 1 \end{vmatrix} = 0. \tag{9}$$

Expand the determinant by minors of elements in the first row of its matrix. Then (9) is

$$12x^1 + 6x^2 + 6x^3 - 78 = 0. \tag{10}$$

The coefficients of x^1, x^2, x^3 in this equation are the three determinants in (4), and they are not all zero exactly because $|\Delta(a_0, a_1, a_2)| > 0$. An equation which is equivalent to (10) is

$$2x^1 + x^2 + x^3 - 13 = 0. \tag{11}$$

Equations (10) and (11) are equations of the form $(a, x) + c = 0$ which was used to describe $(n-1)$-dimensional planes in $\mathbb{R}^n$ in Section 13. Figure 14.1 contains a schematic drawing which shows the points a_0, a_1, a_2 and the plane which these points determine.

Return to the general case in equations (1), (2), (3). If $m = n$, then $P(a_0, a_1, \cdots, a_n)$ is the n-dimensional plane which is the entire space $\mathbb{R}^n$. To prove this statement, it is necessary to show that every x in $\mathbb{R}^n$ is in

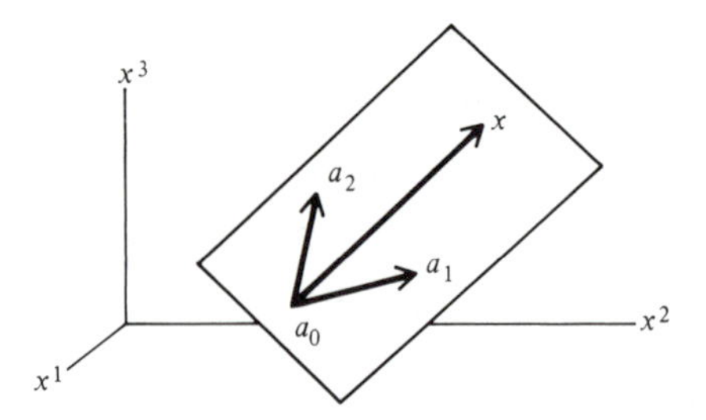

Figure 14.1. The plane $P(a_0, a_1, a_2)$ in Example 14.1.

$P(a_0, a_1, \cdots, a_n)$. Then let x be an arbitrary point in $\mathbb{R}^n$; it is necessary to show that the equation

$$\sum_{i=1}^{n} u^i(a_i - a_0) = x - a_0 \tag{12}$$

has a solution for $(u^1, \cdots, u^n)$. This equation, in coordinate form, is a system of n linear equations in the n unknowns $u^1, \cdots, u^n$. The absolute value of the determinant of this system of equations equals $|\Delta(a_0, a_1, \cdots, a_n)|$. Since the points $a_0, a_1, \cdots, a_n$ are linearly independent, Theorem 13.1 shows that $\Delta(a_0, a_1, \cdots, a_n) \neq 0$. Then Cramer's rule shows that (12) has a unique solution $(u^1, \cdots, u^n)$ for every x in $\mathbb{R}^n$, and $P(a_0, a_1, \cdots, a_n)$ is the entire space as stated. Equation (3) shows that every point in $P(a_0, a_1, \cdots, a_n)$ has barycentric coordinates $(t^0, t^1, \cdots, t^n)$. Thus, if $a_0, a_1, \cdots, a_n$ are points in $\mathbb{R}^n$ such that $\Delta(a_0, a_1, \cdots, a_n) \neq 0$, then every point x in $\mathbb{R}^n$ has barycentric coordinates $(t^0, t^1, \cdots, t^n)$ with respect to $a_0, a_1, \cdots, a_n$ as follows:

$$x = \sum_{i=0}^{n} t^i a_i, \qquad \sum_{i=0}^{n} t^i = 1, \qquad (t^0, t^1, \cdots, t^n) \in \mathbb{R}^{n+1}, \qquad x \in \mathbb{R}^n. \tag{13}$$

If $m = n - 1$, then equations (1), (2), (3) describe an $(n-1)$-dimensional plane in $\mathbb{R}^n$. This plane has an equation which corresponds to (9) and (10) in Example 14.1. Equation (1) shows that $x:(x^1, \cdots, x^n)$ is in $P(a_0, a_1, \cdots, a_{n-1})$ if and only if

$$(x - a_0) - \sum_{i=1}^{n-1} u^i(a_i - a_0) = 0, \tag{14}$$

that is, if and only if the n vectors $(x - a_0), (a_1 - a_0), \cdots, (a_{n-1} - a_0)$ are linearly dependent. But Theorem 13.1 states that these vectors are linearly dependent if and only if $\Delta(x, a_0, a_1, \cdots, a_{n-1}) = 0$. Thus the equation of $P(a_0, a_1, \cdots, a_{n-1})$ is

$$\begin{vmatrix} x^1 & x^2 & \cdots & x^n & 1 \\ a_0^1 & a_0^2 & \cdots & a_0^n & 1 \\ \cdots & \cdots & \cdots & \cdots & \cdots \\ a_{n-1}^1 & a_{n-1}^2 & \cdots & a_{n-1}^n & 1 \end{vmatrix} = 0. \tag{15}$$

This equation can be written in the form $(a, x) + c = 0$ by expanding the determinant by minors of elements in the first row of its matrix. Since $a_0, a_1, \cdots, a_{n-1}$ are linearly independent points, then

$$|\Delta(a_0, a_1, \cdots, a_{n-1})| = \left\{ \sum_{(j_1, \cdots, j_{n-1})} [\Delta[(a_0, a_1, \cdots, a_{n-1})^{(j_1, \cdots, j_{n-1})}]]^2 \right\}^{1/2} > 0, \quad (16)$$

$$(j_1, \cdots, j_{n-1}) \in (n - 1/n).$$

Thus at least one of the coefficients of $x^1, \cdots, x^n$ in (15) is different from zero.

14.2 Definition. Let k be a positive integer, and let $a_1, \cdots, a_k$ be points in $\mathbb{R}^n$. The symbol $[a_1, \cdots, a_k]$ denotes the set of points defined by the following equation:

$$[a_1, \cdots, a_k] = \left\{ x : x = \sum_{i=1}^{k} t^i a_i,\ t^i \geqq 0,\ \sum_{i=1}^{k} t^i = 1 \right\}. \quad (17)$$

The set $[a_1, \cdots, a_k]$ is called the *barycentric extension* of the set $\{a_1, \cdots, a_k\}$.

In this notation, $[a_1, a_2]$ is the closed segment $\{x : x = ta_1 + (1 - t)a_2, 0 \leqq t \leqq 1\}$. It is a trivial exercise to show that, if x_1 and x_2 are in $[a_1, \cdots, a_k]$, then $[x_1, x_2]$ is in $[a_1, \cdots, a_k]$. Since $[a_1, \cdots, a_k]$ contains $\{a_1, \cdots, a_k\}$, then $[a_1, \cdots, a_k]$ is a convex set which contains $\{a_1, \cdots, a_k\}$.

14.3 Definition. The *convex entension* $C(A)$ of a set A in $\mathbb{R}^n$ is the intersection of all convex sets which contain A.

If A is a convex set, then $C(A) = A$. The following argument shows that the convex extension $C(A)$ of every set A is a convex set which contains A. Since A is in every convex set used to define $C(A)$, then A is in the intersection $C(A)$ of these sets. Also, if x_1 and x_2 are two points in $C(A)$, then x_1 and x_2 are contained in every convex set which contains A. Then $[x_1, x_2]$ is contained in every convex set which contains A; hence, $[x_1, x_2]$ is contained in $C(A)$, and $C(A)$ is convex by definition.

14.4 Definition. Let A be a set in $\mathbb{R}^n$, and let $[A]$ denote the set defined by the following equation:

$$[A] = \bigcup \{[a_1, \cdots, a_k] : a_1, \cdots, a_k \text{ are in } A,\ k = 1, 2, \cdots\}. \quad (18)$$

14.5 Theorem. *If A is a set in $\mathbb{R}^n$, then $C(A) = [A]$.*

The proof of this theorem is based on the following three lemmas.

14.6 Lemma. *If A is a set in $\mathbb{R}^n$, then $[A]$ is a convex set which contains A.*

PROOF. By definition 14.4, the set $[A]$ contains $\{[a] : a \in A\}$, which is A itself. Thus $A \subset [A]$. Next, to prove that $[A]$ is convex, it is necessary and sufficient to prove that every segment with end points in $[A]$ is in $[A]$. Let x_1 and x_2 be two points in $[A]$. Then there are points $a_1, \cdots, a_u$ and $b_1, \cdots, b_v$ in A such that $x_1 \in [a_1, \cdots, a_u]$ and $x_2 \in [b_1, \cdots, b_v]$. Then

$$\begin{aligned} x_1 &= \sum_{i=1}^{u} t^i a_i, \qquad 0 \leqq t^i \leqq 1, \qquad \sum_{i=1}^{u} t^i = 1; \\ x_2 &= \sum_{i=1}^{v} s^i b_i, \qquad 0 \leqq s^i \leqq 1, \qquad \sum_{i=1}^{v} s^i = 1. \end{aligned} \tag{19}$$

Thus

$$tx_1 + (1-t)x_2 = \sum_{i=1}^{u} tt^i a_i + \sum_{i=1}^{v} (1-t)s^i b_i. \tag{20}$$

For each t such that $0 \leqq t \leqq 1$,

$$\begin{gathered} 0 \leqq tt^i \leqq 1, \qquad 0 \leqq (1-t)s^i \leqq 1, \\ \sum_{i=1}^{u} tt^i + \sum_{i=1}^{v} (1-t)s^i = t\sum_{i=1}^{u} t^i + (1-t)\sum_{i=1}^{v} s^i = t + (1-t) = 1. \end{gathered} \tag{21}$$

Then $[x_1, x_2] = \{x : x = tx_1 + (1-t)x_2, 0 \leqq t \leqq 1\}$, and (20) and (21) show that

$$[x_1, x_2] \subset [a_1, \cdots, a_u, b_1, \cdots, b_v] \subset [A]. \tag{22}$$

Therefore, by definition, $[A]$ is convex, and the proof of Lemma 14.6 is complete. □

14.7 Lemma. *If b is a point in $[a_1, \cdots, a_k]$, then there is a point b_1 in $[a_1, \cdots, a_r]$ and a point b_2 in $[a_{r+1}, \cdots, a_k]$ such that b is in $[b_1, b_2]$.*

PROOF. Since b is in $[a_1, \cdots, a_k]$, then $b = \Sigma_1^k t^i a_i$. Set

$$s^1 = \sum_{i=1}^{r} t^i, \qquad s^2 = \sum_{i=r+1}^{k} t^i. \tag{23}$$

If $s^1 = 0$, then b is in $[a_{r+1}, \cdots, a_k]$ and the lemma is trivially true; similarly, the lemma is true if $s^2 = 0$. In the remaining cases, s^1 and s^2 are positive, $s^1 + s^2 = 1$, and

$$\sum_{i=1}^{r} (t^i/s^1) = 1, \qquad \sum_{i=r+1}^{k} (t^i/s^2) = 1. \tag{24}$$

Let b_1 and b_2 be points such that

$$b_1 = \sum_{i=1}^{r} (t^i/s^1)a_i, \qquad b_2 = \sum_{i=r+1}^{k} (t^i/s^2)a_i. \tag{25}$$

Then b_1 and b_2 are in $[a_1, \cdots, a_r]$ and $[a_{r+1}, \cdots, a_k]$ respectively, and

$$b = s^1 \sum_{i=1}^{r} (t^i/s^1)a_i + s^2 \sum_{i=r+1}^{k} (t^i/s^2)a_i = s^1 b_1 + s^2 b_2. \tag{26}$$

Thus b is in $[b_1, b_2]$, and the proof of Lemma 14.7 is complete. □

14.8 Lemma. *If $a_1, \cdots, a_k$ are points in a convex set K, then $[a_1, \cdots, a_k]$ is contained in K.*

PROOF. The proof is by induction on k, the number of points. Since the points $a_1, \cdots, a_k$ are in K and K is convex, the lemma is obviously true for $k = 1$ and $k = 2$. Assume that it is true for $q - 1$ points and prove that it is true for q points. The induction hypothesis is that $[a_2, \cdots, a_q] \subset K$. The proof can be completed by showing that, if b is in $[a_1, a_2, \cdots, a_q]$, then b is in K. Lemma 14.7, with $r = 1$, shows that there is a point b_2 in $[a_2, \cdots, a_q]$, and thus in K by the induction hypothesis, such that $b = s^1 a_1 + s^2 b_2$, $s^1 + s^2 = 1$. Since a_1 and b_2 are in K and K is convex, then b is in K. Induction completes the proof. □

PROOF OF THEOREM 14.5. Since $[A]$ is a convex set by Lemma 14.6, then $C(A) \subset [A]$ by the definition of $C(A)$. The proof is completed by proving that $[A] \subset C(A)$. Equation (18) contains the definition of $[A]$. If $a_1, \cdots, a_k$ are points in A, they are contained in the convex set $C(A)$. Then Lemma 14.8 shows that $[a_1, \cdots, a_k]$ is contained in $C(A)$. Since this statement is true for every positive integer k and for every set of points $a_1, \cdots, a_k$ in A, then (18) shows that $[A] \subset C(A)$. The proof of Theorem 14.5 is complete. □

14.9 Corollary. *If $C(a_1, \cdots, a_k)$ denotes the convex extension of the set $\{a_1, \cdots, a_k\}$ in $\mathbb{R}^n$, then $C(a_1, \cdots, a_k) = [a_1, \cdots, a_k]$.*

PROOF. As stated above, $[a_1, \cdots, a_k]$ is a convex set which contains $\{a_1, \cdots, a_k\}$. Then $C(a_1, \cdots, a_k) \subset [a_1, \cdots, a_k]$. Since $a_1, \cdots, a_k$ are points in the convex set $C(a_1, \cdots, a_k)$, then Lemma 14.8 shows that $[a_1, \cdots, a_k] \subset C(a_1, \cdots, a_k)$. Since each set is contained in the other, $C(a_1, \cdots, a_k) = [a_1, \cdots, a_k]$, and the proof is complete. □

14.10 Definition. The convex extension $C(a_0, a_1, \cdots, a_m)$ of a set of linearly independent points $a_0, a_1, \cdots, a_m$ is called an *m-dimensional Euclidean simplex*.

The dimension m of a Euclidean simplex $C(a_0, a_1, \cdots, a_m)$ is the dimension of the plane determined by the points $a_0, a_1, \cdots, a_m$, and it is the dimension of the plane of minimum dimension which contains the simplex. In $\mathbb{R}^n$ there are Euclidean simplexes of every dimension m such that $-1 \leqq m \leqq n$; there are no Euclidean simplexes in $\mathbb{R}^n$ of dimension greater than n. A zero-dimensional simplex consists of a single point. There is only a single simplex of dimension -1; it is the empty set.

Let $\{a_0, a_1, \cdots, a_m\}$ be a set of linearly independent points in $\mathbb{R}^n$. Then the definition given at the beginning of this section shows that each subset of $\{a_0, a_1, \cdots, a_m\}$ which contains at least two points is also linearly independent. Consider, for example, the subset $\{a_1, \cdots, a_k\}$. The vectors whose initial point is a_1 and whose terminal points are $a_0, a_2, \cdots, a_m$ are linearly independent since the points $a_0, a_1, \cdots, a_m$ are linearly independent by hypothesis [compare Exercise 11.6]. Then the vectors with initial point a_1 and terminal points $a_2, \cdots, a_k$ are linearly independent since every nonempty subset of a set of linearly independent vectors is linearly independent [see Exercise 13.14]. Then $a_1, \cdots, a_k$ are linearly independent and $[a_1, \cdots, a_k]$ is a Euclidean simplex; since $a_1, \cdots, a_k$ are contained in $[a_0, a_1, \cdots, a_m]$, which is a convex set, then Lemma 14.8 shows that $[a_1, \cdots, a_k] \subset [a_0, a_1, \cdots, a_m]$. The Euclidean simplex $[a_1, \cdots, a_k]$ is called a *side* of the Euclidean simplex $[a_0, a_1, \cdots, a_m]$. Similar arguments show that, if $\{a_{i_1}, \cdots, a_{i_r}\}$ is a subset of $\{a_0, a_1, \cdots, a_m\}$ which contains at least two points, then $[a_{i_1}, \cdots, a_{i_r}]$ is a Euclidean simplex which is contained in $[a_0, a_1, \cdots, a_m]$. Each such simplex $[a_{i_1}, \cdots, a_{i_r}]$ is called a side of $[a_0, a_1, \cdots, a_m]$. Furthermore, the convex extension of a set consisting of a single point is the set consisting of the single point. Then $[a_0], [a_1], \cdots, [a_m]$ are also called Euclidean simplexes and sides of $[a_0, a_1, \cdots, a_m]$. Finally, the empty set and the entire simplex $[a_0, a_1, \cdots, a_m]$ are *improper sides* of the simplex $[a_0, a_1, \cdots, a_m]$. Thus, if $a_0, a_1, \cdots, a_m$ are linearly independent points, then the convex extension of every subset of these points is a Euclidean simplex which is a side of $[a_0, a_1, \cdots, a_m]$. The points $a_0, a_1, \cdots, a_m$ are called *vertices* of $[a_0, a_1, \cdots, a_m]$.

The linearly independent points $a_0, a_1, \cdots, a_m$ determine a Euclidean simplex which has been characterized as $C(a_0, a_1, \cdots, a_m)$, the intersection of all convex sets which contain $\{a_0, a_1, \cdots, a_m\}$, and also as $[a_0, a_1, \cdots, a_m]$, the barycentric extension of the set $\{a_0, a_1, \cdots, a_m\}$. If $m = n$, then the points $a_0, a_1, \cdots, a_m$ determine an n-dimensional Euclidean simplex, and there is a third important characterization of the simplex in this special case.

Exercise 13.1 states that a closed half-space $\{x : (a, x) + c \geqq 0\}$ is a convex set, and the intersection of any number of convex sets is a convex set [a proof follows Definition 14.3]. These facts will now be used to obtain the third characterization of the n-dimensional Euclidean simplex. Let $a_0, a_1, \cdots, a_n$ be linearly independent points in $\mathbb{R}^n$, $n \geqq 1$; then $\Delta(a_0, a_1, \cdots, a_n) \neq 0$ [see Theorem 13.1]. Assume that the notation has been chosen (the points have been ordered and numbered) so that

$$\Delta(a_0, a_1, \cdots, a_n) > 0. \tag{27}$$

Let $\widehat{\ }$, when placed above a term in a sequence, mean that the term is omitted from the sequence. Thus $\{a_0, \cdots, \widehat{a}_r, \cdots, a_n\}$ is the set obtained by deleting the point a_r from the set $\{a_0, a_1, \cdots, a_n\}$. Then each of the sets $\{a_0, a_1, \cdots, \widehat{a_r}, \cdots, a_n\}$, $r = 0, 1, \cdots, n$, is linearly independent; each of these sets of points determines an $(n-1)$-dimensional plane $P(a_0, a_1, \cdots, \widehat{a_r}, \cdots, a_n)$, one of whose equations is $\Delta(a_0, a_1, \cdots, x/a_r, \cdots, a_n) = 0$ [see equation (15) above; the notation x/a_r means that a_r has been replaced by x]. Let H_r denote, as follows, the positive half-space defined by this plane.

$$H_r = \{x : \Delta(a_0, a_1, \cdots, x/a_r, \cdots, a_n) \geqq 0\}, \qquad r = 0, 1, \cdots, n. \tag{28}$$

14.11 Theorem. *If $\{a_0, a_1, \cdots, a_n\}$ is a set of linearly independent points in $\mathbb{R}^n$, $n \geqq 1$, then its convex extension $C(a_0, a_1, \cdots, a_n)$ is a Euclidean simplex, and*

$$C(a_0, a_1, \cdots a_n) = [a_0, a_1, \cdots, a_n] = \bigcap\{H_r : r = 0, 1, \cdots, n\}. \tag{29}$$

PROOF. Since the points in $\{a_0, a_1, \cdots, a_n\}$ are linearly independent by hypothesis, then $C(a_0, a_1, \cdots, a_n)$ is a Euclidean simplex by Definition 14.10. Furthermore, $C(a_0, a_1, \cdots, a_n) = [a_0, a_1, \cdots, a_n]$ by Corollary 14.9, and the proof of Theorem 14.11 can be completed by showing that

$$[a_0, a_1, \cdots, a_n] = \bigcap\{H_r : r = 0, 1, \cdots, n\}. \tag{30}$$

Let x be a point in $\mathbb{R}^n$. Then x has unique barycentric coordinates $(t^0, t^1, \cdots, t^n)$ with respect to $a_0, a_1, \cdots, a_n$, and

$$x = \sum\{t^i a_i : i = 0, 1, \cdots, n, t^0 + t^1 + \cdots + t^n = 1\}. \tag{31}$$

Theorems 77.1 and 77.3 in Appendix 1 show that, for this x,

$$\Delta(a_0, a_1, \cdots, x/a_r, \cdots, a_n) = t^r \Delta(a_0, a_1, \cdots, a_n), \qquad r = 0, 1, \cdots, n. \tag{32}$$

If x is in $[a_0, a_1, \cdots, a_n]$, then $t^i \geqq 0$ for $i = 0, 1, \cdots, n$ by Definition 14.2, and equations (32) and (27) show that

$$\Delta(a_0, a_1, \cdots, x/a_r, \cdots, a_n) \geqq 0, \qquad r = 0, 1, \cdots, n. \tag{33}$$

Then x is in each H_r and x is in the intersection of these half-spaces. Thus

$$[a_0, a_1, \cdots, a_n] \subset \bigcap\{H_r : r = 0, 1, \cdots, n\}. \tag{34}$$

Assume next that the point x in (31) is in $\bigcap\{H_r : r = 0, 1, \cdots, n\}$. Then for each r the determinant on the left in (32) is non-negative, and (27) shows that $t^r \geqq 0$ for $r = 0, 1, \cdots, n$. Then (31) and Definition 14.2 show that x is in $[a_0, a_1, \cdots, a_n]$; hence,

$$\bigcap\{H_r : r = 0, 1, \cdots, n\} \subset [a_0, a_1, \cdots, a_n]. \tag{35}$$

Equations (34) and (35) complete the proof of (30) and of Theorem 14.11. □

14.12 Example. Let a_0 and a_1 be two points in $\mathbb{R}$ such that $a_0 > a_1$. Then $\{a_0, a_1\}$ is a set of linearly independent points, and their convex extension

Figure 14.2. A Euclidean simplex in $\mathbb{R}$.

$C(a_0, a_1)$ is a Euclidean simplex. Also, $[a_0, a_1] = \{x : x = ta_1 + (1-t)a_0, 0 \leqq t \leqq 1\}$, and $C(a_0, a_1)$ is the closed interval $\{x : a_1 \leqq x \leqq a_0\}$ [see Figure 14.2]. Since $C(a_0, a_1)$ is a 1-dimensional Euclidean simplex in $\mathbb{R}$, then it can be defined as the intersection of closed half-spaces [see Theorem 14.11].

$$\Delta(a_0, a_1) = \begin{vmatrix} a_0 & 1 \\ a_1 & 1 \end{vmatrix} > 0, \quad H_0 = \{x : \Delta(x, a_1) \geqq 0\}, \quad H_1 = \{x : \Delta(a_0, x) \geqq 0\}.$$

Then

$$H_0 = \{x : x \geqq a_1\}, \quad H_1 = \{x : x \leqq a_0\}, \quad \text{and} \quad H_1 \cap H_2 = \{x : a_1 \leqq x \leqq a_0\}.$$

14.13 Example. Let $a_i : (a_i^1, a_i^2)$, $i = 0, 1, 2$, be three points in $\mathbb{R}^2$ such that $\Delta(a_0, a_1, a_2) > 0$. Then $\{a_0, a_1, a_2\}$ is a set of linearly independent points, and their convex extension $C(a_0, a_1, a_2)$ is a Euclidean simplex.

$$\Delta(a_0, a_1, a_2) = \begin{vmatrix} a_0^1 & a_0^2 & 1 \\ a_1^1 & a_1^2 & 1 \\ a_2^1 & a_2^2 & 1 \end{vmatrix}, \qquad \Delta(x, a_1, a_2) = \begin{vmatrix} x^1 & x^2 & 1 \\ a_1^1 & a_1^2 & 1 \\ a_2^1 & a_2^2 & 1 \end{vmatrix},$$

$$\Delta(a_0, x, a_2) = \begin{vmatrix} a_0^1 & a_0^2 & 1 \\ x^1 & x^2 & 1 \\ a_2^1 & a_2^2 & 1 \end{vmatrix}, \qquad \Delta(a_0, a_1, x) = \begin{vmatrix} a_0^1 & a_0^2 & 1 \\ a_1^1 & a_1^2 & 1 \\ x^1 & x^2 & 1 \end{vmatrix}.$$

Then

$$H_0 = \{x : \Delta(x, a_1, a_2) \geqq 0\}, \qquad H_1 = \{x : \Delta(a_0, x, a_2) \geqq 0\},$$
$$H_2 = \{x : \Delta(a_0, a_1, x) \geqq 0\}.$$

Now H_0 is the half-space whose boundary is the line through a_1 and a_2, and which contains the opposite vertex a_0; similarly for H_1 and H_2. Then $C(a_0, a_1, a_2)$ is $H_0 \cap H_1 \cap H_2$, and Figure 14.3 shows that this Euclidean simplex is the triangle whose vertices are a_0, a_1, a_2. Other examples of Euclidean simplexes are shown in Figure 14.4.

14.14 Definition. Let $\{a_0, a_1, \cdots, a_m\}$ be a set of $m+1$ linearly dependent points in $\mathbb{R}^n$. Then the convex extension $C(a_0, a_1, \cdots, a_m)$ of these points is called a *degenerate m-dimensional Euclidean simplex*.

An m-dimensional simplex is more than just a set of points; it is also a collection of sides of all dimensions from -1 to m. As a set of points in $\mathbb{R}^n$, a degenerate m-dimensional Euclidean simplex may or may not be a (non-degenerate) Euclidean simplex of dimension less than m. The convex extension of each subset of $\{a_0, a_1, \cdots, a_m\}$ is a side of the simplex determined

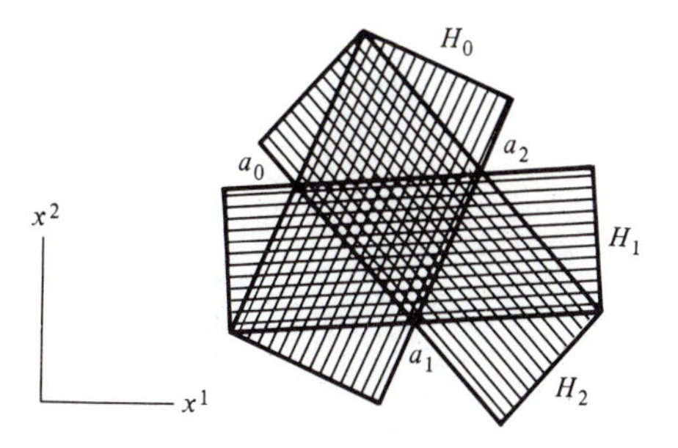

Figure 14.3. $C(a_0, a_1, a_2)$ is $H_0 \cap H_1 \cap H_2$, the triangle $a_0a_1a_2$.

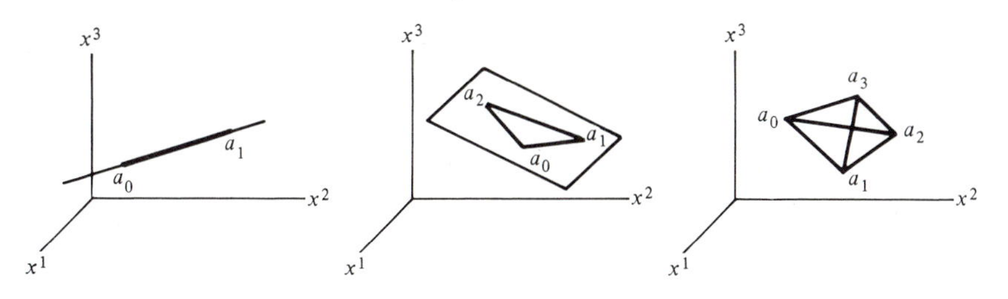

Figure 14.4. Examples of Euclidean simplexes in $\mathbb{R}^3$.

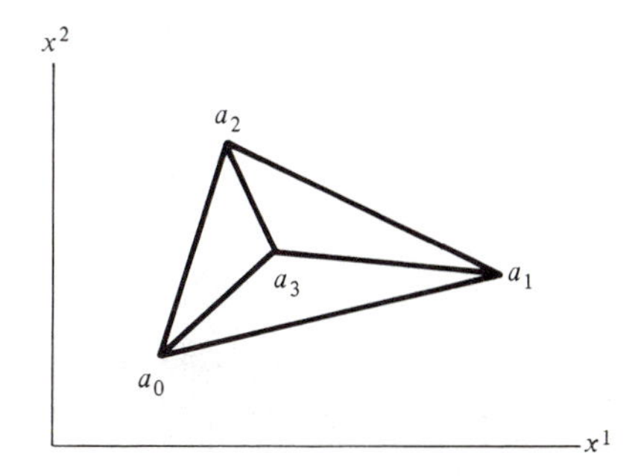

Figure 14.5. A degenerate Euclidean simplex in $\mathbb{R}^2$.

by these points. If each of these sides is a (non-degenerate) Euclidean simplex, then $C(a_0, a_1, \cdots, a_m)$ is an m-dimensional Euclidean simplex. If the points $a_0, a_1, \cdots, a_m$ are linearly dependent, then $C(a_0, a_1, \cdots, a_m)$ has no (improper) side which is a (non-degenerate) Euclidean simplex of dimension m, and $C(a_0, a_1, \cdots, a_m)$ is a degenerate m-dimensional Euclidean simplex.

14.15 Example. The four points $a_0 : (1, 1)$, $a_1 : (5, 3)$, $a_2 : (3, 8)$, and $a_3 : (3, 4)$ in $\mathbb{R}^2$ are linearly dependent; their convex extension $C(a_0, a_1, \cdots, a_3)$ is a degenerate 3-dimensional Euclidean simplex. As a set of points, $C(a_0, a_1, \cdots, a_3)$ is the 2-dimensional Euclidean simplex $[a_0, a_1, a_2]$ as shown in Figure 14.5. The improper side $C(a_0, a_1, \cdots, a_3)$ is a degenerate 3-

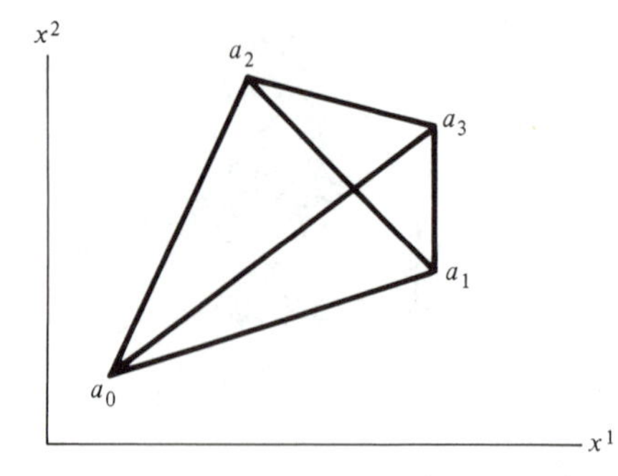

Figure 14.6. A degenerate Euclidean simplex in $\mathbb{R}^2$.

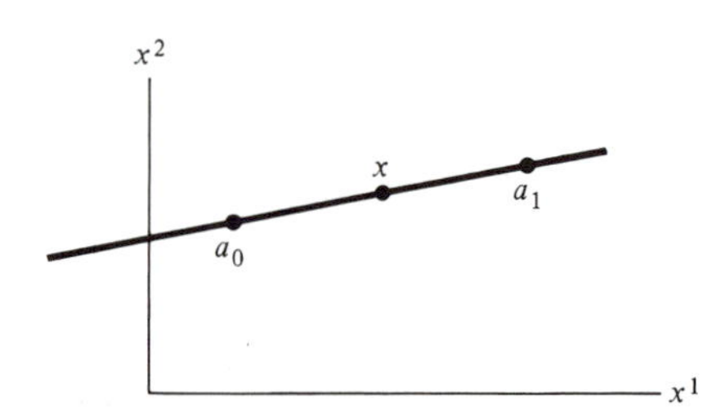

Figure 14.7. The line: $C(x, a_0, a_1)$ is degenerate.

dimensional Euclidean simplex. The other sides of $C(a_0, a_1, \cdots, a_3)$ are the empty set and the following simplexes; they are all (non-degenerate) Euclidean simplexes.

$$[a_0, a_1, a_2], [a_3, a_0, a_1], [a_3, a_1, a_2], [a_3, a_2, a_0];$$
$$[a_0, a_1], [a_1, a_2], [a_2, a_0], [a_0, a_3], [a_1, a_3], [a_2, a_3];$$
$$[a_0], [a_1], [a_2], [a_3].$$

14.16 Example. The four points $a_0 : (1, 1)$, $a_1 : (5, 3)$, $a_2 : (3, 8)$, $a_3 : (5, 7)$ are linearly dependent; they are the vertices of a degenerate 3-dimensional Euclidean simplex [see Figure 14.6]. The point a_3 lies outside the triangle whose vertices are a_0, a_1, a_2; as a result, the convex extension $C(a_0, a_1, \cdots, a_3)$, as a set of points, is a quadrilateral and not a 2-dimensional Euclidean simplex. Nevertheless, all of the proper sides of $C(a_0, a_1, \cdots, a_3)$ are non-degenerate Euclidean simplexes.

14.17 Example. Let $a_0, a_1, \cdots, a_{n-1}$ be n linearly independent points in $\mathbb{R}^n$; as shown in equations (14), $\cdots$, (16), these points determine an $(n-1)$-dimensional plane in $\mathbb{R}^n$. This plane can be described as the set of points x such that the points $x, a_0, a_1, \cdots, a_{n-1}$ are the vertices of a degenerate n-dimensional Euclidean simplex in $\mathbb{R}^n$ [see Figure 14.7].

14.18 Definition. A finite set of points $\{p_0, p_1, \cdots, p_r\}$ is called an *r-dimensional abstract simplex*; each point p_i is called a *vertex* of the simplex.

There is no set of points associated with an abstract simplex other than the points $p_0, p_1, \cdots, p_r$ themselves. Each subset of $\{p_0, p_1, \cdots, p_r\}$ is an abstract simplex which is called a side of $\{p_0, p_1, \cdots, p_r\}$. The empty set and the entire set $\{p_0, p_1, \cdots, p_r\}$ are called improper sides of $\{p_0, p_1, \cdots, p_r\}$. Since the empty set is contained in every set, two abstract simplexes always have a side in common, namely, the improper side consisting of the empty set. They may have other sides in common also; the intersection of the two sets is the common side of maximum dimension. The following theorem suggests how many abstract simplexes arise; the proof of this theorem follows directly from Definition 14.18.

14.19 Theorem. *Let E be a convex set in $\mathbb{R}^n$, and let $a_0, a_1, \cdots, a_r$ be the vertices of an r-dimensional Euclidean simplex in E. Then a function $f: E \to \mathbb{R}^m$ maps the vertices $a_0, a_1, \cdots, a_r$ of $C(a_0, a_1, \cdots, a_r)$ into an r-dimensional abstract simplex whose vertices are $f(a_0), f(a_1), \cdots, f(a_r)$.*

Exercises

14.1. Show that the points $a_0:(3, 6)$, $a_1:(2, 3)$, $a_2:(5, 5)$ in $\mathbb{R}^2$ are linearly independent and that $C(a_0, a_1, a_2)$ is a 2-dimensional Euclidean simplex.

(a) Draw a sketch of $C(a_0, a_1, a_2)$.

(b) Find the barycentric coordinates of the point $b:(4, 5)$ with respect to a_0, a_1, a_2. [Hint. Solve the equations

$$\sum_0^2 t^i a_i = b, \qquad \sum_0^2 t^i = 1 \quad \text{for} \quad t^0, t^1, t^2.]$$

(c) Use (b) to show that $b:(4, 5)$ is in $C(a_0, a_1, a_2)$.

(d) Find H_0, H_1, H_2 and sketch these half-spaces.

(e) Use (d) to show that $b:(4, 5)$ is in $C(a_0, a_1, a_2)$.

14.2. Show that the points $a_0:(1, 3, 9)$, $a_1:(5, 1, 2)$, $a_2:(3, 5, 4)$ are linearly independent and therefore determine a two-dimensional plane $P(a_0, a_1, a_2)$ in $\mathbb{R}^3$.

(a) Show that $P(a_0, a_1, a_2)$ is the set of points $x:(x^1, x^2, x^3)$ such that

$$\begin{aligned} x^1 &= 1 + 4u^1 + 2u^2, \\ x^2 &= 3 - 2u^1 + 2u^2, \qquad (u^1, u^2) \in \mathbb{R}^2. \\ x^3 &= 9 - 7u^1 - 5u^2, \end{aligned}$$

(b) Show that $P(a_0, a_1, a_2)$ is the set of points $x:(x^1, x^2, x^3)$ such that

$$\begin{aligned} x^1 &= t^0 + 5t^1 + 3t^2, \qquad (t^0, t^1, t^2) \in \mathbb{R}^3, \\ x^2 &= 3t^0 + t^1 + 5t^2, \qquad \sum_{i=0}^2 t^i = 1. \\ x^3 &= 9t^0 + 2t^1 + 4t^2, \end{aligned}$$

(c) Show that $P(a_0, a_1, a_2)$ is the set of points $x:(x^1, x^2, x^3)$ whose coordinates satisfy the equation $\Delta(x, a_0, a_1, a_2) = 0$. Show that this equation can be reduced to the form $4x^1 + x^2 + 2x^3 - 25 = 0$.

(d) Find the point whose barycentric coordinates with respect to a_0, a_1, a_2 are $t^0 = 1/3$, $t^1 = 1/3$, $t^2 = 1/3$. Use each of (a), (b), (c) to show that this point is on the plane $P(a_0, a_1, a_2)$.

(e) Explain why $C(a_0, a_1, a_2)$ is a 2-dimensional (non-degenerate) Euclidean simplex. Is (3, 3, 5) a point in this simplex? Why?

14.3. Show that the points $a_0:(1, 3, 9, 3)$, $a_1:(5, 1, 2, 6)$, $a_2 = (3, 5, 4, 12)$ are linearly independent in $\mathbb{R}^4$ and therefore determine a 2-dimensional plane $P(a_0, a_1, a_2)$.

(a) Find the equations of $P(a_0, a_1, a_2)$ which correspond to those in Exercise 14.2(a).

(b) Find the equations of $P(a_0, a_1, a_2)$ which correspond to those in Exercise 14.2(b).

(c) Show that $P(a_0, a_1, a_2)$ is the set of points $x:(x^1, \cdots, x^4)$ in $\mathbb{R}^4$ whose coordinates satisfy the equation $|\Delta(x, a_0, a_1, a_2)| = 0$, which is equivalent to the following four equations:

$$\Delta[(x, a_0, a_1, a_2)^{(j_1, j_2, j_3)}] = 0, \qquad (j_1, j_2, j_3) \in (3/4).$$

[Compare Exercise 14.2(c); see Section 11 for the notation.] The four equations of $P(a_0, a_1, a_2)$ can be simplified to the following:

$$\begin{aligned} 4x^1 + x^2 + 2x^3 \qquad\quad - 25 &= 0,\\ 4x^1 + 5x^2 \qquad\quad - 2x^4 - 13 &= 0,\\ 8x^1 \qquad + 5x^3 + x^4 - 56 &= 0,\\ 4x^2 - 2x^3 - 2x^4 + 12 &= 0. \end{aligned}$$

(d) Use the equations in (c) to show that each of the points a_0, a_1, a_2 is on the plane $P(a_0, a_1, a_2)$.

14.4. Let $a_0, a_1, \cdots, a_m$ be linearly independent points in $\mathbb{R}^n$. Show that $1 \leqq m \leqq n$. The points $a_0, a_1, \cdots, a_m$ determine a plane $P(a_0. a_1, \cdots, a_m)$. Show that, in all cases, $P(a_0, a_1, \cdots, a_m)$ can be described by equations (1) and (3) in this section.

(a) If $m = n - 1$, then $P(a_0, a_1, \cdots, a_{n-1})$ is an $(n-1)$-dimensional plane. Show that this plane is the set of points $x:(x^1, \cdots, x^n)$ in $\mathbb{R}^n$ which satisfy the equation $\Delta(x, a_0, a_1, \cdots, a_{n-1}) = 0$. [Hint. Equation (15).]

(b) If $1 \leqq m \leqq n - 1$, show that $P(a_0, a_1, \cdots, a_m)$ is the set of points $x:(x^1, \cdots, x^n)$ whose coordinates satisfy the equations

$$\Delta[(x, a_0, a_1, \cdots, a_m)^{(j_1, \cdots, j_{m+1})}] = 0, \qquad (j_1, \cdots, j_{m+1}) \in (m + 1/n).$$

Show that, if $m = n - 1$, there is a single equation and it is the equation in (a).

(c) Show that each of the equations in (b) is a linear equation in $x^1, \cdots, x^n$, and that not all of the coefficients of $x^1, \cdots, x^n$ in these equations are zero.

(d) If $m = n$, explain why $P(a_0, a_1, \cdots, a_n)$ is not characterized by an equation such as those in (a) and (b).

14.5. Use each of the descriptions of $P(a_0, a_1, \cdots, a_m)$ in equations (1) and (3) to show that the points $a_0, a_1, \cdots, a_m$ are in $P(a_0, a_1, \cdots, a_m)$. Also, use the equations of $P(a_0, a_1, \cdots, a_m)$ in Exercises 14.4(a) and (b) to show that $a_0, a_1, \cdots, a_m$ are in $P(a_0, a_1, \cdots, a_m)$.

14.6. Prove the following theorem. If A is a bounded set in $\mathbb{R}^n$, then $C(A)$ is a bounded set. [Hint. Definition 92.8 in Appendix 2.]

14.7. Let $(a, x) + c = 0$ be an equation of an $(n-1)$-dimensional plane in $\mathbb{R}^n$. Prove that the set $\{x : (a, x) + c < 0\}$ is an open set and that $\{x : (a, x) + c \geqq 0\}$ is a closed set. [Hint. Definition 92.4, Theorem 96.9, and Theorem 92.5.]
(b) Prove the following theorem. If $C(a_0, a_1, \cdots, a_n)$ is an n-dimensional Euclidean simplex in $\mathbb{R}^n$, then $C(a_0, a_1, \cdots, a_n)$ is a closed, bounded set. [Hint. Theorem 92.6.]

14.8. Let $C(a_0, a_1, \cdots, a_m)$ be a Euclidean simplex. If x is a point in the simplex, then prove that

$$x = \sum_{i=0}^{m} t^i a_i, \qquad 0 \leqq t^i \leqq 1, \qquad \sum_{i=0}^{m} t^i = 1.$$

Find necessary and sufficient conditions that x be a vertex of the simplex. For each side of $C(a_0, a_1, \cdots, a_m)$, find the necessary and sufficient condition that x belong to the side.

14.9. Two sides of the Euclidean simplex $C(a_0, a_1, \cdots, a_m)$ are $C(a_0, a_1, \cdots, a_r)$ and $C(a_{r+1}, \cdots, a_m)$. Prove that every point x in $C(a_0, a_1, \cdots, a_m)$ is contained in a segment which has one end-point in $C(a_0, a_1, \cdots, a_r)$ and the other in $C(a_{r+1}, \cdots, a_m)$. Sketch one figure in $\mathbb{R}^2$ and two figures in $\mathbb{R}^3$ to illustrate this exercise.

14.10. A point in a convex set is called an extreme point of the set if and only if it is not an interior point of a segment connecting two points in the set. Show that the vertices $a_0, a_1, \cdots, a_m$ are extreme points of the Euclidean simplex $C(a_0, a_1, \cdots, a_m)$ and that this simplex has no other extreme points.

14.11. (a) Sketch an example of a bounded convex set which is an open set; of an unbounded convex set which is an open set.
(b) Sketch an example of a bounded convex set which is a closed set; of an unbounded convex set which is a closed set.
(c) Sketch an example of a convex set which is neither open nor closed.
(d) Sketch an example of a convex set which has extreme points; which has no extreme points.
(e) Sketch an example of a convex set which is neither open nor closed, but which has extreme points.

14.12. Prove the following theorem. If $A \subset B$, then $C(A) \subset C(B)$.

14.13. Prove the following theorem. The total number of sides, proper and improper, of an m-dimensional abstract or Euclidean simplex is 2^{m+1}.

14.14. (a) The points $e_0 : (0, 0, 0)$, $e_1 : (1, 0, 0)$, $e_2 : (0, 1, 0)$, $e_3 : (0, 0, 1)$ are linearly independent and their convex extension is a convex set $C(e_0, e_1, \cdots, e_3)$. Make a sketch of this set.

(b) Show that $C(e_0, e_1, \cdots, e_3)$ is the set of points $x: (x^1, x^2, x^3)$ in $\mathbb{R}^3$ such that

$$x^1 = t^1, \qquad (t^0, t^1, \cdots, t^3) \in \mathbb{R}^4,$$
$$x^2 = t^2, \qquad 0 \leqq t^i \leqq 1, \qquad i = 0, 1, \cdots, 3,$$
$$x^3 = t^3, \qquad \sum_{i=0}^{3} t^i = 1.$$

(c) Find the half-spaces H_r [see equation (28) and Theorem 14.11] such that $C(e_0, e_1, \cdots, e_3) = \bigcap\{H_r : r = 0, 1, \cdots, 3\}$.

(d) Repeat (b) and (c) for the convex extension $C(e_0, e_1, \cdots, e_n)$ of the set of points $e_0 : (0, \cdots, 0)$, $e_1 : (1, \cdots, 0)$, $\cdots$, $e_n : (0, \cdots, 1)$ in $\mathbb{R}^n$.

14.15. Let ε be a positive number, let A be a set in $\mathbb{R}^n$, and let $N(A, \varepsilon) = \bigcup\{N(x, \varepsilon) : x \in A\}$. Then $N(A, \varepsilon)$ is the set of points in $\mathbb{R}^n$ whose distance from A is less than ε. Prove the following theorem. If A is a convex set, then $N(A, \varepsilon)$ is a convex set.

14.16. Prove the following theorem. If A is a set in $\mathbb{R}^n$, then the diameter of its convex extension $C(A)$ is equal to the diameter of A. [Outline of the proof. The diameter of a set A in $\mathbb{R}^n$ is defined in Definition 92.9 in Appendix 2. Since $A \subset C(A)$, then $\operatorname{diam}(A) = \operatorname{diam}[C(A)] = \infty$ if A is unbounded, and $\operatorname{diam}(A) \leqq \operatorname{diam}[C(A)]$ if A is bounded. The proof can be completed by showing that $\operatorname{diam}[C(A)] \leqq \operatorname{diam}(A)$ if A is bounded. It is sufficient to prove the following: if $|x_2 - x_1| < d$ for every pair of points x_1, x_2 in A, then $|y_2 - y_1| < d$ for every pair of points y_1, y_2 in $C(A)$. As usual, let $N(x_0, d) = \{x : x \in \mathbb{R}^n, |x - x_0| < d\}$. Let x_1 be a point in A; then by hypothesis, $A \subset N(x_1, d)$. Since $N(x_1, d)$ is convex, the definition of $C(A)$ shows that $C(A) \subset N(x_1, d)$. If y_1 is a point in $C(A)$, then $y_1 \in N(x_1, d)$ and $x_1 \in N(y_1, d)$. Since this statement is true for every x_1 in A, then $A \subset N(y_1, d)$ and $C(A) \subset N(y_1, d)$. If y_2 is in $C(A)$, then $y_2 \in N(y_1, d)$ and $|y_2 - y_1| < d$. The proof of the theorem follows from these arguments.]

14.17. Prove the following theorem. The diameter of a Euclidean simplex is the maximum distance between two of its vertices.

14.18. Let $\{a_0, a_1, \cdots, a_n\}$ be a set of linearly independent points in $\mathbb{R}^n$, and let x be an arbitrary point in $\mathbb{R}^n$. Prove the following identity for determinants:

$$\sum_{r=0}^{n} \Delta(a_0, a_1, \cdots, x/a_r, \cdots, a_n) = \Delta(a_0, a_1, \cdots, a_n).$$

14.19. The points $a_0 : (2, 2)$, $a_1 : (11, 3)$, $a_2 : (5, 7)$ are the vertices of a Euclidean simplex $C(a_0, a_1, a_2)$ in $\mathbb{R}^2$. Write the identity in Exercise 14.18 in the following form:

$$\frac{\Delta(x, a_1, a_2)}{2!} + \frac{\Delta(a_0, x, a_2)}{2!} + \frac{\Delta(a_0, a_1, x)}{2!} = \frac{\Delta(a_0, a_1, a_2)}{2!}.$$

Equation (25) in Section 89 in Appendix 2 shows that each term in this identity represents the signed area of a triangle. With the point x given in each of parts (a), $\cdots$, (d) of this exercise, do the following things: (i) make a sketch which shows the simplexes $C(a_0, a_1, a_2)$, $C(x, a_1, a_2)$, $C(a_0, x, a_2)$, and $C(a_0, a_1, x)$; (ii) evaluate each term in the identity and verify that the identity is satisfied; (iii) explain the geometric significance of the identity.

(a) $x = (6, 4)$; show that x is an interior point of $C(a_0, a_1, a_2)$.
(b) $x = (8, 5)$; show that x is on the side $[a_1, a_2]$ of $C(a_0, a_1, a_2)$.
(c) $x = (2, 9)$; show that x is on the line through a_1, a_2 but not in $C(a_0, a_1, a_2)$.
(d) $x = (14, 8)$; show that x is not in $C(a_0, a_1, a_2)$ nor on a line through two of the points a_0, a_1, a_2.

15. Orientation of Simplexes

Orientation is one of those concepts which the student encounters early in his study of mathematics. A positive angle is generated by rotating the initial side counterclockwise into the terminal side; clockwise rotation generates a negative angle. There are two orientations of the interval $[a, b]$ and two corresponding integrals $\int_a^b f(x)\,dx$ and $\int_b^a f(x)\,dx$; furthermore, $\int_b^a f(x)\,dx = -\int_a^b f(x)\,dx$. The correct orientation in the plane is important in the statement of Green's theorem. Let R be a region in $\mathbb{R}^2$, and let C, with the counterclockwise orientation, be the curve which forms the boundary of R [see Figure 15.1]. Then Green's theorem states, in classical notation, that

$$\iint_R \left(\frac{\partial Q}{\partial x} - \frac{\partial P}{\partial y}\right) dx\,dy = \int_C P\,dx + Q\,dy; \tag{1}$$

and the correct orientation of C in the line integral on the right is essential in the statement of this theorem. The "right-hand rule" and the "left-hand rule" are used to determine orientations in $\mathbb{R}^3$.

The concept of orientation is relatively simple on the line and in the plane because there it is an intuitive concept based on visual inspection of figures. However, later chapters of this book study integrals and integration theory in $\mathbb{R}^n$. If $n > 3$, there are no intuitive ideas and no visual interpretations to assist the study. The purpose of this section is to examine the intuitive and visual notions of orientation in $\mathbb{R}^1$ and $\mathbb{R}^2$ in order to formulate them as definitions in analytical form which can be generalized to $\mathbb{R}^n$.

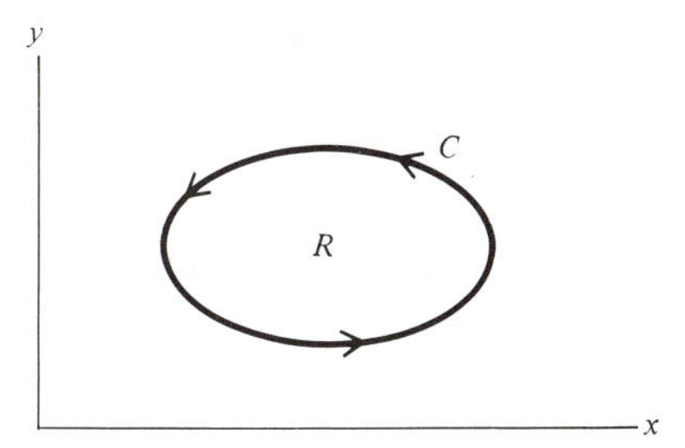

Figure 15.1. Figure for Green's theorem.

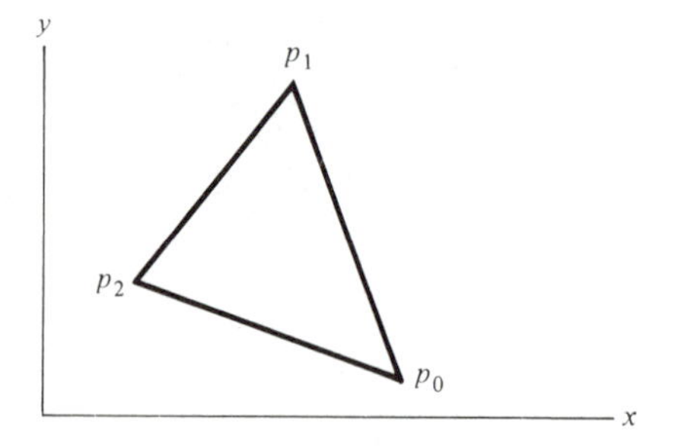

Figure 15.2. A 2-simplex in $\mathbb{R}^2$.

The study of orientation begins with the orientation of simplexes. In $\mathbb{R}^1$, a simplex has two vertices p_0, p_1, and the two orientations of the simplex with these vertices are denoted by (p_0, p_1) and (p_1, p_0). Figure 15.2 shows a simplex in $\mathbb{R}^2$, and (p_0, p_1, p_2) suggests a circuit of the vertices in the counterclockwise direction, starting at the vertex p_0. Also, (p_1, p_2, p_0) and (p_2, p_0, p_1) indicate circuits of the vertices in the same counterclockwise direction but starting at the vertices p_1 and p_2 rather than p_0. In the same way, (p_0, p_2, p_1), (p_2, p_1, p_0), and (p_1, p_0, p_2) indicate the opposite orientation, namely, circuits of the vertices in the clockwise direction. Thus the two orientations of the simplex are indicated by the following two sets of ordered triples of vertices:

$$\{(p_0, p_1, p_2), (p_1, p_2, p_0), (p_2, p_0, p_1)\}, \tag{2}$$

$$\{(p_0, p_2, p_1), (p_2, p_1, p_0), (p_1, p_0, p_2)\}. \tag{3}$$

These two orientations contain all of the permutations (arrangements) of the three letters p_0, p_1, p_2. One-half of the permutations [those in (2)] forms one of the orientations, and the other half [those in (3)] forms the other orientation. What characterizes the permutations in each orientation? An examination shows that each permutation in (2) can be obtained from the others in (2) by an even number of interchanges of adjacent letters. Similarly, an even number of interchanges of adjacent letters converts a specified permutation in (3) into any other permutation in (3). An odd number of interchanges of adjacent letters is required to change a permutation in (2) into a permutation in (3), or to change a permutation in (3) into a permutation in (2). It is customary to write equations of the following types as statements about orientations.

$$\begin{gathered}(p_0, p_1, p_2) = (p_1, p_2, p_0) = (p_2, p_0, p_1);\\ (p_0, p_2, p_1) = (p_1, p_0, p_2) = (p_2, p_1, p_0);\\ (p_0, p_2, p_1) = -(p_0, p_1, p_2), \quad (p_1, p_0, p_2) = -(p_1, p_2, p_0),\\ (p_2, p_1, p_0) = -(p_2, p_0, p_1).\end{gathered} \tag{4}$$

These observations about the orientation of the simplex (p_0, p_1, p_2) in Figure 15.2 suggest how the orientation of the m-dimensional simplex is to be defined.

The simplex of dimension -1 is the empty set, and it is not necessary to define an orientation for it. The simplex of dimension 0 contains only a single point, and it has only a single orientation. All other simplexes have two orientations, and they will now be described. Let $\{p_0, p_1, \cdots, p_m\}$ be an abstract m-dimensional simplex with vertices $p_0, p_1, \cdots, p_m$. If $m \geqq 1$, then there are $(m+1)!$ permutations (or arrangements, or orderings) of these vertices, and they divide into two classes which can be described as follows. The first class consists of the permutation $(p_0, p_1, \cdots, p_m)$ and all of those permutations which can be obtained from it by performing an even number of interchanges of two adjacent letters; it is called the class of even permutations of $(p_0, p_1, \cdots, p_m)$. The second class consists of all those permuations which can be obtained from $(p_0, p_1, \cdots, p_m)$ by performing an odd number of interchanges of two adjacent letters; it is called the class of odd permutations of $(p_0, p_1, \cdots, p_m)$. Each class contains $(m+1)!/2$ permutations, and each class is called an orientation of $\{p_0, p_1, \cdots, p_m\}$. Each permutation in one of the two classes is a representation of the orientation specified by the class, and each permutation in a class describes the same orientation of the simplex. Each permutation in a class is the opposite or the negative of a permutation in the other class. If $(p_0, p_1, \cdots, p_m)$ is an oriented simplex, then $-(p_0, p_1, \cdots, p_m)$ denotes the simplex with the same vertices, but with the opposite orientation; for example, $(p_1, p_0, p_2, \cdots, p_m) = -(p_0, p_1, p_2, \cdots, p_m)$. The symbol $(p_0, p_1, \cdots, p_m)$ denotes the oriented simplex with vertices $p_0, p_1, \cdots, p_m$ and the orientation to which the permutation belongs.

15.1 Definition. The simplex with vertices $p_0, p_1, \cdots, p_m$ has two orientations; one *orientation* is the class of even permutations of $(p_0, p_1, \cdots, p_m)$, and the other is the class of odd permutations of $(p_0, p_1, \cdots, p_m)$.

Determinants can be used to characterize the two orientations of a Euclidean n-simplex in $\mathbb{R}^n$. If $a_0, a_1, \cdots, a_n$ are the vertices of a Euclidean simplex in $\mathbb{R}^n$, then $\{a_0, a_1, \cdots, a_n\}$ is a set of linearly independent points $a_i : (a_i^1, \cdots, a_i^n)$, $i = 0, 1, \cdots, n$, and

$$\Delta(a_0, a_1, \cdots, a_n) = \begin{vmatrix} a_0^1 & a_0^2 & \cdots & a_0^n & 1 \\ a_1^1 & a_1^2 & \cdots & a_1^n & 1 \\ \cdots & \cdots & \cdots & \cdots & \cdots \\ a_n^1 & a_n^2 & \cdots & a_n^n & 1 \end{vmatrix} \neq 0. \tag{5}$$

A determinant changes sign if two adjacent rows in its matrix are interchanged [see Theorem 77.5 in Appendix 2]. Let $r : \{0, 1, \cdots, n\} \to \{0, 1, \cdots, n\}$ be a permutation of $(0, 1, \cdots, n)$. Then

$$\Delta(a_{r(0)}, a_{r(1)}, \cdots, a_{r(n)}) = \pm\Delta(a_0, a_1, \cdots, a_n). \tag{6}$$

15.2 Theorem. *The class of permutations* $(a_{r(0)}, a_{r(1)}, \cdots, a_{r(n)})$ *for which the sign in* (6) *is* $+$ *is one of the orientations of* $(a_0, a_1, \cdots, a_n)$, *and the class of those permutations for which the sign in* (6) *is* $-$ *is the other orientation.*

PROOF. If $(a_{r(0)}, a_{r(1)}, \cdots, a_{r(n)})$ can be obtained from $(a_0, a_1, \cdots, a_n)$ by an even number of interchanges of adjacent letters [rows in the matrix in (5)], then the sign in (6) is $+$, and the permutations $(a_{r(0)}, a_{r(1)}, \cdots, a_{r(n)})$ and $(a_0, a_1, \cdots, a_n)$ belong to the same orientation by Definition 15.1. Similarly, if $(a_{r(0)}, a_{r(1)}, \cdots, a_{r(n)})$ can be obtained from $(a_0, a_1, \cdots, a_n)$ by an odd number of interchanges of adjacent letters, then the sign in (6) is $-$, and the permutations $(a_{r(0)}, a_{r(1)}, \cdots, a_{r(n)})$ and $(a_0, a_1, \cdots, a_n)$ belong to opposite orientations. Thus equation (6) divides the permutations of $(a_0, a_1, \cdots, a_n)$ into the two classes which are the two orientations of the Euclidean simplex $(a_0, a_1, \cdots, a_n)$. □

If $(a_0, a_1, \cdots, a_n)$ is a degenerate Euclidean simplex, then $\Delta(a_0, a_1, \cdots, a_n) = 0$ and determinants cannot be used to define an orientation for the simplex. Determinants provide a means of comparing the orientations of two different Euclidean simplexes as stated in the next definition.

15.3 Definition. Let $(a_0, a_1, \cdots, a_n)$ and $(b_0, b_1, \cdots, b_n)$ be two oriented Euclidean simplexes in $\mathbb{R}^n$. Then these simplexes have the *same orientation* in $\mathbb{R}^n$ if and only if $\Delta(a_0, a_1, \cdots, a_n)\Delta(b_0, b_1, \cdots, b_n) > 0$, and they have *opposite orientations* if and only if $\Delta(a_0, a_1, \cdots, a_n)\Delta(b_0, b_1, \cdots, b_n) < 0$.

15.4 Example. Let (a_0, a_1, a_2), (b_0, b_1, b_2), (c_0, c_1, c_2), and (d_0, d_1, d_2) be four oriented simplexes whose vertices are the following points:

$$\begin{array}{llll} a_0 : (1, 1) & b_0 : (10, 9) & c_0 : (10, 7) & d_0 : (13, 9) \\ a_1 : (4, 3) & b_1 : (3, 10) & c_1 : (11, 2) & d_1 : (17, 10) \\ a_2 : (2, 5) & b_2 : (5, 6) & c_2 : (6, 4) & d_2 : (15, 4) \end{array}$$

Then

$$\Delta(a_0, a_1, a_2) = 10, \qquad \Delta(c_0, c_1, c_2) = -23,$$
$$\Delta(b_0, b_1, b_2) = 26, \qquad \Delta(d_0, d_1, d_2) = -22.$$

The values of the four determinants and Definition 15.3 show that (a_0, a_1, a_2) and (b_0, b_1, b_2) have the same orientation, and that (c_0, c_1, c_2) and (d_0, d_1, d_2) have the same orientation. Also, the orientation of (c_0, c_1, c_2) and (d_0, d_1, d_2) is opposite that of (a_0, a_1, a_2) and (b_0, b_1, b_2). Furthermore, these results correspond to our intuitive feelings about orientation in the plane, because a_0, a_1, a_2 and b_0, b_1, b_2 are counterclockwise circuits of their simplexes, but c_0, c_1, c_2 and d_0, d_1, d_2 are clockwise circuits of their simplexes [see Figure 15.3].

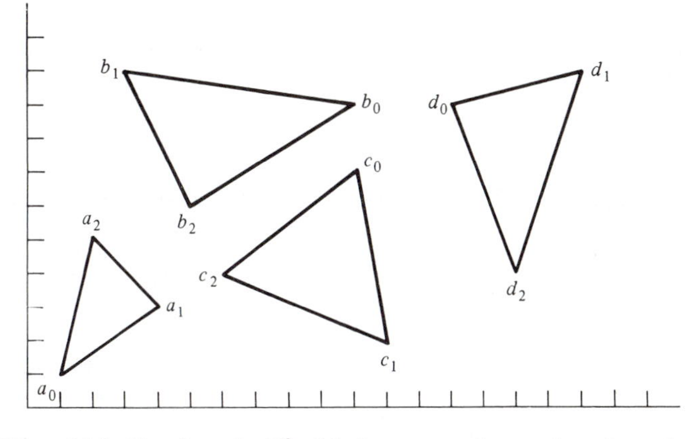

Figure 15.3. Simplexes in $\mathbb{R}^2$ with the same and opposite orientations.

15.5 Theorem. *If translation by the vector $v:(v^1, \cdots, v^n)$ translates the oriented Euclidean simplex $(a_0, a_1, \cdots, a_n)$ into the oriented Euclidean simplex $(b_0, b_1, \cdots, b_n)$, then $(b_0, b_1, \cdots, b_n)$ and $(a_0, a_1, \cdots, a_n)$ have the same orientation in $\mathbb{R}^n$.*

PROOF. Translation by the vector v translates $(a_0, a_1, \cdots, a_n)$ into $(a_0 + v, a_1 + v, \cdots, a_n + v)$. Then

$$\Delta(b_0, b_1, \cdots, b_n)$$

$$= \begin{vmatrix} a_0^1 + v^1 & a_0^2 + v^2 & \cdots & a_0^n + v^n & 1 \\ a_1^1 + v^1 & a_1^2 + v^2 & \cdots & a_1^n + v^n & 1 \\ \cdots & \cdots & \cdots & \cdots & \cdots \\ a_n^1 + v^1 & a_n^2 + v^2 & \cdots & a_n^n + v^n & 1 \end{vmatrix} = \begin{vmatrix} a_0^1 & a_0^2 & \cdots & a_0^n & 1 \\ a_1^1 & a_1^2 & \cdots & a_1^n & 1 \\ \cdots & \cdots & \cdots & \cdots & \cdots \\ a_n^1 & a_n^2 & \cdots & a_n^n & 1 \end{vmatrix}$$

$$= \Delta(a_0, a_1, \cdots, a_n).$$

To prove that the two displayed determinants are equal, multiply the last column of the first matrix in succession by $v^1, v^2, \cdots, v^n$ and then subtract it from columns $1, 2, \cdots, n$. These transformations leave the value of the determinant unchanged [see Theorems 77.11 and 77.17 in Appendix 1]. Then since $\Delta(b_0, b_1, \cdots, b_n) = \Delta(a_0, a_1, \cdots, a_n)$, Definition 15.3 states that the two Euclidean simplexes have the same orientation in $\mathbb{R}^n$. □

15.6 Theorem. *Let $y = Ax$ be an orthogonal linear transformation of $\mathbb{R}^n$ into $\mathbb{R}^n$ such that* $\det A = +1$. *If $y = Ax$ transforms the Euclidean simplex $(x_0, x_1, \cdots, x_n)$ into the simplex $(y_0, y_1, \cdots, y_n)$, then $(y_0, y_1, \cdots, y_n)$ is a Euclidean simplex which has the same orientation as $(x_0, x_1, \cdots, x_n)$.*

PROOF. Exercise 5.3 outlines the proof that

$$\Delta(y_0, y_1, \cdots, y_n) = \Delta(x_0, x_1, \cdots, x_n). \tag{7}$$

Then $\Delta(y_0, y_1, \cdots, y_n) \neq 0$ since $(x_0, x_1, \cdots, x_n)$ is a Euclidean simplex, and $(y_0, y_1, \cdots, y_n)$ is a Euclidean simplex by Definition 14.10. Finally, equation (7) and Definition 15.3 show that $(y_0, y_1, \cdots, y_n)$ and $(x_0, x_1, \cdots, x_n)$ have the same orientation in $\mathbb{R}^n$. □

Definition 15.3 answers the question of which Euclidean simplexes have the same orientation in $\mathbb{R}^n$, but we have not yet decided which of the two orientations shall be called the positive orientation and which the negative orientation. As a result of Definition 15.3, the class of positively oriented n-dimensional Euclidean simplexes in $\mathbb{R}^n$ is completely determined by choosing one of these simplexes and calling its orientation the positive orientation. The simplex usually used for this purpose is $(e_0, e_1, \cdots, e_n)$, the simplex whose vertices are the origin $e_0 : (0, 0, \cdots, 0)$ and the unit points $e_1 : (1, 0, \cdots, 0)$, $e_2 : (0, 1, \cdots, 0)$, $\cdots$, $e_n : (0, 0, \cdots, 1)$ on the axes.

15.7 Theorem. *The simplex $(e_0, e_1, \cdots, e_n)$ is an n-dimensional Euclidean simplex in $\mathbb{R}^n$, and $\Delta(e_0, e_1, \cdots, e_n) = (-1)^n$.*

Proof. Since

$$\Delta(e_0, e_1, \cdots, e_n) = \begin{vmatrix} 0 & 0 & \cdots & 0 & 1 \\ 1 & 0 & \cdots & 0 & 1 \\ \cdots & \cdots & \cdots & \cdots & \cdots \\ 0 & 0 & \cdots & 1 & 1 \end{vmatrix} = (-1)^n, \tag{8}$$

the points $e_0, e_1, \cdots, e_n$ are linearly independent, and $(e_0, e_1, \cdots, e_n)$ is a Euclidean simplex in $\mathbb{R}^n$. □

15.8 Definition. The Euclidean simplex $(e_0, e_1, \cdots, e_n)$ is positively oriented in $\mathbb{R}^n$. The Euclidean simplex $(a_0, a_1, \cdots, a_n)$ is *positively oriented* in $\mathbb{R}^n$ if and only if it has the same orientation as $(e_0, e_1, \cdots, e_n)$; and $(a_0, a_1, \cdots, a_n)$ is *negatively oriented* in $\mathbb{R}^n$ if and only if its orientation is the negative, or the opposite, of that of $(e_0, e_1, \cdots, e_n)$.

15.9 Theorem. *The n-dimensional Euclidean simplex $(a_0, a_1, \cdots, a_n)$ is positively oriented in $\mathbb{R}^n$ if and only if $(-1)^n\Delta(a_0, a_1, \cdots, a_n) > 0$; it is negatively oriented in $\mathbb{R}^n$ if and only if $(-1)^n\Delta(a_0, a_1, \cdots, a_n) < 0$.*

Proof. By Definition 15.8, $(e_0, e_1, \cdots, e_n)$ is positively oriented in $\mathbb{R}^n$, and $(a_0, a_1, \cdots, a_n)$ is positively oriented if and only if it has the same orientation as $(e_0, e_1, \cdots, e_n)$. Now Definition 15.3 states that $(e_0, e_1, \cdots, e_n)$ and $(a_0, a_1, \cdots, a_n)$ have the same orientation if and only if $\Delta(e_0, e_1, \cdots, e_n)$ $\Delta(a_0, a_1, \cdots, a_n) > 0$. Thus $(a_0, a_1, \cdots, a_n)$ is positively oriented if and only if $\Delta(e_0, e_1, \cdots, e_n)\Delta(a_0, a_1, \cdots, a_n) > 0$; hence, by Theorem 15.7, $(a_0, a_1, \cdots, a_n)$ is positively oriented in $\mathbb{R}^n$ if and only if $(-1)^n\Delta(a_0, a_1, \cdots,$

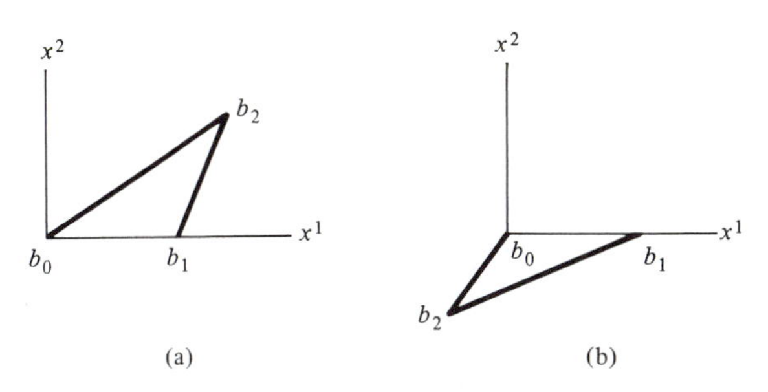

Figure 15.4. Positively and negatively oriented simplexes (b_0, b_1, b_2) in $\mathbb{R}^2$.

$a_n) > 0$. Similar arguments show that $(a_0, a_1, \cdots, a_n)$ is negatively oriented in $\mathbb{R}^n$ if and only if $(-1)^n\Delta(a_0, a_1, \cdots, a_n) < 0$. □

15.10 Example. By Theorem 15.9, the Euclidean simplex (a_0, a_1) is positively oriented in $\mathbb{R}$ if and only if $(-1)\Delta(a_0, a_1) > 0$. Since $\Delta(a_0, a_1) = a_0 - a_1$, then (a_0, a_1) is positively oriented if and only if $(-1)(a_0 - a_1) > 0$ or $a_1 > a_0$. Thus a positively oriented simplex (a_0, a_1) is one for which $a_1 > a_0$, but observe carefully that $\Delta(a_0, a_1) < 0$. Similarly, if (a_0, a_1) is negatively oriented in $\mathbb{R}$, then $a_1 < a_0$ and $\Delta(a_0, a_1) > 0$.

15.11 Theorem. *The Euclidean simplex (a_0, a_1, a_2) is positively oriented (negatively oriented) in $\mathbb{R}^2$ if and only if $\Delta(a_0, a_1, a_2) > 0$ $(\Delta(a_0, a_1, a_2) < 0)$, and also if and only if the vertices a_0, a_1, a_2 are arranged in the counterclockwise order (the clockwise order) around the simplex.*

Proof. By Theorem 15.9, (a_0, a_1, a_2) is positively oriented in $\mathbb{R}^2$ if and only if $(-1)^2\Delta(a_0, a_1, a_2) > 0$, or $\Delta(a_0, a_1, a_2) > 0$. Translate the simplex so that a_0 is at the origin $(0, 0)$; by Theorem 15.5 the value of $\Delta(a_0, a_1, a_2)$ is unchanged by this transformation. Next, use a rotation with positive determinant to rotate the simplex about the origin so that the vertex a_1 is on the positive x^1-axis; by Theorem 15.6, the value of $\Delta(a_0, a_1, a_2)$ again is unchanged by the transformation. Thus, there is a simplex (b_0, b_1, b_2) with vertices $b_0 : (0, 0)$, $b_1 : (b_1^1, 0)$, $b_2 : (b_2^1, b_2^2)$ such that $b_1^1 > 0$ and

$$\Delta(a_0, a_1, a_2) = \Delta(b_0, b_1, b_2) = \begin{vmatrix} 0 & 0 & 1 \\ b_1^1 & 0 & 1 \\ b_2^1 & b_2^2 & 1 \end{vmatrix} = b_1^1 b_2^2.$$

Thus (a_0, a_1, a_2) is positively oriented if and only if $b_2^2 > 0$, and negatively oriented if and only if $b_2^2 < 0$. Figure 15.4(a) shows that the vertices of (b_0, b_1, b_2) are arranged in the counterclockwise order if and only if $b_2^2 > 0$, and Figure 15.4(b) shows that they are arranged in the clockwise order if and only if $b_2^2 < 0$. Since the translation and the rotation do not change the orientation of the vertices of (a_0, a_1, a_2), these arguments show that $(a_0, a_1,$

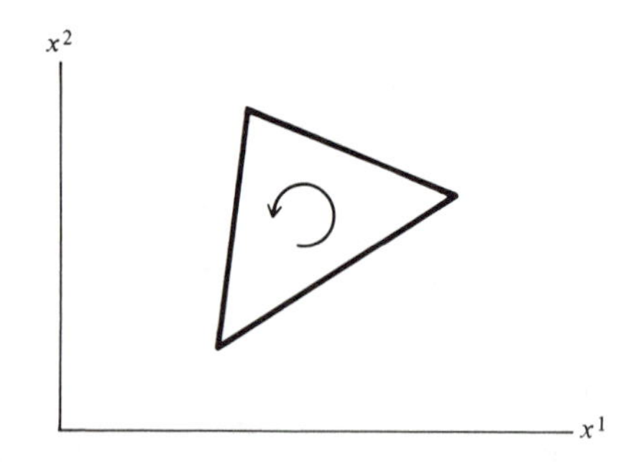

Figure 15.5. The positive orientation of a simplex in $\mathbb{R}^2$.

a_2) is positively oriented in $\mathbb{R}^2$ (negatively oriented in $\mathbb{R}^2$) if and only if its vertices a_0, a_1, a_2 are arranged in the counterclockwise order (the clockwise order) around the simplex. □

The usual graphical indication of the positive orientation of a Euclidean simplex in $\mathbb{R}^2$ is a curved arrow which indicates a counterclockwise circuit of the vertices [see Figure 15.5]. Theorem 15.11 justifies this method of indicating the positive orientation in the plane.

15.12 Example. In Example 15.4, $\Delta(a_0, a_1, a_2) = 10$ and $\Delta(b_0, b_1, b_2) = 26$; thus (a_0, a_1, a_2) and (b_0, b_1, b_2) are positively oriented in $\mathbb{R}^2$ by Theorem 15.9, and Figure 15.3 shows that the vertices of these simplexes are arranged in the counterclockwise order as required by Theorem 15.11. In the same example, $\Delta(c_0, c_1, c_2) = -23$ and $\Delta(d_0, d_1, d_2) = -22$; then (c_0, c_1, c_2) and (d_0, d_1, d_2) are negatively oriented in $\mathbb{R}^2$, and Figure 15.3 shows that their vertices are arranged in the clockwise order.

Finally, it is necessary to consider simplexes of the form $(a_0, a_1, \cdots, a_m)$ in $\mathbb{R}^n$ with $m < n$. The simplest case is the one-dimensional simplex in $\mathbb{R}^2$. Corresponding to the simplex (a_0, a_1) in $\mathbb{R}^2$ with vertices $a_0 : (a_0^1, a_0^2)$ and $a_1 : (a_1^1, a_1^2)$, there are two simplexes (a_0^1, a_1^1) and (a_0^2, a_1^2) in $\mathbb{R}$ whose orientations are determined by the signs of

$$\Delta[(a_0, a_1)^{(1)}] = \begin{vmatrix} a_0^1 & 1 \\ a_1^1 & 1 \end{vmatrix}, \qquad \Delta[(a_0, a_1)^{(2)}] = \begin{vmatrix} a_0^2 & 1 \\ a_1^2 & 1 \end{vmatrix}. \tag{9}$$

Here (a_0^1, a_1^1) and (a_0^2, a_1^2) may have the same or opposite orientations in $\mathbb{R}$; as a result, there is no useful concept of positive and negative simplexes (a_0, a_1) in $\mathbb{R}^2$. What is useful is the concept of "same direction". The oriented simplex (a_0, a_1) in $\mathbb{R}^2$ is usually thought of as a vector, and (a_0, a_1) determines a direction in $\mathbb{R}^2$. If (a_0, a_1) and (b_0, b_1) are two simplexes in $\mathbb{R}^2$, and if r is a positive number such that

$$\Delta[(a_0, a_1)^{(1)}] = r\Delta[(b_0, b_1)^{(1)}], \qquad \Delta[(a_0, a_1)^{(2)}] = r\Delta[(b_0, b_1)^{(2)}], \tag{10}$$

then (a_0, a_1) and (b_0, b_1) determine the same direction in $\mathbb{R}^2$. In the general case of two simplexes $(a_0, a_1, \cdots, a_m)$ and $(b_0, b_1, \cdots, b_m)$ in $\mathbb{R}^n$, $1 \leqq m \leqq n$,

$$\Delta[(a_0, a_1, \cdots, a_m)^{(j_1, \cdots, j_m)}] = \begin{vmatrix} a_0^{j_1} & a_0^{j_2} & \cdots & a_0^{j_m} & 1 \\ \cdots & \cdots & \cdots & \cdots & \cdots \\ a_m^{j_1} & a_m^{j_2} & \cdots & a_m^{j_m} & 1 \end{vmatrix},$$

$$\Delta[(b_0, b_1, \cdots, b_m)^{(j_1, \cdots, j_m)}] = \begin{vmatrix} b_0^{j_1} & b_0^{j_2} & \cdots & b_0^{j_m} & 1 \\ \cdots & \cdots & \cdots & \cdots & \cdots \\ b_m^{j_1} & b_m^{j_2} & \cdots & b_m^{j_m} & 1 \end{vmatrix}, \tag{11}$$

$$(j_1, \cdots, j_m) \in (m/n).$$

15.13 Definition. The Euclidean simplexes $(a_0, a_1, \cdots, a_m)$ and $(b_0, b_1, \cdots, b_m)$ have the same *m-direction in* $\mathbb{R}^n$ if and only if there exists a positive constant r such that for each $(j_1, \cdots, j_m)$ in (m/n),

$$\Delta[(a_0, a_1, \cdots, a_m)^{(j_1, \cdots, j_m)}] = r\Delta[(b_0, b_1, \cdots, b_m)^{(j_1, \cdots, j_m)}]. \tag{12}$$

Observe that, since $(a_0, a_1, \cdots, a_m)$ is a Euclidean simplex, then $\Delta[(a_0, a_1, \cdots, a_m)^{(j_1, \cdots, j_m)}]$ is not zero for every $(j_1, \cdots, j_m)$ in (m/n). The orientation of $(a_0, a_1, \cdots, a_m)^{(j_1, \cdots, j_m)}$ in $\mathbb{R}^m$ is determined if and only if $\Delta[(a_0, a_1, \cdots, a_m)^{(j_1, \cdots, j_m)}] \neq 0$.

15.14 Theorem. *Let $(a_0, a_1, \cdots, a_m)$ and $(b_0, b_1, \cdots, b_m)$ be two Euclidean simplexes which have the same m-direction in $\mathbb{R}^n$, $1 \leqq m \leqq n$. Then either the orientation of $(a_0, a_1, \cdots, a_m)^{(j_1, \cdots, j_m)}$ and $(b_0, b_1, \cdots, b_m)^{(j_1, \cdots, j_m)}$ is undetermined in $\mathbb{R}^m$, or these m-simplexes have the same orientation (positive or negative) in $\mathbb{R}^m$. This statement is true for each $(j_1, \cdots, j_m)$ in (m/n).*

PROOF. If $(a_0, a_1, \cdots, a_m)$ and $(b_0, b_1, \cdots, b_m)$ have the same m-direction in $\mathbb{R}^n$, then by (12) in Definition 15.13, $\Delta[(a_0, a_1, \cdots, a_m)^{(j_1, \cdots, j_m)}]$ and $\Delta[(b_0, b_1, \cdots, b_m)^{(j_1, \cdots, j_m)}]$ have the same sign. If $(-1)^m\Delta[(a_0, a_1, \cdots, a_m)^{(j_1, \cdots, j_m)}] > 0$, then $(a_0, a_1, \cdots, a_m)^{(j_1, \cdots, j_m)}$ and $(b_0, b_1, \cdots, b_m)^{(j_1, \cdots, j_m)}$ are positively oriented in $\mathbb{R}^m$ by Theorem 15.9; if $(-1)^m\Delta[(a_0, a_1, \cdots, a_m)^{(j_1, \cdots, j_m)}] < 0$, they are negatively oriented in $\mathbb{R}^m$. These statements are true for each $(j_1, \cdots, j_m)$ in (m/n). □

15.15 Example. Let (a_0, a_1, a_2) and (b_0, b_1, b_2) be Euclidean simplexes in $\mathbb{R}^3$ with the following vertices:

$$a_0 : (5, 1, 2), \qquad a_1 : (7, 6, 3), \qquad a_2 : (3, 4, 8);$$

$$b_0 : (10, 2, 4), \qquad b_1 : (14, 12, 6), \qquad b_2 : (6, 8, 16).$$

Then

$$\Delta[(a_0, a_1, a_2)^{(1,2)}] = 16, \qquad \Delta[(b_0, b_1, b_2)^{(1,2)}] = 64,$$
$$\Delta[(a_0, a_1, a_2)^{(1,3)}] = 14, \qquad \Delta[(b_0, b_1, b_2)^{(1,3)}] = 56,$$
$$\Delta[(a_0, a_1, a_2)^{(2,3)}] = 27; \qquad \Delta[(b_0, b_1, b_2)^{(2,3)}] = 108.$$

Since each determinant in the second column is 4 times the corresponding determinant in the first column, then (a_0, a_1, a_2) and (b_0, b_1, b_2) have the same 2-direction in $\mathbb{R}^3$. Since all signs are plus, each $(a_0, a_1, a_2)^{(j_1,j_2)}$ and $(b_0, b_1, b_2)^{(j_1,j_2)}$ is positively oriented in $\mathbb{R}^2$ [see Theorems 15.9, 15.11, and 15.14]. In a second example, let (c_0, c_1, c_2) and (d_0, d_1, d_2) be Euclidean simplexes in $\mathbb{R}^3$ with the following vertices:

$$c_0 : (1, -2, 4), \qquad c_1 : (2, -4, 15), \qquad c_2 : (3, -6, 8);$$
$$d_0 : (3, -6, 12), \qquad d_1 : (6, -12, 45), \qquad d_2 : (9, -18, 24).$$

Then

$$\Delta[(c_0, c_1, c_2)^{(1,2)}] = 0, \qquad \Delta[(d_0, d_1, d_2)^{(1,2)}] = 0,$$
$$\Delta[(c_0, c_1, c_2)^{(1,3)}] = -18, \qquad \Delta[(d_0, d_1, d_2)^{(1,3)}] = -162,$$
$$\Delta[(c_0, c_1, c_2)^{(2,3)}] = 36; \qquad \Delta[(d_0, d_1, d_2)^{(2,3)}] = 324.$$

Since each determinant in the second column is 9 times the corresponding determinant in the first column, then (c_0, c_1, c_2) and (d_0, d_1, d_2) have the same 2-direction in $\mathbb{R}^3$. The orientation of $(c_0, c_1, c_2)^{(1,2)}$ and $(d_0, d_1, d_2)^{(1,2)}$ is undetermined in $\mathbb{R}^2$; $(c_0, c_1, c_2)^{(1,3)}$ and $(d_0, d_1, d_2)^{(1,3)}$ are negatively oriented in $\mathbb{R}^2$; and $(c_0, c_1, c_2)^{(2,3)}$ and $(d_0, d_1, d_2)^{(2,3)}$ are positively oriented in $\mathbb{R}^2$.

Exercises

15.1. The points $a_0 : (6, 8)$, $a_1 : (1, 6)$, and $a_2 : (4, 3)$ are the vertices of a simplex (a_0, a_1, a_2) in $\mathbb{R}^2$.
   (a) Prove that the points a_0, a_1, a_2 are linearly independent and that $[a_0, a_1, a_2]$ is a Euclidean simplex.
   (b) Determine from a sketch whether (a_0, a_1, a_2) is positively or negatively oriented in $\mathbb{R}^2$. Verify your answer analytically.

15.2. The points $a_0 : (2, 6, 4)$, $a_1 : (8, 2, 12)$, and $a_2 : (5, 4, 8)$ are vertices of a simplex in $\mathbb{R}^3$.
   (a) Show that a_2 is the mid-point of the segment joining a_0 and a_1. Does this fact indicate that $[a_0, a_1, a_2]$ is a degenerate Euclidean simplex? Explain.
   (b) Prove analytically that $[a_0, a_1, a_2]$ is a degenerate Euclidean simplex.
   (c) Is there a 2-direction assigned to (a_0, a_1, a_2)? Explain.
   (d) Show that the one-dimensional simplexes (a_0, a_2), (a_2, a_1), and (a_0, a_1) have the same direction (that is, 1-direction) in $\mathbb{R}^3$.

15.3. Let (a_0, a_1, a_2) denote the oriented Euclidean simplex $[a_0, a_1, a_2]$ in $\mathbb{R}^3$, and let a_3 be the point such that

$$a_3 = \sum_{i=0}^{2} t^i a_i, \qquad t^i > 0, \qquad \sum_{i=0}^{2} t^i = 1.$$

Explain why a_3 is a point in $[a_0, a_1, a_2]$. Prove that the simplexes (a_3, a_1, a_2), (a_0, a_3, a_2), (a_0, a_1, a_3) determine a 2-direction in $\mathbb{R}^3$ and that they have the same 2-direction as (a_0, a_1, a_2).

15.4. The points $a_0 : (8, 1, 4)$, $a_1 : (1, 10, 3)$, and $a_2 : (-3, 4, 5)$ are the vertices of an oriented Euclidean simplex (a_0, a_1, a_2) in $\mathbb{R}^3$. Show that $a_3 : (2, 5, 4)$ is a point in the interior of $[a_0, a_1, a_2]$, and that the simplexes (a_3, a_1, a_2), (a_0, a_3, a_2), and (a_0, a_1, a_3) have the same 2-direction as (a_0, a_1, a_2). [Hint. Exercise 15.3.]

15.5. Let $(a_0, a_1, \cdots, a_m)$ be an oriented simplex in $\mathbb{R}^n$, $1 \leqq m \leqq n$, and let $(b_0, b_1, \cdots, b_m)$ be the simplex $(a_0 + v, a_1 + v, \cdots, a_m + v)$ obtained by translating $(a_0, a_1, \cdots, a_m)$ by the vector $v : (v^1, \cdots, v^n)$. Show that $(b_0, b_1, \cdots, b_m)$ and $(a_0, a_1, \cdots, a_m)$ have the same m-direction in $\mathbb{R}^n$.

15.6. Let $f : [a, b] \to \mathbb{R}$ be a function which has a continuous derivative f' on the positively oriented Euclidean simplex $[a, b]$. Subdivide $[a, b]$ by equally spaced points $a = x_0 < x_1 < \cdots < x_{i-1} < x_i < \cdots < x_{n-1} < x_n = b$ to form the one-dimensional simplexes $(x_0, x_1), \cdots, (x_{i-1}, x_i), \cdots, (x_{n-1}, x_n)$ in $\mathbb{R}$.

(a) Prove that the simplexes (x_{i-1}, x_i), $i = 1, \cdots, n$, are positively oriented in $\mathbb{R}$.

(b) Prove that

$$\sum_{i=1}^{n} (-1)\Delta f(x_{i-1}, x_i) = \sum_{i=1}^{n} (-1) \begin{vmatrix} f(x_{i-1}) & 1 \\ f(x_i) & 1 \end{vmatrix} = f(x_n) - f(x_0) = f(b) - f(a).$$

(c) Use the mean-value theorem and the definition of the Riemann integral to prove that

$$\lim_{n\to\infty} \sum_{i=1}^{n} (-1)\Delta f(x_{i-1}, x_i) = \lim_{n\to\infty} \sum_{i=1}^{n} f'(x_i^*)(-1)\Delta(x_{i-1}, x_i) = \int_a^b f'(x)\,dx.$$

(d) Prove the following form of the fundamental theorem of the integral calculus.

$$\int_a^b f'(x)\,dx = f(b) - f(a).$$

Section 17 will show that $(b) - (a)$ is the boundary of (a, b), and a later chapter will show that $f(b) - f(a)$ is to be interpreted as the integral of f over this boundary.

15.7. (a) Prove that there are 24 permutations or orderings of the four letters (p_0, p_1, p_2, p_3).

(b) Prove that there are 12 even permutations of (p_0, p_1, p_2, p_3) and also 12 odd permutations.

(c) If $m \geqq 1$, prove that there are $(m + 1)!$ permutations of $(p_0, p_1, \cdots, p_m)$.

(d) Prove that there are $(m + 1)!/2$ even permutations and $(m + 1)!/2$ odd permutations of $(p_0, p_1, \cdots, p_m)$.

15.8. This section has explained how to orient Euclidean n-simplexes in $\mathbb{R}^n$, and the orientation will be essential in many of the applications in the future. Multivectors, which were used in Chapter 1 in the definition of derivatives, were not oriented although from certain points of view multivectors and simplexes are the same. It is true that the initial point x_0 of the multivector $\mathbf{x}$ was usually written last, as in $(x_1, \cdots, x_n, x_0)$, but this ordering was only for convenience in avoiding an unimportant negative sign in certain calculations. Prove that derivatives are independent of orientations by proving the following. If $r: \{0, 1, \cdots, n\} \to \{0, 1, \cdots, n\}$ is a permutation of the integers $0, 1, \cdots, n$, then the quotient

$$\frac{\Delta(f^1, \cdots, f^n)(x_{r(0)}, \cdots, x_{r(n)})}{\Delta(x_{r(0)}, \cdots, x_{r(n)})} = \frac{\begin{vmatrix} f^1(x_{r(0)}) & \cdots & f^n(x_{r(0)}) & 1 \\ \cdots & \cdots & \cdots & \cdots \\ f^1(x_{r(n)}) & \cdots & f^n(x_{r(n)}) & 1 \end{vmatrix}}{\begin{vmatrix} x^1_{r(0)} & \cdots & x^n_{r(0)} & 1 \\ \cdots & \cdots & \cdots & \cdots \\ x^1_{r(n)} & \cdots & x^n_{r(n)} & 1 \end{vmatrix}}$$

has the same value for each of the $(n + 1)!$ permutations r of the points $x_0, x_1, \cdots, x_n$.

16. Complexes and Chains

Thus far our study has emphasized single simplexes rather than sets of simplexes. A set of simplexes is unlikely to hold any special interest, beyond that inherent in the individual simplexes, unless the set has some special structure or organization. For example, Figure 16.1(a) shows a set of three random simplexes $[a_0, a_1, a_2]$, $[b_0, b_1, b_2]$, and $[c_0, c_1, c_2]$, but nothing about this set attracts our interest or suggests that it might have useful applications.

What kind of structure or organization would make a set of simplexes interesting and rich in special properties? The simplexes $[x_{i-1}, x_i]$ in the set $\{[x_{i-1}, x_i]: i = 1, \cdots, n\}$ which forms the subdivision of $[a, b]$ in the proof of the fundamental theorem of the integral calculus in Exercise 15.6 suggest an answer. This set of simplexes $[x_{i-1}, x_i]$ has the property that the intersection of each two of them is a common side of the two simplexes. For example, $[x_{i-1}, x_i] \cap [x_i, x_{i+1}] = [x_i]$, and $[x_i]$ is a 0-dimensional side of $[x_{i-1}, x_i]$ and of $[x_i, x_{i+1}]$. Also, if $j > i$, then $[x_{i-1}, x_i] \cap [x_j, x_{j+1}]$ is

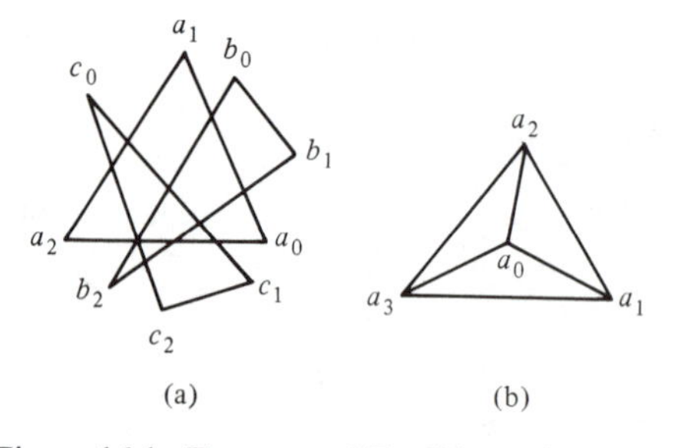

Figure 16.1. Two sets of Euclidean simplexes.

the empty set; but recall from Section 14 that the empty set is an improper side of every simplex. The set of simplexes $[a_0, a_1, a_2]$, $[b_0, b_1, b_2]$, and $[c_0, c_1, c_2]$ in Figure 16.1(a) does not have the property that the intersection of each two simplexes is a side of each simplex. The set $\{[a_0, a_1, a_2], [a_0, a_2, a_3], [a_0, a_3, a_1]\}$ in Figure 16.1(b) does have the intersection property. For example, $[a_0, a_1, a_2] \cap [a_0, a_2, a_3] = [a_0, a_2]$, and $[a_0, a_2]$ is a one-dimensional side of each of the simplexes $[a_0, a_1, a_2]$ and $[a_0, a_2, a_3]$. Since it is desirable that the intersection of each two simplexes in the set be a simplex in the set, an interesting set of Euclidean simplexes is not $\{[a_0, a_1, a_2], [a_0, a_2, a_3], [a_0, a_3, a_1]\}$ but rather the set which contains all of the following simplexes:

$$[a_0, a_1, a_2], [a_0, a_2, a_3], [a_0, a_3, a_1];$$

$$[a_0, a_1], [a_0, a_2], [a_0, a_3], [a_1, a_2], [a_2, a_3], [a_3, a_1];$$

$$[a_0], [a_1], [a_2], [a_3];$$

$\varnothing$ (the empty set).

It is easy to verify that the intersection of each two simplexes in this set is a simplex in the set. These examples suggest the following definition.

16.1 Definition. A finite set K of Euclidean simplexes in $\mathbb{R}^n$ is called a *finite Euclidean simplicial complex* if and only if it has the following two properties:

(1) Each proper or improper side of a simplex in K is a simplex in K.
(2) The intersection of each two simplexes in K is a common side of these two simplexes.

If each simplex in K is a side (proper or improper) of an m-dimensional simplex in K, then K is called a *Euclidean, homogeneous, m-dimensional, simplicial complex* (the modifier "finite" will not be repeated since only finite complexes will be considered in this book). The complex is said to be *oriented* if and only if a definite orientation (which may be chosen arbitrarily) is assigned to each simplex in K.

16.2 Example. The points $a_0 : (0, 0)$, $a_1 : (1, 1)$, $a_2 : (-1, 1)$, $a_3 : (-1, -1)$, and $a_4 : (1, -1)$ are the center and vertices of a square in $\mathbb{R}^2$. The diagonals of the square divide it into four two-dimensional Euclidean simplexes. These simplexes, together with all of their sides, form a Euclidean complex K [see Figure 16.2]. The complex K consists of the following simplexes of dimensions 2, 1, 0, -1:

$$[a_0, a_1, a_2], [a_0, a_2, a_3], [a_0, a_3, a_4], [a_0, a_4, a_1];$$

$$[a_0, a_1], [a_0, a_2], [a_0, a_3], [a_0, a_4], [a_1, a_2], [a_2, a_3], [a_3, a_4], [a_4, a_1];$$

$$[a_0], [a_1], [a_2], [a_3], [a_4];$$

$\varnothing$ (the empty set).

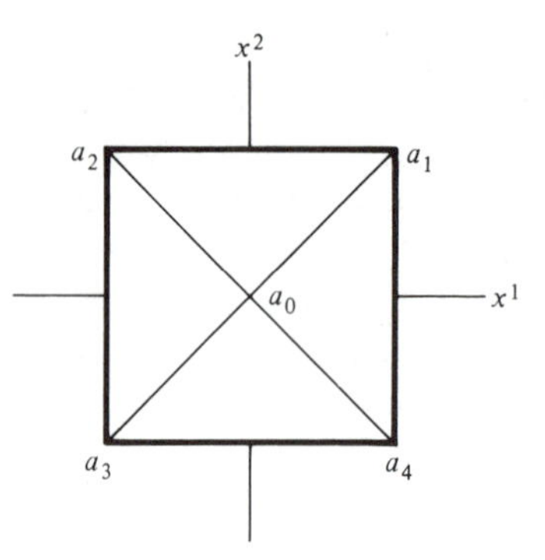

Figure 16.2. The complex in Example 16.2.

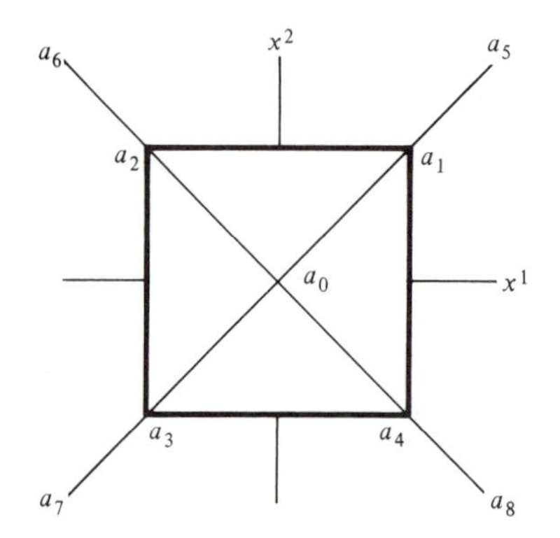

Figure 16.3. The complex in Example 16.3.

This complex K is a homogeneous, 2-dimensional complex since each simplex in it is a side of a 2-dimensional simplex in K.

16.3 Example. Use the points $a_1, \cdots, a_4$ in Example 16.2 and $a_5 : (2, 2)$, $a_6 : (-2, 2)$, $a_7 : (-2, -2)$, $a_8 : (2, -2)$ to construct the one-dimensional Euclidean simplexes $[a_1, a_5]$, $[a_2, a_6]$, $[a_3, a_7]$, $[a_4, a_8]$. Add these four simplexes and the zero-dimensional simplexes $[a_5], \cdots, [a_8]$ to those in Example 16.2 [see Figure 16.3]; the resulting set of 26 simplexes is also a Euclidean complex K, but it is not a homogeneous complex because the added simplexes are not sides of two-dimensional simplexes in the complex.

16.4 Example. Let $a_0, a_1, \cdots, a_4$ be the points in Example 16.2 [see Figure 16.2]. The set consisting of $[a_1, a_2, a_3]$ and $[a_1, a_2, a_4]$ and all of their sides is not a Euclidean complex because the intersection of $[a_1, a_2, a_3]$ and $[a_1, a_2, a_4]$ is $[a_0, a_1, a_2]$, which is not a side of any simplex in the set.

16.5 Example. The eight points $(\pm 1, \pm 1, \pm 1)$ are the vertices of a cube in $\mathbb{R}^3$. Subdivide each face of the cube into four Euclidean simplexes as

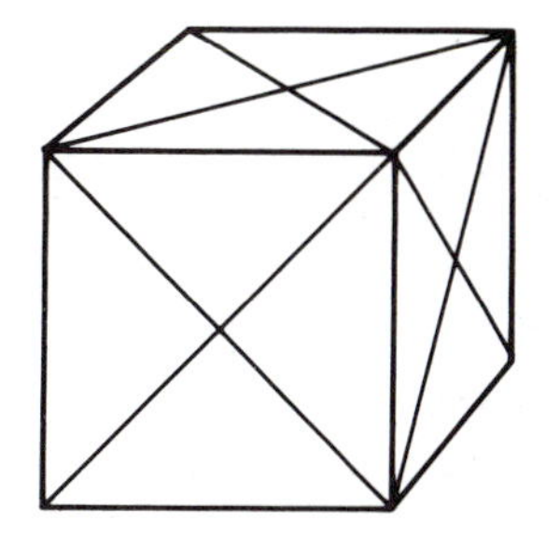

Figure 16.4. The complex in Example 16.5.

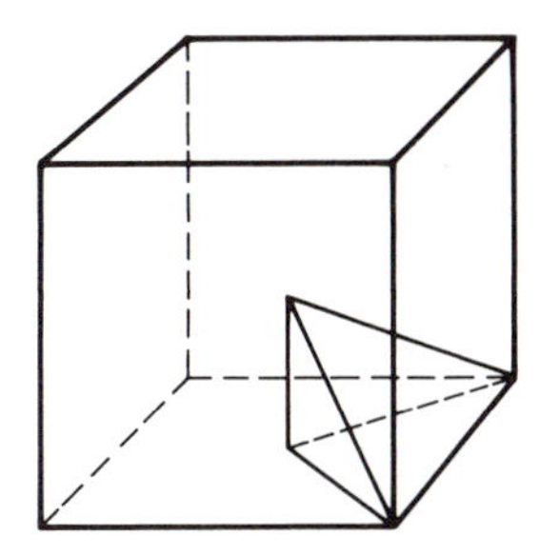

Figure 16.5. The complex in Example 16.7.

in Example 16.2 [see Figure 16.2]; the set consisting of these 24 two-dimensional simplexes and all of their sides of all dimensions is a Euclidean, homogeneous, two-dimensional, simplicial complex K in $\mathbb{R}^3$ [see Figure 16.4].

16.6 Example. The complex in Example 16.5 contains 14 simplexes of dimension zero; they are the vertices of the simplexes of dimension two. The segments which connect these vertices to the point $(0, 0, 0)$ are 14 one-dimensional Euclidean simplexes. The set consisting of the 24 two-dimensional simplexes in Example 16.5 and the 14 one-dimensional simplexes just described, and all of their sides is a Euclidean simplicial complex, but it is not a homogeneous complex.

16.7 Example. Construct 24 three-dimensional Euclidean simplexes as follows. The vertices of each simplex are the point $(0, 0, 0)$ and the three vertices of a two-dimensional simplex in Example 16.5; Figure 16.5 shows one of these three-dimensional simplexes. The set consisting of the 24 three-dimensional Euclidean simplexes and all of their sides is a Euclidean, homogeneous, three-dimensional, simplicial complex K in $\mathbb{R}^3$. The results in Section 15 can be used to give each of the three-dimensional simplexes in K the positive orientation in $\mathbb{R}^3$.

There are also abstract complexes. Let K be a set of abstract simplexes $\{p_0, p_1, \cdots, p_k\}$ such that $k \leqq m$ for some integer m and every simplex in K. The intersection of two simplexes in K is a common side of each of the simplexes; it is the common side of maximum dimension. Then K is an abstract simplicial complex if and only if it satisfies condition (1) in Definition 16.1. The definitions of the terms finite, homogeneous, m-dimensional, and oriented are the same as in the Euclidean complex.

Let K be an oriented (Euclidean or abstract) homogeneous, m-dimensional, simplicial complex in $\mathbb{R}^n$. Let an m-dimensional oriented simplex $(p_0, p_1, \cdots, p_m)$ in K be denoted by s; the simplex with the same vertices but with the opposite orientation is denoted by $-s$. If $s_1, s_2, \cdots, s_k$ are the m-dimensional simplexes in K and $a_1, a_2, \cdots, a_k$ are real numbers, the expression

$$\sum_{i=1}^{k} a_i s_i \tag{3}$$

is called an *m-chain* of simplexes in K. The expression in (3) is a formal sum, and no operations on the simplexes are implied. The numbers a_i are called the coefficients of the chain. It is customary to let 0 denote the zero chain, that is, the chain all of whose coefficients are zero.

Two operations—the addition of two chains and the multiplication of a chain by a real number c—are defined as follows on the set $C_m(K)$ of all m-chains of simplexes in K.

$$\sum_{i=1}^{k} a_i s_i + \sum_{i=1}^{k} b_i s_i = \sum_{i=1}^{k} (a_i + b_i) s_i. \tag{4}$$

$$c \sum_{i=1}^{k} a_i s_i = \sum_{i=1}^{k} (c a_i) s_i. \tag{5}$$

Two chains are equal if and only if they have the same coefficients. Addition is thus commutative. The following equations summarize some of the rules of operation on chains; s, with or without a subscript, denotes an m-simplex in K.

$$1s = s. \tag{6}$$

$$(-1)s = -s. \tag{7}$$

$$0s = 0. \tag{8}$$

$$a(bs) = (ab)s. \tag{9}$$

$$(a + b)s = as + bs. \tag{10}$$

$$\sum_{i=1}^{k} a_i s_i + \sum_{i=1}^{k} b_i s_i = \sum_{i=1}^{k} b_i s_i + \sum_{i=1}^{k} a_i s_i. \tag{11}$$

$$c\left(\sum_{i=1}^{k} a_i s_i + \sum_{i=1}^{k} b_i s_i\right) = c \sum_{i=1}^{k} a_i s_i + c \sum_{i=1}^{k} b_i s_i. \tag{12}$$

As a result of (6), every simplex in K can be considered to be a chain.

As a result of (8), a term with a zero coefficient can be dropped from a chain; alternatively, every chain can be considered to contain a term (perhaps with a zero coefficient) for every m-simplex s in K. The m-chains of simplexes s in K form a linear space $C_m(K)$ and the set of m-simplexes in K forms a basis for this space. If the number of m-simplexes in K is k, then the dimension of $C_m(K)$ is k.

There are important cases in which the coefficients a_i in the chain (3) are restricted to be integers (or elements in other special groups, rings, or fields) rather than real numbers. These chains have many important applications.

16.8 Example. Let K be the Euclidean, homogeneous, one-dimensional, simplicial complex which consists of the simplexes $[a_0, a_1]$, $[a_0, a_2]$, $[a_0, a_3]$, $[a_1, a_2]$, $[a_2, a_3]$, $[a_3, a_1]$ in Figure 16.1(b). If these simplexes are considered to be oriented by the ordering of the simplexes as shown, then K is an oriented complex, and chains of these simplexes are defined as stated above. The sum of the three chains

$$\begin{aligned}&[a_1, a_2] + [a_2, a_0] + [a_0, a_1],\\ &[a_2, a_3] + [a_3, a_0] + [a_0, a_2],\\ &[a_3, a_1] + [a_1, a_0] + [a_0, a_3],\end{aligned} \tag{13}$$

can be found easily by first representing each chain in terms of the oriented basis simplexes as follows:

$$\begin{aligned}&[a_1, a_2] - [a_0, a_2] + [a_0, a_1],\\ &[a_2, a_3] - [a_0, a_3] + [a_0, a_2],\\ &[a_3, a_1] - [a_0, a_1] + [a_0, a_3].\end{aligned} \tag{14}$$

Equation (4) shows that the sum of these three chains is obtained by adding the coefficients of like terms. Thus the sum is

$$\begin{aligned}&(1-1)[a_0, a_1] + (1-1)[a_0, a_2]\\ &\qquad + (1-1)[a_0, a_3] + [a_1, a_2] + [a_2, a_3] + [a_3, a_1].\end{aligned} \tag{15}$$

By the rules given above, this chain is more simply written as $[a_1, a_2] + [a_2, a_3] + [a_3, a_1]$. The next section will explain that the chains in (14) are the boundaries of the simplexes $[a_0, a_1, a_2]$, $[a_0, a_2, a_3]$, $[a_0, a_3, a_1]$, and that (15) is the boundary of the chain formed by the sum of these three simplexes. Compare these statements with the drawing of K in Figure 16.1(b).

Exercises

16.1. Count the number of Euclidean simplexes of dimensions 2, 1, 0, -1 in the Euclidean complex constructed in Example 16.2.

16.2. Show that the Euclidean complex K constructed in Example 16.5 contains 24 simplexes of dimension two, 36 simplexes of dimension one, 14 simplexes of dimension zero, and one simplex of dimension minus one.

16.3. (a) The complex K in Example 16.5 is a Euclidean, homogeneous, two-dimensional simplicial complex in $\mathbb{R}^3$. Prove that $C_2(K)$ is isomorphic to $\mathbb{R}^{24}$.

(b) The complex K in Example 16.7 is a Euclidean, homogeneous, three-dimensional, simplicial complex in $\mathbb{R}^3$. Prove that $C_3(K)$ is also isomorphic to $\mathbb{R}^{24}$.

16.4. Let K be a Euclidean, homogeneous, m-dimensional, complex; assume that the number of m-dimensional simplexes in K is k. Show that the linear space $C_m(K)$ has dimension k.

16.5. Let $a_1 : (1, 0)$, $a_2 : (1, 1)$, $a_3 : (-1, 0)$, $a_4 : (-1, 1)$ be four points in $\mathbb{R}^2$. Form a set of simplexes which consists of $[a_1, a_3, a_4]$, $[a_3, a_1, a_2]$, and all of the sides of these simplexes. Draw a sketch of these simplexes. Does this set of simplexes form a Euclidean complex? Explain your answer.

16.6. Let $a_1, \cdots, a_6$ be the vertices, listed in counterclockwise order, of a regular hexagon in $\mathbb{R}^2$, and let a_0 be the center of this hexagon. Consider the set of Euclidean simplexes $[a_0, a_1, a_2]$, $[a_0, a_2, a_3]$, $\cdots$, $[a_0, a_6, a_1]$ and all of their sides. Draw a sketch which shows these simplexes.

(a) Does the set of simplexes form a Euclidean complex? a homogeneous Euclidean complex K? Give reasons for your answers.

(b) What must be done to make K an oriented complex?

(c) Describe the 2-chains in the oriented, Euclidean, homogeneous, 2-dimensional, simplicial complex K. What is the dimension of the linear space $C_2(K)$ of 2-chains in K?

(d) Describe the linear space $C_1(K)$ of one-chains in K. What is the dimension of $C_1(K)$?

16.7. Show that the abstract simplexes

$$\{p_0, p_1, p_2\}, \{p_0, p_1, p_3\},$$

$$\{p_0, p_1\}, \{p_0, p_2\}, \{p_0, p_3\}, \{p_1, p_2\}, \{p_1, p_3\};$$

$$\{p_0\}, \{p_1\}, \{p_2\}, \{p_3\},$$

$$\varnothing \text{ (the empty set)},$$

form an abstract, homogeneous, 2-dimensional, simplicial complex. Show that this abstract complex can be represented by a Euclidean complex in $\mathbb{R}^2$.

16.8. Let $a_0 : (0, 0)$, $a_1 : (-1, 0)$, $a_2 : (1, 0)$, $a_3 : (0, 1)$, $a_4 : (0, 2)$ be points in $\mathbb{R}^2$. A set of Euclidean simplexes consists of those listed in each part (a), $\cdots$, (d) below and all of their sides. In each part determine whether the set of simplexes forms (i) a Euclidean complex, and (ii) a homogeneous complex. If the set is a homogeneous complex, find its dimension. If the set is a complex, explain how to convert it into an oriented complex. In each case, draw a sketch.

(a) $[a_1, a_2, a_3]$, $[a_3, a_4]$.

(b) $[a_0, a_1, a_3]$, $[a_0, a_2, a_3]$, $[a_0, a_4]$.

(c) $[a_1, a_2, a_3]$, $[a_0, a_3]$, $[a_3, a_4]$.
(d) $[a_0, a_1]$, $[a_0, a_2]$, $[a_0, a_3]$, $[a_1, a_3]$, $[a_2, a_3]$, $[a_3, a_4]$.

16.9. Let A be a square in $\mathbb{R}^2$. Divide A into n^2 equal subsquares by lines parallel to the sides of the square, and divide each subsquare into two triangles by a diagonal of the square. Make a sketch of this subdivision of the square.

(a) Show that each triangle is a two-dimensional Euclidean simplex.
(b) Prove that the set consisting of the $2n^2$ two-dimensional triangles and all of their sides forms a Euclidean, homogeneous, two-dimensional, simplicial complex K. Does this result depend on which diagonal is drawn in each subsquare to divide it into triangles?
(c) Count the number of simplexes of dimensions 2, 1, 0, -1 in the complex K in (b).
(d) Find the dimension of the space $C_2(K)$ of 2-chains in the complex K in (b).
(e) Is the positive orientation in $\mathbb{R}^2$ defined for each of the 2-simplexes in K? If this orientation is defined, explain how to give each 2-simplex in K the positive orientation in $\mathbb{R}^2$.

16.10. (a) Form abstract two-dimensional simplexes in all possible ways from the eight points $(\pm 1, \pm 2)$, $(\pm 2, \pm 1)$, and form an abstract, homogeneous, two-dimensional complex K from these simplexes and all of their sides.
(b) Prove that the set of simplexes specified in (a) really is an abstract, homogeneous, two-dimensional complex. If the simplexes and their sides are taken as Euclidean simplexes, is K a Euclidean complex? Explain by making a sketch.
(c) Count the number of simplexes in K of each dimension 2, 1, 0, -1.
(d) Find the dimension of the linear space of two-dimensional chains in K.

16.11. Let $a_0 : (0, 0, 0)$, $a_1 : (1, 0, 0)$, $a_2 : (0, 1, 0)$, $a_3 : (-1, 0, 0)$, $a_4 : (0, -1, 0)$, and $a_5 : (0, 0, 1)$ be points in $\mathbb{R}^3$.

(a) Form a set of Euclidean simplexes which consists of $[a_0, a_1, a_2, a_5]$, $[a_0, a_2, a_3, a_5]$, $[a_0, a_3, a_4, a_5]$, $[a_0, a_4, a_1, a_5]$ and all of their sides. Make a sketch which shows these simplexes.
(b) Count the number of simplexes of each dimension in this set.
(c) Show that these simplexes form a Euclidean, homogeneous, 3-dimensional, simplicial complex K. Explain how to orient the simplexes so that K is an oriented complex.
(d) Show that the 3-chains in the oriented complex K form a 4-dimensional linear space $C_3(K)$.
(e) Show that, in K, the 2-dimensional simplexes and their sides form a Euclidean, homogeneous, 2-dimensional, simplicial complex, and that the 2-chains in this complex form a 12-dimensional linear space $C_2(K)$.
(f) Show that each 3-simplex in (a), oriented as shown, is positively oriented in $\mathbb{R}^3$. [Hint. Theorem 15.9.] In a right-hand coordinate system, if the right hand is placed at the origin so that the fingers point in the direction which rotates the positive x^1-axis into the positive x^2-axis, then the thumb points in the positive direction on the x^3-axis. Show that the orientation in the four simplexes in (a) follows the right-hand rule; that is, if the curved fingers of the right hand point in the direction of the circuit specified by the first three points, then the thumb points to the fourth point.

17. Boundaries of Simplexes and Chains

There are two kinds of simplexes, Euclidean and abstract, and there are two kinds of boundaries, topological and algebraic. The topological boundary of a simplex is the topological boundary of the set of points which constitute the simplex. Since there is no set of points associated with an abstract simplex $\{p_0, p_1, \cdots, p_m\}$ other than the points $p_0, p_1, \cdots, p_m$ themselves, an abstract simplex does not have a topological boundary.

The topological boundary of a set E in $\mathbb{R}^n$ is defined in Definition 92.7 in Appendix 2. If $\mathrm{cl}(E)$ denotes the closure of E (the union of E and the set E' of its limit points), and if $C(E)$ denotes the complement of E, then the boundary of E is $\mathrm{cl}(E) \cap \mathrm{cl}(C(E))$.

17.1 Theorem. *If $[a_0, a_1, \cdots, a_n]$ is an n-dimensional Euclidean simplex in $\mathbb{R}^n$, then the topological boundary of this simplex is the set*

$$\bigcup_{r=0}^{n} [a_0, a_1, \cdots, \widehat{a_r}, \cdots, a_n]. \tag{1}$$

PROOF. Since $[a_0, a_1, \cdots, a_n]$ is the intersection of closed half-spaces, it is a closed set [see Exercise 14.7], and the closure of $[a_0, a_1, \cdots, a_n]$ is $[a_0, a_1, \cdots, a_n]$ itself. The complement of $[a_0, a_1, \cdots, a_n]$ is an open set whose closure contains the points in the $(n-1)$-dimensional sides $[a_0, a_1, \cdots, \widehat{a_r}, \cdots, a_n]$, $r = 0, 1, \cdots, n$, of $[a_0, a_1, \cdots, a_n]$. Then, by definition, the boundary of $[a_0, a_1, \cdots, a_n]$ is the set in (1). □

If $[a_0, a_1, \cdots, a_m]$ is a Euclidean simplex in $\mathbb{R}^n$, $m < n$, then every point in $[a_0, a_1, \cdots, a_m]$ is in the boundary of this simplex since the closure of $[a_0, a_1, \cdots, a_m]$ is $[a_0, a_1, \cdots, a_m]$ and the closure of the complement of $[a_0, a_1, \cdots, a_m]$ is $\mathbb{R}^n$. Thus the topological boundary of $[a_0, a_1, \cdots, a_m]$, $m < n$, is the set of all points in the simplex. The topological boundary of the simplex does not play a significant role in this case.

As for the algebraic boundary of a simplex, the first problem is to decide on a definition. The definition should be one which leads to an interesting theory which has significant applications in the study of complexes, and especially in the development of the theory of integration. If s is an oriented simplex, then the symbol ∂s denotes the algebraic boundary of s.

17.2 Example. If $\partial[x_{i-1}, x_i]$ is defined to be $(x_i) - (x_{i-1})$ in Exercise 15.6, then the following equations describe operations on chains [see Section 16], and they seem to be related in a significant way to the proof of the fundamental theorem of the integral calculus in that exercise.

$$\partial \sum_{i=1}^{n} [x_{i-1}, x_i] = \sum_{i=1}^{n} \partial[x_{i-1}, x_i] = \sum_{i=1}^{n} [(x_i) - (x_{i-1})] = (x_n) - (x_0). \tag{2}$$

In Example 16.8 [see also Figure 16.1(b)], assume that

$$\begin{aligned} \partial[a_0, a_1, a_2] &= [a_1, a_2] - [a_0, a_2] + [a_0, a_1], \\ \partial[a_0, a_2, a_3] &= [a_2, a_3] - [a_0, a_3] + [a_0, a_2], \\ \partial[a_0, a_3, a_1] &= [a_3, a_1] - [a_0, a_1] + [a_0, a_3], \end{aligned} \tag{3}$$

and that

$$\begin{aligned} &\partial\{[a_0, a_1, a_2] + [a_0, a_2, a_3] + [a_0, a_3, a_1]\} \\ &\qquad = \partial[a_0, a_1, a_2] + \partial[a_0, a_2, a_3] + \partial[a_0, a_3, a_1]. \end{aligned} \tag{4}$$

Then these equations and the calculations in Example 16.8 show that

$$\begin{aligned} &\partial\{[a_0, a_1, a_2] + [a_0, a_2, a_3] + [a_0, a_3, a_1]\} \\ &\qquad = [a_1, a_2] + [a_2, a_3] + [a_3, a_1]. \end{aligned} \tag{5}$$

In these calculations all simplexes are oriented according to the ordering of their vertices as shown. Observe the cancellations which occur in the reduction of (4) to (5); examine the geometrical meaning of (5) in Figure 16.1(b). These examples suggest that the algebraic boundary of a simplex $[a_0, a_1, \cdots, a_n]$ is a chain of the simplexes which form the $(n-1)$-dimensional sides of $[a_0, a_1, \cdots, a_n]$. Thus the examples suggest that the boundary of the simplex in the first column below is a chain of the simplexes in the second column, but it is not yet clear how the coefficients in the chain are determined.

n-simplex	$(n-1)$-dimensional sides
$[a_0, a_1]$	$[a_1], [a_0]$
$[a_0, a_1, a_2]$	$[a_1, a_2], [a_0, a_2], [a_0, a_1]$
$[a_0, a_1, a_2, a_3]$	$[a_1, a_2, a_3], [a_0, a_2, a_3], [a_0, a_1, a_3], [a_0, a_1, a_2]$
....	....
$[a_0, a_1, \cdots, a_r, \cdots, a_n]$	$\{[a_0, a_1, \cdots, \widehat{a_r}, \cdots, a_n] : r = 0, 1, \cdots, n\}$
....	....

(6)

17.3 Example. There is an intimate relationship between Euclidean simplexes and determinants, and determinants suggest how to define the chains which form the boundary of a simplex. Let $[a_0, a_1, \cdots, a_n]$ be a simplex whose vertices are the points $a_i : (a_i^1 \cdots, a_i^n)$, $i = 0, 1, \cdots, n$, in $\mathbb{R}^n$. As convenient notation, set

$$D(a_0, a_1, \cdots, a_n, 1) = \begin{vmatrix} a_0^1 & a_0^2 & \cdots & a_0^n & 1 \\ a_1^1 & a_1^2 & \cdots & a_1^n & 1 \\ \cdots & \cdots & \cdots & \cdots & \cdots \\ a_r^1 & a_r^2 & \cdots & a_r^n & 1 \\ \cdots & \cdots & \cdots & \cdots & \cdots \\ a_n^1 & a_n^2 & \cdots & a_n^n & 1 \end{vmatrix},$$

$$D(a_0, a_1, \cdots, \widehat{a}_r, \cdots, a_n) = \begin{vmatrix} a_0^1 & a_0^2 & \cdots & a_0^n \\ a_1^1 & a_1^2 & \cdots & a_1^n \\ \cdots & \cdots & \cdots & \cdots \\ \widehat{a}_r^1 & \widehat{a}_r^2 & \cdots & \widehat{a}_r^n \\ \cdots & \cdots & \cdots & \cdots \\ a_n^1 & a_n^2 & \cdots & a_n^n \end{vmatrix}. \tag{7}$$

The simplex $[a_0, a_1, \cdots, a_n]$ is positively oriented in $\mathbb{R}^n$ if and only if $(-1)^n D(a_0, a_1, \cdots, a_n; 1) > 0$ [see Theorem 15.9]. Expand the determinant $D(a_0, a_1, \cdots, a_n; 1)$ by minors of the elements in the column of 1's in its matrix; then

$$(-1)^n D(a_0, a_1, \cdots, a_n; 1) = (-1)^{2n+2} \sum_{r=0}^{n} (-1)^r D(a_0, a_1, \cdots, \widehat{a}_r, \cdots, a_n). \tag{8}$$

The terms in the sum on the right are functions of the oriented $(n-1)$-dimensional sides $[a_0, a_1, \cdots, \widehat{a}_r, \cdots, a_n]$ of $[a_0, a_1, \cdots, a_n]$, and these terms suggest the definition of the algebraic boundary of $[a_0, a_1, \cdots, a_n]$. Observe the pattern in the following array.

$$\begin{aligned} (-1)D(a_0, a_1; 1) &= D(a_1) - D(a_0) \\ (-)^2 D(a_0, a_1, a_2; 1) &= D(a_1, a_2) - D(a_0, a_2) + D(a_0, a_1) \\ (-1)^3 D(a_0, a_1, \cdots, a_3; 1) &= \\ D(a_1, a_2, a_3) - D(a_0, a_2, a_3) &+ D(a_0, a_1, a_3) - D(a_0, a_1, a_2) \end{aligned} \tag{9}$$

$$\cdots\cdots\cdots\cdots\cdots\cdots\cdots\cdots\cdots\cdots\cdots\cdots$$

Equations (8) and (9) prove nothing—you do not *prove* definitions—but they *suggest* that it might be profitable to *define* the algebraic boundary as follows.

$$\begin{aligned} \partial[a_0, a_1] &= [a_1] - [a_0] \\ \partial[a_0, a_1, a_2] &= [a_1, a_2] - [a_0, a_2] + [a_0, a_1] \\ \partial[a_0, a_1, \cdots, a_3] &= [a_1, a_2, a_3] - [a_0, a_2, a_3] + [a_0, a_1, a_3] - [a_0, a_1, a_2] \end{aligned}$$

$$\cdots\cdots\cdots\cdots\cdots\cdots\cdots\cdots\cdots\cdots\cdots\cdots$$

$$\partial[a_0, a_1, \cdots, a_m] = \sum_{r=0}^{m} (-1)^r [a_0, a_1, \cdots, \widehat{a}_r, \cdots, a_m] \tag{10}$$

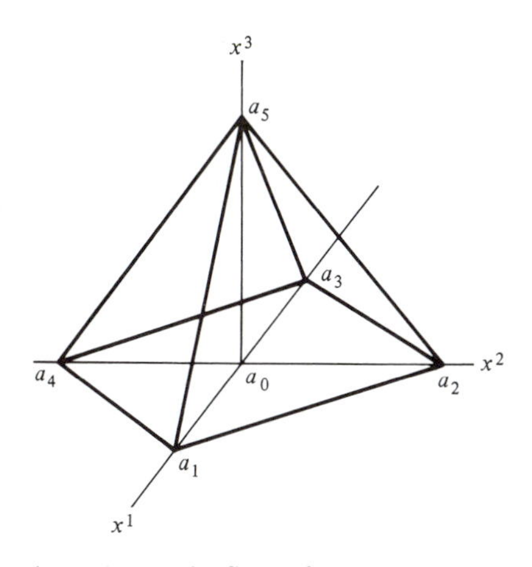

Figure 17.1. The figure for Example 17.6.

Our conviction that the definitions in (10) are appropriate ones is strengthened by the fact that they agree with the suggestions in equations (2) and (3) in Example 17.2.

17.4 Definition. If s is the simplex $(p_0, p_1, \cdots, p_m)$ in an oriented, Euclidean or abstract, homogeneous, m-dimensional, simplicial complex K, then the *boundary* ∂s *of* s is the $(m-1)$-dimensional chain in $C_{(m-1)}(K)$ defined by the following equation.

$$\begin{aligned}\partial s &= \sum_{i=0}^{m} (-1)^i (p_0, p_1, \cdots, p_{i-1}, p_{i+1}, \cdots, p_m) \\ &= \sum_{i=0}^{m} (-1)^i (p_0, p_1, \cdots, \widehat{p_i}, \cdots, p_m).\end{aligned} \tag{11}$$

17.5 Definition. Let K be an oriented, Euclidean or abstract, homogeneous, m-dimensional, simplicial complex, and let $\Sigma_1^k a_i s_i$ be an m-chain in $C_m(K)$. The *boundary* $\partial(\Sigma_1^k a_i s_i)$ of $\Sigma_1^k a_i s_i$ is the $(m-1)$-dimensional chain in $C_{(m-1)}(K)$ defined by the following equation:

$$\partial\left(\sum_{i=1}^{k} a_i s_i\right) = \sum_{i=1}^{k} a_i \partial s_i. \tag{12}$$

The definition in (11) is easily remembered. Each vertex p_i in turn is moved to the left-most position, with an appropriate change in sign to maintain the orientation, and then dropped: $(p_0, p_1, \cdots, p_m) = (-1)^i (p_i, p_0, \cdots, p_{i-1}, p_{i+1}, \cdots, p_m)$; and the boundary is the sum of all terms $(-1)^i(p_0, p_1, \cdots, p_{i-1}, p_{i+1}, \cdots, p_m)$, or $(-1)^i(p_0, p_1, \cdots, \widehat{p_i}, \cdots, p_m)$.

17.6 Example. The 3-dimensional simplexes $s_1 : [a_0, a_1, a_2, a_5]$, $s_2 : [a_0, a_2, a_3, a_5]$, $s_3 : [a_0, a_3, a_4, a_5]$, and $s_4 : [a_0, a_4, a_1, a_5]$ in Exercise 16.11 fit together to make a pyramid with a square base as shown in Figure 17.1.

Observe that the four simplexes have been oriented so that each one is positively oriented in $\mathbb{R}^3$. By Definition 17.5, $\partial(\Sigma_1^4 s_i) = \Sigma_1^4 \partial s_i$, and by Definition 17.4,

$$\begin{aligned}
\partial s_1 &= [a_1, a_2, a_5] - [a_0, a_2, a_5] + [a_0, a_1, a_5] - [a_0, a_1, a_2],\\
\partial s_2 &= [a_2, a_3, a_5] - [a_0, a_3, a_5] + [a_0, a_2, a_5] - [a_0, a_2, a_3],\\
\partial s_3 &= [a_3, a_4, a_5] - [a_0, a_4, a_5] + [a_0, a_3, a_5] - [a_0, a_3, a_4],\\
\partial s_4 &= [a_4, a_1, a_5] - [a_0, a_1, a_5] + [a_0, a_4, a_5] - [a_0, a_4, a_1].
\end{aligned} \tag{13}$$

Then

$$\begin{aligned}
\partial(s_1 + s_2 + s_3 + s_4) &= \partial s_1 + \partial s_2 + \partial s_3 + \partial s_4\\
&= [a_1, a_2, a_5] - [a_0, a_1, a_2] + [a_2, a_3, a_5]\\
&\quad - [a_0, a_2, a_3] + [a_3, a_4, a_5] - [a_0, a_3, a_4]\\
&\quad + [a_4, a_1, a_5] - [a_0, a_4, a_1].
\end{aligned} \tag{14}$$

In $\Sigma_1^4 \partial s_i$, the terms in the second and third columns on the right in (13) cancel, and $\partial(\Sigma_1^4 s_i)$ is the sum of the remaining eight terms. Two terms which cancel arise from the common side of two of the 3-dimensional simplexes. For example, $-[a_0, a_2, a_5]$ is a 2-dimensional side of s_1, and $[a_0, a_2, a_5]$ is also a 2-dimensional side of s_2; they cancel. Similarly, the other pairs which cancel arise from a common side of two simplexes. Each of the 2-dimensional simplexes which remain in $\partial(\Sigma_1^4 s_i)$ belongs to a single one of the simplexes $s_1, \cdots, s_4$. Each of the 2-dimensional simplexes in $\partial(\Sigma_1^4 s_i)$, considered without regard to its orientation, is a set of points; the union of the eight simplexes in $\partial(\Sigma_1^4 s_i)$ is the topological boundary of the union of the sets $s_1, \cdots, s_4$.

17.7 Example. The boundary of the chain $\Sigma_1^4 s_i$ in Example 17.6 is the 2-chain $\Sigma_1^4 \partial s_i$; thus it has a boundary by Definition 17.5, and

$$\partial \sum_{i=1}^{4} \partial s_i = \sum_{i=1}^{4} \partial\partial s_i. \tag{15}$$

Compute $\partial\partial s_1$ by computing the boundary of the chain in the first equation in (13). Thus

$$\begin{aligned}
\partial\partial s_1 = \quad & * & [a_2, a_5] & - [a_1, a_5] & + [a_1, a_2]\\
& - [a_2, a_5] & * & + [a_0, a_5] & - [a_0, a_2]\\
& + [a_1, a_5] & - [a_0, a_5] & * & + [a_0, a_1]\\
& - [a_1, a_2] & + [a_0, a_2] & - [a_0, a_1] & *
\end{aligned} \tag{16}$$

All terms cancel by pairs; thus $\partial\partial s_1 = 0$, and similar calculations show that $\partial\partial s_i = 0$ for $i = 1, 2, 3, 4$. Then (15) shows that

$$\partial\partial\left(\sum_{i=1}^{4} s_i\right) = \partial\left(\sum_{i=1}^{4} \partial s_i\right) = \sum_{i=1}^{4} \partial\partial s_i = \sum_{i=1}^{4} 0 = 0. \tag{17}$$

Geometrically, the eight 2-dimensional Euclidean simplexes in $\partial(\Sigma_1^4 s_i)$ fit together to form a closed surface, the bounding surface of the square pyramid. From a geometric point of view, this surface has no edges or boundary. The eight oriented 2-dimensional simplexes in $\partial(\Sigma_1^4 s_i)$ have oriented one-dimensional sides, each of which belongs to two 2-simplexes. These common one-dimensional sides have opposite orientations in the two 2-simplexes to which they belong, and they cancel in $\partial(\Sigma_1^4 \partial s_i)$. Thus the algebraic boundary of $\partial(\Sigma_1^4 s_i)$ is the zero chain. The geometric boundary of the surface formed by the eight 2-simplexes in $\partial(\Sigma_1^4 s_i)$ is empty; the algebraic boundary of the chain $\partial(\Sigma_1^4 s_i)$ is the zero chain. The two concepts are different, but their statements sound similar.

17.8 Theorem. *Let K be an oriented, Euclidean or abstract, homogeneous, m-dimensional, simplicial complex. If s is a simplex in K, then* $\partial\partial s = 0$. *If* $\Sigma_1^k a_i s_i$ *is an m-chain in* $C_m(K)$, *then*

$$\partial\partial\left(\sum_{i=1}^{k} a_i s_i\right) = \partial\left(\sum_{i=1}^{k} a_i \partial s_i\right) = \sum_{i=1}^{k} a_i \partial\partial s_i = 0. \tag{18}$$

Proof. Equation (18) follows from Definition 17.5 and $\partial\partial s = 0$. Thus the proof of the theorem can be completed by proving that $\partial\partial s = 0$.

Let s be the simplex $(p_0, p_1, \cdots, p_m)$. The term $(p_0, p_1, \cdots, \widehat{p_i}, \cdots, \widehat{p_j}, \cdots, p_m)$ occurs twice in $\partial\partial s$ (the notation has been chosen so that $i < j$). It occurs once when p_i is dropped in computing ∂s and then p_j is dropped in computing $\partial(\partial s)$. The same term arises from dropping p_j in the computation of ∂s and then p_i in the computation of $\partial(\partial s)$. In the first case, dropping p_i gives the term $(-1)^i\,(p_0, p_1, \cdots, \widehat{p_i}, \cdots, p_j, \cdots, p_m)$ in ∂s; dropping p_j then gives the term

$$(-1)^i\,(-1)^{j-1}\,(p_0, p_1, \cdots, \widehat{p_i}, \cdots, \widehat{p_j}, \cdots, p_m) \tag{19}$$

in $\partial(\partial s)$. In the second case, dropping p_j gives the term $(-1)^j\,(p_0, p_1, \cdots, p_i, \cdots, \widehat{p_j}, \cdots, p_m)$; dropping p_i then gives the term

$$(-1)^i\,(-1)^j\,(p_0, p_1, \cdots, \widehat{p_i}, \cdots, \widehat{p_j}, \cdots, p_m) \tag{20}$$

in $\partial(\partial s)$. The terms in (19) and (20) have opposite signs, and they cancel in $\partial\partial s$. Thus all terms in $\partial\partial s$ cancel by pairs, and $\partial\partial s = 0$. □

17.9 Example. The chain $\Sigma_1^4 s_i$ in Example 17.6 is special in some respects because each of the simplexes $s_1, \cdots, s_4$ is positively oriented in $\mathbb{R}^3$. Each chain has a boundary, and it is instructive to examine the boundary of $-s_1 + s_2 + s_3 + s_4$. The boundary $\partial(-s_1)$ of $-s_1$ can be found by changing all signs in the first equation in (13). The cancellations and the final result are now quite different from those in $\partial(\Sigma_1^4 s_i)$. The new set of equations (13) lead to the following result:

$$\begin{aligned}\partial(-s_1+s_2+s_3+s_4) &= \partial(-s_1)+\partial(s_2)+\partial(s_3)+\partial(s_4)\\ &= 2[a_0,a_2,a_5]-2[a_0,a_1,a_5]-[a_1,a_2,a_5]\\ &\quad +[a_0,a_1,a_2]+[a_2,a_3,a_5]-[a_0,a_2,a_3]\\ &\quad +[a_3,a_4,a_5]-[a_0,a_3,a_4]+[a_4,a_1,a_5]\\ &\quad -[a_0,a_4,a_1].\end{aligned} \tag{21}$$

The terms $2[a_0, a_2, a_5]$ and $-2[a_0, a_1, a_5]$ do not appear in (14), and the coefficients of $[a_1, a_2, a_5]$ and $[a_0, a_1, a_2]$ have opposite signs in (14) and (21). The simplexes $[a_0, a_2, a_5]$ and $[a_0, a_1, a_5]$ which appear in the chain in (21) are not, as sets of points, contained in the topological boundary of $s_1 \cup s_2 \cup s_3 \cup s_4$. This example emphasizes that the special relation which exists in Example 17.6 between the algebraic boundary of the chain $\Sigma_1^4 s_i$ and the topological boundary of the union $\bigcup_1^4 s_i$ does not occur in the general case.

Exercises

17.1. Let $a_0, a_1, \cdots, a_4$ be the points defined in Example 16.2. Let K be the oriented Euclidean complex which consists of the simplexes $s_1 : [a_0, a_1, a_2]$, $s_2 : [a_0, a_2, a_3]$, $s_3 : [a_0, a_3, a_4]$, $s_4 : [a_0, a_4, a_1]$, and all of their sides.

(a) Show that each of the simplexes $s_1, \cdots, s_4$ is positively oriented in $\mathbb{R}^2$.

(b) Find $\partial(\Sigma_1^4 s_i)$, and show that each Euclidean simplex, as a set of points, in this chain is contained in the topological boundary of the square whose vertices are $a_1, \cdots, a_4$.

(c) Find $\partial\partial(\Sigma_1^4 s_i)$.

(d) Find $\partial(s_1 + s_2 - s_3 - s_4)$ and $\partial\partial(s_1 + s_2 - s_3 - s_4)$.

17.2. Draw a sketch of the simplexes $[a_0, a_1, a_2]$, $[a_0, a_2, a_3]$, and $[a_0, a_3, a_1]$ in Example 17.2. Use a curved arrow to indicate the positive orientation of each simplex. Use this sketch to show the cancellation of one-simplexes which occurs in the calculation of the algebraic boundary of the 2-chain $[a_0, a_1, a_2] + [a_0, a_2, a_3] + [a_0, a_3, a_1]$.

17.3. Let s_i, $i = 1, \cdots, 4$, be the simplexes defined in Example 17.6. Find $\partial(s_1 - s_2 - s_3 + s_4)$. For each of the simplexes in this boundary, explain why it occurs and does not cancel in the calculation.

17.4. The four points $a_1 : (1, 1)$, $a_2 : (-1, 1)$, $a_3 : (-1, -1)$, $a_4 : (1, -1)$ are vertices of a square in $\mathbb{R}^2$. The simplexes $s_1 : [a_1, a_2, a_3]$, $s_2 : [a_2, a_3, a_4]$, $s_3 : [a_3, a_4, a_1]$, $s_4 : [a_4, a_1, a_2]$ and all of their sides are an oriented, homogeneous, two-dimensional complex K.

(a) Make a sketch and show that each of the simplexes $s_1, s_2, \cdots, s_4$ is positively oriented in $\mathbb{R}^2$.

(b) Is K a Euclidean complex? Give reasons for your answer.

(c) Find $\partial(\Sigma_1^4 s_i)$, and compare the simplexes in this boundary with the topological boundary of the square.

(d) Find $\partial\partial(\Sigma_1^4 s_i)$.

17.5. Let $a_1, a_2, \cdots, a_5$ be five equally spaced points which are arranged in order, counterclockwise, around a circle in $\mathbb{R}^2$. Select three points from $\{a_1, a_2, \cdots, a_5\}$ and order them to form a Euclidean simplex $[a_i, a_j, a_k]$ which is positively oriented in $\mathbb{R}^2$. Select the three points in all possible ways; the resulting two-dimensional simplexes and all of their sides form an oriented complex K.

(a) Show that K contains ten 2-dimensional simplexes $s_1 : [a_1, a_2, a_3], \cdots, s_{10} : [a_5, a_2, a_3]$. Find the remaining 2-simplexes in K.

(b) Is K a Euclidean complex? Give reasons for your answer.

(c) Use the definition of the algebraic boundary to calculate $\partial(\Sigma_1^{10} s_i)$. Then draw a sketch and use a geometric argument to obtain the same result.

(d) Show in two ways that $\partial\partial(\Sigma_1^{10} s_i) = 0$. [Hint. First, $\partial\partial(\Sigma_1^{10} s_i) = \Sigma_1^{10} \partial\partial s_i = 0$. Second, $\partial\partial(\Sigma_1^{10} s_i) = \partial(\Sigma_1^{10} \partial s_i)$. Use your sketch to show that the boundary of the one-chain $\Sigma_1^{10} \partial s_i$ is the zero chain.]

17.6. Repeat Exercise 17.5 for n equally spaced points $a_1, a_2, \cdots, a_n$ arranged counterclockwise around a circle in $\mathbb{R}^2$.

18. Boundaries in a Euclidean Complex

The simplexes in a Euclidean complex are Euclidean simplexes; they are thus sets of points $[a_0, a_1, \cdots, a_n]$ and an associated orientation $(a_0, a_1, \cdots, a_n)$. The symbol $[a_0, a_1, \cdots, a_n]$ emphasizes the set of points, and the symbol $(a_0, a_1, \cdots, a_n)$ emphasizes the orientation; in most cases, however, the symbol $(a_0, a_1, \cdots, a_n)$ will indicate both the set and the orientation of a Euclidean simplex.

Some of the examples in earlier sections have illustrated the fact that special relations may exist between algebraic boundaries and topological boundaries in a Euclidean complex. In some cases the Euclidean simplexes (as sets of points) in the chain $\partial(\Sigma_1^k s_i)$ are contained in the topological boundary of the set $\bigcup_1^k s_i$. This section establishes the conditions under which this relation exists in $\mathbb{R}^n$. Several examples illustrate the relationship under study.

18.1 Example. The proof of the fundamental theorem of the integral calculus [see Exercise 15.6 and equation (2) in Example 17.2] contains the following calculation of the boundary of a chain.

$$\partial \sum_{i=1}^{n} (x_{i-1}, x_i) = \sum_{i=1}^{n} \partial(x_{i-1}, x_i) = \sum_{i=1}^{n} [(x_i) - (x_{i-1})] = (x_n) - (x_0). \quad (1)$$

For $i = 1, \cdots, n-1$, the boundary simplex (x_i) occurs twice, once as (x_i) in $\partial(x_{i-1}, x_i)$ and once as $-(x_i)$ in $\partial(x_i, x_{i+1})$; these cancel and (x_i) does

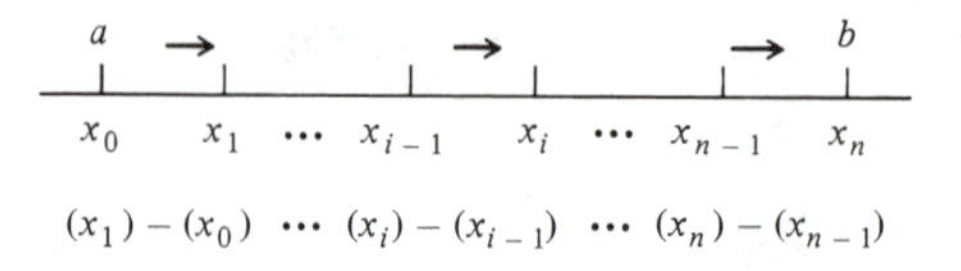

Figure 18.1. The boundary of a one-chain in a Euclidean complex in $\mathbb{R}$.

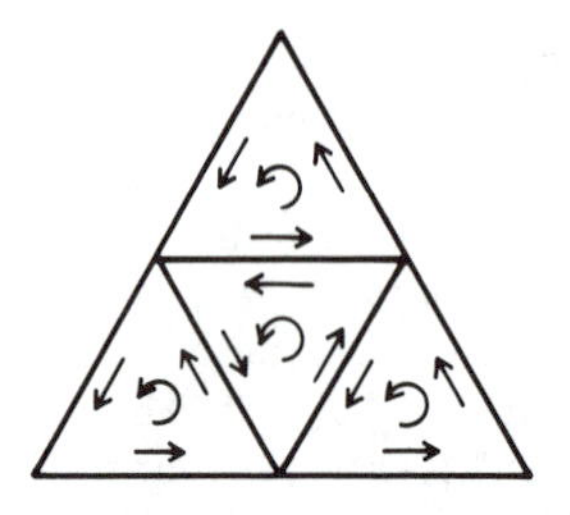

Figure 18.2. A chain of Euclidean 2-simplexes in $\mathbb{R}^2$.

not occur in the boundary $(x_n)-(x_0)$ of the chain [see Figure 18.1]. The union $\bigcup_1^n[x_{i-1}, x_i]$ of the simplexes in the chain is the interval $[a, b]$, and the topological boundary of $[a, b]$ is the set $\{a, b\}$. Since $x_0 = a$ and $x_n = b$, the union of the Euclidean simplexes in $\partial\Sigma_1^n(x_{i-1}, x_i)$ is the topological boundary of $\bigcup_1^n[x_{i-1}, x_i]$.

18.2 Example. Figure 18.2 contains a second example. The 2-simplexes are positively oriented in $\mathbb{R}^2$. A side which belongs to two of the 2-simplexes has opposite orientations (as indicated by the arrows) in the two 2-simplexes, and the side cancels in finding the boundary of the chain of 2-simplexes. The union of the six one-simplexes in the boundary of the chain is the topological boundary of the union of the four Euclidean 2-simplexes.

18.3 Example. As a third example, consider Example 17.6 again. The four 3-simplexes $s_1, \cdots, s_4$ in that example are positively oriented in $\mathbb{R}^3$. If two of these 3-simplexes have a 2-simplex as a side in common, this side has opposite orientations in the boundaries of the two 3-simplexes [see equations (13) in Section 17]. Thus the 2-simplexes which arise from this common side cancel in $\partial(\Sigma_1^4 s_i)$. As stated in that example, each of the 2-dimensional simplexes in $\partial(\Sigma_1^4 s_i)$, considered without regard to its orientation, is a set of points; the union of the eight 2-simplexes in $\partial(\Sigma_1^4 s_i)$ is the topological boundary of the union of the sets $s_1, \cdots, s_4$. Example 17.9 shows that this result depends on the orientation of $s_1, \cdots, s_4$.

18.4 Theorem. *Let K be an oriented, homogeneous, n-dimensional, Euclidean, simplicial complex in $\mathbb{R}^n$, and let s_1 and s_2 be two n-dimensional simplexes in K which have an $(n-1)$-dimensional side in common. Then s_1 lies in one of the closed half-spaces bounded by the plane which contains the common side of s_1 and s_2, and s_2 lies in the other half-space bounded by this plane.*

PROOF. Assume that the vertices in the common side of s_1 and s_2 are $a_1, \cdots, a_n$, and that the other vertices of s_1 and s_2 are a_0 and a_0' respectively. Since $s_1 : (a_0, a_1, \cdots, a_n)$ and $s_2 : (a_0', a_1, \cdots, a_n)$ are Euclidean simplexes, then by Definition 14.10 and Theorem 13.1,

$$\Delta(a_0, a_1, \cdots, a_n) \neq 0, \qquad \Delta(a_0', a_1, \cdots, a_n) \neq 0. \tag{2}$$

The points $a_1, \cdots, a_n$ determine a plane $P(a_1, \cdots, a_n)$, and (15) in Section 14 shows that an equation of $P(a_1, \cdots, a_n)$ is

$$\Delta(x, a_1, \cdots, a_n) = 0. \tag{3}$$

There is no loss of generality in assuming that the vertices have been numbered so that

$$\Delta(a_0, a_1, \cdots, a_n) > 0. \tag{4}$$

Let H_0 and H_0' be the two half-spaces bounded by $P(a_0, a_1, \cdots, a_n)$ as follows:

$$H_0 = \{x : \Delta(x, a_1, \cdots, a_n) \geqq 0\}, \tag{5}$$

$$H_0' = \{x : \Delta(x, a_1, \cdots, a_n) \leqq 0\}. \tag{6}$$

Because (3) is an equation of $P(a_1, \cdots, a_n)$, equation (2) shows that neither a_0 nor a_0' is in $P(a_1, \cdots, a_n)$. Furthermore, (4) and (5) show that a_0 is in H_0; therefore, $(a_0, a_1, \cdots, a_n) \subset H_0$ by Theorem 14.11. To complete the proof of the theorem, it is necessary to show that $(a_0', a_1, \cdots, a_n) \subset H_0'$, for then s_1 lies in one of the half-spaces bounded by $P(a_1, \cdots, a_n)$ and s_2 lies in the other.

If a_0' is in H_0, then $(a_0', a_1, \cdots, a_n)$ is in H_0 by Theorem 14.11. By the same argument, if a_0' is in H_0', then $(a_0', a_1, \cdots, a_n)$ is in H_0'. Thus to complete the proof, it is sufficient to prove that a_0' is not contained in H_0. Assume, then, that a_0' is in H_0 and show that a contradiction follows. If a_0 and a_0' are in H_0, then as already stated, $(a_0, a_1, \cdots, a_n)$ and $(a_0', a_1, \cdots, a_n)$ are in H_0. The point

$$b = (1/n)a_1 + \cdots + (1/n)a_n, \tag{7}$$

whose barycentric coordinates are $(1/n, \cdots, 1/n)$ with respect to $a_1, \cdots, a_n$, is a point in the common side $(a_1, \cdots, a_n)$ of s_1 and s_2. The equations of the sides of $(a_0, a_1, \cdots, a_n)$ and $(a_0', a_1, \cdots, a_n)$, other than the common side $(a_1, \cdots, a_n)$, are respectively

$$\Delta(a_0, a_1, \cdots, x/a_r, \cdots, a_n) = 0, \qquad r = 1, \cdots, n, \tag{8}$$

$$\Delta(a_0', a_1, \cdots, x/a_r, \cdots, a_n) = 0, \qquad r = 1, \cdots, n. \tag{9}$$

Now b is not on any of these planes since, by elementary properties of determinants and (2) above,

$$\Delta(a_0, a_1, \cdots, b/a_r, \cdots, a_n) = (1/n)\Delta(a_0, a_1, \cdots, a_r, \cdots, a_n) \neq 0, \tag{10}$$

$$\Delta(a_0', a_1, \cdots, b/a_r, \cdots, a_n) = (1/n)\Delta(a_0', a_1, \cdots, a_r, \cdots, a_n) \neq 0, \tag{11}$$

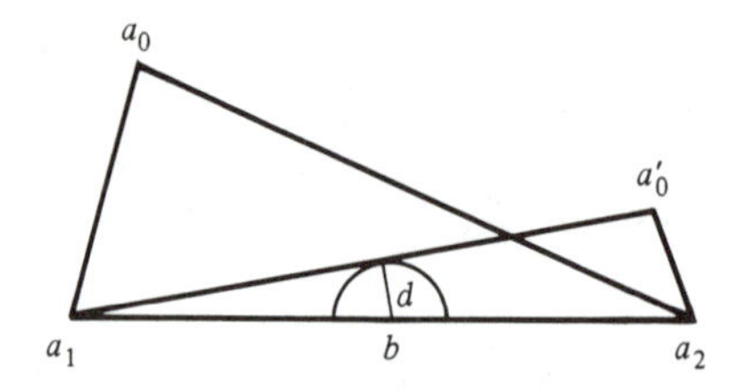

Figure 18.3. $[a_0, a_1, a_2] \cap [a'_0, a_1, a_2]$ contains $N(b, d)$.

for $r = 1, \cdots, n$ in each case. Let d_1 be the minimum of the distances from b to the planes in (10); then $d_1 > 0$ since b is not on any of these planes. Similarly, let d_2 be the minimum of the distances from b to the planes in (11); then $d_2 > 0$ for the same reason. Set $d = \min\{d_1, d_2\}$; then $d > 0$. The definition of d shows that $N(b, d) \cap H_0$ is contained in $[a_0, a_1, \cdots, a_n]$ and in $[a'_0, a_1, \cdots, a_n]$. Figure 18.3 shows the situation in $\mathbb{R}^2$. Since $a_0 \neq a'_0$, the intersection of $[a_0, a_1, \cdots, a_n]$ and $[a'_0, a_1, \cdots, a_n]$ is a proper subset of each Euclidean simplex. Since the intersection of $[a_0, a_1, \cdots, a_n]$ and $[a'_0, a_1, \cdots, a_n]$ contains the n-dimensional set $N(b, d) \cap H_0$, the common side $[a_1, \cdots, a_n]$ of the two simplexes is a proper subset of the intersection of $[a_0, a_1, \cdots, a_n]$ and $[a'_0, a_1, \cdots, a_n]$. Thus the intersection of $[a_0, a_1, \cdots, a_n]$ and $[a'_0, a_1, \cdots, a_n]$ is not a side, proper or improper, of these simplexes. But this statement contradicts the hypothesis that s_1 and s_2 are simplexes in a Euclidean complex. Since the assumption that a'_0 is in H_0 has led to a contradiction, then a'_0 is in H'_0. Thus, as shown above, $s_1 : (a_0, a_1, \cdots, a_n)$ is in H_0 and $s_2 : (a'_0, a_1, \cdots, a_n)$ is in H'_0, and the proof of Theorem 18.4 is complete. □

18.5 Theorem. *Let K be an oriented, homogeneous, n-dimensional, Euclidean, simplicial complex in $\mathbb{R}^n$, and let s_1 and s_2 be two n-dimensional simplexes in K which have an $(n-1)$-dimensional side in common. If s_1 and s_2 have the same orientation (positive or negative) in $\mathbb{R}^n$, then the common side of s_1 and s_2 has opposite orientations in ∂s_1 and ∂s_2.*

PROOF. Assume that the vertices in the common side of s_1 and s_2 are $a_1, \cdots, a_n$, and that the other vertices of s_1 and s_2 are a_0 and a'_0 respectively. Then Theorem 18.4 shows that $(a_0, a_1, \cdots, a_n)$ and $(a'_0, a_1, \cdots, a_n)$ lie in opposite half-spaces bounded by the plane $\Delta(x, a_1, \cdots, a_n) = 0$. Therefore,

$$\Delta(a_0, a_1, \cdots, a_n)\Delta(a'_0, a_1, \cdots, a_n) < 0; \tag{12}$$

if the product (12) were positive, a_0 and a'_0 would belong to the same half-space bounded by $\Delta(x, a_1, \cdots, a_n) = 0$. Thus, (12) and Definition 15.3 show that $(a_0, a_1, \cdots, a_n)$ and $(a'_0, a_1, \cdots, a_n)$ have opposite orientations in $\mathbb{R}^n$, and $(a_0, a_1, \cdots, a_n)$ and $-(a'_0, a_1, \cdots, a_n)$ have the same orientation. But

$$\begin{aligned} \partial(a_0, a_1, \cdots, a_n) &= (a_1, \cdots, a_n) + \cdots, \\ \partial[-(a'_0, a_1, \cdots, a_n)] &= -(a_1, \cdots, a_n) + \cdots. \end{aligned} \tag{13}$$

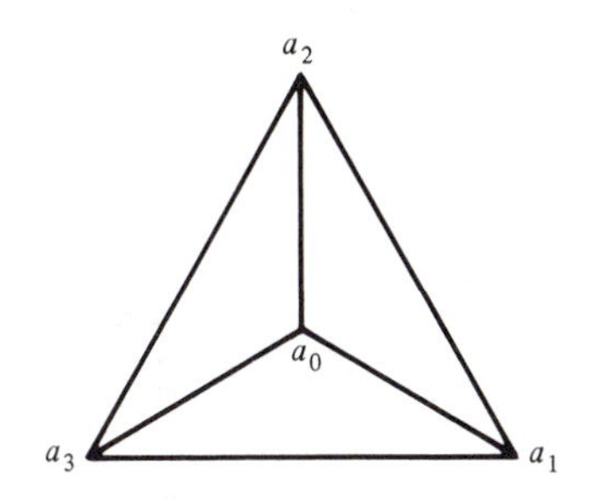

Figure 18.4. A chain of 2-simplexes in $\mathbb{R}^2$.

Hence, if s_1 and s_2 have the same orientation in $\mathbb{R}^n$, then the common side has opposite orientations $(a_1, \cdots, a_n)$ and $-(a_1, \cdots, a_n)$ in ∂s_1 and ∂s_2, respectively. □

18.6 Theorem. *Let K be an oriented, homogeneous, n-dimensional, Euclidean, simplicial complex in $\mathbb{R}^n$, and let $\Sigma_1^k s_i$ be a chain of n-simplexes in K such that the simplexes $s_1, \cdots, s_k$ have the same orientation (positive or negative) in $\mathbb{R}^n$. Then the union of the Euclidean simplexes in $\partial(\Sigma_1^k s_i)$ is the topological boundary of the union $\bigcup_1^k s_i$ of the Euclidean simplexes $s_1, \cdots, s_k$ in $\Sigma_1^k s_i$.*

PROOF. By Definition 17.5, $\partial(\Sigma_1^k s_i) = \Sigma_1^k \partial s_i$. By Definition 17.4, the boundary ∂s_i is a chain of $(n-1)$-dimensional simplexes in the sides of s_i. If two simplexes in $\Sigma_1^k s_i$ have an $(n-1)$-dimensional side in common, then Theorem 18.5 shows that the side has opposite orientations in the boundaries of the two simplexes; hence, it cancels and does not occur in the algebraic boundary of the chain. Furthermore, since K is a Euclidean complex, Theorem 18.4 shows that two simplexes s_i and s_j which have a side in common lie on opposite sides of the plane determined by their common side; therefore, the common side of s_i and s_j is not in the topological boundary of the union $\bigcup_1^k s_i$ of the simplexes $s_1, \cdots, s_k$. Thus, an $(n-1)$-simplex in K belongs to $\partial(\Sigma_1^k s_i)$ if and only if it is a side of a single simplex $s_1, \cdots, s_k$; also, an $(n-1)$-simplex, as a set of points, is contained in the topological boundary of $\bigcup_1^k s_i$ if and only if it is a side of a single simplex $s_1, \cdots, s_k$. Hence, the union of the simplexes in $\partial(\Sigma_1^k s_i)$ is the topological boundary of $\bigcup_1^k s_i$. □

18.7 Example. Theorem 18.6 describes the cancellation which occurs in calculating the boundary of a chain of similarly oriented simplexes in a Euclidean complex. This cancellation has been used already in Exercise 15.6 and Examples 17.2 and 18.1 to prove that $\int_a^b f'(x)\,dx = f(b) - f(a)$, the fundamental theorem of the integral calculus. Another example will suggest how the cancellation in Theorem 18.6 will be used in Chapter 6 to prove more general forms of the fundamental theorem.

Figure 18.4 shows a chain of three positively oriented 2-simplexes (a_0, a_1, a_2), (a_0, a_2, a_3), (a_0, a_3, a_1) in $\mathbb{R}^2$. Let $f: E \to \mathbb{R}^2$ be a function whose components (f^1, f^2) are defined on an open set E which contains $[a_1, a_2, a_3]$. The following formulas are obtained by expanding the determinants by minors of elements in the column of 1's.

$$\Delta f(a_0, a_1, a_2) = \begin{vmatrix} f^1(a_0) & f^2(a_0) & 1 \\ f^1(a_1) & f^2(a_1) & 1 \\ f^1(a_2) & f^2(a_2) & 1 \end{vmatrix}$$

$$= \begin{vmatrix} f^1(a_1) & f^2(a_1) \\ f^1(a_2) & f^2(a_2) \end{vmatrix} - \begin{vmatrix} f^1(a_0) & f^2(a_0) \\ f^1(a_2) & f^2(a_2) \end{vmatrix} + \begin{vmatrix} f^1(a_0) & f^2(a_0) \\ f^1(a_1) & f^2(a_1) \end{vmatrix}.$$

$$\Delta f(a_0, a_2, a_3) = \begin{vmatrix} f^1(a_0) & f^2(a_0) & 1 \\ f^1(a_2) & f^2(a_2) & 1 \\ f^1(a_3) & f^2(a_3) & 1 \end{vmatrix}$$

$$= \begin{vmatrix} f^1(a_2) & f^2(a_2) \\ f^1(a_3) & f^2(a_3) \end{vmatrix} - \begin{vmatrix} f^1(a_0) & f^2(a_0) \\ f^1(a_3) & f^2(a_3) \end{vmatrix} + \begin{vmatrix} f^1(a_0) & f^2(a_0) \\ f^1(a_2) & f^2(a_2) \end{vmatrix}.$$

$$\Delta f(a_0, a_3, a_1) = \begin{vmatrix} f^1(a_0) & f^2(a_0) & 1 \\ f^1(a_3) & f^2(a_3) & 1 \\ f^1(a_1) & f^2(a_1) & 1 \end{vmatrix}$$

$$= \begin{vmatrix} f^1(a_3) & f^2(a_3) \\ f^1(a_1) & f^2(a_1) \end{vmatrix} - \begin{vmatrix} f^1(a_0) & f^2(a_0) \\ f^1(a_1) & f^2(a_1) \end{vmatrix} + \begin{vmatrix} f^1(a_0) & f^2(a_0) \\ f^1(a_3) & f^2(a_3) \end{vmatrix}.$$

Then

$$\Delta f(a_0, a_1, a_2) + \Delta f(a_0, a_2, a_3) + \Delta f(a_0, a_3, a_1)$$

$$= \begin{vmatrix} f^1(a_1) & f^2(a_1) \\ f^1(a_2) & f^2(a_2) \end{vmatrix} + \begin{vmatrix} f^1(a_2) & f^2(a_2) \\ f^1(a_3) & f^2(a_3) \end{vmatrix} + \begin{vmatrix} f^1(a_3) & f^2(a_3) \\ f^1(a_1) & f^2(a_1) \end{vmatrix}. \tag{14}$$

It is easy to verify that the cancellations which occur in deriving this equation from the three above are the same as those in the calculation of $\partial[(a_0, a_1, a_2) + (a_0, a_2, a_3) + (a_0, a_3, a_1)]$. If two simplexes have a side in common, the two terms which arise from this common side have opposite signs and they cancel. Equation (14) expresses a sum of terms defined on the 2-simplexes in a 2-chain as a sum of terms defined on the one-simplexes in the boundary of the 2-chain. Sums convert into integrals; formulas such as the one in (14) can be used to prove one form of the fundamental theorem of the integral calculus.

Exercises

18.1. Let $a_1 : (2, 1)$, $a_2 : (6, 2)$, $a_3 : (1, 4)$, and $a_4 : (4, 6)$ be four points in $\mathbb{R}^2$.

(a) If $s_1 = (a_1, a_2, a_3)$ and $s_2 = (a_3, a_2, a_4)$, show that s_1 and s_2 are positively oriented in $\mathbb{R}^2$. Find $\partial(s_1 + s_2)$. Make a sketch and indicate on it the algebraic boundary of the chain $s_1 + s_2$ and the topological boundary of $s_1 \cup s_2$.

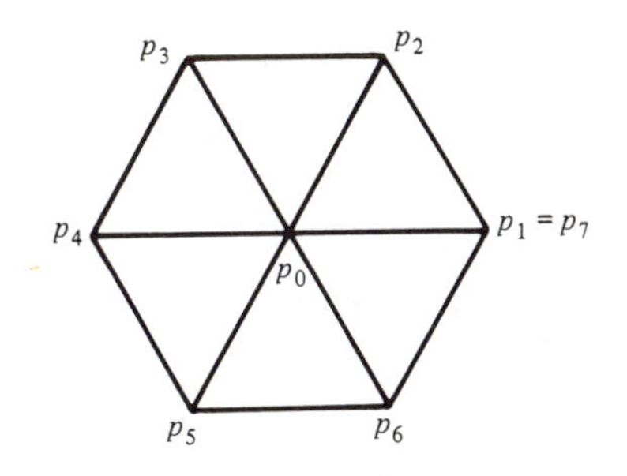

Figure 18.5. The figure for Exercise 18.6.

(b) If $s_1 = (a_1, a_2, a_3)$ and $s_2 = (a_2, a_3, a_4)$, determine the orientation of s_1 and s_2. Find $\partial(s_1 + s_2)$. Make a sketch which shows $\partial(s_1 + s_2)$ and the topological boundary of $s_1 \cup s_2$.

18.2. There are four 2-dimensional simplexes in Example 16.2; give each simplex the positive orientation in $\mathbb{R}^2$ and denote them by $s_1, \cdots, s_4$. Make a sketch of these simplexes, indicate the orientations of the simplexes on it, and use it to find $\partial(s_1 + \cdots + s_4)$. Check your result by an algebraic calculation. Find the topological boundary of $\bigcup_1^4 s_i$.

18.3. Let A be a square in $\mathbb{R}^2$. Divide A into n^2 equal subsquares by lines parallel to the sides of the square, and then divide each subsquare into two 2-simplexes by a diagonal of the subsquare. Let K be the complex which consists of the $2n^2$ 2-simplexes and all of their sides. Give each 2-simplex the positive orientation in $\mathbb{R}^2$. Sketch K for $n = 6$.

(a) Sketch an example of a connected 2-chain in K whose boundary consists of more than one connected set of one-simplexes [each connected set of one-simplexes in this boundary is called a one-cycle].

(b) Sketch an example of a 2-chain in K, the union of whose 2-simplexes is connected, but whose boundary consists of five cycles.

(c) Let c_1 be a 2-chain consisting of a proper subset of the 2-simplexes in K; let c_2 be the chain in K which contains all 2-simplexes not in c_1; and let c denote the chain of all 2-simplexes in K. Investigate the relations among ∂c_1, ∂c_2, $\partial(c_1 + c_2)$, and ∂c.

18.4. Let $f: E \to \mathbb{R}^2$ be a function whose components are defined on an open set E which contains the union of the 2-simplexes shown in Figure 18.2. For the chain of four 2-simplexes in Figure 18.2 derive a formula similar to the one in (14) in Example 18.7.

18.5. Let $s_1 : (a_0, a_1, a_2, a_5)$, $s_2 : (a_0, a_2, a_3, a_5)$, $s_3 : (a_0, a_3, a_4, a_5)$, and $s_4 : (a_0, a_4, a_1, a_5)$ be the 3-simplexes in the complex in Figure 17.1, and let $f: E \to \mathbb{R}^3$ be a function whose components (f^1, f^2, f^3) are defined on an open set E which contains $\bigcup_1^4 s_i$. Derive a formula, similar to the one in (14), for $\Delta f(s_1) + \Delta f(s_2) + \Delta f(s_3) + \Delta f(s_4)$. Compare the terms in this formula with the terms in $\partial(s_1 + \cdots + s_4)$ found in Example 17.6.

18.6. (a) Figure 18.5 shows a hexagon whose center is $p_0 : (x_0, y_0)$, and whose vertices are $p_i : (x_i, y_i)$, $i = 1, \cdots, 6$. For convenience, let $p_7 : (x_7, y_7)$ denote also the point $p_1 : (x_1, y_1)$. Prove that the area of the hexagon is given by the following formula:

$$\frac{1}{2!}\sum_{i=1}^{6}\begin{vmatrix} x_i & y_i \\ x_{i+1} & y_{i+1} \end{vmatrix}, \quad \text{or} \ \frac{1}{2!}\sum_{i=1}^{6}(x_i y_{i+1} - x_{i+1} y_i).$$

[Hint. The area of the hexagon is the sum of the areas of the six 2-simplexes (p_0, p_i, p_{i+1}), $i = 1, \cdots, 6$. By (25) in Section 89 in Appendix 2, the area of the simplex (p_0, p_1, p_2) is

$$\frac{1}{2!}\begin{vmatrix} x_0 & y_0 & 1 \\ x_1 & y_1 & 1 \\ x_2 & y_2 & 1 \end{vmatrix}.$$

Use the method employed in Example 18.7.]

(b) Let K be an oriented, homogeneous, 2-dimensional, Euclidean complex in $\mathbb{R}^2$, and let c be a chain of positively oriented 2-simplexes in K. Find a formula, similar to the one in (a), for the sum of the areas of the simplexes in c. Observe that ∂c may contain more than one cycle.

(c) Let K be an oriented, homogeneous, 3-dimensional, Euclidean complex in $\mathbb{R}^3$, and let c be a chain of positively oriented 3-simplexes in K. Find a formula, similar to the one in (b), for the sum of the volumes of the 3-simplexes in c. Observe that ∂c may contain more than one cycle.

18.7. Let $x_i : (x_i^1, x_i^2, x_i^3)$, $i = 0, 1, \cdots, 3$, be the vertices of a 3-simplex in $\mathbb{R}^3$. Use the fact that

$$\text{(i)} \qquad \begin{vmatrix} 1 & x_0^1 & x_0^2 & 1 \\ 1 & x_1^1 & x_1^2 & 1 \\ \cdots & \cdots & \cdots & \cdots \\ 1 & x_3^1 & x_3^2 & 1 \end{vmatrix} = 0$$

to prove that

$$\text{(ii)} \qquad \begin{vmatrix} x_1^1 & x_1^2 & 1 \\ x_2^1 & x_2^2 & 1 \\ x_3^1 & x_3^2 & 1 \end{vmatrix} - \begin{vmatrix} x_0^1 & x_0^2 & 1 \\ x_2^1 & x_2^2 & 1 \\ x_3^1 & x_3^2 & 1 \end{vmatrix} + \begin{vmatrix} x_0^1 & x_0^2 & 1 \\ x_1^1 & x_1^2 & 1 \\ x_3^1 & x_3^2 & 1 \end{vmatrix} - \begin{vmatrix} x_0^1 & x_0^2 & 1 \\ x_1^1 & x_1^2 & 1 \\ x_2^1 & x_2^2 & 1 \end{vmatrix} = 0.$$

[Hint. Expand the determinant in (i) by minors of the elements in the first column of its matrix.]

(b) Let K be an oriented, homogeneous, 3-dimensional, Euclidean, complex in $\mathbb{R}^3$. Let s_1 and s_2 be two 3-simplexes in K which have the same orientation in $\mathbb{R}^3$ and which have a 2-dimensional side in common. Write formula (ii) for each simplex s_1 and s_2, and then add the formulas. Which terms, if any, cancel because s_1 and s_2 have a 2-dimensional side in common?

(c) Let $\Sigma_1^k s_i$ be a 3-chain c in K such that the simplexes $s_1, \cdots, s_k$ have the same orientation (positive or negative) in $\mathbb{R}^3$. Let $z_i : (z_i^1, z_i^2, z_i^3)$, $i = 1, 2, 3$, be the vertices of a 2-simplex $w : (z_1, z_2, z_3)$ in ∂c. Prove the following formula.

$$\text{(iii)} \qquad \sum_{w \in \partial c}\begin{vmatrix} z_1^1 & z_1^2 & 1 \\ z_2^1 & z_2^2 & 1 \\ z_3^1 & z_3^2 & 1 \end{vmatrix} = 0.$$

(d) Explain the geometric significance of the formula in (iii).

(e) Prove the following formulas and explain their geometric significance.

$$\text{(iv)} \qquad \sum_{w \in \partial c} \begin{vmatrix} z_1^1 & z_1^3 & 1 \\ z_2^1 & z_2^3 & 1 \\ z_3^1 & z_3^3 & 1 \end{vmatrix} = 0, \qquad \sum_{w \in \partial c} \begin{vmatrix} z_1^2 & z_1^3 & 1 \\ z_2^2 & z_2^3 & 1 \\ z_3^2 & z_3^3 & 1 \end{vmatrix} = 0.$$

19. Affine and Barycentric Transformations

This section defines affine and barycentric transformations and establishes some of their important properties. A given Euclidean n-simplex $(a_0, a_1, \cdots, a_n)$ in $\mathbb{R}^n$ can be transformed in a one-to-one manner into an arbitrary Euclidean n-simplex $(b_0, b_1, \cdots, b_n)$ by an affine transformation and, under certain conditions, also by a barycentric transformation. These transformations are thus useful tools in studying a number of problems concerning simplexes. The problems include the following: (i) a comparison of the orientations of two n-simplexes in $\mathbb{R}^n$, and a comparison of the n-directions determined by two n-simplexes in $\mathbb{R}^m$, $n < m$ [see Definition 15.13]; (ii) the mapping of a simplicial subdivision of one n-simplex into another n-simplex; and (iii) the proof that the sum of the measures of the sub-simplexes in a subdivision of a simplex equals the measure of the simplex.

19.1 Example. The equations

$$\begin{aligned} z^1 &= 2x^1 - x^2, \\ z^2 &= -x^1 + 2x^2, \end{aligned} \tag{1}$$

define a linear transformation from (x^1, x^2)-space into (z^1, z^2)-space. This transformation transforms the points $a_0:(1, 1)$, $a_1:(2, 1)$, $a_2:(3, 4)$ in Figure 19.1(a) into the points $c_0:(1, 1)$, $c_1:(3, 0)$, $c_2:(2, 5)$ in Figure 19.1(b). Furthermore, since the transformation (1) is linear and non-singular, it transforms the Euclidean simplex (a_0, a_1, a_2) in a one-to-one manner into the Euclidean simplex (c_0, c_1, c_2). Also, the equations

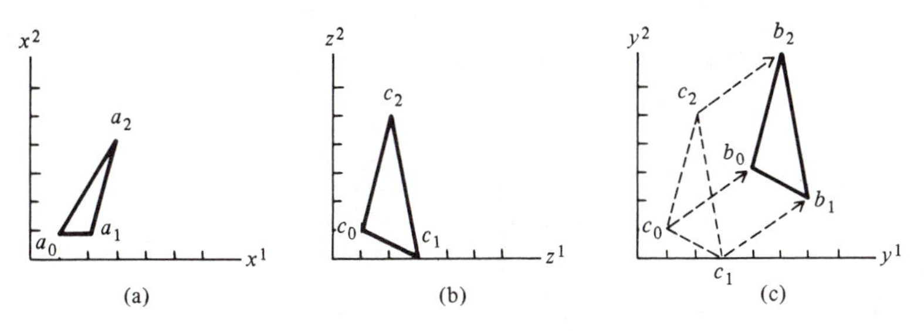

Figure 19.1. The transformations in Example 19.1.

$$\begin{aligned} y^1 &= z^1 + 3, \\ y^2 &= z^2 + 2, \end{aligned} \tag{2}$$

define a translation of $\mathbb{R}^2$ into itself. This translation transforms the points $c_0:(1, 1)$, $c_1:(3, 0)$, $c_2:(2, 5)$ into the points $b_0:(4, 3)$, $b_1:(6, 2)$, $b_2:(5, 7)$ in Figure 19.1(c). This translation transforms the Euclidean simplex (c_0, c_1, c_2) into the Euclidean simplex (b_0, b_1, b_2) by the rigid motion which adds the vector (3, 2) to each point (z^1, z^2) to obtain (y^1, y^2). The result of performing in succession the linear transformation (1) and then the translation (2) is to transform the simplex (a_0, a_1, a_2) into the simplex (b_0, b_1, b_2). The composition of the transformations in (1) and (2) is a single transformation which takes (a_0, a_1, a_2) into (b_0, b_1, b_2). The composition of these transformations is described by the following equations.

$$\begin{aligned} y^1 &= 2x^1 - x^2 + 3, \\ y^2 &= -x^1 + 2x^2 + 2. \end{aligned} \tag{3}$$

The transformation defined by these equations is an example of an affine transformation of $\mathbb{R}^2$ into itself. Definition 19.2 shows that an affine transformation of $\mathbb{R}^n$ into $\mathbb{R}^m$ is a linear transformation followed by a translation. Finally, the example pictured in Figure 19.1 poses the following question: does there exist an affine transformation which transforms a given Euclidean simplex $(a_0, a_1, \cdots, a_n)$ in $\mathbb{R}^n$ into a given n-simplex in $\mathbb{R}^m$? Theorem 19.3 answers this question.

19.2 Definition. Let $x:(x^1, \cdots, x^n)$ and $y:(y^1, \cdots, y^m)$ be points in $\mathbb{R}^n$ and $\mathbb{R}^m$ respectively, and let the c_{ij} denote constants. Then the linear, nonhomogeneous equations

$$y^j = \sum_{i=1}^{n} c_{ij}x^i + c_{n+1,j}, \qquad j = 1, \cdots, m, \tag{4}$$

define a transformation $L:\mathbb{R}^n \to \mathbb{R}^m$ such that $L(x) = y$. The transformation L is called an *affine transformation.*

In matrix form the equations (4) are the following:

$$[y^1 \quad \cdots \quad y^m] = [x^1 \quad \cdots \quad x^n \quad 1]\begin{bmatrix} c_{11} & c_{12} & \cdots & c_{1m} \\ \cdots & \cdots & \cdots & \cdots \\ c_{n1} & c_{n2} & \cdots & c_{nm} \\ c_{n+1,1} & c_{n+1,2} & \cdots & c_{n+1,m} \end{bmatrix}. \tag{5}$$

The notation in (4) has been chosen in an unconventional manner in order to make the notation convenient in (5).

19.3 Theorem. *If $(a_0, a_1, \cdots, a_n)$ is a simplex in $\mathbb{R}^n$ such that $\Delta(a_0, a_1, \cdots, a_n) \neq 0$, and if $(b_0, b_1, \cdots, b_n)$ is a simplex in $\mathbb{R}^m$, then there exists a unique affine transformation $L:\mathbb{R}^n \to \mathbb{R}^m$ such that $L(a_i) = b_i$, $i = 0, 1, \cdots, n$.*

PROOF. The vertices of the two simplexes are the points $a_i:(a_i^1, \cdots, a_i^n)$ and $b_i:(b_i^1, \cdots, b_i^m)$, $i = 0, 1, \cdots, n$. To prove the theorem, it is necessary to show that there exists a unique affine transformation (4) or (5) such that

$$\begin{bmatrix} a_0^1 & a_0^2 & \cdots & a_0^n & 1 \\ a_1^1 & a_1^2 & \cdots & a_1^n & 1 \\ \cdots & \cdots & \cdots & \cdots & \cdots \\ a_n^1 & a_n^2 & \cdots & a_n^n & 1 \end{bmatrix} \begin{bmatrix} c_{11} & c_{12} & \cdots & c_{1m} \\ \cdots & \cdots & \cdots & \cdots \\ c_{n1} & c_{n2} & \cdots & c_{nm} \\ c_{n+1,1} & c_{n+1,2} & \cdots & c_{n+1,m} \end{bmatrix} = \begin{bmatrix} b_0^1 & b_0^2 & \cdots & b_0^m \\ b_1^1 & b_1^2 & \cdots & b_1^m \\ \cdots & \cdots & \cdots & \cdots \\ b_n^1 & b_n^2 & \cdots & b_n^m \end{bmatrix}. \tag{6}$$

The product of the two matrices on the left is an $n \times m$ matrix; the c_{ij} are to be determined so that the product matrix equals the matrix on the right. Equating the elements in the j-th column of the product matrix to the corresponding elements in the j-th column of the matrix on the right yields a system of $n + 1$ equations in the $n + 1$ unknowns $c_{1j}, c_{2j}, \cdots, c_{n+1,j}$. This system of equations has a unique solution by Cramer's rule because $\Delta(a_0, a_1, \cdots, a_n)$, the determinant of the matrix of coefficients in the system of equations, is not zero by hypothesis. Thus the matrix equation (6) has a unique solution for the c_{ij}, and the proof is complete. □

19.4 Theorem. *Let $(a_0, a_1, \cdots, a_n)$ be a simplex in $\mathbb{R}^n$ such that $\Delta(a_0, a_1, \cdots, a_n) \neq 0$; let $(b_0, b_1, \cdots, b_n)$ be a simplex in $\mathbb{R}^m$; and let $L: \mathbb{R}^n \to \mathbb{R}^m$ be an affine transformation such that $L(a_i) = b_i$, $i = 0, 1, \cdots, n$. If x is a point in $\mathbb{R}^n$ such that $x = \Sigma_0^n t^i a_i$, $\Sigma_0^n t^i = 1$, and if $y = L(x)$, then $y = \Sigma_0^n t^i b_i$. Also, L maps each side of the simplex $(a_0, a_1, \cdots, a_n)$ into the corresponding side of $(b_0, b_1, \cdots, b_n)$.*

PROOF. The affine transformation L is described by the matrix equation in (6). It is clear that the matrix multiplication on the left side of (6) is an operation which is linear in the rows of the a_i^j matrix. Thus a linear combination of the rows of the a_i^j matrix is transformed into the same linear combination of the b_i^j matrix. Then multiplication of the row

$$\begin{bmatrix} \sum_{i=0}^n t^i a_i^1 & \sum_{i=0}^n t^i a_i^2 & \cdots & \sum_{i=0}^n t^i a_i^n & \sum_{i=0}^n t^i \end{bmatrix} \tag{7}$$

on the right by the c_{ij} matrix yields the row

$$\begin{bmatrix} \sum_{i=0}^n t^i b_i^1 & \sum_{i=0}^n t^i b_i^2 & \cdots & \sum_{i=0}^n t^i b_i^m \end{bmatrix}. \tag{8}$$

Since $\Sigma_0^n t^i = 1$ in (7), then

$$\begin{bmatrix} \sum_{i=0}^n t^i a_i^1 & \cdots & \sum_{i=0}^n t^i a_i^n & 1 \end{bmatrix} \begin{bmatrix} c_{11} & c_{12} & \cdots & c_{1m} \\ \cdots & \cdots & \cdots & \cdots \\ c_{n1} & c_{n2} & \cdots & c_{nm} \\ c_{n+1,1} & c_{n+1,2} & \cdots & c_{n+1,m} \end{bmatrix} = \begin{bmatrix} \sum_{i=0}^n t^i b_i^1 & \sum_{i=0}^n t^i b_i^2 & \cdots & \sum_{i=0}^n t^i b_i^m \end{bmatrix}. \tag{9}$$

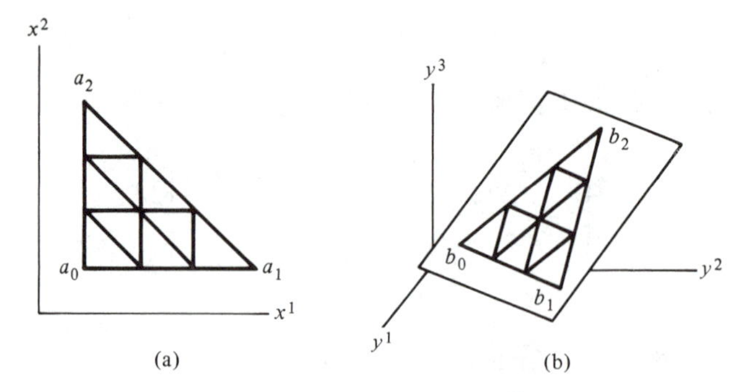

Figure 19.2. The figure for Example 19.5.

Equation (9) states that L transforms the point $\Sigma_0^n t^i a_i$ into the point $\Sigma_0^n t^i b_i$. The side $(a_1, \cdots, a_n)$ of $(a_0, a_1, \cdots, a_n)$ consists of the points $\Sigma_1^n t^i a_i$, $\Sigma_1^n t^i = 1$. By the part of the theorem already proved, each point $\Sigma_1^n t^i a_i$, $\Sigma_1^n t^i = 1$, is transformed into the point $\Sigma_1^n t^i b_i$. Thus the side $(a_1, \cdots, a_n)$ of $(a_0, a_1, \cdots, a_n)$ is transformed into the side $(b_1, \cdots, b_n)$ of $(b_0, b_1, \cdots, b_n)$. In the same way it can be shown that L transforms each side of $(a_0, a_1, \cdots, a_n)$ into the corresponding side of $(b_0, b_1, \cdots, b_n)$. □

19.5 Example. Let (a_0, a_1, a_2) be the Euclidean 2-simplex in $\mathbb{R}^2$ shown in Figure 19.2(a), and let (b_0, b_1, b_2) be the Euclidean 2-simplex in $\mathbb{R}^3$ shown in Figure 19.2(b). By Theorem 19.3, there exists an affine transformation L such that $L(a_i) = b_i$, $i = 0, 1, 2$. Figure 19.2(a) shows a subdivision of (a_0, a_1, a_2) into nine sub-simplexes. A vertex in a sub-simplex has barycentric coordinates (t^0, t^1, t^2) with respect to a_0, a_1, a_2. Theorem 19.4 shows that this vertex is mapped by L into a point in (b_0, b_1, b_2) whose barycentric coordinates with respect to b_0, b_1, b_2 are (t^0, t^1, t^2). The affine transformation L maps the subdivision of (a_0, a_1, a_2) shown in Figure 19.2(a) into the subdivision of (b_0, b_1, b_2) shown in (b).

If (6) is a true equation, then the following matrix equation is also true.

$$\begin{bmatrix} a_0^1 & a_0^2 & \cdots & a_0^n & 1 \\ a_1^1 & a_1^2 & \cdots & a_1^n & 1 \\ \cdots & \cdots & \cdots & \cdots & \cdots \\ a_n^1 & a_n^2 & \cdots & a_n^n & 1 \end{bmatrix} \begin{bmatrix} c_{11} & c_{12} & \cdots & c_{1m} & 0 \\ \cdots & \cdots & \cdots & \cdots & \cdots \\ c_{n1} & c_{n2} & \cdots & c_{nm} & 0 \\ c_{n+1,1} & c_{n+1,2} & \cdots & c_{n+1,m} & 1 \end{bmatrix} = \begin{bmatrix} b_0^1 & b_0^2 & \cdots & b_0^m & 1 \\ b_1^1 & b_1^2 & \cdots & b_1^m & 1 \\ \cdots & \cdots & \cdots & \cdots & \cdots \\ b_n^1 & b_n^2 & \cdots & b_n^m & 1 \end{bmatrix}. \tag{10}$$

If $m = n$, let $\det[c_{ij}]_1^n$ denote $\det[c_{ij}]$, $i, j = 1, \cdots, n$.

19.6 Theorem. *Let $(a_0, a_1, \cdots, a_n)$ be a simplex in $\mathbb{R}^n$ such that $\Delta(a_0, a_1, \cdots, a_n) \neq 0$, and let* (4) *and* (6), *with $m = n$, be an affine transformation*

$L: \mathbb{R}^n \to \mathbb{R}^n$ such that $L(a_i) = b_i$, $i = 0, 1, \cdots, n$. Then

$$\Delta(a_0, a_1, \cdots, a_n) \det[c_{ij}]_1^n = \Delta(b_0, b_1, \cdots, b_n). \tag{11}$$

If $\Delta(b_0, b_1, \cdots, b_n) \neq 0$, then $\det[c_{ij}]_1^n \neq 0$ and L is a non-singular transformation. If $\det[c_{ij}]_1^n > 0$, then $(a_0, a_1, \cdots, a_n)$ and $(b_0, b_1, \cdots, b_n)$ have the same orientation in $\mathbb{R}^n$; if $\det[c_{ij}]_1^n < 0$, these simplexes have opposite orientations in $\mathbb{R}^n$. Finally, $(b_0, b_1, \cdots, b_n)$ is a degenerate Euclidean simplex if and only if $\det[c_{ij}]_1^n = 0$.

PROOF. If $m = n$, then each matrix in (10) is an $(n+1) \times (n+1)$ matrix. The Binet–Cauchy multiplication theorem [see Theorem 80.1(2) in Appendix 1] shows that the product of the determinants of the matrices on the left equals the determinant of the matrix on the right. The determinant of the c_{ij} matrix simplifies to $\det[c_{ij}]_1^n$. These facts establish (11), and the remaining statements in the theorem follows from this equation. □

19.7 Example. Theorem 19.6 shows that the sign of $\det[c_{ij}]_1^n$ determines whether $(a_0, a_1, \cdots, a_n)$ and $(b_0, b_1, \cdots, b_n)$ have the same or opposite orientations in $\mathbb{R}^n$. Furthermore, $|\det[c_{ij}]_1^n|$ has geometric significance. Equation (11) shows that

$$(1/n!)|\Delta(a_0, a_1, \cdots, a_n)|\,|\det[c_{ij}]_1^n| = (1/n!)|\Delta(b_0, b_1, \cdots, b_n)|. \tag{12}$$

Since $(1/n!)|\Delta(a_0, a_1, \cdots, a_n)|$ and $(1/n!)|\Delta(b_0, b_1, \cdots, b_n)|$ are the volumes of $(a_0, a_1, \cdots, a_n)$ and $(b_0, b_1, \cdots, b_n)$ [see Section 89 in Appendix 2 for the cases $n = 2$ and $n = 3$, and Section 20 for the general case], then $|\det[c_{ij}]_1^n|$ is the ratio of the volume of $(b_0, b_1, \cdots, b_n)$ to that of $(a_0, a_1, \cdots, a_n)$. For example, in Example 19.1,

$$\frac{1}{2!}\Delta(a_0, a_1, a_2) = \frac{1}{2!}\begin{vmatrix} 1 & 1 & 1 \\ 2 & 1 & 1 \\ 3 & 4 & 1 \end{vmatrix} = \frac{3}{2}, \qquad \frac{1}{2!}\Delta(b_0, b_1, b_2) = \frac{1}{2!}\begin{vmatrix} 4 & 3 & 1 \\ 6 & 2 & 1 \\ 5 & 7 & 1 \end{vmatrix} = \frac{9}{2}.$$

Since

$$\det[c_{ij}]_1^2 = \begin{vmatrix} 2 & -1 \\ -1 & 2 \end{vmatrix} = 3,$$

then (a_0, a_1, a_2) and (b_0, b_1, b_2) have the same orientation in $\mathbb{R}^2$ [both are positively oriented, as shown already], and the area of (b_0, b_1, b_2) is three times that of (a_0, a_1, a_2). If $m > n$ in (4) and (6), then the relation between the volumes of (a_0, a_1, a_2) and $(b_0, b_1, \cdots, b_n)$ is not as simple as equation (12), but the relation can be found. The next theorem begins the study of this case.

Consider equation (10) again and assume that $m > n$. Let $(j_1, \cdots, j_n)$ be an index set in (n/m). Since the $(j_1, \cdots, j_n)$ columns of the b_{ij} matrix are derived from the $(j_1, \cdots, j_n)$ columns of the c_{ij} matrix in (10), then

$$\begin{bmatrix} a_0^1 & a_0^2 & \cdots & a_0^n & 1 \\ a_1^1 & a_1^2 & \cdots & a_1^n & 1 \\ \cdots & \cdots & \cdots & \cdots & \cdots \\ a_n^1 & a_n^2 & \cdots & a_n^n & 1 \end{bmatrix} \begin{bmatrix} c_{1j_1} & c_{1j_2} & \cdots & c_{1j_n} & 0 \\ \cdots & \cdots & \cdots & \cdots & \cdots \\ c_{nj_1} & c_{nj_2} & \cdots & c_{nj_n} & 0 \\ c_{n+1,j_1} & c_{n+1,j_2} & \cdots & c_{n+1,j_n} & 1 \end{bmatrix} = \begin{bmatrix} b_0^{j_1} & b_0^{j_2} & \cdots & b_0^{j_n} & 1 \\ b_1^{j_1} & b_1^{j_2} & \cdots & b_1^{j_n} & 1 \\ \cdots & \cdots & \cdots & \cdots & \cdots \\ b_n^{j_1} & b_n^{j_2} & \cdots & b_n^{j_n} & 1 \end{bmatrix} \tag{13}$$

The Binet–Cauchy multiplication theorem shows that the product of the determinants of the matrices on the left equals the determinant of the matrix on the right. The determinant of the a_{ij} matrix is denoted by $\Delta(a_0, a_1, \cdots, a_n)$. The determinant of the c_{ij} matrix equals $\det[c_{ij}]$, $i = 1, \cdots, n$ and $j = j_1, \cdots, j_n$; denote this determinant by $\det[c_{ij}]^{(j_1, \cdots, j_n)}$. The determinant of the b_{ij} matrix is denoted by $\Delta[(b_0, b_1, \cdots, b_n)^{(j_1, \cdots, j_n)}]$; see Section 11 for this notation. Then in this condensed notation, the equation derived from (13) is

$$\Delta(a_0, a_1, \cdots, a_n) \det[c_{ij}]^{(j_1, \cdots, j_n)} = \Delta[(b_0, b_1, \cdots, b_n)^{(j_1, \cdots, j_n)}]. \tag{14}$$

By equations (35), (36), (37) and the explanation which accompanies them in Section 11, the volume of $(b_0, b_1, \cdots, b_n)$ is $(1/n!)|\Delta(b_0, b_1, \cdots, b_n)|$, and this volume is given by the following formula:

$$\frac{1}{n!}|\Delta(b_0, b_1, \cdots, b_n)| = \frac{1}{n!}\left\{ \sum_{(j_1, \cdots, j_n)} [\Delta[(b_0, b_1, \cdots, b_n)^{(j_1, \cdots, j_n)}]]^2 \right\}^{1/2}. \tag{15}$$

19.8 Theorem. *Let $L: \mathbb{R}^n \to \mathbb{R}^m$, $m \geqq n$, be the affine transformation which transforms the Euclidean simplex $(a_0, a_1, \cdots, a_n)$ in $\mathbb{R}^n$ into $(b_0, b_1, \cdots, b_n)$ in $\mathbb{R}^m$ so that $L(a_i) = b_i$, $i = 0, 1, \cdots, n$. Then* (14) *holds for each $(j_1, \cdots, j_n)$ in (n/m). Furthermore, the component $\Delta[(b_0, b_1, \cdots, b_n)^{(j_1, \cdots, j_n)}]$ of $\Delta(b_0, b_1, \cdots, b_n)$ has the same sign as $\Delta(a_0, a_1, \cdots, a_n)$ if and only if $\det[c_{ij}]^{(j_1, \cdots, j_n)} > 0$, and the opposite sign if and only if this determinant is negative. Finally,*

$$\frac{1}{n!}|\Delta(b_0, b_1, \cdots, b_n)| = \frac{1}{n!}|\Delta(a_0, a_1, \cdots, a_n)| \left\{ \sum_{(j_1, \cdots, j_n)} [\det[c_{ij}]^{(j_1, \cdots, j_n)}]^2 \right\}^{1/2}. \tag{16}$$

Proof. For each $(j_1, \cdots, j_n)$ in (n/m), equation (13) follows from (10), and (14) follows from (13) as explained above. The relations between the signs of $\Delta[(b_0, b_1, \cdots, b_n)^{(j_1, \cdots, j_n)}]$ and $\Delta(a_0, a_1, \cdots, a_n)$ follows from (14). Finally, the formula in (16) for the volume of $\Delta(b_0, b_1, \cdots, b_n)$ is obtained by substituting the value of $\Delta[(b_0, b_1, \cdots, b_n)^{(j_1, \cdots, j_n)}]$ in (14) for this term in (15) and then simplifying the result. □

19.9 Example. The affine transformation $L: \mathbb{R}^2 \to \mathbb{R}^3$ such that

$$\begin{aligned} y^1 &= 2x^1 - 4x^2 + 5, \\ y^2 &= 3x^1 + 5x^2 - 4, \\ y^3 &= x^1 - 2x^2 + 3, \end{aligned} \tag{17}$$

maps the simplex (a_0, a_1, a_2) in $\mathbb{R}^2$ with vertices $a_0 : (3, 2)$, $a_1 : (4, -5)$, $a_2 : (8, 1)$ into the simplex (b_0, b_1, b_2) in $\mathbb{R}^3$ with vertices $b_0 : (3, 15, 2)$, $b_1 : (33, -17, 17)$, $b_2 : (17, 25, 9)$. A straightforward evaluation of the determinant shows that $\Delta(a_0, a_1, a_2) = 34$; hence the area of (a_0, a_1, a_2) is 17. The formula in (15) shows that the area of (b_0, b_1, b_2) is $187\sqrt{5}$. The affine transformation of (a_0, a_1, a_2) into (b_0, b_1, b_2), written in the matrix form (10), is the following:

$$\begin{bmatrix} a_0^1 & a_0^2 & 1 \\ a_1^1 & a_1^2 & 1 \\ a_2^1 & a_2^2 & 1 \end{bmatrix} \begin{bmatrix} 2 & 3 & 1 & 0 \\ -4 & 5 & -2 & 0 \\ 5 & -4 & 3 & 1 \end{bmatrix} = \begin{bmatrix} b_0^1 & b_0^2 & b_0^3 & 1 \\ b_1^1 & b_1^2 & b_1^3 & 1 \\ b_2^1 & b_2^2 & b_2^3 & 1 \end{bmatrix}. \tag{18}$$

Then the three equations which correspond to (14) are

$$\begin{aligned} \Delta(a_0, a_1, a_2) \begin{vmatrix} 2 & 3 \\ -4 & 5 \end{vmatrix} &= \Delta[(b_0, b_1, b_2)^{(1,2)}], \\ \Delta(a_0, a_1, a_2) \begin{vmatrix} 2 & 1 \\ -4 & -2 \end{vmatrix} &= \Delta[(b_0, b_1, b_2)^{(1,3)}], \\ \Delta(a_0, a_1, a_2) \begin{vmatrix} 3 & 1 \\ 5 & -2 \end{vmatrix} &= \Delta[(b_0, b_1, b_2)^{(2,3)}]. \end{aligned} \tag{19}$$

Then

$$\begin{aligned} \Delta[(b_0, b_1, b_2)^{(1,2)}] &= 22\Delta(a_0, a_1, a_2), \\ \Delta[(b_0, b_1, b_2)^{(1,3)}] &= 0\Delta(a_0, a_1, a_2), \\ \Delta[(b_0, b_1, b_2)^{(2,3)}] &= -11\Delta(a_0, a_1, a_2). \end{aligned} \tag{20}$$

Here $(b_0, b_1, b_2)^{(j_1, j_2)}$, (j_1, j_2) in $(2/3)$, is the projection of (b_0, b_1, b_2) into the (y^{j_1}, y^{j_2}) coordinate plane in $\mathbb{R}^3$, and the equations in (20) describe the relations, with respect to area and orientation, between (a_0, a_1, a_2) and these projections of (b_0, b_1, b_2). Finally, equations (20) show that

$$(1/2!)|\Delta(b_0, b_1, b_2)| = 11\sqrt{5}(1/2!)|\Delta(a_0, a_1, a_2)|. \tag{21}$$

This equation is consistent with the earlier calculation that the areas of (a_0, a_1, a_2) and (b_0, b_1, b_2) are 17 and $187\sqrt{5}$ respectively.

Thus far affine transformations have been used to transform a Euclidean n-simplex $(a_0, a_1, \cdots, a_n)$ into an n-simplex $(b_0, b_1, \cdots, b_n)$. If the two simplexes lie in the same plane, there is a second transformation, called a barycentric transformation, which can be used for this purpose.

19.10 Definition. If there exists a matrix $[t_i^j]$, $i, j = 0, 1, \cdots, m$, such that

$$\begin{bmatrix} b_0^1 & b_0^2 & \cdots & b_0^n & 1 \\ \cdots & \cdots & \cdots & \cdots & \cdots \\ b_m^1 & b_m^2 & \cdots & b_m^n & 1 \end{bmatrix} = \begin{bmatrix} t_0^0 & t_0^1 & \cdots & t_0^m \\ \cdots & \cdots & \cdots & \cdots \\ t_m^0 & t_m^1 & \cdots & t_m^m \end{bmatrix} \begin{bmatrix} a_0^1 & a_0^2 & \cdots & a_0^n & 1 \\ \cdots & \cdots & \cdots & \cdots & \cdots \\ a_m^1 & a_m^2 & \cdots & a_m^n & 1 \end{bmatrix}, \tag{22}$$

$$\sum_{j=0}^{m} t_i^j = 1, \qquad i = 0, 1, \cdots, m, \tag{23}$$

then the transformation of $(a_0, a_1, \cdots, a_m)$ into $(b_0, b_1, \cdots, b_m)$ by the matrix $[t_i^j]_0^m$ is called a *barycentric* transformation.

19.11 Theorem. *If $(a_0, a_1, \cdots, a_n)$ is a simplex in $\mathbb{R}^n$ such that $\Delta(a_0, a_1, \cdots, a_n) \neq 0$, and if $(b_0, b_1, \cdots, b_n)$ is an n-simplex in $\mathbb{R}^n$, then there exists a unique barycentric transformation, with matrix $[t_i^j]$, $i, j = 0, 1 \cdots, n$, which transforms the vertices of $(a_0, a_1, \cdots, a_n)$ into the vertices of $(b_0, b_1, \cdots, b_n)$; furthermore,*

$$\Delta(b_0, b_1, \cdots, b_n) = \det[t_i^j]_0^n \Delta(a_0, a_1, \cdots, a_n). \tag{24}$$

PROOF. Since $\Delta(a_0, a_1, \cdots, a_n) \neq 0$ by hypothesis, equation (13) in Section 14 shows that there is a unique matrix $[t_i^0\ t_i^1\ \cdots\ t_i^n]$, the sum of whose elements is 1, such that

$$[b_i^1 \quad b_i^2 \quad \cdots \quad b_i^n \quad 1] = [t_i^0 \quad t_i^1 \quad \cdots \quad t_i^n] \begin{bmatrix} a_0^1 & a_0^2 & \cdots & a_0^n & 1 \\ \cdots & \cdots & \cdots & \cdots & \cdots \\ a_n^1 & a_n^2 & \cdots & a_n^n & 1 \end{bmatrix}. \tag{25}$$

Because this relation holds for $i = 0, 1, \cdots, n$, then

$$\begin{bmatrix} b_0^1 & b_0^2 & \cdots & b_0^n & 1 \\ \cdots & \cdots & \cdots & \cdots & \cdots \\ b_n^1 & b_n^2 & \cdots & b_n^n & 1 \end{bmatrix} = \begin{bmatrix} t_0^0 & t_0^1 & \cdots & t_0^n \\ \cdots & \cdots & \cdots & \cdots \\ t_n^0 & t_n^1 & \cdots & t_n^n \end{bmatrix} \begin{bmatrix} a_0^1 & a_0^2 & \cdots & a_0^n & 1 \\ \cdots & \cdots & \cdots & \cdots & \cdots \\ a_n^1 & a_n^2 & \cdots & a_n^n & 1 \end{bmatrix}, \tag{26}$$

and $[t_i^j]_0^n$ is the matrix of a barycentric transformation which transforms $(a_0, a_1, \cdots, a_n)$ into $(b_0, b_1, \cdots, b_n)$. Each matrix in (26) has $(n + 1)$ rows and columns. Then (26) and the Binet–Cauchy multiplication theorem [see Theorem 80.1(2) in Appendix 1] show that (24) is true. □

Equation (24) has applications similar to those of equation (11). For example, $(b_0, b_1, \cdots, b_n)$ is a Euclidean simplex if and only if $\det[t_i^j]_0^n \neq 0$; $(b_0, b_1, \cdots, b_n)$ and $(a_0, a_1, \cdots, a_n)$ have the same orientation in $\mathbb{R}^n$ if and only if $\det[t_i^j]_0^n > 0$; and $|\det[t_i^j]_0^n|$ equals the ratio of the volume of $(b_0, b_1, \cdots, b_n)$ to that of $(a_0, a_1, \cdots, a_n)$.

19.12 Theorem. *Let $(a_0, a_1, \cdots, a_m)$ be a Euclidean m-simplex in $\mathbb{R}^n$, $m \leqq n$, and let $P(a_0, a_1, \cdots, a_m)$ be the plane determined by the linearly independent points $a_0, a_1, \cdots, a_m$. If $(b_0, b_1, \cdots, b_m)$ is an m-simplex in $P(a_0, a_1, \cdots, a_m)$, then there exists a barycentric transformation with matrix $[t_i^j]$, $i, j = 0, 1, \cdots, m$, which transforms $(a_0, a_1, \cdots, a_m)$ into $(b_0, b_1, \cdots, b_m)$ as in (22) and (23). Furthermore, for each $(j_1, \cdots, j_m)$ in (m/n),*

$$\begin{bmatrix} b_0^{j_1} & b_0^{j_2} & \cdots & b_0^{j_m} & 1 \\ \cdots & \cdots & \cdots & \cdots & \cdots \\ b_m^{j_1} & b_m^{j_2} & \cdots & b_m^{j_m} & 1 \end{bmatrix} = \begin{bmatrix} t_0^0 & t_0^1 & \cdots & t_0^m \\ \cdots & \cdots & \cdots & \cdots \\ t_m^0 & t_m^1 & \cdots & t_m^m \end{bmatrix} \begin{bmatrix} a_0^{j_1} & a_0^{j_2} & \cdots & a_0^{j_m} & 1 \\ \cdots & \cdots & \cdots & \cdots & \cdots \\ a_m^{j_1} & a_m^{j_2} & \cdots & a_m^{j_m} & 1 \end{bmatrix}, \tag{27}$$

$$\Delta[(b_0, b_1, \cdots, b_m)^{(j_1, \cdots, j_m)}] = \det[t_i^j]_0^m \Delta[(a_0, a_1, \cdots, a_m)^{(j_1, \cdots, j_m)}]. \tag{28}$$

PROOF. Since $(a_0, a_1, \cdots, a_m)$ is a Euclidean simplex, then $|\Delta(a_0, a_1, \cdots, a_m)| > 0$ and the points $a_0, a_1, \cdots, a_m$ are linearly independent by Theorem 13.1. Thus the points $a_0, a_1, \cdots, a_m$ determine a plane $P(a_0, a_1, \cdots, a_m)$, and equations (2) and (3) in Section (14) show that every point x in this plane has barycentric coordinates $(t^0, t^1, \cdots, t^m)$ with respect to $a_0, a_1, \cdots, a_m$. Therefore, as in (25), there exists a unique matrix $[t_i^0 \; t_i^1 \; \cdots \; t_i^m]$ such that

$$[b_i^1 \quad b_i^2 \quad \cdots \quad b_i^n \quad 1] = [t_i^0 \quad t_i^1 \quad \cdots \quad t_i^m] \begin{bmatrix} a_0^1 & a_0^2 & \cdots & a_0^n & 1 \\ \cdots & \cdots & \cdots & \cdots & \cdots \\ a_m^1 & a_m^2 & \cdots & a_m^n & 1 \end{bmatrix},$$

$$\sum_{j=0}^m t_i^j = 1, \qquad i = 0, 1, \cdots, m. \tag{29}$$

Equations (22) and (23) follow from (29), and the unique barycentric transformation of $(a_0, a_1, \cdots, a_m)$ into $(b_0, b_1, \cdots, b_m)$ exists as stated. The definition of matrix multiplication shows that, for each $(j_1, \cdots, j_m)$ in (m/n), equation (27) follows from (22). Then (28) follows from (27) and the Binet–Cauchy multiplication theorem. The proof of Theorem 19.12 is complete. □

19.13 Example. Let $(a_0, a_1, \cdots, a_m)$ be a Euclidean m-simplex, and let $(b_0, b_1, \cdots, b_m)$ be a simplex in $P(a_0, a_1, \cdots, a_m)$. Observe that $(a_0, a_1, \cdots, a_m)^{(j_1, \cdots, j_m)}$ and $(b_0, b_1, \cdots, b_m)^{(j_1, \cdots, j_m)}$ are the projections of $(a_0, a_1, \cdots, a_m)$ and $(b_0, b_1, \cdots, b_m)$ into the $(x^{j_1}, \cdots, x^{j_m})$ coordinate plane in $\mathbb{R}^n$. Equation (27) shows that the same barycentric transformation which transforms $(a_0, a_1, \cdots, a_m)$ into $(b_0, b_1, \cdots, b_m)$ also transforms $(a_0, a_1, \cdots, a_m)^{(j_1, \cdots, j_m)}$ into $(b_0, b_1, \cdots, b_m)^{(j_1, \cdots, j_m)}$. If $(b_0, b_1, \cdots, b_m)$ is a degenerate Euclidean simplex, then

$$\Delta[(b_0, b_1, \cdots, b_m)^{(j_1, \cdots, j_m)}] = 0, \qquad (j_1, \cdots, j_m) \in (m/n); \tag{30}$$

equation (28) shows that $(b_0, b_1, \cdots, b_m)$ is a degenerate Euclidean simplex if and only if $\det[t_i^j]_0^m = 0$. Then (28) and Definition 15.13 show that the Euclidean simplexes $(a_0, a_1, \cdots, a_m)$ and $(b_0, b_1, \cdots, b_m)$ have the same m-direction in $\mathbb{R}^n$ if and only if $\det[t_i^j]_0^m > 0$, and opposite m-directions in $\mathbb{R}^n$

if and only if $\det[t_i^j] < 0$. Also, $|\det[t_i^j]_0^m|$ equals the ratio of the volume of $(b_0, b_1, \cdots, b_m)^{(j_1, \cdots, j_m)}$ to that of $(a_0, a_1, \cdots, a_m)^{(j_1, \cdots, j_m)}$ for $(j_1, \cdots, j_m)$ in (m/n), and therefore equals the ratio of the volume of $(b_0, b_1, \cdots, b_m)$ to that of $(a_0, a_1, \cdots, a_m)$.

Exercises

19.1. Prove the following theorem. If $L : \mathbb{R}^n \to \mathbb{R}^m$, $m \geqq n$, is the affine transformation which transforms the Euclidean simplex $(a_0, a_1, \cdots, a_n)$ into $(b_0, b_1, \cdots, b_n)$ in $\mathbb{R}^m$, and if $|\Delta(b_0, b_1, \cdots, b_m)| > 0$, then the transformation L is one-to-one. [Hints. Equation (16) above and Exercise 12.5.]

19.2. An affine transformation $L : \mathbb{R}^n \to \mathbb{R}^n$ [see (4) and (6)] such that $[c_{ij}]_1^n = I$, the $n \times n$ identity matrix, is called a translation of $\mathbb{R}^n$ into itself. If L is a translation of $\mathbb{R}^n$ into itself, prove that $\det[c_{ij}]_1^n = 1$, and that L preserves all orientations in $\mathbb{R}^n$ [see Theorem 19.6].

19.3. The equations $y^j = \pm x^j$, $j = 1, \cdots, n$, for each of the 2^n choices of signs, define an affine transformation of $\mathbb{R}^n$ into itself. Which ones of these transformations preserve orientations and which ones reverse it?

19.4. (a) Show that each of the affine transformations

$$\text{(i)}\quad \begin{aligned} y^1 &= x^1 \cos\theta - x^2 \sin\theta, \\ y^2 &= x^1 \sin\theta + x^2 \cos\theta, \end{aligned} \qquad \text{(ii)}\quad \begin{aligned} y^1 &= x^1 \cos\theta - x^2 \sin\theta, \\ y^2 &= -x^1 \sin\theta - x^2 \cos\theta, \end{aligned}$$

is an orthogonal (distance preserving) transformation of $\mathbb{R}^2$ into itself [see Exercise 5.3]. Find $\det[c_{ij}]_1^n$ for each transformation.

(b) Show that (i) is a rotation of the plane through an angle θ, and that it preserves all orientations.

(c) Show that (ii) reverses all orientations. Show also that (ii) is a rotation through an angle θ followed by a reflection in the x^1-axis.

19.5. Let $e_0 : (0, 0, \cdots, 0)$, $e_1 : (1, 0, \cdots, 0)$, $\cdots$, $e_n : (0, 0, \cdots, 1)$ be the origin and the unit points on the axes in $\mathbb{R}^n$. Then $(e_0, e_1, \cdots, e_n)$ is a Euclidean simplex. Find the equations (4) of the affine transformation $L : \mathbb{R}^n \to \mathbb{R}^m$ such that $L(e_i) = b_i$, $i = 0, 1, \cdots, n$. [Hint. Solve the matrix equation (6).]

20. Three Theorems on Determinants

The first two theorems in this section contain identities for determinants. The first of these has been stated already—with stronger hypotheses—in Exercise 14.18, and Exercise 14.19 contains several examples which explain its geometric significance. The identity in the second theorem will be used in Chapter 6 to prove a basic form of the fundamental theorem of the integral calculus. The third theorem in this section establishes a determinant formula for the measure of an m-simplex in $\mathbb{R}^n$, $1 \leqq m \leqq n$. The derivation of the formula in the third theorem includes a proof of Hadamard's determinant theorem; a related proof of this theorem can be found in Section 87 in Appendix 1. The results in this section provide important tools which find applications in later chapters; the exercises begin the development of some of these applications.

20.1 Theorem. *If $a_0, a_1, \cdots, a_n, x$ are points in $\mathbb{R}^n$, then*

$$\sum_{r=0}^{n} \Delta(a_0, \cdots, a_{r-1}, x, a_{r+1}, \cdots, a_n) = \Delta(a_0, a_1, \cdots, a_n). \tag{1}$$

PROOF. The proof of this theorem suggested in Exercise 14.18 requires the additional hypothesis that $a_0, a_1, \cdots, a_n$ are the vertices of a Euclidean simplex. To prove the theorem without this hypothesis, expand each determinant on the left in (1) by minors of elements in the row which contains $x^1, \cdots, x^n$. The results show that the sum in (1) is a linear function in $x^1, \cdots, x^n$. The coefficient of x^1 is a sum of cofactors; this sum is the same as the determinant of the matrix obtained from

$$\begin{bmatrix} a_0^1 & \cdots & a_0^n & 1 \\ \cdots & \cdots & \cdots & \cdots \\ a_n^1 & \cdots & a_n^n & 1 \end{bmatrix} \tag{2}$$

by replacing the first column with a column of 1's; thus this coefficient is zero. Similar considerations show that the coefficients of $x^2, \cdots, x^n$ also are zero. The sum of the constant terms in the expansions is the expansion, by minors of elements in the last column of (2), of the determinant $\Delta(a_0, a_1, \cdots, a_n)$. □

20.2 Example. Let a, b, c be three points in $\mathbb{R}$. Then formula (1) states that

$$\begin{vmatrix} c & 1 \\ b & 1 \end{vmatrix} + \begin{vmatrix} a & 1 \\ c & 1 \end{vmatrix} = \begin{vmatrix} a & 1 \\ b & 1 \end{vmatrix}, \tag{3}$$

or $(a - c) + (c - b) = (a - b)$. This formula is related to the following additivity property of the Riemann integral.

$$\int_a^c f(x)\,dx + \int_c^b f(x)\,dx = \int_a^b f(x)\,dx. \tag{4}$$

Theorem 20.1 will be used in Chapter 6 to establish a similar formula for the additivity of the Riemann integral in $\mathbb{R}^n$.

Recall the convention in Section 14 that $\widehat{\ }$, when placed above a term in a sequence, means that the term is omitted from the sequence.

20.3 Theorem. *If $a_i: (a_i^1, a_i^2, \cdots, a_i^n)$, $i = 0, 1, \cdots, n$, are points in $\mathbb{R}^n$, then*

$$\begin{vmatrix} a_0^1 & a_0^2 & \cdots & a_0^n & 1 \\ \cdots & \cdots & \cdots & \cdots & \cdots \\ a_n^1 & a_n^2 & \cdots & a_n^n & 1 \end{vmatrix} = - \begin{vmatrix} \widehat{a}_0^1 + a_1^1 + \cdots + a_n^1 & a_0^2 & \cdots & a_0^n & 1 \\ \cdots & \cdots & \cdots & \cdots & \cdots \\ a_0^1 + a_1^1 + \cdots + \widehat{a}_n^1 & a_n^2 & \cdots & a_n^n & 1 \end{vmatrix}. \tag{5}$$

Proof. In the determinant on the right in (5), multiply the last column of its matrix by $(a_0^1 + a_1^1 + \cdots + a_n^1)$ and subtract it from the first column. Then elementary properties of determinants [see Section 77 in Appendix 1] show that the determinant is unchanged, and that equation (5) is true. □

20.4 Example. Theorems 20.3 and 79.1 show that

$$\begin{vmatrix} f^1(a_0) & f^2(a_0) & 1 \\ f^1(a_1) & f^2(a_1) & 1 \\ f^1(a_2) & f^2(a_2) & 1 \end{vmatrix} = -\begin{vmatrix} f^1(a_1)+f^1(a_2) & f^2(a_0) & 1 \\ f^1(a_0)+f^1(a_2) & f^2(a_1) & 1 \\ f^1(a_0)+f^1(a_1) & f^2(a_2) & 1 \end{vmatrix}$$
$$= -\left\{[f^1(a_1)+f^1(a_2)]\begin{vmatrix} f^2(a_1) & 1 \\ f^2(a_2) & 1 \end{vmatrix} \right. \tag{6}$$
$$- [f^1(a_0)+f^1(a_2)]\begin{vmatrix} f^2(a_0) & 1 \\ f^2(a_2) & 1 \end{vmatrix}$$
$$\left. + [f^1(a_0)+f^1(a_1)]\begin{vmatrix} f^2(a_0) & 1 \\ f^2(a_1) & 1 \end{vmatrix}\right\}.$$

Let (a_0, a_1, a_2) be an oriented Euclidean simplex in $\mathbb{R}^2$. An examination of the terms on the right in (6) shows that each of the three terms is evaluated at the vertices of a simplex in the boundary of (a_0, a_1, a_n). Thus if formula (6) is applied to each simplex (a_0, a_1, a_2) in a chain in an oriented Euclidean complex K, the terms which arise from a common side of two simplexes cancel as in Example 18.7. Hence the sum, over the simplexes in a chain in K, of the determinants on the left in (6) equals the sum, over the simplexes in the boundary of the chain, of the expression on the right in (6). In Chapter 6, Theorem 20.3 is an important tool in the proof of one form of the fundamental theorem of the integral calculus.

The next theorem concerns the measure of an m-simplex $[a_0, a_1, \cdots, a_m]$ in $\mathbb{R}^n$. Here "measure" means the length of a segment, the area of a triangle, the volume of a tetrahedron, $\cdots$, and their generalization for an m-simplex $[a_0, a_1, \cdots, a_m]$ in $\mathbb{R}^n$. The methods of the integral calculus are used to calculate the measure and to derive a determinant formula for the measure of $[a_0, a_1, \cdots, a_m]$. A summary of notation will be helpful. The simplex $[a_0, a_1, \cdots, a_m]$, or $(a_0, a_1, \cdots, a_m)$, is conveniently represented by a matrix as follows:

$$(a_0, a_1, \cdots, a_m) = \begin{bmatrix} a_0^1 & a_0^2 & \cdots & a_0^n \\ \cdots & \cdots & \cdots & \cdots \\ a_m^1 & a_m^2 & \cdots & a_m^n \end{bmatrix} \tag{7}$$

Let $(j_1, \cdots, j_m)$ be an index set in (m/n). Then, as in Section 11,

$$(a_0, a_1, \cdots, a_m)^{(j_1, \cdots, j_m)} = \begin{bmatrix} a_0^{j_1} & \cdots & a_0^{j_m} \\ \cdots & \cdots & \cdots \\ a_m^{j_1} & \cdots & a_m^{j_m} \end{bmatrix}, \qquad (j_1, \cdots, j_m) \in (m/n); \tag{8}$$

$$\Delta[(a_0, a_1, \cdots, a_m)^{(j_1, \cdots, j_m)}] = \begin{vmatrix} a_0^{j_1} & \cdots & a_0^{j_m} & 1 \\ \cdots & \cdots & \cdots & \cdots \\ a_m^{j_1} & \cdots & a_m^{j_m} & 1 \end{vmatrix}, \qquad (j_1, \cdots, j_m) \in (m/n). \tag{9}$$

$$|\Delta(a_0, a_1, \cdots, a_m)| = \left\{ \sum_{(j_1, \cdots, j_m)} [\Delta[(a_0, a_1, \cdots, a_m)^{(j_1, \cdots, j_m)}]]^2 \right\}^{1/2}. \tag{10}$$

Let $v_i = a_i - a_0$, $i = 1, \cdots, m$. Then the Binet–Cauchy multiplication theorem [see Theorem 80.1(3)] shows that

$$\det[(v_i, v_j)]_1^m = \sum_{(j_1, \cdots, j_m)} \begin{vmatrix} a_1^{j_1} - a_0^{j_1} & \cdots & a_1^{j_m} - a_0^{j_m} \\ \cdots & \cdots & \cdots \\ a_m^{j_1} - a_0^{j_1} & \cdots & a_m^{j_m} - a_0^{j_m} \end{vmatrix}^2. \tag{11}$$

Since

$$\begin{vmatrix} a_0^{j_1} & \cdots & a_0^{j_m} & 1 \\ \cdots & \cdots & \cdots & \cdots \\ a_m^{j_1} & \cdots & a_m^{j_m} & 1 \end{vmatrix} = (-1)^m \begin{vmatrix} a_1^{j_1} - a_0^{j_1} & \cdots & a_1^{j_m} - a_0^{j_m} \\ \cdots & \cdots & \cdots \\ a_m^{j_1} - a_0^{j_1} & \cdots & a_m^{j_m} - a_0^{j_m} \end{vmatrix}, \tag{12}$$

equations (9), (10), and (11) show that

$$|\Delta(a_0, a_1, \cdots, a_m)| = \{\det[(v_i, v_j)]_1^m\}^{1/2}. \tag{13}$$

Let $M(a_0, a_1, \cdots, a_m)$ denote the measure of the simplex $(a_0, a_1, \cdots, a_m)$.

20.5 Theorem. *If $a_0, a_1, \cdots, a_m$ are linearly independent points in $\mathbb{R}^n$, then*

$$M(a_0, a_1, \cdots, a_m) = (1/m!)\{\det[(v_i, v_j)]_1^m\}^{1/2} > 0; \tag{14}$$

$$0 < \det[(v_i, v_j)]_1^m \leqq (v_1, v_1) \quad \cdots \quad (v_m, v_m); \tag{15}$$

that is,

$$0 < \det[(v_i, v_j)]_1^m \leqq |a_1 - a_0|^2 \quad \cdots \quad |a_m - a_0|^2. \tag{16}$$

If $m = 1$, the equality holds in (15) *and* (16); *if $m > 1$, the equality holds in* (15) *and* (16) *if and only if $v_1, \cdots, v_m$ are mutually orthogonal.*

Proof. Since the points $a_0, a_1, \cdots, a_m$ are linearly independent, the vectors $v_1, \cdots, v_m$ are linearly independent [see Section 14], and the determinant in (14) is greater than zero by (11) or Theorem 13.1 or Theorem 86.1. If $m = 1$, then

$$M(a_0, a_1) = \{(v_1, v_1)\}^{1/2} = \left\{ \sum_{j=1}^n (a_1^j - a_0^j)^2 \right\}^{1/2} = |a_1 - a_0| > 0, \tag{17}$$

and the theorem is true as stated. From this beginning, the theorem is proved by induction, using the methods of the integral calculus to calculate $M(a_0, a_1, \cdots, a_m)$.

Assume that the theorem is true if $m = r$; the proof by induction is com-

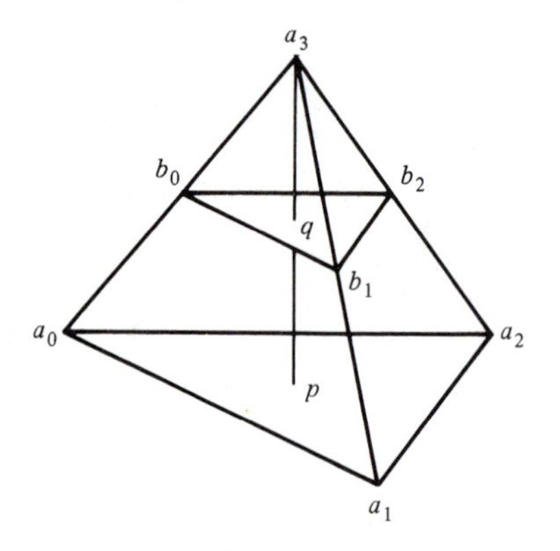

Figure 20.1. Figure for Theorem 20.5 in the case $r + 1 = 3$.

pleted by showing that the theorem is true also if $m = r + 1$. By hypothesis, $v_1, \cdots, v_r, v_{r+1}$ are linearly independent vectors. The vector

$$h = (a_0 + v_{r+1}) - (a_0 + t_1 v_1 + \cdots + t_r v_r), \tag{18}$$

$$h = v_{r+1} - (t_1 v_1 + \cdots + t_r v_r), \tag{19}$$

from the point $p = a_0 + t_1 v_1 + \cdots + t_r v_r$ in the plane $P(a_0, a_1, \cdots, a_r)$ to a_{r+1} is orthogonal to $P(a_0, a_1, \cdots, a_r)$ if and only if it is orthogonal to each of the vectors $v_1, \cdots, v_r$ [Figure 20.1 shows the case $r + 1 = 3$]. Thus, h is an altitude of $(a_0, a_1, \cdots, a_{r+1})$ if and only if

$$(v_i, h) = (v_i, v_{r+1} - t_1 v_1 - \cdots - t_r v_r) = 0, \qquad i = 1, \cdots, r, \tag{20}$$

$$(v_i, v_1)t_1 + \cdots + (v_i, v_r)t_r = (v_i, v_{r+1}), \qquad i = 1, \cdots, r. \tag{21}$$

These equations have a solution for $t_1, \cdots, t_r$ since the vectors $v_1, \cdots, v_r$ are linearly independent by hypothesis and $\det[(v_i, v_j)]_1^r > 0$ as stated above [or by the induction hypothesis that (14) is true for $m = r$]. Henceforth, let $t_1, \cdots, t_r$ denote the unique solution of (21).

Equation (18) shows that the altitude h is the vector (19) from the point $p = a_0 + t_1 v_1 + \cdots + t_r v_r$ in $P(a_0, a_1, \cdots, a_r)$ to the vertex $a_{r+1} = a_0 + v_{r+1}$. For each s in $[0, 1]$ define $(b_0, b_1, \cdots, b_r)$ and q by the following equations:

$$\begin{aligned} b_0 &= a_0 + s v_{r+1}; \\ b_i &= a_0 + v_i + s[(a_{r+1} - a_0) - (a_i - a_0)] \\ &= a_0 + v_i + s(v_{r+1} - v_i), \qquad i = 1, \cdots, r; \\ q &= p + sh. \end{aligned} \tag{22}$$

Since

$$b_i - b_0 = (1 - s)v_i, \qquad i = 1, \cdots, r, \tag{23}$$

the simplex $(b_0, b_1, \cdots, b_r)$ is parallel to $(a_0, a_1, \cdots, a_r)$ and orthogonal to h. The point q is the intersection of h and the plane $P(b_0, b_1, \cdots, b_r)$

because

$$q = b_0 + t_1(b_1 - b_0) + \cdots + t_r(b_r - b_0). \tag{24}$$

Then $|q - p| = s|h|$ is the distance from $P(a_0, a_1, \cdots, a_r)$ to $P(b_0, b_1, \cdots, b_r)$. By equation (23) and the induction hypothesis in (14),

$$\begin{aligned} M(b_0, b_1, \cdots, b_r) &= (1-s)^r M(a_0, a_1, \cdots, a_r), \\ M(a_0, a_1, \cdots, a_r) &= (1/r!)\{\det[(v_i, v_j)]_1^r\}^{1/2}. \end{aligned} \tag{25}$$

Then by the usual calculus definition of measure (volume),

$$M(a_0, a_1, \cdots, a_r, a_{r+1}) = \int_0^1 M(b_0, b_1, \cdots, b_r)|h|\, ds. \tag{26}$$

In the integral, replace $M(b_0, b_1, \cdots, b_r)$ by its value from (25) and integrate; then

$$\begin{aligned} M(a_0, a_1, \cdots, a_r, a_{r+1}) &= M(a_0, a_1, \cdots, a_r)|h| \int_0^1 (1-s)^r\, ds \\ &= \frac{M(a_0, a_1, \cdots, a_r)|h|}{r+1}. \end{aligned} \tag{27}$$

By the induction hypothesis, equation (25) gives the value of $M(a_0, a_1, \cdots, a_r)$; thus

$$M(a_0, a_1, \cdots, a_r, a_{r+1}) = [1/(r+1)!]\{\det[(v_i, v_j)]_1^r |h|^2\}^{1/2}. \tag{28}$$

To complete the proof, it is necessary and sufficient to show that

$$\det[(v_i, v_j)]_1^r |h|^2 = \det[(v_i, v_j)]_1^{r+1}. \tag{29}$$

In the matrix of the determinant on the right, multiply the first r rows in order by $t_1, t_2, \cdots, t_r$ and subtract them from the $(r+1)$-st row; then multiply the first r columns in order by $t_1, t_2, \cdots, t_r$ and subtract them from the $(r+1)$-st column. The resulting determinant, simplified by using equations (19) and (20) and properties of the inner product [see Section 84 in Appendix 1], equals the expression on the left in (29). Thus equations (28) and (29) show that

$$M(a_0, a_1, \cdots, a_r, a_{r+1}) = [1/(r+1)!]\{\det[(v_i, v_j)]_1^{r+1}\}^{1/2}. \tag{30}$$

The proof of equation (29) has used only the assumption that $v_1, \cdots, v_r$ are linearly independent. If $v_1, \cdots, v_r, v_{r+1}$ are linearly independent, then $\det[(v_i, v_j)]_1^{r+1} > 0$ as stated at the beginning of this proof, and equation (29) shows that $|h| > 0$ and (30) shows that $M(a_0, a_1, \cdots, a_r, a_{r+1}) > 0$. Thus formula (14) is true for $m = r + 1$ if it is true for $m = r$, and the induction proof of formula (14) is complete.

Induction proves inequalities (15) and (16) also; assume that (15) and (16) are true if $m = r$ and that $v_1, \cdots, v_r, v_{r+1}$ are linearly independent. By equation (19),

$$v_{r+1} = h + (t_1 v_1 + \cdots + t_r v_r). \tag{31}$$

Then by equation (20),

$$(v_{r+1}, v_{r+1}) = (h, h) + (t_1 v_1 + \cdots + t_r v_r, t_1 v_1 + \cdots + t_r v_r). \tag{32}$$

Thus $0 < |h| \leqq |v_{r+1}|$, and $|h| = |v_{r+1}|$ if and only if $t_1 v_1 + \cdots + t_r v_r = 0$; that is, if and only if v_{r+1} is orthogonal to $v_1, \cdots, v_r$ [see (20) and (31)]. If the inequalities in (15) and (16) hold for $m = r$, then (29) shows that the same inequalities hold if $m = r + 1$. The equality holds in (15) and (16) for $r = 1$. If the equality holds in (15) and (16) for $m = r$, then the equality in (15) and (16) holds for $m = r + 1$ if and only if v_{r+1} is orthogonal to $v_1, \cdots, v_r$. The proof of the entire theorem is complete. □

20.6 Example. For the 3-simplex $(a_0, a_1, \cdots, a_3)$ in $\mathbb{R}^3$ whose vertices are $a_0 : (1, 3, 5)$, $a_1 : (4, 5, 8)$, $a_2 : (3, -4, 5)$, $a_3 : (6, 8, 10)$,

$$\begin{aligned} v_1 &= a_1 - a_0 = (3, 2, 3), & |a_1 - a_0|^2 &= (v_1, v_1) = 22, \\ v_2 &= a_2 - a_0 = (2, -7, 0), & |a_2 - a_0|^2 &= (v_2, v_2) = 53, \\ v_3 &= a_3 - a_0 = (5, 5, 5), & |a_3 - a_0|^2 &= (v_3, v_3) = 75. \end{aligned}$$

Then by Theorem 20.5,

$$\begin{aligned} M(a_0, a_1, \cdots, a_3) &= \frac{1}{3!}\{\det[(v_i, v_j)]_1^3\}^{1/2} \\ &= \frac{1}{3!}\left\{\begin{vmatrix} 22 & -8 & 40 \\ -8 & 53 & -25 \\ 40 & -25 & 75 \end{vmatrix}\right\}^{1/2} = \frac{5}{3}. \end{aligned}$$

Since $\det[(v_i, v_j)]_1^3 = 100$, it is clear that the inequalities in (15) and (16) are satisfied. Since $(a_0, a_1, \cdots, a_3)$ is a 3-simplex in $\mathbb{R}^3$, $M(a_0, a_1, \cdots, a_3) = (1/3!)|\Delta(a_0, a_1, \cdots, a_3)|$ [see (26) in Section 89 in Appendix 2]. A straightforward calculation shows that $\Delta(a_0, a_1, \cdots, a_3) = 10$; thus $M(a_0, a_1, \cdots, a_3) = 5/3$ as before.

20.7 Example. For the 3-simplex $(a_0, a_1, \cdots, a_3)$ in $\mathbb{R}^3$ whose vertices are $a_0 : (3, 5, 4)$, $a_1 : (7, 7, -2)$, $a_2 : (6, 5, 6)$, $a_3 : (7, -21, -2)$,

$$\begin{aligned} v_1 &= a_1 - a_0 = (4, 2, -6), & |a_1 - a_0|^2 &= (v_1, v_1) = 56, \\ v_2 &= a_2 - a_0 = (3, 0, 2), & |a_2 - a_0|^2 &= (v_2, v_2) = 13, \\ v_3 &= a_3 - a_0 = (4, -26, -6), & |a_3 - a_0|^2 &= (v_3, v_3) = 728. \end{aligned}$$

Then by Theorem 20.5,

$$M(a_0, a_1, \cdots, a_3) = \frac{1}{3!}\{\det[(v_i, v_j)]_1^3\}^{1/2} = \frac{1}{3!}\left\{\begin{vmatrix} 56 & 0 & 0 \\ 0 & 13 & 0 \\ 0 & 0 & 728 \end{vmatrix}\right\}^{1/2} = \frac{364}{3}.$$

Since

$$\det[(v_i, v_j)]_1^3 = 56 \times 13 \times 728, \qquad (v_1, v_1)(v_2, v_2)(v_3, v_3) = 56 \times 13 \times 728,$$

the equality holds in (15) and (16). Since $(v_i, v_j) = 0$ if $i \neq j$, then v_1, v_2, v_3 are three mutually orthogonal vectors. The fact that the equality holds in (15) and (16) if and only if the vectors are mutually orthogonal proves that, for fixed lengths of $v_1, \cdots, v_m$, the measure $M(a_0, a_1, \cdots, a_m)$ of the simplex is maximum when the vectors are mutually orthogonal.

Exercises

20.1. Let $a_0 : (1, 2)$, $a_1 : (5, 3)$, $a_2 : (3, 7)$, and $x : (8, 4)$ be four points in $\mathbb{R}^2$.
(a) Write out the determinants in (1) in Theorem 20.1, and verify that the theorem is true in this case.
(b) Make a sketch which shows the four simplexes in the statement of the theorem in this case; find the orientation in $\mathbb{R}^2$ of each of these simplexes.

20.2. Let a_0, a_1, a_2, and x be four points on a line in $\mathbb{R}^2$. Is Theorem 20.1 true for these points? Make a sketch which shows the four points. Use a geometric argument to show why the theorem is trivially true in this case.

20.3. Let $a_0 : (4, 3)$, $a_1 : (6, -4)$, $a_2 : (8, 5)$ be three points in $\mathbb{R}^2$. Write out the two determinants in equation (5) in Theorem 20.3, and verify that the theorem is true in this case.

20.4. (a) Use Theorem 20.5 to find $M(a_0, a_1, \cdots, a_3)$ for the simplex $(a_0, a_1, \cdots, a_3)$ whose vertices are $a_0 : (1, 1, 2)$, $a_1 : (3, 2, 2)$, $a_2 : (2, 4, 5)$, $a_3 : (5, 6, 8)$. Verify that the inequalities in (15) and (16) are satisfied.
(b) Find the length of the altitude h from a_3 to the base (a_0, a_1, a_2) of the simplex in (a).
(c) Find the area of the triangle (a_0, a_1, a_2). Then check your answer in (a) by using the area of (a_0, a_1, a_2) and the length of the altitude h on this side found in (b).

20.5. Use Theorem 20.5 to find $M(a_0, a_1, \cdots, a_3)$ for the simplex whose vertices are $a_0 : (3, 1, 2)$, $a_1 : (1, 4, 7)$, $a_2 : (6, -2, 5)$, $a_3 : (11, 8, 1)$. Verify that $\det[(v_i, v_j)]_1^3 = \prod_1^3 (v_i, v_i)$; explain why the equality rather than the inequality holds in this case.

20.6. Let $(a_0, a_1, \cdots, a_m)$ be an m-simplex in $\mathbb{R}^n$. Show that

$$\begin{aligned} M(a_0, a_1, \cdots, a_m) &= (1/m!)|\Delta(a_0, a_1, \cdots, a_m)| \\ &= (1/m!)\left\{\sum_{(j_1, \cdots, j_m)} |\Delta[(a_0, a_1, \cdots, a_m)^{(j_1, \cdots, j_m)}]|^2\right\}^{1/2}, \\ &\qquad\qquad (j_1, \cdots, j_m) \in (m/n), \\ &= (1/m!)\{\det[(v_i, v_j)]_1^m\}^{1/2}. \end{aligned}$$

20.7. Let (a_0, a_1, a_2) be a simplex in $\mathbb{R}^4$ whose vertices are $a_0 : (2, 3, -4, 3)$, $a_1 : (5, 4, 2, 5)$, $a_2 : (6, 3, 4, 7)$.
(a) Use each of the formulas in Exercise 20.6 to find $M(a_0, a_1, a_2)$.

(b) Verify that the inequalities (15) and (16) are satisfied.

(c) Find the length of the altitude h from the vertex a_2 to the side (a_0, a_1). Use $|h|$ and the length of (a_0, a_1) to check your answers in (a).

20.8. (a) Assume that $v_1, \cdots, v_r$ are linearly independent and that v_{r+1} is a linear combination of these vectors. Use equation (29) to show that $|h| = 0$.

(b) Let $(a_0, a_1, \cdots, a_m)$ be an m-simplex in $\mathbb{R}^n$ such that $v_1, \cdots, v_m$ are linearly dependent. Explain why $M(a_0, a_1, \cdots, a_m)$ should be defined to be zero in this case.

(c) Show that formula (14) gives the correct value for $M(a_0, a_1, \cdots, a_m)$ for every m-simplex $(a_0, a_1, \cdots, a_m)$ in $\mathbb{R}^n$.

20.9. Use the inequality in (15) to prove the following statement of Hadamard's determinant theorem. Let $[a_{ij}]$, $i, j = 1, \cdots, n$, be a matrix of real numbers. Then $|\det[a_{ij}]|$ is equal to or less than the product of the lengths of the row vectors in $[a_{ij}]$, and also equal to or less than the product of the lengths of the column vectors in $[a_{ij}]$. Thus

$$|\det[a_{ij}]| \leqq \prod_{i=1}^{n} \left[\sum_{j=1}^{n} a_{ij}^2 \right]^{1/2}, \qquad |\det[a_{ij}]| \leqq \prod_{j=1}^{n} \left[\sum_{i=1}^{n} a_{ij}^1 \right]^{1/2}.$$

20.10. Show that Theorem 20.5 contains the following formulas: the area of a triangle is one-half the base times the altitude; the volume of a tetrahedron is one-third the area of the base times the altitude; $\cdots$; the measure $M(a_0, a_1, \cdots, a_m)$ is $(1/m)$ times the product of the measure $M(a_0, a_1, \cdots, a_{m-1})$ of the base $(a_0, a_1, \cdots, a_{m-1})$ and the altitude $|h|$ from a_m to the base.

20.11. Show that the measure of the n-simplex $(e_0, e_1, \cdots, e_n)$ in $\mathbb{R}^n$ is $(1/n!)$ for $n = 1, 2, \cdots$.

20.12. Let $(x, a_0, a_1, \cdots, a_n)$ be an oriented simplex whose vertices are points in some space $\mathbb{R}^m$. Find the boundary chain of $(x, a_0, a_1, \cdots, a_n)$ and show that one representation of it is the following:

$$\partial(x, a_0, a_1, \cdots, a_n) = (a_0, a_1, \cdots, a_n) - \sum_{r=0}^{n} (a_0, \cdots, a_{r-1}, x, a_{r+1}, \cdots, a_n).$$

20.13. Let s be the simplex $(a_0, a_1, \cdots, a_n)$ in $\mathbb{R}^n$, and let $\partial s = \Sigma_0^n w_j$. The determinant Δ is a real-valued function which is defined on the set of all n-simplexes in $\mathbb{R}^n$. Show that there is a function $u : \{w_0, w_1, \cdots, w_n\} \to \mathbb{R}$ such that $\Delta(s) = \Sigma_0^n u(w_j)$. [Hint. Theorem 20.3, Example 20.4.]

20.14. Let $x, a_0, a_1, \cdots, a_n$ be points in $\mathbb{R}^n$.

(a) Show that the chain $\partial(x, a_0, a_1, \cdots, a_n)$ is not a chain in a Euclidean complex. Make sketches for $n = 1$, 2, and 3.

(b) Evaluate each of the determinants in the following expression by the method explained in Exercise 20.13.

$$\Delta(a_0, a_1, \cdots, a_n) - \sum_{r=0}^{n} \Delta(a_0, \cdots, a_{r-1}, x, a_{r+1}, \cdots, a_n)$$

(c) The expression in (b) is defined on a chain which is the boundary of $(x, a_0, a_1, \cdots, a_n)$. Use this fact and part (b) to show that the expression in (b) equals zero and thus give a second proof of Theorem 20.1.

20.15. Let $x, a_0, a_1, \cdots, a_n$ be points in $\mathbb{R}^n$, and let f be a real-valued function whose components $f^1, \cdots, f^n$ are defined on a set E which contains these points. Let

$$\Delta f(a_0, a_1, \cdots, a_n) = \begin{vmatrix} f^1(a_0) & f^2(a_0) & \cdots & f^n(a_0) & 1 \\ \cdots & \cdots & \cdots & \cdots & \cdots \\ f^1(a_n) & f^2(a_n) & \cdots & f^n(a_n) & 1 \end{vmatrix},$$

and let $\Delta f(a_0, \cdots, a_{r-1}, x, a_{r+1}, \cdots, a_n)$ have a similar meaning for the other simplexes $(a_0, \cdots, a_{r-1}, x, a_{r+1}, \cdots, a_n)$ in $\partial(x, a_0, a_1, \cdots, a_n)$ [see Exercise 20.12]. Use Theorem 20.3 and Exercise 20.12 to prove that

$$\Delta f(a_0, a_1, \cdots, a_n) = \sum_{r=0}^{n} \Delta f(a_0, \cdots, a_{r-1}, x, a_{r+1}, \cdots, a_n).$$

Show that Exercise 20.14 is a special case of this exercise.

20.16. The points $a_i : (a_i^1, a_i^2, a_i^3)$, $i = 0, 1, 2$, are the vertices of a 2-simplex in $\mathbb{R}^3$. Set $b = \Sigma_0^3 t^i a_i$, $t^i \geqq 0$, $\Sigma_0^3 t^i = 1$. Then (b, a_1, a_2), (a_0, b, a_2), and (a_0, a_1, b) are three simplexes which form a subdivision of (a_0, a_1, a_2). Prove that the area of (a_0, a_1, a_2) is equal to the sum of the areas of the three simplexes in the subdivision. [Outline of solution. By equation (22) and Theorem 19.12 in Section 19,

$$\begin{bmatrix} b^1 & b^2 & b^3 & 1 \\ a_1^1 & a_1^2 & a_1^3 & 1 \\ a_2^1 & a_2^2 & a_2^3 & 1 \end{bmatrix} = \begin{bmatrix} t^1 & t^2 & t^3 \\ 0 & 1 & 0 \\ 0 & 0 & 1 \end{bmatrix} \begin{bmatrix} a_0^1 & a_0^2 & a_0^3 & 1 \\ a_1^1 & a_1^2 & a_1^3 & 1 \\ a_2^1 & a_2^2 & a_2^3 & 1 \end{bmatrix},$$

and there are similar matrix equations which relate (a_0, a_1, a_2) to (a_0, b, a_2) and (a_0, a_1, b). Use Theorem 19.12 and the formula

$$M(a_0, a_1, \cdots, a_m) = (1/m!) \left\{ \sum_{(j_1, \cdots, j_m)} |\Delta[(a_0, a_1, \cdots, a_m)^{(j_1, \cdots, j_m)}]|^2 \right\}^{1/2},$$

$$(j_1, \cdots, j_m) \in (m/n).$$

The sum of the areas of the three simplexes in the subdivision equals the area of (a_0, a_1, a_2) multiplied by

$$\begin{vmatrix} t^1 & t^2 & t^3 \\ 0 & 1 & 0 \\ 0 & 0 & 1 \end{vmatrix} + \begin{vmatrix} 1 & 0 & 0 \\ t^1 & t^2 & t^3 \\ 0 & 0 & 1 \end{vmatrix} + \begin{vmatrix} 1 & 0 & 0 \\ 0 & 1 & 0 \\ t^1 & t^2 & t^3 \end{vmatrix}.$$

In each determinant add the first two columns to the third; the sum of the three determinants equals

$$\begin{vmatrix} t^1 & t^2 & 1 \\ 0 & 1 & 1 \\ 0 & 0 & 1 \end{vmatrix} + \begin{vmatrix} 1 & 0 & 1 \\ t^1 & t^2 & 1 \\ 0 & 0 & 1 \end{vmatrix} + \begin{vmatrix} 1 & 0 & 1 \\ 0 & 1 & 1 \\ t^1 & t^2 & 1 \end{vmatrix}.$$

By Theorem 20.1 this sum has the value 1, and the sum of the areas of (b, a_1, a_2), (a_0, b, a_2), and (a_0, a_1, b) equals the area of (a_0, a_1, a_2).]

20.17. Let $a_0, a_1, \cdots, a_m$ be linearly independent points which form the vertices of an m-simplex in $\mathbb{R}^n$, $m < n$. Let $b = \Sigma_0^m t^i a_i$, $t^i \geqq 0$, $\Sigma_0^m t^i = 1$. Show that the sum of the measures of the simplexes $(a_0, \cdots, a_{r-1}, b, a_{r+1}, \cdots, a_m)$, $r = 0, 1, \cdots, m$, equals the measure of $(a_0, a_1, \cdots, a_m)$.

20.18. This exercise derives the Cayley forms of the multiplication theorem from the Binet–Cauchy multiplication theorem; the special case in this exercise suggests the general case in Exercise 20.19. Let $a_i : (a_i^1, a_i^2, a_i^3)$ and $b_i : (b_i^1, b_i^2, b_i^3)$, $i = 0, 1, 2$, be points in $\mathbb{R}^3$. Set $v_i = a_i - a_0$ and $w_i = b_i - b_0$ for $i = 1, 2$. Let (a_0, a_1, a_2) and (b_0, b_1, b_2) be 2-simplexes in $\mathbb{R}^3$ with the vertices indicated, and let (j_1, j_2) denote an index set in (2/3).

(a) Use the Binet–Cauchy multiplication theorem to show that

$$\sum_{(j_1, j_2)} \Delta[(a_0, a_1, a_2)^{(j_1, j_2)}]\Delta[(b_0, b_1, b_2)^{(j_1, j_2)}] = \det[(v_i, w_j)]_{i,j=1}^2.$$

(b) Two expressions result from applying the Binet–Cauchy multiplication theorem to the following matrices:

$$\begin{bmatrix} \Sigma_1^3 (a_0^j)^2 & -2a_0^1 & -2a_0^2 & -2a_0^3 & 1 \\ \Sigma_1^3 (a_1^j)^2 & -2a_1^1 & -2a_1^2 & -2a_1^3 & 1 \\ \Sigma_1^3 (a_2^j)^2 & -2a_2^1 & -2a_2^2 & -2a_2^3 & 1 \\ 1 & 0 & 0 & 0 & 0 \end{bmatrix}$$

$$\begin{bmatrix} 1 & b_0^1 & b_0^2 & b_0^3 & \Sigma_1^3 (b_0^j)^2 \\ 1 & b_1^1 & b_1^2 & b_1^3 & \Sigma_1^3 (b_1^j)^2 \\ 1 & b_2^1 & b_2^2 & b_2^3 & \Sigma_1^3 (b_2^j)^2 \\ 0 & 0 & 0 & 0 & 1 \end{bmatrix}$$

Show that the sum of products of determinants in the multiplication theorem can be simplified to

$$-2^2 \sum_{(j_1, j_2)} \Delta[(a_0, a_1, a_2)^{(j_1, j_2)}]\Delta[(b_0, b_1, b_2)^{(j_1, j_2)}], \qquad (j_1, j_2) \in (2/3).$$

The determinant of the product of the first matrix and the transpose of the second is the determinant of a 4×4 matrix; find this determinant, simplify it, and complete the proof of the following identities:

$$\det[(v_i, w_j)]_{i,j=1}^2 = \sum_{(j_1, j_2)} \Delta[(a_0, a_1, a_2)^{(j_1, j_2)}]\Delta[(b_0, b_1, b_2)^{(j_1, j_2)}]$$

$$= -\frac{1}{2^2} \begin{vmatrix} |a_0 - b_0|^2 & |a_0 - b_1|^2 & |a_0 - b_2|^2 & 1 \\ |a_1 - b_0|^2 & |a_1 - b_1|^2 & |a_1 - b_2|^2 & 1 \\ |a_2 - b_0|^2 & |a_2 - b_1|^2 & |a_2 - b_2|^2 & 1 \\ 1 & 1 & 1 & 0 \end{vmatrix}$$

$$= -\begin{vmatrix} (a_0, b_0) & (a_0, b_1) & (a_0, b_2) & 1 \\ (a_1, b_0) & (a_1, b_1) & (a_1, b_2) & 1 \\ (a_2, b_0) & (a_2, b_1) & (a_2, b_2) & 1 \\ 1 & 1 & 1 & 0 \end{vmatrix}.$$

20.19. Let $a_i : (a_i^1, a_i^1, \cdots, a_i^n)$ and $b_i : (b_i^1, b_i^2, \cdots, b_i^n)$, $i = 0, 1, \cdots, m$, be the vertices of the m-simplexes $(a_0, a_1, \cdots, a_m)$ and $(b_0, b_1, \cdots, b_m)$ in $\mathbb{R}^n$, $m \leqq n$. Set $v_i = a_i - a_0$ and $w_i = b_i - b_0$, $i = 1, \cdots, m$. State and prove the Cayley additions to the Binet–Cauchy multiplication theorem in the general case; Exercise 20.18 outlines the proof in the special case $n = 3$, $m = 2$.

20.20. (a) Let $(a_0, a_1, \cdots, a_m)$ be an m-simplex in $\mathbb{R}^n$, $m \leqq n$. Prove that

$$M(a_0, a_1, \cdots, a_m)$$
$$= \frac{1}{m!}\left\{\frac{(-1)^{m+1}}{2^m}\begin{vmatrix} 0 & |a_0 - a_1|^2 & \cdots & |a_0 - a_m|^2 & 1 \\ |a_1 - a_0|^2 & 0 & \cdots & |a_1 - a_m|^2 & 1 \\ \cdots & \cdots & \cdots & \cdots & \cdots \\ |a_m - a_0|^2 & |a_m - a_1|^2 & \cdots & 0 & 1 \\ 1 & 1 & \cdots & 1 & 0 \end{vmatrix}\right\}^{1/2}.$$

(b) Use the formula in (a) to find the area of the simplex (triangle) (a_0, a_1, a_2) in Exercise 20.7.

(c) Use the formula in (a) to find the volume of the simplex (tetrahedron) $(a_0, a_1, \cdots, a_3)$ in Exercise 20.4.

20.21. (a) Let $(a_0, a_1, \cdots, a_m)$ be an m-simplex in $\mathbb{R}^n$, $m \leqq n$, each of whose edges has length a; that is, $|a_i - a_j| = a$, $i \neq j$, $i, j = 0, 1, \cdots, m$. Prove that

$$M(a_0, a_1, \cdots, a_m) = \frac{a^m}{m!}\left[\frac{m+1}{2^m}\right]^{1/2}, \qquad m = 1, 2, \cdots, n.$$

(b) A triangle in $\mathbb{R}^2$ has vertices $a_0 : (0, 0)$, $a_1 : (10, 0)$, $a_2 : (5, 5\sqrt{3})$. Show that (a_0, a_1, a_2) is an equilateral triangle, and that the length of each side is 10. Find the area of (a_0, a_1, a_2). Check your answer by using the formula in (a) to find the area.

20.22. Prove that the measure of every simplex in $\mathbb{R}^n$ is invariant under a transformation which preserves distances.

20.23. Let $v = a_1 - a_0$ and $w = b_1 - b_0$ be two vectors in $\mathbb{R}^n$. Show that $(v - w, v - w) = (v, v) - 2(v, w) + (w, w)$, and thus that $(v, w) = (1/2)\{|v|^2 + |w|^2 - |v - w|^2\}$. Use this relation to prove that (v, w) is invariant under a transformation which preserves distances. Describe a generalization of this result which follows from Exercises 20.18 and 20.19.

20.24. Let $(a_0, a_1, \cdots, a_m)$ be an m-simplex in $\mathbb{R}^n$, and let $v_i = a_i - a_0$. The formula

$$M(a_0, a_1, \cdots, a_m) = (1/m!)\{\det[(v_i, v_j)]_1^m\}^{1/2}$$

in Exercise 20.6 for $M(a_0, a_1, \cdots, a_m)$ and $(1/m!)|\Delta(a_0, a_1, \cdots, a_m)|$ suggests that the vertex a_0 plays a preferred role and that $M(a_0, a_1, \cdots, a_m)$ and $|\Delta(a_0, a_1, \cdots, a_m)|$ are not symmetrical functions of $a_0, a_1, \cdots, a_m$. Give three formulas for $M(a_0, a_1, \cdots, a_m)$ and $|\Delta(a_0, a_1, \cdots, a_m)|$ which show that they are in fact symmetrical functions of the vertices of the simplex.

20.25. Let $f: E \to \mathbb{R}$ be a function which is defined on an open set E in $\mathbb{R}^2$ and which is differentiable at x_0 in E. Let $X^*(x_0, \rho)$ be a class of 2-vectors $\mathbf{x} : (x_1, x_2, x_3)$ in E which has the following properties: if $\mathbf{x}$ is in $X^*(x_0, \rho)$, then

(i) x_0 is in the interior of (x_1, x_2, x_3);
(ii) $\mathbf{x}^1 : (x_0, x_2, x_3)$, $\mathbf{x}^2 : (x_1, x_0, x_3)$, and $\mathbf{x}^3 : (x_1, x_2, x_0)$ are in $X(x_0, \rho)$.

(a) Prove that the four determinants $\Delta(\mathbf{x})$, $\Delta(\mathbf{x}^1)$, $\Delta(\mathbf{x}^2)$, $\Delta(\mathbf{x}^3)$ are different from zero and have the same sign.

(b) Prove the following statements for $j = 1, 2$.

$$\Delta_j f(\mathbf{x}) = \sum_{k=1}^{3} \Delta_j f(\mathbf{x}^k);$$

$$\frac{\Delta_j f(\mathbf{x})}{\Delta(\mathbf{x})} - D_j f(x_0) = \sum_{k=1}^{3} \left\{ \frac{\Delta_j f(\mathbf{x}^k)}{\Delta(\mathbf{x}^k)} - D_j f(x_0) \right\} \frac{\Delta(\mathbf{x}^k)}{\Delta(\mathbf{x})};$$

$$\frac{\Delta(\mathbf{x}^k)}{\Delta(\mathbf{x})} < 1, \qquad k = 1, 2, 3.$$

(c) Prove that

$$\lim_{\mathbf{x} \to x_0} \frac{\Delta_j f(\mathbf{x})}{\Delta(\mathbf{x})} = D_j f(x_0), \qquad \mathbf{x} \in X^*(x_0, \rho), \qquad j = 1, 2.$$

20.26. Exercise 20.25 contains a special case of a general theorem. State and prove the general theorem.

21. Simplicial Subdivisions

If $x_0 = a$ and $x_n = b$, the set of intervals $[x_{i-1}, x_i]$, $i = 1, \cdots, n$, is a subdivision $\mathscr{P}$ of $[a, b]$. The theory of Riemann integration employs sequences of such subdivisions of $[a, b]$ in which the maximum length of a subinterval $[x_{i-1}, x_i]$ approaches zero. If $\mathscr{P}_1$ and $\mathscr{P}_2$ are two subdivisions of $[a, b]$, and if each subinterval in $\mathscr{P}_2$ is contained in a subinterval of $\mathscr{P}_1$, then $\mathscr{P}_2$ is called a refinement of $\mathscr{P}_1$. A generalization of these ideas is concerned with the simplicial subdivision of simplexes and cubes in $\mathbb{R}^n$. Sequences of simplicial subdivisions of a given simplex $(a_0, a_1, \cdots, a_m)$ in $\mathbb{R}^n$ have important applications in later chapters of this book, including the chapters on integration. This section gives a constructive proof of the existence of sequences of simplicial subdivisions and establishes several of their properties for later use; it begins with the definition of a simplicial subdivision.

21.1 Definition. Let K be a (finite) Euclidean, homogeneous, m-dimensional, simplicial complex in $\mathbb{R}^n$, $0 < m \leqq n$. Let $\mathscr{P}$ be a complex of the same type which has the following properties: (a) the number of m-dimensional simplexes in $\mathscr{P}$ is greater than the number of m-simplexes in K; (b) each Euclidean simplex in $\mathscr{P}$ is a subset of a Euclidean simplex in K; and (c) the union of the simplexes in $\mathscr{P}$ equals the union of the simplexes in K. Then $\mathscr{P}$ is a *simplicial subdivision* of K. If $\mathscr{P}_1$ and $\mathscr{P}_2$ are two simplicial subdivisions of K, and if $\mathscr{P}_2$ is a simplicial subdivision of $\mathscr{P}_1$, then $\mathscr{P}_2$ is called a *refinement* of $\mathscr{P}_1$.

21.2 Example. The simplexes

$$s_1 : (a_0, a_3, a_5), \qquad s_2 : (a_1, a_4, a_3),$$

$$s_3 : (a_2, a_5, a_4), \qquad s_4 : (a_3, a_4, a_5),$$

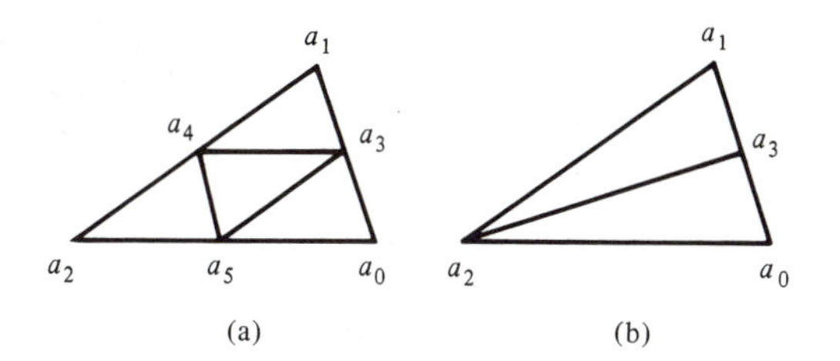

Figure 21.1. Figure for Example 21.2.

in Figure 21.1(a) form a simplicial subdivision of $s:(a_0, a_1, a_2)$. The Δ function is positive on each of these simplexes; in fact, in this case each simplex is positively oriented in $\mathbb{R}^2$. This example will show that $\Delta(s) = \Sigma_1^4 \Delta(s_i)$ and hence that $M(s) = \Sigma_1^4 M(s_i)$. Use Theorem 20.1 to obtain a representation of each $\Delta(s_i)$ in terms of the vertex a_2 as follows.

$$\begin{aligned}
\Delta(s_1) &= \Delta(a_0, a_3, a_5) = \Delta(a_2, a_3, a_5) + \Delta(a_0, a_2, a_5) + \Delta(a_0, a_3, a_2),\\
\Delta(s_2) &= \Delta(a_1, a_4, a_3) = \Delta(a_2, a_4, a_3) + \Delta(a_1, a_2, a_3) + \Delta(a_1, a_4, a_2),\\
\Delta(s_3) &= \Delta(a_2, a_5, a_4) = \Delta(a_2, a_5, a_4) + \Delta(a_2, a_2, a_4) + \Delta(a_2, a_5, a_2),\\
\Delta(s_4) &= \Delta(a_3, a_4, a_5) = \Delta(a_2, a_4, a_5) + \Delta(a_3, a_2, a_5) + \Delta(a_3, a_4, a_2).
\end{aligned} \tag{1}$$

Now

$$\begin{aligned}
\Delta(a_0, a_2, a_5) &= 0, \qquad \Delta(a_1, a_4, a_2) = 0,\\
\Delta(a_2, a_2, a_4) &= 0, \qquad \Delta(a_2, a_5, a_2) = 0,
\end{aligned} \tag{2}$$

because in each case the three vertices of the simplex lie on a line. Also,

$$\begin{aligned}
\Delta(a_2, a_3, a_5) &= -\Delta(a_3, a_2, a_5),\\
\Delta(a_2, a_4, a_3) &= -\Delta(a_3, a_4, a_2),\\
\Delta(a_2, a_5, a_4) &= -\Delta(a_2, a_4, a_5),
\end{aligned} \tag{3}$$

because in each case the two simplexes have the same vertices but opposite orientations. In the first equation, for example, the two terms arise from the simplexes (a_0, a_3, a_5) and (a_3, a_4, a_5); these simplexes have the side (a_3, a_5) in common, but this common side has opposite orientations in the boundaries of the two simplexes since

$$\begin{aligned}
\partial(a_0, a_3, a_5) &= (a_3, a_5) + \cdots,\\
\partial(a_3, a_4, a_5) &= -\partial(a_4, a_3, a_5) = -(a_3, a_5) + \cdots.
\end{aligned} \tag{4}$$

The fact that the common side (a_3, a_5) has opposite orientations in (a_0, a_3, a_5) and (a_3, a_4, a_5) is a consequence of Theorem 18.5 since a_0 and a_4 lie on opposite sides of the common side (a_3, a_5). Thus equations (1), (2), (3) show that

$$\sum_{i=1}^{4} \Delta(s_i) = \Delta(a_0, a_3, a_2) + \Delta(a_1, a_2, a_3). \tag{5}$$

The two simplexes (a_0, a_3, a_2) and (a_1, a_2, a_3) on the right in this equation are shown in Figure 21.1(b). Now use Theorem 20.1 again to obtain a representation of the two terms on the right in (5) in terms of the vertex a_1 as follows.

$$\begin{aligned} \Delta(a_0, a_3, a_2) &= \Delta(a_1, a_3, a_2) + \Delta(a_0, a_1, a_2) + \Delta(a_0, a_3, a_1), \\ \Delta(a_1, a_2, a_3) &= \Delta(a_1, a_2, a_3) + \Delta(a_1, a_1, a_3) + \Delta(a_1, a_2, a_1). \end{aligned} \tag{6}$$

In these equations

$$\begin{gathered} \Delta(a_0, a_3, a_1) = 0, \qquad \Delta(a_1, a_1, a_3) = 0, \qquad \Delta(a_1, a_2, a_1) = 0, \\ \Delta(a_1, a_3, a_2) = -\Delta(a_1, a_2, a_3), \end{gathered} \tag{7}$$

for reasons similar to those which establish (2) and (3). The equations (5), (6), and (7) show that

$$\sum_{i=1}^{4} \Delta(s_i) = \Delta(a_0, a_1, a_2) = \Delta(s). \tag{8}$$

Since $M(s) = (1/2!)\Delta(s)$, this equation proves also that

$$\sum_{i=1}^{4} M(s_i) = (1/2!)\Delta(a_0, a_1, a_2) = M(s). \tag{9}$$

In this example,

$$\begin{aligned} \partial s_1 &= \partial(a_0, a_3, a_5) = (a_3, a_5) - (a_0, a_5) + (a_0, a_3), \\ \partial s_2 &= \partial(a_1, a_4, a_3) = (a_4, a_3) - (a_1, a_3) + (a_1, a_4), \\ \partial s_3 &= \partial(a_2, a_5, a_4) = (a_5, a_4) - (a_2, a_4) + (a_2, a_5), \\ \partial s_4 &= \partial(a_3, a_4, a_5) = (a_4, a_5) - (a_3, a_5) + (a_3, a_4). \end{aligned} \tag{10}$$

Recall that $\partial \Sigma_1^4 s_i = \Sigma_1^4 \partial s_i$; in $\Sigma_1^4 \partial s_i$, the following terms cancel:

$$\begin{aligned} (a_3, a_5) - (a_3, a_5) &= 0, \\ (a_4, a_3) + (a_3, a_4) &= 0, \\ (a_5, a_4) + (a_4, a_5) &= 0. \end{aligned} \tag{11}$$

These terms correspond to the terms in equation (3). Then (10) and (11) show that

$$\partial \sum_{i=1}^{4} s_i = (a_0, a_3) + (a_3, a_1) + (a_1, a_4) + (a_4, a_2) + (a_2, a_5) + (a_5, a_0). \tag{12}$$

Observe that each Euclidean simplex in $\partial \Sigma_1^4 s_i$ is contained in a Euclidean simplex in $\partial(a_0, a_1, a_2)$, and that the two simplexes determine the same direction in $\mathbb{R}^2$ [see Figure 21.1(a) and Definition 15.13].

21.3 Theorem. *Let $s: (a_0, a_1, \cdots, a_n)$ be a Euclidean simplex in $\mathbb{R}^n$ such that $\Delta(a_0, a_1, \cdots, a_n) > 0$. Let s_i, $i = 1, \cdots, N$, be the simplexes in a subdivision $\mathscr{P}$ of $(a_0, a_1, \cdots, a_n)$, and assume that $\Delta(s_i) > 0$. Then:*

$$\sum_{i=1}^{N} \Delta(s_i) = \Delta(a_0, a_1, \cdots, a_n) = \Delta(s). \tag{13}$$

$$\sum_{i=1}^{N} M(s_i) = (1/n!)\Delta(a_0, a_1, \cdots, a_n) = M(s). \tag{14}$$

If $(p_{i_0}, p_{i_1}, \cdots, p_{i_r})$ is a side of a simplex in $\partial\Sigma_1^N s_i$, and if $(p_{i_0}, p_{i_1}, \cdots, p_{i_r})$ is contained in the side $(a_{i_0}, a_{i_1}, \cdots, a_{i_r})$ of a simplex in ∂s, then $(p_{i_0}, p_{i_1}, \cdots, p_{i_r})$ and $(a_{i_0}, a_{i_1}, \cdots, a_{i_r})$ determine the same r-direction in $\mathbb{R}^n$. (15)

Proof. Example 21.2 contains a proof of this theorem for $n = 2$ in a special case. The proof will now be given for $n = 3$ in the general case. These two examples explain the proof of the theorem in all cases.

Let (p_0, p_1, p_2, p_3) be a simplex s_i in $\mathscr{P}$. Then $\Delta(p_0, p_1, p_2, p_3) > 0$. Let p_0, p_1, p_2, q_3 be the vertices of a second simplex in $\mathscr{P}$ which has a side in common with the first. The two simplexes belong to a Euclidean complex $\mathscr{P}$; therefore, they lie on opposite sides of the plane $P(p_0, p_1, p_2)$, and Theorem 18.5 shows that the second simplex is (p_1, p_0, p_2, q_3). Use Theorem 20.1 to obtain a representation of $\Delta(p_0, p_1, p_2, p_3)$ and $\Delta(p_1, p_0, p_2, q_3)$ in terms of the vertex a_3 as follows:

$$\begin{aligned}\Delta(p_0, p_1, p_2, p_3) &= \Delta(a_3, p_1, p_2, p_3) + \Delta(p_0, a_3, p_2, p_3)\\ &\quad + \Delta(p_0, p_1, a_3, p_3) + \Delta(p_0, p_1, p_2, a_3),\\ \Delta(p_1, p_0, p_2, q_3) &= \Delta(a_3, p_0, p_2, q_3) + \Delta(p_1, a_3, p_2, q_3)\\ &\quad + \Delta(p_1, p_0, a_3, q_3) + \Delta(p_1, p_0, p_2, a_3).\end{aligned} \tag{16}$$

Represent each term $\Delta(s_i)$, s_i in $\mathscr{P}$, as in (16). In the sum $\Sigma_1^N \Delta(s_i)$, the last two terms on the right in (16) cancel since

$$\Delta(p_1, p_0, p_2, a_3) = -\Delta(p_0, p_1, p_2, a_3). \tag{17}$$

Let c_3 denote the chain $\Sigma_1^N s_i$. Then in $\Sigma_1^N \Delta(s_i)$, all terms cancel except terms of the form $\Delta(p_0, p_1, p_2, a_3)$ in which (p_0, p_1, p_2) is a simplex in ∂c_3; these simplexes (p_0, p_1, p_2) are contained in the simplexes which form the sides of (a_0, a_1, a_2, a_3). Thus

$$\sum_{i=1}^{N} \Delta(s_i) = \sum_{(p_0, p_1, p_2)} \Delta(p_0, p_1, p_2, a_3), \qquad (p_0, p_1, p_2) \text{ in } \partial c_3. \tag{18}$$

Since $\mathscr{P}$ is a subdivision of $s:(a_0, a_1, a_2, a_3)$, the simplexes in ∂c_3 are contained in the simplexes in ∂s. Now

$$\partial(a_0, a_1, a_2, a_3) = (a_1, a_2, a_3) - (a_0, a_2, a_3) + (a_0, a_1, a_3) - (a_0, a_1, a_2). \tag{19}$$

If (p_0, p_1, p_2) is in a side of s which contains a_3, then the simplex (p_0, p_1, p_2, a_3) has four vertices in a plane and $\Delta(p_0, p_1, p_2, a_3) = 0$. Then (18) simplifies to the following [see Figure 21.2]:

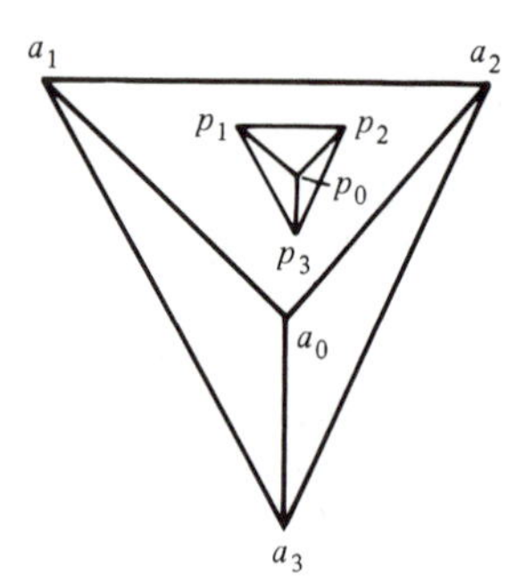

Figure 21.2. The simplex (p_0, p_1, p_2) is in (a_0, a_1, a_2).

$$\sum_{i=1}^{N} \Delta(s_i) = \sum_{(p_0, p_1, p_2)} \Delta(p_0, p_1, p_2, a_3), \tag{20}$$

$$(p_0, p_1, p_2) \quad \text{in} \quad \partial c_3, \qquad (p_0, p_1, p_2) \quad \text{in} \quad (a_0, a_1, a_2). \tag{21}$$

Let c_2 denote the chain which contains the simplexes in (21); then c_2 contains only those simplexes in ∂c_3 which are in (a_0, a_1, a_2). The derivation of (20) has shown that (p_0, p_1, p_2) is a side of a simplex $s_i : (p_0, p_1, p_2, p_3)$ for which $\Delta(s_i) > 0$ by hypothesis. Since p_3 and a_3 are on the same side of the plane $P(p_0, p_1, p_2)$ [see Figure 21.2], then $\Delta(p_0, p_1, p_2, p_3) > 0$ implies that

$$\Delta(p_0, p_1, p_2, a_3) > 0, \qquad (p_0, p_1, p_2) \quad \text{in} \quad c_2. \tag{22}$$

The next step in the proof will show that (p_0, p_1, p_2) and (a_0, a_1, a_2) determine the same 2-direction in $\mathbb{R}^3$. By Theorem 19.12 there exists a barycentric transformation which transforms (a_0, a_1, a_2, a_3) into (p_0, p_1, p_2, a_3). In matrix form this transformation is

$$\begin{bmatrix} p_0^1 & p_0^2 & p_0^3 & 1 \\ p_1^1 & p_1^2 & p_1^3 & 1 \\ p_2^1 & p_2^2 & p_2^3 & 1 \\ a_3^1 & a_3^2 & a_3^3 & 1 \end{bmatrix} = \begin{bmatrix} t_0^1 & t_0^2 & t_0^3 & 0 \\ t_1^1 & t_1^2 & t_1^3 & 0 \\ t_2^1 & t_2^2 & t_2^3 & 0 \\ 0 & 0 & 0 & 1 \end{bmatrix} \begin{bmatrix} a_0^1 & a_0^2 & a_0^3 & 1 \\ a_1^1 & a_1^2 & a_1^3 & 1 \\ a_2^1 & a_2^2 & a_2^3 & 1 \\ a_3^1 & a_3^2 & a_3^3 & 1 \end{bmatrix}. \tag{23}$$

Since $\Delta(a_0, a_1, a_2, a_3) > 0$ by hypothesis and $\Delta(p_0, p_1, p_2, a_3) > 0$ by (22), equation (23) and the Binet–Cauchy multiplication theorem show that

$$\det \begin{bmatrix} t_0^1 & t_0^2 & t_0^3 & 0 \\ t_1^1 & t_1^2 & t_1^3 & 0 \\ t_2^1 & t_2^2 & t_2^3 & 0 \\ 0 & 0 & 0 & 1 \end{bmatrix} > 0, \qquad \det \begin{bmatrix} t_0^1 & t_0^2 & t_0^3 \\ t_1^1 & t_1^2 & t_1^3 \\ t_2^1 & t_2^2 & t_2^3 \end{bmatrix} > 0. \tag{24}$$

Then (23) and (24) show that

$$\begin{bmatrix} p_0^1 & p_0^2 & p_0^3 & 1 \\ p_1^1 & p_1^2 & p_1^3 & 1 \\ p_2^1 & p_2^2 & p_2^3 & 1 \end{bmatrix} = \begin{bmatrix} t_0^1 & t_0^2 & t_0^3 \\ t_1^1 & t_1^2 & t_1^3 \\ t_2^1 & t_2^2 & t_2^3 \end{bmatrix} \begin{bmatrix} a_0^1 & a_0^2 & a_0^3 & 1 \\ a_1^1 & a_1^2 & a_1^3 & 1 \\ a_2^1 & a_2^2 & a_2^3 & 1 \end{bmatrix}, \tag{25}$$

and that

$$\Delta[(p_0, p_1, p_2)^{(j_1,j_2)'}] = \det[t_i^j]\Delta[(a_0, a_1, a_2)^{(j_1,j_2)'}], \qquad (j_1, j_2) \in (2/3). \quad (26)$$

Equations (24) and (26) show that (p_0, p_1, p_2) and (a_0, a_1, a_2) determine the same 2-direction in $\mathbb{R}^3$ [see Definition 15.13, Theorem 19.12, and Example 19.13]. This statement proves one part of conclusion (15) of Theorem 21.3.

Return now to equation (20). Let (p_0, p_1, p_2, a_3) be the simplex of one term in the sum, and let p_0, p_1, q_2, a_3 be the vertices of a simplex which has the vertices p_0, p_1, a_3 in common with the first. Then p_2 and q_2 lie on opposite sides of the plane $P(p_0, p_1, a_3)$, through the common side of the two simplexes. Since all simplexes are oriented so that the Δ function is positive on them by (22), the first simplex is (p_0, p_1, p_2, a_3) by (22) and the second is (p_1, p_0, q_2, a_3). Use Theorem 20.1 to obtain representations of $\Delta(p_0, p_1, p_2, a_3)$ and $\Delta(p_1, p_0, q_2, a_3)$ in terms of the vertex a_2 as follows:

$$\begin{aligned}\Delta(p_0, p_1, p_2, a_3) &= \Delta(a_2, p_1, p_2, a_3) + \Delta(p_0, a_2, p_2, a_3)\\ &\quad + \Delta(p_0, p_1, a_2, a_3) + \Delta(p_0, p_1, p_2, a_2),\\ \Delta(p_1, p_0, q_2, a_3) &= \Delta(a_2, p_0, q_2, a_3) + \Delta(p_1, a_2, q_2, a_3)\\ &\quad + \Delta(p_1, p_0, a_2, a_3) + \Delta(p_1, p_0, q_2, a_2).\end{aligned} \quad (27)$$

The simplexes (p_0, p_1, p_2, a_2) and (p_1, p_0, q_2, a_2) have four vertices in a plane; hence, the last terms on the right in (27) equal zero:

$$\Delta(p_0, p_1, p_2, a_2) = 0, \qquad \Delta(p_1, p_0, q_2, a_2) = 0. \quad (28)$$

If all terms in the sum on the right in (20) are represented as in (27), then (27) shows that two simplexes (p_0, p_1, p_2, a_3) and (p_1, p_0, q_2, a_3), which have a side in common, contribute terms which cancel as follows:

$$\Delta(p_0, p_1, a_2, a_3) + \Delta(p_1, p_0, a_2, a_3) = 0. \quad (29)$$

Thus equations (20), (27), (28), (29) show that

$$\sum_{i=1}^{N} \Delta(s_i) = \sum_{(p_0, p_1)} \Delta(p_0, p_1, a_2, a_3), \qquad (p_0, p_1) \text{ in } \partial c_2. \quad (30)$$

This statement means that the Euclidean simplex (p_0, p_1) is contained in the boundary of a Euclidean simplex (p_0, p_1, p_2) in the chain c_2; the simplex (p_0, p_1) itself is contained in the chain which forms the boundary ∂c_2 of c_2. Now if (p_0, p_1) is in a side of (a_0, a_1, a_2) which contains a_2, then $\Delta(p_0, p_1, a_2, a_3) = 0$ because the simplex (p_0, p_1, a_2, a_3) has four vertices in a plane. As a result, (30) simplifies to the following:

$$\sum_{i=1}^{N} \Delta(s_i) = \sum_{(p_1, p_2)} \Delta(p_0, p_1, a_2, a_3), \quad (31)$$

$$(p_0, p_1) \text{ in } \partial c_2, \qquad (p_0, p_1) \text{ in } (a_0, a_1). \quad (32)$$

Let c_1 denote the chain which contains the simplexes in (32); then c_1 contains only those simplexes (p_0, p_1) in ∂c_2 which are in (a_0, a_1). The derivation of (31) has shown that (p_0, p_1, a_3) is a side of a simplex (p_0, p_1, p_2, a_3) for which $\Delta(p_0, p_1, p_2, a_3) > 0$ by (22). Since p_2 and a_2 are on the same side of the plane $P(p_0, p_1, a_3)$ [compare Figure 21.2], then $\Delta(p_0, p_1, p_2, a_3) > 0$ implies that

$$\Delta(p_0, p_1, a_2, a_3) > 0, \qquad (p_0, p_1) \text{ in } c_1. \tag{33}$$

The next step in the proof will show that (p_0, p_1) and (a_0, a_1) determine the same direction in $\mathbb{R}^3$. By Theorem 19.12 there exists a barycentric transformation which transforms (a_0, a_1, a_2, a_3) into (p_0, p_1, a_2, a_3). In matrix form this transformation is

$$\begin{bmatrix} p_0^1 & p_0^2 & p_0^3 & 1 \\ p_1^1 & p_1^2 & p_1^3 & 1 \\ a_2^1 & a_2^2 & a_2^3 & 1 \\ a_3^1 & a_3^2 & a_3^3 & 1 \end{bmatrix} = \begin{bmatrix} t_0^1 & t_0^2 & 0 & 0 \\ t_1^1 & t_1^2 & 0 & 0 \\ 0 & 0 & 1 & 0 \\ 0 & 0 & 0 & 1 \end{bmatrix} \begin{bmatrix} a_0^1 & a_0^2 & a_0^3 & 1 \\ a_1^1 & a_1^2 & a_1^3 & 1 \\ a_2^1 & a_2^2 & a_2^3 & 1 \\ a_3^1 & a_3^2 & a_3^3 & 1 \end{bmatrix}. \tag{34}$$

Since $\Delta(a_0, a_1, a_2, a_3) > 0$ by hypothesis and $\Delta(p_0, p_1, a_2, a_3) > 0$ by (33), equation (34) and the Binet–Cauchy multiplication theorem show that

$$\det \begin{bmatrix} t_0^1 & t_0^2 & 0 & 0 \\ t_1^1 & t_1^2 & 0 & 0 \\ 0 & 0 & 1 & 0 \\ 0 & 0 & 0 & 1 \end{bmatrix} > 0, \qquad \det \begin{bmatrix} t_0^1 & t_0^2 \\ t_1^1 & t_1^2 \end{bmatrix} > 0. \tag{35}$$

Then (34) and (35) show that

$$\begin{bmatrix} p_0^1 & p_0^2 & p_0^3 & 1 \\ p_1^1 & p_1^2 & p_1^3 & 1 \end{bmatrix} = \begin{bmatrix} t_0^1 & t_0^2 \\ t_1^1 & t_1^2 \end{bmatrix} \begin{bmatrix} a_0^1 & a_0^2 & a_0^3 & 1 \\ a_1^1 & a_1^2 & a_1^3 & 1 \end{bmatrix}, \tag{36}$$

and that

$$\begin{vmatrix} p_0^j & 1 \\ p_1^j & 1 \end{vmatrix} = \begin{vmatrix} t_0^1 & t_0^2 \\ t_1^1 & t_1^2 \end{vmatrix} \begin{vmatrix} a_0^j & 1 \\ a_1^j & 1 \end{vmatrix}, \qquad j = 1, 2, 3. \tag{37}$$

Equations (35), (36), (37) show that (p_0, p_1) and (a_0, a_1) determine the same direction in $\mathbb{R}^3$; this statement is true for each (p_0, p_1) in c_1 [see Definition 15.13, Theorem 19.12, and Example 19.13]. This statement proves one part of conclusion (15) of Theorem 21.3.

Return now to equations (31) and (32). Let the complete set of simplexes (p_0, p_1) in (a_0, a_1) be denoted by

$$(b_0, b_1), (b_1, b_2), \cdots, (b_{k-1}, b_k). \tag{38}$$

Then by (37) each of them has the same direction as (a_0, a_1) [see Figure 21.3]. Then

$$b_0 = a_0, \qquad b_k = a_1, \tag{39}$$

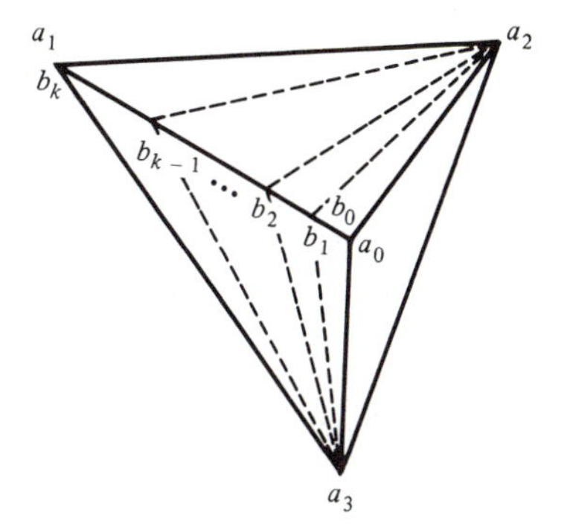

Figure 21.3. The simplexes (p_0, p_1) in (a_0, a_1).

and equation (31) can be written as follows:

$$\sum_{i=1}^{N} \Delta(s_i) = \sum_{r=1}^{k} \Delta(b_{r-1}, b_r, a_2, a_3). \tag{40}$$

Use Theorem 20.1 to obtain a representation of $\Delta(b_{r-1}, b_r, a_2, a_3)$ in terms of the vertex a_1 as follows:

$$\begin{aligned}\Delta(b_{r-1}, b_r, a_2, a_3) = \Delta(a_1, b_r, a_2, a_3) + \Delta(b_{r-1}, a_1, a_2, a_3) \\ + \Delta(b_{r-1}, b_r, a_1, a_3) + \Delta(b_{r-1}, b_r, a_2, a_1).\end{aligned} \tag{41}$$

Now the last two terms on the right equal zero because b_{r-1}, b_r, and a_1 are three points on a line [see Figure 21.3]. Then by (40) and (41),

$$\sum_{i=1}^{N} \Delta(s_i) = \sum_{r=1}^{k} [\Delta(b_{r-1}, a_1, a_2, a_3) - \Delta(b_r, a_1, a_2, a_3)]. \tag{42}$$

The sum on the right is a telescoping sum; thus

$$\sum_{i=1}^{N} \Delta(s_i) = \Delta(b_0, a_1, a_2, a_3) - \Delta(b_k, a_1, a_2, a_3). \tag{43}$$

Since $b_0 = a_0$ and $b_k = a_1$ by (39), this equation simplifies to

$$\sum_{i=1}^{N} \Delta(s_i) = \Delta(a_0, a_1, a_2, a_3). \tag{44}$$

This statement completes the proof of (13), and (14) follows from (13). The statement in (15) has been proved only for simplexes in the sides (a_0, a_1, a_2) and (a_0, a_1), but by changing the order in the evaluation process, (15) can be proved for each 2-dimensional side and each one-dimensional side of (a_0, a_1, a_2, a_3). The proof of Theorem 21.3 is thus complete for a simplicial subdivision of a 3-dimensional simplex (a_0, a_1, a_2, a_3) in $\mathbb{R}^3$. The methods which have been used in this proof can be used to prove the theorem in the general case. □

Let $\mathscr{P}:\{s_i : i = 1, \cdots, N\}$ be a simplicial subdivision of an n-simplex s in $\mathbb{R}^n$; then Theorem 21.3 shows that $\Sigma_1^N M(s_i) = M(s)$. The next theorem shows that the same relation holds if $\mathscr{P}$ is a simplicial subdivision of an n-simplex in $\mathbb{R}^m$, $m > n$. This result can be considered obvious. Measure is invariant under any change of axes for which distance is unchanged [compare Exercise 20.22]. Then choose axes so that the n-simplex is contained in $\mathbb{R}^n$ considered as a subspace of $\mathbb{R}^m$, and apply Theorem 21.3 to show that measure is additive on the simplicial subdivision. Thus, although the result is obvious, the method used to prove the next theorem is of interest.

21.4 Theorem. *If $\mathscr{P}:\{s_i : i = 1, \cdots, N\}$ is a simplicial subdivision of an n-simplex s in $\mathbb{R}^m$, $m \geqq n$, then $\Sigma_1^N M(s_i) = M(s)$.*

PROOF. Let $s:(b_0, b_1, \cdots, b_n)$ be an n-simplex in $\mathbb{R}^m$, $m \geqq n$, and let $\mathscr{P}:\{s_i : i = 1, \cdots, N\}$ be a simplicial subdivision of s. Let points in $\mathbb{R}^m$ be denoted by $y:(y^1, \cdots, y^m)$. Next, let $(e_0, e_1, \cdots, e_n)$ be the unit simplex on the axes in $\mathbb{R}^n$, and let $L:\mathbb{R}^n \to \mathbb{R}^m$ be the affine transformation [see Theorem 19.3] such that $L(e_i) = b_i$, $i = 0, 1, \cdots, n$; it is easy to verify that this transformation is

$$
\begin{aligned}
y^1 &= (b_1^1 - b_0^1)x^1 + (b_2^1 - b_0^1)x^2 + \cdots + (b_n^1 - b_0^1)x^n + b_0^1, \\
&\cdots\cdots\cdots\cdots\cdots\cdots\cdots\cdots\cdots\cdots\cdots\cdots \\
y^m &= (b_1^m - b_0^m)x^1 + (b_2^m - b_0^m)x^2 + \cdots + (b_n^m - b_0^m)x^n + b_0^m.
\end{aligned}
\tag{45}
$$

If s_i is a simplex in $\mathscr{P}$, then by Theorem 19.12 there is a barycentric transformation which transforms $(b_0, b_1, \cdots, b_n)$ into s_i. The same barycentric transformation transforms $(e_0, e_1, \cdots, e_n)$ into a simplex w_i. Thus the collection of barycentric transformations which transforms $(b_0, b_1, \cdots, b_n)$ into the simplicial subdivision $\mathscr{P}:\{s_i : i = 1, \cdots, N\}$ can be used to construct the corresponding simplicial subdivision $\mathscr{Q}:\{w_i : i = 1, \cdots, N\}$ of $(e_0, e_1, \cdots, e_n)$. If w_i is $(x_0, x_1, \cdots, x_n)$ and s_i is $(y_0, y_1, \cdots, y_n)$, then the mapping of w_i into s_i by L, in matrix form, is the following [see equation (10) in Section 19 and Example 19.5]:

$$
\begin{bmatrix}
x_0^1 & x_0^2 & \cdots & x_0^n & 1 \\
x_1^1 & x_1^2 & \cdots & x_1^n & 1 \\
\cdots & \cdots & \cdots & \cdots & \cdots \\
x_n^1 & x_n^2 & \cdots & x_n^n & 1
\end{bmatrix}
\begin{bmatrix}
b_1^1 - b_0^1 & b_1^2 - b_0^2 & \cdots & b_1^m - b_0^m & 0 \\
\cdots & \cdots & \cdots & \cdots & \cdots \\
b_n^1 - b_0^1 & b_n^2 - b_0^2 & \cdots & b_n^m - b_0^m & 0 \\
b_0^1 & b_0^2 & \cdots & b_0^m & 1
\end{bmatrix}
=
\begin{bmatrix}
y_0^1 & y_0^2 & \cdots & y_0^m & 1 \\
y_1^1 & y_1^2 & \cdots & y_1^m & 1 \\
\cdots & \cdots & \cdots & \cdots & \cdots \\
y_n^1 & y_n^2 & \cdots & y_n^m & 1
\end{bmatrix}.
\tag{46}
$$

Let $(j_1, \cdots, j_n)$ be an index set in (n/m). Then the minor, in columns $j_1, \cdots, j_n$, $(m + 1)$ of the matrix on the right in (46), is derived from the minor, in columns $j_1, \cdots, j_n, (m+1)$, of the second matrix on the left. Thus

$$\begin{bmatrix} x_0^1 & x_0^2 & \cdots & x_0^n & 1 \\ x_1^1 & x_1^2 & \cdots & x_1^n & 1 \\ \cdots & \cdots & \cdots & \cdots & \cdots \\ x_n^1 & x_n^2 & \cdots & x_n^n & 1 \end{bmatrix} \begin{bmatrix} b_1^{j_1} - b_0^{j_1} & \cdots & b_1^{j_n} - b_0^{j_n} & 0 \\ \cdots & \cdots & \cdots & \cdots \\ b_n^{j_1} - b_0^{j_1} & \cdots & b_n^{j_n} - b_0^{j_n} & 0 \\ b_0^{j_1} & \cdots & b_0^{j_n} & 1 \end{bmatrix} = \begin{bmatrix} y_0^{j_1} & y_0^{j_2} & \cdots & y_0^{j_n} & 1 \\ y_1^{j_1} & y_1^{j_2} & \cdots & y_1^{j_n} & 1 \\ \cdots & \cdots & \cdots & \cdots & \cdots \\ y_n^{j_1} & y_n^{j_2} & \cdots & y_n^{j_n} & 1 \end{bmatrix}. \tag{47}$$

By the Binet–Cauchy multiplication theorem, the product of the determinants of the matrices on the left is equal to the determinant of the matrix on the right. The result can be simplified to the following.

$$\det \begin{bmatrix} x_0^1 & x_0^2 & \cdots & x_0^n & 1 \\ \cdots & \cdots & \cdots & \cdots & \cdots \\ x_n^1 & x_n^2 & \cdots & x_n^n & 1 \end{bmatrix} (-1)^n \det \begin{bmatrix} b_0^{j_1} & b_0^{j_2} & \cdots & b_0^{j_n} & 1 \\ \cdots & \cdots & \cdots & \cdots & \cdots \\ b_n^{j_1} & b_n^{j_2} & \cdots & b_n^{j_n} & 1 \end{bmatrix} = \det \begin{bmatrix} y_0^{j_1} & y_0^{j_2} & \cdots & y_0^{j_n} & 1 \\ \cdots & \cdots & \cdots & \cdots & \cdots \\ y_n^{j_1} & y_n^{j_2} & \cdots & y_n^{j_n} & 1 \end{bmatrix}. \tag{48}$$

Use the formula in equation (10) in Section 20 to find $|\Delta(y_0, y_1, \cdots, y_n)|$; the result is

$$|\Delta(y_0, y_1, \cdots, y_n)| = |\Delta(x_0, x_1, \cdots, x_n)|\,|\Delta(b_0, b_1, \cdots, b_n)|, \tag{49}$$

$$M(s_i) = |\Delta(w_i)| M(b_0, b_1, \cdots, b_n). \tag{50}$$

Then

$$\sum_{i=1}^{N} M(s_i) = M(b_0, b_1, \cdots, b_n) \sum_{i=1}^{N} |\Delta(w_i)|. \tag{51}$$

Since $\mathscr{Q}: \{w_i : i = 1, \cdots, N\}$ is a simplicial subdivision of $(e_0, e_1, \cdots, e_n)$ in $\mathbb{R}^n$, Theorem 21.3 shows that

$$\sum_{i=1}^{N} |\Delta(w_i)| = |\Delta(e_0, e_1, \cdots, e_n)| = 1. \tag{52}$$

Equations (51) and (52) show that

$$\sum_{i=1}^{N} M(s_i) = M(b_0, b_1, \cdots, b_n) = M(s), \tag{53}$$

and the proof of Theorem 21.4 is complete. □

Definition 21.1 defines a subdivision of a Euclidean complex K, and the next definition defines sequences of oriented simplicial subdivisions of a complex K of n-simplexes in $\mathbb{R}^n$. The importance of these sequences of simplicial subdivisions results from their applications in the theory of integration. The proof of the fundamental theorem of the integral calculus for functions of a single variable which is outlined in Exercise 15.6 employs a sequence of subdivisions of $[a, b]$. Subdivide $[a, b]$ by equally spaced points $a = x_0 < x_1 < \cdots < x_{i-1} < x_i < \cdots < x_{n-1} < x_n = b$ to form the n one-dimensional simplexes $(x_0, x_1), \cdots, (x_{i-1}, x_i), \cdots, (x_{n-1}, x_n)$ which constitute the subdivision $\mathscr{P}_n$ of $[a, b]$. Then

$$f(b) - f(a) = \sum_{i=1}^{n} (-1)\begin{vmatrix} f(x_{i-1}) & 1 \\ f(x_i) & 1 \end{vmatrix} = \sum_{i=1}^{n} f'(x_i^*)(-1)\begin{vmatrix} x_{i-1} & 1 \\ x_i & 1 \end{vmatrix} \to \int_a^b f'(x)\,dx.$$

The theory of integration for functions of n variables requires sequences of subdivisions of n-cubes, n-simplexes, and, more generally, Euclidean complexes K of n-simplexes in $\mathbb{R}^n$. The remainder of this section is devoted to the definition and construction of these sequences of subdivisions.

21.5 Definition. Let K be an oriented, Euclidean, homogeneous, n-dimensional, complex in $\mathbb{R}^n$ in which all n-simplexes have the same (positive or negative) orientation in $\mathbb{R}^n$. Then $\{\mathscr{P}_k : k = 1, 2, \cdots\}$ is called a *sequence of oriented simplicial subdivisions of* K if and only if it has the following properties:

Each $\mathscr{P}_k$ is a simplicial subdivision of K. (54)

The simplexes in each $\mathscr{P}_k$ have the same orientation in $\mathbb{R}^n$ as the simplexes in K. (55)

$\mathscr{P}_{k+1}$ is a refinement of $\mathscr{P}_k$ for $k = 1, 2, \cdots$. (56)

There exists a constant ρ, $0 < \rho \leqq 1$, such that each simplex in each $\mathscr{P}_k$ satisfies the regularity condition with constant ρ [see Definition 2.4] at each of its vertices. (57)

The norm of $\mathscr{P}_k$, that is, the maximum diameter of a simplex in $\mathscr{P}_k$, approaches zero as k tends to infinity. (58)

21.6 Example. In Figure 21.4(a), K is a single simplex; and in Figure 21.4(b), K is a complex which consists of four simplexes. Also, $\mathscr{P}_1$ and $\mathscr{P}_2$ are simplicial subdivisions of K [see Definition 21.1], and $\mathscr{P}_2$ is a refinement of $\mathscr{P}_1$. Orient the simplexes in K and then give each simplex in $\mathscr{P}_1$, $\mathscr{P}_2$, $\cdots$ the same orientation in $\mathbb{R}^2$ as the simplexes in K. If the subdivisions of K are continued in the manner indicated in Figure 21.4, it is clear that each of the sequences $\{\mathscr{P}_k : k = 1, 2, \cdots\}$ has all of the properties stated in Definition 21.5 and is therefore a sequence of oriented simplicial subdivisions of K.

The construction of sequences $\{\mathscr{P}_k : k = 1, 2, \cdots\}$ in certain special complexes will now be described; the construction of sequences of simplicial subdivisions of an arbitrary simplex will be treated afterward. The unit interval (cube A_1 or simplex) $[0, 1]$ and its sides form a complex K in $\mathbb{R}$

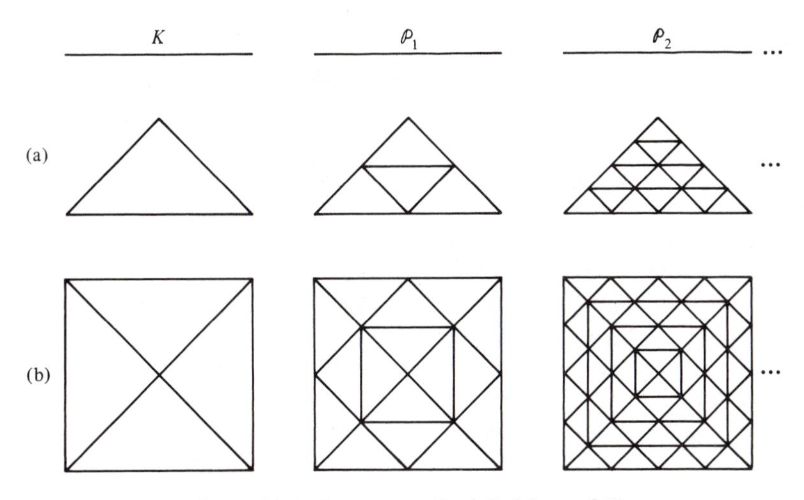

Figure 21.4. Sequences of subdivisions of K.

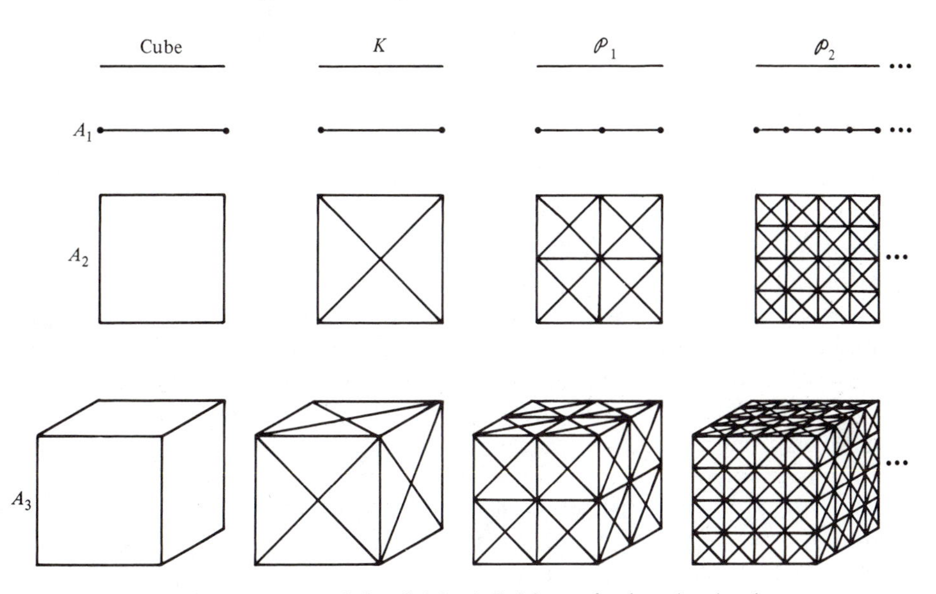

Figure 21.5. Sequences of simplicial subdivisions of cubes $A_1, A_2, A_3, \cdots$.

[see Figure 21.5]. The mid-point of this interval subdivides it into two intervals $[0, \frac{1}{2}]$ and $[\frac{1}{2}, 1]$ which form $\mathscr{P}_1$. Subdivide each interval in $\mathscr{P}_1$ at its mid-point to form four intervals in $\mathscr{P}_2$. Continue this process indefinitely to form the sequence $\{\mathscr{P}_k : k = 1, 2, \cdots\}$ of subdivisions of $[0, 1]$. A similar sequence of subdivisions can be constructed in every interval $[a, b]$ in $\mathbb{R}$.

Next, construct a sequence of subdivisions of the square A_2 in $\mathbb{R}^2$ with vertices $a_1 : (0, 0)$, $a_2 : (1, 0)$, $a_3 : (1, 1)$, $a_4 : (0, 1)$, and center $a_0 : (\frac{1}{2}, \frac{1}{2})$. Let K consist of the following positively oriented simplexes [see Figure 21.5]: (a_0, a_1, a_2), (a_0, a_2, a_3), (a_0, a_3, a_4), (a_0, a_4, a_1). Next, subdivide A_2 into four equal squares by lines through a_0 parallel to the sides of A_2. Subdivide each of the four sub-squares of A_2 in the same way that A_2 was subdivided to form K. The sixteen 2-dimensional simplexes (triangles) obtained in this manner form $\mathscr{P}_1$. Continue this process indefinitely to form $\{\mathscr{P}_k : k = 1, 2, \cdots\}$. Then each $\mathscr{P}_k$ is a simplicial subdivision of A_2, and the simplexes in each $\mathscr{P}_k$ are positively oriented in $\mathbb{R}^2$. Also, each $\mathscr{P}_{k+1}$ is a refinement of $\mathscr{P}_k$. Each simplex in each $\mathscr{P}_k$ is an isosceles right triangle; a simple calculation shows that each simplex satisfies the regularity condition, with $\rho = \sqrt{2}/2$, at each of its vertices. The norm of $\mathscr{P}_k$ clearly approaches zero as $k \to \infty$. Thus $\{\mathscr{P}_k : k = 1, 2, \cdots\}$ satisfies (54), $\cdots$, (58) and is therefore a sequence of oriented simplicial subdivisions of K and of A_2.

Observe that (a_0, a_1, a_2) is a 2-simplex in A_2. Since each subdivision in $\{\mathscr{P}_k : k = 1, 2, \cdots\}$ is a refinement of the preceding by (56), the simplexes in $\mathscr{P}_k$ which are in (a_0, a_1, a_2) form a simplicial subdivision of (a_0, a_1, a_2). Thus, the sequence of subdivisions $\{\mathscr{P}_k : k = 1, 2, \cdots\}$, restricted to (a_0, a_1, a_2), forms a sequence of oriented simplicial subdivisions of (a_0, a_1, a_2).

Next, consider the cube A_3 in $\mathbb{R}^3$ whose vertices are

$$a_1 : (0, 0, 0), \quad a_2 : (1, 0, 0), \quad a_3 : (1, 1, 0), \quad a_4 : (0, 1, 0),$$
$$a_5 : (0, 0, 1), \quad a_6 : (1, 0, 1), \quad a_7 : (1, 1, 1), \quad a_8 : (0, 1, 1),$$

and whose center is $a_0 : (\frac{1}{2}, \frac{1}{2}, \frac{1}{2})$. In each of the six faces of A_3 construct the complex K in A_2 shown in Figure 21.5; then there are four isosceles right triangles in each of the six faces of A_3. Connect a_0 to the vertices of the right triangles in the faces of A_3 to form 3-simplexes; give each 3-simplex the positive orientation in $\mathbb{R}^3$. Each of the 3-simplexes has three mutually perpendicular edges. The twenty-four 3-simplexes form the complex K; it is a simplicial subdivision of the cube A_3. Next, subdivide A_3 into eight congruent cubes by planes through a_0 parallel to the coordinate planes. Construct a subdivision of each of the eight sub-cubes of A_3 in the same way that K was constructed in A_3. The 192 three-dimensional simplexes thus obtained form the simplicial subdivision $\mathscr{P}_1$ of K and of A_3. Continue this process indefinitely to form the sequence $\{\mathscr{P}_k : k = 1, 2, \cdots\}$ of oriented simplicial subdivisions of K and of A_3.

Let s be an arbitrary, but fixed, simplex in the subdivision K of A_3. Since each subdivision in $\{\mathscr{P}_k : k = 1, 2, \cdots\}$ is a refinement of the preceding subdivision, the sequence of subdivisions $\{\mathscr{P}_k : k = 1, 2, \cdots\}$, restricted to s, forms a sequence of oriented simplicial subdivisions of s.

This process can be continued inductively to form sequences of oriented simplicial subdivisions of the unit cube A_n in $\mathbb{R}^n$ and of an n-simplex s in $\mathbb{R}^n$. Thus there exist sequences of subdivisions for all cubes A_n in $\mathbb{R}^n$ and for *some* n-simplexes s in $\mathbb{R}^n$, but there remains the problem of showing that *every* non-degenerate simplex in $\mathbb{R}^n$ has a sequence of simplicial subdivisions. This problem is solved by the following theorem.

21.7 Theorem. *Let s be a simplex $(b_0, b_1, \cdots, b_n)$ in $\mathbb{R}^n$ such that $\Delta(b_0, b_1, \cdots, b_n) \neq 0$. Then there exists a sequence $\{\mathcal{Q}_k : k = 1, 2, \cdots\}$ of oriented simplicial subdivisions of s with the properties stated in Definition 21.5.*

PROOF. The construction just described [see Figure 21.5] shows that there is a simplex $(a_0, a_1, \cdots, a_n)$ in $\mathbb{R}^n$ which has n mutually orthogonal edges and which has a sequence $\{\mathcal{P}_k : k = 1, 2, \cdots\}$ of oriented simplicial subdivisions. Since $(a_0, a_1, \cdots, a_n)$ has n mutually orthogonal edges, the formula in (13) in Section 20 can be used to show that $\Delta(a_0, a_1, \cdots, a_n) \neq 0$. There is no loss of generality in assuming that $\Delta(a_0, a_1, \cdots, a_n) > 0$ and $\Delta(b_0, b_1, \cdots, b_n) > 0$. By Theorem 19.3 there is a unique affine transformation $L : \mathbb{R}^n \to \mathbb{R}^n$ such that $L(a_i) = b_i$, $i = 0, 1, \cdots, n$. The equations for this affine transformation are [see (4) and (6) in Section 19]

$$y^j = \sum_{i=1}^{n} c_{ij} x^i + c_{n+1,j}, \qquad j = 1, 2, \cdots, n. \tag{59}$$

Let $\{\mathcal{P}_k : k = 1, 2, \cdots\}$ be a sequence of simplicial subdivisions of $(a_0, a_1, \cdots, a_n)$ whose orientation is the same as that of $(a_0, a_1, \cdots, a_n)$; the construction of such a sequence has been described just preceding Theorem 21.7. Then Theorems 19.4 and 19.12 [see also Example 19.5] show that L maps each subdivision $\mathcal{P}_k$ of $(a_0, a_1, \cdots, a_n)$ into a simplicial subdivision $\mathcal{Q}_k$ of $(b_0, b_1, \cdots, b_n)$. Thus the sequence $\{\mathcal{Q}_k : k = 1, 2, \cdots\}$ has the property stated in (54) of Definition 21.5. In order to complete the proof of the theorem, it is necessary to show that $\{\mathcal{Q}_k : k = 1, 2, \cdots\}$ has the properties stated in (55), $\cdots$, (58).

By (11) in Theorem 19.6,

$$\Delta(b_0, b_1, \cdots, b_n) = \det[c_{ij}]_1^n \Delta(a_0, a_1, \cdots, a_n). \tag{60}$$

Since $\Delta(a_0, a_1, \cdots, a_n) > 0$ and $\Delta(b_0, b_1, \cdots, b_n) > 0$ by hypothesis, then

$$\det[c_{ij}]_1^n > 0. \tag{61}$$

If $(p_0, p_1, \cdots, p_n)$ is a simplex in $\mathcal{P}_k$ and $(q_0, q_1, \cdots, q_n)$ is the simplex in $\mathcal{Q}_k$ into which it is transformed by L, then

$$\Delta(q_0, q_1, \cdots, q_n) = \det[c_{ij}]_1^n \Delta(p_0, p_1, \cdots, p_n). \tag{62}$$

This equation shows that each simplex $(q_0, q_1, \cdots, q_n)$ in $\mathcal{Q}_k$ has the same orientation as $(p_0, p_1, \cdots, p_n)$, $(a_0, a_1, \cdots, a_n)$, and $(b_0, b_1, \cdots, b_n)$; thus the sequence $\{\mathcal{Q}_k : k = 1, 2, \cdots\}$ has property (55). Since $\mathcal{P}_{k+1}$ is a refinement of $\mathcal{P}_k$, then $\mathcal{Q}_{k+1}$ is a refinement of $\mathcal{Q}_k$ and $\{\mathcal{Q}_k : k = 1, 2, \cdots\}$ has property (56).

Since $\{\mathcal{P}_k : k = 1, 2, \cdots\}$ has property (57), equation (62) shows that

$$\begin{aligned} |\Delta(q_0, q_1, \cdots, q_n)| &= \det[c_{ij}]_1^n |\Delta(p_0, p_1, \cdots, p_n)| \\ &\geq \rho \det[c_{ij}]_1^n \prod_{r=1}^{n} |p_r - p_0|. \end{aligned} \tag{63}$$

Equation (59) and Schwarz's inequality [see Corollary 86.2 and (7) in Section 86, also Example 21.8 below] show that there exists a constant $c > 0$ such that

$$|q_r - q_0| \leqq c|p_r - p_0|, \qquad r = 1, 2, \cdots, n. \tag{64}$$

Then by (63) and (64),

$$|\Delta(q_0, q_1, \cdots, q_n)| \geqq \rho c^{-n} \det[c_{ij}]_1^n \prod_{r=1}^{n} |q_r - q_0|, \tag{65}$$

and $(q_0, q_1, \cdots, q_n)$ satisfies, at the vertex q_0, the regularity condition with parameter $\rho c^{-n}\det[c_{ij}]_1^n$. Since $\mathscr{P}_k$ satisfies (57), the simplex $(p_0, p_1, \cdots, p_n)$ satisfies the regularity condition with parameter ρ at each of its vertices. Then the methods used to prove (65) can be employed to show that $(q_0, q_1, \cdots, q_n)$ satisfies the regularity condition with parameter $\rho c^{-n}\det[c_{ij}]_1^n$ at each of its vertices. Thus $\{\mathscr{Q}_k : k = 1, 2, \cdots\}$ has property (57). Finally, (64) and Exercises 14.16 and 14.17 show that the sequence has property (58). Then $\{\mathscr{Q}_k : k = 1, 2, \cdots\}$ satisfies all of the conditions in Definition 21.5, and it is a sequence of oriented simplicial subdivisions of $(b_0, b_1, \cdots, b_n)$. □

21.8 Example. This is an example to show how Schwarz's inequality is used to prove that there exists a constant c which satisfies (64). Let $x_0 : (x_0^1, x_0^2)$ and $x_1 : (x_1^1, x_1^2)$ be two points in $\mathbb{R}^2$ which are transformed by the affine transformation

$$\begin{aligned} y^1 &= a_{11}x^1 + a_{12}x^2 + a_{13}, \\ y^2 &= a_{21}x^1 + a_{22}x^2 + a_{23}, \end{aligned} \tag{66}$$

into the points $y_0 : (y_0^1, y_0^2)$ and $y_1 : (y_1^1, y_1^2)$. Then

$$\begin{aligned} y_0^1 &= a_{11}x_0^1 + a_{12}x_0^2 + a_{13}, \qquad & y_1^1 &= a_{11}x_1^1 + a_{12}x_1^2 + a_{13}, \\ y_0^2 &= a_{21}x_0^1 + a_{22}x_0^2 + a_{23}, \qquad & y_1^2 &= a_{21}x_1^1 + a_{22}x_1^2 + a_{23}. \end{aligned} \tag{67}$$

From these equations it follows that

$$\begin{aligned} |y_1 - y_0| &= \{(y_1^1 - y_0^1)^2 + (y_1^2 - y_0^2)^2\}^{1/2} \\ &= \{[a_{11}(x_1^1 - x_0^1) + a_{12}(x_1^2 - x_0^2)]^2 \\ &\quad + [a_{21}(x_1^1 - x_0^1) + a_{22}(x_1^2 - x_0^2)]^2\}^{1/2}. \end{aligned} \tag{68}$$

By Schwarz's inequality in Corollary 86.2,

$$\begin{aligned} &[a_{11}(x_1^1 - x_0^1) + a_{12}(x_1^2 - x_0^2)]^2 \leqq (a_{11}^2 + a_{12}^2)[(x_1^1 - x_0^1)^2 + (x_1^2 - x_0^2)^2], \\ &[a_{21}(x_1^1 - x_0^1) + a_{22}(x_1^2 - x_0^2)]^2 \leqq (a_{21}^2 + a_{22}^2)[(x_1^1 - x_0^1)^2 + (x_1^2 - x_0^2)^2]. \end{aligned} \tag{69}$$

These inequalities and (68) show that

$$|y_1 - y_0| \leqq \left\{ \sum_{i,j=1}^{2} (a_{ij})^2 \right\}^{1/2} \{(x_1^1 - x_0^1)^2 + (x_1^2 - x_0^2)^2\}^{1/2}. \tag{70}$$

Thus, if

$$c = \left\{ \sum_{i,j=1}^{2} (a_{ij})^2 \right\}^{1/2}, \tag{71}$$

then $|y_1 - y_0| \leqq c|x_1 - x_0|$. For the general affine transformation in (59), define c as follows:

$$c = \left\{ \sum_{i,j=1}^{n} (c_{ij})^2 \right\}^{1/2}. \tag{72}$$

Then exactly the same procedure proves that

$$|q_r - q_0| \leqq c|p_r - p_0|, \qquad r = 1, 2, \cdots, n, \tag{73}$$

and this is the inequality in (64).

Exercises

21.1. Let s be the simplex (a_0, a_1, a_2, a_3) in $\mathbb{R}^3$ whose vertices are $a_0 : (0, 0, 0)$, $a_1 : (3, 0, 0)$, $a_2 : (0, 3, 0)$, $a_3 : (0, 0, 3)$. Introduce a new vertex $a_4 : (1, 1, 1)$ in the side (a_1, a_2, a_3).

(a) Show that (a_0, a_1, a_2, a_3) is positively oriented in $\mathbb{R}^3$. [Hint. See Definition 15.8 and Theorem 15.9.]

(b) Show that the simplexes (a_0, a_4, a_1, a_2), (a_0, a_4, a_2, a_3), and (a_0, a_4, a_3, a_1) form a subdivision $\mathscr{P}$ of s which is positively oriented in $\mathbb{R}^3$.

(c) Verify Theorem 21.3 for the subdivision $\mathscr{P}$ of s.

(d) The subdivision $\mathscr{P}$ induces a simplicial subdivision in the simplex (a_1, a_2, a_3). Find the measure of each of the simplexes (a_4, a_1, a_2), (a_4, a_2, a_3), (a_4, a_3, a_1), (a_1, a_2, a_3), and verify Theorem 21.4 for this induced subdivision of the side (a_1, a_2, a_3) of s.

21.2. This exercise outlines a proof of Theorem 21.4 which is complete in itself and does not depend on Theorem 21.3. The method used in part (b) is of some interest.

(a) Let $(e_0, e_1, \cdots, e_n)$ be the unit simplex on the axes in $\mathbb{R}^n$. Use Theorem 20.5 to prove that $M(e_0, e_1, \cdots, e_n) = 1/n!$

(b) Prove the following theorem. If $s_1, \cdots, s_N$ are the simplexes in a simplicial subdivision of $(e_0, e_1, \cdots, e_n)$, then $\Sigma_1^N \Delta(s_i) = 1$ and $\Sigma_1^N M(s_i) = 1/n! = M(e_0, e_1, \cdots, e_n)$. [Outline of proof. Since $\Delta(s_i) = n!M(s_i)$, it is sufficient to prove the statement about measure. The proof is by induction. The proof for $n = 1$ is straightforward. In the case $n = 2$, use Theorem 20.1 as in the first step in the proof of Theorem 21.3 to introduce e_2 as a vertex in a representation of $\Delta(s_i)$, $i = 1, \cdots, N$. Then the methods used in the proof of Theorem 21.3 show that $\Sigma_1^N M(s_i) = \Sigma_1^r M(u_j)$. Here each u_j is a 2-simplex whose base is a simplex w_j in a simplicial subdivision $\{w_j : j = 1, \cdots, r\}$ of (e_0, e_1) and whose other vertex is e_2. Draw a sketch to show these simplexes u_j and w_j. Each u_j has altitude 1. The proof of Theorem 20.5 shows that $M(u_j) = (1/2)\cdot 1 \cdot M(w_j)$. Since $\Sigma_1^r M(w_j) = 1$ by the first step in the proof, then $\Sigma_1^N M(s_i) = \Sigma_1^r M(u_j) = (1/2)\Sigma_1^r M(w_j) = 1/(2!) = M(e_0, e_1, e_2)$. Thus the theorem is true for $n = 1$ and $n = 2$. To prove the inductive step, assume that it is true for $(e_0, e_1, \cdots, e_n)$: if $\{w_j : j = 1, \cdots, r\}$ is a simplicial subdivision of $(e_0, e_1, \cdots, e_n)$, then $\Sigma_1^r M(w_j) = 1/n! = M(e_0, e_1, \cdots, e_n)$. If $\{s_i : i = 1, \cdots, N\}$ is a simplicial subdivision of $(e_0, e_1, \cdots, e_n, e_{n+1})$, then Theorem 20.1 can be used as in the proof of Theorem 21.3 to show that $\Sigma_1^N M(s_i) = \Sigma_1^r M(u_j)$. Here u_j is an $(n+1)$-simplex whose base is an n-simplex w_j in a simplicial subdivision of $(e_0, e_1, \cdots, e_n)$ and whose other vertex is e_{n+1}. Draw a sketch to show u_j and w_j in the case $n + 1 = 3$. The altitude of each u_j is 1, and the proof of Theorem 20.5 shows that $M(u_j) = (1/n+1)\cdot 1 \cdot M(w_j)$. Then $\Sigma_1^N M(s_i) = \Sigma_1^r M(u_j) = (1/n+1)\Sigma_1^r M(w_j) = (1/n+1)(1/n!) = 1/(n+1)! = M(e_0, e_1, \cdots, e_n, e_{n+1})$. Therefore, the theorem is true for $(e_0, e_1, \cdots, e_n, e_{n+1})$ if it is true for $(e_0, e_1, \cdots, e_n)$. Since the theorem is true for $n = 1$ and $n = 2$, it is true for $n = 1, 2, \cdots$.]

(c) Prove the following theorem. If $\mathscr{P} : \{s_i : i = 1, \cdots, N\}$ is a simplicial subdivision of a Euclidean simplex $(a_0, a_1, \cdots, a_n)$ in $\mathbb{R}^m$, $m \geqq n$, then $\Sigma_1^N M(s_i) = M(a_0, a_1, \cdots, a_n)$. [Hint. The proof of this theorem is the same as the proof of Theorem 21.4 except that part (b) of this exercise is used in place of Theorem 21.3.]

21.3. Prove that there exists a sequence of oriented simplicial subdivisions $\{\mathscr{P}_k : k = 1, 2, \cdots\}$ of the unit simplex $(e_0, e_1, \cdots, e_n)$ on the axes which has the properties stated in Definition 21.5.

21.4. Use Exercise 21.3 and the affine transformation in equation (45) of this section to construct a sequence $\{\mathscr{Q}_k : k = 1, 2, \cdots\}$ of simplicial subdivisions of the Euclidean simplex $(b_0, b_1, \cdots, b_n)$ in $\mathbb{R}^m$, $m \geqq n$.

21.5. The set $[i, i+1] \times [j, j+1]$, $i, j = 0, \pm 1, \pm 2, \cdots$, subdivides the plane $\mathbb{R}^2$ into a set of unit squares. Subdivide each square into four isosceles right triangles (simplexes) by lines from its center to its four corners; give each simplex the positive orientation in $\mathbb{R}^2$.

(a) Let K denote a connected set of oriented 2-simplexes in the subdivision of the plane just described. Show that K is an oriented, Euclidean, homogeneous, 2-dimensional complex in $\mathbb{R}^2$.

(b) Construct a sequence $\{\mathscr{P}_k : k = 1, 2, \cdots\}$ of oriented simplicial subdivisions of K which has the properties stated in Definition 21.5.

21.6. State and solve a generalization of Exercise 21.5 in $\mathbb{R}^n$. The steps are these: (a) subdivide $\mathbb{R}^n$ into unit cubes; (b) subdivide each unit cube into positively oriented n-simplexes; (c) form a complex K of n-simplexes; and (d) construct a sequence $\{\mathscr{P}_k : k = 1, 2, \cdots\}$ of subdivisions of K.

21.7. Subdivide the cube $[0, 3] \times [0, 3] \times [0, 3]$ in $\mathbb{R}^3$ into unit cubes by planes parallel to the coordinate planes and remove the center cube to form a hollow cube. Subdivide each of the 26 remaining cubes into twenty-four 3-simplexes as in K in Figure 21.5. The 624 3-simplexes, positively oriented in $\mathbb{R}^3$, form a connected, oriented, Euclidean, homogeneous, 3-dimensional, simplicial complex K_3 in $\mathbb{R}^3$. The boundary of the chain formed by the 3-simplexes in K_3 is a chain whose 2-simplexes form an oriented, Euclidean, homogeneous, 2-

dimensional, simplicial complex K_2 in $\mathbb{R}^3$; each of the two components of K_2 is connected. Construct a sequence $\{\mathscr{P}_k : k = 1, 2, \cdots\}$ of simplicial subdivisions of K_2 which has the properties stated in Definition 21.5.

21.8. Let $[a, b] \times [c, d]$ be a rectangle in $\mathbb{R}^2$, and let $\mathscr{P}$ be a simplicial subdivision $\{s_i : i = 1, \cdots, N\}$ of the rectangle whose simplexes form a complex. Prove that $\Sigma_1^N M(s_i) = (b - a)(d - c)$, which is the area of the rectangle.

21.9. State and prove the generalization of Exercise 21.8 for a rectangular parallelepiped in $\mathbb{R}^3$.

21.10. Let $\{\mathscr{P}_k : k = 1, 2, \cdots\}$ be the sequence of simplicial subdivisions of the cube A_3 shown in Figure 21.5.

(a) Show that the simplexes in each $\mathscr{P}_k$ are congruent.

(b) Show that the simplexes in each two subdivisions in $\{\mathscr{P}_k : k = 1, 2, \cdots\}$ are similar.

(c) Find the best constant of regularity ρ at each vertex of a simplex in the subdivision $\mathscr{P}_1$ of A_3. Find the best constant of regularity ρ for all vertices of all simplexes in all subdivisions $\mathscr{P}_k : k = 1, 2, \cdots$, of A_3.

21.11. Solve the generalization of Exercise 21.10 for the corresponding sequence $\{\mathscr{P}_k : k = 1, 2, \cdots\}$ of subdivisions of the cube A_n in $\mathbb{R}^n$.

21.12. Let $(a_0, a_1, \cdots, a_n)$ be an n-simplex in $\mathbb{R}^n$ such that $\Delta(a_0, a_1, \cdots, a_n) > 0$, and let s denote the side $(a_1, \cdots, a_n)$ of $(a_0, a_1, \cdots, a_n)$. Then $(a_0, a_1, \cdots, a_n)$ and s are represented by matrices $[a_i^j]_0^n$ and $[a_i^j]_1^n$ respectively. Let $\mathscr{P}$ be a subdivision of $(a_0, a_1, \cdots, a_n)$, each of whose simplexes has the same orientation in $\mathbb{R}^n$ as $(a_0, a_1, \cdots, a_n)$. Then $\mathscr{P}$ induces a subdivision in the sides of $(a_0, a_1, \cdots, a_n)$; let $s_1, \cdots, s_r$ be the $(n-1)$-simplexes induced by $\mathscr{P}$ in the side s of $(a_0, a_1, \cdots, a_n)$. Each simplex $s_1, \cdots, s_r$ has a matrix representation. Let $k_1, \cdots, k_j, \cdots, k_n$ be the index sets in $(n - 1/n)$; thus $k_j = (1, 2, \cdots, n-j, n-j+2, \cdots, n)$, $j = 1, \cdots, n$. The following table displays the component matrices [see (14) and (33) in Section 11] of $s, s_1, \cdots, s_r$.

$$
\begin{array}{l}
s : s^{k_1}, \cdots, s^{kj}, \cdots, s^{kn};\\
s_1 : s_1^{k_1}, \cdots, s_1^{kj}, \cdots, s_1^{kn};\\
\cdots\cdots\cdots\cdots\cdots\cdots\cdots\\
s_r : s_r^{k_1}, \cdots, s_r^{kj}, \cdots, s_r^{kn}.
\end{array}
$$

(a) Show that there exist $n \times n$ matrices $T_1, \cdots, T_r$ which define barycentric transformations such that $s_i = T_i s$ and $\det T_i > 0$ for $i = 1, \cdots, r$. [Hint. See the proof of Theorem 21.3.]

(b) The component matrices of s are the projections of s into the $(n-1)$-dimensional coordinate planes of $\mathbb{R}^n$; thus each component matrix of s can be interpreted as an $(n-1)$-dimensional simplex in $\mathbb{R}^{n-1}$. Similar statements hold for each of the matrices $s_1, \cdots, s_r$. Prove that $s_1^{kj}, \cdots, s_r^{kj}$ have the same orientation in $\mathbb{R}^{n-1}$ as s^{kj} for $j = 1, \cdots, n$. [Hint. See Theorem 19.12 and the proof of Theorem 21.3.]

(c) Prove that each of the simplexes $s_1, \cdots, s_r$ has the same $(n-1)$-direction in $\mathbb{R}^n$ as s. [Hint. Example 19.13.]

(d) Prove that $\{s_1^{kj}, \cdots, s_r^{kj}\}$ is a subdivision of s^{kj} for $j = 1, \cdots, n$.

(e) Prove the following:

$$\sum_{i=1}^{r} \Delta(s_i^{kj}) = \Delta(s^{kj}), \qquad \sum_{i=1}^{r} M(s_i^{kj}) = M(s^{kj}), \qquad j = 1, \cdots, n;$$

$$\sum_{i=1}^{r} |\Delta(s_i)| = |\Delta(s)|, \qquad \sum_{i=1}^{r} M(s_i) = M(s).$$

21.13. Let E be an open set in $\mathbb{R}^2$ which contains the unit square $A_2 = [0, 1] \times [0, 1]$, and let $f: E \to \mathbb{R}^2$ be a function whose components (f^1, f^2) have continuous derivatives in E. Let $\{\mathscr{P}_k : k = 1, 2, \cdots\}$ be a sequence of oriented simplicial subdivisions of A_2 [see Definition 21.5]; assume that each simplex (p_0, p_1, p_2) in each subdivision $\mathscr{P}_k$ is positively oriented in $\mathbb{R}^2$, that is, $\Delta(p_0, p_1, p_2) > 0$. If $[a_0, a_1, a_2]$ is a (convex) Euclidean simplex in a subdivision of A_2, then f does not necessarily map $[a_0, a_1, a_2]$ into a (convex) Euclidean simplex, but f does map the points a_0, a_1, a_2 into three points $f(a_0)$, $f(a_1)$, $f(a_2)$ which form the vertices of an abstract simplex. Prove the following. If $D_{(1,2)}(f_1, f_2)(x) > 0$ for every x in A_2, then there exists a positive integer k_0 and a parameter of regularity ρ_f, $0 < \rho_f \leqq 1$, such that

$$\Delta[f(p_0), f(p_1), f(p_2)] \geqq \rho_f |f(p_1) - f(p_0)| |f(p_2) - f(p_0)| > 0$$

for each simplex (p_0, p_1, p_2) in each subdivision $\mathscr{P}_k$ with $k \geqq k_0$. Thus the (abstract) simplexes in $\mathscr{P}_k$ form an abstract complex whose simplexes are positively oriented in $\mathbb{R}^2$, and f maps this complex, for $k \geqq k_0$, into a complex whose simplexes are positively oriented in $\mathbb{R}^2$ and satisfy the condition of regularity with parameter ρ_f. [Hint. Theorem 12.8.]

21.14. State and prove a generalization in $\mathbb{R}^n$ of Exercise 21.13.

CHAPTER 4

Sperner's Lemma and the Intermediate-Value Theorem

22. Introduction

Let ABC be the triangle, positively oriented in $\mathbb{R}^2$, shown in Figure 22.1(a). Construct a simplicial subdivision $\mathscr{P}$ of ABC; one such subdivision is shown in Figure 22.1(b). Label each new vertex on the side AB with one of the letters A or B, chosen arbitrarily; similarly, label the new vertices on AC with A or C, and those on BC with B or C. Finally, label the vertices in $\mathscr{P}$ which are in the interior of ABC with one of the letters A, B, C, chosen arbitrarily. Then according to the classical Sperner's lemma, there is at least one triangle in the subdivision $\mathscr{P}$ whose three vertices are labeled with the three letters A, B, C. More precisely, the number of triangles in $\mathscr{P}$ which are labeled (A, B, C) exceeds by one the number of those which are labeled (A, C, B) [the letters are ordered as they appear on the positively oriented

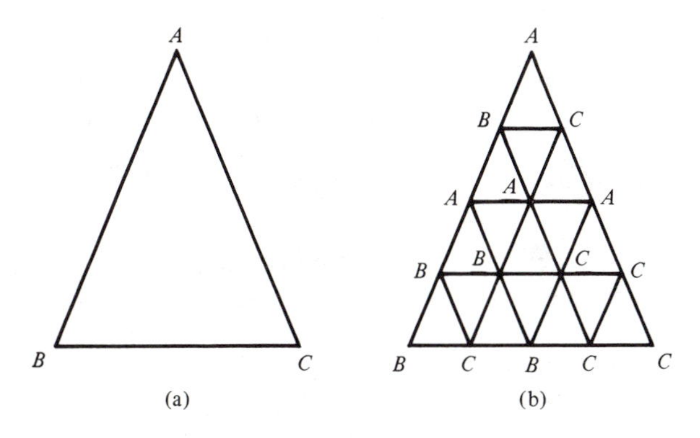

Figure 22.1. Sperner's lemma.

simplex]. In Figure 22.1(b), two triangles are labeled (A, B, C) and one is labeled (A, C, B).

Sperner's lemma is a special case of a general theorem which is proved in this chapter. The classical Sperner's lemma states a property of a special type of function defined on the vertices in simplicial subdivisions $\mathscr{P}$ of a *simplex*; the generalized Sperner's lemma states a property of similar functions defined on the vertices of the 2-simplexes in the simplicial subdivisions $\mathscr{P}$ of a 2-dimensional, homogeneous, oriented, simplicial *complex*. Furthermore, there is a generalized Sperner's lemma for each dimension n, $n = 1, 2, \cdots$; the classical Sperner's lemma is a special case of the two-dimensional theorem. The one-dimensional theorem is so trivial as to be easily overlooked, but the higher dimensional cases are far from obvious, and proofs, although elementary, are complicated in their details.

The methods employed in this chapter are based on results developed in Chapter 3. These methods take advantage of the structure in a homogeneous, oriented, simplicial, Euclidean complex and of the properties of determinants defined on the vertices of simplexes in such a complex. The connection between oriented complexes and determinants is made as follows. In two dimensions, employ $e_0 : (0, 0)$, $e_1 : (1, 0)$, and $e_2 : (0, 1)$—rather than A, B, C—as labels for the vertices of a simplex. Use the points e_i, e_j, e_k at the vertices of an oriented 2-simplex to form the determinant $\Delta(e_i, e_j, e_k)$. For example, if the Sperner function s is defined on the positively oriented simplex (a_0, a_1, a_2) so that $s(a_0) = e_0$, $s(a_1) = e_1$, $s(a_2) = e_2$, then

$$\Delta[s(a_0), s(a_1), s(a_2)] = \Delta[e_0, e_1, e_2] = \begin{vmatrix} 0 & 0 & 1 \\ 1 & 0 & 1 \\ 0 & 1 & 1 \end{vmatrix} = 1. \tag{1}$$

In $\mathbb{R}^n$ use $e_0, e_1, \cdots, e_n$ as labels for the vertices of n-simplexes in a complex; the methods and results generalize easily, although the details become increasingly complicated as n increases. A labeling of vertices in a complex is a function defined on these vertices. A function which is defined on the vertices of simplexes in a homogeneous, oriented, simplicial complex and has its values in the set $\{e_0, e_1, \cdots, e_n\}$ is called a *Sperner function*; this chapter is a study of Sperner functions and some of their applications.

Sperner's lemma and its generalizations and applications are interesting subjects in themselves, but they are of special interest here because they provide tools which are used in this chapter to establish general forms of the intermediate-value theorem [see Exercise 7.9 for the one-dimensional theorem], and in the next chapter to establish the inverse-function theorem. The following is one form of the intermediate-value theorem which is established in this chapter [see Theorem 27.1 and Example 27.3] as an application of the generalized Sperner's lemma: Let f be a continuous function which maps the oriented Euclidean simplex $[a_0, a_1, a_2]$ into $\mathbb{R}^2$ in such a way that each of the sides $[a_0, a_1]$, $[a_1, a_2]$, $[a_2, a_0]$ is mapped onto the corresponding side $[b_0, b_1]$, $[b_1, b_2]$, $[b_2, b_0]$ of $[b_0, b_1, b_2]$. Then for each

y in $[b_0, b_1, b_2]$ there exists at least one x in $[a_0, a_1, a_2]$ such that $f(x) = y$. Observe that f is assumed to be only continuous—not differentiable.

Exercises

22.1. Give a second example of the classical Sperner's lemma by making a second labeling of the vertices in the simplicial subdivision $\mathscr{P}$ of ABC in Figure 22.1(b). In your labeling verify that the number of triangles with labels (A, B, C) exceeds by one the number of those triangles with labels (A, C, B).

22.2. Let $\mathscr{P}$ denote the simplicial subdivision shown in Figure 22.1(b) of the 2-simplex ABC shown in Figure 22.1(a); here ABC is positively oriented in $\mathbb{R}^2$, and $\mathscr{P}$ is a complex of simplexes (a_0, a_1, a_2) which are positively oriented in $\mathbb{R}^2$. Let $V(\mathscr{P})$ denote the set of vertices of simplexes in $\mathscr{P}$. Let the initial labels of the vertices $V(\mathscr{P})$ be the ones you chose in Exercise 22.1. Introduce new labels by making the following substitution of labels: $A \to e_0$, $B \to e_1$, $C \to e_2$. The new labeling of vertices defines a function $s: V(\mathscr{P}) \to \{e_0, e_1, e_2\}$.

(a) If the initial labels of (a_0, a_1, a_2) are an even permutation of (A, B, C) show that the new labels are an even permutation of (e_0, e_1, e_2), and hence that $\Delta[s(a_0), s(a_1), s(a_2)] = \Delta(e_0, e_1, e_2) = +1$.

(b) If the initial labels of (a_0, a_1, a_2) are an odd permutation of (A, B, C), show that the new labels are an odd permutation of (e_0, e_1, e_2), and hence that $\Delta[s(a_0), s(a_1), s(a_2)] = -\Delta(e_0, e_1, e_2) = -1$.

(c) If (a_0, a_1, a_2) is a simplex in $\mathscr{P}$ whose vertices do not have three different (initial or new) labels, show that $\Delta[s(a_0), s(a_1), s(a_2)] = 0$.

(d) Find the value of $\Delta[s(a_0), s(a_1), s(a_2)]$ for each of the 16 simplexes (a_0, a_1, a_2) in $\mathscr{P}$, and show that the sum of these 16 determinants is 1.

22.3. In the triangle ABC shown in Figure 22.1(a) construct a simplicial subdivision $\mathscr{P}$ which has at least 25 simplexes (a_0, a_1, a_2), each of which is positively oriented in $\mathbb{R}^2$. Label the vertices $V(\mathscr{P})$ with A, B, C in the manner described in the first paragraph of Section 22. Then make the substitution $A \to e_0$, $B \to e_1$, $C \to e_2$ of labels described in Exercise 22.2. The labeling of vertices is a Sperner function $s: V(\mathscr{P}) \to \{e_0, e_1, e_2\}$. For each simplex (a_0, a_1, a_2) in $\mathscr{P}$, find the value of $\Delta[s(a_0), s(a_1), s(a_2)]$; then show that $\Sigma\{\Delta[s(a_0), s(a_1), s(a_2)] : (a_0, a_1, a_2) \in \mathscr{P}\} = 1$.

22.4. What is your conjecture for the statement of Sperner's lemma in $\mathbb{R}^3$? Construct an example to illustrate your conjecture.

22.5. Is there a Spermer's lemma for a one-dimensional simplex in $\mathbb{R}$? If there is, state it and construct two examples to illustrate it.

23. Sperner Functions; Sperner's Lemma

This section defines a class of Sperner functions and establishes a form of Sperner's lemma for these functions. These Sperner functions are defined on the vertices of certain Euclidean and abstract complexes in $\mathbb{R}^n$, and the section begins with a description of these complexes.

Let K be an oriented, Euclidean, homogeneous, 2-dimensional complex in $\mathbb{R}^n$, $n \geqq 2$. The simplexes in K are Euclidean simplexes $[a_0, a_1, a_2]$ in $\mathbb{R}^n$; the union of these Euclidean simplexes is a closed set, denoted by A, in $\mathbb{R}^n$. If $[a_0, a_1]$ is a one-dimensional simplex in K, then by special hypothesis in this chapter it is a side of at most two 2-dimensional simplexes in K; if K is in $\mathbb{R}^2$, then this hypothesis is satisfied automatically since K is a Euclidean complex. It is assumed that the simplexes in K are oriented so that the boundary of the chain which contains all 2-simplexes in K consists of all one-dimensional sides $[a_0, a_1]$ which belong to a single simplex in K. This condition also is satisfied automatically if K is in $\mathbb{R}^2$ and all 2-simplexes in K have the same orientation in $\mathbb{R}^2$.

23.1 Examples. Let A be a square in a 2-dimensional plane in $\mathbb{R}^3$; subdivide A into 2-dimensional simplexes and give each one the same orientation. If K consists of these oriented 2-dimensional simplexes and all of their sides, then K has all of the properties specified in the last paragraph. It is not necessary, however, that K lie in a 2-dimensional plane in $\mathbb{R}^n$. For example, let A_3 be a cube in $\mathbb{R}^3$; an oriented subdivision of A_3 into a homogeneous, 3-dimensional, Euclidean complex induces a simplicial subdivision into the topological boundary of A_3 which forms an oriented, homogeneous, 2-dimensional Euclidean complex [see the subdivision of A_3 in Figure 21.5]. Any connected subset of 2-simplexes in this boundary complex is a complex K with the properties specified in the last paragraph.

Let $\mathscr{P}$ be a simplicial subdivision of K; then $\mathscr{P} = \{\mathbf{a}_r : r = 1, \cdots, N\}$, and $\mathbf{a}_r$ is a 2-dimensional Euclidean simplex with vertices (p_0^r, p_1^r, p_2^r) which determines the same 2-direction in $\mathbb{R}^n$ as the simplex in K which contains it [if K is in $\mathbb{R}^2$, then $\mathbf{a}_r$ has the same orientation in $\mathbb{R}^2$ as the simplexes in K]. Let $f: A \to \mathbb{R}^2$ be a function which is defined and continuous on A. Then f maps (the vertices of) $\mathscr{P}$ into an abstract, oriented, 2-dimensional, simplicial complex $\mathscr{Q} = \{\mathbf{b}_r : r = 1, \cdots, N\}$ in which

$$\mathbf{b}_r = (f(p_0^r), f(p_1^r), f(p_2^r)) = (q_0^r, q_1^r, q_2^r). \tag{1}$$

Let $V(\mathscr{Q})$ denote the set of vertices of simplexes in $\mathscr{Q}$.

23.2 Definition. A function $s: V(\mathscr{Q}) \to \{e_0, e_1, e_2\}$ whose domain is $V(\mathscr{Q})$ and whose range is contained in $\{e_0, e_1, e_2\}$ is called a *Sperner function* on $\mathscr{Q}$. The *Sperner number* $S(s)$ of s is defined by the following equation:

$$S(s) = \sum_{r=1}^{N} \Delta[s(q_0^r), s(q_1^r), s(q_2^r)]. \tag{2}$$

Observe that $S(s)$ is an integer since each term $\Delta[s(q_0^r), s(q_1^r), s(q_2^r)]$ in the sum in (2) has the value $+1$, -1, or 0. Let $P(s)$ and $N(s)$ be the number of simplexes (q_0^r, q_1^r, q_2^r) in $\mathscr{Q}$ for which $\Delta[s(q_0^r), s(q_1^r), s(q_2^r)]$ is $+1$ and -1 respectively; then

$$S(s) = P(s) - N(s). \tag{3}$$

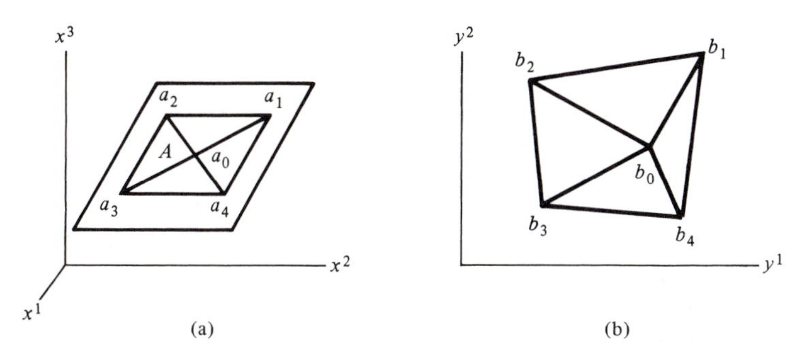

Figure 23.1. Figure for Example 23.3.

23.3 Example. Let $\mathscr{P}$ be the Euclidean complex in $\mathbb{R}^3$ which has the oriented 2-dimensional simplexes (a_0, a_1, a_2), (a_0, a_2, a_3), (a_0, a_3, a_4), (a_0, a_4, a_1) shown in Figure 23.1(a). Then $\mathscr{P}$ is a simplicial subdivision of the square A in a plane in $\mathbb{R}^3$. Let $f: A \to \mathbb{R}^2$ be a continuous function which maps the oriented simplexes in $\mathscr{P}$ into the oriented abstract simplexes in $\mathscr{Q}$ shown in Figure 23.1(b) such that $b_i = f(a_i)$, $i = 0, 1, \cdots, 4$. Let $s: V(\mathscr{Q}) \to \{e_0, e_1, e_2\}$ be a Sperner function which is defined on $\mathscr{Q}$ as follows:

$$s(b_0) = e_0, \qquad s(b_1) = e_1, \qquad s(b_2) = e_2, \qquad s(b_3) = e_1, \qquad s(b_4) = e_0. \tag{4}$$

Then

$$\begin{aligned}
&(p_0^1, p_1^1, p_2^1) = (a_0, a_1, a_2),\\
&(q_0^1, q_1^1, q_2^1) = [f(a_0), f(a_1), f(a_2)] = (b_0, b_1, b_2);\\
&(p_0^2, p_1^2, p_2^2) = (a_0, a_2, a_3),\\
&(q_0^2, q_1^2, q_2^2) = [f(a_0), f(a_2), f(a_3)] = (b_0, b_2, b_3);\\
&(p_0^3, p_1^3, p_2^3) = (a_0, a_3, a_4),\\
&(q_0^3, q_1^3, q_2^3) = [f(a_0), f(a_3), f(a_4)] = (b_0, b_3, b_4);\\
&(p_0^4, p_1^4, p_2^4) = (a_0, a_4, a_1),\\
&(q_0^4, q_1^4, q_2^4) = [f(a_0), f(a_4), f(a_1)] = (b_0, b_4, b_1).
\end{aligned} \tag{5}$$

The Sperner number $S(s)$ of the Sperner function s is calculated by the formula in (2) as follows:

$$\begin{aligned}
&\Delta[s(q_0^1), s(q_1^1), s(q_2^1)] = \Delta[s(b_0), s(b_1), s(b_2)] = \Delta(e_0, e_1, e_2) = 1;\\
&\Delta[s(q_0^2), s(q_1^2), s(q_2^2)] = \Delta[s(b_0), s(b_2), s(b_3)] = \Delta(e_0, e_2, e_1) = -1;\\
&\Delta[s(q_0^3), s(q_1^3), s(q_2^3)] = \Delta[s(b_0), s(b_3), s(b_4)] = \Delta(e_0, e_1, e_0) = 0;\\
&\Delta[s(q_0^4), s(q_1^4), s(q_2^4)] = \Delta[s(b_0), s(b_4), s(b_1)] = \Delta(e_0, e_0, e_1) = 0.
\end{aligned} \tag{6}$$

Then by (2), the Sperner number of the Sperner function defined in (4) is zero, or $S(s) = 0$. Also, $P(s) = N(s) = 1$, and $S(s) = 0$ by equation (3).

The boundary $\partial\mathcal{Q}$ of $\Sigma\{\mathbf{b} : \mathbf{b} \in \mathcal{Q}\}$ is a chain of one-dimensional simplexes (q_i, q_{i+1}) as follows:

$$\partial\sum\{\mathbf{b} : \mathbf{b} \in \mathcal{Q}\} = \sum\{(q_i, q_{i+1}) : i \in I\}. \tag{7}$$

Simple examples show that $\partial\mathcal{Q}$ may be the zero-chain. For example, the 2-simplexes in the surface of the cube A_3 in Figure 21.5 form a chain whose boundary is the zero-chain. The next theorem shows that $S(s)$ is completely determined by the values of s on $\partial\mathcal{Q}$.

23.4 Theorem. *Let s be a Sperner function on $\mathcal{Q}$, and let e be an arbitrary, but fixed, point in $\{e_0, e_1, e_2\}$. Then*

$$S(s) = \sum_{i \in I} \Delta[e, s(q_i), s(q_{i+1})]. \tag{8}$$

PROOF. Let e be one of the points in $\{e_0, e_1, e_2\}$. Evaluate each of the determinants in (2) by Theorem 20.1, using e as the point x in that theorem. Thus

$$S(s) = \sum_{r=1}^{N} \{\Delta[e, s(q_1^r), s(q_2^r)] + \Delta[s(q_0^r), e, s(q_2^r)] + \Delta[s(q_0^r), s(q_1^r), e]\}. \tag{9}$$

If two simplexes $\mathbf{b}$ in $\mathcal{Q}$ have a side in common, then the two terms in (9) which arise from this common side cancel in the sum because the side has opposite orientations in the two simplexes to which it belongs. If (q_i, q_{i+1}) is a one-simplex in $\partial\mathcal{Q}$, then there is a simplex of the form (q, q_i, q_{i+1}) in $\mathcal{Q}$, and (9) shows that the term $\Delta[e, s(q_i), s(q_{i+1})]$ remains in the sum after all cancellations have been made. Thus the sum in (9) simplifies to that in (8). □

23.5 Example. Use Theorem 23.4 to evaluate $S(s)$ in Example 23.3. Choose e to be e_1. Then by the equations in (6),

$$\begin{aligned}
&\Delta[s(b_0), s(b_1), s(b_2)] \\
&\quad = \Delta[e_1, s(b_1), s(b_2)] + \Delta[s(b_0), e_1, s(b_2)] + \Delta[s(b_0), s(b_1), e_1], \\
&\Delta[s(b_0), s(b_2), s(b_3)] \\
&\quad = \Delta[e_1, s(b_2), s(b_3)] + \Delta[s(b_0), e_1, s(b_3)] + \Delta[s(b_0), s(b_2), e_1], \\
&\Delta[s(b_0), s(b_3), s(b_4)] \\
&\quad = \Delta[e_1, s(b_3), s(b_4)] + \Delta[s(b_0), e_1, s(b_4)] + \Delta[s(b_0), s(b_3), e_1], \\
&\Delta[s(b_0), s(b_4), s(b_1)] \\
&\quad = \Delta[e_1, s(b_4), s(b_1)] + \Delta[s(b_0), e_1, s(b_1)] + \Delta[s(b_0), s(b_4), e_1].
\end{aligned} \tag{10}$$

Add these four equations; the sum on the left is $S(s)$ by (2). On the right, eight terms cancel and the result is

$$S(s) = \Delta[e_1, s(b_1), s(b_2)] + \Delta[e_1, s(b_2), s(b_3)] \\ + \Delta[e_1, s(b_3), s(b_4)] + \Delta[e_1, s(b_4), s(b_1)]. \tag{11}$$

Thus $S(s)$ is completely determined by the values of s on vertices in $\partial\mathcal{Q}$ as stated in Theorem 23.4. Each of the terms on the right in (11) has the value zero by (4); hence $S(s) = 0$ as in Example 23.3. The results are similar if e is chosen to be e_0 or e_2.

Return to the general case: $\mathcal{P}$ is a simplicial subdivision of the complex K, $f: A \to \mathbb{R}^2$ is a continuous function which maps $\mathcal{P}$ into $\mathcal{Q}$, and $s: V(\mathcal{Q}) \to \{e_0, e_1, e_2\}$ is a Sperner function on $\mathcal{Q}$. Let $\mathcal{P}'$ be a refinement of $\mathcal{P}$; then f maps $\mathcal{P}$ and $\mathcal{P}'$ into abstract complexes $\mathcal{Q}$ and $\mathcal{Q}'$, and $\mathcal{Q}'$ is called a refinement of $\mathcal{Q}$. Then corresponding to a simplex (q_i, q_{i+1}) in $\partial\mathcal{Q}$, there are simplexes (b_j, b_{j+1}) of the following kind in $\partial\mathcal{Q}'$:

$$(b_0, b_1), (b_1, b_2), \cdots, (b_m, b_{m+1}), \\ b_0 = q_i, \quad b_{m+1} = q_{i+1}. \tag{12}$$

Let s and s' be Sperner functions on $\mathcal{Q}$ and $\mathcal{Q}'$ respectively such that (a) s and s' have the same values on $V(\mathcal{Q})$; and (b) for each (q_i, q_{i+1}) in $\partial\mathcal{Q}$, the function s' takes on at most two distinct values on the associated vertices $b_0, \cdots, b_{m+1}$ in $V(\mathcal{Q}')$. Thus

$$s'(b_0) = s(q_i), s'(b_{m+1}) = s(q_{i+1}); \tag{13}$$

$$\{s'(b_j); j = 0, 1, \cdots, m+1\} \text{ contains at most two distinct elements in } \{e_0, e_1, e_2\}. \tag{14}$$

23.6 Definition. Let $\mathcal{Q}'$ be a refinement of $\mathcal{Q}$, and s and s' be Sperner functions on $\mathcal{Q}$ and $\mathcal{Q}'$, respectively, which have properties (13) and (14). Then s' is said to be a *refinement* of s.

23.7 Example. Let K and $\mathcal{P}$ consist of a single positively oriented simplex (a_0, a_1, a_2) in $\mathbb{R}^2$, and let $A = [a_0, a_1, a_2]$. Then the function $f: A \to \mathbb{R}^2$ maps (a_0, a_1, a_2) into the single abstract simplex $(f(a_0), f(a_1), f(a_2))$ in $\mathcal{Q}$. Define $s: V(\mathcal{Q}) \to \{e_0, e_1, e_2\}$ as follows: $s[f(a_i)] = e_i$, $i = 0, 1, 2$. Let $\mathcal{P}'$ be a refinement of $\mathcal{P}$; then f maps $\mathcal{P}'$ into $\mathcal{Q}'$. Define the Sperner function $s': V(\mathcal{Q}') \to \{e_0, e_1, e_2\}$ as follows: (i) $s'[f(a_i)] = e_i$, $i = 0, 1, 2$; (ii) at each new vertex in $\mathcal{Q}'$ associated with the side $(f(a_0), f(a_1))$ in $\mathcal{Q}$ let s' have either the value e_0 or the value e_1; at each vertex of $\mathcal{Q}'$ on $(f(a_1), f(a_2))$ let s' have one of the values e_1 or e_2; and at each vertex of $\mathcal{Q}'$ on $(f(a_2), f(a_0))$ let s' have one of the values e_2 or e_0. Then s' is a refinement of s. If f is the identity function, then s and s' can be considered to be defined on the vertices of $\mathcal{P}$ and $\mathcal{P}'$ in the complex K.

23.8 Theorem (Sperner's Lemma). *Let s and s' be Sperner functions on $\mathcal{Q}$ and $\mathcal{Q}'$ respectively. If $\mathcal{Q}'$ and s' are refinements of $\mathcal{Q}$ and s, respectively, then*

$$S(s') = S(s); \tag{15}$$

if $S(s) \neq 0$, *then there is a simplex* (q_0', q_1', q_2') *in* $\mathscr{Q}'$ *such that* $\{s'(q_0'), s'(q_1'), s'(q_2')\} = \{e_0, e_1, e_2\}$. (16)

PROOF. Use (8) in Theorem 23.4 to evaluate $S(s')$. Recall the relations in (12). Then by Theorem 20.1, with $s(q_i)$ as the point x in that theorem,

$$\begin{aligned} &\Delta[e, s'(b_j), s'(b_{j+1})] \\ &\quad = \Delta[s(q_i), s'(b_j), s'(b_{j+1})] + \Delta[e, s(q_i), s'(b_{j+1})] + \Delta[e, s'(b_j), s(q_i)]. \end{aligned} \tag{17}$$

Since $s(q_i) = s'(b_0)$ by (13) and $\Delta[s'(b_0), s'(b_j), s'(b_{j+1})] = 0$ by (14), then the first term on the right in (17) is zero and

$$\Delta[e, s'(b_j), s'(b_{j+1})] = \Delta[e, s(q_i), s'(b_{j+1})] - \Delta[e, s(q_i), s'(b_j)]. \tag{18}$$

This equation is true for $j = 0, 1, \cdots, m$. Sum equation (18) for $j = 0, 1, \cdots, m$ and observe that $\Delta[e, s(q_i), s'(b_0)] = 0$ and $\Delta[e, s(q_i), s'(b_{m+1})] = \Delta[e, s(q_i), s(q_{i+1})]$ by (13). Then

$$\begin{aligned} \sum_{j=0}^{m} \Delta[e, s'(b_j), s'(b_{j+1})] &= \sum_{j=0}^{m} \{\Delta[e, s(q_i), s'(b_{j+1})] - \Delta[e, s(q_i), s'(b_j)]\} \\ &= \Delta[e, s(q_i), s'(b_{m+1})] - \Delta[e, s(q_i), s'(b_0)] \\ &= \Delta[e, s(q_i), s(q_{i+1})]. \end{aligned} \tag{19}$$

To complete the proof, sum equation (19) over all simplexes (q_i, q_{i+1}) in $\partial\mathscr{Q}$. The sum of terms on the left is $S(s')$ by Theorem 23.4, and the sum of the terms on the right is $S(s)$ by the same theorem. The proof of (15) is complete. If $S(s) \neq 0$, then $S(s') \neq 0$ by (15), and $P(s') - N(s') \neq 0$ by (3). Then at least one of the numbers $P(s')$, $N(s')$ is different from zero and (16) is true. The entire proof is complete. □

EXERCISES

23.1. Let K consist of six positively oriented equilateral triangles in $\mathbb{R}^2$ which form a regular hexagon, and let $\mathscr{P}$ be these six triangles. Let f be the identity transformation; then $\mathscr{Q} = \mathscr{P}$.

(a) Define a Sperner function s on $\mathscr{Q}$, and use equation (2) to find $S(s)$. Verify (8) in Theorem 23.4 first with $e = e_0$, then with $e = e_1$, and finally with $e = e_2$.

(b) Construct $\mathscr{Q}'$ by subdividing each triangle in $\mathscr{Q}$ into four equilateral triangles, and define a Sperner function s' on $\mathscr{Q}'$ which is a refinement of s in (a). Find $S(s')$ by equation (2); then find $S(s')$ by (8) in Theorem 23.4, using each of the values e_0, e_1, e_2 for e in succession. Then verify Theorem 23.8 (Sperner's lemma) for s and s'.

23.2. Show that Theorem 23.8 contains, as a special case, the classical Sperner's lemma for the triangle ABC as stated in Section 22.

23.3. (a) Show that $S(s) = 0$ if $\partial\mathcal{Q} = 0$.
(b) Give an example of a complex K in $\mathbb{R}^3$ for which $\partial\mathcal{Q} = 0$. Explain why there does not exist a complex K of this type in $\mathbb{R}^2$.
(c) Let K be a complex for which $\partial\mathcal{Q} = 0$, and let $s: V(\mathcal{Q}) \to \{e_0, e_1, e_2\}$ be a Sperner function on $\mathcal{Q}$. Prove that $P(s) = N(s)$.

24. A Special Class of Sperner Functions

This section defines a class of Sperner functions which has special significance for the mapping $f: A \to \mathbb{R}^2$. In particular, these Sperner functions will be used to prove a form of the intermediate-value theorem for the mapping $f: A \to \mathbb{R}^2$.

Let A be the set described in the second paragraph of Section 23, and let $\{\mathcal{P}_k : k = 1, 2, \cdots\}$ be a sequence of simplicial subdivisions of A which has the following properties:

Each simplex in $\mathcal{P}_k$ determines the same 2-direction in $\mathbb{R}^n$ as the simplex in K which contains it. If A is a set in $\mathbb{R}^2$, this assumption means that all simplexes in K and in $\mathcal{P}_k$ have the same orientation in $\mathbb{R}^2$. (1)

$$\mathcal{P}_{k+1} \text{ is a refinement of } \mathcal{P}_k. \tag{2}$$

$$\text{The norm of } \mathcal{P}_k \text{ approaches zero as } k \text{ tends to infinity.} \tag{3}$$

Choose notation so that

$$\mathcal{P}_k = \{\mathbf{a}_r^k : r = 1, \cdots, N_k\}; \tag{4}$$

$$\partial\mathcal{P}_k = \sum\{(p_i^k, p_{i+1}^k) : i \in I_k\}; \tag{5}$$

$$\partial A = \bigcup\{[p_i^k, p_{i+1}^k] : i \in I_k\}. \tag{6}$$

In (5), $\partial\mathcal{P}_k$ denotes the boundary of the chain which contains all 2-simplexes in the complex $\mathcal{P}_k$; similarly, in (8) below, $\partial\mathcal{Q}_k$ denotes the boundary of the chain of abstract 2-simplexes in $\mathcal{Q}_k$. Let $f: A \to \mathbb{R}^2$ be a continuous function. Then f maps the vertices of $\mathbf{a}_r^k$ into the vertices of an abstract simplex $\mathbf{b}_r^k$, and f maps the simplicial complex $\mathcal{P}_k$ into an abstract simplicial complex $\mathcal{Q}_k$ such that

$$\mathcal{Q}_k = \{\mathbf{b}_r^k : r = 1, \cdots, N_k\}, \qquad \mathbf{b}_r^k = f(\mathbf{a}_r^k); \tag{7}$$

$$\partial\mathcal{Q}_k = \sum\{(q_i^k, q_{i+1}^k) : i \in I_k\}, \qquad q_i^k = f(p_i^k). \tag{8}$$

Assume that $\partial\mathcal{P}_1 \neq 0$; then $\partial\mathcal{P}_k \neq 0$ for $k = 1, 2, \cdots$ and $\partial A \neq \varnothing$.

Let b be any point in $\mathbb{R}^2$ such that $b \notin f(\partial A)$. Let r_0, r_1, r_2 be three rays which emanate from b as follows: r_0 is directed horizontally to the right, r_1 makes an angle of $2\pi/3$ with r_0, and r_2 makes an angle of $2\pi/3$ with r_1 [see

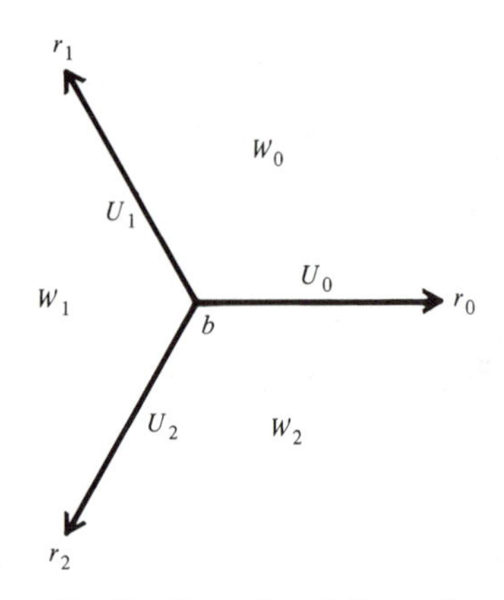

Figure 24.1. The regions U_0, U_1, U_2 used to define a class of Sperner functions.

Figure 24.1]. Then r_0 and r_1, r_1 and r_2, and r_2 and r_0 bound the closed regions W_0, W_1, and W_2 respectively; each point in $\mathbb{R}^2$ belongs to at least one of these regions, and the intersection of these regions is the set whose only point is b. Let $\text{int}(W_i)$ and $\text{int}(r_i)$ denote the interior of W_i and the interior of r_i (the ray which is open at b), respectively, for $i = 0, 1, 2$. Define sets U_0, U_1, U_2 as follows.

$$\begin{aligned} U_0 &= \text{int}(W_0) \cup \text{int}(r_0), \\ U_1 &= \text{int}(W_1) \cup \text{int}(r_1), \\ U_2 &= \text{int}(W_2) \cup \text{int}(r_2) \cup \{b\}. \end{aligned} \tag{9}$$

The sets U_i, $i = 0, 1, 2$, form a disjoint decomposition of $\mathbb{R}^2$.

Corresponding to the point b and to each subdivision $\mathscr{P}_k$ of A, define a Sperner function $s_k : V(\mathscr{Q}_k) \to \{e_0, e_1, e_2\}$ as follows:

$$s_k(q) = e_i \quad \text{if} \quad q \in U_i, \qquad q \in V(\mathscr{Q}_k), \qquad i = 0, 1, 2. \tag{10}$$

Then s_k has a Sperner number as defined in Definition 23.2. If (q_0, q_1, q_2) are the vertices of $\mathbf{b}_r^k$, let $s_k(\mathbf{b}_r^k)$ denote $(s(q_0), s(q_1), s(q_2))$; then by (2) in Definition 23.2,

$$S(s_k) = \sum_{r=1}^{N_k} \Delta[s_k(\mathbf{b}_r^k)], \qquad \mathbf{b}_r^k \text{ in } \mathscr{Q}_k. \tag{11}$$

Also, $S(s_k)$ can be evaluated by (8) in Theorem 23.4; thus by (8) in Section 23 and (8) in Section 24,

$$S(s_k) = \sum_{i \in I_k} \Delta[e, s_k(q_i^k), s_k(q_{i+1}^k)]. \tag{12}$$

24.1 Example. Let A be the oriented Euclidean simplex $[a_0, a_1, a_2]$ shown in Figure 24.2, and let $f: A \to \mathbb{R}^2$ be a continuous function. Let b be a point in $\mathbb{R}^2$ which is not on $f(\partial A)$; here $f(\partial A)$ is the curve in $\mathbb{R}^2$ into which f maps

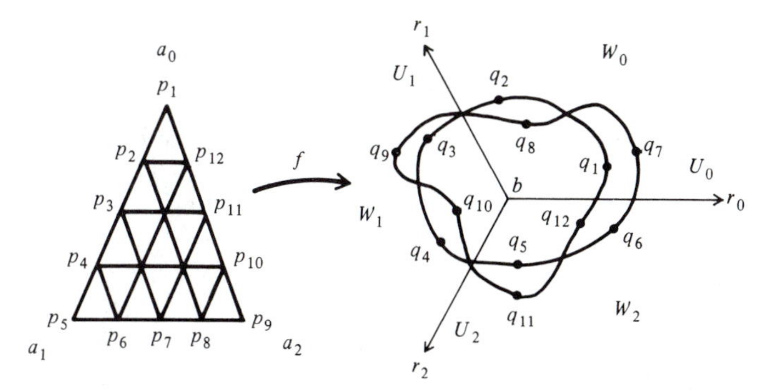

Figure 24.2. The point b, the simplicial subdivision $\mathscr{P}_k$, and the Sperner number $S(s)$ for a mapping $f: [a_0, a_1, a_2] \to \mathbb{R}^2$.

the three sides of $[a_0, a_1, a_2]$, or A [see (6)]. A subdivision $\mathscr{P}$ of A is shown in the figure; the function f maps the vertices in $\mathscr{P}$ into an abstract complex $\mathscr{Q}$ in $\mathbb{R}^2$. Construct the regions U_0, U_1, U_2 as stated in (9), and use these regions to define the Sperner function $s: V(\mathscr{Q}) \to \{e_0, e_1, e_2\}$ as stated in (10). The Sperner number $S(s)$ of s can be found by using either the formula in (11) or the one in (12), but the latter is simpler. Let $p_1, p_2, \cdots, p_{12}$ be the vertices of $\mathscr{P}$ in ∂A, and let $q_i = f(p_i)$, $i = 1, 2, \cdots, 12$. Assume that f maps the boundary ∂A of A into the curve shown in Figure 24.2, and assume that the points $q_1, q_2, \cdots, q_{12}$ are located as shown in the figure. Then by (10),

$$\begin{aligned} &s(q_1) = s(q_2) = s(q_7) = s(q_8) = e_0, \\ &s(q_3) = s(q_4) = s(q_9) = s(q_{10}) = e_1, \\ &s(q_5) = s(q_6) = s(q_{11}) = s(q_{12}) = e_2. \end{aligned} \tag{13}$$

Choose e to be e_2; the final result would be the same if e were chosen to be e_0 or e_1. The following is the calculation of the terms in the formula for $S(s)$ in (12).

$$\Delta[e_2, s(q_1), s(q_2)] = \Delta[e_2, e_0, e_0] = 0.$$

$$\Delta[e_2, s(q_2), s(q_3)] = \Delta[e_2, e_0, e_1] = 1.$$

$$\Delta[e_2, s(q_3), s(q_4)] = \Delta[e_2, e_1, e_1] = 0.$$

$$\Delta[e_2, s(q_4), s(q_5)] = \Delta[e_2, e_1, e_2] = 0.$$

$$\Delta[e_2, s(q_5), s(q_6)] = \Delta[e_2, e_2, e_2] = 0.$$

$$\Delta[e_2, s(q_6), s(q_7)] = \Delta[e_2, e_2, e_0] = 0.$$

$$\Delta[e_2, s(q_7), s(q_8)] = \Delta[e_2, e_0, e_0] = 0.$$

$$\Delta[e_2, s(q_8), s(q_9)] = \Delta[e_2, e_0, e_1] = 1.$$

$$\Delta[e_2, s(q_9), s(q_{10})] = \Delta[e_2, e_1, e_1] = 0.$$

$$\Delta[e_2, s(q_{10}), s(q_{11})] = \Delta[e_2, e_1, e_2] = 0.$$

$$\Delta[e_2, s(q_{11}), s(q_{12})] = \Delta[e_2, e_2, e_2] = 0.$$

$$\Delta[e_2, s(q_{12}), s(q_1)] = \Delta[e_2, e_2, e_0] = 0.$$

Then $S(s)$ is the sum of the column of numbers at the right; hence, $S(s) = 2$. The calculation of $S(s)$ could have been greatly shortened; inspection of the term $\Delta[e_2, s(q_i), s(q_{i+1})]$ [see (12) with $e = e_2$, and (10)] shows that it is zero except when one of the points q_i, q_{i+1} is in U_0 and the other is in U_1. As the table above shows [see also Figure 24.2], there are only two terms $\Delta[e_2, s(q_i), s(q_{i+1})]$ which are different from zero. The fact that $S(s) = 2$ has geometric significance: if p traces the boundary of $[a_0, a_1, a_2]$, or A, once in the positive direction, then $f(p)$ traces a curve which winds twice around the point b in the positive direction.

Return now to the general problem described at the beginning of this section. The function $f: A \to \mathbb{R}^2$ maps the sequence of subdivisions $\mathscr{P}_k$ of A into a sequence of abstract complexes $\mathscr{Q}_k$; there is a Sperner function $s_k: V(\mathscr{Q}_k) \to \{e_0, e_1, e_2\}$ and a Sperner number $S(s_k)$ for $k = 1, 2, \cdots$. For some functions f it is not true that $S(s_1) = S(s_2) = \cdots$; however, for every f it is true that $S(s_k) = S(s_{k+1}) = \cdots$ if k is sufficiently large. The following lemma is needed to establish this result in Theorem 24.3 below.

24.2 Lemma. *Let $f: A \to \mathbb{R}^2$ be the continuous function described at the beginning of this section, and let b be a point in $\mathbb{R}^2$ such that $b \notin f(\partial A)$. Then there exists a positive integer K such that, for $k \geqq K$, the function f maps each Euclidean simplex $[p_i^k, p_{i+1}^k]$ in $\partial\mathscr{P}_k$ into at most two of the regions W_0, W_1, W_2.*

PROOF. Assume that the lemma is false, that is, assume that there is no integer K with the stated property. If f has the stated property with respect to $\mathscr{P}_k$, then it has this property with respect to $\mathscr{P}_{k+1}$ since $\mathscr{P}_{k+1}$ is a refinement of $\mathscr{P}_k$ by (2). Then since the lemma is false, for each $\mathscr{P}_k$, $k = 1, 2, \cdots$, there is at least one Euclidean simplex $[p_i^k, p_{i+1}^k]$ in $\partial\mathscr{P}_k$ which contains three points x_0^k, x_1^k, x_2^k such that $f(x_i^k) \in W_i$, $i = 0, 1, 2$. The infinite set $\{x_0^k : k = 1, 2, \cdots\}$ belongs to the compact set ∂A [see Definition 92.8 in Appendix 2]; hence, this set has a point of accumulation x which belongs to ∂A [see Theorem 94.1 and Definition 92.3 in Appendix 2]. Then the sequence $\{x_0^k : k = 1, 2, \cdots\}$ contains a subsequence $\{x_0^{k_j} : j = 1, 2, \cdots\}$ such that

$$\lim_{j\to\infty} x_0^{k_j} = x. \tag{14}$$

Hence, since f is continuous on A,

$$\lim_{j\to\infty} f(x_0^{k_j}) = f(x). \tag{15}$$

Since the $x_i^{k_j}$, $i = 0, 1, 2$, belong to the Euclidean simplex $[p_i^{k_j}, p_{i+1}^{k_j}]$ in $\partial\mathscr{P}_{k_j}$, and since the norms of the subdivisions $\mathscr{P}_{k_j}$ tend to zero by (3),

$$\lim_{j\to\infty} x_i^{k_j} = x, \qquad \lim_{j\to\infty} f(x_i^{k_j}) = f(x), \qquad i = 0, 1, 2. \tag{16}$$

Since W_0, W_1, W_2 are closed sets, and since $f(x_i^k) \in W_i$ for $i = 0, 1, 2$ and $k = 1, 2, \cdots$, equation (16) shows that $f(x) \in W_i$ for $i = 0, 1, 2$. But the only point in $W_0 \cap W_1 \cap W_2$ is b; hence, $f(x) = b$. However, $f(x) \neq b$ since $x \in \partial A$ and, by hypothesis, $b \notin f(\partial A)$. This contradiction establishes the lemma. □

24.3 Theorem. *Let s_k, $k = 1, 2, \cdots$, be the set of Sperner functions defined in* (10), *and let K be the integer defined in Lemma 24.2. Then there exists an integer $d(f, b)$ such that*

$$S(s_k) = d(f, b), \qquad k \geqq K. \tag{17}$$

PROOF. Let k be an integer such that $k \geqq K$. Since $\mathscr{P}_{k+1}$ is a refinement of $\mathscr{P}_k$ by the hypothesis in (2), $\mathscr{P}_{k+1}$ induces a subdivision in each simplex $[p_i^k, p_{i+1}^k]$ in $\partial\mathscr{P}_k$ [see (5)]; assume that this subdivision is

$$\begin{gathered}(a_0, a_1), (a_1, a_2), \cdots, (a_m, a_{m+1}),\\ a_0 = p_i^k, \qquad a_{m+1} = p_{i+1}^k.\end{gathered} \tag{18}$$

Then f maps the vertices of (p_i^k, p_{i+1}^k) into the vertices of (q_i^k, q_{i+1}^k), and f maps the vertices in (18) into the vertices of

$$\begin{gathered}(b_0, b_1), (b_1, b_2), \cdots, (b_m, b_{m+1}),\\ b_0 = q_i^k, \qquad b_{m+1} = q_{i+1}^k.\end{gathered} \tag{19}$$

The points $b_0, b_1, \cdots, b_{m+1}$ lie in at most two of the regions W_i by Lemma 24.2, and the definition of the Sperner functions in (10) shows that

$$s_{k+1}(b_0) = s_k(q_i^k), \qquad s_{k+1}(b_{m+1}) = s_k(q_{i+1}^k); \tag{20}$$

the set $\{s_{k+1}(b_j) : j = 0, 1, \cdots, m+1\}$ contains at most two distinct elements in $\{e_0, e_1, e_2\}$. (21)

Then $\mathscr{Q}_{k+1}$ is a refinement of $\mathscr{Q}_k$, and s_{k+1} is a refinement of s_k. Hence, by Theorem 23.8 [Sperner's lemma], $S(s_{k+1}) = S(s_k)$. A complete induction and the definition $d(f, b) = S(s_k)$ complete the proof of (17). □

24.4 Definition. The integer $S(s_k)$ is called the *degree of f at b* and denoted by $d(f, b)$.

Equation (17) shows that $\lim_{k\to\infty} S(s_k)$ exists, and that

$$d(f, b) = \lim_{k\to\infty} S(s_k). \tag{22}$$

EXERCISES

24.1. Let (a_0, a_1, a_2) and (b_0, b_1, b_2) be two Euclidean simplexes which are positively oriented in $\mathbb{R}^2$. Let $f: [a_0, a_1, a_2] \to \mathbb{R}^2$ be the affine transformation $L: \mathbb{R}^2 \to \mathbb{R}^2$ such that $L(a_i) = b_i$, $i = 0, 1, 2$. Find all of the points b at which $d(f, b)$ is defined.

Show that $d(f, b) = 1$ if b is in the interior of (b_0, b_1, b_2), and that $d(f, b) = 0$ if b is in the exterior of (b_0, b_1, b_2).

24.2. In Exercise 24.1 rotate the rays r_0, r_1, r_2 through 45° and then find $d(f, b)$ with respect to the new regions U_0, U_1, U_2 determined by the new location of the rays. Show that the change in the location of the rays does not affect the value of $d(f, b)$.

24.3. Let A be a set in $\mathbb{R}^2$ which consists of six equilateral triangles whose sides have length 4 and which form a regular hexagon with center at the origin. Map A into $\mathbb{R}^2$ by a function f which in polar coordinates is $r' = r$, $\theta' = 6\theta$. Find the degree of f at each of the following points: $(0, 0)$, $(1, 1)$, $(2, 0)$, $(0, 12)$, $(-20, -20)$.

24.4. (a) Let A be a regular hexagon in $\mathbb{R}^2$, and let f be a continuous function which maps each vertex of the hexagon into itself, each one-dimensional side of the hexagon into itself, and the hexagon into itself.
(b) Show that the hypotheses in (a) are satisfied if f is the identity function, that is, if f maps each point of the hexagon into itself. If f is the identity function, show that the degree of f is defined at each point b which is not on the topological boundary of the hexagon, and that $d(f, b) = 1$ if b is on the interior of the hexagon and $d(f, b) = 0$ if b is on the exterior of the hexagon.
(c) Show that $d(f, b)$ is defined and has the values stated in (b) if f is any function which satisfies the hypotheses stated in (a).

24.5. In equation (9), the point b is included in U_2. Show that the same value is obtained for the degree of f at b if b is assigned to U_0 or to U_1.

25. Properties of the Degree of a Function

The notation $d(f, b)$ suggests that the degree of a function depends only on f and b, but the definition of $d(f, b)$ in Section 24 apparently depends also on the position of the rays r_0, r_1, r_2. The first theorem in this section shows that the degree is independent of the location of these rays [see Exercise 24.2], and the second theorem establishes the properties of the degree as a function of b.

25.1 Theorem. *Let $d(f, b)$ be the degree at b of f as defined in Section 24. Let $d'(f, b)$ be defined in a similar manner with respect to three rays which, in the counterclockwise order, are denoted by r'_0, r'_1, r'_2. Then $d'(f, b) = d(f, b)$.*

PROOF. First consider the case in which r_0 is replaced by a ray r'_0 obtained by rotating r_0 through a small angle, and r_1 and r_2 are unchanged. Then W_0, W_1, W_2 are replaced by regions W'_0, W'_1, W'_2, but $W'_1 = W_1$. Also, U_0, U_1, U_2 are replaced by U'_0, U'_1, U'_2, but $U'_1 = U_1$. Let s_k and s'_k be the Sperner functions defined with respect to $\mathscr{P}_k$ and the rays r_0, r_1, r_2 and r'_0, r'_1, r'_2, respectively. Then, by Theorem 23.4,

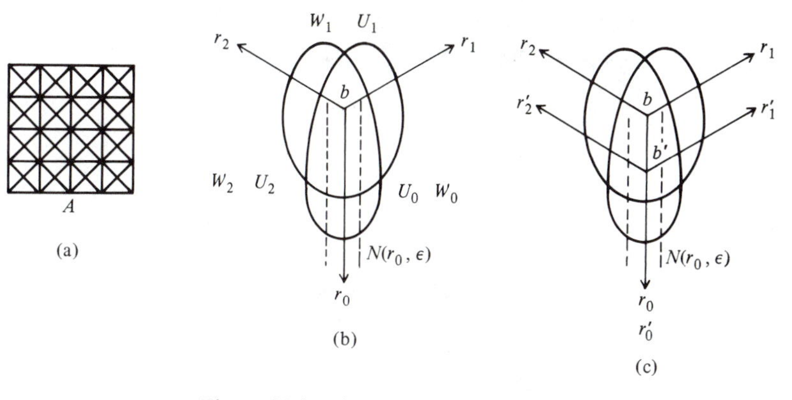

Figure 25.1. Figures for Example 25.2.

$$S(s_k) = \sum_{i \in I_k} \Delta[e_0, s_k(q_i^k), s_k(q_{i+1}^k)], \tag{1}$$

$$S(s_k') = \sum_{i \in I_k} \Delta[e_0, s_k'(q_i^k), s_k'(q_{i+1}^k)]. \tag{2}$$

Choose k so large that f maps each simplex $[p_i^k, p_{i+1}^k]$ [see equation (6) in Section 24] into at most two of the regions U_0, U_1, U_2 and at most two of the regions U_0', U_1', U_2'; this choice of k is possible by Lemma 24.2. Let $\Delta[e_0, s_k(q_i^k), s_k(q_{i+1}^k)]$ be a term in (1) which is not zero. Then one of the points q_i^k, q_{i+1}^k is in U_1 and the other is in U_2. The point which is in U_1 is also in U_1' since these sets are identical, and the point in U_2 is also in U_2' because of the choice of k. Thus,

$$s_k'(q_i^k) = s_k(q_i^k), \qquad s_k'(q_{i+1}^k) = s_k(q_{i+1}^k), \tag{3}$$

$$\Delta[e_0, s_k'(q_i^k), s_k'(q_{i+1}^k)] = \Delta[e_0, s_k(q_i^k), s_k(q_{i+1}^k)]. \tag{4}$$

Conversely, if $\Delta[e_0, s_k'(q_i^k), s_k'(q_{i+1}^k)]$ is a term in (2) which is not zero, then similar arguments show that (3) and (4) are true. Then the sums in (1) and (2) are equal for all sufficiently large k, and

$$d'(f, b) = \lim_{k \to \infty} S(s_k') = \lim_{k \to \infty} S(s_k) = d(f, b). \tag{5}$$

By a succession of steps of the kind described, the rays r_0, r_1, r_2 can be replaced by any other set r_0', r_1', r_2' of rays which have the counterclockwise order about b. Arguments similar to those just given show that the degree is unchanged at each step; hence $d'(f, b) = d(f, b)$. □

25.2 Example. Let $\{\mathscr{P}_k : k = 1, 2, \cdots\}$ be a sequence of simplicial subdivisions of the square A in Figure 25.1 (a), and let $f: A \to \mathbb{R}^2$ be a continuous function which maps the boundary of A into a curve as shown in Figure 25.1 (b) and (c). Let ε be a small positive number, and let $N(r_0, \varepsilon)$ be the set of points in $W_0 \cup W_2$ whose distance from r_0 is less than ε; this strip is shown in Figure 25.1 (b) and (c). The proof of Theorem 25.1 has shown that

$d(f, b)$ is completely determined by the part of $f(\partial A)$ which is in $N(r_0, \varepsilon)$ if ε is sufficiently small, for Theorem 23.4 shows that

$$S(s_k) = \sum_{i \in I_k} \Delta[e_1, s_k(q_i^k), s_k(q_{i+1}^k)]. \tag{6}$$

A term in the sum is different from zero if and only if one of the points q_i^k, q_{i+1}^k is in U_0 and the other is in U_2. Thus for all sufficiently large k, the terms $\Delta[e_1, s_k(q_i^k), s_k(q_{i+1}^k)]$ which are different from zero arise from points q_i^k, q_{i+1}^k, one of which is in $U_0 \cap N(r_0, \varepsilon)$ and the other of which is in $U_2 \cap N(r_0, \varepsilon)$. Thus the position of the rays r_1 and r_2 can be changed without changing the value of the sum in (6) and thus the value of $S(s_k)$ and $d(f, b)$. This example illustrates the argument used in proving Theorem 25.1. In Figure 25.1 (c), let b and b' be two points such that the segment $[b, b']$ contains no point in $f(\partial A)$, that is, in the curve into which f maps the boundary of A. Choose rays r_0, r_1, r_2 at b and r_0', r_1', r_2' at b' as shown in the figure. Then the figure shows that the intersections of $f(\partial A)$ with $N(r_0, \varepsilon)$ are the same for b and b', and hence that $d(f, b') = d(f, b)$. This example is an illustration of the next theorem.

Recall that a polygonal curve is a curve which consists of a finite number of straight line segments.

25.3 Theorem. *Let b and b' be two points in the complement of $f(\partial A)$ in $\mathbb{R}^2$ which can be connected by a polygonal curve which contains no point in $f(\partial A)$. Then $d(f, b') = d(f, b)$.*

PROOF. Consider first the special case in which the points b, b' can be connected by a polygonal curve consisting of the single segment $[b, b']$. Theorem 25.1 shows that any convenient set of rays r_0, r_1, r_2, ordered counterclockwise, can be used to determine the degrees of f at b and b'. Accordingly, choose r_0 as the ray from b through b', and choose r_0' as the ray from b' which is contained in r_0; choose additional rays r_1, r_2 and r_1', r_2' so that r_1' and r_2' are parallel to r_1 and r_2, respectively [see Figure 25.1 (c)]. The rays r_0, r_1, r_2 and r_0', r_1', r_2' determine regions U_0, U_1, U_2 and U_0', U_1', U_2' as in Section 24. Then U_0' and U_2' are proper subsets of U_0 and U_2, respectively, and U_1 is a proper subset of U_1'. As shown in the proof of Theorem 25.1 and in Example 25.2, for all sufficiently large k,

$$d(f, b) = S(s_k) = \sum_{i \in I_k} \Delta[e_1, s_k(q_i^k), s_k(q_{i+1}^k)], \tag{7}$$

$$d(f, b') = S(s_k') = \sum_{i \in I_k} \Delta[e_1, s_k'(q_i^k), s_k'(q_{i+1}^k)]. \tag{8}$$

For k sufficiently large, the non-zero terms in the sum in (7) are those for which one of the points q_i^k, q_{i+1}^k is in $U_0 \cap N(r_0, \varepsilon)$ and the other is in $U_2 \cap N(r_0, \varepsilon)$. Likewise, the non-zero terms in the sum in (8) are those for which one of the points q_i^k, q_{i+1}^k is in $U_0' \cap N(r_0, \varepsilon)$ and the other is in $U_2' \cap N(r_0, \varepsilon)$. But since $[b, b']$ contains no point in $f(\partial A)$, the intersections of $f(\partial A)$ with $U_0 \cap N(r_0, \varepsilon)$ and $U_2 \cap N(r_0, \varepsilon)$ are exactly the same as the intersections of $f(\partial A)$ with $U_0' \cap N(r_0, \varepsilon)$ and $U_2' \cap N(r_0, \varepsilon)$. Then equations (7) and (8) show that $S(s_k) = S(s_k')$ for all sufficiently large k; therefore, $d(f, b') = d(f, b)$. Thus, the proof of Theorem 25.3 is complete in the special case in which b and b' can be connected by a polygonal curve in $\mathbb{R}^2 - f(\partial A)$ consisting of the single segment $[b, b']$. In the general case, b and b' are connected by a polygonal curve with segments $[b, b_1]$, $[b_1, b_2]$, $\cdots$, $[b_n, b']$. By the special case already proved,

$$d(f, b) = d(f, b_1) = d(f, b_2) = \cdots = d(f, b_n) = d(f, b'). \tag{9}$$

The proof of the entire theorem is complete. □

EXERCISES

25.1. The degree $d(f, b)$ has been defined relative to positively oriented rays r_0, r_1, r_2 [for positively oriented rays, the counterclockwise order is r_0, r_1, r_2]. A second degree $d'(f, g)$ can be defined relative to negatively oriented rays [the counterclockwise order is r_0, r_2, r_1]. Establish the properties of $d'(f, b)$, and, in particular, prove that $d'(f, b) = -d(f, b)$.

25.2. Prove the following theorem. If the boundary of the chain of all 2-simplexes in K is empty, then $d(f, b) = 0$ for every continuous function $f: A \to \mathbb{R}^2$ and every b in $\mathbb{R}^2$. [Hint. Exercise 23.3 and the paragraph which contains equation (7) in Section 23.]

25.3. Let (p_0, p_1, p_2) and (q_0, q_1, q_2) be two non-degenerate simplexes in $\mathbb{R}^2$, and assume that each one is positively oriented in $\mathbb{R}^2$. Let A and B denote the closed, Euclidean simplexes $[p_0, p_1, p_2]$ and $[q_0, q_1, q_2]$, and let $f: A \to \mathbb{R}^2$ be a continuous function with the following properties: (i) $f(p_i) = q_i$, $i = 0, 1, 2$; and (ii) f maps the sides $[p_0, p_1]$, $[p_1, p_2]$, $[p_2, p_0]$ of A into the corresponding sides $[q_0, q_1]$, $[q_1, q_2]$, $[q_2, q_0]$ of B. Observe that f may map points in the interior of A into points in the exterior of B.

(a) Show that $\partial A = [p_0, p_1] \cup [p_1, p_2] \cup [p_2, p_0]$ and that $f(\partial A) = [q_0, q_1] \cup [q_1, q_2] \cup [q_2, q_0]$. Show also that $\mathbb{R}^2 - f(\partial A)$ consists of two open sets which are the interior and exterior of B. Finally, show that, in each of these open sets, two points can be connected by a polygonal curve.

(b) If b is a point in the exterior of B, show that the degree of f is defined at b and that $d(f, b) = 0$. [Hint. Consider first a point b at a great distance from B.]

(c) Let $\mathscr{P}_k$, $k = 1, 2, \cdots$, be subdivisions of A, and assume that $\mathscr{P}_1$ consists of the single simplex (p_0, p_1, p_2). Let b be a point in the interior of B. Choose rays r_0, r_1, r_2 from b through the mid-points of the sides $[q_2, q_0]$, $[q_0, q_1]$, $[q_1, q_2]$ respectively of B. Let s_k be the Sperner function on $\mathscr{Q}_k$, $k = 1, 2, \cdots$, with respect to r_0, r_1, r_2. Show that $S(s_k) = S(s_1) = 1$ for $k \geqq 1$. Compare this result with Theorem 24.3.

(d) Part (c) has shown that $d(f, b) = 1$ for one point b on the interior of B; use Theorem 25.3 to prove that $d(f, b) = 1$ at every point b on the interior of B. Finally, give a direct proof, without using Theorem 25.3, that $d(f, b) = 1$ at each b in the interior of B.

26. The Degree of a Curve

Equations (12) and (22) in Section 24 show that $d(f, b)$ is completely determined by the mapping of ∂A by f. This situation suggests that the concept of degree can be extended to curves. This section contains the basic results in this extension.

Let A be a closed interval $[a_0, a_1]$, and let $\mathscr{P}_k$, $k = 1, 2, \cdots$, be a set of oriented simplicial subdivisions $\{(p_i^k, p_{i+1}^k) : i = 1, 2, \cdots, N_k\}$ of A such that $\mathscr{P}_{k+1}$ is a refinement of $\mathscr{P}_k$, and such that the norm of $\mathscr{P}_k$ approaches zero as $k \to \infty$. Next, let $f: A \to \mathbb{R}^2$ be a continuous function on A such that $f(a_1) = f(a_0)$. Then f defines a continuous closed curve C in $\mathbb{R}^2$, and f maps $\mathscr{P}_k$, $k = 1, 2, \cdots$, into $\mathscr{Q}_k$ as follows:

$$\mathscr{Q} = \{(q_i^k, q_{i+1}^k) : q_i^k = f(p_i^k), q_{i+1}^k = f(p_{i+1}^k), i = 1, 2, \cdots, N_k\}.$$

The results in Sections 24 and 25 will be used to define the degree $d(C, b)$ of C with respect to each point b in $\mathbb{R}^2 - f(A)$.

It is convenient, although not necessary, to think of A as a segment in the plane. Let p^* be an arbitrary point in the plane but not in A, and define $f(p^*)$ to be a point q^* chosen arbitrarily in the plane of C. Then f maps the oriented simplexes (p_i^k, p_{i+1}^k, p^*) into the oriented simplexes (q_i^k, q_{i+1}^k, q^*). Let b be a point in $\mathbb{R}^2$ which is not on C. Choose rays r_0, r_1, r_2 at b and define the regions W_0, W_1, W_2 and U_0, U_1, U_2 as in Section 24. Define a Sperner function s_k on q^* and the vertices of $\mathscr{Q}_k$ as in equation (10) in Section 24. The Sperner number $S(s_k)$ of s_k is defined as in (11) of that section; thus

$$S(s_k) = \sum_{i=1}^{N_k} \Delta[s_k(q_i^k), s_k(q_{i+1}^k), s_k(q^*)]. \tag{1}$$

Then the methods used in Section 23 can be employed again to show that

$$S(s_k) = \sum_{i=1}^{N_k} \Delta[e, s_k(q_i^k), s_k(q_{i+1}^k)]. \tag{2}$$

Here e is an arbitrary, but fixed, point in the set $\{e_0, e_1, e_2\}$. Now f is continuous, and Lemma 24.2 shows that there exists an integer K such that, if $k \geqq K$, then f maps each Euclidean simplex (p_i^k, p_{i+1}^k) in $\mathscr{P}_k$ into at most two of the regions W_0, W_1, W_2. Then, for $k \geqq K$, each $\mathscr{Q}_{k+1}$ is a refinement of $\mathscr{Q}_k$ and each s_{k+1} is a refinement of s_k [see Definition 23.6]. Finally, Sperner's lemma [Theorem 23.8] shows that $S(s_k)$ is constant for $k \geqq K$. Define this constant value to be the degree of C at b and denote it by $d(C, b)$. Then

$$d(C, b) = S(s_k) = \sum_{i=1}^{N_k} \Delta[e, s_k(q_i^k), s_k(q_{i+1}^k)], \qquad k \geqq K, \tag{3}$$

$$d(C, b) = \lim_{k \to \infty} S(s_k). \tag{4}$$

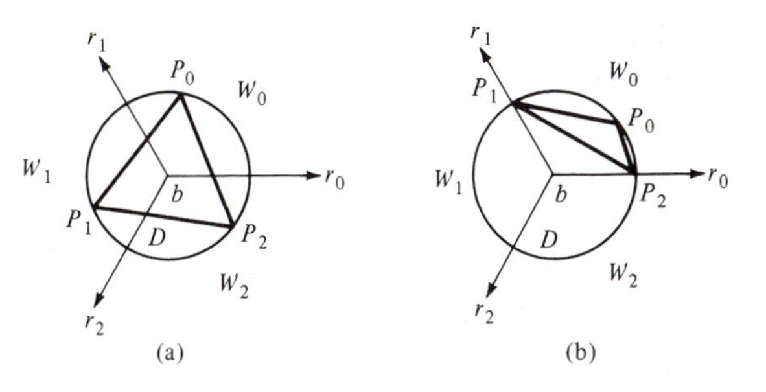

Figure 26.1. Figures for Lemma 26.1.

These equations and equations (12) and (22) in Section 24 show that $d(f, b)$ and $d(C, b)$ are basically the same.

The next theorem proves that, if two curves C_1 and C_2 are sufficiently near each other (in a sense to be made precise), then $d(C_1, b) = d(C_2, b)$. The proof of this theorem requires a lemma, and the following explanation is needed for the statement of the lemma.

Let b be a point in $\mathbb{R}^2$, and let r_0, r_1, r_2 be three rays from b which make angles of $2\pi/3$ radians with one another and have the counterclockwise order. Define the closed regions W_0, W_1, W_2 as in Section 24. Let D denote the circle with center b and radius r.

26.1 Lemma. *The minimum diameter of all sets $\{P_0, P_1, P_2\}$ such that P_i is in W_i but not in the interior of D, $i = 0, 1, 2$, is $\sqrt{3}r$.*

PROOF. Let P_0, P_1, P_2 be three points such that $P_i \in W_i$, $i = 0, 1, 2$, but such that no one of the points is in the interior of D. If the points P_i are not on the circle D, then project them from b into points P_0', P_1', P_2' on the circle D. Then the cosine law in trigonometry can be used to show that the diameter of $\{P_0', P_1', P_2'\}$ is equal to or less than that of $\{P_0, P_1, P_2\}$. Thus in seeking the minimum diameter of $\{P_0, P_1, P_2\}$, there is no loss of generality in assuming that P_0, P_1, P_2 are points on D. It is easy to identify two types of sets $\{P_0, P_1, P_2\}$ with P_i on D whose diameter is $\sqrt{3}r$. In the first type, P_0, P_1, P_2 are the vertices of an equilateral triangle inscribed in D [see Figure 26.1(a)]. An example of the second type is shown in Figure 26.1(b); P_2 and P_1 are the points at which r_0 and r_1 intersect D, and P_0 is a point on the shorter arc bounded by P_2 and P_1. If two points, say P_i and P_j, are the end-points of an arc of D which subtends an angle θ at b such that $(2\pi/3) < \theta < \pi$, then the distance between P_i and P_j is greater than $\sqrt{3}r$ and the diameter of $\{P_0, P_1, P_2\}$ is greater than $\sqrt{3}r$. Therefore, the minimum diameter of all sets $\{P_0, P_1, P_2\}$ such that P_i is in W_i but not in the interior of D, $i = 0, 1, 2$, is $\sqrt{3}r$. □

The following explanation is needed for the statement of the theorem. Let $A = [a_0, a_1]$, and let $\{\mathscr{P}_k : k = 1, 2, \cdots\}$ be the simplicial subdivisions of A described at the beginning of this section. Let $f: A \to \mathbb{R}^2$ and $g: A \to \mathbb{R}^2$ be two continuous functions which define closed curves C_f and C_g in $\mathbb{R}^2$. As before, set

$$\mathscr{P}_k = \{(p_i^k, p_{i+1}^k): i = 1, 2, \cdots, N_k\}, \qquad k = 1, 2, \cdots.$$

Then f and g map $\mathscr{P}_k$ into

$$\mathscr{Q}_k = \{(q_i^k, q_{i+1}^k): i = 1, 2, \cdots, N_k\}, \qquad k = 1, 2, \cdots,$$

$$\mathscr{R}_k = \{(r_i^k, r_{i+1}^k): i = 1, 2, \cdots, N_k\}, \qquad k = 1, 2, \cdots.$$

26.2 Theorem. *Assume that the curves C_f and C_g just described satisfy the following hypotheses:*

The closed disk D with center b and radius r contains no points in the sets $f(A)$ and $g(A)$; that is, the curves C_f and C_g lie outside D. (5)

$$|f(p) - g(p)| < \sqrt{3}r/2 \text{ for every } p \text{ in } A. \tag{6}$$

Then $d(C_f, b) = d(C_g, b)$.

Proof. For the calculation of $d(C_f, b)$ and $d(C_g, b)$, use rays r_0, r_1, r_2 which make angles of $2\pi/3$ radians with one another and are arranged in the counterclockwise order. There is a Sperner function s_k on the vertices of $\mathscr{Q}_k$ such that

$$d(C_f, b) = S(s_k) = \sum_{i=1}^{N_k} \Delta[e, s_k(q_i^k), s_k(q_{i+1}^k)], \qquad k \geqq K_f, \tag{7}$$

and also a Sperner function t_k defined on the vertices of $\mathscr{R}_k$ such that

$$d(C_g, b) = S(t_k) = \sum_{i=1}^{N_k} \Delta[e, t_k(r_i^k), t_k(r_{i+1}^k)], \qquad k \geqq K_g. \tag{8}$$

Choose k so large that $k \geqq K_f$, $k \geqq K_g$, and also so that

$$|f(p_{i+1}^k) - f(p_i^k)| < \sqrt{3}r/2, \quad |g(p_{i+1}^k) - g(p_i^k)| < \sqrt{3}r/2, \quad i = 1, 2, \cdots, N_k. \tag{9}$$

The last choice is possible because f and g are uniformly continuous on the compact set A and the norm of $\mathscr{P}_k$ tends to zero as $k \to \infty$ [see Definition 96.16 and Theorem 96.18 in Appendix 2].

Corresponding to the simplexes (q_i^k, q_{i+1}^k) and (r_i^k, r_{i+1}^k), in $\mathscr{Q}_k$ and $\mathscr{R}_k$ respectively, construct the following abstract simplexes [see Figure 26.2].

$$(q_i^k, q_{i+1}^k, r_i^k), \qquad (r_{i+1}^k, r_i^k, q_{i+1}^k), \qquad i = 1, 2, \cdots, N_k. \tag{10}$$

Now because of the choice of k [see (9)] and the hypothesis in (6), the diameter of each of these abstract simplexes is less than $\sqrt{3}r$. Then by Lemma 26.1, each simplex in (10) has vertices in at most two of the regions

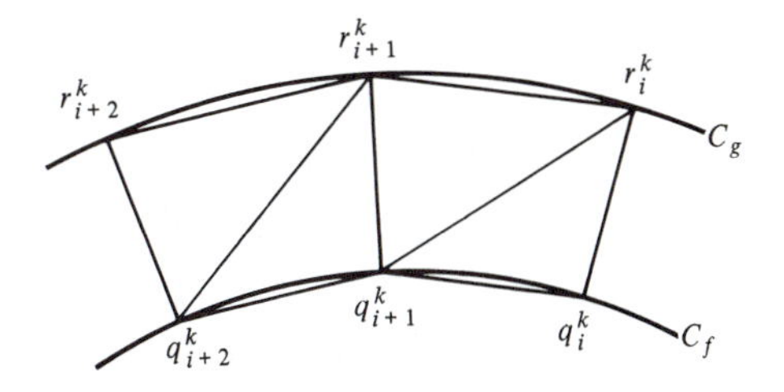

Figure 26.2. The abstract simplexes in (10).

W_0, W_1, W_2, and hence in at most two of the regions U_0, U_1, U_2. Let u_k be the Sperner function defined on the union of the vertices in $\mathscr{Q}_k$ and $\mathscr{R}_k$ as follows:

$$u_k(v) = e_i \quad \text{if} \quad v \in U_i, \qquad v \in V(\mathscr{Q}_k) \cup V(\mathscr{R}_k). \tag{11}$$

Then

$$u_k(v) = \begin{cases} s_k(v), & v \in V(\mathscr{Q}_k), \\ t_k(v), & v \in V(\mathscr{R}_k). \end{cases} \tag{12}$$

Since each simplex in (10) has vertices in at most two of the regions by Lemma 26.1, the definition of u_k in (11) shows that each of the sets

$$\{u_k(q_i^k), u_k(q_{i+1}^k), u_k(r_i^k)\}, \qquad \{u_k(r_{i+1}^k), u_k(r_i^k), u_k(q_{i+1}^k)\}, \tag{13}$$

contains at most two distinct elements in $\{e_0, e_1, e_2\}$. Then since a determinant with two rows identical is zero,

$$\sum_{i=1}^{N_k} \Delta[u_k(q_i^k), u_k(q_{i+1}^k), u_k(r_i^k)] + \sum_{i=1}^{N_k} \Delta[u_k(r_{i+1}^k), u_k(r_i^k), u_k(q_{i+1}^k)] = 0, \tag{14}$$

because each term in each of the sums is zero. Let e denote an arbitrary, but fixed, element in $\{e_0, e_1, e_2\}$. Then by Theorem 20.1 as in the proof of Theorem 23.4,

$$\Delta[u_k(q_i^k), u_k(q_{i+1}^k), u_k(r_i^k)] = \Delta[e, u_k(q_{i+1}^k), u_k(r_i^k)] + \cdots,$$
$$\Delta[u_k(r_{i+1}^k), u_k(r_i^k), u_k(q_{i+1})] = \Delta[e, u_k(r_i^k), u_k(q_{i+1}^k)] + \cdots.$$

The two terms shown on the right in these equations cancel in the sum (14) because each is the negative of the other. The side with vertices r_i^k and q_{i+1}^k belongs to each of the simplexes in (10), and this side has opposite orientations in the two simplexes. In the same way,

$$\Delta[u_k(r_{i+1}^k), u_k(r_i^k), u_k(q_{i+1}^k)] = \cdots + \Delta[u_k(r_{i+1}^k), e, u_k(q_{i+1}^k)] + \cdots,$$
$$\Delta[u_k(q_{i+1}^k), u_k(q_{i+2}^k), u_k(r_{i+1}^k)] = \cdots + \Delta[u_k(q_{i+1}^k), e, u_k(r_{i+1}^k)] + \cdots,$$

and again the terms shown on the right cancel in the sum in (14). Transform each term in (14) by Theorem 20.1 as in these examples, replacing x in Theorem 20.1 by e. If a one-dimensional side belongs to two simplexes in

(10), there are two corresponding terms in the transformed sum in (14) and they cancel. Recall that the curves C_f and C_g are closed. Then all terms cancel except those formed from sides (q_i^k, q_{i+1}^k) and (r_i^k, r_{i+1}^k), and (14) simplifies to

$$\sum_{i=1}^{N_k} \Delta[e, u_k(q_i^k), u_k(q_{i+1}^k)] + \sum_{i=1}^{N_k} \Delta[e, u_k(r_{i+1}^k), u_k(r_i^k)] = 0. \tag{15}$$

Then (12) and (15) and elementary properties of determinants show that

$$\sum_{i=1}^{N_k} \Delta[e, s_k(q_i^k), s_k(q_{i+1}^k)] - \sum_{i=1}^{N_k} \Delta[e, t_k(r_i^k), t_k(r_{i+1}^k)] = 0. \tag{16}$$

Then by equations (7) and (8), this equation asserts that $d(C_f, b) = d(C_g, b)$. □

26.3 Example. Let A be the circumference of a circle with center b and radius $2r$, and let $f: A \to \mathbb{R}^2$ be the identity function on A. Then $f(p) = p$ for p in A; C_f is the circle; and $d(C_f, b) = 1$. Let $g: A \to \mathbb{R}^2$ be a continuous function such that $|f(p) - g(p)| < \sqrt{3}r/2$. Then the curves C_f and C_g lie outside the circle D with center b and radius r, and Theorem 26.2 shows that $d(C_g, b) = 1$. The methods used in proving Theorem 25.3 [see also Exercise 26.7] can be employed to show that the degree of C_g is 1 at each point in D.

Exercises

26.1. Let A be the closed interval $[a_0, a_1]$, and let $f: A \to \mathbb{R}^2$ be a continuous function such that $f(a_0) = f(a_1)$. Then f defines a closed curve C. Prove that $d(C, b)$ is defined at each point b such that $b \notin f(A)$, and that $d(C, b)$ is a positive, zero, or negative integer.

26.2. Let C be an ellipse E. Prove that $d(C, b) = 1$ for each point b on the interior of the ellipse, and that $d(C, b) = 0$ for each point b on the exterior of the ellipse. Here it is assumed that C is positively oriented in $\mathbb{R}^2$, and that r_0, r_1, r_2 are ordered counterclockwise.

26.3. Repeat Exercise 26.2 for the following curves: (a) C is the polygonal curve which forms the sides of a triangle; and (b) C is the polygonal curve which forms the sides of a rectangle.

26.4. Let $f_j: A \to \mathbb{R}^2$, $j = 0, 1, \cdots, m$, be continuous functions which define closed curves $C_0, C_1, \cdots, C_m$. If C_{j-1} is near C_j in the sense of Theorem 26.2 for $j = 1, \cdots, m$, then prove that $d(C_0, b) = d(C_m, b)$.

26.5. Let $f: A \to \mathbb{R}^2$ be a constant function such that $f(p) = q_0$ for all p in A, and let C be the curve (which consists of the single point q_0) defined by f. If $b \neq q_0$, prove that $d(C, b)$ is defined and that $d(C, b) = 0$.

26.6. Let C be a curve which lies on an open polygonal curve or on an arc (an open curve) which have no self intersections. Prove that $d(C, b)$ is defined at each point b not on C and that $d(C, b) = 0$.

26.7. Let C be a closed curve defined by a continuous function $f: A \to \mathbb{R}^2$. The set $f(A)$ is closed and its complement $\mathbb{R}^2 - f(A)$ is an open set [see Section 92 in Appendix 2]. Prove the following: if b and b' are points in $\mathbb{R}^2 - f(A)$ which can be connected by a polygonal curve which contains no point in $f(A)$, then $d(C, b) = d(C, b')$.

26.8. Consider again the curve C in Exercise 26.7, and prove the following. If b is a point in $\mathbb{R}^2 - f(A)$ which can be connected to the point at infinity by a polygonal curve in $\mathbb{R}^2 - f(A)$, then $d(C, b) = 0$.

26.9. The degree is a property of an oriented curve [all of the simplexes (p_i^k, p_{i+1}^k) in $\mathscr{P}_k$, $k = 1, 2, \cdots$, have the same orientation in A]. If this orientation is reversed, the resulting curve is denoted by $-C$. Prove the following: if b is not on C, then the degrees of both C and $-C$ are defined at b, and $d(-C, b) = -d(C, b)$.

26.10. Let C be the curve in $\mathbb{R}^2$ defined by the following equations:

$$y_1 = r\cos nx, \qquad y_2 = r\sin nx, \qquad 0 \leqq x \leqq 2\pi.$$

(a) If $n = 0$, then C consists of the single point $(r, 0)$. If $b \neq (r, 0)$, then prove that $d(C, b) = 0$.

(b) If n is an integer and $n \neq 0$, then C lies on the circle $y_1^2 + y_2^2 = r^2$. Let $\mathscr{P}_k$ be a subdivision of $[0, 2\pi]$ such that $0 = p_1^k < p_2^k < \cdots < p_{N_k+1}^k = 2\pi$, and let the rays r_0, r_1, r_2 be ordered in the counterclockwise direction. Prove that $d(C, b) = n$ if b is inside the circle and $d(C, b) = 0$ if b is outside the circle.

27. The Intermediate-Value Theorem

The intermediate-value theorem states that a real-valued function which is continuous on an interval in $\mathbb{R}$ assumes every value between two of its values. More precisely, if $f: [a_0, a_1] \to \mathbb{R}$ is continuous, if $f(a_0) \neq f(a_1)$, and if c is any number between $f(a_0)$ and $f(a_1)$, then there is an x such that $a_0 < x < a_1$ and $f(x) = c$. The purpose of this section is to prove, for functions of the form $f: A \to \mathbb{R}^2$, a generalization of this elementary theorem. As in the one-dimensional case, a point b must satisfy certain conditions relative to $f(\partial A)$ before it is possible to assert that there exists a point a in A such that $f(a) = b$. The conditions that b must satisfy are described in terms of the degree $d(f, b)$ of f at b.

The theorem in this section is an interesting result in itself. The proof is an unexpected application of Sperner's lemma, and the result is the best possible in the sense that the theorem assumes only the minimum hypotheses: the only assumption on f is that it is continuous. Finally, the theorem in this section has significant applications; in Chapter 5 it will be used to prove a good form of the inverse-function theorem.

Let K be a Euclidean, homogeneous, 2-dimensional complex in $\mathbb{R}^n$, $n \geqq 2$. The simplexes in K are Euclidean simplexes $[a_0, a_1, a_2]$; the union of these

Euclidean simplexes is a closed set, denoted by A, in $\mathbb{R}^n$. The simplexes in K are oriented so that the boundary of the chain which contains all 2-simplexes in K is the chain which consists of all one-dimensional sides $[a_0, a_1]$ which belong to a single simplex in K [see the second paragraph of Section 23]. Finally, $\mathscr{P}_k$, $k = 1, 2, \cdots$, is a set of simplicial subdivisions of K, each of which is a Euclidean complex which has the same properties as K [see the further description of these subdivisions in equations (1), $\cdots$, (8) in Section 24].

27.1 Theorem (Intermediate-Value Theorem). *If the function $f: A \to \mathbb{R}^2$ is continuous, if b is in $\mathbb{R}^2$ but not in $f(\partial A)$, and if $d(f, b) \neq 0$, then there is a point a in A such that $f(a) = b$.*

Proof. Construct the Sperner functions s_k corresponding to the subdivisions $\mathscr{P}_k$ and the point b as in Section 24. If $k \geqq K$ [see Lemma 24.2], then $S(s_k) = d(f, b)$ by Theorem 24.3, and, since $d(f, b) \neq 0$ by hypothesis, there is at least one simplex (q_0^k, q_1^k, q_2^k) in $\mathscr{Q}_k$ on which s_k has the three values e_0, e_1, e_2 [see Sperner's lemma in Theorem 23.8]. This statement means that there is a simplex (p_0^k, p_1^k, p_2^k) in $\mathscr{P}_k$ such that one element in $\{f(p_0^k), f(p_1^k), f(p_2^k)\}$ is in U_0, one is in U_1, and one is in U_2. Assume that the notation is such that $f(p_i^k) \in U_i$; then $f(p_i^k) \in W_i$, $i = 0, 1, 2$. The set $\{p_0^k : k \geqq K\}$ is contained in the compact set A; hence, it has at least one point of accumulation a in A [see the Bolzano–Weierstrass theorem in Theorem 94.1 in Appendix 2], and there is a sequence $\{p_0^{k_j} : j = 1, 2, \cdots\}$ which converges to a. Since f is continuous at a, then f converges on this sequence to $f(a)$. Again, since the norm of $\mathscr{P}_k$ tends to zero as $k \to \infty$,

$$\lim_{j\to\infty} p_i^{k_j} = a, \qquad \lim_{j\to\infty} f(p_i^{k_j}) = f(a), \qquad i = 0, 1, 2.$$

Since $f(p_i^{k_j}) \in W_i$ and W_i is closed, then $f(a) \in W_i$ for $i = 0, 1, 2$. Since the only point in $W_0 \cap W_1 \cap W_2$ is b, then $f(a) = b$. □

27.2 Example. Let A and B be the Euclidean simplexes $[a_0, a_1, a_2]$ and $[b_0, b_1, b_2]$, respectively, and assume that A and B are positively oriented in $\mathbb{R}^2$. Let $f: A \to \mathbb{R}^2$ be the affine transformation $L: \mathbb{R}^2 \to \mathbb{R}^2$ such that $L(a_i) = b_i$, $i = 0, 1, 2$. If b is in the interior of $[b_0, b_1, b_2]$, then $d(f, b) = 1$ and Theorem 27.1 shows that there is a point a in $[a_0, a_1, a_2]$ such that $f(a) = b$. There are, of course, other proofs of this fact.

27.3 Example. Let A and B be the 2-simplexes in $\mathbb{R}^2$ in Example 27.2, and let $f: A \to \mathbb{R}^2$ be a continuous function with the following properties [see Exercise 25.3]:

(i) $f(a_i) = b_i$, $i = 0, 1, 2$;
(ii) f maps the sides $[a_0, a_1]$, $[a_1, a_2]$, $[a_2, a_0]$ of A into the sides $[b_0, b_1]$, $[b_1, b_2]$, $[b_2, b_0]$, respectively, of B.

It is not assumed, however, that f maps A into B; points in A may be mapped into points in the exterior of B. If b is in the interior of $[b_0, b_1, b_2]$, then it is easy to use the formulas in equations (12) and (22) in Section 24 [see also Exercise 25.3] to show that $d(f, b) = 1$, and Theorem 27.1 shows that there is a point a in the interior of $[a_0, a_1, a_2]$ such that $f(a) = b$. If b is in the exterior of $[b_0, b_1, b_2]$, then $d(f, b) = 0$, and there may or may not be a point a such that $f(a) = b$.

27.4 Example. Let A_3 be the unit cube in $\mathbb{R}^3$. Construct a simplicial subdivision of A_3; a sequence of such subdivisions is shown in Figure 21.5. The boundary of the chain of simplexes in the subdivision of A_3 is a chain C_2 of 2-dimensional simplexes in the surface of A_3. Let A be the union of the Euclidean simplexes in C_2, and let $f: A \to \mathbb{R}^2$ be a continuous function. Then $d(f, b) = 0$ for every point b in $\mathbb{R}^2$ [see Exercise 25.2]; and for every b in $\mathbb{R}^2$, it is impossible to use Theorem 27.1 to show that there is a point a in A such that $f(a) = b$.

27.5 Example. Delete one or more 2-simplexes from the chain C_2 in Example 27.4 to form a chain C_2' whose boundary contains one or more cycles. Let A be the union of the 2-simplexes in C_2', and let $f: A \to \mathbb{R}^2$ be a continuous function. Then since the boundary of C_2' is not empty, there may be points b for which $d(f, b) \neq 0$; these points are the images under f of points in A. For example, suppose that C_2' is formed by deleting from C_2 all 2-simplexes in the top face of A_3, and suppose that $f: A \to \mathbb{R}^2$ is the projection of A into the (x^1, x^2)-plane. Then f maps A onto the square in $\mathbb{R}^2$ whose vertices are $(0, 0)$, $(1, 0)$, $(1, 1)$, $(0, 1)$ [see the description in Section 21 of the subdivisions of A_3 in Figure 21.5]. If b is in the interior of the square, then $d(f, b) \neq 0$ and Theorem 27.1 shows that there is a point a in A such that $f(a) = b$.

27.6 Example. Let $A: [a_0, a_1, a_2]$ and $B: [b_0, b_1, b_2]$ be two Euclidean simplexes which are positively oriented in $\mathbb{R}^2$, and let $f: A \to B$ be the affine transformation such that $f(a_i) = b_i$, $i = 0, 1, 2$. Then the intermediate-value theorem shows [see Example 27.2] that every point on the interior of B is in $f(A)$. The purpose of the present example is to show that, if $g: A \to \mathbb{R}^2$ is a function whose values on ∂A are near those of $f: A \to \mathbb{R}^2$ on ∂A [in a sense to be made precise], then there is at least a disk in the interior of B which is contained in $g(A)$.

Assume that the inscribed circle in B has center b and radius $2r$, and let D be the disk with center b and radius r. Let $g: A \to \mathbb{R}^2$ be a continuous function such that

$$|g(x) - f(x)| < \sqrt{3}r/2, \qquad x \in \partial A. \tag{1}$$

If C_f and C_g are the curves $f: \partial A \to \mathbb{R}^2$ and $g: \partial A \to \mathbb{R}^2$, then C_f is the polygonal curve which bounds B and C_g is a curve which lies outside D because of

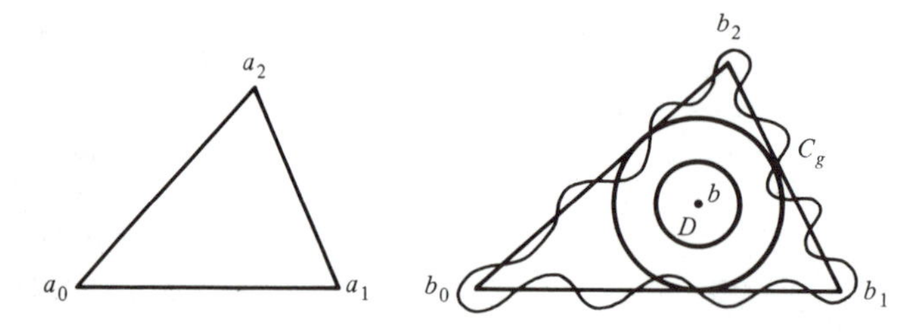

Figure 27.1. Figure for Example 27.6.

(1) [see Figure 27.1]. Now $d(f, b) = d(C_f, b) = 1$, and $d(C_g, b) = d(C_f, b)$ by Theorem 26.2. Thus $d(g, b) = d(C_g, b) = d(C_f, b) = 1$. Now because of (1), every point x in D can be connected to b by a segment $[b, x]$ which contains no point in $g(\partial A)$. Then since $d(g, b) = 1$, Theorem 25.3 shows that $d(g, x) = 1$ for all x in D. Finally, Theorem 27.1 shows that the disk D is contained in $g(A)$. Of course there may be other points x in $g(A)$, some at which $d(g, x) \neq 0$ and some at which $d(g, x) = 0$.

27.7 Example. Let A be a closed disk in $\mathbb{R}^2$, and let $f: A \to \mathbb{R}^2$ be a continuous function. There ought to be an intermediate-value theorem for this function, but some explanations are required in order to show that the theory which has been developed can be applied in this case also. Since A is a disk, it does not have a sequence of simplicial subdivisions $\mathscr{P}_k$, $k = 1, 2, \cdots$, of the kind used in proving Sperner's lemma and the intermediate-value theorem. This example forces us to look more carefully at the details in the proof of the lemma and theorem.

Begin by constructing a complex $\mathscr{P}$; it will be almost a subdivision of A. Let b be a point in the plane of $f(A)$ but not a point in $f(\partial A)$. Construct the rays r_0, r_1, r_2 and the regions W_0, W_1, W_2 and U_0, U_1, U_2 as in Section 24. Subdivide the disk A into k equal sectors [see Figure 27.2 (a)]. Let the arcs of these sectors be $(p_0, p_1), \cdots, (p_i, p_{i+1}), \cdots, (p_{k-1}, p_k), p_k = p_0$. Choose k so large that f maps each arc (p_i, p_{i+1}) into at most two of the regions W_0, W_1, W_2; this choice of k is possible by the same arguments used to prove Lemma 24.2. Subdivide each radius of A in the boundary of a sector

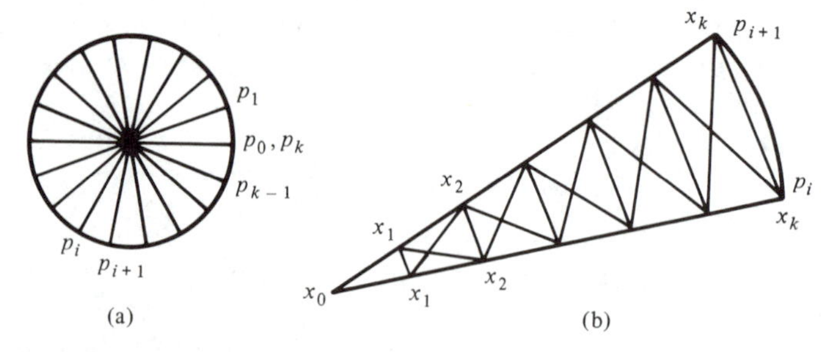

Figure 27.2. Construction of a complex $\mathscr{P}$ for the disk.

into k equal subintervals by points $x_0, x_1, \cdots, x_k$. On the intervals $[x_0, x_1]$, $[x_1, x_2], \cdots, [x_{k-1}, x_k]$ construct simplexes as shown in Figure 27.2(b). Give each simplex the positive orientation in $\mathbb{R}^2$. The collection of all the simplexes constructed in the k sectors is the oriented, 2-dimensional, simplicial complex $\mathscr{P}$. The boundary of the chain of all 2-simplexes in $\mathscr{P}$ is the chain $\Sigma_0^{k-1}[p_i, p_{i+1}]$ of one-dimensional simplexes. Observe that $\mathscr{P}$ is a Euclidean complex, but that it is not a subdivision of A since there are points along the boundary of the disk which are not in any simplex in $\mathscr{P}$. As in Section 23, the function f maps (the vertices of) $\mathscr{P}$ into an abstract, oriented, 2-dimensional, simplicial complex $\mathscr{Q}$ with simplexes (q_0^r, q_1^r, q_2^r), $r = 1, \cdots, N$. Define the Sperner function $s: V(\mathscr{Q}) \to \{e_0, e_1, e_2\}$ as in (10) in Section 24. Then

$$S(s) = \sum_{r=1}^{N} \Delta[s(q_0^r), s(q_1^r), s(q_2^r)]. \tag{2}$$

Then by using Theorem 20.1 and the cancellation properties in the oriented simplicial complex $\mathscr{Q}$, it can be shown that

$$S(s) = \sum_{i=0}^{k-1} \Delta[e, s(q_i), s(q_{i+1})], \qquad q_i = f(p_i), \qquad i = 0, 1, \cdots, k. \tag{3}$$

Return now to the beginning to construct a complex $\mathscr{P}'$; it will be almost a refinement of $\mathscr{P}$. Subdivide each of the k sectors of A used in constructing $\mathscr{P}$ into $m+1$ equal sectors, $m \geqq 1$. Then subdivide each radius which bounds a sector into $k(m+1)$ equal intervals and in each interval construct simplexes as before [see Figure 27.2(b)]; give each simplex the positive orientation in $\mathbb{R}^2$. The collection of all these oriented simplexes is the Euclidean complex $\mathscr{P}'$. It is not true that $\mathscr{P}'$ is a refinement of $\mathscr{P}$, but it has all of the properties needed to prove Sperner's lemma.

Now f maps $\mathscr{P}'$ into the abstract complex $\mathscr{Q}'$ with vertices $V(\mathscr{Q}')$. Define the Sperner function $s': V(\mathscr{Q}') \to \{e_0, e_1, e_2\}$ as in (10) in Section 24. We shall now prove that $S(s') = S(s)$ for all m such that $m \geqq 1$. Figure 27.2(b) shows that $[p_i, p_{i+1}]$ is a simplex in $\mathscr{P}$; the complex $\mathscr{P}'$ has the following simplexes associated with the arc (p_i, p_{i+1}).

$$\begin{gathered}[a_0, a_1], [a_1, a_2], \cdots, [a_m, a_{m+1}], \\ a_0 = p_i, \qquad a_{m+1} = p_{i+1}.\end{gathered} \tag{4}$$

Observe that each vertex of these simplexes is a point on the arc (p_i, p_{i+1}). Then f maps these simplexes into the following simplexes in $\mathscr{Q}'$.

$$\begin{gathered}[b_0, b_1], [b_1, b_2], \cdots, [b_m, b_{m+1}], \qquad b_j = f(a_j), \qquad j = 0, 1, \cdots, m+1, \\ b_0 = q_i, \qquad b_{m+1} = q_{i+1}.\end{gathered} \tag{5}$$

Because of the choice of k, each of the arcs (p_i, p_{i+1}), $i = 0, 1, \cdots, k-1$, is mapped by f into at most two of the regions W_0, W_1, W_2. Then the set $\{s'(b_j): j = 0, 1, \cdots, m+1\}$ contains at most two distinct elements in the

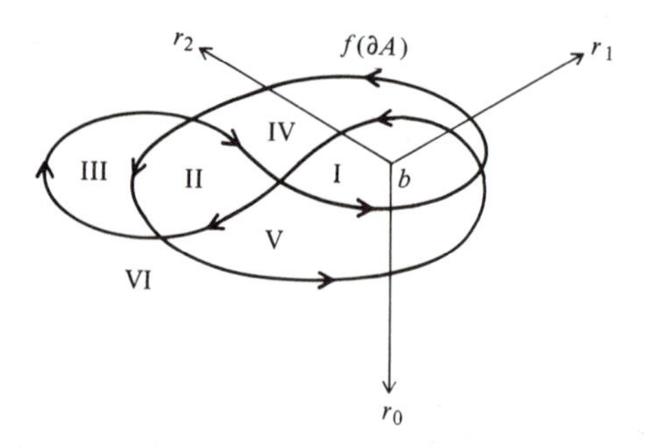

Figure 27.3. Calculation of $d(f, b)$ in Example 27.7.

set $\{e_0, e_1, e_2\}$. Then

$$\sum_{j=1}^{m} \Delta[e, s'(b_j), s'(b_{j+1})] = \Delta[e, s(q_i), s(q_{i+1})]. \tag{6}$$

The proof of this equation is exactly the same as in the proof of Sperner's lemma [see Theorem 23.8]. Sum equation (6) over all simplexes $[p_i, p_{i+1}]$ in $\mathscr{P}$; the sum on the left is $S(s')$ and the sum on the right is $S(s)$ [see (3)]. Therefore $S(s') = S(s)$ for all m such that $m \geqq 1$, and $d(f, b)$, the degree of f at b, is defined as before. The degree is defined at each point b which is not in $f(\partial A)$; if $d(f, b) \neq 0$, then the intermediate-value theorem shows that there is a point a in A such that $f(a) = b$.

The boundary ∂A is the circle which is the boundary of the disk A. Assume that $f\colon A \to \mathbb{R}^2$ maps the circle ∂A into the curve shown in Figure 27.3. Then $f(\partial A)$ divides the plane into six regions as shown, and the degree of f can be found in each of these regions. Let b be a point in region I. Choose rays r_0, r_1, r_2 at b as shown in Figure 27.3. Let e in the formula in (8) in Theorem 23.4 be e_1; then

$$S(s) = \sum_{i \in I} \Delta[e_1, s(q_i), s(q_{i+1})]. \tag{7}$$

For all sufficiently large values of m there are exactly two terms in the sum which are different from zero; they arise from the two crossings of r_0 by $f(\partial A)$. At these crossings, q_i is in U_2 and q_{i+1} is in U_0; then $s(q_i) = e_2$ and $s(q_{i+1}) = e_0$. Therefore, for two terms, $\Delta[e_1, s(q_i), s(q_{i+1})] = \Delta[e_1, e_2, e_0] = \Delta[e_0, e_1, e_2] = 1$, and (7) shows that $d(f, b) = 2$. The value of the degree of f in the other five regions can be found by similar methods.

Exercises

27.1. Let A be a regular hexagon in $\mathbb{R}^2$, and let $f(A) \to \mathbb{R}^2$ be a continuous function which maps each vertex of the hexagon into itself and each one-dimensional side into itself. However, these one-dimensional sides are not assumed to be point-wise fixed, and f may map points in the hexagon into points outside the hexagon. If b is any point inside the hexagon, prove that there is a point a in the hexagon such that $f(a) = b$. What can you say if b is outside the hexagon?

27.2. Let A be the regular hexagon in Exercise 27.1. Prove that there exists no continuous function $f: A \to \mathbb{R}^2$ with the following properties:

(a) if $p \in \partial A$, then $f(p) = p$;
(b) if $p \in A$, then $f(p) \in \partial A$.

27.3. Prove the following theorem. If f is a continuous function which maps the regular hexagon A into the interior of A, then there is at least one point p_0 in A such that $f(p_0) = p_0$. [Outline of solution. Assume the theorem false. Define a function $g: A \to \mathbb{R}^2$ as follows: if p is a point in A, then $g(p)$ is the point in which the ray from $f(p)$ through p intersects ∂A. Then g is continuous since f is continuous and $f(p) \neq p$. Show that a contradiction with Exercise 27.2 results.]

27.4. If A is a disk (solid circle), prove that there exists no continuous function $f: A \to \mathbb{R}^2$ which has properties (a) and (b) in Exercise 27.2.

27.5. Prove the following theorem. If f is a continuous function which maps the disk A (solid circle) either into a subset of A or onto all of A, then there is at least one point p_0 in A such that $f(p_0) = p_0$.

27.6. For the mapping $f: A \to \mathbb{R}^2$ described in Example 27.7 and pictured in Figure 27.3, prove that $d(f, b)$ has the following values in regions I, $\cdots$, VI: in I, $d(f, b) = 2$; in II, $d(f, b) = 0$; in III, $d(f, b) = -1$; in IV and V, $d(f, b) = 1$; and in VI, $d(f, b) = 0$. If b is a point in regions I, III, IV, or V, prove that there is a point a in A such that $f(a) = b$. Explain what can be said about points b in regions II and VI.

28. Sperner's Lemma Generalized

The intermediate-value theorem in Section 27 for functions $f: A \to \mathbb{R}^2$ is a good theorem, but it has severe limitations on the set A. Theorem 27.1 assumes that A is the union of the Euclidean simplexes in a Euclidean complex K which has a sequence of simplicial subdivisions $\mathscr{P}_k$, $k = 1, 2, \cdots$, with the properties described in Sections 23 and 24. These hypotheses are very restrictive. The construction in Example 27.7 shows that some extensions of Theorem 27.1 are possible, but these extensions are quite limited because they require a special construction for each different type of region A. The purpose of this section is to generalize the results established in earlier sections of this chapter. The generalizations include an extended form of Sperner's lemma which can be used to prove an intermediate-value theorem for a large class of continuous functions.

Let A be a set in $\mathbb{R}^n$, $n \geqq 2$, which is the union of the 2-dimensional Euclidean simplexes in a complex K as described in Section 23, and let $\mathscr{P}_k$, $k = 1, 2, \cdots$, be oriented simplicial subdivisions of A which have

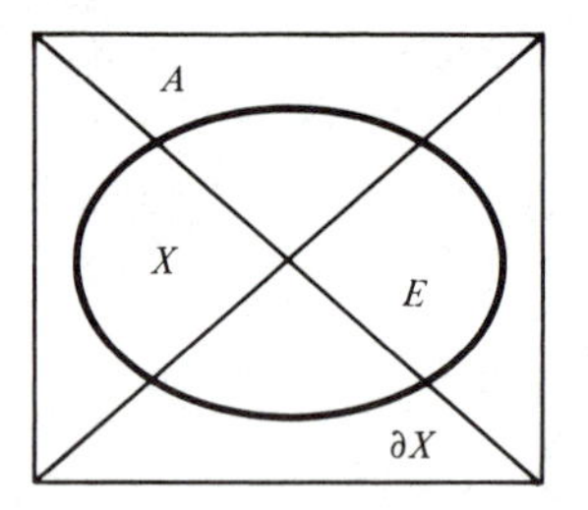

Figure 28.1. An example of a set X.

properties (1), (2), (3) in Section 24. As an example in a special case, A is the union of a set of 2-dimensional simplexes which form a Euclidean complex K in $\mathbb{R}^2$. Let X be a closed subset of A whose interior relative to A is a non-empty open set E, any two points of which can be connected by a polygonal curve in the set. Let ∂X denote the boundary of X relative to A [see Definition 92.7], and assume that $\partial X \neq \varnothing$. This section will establish an intermediate-value theorem for continuous functions of the form $f: X \to \mathbb{R}^2$.

28.1 Example. Let K be the Euclidean complex which consists of the four oriented 2-simplexes whose union is the rectangle A in Figure 28.1. Let $\mathscr{P}_k$, $k = 1, 2, \cdots$, be a sequence of simplicial subdivisions of A [see Figure 21.5]. Let X be a closed oval region in A with interior E and boundary ∂X. As a second example, let K be an oriented Euclidean complex of two-dimensional simplexes in the surface of the cube in $\mathbb{R}^3$ [see Figure 21.5], and let $\mathscr{P}_k$, $k = 1, 2, \cdots$, be a sequence of simplicial subdivisions of K. Then A is the surface of the cube. Let X be a closed set in A which is bounded by a finite number of non-intersecting curves, and whose interior is an open set E, relative to A, whose boundary ∂X consists of the bounding curves.

Let $\mathscr{R}_k$ denote, for $k = 1, 2, \cdots$, the set of simplexes $\mathbf{a}$ in $\mathscr{P}_k$ such that the Euclidean simplex $\mathbf{a}$ is contained in E. Since E is open and since the norm of $\mathscr{P}_k$ tends to zero as $k \to \infty$, then for all sufficiently large k the set $\mathscr{R}_k$ is not empty. For each subdivision $\mathscr{P}_{k+1}$, $k = 1, 2, \cdots$, define sets $\mathscr{A}_{k+1}$ and $\mathscr{B}_{k+1}$ as follows.

$$\mathscr{A}_{k+1} = \{\mathbf{a} : \mathbf{a} \in \mathscr{R}_{k+1}, \mathbf{a} \text{ is contained in a simplex in } \mathscr{R}_k\};$$

$$\mathscr{B}_{k+1} = \{\mathbf{a} : \mathbf{a} \in \mathscr{R}_{k+1}, \mathbf{a} \text{ is not contained in a simplex in } \mathscr{R}_k\}.$$

Then

$$\begin{aligned} \mathscr{R}_1 &= \mathscr{A}_1, \\ \mathscr{R}_2 &= \mathscr{A}_2 \cup \mathscr{B}_2, \\ &\cdots\cdots\cdots\cdots \\ \mathscr{R}_k &= \mathscr{A}_k \cup \mathscr{B}_k, \\ &\cdots\cdots\cdots\cdots \end{aligned} \tag{1}$$

Thus $\mathscr{A}_{k+1}$ is a subdivision of $\mathscr{R}_k$ [it consists of Euclidean simplexes in $\mathscr{R}_{k+1}$ which are in E and in a simplex in $\mathscr{R}_k$]; and $\mathscr{B}_{k+1}$ consists of all other simplexes in $\mathscr{P}_{k+1}$ whose Euclidean simplexes are in E. If $[\mathscr{R}_k]$ denotes the union of the Euclidean simplexes in $\mathscr{R}_k$, then

$$[\mathscr{R}_1] \subset [\mathscr{R}_2] \subset \cdots \subset [\mathscr{R}_k] \subset \cdots, \tag{2}$$

and the boundary of $[\mathscr{R}_k]$ approaches the boundary of X. The set $[\mathscr{R}_k]$ may not be connected in some cases.

Let $f: X \to \mathbb{R}^n$ be a continuous function which maps X into $\mathbb{R}^n$, $n \geqq 2$. Then f maps the three vertices of a simplex $\mathbf{a}$ in $\mathscr{R}_k$ into three points which form the vertices of an abstract simplex $\mathbf{b}$, and f maps $\mathscr{R}_k$ into an oriented abstract complex; its set of simplexes is $\{\mathbf{b} : \mathbf{b} = f(\mathbf{a}), \mathbf{a} \in \mathscr{R}_k\}$, which will be denoted by $f(\mathscr{R}_k)$. Let V_k denote the set of all vertices of simplexes $\mathbf{b}$ in $f(\mathscr{R}_k)$, and let $s_k : V_k \to \{e_0, e_1, e_2\}$ be a Sperner function defined on V_k. Also, let $S(s_k)$ denote the Sperner number of s_k [see Definition 23.2].

The function f maps a simplex (p_i^k, p_{i+1}^k) in the boundary of the chain $\mathscr{R}_k$ into a simplex (q_i^k, q_{i+1}^k) in the boundary of (the abstract chain) $f(\mathscr{R}_k)$. Then $\mathscr{A}_{k+1}$ induces a subdivision in (p_i^k, p_{i+1}^k) as follows:

$$\begin{gathered}(a_0, a_1), (a_1, a_2), \cdots, (a_m, a_{m+1}),\\ a_0 = p_i^k, \qquad a_{m+1} = p_{i+1}^k.\end{gathered} \tag{3}$$

The function f maps this subdivision of (p_i^k, p_{i+1}^k) into the following subdivision of (q_i^k, q_{i+1}^k):

$$\begin{gathered}(b_0, b_1), (b_1, b_2), \cdots, (b_m, b_{m+1}), \qquad b_j = f(a_j),\\ b_0 = q_i^k, \qquad b_{m+1} = q_{i+1}^k.\end{gathered} \tag{4}$$

In the next theorem, the following statements (a) and (b) will be hypotheses concerning the Sperner functions s_k, $k = 1, 2, \cdots$.

(a) For each simplex (q_i^k, q_{i+1}^k) in the boundary of $f(\mathscr{R}_k)$, the Sperner function s_{k+1} has the following properties:

$$s_{k+1}(b_0) = s_k(q_i^k), \qquad s_{k+1}(b_{m+1}) = s_k(q_{i+1}^k); \tag{5}$$

the set $\{s_{k+1}(b_j) : j = 0, 1, \cdots, m+1\}$ contains at most two distinct elements in $\{e_0, e_1, e_2\}$. (6)

(b) The Sperner function s_{k+1} maps the three vertices of each simplex $\mathbf{b}$ in $f(\mathscr{B}_{k+1})$ into at most two distinct points in $\{e_0, e_1, e_2\}$.

28.2 Theorem (Sperner's Lemma). *Let* $s_k : V_k \to \{e_0, e_1, e_2\}$, $k = 1, 2, \cdots$, *be Sperner functions, defined on the vertices* V_k *of* $f(\mathscr{R}_k)$, *which have the following property: there is an integer* K *such that, for* $k \geqq K$, *the statements in* (a) *and* (b) *are true. Then*

$$S(s_{k+1}) = S(s_k), \qquad k \geqq K. \tag{7}$$

PROOF. Let $\mathbf{b}$ denote a simplex (q_0, q_1, q_2) in $f(\mathscr{R}_{k+1})$, and let $\Delta[s_{k+1}(\mathbf{b})]$ be an abbreviation for $\Delta[s_{k+1}(q_0), s_{k+1}(q_1), s_{k+1}(q_2)]$. Then since $\mathscr{R}_{k+1} = \mathscr{A}_{k+1} \cup \mathscr{B}_{k+1}$ by (1),

$$\sum_{f(\mathscr{R}_{k+1})} \Delta[s_{k+1}(\mathbf{b})] = \sum_{f(\mathscr{A}_{k+1})} \Delta[s_{k+1}(\mathbf{b})] + \sum_{f(\mathscr{B}_{k+1})} \Delta[s_{k+1}(\mathbf{b})], \tag{8}$$

$$= \sum_{f(\mathscr{A}_{k+1})} \Delta[s_{k+1}(\mathbf{b})], \qquad k \geqq K, \tag{9}$$

because, for $k \geqq K$, each term in the sum over $f(\mathscr{B}_{k+1})$ is zero as a result of the hypothesis in (b). In the sums in (8) and (9), $\mathbf{b}$ varies over the sets indicated. The sum on the left in (8) is the Sperner number $S(s_{k+1})$ of s_{k+1} by Definition 23.2. Now $\mathscr{A}_{k+1}$ is a refinement of $\mathscr{R}_k$, and s_{k+1}, restricted to the vertices of $f(\mathscr{A}_{k+1})$, is a refinement of s_k [see hypothesis (a) and Definition 23.6]. Then by the original form of Sperner's lemma in Theorem 23.8, the sum in (9) equals $S(s_k)$. Then (8) and (9) show that $S(s_{k+1}) = S(s_k)$ for $k \geqq K$. □

Assume now that $f: X \to \mathbb{R}^2$ is a continuous function which maps the closed set X into the plane. Let b be a point in the plane of $f(X)$ but not in $f(\partial X)$. Choose rays r_0, r_1, r_2 from b which make equal angles with one another and are located in the counterclockwise order, and define regions W_i and U_i, $i = 0, 1, 2$, as in Section 24. Let V_k denote the vertices in $f(\mathscr{R}_k)$ as before, and define Sperner functions $s_k: V_k \to \{e_0, e_1, e_2\}$ as follows [compare (10) in Section 24]:

$$s_k(q) = e_i, \qquad q \in U_i, \qquad q \in V_k, \qquad k = 1, 2, \cdots. \tag{10}$$

28.3 Theorem. *Let s_k, $k = 1, 2, \cdots$, be the Sperner functions defined in* (10). *Then there exists an integer K and an integer $d(f, b)$ such that*

$$S(s_k) = d(f, b), \qquad k \geqq K. \tag{11}$$

This theorem is analogous to Theorem 24.3, and their proofs are similar. The proof of Theorem 28.3 requires the following lemma, which is similar to Lemma 24.2.

28.4 Lemma. *Let $f: X \to \mathbb{R}^2$ be the continuous function described above. Then there exists a positive integer K such that, for $k \geqq K$, the function f maps each Euclidean simplex $[p_i^k, p_{i+1}^k]$ in the boundary of $\mathscr{R}_k$ and each Euclidean simplex $\mathbf{a}$ in $\mathscr{B}_{k+1}$ into at most two of the regions W_i, $i = 0, 1, 2$.*

PROOF OF THE LEMMA. Assume that the lemma is false. Then for an infinite number of the sets $\mathscr{R}_1, \mathscr{R}_2, \cdots, \mathscr{R}_k, \cdots, f$ fails to have the property stated in the lemma. Thus for an infinite number of values of k, there is either a simplex $[p_i^k, p_{i+1}^k]$ in $\partial\mathscr{R}_k$ or a Euclidean simplex $\mathbf{a}$ in $\mathscr{B}_{k+1}$ which contains three points x_0^k, x_1^k, x_2^k such that $f(x_i^k) \in W_i$, $i = 0, 1, 2$. Then by the Bolzano–

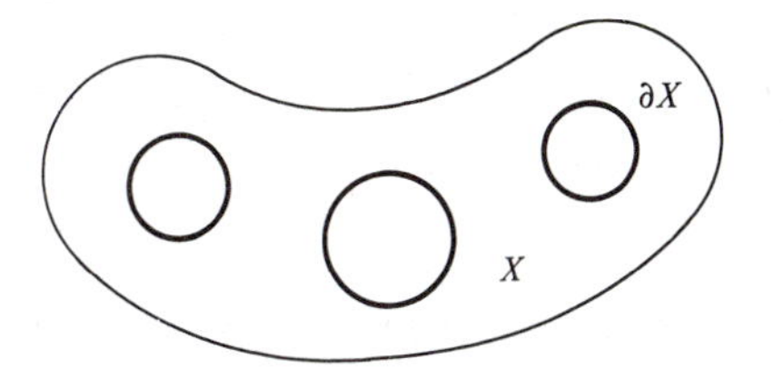

Figure 28.2. A region X and its boundary ∂X.

Weierstrass theorem [see Theorem 94.1] the infinite set of points x_0^k in the compact set X has a point of accumulation x in X. Then there is a sequence $\{x_0^{k_j}: j = 1, 2, \cdots\}$ such that $\lim_{j\to\infty} x_0^{k_j} = x$. Since the norm of $\mathscr{P}_k$ approaches zero as $k \to \infty$, and since f is continuous at x, then

$$\lim_{j\to\infty} x_i^{k_j} = x, \qquad \lim_{j\to\infty} f(x_i^{k_j}) = f(x), \qquad i = 0, 1, 2. \tag{12}$$

Also, $f(x) \in W_i$, $i = 0, 1, 2$, since $f(x_i^k) \in W_i$ and W_i is closed. Therefore $f(x) = b$ since $W_0 \cap W_1 \cap W_2$ contains the single point b. But the boundary of $\mathscr{R}_k$ approaches the boundary of X as $k \to \infty$, and thus the distance from x_i^k, $i = 0, 1, 2$, to ∂X approaches zero as $k \to \infty$. Then (12) shows that $x \in \partial X$ and $f(x) \in f(\partial X)$; also, $b \in f(\partial X)$ since $f(x) = b$. But $b \notin f(\partial X)$ by hypothesis, and this contradiction establishes the lemma. □

Proof of Theorem 28.3. Let K be the integer which exists by Lemma 28.4. Then hypotheses (a) and (b) of Theorem 28.2 are satisfied for $k \geqq K$, and the Sperner number $S(s_k)$ is constant for $k \geqq K$ by (7). This constant is denoted by $d(f, b)$ and called the degree of f at b, and the proof of (11) and Theorem 28.3 is complete. □

Thus the degree $d(f, b)$ of $f: X \to \mathbb{R}^2$ is defined at every point b in the plane of $f(X)$ which is not in $f(\partial X)$; the degree is a positive, zero, or negative integer. The properties and applications of this degree are entirely similar to those in the earlier case, and their investigation is left to the reader [see Example 28.5 and the exercises].

28.5 Example. Let X be the region shown in Figure 28.2. Then X has an open interior E which is not empty, and its boundary consists of four curves as shown. The pieces of ∂X are shown as curves, but the fact that the boundary consists of curves is of no consequence. Let $f: X \to \mathbb{R}^2$ be a continuous function; the purpose of this example is to indicate how an intermediate-value theorem can be proved for f. First it is necessary to construct a sequence of simplicial subdivisions $\mathscr{P}_k$, $k = 1, 2, \cdots$, which can be used to construct a sequence of Sperner functions s_k, $k = 1, 2, \cdots$. The squares $[i, i+1] \times [j, j+1]$, $i, j = 0, \pm 1, \pm 2, \cdots$, subdivide the plane into unit squares [see Exercise 21.5]. Since X is bounded, X is contained in a finite number of these unit squares; let A denote the union of the squares which contain X.

In each square construct a sequence of positively oriented simplicial subdivisions of the kind shown in Figure 21.5. Combine corresponding subdivisions of the squares to form the subdivisions $\mathscr{P}_k$, $k = 1, 2, \cdots$, of A. Let b be a point in the plane of $f(X)$ which is not in $f(\partial X)$. Choose rays r_0, r_1, r_2 at b, and define the Sperner function $s_k: V_k \to \{e_0, e_1, e_2\}$; here V_k is the set of vertices of $f(\mathscr{R}_k)$. Then the Sperner number $S(s_k)$ is constant for $k \geqq K$, and this constant value is $d(f, b)$. If $d(f, b) \neq 0$, then the properties of the Sperner number $S(s_k)$ and the arguments used in proving Theorem 27.1 can be employed to show that there is a point a in X such that $f(a) = b$.

Exercises

28.1. In Example 28.5, let b be a point in the plane of $f(X)$ which is not in $f(\partial X)$, and assume that $d(f, b) \neq 0$. Write out the details in the proof that there exists a point a in X such that $f(a) = b$.

28.2. Let K be a Euclidean 2-complex [see Section 23] whose 2-dimensional simplexes are $[p_3, p_0, p_1]$, $[p_3, p_1, p_2]$, and $[p_3, p_2, p_0]$; let the union of these Euclidean simplexes be denoted by A. Assume that the 2-dimensional simplexes in K are three sides of a regular tetrahedron in $\mathbb{R}^n$, $n \geqq 3$. Let $[q_0, q_1, q_2]$ be a Euclidean simplex in $\mathbb{R}^2$ which is positively oriented in $\mathbb{R}^2$. Finally, let $f: A \to \mathbb{R}^2$ be a continuous function which maps $[p_0, p_1]$, $[p_1, p_2]$, $[p_2, p_0]$ into $[q_0, q_1]$, $[q_1, q_2]$, $[q_2, q_0]$, respectively, so that $f(p_i) = q_i$, $i = 0, 1, 2$. Observe that points in A may be mapped into points in the exterior of $[q_0, q_1, q_2]$.

(a) Construct a set $\mathscr{P}_k$, $k = 1, 2, \cdots$, of simplicial subdivisions of K which have properties (1), (2), (3) in Section 24.

(b) If b is a point in the plane of $f(\partial A)$ but in the exterior of $[q_0, q_1, q_2]$, show that $d(f, b)$ is defined and that $d(f, b) = 0$.

(c) If b is a point in the interior of $[q_0, q_1, q_2]$, show that $d(f, b) = 1$ and that there is a point a in A such that $f(a) = b$.

28.3. Let X be a set of the kind described in the second paragraph of Section 28. Then A is a set in $\mathbb{R}^n$ which is the union of the simplexes in a Euclidean complex K; $\mathscr{P}_k$, $k = 1, 2, \cdots$, is a sequence of simplicial subdivisions of A; and X is a closed set whose interior relative to A is not empty. Prove the following theorem. There exists no function f which has the following properties:

(a) f is a continuous function which maps X into $\mathbb{R}^2$.

(b) There exists a point b in $\mathbb{R}^2$ such that $b \notin f(\partial X)$ and $d(f, b) \neq 0$.

(c) f maps X into $f(\partial X)$.

28.4. Prove the following theorem. If X is a closed, bounded, convex set in $\mathbb{R}^2$ whose interior E is not empty, then there exists no function f which has the following properties:

(a) f is a continuous function which maps X into $\mathbb{R}^2$.

(b) If x is in ∂X, then $f(x) = x$.

(c) If $x \in X$, then $f(x) \in \partial X$.

28.5. Prove the following theorem. Let X be a closed, bounded, convex set in $\mathbb{R}^2$ whose interior E is not empty, and let $f: X \to \mathbb{R}^2$ be a continuous function which maps the convex set into its interior. Then f has a fixed point; that is, there exists an x in X such that $f(x) = x$.

28.6. (Coincidences). Let X be a set of the kind described in the second paragraph of Section 28, and let $g_i : X \to \mathbb{R}^2$, $i = 1, 2$, be two continuous functions. Let $f: X \to \mathbb{R}^2$ be the function such that $f(x) = g_1(x) - g_2(x)$ for x in X. If the origin e_0 in $\mathbb{R}^2$ belongs to $f(\partial X)$, then there exists an x in ∂X such that $g_1(x) = g_2(x)$. Prove the following: if $e_0 \notin f(\partial X)$ and $d(f, e_0) \neq 0$, then there exists an x in the interior of X such that $g_1(x) = g_2(x)$. [Remark. A point x such that $g_1(x) = g_2(x)$ is called a *coincidence* of g_1 and g_2. If $g_2(x) \equiv x$, then a coincidence of g_1 and g_2 is a fixed point of g_1.]

28.7. Let $X = \{x : |x| \leqq a\}$ in $\mathbb{R}^2$, and let $g_i : X \to \mathbb{R}^2$, $i = 1, 2$, be continuous functions. Assume the following: (a) $d(g_1, e_0) \neq 0$; (b) $|g_1(x)| \geqq 2r > 0$ if $x \in \partial X$; and (c) $|g_2(x)| < \sqrt{3}r/2$ if $x \in \partial X$. Prove that there exists an x on the interior of X such that $g_1(x) = g_2(x)$. [Hint. Set $f(x) = g_1(x) - g_2(x)$. Then $|f(x)| > r$ if $x \in \partial X$ and $|g_1(x) - f(x)| = |g_2(x)| < \sqrt{3}r/2$ if $x \in \partial X$. Use Theorem 26.2 to show that $d(f, e_0) \neq 0$.]

28.8. Prove the following theorem. Let $X = \{x : |x| \leqq \sqrt{3}r/2\}$ in $\mathbb{R}^2$, and let $f: X \to \mathbb{R}^2$ be a continuous function such that (a) $d(f, e_0) \neq 0$ and (b) $|f(x)| \geqq 2r$ if $x \in \partial X$. Then f has a fixed point on the interior of X.

28.9. Prove the following theorem. Let $X = \{x : |x| \leqq 2r\}$ in $\mathbb{R}^2$, and let $f: X \to \mathbb{R}^2$ be a continuous function such that $|f(x)| < \sqrt{3}r/2$ for x in ∂X. Then f has a fixed point on the interior of X.

28.10. Prove the following theorem. Let X be the closed square in $\mathbb{R}^2$ whose four vertices are the points $(\pm 2r, \pm 2r)$, and let $f: X \to \mathbb{R}^2$ be a continuous function such that $|f(x)| < \sqrt{3}r/2$ for x in ∂X. Then f has a fixed point on the interior of X.

28.11. (Critical points). Let X be a closed set in $\mathbb{R}^2$ whose interior E is not empty. Let $g : X \to \mathbb{R}$ be a function which has continuous derivatives $D_1 g$ and $D_2 g$ on X. Let $f: X \to \mathbb{R}^2$ be a function with components (f_1, f_2) such that $f_i(x) = D_i g(x)$, $i = 1, 2$, for x in X. Assume that $e_0 \notin f(\partial X)$ and that $d(f, e_0) \neq 0$. Prove that there exists a point x in the interior of X such that $D_1 g(x) = 0$ and $D_2 g(x) = 0$. [A point at which the two derivatives of g vanish is called a *critical point* of g.]

28.12. Let $g(x_1, x_2) = 1 - (x_1^2 + x_2^2)$. Use Exercise 28.11 to prove that g has a critical point in $\{(x_1, x_2) : x_1^2 + x_2^2 < 1\}$.

29. Generalizations to Higher Dimensions

The purpose of this section is to outline the extension to other dimensions of the results in the preceding sections of this chapter.

Let $\{\mathbf{a}_r : r = 1, \cdots, N\}$ be a set of oriented, m-dimensional, Euclidean simplexes $(p_{r0}, p_{r1}, \cdots, p_{rm})$ with vertices $p_{r0}, p_{r1}, \cdots, p_{rm}$ which are points in some space $\mathbb{R}^n, n \geqq m$; assume that these simplexes form a Euclidean complex K in which at most two m-simplexes have an $(m-1)$-dimensional side in common. Assume that the simplexes in K are oriented so that $\partial\Sigma\{\mathbf{a}_r : r = 1, \cdots, N\}$ contains those $(m-1)$-dimensional simplexes in K which are a side of a single m-dimensional simplex. These hypotheses are satisfied automatically if K is a Euclidean m-complex of m-simplexes which are positively oriented in $\mathbb{R}^m$. Let the union of the Euclidean simplexes in K be denoted by A, and let $\mathscr{P}_k$, $k = 1, 2, \cdots$, be a set of simplicial subdivisions of A which has the following properties [see (1), (2), (3) in Section 24]:

(1) Each simplex in $\mathscr{P}_k$ determines the same m-direction in $\mathbb{R}^n$ as the simplex in K which contains it. If A is a set in $\mathbb{R}^m$, this assumption means that all simplexes in K and in $\mathscr{P}_k$ have the same orientation in $\mathbb{R}^m$.
(2) $\mathscr{P}_{k+1}$ is a refinement of $\mathscr{P}_k$.
(3) The norm of $\mathscr{P}_k$ approaches zero as k tends to infinity.

Let X be a closed subset of A whose interior relative to A is a non-empty, open set E, each two points of which can be connected by a polygonal curve in E, and whose boundary ∂X is not empty. Define sets $\mathscr{R}_k$, $\mathscr{A}_k$, $\mathscr{B}_k$ for $k = 1, 2, \cdots$, in the manner suggested by the definition of similar sets in Section 28.

Let $(e_0, e_1, \cdots, e_m)$ be the simplex whose vertices $e_0, e_1, \cdots, e_m$ are, respectively, the origin $(0, 0, \cdots, 0)$ and the unit points $(1, 0, \cdots, 0)$, $(0, 1, \cdots, 0)$, $\cdots$, $(0, 0, \cdots, 1)$ on the axes in $\mathbb{R}^m$; then $(e_0, e_1, \cdots, e_m)$ is positively oriented in $\mathbb{R}^m$ and $\Delta(e_0, e_1, \cdots, e_m) = (-1)^m$.

Let $f: X \to \mathbb{R}^m$ be a continuous function on X. Then f maps the $m+1$ vertices of a simplex $\mathbf{a}$ in $\mathscr{R}_k$ into $m+1$ points which form the vertices of an abstract simplex $\mathbf{b}$, and f maps $\mathscr{R}_k$ into an oriented abstract complex $\{\mathbf{b} : \mathbf{b} = f(\mathbf{a}), \mathbf{a} \in \mathscr{R}_k\}$ which will be denoted also by $f(\mathscr{R}_k)$. Let V_k denote the set of all vertices of simplexes $\mathbf{b}$ in $f(\mathscr{R}_k)$; then a function $s_k : V_k \to \{e_0, e_1, \cdots, e_m\}$ is called a Sperner function on V_k [compare Definition 23.2].

Introduce the following notation for $k = 1, 2, \cdots$:

$$\mathscr{R}_k = \{(p_{r0}^k, p_{r1}^k, \cdots, p_{rm}^k) : r = 1, 2, \cdots, N_k\}; \tag{4}$$

$$\partial\mathscr{R}_k = \sum\{(u_{i1}^k, u_{i2}^k, \cdots, u_{im}^k) : i = 1, 2, \cdots, I_k\}; \tag{5}$$

$$f(\mathscr{R}_k) = \{(q_{r0}^k, q_{r1}^k, \cdots, q_{rm}^k) : r = 1, 2, \cdots, N_k\}; \tag{6}$$

$$\partial f(\mathscr{R}_k) = \sum\{(v_{i1}^k, v_{i2}^k, \cdots, v_{im}^k) : i = 1, 2, \cdots, I_k\}. \tag{7}$$

If s_k is a Sperner function on V_k, then each of the determinants $\Delta[s_k(q_{r0}^k), s_k(q_{r1}^k), \cdots, s_k(q_{rm}^k)]$ has the value 1, 0, or -1, and the Sperner number $S(s_k)$ of s_k is defined by the following equation:

$$S(s_k) = \sum_{r=1}^{N_k} \Delta[s_k(q_{r0}^k), s_k(q_{r1}^k), \cdots, s_k(q_{rm}^k)]. \tag{8}$$

If e is an arbitrary but fixed element in $\{e_0, e_1, \cdots, e_m\}$, then Theorem 20.1 and the cancellation properties in the oriented complex K can be used to prove that

$$S(s_k) = \sum_{i=1}^{I_k} \Delta[e, s_k(v_{i1}^k), \cdots, s_k(v_{im}^k)]. \tag{9}$$

Now $\mathscr{A}_{k+1}$ is a refinement of $\mathscr{R}_k$, and $\mathscr{A}_{k+1}$ induces a subdivision in $\partial\mathscr{R}_k$. The function $f: X \to \mathbb{R}^m$ maps this subdivision of $\partial\mathscr{R}_k$ into a subdivision of $\partial f(\mathscr{R}_k)$. The vertices in a subdivision of a simplex $(v_{i1}^k, v_{i2}^k, \cdots, v_{im}^k)$ in $\partial f(\mathscr{R}_k)$ are said to lie in $(v_{i1}^k, v_{i2}^k, \cdots, v_{im}^k)$ although these simplexes are only abstract simplexes. These statements about subdivisions of $f(\mathscr{R}_k)$ derive their meaning from relationships in the Euclidean simplexes in $\mathscr{R}_k$. Consider the following two statements about the Sperner function $s_{k+1}: V_{k+1} \to \{e_0, e_1, \cdots, e_m\}$; these statements are hypotheses in the next theorem.

(a) For each simplex $(v_{i1}^k, v_{i2}^k, \cdots, v_{im}^k)$ in $\partial f(\mathscr{R}_k)$, the function s_{k+1} maps those vertices in V_{k+1} which lie in $(v_{i1}^k, v_{i2}^k, \cdots, v_{im}^k)$ into at most m distinct points in $\{e_0, e_1, \cdots, e_m\}$, and s_k and s_{k+1} have the same values on those vertices in V_{k+1} which are also in V_k.
(b) The function s_{k+1} maps the $m+1$ vertices of each simplex in $f(\mathscr{B}_{k+1})$ into at most m distinct points in $\{e_0, e_1, \cdots, e_m\}$.

29.1 Theorem (Sperner's Lemma). *Let* $s_k: V_k \to \{e_0, e_1, \cdots, e_m\}$, $k = 1, 2, \cdots$, *be Sperner functions which satisfy the following hypotheses: there is an integer* K *such that, for* $k \geqq K$, *the statements in* (a) *and* (b) *are true. Then*

$$S(s_{k+1}) = S(s_k), \qquad k \geqq K. \tag{10}$$

Proof. The proof is similar to the proof of Theorem 28.2. For convenience in notation, if $\mathbf{b} = (q_0, q_1, \cdots, q_m)$, then let $\Delta[s_{k+1}(\mathbf{b})]$ denote the determinant

$$\Delta[s_{k+1}(q_0), s_{k+1}(q_1), \cdots, s_{k+1}(q_m)]. \tag{11}$$

Then since $\mathscr{R}_{k+1} = \mathscr{A}_{k+1} \cup \mathscr{B}_{k+1}$,

$$\sum_{f(\mathscr{R}_{k+1})} \Delta[s_{k+1}(\mathbf{b})] = \sum_{f(\mathscr{A}_{k+1})} \Delta[s_{k+1}(\mathbf{b})] + \sum_{f(\mathscr{B}_{k+1})} \Delta[s_{k+1}(\mathbf{b})], \tag{12}$$

$$= \sum_{f(\mathscr{A}_{k+1})} \Delta[s_{k+1}(\mathbf{b})], \qquad k \geqq K, \tag{13}$$

because each term in the sum over $f(\mathscr{B}_{k+1})$ is zero since s_{k+1} satisfies the hypothesis in (b) for $k \geqq K$. The sum on the left in (12) is $S(s_{k+1})$ by (8), and the proof can be completed by showing that the sum in (13) is equal to $S(s_k)$. The proof is similar to the proof of (15) in Theorem 23.8, but it is not obvious. The proof will be given in the special case $m = 3$ in order to simplify the notation and the exposition; this special case exhibits the essential features of the proof in the general case.

Assume that $k \geqq K$ so that the hypotheses in (a) and (b) are satisfied. Let

$$\partial f(\mathscr{A}_{k+1}) = \sum \{(a_{m1}^{k+1}, a_{m2}^{k+1}, a_{m3}^{k+1}) : m = 1, 2, \cdots, M_{k+1}\}. \tag{14}$$

Then Theorem 20.1 and the cancellation properties in the oriented complex K show that

$$\sum_{f(\mathscr{A}_{k+1})} \Delta[s_{k+1}(\mathbf{b})] = \sum_{m=1}^{M_{k+1}} \Delta[e, s_{k+1}(a_{m1}^{k+1}), \cdots, s_{k+1}(a_{m3}^{k+1})]. \tag{15}$$

Here e is an arbitrary but fixed element in $\{e_0, e_1, \cdots, e_3\}$. Let

$$\{(a_{t1}^i, a_{t2}^i, a_{t3}^i) : t = 1, 2, \cdots, T_i\} \tag{16}$$

be the set of simplexes in $\partial f(\mathscr{A}_{k+1})$ which are contained in the simplex $(v_{i1}^k, v_{i2}^k, v_{i3}^k)$ in $\partial f(\mathscr{R}_k)$. Then [see (7)]

$$\partial f(\mathscr{A}_{k+1}) = \sum_{i=1}^{I_k} \sum_{t=1}^{T_i} (a_{t1}^i, a_{t2}^i, a_{t3}^i), \tag{17}$$

and equation (15) can be written as follows:

$$S(s_{k+1}) = \sum_{i=1}^{I_k} \sum_{t=1}^{T_i} \Delta[e, s_{k+1}(a_{t1}^i), \cdots, s_{k+1}(a_{t3})]. \tag{18}$$

Here the sum on the left in (15) has been replaced by its value $S(s_{k+1})$ [see (12) and (13)]. The proof will be completed by showing that the sum on the right in (18) is equal to $S(s_k)$ for $k \geqq K$.

Let (a, b, c) and (a', c, b) be two oriented simplexes in (16) which have the vertices b, c in common. Then by Theorem 20.1, with x in that theorem chosen to be $s_{k+1}(v_{i1}^k)$,

$$\begin{aligned}
\Delta[e, s_{k+1}(a), s_{k+1}(b), s_{k+1}(c)] = {} & \Delta[s_{k+1}(v_{i1}^k), s_{k+1}(a), s_{k+1}(b), s_{k+1}(c)] \\
& + \Delta[e, s_{k+1}(v_{i1}^k), s_{k+1}(b), s_{k+1}(c)] \\
& + \Delta[e, s_{k+1}(a), s_{k+1}(v_{i1}^k), s_{k+1}(c)] \\
& + \Delta[e, s_{k+1}(a), s_{k+1}(b), s_{k+1}(v_{i1}^k)].
\end{aligned} \tag{19}$$

The first term on the right is zero because of the hypothesis in (a). Evaluate $\Delta[e, s_{k+1}(a'), s_{k+1}(c), s_{k+1}(b)]$ in the same way and add the two equations to obtain the following [observe that two terms cancel in the sum]:

$$\begin{aligned}
& \Delta[e, s_{k+1}(a), s_{k+1}(b), s_{k+1}(c)] + \Delta[e, s_{k+1}(a'), s_{k+1}(c), s_{k+1}(b)] \\
& \quad = \Delta[e, s_{k+1}(v_{i1}^k), s_{k+1}(a), s_{k+1}(b)] + \Delta[e, s_{k+1}(v_{i1}^k), s_{k+1}(b), s_{k+1}(a')] \\
& \qquad + \Delta[e, s_{k+1}(v_{i1}^k), s_{k+1}(a'), s_{k+1}(c)] + \Delta[e, s_{k+1}(v_{i1}^k), s_{k+1}(c), s_{k+1}(a)].
\end{aligned} \tag{20}$$

Introduce simplified notation as follows:

$$\begin{gathered}
\partial \sum_{t=1}^{T_i} (a_{t1}^i, a_{t2}^i, a_{t3}^i) = \sum_{j=0}^{J(i,a)} (a_j, a_{j+1}) + \sum_{j=0}^{J(i,b)} (b_j, b_{j+1}) + \sum_{j=0}^{J(i,c)} (c_j, c_{j+1}). \\
a_0 = c_{J(i,c)} = v_{i1}^k, \qquad b_0 = a_{J(i,a)} = v_{i2}^k, \qquad c_0 = b_{J(i,b)} = v_{i3}^k.
\end{gathered} \tag{21}$$

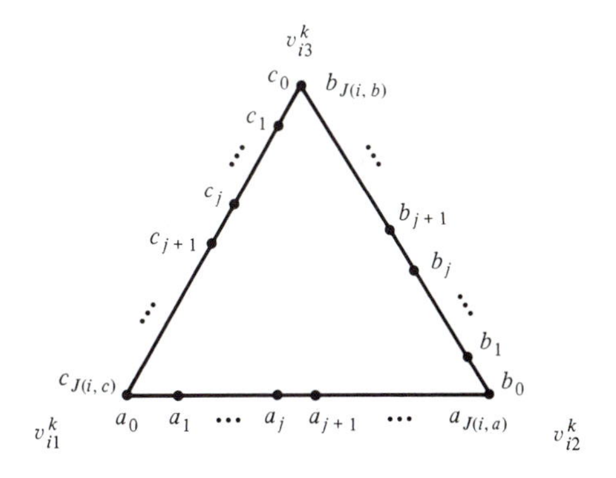

Figure 29.1. The notation in (21).

Figure 29.1 is designed to explain this notation. The term on the left in the first equation in (21) is the boundary of the chain of simplexes $(a_{t1}^i, a_{t2}^i, a_{t3}^i)$ in $\partial f(\mathscr{A}_{k+1})$ which are in the simplex $(v_{i1}^k, v_{i2}^k, v_{i3}^k)$ in $\partial f(\mathscr{R}_k)$ [see (17)]. The boundary of the simplexes in $(v_{i1}^k, v_{i2}^k, v_{i3}^k)$ is a one-chain which consists of one-simplexes (a_j, a_{j+1}), (b_j, b_{j+1}), (c_j, c_{j+1}) on the sides (v_{i1}^k, v_{i2}^k), (v_{i2}^k, v_{i3}^k), (v_{i3}^k, v_{i1}^k) respectively [see Figure 29.1]. Evaluate each term in the inner sum on the right in (18) by the method used in establishing (20); the result is the following:

$$\begin{aligned}
\sum_{t=1}^{T_i} \Delta[e, s_{k+1}(a_{t1}^i), \cdots, s_{k+1}(a_{t3}^i)]
&= \sum_{j=0}^{J(i,a)-1} \Delta[e, s_{k+1}(v_{i1}^k), s_{k+1}(a_j), s_{k+1}(a_{j+1})] \\
&\quad + \sum_{j=0}^{J(i,b)-1} \Delta[e, s_{k+1}(v_{i1}^k), s_{k+1}(b_j), s_{k+1}(b_{j+1})] \\
&\quad + \sum_{j=0}^{J(i,c)-1} \Delta[e, s_{k+1}(v_{i1}^k), s_{k+1}(c_j), s_{k+1}(c_{j+1})].
\end{aligned} \tag{22}$$

Use Theorem 20.1, with x in that theorem chosen to be $s_{k+1}(v_{i2}^k)$, to evaluate the terms in the middle sum on the right in (22) as follows:

$$\begin{aligned}
\Delta[e, s_{k+1}(v_{i1}^k), s_{k+1}(b_j), s_{k+1}(b_{j+1})]
&= \Delta[s_{k+1}(v_{i2}^k), s_{k+1}(v_{i1}^k), s_{k+1}(b_j), s_{k+1}(b_{j+1})] \\
&\quad + \Delta[e, s_{k+1}(v_{i2}^k), s_{k+1}(b_j), s_{k+1}(b_{j+1})] \\
&\quad + \Delta[e, s_{k+1}(v_{i1}^k), s_{k+1}(v_{i2}^k), s_{k+1}(b_{j+1})] \\
&\quad + \Delta[e, s_{k+1}(v_{i1}^k), s_{k+1}(b_j), s_{k+1}(v_{i2}^k)].
\end{aligned} \tag{23}$$

The first term on the right is zero because of hypothesis (a). Thus

$$\sum_{j=0}^{J(i,b)-1} \Delta[e, s_{k+1}(v_{i1}^k), s_{k+1}(b_j), s_{k+1}(b_{j+1})]$$
$$= \sum_{j=0}^{J(i,b)-1} \{\Delta[e, s_{k+1}(v_{i1}^k), s_{k+1}(v_{i2}^k), s_{k+1}(b_{j+1})] - \Delta[e, s_{k+1}(v_{i1}^k), s_{k+1}(v_{i2}^k), s_{k+1}(b_j)]\} \tag{24}$$
$$+ \sum_{j=0}^{J(i,b)-1} \Delta[e, s_{k+1}(v_{i2}^k), s_{k+1}(b_j), s_{k+1}(b_{j+1})].$$

The first sum on the right is a telescoping sum whose value, by (21), elementary properties of determinants, and hypothesis (a), is this:

$$\Delta[e, s_{k+1}(v_{i1}^k), s_{k+1}(v_{i2}^k), s_{k+1}(b_{J(i,b)})] - \Delta[e, s_{k+1}(v_{i1}^k), s_{k+1}(v_{i2}^k), s_{k+1}(b_0)]$$
$$= \Delta[e, s_{k+1}(v_{i1}^k), s_{k+1}(v_{i2}^k), s_{k+1}(v_{i3}^k)] - \Delta[e, s_{k+1}(v_{i1}^k), s_{k+1}(v_{i2}^k), s_{k+1}(v_{i2}^k)]$$
$$= \Delta[e, s_k(v_{i1}^k), s_k(v_{i2}^k), s_k(v_{i3}^k)]. \tag{25}$$

Substitute from (25) in (24); the result is

$$\sum_{j=0}^{J(i,b)-1} \Delta[e, s_{k+1}(v_{i1}^k), s_{k+1}(b_j), s_{k+1}(b_{j+1})]$$
$$= \Delta[e, s_k(v_{i1}^k), s_k(v_{i2}^k), s_k(v_{i3}^k)] \tag{26}$$
$$+ \sum_{j=0}^{J(i,b)-1} \Delta[e, s_{k+1}(v_{i2}^k), s_{k+1}(b_j), s_{k+1}(b_{j+1})].$$

Next use Theorem 20.1 to evaluate the third sum on the right in (22); if the point x in Theorem 20.1 is chosen to be $s_{k+1}(v_{i3}^k)$, then simplifications similar to those used in proving (26) show that

$$\sum_{j=0}^{J(i,c)-1} \Delta[e, s_{k+1}(v_{i1}^k), s_{k+1}(c_j), s_{k+1}(c_{j+1})]$$
$$= \sum_{j=0}^{J(i,c)-1} \Delta[e, s_{k+1}(v_{i3}^k), s_{k+1}(c_j), s_{k+1}(c_{j+1})]. \tag{27}$$

Substitute from (26) and (27) in (22), and then from (22) in (18); the result is this:

$$S(s_{k+1}) = \sum_{i=1}^{I_k} \Delta[e, s_k(v_{i1}^k), s_k(v_{i2}^k), s_k(v_{i3}^k)] \tag{28}$$
$$+ \sum_{i=1}^{I_k} \left\{ \sum_{j=0}^{J(i,a)-1} \Delta[e, s_{k+1}(v_{i1}^k), s_{k+1}(a_j), s_{k+1}(a_{j+1})] \right.$$
$$+ \sum_{j=0}^{J(i,b)-1} \Delta[e, s_{k+1}(v_{i2}^k), s_{k+1}(b_j), s_{k+1}(b_{j+1})] \tag{29}$$
$$\left. + \sum_{j=0}^{J(i,c)-1} \Delta[e, s_{k+1}(v_{i3}^k), s_{k+1}(c_j), s_{k+1}(c_{j+1})] \right\}.$$

The sum on the right in (28) is $S(s_k)$ by (9), and the proof can be completed by showing that the sum in (29) is zero. Since $(v_{i1}^k, v_{i2}^k, v_{i3}^k)$ belongs to $\partial f(\mathscr{R}_k)$ by (7), and since the boundary of a boundary is the zero chain [see Theorem 17.8], the edge (v_{i1}^k, v_{i2}^k) belongs to two simplexes in $\partial f(\mathscr{R}_k)$, but it has opposite orientations in the boundaries of the two simplexes. Thus these two simplexes contribute the following two terms to the sums in (29).

$$\begin{aligned}&\sum_{j=0}^{J(i,a)-1} \Delta[e, s_{k+1}(v_{i1}^k), s_{k+1}(a_j), s_{k+1}(a_{j+1})]\\ &\sum_{j=0}^{J(i,a)-1} \Delta[e, s_{k+1}(v_{i2}^k), s_{k+1}(a_{j+1}), s_{k+1}(a_j)]\end{aligned} \tag{30}$$

Since by a proof similar to that of (27),

$$\begin{aligned}\sum_{j=0}^{J(i,a)-1} &\Delta[e, s_{k+1}(v_{i1}^k), s_{k+1}(a_j), s_{k+1}(a_{j+1})]\\ &= \sum_{j=0}^{J(i,a)-1} \Delta[e, s_{k+1}(v_{i2}^k), s_{k+1}(a_j), s_{k+1}(a_{j+1})]\\ &= -\sum_{j=0}^{J(i,a)-1} \Delta[e, s_{k+1}(v_{i2}^k), s_{k+1}(a_{j+1}), s_{k+1}(a_j)],\end{aligned}$$

then the sum of the two terms in (29) which arise from the edge (v_{i1}^k, v_{i2}^k) is zero. In the same way, the sum of the two terms which arise from every edge is zero, and the sum in (29) is zero. The proof of Theorem 29.1 (Sperner's lemma) is complete in the case $m = 3$. □

Return to the general case described at the beginning of this section. Let $f: A \to \mathbb{R}^m$ be a continuous function on the closed set A, and let b be a point in $\mathbb{R}^m$ but not in $f(\partial A)$. Let $(a_0, a_1, \cdots, a_m)$ be an m-dimensional equilateral simplex (all edges equal) in $\mathbb{R}^m$ whose medians intersect at b. Define regions W_i, $i = 0, 1, \cdots, m$, in $\mathbb{R}^m$ as follows: W_i is the set of points q in $\mathbb{R}_m$ which lie on a closed ray from b through a point in the side $(a_0, a_1, \cdots, \widehat{a_i}, \cdots, a_m)$ of $(a_0, a_1, \cdots, a_m)$. These regions W_i are closed, their union is $\mathbb{R}^m$, and their intersection is $\{b\}$. Define sets U_i, $i = 0, 1, \cdots, m$, as follows.

$$U_0 = W_0, \qquad U_i = W_i - \bigcup_{j=0}^{i-1} U_j, \qquad i = 1, 2, \cdots, m.$$

The sets U_i form a disjoint decomposition of $\mathbb{R}^m$. Let V_k be the set of all vertices in $f(\mathscr{P}_k)$. Define the Sperner function $s_k : V_k \to \{e_0, e_1, \cdots, e_m\}$ as follows [compare (10) in Section 24].

$$s_k(q) = e_i \quad \text{if} \quad q \in U_i, \qquad q \in V_k, \qquad k = 1, 2, \cdots.$$

The reader should now be able to repeat the results of this chapter for the function $f: A \to \mathbb{R}^m$ and its associated Sperner functions s_k, and, more generally, for functions $f: X \to \mathbb{R}^m$ and their Sperner functions.

EXERCISES

29.1. There is a one-dimensional Sperner's lemma, and it can be used to prove the elementary intermediate-value theorem; this exercise outlines the treatment of this case. Let $\mathscr{Q}$ be an abstract, oriented, one-dimensional complex $\{(q_{i-1}, q_i): i = 1, 2, \cdots, N\}$ in some space $\mathbb{R}^m$, $m \geqq 1$. Then (considering $\mathscr{Q}$ as a chain), $\partial\mathscr{Q} = (q_N) - (q_0)$. Let e_0 and e_1 be the origin (0) and the unit point (1) in $\mathbb{R}$; then $\Delta(e_0, e_1) = -1$. Let $V(\mathscr{Q})$ be the set of vertices of $\mathscr{Q}$, and let $s: V(\mathscr{Q}) \to \{e_0, e_1\}$ be a Sperner function. The Sperner number $S(s)$ of s is defined as follows:

$$S(s) = \sum_{i=1}^{N} \Delta[s(q_{i-1}), s(q_i)].$$

(a) Let e be one of the points in $\{e_0, e_1\}$. Prove that

$$S(s) = \Delta[e, s(q_N)] - \Delta[e, s(q_0)].$$

(b) Prove that $S(s)$ equals 0, +1, or −1.
(c) If $\partial\mathscr{Q} = 0$, prove that $S(s) = 0$.
(d) If $\mathscr{Q}'$ is a simplicial subdivision of $\mathscr{Q}$, then $\partial\mathscr{Q}' = \partial\mathscr{Q}$. Let $s': V(\mathscr{Q}') \to \{e_0, e_1\}$ be a Sperner function such that, if q is a point in the common boundary of $\mathscr{Q}$ and $\mathscr{Q}'$, then $s'(q) = s(q)$. Prove that $S(s') = S(s)$. This statement is Sperner's lemma in the one-dimensional case.
(e) Let (a_{i-1}, a_i), $i = 1, 2, \cdots, N$, be one-dimensional simplexes which form a Euclidean complex K in some space $\mathbb{R}^n$, $n \geqq 1$. Let $A = \bigcup_1^N [a_{i-1}, a_i]$, and let $\mathscr{P}_k$, $k = 1, 2, \cdots$, be a set of simplicial subdivisions of A which satisfy the statements in equations (1), (2), (3). Let f be a continuous function which maps A into a line L in some space $\mathbb{R}^m$. A point b on L divides L into two closed sets W_0, W_1. Set $U_0 = W_0$ and $U_1 = W_1 - W_0$; then U_0, U_1 are the sets in a disjoint decomposition of L. Then f maps (p_{i-1}^k, p_i^k) in $\mathscr{P}_k$ into (q_{i-1}^k, q_i^k) in $\mathscr{Q}_k$. Define $s_k: V(\mathscr{Q}_k) \to \{e_0, e_1\}$ as follows: $s_k(q) = e_i$ if q is in $V(\mathscr{Q}_k)$ and in U_i. Show that $S(s_k) = S(s_1)$ for $k = 1, 2, \cdots$. Define $d(f, b)$ to be $S(s_1)$.
(f) Prove the following form of the elementary intermediate-value theorem: if b is a point on L such that $d(f, b) = \pm 1$, then there is a point a in A such that $f(a) = b$.

29.2. Let A, with simplicial subdivisions $\mathscr{P}_1, \mathscr{P}_2, \cdots$, be the set described in Sections 23 and 24; let H be an open hemisphere in $\mathbb{R}^3$; and let $f: A \to H$ be a continuous function which maps A into H. Let b be a point on H such that $b \notin f(\partial A)$. Subdivide H into three regions W_i, $i = 0, 1, 2$, by suitably chosen "rays" r_i, $i = 0, 1, 2$, which are arcs of great circles with one end at b and the other in the great circle which bounds H. Then every point in H is contained in at least one of the regions W_i, $i = 0, 1, 2$, and, since H is an open hemisphere, b is the only point in $W_0 \cap W_1 \cap W_2$. Let the regions U_i, $i = 0, 1, 2$, be defined as in (9) in Section 24; let the Sperner functions $s_k: V_k \to \{e_0, e_1, e_2\}$ be defined as in (10) in Section 24; then the results in Sections 24 and 25 hold without change. State and prove the intermediate-value theorem for the function $f: A \to H$.

CHAPTER 5

The Inverse-Function Theorem

30. Introduction

A function f with domain A and range B is a set $\{(x, y) : x \in A, y \in B\}$ of ordered pairs (x, y) in $A \times B$ which satisfies the following condition [see Section 96 in Appendix 2]: if (x, y) and (x, y') are in f, then $y' = y$. We say f maps A into B and write $f: A \to B$. If (x, y) is in f, we say y is the value of f at x and write $y = f(x)$. Then f is the set $\{(x, f(x)) : x \in A\}$, and $B = \{f(x) : x \in A\}$.

30.1 Definition. If f is a function $\{(x, y) : x \in A, y \in B\}$, and if $\{(y, x) : (x, y) \in f\}$ is a function g, then g is called the *inverse function* of the function f, and it is denoted by f^{-1}.

If f is the function $f: A \to B$, and if f has the inverse $f^{-1} : B \to A$, then $y = f(x)$ and $x = f^{-1}(y)$, and

$$\begin{aligned} x &= f^{-1}[f(x)], \qquad x \in A, \\ y &= f[f^{-1}(y)], \qquad y \in B. \end{aligned} \tag{1}$$

If f^{-1} is the inverse of f, then f is the inverse of f^{-1}. Some functions do not have an inverse. For example, $\{(1, 6), (2, 7), (3, 6), (4, 8), (5, 10)\}$ is a function, but $\{(6, 1), (7, 2), (6, 3), (8, 4), (10, 5)\}$ is not a function because the set contains the pairs $(6, 1)$ and $(6, 3)$, which do not satisfy the condition for a function. A function $f: A \to B$ has an inverse $f^{-1} : B \to A$ if and only if, for each y in B, the equation $f(x) = y$ has a unique solution for x. A function f has an inverse f^{-1} if and only if f maps every pair of distinct points x_1, x_2 in A into distinct points $f(x_1), f(x_2)$ in B; if f maps distinct points in A into distinct points in B, then $f: A \to B$ is called a one-to-one mapping of A onto B.

30.2 Examples. The student is familiar with inverse functions from his study of elementary calculus. If A and B are sets in $\mathbb{R}$, and if $f: A \to B$, with domain A and range B, is strictly monotonically increasing (or strictly monotonically decreasing), then f has an inverse $f^{-1}: B \to A$ since f maps distinct points x_1, x_2 in its domain into distinct points $f(x_1), f(x_2)$ in its range. For example, the natural logarithm function is defined and strictly monotonically increasing on its domain $x > 0$; its range is $-\infty < x < \infty$. Then the natural logarithm function $\log_e$ has an inverse; this inverse is the exponential function exp, and its domain is $-\infty < x < \infty$. For these functions, the relations in (1) are the following:

$$\begin{aligned} \exp[\log_e x] &= x, \qquad x > 0, \\ \log_e[\exp y] &= y, \qquad -\infty < y < \infty. \end{aligned} \tag{2}$$

The sine function does not have an inverse because $\sin(x + 2\pi) = \sin x$ for every x. However, the sine function, restricted to the interval $-\frac{\pi}{2} \leqq x \leqq \frac{\pi}{2}$, is strictly monotonically increasing; the inverse of this function is the Arc sine function, defined on $-1 \leqq y \leqq 1$. Finally, let $f: \mathbb{R}^2 \to \mathbb{R}^2$ be the function which is defined by the following equations:

$$\begin{aligned} y_1 &= a_{11}x_1 + a_{12}x_2 + b_1, \\ y_2 &= a_{21}x_1 + a_{22}x_2 + b_2, \end{aligned} \qquad \begin{vmatrix} a_{11} & a_{12} \\ a_{21} & a_{22} \end{vmatrix} \neq 0. \tag{3}$$

Then f has an inverse f^{-1} since the equations in (3) have a unique solution for (x_1, x_2) for each point (y_1, y_2) in $\mathbb{R}^2$; this inverse can be found explicitly by solving equations (3) for x_1, x_2 by Cramer's rule. In most cases the equation $f(x) = y$ cannot be solved explicitly to prove the existence of an inverse f^{-1} of f; instead, special methods are used to prove the existence of a solution. The intermediate-value theorem [see Theorem 27.1] is a tool which is frequently used for this purpose.

30.3 Example. Let $A = \mathbb{R}^2$ and $B = \{(y_1, y_2) : (y_1, y_2) \in \mathbb{R}^2, y_1 \geqq 0\}$, and let $f: A \to B$ be the function defined by the following equations:

$$\begin{aligned} y_1 &= x_1^2, \\ y_2 &= x_2, \end{aligned} \qquad (x_1, x_2) \in \mathbb{R}^2. \tag{4}$$

This function does not have an inverse since $f(-x_1, x_2) = f(x_1, x_2)$ for every (x_1, x_2) in $\mathbb{R}^2$. However, the function f, restricted to the set $x_1 \geqq 0$ (or to the set $x_1 \leqq 0$) has an inverse. Thus f does not have an inverse, but f—restricted to a neighborhood $N(p_0, r)$ which contains no point on the line $x_1 = 0$—does have an inverse. The inverse-function theorem is concerned with local inverses in a neighborhood of a point rather than with inverses in the large.

Let $U = N(x_0, a) = \{x : x \in \mathbb{R}^n, |x - x_0| < a\}$ for some $a > 0$, and let $f^1, \cdots, f^n$ be the components of a function $f: U \to \mathbb{R}^n$. Then f maps U onto a set V in $\mathbb{R}^n$. Set $f(x_0) = y_0$. The inverse-function theorem states sufficient conditions under which (a) there exists a $\delta > 0$ such that f, restricted to a sufficiently small neighborhood of x_0, has an inverse $g: N(y_0, \delta) \to U$, and (b) the components $g^1, \cdots, g^n$ of g are differentiable at y_0. Section 31 establishes the inverse-function theorem in the one-dimensional case under three different sets of hypotheses. The purpose of this chapter is to prove the inverse-function theorem in the n-dimensional case under three corresponding sets of hypotheses. There are three steps in the proofs. The first establishes conditions on f under which V contains a neighborhood $N(y_0, \delta)$ of $f(x_0)$. The intermediate-value theorem in Theorem 27.1 is the fundamental tool in this step. The second step establishes conditions, using Theorems 12.2 and 12.4, under which f is one-to-one in a neighborhood of x_0 so that an inverse $g: N(y_0, \delta) \to U$ of f exists in a neighborhood of y_0. The third step establishes conditions on f under which the components $g^1, \cdots, g^n$ of g are differentiable at y_0; the theory of differentiation developed in Chapter 1 is an essential tool in this step. In particular, the third step uses the definition of the derivative in Definition 2.8 and also the fact that a differentiable function satisfies the Stolz condition [see Definition 3.1]. The emphasis throughout the chapter is on theorems with the weakest possible hypotheses. Theorem 31.1, which states the inverse-function theorem in the one-dimensional case, indicates the nature of the theorems which are proved in the n-dimensional case, but it gives only a small hint about the methods of proof which are required in the general case.

Exercises

30.1. Let A and B be, respectively, the (non-degenerate) Euclidean simplexes $[a_0, a_1, a_2]$ and $[b_0, b_1, b_2]$ in $\mathbb{R}^2$, and let $f:(f^1, f^2)$ be the affine transformation which maps points $x:(x^1, x^2)$ into points $y:(y^1, y^2)$, and which transforms A into B so that $f(a_i) = b_i$, $i = 0, 1, 2$. Then there are constants c_{ij} such that [see (4) in Definition 19.2, and Theorem 19.3]

$$y^1 = f^1(x) = c_{11}x^1 + c_{21}x^2 + c_{31},$$

$$y^2 = f^2(x) = c_{12}x^1 + c_{22}x^2 + c_{32}.$$

(a) Find $D_{(1,2)}(f^1, f^2)(x)$ and show that $D_{(1,2)}(f^1, f^2)(x) \neq 0$ for x in A. [Hint. Theorem 19.6.]

(b) Use the intermediate-value theorem in Section 27 to show that every point y in the interior of B is the image $f(x)$ of a point x in the interior of A. [Hint. Example 27.2.]

(c) Use Theorem 12.2 to prove that the mapping $f: A \to B$ is one-to-one. Then prove that the function $f: A \to B$ has an inverse $g: B \to A$.

(d) Find explicit formulas for g^1 and g^2, the components of g, by solving for $x:(x^1, x^2)$ the system of equations $y = f(x)$ in the introduction of this exercise. Use the results to show that g is continuous and differentiable.

(e) Find $D_{(1,2)}(g^1, g^2)(y)$ and show that

$$D_{(1,2)}(f^1, f^2)(x) D_{(1,2)}(g^1, g^2)(y) = 1$$

for all x and y such that $y = f(x)$.

30.2. Consider Exercise 30.1 again. Let y_0 and $2r$ be the center and radius, respectively, of the inscribed circle of the simplex B, and let $h: A \to \mathbb{R}^2$ be a continuous function such that $|h(x)| < \sqrt{3}r/2$ for x in ∂A. Prove that $N(y_0, r) \subset (f+h)(A)$. [Hint. Example 27.6.]

31. The One-Dimensional Case

This section treats the one-dimensional case of the inverse-function theorem. The student is already somewhat familiar with this case from the study of elementary calculus. A real-valued function f which is defined on an interval in $\mathbb{R}$ has an inverse in the neighborhood of a point if and only if f is a strictly increasing or strictly decreasing function in the neighborhood. This statement is used to show that the logarithm function $\log_e$, the exponential function exp, and the sine function restricted to $-\frac{\pi}{2} \leqq x \leqq \frac{\pi}{2}$, have inverses. Also, if f has a derivative at x_0, then the inverse function g has a derivative at the corresponding point y_0 and $f'(x_0)g'(y_0) = 1$. This section establishes three sufficient conditions for the existence of the inverse g of f in a neighborhood of x_0, and a sufficient condition for the differentiability of g at y_0.

31.1 Theorem (Inverse-Function Theorem). *Let U be the interval $(x_0 - a, x_0 + a)$ in $\mathbb{R}$, and let $f: U \to V$, $V \subset \mathbb{R}$, be a function which satisfies at least one of the following conditions:*

$$f \text{ is continuous and strictly increasing (or strictly decreasing) on } U;\ f \text{ has a derivative at the single point } x_0 \text{ and } f'(x_0) \neq 0. \tag{1}$$

$$f \text{ has a derivative } f' \text{ on } U, \text{ and } f'(x) \neq 0 \text{ for } x \text{ in } U. \tag{2}$$

$$f \text{ has a derivative } f' \text{ on } U;\ f'(x_0) \neq 0;\ \text{and } f' \text{ is continuous at } x_0. \tag{3}$$

Let $y_0 = f(x_0)$. Then there is an interval $I = (x_0 - \varepsilon, x_0 + \varepsilon)$, $I \subset U$, on which $f: I \to V$ has an inverse $g: f(I) \to U$, and $f(I)$ is an interval $(y_0 - \delta, y_0 + \delta)$. Finally, $g'(y_0)$ exists, and $f'(x_0)g'(y_0) = 1$.

Proof. Assume first that f satisfies (1). Since f is a strictly increasing (or decreasing) function, I can be taken as the interval U. Because f is continuous on U by hypothesis, the elementary intermediate-value theorem shows that $f(I)$ is an interval (which may be infinite in one or both directions). Then $f: U \to V$ has a well-defined inverse $g: f(I) \to U$ on the interval $f(I)$.

An elementary proof shows that $g: f(I) \to U$ is continuous because $f: I \to V$ is continuous and strictly monotonic. The following special proof that the differentiability of f at x_0 implies that g is continuous at y_0, where $y_0 = f(x_0)$, is similar to a proof in Section 33 in the general case. Since $f'(x_0)$ exists by hypothesis,

$$f(x) - f(x_0) = f'(x_0)(x - x_0) + r(f; x_0, x)|x - x_0|, \tag{4}$$

$$\big||f'(x_0)| - |r(f; x_0, x)|\big|\,|x - x_0| \leqq |f(x) - f(x_0)|. \tag{5}$$

Now $|f'(x_0)| > 0$ by hypothesis; choose $\eta > 0$ so that

$$|r(f; x_0, x)| < (1/2)|f'(x_0)|, \qquad |x - x_0| < \eta. \tag{6}$$

Then by (5) and (6),

$$|x - x_0| \leqq [2/f'(x_0)]|f(x) - f(x_0)|, \qquad |x - x_0| < \eta. \tag{7}$$

Since $y = f(x)$ and $x = g(y)$, the same relationship can be stated as follows:

$$|g(y) - g(y_0)| \leqq [2/f'(x_0)]|y - y_0| \tag{8}$$

for all y between $f(x_0 - \eta)$ and $f(x_0 + \eta)$. This statement implies that g is continuous at y_0.

Since g is continuous at y_0, the equations $x_0 = g(y_0)$ and $x = g(y)$ show that x approaches x_0 as y approaches y_0. Then since each of the equations $y = f(x)$ and $x = g(y)$ implies the other,

$$g'(y_0) = \lim_{y \to y_0} \frac{g(y) - g(y_0)}{y - y_0} = \lim_{x \to x_0} \frac{x - x_0}{f(x) - f(x_0)} = \frac{1}{f'(x_0)}. \tag{9}$$

Thus $g'(y_0)$ exists, and $f'(x_0)g'(y_0) = 1$. If $g'(y_0)$ is known to exist, a second proof of the equation $f'(x_0)g'(y_0) = 1$ follows from an application of the chain rule to the identity $g[f(x)] = x$ [see (1) in Section 30] to obtain $g'[f(x_0)]f'(x_0) = 1$. The proof of Theorem 31.1 in case (1) is satisfied is complete.

Assume next that f satisfies (2). Since f is differentiable on U, then f is continuous on U. Since $f'(x) \neq 0$ for x in U by hypothesis, either $f'(x) > 0$ for all x in U or $f'(x) < 0$ for all x in U; for, by Darboux's theorem in Exercise 7.10, if $f'(x_1)$ and $f'(x_2)$ had opposite signs, then $f'(x)$ would equal zero—contrary to hypothesis—at some point between x_1 and x_2. Thus f' is either positive in U or it is negative in U, and f is either strictly increasing in U or strictly decreasing in U by the mean-value theorem. Thus condition (2) implies (1), and the proof of Theorem 31.1 in this case follows from the first case.

Finally, if f' is continuous at x_0 and $f'(x_0) \neq 0$, there is an interval $(x_0 - \varepsilon, x_0 + \varepsilon)$ in U in which $f'(x_0)$ is not zero [see Theorem 96.9 in Appendix 2]. Then the proof of Theorem 31.1, with the hypothesis stated in (3), follows from the second case, and the proof of the entire theorem is complete. □

31.2 Example. The integral $\int_1^x (1/t)\,dt$ exists for every $x > 0$ since its integrand is continuous. This integral defines a function f as follows:

$$f(x) = \int_1^x \frac{dt}{t}, \qquad x > 0. \tag{10}$$

The domain of this function is the set $\{x : x > 0\}$, and its range is the set $\{y : -\infty < y < \infty\}$. Then

$$f'(x) = \frac{1}{x} > 0, \qquad x > 0, \tag{11}$$

and f is a continuous function which is strictly increasing on its domain. Then f has an inverse g whose domain is the set $\{y : -\infty < y < \infty\}$ and whose range is the set $\{x : x > 0\}$. Since f is differentiable at each point in its domain, Theorem 31.1 shows that g is differentiable at each point in its domain and that $f'(x_0)g'(y_0) = 1$; hence, by (11),

$$g'(y_0) = \frac{1}{f'(x_0)} = x_0. \tag{12}$$

Since g is the inverse of f, then $y_0 = f(x_0)$ and $x_0 = g(y_0)$. Thus (12) can be written as follows:

$$g'(y_0) = g(y_0). \tag{13}$$

Now f is usually called the natural logarithm function and denoted by $\log_e$; and g, the inverse of f, is called the exponential function and denoted by exp. Thus

$$\log_e x = \int_0^x \frac{dt}{t}, \qquad x > 0, \tag{14}$$

and the formulas in (11) and (13) are the following familiar differentiation formulas.

$$\frac{d}{dx}\log_e x = \frac{1}{x}, \qquad x > 0, \tag{15}$$

$$\frac{d}{dx}\exp x = \exp x, \qquad -\infty < x < \infty. \tag{16}$$

The reader knows from elementary calculus that $\exp x = e^x$.

Exercises

31.1. Let $f: \mathbb{R} \to \mathbb{R}$ be a function such that $f(x) = (1/4)x + x^2 \sin(1/x)$ if $x \neq 0$ and $f(0) = 0$.
- (a) Show that (i) f is continuous for every x in $\mathbb{R}$; (ii) f has a derivative f' for every x; and (iii) $f'(0) = 1/4$.
- (b) Prove that there is no neighborhood of the origin in which f has an inverse function.
- (c) Explain why the result in (b) does not contradict Theorem 31.1.

31.2. Let $f: \mathbb{R} \to \mathbb{R}$ be a function such that $f(x) = 2x + x^2 \sin(1/x)$ if $x \neq 0$, and $f(0) = 0$. Prove the following: (i) f is a strictly increasing function on the interval $(-1/4, 1/4)$; (ii) f has an inverse g on $(-1/4, 1/4)$; and (iii) g has a derivative g' and $g'(0) = 1/2$.

31.3. The function $f: \mathbb{R} \to \mathbb{R}$ such that $f(x) = x^3$ is a strictly increasing function on $\mathbb{R}$, and it therefore has an inverse function $g: \mathbb{R} \to \mathbb{R}$. Show that $g'(0)$ does not exist and explain why. Find an algebraic representation—a formula—for g.

31.4. Let $f: I \to V$ be the function described in Theorem 31.1, and let $g: f(I) \to U$ be the inverse of f. Then $y = f(x)$, $y_0 = f(x_0)$, and $x = g(y)$, $x_0 = g(y_0)$. The following is an outline of another proof that g is differentiable at y_0 and that $f'(x_0)g'(y_0) = 1$. The method explained here will be used in Section 33 to prove the differentiability of the inverse function in the n-dimensional case.

(a) Use equation (8) in Section 31 to prove that g is continuous at y_0.

(b) Since $f'(x_0)$ exists, show that f satisfies the following Stolz condition:

$$f(x) - f(x_0) = f'(x_0)(x - x_0) + r(f; x_0, x)|x - x_0|, \qquad \lim_{x \to x_0} r(f; x_0, x) = 0.$$

(c) In the Stolz condition in (b), replace x and x_0 by $g(y)$ and $g(y_0)$ respectively to obtain

$$f[g(y)] - f[g(y_0)]$$
$$= f'[g(y_0)][g(y) - g(y_0)] + r[f; g(y_0), g(y)]|g(y) - g(y_0)|,$$
$$\lim_{y \to y_0} r[f; g(y_0), g(y)] = 0.$$

(d) In (c), replace $f[g(y)]$ and $f[g(y_0)]$ by y and y_0 respectively [see equation (1) in Section 30]. Then divide the equation by $y - y_0$ and take the limit as $y \to y_0$.

(e) Use (c), equation (8), and (d) to show that

$$\lim_{y \to y_0} r[f; g(y_0), g(y)] \frac{|g(y) - g(y_0)|}{y - y_0} = 0,$$

and hence that

$$\lim_{y \to y_0} \frac{g(y) - g(y_0)}{y - y_0} \text{ exists}, \qquad f'(x_0)g'(y_0) = 1.$$

32. The First Step: A Neighborhood is Covered

Let $U = N(x_0, a) = \{x : x \in \mathbb{R}^n, |x - x_0| < a\}$ for some $n \geqq 2$ and some $a > 0$, and let $f^1, \cdots, f^n$ be the components of a function $f: U \to \mathbb{R}^n$. Then f maps U onto a set V in $\mathbb{R}^n$; let $f(x_0) = y_0$. The purpose of this section is to use the intermediate-value theorem and other results in Chapter 4 to prove that $f(U)$, under certain hypotheses on f, contains a neighborhood $N(y_0, \delta)$ of y_0. A lemma required for the proof of the principal theorem begins the section.

Let $[a_{ij}]$, $i, j = 1, \cdots, n$, be a matrix of constants such that

$$\det [a_{ij}] \neq 0. \tag{1}$$

The equations

$$y^i = \sum_{j=1}^{n} a_{ij} x^j, \qquad i = 1, \cdots, n, \tag{2}$$

define a linear transformation $L: \mathbb{R}^n \to \mathbb{R}^n$ which has the following properties.

$$L(tx) = tL(x), \qquad x \in \mathbb{R}^n, \qquad t \in \mathbb{R}. \tag{3}$$

$$L(x_1 + x_2) = L(x_1) + L(x_2), \qquad x_1, x_2 \quad \text{in} \quad \mathbb{R}^n. \tag{4}$$

Intuitively, the transformation described by (1) and (2) transforms the unit sphere in $\mathbb{R}^n$ into an ellipsoid in $\mathbb{R}^n$, and this fact can be used to obtain useful bounds for $|L(x)|$, $x \in \mathbb{R}^n$. The precise results are stated in the following lemma.

32.1 Lemma. *If $L: \mathbb{R}^n \to \mathbb{R}^n$ is the linear function defined by* (1) *and* (2), *then there exist constants $m > 0$ and $M > 0$ such that*

$$m|x| \leqq |L(x)| \leqq M|x| \tag{5}$$

for every x in $\mathbb{R}^n$.

Proof. Let S denote the sphere $\{x : x \in \mathbb{R}^n, |x| = 1\}$ with radius 1 and center at the origin in $\mathbb{R}^n$. The sphere S is closed and bounded and therefore compact by Definition 92.8. Next, the following arguments show that the norm of L is continuous on S. By (4) and the triangle inequality for the norm in $\mathbb{R}^n$,

$$\big||L(x_2)| - |L(x_1)|\big| \leqq |L(x_2 - x_1)|. \tag{6}$$

Schwarz's inequality can be used as in Example 21.8 to show that there exists a constant C such that $|L(x_2 - x_1)| \leqq C|x_2 - x_1|$. Then by (6),

$$\big||L(x_2)| - |L(x_1)|\big| \leqq C|x_2 - x_1|, \tag{7}$$

and this inequality shows that $|L|: S \to \mathbb{R}$ is a continuous function. Now by Theorem 96.14, a continuous function on a compact set has a maximum and a minimum value; hence, there exist (i) constants $m \geqq 0$ and $M \geqq 0$ such that $m \leqq M$ and

$$m \leqq |L(x)| \leqq M, \qquad x \in S, \tag{8}$$

and (ii) points x_1, x_2 in S such that

$$|L(x_1)| = m, \qquad |L(x_2)| = M. \tag{9}$$

To show that $m > 0$, assume that $m = 0$ and show that this assumption leads to a contradiction. If $m = 0$, then by (9) there is a point x in S such that $|L(x)| = 0$ and $L(x) = 0$. But if $y^1 = \cdots = y^n = 0$ in (2), then the hypothesis $\det [a_{ij}] \neq 0$ in (1) shows that $x^1 = \cdots = x^n = 0$, or $x = 0$. This statement, however, contradicts the assumption that $x \in S$ and hence that $|x| = 1$. Thus $m > 0$ since the assumption that $m = 0$ leads to a contradiction.

Next, let x be an arbitrary point such that $|x| > 0$ in $\mathbb{R}^n$. Then $x/|x|$ is a point in S, and

$$m \leqq \left| L\left(\frac{x}{|x|}\right) \right| \leqq M \tag{10}$$

by (8). But L is homogeneous by (3), and $L(x/|x|) = L(x)/|x|$. Then (10) is equivalent to the following equation.

$$m|x| \leqq |L(x)| \leqq M|x|. \tag{11}$$

Finally, if $x = 0$, then $L(x) = 0$ by (2), and (11) is true in this case also. The proof of the lemma is complete. □

32.2 Theorem. *If the function $f: U \to V$ is continuous on U, if the functions $f^1, \cdots, f^n$ are differentiable at x_0, and if $D_{(1,\cdots,n)} f(x_0) \neq 0$, then there exists a $\delta > 0$ such that $N(y_0, \delta) \subset f(U)$.*

Proof. Example 27.6 is a model for the proof of this theorem and should be reviewed at this time. Since f has a derivative at x_0, the Stolz condition can be used to show that the mapping $f: U \to V$ differs only slightly from an affine transformation in a small neighborhood of x_0. Then, as in Example 27.6, the intermediate-value theorem and other results in Chapter 4 are used to show that $f(U)$ contains a small neighborhood of $f(x_0)$.

Let $x = (x^1, \cdots, x^n)$ and $x_0 = (x_0^1, \cdots, x_0^n)$; then Theorem 3.6 shows that f^i satisfies the following Stolz condition:

$$f^i(x) - f^i(x_0) = \sum_{j=1}^{n} D_j f^i(x_0)(x^j - x_0^j) + r(f^i; x_0, x)|x - x_0|, \qquad i = 1, \cdots, n. \tag{12}$$

This system of equations in briefer notation is

$$f(x) - f(x_0) = H(x - x_0) + R(x_0, x). \tag{13}$$

Here $H: \mathbb{R}^n \to \mathbb{R}^n$ is a transformation which is linear and homogeneous in $x - x_0$ such that

$$H(x - x_0) = \left(\sum_{j=1}^{n} D_j f^1(x_0)(x^j - x_0^j), \cdots, \sum_{j=1}^{n} D_j f^n(x_0)(x^j - x_0^j) \right), \tag{14}$$

and

$$R(x_0, x) = (r(f^1; x_0, x)|x - x_0|, \cdots, r(f^n; x_0, x)|x - x_0|). \tag{15}$$

Since by Theorem 3.12 the determinant of H in (14) is $D_{(1,\cdots,n)} f(x_0)$, which is not zero by hypothesis, then Lemma 32.1 shows that there exist positive constants m and M such that

$$m|x - x_0| \leqq |H(x - x_0)| \leqq M|x - x_0|, \qquad x \in U. \tag{16}$$

Now

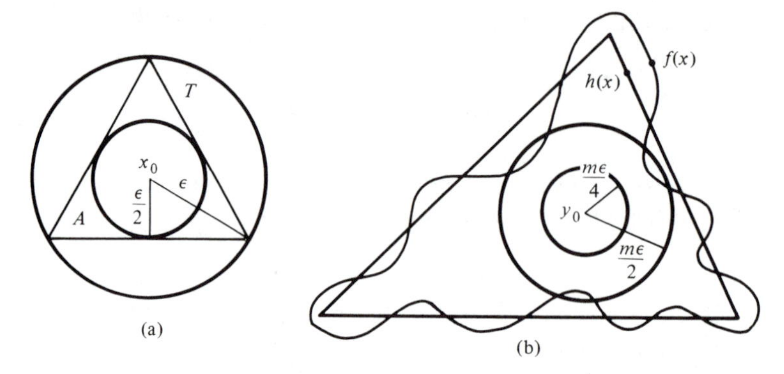

Figure 32.1. The images of T under h and f.

$$|R(x_0, x)| = \left\{ \sum_{i=1}^{n} [r(f^i; x_0, x)]^2 \right\}^{1/2} |x - x_0|, \qquad x \in U, \tag{17}$$

and Theorem 3.6 and Definition 3.1 show that

$$\lim_{x \to x_0} \left\{ \sum_{i=1}^{n} [r(f^i; x_0, x)]^2 \right\}^{1/2} = 0. \tag{18}$$

Consider the transformations $f: U \to V$ and $h: U \to \mathbb{R}^n$ such that

$$\begin{aligned} f(x) &= f(x_0) + H(x - x_0) + R(x_0, x), \\ h(x) &= f(x_0) + H(x - x_0). \qquad x \in U, \end{aligned} \tag{19}$$

Here $h: \mathbb{R}^n \to \mathbb{R}^n$ is an affine transformation, and the first equation shows that f differs only slightly from the affine transformation h in the neighborhood of x_0 since $f(x) - h(x) = R(x_0, x)$.

The remainder of the proof of Theorem 32.2 will be given in the special case in which $n = 2$. The methods are general, however, and they can be used to prove the theorem for any value of n for which the results of Chapter 4 have been established.

Choose ε so that $0 < \varepsilon < a$ and so that

$$\left\{ \sum_{i=1}^{2} [r(f^i; x_0, x)]^2 \right\}^{1/2} < \frac{m\sqrt{3}}{8}, \qquad |x - x_0| \leqq \varepsilon. \tag{20}$$

This choice of ε is possible by (18). Let T denote (the boundary of) an equilateral triangle inscribed in the circle with center x_0 and radius ε [see Figure 32.1(a)]. Then the circle inscribed in T has center x_0 and radius $\varepsilon/2$, and

$$\frac{\varepsilon}{2} \leqq |x - x_0| \leqq \varepsilon, \qquad x \in T. \tag{21}$$

Then

$$|R(x_0, x)| < \frac{m\varepsilon\sqrt{3}}{8}, \qquad x \in T, \tag{22}$$

by (17), (20), and (21). Also,

$$|f(x) - h(x)| = |R(x_0, x)| < \frac{m\varepsilon\sqrt{3}}{8}, \qquad x \in T, \tag{23}$$

by (19) and (22). Next,

$$\begin{aligned} |h(x) - f(x_0)| = |H(x - x_0)| &\geqq m|x - x_0|, && x \in U, \\ &\geqq \frac{m\varepsilon}{2}, && x \in T, \end{aligned} \tag{24}$$

by (16) and (21). Finally,

$$\begin{aligned} |f(x) - f(x_0)| &\geqq \big||h(x) - f(x_0)| - |f(x) - h(x)|\big|, \\ &> \frac{m\varepsilon}{2} - \frac{m\varepsilon\sqrt{3}}{8}, && x \in T, \\ &> \frac{m\varepsilon}{4}, && x \in T, \end{aligned} \tag{25}$$

by (24) and (23). These results can be summarized as follows [see Figure 32.1(b)]: the affine transformation h maps the triangle T into a triangle which lies outside the circle with center y_0 and radius $m\varepsilon/2$ [see (24)]; f maps T into a curve which lies outside the circle with center y_0 and radius $m\varepsilon/4$ [see (25)]; and the distance between $f(x)$ and $h(x)$ is less than $m\varepsilon\sqrt{3}/8$ for all x in T [see (23)]. Set $\delta = m\varepsilon/4$; the facts just summarized and the results in Chapter 4 will now be used to show that $N(y_0, \delta) \subset f(U)$.

Let A be the closed Euclidean simplex whose boundary is T, and consider the mappings $f: A \to \mathbb{R}^2$ and $h: A \to \mathbb{R}^2$. Let $\mathscr{P}_1, \mathscr{P}_2, \cdots$ be a set of simplicial subdivisions of A which has properties (1), (2), (3) in Section 24. Use these subdivisions to find $d(h, y_0)$ and $d(f, y_0)$ as in Chapter 4 [see especially Exercise 25.3, Theorem 26.2, and also Exercise 30.2]. Chapter 4 has shown that $d(h, y_0) = +1$ if $D_{(1,2)}f(x_0) > 0$ and $d(h, y_0) = -1$ if $D_{(1,2)}f(x_0) < 0$ [see equation (14) in Section 19, Exercise 25.3, and Example 27.6]. Thus $d(h, y_0) \neq 0$ since $D_{(1,2)}f(x_0) \neq 0$ by hypothesis. Finally, (23), (24), (25), and Theorem 26.2 show that $d(f, y_0) = d(h, y_0)$. Thus $d(f, y_0) \neq 0$ for the mapping $f: A \to \mathbb{R}^2$.

The complement in $\mathbb{R}^2$ of the closed set $f(T)$ is an open set. Theorem 25.3 shows that, if two points can be connected by a polygonal curve which contains no point in $f(T)$, then the degree of f is the same at these two points. In particular, if y can be connected to y_0 by a polygonal curve which does not intersect $f(T)$, then $d(f, y) = d(f, y_0)$. But if $\delta = m\varepsilon/4$, and if y is in $N(y_0, \delta)$, then (25) shows that the segment $[y_0, y]$ does not intersect $f(T)$. Therefore, $d(f, y) \neq 0$ at every point y in $N(y_0, \delta)$, and Theorem 27.1, the intermediate-value theorem, shows that there exists an x in A such that $f(x) = y$ for every y in $N(y_0, \delta)$. Thus $N(y_0, \delta) \subset f(U)$, and the proof of Theorem 32.2 is complete in the case $n = 2$. □

32.3 Corollary. *Let $f: U \to \mathbb{R}^n$ be defined and continuous on $U = N(x_0, a)$, and let $f^1, \cdots, f^n$ be differentiable at x_0. If $f(x_0)$ is a boundary point of $f(U)$, then $D_{(1,\cdots,n)} f(x_0) = 0$.*

Proof. Assume that $D_{(1,\cdots,n)} f(x_0) \neq 0$. Then Theorem 32.2 shows that there is a neighborhood $N[f(x_0), \delta]$ of $f(x_0)$ which is contained in $f(U)$. Then $f(x_0)$ is not a boundary point of $f(U)$ [see Definition 92.7]. This contradiction of the hypothesis shows that $D_{(1,\cdots,n)} f(x_0) = 0$ and completes the proof. □

32.4 Example. Let $f: \mathbb{R}^2 \to \mathbb{R}^2$ be the function defined by the following equations:

$$\begin{aligned} y^1 &= (x^1 - 1)^2, \\ y^2 &= (x^2 - 1)^2, \qquad (x^1, x^2) \in R^2. \end{aligned} \tag{26}$$

Let x_0 be the point (2,2). Then f maps x_0 into $y_0 : (1,1)$. Since

$$D_{(1,2)} f(x_0) \doteq \begin{vmatrix} 2 & 0 \\ 0 & 2 \end{vmatrix} = 4 \neq 0,$$

then Theorem 32.2 asserts that there is a neighborhood of y_0 which is covered by $f(\mathbb{R}^2)$. One such neighborhood is $N(y_0, 1)$, because, if y is in $N(y_0, 1)$, then the equations in (26) can be solved explicitly to find the four points $(1 \pm \sqrt{y^1}, 1 \pm \sqrt{y^2})$ which are mapped by f into $y : (y^1, y^2)$. The boundary of $f(\mathbb{R}^2)$ is the set $\{(y^1, 0) : y^1 \geqq 0\} \cup \{(0, y^2) : y^2 \geqq 0\}$ in $\mathbb{R}^2$. Then $f(x^1, x^2)$ is a boundary point of $f(\mathbb{R}^2)$ if and only if $x^1 = 1$ or $x^2 = 1$. Now $D_{(1,2)} f(x) = 4(x^1 - 1)(x^2 - 1)$. Then, as required by Corollary 32.3, the derivative $D_{(1,2)} f(x)$ is zero at each point x such that $f(x)$ is a boundary point of $f(\mathbb{R}^2)$.

Exercises

32.1. (a) Prove Theorem 32.2 in the special case $n = 1$; that is, prove the following theorem. Let $U = \{x : x \in R, |x - x_0| < a\}$ and $V = \{y : y \in R, |y - y_0| < b\}$. Let $f: U \to V$ be a continuous function such that $y_0 = f(x_0)$. If f has a derivative $D_1 f(x_0)$ at x_0 and $D_1 f(x_0) \neq 0$, then there is a $\delta > 0$ such that $\{y : y \in R, |y - y_0| < \delta\}$ is contained in $f(U)$.

(b) Prove Corollary 32.3 in the special case $n = 1$; that is, prove the following corollary of the theorem in (a). Let $f: U \to V$ be a continuous function on U which has a derivative at x_0. If $f(x_0)$ is a boundary point of $f(U)$ (a maximum or minimum value of f), then $D_1 f(x_0) = 0$.

32.2. Show that the corollary in Exercise 32.1(b) contains the following elementary theorem. Let $f: U \to R$ be defined and continuous on $U = (x_0 - a, x_0 + a)$, and assume that f has a derivative $D_1 f(x_0)$ at x_0. If f has a relative maximum or minimum at x_0, then $D_1 f(x_0) = 0$.

32.3. Let $f: U \to \mathbb{R}^2$ be a function with components (f^1, f^2) such that, for some constants a_1, a_2, a_3,

$$f^1(x^1, x^2) = (a_1 x^1 + a_2 x^2 + a_3)^2,$$
$$f^2(x^1, x^2) = 2x^1 x^2.$$

(a) Show that f is differentiable in U, and find $D_{(1,2)}f(x)$.
(b) If $x:(x^1, x^2)$ is a point in U such that $a_1 x^1 + a_2 x^2 + a_3 = 0$, show that $f(x)$ is a boundary point of $f(U)$ and verify Corollary 32.3 by showing that $D_{(1,2)}f(x) = 0$.

32.4. Prove the following corollary of Theorem 32.2. If $f: U \to V$ is continuous on U and differentiable at x_0, and if $D_{(1,\cdots,n)}f(x_0) \neq 0$, then there exists an ε such that (i) $0 < \varepsilon < a$ and (ii) $f(x) \neq f(x_0)$ for $0 < |x - x_0| \leqq \varepsilon$. [Hint. In the special case $n = 2$, choose ε as in (20) and prove that

$$|h(x) - f(x_0)| \geqq m|x - x_0|, \qquad x \in U,$$

$$|f(x) - h(x)| < \frac{m\sqrt{3}}{8}|x - x_0|, \qquad 0 < |x - x_0| \leqq \varepsilon.$$

Then

$$|f(x) - f(x_0)| > \frac{3m}{4}|x - x_0|, \qquad 0 < |x - x_0| \leqq \varepsilon.]$$

32.5. Give a second proof of the corollary in Exercise 32.4 as follows. Assume the corollary false; then in every neighborhood of x_0 there exists a point x such that $f(x) = f(x_0)$. Use the definition of $D_{(1,\cdots,n)}f(x_0)$ in Definition 2.8 to show that $D_{(1,\cdots,n)}f(x_0) = 0$. The contradiction establishes the corollary.

32.6. Let $r = [(x^1)^2 + (x^2)^2]^{1/2}$, let x_0 denote the origin $(0, 0)$ in $\mathbb{R}^2$, and let $A = \{x : x \in \mathbb{R}^2, |x| \leqq 1\}$. Let $f: A \to \mathbb{R}^2$ be a function with components f^i, $i = 1, 2$, such that

$$f^i(x) = r^2 \sin(1/r), \qquad r \neq 0,$$
$$= 0, \qquad r = 0.$$

(a) Show by inspection that f^1 and f^2 satisfy the Stolz condition at x_0 and are therefore differentiable at x_0 [compare Exercise 3.5]. Show that $D_1 f^i(x_0)$ and $D_2 f^i(x_0)$, $i = 1, 2$, equal zero and hence that $D_{(1,2)}f(x_0) = 0$.
(b) Show that f maps all of the points on the circle $r = r_0$ into the same point. Show also that $f^1(x) = f^2(x)$ for every x in A and hence that f maps A into a segment of a straight line. Show that no neighborhood of $f(x_0)$ is covered by $f(A)$.
(c) Is $d(f, x_0)$ defined? If it is, what is its value?

32.7. Let r, A, and x_0 have the meanings given them in Exercise 32.6. Let $f: A \to \mathbb{R}^2$ have the following components:

$$f^1(x) = r^2 \cos(1/r), \qquad f^2(x) = r^2 \sin(1/r), \qquad r \neq 0,$$
$$f^1(x_0) = 0, \qquad f^2(x_0) = 0.$$

(a) Show that $f(A)$ is a curve which spirals, with infinitely many turns, into the origin. Show that there is no neighborhood of $f(x_0)$ which is contained in $f(A)$.
(b) Show that the following derivatives exist and have the value zero: $D_1 f^i(x_0)$, $D_2 f^i(x_0)$, $i = 1, 2$, and $D_{(1,2)}f(x_0)$.
(c) Show that $d(f, x_0)$ exists and has the value zero.

32.8. Define r, x_0, and A as in Exercise 32.6, and let $f: A \to \mathbb{R}^2$ be the function which has the following components:

$$f^1(x) = x^1 r \sin(1/r), \qquad f^2(x) = x^2 r \sin(1/r), \qquad r \neq 0,$$

$$f^1(x_0) = 0, \qquad f^2(x_0) = 0.$$

(a) Show that f is continuous on A, and that the derivatives $D_1 f^i(x_0)$, $D_2 f^i(x_0)$, and $D_{(1,2)}f(x_0)$ exist and have the value zero.
(b) Show that $d(f, x_0) = 1$ and hence that $f(A)$ contains the neighborhood of $f(x_0)$ which has radius $\sin(1)$. Use this example to show that the conditions in Theorem 32.2 are sufficient but not necessary.
(c) Show that there is no neighborhood of x_0 in which the mapping $f: A \to \mathbb{R}^2$ is one-to-one.

32.9. Let $A = \{x : x \in \mathbb{R}^2, |x| \leqq 1\}$, and let $f: A \to \mathbb{R}^2$ be the function whose components are defined as follows:

$$f^1(x) = (x^1)^2 - (x^2)^2, \qquad f^2(x) = 2x^1 x^2.$$

Here $f: A \to \mathbb{R}^2$ is the transformation $w = z^2$ in complex variables.
(a) Show that $f(A)$ contains the neighborhood $N[f(x_0), 1]$.
(b) Show that $D_{(1,2)}f(x)$ exists for all x in A and that its value is zero at $x_0 : (0, 0)$.
(c) Show that $d(f, x_0) = 2$, and that every point except $f(x_0)$ in $N[f(x_0), 1]$ is the image of two distinct points in A. Use this example to show that the conditions in Theorem 32.2 are sufficient but not necessary.

32.10. Let $A = [-1, 1] \times [-1, 1]$, and let $f: A \to \mathbb{R}^2$ be the function whose components are

$$f^1(x) = (x^1)^3, \qquad f^2(x) = (x^2)^3, \qquad x : (x^1, x^2) \text{ in } A.$$

(a) Show that $f: A \to \mathbb{R}^2$ is a one-to-one mapping of A into A. Find the inverse $g: A \to A$ of $f: A \to A$.
(b) Show that $D_{(1,2)}f(0, 0) = 0$ and that $D_{(1,2)}g(0, 0)$ does not exist.
(c) Consider Theorem 32.2. Use this example to show that (i) $f(U)$ may contain a neighborhood of $N(y_0, \delta)$ and (ii) f may have an inverse which is defined in a neighborhood of y_0, both in spite of the fact that $D_{(1,\cdots,n)}f(x_0) = 0$.

33. The Inverse-Function Theorem

The first theorem to be proved in this section is a generalization of the one-dimensional inverse-function theorem in Theorem 31.1 with the hypothesis stated in (1).

33.1 Theorem (Inverse-Function Theorem). *Let $f:(f^1, \cdots f^n)$ be a function $f: U \to V$ which maps $U = N(x_0, a)$ in $\mathbb{R}^n$ onto $V = f(U)$ in $\mathbb{R}^n$ and which has the following properties:*

$$f \text{ is continuous and one-to-one on } U. \tag{1}$$

$$f^1, \cdots, f^n \text{ are differentiable at the single point } x_0. \tag{2}$$

$$D_{(1,\cdots,n)}f(x_0) \neq 0. \tag{3}$$

Then $f: U \to V$ has an inverse $g: V \to U$; g is defined at least in some neighborhood $N(y_0, \delta)$ of $y_0 = f(x_0)$; the components $g^i: V \to R$, $i = 1, \cdots, n$, of g are differentiable at y_0, and

$$D_{(1,\cdots,n)}f(x_0)D_{(1,\cdots,n)}g(y_0) = 1. \tag{4}$$

Proof. The inverse function $g: V \to U$ exists because hypothesis (1) assumes that $f: U \to V$ is one-to-one; by Theorem 32.2 there is a neighborhood $N(y_0, \delta)$ on which g is defined. The first step in the proof that $g^1, \cdots, g^n$ are differentiable at y_0 is to show that $g: V \to U$ is continuous at y_0.

Let m and M be the constants in (16) in Section 32. Choose ε so that $0 < \varepsilon < a$ and so that

$$\left\{\sum_{i=1}^{n} [r(f^i; x_0, x)]^2\right\}^{1/2} \leqq \frac{m}{4n}, \qquad |x - x_0| \leqq \varepsilon. \tag{5}$$

This choice is possible by (18) in Section 32. Then

$$|R(x_0, x)| \leqq \frac{m}{4n}|x - x_0|, \qquad |x - x_0| \leqq \varepsilon, \tag{6}$$

by (17) in Section 32 and (5) in Section 33. Also, the methods used in Section 32 show that

$$|f(x) - h(x)| = |R(x_0, x)| \leqq \frac{m}{4n}|x - x_0|, \qquad |x - x_0| \leqq \varepsilon, \tag{7}$$

$$m|x - x_0| \leqq |h(x) - f(x_0)| = |H(x - x_0)| \leqq M|x - x_0|, \quad |x - x_0| \leqq \varepsilon. \tag{8}$$

Then

$$|f(x) - f(x_0)| = |[h(x) - f(x_0)] + [f(x) - h(x)]|,$$

$$||h(x) - f(x_0)| - |f(x) - h(x)|| \leqq |f(x) - f(x_0)|$$

$$\leqq |h(x) - f(x_0)| + |f(x) - h(x)|.$$

This inequality, together with those in (7) and (8), show that

$$\left(m - \frac{m}{4n}\right)|x - x_0| \leqq |f(x) - f(x_0)| \leqq \left(M + \frac{m}{4n}\right)|x - x_0|, \qquad |x - x_0| \leqq \varepsilon. \tag{9}$$

These inequalities will be used to show that $g: V \to U$ is continuous at y_0.

Theorem 32.2 shows that $f[N(x_0, \varepsilon)]$ contains a neighborhood $N(y_0, \eta)$ for some $\eta > 0$. Since $f: U \to V$ has an inverse $g: V \to U$, for each y in $N(y_0, \eta)$ there is an x in $N(x_0, \varepsilon)$ such that $f(x) = y$ and $x = g(y)$. Also, $f(x_0) = y_0$ and $x_0 = g(y_0)$. Then (9) shows that, for all y in $N(y_0, \eta)$,

$$\left(m - \frac{m}{4n}\right)|g(y) - g(y_0)| \leqq |y - y_0| \leqq \left(M + \frac{m}{4n}\right)|g(y) - g(y_0)|,$$

$$\left(M + \frac{m}{4n}\right)^{-1}|y - y_0| \leqq |g(y) - g(y_0)| \leqq \left(m - \frac{m}{4n}\right)^{-1}|y - y_0|. \quad (10)$$

These inequalities show that g is continuous at y_0.

Now that the proof of this preliminary result has been completed, we proceed to the proof that $g^1, \cdots, g^n$ are differentiable. The proof employs the definition of $D_i g^j(y_0)$, $i, j = 1, \cdots, n$, in Definition 2.8 to prove the existence of these derivatives, and thus of $D_{(1,\cdots,n)}g(y_0)$ by Theorem 3.12, and simultaneously to calculate their values.

Let $Y(y_0, \rho)$ be the class of ρ-increments $\mathbf{y}$ at y_0 and in $N(y_0, \delta)$ [see Definition 2.4]. Thus if $\mathbf{y} \in Y(y_0, \rho)$, then $\mathbf{y} = (y_1, \cdots, y_n, y_0)$ and $y_i \in N(y_0, \delta)$, $i = 1, \cdots, n, 0$. Let $y_i = (y_i^1, \cdots, y_i^n)$; then $x_i = g(y_i)$, $i = 1, \cdots, n$, 0, and $x_i = (x_i^1, \cdots, x_i^n)$. Since f and g are inverses of each other,

$$\begin{array}{ll} x_i^1 = g^1(y_i^1, \cdots, y_i^n), & y_i^1 = f^1(x_i^1, \cdots, x_i^n), \\ \cdots\cdots\cdots\cdots & \cdots\cdots\cdots\cdots \\ x_i^n = g^n(y_i^1, \cdots, y_i^n); & y_i^n = f^n(x_i^1, \cdots, x_i^n). \end{array} \quad (11)$$

These equations and the relations in (1) in Section 30 show that

$$\begin{vmatrix} f^t[g(y_1)] & y_1^2 & \cdots & y_1^n & 1 \\ \cdots & \cdots & \cdots & \cdots & \cdots \\ f^t[g(y_n)] & y_n^2 & \cdots & y_n^n & 1 \\ f^t[g(y_0)] & y_0^2 & \cdots & y_0^n & 1 \end{vmatrix} = \begin{cases} \Delta(y_1, \cdots, y_n, y_0) & \text{if} \quad t = 1, \\ 0 \quad \text{if} \quad t = 2, \cdots, n. \end{cases} \quad (12)$$

The determinant on the left in (12), in the case $t = 1$, equals

$$\begin{vmatrix} f^1[g(y_1)] - f^1[g(y_0)] & y_1^2 - y_0^2 & \cdots & y_1^n - y_0^n \\ \cdots & \cdots & \cdots & \cdots \\ f^1[g(y_n)] - f^1[g(y_0)] & y_n^2 - y_0^2 & \cdots & y_n^n - y_0^n \end{vmatrix} \quad (13)$$

Each of the functions $f^1, \cdots, f^n$ is differentiable at x_0 by the hypothesis in (2) and therefore satisfies the Stolz condition at x_0, that is, at $g(y_0)$, by Theorem 3.6. Then

$$f^1[g(y_k)] - f^1[g(y_0)]$$
$$= \sum_{j=1}^{n} D_j f^1(x_0)[g^j(y_k) - g^j(y_0)] + r(f^1; g(y_0), g(y_k))|g(y_k) - g(y_0)| \quad (14)$$

for $k = 1, \cdots, n$. Use these equations to evaluate the terms in the first column of the matrix in (13); the result can be written in the following two forms [see the properties of determinants in Section 77 in Appendix 1]:

$$\sum_{j=1}^{n} D_j f^1(x_0)\Delta_1 g^j(y_1, \cdots, y_n, y_0) + \det R(f^1) = \Delta(y_1, \cdots, y_n, y_0); \quad (15)$$

$$\sum_{j=1}^{n} D_j f^1(x_0)\frac{\Delta_1 g^j(\mathbf{y})}{\Delta(\mathbf{y})} = 1 - \frac{\det R(f^1)}{\Delta(\mathbf{y})}. \quad (16)$$

Here $R(f^1)$ is the $n \times n$ matrix which is written out explicitly in equation (18) below. The result of transforming each of the equations in (12) in the same manner is the following system of linear equations in the unknowns $\Delta_1 g^j(\mathbf{y})/\Delta(\mathbf{y}), j = 1, \cdots, n$.

$$\begin{aligned} \sum_{j=1}^{n} D_j f^1(x_0)\frac{\Delta_1 g^j(\mathbf{y})}{\Delta(\mathbf{y})} &= 1 - \frac{\det R(f^1)}{\Delta(\mathbf{y})}, \\ \sum_{j=1}^{n} D_j f^2(x_0)\frac{\Delta_1 g^j(\mathbf{y})}{\Delta(\mathbf{y})} &= 0 - \frac{\det R(f^2)}{\Delta(\mathbf{y})}, \\ &\cdots\cdots\cdots \\ \sum_{j=1}^{n} D_j f^n(x_0)\frac{\Delta_1 g^j(\mathbf{y})}{\Delta(\mathbf{y})} &= 0 - \frac{\det R(f^n)}{\Delta(\mathbf{y})}. \end{aligned} \quad (17)$$

In these equations,

$$\det R(f^i) = \begin{vmatrix} r[f^i; g(y_0), g(y_1)]|g(y_1) - g(y_0)| & y_1^2 - y_0^2 & \cdots & y_1^n - y_0^n \\ r[f^i; g(y_0), g(y_2)]|g(y_2) - g(y_0)| & y_2^2 - y_0^2 & \cdots & y_2^n - y_0^n \\ \cdots & \cdots & \cdots & \cdots \\ r[f^i; g(y_0), g(y_n)]|g(y_n) - g(y_0)| & y_n^2 - y_0^2 & \cdots & y_n^n - y_0^n \end{vmatrix}. \quad (18)$$

Since $\mathbf{y}$ is in $Y(y_0, \rho)$,

$$\frac{|\det R(f^i)|}{|\Delta(y)|} \leqq \frac{|\det R(f^i)|}{\rho|y_1 - y_0||y_2 - y_0| \cdots |y_n - y_0|}, \qquad i = 1, \cdots, n. \quad (19)$$

Divide the successive rows of the matrix in the numerator on the right by $|y_1 - y_0|, |y_2 - y_0|, \cdots, |y_n - y_0|$, respectively, and divide the denominator by their product. The absolute value of each term in columns $2, 3, \cdots, n$ in the numerator is now equal to or less than 1; the first column contains the terms

$$r[f^i; g(y_0), g(y_k)]\frac{|g(y_k) - g(y_0)|}{|y_k - y_0|}, \qquad k = 1, 2, \cdots, n. \quad (20)$$

These terms approach zero as $\mathbf{y} \to y_0$; the proof follows. First,

$$\frac{|g(y_k) - g(y_0)|}{|y_k - y_0|} \leqq \left(m - \frac{m}{4n}\right)^{-1}, \qquad k = 1, 2, \cdots, n, \quad (21)$$

by (10) for y_k in $N(y_0, \eta)$; observe that $y_k \neq y_0$ since $\mathbf{y} \in Y(y_0, \rho)$. Next,

$$\lim_{\mathbf{y} \to y_0} r[f^i; g(y_0), g(y_k)] = 0, \qquad k = 1, 2, \cdots, n, \tag{22}$$

by Theorem 3.6 since (10) shows that

$$\lim_{\mathbf{y} \to y_0} g(y_k) = g(y_0)$$

for each k. These facts prove that the limit of each term (20) is zero, and Hadamard's determinant theorem [see Corollary 87.2] shows that

$$\lim_{\mathbf{y} \to y_0} \frac{\det R(f^i)}{\Delta(\mathbf{y})} = 0, \qquad i = 1, 2, \cdots, n. \tag{23}$$

The system of linear equations in (17) has a unique solution for

$$\frac{\Delta_1 g^1(\mathbf{y})}{\Delta(\mathbf{y})}, \frac{\Delta_1 g^2(\mathbf{y})}{\Delta(\mathbf{y})}, \cdots, \frac{\Delta_1 g^n(\mathbf{y})}{\Delta(\mathbf{y})}, \tag{24}$$

since the determinant of the system is $D_{(1,\cdots,n)}f(x_0)$, which is not zero by hypothesis (3). Solve the system (17) for the terms in (24) by Cramer's rule. Then (23) shows that the limit as $\mathbf{y} \to y_0$ of each term in (24) exists and that

$$\begin{aligned} D_1 g^j(y_0) &= \lim_{\mathbf{y} \to y_0} \frac{\Delta_1 g^j(\mathbf{y})}{\Delta(\mathbf{y})} \\ &= \frac{1}{D_{(1,\cdots,n)}f(x_0)} \begin{vmatrix} D_1 f^1(x_0) & \cdots & D_{j-1} f^1(x_0) & 1 & D_{j+1} f^1(x_0) & \cdots & D_n f^1(x_0) \\ D_1 f^2(x_0) & \cdots & D_{j-1} f^2(x_0) & 0 & D_{j+1} f^2(x_0) & \cdots & D_n f^2(x_0) \\ \cdots & \cdots & \cdots & \cdots & \cdots & \cdots & \cdots \\ D_1 f^n(x_0) & \cdots & D_{j-1} f^n(x_0) & 0 & D_{j+1} f^n(x_0) & \cdots & D_n f^n(x_0) \end{vmatrix} \end{aligned} \tag{25}$$

for $j = 1, 2, \cdots, n$. The equations (25) establish these two results: they prove that the derivatives $D_1 g^j(y_0)$ exist, and they give also the values of these derivatives.

To find $D_2 g^j(y_0)$, $j = 1, 2, \cdots, n$, the calculation must be repeated from the beginning in (12). Substitute the column of terms $f^i[g(y_k)]$ in the second column of the matrix of $\Delta(\mathbf{y})$ to obtain $D_2 g^j(y_0)$, $j = 1, 2, \cdots, n$, $\cdots$, and in the n-th column to obtain $D_n g^j(y_0)$, $j = 1, 2, \cdots, n$. These calculations have found the values of the derivatives $D_i g^j(y_0)$, and the next step is to find the value of $D_{(1,\cdots,n)}g(y_0)$. Replace each term $D_i g^j(y_0)$ in the matrix $[D_i g^j(y_0)]$, $i, j = 1, \cdots, n$, by the values found in (25) and in the remaining calculations. Each element in $[D_i g^j(y_0)]_1^n$ now contains the determinant of an $n \times n$ matrix such as the one shown in (25). Expand this determinant by the elements in the column which contains the $n - 1$ zeros. Then properties of determinants [see especially Exercise 5.1] show that

$$D_{(1,\cdots,n)}g(y_0) = \det[D_i g^j(y_0)]_1^n = \frac{\det\{\operatorname{adj}[D_i f^j(x_0)]_1^n\}}{\{D_{(1,\cdots,n)}f(x_0)\}^n}. \tag{26}$$

Since $(\operatorname{adj} A)A = (\det A)I$ as stated in Exercise 5.1, then the Binet–Cauchy multiplication theorem [see Theorem 80.1] shows that

$$\det(\operatorname{adj} A)\det A = (\det A)^n,$$
$$\det(\operatorname{adj} A) = (\det A)^{n-1}.$$

This formula shows that

$$\det\{\operatorname{adj}[D_i f^j(x_0)]_1^n\} = \{\det[D_i f^j(x_0)]_1^n\}^{n-1} = \{D_{(1,\cdots,n)} f(x_0)\}^{n-1}.$$

Use this formula to evaluate the expression on the right in (26). The result is

$$D_{(1,\cdots,n)} g(y_0) = \frac{1}{D_{(1,\cdots,n)} f(x_0)}, \tag{27}$$

and this statement establishes (4) in Theorem 33.1 and completes the proof of that theorem. □

33.2 Example. Once the existence of the derivatives $D_i g^j(y_0)$, $i = 1, \cdots, n$, has been established in the proof of Theorem 33.1, there is a simple proof of the relation in (27), which is equivalent to the conclusion in (4) in Theorem 33.1. By (1) in Section 30,

$$\begin{aligned} g^1[f(x)] &= x^1, \\ g^2[f(x)] &= x^2, \\ &\cdots\cdots\cdots \\ g^n[f(x)] &= x^n. \end{aligned} \tag{28}$$

The chain rule in Theorem 4.5 shows that the derivative at x_0 of the composite function $g \circ f$ whose components are shown on the left in (28) is $D_{(1,\cdots,n)} g(y_0) D_{(1,\cdots,n)} f(x_0)$, and the derivative at x_0 of the function whose components appear on the right in (28) is the determinant of the identity matrix, which is 1. For example, if $n = 2$ [see Example 4.2],

$$\begin{vmatrix} \sum_{i=1}^{2} D_i g^1(y_0) D_1 f^i(x_0) & \sum_{i=1}^{2} D_i g^1(y_0) D_2 f^i(x_0) \\ \sum_{i=1}^{2} D_i g^2(y_0) D_1 f^i(x_0) & \sum_{i=1}^{2} D_i g^2(y_0) D_2 f^i(x_0) \end{vmatrix} = \begin{vmatrix} 1 & 0 \\ 0 & 1 \end{vmatrix},$$

$$\begin{vmatrix} D_1 g^1(y_0) & D_2 g^1(y_0) \\ D_1 g^2(y_0) & D_2 g^2(y_0) \end{vmatrix} \begin{vmatrix} D_1 f^1(x_0) & D_2 f^1(x_0) \\ D_1 f^2(x_0) & D_2 f^2(x_0) \end{vmatrix} = 1,$$

$$D_{(1,2)} g(y_0) D_{(1,2)} f(x_0) = 1.$$

The next theorem is the generalization of the one-dimensional inverse-function theorem in Theorem 31.1 with the hypothesis stated in (2).

33.3 Theorem (Inverse-Function Theorem). *Let $f: (f^1, \cdots, f^n)$ be a function $f: U \to V$ which maps $U = N(x_0, a)$ in $\mathbb{R}^n$ onto $V = f(U)$ in $\mathbb{R}^n$ and which has the following properties:*

The functions $f^i : U \to \mathbb{R}$, $i = 1, \cdots, n$, *are differentiable on* U. (29)

If $x_1^*, x_2^*, \cdots, x_n^*$ *are n points on an open segment joining two points in* U, *then* $\det[D_j f^i(x_i^*)] \neq 0$, $i, j = 1, \cdots, n$. (30)

Then $f: U \to V$ *has an inverse* $g : V \to U$; *g is defined at least in some neighborhood* $N(y_0, \delta)$ *of* $y_0 = f(x_0)$; *the components* $g^i : V \to \mathbb{R}$, $i = 1, \cdots, n$, *of g are differentiable at* y_0, *and*

$$D_{(1,\cdots,n)}f(x_0)D_{(1,\cdots,n)}g(y_0) = 1. \tag{31}$$

PROOF. The function $f: U \to V$ is continuous in U because $f^1, \cdots, f^n$ are differentiable in U. The hypothesis (30) and Theorem 12.2 show that the mapping $f: U \to V$ is one-to-one. Finally, if x is in U and $x_1^* = x_2^* = \cdots = x_n^* = x$, hypothesis (30) shows that $D_{(1,\cdots,n)}f(x) \neq 0$. Thus all of the hypotheses of Theorem 33.1 are satisfied, and Theorem 33.3 follows from Theorem 33.1. □

The final theorem in this section is the generalization of the one-dimensional inverse-function theorem in Theorem 31.1 with the hypothesis stated in (3).

33.4 Theorem (Inverse-Function Theorem). *Let* $f: (f^1, \cdots, f^n)$ *be a function* $f: U \to V$ *which maps* $U = N(x_0, a)$ *in* $\mathbb{R}^n$ *onto* $V = f(U)$ *in* $\mathbb{R}^n$ *and which has the following properties:*

The functions $f^i : U \to \mathbb{R}$, $i = 1, \cdots, n$, *are differentiable on* U. (32)

The derivatives $D_j f^i$, $i, j = 1, \cdots, n$, *are continuous at* x_0 *in* U. (33)

$D_{(1,\cdots,n)}f(x_0) \neq 0.$ (34)

Then in a sufficiently small neighborhood $U' = N(x_0, a')$, $U' \subset U$, *the function* $f: U' \to V' = f(U')$ *has an inverse* $g : V' \to U'$ *with components* $(g', \cdots, g^n)$; *there is a* $\delta' > 0$ *such that g is defined in the neighborhood* $N(y_0, \delta')$ *of* $y_0 = f(x_0)$; *the functions* $g^i : V' \to R$, $i = 1, \cdots, n$, *are differentiable at* y_0; *and*

$$D_{(1,\cdots,n)}f(x_0)D_{(1,\cdots,n)}g(y_0) = 1. \tag{35}$$

PROOF. Since $D_{(1,\cdots,n)}f(x_0) \neq 0$ by (34), since the derivatives $D_j f^i$, $i, j = 1, \cdots, n$, are continuous at x_0 by (33), and since a determinant is a continuous function of its elements, then in a sufficiently small neighborhood $U' = N(x_0, a')$ in U the conditions stated in (29) and (30) are satisfied. Thus the function $f: U' \to V'$ satisfies all of the hypotheses of Theorem 33.3, and Theorem 33.4 follows from Theorem 33.3. □

33.5 Example. The equations

$$\begin{aligned} y^1 &= f^1(x) = (x^1 - x^2 - 1)^2, \\ y^2 &= f^2(x) = (x^1 + 2x^2 - 4)^2, \end{aligned} \tag{36}$$

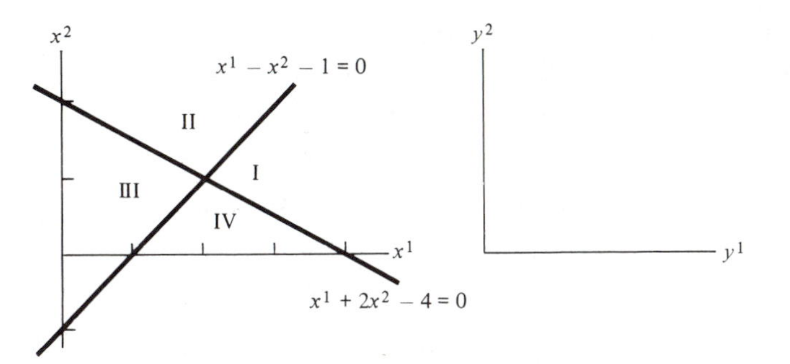

Figure 33.1. The regions I, ..., IV and the quadrant into which they are mapped.

define a function $f\colon \mathbb{R}^2 \to \mathbb{R}^2$ for which

$$D_{(1,2)}(f^1, f^2)(x^1, x^2) = 12(x^1 - x^2 - 1)(x^1 + 2x^2 - 4). \tag{37}$$

This derivative equals zero at each point on each of the lines $x^1 - x^2 - 1 = 0$ and $x^1 + 2x^2 - 4 = 0$, but at no other points. Now it is easy to verify that f maps each of the points

$$(3, 1), \quad (5/3, 5/3), \quad (1, 1), \quad (7/3, 1/3) \tag{38}$$

into the same point $(1, 1)$; thus f is not one-to-one in the entire plane. Observe, however, that

$$\begin{aligned} &D_{(1,2)}(f^1, f^2)(3, 1) = 12, \qquad D_{(1,2)}(f^1, f^2)(5/3, 5/3) = -12, \\ &D_{(1,2)}(f^1, f^2)(1, 1) = 12, \qquad D_{(1,2)}(f^1, f^2)(7/3, 1/3) = -12. \end{aligned} \tag{39}$$

Then (37) and (39) show that the hypotheses in (32), (33), (34) are satisfied at the four points in (38), and therefore Theorem 33.4 shows that, in sufficiently small neighborhoods of these points, the function $f\colon (f^1, f^2)$ has an inverse $g\colon (g^1, g^2)$. But much more can be proved in this example. The lines $x^1 - x^2 - 1 = 0$ and $x^1 + 2x^2 - 4 = 0$ divide the plane into four regions I, $\cdots$, IV as shown in Figure 33.1. The line segments which bound region I are mapped by f into the non-negative y^1-axis and the non-negative y^2-axis, and the same statement is true also for each of the regions II, III, and IV. Also, if $y\colon (y^1, y^2)$ is an arbitrary point in the open first quadrant in the (y^1, y^2)-plane, then the intermediate-value theorem [Theorem 27.1] and other results in Chapter 4 can be used to show that there are points $x_1, \cdots, x_4$ in regions I, $\cdots$, IV such that $f(x_i) = y$, $i = 1, \cdots, 4$. As shown below, however, more can be proved in this special case quite simply without using the results in Chapter 4. If $x_i^* = (x_i^1, x_i^2)$, $i = 1, 2$, then

$$\begin{aligned} \det[D_j f^i(x_i^*)]_1^2 &= \begin{vmatrix} 2(x_1^1 - x_1^2 - 1) & -2(x_1^1 - x_1^2 - 1) \\ 2(x_2^1 + 2x_2^2 - 4) & 4(x_2^1 + 2x_2^2 - 4) \end{vmatrix} \\ &= 12(x_1^1 - x_1^2 - 1)(x_2^1 + 2x_2^2 - 4); \end{aligned} \tag{40}$$

and for every pair of points x_1^*, x_2^* on a segment contained in one of the four regions, the determinant $\det[D_j f^i(x_i^*)]_1^2$ is not zero. Thus, Theorem 12.2

asserts that f defines a one-to-one mapping in each open region I, $\cdots$, IV; therefore, f has an inverse in each of these regions. These inverses can be found by solving the equations in (36) for x^1, x^2. The results are the following.

$$\begin{aligned}
&\text{Region I:} && x^1 = g^1(y) = (6 + 2\sqrt{y^1} + \sqrt{y^2})/3, \\
& && x^2 = g^2(y) = (3 - \ \sqrt{y^1} + \sqrt{y^2})/3. \\
&\text{Region II:} && x^1 = g^1(y) = (6 - 2\sqrt{y^1} + \sqrt{y^2})/3, \\
& && x^2 = g^2(y) = (3 + \ \sqrt{y^1} + \sqrt{y^2})/3. \\
&\text{Region III:} && x^1 = g^1(y) = (6 - 2\sqrt{y^1} - \sqrt{y^2})/3, \\
& && x^2 = g^2(y) = (3 + \ \sqrt{y^1} - \sqrt{y^2})/3. \\
&\text{Region IV:} && x^1 = g^1(y) = (6 + 2\sqrt{y^1} - \sqrt{y^2})/3, \\
& && x^2 = g^2(y) = (3 - \ \sqrt{y^1} - \sqrt{y^2})/3.
\end{aligned} \tag{41}$$

These functions map $y:(1, 1)$ into the points $(3, 1)$, $(5/3, 5/3)$, $(1, 1)$, and $(7/3, 1/3)$, in regions I, II, III, IV, respectively, and the functions in (41) are the inverses of f, in neighborhoods of these points, which exist by (39) and Theorem 33.4. The equations in (36) show that f maps points in each of the closed regions into points in the closed first quadrant in the (y^1, y^2)-plane; and the equations (41) show that each point in the closed first quadrant in the (y^1, y^2)-plane is the image under f of a point in each of the closed regions I, $\cdots$, IV. The inverse-function theorem [Theorem 33.4] shows that the inverse functions in (41) are differentiable at each point (y^1, y^2) in the open first quadrant, and this fact can be verified directly from the analytic expressions which define these functions in (41).

33.6 Example. The equations

$$\begin{aligned}
y^1 &= f^1(x) = (2x^1 - 4x^2 + 6)^2, \\
y^2 &= f^2(x) = (x^1 - 2x^2 + 2)^2,
\end{aligned} \tag{42}$$

define a function $f: \mathbb{R}^2 \to \mathbb{R}^2$ for which

$$D_{(1,2)}(f^1, f^2)(x^1, x^2) = 0, \qquad (x^1, x^2) \quad \text{in} \quad \mathbb{R}^2. \tag{43}$$

Thus f does not satisfy the hypotheses of any of the inverse-function theorems in this section, and we shall now show that f does not have an inverse in the neighborhood of any point (x^1, x^2) in $\mathbb{R}^2$. Since

$$\begin{aligned}
y^1 &= [2(x^1 - 2x^2 + 2) + 2]^2 \\
&= 4(x^1 - 2x^2 + 2)^2 + 8(x^1 - 2x^2 + 2) + 4,
\end{aligned} \tag{44}$$

the second equation in (42) shows that

$$\begin{aligned}
y^1 &= 4y^2 \pm 8\sqrt{y^2} + 4, \\
(y^1 - 4y^2 - 4)^2 &= 64y^2.
\end{aligned} \tag{45}$$

This equation represents a curve which lies in the first quadrant; the curve is one of the conic sections. Thus f maps every point in $\mathbb{R}^2$ onto a point on the curve (45), and the equations in (42) show that f fails to be a one-to-one mapping in every neighborhood of every point (x^1, x^2) in $\mathbb{R}^2$ since all points (x^1, x^2) on a line $x^1 - 2x^2 + 2 = \text{constant}$ are mapped into the same point (y^1, y^2). Therefore, in every neighborhood of every point in $\mathbb{R}^2$ the function f fails to have an inverse.

Exercises

33.1. The equations

$$y^1 = f^1(x) = (4x^1 - 6x^2 + 10)^2,$$
$$y^2 = f^2(x) = (6x^1 - 9x^2 + 18)^2,$$

define a function $f: \mathbb{R}^2 \to \mathbb{R}^2$.
(a) Show that $D_{(1,2)}(f^1, f^2)(x^1, x^2) = 0$ for every (x^1, x^2) in $\mathbb{R}^2$.
(b) Show that f maps all the points on the line $2x^1 - 3x^2 + 5 = c$ into the same point $[4c^2, 9(c+1)^2]$.
(c) Find the equation of the curve into which f maps the entire plane $\mathbb{R}^2$.
(d) Show that f fails to have an inverse in every neighborhood of every point in $\mathbb{R}^2$.

33.2. The equations

$$y^1 = f^1(x) = (x^1 - x^2 + 1)^2,$$
$$y^2 = f^2(x) = (x^1 + x^2 + 3)^2,$$

define a function $f: \mathbb{R}^2 \to \mathbb{R}^2$.
(a) Show that f maps each of the points $(3, 2)$, $(1, 4)$, $(-5, -6)$, and $(-7, -4)$ into the same point and that f is therefore not one-to-one in the entire plane.
(b) Show that f has an inverse in sufficiently small neighborhoods of the points $(3, 2)$, $(1, 4)$, $(-5, -6)$, $(-7, -4)$.
(c) Show that f defines a one-to-one mapping in each of the four regions into which the lines $x^1 - x^2 + 1 = 0$ and $x^1 + x^2 + 3 = 0$ divide the plane, and find the inverses of these one-to-one mappings.

33.3. Consider again the functions $g^i: V' \to \mathbb{R}$, $i = 1, \cdots, n$, in Theorem 33.4. Prove that the derivatives $D_j g^i$, $i, j = 1, \cdots, n$, exist in a sufficiently small neighborhood of y_0 and are continuous at y_0.

33.4. Consider again the functions $f: U \to V$ and $g: V \to U$ in Theorem 33.3. Prove that the functions $g^i: V \to \mathbb{R}$, $i = 1, \cdots, n$, are differentiable at each point y in V.

33.5. Consider again the functions $f: U \to V$ and $g: V \to U$ in Theorem 33.1. Assume that the functions $f^i: U \to \mathbb{R}$, $i = 1, \cdots, n$, are differentiable at x_0 and that the functions $g^i: V \to \mathbb{R}$, $i = 1, \cdots, n$, are differentiable at y_0. Use the chain rule and the identities in (28) to derive formulas for $D_j g^i(y_0)$, $i, j = 1, \cdots, n$, similar to the one given in (25).

33.6. Let U be the neighborhood $N(x_0, a)$ in $\mathbb{R}^2$; let $f^i : U \to \mathbb{R}$, $i = 1, 2, 3$, be functions which are defined and differentiable on U, and let S be the surface defined by the equations $y^i = f^i(x^1, x^2)$, (x^1, x^2) in U, $i = 1, 2, 3$.

(a) If the matrix $[D_j f^i(x_i^*)]$, $i = 1, 2, 3$ and $j = 1, 2$, has rank 2 [that is, if the determinant of at least one 2 by 2 submatrix is different from zero] for each set of points x_i^*, $i = 1, 2, 3$, on a line segment in U, show that the mapping $(f^1, f^2, f^3) : U \to \mathbb{R}^3$ is one-to-one; that is, show that the surface S does not intersect itself. [Hint. Exercise 12.3.]

(b) Set $V = (f^1, f^2)(U)$. If for every pair of points x_1^*, x_2^* on a line segment in U the determinant $\det[D_j f^i(x_i^*)]$, $i, j = 1, 2$, is not zero, show that the equations

$$y^1 = f^1(x^1, x^2),$$
$$y^2 = f^2(x^1, x^2), \qquad (x^1, x^2) \text{ in } U,$$

can be solved for x^1, x^2 to obtain

$$x^1 = g^1(y^1, y^2),$$
$$x^2 = g^2(y^1, y^2), \qquad (y^1, y^2) \text{ in } V.$$

Show also that the functions $g^i : V \to \mathbb{R}$, $i = 1, 2$, are differentiable in V.

(c) If the hypotheses in (b) are satisfied, show that the surface S is represented by an equation of the form $y^3 = F(y^1, y^2)$, (y^1, y^2) in V, and that F is differentiable in V.

(d) Under what conditions can S be represented by an equation of the form $y^1 = G(y^2, y^3)$? by an equation of the form $y^2 = H(y^1, y^3)$?

33.7. The purpose of this exercise is to outline the proof of the implicit-function theorem in a special case. If f^1 and f^2 are two functions which are defined in a neighborhood of $x_0 : (x_0^1, \cdots, x_0^4)$ in $\mathbb{R}^4$, and if

$$f^1(x_0^1, \cdots, x_0^4) = 0,$$
$$f^2(x_0^1, \cdots, x_0^4) = 0,$$

then, under suitable hypotheses on f^1, f^2, the implicit-function theorem asserts that the equations

$$f^1(x^1, x^2, x^3, x^4) = 0,$$
$$f^2(x^1, x^2, x^3, x^4) = 0,$$

can be solved for x^1, x^2 to obtain x^1, x^2 as functions of (x^3, x^4) which satisfy the given equations. An application of the inverse-function theorem provides the proof of this special case and of the general case of the implicit-function theorem. Begin with the notation and the hypotheses. Let $x : (x^1, \cdots, x^4)$ and $x_0 : (x_0^1, \cdots, x_0^4)$ be points in $\mathbb{R}^4$, and let U be a neighborhood $N(x_0, a)$ of x_0 in $\mathbb{R}^4$. Let $f^i : U \to R$, $i = 1, 2$, be functions which are differentiable on U. Assume also that $f^i(x_0) = 0$, $i = 1, 2$, as stated above, and that

$$\begin{vmatrix} D_1 f^1(x_1^*) & D_2 f^1(x_1^*) \\ D_1 f^2(x_2^*) & D_2 f^2(x_2^*) \end{vmatrix} \neq 0$$

for every pair of points x_1^*, x_2^* on a line segment in U.

(a) Show that the transformation $f: U \to R^4$ defined by the equations

$$\begin{aligned} y^1 &= f^1(x^1, x^2, x^3, x^4), \\ y^2 &= f^2(x^1, x^2, x^3, x^4), \\ y^3 &= x^3, \\ y^4 &= x^4, \end{aligned} \qquad (x^1, \cdots, x^4) \text{ in } U,$$

is one-to-one on U. If $y_0 = f(x_0)$, show that $y_0 = (0, 0, x_0^3, x_0^4)$.

(b) Let $V = f(U)$, and let y be the point with coordinates $(y^1, \cdots, y^4)$. Show that $f: U \to V$ has an inverse $g: V \to U$ with the following properties:

$$\begin{aligned} x^1 &= g^1(y^1, \cdots, y^4), \\ x^2 &= g^2(y^1, \cdots, y^4), \\ x^3 &= g^3(y^1, \cdots, y^4) \equiv y^3, \\ x^4 &= g^4(y^1, \cdots, y^4) \equiv y^4. \end{aligned} \qquad (y^1, \cdots, y^4) \text{ in } V,$$

The functions $g^i: V \to R$, $i = 1, \cdots, 4$, are differentiable in V. If $\delta > 0$ is sufficiently small, then $N(y_0, \delta) \subset V$. Finally, for every y in V,

$$\begin{aligned} y^1 &= f^1[g^1(y), \cdots, g^4(y)], \\ y^2 &= f^2[g^1(y), \cdots, g^4(y)]. \end{aligned}$$

(c) Let y be the point $(0, 0, y^3, y^4)$ in $N(y_0, \delta)$. Use the equations in (b) to show that

$$\begin{aligned} f^1[g^1(0, 0, y^3, y^4), g^2(0, 0, y^3, y^4), y^3, y^4] &= 0, \\ f^2[g^1(0, 0, y^3, y^4), g^2(0, 0, y^3, y^4), y^3, y^4] &= 0. \end{aligned}$$

Use these equations and the fact that $x^3 = y^3$, $x^4 = y^4$ to show that

$$\begin{aligned} f^1[g^1(0, 0, x^3, x^4), g^2(0, 0, x^3, x^4), x^3, x^4] &= 0, \\ f^2[g^1(0, 0, x^3, x^4), g^2(0, 0, x^3, x^4), x^3, x^4] &= 0, \end{aligned}$$

for all x^3, x^4 such that $[(x^3 - x_0^3)^2 + (x^4 - x_0^4)^2]^{1/2} < \delta$. Thus show that

$$\begin{aligned} x^1 &= g^1(0, 0, x^3, x^4), \\ x^2 &= g^2(0, 0, x^3, x^4), \end{aligned}$$

are the solution of the system of equations

$$\begin{aligned} f^1(x^1, \cdots, x^4) &= 0, \\ f^2(x^1, \cdots, x^4) &= 0, \end{aligned}$$

for all (x^3, x^4) in $N[(x_0^3, x_0^4), \delta]$.

(d) By differentiating the identities in (c), show that

$$D_3 g^1(0, 0, x_0^3, x_0^4) = -\frac{\begin{vmatrix} D_3 f^1(x_0) & D_2 f^1(x_0) \\ D_3 f^2(x_0) & D_2 f^2(x_0) \end{vmatrix}}{\begin{vmatrix} D_1 f^1(x_0) & D_2 f^1(x_0) \\ D_1 f^2(x_0) & D_2 f^2(x_0) \end{vmatrix}},$$

and find similar formulas for $D_4g^1(0, 0, x_0^3, x_0^4)$, $D_3g^2(0, 0, x_0^3, x_0^4)$, and $D_4g^2(0, 0, x_0^3, x_0^4)$.

(e) If the equations $f^1(x^1, \cdots, x^4) = 0, f^2(x^1, \cdots, x^4) = 0$ are linear in x^1 and x^2, they have the form

$$a_{11}(x^3, x^4)x^1 + a_{12}(x^3, x^4)x^2 + a_{13}(x^3, x^4) = 0,$$

$$a_{21}(x^3, x^4)x^1 + a_{22}(x^3, x^4)x^2 + a_{23}(x^3, x^4) = 0.$$

Compare the solution of these equations for x^1, x^2 by Cramer's rule with the solution obtained by the method used in this exercise. In particular, compare the conditions for the existence of a solution and the formulas obtained in the two cases for the derivatives in (d).

33.8. Exercise 33.7 contains a special case of the implicit-function theorem. State and prove two versions of the general implicit-function theorem corresponding to two versions of the inverse-function theorem.

CHAPTER 6

Integrals and the Fundamental Theorem of the Integral Calculus

34. Introduction

The purpose of this chapter is to define integrals, to establish their principal properties, and to describe some of their applications. Integrals are the limits of certain types of finite sums. As shown in elementary calculus, the problem of finding the area under a curve leads to these finite sums and to integrals. Intuitively, the area under the curve $y=f(x)$ in Figure 34.1 is approximately

$$\begin{aligned} &\sum_{i=1}^{n} f(x_{i-1})(x_i - x_{i-1}) \\ \text{or} \quad &\sum_{i=1}^{n} (-1) f(x_{i-1}) \begin{vmatrix} x_{i-1} & 1 \\ x_i & 1 \end{vmatrix}, \end{aligned} \tag{1}$$

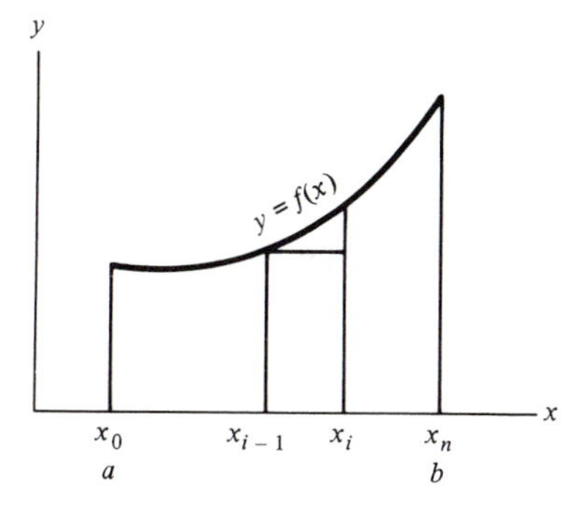

Figure 34.1. The approximate area under a curve.

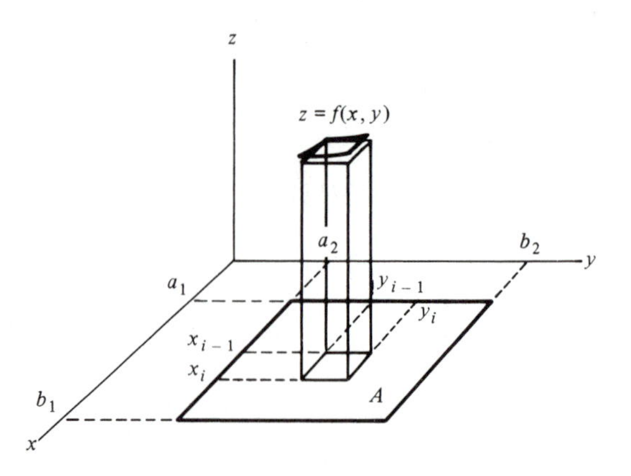

Figure 34.2. Approximate volume under a surface.

and the area is the limit of this sum as n tends to infinity and the maximum length of the intervals (x_{i-1}, x_i) tends to zero. This limit of the sums (1), if it exists, is the integral

$$\int_a^b f(x)\, dx \tag{2}$$

of the function $f\colon [a, b] \to \mathbb{R}$; the integral exists at least if f is continuous.

If $A = [a_1, b_1] \times [a_2, b_2]$, then the graph of $f\colon A \to \mathbb{R}$ is a surface, and the corresponding problem is to find the volume under the surface $z = f(x, y)$. The usual procedure [see Figure 34.2] is to subdivide A into rectangles $[x_{i-1}, x_i] \times [y_{j-1}, y_j]$, $i = 1, \cdots, n$, $j = 1, \cdots, m$, and to approximate the volume under the surface $z = f(x, y)$ by the sum

$$\sum_{i=1}^{n} \sum_{j=1}^{m} f(x_{i-1}, y_{j-1})(x_i - x_{i-1})(y_j - y_{j-1}). \tag{3}$$

The limit of this sum is the double integral

$$\iint_A f(x, y)\, dx\, dy. \tag{4}$$

Once more the integral exists at least if f is continuous.

But there are other possible ways to proceed. The same integral (4) can be obtained by subdividing A into triangular subregions rather than rectangular subregions. In particular, simplicial subdivisions of A could be used to define (4); a simplicial subdivision of A is a set of triangular subregions of A which form a simplicial complex [see Section 21 and Figure 21.5]. Since a choice of type of subdivision of A is available, the choice should be made in a way which facilitates the development of the theory of the integral. For some parts of the theory, rectangular subdivisions are the most suitable, but there are good reasons why simplicial subdivisions of the region of integration will be employed in most cases in this book. Let us now explain why.

The fundamental theorem of the integral calculus is one of the most important theorems in the theory of integration. This theorem represents the integral of certain functions (derivatives) over a region A as the integral of a related function over the boundary of A. The cancellation properties in an oriented simplicial complex have been employed already to establish important results [see, for example, Theorem 18.6, Exercise 18.7, and Theorem 23.4], and Exercise 15.6 outlines the following proof, based on these cancellation properties, of the fundamental theorem in the one-dimensional case. Assume that f has a continuous derivative f' on $[a, b]$. Then

$$\lim_{n\to\infty}\sum_{i=1}^{n}(-1)\begin{vmatrix} f(x_{i-1}) & 1 \\ f(x_i) & 1 \end{vmatrix} = \lim_{n\to\infty}\sum_{i=1}^{n}[f(x_i)-f(x_{i-1})] \\ = f(x_n)-f(x_0)=f(b)-f(a). \tag{5}$$

By the mean-value theorem,

$$\lim_{n\to\infty}\sum_{i=1}^{n}(-1)\begin{vmatrix} f(x_{i-1}) & 1 \\ f(x_i) & 1 \end{vmatrix} = \lim_{n\to\infty}\sum_{i=1}^{n}(-1)f'(x_i^*)\begin{vmatrix} x_{i-1} & 1 \\ x_i & 1 \end{vmatrix} = \int_a^b f'(x)\,dx. \tag{6}$$

Then equations (5) and (6) show that

$$\int_a^b f'(x)\,dx = f(b)-f(a), \tag{7}$$

and this statement is the fundamental theorem of the integral calculus in the one-dimensional case. The cancellations which occur in (5) result from the fact that the one-dimensional simplexes (x_{i-1}, x_i) always form an oriented simplicial complex. The proof of the fundamental theorem given in (5) and (6) does not have an exact generalization in dimensions greater than one; the reason is that there is no analog of the mean-value theorem, used in (6), which is available at this time (the desired mean-value theorem will be derived, later in Section 42, from the fundamental theorem of the integral calculus). Nevertheless, only a small change needs to be made in (6) to obtain a proof of (7) which has an exact generalization to higher dimensions; the following proof, with more detail, can be found in Example 11.1. Since f has a continuous derivative f' on $[a, b]$, it satisfies the Stolz condition and

$$\begin{vmatrix} f(x_{i-1}) & 1 \\ f(x_i) & 1 \end{vmatrix} = f'(x_{i-1})\begin{vmatrix} x_{i-1} & 1 \\ x_i & 1 \end{vmatrix} - r(f;x_{i-1},x_i)|x_{i-1}-x_i|. \tag{8}$$

Furthermore, for each $\varepsilon > 0$ there exists a $\delta > 0$ such that

$$|r(f;x_{i-1},x_i)| < \varepsilon, \qquad i=1,\cdots,n, \tag{9}$$

if $|x_{i-1}-x_i| < \delta$ for $i = 1, \cdots, n$. Then (8) and (9) show that (6) can be replaced by

$$\lim_{n\to\infty}\sum_{i=1}^{n}(-1)\begin{vmatrix} f(x_{i-1}) & 1 \\ f(x_i) & 1 \end{vmatrix}$$
$$= \lim_{n\to\infty}\left\{\sum_{i=1}^{n}(-1)f'(x_{i-1})\begin{vmatrix} x_{i-1} & 1 \\ x_i & 1 \end{vmatrix} + \sum_{i=1}^{n} r(f; x_{i-1}, x_i)|x_{i-1} - x_i|\right\}. \tag{10}$$

If $|x_{i-1} - x_i| < \delta$ for $i = 1, \cdots, n$, then

$$\left|\sum_{i=1}^{n} r(f; x_{i-1}, x_i)|x_{i-1} - x_i|\right| \leqq \varepsilon \sum_{i=1}^{n} |x_{i-1} - x_i| = \varepsilon|b - a|. \tag{11}$$

This equation shows that the limit of the second sum on the right in (10) is zero; since the limit of the first sum is $\int_a^b f'(x)\,dx$, equation (10) simplifies to

$$\lim_{n\to\infty}\sum_{i=1}^{n}(-1)\begin{vmatrix} f(x_{i-1}) & 1 \\ f(x_i) & 1 \end{vmatrix} = \int_a^b f'(x)\,dx. \tag{12}$$

Then the fundamental theorem in (7) follows from (12) and (5). The virtue of the second proof is that it has an exact generalization in dimensions greater than one.

Let A be a Euclidean simplex in $\mathbb{R}^2$, and let $f: A \to \mathbb{R}^2$ be a function whose components (f^1, f^2) have continuous derivatives on A. Also, let $p_i : (x_i, y_i)$, $i = 0, 1, 2$, be the vertices of a simplex in an oriented simplicial subdivision $\mathscr{P}$ of A. Then by Theorem 11.4,

$$\frac{1}{2!}\begin{vmatrix} f^1(x_0, y_0) & f^2(x_0, y_0) & 1 \\ f^1(x_1, y_1) & f^2(x_1, y_1) & 1 \\ f^1(x_2, y_2) & f^2(x_2, y_2) & 1 \end{vmatrix} = \frac{D_{(1,2)}(f^1, f^2)(x_0, y_0)}{2!}\begin{vmatrix} x_0 & y_0 & 1 \\ x_1 & y_1 & 1 \\ x_2 & y_2 & 1 \end{vmatrix}$$
$$+ r(f; p_0, p_1, p_2)\frac{|\Delta(p_0, p_1, p_2)|}{2!}. \tag{13}$$

This equation in condensed notation is

$$\frac{\Delta_{(1,2)}(f^1, f^2)(p_0, p_1, p_2)}{2!} = D_{(1,2)}(f^1, f^2)(p_0)\frac{\Delta(p_0, p_1, p_2)}{2!}$$
$$+ r(f; p_0, p_1, p_2)\frac{|\Delta(p_0, p_1, p_2)|}{2!}. \tag{14}$$

Let $\mathscr{P}_k$, $k = 1, 2, \cdots$, be a sequence of simplicial subdivisions of A [see Definition 21.5]. Write the identity (14) for each simplex (p_0, p_1, p_2) in $\mathscr{P}_k$ and then add these identities to obtain

$$\sum_{\mathscr{P}_k}\frac{\Delta_{(1,2)}(f^1, f^2)(p_0, p_1, p_2)}{2!} = \sum_{\mathscr{P}_k} D_{(1,2)}(f^1, f^2)(p_0)\frac{\Delta(p_0, p_1, p_2)}{2!}$$
$$+ \sum_{\mathscr{P}_k} r(f; p_0, p_1, p_2)\frac{|\Delta(p_0, p_1, p_2)|}{2!}. \tag{15}$$

The limit, as k tends to infinity, of the first sum on the right is an integral, and the limit of the second sum is zero. Thus

$$\lim_{k\to\infty} \sum_{\mathscr{P}_k} \frac{\Delta_{(1,2)}(f^1, f^2)(p_0, p_1, p_2)}{2!} = \iint_A D_{(1,2)}(f^1, f^2)(x, y)\, dx\, dy. \tag{16}$$

This equation corresponds to (12) in the one-dimensional case. But there is a second way to evaluate the limit in (16). By Theorem 20.3 and elementary properties of determinants,

$$\begin{aligned} \frac{1}{2!}\begin{vmatrix} f^1(x_0, y_0) & f^2(x_0, y_0) & 1 \\ f^1(x_1, y_1) & f^2(x_1, y_1) & 1 \\ f^1(x_2, y_2) & f^2(x_2, y_2) & 1 \end{vmatrix} &= -\frac{[f^1(x_1, y_1) + f^1(x_2, y_2)]}{2}\begin{vmatrix} f^2(x_1, y_1) & 1 \\ f^2(x_2, y_2) & 1 \end{vmatrix} \\ &\quad + \frac{[f^1(x_0, y_0) + f^1(x_2, y_2)]}{2}\begin{vmatrix} f^2(x_0, y_0) & 1 \\ f^2(x_2, y_2) & 1 \end{vmatrix} \\ &\quad - \frac{[f^1(x_0, y_0) + f^1(x_1, y_1)]}{2}\begin{vmatrix} f^2(x_0, y_0) & 1 \\ f^2(x_1, y_1) & 1 \end{vmatrix}. \end{aligned} \tag{17}$$

Each term on the right contains values of f^1 and f^2 on a single simplex in the boundary of (p_0, p_1, p_2); each of these terms changes sign if the orientation of the boundary simplex is reversed. Write the identity (17) for each simplex (p_0, p_1, p_2) in $\mathscr{P}_k$, and then add these identities. If two simplexes in $\mathscr{P}_k$ have a side in common, the two terms in the sum which arise from this common side cancel because the common side has opposite orientations in the two simplexes [see Theorem 18.5]. Thus, as a result of the cancellation properties in an oriented simplicial complex,

$$\begin{aligned} &\sum_{\mathscr{P}_k} \frac{\Delta_{(1,2)}(f^1, f^2)(p_0, p_1, p_2)}{2!} \\ &\qquad = \sum_{\partial\mathscr{P}_k} (-1)\frac{[f^1(x_1, y_1) + f^1(x_2, y_2)]}{2}\begin{vmatrix} f^2(x_1, y_1) & 1 \\ f^2(x_2, y_2) & 1 \end{vmatrix}. \end{aligned} \tag{18}$$

Here $\partial\mathscr{P}_k$ denotes the chain of simplexes in the boundary of the chain of all simplexes in $\mathscr{P}_k$; the simplexes in $\partial\mathscr{P}_k$ are in the topological boundary of the simplex A. It will be shown later that the limit of the sum on the right in (18) is an integral around the boundary of A in the positive direction. Thus

$$\begin{aligned} &\lim_{k\to\infty} \sum_{\mathscr{P}_k} \frac{\Delta_{(1,2)}(f^1, f^2)(p_0, p_1, p_2)}{2!} \\ &\qquad = \int_{\partial A} f^1(x, y) D_1 f^2(x, y)\, dx + f^1(x, y) D_2 f^2(x, y)\, dy. \end{aligned} \tag{19}$$

This equation corresponds to (5) in the one-dimensional case although their appearances are somewhat different. The following statement of the fundamental theorem of the integral calculus results from the two evaluations of the same limit in (16) and (19).

$$\iint_A D_{(1,2)}(f^1, f^2)(x, y)\,dx\,dy = \int_{\partial A} f^1(x, y) D_1 f^2(x, y)\,dx + f^1(x, y) D_2 f^2(x, y)\,dy. \tag{20}$$

The complete details of the proof of this formula are given in Section 38. The proof depends in an essential way on the fact that a sequence $\mathscr{P}_k$, $k = 1, 2, \cdots$, of simplicial subdivisions of A, with the properties described in Definition 21.5, is used to define the integral in (16).

Exercises

34.1. Let A be the positively oriented Euclidean simplex $[p_0, p_2, p_4]$ shown in Figure 34.3, and let $f: A \to \mathbb{R}^2$ be a function whose components (f^1, f^2) have continuous derivatives on A.

(a) Let $\mathscr{P}$ be the oriented simplicial subdivision of A which consists of the simplexes (p_0, p_1, p_5), (p_1, p_2, p_3), (p_3, p_4, p_5), (p_1, p_3, p_5). Find the boundary of the chain of these four simplexes.

(b) Use Theorem 20.3 to prove the following identity [see equation (17) in this section].

$$(1/2!)\{\Delta_{(1,2)}f(p_0, p_1, p_5) + \Delta_{(1,2)}f(p_1, p_2, p_3) + \Delta_{(1,2)}f(p_3, p_4, p_5) + \Delta_{(1,2)}f(p_1, p_3, p_5)\}$$

$$= -\left\{\frac{f^1(p_0) + f^1(p_1)}{2}\begin{vmatrix} f^2(p_0) & 1 \\ f^2(p_1) & 1 \end{vmatrix} + \frac{f^1(p_1) + f^1(p_2)}{2}\begin{vmatrix} f^2(p_1) & 1 \\ f^2(p_2) & 1 \end{vmatrix}\right.$$

$$+ \frac{f^1(p_2) + f^1(p_3)}{2}\begin{vmatrix} f^2(p_2) & 1 \\ f^2(p_3) & 1 \end{vmatrix} + \frac{f^1(p_3) + f^1(p_4)}{2}\begin{vmatrix} f^2(p_3) & 1 \\ f^2(p_4) & 1 \end{vmatrix}$$

$$\left.+ \frac{f^1(p_4) + f^1(p_5)}{2}\begin{vmatrix} f^2(p_4) & 1 \\ f^2(p_5) & 1 \end{vmatrix} + \frac{f^1(p_5) + f^1(p_0)}{2}\begin{vmatrix} f^2(p_5) & 1 \\ f^2(p_0) & 1 \end{vmatrix}\right\}.$$

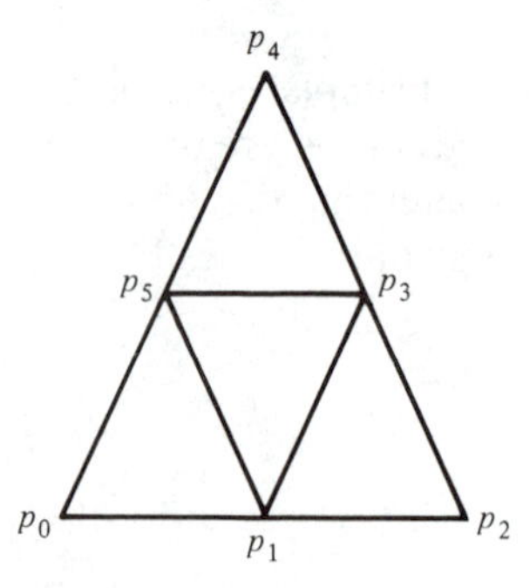

Figure 34.3. A simplicial subdivision of $[p_0, p_2, p_4]$.

Compare the terms on the right in this identity with the boundary of the chain found in (a).

(c) Describe briefly the relation of this identity and its extensions, for a sequence $\mathscr{P}_k$, $k = 1, 2, \cdots$, of simplicial subdivisions of (p_0, p_2, p_4), to the statement of the fundamental theorem of the integral calculus in (20).

34.2. Prove the following theorem. Let E be an open interval which contains $[a, b]$. If $f: E \to \mathbb{R}$ is a continuous function, then there exists a function $F: E \to \mathbb{R}$ such that $F'(x) = f(x)$ and

$$\int_a^b f(x)\,dx = \int_a^b F'(x)\,dx = F(b) - F(a).$$

[Hint. Set $F(x) = \int_a^x f(t)\,dt$.]

34.3. Let E be an open interval which contains $[a, b]$. Let $g: [a, b] \to \mathbb{R}^3$ be a function whose components (g_1, g_2, g_3) have continuous derivatives on E. Then g defines a curve in $\mathbb{R}^3$; assume that the trace of this curve is contained in an open set G in $\mathbb{R}^3$. Let $f: G \to \mathbb{R}^3$ have continuous components (f_1, f_2, f_3).

(a) Show that the following line integral exists as a Riemann integral.

$$\int_a^b \sum_{i=1}^3 f_i[g(x)]g_i'(x)\,dx.$$

(b) Let $F: G \to \mathbb{R}$ be a function which has continuous derivatives D_iF, $i = 1, 2, 3$, such that $f_i(y) = D_iF(y)$ for y in G and $i = 1, 2, 3$. Prove that

$$\begin{aligned}\int_a^b \sum_{i=1}^3 f_i[g(x)]g_i'(x)\,dx &= \int_a^b \sum_{i=1}^3 D_iF[g(x)]g_i'(x)\,dx \\ &= \int_a^b (F \circ g)'(x)\,dx = F[g(b)] - F[g(a)].\end{aligned}$$

34.4. Let E be an open interval in $\mathbb{R}$, let $f: E \to \mathbb{R}$ be continuous, and let $g: [a, b] \to \mathbb{R}$ be a function which maps $[a, b]$ into E and which has a continuous derivative g'. Prove the following change-of-variable theorem.

$$\int_{g(a)}^{g(b)} f(y)\,dy = \int_a^b (f \circ g)(x)g'(x)\,dx.$$

[Outline of the proof. Observe that g is not assumed to be a monotonic function, nor is it assumed that g maps $[a, b]$ into an interval bounded by $g(a)$ and $g(b)$. In reality, $g: [a, b] \to \mathbb{R}$ defines a "curve" whose trace lies in the one-dimensional set E and has end points $g(a)$ and $g(b)$. Let $F: E \to \mathbb{R}$ be a function such that $F'(y) = f(y)$ for y in E [see Exercise 34.2]; then

$$\int_{g(a)}^{g(b)} f(y)\,dy = \int_{g(a)}^{g(b)} F'(y)\,dy = F[g(b)] - F[g(a)];$$

$$\begin{aligned}\int_a^b (f \circ g)(x)g'(x)\,dx &= \int_a^b (F' \circ g)(x)g'(x)\,dx = \int_a^b (F \circ g)'(x)\,dx \\ &= (F \circ g)(b) - (F \circ g)(a) = F[g(b)] - F[g(a)].\end{aligned}$$

34.5. Use Exercise 34.4 to show that $\int_0^1 (1 + x^2)^2 2x\,dx = 7/3$.

34.6. Find the value of $\int_0^1 x^2\,dx$ by using only the following definition of the integral.

$$\int_a^b f(x)\,dx = \lim_{n\to\infty} \sum_{i=1}^{n} f(x_i)(x_i - x_{i-1}).$$

Check your answer by Exercise 34.2. [Hint. Subdivide $[0, 1]$ into n equal subintervals (x_{i-1}, x_i) and use the formula $1^2 + 2^2 + \cdots + n^2 = n(n+1)(2n+1)/6$.]

35. The Riemann Integral in $\mathbb{R}^n$

Section 34 has provided an introduction to the Riemann integral, and it has described some of the problems to be investigated in this chapter and some of the methods to be employed in solving them. This section gives the formal definition of the Riemann integral and establishes its fundamental properties.

In the usual treatment of the Riemann integral, a function is integrated over a rectangular region and the integral is defined by rectangular subdivisions of the region. Section 34 has shown that, in some situations, there are advantages to be gained from using sequences of simplicial subdivisions. The simplexes in a simplicial subdivision form a simplicial complex. It is necessary to begin the study, however, with an even more general type of subdivision to be called a simplicial partition. The simplicial partition will be defined after a summary of notation and previous results.

As heretofore, let x denote a point with coordinates $(x^1, \cdots, x^n)$ in $\mathbb{R}^n$. The points $a_i : (a_i^1, \cdots, a_i^n)$, $i = 0, 1, \cdots, n$, are the vertices of a Euclidean simplex $[a_0, a_1, \cdots, a_n]$ which will be denoted also by $\mathbf{a}$. The symbol $\mathbf{x}$ will have several interpretations, all consistent and easily understood from the context. Let $x_i : (x_i^1, \cdots, x_i^n)$, $i = 0, 1, \cdots, n$, be $n+1$ points in $\mathbb{R}^n$ which form the vertices of a Euclidean simplex. Then $\mathbf{x}$ denotes the oriented simplex $(x_0, x_1, \cdots, x_n)$ or the oriented solid simplex $[x_0, x_1, \cdots, x_n]$. Also, $\mathbf{x}$ or $(x_0, x_1, \cdots, x_n)$ will denote the n-vector whose initial point is x_0 and whose terminal points are $x_1, \cdots, x_n$. The matrix $[x_i^j]$, $i = 0, 1, \cdots, n$, $j = 1, \cdots, n$, is a convenient representation of $\mathbf{x}$, and $\Delta(\mathbf{x})$ is the determinant $\Delta(x_0, x_1, \cdots, x_n)$ or $\det[x_i^j 1]$:

$$\Delta(\mathbf{x}) = \Delta(x_0, x_1, \cdots, x_n) = \det \begin{bmatrix} x_0^1 & \cdots & x_0^n & 1 \\ \cdots & \cdots & \cdots & \cdots \\ x_n^1 & \cdots & x_n^n & 1 \end{bmatrix}. \tag{1}$$

By Definition 14.10 and Theorem 13.1, the simplexes $\mathbf{a}$ and $\mathbf{x}$ are (nondegenerate) n-dimensional Euclidean simplexes if and only if $\Delta(\mathbf{a}) \neq 0$ and $\Delta(\mathbf{x}) \neq 0$. Theorem 15.9 shows that $\mathbf{a}$ is positively oriented in $\mathbb{R}^n$ if and only if $(-1)^n\Delta(\mathbf{a}) > 0$, and negatively oriented in $\mathbb{R}^n$ if and only if $(-1)^n\Delta(\mathbf{a}) < 0$. Finally, Definition 15.3 states that $\mathbf{x}$ and $\mathbf{a}$ have the same orientation if and only if $\Delta(\mathbf{x})\Delta(\mathbf{a}) > 0$, and the opposite orientation in $\mathbb{R}^n$ if and only if $\Delta(\mathbf{x})\Delta(\mathbf{a}) < 0$.

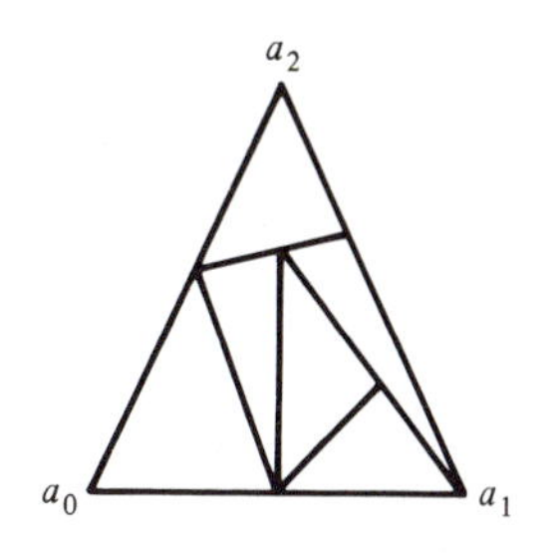

Figure 35.1. A simplicial partition P of $[a_0, a_1, a_2]$.

35.1 Definition. Let $\mathbf{a}$ be the n-dimensional Euclidean simplex $[a_0, a_1, \cdots, a_n]$ in $\mathbb{R}^n$. An *oriented simplicial partition* P of $\mathbf{a}$ is a finite set of n-dimensional Euclidean simplexes $\mathbf{x}$ with the following properties:

If x is a point in $\mathbf{a}$, then x belongs to at least one simplex $\mathbf{x}$ in P. (2)

Each $\mathbf{x}$ in P is contained in $\mathbf{a}$. (3)

The simplexes in P do not overlap; that is, the intersection of two distinct simplexes in P is either empty or it is contained in a side of each of the simplexes. (4)

Each simplex $\mathbf{x}$ in P has the same orientation as $\mathbf{a}$; that is, $\Delta(\mathbf{a})\Delta(\mathbf{x}) > 0$ for every $\mathbf{x}$ in P. (5)

35.2 Example. Figure 35.1 shows a simplicial partition P of the Euclidean simplex $\mathbf{a} : [a_0, a_1, a_2]$. The simplexes in P do not form a Euclidean complex [see Definition 16.1] and hence they do not form a simplicial subdivision of $\mathbf{a}$ as defined in Definition 21.1. Observe that a simplicial subdivision $\mathscr{P}$ of $\mathbf{a}$ is a simplicial partition P of $\mathbf{a}$, but that a simplicial partition P of $\mathbf{a}$ need not be a simplicial subdivision $\mathscr{P}$ of $\mathbf{a}$.

Let P and P' be two simplicial partitions of $\mathbf{a}$. If (i) each simplex in P' is contained in a simplex in P, and (ii) the number of simplexes in P' is greater than the number in P, then P' is called a *refinement* of P. For example, if at least one of the simplexes in the partition P of $[a_0, a_1, a_2]$ in Figure 35.1 is divided into two or more simplexes, the resulting simplicial partition of $[a_0, a_1, a_2]$ is a refinement of P.

35.3 Lemma. *Let* $\mathbf{a}$ *be an n-dimensional Euclidean simplex in* $\mathbb{R}^n$, *and let* P_1 *and* P_2 *be two simplicial partitions of* $\mathbf{a}$. *Then there exists a simplicial subdivision* $\mathscr{P}$ *of* $\mathbf{a}$ *which is a refinement of* P_1 *and also of* P_2.

PROOF. Consider first the special case shown in Figure 35.2; (a) and (b) show the partitions P_1 and P_2 of $[a_0, a_1, a_2]$. Figure 35.2(c) shows P_1 and P_2 constructed in the same simplex. Extend each side of a simplex in (c) until it intersects the boundary of $[a_0, a_1, a_2]$; those extensions are shown

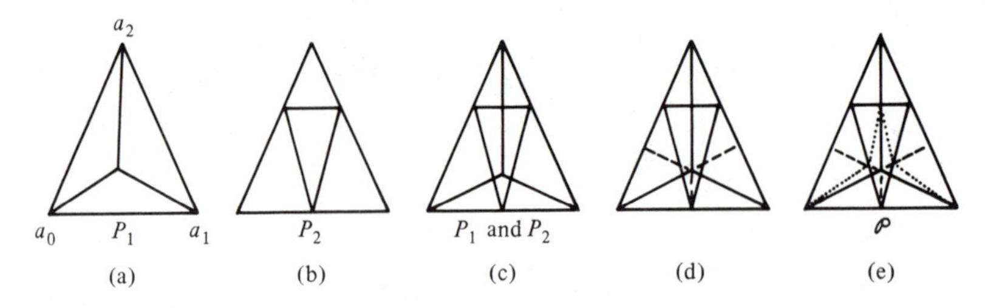

Figure 35.2. Construction of the refinement $\mathscr{P}$ of P_1 and P_2.

in dashed lines in (d). At this point the simplex $[a_0, a_1, a_2]$ is divided into a collection of several triangles and four convex quadrilaterals. Draw lines to divide the quadrilaterals into triangles. This can be done in several ways, one of which is shown by the dotted lines in Figure 35.2(e). The collection of triangles (2-simplexes) in (e) form a simplicial subdivision $\mathscr{P}$ of $[a_0, a_1, a_2]$ which is a refinement of P_1 and P_2. Proceed in the same way if $\mathbf{a}$ is an n-simplex in $\mathbb{R}^n$. Extend each $(n-1)$-dimensional side of the simplexes in P_1 and P_2 until it intersects the boundary of $\mathbf{a}$. These planes subdivide $\mathbf{a}$ into simplexes and convex polygonal or polyhedral regions. Subdivide each of the latter into simplexes. The result is a simplicial subdivision $\mathscr{P}$ of $\mathbf{a}$ which is a refinement of both P_1 and P_2. □

35.4 Corollary. *Let P be a positively oriented, simplicial partition of* $\mathbf{a}$, *and let $M(\mathbf{x})$ denote $(-1)^n\Delta(\mathbf{x})/n!$, the measure of* $\mathbf{x}$. *Then*

$$\sum_{\mathbf{x}\in P} M(\mathbf{x}) = \sum_{\mathbf{x}\in P} \frac{(-1)^n}{n!}\Delta(\mathbf{x}) = \frac{(-1)^n}{n!}\Delta(\mathbf{a}) = M(\mathbf{a}). \tag{6}$$

Proof. Let $\mathscr{P}$ be a simplicial subdivision of $\mathbf{a}$ which is a refinement of P; a $\mathscr{P}$ of this kind exists by Lemma 35.3. Let $\mathbf{x}$ and $\mathbf{y}$ denote simplexes in P and $\mathscr{P}$ respectively. Now the simplexes in the set $\{\mathbf{y}:\mathbf{y}\subset\mathbf{x}\}$ form a simplicial subdivision of $\mathbf{x}$ since $\mathscr{P}$ is a simplicial subdivision of $\mathbf{a}$ which is a refinement of P. Then

$$M(\mathbf{x}) = \sum\{M(\mathbf{y}):\mathbf{y}\subset\mathbf{x}\} \tag{7}$$

by Theorem 21.3, and

$$\sum_{\mathbf{x}\in P} M(\mathbf{x}) = \sum_{\mathbf{x}\in P}\sum\{M(\mathbf{y}):\mathbf{y}\subset\mathbf{x}\} = \sum_{\mathbf{y}\in\mathscr{P}} M(\mathbf{y}). \tag{8}$$

Since $\mathscr{P}$ is a simplicial subdivision of $\mathbf{a}$, the sum on the right in (8) is $M(\mathbf{a})$ by Theorem 21.3; therefore, (8) shows that the sum on the left in (8) is $M(\mathbf{a})$ and thus that (6) is true. □

Let $f:\mathbf{a}\to\mathbb{R}$ be a bounded function such that

$$m \leqq f(x) \leqq M, \qquad x\in\mathbf{a}, \tag{9}$$

for some constants m and M. Assume that $\mathbf{a}$ is positively oriented in

$\mathbb{R}^n : (-1)^n \Delta(\mathbf{a}) > 0$. Let P be an oriented simplicial partition of $\mathbf{a}$; then $(-1)^n \Delta(\mathbf{x}) > 0$ for each $\mathbf{x}$ in P. For each such partition P and each $\mathbf{x}$ in P set

$$m(f, \mathbf{x}) = \inf\{f(x) : x \in \mathbf{x}\}, \; M(f, \mathbf{x}) = \sup\{f(x) : x \in \mathbf{x}\}. \tag{10}$$

The infimum and supremum in (10) exist by Axiom 88.2 [see Section 88 in Appendix 2] since f is bounded by (9). Furthermore,

$$m \leqq m(f, \mathbf{x}) \leqq M(f, \mathbf{x}) \leqq M, \qquad \mathbf{x} \text{ in } P. \tag{11}$$

Define $m(f, P)$ and $M(f, P)$ by the following equations:

$$\begin{aligned} m(f, P) &= \sum_{\mathbf{x} \in P} m(f, \mathbf{x}) \frac{(-1)^n}{n!} \Delta(\mathbf{x}), \\ M(f, P) &= \sum_{\mathbf{x} \in P} M(f, \mathbf{x}) \frac{(-1)^n}{n!} \Delta(\mathbf{x}). \end{aligned} \tag{12}$$

Multiply the inequalities in (11) by $(-1)^n \Delta(\mathbf{x})/n!$, which is positive by hypothesis and which equals $M(\mathbf{x})$. Then

$$mM(\mathbf{x}) \leqq m(f, \mathbf{x}) \frac{(-1)^n}{n!} \Delta(\mathbf{x}) \leqq M(f, \mathbf{x}) \frac{(-1)^n}{n!} \Delta(\mathbf{x}) \leqq MM(\mathbf{x}). \tag{13}$$

Write these inequalities for each $\mathbf{x}$ in P and then add them; by (6) in Corollary 35.4, the sum can be simplified to this:

$$mM(\mathbf{a}) \leqq \sum_{\mathbf{x} \in P} m(f, \mathbf{x}) \frac{(-1)^n}{n!} \Delta(\mathbf{x}) \leqq \sum_{\mathbf{x} \in P} M(f, \mathbf{x}) \frac{(-1)^n}{n!} \Delta(\mathbf{x}) \leqq MM(\mathbf{a}). \tag{14}$$

In the notation of equations (12), the inequalities in (14) assert that

$$mM(\mathbf{a}) \leqq m(f, P) \leqq M(f, P) \leqq MM(\mathbf{a}) \tag{15}$$

for every oriented simplicial partition P of $\mathbf{a}$. The sums $m(f, P)$ and $M(f, P)$ are called the lower and upper Riemann sums for f corresponding to the partition P.

35.5 Lemma. *Let P be an oriented simplicial partition of $\mathbf{a}$, and let $\mathscr{P}$ be an oriented simplicial subdivision of $\mathbf{a}$ which is a refinement of P. Then*

$$m(f, P) \leqq m(f, \mathscr{P}) \leqq M(f, \mathscr{P}) \leqq M(f, P). \tag{16}$$

Proof. The center inequality in (16) follows from (15); thus only the first and third inequalities remain to be proved. As in the proof of Corollary 35.4, let $\mathbf{x}$ and $\mathbf{y}$ denote simplexes in P and $\mathscr{P}$, respectively. Now the simplexes in the set $\{\mathbf{y} : \mathbf{y} \subset \mathbf{x}\}$ form a simplicial subdivision of $\mathbf{x}$ since $\mathscr{P}$ is a simplicial subdivision of $\mathbf{a}$ which is a refinement of P. Then

$$\frac{(-1)^n}{n!} \Delta(\mathbf{x}) = \sum \left\{ \frac{(-1)^n}{n!} \Delta(\mathbf{y}) : \mathbf{y} \subset \mathbf{x} \right\} \tag{17}$$

by equation (7) above. Multiply the two sides of (17) by $m(f; \mathbf{x})$; then since $m(f, \mathbf{x}) \leqq m(f, y)$ for each $\mathbf{y} \subset \mathbf{x}$,

$$m(f, \mathbf{x})\frac{(-1)^n}{n!}\Delta(\mathbf{x}) = \sum\left\{m(f, \mathbf{x})\frac{(-1)^n}{n!}\Delta(\mathbf{y}) : \mathbf{y} \subset \mathbf{x}\right\} \leqq \sum\left\{m(f, \mathbf{y})\frac{(-1)^n}{n!}\Delta(\mathbf{y}) : \mathbf{y} \subset \mathbf{x}\right\}. \tag{18}$$

Now sum these inequalities over all simplexes $\mathbf{x}$ in P. Since $\mathscr{P}$ is a refinement of P, the resulting inequality and (12) show that $m(f, P) \leqq m(f, \mathscr{P})$, which is the inequality on the left in (16). To prove the inequality on the right in (16), multiply the two sides of (17) by $M(f, \mathbf{x})$ and use the fact that $M(f, \mathbf{y}) \leqq M(f, \mathbf{x})$ for each $\mathbf{y} \subset \mathbf{x}$ to obtain

$$\sum\left\{M(f, \mathbf{y})\frac{(-1)^n}{n!}\Delta(\mathbf{y}) : \mathbf{y} \subset \mathbf{x}\right\} \leqq M(f, \mathbf{x})\frac{(-1)^n}{n!}\Delta(\mathbf{x}). \tag{19}$$

Sum this inequality over all $\mathbf{x}$ in P; since $\mathscr{P}$ is a refinement of P, the resulting inequality, by (12), is $M(f, \mathscr{P}) \leqq M(f, P)$. The proof of all parts of (16) and of Lemma 35.5 is complete. □

35.6 Lemma. *Let $f : \mathbf{a} \to \mathbb{R}$ be a bounded function which satisfies (9). Then every lower Riemann sum of f is equal to or less than every upper Riemann sum of f. Furthermore,* $\sup\{m(f, P) : P$ *is a partition of* $\mathbf{a}\}$ *and* $\inf\{M(f, P) : P$ *is a partition of* $\mathbf{a}\}$ *exist, and*

$$\sup_P m(f, P) \leqq \inf_P M(f, P). \tag{20}$$

Proof. Compare the lower and upper Riemann sums of f corresponding to two arbitrary simplicial partitions P_1 and P_2 of $\mathbf{a}$. Let $\mathscr{P}$ be a simplicial subdivision of $\mathbf{a}$ which is a refinement of both P_1 and P_2; a $\mathscr{P}$ of this kind exists by Lemma 35.3. Then since $\mathscr{P}$ is a refinement of both P_1 and P_2, Lemma 35.5 shows that

$$\left.\begin{matrix} m(f, P_1) \\ m(f, P_2) \end{matrix}\right\} \leqq m(f, \mathscr{P}) \leqq M(f, \mathscr{P}) \leqq \left\{\begin{matrix} M(f, P_1), \\ M(f, P_2). \end{matrix}\right. \tag{21}$$

Here $m(f, P_1)$ [or $m(f, P_2)$] represents an arbitary lower Riemann sum of f and $M(f, P_2)$ [or $M(f, P_1)$] represents an arbitrary upper Riemann sum of f. Thus (21) shows that every lower sum is equal to or less than every upper sum, and the proof of the first conclusion in the lemma is complete.

The set $\{m(f, P) : P$ is a partition of $\mathbf{a}\}$ is bounded above, and the set $\{M(f, P) : P$ is a partition of $\mathbf{a}\}$ is bounded below, both by (15). Then the first of these sets has a supremum and the second has an infimum, both by Axiom 88.2 in Appendix 2. This statement completes the proof of the second conclusion in the lemma, and only (20) remains to be proved. To prove (20), assume the statement false and show that a contradiction results. If (20) is false, then by (2) in Section 88,

$$\sup_P m(f, P) > \inf_P M(f, P). \tag{22}$$

Set

$$d = \sup_P m(f, P) - \inf_P M(f, P); \tag{23}$$

then $d > 0$ by (22). Let ε be a number such that $0 < \varepsilon < d/4$. Then by Theorem 88.3 there exist simplicial partitions P_1 and P_2 of $\mathbf{a}$ such that

$$m(f, P_1) > \sup_P m(f, P) - \varepsilon, \qquad M(f, P_2) < \inf_P M(f, P) + \varepsilon. \tag{24}$$

Subtract the second of these inequalities from the first; then by (23),

$$m(f, P_1) - M(f, P_2) > \sup_P m(f, P) - \inf_P M(f, P) - 2\varepsilon > (d/2) > 0. \tag{25}$$

Therefore $m(f, P_1) > M(f, P_2)$. But this statement contradicts (21); hence, the assumption in (22) has led to a contradiction and (20) is true. This statement completes the proof of Lemma 35.6. □

35.7 Definition. Let $f : \mathbf{a} \to \mathbb{R}$ be a bounded function. If

$$\sup_P m(f, P) = \inf_P M(f, P), \tag{26}$$

then f is *Riemann integrable on* $\mathbf{a}$, the Riemann integral $\int_{\mathbf{a}} f(x)\, d(x^1, \cdots, x^n)$ *exists*, and

$$\int_{\mathbf{a}} f(x)\, d(x^1, \cdots, x^n) = \sup_P m(f, P) = \inf_P M(f, P). \tag{27}$$

35.8 Theorem. *If $f : \mathbf{a} \to \mathbb{R}$ is continuous on the closed simplex $\mathbf{a} : [a_0, a_1, \cdots, a_n]$, then f is Riemann integrable on $\mathbf{a}$.*

Proof. Since $\mathbf{a}$ is closed and bounded, $\mathbf{a}$ is compact by Definition 92.8. Then f is continuous on a compact set, and f has a minimum and a maximum value by Theorem 96.14. Therefore f is bounded on $\mathbf{a}$, and the bounds m and M can be chosen as follows.

$$m = \min\{f(x) : x \in \mathbf{a}\}, \qquad M = \max\{f(x) : x \notin \mathbf{a}\}. \tag{28}$$

Then by Lemma 35.6, $\sup\{m(f, P) : P \text{ is a partition of } \mathbf{a}\}$ and $\inf\{M(f, P) : P$ is a partition of $\mathbf{a}\}$ exist and

$$\sup_P m(f, P) \leqq \inf_P M(f, P). \tag{29}$$

By Definition 35.7, the proof can be completed by showing that the equality holds in (29). Since f is continuous on the compact set $\mathbf{a}$, it is uniformly continuous by Theorem 96.18. Thus for each $\varepsilon > 0$ there exists a $\delta(\varepsilon)$ such that

$$|f(x_1) - f(x_2)| < \varepsilon, \qquad x_1, x_2 \text{ in } \mathbf{a}, \qquad |x_1 - x_2| < \delta(\varepsilon). \tag{30}$$

Let P be a simplicial partition of $\mathbf{a}$ whose norm is less than $\delta(\varepsilon)$; the norm of P is the maximum diameter of a simplex $\mathbf{x}$ in P [see Definition 92.9]. There exists a P with norm less than $\delta(\varepsilon)$ by Definition 2.5 and Theorem 21.7. Now

$$M(f, P) - m(f, P) = \sum_{\mathbf{x} \in P} [M(f, \mathbf{x}) - m(f, \mathbf{x})] \frac{(-1)^n}{n!} \Delta(\mathbf{x}) \tag{31}$$

by the definitions in (10) and (12). Thus

$$|M(f, P) - m(f, P)| \leqq \sum_{\mathbf{x} \in P} |M(f, \mathbf{x}) - m(f, \mathbf{x})| \frac{|\Delta(\mathbf{x})|}{n!}. \tag{32}$$

Now f is uniformly continuous on $\mathbf{a}$, and the diameter of each $\mathbf{x}$ in P is less than $\delta(\varepsilon)$; then

$$|M(f, \mathbf{x}) - m(f, \mathbf{x})| \leqq \varepsilon, \qquad \mathbf{x} \text{ in } P. \tag{33}$$

If this statement were not true, then in some $\mathbf{x}$ in P there would be two points x_1, x_2 such that

$$|f(x_1) - f(x_2)| > \varepsilon, \qquad |x_1 - x_2| < \delta(\varepsilon). \tag{34}$$

Since (34) contradicts (30), then (33) is true. Then since

$$\sum_{\mathbf{x} \in P} \frac{|\Delta(\mathbf{x})|}{n!} = M(\mathbf{a}) \tag{35}$$

by Corollary 35.4, equations (32) and (33) show that

$$|M(f, P) - m(f, P)| \leqq \varepsilon M(\mathbf{a}). \tag{36}$$

Thus for each $\varepsilon > 0$ there is a partition P for which (36) is true; this statement proves that the equality holds in (29). Then $\int_{\mathbf{a}} f(x)\,d(x^1, \cdots, x^n)$ exists by Definition 35.7, and the proof of Theorem 35.8 is complete. □

35.9 Corollary. *Let $f : \mathbf{a} \to \mathbb{R}$ be a function which is continuous on $\mathbf{a}$, and let P_k, $k = 1, 2, \cdots$, be a sequence of simplicial partitions of $\mathbf{a}$ for which the norm of P_k tends to zero as k tends to infinity. Then*

$$\int_{\mathbf{a}} f(x)\,d(x^1, \cdots, x^n) = \lim_{k \to \infty} m(f, P_k) = \lim_{k \to \infty} M(f, P_k). \tag{37}$$

PROOF. Since f is Riemann integrable on $\mathbf{a}$ by Theorem 35.8, then

$$m(f, P_k) \leqq \sup_P m(f, P) = \inf_P M(f, P) \leqq M(f, P_k), \qquad k = 1, 2, \cdots. \tag{38}$$

By (27) and (38),

$$\begin{aligned}\left|m(f, P_k) - \int_{\mathbf{a}} f(x)\,d(x^1, \cdots, x^n)\right| &= \left|\inf_P M(f, P) - m(f, P_k)\right| \\ &\leqq \left|M(f, P_k) - m(f, P_k)\right|.\end{aligned} \tag{39}$$

Let $\varepsilon > 0$ be given and let $\delta(\varepsilon)$ be the number defined in (30). Since the norm of P_k tends to zero as $k \to \infty$, there exists a k_0 such that the norm of P_k is less than $\delta(\varepsilon)$ for $k \geqq k_0$. Then by (39) and (36),

$$\left|m(f, P_k) - \int_{\mathbf{a}} f(x)\,d(x^1, \cdots, x^n)\right| < \varepsilon M(\mathbf{a}), \qquad k \geqq k_0. \tag{40}$$

This statement proves the first equality in (37). Similarly, (27) and (38) show that

$$\begin{aligned}\left|M(f, P_k) - \int_{\mathbf{a}} f(x)\,d(x^1, \cdots, x^n)\right| &\leqq \left|M(f, P_k) - m(f, P_k)\right| \\ &< \varepsilon M(\mathbf{a}), \qquad k \geqq k_0,\end{aligned} \tag{41}$$

and this statement establishes the second equality in (37). □

35.10 Corollary. *Let $f: \mathbf{a} \to \mathbb{R}$ be a function which is continuous on the positively oriented n-dimensional simplex $\mathbf{a}$ in $\mathbb{R}^n$, and let P_k, $k = 1, 2, \cdots$, be a sequence of simplicial partitions of $\mathbf{a}$ whose norms tend to zero as $k \to \infty$. Finally, for each $\mathbf{x}$ in P_k, let ξ denote a point in $\mathbf{x}$. Then*

$$\lim_{k\to\infty} \sum_{\mathbf{x} \in P_k} f(\xi)\frac{(-1)^n}{n!}\Delta(\mathbf{x}) = \int_{\mathbf{a}} f(x)\,d(x^1, \cdots, x^n). \tag{42}$$

PROOF. The definitions of $m(f, P_k)$ and $M(f, P_k)$ in equations (10) and (12) show that

$$m(f, P_k) \leqq \sum_{\mathbf{x} \in P_k} f(\xi)\frac{(-1)^n}{n!}\Delta(\mathbf{x}) \leqq M(f, P_k), \qquad k = 1, 2, \cdots. \tag{43}$$

Since

$$m(f, P_k) \leqq \int_{\mathbf{a}} f(x)\,d(x^1, \cdots, x^n) \leqq M(f, P_k), \qquad k = 1, 2, \cdots, \tag{44}$$

then (43) and (44) show that

$$\left|\sum_{\mathbf{x} \in P_k} f(\xi)\frac{(-1)^n}{n!}\Delta(\mathbf{x}) - \int_{\mathbf{a}} f(x)\,d(x^1, \cdots, x^n)\right| \leqq M(f, P_k) - m(f, P_k) \tag{45}$$

for $k = 1, 2, \cdots$. Since the expression on the right in this inequality approaches zero as $k \to \infty$ by (37) in Corollary 35.9, the statement in (42) follows. □

Thus far the integral of f over a negatively oriented simplex has not been defined, but Corollary 35.10 now suggests how this definition should be made. If $\mathbf{a}$ and $\mathbf{x}$ are positively oriented simplexes, then $-\mathbf{a}$ and $-\mathbf{x}$ denote the same simplexes with the negative orientation in $\mathbb{R}^n$. Let $f: \mathbf{a} \to \mathbb{R}$ be a continuous function, and let P_k, $k = 1, 2, \cdots$, be a sequence of simplicial partitions of $\mathbf{a}$ whose norms tend to zero. Multiply the inequalities in (43) by -1; since $-\Delta(\mathbf{x}) = \Delta(-\mathbf{x})$, the result can be written in the following form:

$$-M(f, P_k) \leqq \sum_{\mathbf{x} \in P_k} f(\xi)\frac{(-1)^n}{n!}\Delta(-\mathbf{x}) \leqq -m(f, P_k). \tag{46}$$

Now $\{-\mathbf{x} : \mathbf{x} \in P_k\}$ is a simplicial partition of the negatively oriented simplex $-\mathbf{a}$ by Definition 35.1, and by (37) in Corollary 35.9,

$$\lim_{k\to\infty} [-M(f, P_k)] = \lim_{k\to\infty} [-m(f, P_k)] = -\int_{\mathbf{a}} f(x)\,d(x^1, \cdots, x^n). \tag{47}$$

Then (47) and the inequalities in (46) show that

$$\lim_{k\to\infty} \sum_{\mathbf{x} \in P_k} f(\xi)\frac{(-1)^n}{n!}\Delta(-\mathbf{x}) = -\int_{\mathbf{a}} f(x)\,d(x^1, \cdots, x^n). \tag{48}$$

This equation suggests the following definition.

35.11 Definition. If $\mathbf{a}$ is positively oriented in $\mathbb{R}^n$, then $-\mathbf{a}$ is negatively oriented in $\mathbb{R}^n$, and the Riemann integral of f over $-\mathbf{a}$, denoted by $\int_{-\mathbf{a}} f(x)\,d(x^1, \cdots, x^n)$, is defined as follows. Let P_k, $k = 1, 2, \cdots$, be a sequence of simplicial partitions of $\mathbf{a}$ whose norms approach zero. Then

$$\begin{aligned}\int_{-\mathbf{a}} f(x)\,d(x^1, \cdots, x^n) &= \lim_{k\to\infty} \sum_{\mathbf{x} \in P_k} f(\xi)\frac{(-1)^n}{n!}\Delta(-\mathbf{x}) \\ &= -\int_{\mathbf{a}} f(x)\,d(x^1, \cdots, x^n).\end{aligned} \tag{49}$$

Observe that both $\int_{\mathbf{a}} f(x)\,d(x^1, \cdots, x^n)$ and $\int_{-\mathbf{a}} f(x)\,d(x^1, \cdots, x^n)$ exist if f is continuous on $\mathbf{a}$. Finally, let $\mathbf{a}$ be an n-dimensional Euclidean simplex in $\mathbb{R}^n$ which is either positively oriented or negatively oriented in $\mathbb{R}^n$. Recall (5), and let P_k, $k = 1, 2, \cdots$, be a sequence of simplicial partitions of $\mathbf{a}$ whose norms approach zero. Then (42) and (49) show that in all cases

$$\int_{\mathbf{a}} f(x)\,d(x^1, \cdots, x^n) = \lim_{k\to\infty} \sum_{\mathbf{x} \in P_k} f(\xi)\frac{(-1)^n}{n!}\Delta(\mathbf{x}). \tag{50}$$

Thus far the Riemann integral has been defined only over simplicial regions $\mathbf{a}$. The same methods which have been used in this case, however, can be used to define the Riemann integral over cubes and rectangular regions in $\mathbb{R}^n$, and, in fact, over any region in $\mathbb{R}^n$, bounded by planes, for which sequences of oriented partitions P_k, $k = 1, 2, \cdots$, can be constructed. There is nothing essentially new in these cases, and the details are left to the reader. The remainder of this section establishes some of the fundamental properties of the Riemann integral.

35.12 Theorem. *Let* **a** *be an oriented Euclidean simplex in* $\mathbb{R}^n$*; let* c, c_1, c_2 *be constants, and let* $f_i: \mathbf{a} \to \mathbb{R}$, $i = 1, 2$, *be continuous functions. Then the constant function* c *and the function* $c_1 f_1 + c_2 f_2$ *are continuous and therefore integrable, and*

$$\int_{\mathbf{a}} c\, d(x^1, \cdots, x^n) = c\frac{(-1)^n}{n!}\Delta(\mathbf{a}), \tag{51}$$

$$\begin{aligned}\int_{\mathbf{a}} [c_1 f_1(x) + c_2 f_2(x)]\, d(x^1, \cdots, x^n) &= c_1 \int_{\mathbf{a}} f_1(x)\, d(x^1, \cdots, x^n) \\ &\quad + c_2 \int_{\mathbf{a}} f_2(x)\, d(x^1, \cdots, x^n).\end{aligned} \tag{52}$$

PROOF. By (50) and (6),

$$\begin{aligned}\int_{\mathbf{a}} c\, d(x^1, \cdots, x^n) &= \lim_{k\to\infty} \sum_{\mathbf{x}\in P_k} c\frac{(-1)^n}{n!}\Delta(\mathbf{x}) \\ &= c \lim_{k\to\infty} \sum_{\mathbf{x}\in P_k} \frac{(-1)^n}{n!}\Delta(\mathbf{x}) = c\frac{(-1)^n}{n!}\Delta(\mathbf{a}).\end{aligned}$$

These statements establish (51). By Example 96.5, the function $c_1 f_1 + c_2 f_2$ is continuous since f_1 and f_2 are continuous by hypothesis; its integral exists by Theorem 35.8, and

$$\begin{aligned}&\int_{\mathbf{a}} [c_1 f_1(x) + c_2 f_2(x)]\, d(x^1, \cdots, x^n) \\ &\qquad = \lim_{k\to\infty} \sum_{\mathbf{x}\in P_k} [c_1 f_1(\xi) + c_2 f_2(\xi)]\frac{(-1)^n}{n!}\Delta(\mathbf{x}) \\ &\qquad = c_1 \lim_{k\to\infty} \sum_{\mathbf{x}\in P_k} f_1(\xi)\frac{(-1)^n}{n!}\Delta(\mathbf{x}) + c_2 \lim_{k\to\infty} \sum_{\mathbf{x}\in P_k} f_2(\xi)\frac{(-1)^n}{n!}\Delta(\mathbf{x}) \\ &\qquad = c_1 \int_{\mathbf{a}} f_1(x)\, d(x^1, \cdots, x^n) + c_2 \int_{\mathbf{a}} f_2(x)\, d(x^1, \cdots, x^n).\end{aligned}$$

These statements establish (52) and complete the proof of Theorem 35.12. □

35.13 Theorem. *Let* **a** *be an oriented Euclidean simplex in* $\mathbb{R}^n$*, let* $f: \mathbf{a} \to \mathbb{R}$ *be a continuous function, and let* $\{\mathbf{a}_i : i = 1, \cdots, N\}$ *be a simplicial partition of* **a**. *Then* f *is Riemann integrable on* **a** *and on each of the simplexes* $\mathbf{a}_i$, $i = 1, \cdots, n$, *and*

$$\int_{\mathbf{a}} f(x)\, d(x^1, \cdots, x^n) = \sum_{i=1}^{N} \int_{\mathbf{a}_i} f(x)\, d(x^1, \cdots, x^n). \tag{53}$$

PROOF. Since f is continuous on **a**, the integrals in (53) exist by Theorem 35.8. In each simplex $\mathbf{a}_i$ construct a sequence of simplicial partitions P_k^i, $k = 1, 2,$

$\cdots$, whose norms approach zero as $k \to \infty$. For each k, combine the partitions P_k^i of $\mathbf{a}_i$, $i = 1, \cdots, N$, to form the partition P_k of $\mathbf{a}$. By (50), the definition of P_k, and the properties of sums and limits,

$$\begin{aligned}\int_{\mathbf{a}} f(x)\,d(x^1, \cdots, x^n) &= \lim_{k\to\infty} \sum_{\mathbf{x}\in P_k} f(\xi)\frac{(-1)^n}{n!}\Delta(\mathbf{x}) \\ &= \lim_{k\to\infty} \sum_{i=1}^{N} \sum_{\mathbf{x}\in P_k^i} f(\xi)\frac{(-1)^n}{n!}\Delta(\mathbf{x}) \\ &= \sum_{i=1}^{N} \left\{ \lim_{k\to\infty} \sum_{\mathbf{x}\in P_k^i} f(\xi)\frac{(-1)^n}{n!}\Delta(\mathbf{x}) \right\} \\ &= \sum_{i=1}^{N} \int_{\mathbf{a}_i} f(x)\,d(x^1, \cdots, x^n).\end{aligned}$$

The proof of (53) and of Theorem 35.13 is complete. □

35.14 Theorem. *If* $\mathbf{a}$ *is positively oriented in* $\mathbb{R}^n$, *and if* $f: \mathbf{a} \to \mathbb{R}$ *is a continuous function such that* $f(x) \geqq 0$ *for every* x *in* $\mathbf{a}$, *then*

$$\int_{\mathbf{a}} f(x)\,d(x^1, \cdots, x^n) \geqq 0, \tag{54}$$

and the equality holds if and only if $f(x) = 0$ *for every* x *in* $\mathbf{a}$.

Proof. Let P_k, $k = 1, 2, \cdots$, be a sequence of simplicial partitions of $\mathbf{a}$. If f is non-negative in $\mathbf{a}$, and if $\mathbf{a}$ is positively oriented in $\mathbb{R}^n$, then

$$\sum_{\mathbf{x}\in P_k} f(\xi)\frac{(-1)^n}{n!}\Delta(\mathbf{x}) \geqq 0, \qquad k = 1, 2, \cdots. \tag{55}$$

By (42) in Corollary 35.10,

$$\int_{\mathbf{a}} f(x)\,d(x^1, \cdots, x^n) = \lim_{k\to\infty} \sum_{\mathbf{x}\in P_k} f(\xi)\frac{(-1)^n}{n!}\Delta(\mathbf{x}) \geqq 0. \tag{56}$$

Thus (54) is true. Also, if $f(x) = 0$ for every x in $\mathbf{a}$, then the equality holds in (55) and (56). Finally, to complete the proof of the theorem, it is necessary to show that $f(x) = 0$ for every x in $\mathbf{a}$ if the equality holds in (54). Assume that this statement is false and show that a contradiction results; that is, assume that

$$\int_{\mathbf{a}} f(x)\,d(x^1, \cdots, x^n) = 0, \qquad f(x) \geqq 0 \text{ for every } x \text{ in } \mathbf{a}, \tag{57}$$

and that there is a point x_0 in $\mathbf{a}$ such that $f(x_0) > 0$. Then there is a neighborhood $N(x_0, r)$ of x_0 such that $f(x) > f(x_0)/2$ for every x in $N(x_0, r) \cap \mathbf{a}$ [see equation (9) in the proof of Theorem 96.9 in Appendix 2]. Now by Section 21 there exists a simplicial subdivision $\{\mathbf{a}_i : i = 1, \cdots, N\}$ of $\mathbf{a}$ in which at least one simplex—call it $\mathbf{a}_1$—is contained in $N(x_0, r) \cap \mathbf{a}$. Then (42) in Corollary 35.10 can be used to show that

$$\int_{\mathbf{a}_1} f(x)\,d(x^1, \cdots, x^n) \geqq \frac{f(x_0)}{2}\frac{(-1)^n}{n!}\Delta(\mathbf{a}_1) > 0. \tag{58}$$

The part of the theorem already proved shows that

$$\int_{\mathbf{a}_i} f(x)\,d(x^1, \cdots, x^n) \geqq 0, \qquad i = 2, \cdots, N. \tag{59}$$

Then (53), (58), and (59) show that

$$\int_{\mathbf{a}} f(x)\,d(x^1, \cdots, x^n) = \sum_{i=1}^{N} \int_{\mathbf{a}_i} f(x)\,d(x^1, \cdots, x^n) \geqq \frac{f(x_0)}{2}\frac{(-1)^n}{n!}\Delta(\mathbf{a}_1) > 0. \tag{60}$$

This statement contradicts the assumption in (57). Thus the assumption $f(x_0) > 0$ has led to a contradiction; therefore, if f is continuous and $f(x) \geqq 0$ in $\mathbf{a}$, and if $\int_{\mathbf{a}} f(x)\,d(x^1, \cdots, x^n) = 0$, then $f(x) = 0$ at every point x in $\mathbf{a}$. This statement completes the proof of Theorem 35.14. □

35.15 Corollary. *Let* $\mathbf{a}$ *be a Euclidean simplex which is positively oriented in* $\mathbb{R}^n$, *and let* $f_i : \mathbf{a} \to \mathbb{R}$, $i = 1, 2$, *be functions which are continuous on* $\mathbf{a}$. *If* $f_1(x) \leqq f_2(x)$ *for every* x *in* $\mathbf{a}$, *then*

$$\int_{\mathbf{a}} f_1(x)\,d(x^1, \cdots, x^n) \leqq \int_{\mathbf{a}} f_2(x)\,d(x^1, \cdots, x^n). \tag{61}$$

and the equality holds if and only if $f_1(x) = f_2(x)$ *for every* x *in* $\mathbf{a}$.

Proof. The function $f_2 - f_1$ is continuous and integrable by Theorem 35.12, and $f_2(x) - f_1(x) \geqq 0$ in $\mathbf{a}$ since $f_1(x) \leqq f_2(x)$ by hypothesis. Then, by Theorem 35.14,

$$\int_{\mathbf{a}} [f_2(x) - f_1(x)]\,d(x^1, \cdots, x^n) \geqq 0. \tag{62}$$

Since

$$\begin{aligned}\int_{\mathbf{a}} [f_2(x) - f_1(x)]\,d(x^1, \cdots, x^n) \\ = \int_{\mathbf{a}} f_2(x)\,d(x^1, \cdots, x^n) - \int_{\mathbf{a}} f_1(x)\,d(x^1, \cdots, x^n)\end{aligned} \tag{63}$$

by Theorem 35.12, inequality (62) is equivalent to (61). Since, by Theorem 35.14, the equality holds in (62) if and only if $f_2(x) - f_1(x) = 0$ for every x in $\mathbf{a}$, the equality holds in (61) if and only if $f_1(x) = f_2(x)$ for every x in $\mathbf{a}$. □

35.16 Corollary. *If* $\mathbf{a}$ *is an n-dimensional Euclidean simplex which is positively oriented in* $\mathbb{R}^n$, *and if* $f : \mathbf{a} \to \mathbb{R}$ *is continuous, then*

$$\left|\int_{\mathbf{a}} f(x)\,d(x^1, \cdots, x^n)\right| \leqq \int_{\mathbf{a}} |f(x)|\,d(x^1, \cdots, x^n). \tag{64}$$

The equality holds if and only if either $f(x) = |f(x)|$ *for every* x *in* $\mathbf{a}$ *or* $f(x) = -|f(x)|$ *for every* x *in* $\mathbf{a}$.

PROOF. Now $\int_{\mathbf{a}} -f(x)\,d(x^1, \cdots, x^n) = -\int_{\mathbf{a}} f(x)\,d(x^1, \cdots, x^n)$ by (52) in Theorem 35.12, and $|f|$ is continuous since f is continuous. Since $f(x) \leqq |f(x)|$ and $-f(x) \leqq |f(x)|$, then by Corollary 35.15,

$$\begin{aligned} \int_{\mathbf{a}} f(x)\,d(x^1, \cdots, x^n) &\leqq \int_{\mathbf{a}} |f(x)|\,d(x^1, \cdots, x^n), \\ -\int_{\mathbf{a}} f(x)\,d(x^1, \cdots, x^n) &\leqq \int_{\mathbf{a}} |f(x)|\,d(x^1, \cdots, x^n). \end{aligned} \tag{65}$$

Inequality (64) follows from the two inequalities in (65). Next, if $f(x) = |f(x)|$ for every x in $\mathbf{a}$, then the equality holds in (64); also, if $f(x) = -|f(x)|$, then

$$\left|\int_{\mathbf{a}} f(x)\,d(x^1, \cdots, x^n)\right| = \left|-\int_{\mathbf{a}} |f(x)|\,d(x^1, \cdots, x^n)\right| = \int_{\mathbf{a}} |f(x)|\,d(x^1, \cdots, x^n),$$

and the equality holds in (64). Finally, if the equality holds in (64), then the equality holds in one of the inequalities in (65): if in the first one, then $f(x) = |f(x)|$ for every x in $\mathbf{a}$ by Corollary 35.15; if in the second one, then $f(x) = -|f(x)|$ for every x in $\mathbf{a}$ by the same corollary. □

35.17 Corollary. *If* $\mathbf{a}$ *is an n-dimensional Euclidean simplex which is positively oriented in* $\mathbb{R}^n$, *if* m *and* M *are constants, and if* $f: \mathbf{a} \to \mathbb{R}$ *is a continuous function such that*

$$m \leqq f(x) \leqq M, \qquad x \in \mathbf{a}, \tag{66}$$

then

$$m\frac{(-1)^n}{n!}\Delta(\mathbf{a}) \leqq \int_{\mathbf{a}} f(x)\,d(x^1, \cdots, x^n) \leqq M\frac{(-1)^n}{n!}\Delta(\mathbf{a}). \tag{67}$$

If f *is constant in* $\mathbf{a}$ *and* $f(x) = m$ [*or* $f(x) = M$] *for every* x *in* $\mathbf{a}$, *then in* (67) *the equality holds on the left* [*or on the right*]. *Also, if* (66) *is true and the equality in* (67) *holds on the left* [*or on the right*], *then* $f(x) = m$ [*or* $f(x) = M$] *for every* x *in* $\mathbf{a}$.

PROOF. A constant function is continuous and hence Riemann integrable. Then (67) follows from (66), (61), and (51). Next, if $f(x) = m$ [or $f(x) = M$] for every x in $\mathbf{a}$, then (51) shows that in (67) the equality holds on the left [or on the right]. Finally, assume (66) and assume that the equality holds on the left in (67). Then

$$\int_{\mathbf{a}} f(x)\,d(x^1, \cdots, x^n) = \int_{\mathbf{a}} m\,d(x^1, \cdots, x^n) \tag{68}$$

and $f(x) = m$ for every x in $\mathbf{a}$ by Corollary 35.15. If the equality holds on the right in (67), similar arguments show that $f(x) = M$ for every x in $\mathbf{a}$. □

35.18 Theorem (The Mean-Value Theorem for Integrals). *Let* $\mathbf{a}$ *be an oriented Euclidean n-simplex in* $\mathbb{R}^n$, *and let* $f: \mathbf{a} \to \mathbb{R}$ *be a continuous function. Then there exists a point* x^* *on the interior of* $\mathbf{a}$ *such that*

$$\int_{\mathbf{a}} f(x)\, d(x^1, \cdots, x^n) = f(x^*) \frac{(-1)^n}{n!} \Delta(\mathbf{a}). \tag{69}$$

Proof. Consider first the case in which $\mathbf{a}$ is positively oriented in $\mathbb{R}^n$; then

$$(-1)^n \Delta(\mathbf{a}) > 0. \tag{70}$$

In (67), let m and M denote respectively the minimum and maximum of f on $\mathbf{a}$; this minimum and maximum exist by Theorem 96.14 since f is continuous on the compact set $\mathbf{a}$. Then (67) and (70) show that

$$m \leqq \frac{n!}{(-1)^n \Delta(\mathbf{a})} \int_{\mathbf{a}} f(x)\, d(x^1, \cdots, x^n) \leqq M. \tag{71}$$

If either equality holds in (71), then Corollary 33.17 shows that f is constant on $\mathbf{a}$ and that $f(x) = m = M$ for every x in $\mathbf{a}$. Let x^* be a point on the interior of $\mathbf{a}$; then $f(x^*) = m = M$ and (71) shows that (69) is true. In the remaining cases (71) has the form

$$m < \frac{n!}{(-1)^n \Delta(\mathbf{a})} \int_{\mathbf{a}} f(x)\, d(x^1, \cdots, x^n) < M. \tag{72}$$

Let x_m and x_M be points in $\mathbf{a}$ at which f assumes its minimum and maximum values respectively; then $f(x_m) = m$ and $f(x_M) = M$. Define the function $g: [0, 1] \to \mathbb{R}$ as follows:

$$g(t) = f[tx_M + (1 - t)x_m], \qquad 0 \leqq t \leqq 1. \tag{73}$$

Then g is a continuous function since it is the composition of two continuous functions. Also

$$g(0) = f(x_m) = m, \qquad g(1) = f(x_M) = M. \tag{74}$$

Then by the elementary intermediate-value theorem [see Exercises 7.9 and 29.1], g takes on every value between m and M. One such value is the middle term in (72). Thus there is a value of t, call it t^*, such that $0 < t^* < 1$ and

$$g(t^*) = f[t^* x_M + (1 - t^*)x_m] = \frac{n!}{(-1)^n \Delta(\mathbf{a})} \int_{\mathbf{a}} f(x)\, d(x^1, \cdots, x^n). \tag{75}$$

Define x^* to be $t^* x_M + (1 - t^*)x_m$; then (75) states that

$$f(x^*) = \frac{n!}{(-1)^n \Delta(\mathbf{a})} \int_{\mathbf{a}} f(x)\, d(x^1, \cdots, x^n), \tag{76}$$

and this equation is equivalent to (69) in Theorem 35.18 if x^* is an interior point of $\mathbf{a}$. But if x_m and x_M are points in the same side of $\mathbf{a}$, then all of the points $tx_M + (1 - t)x_m$ for $0 \leqq t \leqq 1$ are in the boundary of $\mathbf{a}$ and not in the interior. To prove that a point x^* with the desired properties exists in this case, proceed as follows. Let $\mathbf{a}$ be the simplex $(a_0, a_1, \cdots, a_n)$, and define y to be the point $[1/(n+1)](a_0 + a_1 + \cdots + a_n)$. Here y is a point in $\mathbf{a}$ [see Section 14], and it is not on a side of $\mathbf{a}$ since

$$\begin{aligned}(-1)^n\Delta(a_0, a_1, \cdots, y/a_i, \cdots, a_n)\\ = [1/(n+1)](-1)^n\Delta(\mathbf{a}) > 0, \qquad i = 0, 1, \cdots, n.\end{aligned} \tag{77}$$

Define $g: [0, 2] \to R$ as follows:

$$g(t) = \begin{cases} f[ty + (1-t)x_m], & 0 \leqq t \leqq 1, \\ f[(t-1)x_M + (2-t)y], & 1 \leqq t \leqq 2. \end{cases} \tag{78}$$

Now g is continuous and

$$g(0) = f(x_m) = m, \qquad g(2) = f(x_M) = M. \tag{79}$$

Also, since y is in the interior of the convex set $\mathbf{a}$, the segments

$$\begin{aligned} &ty + (1-t)x_m, && 0 < t \leqq 1, \\ &(t-1)x_M + (2-t)y, && 1 \leqq t < 2, \end{aligned} \tag{80}$$

are in the interior of $\mathbf{a}$ even if x_m and x_M are boundary points of $\mathbf{a}$. Then by (79), (72), and the elementary intermediate-value theorem, there is a t^* such that $0 < t^* < 2$ and

$$g(t^*) = \frac{n!}{(-1)^n\Delta(\mathbf{a})} \int_{\mathbf{a}} f(x)\, d(x^1, \cdots, x^n). \tag{81}$$

The equations (80) show that t^* determines a point x^* on the interior of $\mathbf{a}$. Then (81) and (78) show that

$$f(x^*) = \frac{n!}{(-1)^n\Delta(\mathbf{a})} \int_{\mathbf{a}} f(x)\, d(x^1, \cdots, x^n), \tag{82}$$

and this equation is equivalent to (69). The proof of Theorem 35.17 is complete if $\mathbf{a}$ is positively oriented in $\mathbb{R}^n$. To prove the theorem in the case in which $\mathbf{a}$ is negatively oriented in $\mathbb{R}^n$, observe that $-\mathbf{a}$ is positively oriented in $\mathbb{R}^n$, and that, by the part of the theorem already proved, there exists a point x^* on the interior of $\mathbf{a}$ such that

$$\int_{-\mathbf{a}} f(x)\, d(x^1, \cdots, x^n) = f(x^*)\frac{(-1)^n}{n!}\Delta(-\mathbf{a}). \tag{83}$$

Since $\int_{-\mathbf{a}} f(x)\, d(x^1, \cdots, x^n) = -\int_{\mathbf{a}} f(x)\, d(x^1, \cdots, x^n)$ by Definition 35.11, and since $\Delta(-\mathbf{a}) = -\Delta(\mathbf{a})$, equation (83) is equivalent to (69). The proof of Theorem 35.18 is now complete in all cases. □

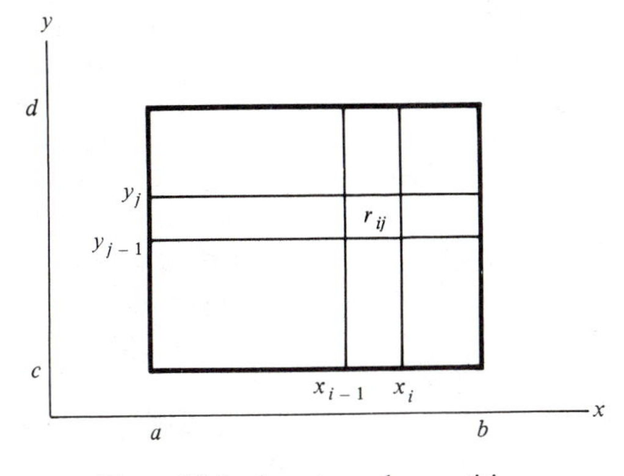

Figure 35.3. A rectangular partition.

35.19 Example. This example contains a brief introduction to the Riemann integral based on rectangular partitions of a rectangle. Let $f: A \to \mathbb{R}$ be a function which is defined and bounded on the rectangle $A = [a, b] \times [c, d]$; thus $m \leqq f(x, y) \leqq M$ for (x, y) in A. Introduce subdivisions of $[a, b]$ and $[c, d]$ as follows [see Figure 35.3]:

$$a = x_0 < x_1 < \cdots < x_{i-1} < x_i < \cdots < x_n = b,$$

$$c = y_0 < y_1 < \cdots < y_{j-1} < y_j < \cdots < y_m = d.$$

Let the subrectangle $[x_{i-1}, x_i] \times [y_{j-1}, y_j]$ of A be denoted by r_{ij}. The set $\{r_{ij} : i = 1, \cdots, n, j = 1, \cdots, m\}$ of subrectangles r_{ij} is called a rectangular partition of A and denoted by $\mathscr{R}$. Let $|r_{ij}|$ denote the area $(x_i - x_{i-1})(y_j - y_{j-1})$ of r_{ij}. Introduce the following definitions corresponding to those in (10) and (12).

$$\begin{aligned} m(f, r_{ij}) &= \inf\{f(x, y) : (x, y) \text{ in } r_{ij}\}, \\ M(f, r_{ij}) &= \sup\{f(x, y) : (x, y) \text{ in } r_{ij}\}. \end{aligned} \tag{84}$$

$$\begin{aligned} m(f, \mathscr{R}) &= \sum_{r_{ij} \in \mathscr{R}} m(f, r_{ij})|r_{ij}|, \\ M(f, \mathscr{R}) &= \sum_{r_{ij} \in \mathscr{R}} M(f, r_{ij})|r_{ij}|. \end{aligned} \tag{85}$$

Then $m(f, \mathscr{R}) \leqq M(f, \mathscr{R})$. If $\mathscr{R}$ is a refinement of $\mathscr{R}_1$ and $\mathscr{R}_2$, then the proof of the following inequalities is similar to the proof of (21).

$$\left.\begin{matrix} m(f, \mathscr{R}_1) \\ m(f, \mathscr{R}_2) \end{matrix}\right\} \leqq m(f, \mathscr{R}) \leqq M(f, \mathscr{R}) \leqq \left\{\begin{matrix} M(f, \mathscr{R}_1), \\ M(f, \mathscr{R}_2). \end{matrix}\right. \tag{86}$$

Also

$$\sup_{\mathscr{R}} m(f, \mathscr{R}) \leqq \inf_{\mathscr{R}} M(f, \mathscr{R}), \tag{87}$$

and f is integrable over A with respect to rectangular partitions if and only if the equality holds in (87). If f is continuous on A, a proof similar to the proof of Theorem 35.8 shows that the equality holds and the integral, denoted by $\iint_A f(x, y)\,dx\,dy$, exists. If f is continuous on A, then f also has an integral $\int_A f(x, y)\,d(x, y)$ defined with respect to simplicial partitions P as explained earlier in this section. To show that these two integrals are equal, proceed as follows.

In each subrectangle r_{ij} of $\mathscr{R}$, draw the two diagonals to form four triangles (2-simplexes) in r_{ij}. This construction yields a simplicial partition $\mathscr{P}$ which is a refinement of $\mathscr{R}$. Then the methods used earlier can be employed again to show that

$$m(f, \mathscr{R}) \leqq m(f, \mathscr{P}) \leqq \sup_P m(f, P) \leqq \inf_P M(f, P) \leqq M(f, \mathscr{P}) \leqq M(f, \mathscr{R}). \tag{88}$$

These inequalities show that, if f is integrable with respect to rectangular partitions $\mathscr{R}$, it is also integrable with respect to simplicial partitions P, and that

$$\iint_A f(x, y)\,dx\,dy = \int_A f(x, y)\,d(x, y). \tag{89}$$

Finally, let ξ_{ij} denote a point in r_{ij}. Then (84) and (85) show that

$$m(f, \mathscr{R}) \leqq \sum_{r_{ij} \in \mathscr{R}} f(\xi_{ij})|r_{ij}| \leqq M(f, \mathscr{R}). \tag{90}$$

If $\mathscr{R}_k$, $k = 1, 2, \cdots$, is a sequence of rectangular partitions of A whose norms tend to zero, then

$$\lim_{k\to\infty} m(f, \mathscr{R}_k) = \lim_{k\to\infty} M(f, \mathscr{R}_k) = \iint_A f(x, y)\,dx\,dy. \tag{91}$$

Then (90) shows that

$$\lim_{k\to\infty} \sum_{r_{ij} \in \mathscr{R}_k} f(\xi_{ij})|r_{ij}| = \iint_A f(x, y)\,dx\,dy. \tag{92}$$

35.20 Example. This example contains a proof that, if the function $f: [a, b] \times [c, d] \to \mathbb{R}$ in Example 35.19 is continuous, then

$$\iint_A f(x, y)\,dx\,dy = \int_a^b \left(\int_c^d f(x, y)\,dy \right) dx = \int_c^d \left(\int_a^b f(x, y)\,dx \right) dy. \tag{93}$$

The integral on the left in (93) is called a double integral, and the other two integrals are called iterated integrals. The proof to be given shows that, if f is continuous, then the iterated integrals exist and equal the double integral. The proof has a simple geometric motivation. The double integral $\iint_A f(x, y)\,dx\,dy$ is the volume of the solid under the surface $z = f(x, y)$ and over the region A [see Figure 35.4]. To find this volume, slice the solid by planes, parallel to the (y, z)-plane, at the points $x_0, x_1, \cdots, x_n$. The area of

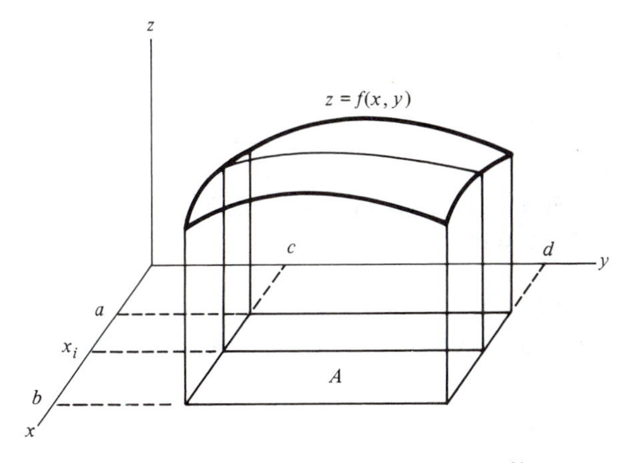

Figure 35.4. The area of the cross section at x_i is $\int_c^d f(x_i, y)\,dy$.

the cross section at x_i is $\int_c^d f(x_i, y)\,dy$, and the volume of the slice bounded by the planes at x_{i-1} and x_i is approximately $(\int_c^d f(x_i, y)\,dy)(x_i - x_{i-1})$. Then

$$\iint_A f(x, y)\,dx\,dy \approx \sum_{i=1}^{n}\left(\int_c^d f(x_i, y)\,dy\right)(x_i - x_{i-1}), \tag{94}$$

and geometric intuition leads one to expect that

$$\iint_A f(x, y)\,dx\,dy = \lim_{n\to\infty}\sum_{i=1}^{n}\left(\int_c^d f(x_i, y)\,dy\right)(x_i - x_{i-1}). \tag{95}$$

If $\int_c^d f(x, y)\,dy$ is a continuous function of x, then the limit of the sum is an integral and

$$\iint_A f(x, y)\,dx\,dy = \int_a^b\left(\int_c^d f(x, y)\,dy\right)dx. \tag{96}$$

The analytic proof merely fills in the details to justify the steps in this geometric description of the proof.

Since $f: A \to \mathbb{R}$ is a continuous function on $[a, b] \times [c, d]$, then f is a continuous function of y for each fixed x in $[a, b]$. Therefore the integral

$$\int_c^d f(x, y)\,dy, \qquad a \leqq x \leqq b, \tag{97}$$

exists for each x in $[a, b]$ and is a function of x. The value of this integral is the area of the cross section at x of the solid shown in Figure 35.4; denote it by $g(x)$. Thus

$$g(x) = \int_c^d f(x, y)\,dy, \qquad a \leqq x \leqq b. \tag{98}$$

We shall now show that g is continuous on $[a, b]$. By the definition of g in (98) and the linear property of integrals in (52) of Theorem 35.12,

$$\begin{aligned} g(x_i) - g(x_{i-1}) &= \int_c^d f(x_i, y)\,dy - \int_c^d f(x_{i-1}, y)\,dy \\ &= \int_c^d [f(x_i, y) - f(x_{i-1}, y)]\,dy. \end{aligned} \tag{99}$$

Since f is continuous on the compact set A, it is uniformly continuous on A. Then to each $\varepsilon > 0$ there corresponds a $\delta(\varepsilon)$ such that

$$|f(x_i, y) - f(x_{i-1}, y)| < \varepsilon \tag{100}$$

for each y in $[c, d]$ provided $|x_i - x_{i-1}| < \delta(\varepsilon)$. Now by (64) in Corollary 35.16, and (99),

$$\begin{aligned} |g(x_i) - g(x_{i-1})| &= \left| \int_c^d [f(x_i, y) - f(x_{i-1}, y)]\,dy \right| \\ &\leqq \int_c^d |f(x_i, y) - f(x_{i-1}, y)|\,dy. \end{aligned} \tag{101}$$

Thus if $|x_i - x_{i-1}| < \delta(\varepsilon)$, then (100) and (101) show that

$$|g(x_i) - g(x_{i-1})| < \int_c^d \varepsilon\,dy = \varepsilon(d - c). \tag{102}$$

This statement proves that g is continuous at each point x in $[a, b]$. Since g is continuous, it has an integral on $[a, b]$, and

$$\int_a^b g(x)\,dx = \lim_{n\to\infty} \sum_{i=1}^n g(x_i)(x_i - x_{i-1}). \tag{103}$$

In this limit, it is understood that the maximum length of a subinterval (x_{i-1}, x_i) tends to zero as $n \to \infty$. In (103) replace $g(x_i)$ and $g(x)$ by their values from (98). Then

$$\int_a^b \left(\int_c^d f(x, y)\,dy \right) dx = \lim_{n\to\infty} \sum_{i=1}^n \left(\int_c^d f(x_i, y)\,dy \right)(x_i - x_{i-1}). \tag{104}$$

Thus the iterated integral on the left in (104) has been shown to exist; to complete the proof it is necessary to show that this iterated integral is equal to the double integral $\iint_A f(x, y)\,dx\,dy$. The proof will be completed by showing that the limit on the right in (104) equals the limit in (92) which defines the double integral in Example 35.19.

Let $\mathscr{R}_k$, $k = 1, 2, \cdots$, be a sequence of rectangular partitions of A as described in Example 35.19; then for each x_i the interval $c \leqq y \leqq d$ is subdivided by points $c = y_0 < y_1 < \cdots < y_j < \cdots < y_m = d$. By the additive property of the integral with respect to intervals [see Theorem 35.13] and the mean-value theorem for integrals [see Theorem 35.18],

$$\int_c^d f(x_i, y)\,dy = \sum_{j=1}^m \int_{y_{j-1}}^{y_j} f(x_i, y)\,dy = \sum_{j=1}^m f(x_i, \eta_j)(y_j - y_{j-1}). \tag{105}$$

Here η_j is a point between y_{j-1} and y_j on the side x_i of r_{ij}; let ξ_{ij} denote the point (x_i, η_j). Then (105) shows that

$$\begin{aligned} \sum_{i=1}^{n}\left(\int_c^d f(x_i, y)\,dy\right)(x_i - x_{i-1}) &= \sum_{i=1}^{n}\left(\sum_{j=1}^{m} f(\xi_{ij})(y_j - y_{j-1})\right)(x_i - x_{i-1}) \\ &= \sum_{r_{ij}\in\mathscr{R}_k} f(\xi_{ij})|r_{ij}|. \end{aligned} \tag{106}$$

Observe that $n \to \infty$ as $k \to \infty$; equation (106) shows that

$$\lim_{k\to\infty} \sum_{i=1}^{n}\left(\int_c^d f(x_i, y)\,dy\right)(x_i - x_{i-1}) = \lim_{k\to\infty} \sum_{r_{ij}\in\mathscr{R}_k} f(\xi_{ij})|r_{ij}|. \tag{107}$$

By (104) the limit on the left is the iterated integral $\int_a^b(\int_c^d f(x, y)\,dy)\,dx$; and by (92) in Exercise 35.19, the limit on the right is the double integral $\iint_A f(x, y)\,dx\,dy$. Therefore,

$$\int_a^b\left(\int_c^d f(x, y)\,dy\right)dx = \iint_A f(x, y)\,dx\,dy, \tag{108}$$

and the proof of the first equality in (93) is complete. A similar proof can be constructed to show that

$$\int_c^d\left(\int_a^b f(x, y)\,dx\right)dy = \iint_A f(x, y)\,dx\,dy, \tag{109}$$

which is the other statement in (93).

Example 35.19 has shown that $f: A \to \mathbb{R}$ has a Riemann integral $\iint_A f(x, y)\,dx\,dy$ if f is continuous on $A = [a, b] \times [c, d]$. It is almost obvious that the integral exists if there are only a finite number of points at which f is discontinuous, but much more can be proved as stated in the following theorem.

35.21 Theorem. *Let $f: A \to \mathbb{R}$ be a bounded function on the rectangle $A = [a, b] \times [c, d]$, and let E be the set of points in A at which f is not continuous. Then f is Riemann integrable on A if and only if E is a set whose Lebesgue measure is zero.*

A set E has Lebesgue measure zero if and only if, for each $\varepsilon > 0$, the set E can be enclosed in a finite or denumerably infinite number of rectangles the sum of whose areas is less than ε. Every finite set and every denumerably infinite set has measure zero; the set of points on a line segment has measure zero; and there are many other sets of measure zero. The proof of Theorem 35.21 is omitted since it is beyond the scope of this book.

If $f: A \to \mathbb{R}$ is continuous, then $\iint_A f(x, y)\,dx\,dy$ can be evaluated by iterated integrals as shown in Example 35.20. Fubini's theorem treats the evaluation of the integral by iterated integrals in the most general case [f is merely integrable], but this theorem also must be omitted.

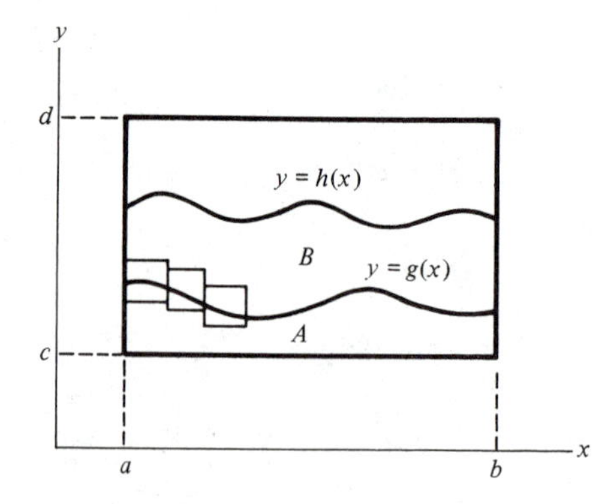

Figure 35.5. Figure for Example 35.23.

Thus far integrals have been defined only on rectangles and simplexes, but integrals must be defined on more general sets in order to complete the theory. Let B be a bounded set in the plane; then B is contained in a rectangle $A = [a, b] \times [c, d]$. Let $f: B \to \mathbb{R}$ be a bounded function. Define a new function $F: A \to \mathbb{R}$ as follows:

$$F(x, y) = \begin{cases} f(x, y), & (x, y) \text{ in } B, \\ 0, & (x, y) \text{ in } A \text{ but not in } B. \end{cases} \tag{110}$$

35.22 Definition. The function f is *Riemann integrable on B* if and only if F is Riemann integrable on A; if $\iint_A F(x, y)\,dx\,dy$ exists, then $\iint_B f(x, y)\,dx\,dy$ exists, and

$$\iint_B f(x, y)\,dx\,dy = \iint_A F(x, y)\,dx\,dy. \tag{111}$$

35.23 Example. This example describes an important special case in which an integral, defined as in Definition 35.22, not only exists but also can be evaluated easily by an iterated integral. Let $g: [a, b] \to \mathbb{R}$ and $h: [a, b] \to \mathbb{R}$ be continuous functions such that $g(x) \leqq h(x)$ for x in $[a, b]$, and let B be the closed region in the plane bounded by the vertical lines $x = a$, $x = b$ and by the curves $y = g(x)$, $y = h(x)$ [see Figure 35.5].

Let $f: B \to \mathbb{R}$ be a continuous function. In order to define the integral of f on B, first enclose B in a rectangle $A = [a, b] \times [c, d]$ as shown in Figure 35.5, and then define a function $F: A \to \mathbb{R}$ as in (110). Since $f: B \to \mathbb{R}$ is continuous by hypothesis, the definition of F in (110) shows that F is continuous except perhaps at points on the curves $y = g(x)$, $y = h(x)$. Theorem 35.21 shows that, in order to prove the existence of $\iint_A F(x, y)\,dx\,dy$, it is necessary to prove that the points on these two curves form a set of measure zero. Now g is continuous on the compact set $[a, b]$; therefore, g is uniformly continuous on $[a, b]$ by Theorem 96.18. Thus to each $\varepsilon > 0$ there corresponds a $\delta = \delta(\varepsilon)$ such that $|f(x_2) - f(x_1)| < \varepsilon$ if $|x_2 - x_1| = \delta$. Then

$$\begin{aligned} |f(x) - f(a)| &< \varepsilon, \qquad a \leqq x \leqq a + \delta, \\ f(a) - \varepsilon < f(x) &< f(a) + \varepsilon, \qquad a \leqq x \leqq a + \delta. \end{aligned} \tag{112}$$

Thus on the interval $[a, a + \delta]$, the points (x, y) on the curve $y = f(x)$ are contained in a rectangle of length δ and altitude 2ε. A similar statement holds for the interval $[a + \delta, a + 2\delta]$ and for every interval whose length is equal to or less than δ. Then by a succession of steps as indicated in Figure 35.5, the points (x, y) on the curve $y = g(x)$ can be enclosed in a finite number of rectangles, the sum of whose lengths is $b - a$ and whose altitudes are 2ε. The sum of the areas of these rectangles is $2\varepsilon(b - a)$. Since this statement holds for every $\varepsilon > 0$, the measure of $\{(x, y): x \in [a, b], y = g(x)\}$ is zero. Similar arguments show that the measure of $\{(x, y): x \in [a, b], y = h(x)\}$ is zero. The points of discontinuity of F are contained in the union of these two sets and thus they form a set of measure zero. Then $\iint_A F(x, y)\,dx\,dy$ exists by Theorem 35.21, and $\iint_B f(x, y)\,dx\,dy$ exists by Definition 35.22; equation (111) is true.

This example will be completed by showing that the integral of F on A, and hence the integral of f on B, can be evaluated by an iterated integral; the proof is similar to the proof given in Example 35.20. First, it is necessary to show that $\int_c^d F(x, y)\,dy$ is a continuous function of x. Now

$$\begin{aligned} \int_c^d F(x, y)\,dy &= \int_c^{g(x)} F(x, y)\,dy + \int_{g(x)}^{h(x)} F(x, y)\,dy + \int_{h(x)}^d F(x, y)\,dy \\ &= \int_{g(x)}^{h(x)} f(x, y)\,dy \end{aligned} \tag{113}$$

since $F(x, y) = f(x, y)$ in B and $F(x, y) = 0$ outside B. Thus to prove that $\int_c^d F(x, y)\,dy$ is continuous, it is sufficient to prove that $\int_{g(x)}^{h(x)} f(x, y)\,dy$ is a continuous function of x. This proof is similar to that given in equations (100), $\cdots$, (102), but there are some additional details because the limits of the integral are the continuous functions $g(x)$ and $h(x)$. Thus $\int_c^d F(x, y)\,dy$ is a continuous function of x, and the iterated integral $\int_a^b (\int_c^d F(x, y)\,dy)\,dx$ exists. By a proof which differs only in some small details from that given in Example 35.20, it can be shown that

$$\int_a^b \left(\int_c^d F(x, y)\,dy \right) dx = \iint_A F(x, y)\,dx\,dy. \tag{114}$$

Then (111), (113), and (114) show that

$$\begin{aligned} \iint_B f(x, y)\,dx\,dy &= \iint_A F(x, y)\,dx\,dy \\ &= \int_a^b \left(\int_c^d F(x, y)\,dy \right) dx = \int_a^b \left(\int_{g(x)}^{h(x)} f(x, y)\,dy \right) dx. \end{aligned} \tag{115}$$

35.24 Example. The purpose of this example is to show that iterated integrals can be used to evaluate the integrals over simplexes which were defined earlier in this section. Let $\mathbf{a}:(a_0, a_1, a_2)$ be a positively oriented simplex in $\mathbb{R}^2$, and let $f:\mathbf{a}\to\mathbb{R}$ be a continuous function. Then the integral $\int_{\mathbf{a}} f(x, y)\,d(x, y)$ exists by Theorem 35.8. Example 35.23 has defined a second integral of $f:\mathbf{a}\to\mathbb{R}$ as follows. Enclose $\mathbf{a}$ in a rectangle $A:[a, b]\times[c, d]$, and then define a function $F:A\to\mathbb{R}$ as in (110). Then the integral $\iint_{\mathbf{a}} f(x, y)\,dx\,dy$ equals $\iint_A F(x, y)\,dx\,dy$ by definition. A proof will now be given that $\int_{\mathbf{a}} f(x, y)\,d(x, y)$ and $\iint_{\mathbf{a}} f(x, y)\,dx\,dy$ are equal, and thus that $\int_{\mathbf{a}} f(x, y)\,d(x, y)$ can be evaluated by an iterated integral as explained in Example 35.23. Construct a sequence $\mathscr{R}_k$, $k = 1, 2, \cdots$, of rectangular partitions, as in Example 35.19, whose norms approach zero. For each $\mathscr{R}_k$ construct a refinement P_k which is a simplicial partition of A, and which contains a simplicial partition P_k' of $\mathbf{a}$. For each $\mathbf{x}$ in P_k let ξ denote an arbitrary, but fixed, point in the interior of $\mathbf{x}$. Then

$$\lim_{k\to\infty}\sum_{\mathbf{x}\in P_k} F(\xi)\frac{(-1)^2}{2!}\Delta(\mathbf{x}), \qquad \xi\in\mathbf{x}, \tag{116}$$

equals $\iint_A F(x, y)\,dx\,dy$ since P_k is a refinement of $\mathscr{R}_k$ for k = 1, 2, $\cdots$ [see Example 35.19]. But, by the definition of F, the construction of the refinement P_k of $\mathscr{R}_k$, and the choice of ξ in each $\mathbf{x}$ in P_k,

$$F(\xi) = \begin{cases} f(\xi) & \text{if } \mathbf{x} \text{ is in } \mathbf{a}, \\ 0 & \text{if } \mathbf{x} \text{ is not in } \mathbf{a}. \end{cases} \tag{117}$$

Then (117) shows that (116) equals

$$\lim_{k\to\infty}\sum_{\mathbf{x}\in P_k'} f(\xi)\frac{(-1)^2}{2!}\Delta(\mathbf{x}), \tag{118}$$

and this limit is $\int_{\mathbf{a}} f(x, y)\,d(x, y)$ by (42) in Corollary 35.10. Thus the limit in (116) is both $\iint_A F(x, y)\,dx\,dy$, which is $\iint_{\mathbf{a}} f(x, y)\,dx\,dy$, and also $\int_{\mathbf{a}} f(x, y)\,d(x, y)$. Since Example 35.23 shows that $\iint_{\mathbf{a}} f(x, y)\,dx\,dy$ can be evaluated by an iterated integral, the integral $\int_{\mathbf{a}} f(x, y)\,d(x, y)$ can be evaluated by the same iterated integral. A numerical example illustrates this procedure.

Let $\mathbf{a}:[a_0, a_1, a_2]$ be the simplex whose vertices are $a_0:(2, 1)$, $a_1:(8, 7)$, and $a_2:(5, 7)$, and let $f:\mathbf{a}\to\mathbb{R}$ be the function such that $f(x, y) = 2x + 4y + 5$. Use the formula

$$\int_{\mathbf{a}} f(x, y)\,d(x, y) = \int_a^b\left(\int_{g(x)}^{h(x)} f(x, y)\,dy\right)dx \tag{119}$$

to evaluate the integral on the left. A sketch of $\mathbf{a}$ shows that

$$\begin{aligned} g(x) &= x - 1, \qquad 2 \leqq x \leqq 8, \\ h(x) &= \begin{cases} 2x - 3, & 2 \leqq x \leqq 5, \\ 7, & 5 \leqq x \leqq 8. \end{cases} \end{aligned} \tag{120}$$

Then

$$\int_{\mathbf{a}} f(x, y)\,d(x, y)$$

$$= \int_2^5 \left(\int_{x-1}^{2x-3} (2x + 4y + 5)\,dy \right) dx + \int_5^8 \left(\int_{x-1}^{7} (2x + 4y + 5)\,dy \right) dx \tag{121}$$

$$= \frac{783}{6} + \frac{1107}{6} = 315.$$

The material beginning with Example 35.19 in this section has dealt with integrals over regions in the plane. The reader will readily understand that this material has generalizations for integrals over rectangular regions, simplexes, and other regions in $\mathbb{R}^n$, $n \geqq 3$, which are analogous to those in the theorems and examples in this part of the section.

Exercises

35.1. Let $\mathbf{a}: [a_0, a_1, a_2]$ be the simplex whose vertices are $a_0: (0, 0)$, $a_1: (1, 0)$, and $a_2: (0, 1)$, and let $f: \mathbf{a} \to \mathbb{R}$ be the function such that $f(x, y) = x^2 + y^2$. Use the iterated integral in the formula in (119) to find the value of $\int_{\mathbf{a}} f(x, y)\,d(x, y)$.

35.2. Repeat Exercise 35.1 for the simplex $\mathbf{a}: [a_0, a_1, a_2]$ with vertices $a_0: (-1, 0)$, $a_1: (1, 0)$, $a_2: (0, 1)$ and the function $f: \mathbf{a} \to \mathbb{R}$ such that $f(x, y) = x^2 - y^2$.

35.3. Let $\mathbf{a}: [a_0, a_1, a_2]$ have the vertices $a_0: (2, 3)$, $a_1: (10, 7)$, and $a_2: (8, 9)$, and let $f: \mathbf{a} \to \mathbb{R}$ be the function such that $f(x, y) = x + y$.
(a) Make a sketch of the simplex $\mathbf{a}$, and use an iterated integral to find the value of $\int_{\mathbf{a}} f(x, y)\,d(x, y)$.
(b) Use a second iterated integral to find the value of $\int_{\mathbf{a}} f(x, y)\,d(x, y)$, and show that the two iterated integrals give the same value for $\int_{\mathbf{a}} f(x, y)\,d(x, y)$.

35.4. Evaluate the following iterated integrals and show that they are equal:

$$\int_3^5 \left(\int_2^4 (2x + 6y)\,dy \right) dx, \qquad \int_2^4 \left(\int_3^5 (2x + 6y)\,dx \right) dy.$$

35.5. Let $f: [0, 1] \to \mathbb{R}$ be the function which is defined as follows: $f(x) = 0$ if $x \in [0, 1]$ and x is rational; $f(x) = 1$ if $x \in [0, 1]$ and x is irrational. Show that f does not have a Riemann integral. [Hint. Find $m(f, P)$ and $M(f, P)$.]

35.6. Let $f: [a, b] \to \mathbb{R}$ be a monotonically increasing function on the closed interval $[a, b]$.
(a) Show that f is a bounded function, and find the best values for its bounds.
(b) Let P_n be a partition of $[a, b]$ into n equal subintervals. Show that

$$M(f, P_n) - m(f, P_n) = \frac{[f(b) - f(a)][b - a]}{n}.$$

Make a sketch to show geometrically why this statement is true.
(c) Use the result in (b) to show that the Riemann integral $\int_a^b f(x)\,dx$ exists. Observe that this integral exists although f is not assumed to be continuous.

35.7. Let $\mathbf{a}$ be a 2-simplex (triangle) in the (x^1, x^2)-plane, and let P_1 and P_2 be two simplicial partitions of $\mathbf{a}$. Make a sketch which shows these two partitions in the same copy of $\mathbf{a}$. Then use the method explained in the proof of Lemma 35.3 to construct a simplicial subdivision $\mathscr{P}$ of $\mathbf{a}$ which is a refinement of both P_1 and P_2.

35.8. Let $\mathbf{a}$ be a Euclidean 3-simplex in $\mathbb{R}^3$; let $f: \mathbf{a} \to \mathbb{R}$ be a continuous function; and let P_k, $k = 1, 2, \cdots$, be a sequence of simplicial partitions of $\mathbf{a}$ whose norms approach zero as $k \to \infty$. For each $\mathbf{x}$ in P_k let p_0, p_1, p_2 be three points in $\mathbf{x}$; for example, p_0, p_1, p_2 might be three vertices of $\mathbf{x}$. Prove that

$$\lim_{k\to\infty} \sum_{\mathbf{x}\in P_k} \frac{f(p_0) + f(p_1) + f(p_2)}{3} \frac{(-1)^3}{3!} \Delta(\mathbf{x}) = \int_{\mathbf{a}} f(x)\, d(x^1, \cdots, x^3).$$

35.9. Let $\mathbf{a}$ be a 2-simplex (a_0, a_1, a_2) in the (x, y)-plane whose vertices are $a_0: (0, 0)$, $a_1: (4, 0)$, and $a_2: (0, 4)$. Let $f: \mathbf{a} \to \mathbb{R}$ be the function such that $f(x, y) = 12 - 3x - 3y$.

(a) Show that the solid under the surface $z = f(x, y)$ for (x, y) in $\mathbf{a}$ is a triangular pyramid (tetrahedron) whose vertices are $(0, 0, 0)$, $(4, 0, 0)$, $(0, 4, 0)$, and $(0, 0, 12)$.

(b) Show that $\int_{\mathbf{a}} f(x, y)\, d(x, y)$ exists and has the value 32. [Hint. Recall the formula for the volume of a pyramid; see also (26) in Section 89 and (119) above.]

(c) Show that $(-1)^2\Delta(\mathbf{a})/2! = 8$. Then Theorem 35.18 asserts that there is a point (x^*, y^*) in the interior of $\mathbf{a}$ such that

$$\int_{\mathbf{a}} (12 - 3x - 3y)\, d(x, y) = 8f(x^*, y^*).$$

Show that every point (x^*, y^*) on the segment of the line $3x - 3y - 8 = 0$ which is in $\mathbf{a}$ satisfies this statement of the mean-value theorem for integrals. Make a sketch to show that there are two of these points on the boundary of $\mathbf{a}$, and an open segment of these points which are on the interior of $\mathbf{a}$. [Hint. $f(x^*, y^*) = 12 - 3x^* - 3y^*$, and $\int_{\mathbf{a}} (12 - 3x - 3y)\, d(x, y) = 32$. Solve the equation $8(12 - 3x^* - 3y^*) = 32$.]

35.10. Let A be the rectangular region $[a_1, b_1] \times [a_2, b_2] \times [a_3, b_3]$ in $\mathbb{R}^3$, and let $f: A \to \mathbb{R}$ be a continuous function.

(a) Use rectangular partitions of A to show that f has a Riemann integral $\iiint_A f(x, y, z)\, dx\, dy\, dz$.

(b) Prove the following formula:

$$\iiint_A f(x, y, z)\, dx\, dy\, dz = \int_{a_1}^{b_1} \left(\int_{a_2}^{b_2} \left(\int_{a_3}^{b_3} f(x, y, z)\, dz \right) dy \right) dx.$$

(c) Show that there are six iterated integrals of f on A. Write out the expressions for these six iterated integrals.

35.11. Let $A = [a, b] \times [c, d]$, and let $P: A \to \mathbb{R}$ and $Q: A \to \mathbb{R}$ be continuous functions. Assume also that the partial derivatives $\partial P/\partial y$ and $\partial Q/\partial x$ exist and are continuous on A.

(a) Use the properties of integrals and iterated integrals to prove the following formulas:

$$\iint_A \frac{\partial P}{\partial y} dx\,dy = \int_a^b P(x, d)\,dx - \int_a^b P(x, c)\,dx,$$

$$\iint_A \frac{\partial Q}{\partial x} dx\,dy = \int_c^d Q(b, y)\,dy - \int_c^d Q(a, y)\,dy.$$

(b) Use the properties of integrals and the formulas in (a) to prove the following formula.

$$\iint_A \left[\frac{\partial Q}{\partial x} - \frac{\partial P}{\partial y}\right] dx\,dy$$

$$= \int_a^b P(x, c)\,dx + \int_c^d Q(b, y)\,dy + \int_b^a P(x, d)\,dx + \int_d^c Q(a, y)\,dy.$$

This formula is a special case of Green's theorem; it evaluates the double integral on the left by means of the integral $\int_{\partial A} P(x, y)\,dx + Q(x, y)\,dy$ around the boundary of A in the positive direction. This statement of Green's theorem is a special case of one form of the fundamental theorem of the integral calculus; the proof of the general form of this theorem is one of the major objectives in this chapter.

36. Surface Integrals in $\mathbb{R}^n$

Section 35 has shown that the problem of finding the area under the curve $y = f(x)$, and the volume under the surface $z = f(x, y)$, leads to the integrals $\int_a^b f(x)\,dx$ and $\int_A f(x, y)\,d(x, y)$, respectively. This section begins with two examples which show that some problems lead to integrals of a type known as line and surface integrals. After these examples the section defines the surface integral of a function on an m-dimensional surface in $\mathbb{R}^n$ and proves that it exists under appropriate hypotheses.

36.1 Example. The work w performed by a force f which acts through a distance d is calculated by the formula $w = fd$. Using this elementary formula as the point of departure, we wish to find a formula for the work done by a force which acts along a curve in $\mathbb{R}^3$. Proceed as follows. Let $\mathbf{a}: [a_0, a_1]$ be a positively oriented simplex in $\mathbb{R}$, and let $g: \mathbf{a} \to \mathbb{R}^3$ be a function whose components (g^1, g^2, g^3) have continuous derivatives. Then the equations

$$y^j = g^j(x), \qquad j = 1, 2, 3, \qquad x \in \mathbf{a}, \tag{1}$$

define a curve G_1 in $\mathbb{R}^3$ which has the trace $T(G_1) = \{y: y = g(x), x \in \mathbf{a}\}$. The force which acts along this curve is represented by a function $f: T(G_1) \to \mathbb{R}^3$ whose components (f^1, f^2, f^3) are continuous. Let $\mathscr{P}_k$, $k = 1, 2, \cdots$, be a sequence of subdivisions of $\mathbf{a}$ of the form $a_0 = x_0 < x_1 < \cdots < x_n = a_1$. Begin by approximating the work done by the force on the interval (x_{i-1}, x_i). The expression

$$f^1[g(x_{i-1})](-1)\begin{vmatrix} g^1(x_{i-1}) & 1 \\ g^1(x_i) & 1 \end{vmatrix} + \cdots + f^3[g(x_{i-1})](-1)\begin{vmatrix} g^3(x_{i-1}) & 1 \\ g^3(x_i) & 1 \end{vmatrix} \tag{2}$$

is the inner product of two vectors; here

$$f^1[g(x_{i-1})], \cdots, f^3[g(x_{i-1})] \tag{3}$$

are the components of the force vector at $g(x_{i-1})$, and

$$(-1)\begin{vmatrix} g^1(x_{i-1}) & 1 \\ g^1(x_i) & 1 \end{vmatrix}, \cdots, (-1)\begin{vmatrix} g^3(x_{i-1}) & 1 \\ g^3(x_i) & 1 \end{vmatrix} \tag{4}$$

are the components of a vector which is approximately the distance through which the force acts on the i-th interval of the curve. Let θ_i denote the angle between the vectors in (3) and (4). Then by (7) in Theorem 84.2, the inner product in (2) equals the product of the lengths of the vectors multiplied by the cosine of the angle between them, or

$$\left\{\sum_{j=1}^{3} [f^j[g(x_{i-1})]]^2\right\}^{1/2} \cos\theta_i \cdot \left\{\sum_{j=1}^{3} [g^j(x_i) - g^j(x_{i-1})]^2\right\}^{1/2}. \tag{5}$$

The first of the two factors in (5) is the component, in the direction of the distance vector in (4), of the force vector in (3); the second factor is the length of the vector in (4). Then (5) and the elementary formula $w = fd$ suggest that (2) is approximately the work done by the force on the i-th subinterval, and that the work done by f on the curve G_1 should be defined to be

$$\lim_{k\to\infty} \sum_{\mathscr{P}_k} \left\{ f^1[g(x_{i-1})](-1)\begin{vmatrix} g^1(x_{i-1}) & 1 \\ g^1(x_i) & 1 \end{vmatrix} + \cdots + f^3[g(x_{i-1})](-1)\begin{vmatrix} g^3(x_{i-1}) & 1 \\ g^3(x_i) & 1 \end{vmatrix} \right\}. \tag{6}$$

In order for this definition to be meaningful, it must be shown that this limit exists. Since the functions $g^1, \cdots, g^3$ have continuous derivatives, they satisfy the Stolz condition, and

$$\begin{gathered} (-1)\begin{vmatrix} g^j(x_{i-1}) & 1 \\ g^j(x_i) & 1 \end{vmatrix} = D_1 g^j(x_{i-1})(-1)\begin{vmatrix} x_{i-1} & 1 \\ x_i & 1 \end{vmatrix} + r(g^j; x_{i-1}, x_i)|x_i - x_{i-1}|, \\ \lim_{k\to\infty} r(g^j; x_{i-1}, x_i) = 0, \qquad i = 1, \cdots, n, \qquad j = 1, 2, 3. \end{gathered} \tag{7}$$

Thus (6) can be written in the following form:

$$\begin{aligned} &\lim_{k\to\infty} \sum_{\mathscr{P}_k} \{f^1[g(x_{i-1})]D_1 g^1(x_{i-1}) + \cdots \\ &\qquad + f^3[g(x_{i-1})]D_1 g^3(x_{i-1})\}(-1)\begin{vmatrix} x_{i-1} & 1 \\ x_i & 1 \end{vmatrix} \\ &\qquad + \lim_{k\to\infty} \sum_{\mathscr{P}_k} \{f^1[g(x_{i-1})]r(g^1; x_{i-1}, x_i) + \cdots \\ &\qquad + f^3[g(x_{i-1})]r(g^3; x_{i-1}, x_i)\}|x_i - x_{i-1}|. \end{aligned} \tag{8}$$

Since the components of $f\colon T(G_1) \to \mathbb{R}^3$ are continuous, and since the components of $g\colon \mathbf{a} \to \mathbb{R}^3$ have continuous derivatives, then

$$f^1 \circ gD_1g^1 + f^2 \circ gD_1g^2 + f^3 \circ gD_1g^3 \tag{9}$$

is a continuous function on the compact set $\mathbf{a}$. Then the function in (9) has a Riemann integral in $\mathbf{a}$ by Theorem 35.8, and (50) in Section 35 shows that the first limit in (8) is

$$\int_{\mathbf{a}} \{f^1 \circ g(x)D_1g^1(x) + f^2 \circ g(x)D_1g^2(x) + f^3 \circ g(x)D_1g^3(x)\}\,dx. \tag{10}$$

Furthermore, the second limit in (8) exists and equals zero. To prove this statement, observe first that $f^1, \cdots, f^3$ are continuous on the compact set $T(G_1)$; then there is a constant M such that

$$|f^j[g(x)]| \leqq M, \qquad j = 1, 2, 3. \tag{11}$$

Next, let $\varepsilon > 0$ be given; then since $g^1, \cdots, g^3$ have continuous derivatives, Theorem 9.9 shows that

$$|r(g^j; x_{i-1}, x_i)| < \varepsilon, \qquad i = 1, \cdots, n, \qquad j = 1, 2, 3, \tag{12}$$

for all sufficiently large k. Then (11) and (12) show that the second sum in (8) is equal to or less than

$$3M\varepsilon \sum_{\mathscr{P}_k} |x_i - x_{i-1}| = 3M\varepsilon|a_1 - a_0|. \tag{13}$$

This statement shows that the second limit in (8) exists and has the value zero. Then the limit in (6) exists and is the Riemann integral in (10), and the work done by the force $f\colon T(G_1) \to \mathbb{R}^3$ acting along the curve $G_1 : y = g(x)$ is given by the integral (10). Because of the special relation of the integrand to the curve G_1, the special Riemann integral in (10) is called a line integral. It is frequently denoted by the symbol

$$\int_{G_1} f^1(y)\,dy^1 + f^2(y)\,dy^2 + f^3(y)\,dy^3,$$

but in all cases it is evaluated by the formula in (10). It is a one-dimensional example of a class of integrals to be known collectively as surface integrals.

36.2 Example. Let $g\colon \mathbf{a} \to \mathbb{R}^3$ be a function whose components (g^1, g^2, g^3) have continuous derivatives on $\mathbf{a} : [a_0, a_1, a_2]$ in $\mathbb{R}^2$. The equations

$$y^j = g^j(x), \qquad j = 1, 2, 3, \qquad x \in \mathbf{a}, \tag{14}$$

define a surface G_2 in $\mathbb{R}^3$ whose trace $T(G_2)$ is $\{y : y = g(x),\ x \in \mathbf{a}\}$. Let $f\colon T(G_2) \to \mathbb{R}^3$ be a function whose components (f^1, f^2, f^3) are continuous. Let $f(y)$ denote the velocity with which a fluid flows across the surface G_2 at the point y on G_2. The problem is to find a formula for the rate of flow of the fluid across the entire surface G_2. The procedure is the usual one: approximate the flow across each simplex in a piecewise linear surface inscribed in G_2 and then take the limit. The details follow.

Let P_k, $k = 1, 2, \cdots$, be a sequence of simplicial partitions of $\mathbf{a}$; later it will be necessary to impose a restriction on these partitions. Let $\mathbf{x}:(x_0, x_1, x_2)$ be a simplex in P_k; then g maps the vertices of $\mathbf{x}$ into $g(\mathbf{x})$, where

$$\mathbf{x} = \begin{bmatrix} x_0^1 & x_0^2 \\ \cdots & \cdots \\ x_2^1 & x_2^2 \end{bmatrix}, \qquad g(\mathbf{x}) = \begin{bmatrix} g^1(x_0) & g^2(x_0) & g^3(x_0) \\ \cdots & \cdots & \cdots \\ g^1(x_2) & g^2(x_2) & g^3(x_2) \end{bmatrix}. \tag{15}$$

In the usual notation,

$$\Delta(g^{j_1}, g^{j_2})(\mathbf{x}) = \begin{vmatrix} g^{j_1}(x_0) & g^{j_2}(x_0) & 1 \\ g^{j_1}(x_1) & g^{j_2}(x_1) & 1 \\ g^{j_1}(x_2) & g^{j_2}(x_2) & 1 \end{vmatrix}, \qquad (j_1, j_2) \in (2/3). \tag{16}$$

Then

$$\frac{(-1)^2}{2!}\Delta(g^2, g^3)(\mathbf{x}), \qquad -\frac{(-1)^2}{2!}\Delta(g^1, g^3)(\mathbf{x}), \qquad \frac{(-1)^2}{2!}\Delta(g^1, g^2)(\mathbf{x}) \tag{17}$$

are the components of the vector product of $g(x_1) - g(x_0)$ and $g(x_2) - g(x_0)$; thus (17) is a vector which is orthogonal to the plane of the simplex $g(\mathbf{x})$. The expression

$$\begin{aligned} f^1[g(x_0)]\frac{(-1)^2}{2!}\Delta(g^2, g^3)(\mathbf{x}) &- f^2[g(x_0)]\frac{(-1)^2}{2!}\Delta(g^1, g^3)(\mathbf{x}) \\ &+ f^3[g(x_0)]\frac{(-1)^2}{2!}\Delta(g^1, g^2)(\mathbf{x}) \end{aligned} \tag{18}$$

is the inner product of the velocity vector $f:(f^1, f^2, f^3)$ at $g(x_0)$ and the vector in (17). Let θ denote the angle between these two vectors. Then the inner product (18) equals $\cos\theta$ multiplied by the product of the lengths of these vectors. The length of (17) is the area of the simplex (triangle) $g(\mathbf{x})$. The length of f multiplied by $\cos\theta$ is the component of the velocity normal to the plane of $g(\mathbf{x})$. Thus (18) is the area of $g(\mathbf{x})$ multiplied by the velocity, normal to $g(\mathbf{x})$, at $g(x_0)$; hence, (18) is approximately the flow across $g(\mathbf{x})$, and

$$\begin{aligned} \sum_{\mathbf{x} \in P_k} \Big\{ f^1[g(x_0)]\frac{(-1)^2}{2!}\Delta(g^2, g^3)(\mathbf{x}) &- f^2[g(x_0)]\frac{(-1)^2}{2!}\Delta(g^1, g^3)(\mathbf{x}) \\ &+ f^3[g(x_0)]\frac{(-1)^2}{2!}\Delta(g^1, g^2)(\mathbf{x}) \Big\} \end{aligned} \tag{19}$$

is approximately the rate at which the fluid is flowing across the entire surface. The limit of (19) as $k \to \infty$ and the norm of P_k tends to zero is defined to be the rate of flow across G_2 if this limit exists. This limit is similar to the one in (6), but this time it is necessary to impose a special restriction on the sequence P_k, $k = 1, 2, \cdots$, of simplicial partitions of $\mathbf{a}$.

All partitions of a one-dimensional simplex $[a_0, a_1]$ in $\mathbb{R}$ form a simplicial subdivision $\mathscr{P}$, and the simplexes in $\mathscr{P}$ form a Euclidean complex. Furthermore, in one dimension there is no troublesome regularity condition to be considered. There are many types of partitions of simplexes $[a_0, a_1, \cdots, a_n]$ for $n \geqq 2$, and different types of partitions can be used to establish different types of results. For example, rectangular partitions were used in Section 35 to establish the evaluation of Riemann integrals by iterated integrals. In the present section, partitions P_k whose simplexes satisfy a regularity condition [see Definition 2.4] will be used to show that the limit of (19) is a certain Riemann integral. A later section will show that sequences of simplicial subdivisions $\mathscr{P}_k$, $k = 1, 2, \cdots$, are required to establish the fundamental theorem of the integral calculus. In $\mathscr{P}_k$, the simplexes satisy the regularity condition and in addition they form a Euclidean complex.

Return to the evaluation of the limit of (19). Comparison of (19) with (6) and (10) suggests that derivatives will appear in the limit. As a result, it is necessary to assume that the simplexes in P_k, $k = 1, 2, \cdots$, satisfy the regularity condition [see Definition 2.4]; then Theorem 11.4 shows that

$$\frac{(-1)^2}{2!}\Delta(g^{j_1}, g^{j_2})(\mathbf{x}) = D_{(1,2)}(g^{j_1}, g^{j_2})(x_0)\frac{(-1)^2}{2!}\Delta(\mathbf{x}) + r(g^{j_1}, g^{j_2}; \mathbf{x})\frac{|\Delta(\mathbf{x})|}{2!},$$

$$\lim_{k\to\infty} r(g^{j_1}, g^{j_2}; \mathbf{x}) = 0, \qquad \mathbf{x} \text{ in } P_k, \qquad (j_1, j_2) \text{ in } (2/3). \tag{20}$$

Furthermore, Theorem 11.4 shows that the limit in (20) is uniform with respect to the simplexes $\mathbf{x}$ in P_k. This statement means the following: for each $\eta > 0$ there exists a $k(\eta)$, which depends only on η, such that

$$|r(g^{j_1}, g^{j_2}; \mathbf{x})| < \eta, \qquad (j_1, j_2) \text{ in } (2/3), \tag{21}$$

for every $\mathbf{x}$ in P_k and for every P_k provided $k \geqq k(\eta)$. Since f is continuous on the compact set $T(G_2)$, and since

$$\sum_{\mathbf{x}\in P_k} \frac{|\Delta(\mathbf{x})|}{2!} = \frac{|\Delta(a_0, a_1, a_2)|}{2!} \tag{22}$$

the relations in (20), (21), (22) can be used as in Example 36.1 to show that the limit of the sum in (19) is

$$\begin{aligned}\int_{\mathbf{a}} \{f^1 \circ g(x) D_{(1,2)}(g^2, g^3)(x) - f^2 \circ g(x) D_{(1,2)}(g^1, g^3)(x) \\ + f^3 \circ g(x) D_{(1,2)}(g^1, g^2)(x)\}\, d(x^1, x^2).\end{aligned} \tag{23}$$

Because of the special relation of the integrand to the surface G_2 defined by the function $g : \mathbf{a} \to \mathbb{R}^3$, the special Riemann integral in (23) is called a surface integral; it is often denoted by the following symbol:

$$\int_{G_2} f^1(y)\,d(y^2, y^3) - f^2(y)\,d(y^1, y^3) + f^3(y)\,d(y^1, y^2). \tag{24}$$

In all cases, however, it is evaluated by the formula in (23). There are strong similarities between the integrals in (10) and (23); they suggest the definition and study of a large class of integrals to be known as surface integrals.

With these two examples as an introduction, we turn to the definition of the surface integral in the general case. Let $\mathbf{a}$ be the m-dimensional Euclidean simplex $[a_0, a_1, \cdots, a_m]$ in $\mathbb{R}^m$, and let $g : \mathbf{a} \to \mathbb{R}^n$, $n \geqq m$, be a function whose components $(g^1, \cdots, g^n)$ have continuous derivatives on $\mathbf{a}$. The equations

$$y^j = g^j(x), \qquad j = 1, \cdots, n, \qquad x \in \mathbf{a}, \qquad x = (x^1, \cdots, x^m), \tag{25}$$

define an m-dimensional surface G_m in $\mathbb{R}^n$ which has the trace $T(G_m) = \{y : y = g(x), x \in \mathbf{a}\}$. Let f: $T(G_m) \to \mathbb{R}^{\binom{n}{m}}$ be a function which has continuous components $f^{(j_1, \cdots, j_m)}$, $(j_1, \cdots, j_m) \in (m/n)$. Here it is in order to recall [see Section 2] that $(m/n) = \{(j_1, \cdots, j_m) : 1 \leqq j_1 < j_2 < \cdots < j_m \leqq n\}$, and that the index sets in (m/n) are ordered lexicographically. The following methods are used to indicate summations over the set (m/n):

$$\sum_{(m/n)}; \quad \sum_{(j_1, \cdots, j_m)}, \qquad (j_1, \cdots, j_m) \in (m/n). \tag{26}$$

Let P_k, $k = 1, 2, \cdots$, be a sequence of oriented simplicial partitions of $\mathbf{a}$ which has the properties stated in Definition 35.1. Then $P_k = \{\mathbf{x} : \mathbf{x} \in P_k\}$, and $\mathbf{x}$ is an m-dimensional Euclidean simplex $(x_0, x_1, \cdots, x_m)$ which is contained in $\mathbf{a}$ and which has the same orientation in $\mathbb{R}^m$ as $\mathbf{a}$. Then g maps each $\mathbf{x}$ in P_k into an abstract simplex $g(\mathbf{x})$ whose vertices are $g(x_0), g(x_1), \cdots, g(x_m)$. Write the coordinates of each vertex of $g(\mathbf{x})$ as a row in a matrix as in (15); then

$$g(\mathbf{x}) = [g^j(x_i)], \qquad i = 0, 1, \cdots, m; j = 1, \cdots, n. \tag{27}$$

Here $g(\mathbf{x})$ is an m-vector in $\mathbb{R}^n$; it has $\binom{n}{m}$ components, and the index sets in (m/n) can be used to index them. Then, employing the notation introduced in Sections 2 and 3,

$$\Delta_{(1, \cdots, m)}(g^{j_1}, \cdots, g^{j_m})(\mathbf{x}) = \det \begin{bmatrix} g^{j_1}(x_0) & \cdots & g^{j_m}(x_0) & 1 \\ \cdots & \cdots & \cdots & \cdots \\ g^{j_1}(x_m) & \cdots & g^{j_m}(x_m) & 1 \end{bmatrix}, \tag{28}$$

$$(j_1, \cdots, j_m) \in (m/n).$$

This determinant can be evaluated by (41) in Theorem 11.4. The approximate value of the integral of f on G_m is

$$\sum_{\mathbf{x} \in P_k} \sum_{(m/n)} f^{(j_1, \cdots, j_m)}[g(x_0)] \frac{(-1)^m}{m!} \Delta_{(1, \cdots, m)}(g^{j_1}, \cdots, g^{j_m})(\mathbf{x}). \tag{29}$$

36.3 Definition. The *surface integral of* $f\colon T(G_m) \to \mathbb{R}^{\binom{n}{m}}$ *on* G_m, denoted by

$$\int_{G_m} \sum_{(m/n)} f^{(j_1,\cdots,j_m)}(y)\, d(y^{j_1}, \cdots, y^{j_m}), \tag{30}$$

is the limit, as $k \to \infty$, of the expression in (29) if this limit exists. If the limit does not exist, then the integral of f on G_m is not defined.

36.4 Theorem. *If the components of* $f\colon T(G_m) \to \mathbb{R}^{\binom{n}{m}}$ *are continuous, and if the components of* $g\colon \mathbf{a} \to \mathbb{R}^n$ *have continuous derivatives, then the surface integral* (30) *exists and*

$$\begin{aligned}\int_{G_m} \sum_{(m/n)} & f^{(j_1,\cdots,j_m)}(y)\, d(y^{j_1}, \cdots, y^{j_m}) \\ &= \int_{\mathbf{a}} \sum_{(m/n)} f^{(j_1,\cdots,j_m)}[g(x)] D_{(1,\cdots,m)}(g^{j_1}, \cdots, g^{j_m})(x)\, d(x^1, \cdots, x^m).\end{aligned} \tag{31}$$

PROOF. In order to prove this theorem, it is necessary to assume that, for some $\rho > 0$, each simplex $\mathbf{x}$ in each P_k satisfies the regularity condition [see Definition 2.4] with parameter ρ at each vertex. The construction of sequences of subdivisions which have this property was explained in Section 21. The proof of the theorem consists of showing that the limit, as $k \to \infty$, of the sum in (29) is the Riemann integral in (31). Observe that $(g^{j_1}, \cdots, g^{j_m})\colon \mathbf{a} \to \mathbb{R}^m$ is a function with m components which are defined and have continuous derivatives on the m-dimensional Euclidean simplex $\mathbf{a}$ in $\mathbb{R}^m$. Then, by Theorem 11.4, for each $\mathbf{x}$ in P_k,

$$\begin{aligned}&\frac{(-1)^m}{m!}\Delta_{(1,\cdots,m)}(g^{j_1}, \cdots, g^{j_m})(\mathbf{x}) \\ &\quad = D_{(1,\cdots,m)}(g^{j_1}, \cdots, g^{j_m})(x_0)\frac{(-1)^m}{m!}\Delta(\mathbf{x}) + \frac{(-1)^m}{m!} r(g^{j_1}, \cdots, g^{j_m}; x)|\Delta(\mathbf{x})|.\end{aligned} \tag{32}$$

Substitute the expression on the right in (32) for $[(-1)^m/m!]\Delta_{(1,\cdots,m)}(g^{j_1}, \cdots, g^{j_m})(\mathbf{x})$ in (29); then (29) becomes

$$\sum_{\mathbf{x} \in P_k} \sum_{(m/n)} f^{(j_1,\cdots,j_m)}[g(x_0)] D_{(1,\cdots,m)}(g^{j_1}, \cdots, g^{j_m})(x_0)\frac{(-1)^m}{m!}\Delta(\mathbf{x}) \tag{33}$$

$$+ \sum_{\mathbf{x} \in P_k} \sum_{(m/n)} \frac{(-1)^m}{m!} f^{(j_1,\cdots,j_m)}[g(x_0)] r(g^{j_1}, \cdots, g^{j_m}; \mathbf{x})|\Delta(\mathbf{x})| \tag{34}$$

Now f is continuous on $T(G_m)$ and g has continuous derivatives on $\mathbf{a}$; then

$$\sum_{(m/n)} [f^{(j_1,\cdots,j_m)} \circ g] D_{(1,\cdots,m)}(g^{j_1}, \cdots, g^{j_m}) \tag{35}$$

is a continuous function on $\mathbf{a}$, and the limit of (33) as $k \to \infty$ is the Riemann integral in (31) by equation (50) in Section 35. Thus the proof can be completed by showing that the limit of the sum in (34) is zero. Since f is continuous on the compact set $T(G_m)$, then f is bounded and there is a constant M such that

$$|f^{(j_1,\cdots,j_m)}[g(x)]| \leqq M, \qquad x \in \mathbf{a}, \qquad (j_1, \cdots, j_m) \in (m/n). \tag{36}$$

Also, since $g^1, \cdots, g^n$ have continuous derivatives, Theorem 11.4 shows that for each $\eta > 0$ there exists a $k(\eta)$, which depends only on η, such that

$$|r(g^{j_1}, \cdots, g^{j_m}; \mathbf{x})| < \eta, \qquad \mathbf{x} \in P_k, \qquad (j_1, \cdots, j_m) \in (m/n), \tag{37}$$

if $k \geqq k(\eta)$. Then for $k \geqq k(\eta)$, the absolute value of the expression in (34) is less than

$$\sum_{\mathbf{x} \in P_k} \sum_{(m/n)} M\eta \frac{|\Delta(\mathbf{x})|}{m!} = \binom{n}{m} M\eta \sum_{\mathbf{x} \in P_k} \frac{|\Delta(\mathbf{x})|}{m!}. \tag{38}$$

Since P_k is a simplicial partition of $\mathbf{a}$, and since $\mathbf{a}$ and its partitions have the same orientation in $\mathbb{R}^m$, then Corollary 35.4 shows that

$$\sum_{\mathbf{x} \in P_k} \frac{|\Delta(\mathbf{x})|}{m!} = \frac{|\Delta(\mathbf{a})|}{m!}. \tag{39}$$

Then (38) and (39) show that the sum in (34) is less than $\binom{n}{m} M\eta |\Delta(\mathbf{a})|/m!$ for $k \geqq k(\eta)$; hence, the limit, as $k \to \infty$, of (34) is zero, and the proof of Theorem 36.4 is complete. □

36.5 Example. In Example 36.1, assume that $T(G_1)$ is contained in an open set E in $\mathbb{R}^3$, and that there is a function $F: E \to \mathbb{R}$ which has continuous derivatives D_jF such that

$$f^j(y) = D_jF(y), \qquad y \in E, \qquad j = 1, 2, 3. \tag{40}$$

Then (10) shows that the work done by the force f acting along the curve G_1 is

$$\int_{\mathbf{a}} \{D_1F[g(x)]D_1g^1(x) + D_2F[g(x)]D_1g^2(x) + D_3F[g(x)]D_1g^3(x)\}\,dx. \tag{41}$$

By the chain rule in Theorem 4.1, the integrand of this integral is $D_x[F \circ g](x)$, and by the fundamental theorem of the integral calculus [see (7) in Section 34],

$$\int_{a_0}^{a_1} D_x[F \circ g](x)\,dx = F[g(a_1)] - F[g(a_0)]. \tag{42}$$

In this case the work done by the force depends only on the end points $g(a_0)$, $g(a_1)$ of the curve but not on the curve itself. For example, if $h: \mathbf{a} \to \mathbb{R}^3$ is another curve H_1 whose trace $T(H_1)$ is in E and $h(a_0) = g(a_0)$, $h(a_1) = g(a_1)$, then the work done by f acting along H_1 is $\int_{a_0}^{a_1} D_x[F \circ h](x)\,dx$, and

$$\int_{a_0}^{a_1} D_x[F \circ h](x)\,dx = F[h(a_1)] - F[h(a_0)]$$
$$= F[g(a_1)] - F[g(a_0)] = \int_{a_0}^{a_1} D_x[F \circ g](x)\,dx. \tag{43}$$

Forces for which the work is independent of the path have important applications in physics.

36.6 Example. Let E be an open set in $\mathbb{R}^3$, and let $\mathbf{a}: [a_0, a_1, a_2]$ be a Euclidean simplex in $\mathbb{R}^2$. Let $g: \mathbf{a} \to \mathbb{R}^3$, with components (g^1, g^2, g^3), define a surface G_2 whose trace $T(G_2)$ is in E. By (31), the surface integral of $f: E \to \mathbb{R}^3$, with components $(f^{(1,2)}, f^{(1,3)}, f^{(2,3)})$, on G_2 is

$$\int_{\mathbf{a}} \{f^{(1,2)}[g(x)]D_{(1,2)}(g^1, g^2)(x) + f^{(1,3)}[g(x)]D_{(1,2)}(g^1, g^3)(x)$$
$$+ f^{(2,3)}[g(x)]D_{(1,2)}(g^2, g^3)(x)\}\,d(x^1, x^2). \tag{44}$$

Let $F: E \to \mathbb{R}^2$ be a function whose components (F^1, F^2) have continuous derivatives such that

$$f^{(j_1,j_2)}(y) = D_{(j_1,j_2)}(F^1, F^2)(y), \qquad y \in E, \qquad (j_1, j_2) \in (2/3). \tag{45}$$

Then the surface integral (44) is

$$\int_{\mathbf{a}} \sum_{(j_1,j_2)} D_{(j_1,j_2)}(F^1, F^2)[g(x)]D_{(1,2)}(g^{j_1}, g^{j_2})(x)\,d(x^1, x^2). \tag{46}$$

Now by the chain rule [see Example 4.3 and Theorem 4.5] the integrand of the integral in (46) is $D_{(1,2)}(F^1 \circ g, F^2 \circ g)(x)$; therefore, the integral (44) in this special case equals (46) and also

$$\int_{\mathbf{a}} D_{(1,2)}(F^1 \circ g, F^2 \circ g)(x)\,d(x^1, x^2). \tag{47}$$

Example 36.5 suggests that there should be a theorem which evaluates this integral by means of an integral over the boundary of $\mathbf{a}$. Section 38 establishes this fundamental theorem of the integral calculus.

36.7 Example. This example uses the formula in (23) to find the rate of flow of a fluid across a sphere. The equations

$$\begin{aligned} y^1 &= r\cos\theta\sin\varphi, \\ y^2 &= r\sin\theta\sin\varphi, \\ y^3 &= r\cos\varphi, \end{aligned} \tag{48}$$

map the rectangle $A: 0 \leqq \theta \leqq 2\pi,\ 0 \leqq \varphi \leqq \pi$ onto the sphere with center at the origin and radius r. Then

$$\begin{aligned} D_{(1,2)}(g^2, g^3)(\theta, \varphi) &= -r^2 \cos\theta \sin^2\varphi, \\ -D_{(1,2)}(g^1, g^3)(\theta, \varphi) &= -r^2 \sin\theta \sin^2\varphi, \\ D_{(1,2)}(g^1, g^2)(\theta, \varphi) &= -r^2 \sin\varphi \cos\varphi. \end{aligned} \tag{49}$$

These formulas give the components of the inner normal of the sphere at the point which is the image of the point (θ, φ). The outer normal is obtained by reversing all signs. First consider the case in which the velocity is constant over the sphere and has components (v_1, v_2, v_3). Then (23) and (49) show that the rate of flow out of the sphere is

$$\int_A r^2\{v_1 \cos\theta \sin^2\varphi + v_2 \sin\theta \sin^2\varphi + v_3 \sin\varphi \cos\varphi\}\, d(\theta, \varphi).$$

This integral equals the sum of three integrals, each of which can be evaluated easily by iterated integrals and shown to have the value zero. This answer means that fluid is flowing into the sphere as fast as it is flowing out. Next, consider the case in which $f^j(y) = y^j/r$, $j = 1, 2, 3$. Then the velocity vector at each point on the sphere is the unit vector which is normal to the sphere and directed outward. Expressed in terms of θ, φ,

$$f^1(y) = \cos\theta \sin\varphi, \qquad f^2(y) = \sin\theta \sin\varphi, \qquad f^3(y) = \cos\varphi. \tag{50}$$

Then using the outer normal [see (49)] and the velocity in (50), formula (23) for the flow across the sphere is

$$\begin{aligned} &\int_A r^2\{\cos^2\theta \sin^3\varphi + \sin^2\theta \sin^3\varphi + \sin\varphi \cos^2\varphi\}\, d(\theta, \varphi) \\ &\qquad = \int_A r^2 \sin\varphi\, d(\theta, \varphi). \end{aligned}$$

Evaluation of this integral by iterated integrals shows that it equals

$$r^2 \int_0^{2\pi} \left(\int_0^{\pi} \sin\varphi\, d\varphi \right) d\theta = r^2 \int_0^{2\pi} 2\, d\theta = 4\pi r^2.$$

The fluid flows outward, normal to the surface, with unit velocity. As is to be expected in this situation, the rate of flow across the surface equals the area of the sphere.

Exercises

36.1. Let $\mathbf{a}$ be $[0, 1]$, and let $g : \mathbf{a} \to \mathbb{R}^3$ be the function whose components are $g^1(x) = x$, $g^2(x) = 2x$, $g^3(x) = 3x$. Then g defines a curve G_1 in $\mathbb{R}^3$. If $f^1(y) = y^1$, $f^2(y) = y^1 + y^2$, $f^3(y) = y^1 + y^2 + y^3$, find the value of $\int_{G_1} f^1(y)\, dy^1 + f^2(y)\, dy^2 + f^3(y)\, dy^3$.

36.2. If $f^j(y) = D_j F(y)$, $j = 1, 2, 3$, show that the work done by the force f acting along a closed curve G_1 is zero. The complete assumptions are these: (i) $\mathbf{a}$ is the interval

$[a_0, a_1]$ in $\mathbb{R}$; (ii) E is an open set in $\mathbb{R}^3$; (iii) $g : \mathbf{a} \to \mathbb{R}^3$ has continuous derivatives and defines a curve G_1 whose trace $T(G_1)$ is in E; (iv) $g(a_1) = g(a_0)$; (v) $F : E \to \mathbb{R}$ has continuous derivatives and $f^j(y) = D_jF(y)$ for $j = 1, 2, 3$. Then the conclusion is $\int_{G_1} \Sigma_1^3 f^j(y)\,dy^j = 0$.

36.3. Let $A = [0, 1] \times [0, 1]$. The functions

$$g^1(x) = 2x^1 - x^2,$$
$$g^2(x) = x^1 + 2x^2, \qquad (x^1, x^2) \in A,$$
$$g^3(x) = x^1 + x^2,$$

define a surface G_2. If $f^j(y) = y^j$, $j = 1, 2, 3$, show that

$$\int_{G_2} f^1(y)\,d(y^1, y^2) + f^2(y)\,d(y^1, y^3) + f^3(y)\,d(y^2, y^3) = 6.$$

36.4. Let E be an open set in $\mathbb{R}^n$ which contains $T(G_m)$ in Theorem 36.4. Assume that f is defined on E and that f is a derivative; that is, assume that there is a function $F : E \to \mathbb{R}^m$ whose components $(F^1, \cdots, F^m)$ have continuous derivatives on E such that

$$f^{(j_1, \cdots, j_m)}(y) = D_{(j_1, \cdots, j_m)}(F^1, \cdots, F^m)(y), \qquad y \in E, \qquad (j_1, \cdots, j_m) \in (m/n).$$

Prove that the surface integral in (31) equals

$$\int_{\mathbf{a}} D_{(1, \cdots, m)}(F^1 \circ g, \cdots, F^m \circ g)(x)\,d(x^1, \cdots, x^m).$$

36.5. Prove the following theorem. If g^1, g^2, g^3 have continuous derivatives on $\mathbf{a} : [a_0, a_1]$ in $\mathbb{R}$, if f^1, f^2, f^3 are defined and continuous on the trace $T(G_1)$ of the curve defined by $g : (g^1, g^2, g^3)$, and if M is the maximum of $\{\Sigma_1^3 [f^j[g(x)]]^2\}^{1/2}$ on $\mathbf{a}$, then

$$\left| \int_{\mathbf{a}} \sum_{j=1}^{3} f^j[g(x)] D_1 g^j(x)\,dx \right| \leqq M \int_{\mathbf{a}} \left\{ \sum_{j=1}^{3} [D_1 g^j(x)]^2 \right\}^{1/2} dx.$$

Observe that the integral on the right in this inequality is the length of G_1. [Hints. Use Schwarz's inequality to show that

$$\left| \sum_{j=1}^{3} f^j[g(x)] D_1 g^j(x) \right| \leqq \left\{ \sum_{j=1}^{3} \left[f^j[g(x)] \right]^2 \right\}^{1/2} \left\{ \sum_{j=1}^{3} [D_1 g^j(x)]^2 \right\}^{1/2}.$$

Then apply Corollary 35.16.]

36.6. Prove the following theorem. If g^1, g^2, g^3 have continuous derivatives on $\mathbf{a} : [a_0, a_1, a_2]$ in $\mathbb{R}^2$, if $f^{(1,2)}$, $f^{(1,3)}$, $f^{(2,3)}$ are defined and continuous on the trace $T(G_2)$ of the surface G_2 defined by $g : (g^1, g^2, g^3)$, and if M is the maximum of $\{\Sigma_{(j_1, j_2)} [f^{(j_1, j_2)}[g(x)]]^2\}^{1/2}$, $(j_1, j_2) \in (2/3)$, on $\mathbf{a}$, then

$$\left| \int_{\mathbf{a}} \sum_{(j_1, j_2)} f^{(j_1, j_2)}[g(x)] D_{(1,2)}(g^{j_1}, g^{j_2})(x)\,d(x^1, x^2) \right|$$
$$\leqq M \int_{\mathbf{a}} \left\{ \sum_{(j_1, j_2)} [D_{(1,2)}(g^{j_1}, g^{j_2})(x)]^2 \right\}^{1/2} d(x^1, x^2).$$

Observe that the integral on the right in this inequality is the area of the surface G_2 [see Section 45].

36.7. Exercises 36.5 and 36.6 contain special cases of a general theorem about the surface integral on G_m in $\mathbb{R}^n$ described in Theorem 36.4. State and prove the general theorem.

37. Integrals on an m-Simplex in $\mathbb{R}^n$

Section 35 contains the definition of the Riemann integral of a function $f:\mathbf{a} \to \mathbb{R}$ on a Euclidean n-simplex $\mathbf{a}:[a_0, a_1, \cdots, a_n]$ in $\mathbb{R}^n$, and Section 36 contains the definition of the integral of a function $f: T(G_m) \to \mathbb{R}^{\binom{n}{m}}$ on a surface G_m in $\mathbb{R}^n$, $m \leq n$. Let $\mathbf{b}$ be an oriented m-dimensional simplex $\mathbf{b}:[b_0, b_1, \cdots, b_m]$ in $\mathbb{R}^n$, and let $f:\mathbf{b} \to \mathbb{R}^{\binom{n}{m}}$ be continuous. Then there is an integral of f on $\mathbf{b}$ which can be treated as a generalization of the Riemann integral in Section 35 and also as a special case of the surface integral in Section 36. This section contains these two treatments of the integral of f on $\mathbf{b}$. In order to simplify the exposition, the explanation will be given in the special case of a 2-dimensional simplex $\mathbf{b}:[b_0, b_1, b_2]$ in $\mathbb{R}^3$.

An outline of this section may help to understand the details which follow. The simplex $\mathbf{b}$ is represented by a matrix $[b_i^j]$, $i = 0, 1, 2$ and $j = 1, 2, 3$. Let $\mathbf{b}^{(1,2)}$, $\mathbf{b}^{(1,3)}$, and $\mathbf{b}^{(2,3)}$ denote the 3×2 minors of $\mathbf{b}$ in columns (1, 2), (1, 3), and (2, 3) respectively; this notation has been used in Section 11. Then $\mathbf{b}^{(1,2)}$, $\mathbf{b}^{(1,3)}$, and $\mathbf{b}^{(2,3)}$ represent 2-dimensional simplexes in the 2-dimensional (y^1, y^2), (y^1, y^3), and (y^2, y^3) spaces. It will be shown that simplicial partitions of $\mathbf{b}^{(j_1,j_2)}$, $(j_1, j_2) \in (2/3)$, can be derived from simplicial partitions of $\mathbf{b}$, and that the simplicial partitions of $\mathbf{b}^{(j_1,j_2)}$ can be used to define a Riemann-type integral of $f^{(j_1,j_2)}:\mathbf{b} \to \mathbb{R}$. The sum of these integrals on $\mathbf{b}^{(j_1,j_2)}$, $(j_1, j_2) \in (2/3)$, is the integral on the simplex $\mathbf{b}$ of $f:\mathbf{b} \to \mathbb{R}^3$ with components $f^{(j_1,j_2)}$. Integrals of this type arise in connection with the proof of the fundamental theorem of the integral calculus in Section 38.

The definition of all of the integrals begins with the construction of a set Q_k, $k = 1, 2, \cdots$, of simplicial partitions of $\mathbf{b}:[b_0, b_1, b_2]$. As usual, the sequence Q_k, $k = 1, 2, \cdots$, is obtained by mapping a sequence P_k, $k = 1, 2, \cdots$, of simplicial partitions of a 2-simplex $\mathbf{a}:[a_0, a_1, a_2]$ in $\mathbb{R}^2$ into $\mathbf{b}$ by means of an affine transformation. Let $\mathbf{a}$ be a simplex in $\mathbb{R}^2$ such that $\Delta(a_0, a_1, a_2) \neq 0$. Then Theorem 19.3 shows that there is a unique affine transformation $L:\mathbb{R}^2 \to \mathbb{R}^3$ such that $L(a_i) = b_i$, $i = 0, 1, 2$. The equations of this transformation are [see Section 19]

$$y^j = g^j(x) = c_{1j}x^1 + c_{2j}x^2 + c_{3j}, \qquad j = 1, 2, 3. \tag{1}$$

Let $a_i = (a_i^1, a_i^2)$ and $b_i = (b_i^1, b_i^2, b_i^3)$ for $i = 0, 1, 2$. The coefficients c_{ij} in (1) are found by solving the following matrix equation [compare equation (6) in Section 19].

$$\begin{bmatrix} a_0^1 & a_0^2 & 1 \\ a_1^1 & a_1^2 & 1 \\ a_2^1 & a_2^2 & 1 \end{bmatrix} \begin{bmatrix} c_{11} & c_{12} & c_{13} \\ c_{21} & c_{22} & c_{23} \\ c_{31} & c_{32} & c_{33} \end{bmatrix} = \begin{bmatrix} b_0^1 & b_0^2 & b_0^3 \\ b_1^1 & b_1^2 & b_1^3 \\ b_2^1 & b_2^2 & b_2^3 \end{bmatrix}. \tag{2}$$

This matrix equation has a unique solution for $[c_{ij}]$ since $\Delta(a_0, a_1, a_2) \neq 0$. Henceforth let c_{ij} denote the constants obtained as solutions of (2). Then

$$\begin{bmatrix} a_0^1 & a_0^2 & 1 \\ a_1^1 & a_1^2 & 1 \\ a_2^1 & a_2^2 & 1 \end{bmatrix} \begin{bmatrix} c_{11} & c_{12} & c_{13} & 0 \\ c_{21} & c_{22} & c_{23} & 0 \\ c_{31} & c_{32} & c_{33} & 1 \end{bmatrix} = \begin{bmatrix} b_0^1 & b_0^2 & b_0^3 & 1 \\ b_1^1 & b_1^2 & b_1^3 & 1 \\ b_2^1 & b_2^2 & b_2^3 & 1 \end{bmatrix}. \tag{3}$$

The components $\Delta(\mathbf{b}^{(j_1, j_2)})$, $(j_1, j_2) \in (2/3)$, are defined as follows:

$$\Delta(\mathbf{b}^{(j_1, j_2)}) = \det[b_i^j 1], \qquad i = 0, 1, 2, \qquad j = j_1, j_2. \tag{4}$$

Also, let $c^{(j_1, j_2)}$ denote the 2×2 minor in columns j_1, j_2 of the following matrix c.

$$c = \begin{bmatrix} c_{11} & c_{12} & c_{13} \\ c_{21} & c_{22} & c_{23} \end{bmatrix}. \tag{5}$$

Then (3) and the multiplication theorem for determinants show that

$$\Delta(\mathbf{b}^{(j_1, j_2)}) = \det[c^{(j_1, j_2)}]\Delta(\mathbf{a}), \qquad (j_1, j_2) \in (2/3). \tag{6}$$

Preparations are now complete for showing that simplicial partitions of $\mathbf{a}$ can be mapped into partitions of $\mathbf{b}$ and then into simplicial partitions of $\mathbf{b}^{(j_1, j_2)}$. Let P_k, $k = 1, 2, \cdots$, be a sequence of simplicial partitions of $\mathbf{a}$ which has the properties stated in Definition 35.1. Then $P_k = \{\mathbf{x} : \mathbf{x} \in P_k\}$, and the affine transformation (1) maps P_k into a simplical partition $Q_k = \{\mathbf{y} : \mathbf{y} \in Q_k\}$ of $\mathbf{b}$. Equations (2) and (3) are true with $\mathbf{a}$ and $\mathbf{b}$ replaced by $\mathbf{x}$ and $\mathbf{y}$ respectively. Also, (6) is true with $\mathbf{a}$ and $\mathbf{b}^{(j_1, j_2)}$ replaced by $\mathbf{x}$ and $\mathbf{y}^{(j_1, j_2)}$ respectively; thus

$$\Delta(\mathbf{y}^{(j_1, j_2)}) = \det[c^{(j_1, j_2)}]\Delta(\mathbf{x}), \qquad (j_1, j_2) \in (2/3). \tag{7}$$

The simplexes $\mathbf{y}^{(j_1, j_2)}$ form a simplicial partition of $\mathbf{b}^{(j_1, j_2)}$; since (7) shows that all simplexes $\mathbf{y}^{(j_1, j_2)}$ have the same orientation in (y^{j_1}, y^{j_2})-space, they form a simplicial partition $Q_k^{(j_1, j_2)}$ of $\mathbf{b}^{(j_1, j_2)}$ in the sense of Definition 35.1.

Let $f : \mathbf{b} \to \mathbb{R}^3$ be a function with bounded components $f^{(j_1, j_2)}$, $(j_1, j_2) \in (2/3)$; the partitions $Q_k^{(j_1, j_2)}$ will now be used to define an integral of $f^{(j_1, j_2)}$ on $\mathbf{b}^{(j_1, j_2)}$. Consider the integral of $f^{(1,2)}$ on $\mathbf{b}^{(1,2)}$; the definition of the other two integrals is similar. Then $\mathbf{b}^{(1,2)}$ is a 2-simplex in the (y^1, y^2)-plane. Consider first the case in which

$$\Delta(\mathbf{a}) > 0, \qquad \det[c^{(1,2)}] > 0; \tag{8}$$

the other cases will be considered later. Then (8) and (6) show that $\mathbf{b}^{(1,2)}$ is a non-degenerate 2-simplex with $\Delta(\mathbf{b}^{(1,2)}) > 0$. Since $\mathbf{x}$ is in P_k and $\Delta(\mathbf{a}) > 0$, then $\Delta(\mathbf{x}) > 0$ for every $\mathbf{x}$. The affine transformation (1) maps each $\mathbf{x}$ in P_k into $\mathbf{y}$ in Q_k, and (8) and (7) show that $\Delta(\mathbf{y}^{(1,2)}) > 0$. Thus, if $Q_k^{(1,2)} = \{\mathbf{y}^{(1,2)} : \mathbf{y} \in Q_k\}$, then $Q_k^{(1,2)}$ is a positively oriented simplicial partition of $\mathbf{b}^{(1,2)}$. Now use the definition of the Riemann integral in Section 35 as a model in defining the integral of $f^{(1,2)}$ over $\mathbf{b}^{(1,2)}$. Let Q be an oriented partition of $\mathbf{b}$ in which $\Delta(\mathbf{y}^{(1,2)}) > 0$ for each $\mathbf{y}$ in Q. Then, for each such partition Q of $\mathbf{b}$ and each

$\mathbf{y}$ in Q, set

$$\begin{aligned} m(f^{(1,2)}, \mathbf{y}) &= \inf\{f^{(1,2)}(y) : y \in \mathbf{y}\}, \\ M(f^{(1,2)}, \mathbf{y}) &= \sup\{f^{(1,2)}(y) : y \in \mathbf{y}\}. \end{aligned} \tag{9}$$

$$\begin{aligned} m(f^{(1,2)}, Q) &= \sum_{\mathbf{y} \in Q} m(f^{(1,2)}, \mathbf{y}) \frac{(-1)^2}{2!} \Delta(\mathbf{y}^{(1,2)}), \\ M(f^{(1,2)}, Q) &= \sum_{\mathbf{y} \in Q} M(f^{(1,2)}, \mathbf{y}) \frac{(-1)^2}{2!} \Delta(\mathbf{y}^{(1,2)}). \end{aligned} \tag{10}$$

Then, as in Section 35, it is easy to prove that, if Q' is a refinement of Q, then

$$m(f^{(1,2)}, Q) \leqq m(f^{(1,2)}, Q') \leqq M(f^{(1,2)}, Q') \leqq M(f^{(1,2)}, Q). \tag{11}$$

If Q_1 and Q_2 are two partitions of $\mathbf{b}$, then there is a partition which is a refinement of each one, and

$$m(f^{(1,2)}, Q_i) \leqq M(f^{(1,2)}, Q_j), \qquad i, j = 1, 2; \tag{12}$$

$$\sup_Q m(f^{(1,2)}, Q) \leqq \inf_Q M(f^{(1,2)}, Q). \tag{13}$$

If the equality holds in (13), then $f^{(1,2)}$ is said to be integrable on $\mathbf{b}^{(1,2)}$, and the common value in (13) is the value of the integral, which is denoted by $\int_{\mathbf{b}^{(1,2)}} f^{(1,2)}(y)\, d(y^1, y^2)$.

If $\det[c^{(1,2)}] = 0$, then (6) shows that $\Delta(\mathbf{b}^{(1,2)}) = 0$ and (7) shows that $\Delta(\mathbf{y}^{(1,2)}) = 0$ for every $\mathbf{y}$ in Q. Then $m(f^{(1,2)}, Q)$ and $M(f^{(1,2)}, Q)$ in (10) equal zero, and the integral of $f^{(1,2)}$ on $\mathbf{b}^{(1,2)}$ exists and has the value zero. If $\Delta(\mathbf{x}) > 0$ and $\det[c^{(1,2)}] < 0$, or if $\Delta(\mathbf{x}) < 0$ and $\det[c^{(1,2)}] > 0$, then $\Delta(\mathbf{y}^{(1,2)}) < 0$ for every $\mathbf{y}$ in Q. The integral which results in this case is the negative of the integral obtained by reversing the orientation of each simplex $\mathbf{y}^{(1,2)}$. The definition of the integral of $f^{(1,3)}$ and $f^{(2,3)}$ on $\mathbf{b}^{(1,3)}$ and $\mathbf{b}^{(2,3)}$, respectively, is similar to that of $f^{(1,2)}$ on $\mathbf{b}^{(1,2)}$. Thus the definition of the following integrals is complete:

$$\int_{\mathbf{b}^{(j_1,j_2)}} f^{(j_1,j_2)}(y)\, d(y^{j_1}, y^{j_2}), \qquad (j_1, j_2) \in (2/3). \tag{14}$$

37.1 Definition. Let $f: \mathbf{b} \to \mathbb{R}^3$ be a function whose components $f^{(1,2)}$, $f^{(1,3)}$, $f^{(2,3)}$ are bounded. If the three integrals in (14) exist, then f *is integrable on the simplex* $\mathbf{b}$; the value of its integral, denoted by

$$\int_{\mathbf{b}} \sum_{(j_1,j_2)} f^{(j_1,j_2)}(y)\, d(y^{j_1}, y^{j_2}), \qquad (j_1, j_2) \in (2/3), \tag{15}$$

is the sum of the integrals in (14).

As in the case of the Riemann integral in Section 35, it is easy to show that the integrals in (14) and (15) exist if the components of $f: \mathbf{b} \to \mathbb{R}^3$ are continuous. Also, let Q_k, $k = 1, 2, \cdots$, be a sequence of simplicial partitions of $\mathbf{b}$ whose norms approach zero as $k \to \infty$. Then the discussion above and the proof of Corollary 35.10 show that

$$\int_{\mathbf{b}} \sum_{(j_1, j_2)} f^{(j_1, j_2)}(y)\, d(y^{j_1}, y^{j_2})$$
$$= \lim_{k\to\infty} \sum_{y\in Q_k} \sum_{(j_1, j_2)} f^{(j_1, j_2)}(y) \frac{(-1)^2}{2!} \Delta(\mathbf{y}^{(j_1, j_2)}), \qquad y \in \mathbf{y}, \qquad (j_1, j_2) \in (2/3). \tag{16}$$

The final object of this section is to show that the integral in (15) is equal to a surface integral of the kind defined in Section 36. The affine transformation (1) is a function $g: \mathbf{a} \to \mathbb{R}^3$ whose components g^j have continuous derivatives. Then g defines a surface G_2 in $\mathbb{R}^3$. Since the affine transformation (1) maps $\mathbf{a}$ into $\mathbf{b}$, the trace $T(G_2)$ of G_2 is $\mathbf{b}$. If $f: \mathbf{b} \to \mathbb{R}^3$ is defined and continuous on $T(G_2)$, then the surface integral

$$\int_{G_2} \sum_{(j_1, j_2)} f^{(j_1, j_2)}(y)\, d(y^{j_1}, y^{j_2}), \qquad (j_1, j_2) \in (2/3), \tag{17}$$

defined in Section 36, exists.

37.2 Theorem. *Let $f: \mathbf{b} \to \mathbb{R}^3$ be a function whose components $f^{(j_1, j_2)}$ are continuous on* $\mathbf{b}$. *Then the integral on the simplex* $\mathbf{b}$ *in* (15) *equals the surface integral in* (17), *and both are equal to the following Riemann integral over* $\mathbf{a}$ *in* $\mathbb{R}^2$.

$$\int_{\mathbf{a}} \left\{ \sum_{(j_1, j_2)} (f^{(j_1, j_2)} \circ g)(x) \det[c^{(j_1, j_2)}] \right\} d(x^1, x^2), \qquad (j_1, j_2) \in (2/3). \tag{18}$$

PROOF. To find the value of the surface integral in (17), it is necessary to find the limit of the sum in equation (29) in Section 36. Let $\mathbf{x} = (x_0, x_1, x_2)$; then

$$\Delta_{(1,2)}(g^{j_1}, g^{j_2})(\mathbf{x}) = \begin{vmatrix} g^{j_1}(x_0) & g^{j_2}(x_0) & 1 \\ g^{j_1}(x_1) & g^{j_2}(x_1) & 1 \\ g^{j_1}(x_2) & g^{j_2}(x_2) & 1 \end{vmatrix} = \begin{vmatrix} y_0^{j_1} & y_0^{j_2} & 1 \\ y_1^{j_1} & y_1^{j_2} & 1 \\ y_2^{j_1} & y_2^{j_2} & 1 \end{vmatrix}$$
$$= \Delta(\mathbf{y}^{(j_1, j_2)}) = \det[c^{(j_1, j_2)}]\Delta(\mathbf{x}). \tag{19}$$

Substitute $\Delta(\mathbf{y}^{(j_1, j_2)})$ for $\Delta_{(1,2)}(g^{j_1}, g^{j_2})(\mathbf{x})$ in (29) in Section 36 with $m = 2$ and $n = 3$; then the integral in (17) is

$$\lim_{k\to\infty} \sum_{\mathbf{x}\in P_k} \sum_{(j_1, j_2)} f^{(j_1, j_2)}[g(x_0)] \frac{(-1)^2}{2!} \Delta(\mathbf{y}^{(j_1, j_2)}). \tag{20}$$

By (16) the value of this limit is (15), the integral of f over the simplex $\mathbf{b}$. Thus the integral in (17) equals the integral in (15). Next, in (20), replace $\Delta(y^{(j_1, j_2)})$ by its value $\det[c^{(j_1, j_2)}]\Delta(\mathbf{x})$ in (19). Then (20) equals

$$\lim_{k\to\infty} \sum_{\mathbf{x}\in P_k} \sum_{(j_1, j_2)} f^{(j_1, j_2)}[g(x_0)] \det[c^{(j_1, j_2)}] \frac{(-1)^2}{2!} \Delta(\mathbf{x}). \tag{21}$$

This limit, by (50) in Section 35, is the integral in (18), and the proof of Theorem 37.2 is complete. □

This section has treated the integral of a function f on a 2-simplex $\mathbf{b}$ in $\mathbb{R}^3$. The treatment of the integral of a function $f: \mathbf{b} \to \mathbb{R}^{\binom{n}{m}}$ on an m-simplex $\mathbf{b}$ in $\mathbb{R}^n$ is entirely similar, and it is left to the reader.

Exercises

37.1. Let $\mathbf{a}: (a_0, a_1, a_2)$ be the simplex in $\mathbb{R}^2$ whose vertices are $a_0: (0, 0)$, $a_1: (1, 0)$, and $a_2: (0, 1)$. Let $\mathbf{b}: (b_0, b_1, b_2)$ be the simplex in $\mathbb{R}^3$ whose vertices are $b_0: (0, 0, 1)$, $b_1: (1, 0, 0)$, and $b_2: (0, 1, 0)$. Show that

$$\begin{aligned} y^1 &= x^1, \\ y^2 &= x^2, \\ y^3 &= 1 - (x^1 + x^2), \end{aligned}$$

is an affine transformation $L: \mathbb{R}^2 \to \mathbb{R}^3$ such that $L(a_i) = b_i$, $i = 0, 1, 2$. Let $y: (y^1, y^2, y^3)$ denote a point in $\mathbf{b}: [b_0, b_1, b_2]$, and let $f: \mathbf{b} \to \mathbb{R}^3$ be a function whose components have the following values:

$$\begin{aligned} f^{(1,2)}(y) &= 4(y^1 + y^2 + y^3), \\ f^{(1,3)}(y) &= -2(y^1 + y^2 + y^3), \\ f^{(2,3)}(y) &= 6(y^1 + y^2 + y^3). \end{aligned}$$

(a) Show that the following integrals exist and have the values shown.

$$\int_{\mathbf{b}^{(1,2)}} f^{(1,2)}(y)\, d(y^1, y^2) = 2, \qquad \int_{\mathbf{b}^{(1,3)}} f^{(1,3)}(y)\, d(y^1, y^3) = 1,$$

$$\int_{\mathbf{b}^{(2,3)}} f^{(2,3)}(y)\, d(y^2, y^3) = 3.$$

(b) Find the value of the integral of f on the simplex $\mathbf{b}$ [see (15)] in two ways and show that this value is 6.

37.2. Show that equation (3) is true if (2) is true.

37.3. If $\mathbf{x}$ is a simplex in a partition P_k of $\mathbf{a}$, prove that equation (7) is true. [Hint. Use barycentric coordinates. If $\mathbf{x} = [x_i^j]$, $i = 0, 1, 2$ and $j = 1, 2$, show that

$$\begin{bmatrix} x_0^1 & x_0^2 & 1 \\ x_1^1 & x_1^2 & 1 \\ x_2^1 & x_2^2 & 1 \end{bmatrix} = \begin{bmatrix} t_{00} & t_{01} & t_{02} \\ t_{10} & t_{11} & t_{12} \\ t_{20} & t_{21} & t_{22} \end{bmatrix} \begin{bmatrix} a_0^1 & a_0^2 & 1 \\ a_1^1 & a_1^2 & 1 \\ a_2^1 & a_2^2 & 1 \end{bmatrix}.$$

Show that (i) the elements in the t_{ij} matrix are non-negative; (ii) the sum of the elements in each row of the t_{ij} matrix is 1; and (iii) the determinant of the t_{ij} matrix is positive since $\mathbf{x}$ has the same orientation as $\mathbf{a}$. To obtain the desired conclusion, multiply equation (3) on the left by the t_{ij} matrix.]

37.4. Let $\mathbf{b}:(b_0, b_1, \cdots, b_m)$ be an m-dimensional simplex in $\mathbb{R}^n$, $m \leqq n$, and let $f: \mathbf{b} \to \mathbb{R}^{\binom{n}{m}}$ be a bounded function. Define the integral of f on $\mathbf{b}$ and show that this integral

$$\int_{\mathbf{b}} \sum_{(j_1, \cdots, j_m)} f^{(j_1, \cdots, j_m)}(y)\, d(y^{j_1}, \cdots, y^{j_m}), \qquad (j_1, \cdots, j_m) \in (m/n),$$

exists if f is continuous. Consider the case in which $\mathbf{b}$ is a degenerate simplex. [Hint. Repeat the entire discussion in Section 37 for this general case.]

37.5. Let $\mathbf{b}:[b_0, b_1, \cdots, b_m]$ be an m-simplex in $\mathbb{R}^n$, $m \leqq n$, and let $f: \mathbf{b} \to \mathbb{R}^{\binom{n}{m}}$ be continuous. Let $\mathbf{a}:[a_0, a_1, \cdots, a_m]$ be a simplex in $\mathbb{R}^m$ such that $\Delta(\mathbf{a}) > 0$, and let $g: \mathbf{a} \to \mathbf{b}$ be an affine transformation such that $g(a_i) = b_i$, $i = 0, 1, \cdots, m$. Then g defines a surface G_m whose trace $T(G_m)$ is $\mathbf{b}$. A second affine transformation $g': \mathbf{a}' \to \mathbf{b}$ defines another surface G'_m whose trace is $\mathbf{b}$. Prove that the surface integrals of f on G_m and G'_m are equal.

37.6. In Exercise 37.5 let

$$y^j = g^j(x) = c_{1j}x^1 + \cdots + c_{mj}x^m + c_{m+1,j}, \qquad j = 1, \cdots, n,$$

be the affine transformation which maps $\mathbf{a}$ in $\mathbb{R}^m$ into $\mathbf{b}$, and let c be the matrix $[c_{ij}]$, $i = 1, \cdots, m$ and $j = 1, \cdots, n$; this matrix is the general case of the matrix c in (5). Let $c^{(j_1, \cdots, j_m)}$ be the $m \times m$ minor of c in columns $j_1, \cdots, j_m$.

(a) Prove that

$$\int_{\mathbf{b}} \sum_{(j_1, \cdots, j_m)} f^{(j_1, \cdots, j_m)}(y)\, d(y^{j_1}, \cdots, y^{j_m})$$

$$= \int_{\mathbf{a}} \sum_{(j_1, \cdots, j_m)} f^{(j_1, \cdots, j_m)}[g(x)] \det[c^{(j_1, \cdots, j_m)}]\, d(x^1, \cdots, x^m),$$

$$(j_1, \cdots, j_m) \in (m/n).$$

(b) Use Theorem 35.18 to prove that there exists a point x^* on the interior of $\mathbf{a}$ such that the integral in (a) equals

$$\sum_{(j_1, \cdots, j_m)} f^{(j_1, \cdots, j_m)}[g(x^*)] \det[c^{(j_1, \cdots, j_m)}] \frac{(-1)^m}{m!} \Delta(\mathbf{a}).$$

(c) Use the generalization of (3) to prove the following generalization of (6):

$$\Delta(\mathbf{b}^{(j_1, \cdots, j_m)}) = \det[c^{(j_1, \cdots, j_m)}] \Delta(\mathbf{a}).$$

Use this result and (b) to prove that the integral in (a) equals

$$\sum_{(j_1, \cdots, j_m)} f^{(j_1, \cdots, j_m)}[g(x^*)] \frac{(-1)^m}{m!} \Delta(\mathbf{b}^{(j_1, \cdots, j_m)}).$$

(d) Show that the mean-value theorem in part (b) can be used to evaluate the integral in Exercise 37.1(b); verify that the value obtained in that exercise is correct.

37.7. Let $\mathbf{b}:[b_0, b_1, \cdots, b_m]$ be a simplex in $\mathbb{R}^n$, $m \leqq n$, and let $\mathbf{e}:[e_0, e_1, \cdots, e_m]$ be the simplex in $\mathbb{R}^m$ whose vertices are the origin and the unit points on the axes. Then $\mathbf{b}$ is represented by the matrix $[b_i^j]$, $i = 0, 1, \cdots, m, j = 1, \cdots, n$.

(a) Use the generalization of equation (3) to find the equations of the affine transformation which maps **e** into **b**. Show that these equations are

$$y^j = g^j(x) = \sum_{i=1}^{m} (b_i^j - b_0^j)x^i + b_0^j, \qquad j = 1, \cdots, n,$$

and that

$$c = [(b_i^j - b_0^j)], \qquad i = 1, \cdots, m, \qquad j = 1, \cdots, n.$$

(b) Show by direct calculation that

$$\begin{aligned}\det[c^{(j_1,\cdots,j_m)}] &= (-1)^m \det[b_i^j \quad 1], \qquad i = 0, 1, \cdots, m, \qquad j = j_1, \cdots, j_m,\\ &= (-1)^m \Delta(\mathbf{b}^{(j_1,\cdots,j_m)}).\end{aligned}$$

(c) Observe that $\Delta(\mathbf{e}) = (-1)^m$, and give a second proof of the second equation in (b) by using the generalization of equation (6).

(d) Use these results to verify the following statement of the mean-value theorem for integrals. If $f: \mathbf{b} \to \mathbb{R}^{\binom{n}{m}}$ is continuous, then there exists a point x^* in the interior of **e** such that

$$\int_{\mathbf{b}} \sum_{(j_1,\cdots,j_m)} f^{(j_1,\cdots,j_m)}(y)\, d(y^{j_1}, \cdots, y^{j_m})$$

$$= \sum_{(j_1,\cdots,j_m)} f^{(j_1,\cdots,j_m)}[g(x^*)] \frac{(-1)^m}{m!} \Delta(\mathbf{b}^{(j_1,\cdots,j_m)}).$$

Observe that $g(x^*)$ is a point y^* which can be described as being on the interior of **b**. [Hint. Exercise 37.6(b) and Exercise 37.7(b) and (c).]

38. The Fundamental Theorem of the Integral Calculus

The formula $\int_a^b f'(x)\,dx = f(b) - f(a)$ is the one-dimensional form of the fundamental theorem of the integral calculus; it expresses the integral of the derivative f' on $[a, b]$ as the integral of f on the boundary of $[a, b]$. The proof of this elementary theorem, by methods which will be generalized in this section, can be found in Example 11.1 and in Section 34. In n-dimensions, the fundamental theorem of the integral calculus represents the integral $\int_{\mathbf{a}} D_{(1,\cdots,n)}(f^1, \cdots, f^n)(x)\, d(x^1, \cdots, x^n)$ as the integral of a certain function over the boundary $\partial\mathbf{a}$ of **a**.

Experience has shown that special properties of the partitions used to define the integral are frequently required to establish special properties of the integral. Section 35 contains a first example of this statement; there rectangular partitions $\mathscr{R}_k$, $k = 1, 2, \cdots$, were used to establish the evaluation of $\int_{\mathbf{a}} f(x)\, d(x^1, \cdots, x^n)$ by iterated integrals. Section 36 contains a second example, because there the definition of the surface integral involves the derivatives of g. The definition and the existence of these derivatives requires, first, that the partitions be simplicial partitions P_k, $k = 1, 2, \cdots$, and, second, that the simplexes in these partitions satisfy the regularity condition. What restrictions must be placed on the partitions in order to prove the fundamental theorem of the integral calculus? The answer seems almost obvious. Since a derivative occurs in $\int_{\mathbf{a}} D_{(1,\cdots,n)}(f^1, \cdots, f^n)(x)\,d(x^1, \cdots, x^n)$, it is to be expected that the partitions will be simplicial partitions and that the simplexes in the partitions will be required to satisfy the regularity condition. Also, previous experience [see, for example, Theorem 18.6, Example 18.7, Exercise 18.7, and Theorem 23.4] has shown that the cancellation properties in an oriented complex can be used frequently to represent a sum over the simplexes in the complex as a sum over the simplexes in the boundary of the complex; this experience suggests that the simplexes in each partition of $\mathbf{a}$ should form an oriented complex. The partitions in a sequence of oriented simplicial subdivisions $\mathscr{P}_k$, $k = 1, 2, \cdots$, have both of the suggested properties [see Definition 21.5]. Thus it should come as no surprise that, in all proofs related to the fundamental theorem of the integral calculus, the sequences of partitions of the region of integration are restricted to be sequences $\mathscr{P}_k$, $k = 1, 2, \cdots$, of oriented simplicial subdivisions as defined in Definition 21.5.

38.1 Example. Let $A = [a_1, b_1] \times [a_2, b_2]$, and let $f: A \to \mathbb{R}^2$ be a function whose components (f^1, f^2) have continuous derivatives on A. The purpose of this example is to prove the following statement of the fundamental theorem of the integral calculus. If A is positively oriented in $\mathbb{R}^2$, then

$$\begin{aligned}\int_A D_{(1,2)}(f^1, f^2)(x)\,d(x^1, x^2) = &\int_{a_1}^{b_1} f^1(x^1, a_2)D_1 f^2(x^1, a_2)\,dx^1\\ &+ \int_{a_2}^{b_2} f^1(b_1, x^2)D_2 f^2(b_1, x^2)\,dx^2\\ &+ \int_{b_1}^{a_1} f^1(x^1, b_2)D_1 f^2(x^1, b_2)\,dx^1\\ &+ \int_{b_2}^{a_2} f^1(a_1, x^2)D_2 f^2(a_1, x^2)\,dx^2.\end{aligned} \tag{1}$$

Let $\mathscr{P}_k$, $k = 1, 2, \cdots$, be a sequence of simplicial subdivisions of A which are positively oriented in $\mathbb{R}^2$, and let $\mathbf{x}: (x_0, x_1, x_2)$ denote a simplex in $\mathscr{P}_k$. The proof of (1) is obtained by making two evaluations of the following limit:

$$\lim_{k\to\infty} \sum_{\mathbf{x}\in\mathscr{P}_k} \frac{(-1)^2}{2!}\Delta_{(1,2)}(f^1, f^2)(\mathbf{x}). \tag{2}$$

The first evaluation uses (a) the fact that the simplexes in $\mathscr{P}_k$ satisfy the regularity condition, and (b) Theorem 11.4 to prove that

$$\lim_{k\to\infty}\sum_{\mathbf{x}\in\mathscr{P}_k}\frac{(-1)^2}{2!}\Delta_{(1,2)}(f^1,f^2)(\mathbf{x})=\int_A D_{(1,2)}(f^1,f^2)(x)\,d(x^1,x^2). \tag{3}$$

Theorem 11.4 shows that the limit in (2) equals

$$\lim_{k\to\infty}\sum_{\mathbf{x}\in\mathscr{P}_k}D_{(1,2)}(f^1,f^2)(x_0)\frac{(-1)^2}{2!}\Delta(\mathbf{x})+\lim_{k\to\infty}\sum_{\mathbf{x}\in\mathscr{P}_k}r(f^1,f^2;\mathbf{x})\frac{(-1)^2}{2!}|\Delta(\mathbf{x})| \tag{4}$$

if the two limits in this expression exist. Since f^1, f^2 have continuous derivatives on A, then $D_{(1,2)}(f^1, f^2)$ is a continuous function on A, and Corollary 35.10 shows that the first limit in (4) is the integral on the right in (3). Then the proof of (3) will be complete if it can be shown that the second limit in (4) is zero. Since the norm of $\mathscr{P}_k$ approaches zero as $k \to \infty$, Theorem 11.4 shows that, to each $\eta > 0$ there corresponds a $k(\eta)$ such that $|r(f^1, f^2; \mathbf{x})| < \eta$ for every $\mathbf{x}$ in $\mathscr{P}_k$ provided $k \geqq k(\eta)$. Then

$$\left|\sum_{\mathbf{x}\in\mathscr{P}_k}r(f^1,f^2;\mathbf{x})\frac{(-1)^2}{2!}|\Delta(\mathbf{x})|\right|<\eta\sum_{\mathbf{x}\in\mathscr{P}_k}\frac{|\Delta(\mathbf{x})|}{2!},\qquad k\geqq k(\eta). \tag{5}$$

Since the sum on the right of this inequality is the area of A, a constant, the statement in (5) proves that the second limit in (4) equals zero. Thus the proof of (3) is complete.

The second evaluation of (2) uses the fact that the simplexes in $\mathscr{P}_k$ form a Euclidean complex, together with Theorems 20.3 and 18.5, to prove that the limit in (2) equals the sum of the four integrals on the right in (1). Theorem 20.3 and elementary properties of determinants show that

$$\begin{aligned}\frac{(-1)^2}{2!}\begin{vmatrix}f^1(x_0)&f^2(x_0)&1\\f^1(x_1)&f^2(x_1)&1\\f^1(x_2)&f^2(x_2)&1\end{vmatrix}&=\frac{(-1)}{2}\begin{vmatrix}f^1(x_1)+f^1(x_2)&f^2(x_0)&1\\f^1(x_0)+f^1(x_2)&f^2(x_1)&1\\f^1(x_0)+f^1(x_1)&f^2(x_2)&1\end{vmatrix}\\&=\frac{[f^1(x_1)+f^1(x_2)]}{2}(-1)\begin{vmatrix}f^2(x_1)&1\\f^2(x_2)&1\end{vmatrix}\\&\quad-\frac{[f^1(x_0)+f^1(x_2)]}{2}(-1)\begin{vmatrix}f^2(x_0)&1\\f^2(x_2)&1\end{vmatrix}\\&\quad+\frac{[f^1(x_0)+f^1(x_1)]}{2}(-1)\begin{vmatrix}f^2(x_0)&1\\f^2(x_1)&1\end{vmatrix}.\end{aligned} \tag{6}$$

The expression

$$\frac{[f^1(x_1)+f^1(x_2)]}{2}(-1)\begin{vmatrix}f^2(x_1)&1\\f^2(x_2)&1\end{vmatrix} \tag{7}$$

is a function $\Phi(x_1, x_2)$ of the oriented side (x_1, x_2) of $\mathbf{x}$, and $\Phi(x_2, x_1) = -\Phi(x_1, x_2)$. Similar statements hold for the terms in (6) which are defined on (x_0, x_2) and (x_0, x_1). Represent each term in the sum in (2) as in (6). If

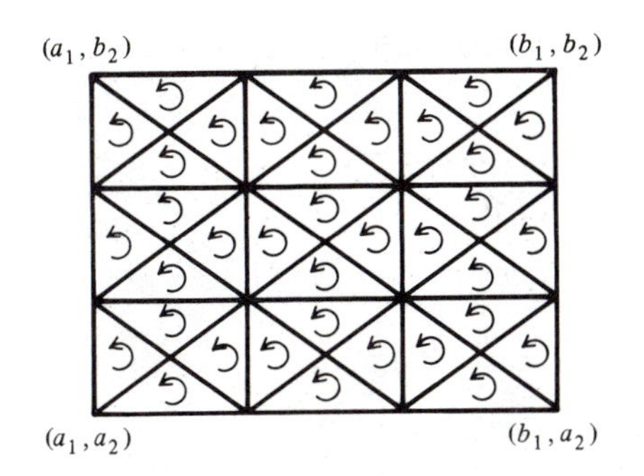

Figure 38.1. Simplicial subdivision of A in Example 38.1.

two simplexes $\mathbf{x}$ and $\mathbf{x}'$ have a one-dimensional side in common, then this side has opposite orientations by Theorem 18.5 because the simplexes in each $\mathscr{P}_k$ form an oriented Euclidean complex. Then two simplexes $\mathbf{x}$ and $\mathbf{x}'$ with a one-dimensional side in common contribute two terms which cancel in the sum in (2). The term in (7) remains after all cancellations have been made if and only if (x_1, x_2) is a side of only a single simplex $\mathbf{x}$. Thus (7) is a term in (2), after cancellations, if and only if (x_1, x_2) is a one-simplex in the boundary of the chain of all simplexes in $\mathscr{P}_k$. There is no loss of generality in assuming that x_0, in the simplex $\mathbf{x}: (x_0, x_1, x_2)$, is in the interior of A, and that the side of $\mathbf{x}$ in the boundary ∂A of A has vertices x_1 and x_2. Then the sum in (2) equals the sum of terms of the form shown in (7). Figure 38.1 indicates graphically the cancellation which occurs in the sum in (2) when the terms in the sum are represented in the form indicated in (6). Figure 38.1 also indicates that the terms (7) correspond to a counterclockwise, or positive, subdivision of the boundary of A.

Group the terms on the four sides of A. The sum on the side $x^2 = a_2$ of A, with vertices (a_1, a_2) and (b_1, a_2), has the form

$$\sum_{i=1}^{n_k} \frac{[f^1(x_{i-1}^1, a_2) + f^1(x_i^1, a_2)]}{2}(-1)\begin{vmatrix} f^2(x_{i-1}^1, a_2) & 1 \\ f^2(x_i^1, a_2) & 1 \end{vmatrix}. \tag{8}$$

As $k \to \infty$, the norm of the subdivisions $\mathscr{P}_k$ tends to zero. Correspondingly, the norm of the subdivisions $a_1 = x_0^1 < x_1^1 < \cdots < x_{i-1}^1 < x_i^1 < \cdots < x_{n_k}^1 = b_1$ of the side $x^2 = a_2$ of A tends to zero. Then the limit, as $k \to \infty$, of the sum in (8) equals

$$\lim_{k\to\infty} \sum_{i=1}^{n_k} f^1(x_{i-1}^1, a_2)(-1)\begin{vmatrix} f^2(x_{i-1}^1, a_2) & 1 \\ f^2(x_i^1, a_2) & 1 \end{vmatrix}. \tag{9}$$

To prove this statement, observe that the difference between the sums in (8) and (9) is

$$\sum_{i=1}^{n_k} \frac{[f^1(x_i^1, a_2) - f^1(x_{i-1}^1, a_2)]}{2}(-1)\begin{vmatrix} f^2(x_{i-1}^1, a_2) & 1 \\ f^2(x_i^1, a_2) & 1 \end{vmatrix}. \tag{10}$$

Now by the elementary mean-value theorem, there is a ξ_i between x_{i-1}^1 and x_i^1 such that

$$(-1)\begin{vmatrix} f^2(x_{i-1}^1, a_2) & 1 \\ f^2(x_i^1, a_2) & 1 \end{vmatrix} = D_1 f^2(\xi_i, a_2)(-1)\begin{vmatrix} x_{i-1}^1 & 1 \\ x_i^1 & 1 \end{vmatrix}. \tag{11}$$

Then the sum in (10) equals

$$\sum_{i=1}^{n_k} \frac{[f^1(x_i^1, a_2) - f^1(x_{i-1}^1, a_2)]}{2} D_1 f^2(\xi_i, a_2)(-1)\begin{vmatrix} x_{i-1}^1 & 1 \\ x_i^1 & 1 \end{vmatrix}. \tag{12}$$

Now since $D_1 f^2$ is continuous on the compact set A, the terms $D_1 f^2(\xi_i, a_2)$ in (12) are uniformly bounded. Since f^1 is continuous on the compact set A, it is uniformly continuous on A, and $(1/2)[f^1(x_i^1, a_2) - f^1(x_{i-1}^1, a_2)]$ tends uniformly to zero as $k \to \infty$. Then for each $\varepsilon > 0$ there is a $k(\varepsilon)$ such that

$$|(1/2)[f^1(x_i^1, a_2) - f^1(x_{i-1}^1, a_2)] D_1 f^2(\xi_i, a_2)| < \varepsilon \tag{13}$$

for $i = 1, \cdots, n_k$ and all k for which $k \geqq k(\varepsilon)$. Thus for $k \geqq k(\varepsilon)$, the absolute value of the sum in (10) and (12) is less than

$$\sum_{i=1}^{n_k} \varepsilon |x_i^1 - x_{i-1}^1| = \varepsilon(b_1 - a_1). \tag{14}$$

This statement completes the proof that the limit of (10) is zero, and hence that the limit of the sum in (8) equals the limit in (9). To evaluate (9), recall that f^2 has a continuous derivative on A and therefore satisfies the Stolz condition on the side $x^2 = a_2$ of A. Thus

$$\begin{aligned}(-1)\begin{vmatrix} f^2(x_{i-1}^1, a_2) & 1 \\ f^2(x_i^1, a_2) & 1 \end{vmatrix} &= D_1 f^2(x_{i-1}^1, a_2)(-1)\begin{vmatrix} x_{i-1}^1 & 1 \\ x_i^1 & 1 \end{vmatrix} \\ &\quad + r(f^2; x_{i-1}^1, x_i^1)|x_i^1 - x_{i-1}^1|,\end{aligned} \tag{15}$$

and $r(f^2; x_{i-1}^1, x_i^1)$ is uniformly small for $i = 1, \cdots, n_k$ [see Theorem 9.9] and approaches zero as $k \to \infty$. Then the limit in (9) equals

$$\begin{aligned}&\lim_{k\to\infty} \sum_{i=1}^{n_k} f^1(x_{i-1}^1, a_2) D_1 f^2(x_{i-1}^1, a_2)(-1)\begin{vmatrix} x_{i-1}^1 & 1 \\ x_i^1 & 1 \end{vmatrix} \\ &\qquad + \lim_{k\to\infty} \sum_{i=1}^{n_k} f^1(x_{i-1}^1, a_2) r(f^2; x_{i-1}^1, x_i^1)|x_i^1 - x_{i-1}^1|.\end{aligned} \tag{16}$$

Now the first limit in (16) is

$$\int_{a_1}^{b_1} f^1(x^1, a_2) D_1 f^2(x^1, a_2)\, dx^1 \tag{17}$$

by (42) in Section 35, and the second limit is zero since f^1 is bounded on A, $r(f^2; x_{i-1}^1, x_i^1)$ is uniformly small for $i = 1, \cdots, n_k$, and $\Sigma_1^{n_k} |x_i^1 - x_{i-1}^1| = b_1 - a_1$. Thus the sum of the terms (7) on the side $x^2 = a_2$ of A is the sum in (8), and the limit of this sum, as $k \to \infty$, is (17), which is the first of the integrals on the right in (1). Similar arguments show that the sums of the terms (7) on the other three sides of A lead to the other three integrals on the right in (1). Observe carefully that the integration around the boundary of A is in the positive, or counterclockwise direction. The proof of (1) is complete.

As a numerical example of the fundamental theorem of the integral calculus in (1), let $A = [2, 4] \times [4, 8]$ and let $f^1(x^1, x^2) = x^1 - x^2$, $f^2(x^1, x^2) = x^1x^2$. Then

$$D_{(1,2)}(f^1, f^2)(x^1, x^2) = \begin{vmatrix} 1 & -1 \\ x^2 & x^1 \end{vmatrix} = x^1 + x^2. \tag{18}$$

The formula for the evaluation of double integrals by iterated integrals [see Example 35.19] shows that

$$\begin{aligned} \int_A D_{(1,2)}(f^1, f^2)(x^1, x^2)\,d(x^1, x^2) &= \int_2^4 \left(\int_4^8 (x^1 + x^2)\,dx^2 \right) dx^1 \\ &= \int_2^4 (x^1x^2 + (1/2)(x^2)^2|_4^8)\,dx^1 \\ &= \int_2^4 (4x^1 + 24)\,dx^1 \\ &= 2(x^1)^2 + 24x^1|_2^4 \\ &= 72. \end{aligned} \tag{19}$$

Also,

$$\begin{aligned} \int_{a_1}^{b_1} f^1(x^1, a_2)D_1f^2(x^1, a_2)\,dx^1 &= \int_2^4 (x^1 - 4)4dx^1 = -8. \\ \int_{a_2}^{b_2} f^1(b_1, x^2)D_2f^2(b_1, x^2)\,dx^2 &= \int_4^8 (4 - x^2)4dx^2 = -32. \\ \int_{b_1}^{a_1} f^1(x^1, b_2)D_1f^2(x^1, b_2)\,dx^1 &= \int_4^2 (x^1 - 8)8dx^1 = 80. \\ \int_{b_2}^{a_2} f^1(a_1, x^2)D_2f^2(a_1, x^2)\,dx^2 &= \int_8^4 (2 - x^2)2dx^2 = 32. \end{aligned} \tag{20}$$

Since the sum of the four integrals in (20) is equal to the integral in (19), this example verifies the statement of the fundamental theorem of the integral calculus in (1).

38.2 Example. Let $\mathbf{a}$ be the simplex $[a_0, a_1, a_2]$ in $\mathbb{R}^2$, and let $f: \mathbf{a} \to \mathbb{R}^2$ be a function whose components (f^1, f^2) have continuous derivatives on

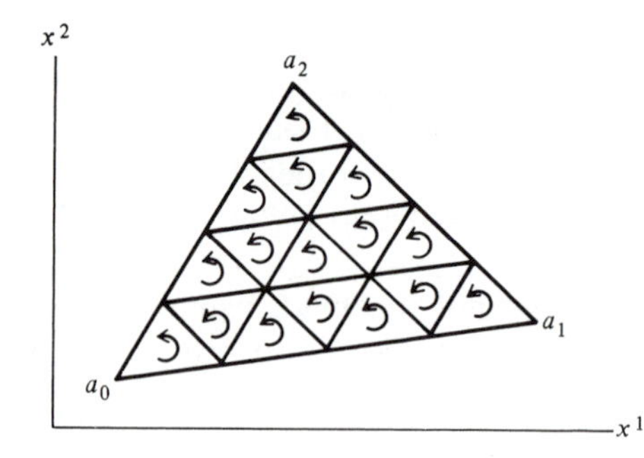

Figure 38.2. A simplicial subdivision of **a** in Example 38.2.

$[a_0, a_1, a_2]$. The purpose of this example is to prove the following statement of the fundamental theorem of the integral calculus.

$$\begin{aligned}\int_{[a_0,a_1,a_2]} & D_{(1,2)}(f^1, f^2)(x^1, x^2)\,d(x^1, x^2) \\ &= \int_{[a_1,a_2]} f^1(x^1, x^2)D_1 f^2(x^1, x^2)\,dx^1 + f^1(x^1, x^2)D_2 f^2(x^1, x^2)\,dx^2 \\ &\quad + \int_{[a_2,a_0]} f^1(x^1, x^2)D_1 f^2(x^1, x^2)\,dx^1 + f^1(x^1, x^2)D_2 f^2(x^1, x^2)\,dx^2 \\ &\quad + \int_{[a_0,a_1]} f^1(x^1, x^2)D_1 f^2(x^1, x^2)\,dx^1 + f^1(x^1, x^2)D_2 f^1(x^1, x^2)\,dx^2.\end{aligned} \tag{21}$$

The integral on the left is an integral on the 2-dimensional simplex **a** in $\mathbb{R}^2$; integrals of this type are treated in Section 35. The integrals on the right in (21) are integrals on the one-dimensional simplexes $[a_1, a_2]$, $[a_2, a_0]$, $[a_0, a_1]$ in $\mathbb{R}^2$; integrals of this type are treated in Section 37. Formula (21) expresses the integral on **a** as an integral around the boundary of **a**. Observe that

$$\begin{aligned}\partial(a_0, a_1, a_2) &= (a_1, a_2) - (a_0, a_2) + (a_0, a_1) \\ &= (a_1, a_2) + (a_2, a_0) + (a_0, a_1).\end{aligned} \tag{22}$$

Figure 38.2 shows **a** as positively oriented in $\mathbb{R}^2$, but (21) is a correct statement in all cases. The proof of (21) is the same as the proof of (1) in Example 38.1, but some of the details are different because the sides of the simplex **a** are not parallel to the axes. Let $\mathscr{P}_k$, $k = 1, 2, \cdots$, be a sequence of simplicial subdivisions of $[a_0, a_1, a_2]$ which have the same orientation in $\mathbb{R}^2$ as **a**. Let $\mathbf{x}: (x_0, x_1, x_2)$ denote a simplex in $\mathscr{P}_k$. Then the proof of (21) is obtained by making two evaluations of the following limit [see (2)]:

$$\lim_{k\to\infty} \sum_{\mathbf{x}\in\mathscr{P}_k} \frac{(-1)^2}{2!}\Delta_{(1,2)}(f^1, f^2)(\mathbf{x}). \tag{23}$$

Now exactly the same proof that was used to prove (3) can be used again to prove that the limit in (23) equals

$$\int_{[a_0, a_1, a_2]} D_{(1,2)}(f^1, f^2)(x^1, x^2)\, d(x^1, x^2). \tag{24}$$

Next, represent each term in the sum in (23) as in (6). Then because the simplexes in $\mathscr{P}_k$ form a Euclidean complex, terms cancel as before, and (7) shows that the sum in (23) simplifies to a sum which can be represented as follows:

$$\sum_{(x_1, x_2) \in \partial\mathscr{P}_k} \frac{[f^1(x_1) + f^1(x_2)]}{2}(-1)\begin{vmatrix} f^2(x_1) & 1 \\ f^2(x_2) & 1 \end{vmatrix}. \tag{25}$$

The limit of this sum is equal to the limit of

$$\sum_{(x_1, x_2) \in \partial\mathscr{P}_k} f^1(x_1)(-1)\begin{vmatrix} f^2(x_1) & 1 \\ f^2(x_2) & 1 \end{vmatrix}. \tag{26}$$

The difference between the sum in (25) and the one in (26) is [compare (10)]

$$\sum_{(x_1, x_2) \in \partial\mathscr{P}_k} \frac{[f^1(x_1) - f^1(x_2)]}{2}(-1)\begin{vmatrix} f^2(x_1) & 1 \\ f^2(x_2) & 1 \end{vmatrix}. \tag{27}$$

The limit of this sum as $k \to \infty$ is zero for the following reasons: (a) since f^1 is continuous on the compact set **a**, $|f^1(x_1) - f^1(x_2)|$ is uniformly less than ε for (x_1, x_2) in $\partial\mathscr{P}_k$ if k is sufficiently large; and (b) by (20) in Corollary 3.9,

$$|f^2(x_2) - f^2(x_1)| \leq \left\{\left[\sum_{j=1}^{2} |D_j f^2(x_1)|^2\right]^{1/2} + |r(f^2; x_1, x_2)|\right\}|x_2 - x_1|. \tag{28}$$

The quantity in the brace is bounded by a constant M since the derivatives $D_j f^2$ are continuous by hypothesis and $|r(f^2; x_1, x_2)|$ is uniformly small by Theorem 9.9; thus the absolute value of the sum in (27) is less than

$$\sum_{(x_1, x_2) \in \partial\mathscr{P}_k} \varepsilon M |x_2 - x_1| = \varepsilon M \sum_{(x_1, x_2) \in \partial\mathscr{P}_k} |x_2 - x_1|. \tag{29}$$

The sum on the right is the perimeter of **a** and therefore a constant. Thus the limit of the sum in (27), as $k \to \infty$, is zero, and the limit of the sum in (25) equals the limit of the sum in (26). To evaluate the latter sum, group the terms on the three sides of **a**. Now f^2 is differentiable on **a** and it satisfies the Stolz condition. Thus

$$(-1)\begin{vmatrix} f^2(x_1) & 1 \\ f^2(x_2) & 1 \end{vmatrix} = D_1 f^2(x_1)(-1)\begin{vmatrix} x_1^1 & 1 \\ x_2^1 & 1 \end{vmatrix} + D_2 f^2(x_1)(-1)\begin{vmatrix} x_1^2 & 1 \\ x_2^2 & 1 \end{vmatrix} + r(f^2; x_1, x_2)|x_2 - x_1|. \tag{30}$$

Then (30) shows that the limit of the sum in (26) equals

$$\lim_{k\to\infty} \sum_{(x_1,x_2)\in\partial\mathscr{P}_k} f^1(x_1)\left\{D_1 f^2(x_1)(-1)\begin{vmatrix} x_1^1 & 1 \\ x_2^1 & 1 \end{vmatrix} + D_2 f^1(x_1)(-1)\begin{vmatrix} x_1^2 & 1 \\ x_2^2 & 1 \end{vmatrix}\right\} \tag{31}$$

$$+\lim_{k\to\infty} \sum_{(x_1,x_2)\in\partial\mathscr{P}_k} f^1(x_1) r(f^2; x_1, x_2)|x_2 - x_1| \tag{32}$$

if these limits exist. In (31) group the terms on the three sides $[a_1, a_2]$, $[a_2, a_0]$, and $[a_0, a_1]$ of $[a_0, a_1, a_2]$. The limit of the terms on each of these sides is an integral on a simplex of the kind treated in Section 37; thus the limit in (31) equals the sum of the three integrals shown on the right in (21). Furthermore, f^1 is continuous on **a** and therefore bounded; $r(f^2; x_1, x_2)$ is uniformly small on the simplexes (x_1, x_2) and approaches zero by Theorem 9.9; and $\Sigma|x_2 - x_1|$ for (x_1, x_2) in $\partial\mathscr{P}_k$ is the perimeter of **a**. These facts prove that the limit in (32) equals zero; hence, the proof of (21) is complete.

As a numerical example of (21), let **a** be the simplex with vertices $a_0 : (0, 0)$, $a_1 : (4, 0)$, and $a_2 : (2, 2)$, and let $f^1(x^1, x^2) = x^1 - x^2$, $f^2(x^1, x^2) = x^1x^2$. Then $D_{(1,2)}(f^1, f^2)(x^1, x^2) = x^1 + x^2$ as in (18). The integral on the left in (21) is a double integral which can be evaluated by iterated integrals as follows:

$$\begin{aligned} \int_{\mathbf{a}} D_{(1,2)}(f^1, f^2)(x^1, x^2)\, d(x^1, x^2) &= \int_{\mathbf{a}} (x^1 + x^2)\, d(x^1, x^2) \\ &= \int_0^2 \left(\int_0^{x^1} (x^1 + x^2)\, dx^2 \right) dx^1 \\ &\quad + \int_2^4 \left(\int_0^{4-x^1} (x^1 + x^2)\, dx^2 \right) dx^1 \\ &= 4 + \frac{20}{3} = \frac{32}{3}. \end{aligned} \tag{33}$$

The integrals on the right in (21) are integrals on the simplexes $[a_1, a_2]$, $[a_2, a_0]$, and $[a_0, a_1]$; they are evaluated as surface integrals by the methods explained in Section 36. The procedure is this: find affine transformations which transform (e_0, e_1) into the three simplexes. Now the affine transformation which transforms (e_0, e_1) in $\mathbb{R}$ into (b_0, b_1) in $\mathbb{R}^2$ is [see Exercises 19.5 and 37.7]

$$\begin{aligned} x^1 &= (b_1^1 - b_0^1)t + b_0^1, \\ x^2 &= (b_1^2 - b_0^2)t + b_0^2. \end{aligned} \tag{34}$$

This formula shows that the three transformations are the following:

$$\begin{aligned} [e_0, e_1] \to [a_1, a_2] : x^1 &= 4 - 2t, \\ x^2 &= 2t, \qquad 0 \leqq t \leqq 1. \end{aligned} \tag{35}$$

$$\begin{aligned} [e_0, e_1] \to [a_2, a_0] : x^1 &= 2 - 2t, \\ x^2 &= 2 - 2t, \qquad 0 \leqq t \leqq 1. \end{aligned} \tag{36}$$

$$[e_0, e_1] \to [a_0, a_1] : x^1 = 4t, \quad x^2 = 0, \qquad 0 \leqq t \leqq 1. \tag{37}$$

Use the transformation in (35) to evaluate, as follows, the first integral on the right in (21).

$$\begin{aligned}
\int_{[a_1,a_2]} & f^1(x^1, x^2)D_1 f^2(x^1, x^2)\,dx^1 + f^1(x^1, x^2)D_2 f^2(x^1, x^2)\,dx^2 \\
&= \int_{[a_1,a_2]} (x^1 - x^2)x^2\,dx^1 + (x^1 - x^2)x^1\,dx^2 \\
&= \int_0^1 \{(4 - 4t)2t(-2) + (4 - 4t)(4 - 2t)(2)\}\,dt \\
&= \int_0^1 (32t^2 - 64t + 32)\,dt = \frac{32}{3}.
\end{aligned} \tag{38}$$

In the same way the transformations in (36) and (37) can be used to show that the second and third integrals on the right in (21) have the value zero. Then (33) and (38) show that the statement of the fundamental theorem of the integral calculus in (21) is verified in this numerical example.

38.3 Example. Let $\mathbf{a}$ be the simplex $[a_0, a_1, a_2, a_3]$ in $\mathbb{R}^3$, and let $f: \mathbf{a} \to \mathbb{R}^3$ be a function whose components (f^1, f^2, f^3) have continuous derivatives on $\mathbf{a}$. The purpose of this example is to prove the following statement of the fundamental theorem of the integral calculus.

$$\begin{aligned}
\int_{\mathbf{a}} & D_{(1,2,3)} f(x)\,d(x^1, x^2, x^3) \\
&= \int_{\partial \mathbf{a}} \sum_{(j_1,j_2)} f^1(x) D_{(j_1,j_2)}(f^2, f^3)(x)\,d(x^{j_1}, x^{j_2}), \qquad (j_1, j_2) \in (2/3).
\end{aligned} \tag{39}$$

The integral on the left is an integral on the 3-dimensional simplex $\mathbf{a}$ in $\mathbb{R}^3$; integrals of this type are treated in Section 35. The integral on the right is the sum of four integrals on the 2-dimensional simplexes $[a_1, a_2, a_3]$, $-[a_0, a_2, a_3]$, $[a_0, a_1, a_3]$, and $-[a_0, a_1, a_2]$ which constitute the boundary of $[a_0, a_1, a_2, a_3]$. Integrals on simplexes of this type are treated in Section 37. Formula (39) expresses the integral on $\mathbf{a}$ as an integral over the boundary of $\mathbf{a}$. Observe that

$$\partial(a_0, a_1, a_2, a_3) = (a_1, a_2, a_3) - (a_0, a_2, a_3) + (a_0, a_1, a_3) - (a_0, a_1, a_2). \tag{40}$$

The proof of (39) is the same as the proof of (21) in Example 38.2, but some of the details need further emphasis. Let $\mathscr{P}_k$, $k = 1, 2, \cdots$, be a sequence of simplicial subdivisions of the Euclidean simplex $[a_0, a_1, a_2, a_3]$. Then the simplexes in each $\mathscr{P}_k$ have the same orientation in $\mathbb{R}^3$ as $\mathbf{a}$, and the norm of

$\mathscr{P}_k$ tends to zero as $k \to \infty$ [see Definition 21.5]. Let $\mathbf{x}:(x_0, x_1, \cdots, x_3)$ denote a simplex in $\mathscr{P}_k$. Then the proof of (39) is obtained by making two evaluations of the following limit [see (23)]:

$$\lim_{k\to\infty} \sum_{\mathbf{x}\in\mathscr{P}_k} \frac{(-1)^3}{3!}\Delta_{(1,2,3)}(f^1, f^2, f^3)(\mathbf{x}). \tag{41}$$

Now the same proof that was used to prove (3) can be used again to prove that the limit in (41) equals

$$\int_{\mathbf{a}} D_{(1,2,3)}f(x)\,d(x^1, x^2, x^3), \tag{42}$$

because Theorem 11.4 shows that

$$\begin{aligned}\lim_{k\to\infty} \sum_{\mathbf{x}\in\mathscr{P}_k} \frac{(-1)^3}{3!}\Delta_{(1,2,3)}f(\mathbf{x}) &= \lim_{k\to\infty} \sum_{\mathbf{x}\in\mathscr{P}_k} D_{(1,2,3)}f(x_0)\frac{(-1)^3}{3!}\Delta(\mathbf{x}) \\ &\quad + \lim_{k\to\infty} \sum_{\mathbf{x}\in\mathscr{P}_k} r(f;\mathbf{x})\frac{(-1)^3}{3!}|\Delta(\mathbf{x})|.\end{aligned} \tag{43}$$

The first limit on the right in (43) is the integral in (42) by Corollary 35.10. Furthermore, the second limit is zero because the terms $r(f;\mathbf{x})$ are uniformly small on the simplexes $\mathbf{x}$ in $\mathscr{P}_k$ by Theorem 11.4 and

$$\sum_{\mathbf{x}\in\mathscr{P}_k} \frac{1}{3!}|\Delta(\mathbf{x})| \tag{44}$$

is the volume of the simplex $\mathbf{a}$ by Theorem 21.3. These facts prove that the limit in (41) equals the integral in (42). Next, Theorem 20.3 shows that

$$\frac{(-1)^3}{3!}\begin{vmatrix} f^1(x_0) & \cdots & f^3(x_0) & 1 \\ f^1(x_1) & \cdots & f^3(x_1) & 1 \\ \cdots & \cdots & \cdots & \cdots \\ f^1(x_3) & \cdots & f^3(x_3) & 1 \end{vmatrix} = \frac{(-1)^2}{3!}\begin{vmatrix} f^1(x_1) + \cdots + f^1(x_3) & f^2(x_0) & f^3(x_0) & 1 \\ f^1(x_0) + \cdots + f^1(x_3) & f^2(x_1) & f^3(x_1) & 1 \\ \cdots & \cdots & \cdots & \cdots \\ f^1(x_0) + \cdots + f^1(x_2) & f^2(x_3) & f^3(x_3) & 1 \end{vmatrix}. \tag{45}$$

Expand the determinant on the right by minors of elements in the first column of its matrix. Then

$$\frac{(-1)^3}{3!}\Delta_{(1,2,3)}f(\mathbf{x}) = \frac{f^1(x_1) + \cdots + f^1(x_3)}{3}\frac{(-1)^2}{2!}\begin{vmatrix} f^2(x_1) & f^3(x_1) & 1 \\ f^2(x_2) & f^3(x_2) & 1 \\ f^2(x_3) & f^3(x_3) & 1 \end{vmatrix} + \cdots. \tag{46}$$

Each term on the right is a function of one of the oriented sides in the boundary of $\mathbf{x}$. Represent every term in the sum in (41) as in (46). If a side belongs to two simplexes in $\mathscr{P}_k$, then these sides have opposite orientations by Theorem 18.5 and the two terms which arise from it cancel in the sum in (41). These considerations show that, after all cancellations have been made,

$$\sum_{\mathbf{x}\in\mathscr{P}_k}\frac{(-1)^3}{3!}\Delta_{(1,2,3)}f(\mathbf{x})$$
$$= \sum_{(x_1,\cdots,x_3)\in\partial\mathscr{P}_k}\frac{f^1(x_1)+\cdots+f^1(x_3)}{3}\frac{(-1)^2}{2!}\begin{vmatrix} f^2(x_1) & f^3(x_1) & 1\\ f^2(x_2) & f^3(x_2) & 1\\ f^2(x_3) & f^3(x_3) & 1\end{vmatrix}. \tag{47}$$

The limit, as $k\to\infty$, of the sum on the right is equal to the limit of the following sum; the proof of this statement is the same as the proof that the limit of (25) equals the limit of (26).

$$\sum_{(x_1,\cdots,x_3)\in\partial\mathscr{P}_k} f^1(x_1)\frac{(-1)^2}{2!}\begin{vmatrix} f^2(x_1) & f^3(x_1) & 1\\ f^2(x_2) & f^3(x_2) & 1\\ f^2(x_3) & f^3(x_3) & 1\end{vmatrix}. \tag{48}$$

Use the following notation [see Section 11]:

$$\Delta_{(1,2)}(f^2,f^3)(x_1,x_2,x_3) = \begin{vmatrix} f^2(x_1) & f^3(x_1) & 1\\ \cdots & \cdots & \cdots\\ f^2(x_3) & f^3(x_3) & 1\end{vmatrix}. \tag{49}$$

$$\Delta[(x_1,x_2,x_3)^{(j_1,j_2)}] = \begin{vmatrix} x_1^{j_1} & x_1^{j_2} & 1\\ \cdots & \cdots & \cdots\\ x_3^{j_1} & x_3^{j_2} & 1\end{vmatrix},\qquad (j_1,j_2)\in(2/3). \tag{50}$$

Theorems 11.2 and 11.3 show that

$$\frac{(-1)^2}{2!}\Delta_{(1,2)}(f^2,f^3)(x_1,\cdots,x_3) \tag{51}$$
$$= \sum_{(j_1,j_2)} D_{(j_1,j_2)}(f^2,f^3)(x_1)\frac{(-1)^2}{2!}\Delta[(x_1,\cdots,x_3)^{(j_1,j_2)}] \tag{52}$$
$$+ r(f^2,f^3;x_1,x_2,x_3)\frac{(-1)^2}{2!}\left\{\sum_{(j_1,j_2)}\left[\Delta[(x_1,x_2,x_3)^{(j_1,j_2)}]\right]^2\right\}^{1/2}. \tag{53}$$

Here the summations extend over all (j_1,j_2) in (2/3). Replace the determinant in (48) by its value from these equations. The next step is to show that

$$\lim_{k\to\infty}\sum_{(x_1,x_2,x_3)\in\partial\mathscr{P}_k} f^1(x_1)r(f^2,f^3;x_1,x_2,x_3)\frac{(-1)^2}{2!}\times\left\{\sum_{(j_1,j_2)}\left[\Delta[(x_1,x_2,x_3)^{(j_1,j_2)}]\right]^2\right\}^{1/2} = 0. \tag{54}$$

Now $f^1:\mathbf{a}\to\mathbb{R}$ is continuous on the compact set $\mathbf{a}$ and is therefore bounded. The remainder terms $r(f^2,f^3;x_1,x_2,x_3)$ are uniformly small over the simplexes (x_1,x_2,x_3) in $\partial\mathscr{P}_k$ by Theorem 11.2 and they approach zero as $k\to\infty$. Next,

$$\frac{1}{2!}\left\{\sum_{(j_1,j_2)}\left[\Delta[(x_1, x_2, x_3)^{(j_1,j_2)}]\right]^2\right\}^{1/2} \tag{55}$$

is the area of the simplex (x_1, x_2, x_3) by Exercise 20.6, and the sum of (55) over all simplexes in $\partial\mathscr{P}_k$ is the area of the surface of **a** by Theorem 21.4. These facts show that the limit in (54) equals zero. The next step is to prove that

$$\lim_{k\to\infty}\sum_{(x_1,x_2,x_3)\in\partial\mathscr{P}_k}\sum_{(j_1,j_2)} f^1(x_1)D_{(j_1,j_2)}(f^2, f^3)(x_1)\frac{(-1)^2}{2!}\Delta[(x_1, x_2, x_3)^{(j_1,j_2)}]$$
$$= \int_{\partial\mathbf{a}}\sum_{(j_1,j_2)} f^1(x)D_{(j_1,j_2)}(f^2, f^3)(x)\,d(x^{j_1}, x^{j_2}), \qquad (j_1, j_2)\in(2/3). \tag{56}$$

The proof of this equation will complete the proof of the statement of the fundamental theorem of the integral calculus in (39) since the limit in (41) equals the integral on the left in (39) by (42) and (43), and the limit in (41) also equals the integral on the right in (39) by equations (47) through (56).

Thus the remaining detail is the proof of (56). Although (56) is almost obvious, there is one detail which requires attention. The subdivisions in the sides of **a** are the subdivisions which are induced by the subdivisions $\mathscr{P}_k$, $k = 1, 2, \cdots$, of **a**. In Examples 38.1 and 38.2 it is geometrically obvious from Figures 38.1 and 38.2 that the subdivisions of the boundaries of A and $[a_0, a_1, a_2]$ lead to integrals, but an analytic proof is needed in dimensions greater than two. Observe that $(x_1, x_2, x_3)^{(j_1,j_2)}$ is the simplex into which (x_1, x_2, x_3) projects in the (x^{j_1}, x^{j_2})-coordinate plane. It is necessary to show that, for all the simplexes (x_1, x_2, x_3) in a given side of **a**, the simplexes $(x_1, x_2, x_3)^{(j_1,j_2)}$ in the (x^{j_1}, x^{j_2})-coordinate plane have the same orientation; that is, it is necessary to show that all of the determinants $\Delta[(x_1, x_2, x_3)^{(j_1,j_2)}]$ have the same sign. The proof can be given by means of barycentric coordinates. Let (x_1, x_2, x_3) be a simplex in $\partial\mathscr{P}_k$ which is in the side (a_1, a_2, a_3) of $\mathbf{a}:(a_0, a_1, a_2, a_3)$. Then (x_1, x_2, x_3) is a side of a simplex $\mathbf{x}:(x_0, x_1, x_2, x_3)$ and $\Delta(\mathbf{a})$ and $\Delta(\mathbf{x})$ have the same sign since $\mathbf{x}$ is a simplex in a simplicial subdivision of **a**. The points $x_0, x_1, \cdots, x_3$ have barycentric coordinates with respect to $a_0, a_1, \cdots, a_3$ such that

$$\begin{bmatrix} x_0^1 & x_0^2 & x_0^3 & 1 \\ x_1^1 & x_1^2 & x_1^3 & 1 \\ x_2^1 & x_2^2 & x_2^3 & 1 \\ x_3^1 & x_3^2 & x_3^3 & 1 \end{bmatrix} = \begin{bmatrix} t_0^0 & t_0^1 & t_0^2 & t_0^3 \\ 0 & t_1^1 & t_1^2 & t_1^3 \\ 0 & t_2^1 & t_2^2 & t_2^3 \\ 0 & t_3^1 & t_3^2 & t_3^3 \end{bmatrix}\begin{bmatrix} a_0^1 & a_0^2 & a_0^3 & 1 \\ a_1^1 & a_1^2 & a_1^3 & 1 \\ a_2^1 & a_2^2 & a_2^3 & 1 \\ a_3^1 & a_3^2 & a_3^3 & 1 \end{bmatrix}, \tag{57}$$

$$t_i^j \geqq 0, \qquad i, j = 0, 1, \cdots, 3, \tag{58}$$

$$\sum_{j=0}^{3} t_i^j = 1, \qquad i = 0, 1, \cdots, 3. \tag{59}$$

Observe that $t_i^0 = 0$, $i = 1, 2, 3$, because x_1, x_2, x_3 are in the side (a_1, a_2, a_3) of $\mathbf{a}$. The Binet–Cauchy multiplication theorem and (57) show that

$$\Delta(\mathbf{x}) = \det[t_i^j]_0^3 \Delta(\mathbf{a}). \tag{60}$$

Then $\det[t_i^j]_0^3 > 0$ because $\Delta(\mathbf{x})$ and $\Delta(\mathbf{a})$ have the same sign. Expand $\det[t_i^j]_0^3$ by minors of elements in the first column of its matrix. Then $\det[t_i^j]_0^3 = t_0^0 \det[t_i^j]_1^3$, and

$$\det[t_i^j] > 0, \qquad i, j = 1, 2, 3, \tag{61}$$

since $t_0^0 > 0$. The definition of matrix multiplication and (57) show that

$$\begin{bmatrix} x_1^1 & x_1^2 & x_1^3 & 1 \\ x_2^1 & x_2^2 & x_2^3 & 1 \\ x_3^1 & x_3^2 & x_3^3 & 1 \end{bmatrix} = \begin{bmatrix} t_1^1 & t_1^2 & t_1^3 \\ t_2^1 & t_2^2 & t_2^3 \\ t_3^1 & t_3^2 & t_3^3 \end{bmatrix} \begin{bmatrix} a_1^1 & a_1^2 & a_1^3 & 1 \\ a_2^1 & a_2^2 & a_2^3 & 1 \\ a_3^1 & a_3^2 & a_3^3 & 1 \end{bmatrix}, \tag{62}$$

$$\begin{vmatrix} x_1^{j_1} & x_1^{j_2} & 1 \\ x_2^{j_1} & x_2^{j_2} & 1 \\ x_3^{j_1} & x_3^{j_2} & 1 \end{vmatrix} = \begin{vmatrix} t_1^1 & t_1^2 & t_1^3 \\ t_2^1 & t_2^2 & t_2^3 \\ t_3^1 & t_3^2 & t_3^3 \end{vmatrix} \begin{vmatrix} a_1^{j_1} & a_1^{j_2} & 1 \\ a_2^{j_1} & a_2^{j_2} & 1 \\ a_3^{j_1} & a_3^{j_2} & 1 \end{vmatrix}. \tag{63}$$

Thus (61) and (63) show that $\Delta[(x_1, x_2, x_3)^{(j_1, j_2)}]$ has the same sign as $\Delta[(a_1, a_2, a_3)^{(j_1, j_2)}]$ for every simplex (x_1, x_2, x_3) in $\partial\mathscr{P}_k$ which is in the side (a_1, a_2, a_3) of $\mathbf{a}$. Hence for each (j_1, j_2) in $(2/3)$, the simplexes $(x_1, x_2, x_3)^{(j_1, j_2)}$ form a simplicial subdivision of $(a_1, a_2, a_3)^{(j_1, j_2)}$ which can be used to define an integral on $(a_1, a_2, a_3)^{(j_1, j_2)}$, and similar statements hold for the other three sides of $(a_0, a_1, \cdots, a_3)$. These statements complete the proof of (56) and of the fundamental theorem of the integral calculus in (39).

38.4 Theorem (Fundamental Theorem of the Integral Calculus). *If the components $(f^1, \cdots, f^n)$ of $f: \mathbf{a} \to \mathbb{R}^n$ have continuous derivatives on the simplex $\mathbf{a}: [a_0, a_1, \cdots, a_n]$ in $\mathbb{R}^n$, then*

$$\int_{\mathbf{a}} D_{(1, \cdots, n)}(f^1, \cdots, f^n)(x)\, d(x^1, \cdots, x^n)$$

$$= \int_{\partial\mathbf{a}} \sum_{(j_1, \cdots, j_{n-1})} f^1(x) D_{(j_1, \cdots, j_{n-1})}(f^2, \cdots, f^n)(x)\, d(x^{j_1}, \cdots, x^{j_{n-1}}), \tag{64}$$

$$(j_1, \cdots, j_{n-1}) \in (n - 1/n).$$

Proof. Let $\mathscr{P}_k$, $k = 1, 2, \cdots$, be a sequence of simplicial subdivisions of $\mathbf{a}$ [see Definition 21.5]. Two lemmas as follows are needed to complete the proof.

38.5 Lemma. *If the components $(f^1, \cdots, f^n)$ of $f: \mathbf{a} \to \mathbb{R}^n$ have continuous derivatives on $\mathbf{a}: [a_0, a_1, \cdots, a_n]$ in $\mathbb{R}^n$, and if $\mathscr{P}_k$, $k = 1, 2, \cdots$, is a sequence* *of simplicial subdivisions of* **a**, *then*

$$\lim_{k\to\infty}\sum_{\mathbf{x}\in\mathscr{P}_k}\frac{(-1)^n}{n!}\Delta_{(1,\cdots,n)}(f^1,\cdots,f^n)(\mathbf{x}) = \int_{\mathbf{a}} D_{(1,\cdots,n)}(f^1,\cdots,f^n)(x)\,d(x^1,\cdots,x^n). \tag{65}$$

Proof of Lemma 38.5. The proof of this lemma is exactly the same as the proof of the special cases of it given in Examples 38.1, $\cdots$, 38.3; see, for example, the proof that (41) equals (42) given in equations (43) and (44). □

38.6 Lemma. *If the components* $(f^1,\cdots,f^n)$ *of* $f:\mathbf{a}\to\mathbb{R}^n$ *have continuous derivatives on* $\mathbf{a}:[a_0,a_1,\cdots,a_n]$ *in* $\mathbb{R}^n$, *and if* $\mathscr{P}_k$, $k=1,2,\cdots$, *is a sequence of simplicial subdivisions of* **a**, *then*

$$\lim_{k\to\infty}\sum_{\mathbf{x}\in\mathscr{P}_k}\frac{(-1)^n}{n!}\Delta_{(1,\cdots,n)}(f^1,\cdots,f^n)(\mathbf{x}) = \int_{\partial\mathbf{a}}\sum_{(j_1,\cdots,j_{n-1})} f^1(x)D_{(j_1,\cdots,j_{n-1})}(f^2,\cdots,f^n)(x)\,d(x^{j_1},\cdots,x^{j_{n-1}}), \tag{66}$$
$$(j_1,\cdots,j_{n-1})\in(n-1/n).$$

Proof of Lemma 38.6. The proof of this lemma is exactly the same as the proof of the special cases of it given in Examples 38.1, $\cdots$, 38.3; see especially the proof that (41) equals the integral on the right in (39) which begins with equation (45) in Example 38.3. No new ideas or methods are required to prove (66). □

To complete the proof of Theorem 38.4, observe that equations (65) and (66) in Lemmas 38.5 and 38.6 show that the two integrals in (64) are equal. □

38.7 Corollary (Fundamental Theorem of the Integral Calculus). *If the components* $(f^1,\cdots,f^n)$ *of* $f:\mathbf{a}\to\mathbb{R}^n$ *have continuous derivatives on the simplex* $\mathbf{a}:[a_0,a_1,\cdots,a_n]$, *then*

$$\int_{\mathbf{a}} D_{(1,\cdots,n)}(f^1,\cdots,f^n)(x)\,d(x^1,\cdots,x^n) = (-1)^{i-1}\int_{\partial\mathbf{a}}\sum_{(j_1,\cdots,j_{n-1})} f^i(x)D_{(j_1,\cdots,j_{n-1})}(f^1,\cdots,\widehat{f^i},\cdots,f^n)\,d(x^{j_1},\cdots,x^{j_{n-1}}),$$
$$(j_1,\cdots,j_{n-1})\in(n-1/n),\qquad i=1,2,\cdots,n. \tag{67}$$

Proof. By (32) in Theorem 3.12,

$$D_{(1,\cdots,n)}(f^1,\cdots,f^n)(x)=\det[D_jf^i(x)],\qquad i,j=1,\cdots,n.$$

Then

$$D_{(1,\cdots,n)}(f^i, f^1, \cdots, \widehat{f^i}, \cdots, f^n)(x) = (-1)^{i-1} D_{(1,\cdots,n)}(f^1, \cdots, f^n)(x) \quad (68)$$

because the determinant on the left is obtained from the one on the right by moving the i-th row across $(i-1)$ rows to the first row. Then by (68) and (64),

$$\begin{aligned}\int_{\mathbf{a}} & D_{(1,\cdots,n)}(f^1, \cdots, f^n)(x)\, d(x^1, \cdots, x^n) \\ &= (-1)^{i-1} \int_{\mathbf{a}} D_{(1,\cdots,n)}(f^i, f^1, \cdots, \widehat{f^i}, \cdots, f^n)(x)\, d(x^1, \cdots, x^n) \\ &= (-1)^{i-1} \int_{\partial\mathbf{a}} \sum_{(j_1,\cdots,j_{n-1})} f^i(x) \\ &\quad \times D_{(j_1,\cdots,j_{n-1})}(f^1, \cdots, \widehat{f^i}, \cdots, f^n)(x)\, d(x^{j_1}, \cdots, x^{j_{n-1}}).\end{aligned} \quad (69)$$

The proof of (67) and of Corollary 38.7 is complete. □

Before the next theorem is stated, it will be helpful to introduce some new notation as follows. Let $f: \mathbf{a} \to \mathbb{R}^n$ be a function whose components $(f^1, \cdots, f^n)$ have continuous derivatives on $\mathbf{a}: [a_0, a_1, \cdots, a_n]$ in $\mathbb{R}^n$. Let $D_j f(x)$ denote the vector $(D_j f^1(x), \cdots, D_j f^n(x))$ for $j = 1, \cdots, n$, and define $\det[f(x), D_{j_1} f(x), \cdots, D_{j_{n-1}} f(x)]$ for $(j_1, \cdots, j_{n-1})$ in $(n-1/n)$ by the following equation:

$$\det[f(x), D_{j_1} f(x), \cdots, D_{j_{n-1}} f(x)] = \begin{vmatrix} f^1(x) & f^2(x) & \cdots & f^n(x) \\ D_{j_1} f^1(x) & D_{j_1} f^2(x) & \cdots & D_{j_1} f^n(x) \\ \cdots & \cdots & \cdots & \cdots \\ D_{j_{n-1}} f^1(x) & D_{j_{n-1}} f^2(x) & \cdots & D_{j_{n-1}} f^n(x) \end{vmatrix}. \quad (70)$$

38.8 Theorem (Fundamental Theorem of the Integral Calculus). *If the components $(f^1, \cdots, f^n)$ of $f: \mathbf{a} \to \mathbb{R}^n$ have continuous derivatives on the simplex $\mathbf{a}: [a_0, a_1, \cdots, a_n]$ in $\mathbb{R}^n$, then*

$$\begin{aligned}\int_{\mathbf{a}} & D_{(1,\cdots,n)}(f^1, \cdots, f^n)(x)\, d(x^1, \cdots, x^n) \\ &= \frac{1}{n} \int_{\partial\mathbf{a}} \sum_{(j_1,\cdots,j_{n-1})} \det[f(x), D_{j_1} f(x), \cdots, D_{j_{n-1}} f(x)]\, d(x^{j_1}, \cdots, x^{j_{n-1}}), \quad (71) \\ &\qquad (j_1, \cdots, j_{n-1}) \in (n-1/n).\end{aligned}$$

PROOF. The proof of this theorem can be explained most clearly and easily by proving it in the special case in which $n = 3$. In this special case, by

Corollary 38.7

$$\int_{\mathbf{a}} D_{(1,2,3)}(f^1, \cdots, f^3)(x)\,d(x^1, \cdots, x^3)$$

$$\begin{aligned} &= \int_{\partial \mathbf{a}} \sum_{(j_1, j_2)} f^1(x) D_{(j_1, j_2)}(f^2, f^3)(x)\,d(x^{j_1}, x^{j_2}) \\ &= -\int_{\partial \mathbf{a}} \sum_{(j_1, j_2)} f^2(x) D_{(j_1, j_2)}(f^1, f^3)(x)\,d(x^{j_1}, x^{j_2}) \\ &= \int_{\partial \mathbf{a}} \sum_{(j_1, j_2)} f^3(x) D_{(j_1, j_2)}(f^1, f^2)(x)\,d(x^{j_1}, x^{j_2}). \end{aligned} \tag{72}$$

Add the three equations in (72); then $\int_{\mathbf{a}} D_{(1,2,3)}(f^1, \cdots, f^3)\,d(x^1, \cdots, x^3)$ is one-third of the sum of the three integrals on the right in (72). Write the sum of the three integrals as a single integral on $\partial \mathbf{a}$. The coefficient of $d(x^1, x^2)$ is

$$f^1(x) D_{(1,2)}(f^2, f^3)(x) - f^2(x) D_{(1,2)}(f^1, f^3)(x) + f^3(x) D_{(1,2)}(f^1, f^2)(x) \tag{73}$$

In this expression, each of the derivatives is a determinant by Corollary 3.20, and it can be written in the following form:

$$\begin{aligned} f^1(x)\begin{vmatrix} D_1 f^2(x) & D_1 f^3(x) \\ D_2 f^2(x) & D_2 f^3(x) \end{vmatrix} &- f^2(x)\begin{vmatrix} D_1 f^1(x) & D_1 f^3(x) \\ D_2 f^1(x) & D_2 f^3(x) \end{vmatrix} \\ &+ f^3(x)\begin{vmatrix} D_1 f^1(x) & D_1 f^2(x) \\ D_2 f^1(x) & D_2 f^2(x) \end{vmatrix}. \end{aligned} \tag{74}$$

This expression is easily seen to be the expansion of the following determinant:

$$\begin{vmatrix} f^1(x) & f^2(x) & f^3(x) \\ D_1 f^1(x) & D_1 f^2(x) & D_1 f^3(x) \\ D_2 f^1(x) & D_2 f^2(x) & D_2 f^3(x) \end{vmatrix}. \tag{75}$$

By (70), this determinant is denoted by the symbol $\det[f(x), D_1 f(x), D_2 f(x)]$. Similar considerations show that, in the sum of the three integrals on the right in (72), the coefficients of $d(x^1, x^3)$ and $d(x^2, x^3)$ are $\det[f(x), D_1 f(x), D_3 f(x)]$ and $\det[f(x), D_2 f(x), D_3 f(x)]$, respectively. Thus

$$\int_{\mathbf{a}} D_{(1,2,3)}(f^1, \cdots, f^3)(x)\,d(x^1, \cdots, x^3)$$

$$= \frac{1}{3} \int_{\partial \mathbf{a}} \sum_{(j_1, j_2)} \det[f(x), D_{j_1} f(x), D_{j_2} f(x)]\,d(x^{j_1}, x^{j_2}), \qquad (j_1, j_2) \in (2/3). \tag{76}$$

Here (76) is formula (71) in the special case $n = 3$. The general case of (71) can be proved in exactly the same way to complete the proof of Theorem 38.8. □

As shown above, Theorem 38.4 implies Theorem 38.8, but it is not clear whether or not Theorem 38.8 implies Theorem 38.4. The proof of Theorem 38.4 uses Theorem 20.3 and Lemmas 38.5 and 38.6; thus the proof of Theorem 38.8 given above also depends on Theorem 20.3. There is a direct proof of Theorem 38.8 which does not use Theorems 20.3 and 38.4; this proof is derived from Lemma 38.5 and the following lemma.

38.9 Lemma. *If the components* $(f^1, \cdots, f^n)$ *of* $f: \mathbf{a} \to \mathbb{R}^n$ *have continuous derivatives on* $\mathbf{a}: [a_0, a_1, \cdots, a_n]$ *in* $\mathbb{R}^n$, *and if* $\mathscr{P}_k$, $k = 1, 2, \cdots$, *is a sequence of simplicial subdivisions of* $\mathbf{a}$, *then*

$$\lim_{k\to\infty} \sum_{\mathbf{x}\in\mathscr{P}_k} \frac{(-1)^n}{n!} \Delta_{(1,\cdots,n)}(f^1, \cdots, f^n)(\mathbf{x})$$

$$= \frac{1}{n} \int_{\partial\mathbf{a}} \sum_{(j_1,\cdots,j_{n-1})} \det[f(x), D_{j_1}f(x), \cdots, D_{j_{n-1}}f(x)]\, d(x^{j_1}, \cdots, x^{j_{n-1}}), \quad (77)$$

$$(j_1, \cdots, j_{n-1}) \in (n-1/n).$$

PROOF. The proof will be outlined in the special case in which $n = 2$. Let $\mathbf{x}: (x_0, x_1, x_2)$ be a simplex in $\mathscr{P}_k$. In order to obtain the cancellations in the Euclidean complex, it is necessary to represent each term $\Delta_{(1,2)}(f^1, f^2)(\mathbf{x})$ as a sum of terms on the sides of $\mathbf{x}$; in the proof of Lemma 38.6 [see equation (45)], this was accomplished by applying Theorem 20.3; in the present case, expand the determinant by minors of elements in the column of 1's as follows:

$$\frac{(-1)^2}{2!}\begin{vmatrix} f^1(x_0) & f^2(x_0) & 1 \\ f^1(x_1) & f^2(x_1) & 1 \\ f^1(x_2) & f^2(x_2) & 1 \end{vmatrix}$$

$$= \frac{(-1)^2}{2!}\begin{vmatrix} f^1(x_1) & f^2(x_1) \\ f^1(x_2) & f^2(x_2) \end{vmatrix} - \frac{(-1)^2}{2!}\begin{vmatrix} f^1(x_0) & f^2(x_0) \\ f^1(x_2) & f^2(x_2) \end{vmatrix} \quad (78)$$

$$+ \frac{(-1)^2}{2!}\begin{vmatrix} f^1(x_0) & f^2(x_0) \\ f^1(x_1) & f^2(x_1) \end{vmatrix}.$$

Represent each term $\Delta_{(1,2)}(f^1, f^2)(\mathbf{x})$ in this manner. The usual cancellations in the Euclidean complex $\mathscr{P}_k$ show that

$$\sum_{\mathbf{x}\in\mathscr{P}_k} \frac{(-1)^2}{2!}\Delta_{(1,2)}(f^1, f^2)(\mathbf{x}) = \sum_{(x_1,x_2)\in\partial\mathscr{P}_k} \frac{(-1)^2}{2!}\begin{vmatrix} f^1(x_1) & f^2(x_1) \\ f^1(x_2) & f^2(x_2) \end{vmatrix}. \quad (79)$$

In the determinant on the right, subtract the first row from each of the other rows. Then expand the determinant by minors of elements in the first row of its matrix. Write the determinants in this expansion as determinants of one higher order with a column of 1's. Thus

$$\begin{aligned}\begin{vmatrix} f^1(x_1) & f^2(x_1) \\ f^1(x_2) & f^2(x_2) \end{vmatrix} &= \begin{vmatrix} f^1(x_1) & f^2(x_1) \\ f^1(x_2) - f^1(x_1) & f^2(x_2) - f^2(x_1) \end{vmatrix} \\ &= -f^1(x_1)\begin{vmatrix} f^2(x_1) & 1 \\ f^2(x_2) & 1 \end{vmatrix} + f^2(x_1)\begin{vmatrix} f^1(x_1) & 1 \\ f^1(x_2) & 1 \end{vmatrix}.\end{aligned} \tag{80}$$

Then

$$\lim_{k\to\infty} \sum_{\mathbf{x}\in\mathscr{P}_k} \frac{(-1)^2}{2!}\Delta_{(1,2)}(f^1, f^2)(\mathbf{x})$$
$$= \frac{1}{2}\lim_{k\to\infty} \sum_{(x_1,x_2)\in\partial\mathscr{P}_k} \left\{ f^1(x_1)(-1)\begin{vmatrix} f^2(x_1) & 1 \\ f^2(x_2) & 1 \end{vmatrix} - f^2(x_1)(-1)\begin{vmatrix} f^1(x_1) & 1 \\ f^1(x_2) & 1 \end{vmatrix} \right\}. \tag{81}$$

To complete the proof, evaluate the limit on the right by the methods used to evaluate the limit of the sum in equation (26) and also the limit of the sum in (48). No new ideas or methods are needed to complete the proof. The methods used in this special case can be used to prove the lemma in the general case. □

38.10 Example. The purpose of this example is to establish a formula for the area of a triangle in the plane. Let $\mathbf{a}: [a_0, a_1, a_2]$ be a simplex which is positively oriented in $\mathbb{R}^2$. Then formula (71) in Theorem 38.8 shows that

$$\int_{\mathbf{a}} D_{(1,2)}(f^1, f^2)(x)\, d(x^1, x^2)$$
$$= \frac{1}{2}\int_{\partial\mathbf{a}} \begin{vmatrix} f^1(x) & f^2(x) \\ D_1 f^1(x) & D_1 f^2(x) \end{vmatrix} dx^1 + \begin{vmatrix} f^1(x) & f^2(x) \\ D_2 f^1(x) & D_2 f^2(x) \end{vmatrix} dx^2. \tag{82}$$

If $f^1(x) = x^1$ and $f^2(x) = x^2$, then $D_{(1,2)}(f^1, f^2)(x) = 1$, and the integral on the left in (82) is $\int_{\mathbf{a}} 1 d(x^1, x^2)$, which is the area of $\mathbf{a}$. Thus the integral on the right in (82) is also the area of $\mathbf{a}$. This integral can be simplified to

$$\frac{1}{2}\int_{\partial\mathbf{a}} x^1\, dx^2 - x^2\, dx^1, \tag{83}$$

which is the desired formula for the area of $\mathbf{a}$.

Exercises

38.1. Let A be the positively oriented square $[0, 6] \times [0, 6]$ in $\mathbb{R}^2$, and let $f^1(x) = x^1 - x^2, f^2(x) = (x^1)^2 + (x^2)^2$.
(a) Show that $D_{(1,2)}(f^1, f^2)(x) = 2(x^1 + x^2)$.
(b) Use iterated integrals to show that $\int_A D_{(1,2)}(f^1, f^2)(x)\, d(x^1, x^2) = 432$.
(c) Find the value of the sum of the four integrals on the right in equation (1) and thus verify the fundamental theorem of the integral calculus in this numerical example.

38.2. Let $\mathbf{a}:[a_0, a_1, a_2]$ be the simplex with vertices $a_0:(0, 0)$, $a_1:(12, 0)$, $a_2:(6, 6)$, and let $f^1(x) = x^1 - x^2, f^2(x) = (x^1)^2 + (x^2)^2$.

(a) Use iterated integrals to show that

$$\int_{\mathbf{a}} D_{(1,2)}(f^1, f^2)\, d(x^1, x^2) = 576.$$

(b) Find the value of the sum of the three integrals on the right in equation (21) and thus verify the statement of the fundamental theorem of the integral calculus in (21) in this numerical example.

38.3. Let $\mathbf{a}:[a_0, a_1, a_2]$ be a simplex whose vertices are $a_0:(-1, 0), a_1:(1, 0), a_2:(0, 1)$, and let $f^1(x) = x^1 - x^2$ and $f^2(x) = x^1 + x^2$.

(a) Show that $\int_{\mathbf{a}} D_{(1,2)}(f^1, f^2)\, d(x^1, x^2) = 2$ by evaluating this integral.

(b) Use the fundamental theorem of the integral calculus in equation (21) to find the value of the integral in (a).

38.4. Let $\mathbf{a}:[a_0, a_1, a_2]$ be a Euclidean simplex in $\mathbb{R}^2$, and let $f:\mathbf{a} \to \mathbb{R}^2$ be a function whose components have continuous derivatives on $\mathbf{a}$.

(a) If $f^1(x) = P(x)$ and $f^2(x) = x^1$, prove the following formulas:

$$\int_{\mathbf{a}} D_{(1,2)}(f^1, f^2)(x)\, d(x^1, x^2) = \int_{\mathbf{a}} - D_2 P(x)\, d(x^1, x^2).$$

$$\int_{\partial\mathbf{a}} f^1(x) D_1 f^2(x)\, dx^1 + f^1(x) D_2 f^2(x)\, dx^2 = \int_{\partial\mathbf{a}} P(x)\, dx^1.$$

$$\int_{\mathbf{a}} - D_2 P(x)\, d(x^1, x^2) = \int_{\partial\mathbf{a}} P(x)\, dx^1.$$

(b) If $f^1(x) = Q(x), f^2(x) = x^2$, then in the same way prove that

$$\int_{\mathbf{a}} D_1 Q(x)\, d(x^1, x^2) = \int_{\partial\mathbf{a}} Q(x)\, dx^2.$$

(c) Use the formulas in (a) and (b) to prove the following formula, which is known as Green's theorem.

$$\int_{\mathbf{a}} [D_1 Q(x) - D_2 P(x)]\, d(x^1, x^2) = \int_{\partial\mathbf{a}} P(x)\, dx^1 + Q(x)\, dx^2.$$

38.5. A simplex $\mathbf{a}:[a_0, a_1, a_2]$ has vertices $a_0:(2, 0)$, $a_1:(6, 0)$, and $a_2:(4, 12)$ in $\mathbb{R}^2$. Use the formula $(1/2)\int_{\partial\mathbf{a}} x^1\, dx^2 - x^2\, dx^1$ in equation (83) to show that the area of $\mathbf{a}$ is 24. [Hint. Evaluate the integral around the boundary of $\mathbf{a}$ by the method explained in equations (34)–(38).]

38.6. A simplex $\mathbf{a}:[a_0, a_1, a_2]$ with vertices $a_i:(a_i^1, a_i^2), i = 0, 1, 2$, is positively oriented in $\mathbb{R}^2$. Use the formula $(1/2)\int_{\partial\mathbf{a}} x^1\, dx^2 - x^2\, dx^1$ in (83) and the methods suggested in Exercise 38.5 to prove that the area of $\mathbf{a}$ is

$$\frac{1}{2!}\begin{vmatrix} a_0^1 & a_0^2 & 1 \\ a_1^1 & a_1^2 & 1 \\ a_2^1 & a_2^2 & 1 \end{vmatrix}.$$

38.7. Let $\mathbf{a}:[a_0, a_1, \cdots, a_3]$ be a simplex which is positively oriented in $\mathbb{R}^3$. By methods similar to those used in Example 38.10, prove that the volume of $\mathbf{a}$ is given by the following formula.

$$\frac{1}{3}\int_{\partial \mathbf{a}} x^3\,d(x^1, x^2) - x^2\,d(x^1, x^3) + x^1\,d(x^2, x^3).$$

38.8. Let $\mathbf{a}: [a_0, a_1, a_2]$ be a simplex in $\mathbb{R}^2$; let $f^1: \mathbf{a} \to \mathbb{R}$ be a function which has a continuous derivative on $\mathbf{a}$; and let $f^2(x) = c$, a constant, on $\mathbf{a}$.

(a) Prove that $D_{(1,2)}(f^1, f^2)(x) = 0$ on $\mathbf{a}$ and hence that $\int_{\mathbf{a}} D_{(1,2)}(f^1, f^2)(x)\,d(x^1, x^2) = 0$.

(b) Use the statement of the fundamental theorem of the integral calculus in equation (82) to prove that

$$\int_{\partial \mathbf{a}} D_1 f^1(x)\,dx^1 + D_2 f^1(x)\,dx^2 = 0.$$

(c) Prove the result in (b) by using the formula in (67) in Corollary 38.7.

38.9. Let E be an open set in $\mathbb{R}^2$ which contains the simplexes $[a_0, a_1]$, $[a_1, a_2]$, $[a_2, a_0]$, and let $f^1: E \to \mathbb{R}$ be a function which has continuous derivatives.

(a) Use the methods outlined in equations (34)–(38) to prove the following statements [they are forms of the fundamental theorem of the integral calculus]:

$$\int_{[a_0, a_1]} D_1 f^1(x)\,dx^1 + D_2 f^1(x)\,dx^2 = f^1(a_1) - f^1(a_0).$$

$$\int_{[a_1, a_2]} D_1 f^1(x)\,dx^1 + D_2 f^1(x)\,dx^2 = f^1(a_2) - f^1(a_1).$$

$$\int_{[a_2, a_0]} D_1 f^1(x)\,dx^1 + D_2 f^1(x)\,dx^2 = f^1(a_0) - f^1(a_2).$$

(b) If $\mathbf{a}$ is the simplex (a_0, a_1, a_2), use the results in (a) to show that

$$\int_{\partial \mathbf{a}} D_1 f^1(x)\,dx^1 + D_2 f^1(x)\,dx^2 = 0.$$

This proof of the equation [compare Exercise 38.8(b)] does not require that f^1 be defined on $[a_0, a_1, a_2]$, but only on the set E which contains the boundary $\partial \mathbf{a}$ of this simplex.

38.10. Let the components (f^1, f^2, f^3) of $f: \mathbf{a} \to \mathbb{R}^3$ have continuous derivatives on the simplex $\mathbf{a}: [a_0, a_1, \cdots, a_3]$ in $\mathbb{R}^3$, and let $f^3(x) = c$, a constant, on $\mathbf{a}$.

(a) Use Corollary 38.7 to prove that

$$\int_{\partial \mathbf{a}} D_{(1,2)}(f^1, f^2)(x)\,d(x^1, x^2) + D_{(1,3)}(f^1, f^2)(x)\,d(x^1, x^3)$$
$$+ D_{(2,3)}(f^1, f^2)(x)\,d(x^2, x^3) = 0.$$

(b) Give a second proof of the result in (a) by using Theorem 38.8.

38.11. Use the fundamental theorem of the integral calculus to prove that

$$\int_a^c D_1 f(x)\,dx = \int_a^b D_1 f(x)\,dx + \int_b^c D_1 f(x)\,dx.$$

[Hint. $\int_a^b D_1 f(x)\,dx = f(b) - f(a)$, etc.]

38.12. Let E be an open convex set which contains the simplex $[a_1, a_2, a_3]$ in $\mathbb{R}^2$, and let the components (f^1, f^2) of $f: E \to \mathbb{R}^2$ have continuous derivatives. Let a_0 be a point on the interior of $[a_1, a_2, a_3]$.

(a) Use the fundamental theorem of the integral calculus in equation (21) to prove the following formula [compare Exercise 38.11].

$$\int_{[a_1,a_2,a_3]} D_{(1,2)}(f^1,f^2)(x)\,d(x^1,x^2)$$
$$= \left\{\int_{[a_0,a_2,a_3]} + \int_{[a_1,a_0,a_3]} + \int_{[a_1,a_2,a_0]}\right\} D_{(1,2)}(f^1,f^2)(x)\,d(x^1,x^2).$$

(b) Make a sketch and show that $[a_0, a_2, a_3]$, $[a_1, a_0, a_3]$, and $[a_1, a_2, a_0]$ have the same orientation in $\mathbb{R}^2$ as $[a_1, a_2, a_3]$ if a_0 is on the interior of $[a_1, a_2, a_3]$.

(c) Show that the formula in (a) is valid even if a_0 is outside $[a_1, a_2, a_3]$ or on a side of this simplex.

(d) Compare the formula in (a) with Exercise 14.19 and Theorem 20.1. In particular, show that the formula reduces to a statement of Theorem 20.1 if $D_{(1,2)}(f^1,f^2)(x)$ is a constant.

(e) If the four vertices a_0, a_1, a_2, a_3 of $\mathbf{a}$ in Exercise 38.10 are in the (x^1, x^2)-coordinate plane, show that the formula in Exercise 38.10(a) reduces to the formula in Exercise 38.12(a).

39. The Fundamental Theorem of the Integral Calculus for Surfaces

Section 38 has shown that the integral of a derivative $D_{(1,\cdots,n)}(f^1, \cdots, f^n)$ over an n-simplex $\mathbf{a}$ in $\mathbb{R}^n$ is equal to certain integrals over the boundary of $\mathbf{a}$. Section 36 has defined the integral of a function on a surface G_m in $\mathbb{R}^n$; the surface integrals of the derivatives of functions are of special interest. Exercise 36.4 suggests that the surface integral of a derivative equals an integral of a derivative on an m-simplex in $\mathbb{R}^m$; thus there is strong reason to conjecture that there is a form of the fundamental theorem of the integral calculus for surface integrals of derivatives. The conjecture is correct. This section states and proves the fundamental theorem for surface integrals.

Let $\mathbf{a}$ be the m-dimensional Euclidean simplex $[a_0, a_1, \cdots, a_m]$ in $\mathbb{R}^m$, and let $g: \mathbf{a} \to \mathbb{R}^n$, $m \leqq n$, be a function whose components $(g^1, \cdots, g^n)$ have continuous derivatives. Then the equations

$$y^j = g^j(x), \qquad j = 1, \cdots, n, \qquad x \in \mathbf{a}, \qquad x = (x^1, \cdots, x^m), \tag{1}$$

define an m-dimensional surface G_m in $\mathbb{R}^n$ whose trace is a compact set $T(G_m)$. Let E be an open set in $\mathbb{R}^n$ which contains $T(G_m)$, and assume that the components $(f^1, \cdots, f^m)$ of $f: E \to \mathbb{R}^m$ have continuous derivatives. Then the derivatives $D_{(j_1,\cdots,j_m)}(f^1, \cdots, f^m)$, $(j_1, \cdots, j_m) \in (m/n)$, exist and are continuous functions on E, and Theorem 36.4 shows that the following integral exists:

$$\int_{G_m} \sum_{(j_1,\cdots,j_m)} D_{(j_1,\cdots,j_m)}(f^1, \cdots, f^m)(y)\,d(y^{j_1}, \cdots, y^{j_m}), \qquad (j_1, \cdots, j_m) \in (m/n). \tag{2}$$

The boundary ∂G_m of G_m is a surface of dimension $(m-1)$ in $\mathbb{R}^n$. The boundary $\partial\mathbf{a}$ of $\mathbf{a}$ is a chain of simplexes, $(m+1)$ in number, each of which has dimension $(m-1)$. The function $g:\mathbf{a}\to\mathbb{R}^n$ maps the set $\partial\mathbf{a}$ into a surface ∂G_m in $\mathbb{R}^n$; this surface, which has dimension $(m-1)$, is called the boundary of G_m. Integrals on ∂G_m are defined in Section 36, but there is one important difference between the surfaces G_m and ∂G_m as follows: the surface G_m is defined by a function g on an m-dimensional simplex $\mathbf{a}$ in $\mathbb{R}^m$, but ∂G_m is defined by a function g on an $(m-1)$-dimensional chain $\partial\mathbf{a}$ in $\mathbb{R}^m$. It is necessary to take account of this difference in proving the next theorem.

39.1 Theorem (Fundamental Theorem of the Integral Calculus). *If G_m is a surface in $\mathbb{R}^n$, $m\leqq n$, which is defined by a function $g:\mathbf{a}\to\mathbb{R}^n$ whose components have continuous derivatives, and if the components $(f^1,\cdots,f^m)$ of $f:E\to\mathbb{R}^m$ have continuous derivatives, then*

$$\int_{G_m}\sum_{(m/n)}D_{(j_1,\cdots,j_m)}(f^1,\cdots,f^m)(y)\,d(y^{j_1},\cdots,y^{j_m}) \tag{3}$$

$$=\int_{\mathbf{a}}D_{(1,\cdots,m)}(f^1\circ g,\cdots,f^m\circ g)(x)\,d(x^1,\cdots,x^m) \tag{4}$$

$$=\int_{\partial\mathbf{a}}\sum_{(m-1/m)}f^1[g(x)]D_{(i_1,\cdots,i_{m-1})}(f^2\circ g,\cdots,f^m\circ g)(x)\,d(x^{i_1},\cdots,x^{i_{m-1}}) \tag{5}$$

$$=\int_{\partial G_m}\sum_{(m-1/n)}f^1(y)D_{(j_1,\cdots,j_{m-1})}(f^2,\cdots,f^m)(y)\,d(y^{j_1},\cdots,y^{j_{m-1}}). \tag{6}$$

PROOF. As stated in (2), the integral (3) exists by Theorem 36.4. By the same theorem, the integral in (3) equals

$$\int_{\mathbf{a}}\sum_{(m/n)}D_{(j_1,\cdots,j_m)}(f^1,\cdots,f^m)[g(x)]D_{(1,\cdots,m)}(g^{j_1},\cdots,g^{j_m})(x)\,d(x^1,\cdots,x^m). \tag{7}$$

The chain rule in Theorem 4.5 shows that the integrand of this integral equals

$$D_{(1,\cdots,m)}(f^1\circ g,\cdots,f^m\circ g)(x). \tag{8}$$

Therefore the integral in (3) equals the integral in (4), and the first conclusion in the theorem has been established. Also, the integral in (4) equals the integral in (5) by the fundamental theorem of the integral calculus in Theorem 38.4, and the proof of the second conclusion is complete. Observe that $(i_1,\cdots,i_{m-1})$ has been used to denote derivatives with respect to variables in the set $(x^1,\cdots,x^m)$, and that $(j_1,\cdots,j_m)$ and $(j_1,\cdots,j_{m-1})$ denote derivatives with respect to variables in the set $(y^1,\cdots,y^n)$.

The proof of Theorem 39.1 will be completed by proving that the integral in (6) equals the integral in (5). For the sake of clarity and ease of understanding, the proof will be given in the special case in which $m=3$ and $n=3$. Thus the proof of the theorem will be completed in this special case by proving that

$$\int_{\partial G_3} \sum_{(j_1,j_2)} f^1(y) D_{(j_1,j_2)}(f^2, f^3)(y)\, d(y^{j_1}, y^{j_2}), \qquad (j_1, j_2) \in (2/3), \tag{9}$$

$$= \int_{\partial \mathbf{a}} \sum_{(i_1,i_2)} f^1[g(x)] D_{(i_1,i_2)}(f^2 \circ g, f^3 \circ g)(x)\, d(x^{i_1}, x^{i_2}), \quad (i_1, i_2) \in (2/3). \tag{10}$$

The integral in (9) will be evaluated by using the definition of this integral in Section 36 to show that (9) equals (10).

Let $\mathscr{P}_k$, $k = 1, 2, \cdots$, be a sequence of simplicial subdivisions of $\mathbf{a}$. These subdivisions induce a sequence $\partial\mathscr{P}_k$, $k = 1, 2, \cdots$, of simplicial subdivisions in the boundary $\partial\mathbf{a}$ of $\mathbf{a}$. Let (x_1, x_2, x_3) denote a simplex in $\partial\mathscr{P}_k$; then there is a simplex $(x_0, x_1, \cdots, x_3)$ in $\mathscr{P}_k$, and x_0 is in the interior of $\mathbf{a}$.

Let the simplexes (x_1, x_2, x_3) in $\partial\mathscr{P}_k$ be denoted by $\mathbf{x}$. Then g maps the vertices of $\mathbf{x}$ into three points which can be denoted by $g(\mathbf{x})$ and the matrix

$$\begin{bmatrix} g^1(x_1) & \cdots & g^3(x_1) \\ \cdots & \cdots & \cdots \\ g^1(x_3) & \cdots & g^3(x_3) \end{bmatrix}. \tag{11}$$

Then the integral in (9) is

$$\begin{aligned} \lim_{k\to\infty} \sum_{\mathbf{x}\in\partial\mathscr{P}_k} \Big\{ & f^1[g(x_1)] D_{(1,2)}(f^2, f^3)[g(x_1)] \frac{(-1)^2}{2!} \Delta_{(1,2)}(g^1, g^2)(\mathbf{x}) \\ & + f^1[g(x_1)] D_{(1,3)}(f^2, f^3)[g(x_1)] \frac{(-1)^2}{2!} \Delta_{(1,2)}(g^1, g^3)(\mathbf{x}) \\ & + f^1[g(x_1)] D_{(2,3)}(f^2, f^3)[g(x_1)] \frac{(-1)^2}{2!} \Delta_{(1,2)}(g^2, g^3)(\mathbf{x}) \Big\}. \end{aligned} \tag{12}$$

To evaluate this limit, replace the terms $\Delta_{(1,2)}(g^{j_1}, g^{j_2})(\mathbf{x})$, $(j_1, j_2) \in (2/3)$, by their values from Theorem 11.2 as follows.

$$\begin{aligned} \Delta_{(1,2)}(g^{j_1}, g^{j_2})(\mathbf{x}) = & \sum_{(i_1,i_2)} D_{(i_1,i_2)}(g^{j_1}, g^{j_2})(x_1) \Delta(\mathbf{x}^{(i_1,i_2)}) \\ & + r(g^{j_1}, g^{j_2}; \mathbf{x}) |\Delta(\mathbf{x})|, \qquad (i_1, i_2) \in (2/3). \end{aligned} \tag{13}$$

Substitute from (13) in (12) and then rearrange the terms to find the coefficients of $[(-1)^2/2!]\Delta(\mathbf{x}^{(i_1,i_2)})$, $(i_1, i_2) \in (2/3)$. The coefficient of $[(-1)^2/2!]$ $\Delta(\mathbf{x}^{(1,2)})$ is

$$\begin{aligned} f^1[g(x_1)]\{ & D_{(1,2)}(f^2, f^3)[g(x_1)] D_{(1,2)}(g^1, g^2)(x_1) \\ & + D_{(1,3)}(f^2, f^3)[g(x_1)] D_{(1,2)}(g^1, g^3)(x_1) \\ & + D_{(2,3)}(f^2, f^3)[g(x_1)] D_{(1,2)}(g^2, g^3)(x_1)\}. \end{aligned} \tag{14}$$

By the chain rule [see Example 4.3 and Theorem 4.5], the expression in (14) is $f^1[g(x_1)] D_{(1,2)}(f^2 \circ g, f^3 \circ g)(x_1)$. In the same way the coefficients of $[(-1)^2/2!]\Delta(\mathbf{x}^{(1,3)})$ and $[(-1)^2/2!]\Delta(\mathbf{x}^{(2,3)})$ can be shown to be, respectively,

$f^1[g(x_1)]D_{(1,3)}(f^2 \circ g, f^3 \circ g)(x_1)$ and $f^1[g(x_1)]D_{(2,3)}(f^2 \circ g, f^3 \circ g)(x_1)$. Thus, after the substitution from (13) in (12), the part of (12) which does not involve the remainder terms reduces to

$$\lim_{k\to\infty} \sum_{\mathbf{x}\in\partial\mathscr{P}_k} \sum_{(i_1,i_2)} f^1[g(x_1)]D_{(i_1,i_2)}(f^2 \circ g, f^3 \circ g)(x_1)\frac{(-1)^2}{2!}\Delta(\mathbf{x}^{(i_1,i_2)}). \quad (15)$$

This limit is the integral in (10). The terms in the sum on each side of **a** lead to an integral on a simplex of the kind treated in Section 37, and the integral in (10) is the sum of these integrals on the sides of **a**.

Thus a part of the terms which result from substituting (13) in (12) leads, in the limit, to the integral (10), which is the desired result. In order to complete the proof, it is necessary to show that the limit of the remaining terms is zero. The remaining limit is

$$\begin{aligned}\lim_{k\to\infty} \sum_{\mathbf{x}\in\partial\mathscr{P}_k} \Big\{ & f^1[g(x_1)]D_{(1,2)}(f^2, f^3)[g(x_1)]\frac{(-1)^2}{2!}r(g^1, g^2; \mathbf{x})|\Delta(\mathbf{x})| \\ & + f^1[g(x_1)]D_{(1,3)}(f^2, f^3)[g(x_1)]\frac{(-1)^2}{2!}r(g^1, g^3; \mathbf{x})|\Delta(\mathbf{x})| \\ & + f^1[g(x_1)]D_{(2,3)}(f^2, f^3)[g(x_1)]\frac{(-1)^2}{2!}r(g^2, g^3; \mathbf{x})|\Delta(\mathbf{x})| \Big\}.\end{aligned} \quad (16)$$

Now f^1 and the derivatives $D_{(j_1,j_2)}(f^2, f^3)$, $(j_1, j_2) \in (2/3)$, are continuous on the compact set $T(G_3)$; hence, there is a constant M such that

$$|f^1[g(x)]D_{(j_1,j_2)}(f^2, f^3)[g(x)]| < M, \qquad x \in \mathbf{a}, \qquad (j_1, j_2) \in (2/3). \quad (17)$$

Also, for each $\varepsilon > 0$ there exists a $k(\varepsilon)$ such that

$$|r(g^{j_1}, g^{j_2}; \mathbf{x})| < \varepsilon, \qquad (j_1, j_2) \in (2/3), \quad (18)$$

for each $\mathbf{x}$ in $\partial\mathscr{P}_k$ if $k \geqq k(\varepsilon)$; this statement follows from Theorem 11.2. Then the absolute value of the sum in (16) is less than

$$3M\varepsilon \sum_{\mathbf{x}\in\partial\mathscr{P}_k} \frac{|\Delta(\mathbf{x})|}{2!} \quad (19)$$

for all k such that $k \geqq k(\varepsilon)$. Now equation (16) in Section 11 and equations (10), (11), (13) and Theorem 20.5 in Section 20 show that $|\Delta(\mathbf{x})|/2!$ is the area of the simplex (triangle) $\mathbf{x}$. Finally, Theorem 21.4 shows that the sum in (19) is the sum of the areas of the four sides of **a**. Thus the sum in (19) is a constant, and (19) proves that the limit in (16) is zero. This statement completes the proof that the integral in (9) equals the integral in (10). This proof, in the special case $m = n = 3$, that the integral in (6) equals the integral in (5) contains all of the ideas and methods to be used in the proof of the general case.

This proof of the last conclusion in Theorem 39.1 completes the proof of the theorem. □

39.2 Example. An oriented simplex $\mathbf{b}:[b_0, b_1, \cdots, b_m]$ in $\mathbb{R}^n$, $m \leqq n$, can always be considered the trace of a surface G_m. If $\mathbf{a}:[a_0, a_1, \cdots, a_m]$ is an oriented simplex in $\mathbb{R}^m$, then there is an affine transformation—call it g—such that $g(a_i) = b_i$, $i = 0, 1, \cdots, m$. Then the components $(g^1, \cdots, g^n)$ of g have continuous derivatives, and the function $g:\mathbf{a} \to \mathbb{R}^n$ defines a surface G_m whose trace $T(G_m)$ is $\mathbf{b}$. The fundamental theorem of the integral calculus in Theorem 39.1 holds without change for integrals defined on this surface. In this case the integrals on G_m and ∂G_m are denoted as integrals on $\mathbf{b}$ and $\partial \mathbf{b}$, and the fundamental theorem of the integral calculus in (3) and (6) can be written as follows:

$$\int_{\mathbf{b}} \sum_{(m/n)} D_{(j_1,\cdots,j_m)}(f^1, \cdots, f^m)(y)\, d(y^{j_1}, \cdots, y^{j_m}) \tag{20}$$

$$= \int_{\partial \mathbf{b}} \sum_{(m-1/n)} f^1(y) D_{(j_1,\cdots,j_{m-1})}(f^2, \cdots, f^m)(y)\, d(y^{j_1}, \cdots, y^{j_{m-1}}). \tag{21}$$

The formulas in equations (3) to (6) are valid for all functions $f:(f^1, \cdots, f^m)$ and all surface G_m. Special formulas result if the function f has a special form, or if the surface G_m has a special form, or if both f and g have special forms simultaneously. The remainder of this section derives some of these special formulas which result from the general form of the fundamental theorem of the integral calculus, and other special formulas are derived in later sections. Some of the most important consequences of the fundamental theorem result from these special formulas; they are treated more fully later in this chapter. The first of these special formulas, stated in the next theorem, results from the assumption that G_m has a special degenerate form.

39.3 Theorem. *Let G_m be a surface in $\mathbb{R}^n$, $m \leqq n$, which is defined by a function $g:\mathbf{a} \to \mathbb{R}^n$, $\mathbf{a}$ in $\mathbb{R}^m$, whose components $(g^1, \cdots, g^n)$ have continuous derivatives, and let the components $(f^1, \cdots, f^m)$ of $f:E \to \mathbb{R}^m$ have continuous derivatives. If the trace $T(G_m)$ of G_m is contained in a plane in $\mathbb{R}^n$ which has dimension d, $d < m$, then*

$$\int_{\partial G_m} \sum_{(m-1/n)} f^1(y) D_{(j_1,\cdots,j_{m-1})}(f^2, \cdots, f^m)(y)\, d(y^{j_1}, \cdots, y^{j_{m-1}}) = 0. \tag{22}$$

PROOF. Theorem 39.1 shows that (22) can be established by proving that

$$\int_{G_m} \sum_{(m/n)} D_{(j_1,\cdots,j_m)}(f^1, \cdots, f^m)(y)\, d(y^{j_1}, \cdots, y^{j_m}) = 0. \tag{23}$$

The proof of (23) employs the definition of the surface integral in Section 36. Let $\mathscr{P}_k$, $k = 1, 2, \cdots$, be a sequence of subdivisions of $\mathbf{a}$, and let $\mathbf{x}:(x_0, x_1, \cdots, x_m)$ be a simplex in $\mathscr{P}_k$. By Definition 36.3, the integral in (23) is

$$\lim_{k\to\infty} \sum_{\mathbf{x}\in\mathscr{P}_k} \sum_{(m/n)} D_{(j_1,\cdots,j_m)}(f^1, \cdots, f^m)[g(x_0)] \frac{(-1)^m}{m!} \Delta_{(1,\cdots,m)}(g^{j_1}, \cdots, g^{j_m})(\mathbf{x}). \tag{24}$$

The $(m+1)$ points $g(x_0), g(x_1), \cdots, g(x_m)$ are in a d-dimensional plane in $\mathbb{R}^n$, $d < m$. Then these points are the vertices of a degenerate simplex [see Sections 13 and 14] and

$$\left\{ \sum_{(j_1, \cdots, j_m)} [\Delta_{(1, \cdots, m)}(g^{j_1}, \cdots, g^{j_m})(\mathbf{x})]^2 \right\}^{1/2} = 0. \tag{25}$$

As a result of (25), for each $\mathbf{x}$ in $\mathscr{P}_k$ the sum in (24) is zero; therefore the limit in (24) exists and has the value zero, the integral in (23) equals zero, and the proof of (22) and of Theorem 39.3 is complete. □

39.4 Example. Let $\mathbf{a}:[a_0, a_1, \cdots, a_3]$ be a Euclidean simplex in $\mathbb{R}^3$, and let $g:\mathbf{a} \to \mathbb{R}^3$ be a function with components $(g^1, \cdots, g^3)$ such that $g^3(x) = 0$ for every x in $\mathbf{a}$. Then the equations

$$\begin{aligned} y^1 &= g^1(x), \\ y^2 &= g^2(x), \qquad x \in \mathbf{a}, \\ y^3 &= 0, \end{aligned} \tag{26}$$

describe a surface G_3 whose trace $T(G_3)$ lies in the coordinate plane $y^3 = 0$. If $f:(f^1, f^2, f^3)$ is a function whose components have continuous derivatives, then Theorem 39.1 states that

$$\begin{aligned} &\int_{G_3} D_{(1, \cdots, 3)}(f^1, \cdots, f^3)(y)\, d(y^1, \cdots, y^3) \\ &\qquad = \int_{\partial G_3} \sum_{(j_1, j_2)} f^1(y) D_{(j_1, j_2)}(f^2, f^3)(y)\, d(y^{j_1}, y^{j_2}). \end{aligned} \tag{27}$$

Now the integral on the left equals zero since its value is

$$\int_{\mathbf{a}} D_{(1, \cdots, 3)}(f^1, \cdots, f^3)[g(x)] D_{(1, \cdots 3)}(g^1, \cdots, g^3)(x)\, d(x^1, \cdots, x^3), \tag{28}$$

and $D_{(1, \cdots, 3)}(g^1, \cdots, g^3)(x) = 0$ for every x in $\mathbf{a}$ by (26). Then

$$\int_{\partial G_3} \sum_{(j_1, j_2)} f^1(y) D_{(j_1, j_2)}(f^2, f^3)(y)\, d(y^{j_1}, y^{j_2}) = 0 \tag{29}$$

by (27) and (28), and also as stated in equation (22) in Theorem 39.3. Because $T(G_3)$ is in the plane $y^3 = 0$ [see (26)], the integral in (29) equals

$$\int_{\partial G_3} f^1(y) D_{(1, 2)}(f^2, f^3)(y)\, d(y^1, y^2). \tag{30}$$

In this case G_3 is a 3-dimensional "surface" whose trace $T(G_3)$ is collapsed into the plane $y^3 = 0$. The boundary ∂G_3 of G_3 is a closed 2-dimensional surface whose trace is also in $y^3 = 0$.

39.5 Theorem. *If G_m is a surface in $\mathbb{R}^n$, $m \leqq n$, which is defined by a function $g:\mathbf{a} \to \mathbb{R}^n$ whose components have continuous derivatives, and if the components*

$(f^2, \cdots, f^m)$ *of* $f: E \to \mathbb{R}^{m-1}$ *have continuous derivatives, then*

$$\int_{\partial G_m} \sum_{(m-1/n)} D_{(j_1,\cdots,j_{m-1})}(f^2, \cdots, f^m)(y)\, d(y^{j_1}, \cdots, y^{j_{m-1}}) = 0. \tag{31}$$

Proof. Define $f^1: E \to \mathbb{R}$ to be the constant function whose value is 1. Then

$$D_{(j_1,\cdots,j_m)}(f^1, \cdots, f^m)(y) = 0, \qquad y \in E, \qquad (j_1, \cdots, j_m) \in (m/n), \tag{32}$$

because this derivative is the determinant of a matrix of derivatives by Corollary 3.20, and the first row of the matrix consists of zeros since f^1 is constant. Thus

$$\int_{G_m} \sum_{(m/n)} D_{(j_1,\cdots,j_m)}(f^1, \cdots, f^m)(y)\, d(y^{j_1}, \cdots, y^{j_m}) = 0. \tag{33}$$

Then by (3) and (6) in Theorem 39.1,

$$\int_{\partial G_m} \sum_{(m-1/n)} f^1(y) D_{(j_1,\cdots,j_{m-1})}(f^2, \cdots, f^m)(y)\, d(y^{j_1}, \cdots, y^{j_{m-1}}) = 0. \tag{34}$$

Since $f^1(y) = 1$ for y in E, equation (34) is the same as (31) and the proof is complete. □

Let c and d denote arbitrary constants. The following elementary formula emphasizes that the constant of integration can be ignored, or chosen arbitrarily, in evaluating a definite integral by the fundamental theorem of the integral calculus:

$$\int_a^b \frac{d}{dx}[f(x) + c]\, dx = [f(b) + d] - [f(a) + d]. \tag{35}$$

The next theorem contains the generalization of this formula for surface integrals of derivatives. It has applications later.

39.6 Theorem. *If* G_m *is a surface in* $\mathbb{R}^n$, $m \leqq n$, *which is defined by a function* $g: \mathbf{a} \to \mathbb{R}^n$, $\mathbf{a}$ *in* $\mathbb{R}^m$, *whose components have continuous derivatives; if the components* $(f^1, \cdots, f^m)$ *of* $f: E \to \mathbb{R}^m$ *have continuous derivatives; and if* $c_1, \cdots, c_m$ *and* $d_1, \cdots, d_m$ *are arbitrary constants, then*

$$\int_{G_m} \sum_{(m/n)} D_{(j_1,\cdots,j_m)}(f^1 + c_1, \cdots, f^m + c_m)(y)\, d(y^{j_1}, \cdots, y^{j_m}) \tag{36}$$

$$= \int_{G_m} \sum_{(m/n)} D_{(j_1,\cdots,j_m)}(f^1, \cdots, f^m)(y)\, d(y^{j_1}, \cdots, y^{j_m}) \tag{37}$$

$$= \int_{\partial G_m} \sum_{(m-1/n)} f^1(y) D_{(j_1,\cdots,j_{m-1})}(f^2, \cdots, f^m)(y)\, d(y^{j_1}, \cdots, y^{j_{m-1}}) \tag{38}$$

$$= \int_{\partial G_m} \sum_{(m-1/n)} [f^1(y) + d_1] D_{(j_1,\cdots,j_{m-1})}(f^2 + d_2, \cdots, f^m + d_m) \times (y)\, d(y^{j_1}, \cdots, y^{j_{m-1}}). \tag{39}$$

PROOF. Since

$$D_{(j_1,\cdots,j_m)}(f^1 + c_1, \cdots, f^m + c_m)(y) = D_{(j_1,\cdots,j_m)}(f^1, \cdots, f^m)(y), \quad (j_1, \cdots, j_m) \in (m/n), \tag{40}$$

the integrals in (36) and (37) are equal. The integrals in (37) and (38) are equal by Theorem 39.1. Since the derivatives of $(f^2 + d_2, \cdots, f^m + d_m)$ are the same as the derivatives of $(f^2, \cdots, f^m)$, the integral in (39) equals

$$\int_{\partial G_m} \sum_{(m-1/n)} [f^1(y) + d_1] D_{(j_1,\cdots,j_{m-1})}(f^2, \cdots, f^m)(y) d(y^{j_1}, \cdots, y^{j_{m-1}}). \tag{41}$$

This integral equals the integral in (38) plus d_1 times the integral in (31), which equals zero by Theorem 39.5. Thus the integrals in (39) and (38) are equal, and the proof of Theorem 39.6 is complete. □

EXERCISES

39.1. Let A be the square $[0, 1] \times [0, 1]$ in $\mathbb{R}^2$, and let the surface G_2 be defined by the function $g: A \to \mathbb{R}^2$ such that

$$g^1(x) = x^1, \qquad g^2(x) = x^2, \qquad g^3(x) = x^1 + x^2. \tag{42}$$

Let E be an open set which contains $T(G_2)$, and let $f: E \to \mathbb{R}^2$ be the function whose components are defined as follows:

$$f^1(y) = (y^1)^2 + (y^2)^2 + (y^3)^2, \qquad f^2(y) = y^1 + y^2 + y^3. \tag{43}$$

(a) Show that the integral in Theorem 39.1, equation (3), for this problem is

$$\int_{G_2} D_{(1,2)}(f^1, f^2)(y)\, d(y^1, y^2) + D_{(1,3)}(f^1, f^2)(y)\, d(y^1, y^3) + D_{(2,3)}(f^1, f^2)(y)\, d(y^2, y^3), \tag{44}$$

and that, for the functions in (43), this integral is

$$\int_{G_2} 2(y^1 - y^2)\, d(y^1, y^2) + 2(y^1 - y^3)\, d(y^1, y^3) + 2(y^2 - y^3)\, d(y^2, y^3). \tag{45}$$

Express this integral as an integral in (x^1, x^2) over A by making the following substitutions:

$$\begin{aligned} y^1 &= g^1(x) = x^1, & d(y^1, y^2) &= D_{(1,2)}(g^1, g^2)(x)\, d(x^1, x^2) = d(x^1, x^2), \\ y^2 &= g^2(x) = x^2, & d(y^1, y^3) &= D_{(1,2)}(g^1, g^3)(x)\, d(x^1, x^2) = d(x^1, x^2), \\ y^3 &= g^3(x) = x^1 + x^2; & d(y^2, y^3) &= D_{(1,2)}(g^2, g^3)(x)\, d(x^1, x^2) = -d(x^1, x^2). \end{aligned} \tag{46}$$

Show that these substitutions reduce the integral in (45) to the following:

$$\int_A 4(x^1 - x^2)\, d(x^1, x^2). \tag{47}$$

Use iterated integrals to evaluate this integral and show that its value is zero.

(b) Show that the integral in Theorem 39.1, equation (4), for this problem is

$$\int_A D_{(1,2)}(f^1 \circ g, f^2 \circ g)(x)\, d(x^1, x^2). \tag{48}$$

Show also that, for the functions in (42) and (43),

$$\begin{aligned}(f^1 \circ g)(x) &= 2[(x^1)^2 + x^1x^2 + (x^2)^2],\\ (f^2 \circ g)(x) &= 2[x^1 + x^2],\\ D_{(1,2)}(f^1 \circ g, f^2 \circ g)(x) &= 4(x^1 - x^2).\end{aligned} \tag{49}$$

Use these values to show that the integral in (48) is the integral in (47), and that it therefore has the value zero.

(c) Show that the integral in Theorem 39.1, equation (5), for this problem is

$$\int_{\partial A} f^1[g(x)]D_1[f^2 \circ g](x)\, dx^1 + f^1[g(x)]D_2[f^2 \circ g](x)\, dx^2. \tag{50}$$

Show also that, for the functions in (42), (43), and (49), this integral is

$$\int_{\partial A} 2[(x^1)^2 + x^1x^2 + (x^2)^2][2\, dx^1 + 2\, dx^2]. \tag{51}$$

Evaluate this integral and show that its value is zero.

(d) Show that the integral in Theorem 39.1, equation (6), for this problem is

$$\int_{\partial G_2} f^1(y)D_1f^2(y)\, dy^1 + f^1(y)D_2f^2(y)\, dy^2 + f^1(y)D_3f^2(y)\, dy^3. \tag{52}$$

Show that this integral equals the integral in (51) and that its value is therefore zero.

(e) Verify the fundamental theorem of the integral calculus in this numerical example by collecting results from parts (a), $\cdots$, (d) and showing that the integrals in equations (3), $\cdots$, (6) in Theorem 39.1 are equal.

39.2. If $g: \mathbf{a} \to \mathbb{R}^n$ is the identity function so that $g(x) = x$, show that the fundamental theorem of the integral calculus in Theorem 39.1 becomes the fundamental theorem of the integral calculus in Theorem 38.4.

39.3. Prove the following form of the fundamental theorem of the integral calculus. If G_m is a surface in $\mathbb{R}^n$, $m \leq n$, which is defined by a function $g: \mathbf{a} \to \mathbb{R}^n$ whose components have continuous derivatives, and if the components $(f^1, \cdots, f^m)$ of $f: E \to \mathbb{R}^m$ have continuous derivatives, then

$$\int_{G_m} \sum_{(m/n)} D_{(j_1, \cdots, j_m)}(f^1, \cdots, f^m)(y)\, d(y^{j_1}, \cdots, y^{j_m}) \tag{53}$$

$$= \int_{\mathbf{a}} D_{(1, \cdots, m)}(f^1 \circ g, \cdots, f^m \circ g)(x)\, d(x^1, \cdots, x^m) \tag{54}$$

$$= \frac{1}{m}\int_{\partial \mathbf{a}} \sum_{(m-1/n)} \det[f \circ g)(x), D_{j_1}(f \circ g)(x), \cdots, D_{j_{m-1}}(f \circ g)(x)]\, d(x^{j_1}, \cdots, x^{j_{m-1}}) \tag{55}$$

$$= \frac{1}{m}\int_{\partial G_m} \sum_{(m-1/n)} \det[f(y), D_{j_1}f(y), \cdots, D_{j_{m-1}}f(y)]\, d(y^{j_1}, \cdots, y^{j_{m-1}}). \tag{56}$$

39.4. Show that the integral in Exercise 39.3, equation (56), for the problem in Exercise 39.1 is

$$\frac{1}{2}\int_{\partial G_2}\begin{vmatrix} f^1(y) & f^2(y) \\ D_1f^1(y) & D_1f^2(y)\end{vmatrix} dy^1 + \begin{vmatrix} f^1(y) & f^2(y) \\ D_2f^1(y) & D_2f^2(y)\end{vmatrix} dy^2 + \begin{vmatrix} f^1(y) & f^2(y) \\ D_3f^1(y) & D_3f^2(y)\end{vmatrix} dy^3.$$

Evaluate this integral for the problem in Exercise 39.1. Show that the value of the integral is zero, which is the same as the value of the integrals in equations (44), (48), (50), and (52) in Exercise 39.1.

39.5. Let G_2 be a surface in $\mathbb{R}^3$ which is defined by a function $g: \mathbf{a} \to \mathbb{R}^3$, $\mathbf{a}$ in $\mathbb{R}^2$, whose components have continuous derivatives, and let the components (f^1, f^2) of $f: E \to \mathbb{R}^2$ have continuous derivatives.

(a) Show that, for this surface G_2 and function f, the following is a statement of the fundamental theorem of the integral calculus.

$$\int_{G_2} D_{(1,2)}(f^1, f^2)(y)\, d(y^1, y^2) + D_{(1,3)}(f^1, f^2)(y)\, d(y^1, y^3)$$
$$+ D_{(2,3)}(f^1, f^2)(y)\, d(y^2, y^3)$$
$$= \int_{\partial G_2} f^1(y) D_1 f^2(y)\, dy^1 + f^1(y) D_2 f^2(y)\, dy^2 + f^1(y) D_3 f^2(y)\, dy^3.$$

(b) If $f: \mathbf{a} \to \mathbb{R}^2$ is a function with components (f^1, f^2) such that $f^1(y) = P(y)$ and $f^2(y) = y^1$, show that the formula in (a) takes the following special form.

$$\int_{G_2} - D_2 P(y)\, d(y^1, y^2) - D_3 P(y)\, d(y^1, y^3) = \int_{\partial G_2} P(y)\, dy^1.$$

(c) Find the special formulas which result from the formula in (a) if $f^1(y) = Q(y)$ and $f^2(y) = y^2$; if $f^1(y) = R(y)$ and $f^2(y) = y^3$.

39.6. If G_2 is a 2-dimensional surface in $\mathbb{R}^4$, write the statement of the fundamental theorem of the integral calculus which corresponds to the one in Exercise 39.5(a). Find the special form which this formula takes if $f^1(y) = P(y)$ and $f^2(y) = y^1$. Find three other similar formulas. [Hint. Exercise 39.5(c).]

40. The Fundamental Theorem on Chains

Example 39.2 has explained that an integral can be defined on every simplex $\mathbf{b}: [b_0, b_1, \cdots, b_m]$ in $\mathbb{R}^n$, and integrals on the chain of simplexes which form the boundary of a simplex occur in the fundamental theorem of the integral calculus for a simplex. This section extends the fundamental theorem from simplexes to chains of simplexes of a certain type. These chains are described in the next definition.

40.1 Definition. A chain $\mathbf{c}$ of simplexes is called an oriented simplicial chain if it has the following properties:

(a) The chain $\mathbf{c}$ is the sum of a finite number of oriented, (degenerate or non-degenerate) m-dimensional Euclidean simplexes in $\mathbb{R}^n$, $m \leqq n$ [see Definitions 14.10 and 14.14].

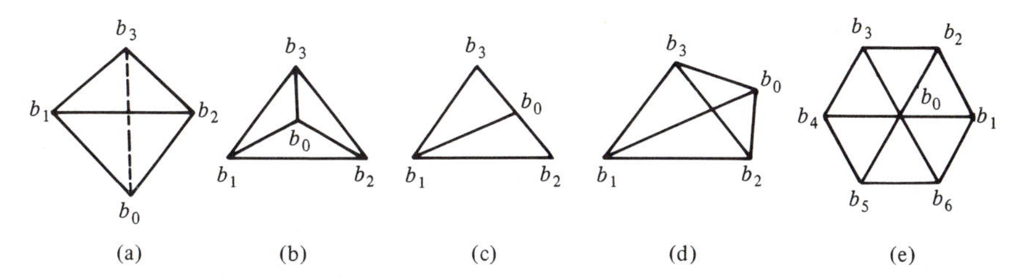

Figure 40.1. The oriented simplicial chains in Example 40.2.

(b) The chain $\mathbf{c}$ is connected; that is, each two vertices in $\mathbf{c}$ are connected by edges in $\mathbf{c}$.
(c) An $(m-1)$-dimensional side of a simplex in $\mathbf{c}$ is a side of at most two simplexes in $\mathbf{c}$.
(d) If an $(m-1)$-simplex is a side of two simplexes in $\mathbf{c}$, then it has opposite orientations in the boundaries of these two simplexes; thus the boundary $\partial\mathbf{c}$ of $\mathbf{c}$ consists of those $(m-1)$-dimensional simplexes which are a side of a single simplex in $\mathbf{c}$.

40.2 Example. If $[b_0, b_1, \cdots, b_3]$ is an oriented Euclidean (non-degenerate) simplex in $\mathbb{R}^3$ [see Figure 40.1(a)], then its boundary is an oriented simplicial chain according to Definition 40.1. Now

$$\begin{aligned}\partial&[b_0, b_1, \cdots, b_3]\\ &= [b_1, b_2, b_3] - [b_0, b_2, b_3] + [b_0, b_1, b_3] - [b_0, b_1, b_2] \qquad (1)\\ &= [b_1, b_2, b_3] - [b_0, b_2, b_3] - [b_1, b_0, b_3] - [b_1, b_2, b_0].\end{aligned}$$

Let

$$\mathbf{c} = [b_1, b_2, b_3] - [b_0, b_2, b_3] - [b_1, b_0, b_3] - [b_1, b_2, b_0]. \qquad (2)$$

To verify that $\mathbf{c}$ is an oriented simplicial chain, observe first that $\mathbf{c}$ is a chain of oriented (degenerate or non-degenerate) 2-dimensional Euclidean simplexes in $\mathbb{R}^3$. Next, the chain is connected. Also, each side (b_{i_1}, b_{i_2}) belongs to exactly two of the simplexes in $\mathbf{c}$. Finally, if a one-dimensional side belongs to two simplexes in $\mathbf{c}$, the side has opposite orientations in the boundaries of these two simplexes. For example, $[b_1, b_2]$ is a side in the boundary of $[b_1, b_2, b_3]$, but $-[b_1, b_2]$ is a side in the boundary of $-[b_1, b_2, b_0]$. Therefore $\mathbf{c}$ is an oriented simplicial chain by Definition 40.1. If the four points $b_0, b_1, \cdots, b_3$ are in the plane $\mathbb{R}^2$ as shown in Figure 40.1(b), then the chain $\mathbf{c}$ defined by (2) is still an oriented simplicial chain. The simplex $[b_0, b_1, \cdots, b_3]$ is now a degenerate Euclidean simplex, but $\mathbf{c}$ is a chain which consists of four non-degenerate Euclidean simplexes. If b_0 is on the side $[b_2, b_3]$ as shown in Figure 40.1(c), the only change is that one of the simplexes in $\mathbf{c}$, namely $-[b_0, b_2, b_3]$, is now a degenerate Euclidean simplex. If b_0 is outside the simplex $[b_1, b_2, b_3]$ as in Figure 40.1(d), then $[b_1, b_2, b_3]$

and $-[b_0, b_2, b_3]$ are positively oriented in $\mathbb{R}^2$, but $-[b_1, b_0, b_3]$ and $-[b_1, b_2, b_0]$ are negatively oriented; all simplexes, however, are non-degenerate. For each of the chains $\mathbf{c}$ in Figures 40.1(a), $\cdots$, (d), the boundary $\partial\mathbf{c}$ of $\mathbf{c}$ is the zero chain by Theorem 17.8 since $\mathbf{c}$ itself is the boundary of the simplex $[b_0, b_1, \cdots, b_3]$. Finally, Figure 40.1(e) shows the oriented simplicial chain

$$\mathbf{c} = [b_0, b_1, b_2] + [b_0, b_2, b_3] + \cdots + [b_0, b_6, b_1], \tag{3}$$

which consists of six non-degenerate, 2-dimensional, oriented, Euclidean simplexes. The boundary is not the zero chain in this case since

$$\partial\mathbf{c} = [b_1, b_2] + [b_2, b_3] + \cdots + [b_6, b_1]. \tag{4}$$

Integrals on chains have been encountered already in the fundamental theorem of the integral calculus; nevertheless, the following definition is needed.

40.3 Definition. Let $\mathbf{c}$ be an m-dimensional oriented simplicial chain $\Sigma_1^r \mathbf{b}_i$ in $\mathbb{R}^n$, $m \leqq n$. Let $f^{(j_1, \cdots, j_m)}$, $(j_1, \cdots, j_m) \in (m/n)$, be the components of a function which is continuous on $\mathbf{c}$. Then

$$\int_{\Sigma_1^r \mathbf{b}_i} \sum_{(m/n)} f^{(j_1, \cdots, j_m)}(y)\, d(y^{j_1}, \cdots, y^{j_m}) = \sum_{i=1}^{r} \int_{\mathbf{b}_i} \sum_{(m/n)} f^{(j_1, \cdots, j_m)}(y)\, d(y^{j_1}, \cdots, y^{j_m}). \tag{5}$$

40.4 Example. Let E be an open set which contains the four simplexes in the boundary of $[b_0, b_1, \cdots, b_3]$ in Figure 40.1(a), and let $f: E \to \mathbb{R}^2$ be a function whose components (f^1, f^2) have continuous derivatives on E. If $\mathbf{c}$ is the chain $\Sigma_1^4 \mathbf{b}_i$ in (2), then by Definition 40.3,

$$\begin{aligned} &\int_{\Sigma_1^4 \mathbf{b}_i} \sum_{(j_1, j_2)} D_{(j_1, j_2)}(f^1, f^2)(y)\, d(y^{j_1}, y^{j_2}) \\ &\quad = \sum_{i=1}^{4} \int_{\mathbf{b}_i} \sum_{(j_1, j_2)} D_{(j_1, j_2)}(f^1, f^2)(y)\, d(y^{j_1}, y^{j_2}), \qquad (j_1, j_2) \in (2/3). \end{aligned} \tag{6}$$

40.5 Theorem (Fundamental Theorem of the Integral Calculus). *Let $\mathbf{c}$ be an m-dimensional oriented simplicial chain $\Sigma_1^r \mathbf{b}_i$ in $\mathbb{R}^n$, $m \leqq n$, and let E be an open set which contains the simplexes $\mathbf{b}_1, \cdots, \mathbf{b}_r$. If the components $(f^1, \cdots, f^m)$ of $f: E \to \mathbb{R}^m$ have continuous derivatives, then*

$$\int_{\mathbf{c}} \sum_{(m/n)} D_{(j_1, \cdots, j_m)}(f^1, \cdots, f^m)(y)\, d(y^{j_1}, \cdots. y^{j_m}) \tag{7}$$

$$= \int_{\partial\mathbf{c}} \sum_{(m-1/n)} f^1(y) D_{(j_1, \cdots, j_{m-1})}(f^2, \cdots, f^m)(y)\, d(y^{j_1}, \cdots, y^{j_m}). \tag{8}$$

If $\partial\mathbf{c} = 0$, then the integral in (7) equals zero.

PROOF. By Definition 40.3, the integral (7) equals

$$\sum_{i=1}^{r} \int_{\mathbf{b}_i} \sum_{(m/n)} D_{(j_1,\cdots,j_m)}(f^1, \cdots, f^m)(y)\, d(y^{j_1}, \cdots, y^{j_m}). \tag{9}$$

By the fundamental theorem of the integral calculus [see Theorems 38.4 and 39.1 and Example 39.2], this sum equals

$$\sum_{i=1}^{r} \int_{\partial \mathbf{b}_i} \sum_{(m-1/n)} f^1(y) D_{(j_1,\cdots,j_{m-1})}(f^2, \cdots, f^m)(y)\, d(y^{j_1}, \cdots, y^{j_{m-1}}). \tag{10}$$

If two simplexes in $\mathbf{c}$ have an $(m-1)$-dimensional side in common, then by the definition of $\mathbf{c}$, these sides have opposite orientations and the corresponding integrals in (10) cancel. The sum (10) reduces to (8) after all cancellations have been made. If $\partial \mathbf{c} = 0$, then all integrals in (10) cancel, and (7) equals zero. □

40.6 Example. Continue Example 40.2. By the fundamental theorem for a simplex,

$$\begin{aligned} &\int_{[b_1,b_2,b_3]} \sum_{(j_1,j_2)} D_{(j_1,j_2)}(f^1, f^2)(y)\, d(y^{j_1}, y^{j_2}) \qquad (j_1, j_2) \in (2/3) \\ &\qquad = \left\{ \int_{[b_1,b_2]} + \int_{[b_2,b_3]} + \int_{[b_3,b_1]} \right\} \sum_{j=1}^{3} f^1(y) D_j f^2(y)\, dy^j. \end{aligned} \tag{11}$$

In the same way, evaluate the integral on the other simplexes $-[b_0, b_2, b_3]$, $-[b_1, b_0, b_3]$, and $-[b_1, b_2, b_0]$ in $\mathbf{c}$ in (2). The results show that the sum of the integrals on the four simplexes in $\mathbf{c}$ is zero. Thus Theorem 40.5 is verified in this special case.

EXERCISES

40.1. Equation (11) evaluates the integral on the left by the fundamental theorem of the integral calculus. In the same way evaluate this integral on $-[b_0, b_2, b_3]$, $-[b_1, b_0, b_3]$, and $-[b_1, b_2, b_0]$ and complete all details in the verification of Theorem 40.5 in Example 40.6.

40.2. Let H be the hexagon in Figure 40.1(e), and let $f: H \to \mathbb{R}^2$ be a function whose components have continuous derivatives.

(a) Prove the following statement of the fundamental theorem of the integral calculus.

$$\int_H D_{(1,2)}(f^1, f^2)(x)\, d(x^1, x^2) = \left\{ \int_{[a_1,a_2]} + \cdots + \int_{[a_6,a_1]} \right\} \sum_{j=1}^{2} f^1(x) D_j f^2(x)\, dx^j.$$

(b) If $f^1(x) = d$, a constant, for every x in the boundary of H, prove that

$$\int_H D_{(1,2)}(f^1, f^2)(x)\, d(x^1, x^2) = 0.$$

41. Stokes' Theorem and Related Results

Section 39 has remarked already that certain important results are obtained by applying the fundamental theorem of the integral calculus to special functions. The formula in Theorem 39.5 is one of these results, and Stokes' theorem, to be proved in this section, is another. Green's theorem and Gauss' theorem are special cases of Stokes' theorem.

A reminder of the notation for certain functions will be helpful. As usual, $(y^1, \cdots, y^n)$ denote the coordinates of a point y in $\mathbb{R}^n$. It is convenient to let y^i, $i = 1, \cdots, n$, denote also the function $y^i: \mathbb{R}^n \to \mathbb{R}$ whose value at $(y^1, \cdots, y^n)$ in $\mathbb{R}^n$ is y^i. The functions $y^i: \mathbb{R}^n \to \mathbb{R}$ are continuous and have continuous derivatives of all orders in $\mathbb{R}^n$. Also, $D_i y^j(y) = 1$ if $i = j$ and $D_i y^j(y) = 0$ if $i \neq j$.

41.1 Example. The proof of Green's theorem illustrates the methods employed in the proof of the general Stokes' theorem. Let $\mathbf{a}$ be an oriented simplex in $\mathbb{R}^2$, and let $g: \mathbf{a} \to \mathbb{R}^2$ be a function whose components (g^1, g^2) have continuous derivatives. Then g defines a surface G_2 whose trace $T(G_2)$ is in $\mathbb{R}^2$. Let E be an open set which contains $T(G_2)$, and let $P: E \to \mathbb{R}$ and $Q: E \to \mathbb{R}$ be functions which have continuous derivatives. Then the functions $(P, y^1): E \to \mathbb{R}^2$ and $(Q, y^2): E \to \mathbb{R}^2$ have continuous derivatives, and the fundamental theorem of the integral calculus in Theorem 39.1 states that

$$\int_{G_2} D_{(1,2)}(P, y^1)(y)\,d(y^1, y^2) = \int_{\partial G_2} P(y)D_1 y^1(y)\,dy^1 + P(y)D_2 y^1(y)\,dy^2, \quad (1)$$

$$\int_{G_2} D_{(1,2)}(Q, y^2)(y)\,d(y^1, y^2) = \int_{\partial G_2} Q(y)D_1 y^2(y)\,dy^1 + Q(y)D_2 y^2(y)\,dy^2. \quad (2)$$

Since

$$D_{(1,2)}(P, y^1)(y) = -D_2 P(y), \qquad D_1 y^1(y) = 1, \qquad D_2 y^1(y) = 0, \quad (3)$$

$$D_{(1,2)}(Q, y^2) = D_1 Q(y), \qquad D_1 y^2(y) = 0, \qquad D_2 y^2(y) = 1, \quad (4)$$

equations (1) and (2) simplify to

$$\int_{G_2} -D_2 P(y)\,d(y^1, y^2) = \int_{\partial G_2} P(y)\,dy^1, \quad (5)$$

$$\int_{G_2} D_1 Q(y)\,d(y^1, y^2) = \int_{\partial G_2} Q(y)\,dy^2. \quad (6)$$

Add equations (5) and (6) to obtain the following statement of Green's theorem:

$$\int_{G_2} [D_1 Q(y) - D_2 P(y)]\,d(y^1, y^2) = \int_{\partial G_2} P(y)\,dy^1 + Q(y)\,dy^2. \quad (7)$$

The function $g: \mathbf{a} \to \mathbb{R}^2$ and the orientation of $\mathbf{a}$ determine the orientation of G_2 and ∂G_2. As stated in Section 39, the surface G_2 can be replaced by a simplex $\mathbf{b}$ in $\mathbb{R}^2$.

41.2 Theorem. *If G_m is a surface in $\mathbb{R}^n$, $m \leqq n$, which is defined by a function $g: \mathbf{a} \to \mathbb{R}^n$, $\mathbf{a}$ in $\mathbb{R}^m$, whose components have continuous derivatives, and if the functions $P^{(j_1, \cdots, j_{m-1})}: E \to \mathbb{R}$, $T(G_m) \subset E$, $(j_1, \cdots, j_{m-1})$ in $(m - 1/n)$, have continuous derivatives, then*

$$\int_{G_m} \sum_{(i_1, \cdots, i_m)} D_{(i_1, \cdots, i_m)}(P^{(j_1, \cdots, j_{m-1})}, y^{j_1}, \cdots, y^{j_{m-1}})(y)\, d(y^{i_1}, \cdots, y^{i_m}) \tag{8}$$

$$= \int_{\partial G_m} P^{(j_1, \cdots, j_{m-1})}(y)\, d(y^{j_1}, \cdots, y^{j_{m-1}}), \qquad (i_1, \cdots, i_m) \in (m/n), \tag{9}$$

$$(j_1, \cdots, j_{m-1}) \in (m - 1/n).$$

PROOF. Apply the fundamental theorem of the integral calculus in Theorem 39.1 to the function whose components are $(P^{(j_1, \cdots, j_{m-1})}, y^{j_1}, \cdots, y^{j_{m-1}})$; the result is this: the integral in (8) equals

$$\int_{\partial G_m} \sum_{(i_1, \cdots, i_{m-1})} P^{(j_1, \cdots, j_{m-1})}(y) D_{(i_1, \cdots, i_{m-1})}(y^{j_1}, \cdots, y^{j_{m-1}})(y) d(y^{i_1}, \cdots, y^{i_{m-1}}),$$
$$(i_1, \cdots, i_{m-1}) \in (m - 1/n), \qquad (j_1, \cdots, j_{m-1}) \in (m - 1/n). \tag{10}$$

It is easy to verify that

$$\begin{aligned} D_{(i_1, \cdots, i_{m-1})}(y^{j_1}, \cdots, y^{j_{m-1}})(y) &= 1, \qquad (i_1, \cdots, i_{m-1}) = (j_1, \cdots, j_{m-1}), \\ &= 0, \qquad (i_1, \cdots, i_{m-1}) \neq (j_1, \cdots, j_{m-1}), \end{aligned} \tag{11}$$

for every y in $\mathbb{R}^n$. As a result of (11), the integral in (10) simplifies to (9). □

41.3 Theorem (Stokes' Theorem). *If the hypotheses are those stated in Theorem 41.2, then*

$$\int_{G_m} \sum_{(i_1, \cdots, i_m)} \left[\sum_{(j_1, \cdots, j_{m-1})} D_{(i_1, \cdots, i_m)}(P^{(j_1, \cdots, j_{m-1})}, y^{j_1}, \cdots, y^{j_{m-1}})(y) \right] d(y^{i_1}, \cdots, y^{i_m})$$

$$= \int_{\partial G_m} \sum_{(j_1, \cdots, j_{m-1})} P^{(j_1, \cdots, j_{m-1})}(y)\, d(y^{j_1}, \cdots, y^{j_{m-1}}), \tag{12}$$
$$(i_1, \cdots, i_m) \in (m/n), \qquad (j_1, \cdots, j_{m-1}) \in (m - 1/n).$$

PROOF. For each index set $(j_1, \cdots, j_{m-1})$ in $(m - 1/n)$, the equation in (8) and (9) is true; the result of adding these equations is (12). □

41.4 Example. The formula in (12) is forbidding in the general case, but it is familiar in some special cases. The first of these is Green's theorem in (7); this is formula (12), with slightly different notation, in the special case $m =$

$n = 2$. If $m = n = 3$, then (12) can be simplified to

$$\begin{aligned}\int_{G_3} & [D_3 P^{(1,2)}(y) - D_2 P^{(1,3)}(y) + D_1 P^{(2,3)}(y)]\, d(y^1, y^2, y^3) \\ &= \int_{\partial G_3} P^{(1,2)}(y)\, d(y^1, y^2) + P^{(1,3)}(y)\, d(y^1, y^3) + P^{(2,3)}(y)\, d(y^2, y^3).\end{aligned} \tag{13}$$

This formula is known as Gauss' theorem. Finally, if $m = 2$ and $n = 3$, Stokes' theorem in formula (12), written out in full detail, is the following:

$$\begin{aligned}\int_{G_2} & \{[D_{(1,2)}(P^1, y^1)(y) + D_{(1,2)}(P^2, y^2)(y) + D_{(1,2)}(P^3, y^3)(y)]\, d(y^1, y^2) \\ &+ [D_{(1,3)}(P^1, y^1)(y) + D_{(1,3)}(P^2, y^2)(y) + D_{(1,3)}(P^3, y^3)(y)]\, d(y^1, y^3) \\ &+ [D_{(2,3)}(P^1, y^1)(y) + D_{(2,3)}(P^2, y^2)(y) \\ &+ D_{(2,3)}(P^3, y^3)(y)]\, d(y^2, y^3)\} \\ &= \int_{\partial G_2} P^1(y)\, dy^1 + P^2(y)\, dy^2 + P^3(y)\, dy^3.\end{aligned} \tag{14}$$

Now

$$D_{(1,2)}(P^1, y^1)(y) = -D_2 P^1(y), \qquad D_{(1,2)}(P^2, y^2)(y) = D_1 P^2(y),$$
$$D_{(1,2)}(P^3, y^3) = 0,$$

and the other derivatives in (14) can be evaluated in a similar manner. Thus the formula in (14) can be simplified to the following:

$$\begin{aligned}\int_{G_2} & \{[D_1 P^2(y) - D_2 P^1(y)]\, d(y^1, y^2) + [D_1 P^3(y) - D_3 P^1(y)]\, d(y^1, y^3) \\ &+ [D_2 P^3(y) - D_3 P^2(y)]\, d(y^2, y^3)\} \\ &= \int_{\partial G_2} P^1(y)\, dy^1 + P^2(y)\, dy^2 + P^3(y)\, dy^3.\end{aligned} \tag{15}$$

This is the formula which is usually called Stokes' theorem; the formula in (12) is the general case of Stokes' theorem which includes the formulas in (7), (13), and (15) as special cases. As the proof has shown, Stokes' theorem is one of the formulas which results from applying the fundamental theorem of the integral calculus to special functions.

41.5 Theorem. *Let E be an open set in $\mathbb{R}^3$, and let $f: E \to \mathbb{R}$ be a function which has two continuous derivatives. If*

$$P^i(y) = D_i f(y), \qquad i = 1, 2, 3, \qquad y \in E, \tag{16}$$

then

$$D_i P^j(y) - D_j P^i(y) = 0, \tag{17}$$

$$D_i(D_j f)(y) = D_j(D_i f)(y), \qquad (i, j) = (1, 2), (1, 3), (2, 3), \qquad y \in E. \tag{18}$$

PROOF. To prove (17), assume the statement false and show that a contradiction results. Thus let y_0 be a point in E at which $D_1P^2(y_0) - D_2P^1(y_0) > 0$. Since $D_1P^2 - D_2P^1$ is continuous by hypothesis, there is a neighborhood of y_0 in which it is positive [see Theorem 96.9 in Appendix 2]. In this neighborhood choose a non-degenerate 2-simplex $\mathbf{b}$ on which y^3 is constant. Then as a result of the hypothesis in (16), the integral on the right of (15), written for the simplex $\mathbf{b}$, is

$$\int_{\partial \mathbf{b}} D_1 f(y)\,dy^1 + D_2 f(y)\,dy^2 + D_3 f(y)\,dy^3. \tag{19}$$

But the integral of a derivative over the boundary of a surface is zero by Theorem 39.5; hence, the integral (19) equals zero. Then the integral on the left in (15), written for the simplex $\mathbf{b}$, also equals zero; thus, since y^3 is constant on $\mathbf{b}$,

$$\int_{\mathbf{b}} [D_1P^2(y) - D_2P^1(y)]\,d(y^1, y^2) = 0. \tag{20}$$

But this equation is impossible since $D_1P^2(y) - D_2P^1(y) > 0$ on $\mathbf{b}$. Thus there is no point y_0 in E at which $D_1P^2(y_0) - D_2P_1(y_0) > 0$, and a similar argument shows that there is no point at which the expression is negative. Similar arguments complete the proof of (17), and (16) shows that (18) is another statement of (17). The proof of Theorem 41.5 is complete. □

Theorem 41.5 results from the special case of Stokes' theorem in (15) by assuming that the integral on the right is the integral of a derivative. It is clear that a similar theorem can be obtained from the general statement of Stokes' theorem by assuming that the integral on the right in (12) is the integral of a derivative. The theorem obtained in this way is the following one.

41.6 Theorem. *Let E be an open set in $\mathbb{R}^n$, and let $(f^1, \cdots, f^{m-1}) : E \to \mathbb{R}^{m-1}$, $m \leqq n$, be a function whose components have two continuous derivatives. If*

$$\begin{aligned} P^{(j_1, \cdots, j_{m-1})}(y) &= D_{(j_1, \cdots, j_{m-1})}(f^1, \cdots, f^{m-1})(y), \qquad y \in E, \\ &(j_1, \cdots, j_{m-1}) \in (m-1/n), \end{aligned} \tag{21}$$

then

$$\begin{aligned} \sum_{(j_1, \cdots, j_{m-1})} D_{(i_1, \cdots, i_m)}(P^{(j_1, \cdots, j_{m-1})}, y^{j_1}, \cdots, y^{j_{m-1}})(y) = 0, \qquad y \in E, \\ (j_1, \cdots, j_{m-1}) \in (m-1/n), \qquad (i_1, \cdots, i_m) \in (m/n). \end{aligned} \tag{22}$$

PROOF. Theorem 39.5 and the hypothesis in (21) show that the integral on the right in (12) equals zero. Then the integral on the left in (12) equals zero, and arguments similar to those used in the proof of Theorem 41.5 show that the function in each square bracket vanishes identically in E. This statement is the conclusion (22) in Theorem 41.6. □

EXERCISES

41.1. Let $A = [0, 1] \times [0, 1]$, and let $P(x) = 2x^1x^2$, $Q(x) = (x^1)^2$.
(a) Prove that $\int_{\partial A} P(x)\,dx^1 + Q(x)\,dx^2 = 0$ by direct evaluation of the integral.
(b) Use Green's theorem to prove that the integral in (a) equals zero.
(c) Let $f(x) = (x^1)^2x^2$, and show that $P(x) = D_1 f(x)$, $Q(x) = D_2 f(x)$. Then use Theorem 39.5 (stated for the square A) to prove that the integral in (a) equals zero.

41.2. Let $A = [0, 1] \times [0, 1]$, and let $P(x) = 2x^1x^2$, $Q(x) = (x^1)^2 - (x^2)^2$.
(a) Prove that $\int_{\partial A} P(x)\,dx^1 + Q(x)\,dx^2 = 0$ by direct evaluation of the integral.
(b) Use Green's theorem to prove that the integral in (a) equals zero.
(c) Let $f(x) = (x^1)^2x^2 - (1/3)(x^2)^3$, and show that $P(x) = D_1 f(x)$, $Q(x) = D_2 f(x)$. Then use Theorem 39.5 to prove that the integral in (a) equals zero.

41.3. Let G_2 be a two-dimensional surface in $\mathbb{R}^3$, and let E be an open set which contains $T(G_2)$. If $P^i: E \to \mathbb{R}$, $i = 1, 2, 3$, are functions which have continuous derivatives on E such that $D_{i_1} P^{i_2}(y) = D_{i_2} P^{i_1}(y)$ for (i_1, i_2) in $(2/3)$ and all y in E, prove that

$$\int_{\partial G_2} P^1(y)\,dy^1 + P^2(y)\,dy^2 + P^3(y)\,dy^3 = 0.$$

Question: would you conjecture that there is a function $f: E \to \mathbb{R}$ such that $P^i(y) = D_i f(y)$ for $i = 1, 2, 3$ and y in E?

41.4. Let E be an open set in $\mathbb{R}^4$, and let $f: E \to \mathbb{R}$ be a function which has two continuous derivatives.
(a) Prove that

$$D_{i_1}(D_{i_2} f)(y) - D_{i_2}(D_{i_1} f)(y) = 0, \qquad (i_1, i_2) \in (2/4), \qquad y \in E.$$

(b) If f has two continuous derivatives on an open set E in $\mathbb{R}^n$, state the generalization of the identity in (a) and outline its proof.

41.5. Let E be an open set in $\mathbb{R}^3$, and let the components (f^1, f^2) of $f: E \to \mathbb{R}^2$ have two continuous derivatives.
(a) Prove that

$$D_1[D_{(2,3)}(f^1, f^2)](y) - D_2[D_{(1,3)}(f^1, f^2)](y) + D_3[D_{(1,2)}(f^1, f^2)](y) = 0, \qquad y \in E.$$

(b) If E is an open set in $\mathbb{R}^4$, state and prove the identities which are satisfied by the derivatives $D_{(j_1, j_2)}(f^1, f^2)$, $(j_1, j_2) \in (2/4)$.
(c) Explain why the identities in this exercise are generalizations of those in Exercise 41.4.

42. The Mean-Value Theorem

Let f be a real-valued function which is defined and continuous for $a_0 \leqq x \leqq a_1$ and which has a derivative on the open interval. Then there is a point x^* such that $a_0 < x^* < a_1$ and $f(a_1) - f(a_0) = f'(x^*)(a_1 - a_0)$. This form of the mean-value theorem is a highly satisfactory result: the hypotheses are weak, the proof is elementary, and the applications are numerous and important. Nevertheless, the result is special, and Theorem 9.1 is its only generalization. There is a weaker theorem, however, which can be generalized: if f has a continuous derivative f' on the closed interval, then there is a point x^* in the open interval such that

$$\int_{a_0}^{a_1} f'(x)\,dx = f'(x^*)(a_1 - a_0) = f(a_1) - f(a_0), \qquad a_0 < x^* < a_1. \tag{1}$$

The next theorem is a generalization of this form of the elementary mean-value theorem.

42.1 Theorem (Mean-Value Theorem). *Let* $\mathbf{a} : [a_0, a_1, \cdots, a_n]$ *be an oriented* n*-simplex in* $\mathbb{R}^n$*, and let* $f: \mathbf{a} \to \mathbb{R}^n$ *be a function whose components* $(f^1, \cdots, f^n)$ *have continuous derivatives. Then there exists a point* x^* *on the interior of* $\mathbf{a}$ *such that*

$$\int_{\mathbf{a}} D_{(1,\cdots,n)}(f^1, \cdots, f^n)(x)\,d(x^1, \cdots, x^n) \tag{2}$$

$$= D_{(1,\cdots,n)}(f^1, \cdots, f^n)(x^*)\frac{(-1)^n}{n!}\Delta(\mathbf{a}) \tag{3}$$

$$= \int_{\partial\mathbf{a}} \sum_{(n-1/n)} f^1(x) D_{(j_1,\cdots,j_{n-1})}(f^2, \cdots, f^n)(x)\,d(x^{j_1}, \cdots, x^{j_{n-1}}). \tag{4}$$

PROOF. The integral in (2) equals (3) by the mean-value theorem for integrals in Theorem 35.18, and the integral in (2) equals (4) by the fundamental theorem of the integral calculus in Theorem 38.4. □

42.2 Corollary. *If* $\mathbf{a} : [a_0, a_1, \cdots, a_m]$ *is an oriented simplex in* $\mathbb{R}^m$*; if* $g : \mathbf{a} \to \mathbb{R}^n$*,* $m \leqq n$*, is a function whose components* $(g^1, \cdots, g^n)$ *have continuous derivatives; if* g *defines a surface* G_m *whose trace* $T(G_m)$ *is contained in an open set* E *in* $\mathbb{R}^n$*; and if* $f: E \to \mathbb{R}^m$ *is a function whose components* $(f^1, \cdots, f^m)$ *have continuous derivatives; then there exists a point* x^* *in the interior of* $\mathbf{a}$ *such that*

$$\int_{G_m} \sum_{(m/n)} D_{(j_1,\cdots,j_m)}(f^1, \cdots, f^m)(y)\,d(y^{j_1}, \cdots, y^{j_m}) \tag{5}$$

$$= \sum_{(m/n)} D_{(j_1,\cdots,j_m)}(f^1, \cdots, f^m)[g(x^*)] D_{(1,\cdots,m)}(g^{j_1}, \cdots, g^{j_m})(x^*)\frac{(-1)^m}{m!}\Delta(\mathbf{a}) \tag{6}$$

$$= \int_{\partial G_m} \sum_{(m-1/n)} f^1(y) D_{(j_1,\cdots,j_{m-1})}(f^2, \cdots, f^m)(y)\,d(y^{j_1}, \cdots, y^{j_{m-1}}). \tag{7}$$

PROOF. The integral in (5) equals the integral in (7) by the fundamental theorem of the integral calculus in Theorem 39.1. Also, by Theorem 36.4, the integral in (5) equals

$$\int_{\mathbf{a}} \sum_{(m/n)} D_{(j_1,\cdots,j_m)}(f^1,\cdots,f^m)[g(x)]D_{(1,\cdots,m)}(g^{j_1},\cdots,g^{j_m})(x)\,d(x^1,\cdots,x^m), \tag{8}$$

and this integral equals the sum in (6) by the mean-value theorem for integrals in Theorem 35.18. Stated another way, the entire corollary follows from the application of Theorem 42.1 to the integral in (8). □

42.3 Corollary. *Let* **b** *be an m-dimensional simplex in* $\mathbb{R}^n$, $m \leqq n$; *let E be an open set in* $\mathbb{R}^n$ *which contains* **b**; *and let the components* $(f^1,\cdots,f^m)$ *of* $f: E \to \mathbb{R}^m$ *have continuous derivatives. Then there exists a point* y^* *on the interior of* **b** *such that*

$$\int_{\mathbf{b}} \sum_{(m/n)} D_{(j_1,\cdots,j_m)}(f^1,\cdots,f^m)(y)\,d(y^{j_1},\cdots,y^{j_m}) \tag{9}$$

$$= \sum_{(m/n)} D_{(j_1,\cdots,j_m)}(f^1,\cdots,f^m)(y^*)\frac{(-1)^m}{m!}\Delta(\mathbf{b}^{(j_1,\cdots,j_m)}) \tag{10}$$

$$= \int_{\partial\mathbf{b}} \sum_{(m-1/n)} f^1(y)D_{(j_1,\cdots,j_{m-1})}(f^2,\cdots,f^m)(y)\,d(y^{j_1},\cdots,y^{j_{m-1}}). \tag{11}$$

PROOF. This corollary is the special case of Corollary 42.2 in which the trace $T(G_m)$ of G_m is the simplex **b**. Thus, (9) equals (11) by the fundamental theorem of the integral calculus in Theorem 39.1. The statement that (9) equals (10) is the mean-value theorem for integrals in this case; the details of the proof are outlined in Exercises 37.6 and 37.7. □

EXERCISES

42.1. Prove the following form of Rolle's theorem. Let $\mathbf{a}: [a_0, a_1, \cdots, a_n]$ be an oriented simplex in $\mathbb{R}^n$ such that $\Delta(\mathbf{a}) \neq 0$, and let $f: \mathbf{a} \to \mathbb{R}^n$ be a function whose components $(f^1,\cdots,f^n)$ have continuous derivatives. If

$$\int_{\partial\mathbf{a}} \sum_{(n-1/n)} f^1(x)D_{(j_1,\cdots,j_{n-1})}(f^2,\cdots,f^n)(x)\,d(x^{j_1},\cdots,x^{j_{n-1}}) = 0,$$

then there is a point x^* in the interior of **a** such that $D_{(1,\cdots,n)}(f^1,\cdots,f^n)(x^*) = 0$.

42.2. In Exercise 42.1 assume that f^1 is constant on the boundary $\partial\mathbf{a}$ of **a**. Prove that the integral in Exercise 42.1 equals zero. Prove also that there is a point x^* in the interior of **a** such that $D_i f^1(x^*) = 0$, $i = 1, \cdots, n$, and hence that $D_{(1,\cdots,n)}(f^1,\cdots,f^n)(x^*) = 0$. Thus verify Rolle's theorem in Exercise 42.1 in this special case.

43. An Addition Theorem for Integrals

The following are two familiar statements that the simple Riemann integral is additive with respect to intervals:

$$\int_a^b f(x)\,dx + \int_b^c f(x)\,dx + \int_c^a f(x)\,dx = 0. \tag{1}$$

$$\int_a^c f(x)\,dx = \int_a^b f(x)\,dx + \int_b^c f(x)\,dx. \tag{2}$$

This section establishes some important generalizations of these formulas. To simplify the notation, let $[b_1, \cdots, b_0/b_i, \cdots, b_m]$ denote the simplex $[b_i, \cdots, b_{i-1}, b_0, b_{i+1}, \cdots, b_m]$.

43.1 Theorem. *Let* $\mathbf{b}: [b_0, b_1, \cdots, b_m]$ *be an oriented, degenerate simplex in* $\mathbb{R}^n$, $m - 1 \leq n$; *let* E *be an open set in* $\mathbb{R}^n$ *which contains* $\mathbf{b}$; *and let the components* $(f^1, \cdots, f^m)$ *of* $f: E \to \mathbb{R}^m$ *have continuous derivatives. Then*

$$\int_{\partial \mathbf{b}} \sum_{(m-1/n)} f^1(y) D_{(j_1, \cdots, j_{m-1})}(f^2, \cdots, f^m)(y)\, d(y^{j_1}, \cdots, y^{j_{m-1}}) = 0. \tag{3}$$

If I *denotes the integrand in* (3), *then*

$$\int_{[b_1, \cdots, b_m]} I = \sum_{i=1}^m \int_{[b_1, \cdots, b_0/b_i, \cdots, b_m]} I. \tag{4}$$

Proof. Since $\mathbf{b}$ is a degenerate simplex, it is contained in a d-dimensional plane in $\mathbb{R}^n$, and $d < m$. Then (3) follows from equation (22) in Theorem 39.3. The following calculation begins with the definition of $\partial\mathbf{b}$ in Definition 17.4.

$$\begin{aligned}
&\partial[b_0, b_1, \cdots, b_m] \\
&\quad = [b_1, \cdots, b_m] + \sum_{i=1}^m (-1)^i [b_0, \cdots, b_{i-1}, b_{i+1}, \cdots, b_m] \\
&\quad = [b_1, \cdots, b_m] + \sum_{i=1}^m (-1)^i (-1)^{i-1} [b_1, \cdots, b_{i-1}, b_0, b_{i+1}, \cdots, b_m] \\
&\quad = [b_1, \cdots, b_m] + \sum_{i=1}^m (-1)[b_1, \cdots, b_0/b_i, \cdots, b_m].
\end{aligned} \tag{5}$$

Then (3) can be written as follows [see Definition 40.3]:

$$\int_{[b_1, \cdots, b_m]} I + \sum_{i=1}^m \int_{(-1)[b_1, \cdots, b_0/b_i, \cdots, b_m]} I = 0. \tag{6}$$

But

$$\int_{(-1)[b_1,\cdots,b_0/b_i,\cdots,b_m]} I = -\int_{[b_1,\cdots,b_0/b_i,\cdots,b_m]} I \tag{7}$$

by Definition 35.11, and therefore (6) is equivalent to (4). □

43.2 Example. If $b_0, b_1, \cdots, b_3$ are four points on a plane in $\mathbb{R}^3$, then the formula in (4) in this case is the following:

$$\begin{aligned}\int_{[b_1,b_2,b_3]} &\sum_{(2/3)} f^1(y)D_{(j_1,j_2)}(f^2,f^3)(y)\,d(y^{j_1}, y^{j_2}) \\ &= \left\{\int_{[b_0,b_2,b_3]} + \int_{[b_1,b_0,b_3]} + \int_{[b_1,b_2,b_0]}\right\} \sum_{(2/3)} f^1(y)D_{(j_1,j_2)}(f^2,f^3)(y)\,d(y^{j_1}, y^{j_2}).\end{aligned} \tag{8}$$

43.3 Corollary. *If the simplex* $\mathbf{b}: [b_0, b_1, \cdots, b_m]$ *in Theorem 43.1 is in the plane* $y^m = 0, \cdots, y^n = 0$ *in* $\mathbb{R}^n$, *then the formula in* (3) *is*

$$\int_{\partial \mathbf{b}} f^1(y)D_{(1,\cdots,m-1)}(f^2, \cdots, f^m)(y)\,d(y^1, \cdots, y^{m-1}) = 0. \tag{9}$$

If I denotes the integrand in (9), *then formula* (4) *in this case is*

$$\int_{[b_1,\cdots,b_m]} I = \sum_{i=1}^{m} \int_{[b_1,\cdots,b_0/b_i,\cdots,b_m]} I. \tag{10}$$

Proof. Recall the procedure for the evaluation of a surface integral as explained in Theorem 36.4. If $(y^{j_1}, \cdots, y^{j_{m-1}})$ contains one or more of the coordinates $y^m, \cdots, y^n$ whose value is zero, then the term which contains $d(y^{j_1}, \cdots, y^{j_{m-1}})$ disappears from the integrand. Thus, only a single term remains; it is the one which contains $d(y^1, \cdots, y^{m-1})$. This simplification is explained in detail in a special case in Example 39.4. For example, if the simplex $[b_0, b_1, \cdots, b_3]$ in Example 43.2 is contained in the plane $y^3 = 0$, then the formula in (8) simplifies to

$$\begin{aligned}\int_{[b_1,b_2,b_3]} &f^1(y)D_{(1,2)}(f^2,f^3)(y)\,d(y^1, y^2) \\ &= \left\{\int_{[b_0,b_2,b_3]} + \int_{[b_1,b_0,b_3]} + \int_{[b_1,b_2,b_0]}\right\} f^1(y)D_{(1,2)}(f^2,f^3)(y)\,d(y^1, y^2).\end{aligned} \tag{11}$$

□

43.4 Corollary. *If* $f^2(y) = y^1, \cdots, f^m(y) = y^{m-1}$ *in Corollary 43.3; then formula* (9) *simplifies to*

$$\int_{\partial \mathbf{b}} f^1(y)\,d(y^1, \cdots, y^{m-1}) = 0. \tag{12}$$

If I denotes the integrand in (12), *the formula* (10) *simplifies to*

$$\int_{[b_1,\cdots,b_m]} I = \sum_{i=1}^{m} \int_{[b_1,\cdots,b_0/b_i,\cdots,b_m]} I. \tag{13}$$

PROOF. If $f^2(y) = y^1, \cdots, f^m(y) = y^{m-1}$, then $D_{(1,\cdots,m-1)}(f^2, \cdots, f^m)(y) = 1$, and the formulas in (9) and (10) simplify immediately to those in (12) and (13). For example, if $f^2(y) = y^1, f^3(y) = y^2$ in (11), then this formula simplifies to

$$\begin{aligned}\int_{[b_1,b_2,b_3]} f^1(y)\,d(y^1, y^2) = &\int_{[b_0,b_2,b_3]} f^1(y)\,d(y^1, y^2)\\ &+ \int_{[b_1,b_0,b_3]} f^1(y)\,d(y^1, y^2) \qquad (14)\\ &+ \int_{[b_1,b_2,b_0]} f^1(y)\,d(y^1, y^2).\end{aligned}$$

□

43.5 Corollary. *Let* $\mathbf{b}: [b_0, b_1, \cdots, b_m]$ *be an oriented, degenerate simplex in* $\mathbb{R}^n$, $m - 1 \leqq n$; *let* E *be an open set in* $\mathbb{R}^n$ *which contains* $\mathbf{b}$; *and let the functions* $f^{(j_1,\cdots,j_{m-1})}: E \to \mathbb{R}$, $(j_1, \cdots, j_{m-1}) \in (m - 1/n)$, *have continuous derivatives. Then*

$$\int_{\partial \mathbf{b}} \sum_{(m-1/n)} f^{(j_1,\cdots,j_{m-1})}(y)\,d(y^{j_1}, \cdots, y^{j_{m-1}}) = 0. \tag{15}$$

If I *denotes the integrand in* (15), *then*

$$\int_{[b_1,\cdots,b_m]} I = \sum_{i=1}^{m} \int_{[b_1,\cdots,b_0/b_i,\cdots,b_m]} I. \tag{16}$$

PROOF. In Theorem 43.1, let

$$f^1(y) = f^{(j_1,\cdots,j_{m-1})}(y), \qquad f^2(y) = y^{j_1}, \cdots, f^m(y) = y^{j_{m-1}}. \tag{17}$$

For these functions, equation (3) can be simplified to

$$\int_{\partial \mathbf{b}} f^{(j_1,\cdots,j_{m-1})}(y)\,d(y^{j_1}, \cdots, y^{j_{m-1}}) = 0, \qquad (j_1, \cdots, j_{m-1}) \in (m - 1/n), \tag{18}$$

and the integrand I in (4) becomes the integrand in (18). Add the equations in (18); the result is (15). Add the equations which correspond to (4); the result is (16). □

The formulas in this section can be used in many cases to assist in the evaluation of an integral on a chain. Let $\mathbf{c}$ be the oriented simplicial chain $\Sigma_1^r \mathbf{a}_i$ of simplexes $\mathbf{a}_i: [a_0, a_1, a_2]$ in $\mathbb{R}^2$ [see Definition 40.1]. Then, as shown in Theorem 40.5, the fundamental theorem of the integral calculus can be used to evaluate the integral of a derivative on $\mathbf{c}$ and to show that

$$\int_{\mathbf{c}} D_{(1,2)}(f^1, f^2)(x)\,d(x^1, x^2) = \int_{\partial \mathbf{c}} f^1(x) D_1 f^2(x)\,dx^1 + f^1(x) D_2 f^2(x)\,dx^2. \tag{19}$$

Thus the integral on $\mathbf{c}$ is completely determined by the integral on $\partial\mathbf{c}$. If $\mathbf{c}_1$ and $\mathbf{c}_2$ are two chains with the same boundary $\partial\mathbf{c}$, then

$$\int_{\mathbf{c}_1} D_{(1,2)}(f^1, f^2)(x)\, d(x^1, x^2) = \int_{\mathbf{c}_2} D_{(1,2)}(f^1, f^2)(x)\, d(x^1, x^2). \tag{20}$$

But the fundamental theorem applies only to the integrals of derivatives; thus these methods cannot be applied to integrals of the form $\int_{\mathbf{c}} f(x)\, d(x^1, x^2)$. Nevertheless, the addition formulas in this section can be used to establish similar results, at least in some cases.

Let E be an open set in $\mathbb{R}^2$, and let $f: E \to \mathbb{R}$ be a function which has continuous derivatives. Also, let $\mathbf{c}$ be an oriented simplicial chain of simplexes $\mathbf{a}: [a_0, a_1, a_2]$ in E. Then the integral of f on $\mathbf{c}$ exists by Definition 40.3, and

$$\int_{\mathbf{c}} f(x)\, d(x^1, x^2) = \sum_{\mathbf{a} \in \mathbf{c}} \int_{\mathbf{a}} f(x)\, d(x^1, x^2). \tag{21}$$

Let the boundary $\partial\mathbf{c}$ of $\mathbf{c}$ be the chain $\Sigma\{[b_1, b_2]: [b_1, b_2] \in \partial\mathbf{c}\}$, and assume that there is a point p such that, for each simplex $[b_1, b_2]$ in $\partial\mathbf{c}$, the simplex $[p, b_1, b_2]$ is in E.

43.6 Theorem. *If $\int_{\mathbf{c}} f(x)\, d(x^1, x^2)$ is the integral on $\mathbf{c}$ in E just described, then*

$$\int_{\mathbf{c}} f(x)\, d(x^1, x^2) = \sum_{[b_1, b_2] \in \partial\mathbf{c}} \int_{[p, b_1, b_2]} f(x)\, d(x^1, x^2). \tag{22}$$

If $\mathbf{c}_1$ and $\mathbf{c}_2$ have the same boundary, then

$$\int_{\mathbf{c}_1} f(x)\, d(x^1, x^2) = \int_{\mathbf{c}_2} f(x)\, d(x^1, x^2). \tag{23}$$

Finally, if $\partial\mathbf{c} = 0$, then

$$\int_{\mathbf{c}} f(x)\, d(x^1, x^2) = 0. \tag{24}$$

PROOF. If $\mathbf{a}: [a_0, a_1, a_2]$ is a simplex in $\mathbf{c}$, and if p is a point in E with the properties described in the hypotheses, then Corollary 43.4 [see also equation (14)] and equation (21) show that

$$\int_{\mathbf{c}} f(x)\, d(x^1, x^2) = \sum_{\mathbf{a} \in \mathbf{c}} \left\{ \int_{[p, a_1, a_2]} + \int_{[a_0, p, a_2]} + \int_{[a_0, a_1, p]} \right\} f(x)\, d(x^1, x^2). \tag{25}$$

If two simplexes in $\mathbf{c}$ have a side in common, then the two terms in the sum in (25) which arise from that side cancel [see Definition 40.1(d)], and the only terms which remain after all cancellations have been made are those which appear in the sum in (22). Thus (22) has been established. If $\mathbf{c}_1$ and $\mathbf{c}_2$ have the same boundary $\partial\mathbf{c}$, then the integrals in (23) are each equal to the sum on the right in (22); therefore, they equal each other, and (23) follows. If $\partial\mathbf{c} = 0$, then the sum on the right equals zero and (24) is true. The proof of the entire theorem is complete. □

43.7 Example. Theorem 43.6 shows that the integrals in (23) are equal if $\partial\mathbf{c}_1 = \partial\mathbf{c}_2$, but examples show that the integrals may be equal under much weaker assumptions. Let $b_0, b_1, \cdots, b_{i-1}, b_i, \cdots, b_r$ be a set of points on the line through the vertices a_1, a_2 of the simplex $[a_0, a_1, a_2]$; assume that $b_0 = a_1$ and $b_r = a_2$. Observe that no special order is assumed for the points $b_0, b_1, \cdots, b_r$; also, some of these points may lie outside the side $[a_1, a_2]$ of $[a_0, a_1, a_2]$. Nevertheless,

$$\int_{[a_0,a_1,a_2]} f(x)\, d(x^1, x^2) = \sum_{i=1}^{r} \int_{[a_0,b_{i-1},b_i]} f(x)\, d(x^1, x^2), \tag{26}$$

although the chains $[a_0, a_1, a_2]$ and $\Sigma\{[a_0, b_{i-1}, b_i] : i = 1, \cdots, r\}$ do not have the same boundary. To prove (26), use (14) to introduce a_1 as a new vertex to evaluate each integral on the right. Thus

$$\begin{aligned} &\int_{[a_0,b_{i-1},b_i]} f(x)\, d(x^1, x^2) \\ &\qquad = \left\{\int_{[a_1,b_{i-1},b_i]} + \int_{[a_0,a_1,b_i]} + \int_{[a_0,b_{i-1},a_1]}\right\} f(x)\, d(x^1, x^2). \end{aligned} \tag{27}$$

Since a_1, b_{i-1}, b_i are three points on a line in $\mathbb{R}^2$ by hypothesis, the integral on $[a_1, b_{i-1}, b_i]$ equals zero. Then

$$\begin{aligned} &\sum_{i=1}^{r} \int_{[a_0,b_{i-1},b_i]} f(x)\, d(x^1, x^2) \\ &\qquad = \sum_{i=1}^{r} \left\{\int_{[a_0,a_1,b_i]} f(x)\, d(x^1, x^2) + \int_{[a_0,b_{i-1},a_1]} f(x)\, d(x^1, x^2)\right\} \\ &\qquad = \sum_{i=1}^{r} \left\{\int_{[a_0,a_1,b_i]} f(x)\, d(x^1, x^2) - \int_{[a_0,a_1,b_{i-1}]} f(x)\, d(x^1, x^2)\right\} \\ &\qquad = \int_{[a_0,a_1,b_r]} f(x)\, d(x^1, x^2) - \int_{[a_0,a_1,b_0]} f(x)\, d(x^1, x^2). \end{aligned} \tag{28}$$

Since $b_0 = a_1$ and $b_r = a_2$ by hypothesis, the last equation simplifies to

$$\sum_{i=1}^{r} \int_{[a_0,b_{i-1},b_i]} f(x)\, d(x^1, x^2) = \int_{[a_0,a_1,a_2]} f(x)\, d(x^1, x^2), \tag{29}$$

and the proof of (26) is complete.

43.8 Example. The formulas in (1) and (2) state that the integral $\int_a^b f(x)\,dx$ is additive with respect to intervals, and this section contains formulas which are generalizations of these elementary addition formulas. Also, these formulas can be viewed as generalizations of certain identities for determinants. For example, if $f^1(y) = 2!$, then equation (14) can be simplified to the following statement:

$$\Delta(b_1, b_2, b_3) = \Delta(b_0, b_2, b_3) + \Delta(b_1, b_0, b_3) + \Delta(b_1, b_2, b_0). \tag{30}$$

This equation is a statement of the identity for determinants given in Theorem 20.1.

EXERCISES

43.1. (a) Show that the following chain is an oriented simplicial chain whose boundary is the zero chain [see Definition 40.1].

$$[a_0, a_1, a_2] + [a_0, a_2, a_3] + [a_0, a_3, a_4] + [a_0, a_4, a_1]$$
$$+ [b_0, a_2, a_1] + [b_0, a_3, a_2] + [b_0, a_4, a_3] + [b_0, a_1, a_4].$$

(b) Make a sketch which shows a model in $\mathbb{R}^3$ of the chain in (a).

(c) Make a second sketch which shows a model in $\mathbb{R}^2$ of the chain in (a). Observe that it is possible to sketch models with markedly different appearances both in $\mathbb{R}^2$ and in $\mathbb{R}^3$.

43.2. (a) Explain why f^1 in Theorem 43.1 and in Corollaries 43.3 and 43.4 is assumed to have continuous derivatives.

(b) Prove the following theorem. If a_0 is a point in the interior of the simplex $\mathbf{a}: [a_1, a_2, a_3]$ in $\mathbb{R}^2$, and if $f: \mathbf{a} \to \mathbb{R}$ is a continuous function, then

$$\int_{[a_1,a_2,a_3]} f(x)\,d(x^1, x^2) = \left\{\int_{[a_0,a_2,a_3]} + \int_{[a_1,a_0,a_3]} + \int_{[a_1,a_2,a_0]}\right\} f(x)\,d(x^1, x^2).$$

43.3. Theorem 43.6 contains the hypothesis that there exists a point P such that the simplex $[p, b_1, b_2]$ is in E for each $[b_1, b_2]$ in $\partial\mathbf{c}$. Make a sketch which shows a set E and a chain $\mathbf{c}$ for which no such p exists.

44. Integrals Which Are Independent of the Path

This section treats another consequence of the fundamental theorem of the integral calculus. The surface integral of a derivative equals an integral over the boundary of the surface; this fact suggests that the integrals of a derivative over two surfaces with the same boundary should be equal. A simple example will serve as an introduction to the precise theorem.

Let E be an open set in $\mathbb{R}^3$, and let $f: E \to \mathbb{R}$ be a function which has continuous derivatives $D_i f$, $i = 1, 2, 3$. Also, let $g: [a, b] \to \mathbb{R}^3$ and $h: [a, b] \to \mathbb{R}^3$ be functions which define two curves C_g and C_h, with traces in E, and which have the following properties: (a) the components of g and h have continuous derivatives, and (b) C_g and C_h have the same end points; that is, $g(a) = h(a)$ and $g(b) = h(b)$. Then by the fundamental theorem of the integral calculus,

$$\begin{aligned}\int_{C_g} \sum_{i=1}^{3} D_i f(y)\,dy^i &= f[g(b)] - f[g(a)] \\ &= f[h(b)] - f[h(a)] = \int_{C_h} \sum_{i=1}^{3} D_i f(y)\,dy^i.\end{aligned} \tag{1}$$

The line integral $\int_C \Sigma_1^3 D_i f(y)\,dy^i$ is said to be "independent of the path" because (1) shows that it has the same value on all curves in E which have the same end points. The following theorem states the general theorem which includes this elementary result.

44.1 Theorem. *Let* $\mathbf{a}:[a_0, a_1, \cdots, a_m]$ *be an oriented simplex in* $\mathbb{R}^m$; *let* E *be an open set in* $\mathbb{R}^n$, $m \leqq n$; *and let* $g:\mathbf{a} \to \mathbb{R}^n$ *and* $h:\mathbf{a} \to \mathbb{R}^n$ *be functions whose components* $(g^1, \cdots, g^n)$ *and* $(h^1, \cdots, h^n)$ *have continuous derivatives on* $\mathbf{a}$ *and define surfaces* G_m *and* H_m *whose traces are in* E. *Let the components* $(f^1, \cdots, f^m)$ *of* $f: E \to \mathbb{R}^m$ *have continuous derivatives. If* $g(x) = h(x)$ *for* x *in* $\partial\mathbf{a}$, *then*

$$\int_{G_m} \sum_{(m/n)} D_{(j_1, \cdots, j_m)}(f^1, \cdots, f^m)(y)\,d(y^{j_1}, \cdots, y^{j_m})$$

$$= \int_{H_m} \sum_{(m/n)} D_{(j_1, \cdots, j_m)}(f^1, \cdots, f^m)(y)\,d(y^{j_1}, \cdots, y^{j_m}). \qquad (2)$$

Proof. By the fundamental theorem of the integral calculus in Theorem 39.1, the integrals in (2) are equal respectively to the following integrals.

$$\int_{\partial G_m} \sum_{(m-1/n)} f^1(y) D_{(j_1, \cdots, j_{m-1})}(f^2, \cdots, f^m)(y)\,d(y^{j_1}, \cdots, y^{j_{m-1}})$$

$$\int_{\partial H_m} \sum_{(m-1/n)} f^1(y) D_{(j_1, \cdots, j_{m-1})}(f^2, \cdots, f^m)(y)\,d(y^{j_1}, \cdots, y^{j_{m-1}}). \qquad (3)$$

These two integrals are equal as a result of the hypothesis that $g(x) = h(x)$ for all x in $\partial\mathbf{a}$. Then the integrals in (2) are equal, and the proof is complete. □

Exercises

44.1. Let $\mathbf{a}:[a_0, a_1]$ be an oriented simplex in $\mathbb{R}$, and let $g:\mathbf{a} \to \mathbb{R}^3$ be a function whose components have continuous derivatives and which defines a curve G_1 in $\mathbb{R}^3$. Let E be an open set in $\mathbb{R}^3$ which contains the trace $T(G_1)$ of G_1. Let $f: T(G_1) \to \mathbb{R}^3$ be a function which has continuous components (f^1, f^2, f^3). If f is interpreted as a force which acts along G_1, then the work performed by f on G_1 is given by the following formula [see Example 36.1]:

$$w = \int_{G_1} f^1(y)\,dy^1 + f^2(y)\,dy^2 + f^3(y)\,dy^3.$$

If there exists a function $F: E \to \mathbb{R}$ such that $f^j(y) = D_j F(y)$, $j = 1, 2, 3$, find the work performed by f and show that it is independent of the path from $g(a_0)$ to $g(a_1)$.

44.2. Let $\mathbf{a}:[a_0, a_1, a_2]$ be an oriented simplex in $\mathbb{R}^2$, and let $g:\mathbf{a} \to \mathbb{R}^3$ and $h:\mathbf{a} \to \mathbb{R}^3$ be functions whose components have continuous derivatives and which define surfaces G_2 and H_2 whose traces are in the open set E in $\mathbb{R}^3$. Let $P_j: E \to \mathbb{R}$, $j =$

1, 2, 3, be functions which have continuous derivatives. If $g(x) = h(x)$ for x in $\partial\mathbf{a}$, prove that

$$\int_{G_2} I = \int_{H_2} I, \qquad I = \sum_{(2/3)} [D_{j_1}P_{j_2}(y) - D_{j_2}P_{j_1}(y)]\, d(y^{j_1}, y^{j_2}),$$

and thus show that Stokes' integral [see equation (15) in Section 41] has the same value on all surfaces which have the same boundary. [Hint. Use the fundamental theorem of the integral calculus to show that

$$\int_{G_2} \sum_{(j_1,j_2)} D_{(j_1,j_2)}(P_1, y^1)(y)\, d(y^{j_1}, y^{j_2}) = \int_{\partial G_2} P_1(y)\, dy^1$$
$$= \int_{\partial H_2} P_1(y)\, dy^1 = \int_{H_2} \sum_{(j_1,j_2)} D_{(j_1,j_2)}(P_1, y^1)(y)\, d(y^{j_1}, y^{j_2}).$$

Evaluate the integrals of the derivatives of (P_2, y^2) and (P_3, y^3) in the same way and then add the three equations.]

45. The Area of a Surface

This section defines the area $|G_m|$ of the m-dimensional surface G_m and derives the classical formula for this area.

Let $\mathbf{a}: [a_0, a_1, \cdots, a_m]$ be a non-degenerate, oriented simplex in $\mathbb{R}^m$ such that $(-1)^m\Delta(\mathbf{a}) > 0$, and let the components $(g^1, \cdots, g^n)$ of $g: \mathbf{a} \to \mathbb{R}^n$, $m \leqq n$, have continuous derivatives. The function g defines a surface G_m in $\mathbb{R}^n$. Let $\mathscr{P}_k$, $k = 1, 2, \cdots$, be a sequence of oriented simplicial subdivisions of $\mathbf{a}$ such that $\mathscr{P}_k = \{\mathbf{x}: \mathbf{x} \in \mathscr{P}_k\}$ and $\mathbf{x} = (x_0, x_1, \cdots, x_m)$. Then g maps the vertices of $\mathbf{x}$ into the vertices of a simplex $g(\mathbf{x})$. Let $M[g(\mathbf{x})]$ denote the measure of $g(\mathbf{x})$ [see Theorem 20.5].

45.1 Definition. The *area* $|G_m|$ of the surface G_m is defined by the following equation:

$$|G_m| = \lim_{k\to\infty} \sum_{\mathbf{x}\in\mathscr{P}_k} M[g(\mathbf{x})]. \tag{1}$$

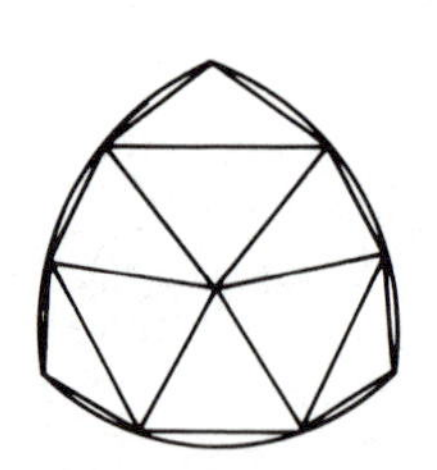

Figure 45.1. An inscribed simplicial surface in G_2.

This definition has a simple geometric interpretation. If $m = 2$ and $n = 3$, then $g(\mathbf{x})$ is a 2-dimensional simplex (triangle) whose vertices are points in an ordinary surface in $\mathbb{R}^3$, and the area of this simplex (triangle) is $M[g(\mathbf{x})]$. The simplexes in the set $\{g(\mathbf{x}) : \mathbf{x} \in \mathscr{P}_k\}$ form a piecewise linear, continuous surface which is inscribed in G_2 [see Figure 45.1], and this simplicial surface approximates the given surface G_2. Also, geometric intuition suggests that

$$\sum_{\mathbf{x} \in \mathscr{P}_k} M[g(\mathbf{x})], \tag{2}$$

which is the sum of the areas of the simplexes in the simplicial surface, should be approximately the area of G_2. Furthermore, the simplicial surface is a better approximation to G_2 when k is large, and it is reasonable to accept (2) as a better approximation to $|G_m|$. Finally, these considerations suggest that the area of G_2 be defined to be the limit of (2) as $k \to \infty$. This interpretation of Definition 45.1 is easily pictured and understood in the case $m = 2$ and $n = 3$, but it is valid for all values of m and n.

45.2 Theorem. *If the components $(g^1, \cdots, g^n)$ of $g : \mathbf{a} \to \mathbb{R}^n$ have continuous derivatives, then*

$$|G_m| = \int_{\mathbf{a}} \left\{ \sum_{(m/n)} [D_{(1,\cdots,m)}(g^{j_1}, \cdots, g^{j_m})(x)]^2 \right\}^{1/2} d(x^1, \cdots, x^m). \tag{3}$$

Proof. The elements in the rows in the following matrix are the coordinates of the vertices of $g(\mathbf{x})$:

$$\begin{bmatrix} g^1(x_0) & g^2(x_0) & \cdots & g^n(x_0) \\ g^1(x_1) & g^2(x_1) & \cdots & g^n(x_1) \\ \cdots & \cdots & \cdots & \cdots \\ g^1(x_m) & g^2(x_m) & \cdots & g^n(x_m) \end{bmatrix}. \tag{4}$$

Form an $(m+1) \times (m+1)$ matrix by bordering the minor of (4) in columns $(j_1, \cdots, j_m)$ with a column of 1's on the right; denote the determinant of this matrix as usual by $\Delta_{(1,\cdots,m)}(g^{j_1}, \cdots, g^{j_m})(\mathbf{x})$. Then by Exercise 20.6,

$$M[g(\mathbf{x})] = \frac{1}{m!} \left\{ \sum_{(m/n)} [\Delta_{(1,\cdots,m)}(g^{j_1}, \cdots, g^{j_m})(\mathbf{x})]^2 \right\}^{1/2}. \tag{5}$$

By Theorem 11.4,

$$\begin{aligned} &\Delta_{(1,\cdots,m)}(g^{j_1}, \cdots, g^{j_m})(\mathbf{x}) \\ &\quad = D_{(1,\cdots,m)}(g^{j_1}, \cdots, g^{j_m})(x_0)\Delta(\mathbf{x}) + r(g^{j_1}, \cdots, g^{j_m}; \mathbf{x})|\Delta(\mathbf{x})|. \end{aligned} \tag{6}$$

Furthermore, the remainder terms $r(g^{j_1}, \cdots, g^{j_m}; \mathbf{x})$ approach zero uniformly with respect to $\mathbf{x}$ as $k \to \infty$ since the diameter of $\mathbf{x}$, $\mathbf{x}$ in $\mathscr{P}_k$, approaches zero uniformly [see Definition 21.5]. Now by (1), (5), and (6),

$$|G_m| = \lim_{k\to\infty} \sum_{\mathbf{x}\in\mathscr{P}_k} \left\{ \sum_{(m/n)} [D_{(1,\cdots,m)}(g^{j_1}, \cdots, g^{j_m})(x_0) + r(g^{j_1}, \cdots, g^{j_m}; \mathbf{x})]^2 \right\}^{1/2} \frac{(-1)^m}{m!} \Delta(\mathbf{x}). \tag{7}$$

Since $g^1, \cdots, g^n$ have continuous derivatives, then the derivatives $D_{(1,\cdots,m)}(g^{j_1}, \cdots, g^{j_m})$ are continuous on $\mathbf{a}$ and, by (42) in Corollary 35.10,

$$\lim_{k\to\infty} \sum_{\mathbf{x}\in\mathscr{P}_k} \left\{ \sum_{(m/n)} [D_{(1,\cdots,m)}(g^{j_1}, \cdots, g^{j_m})(x_0)]^2 \right\}^{1/2} \frac{(-1)^m}{m!} \Delta(\mathbf{x}) = \int_{\mathbf{a}} \left\{ \sum_{(m/n)} [D_{(1,\cdots,m)}(g^{j_1}, \cdots, g^{j_m})(x)]^2 \right\}^{1/2} d(x^1, \cdots, x^m). \tag{8}$$

Thus to complete the proof, it is sufficient to prove that the limits in (7) and (8) are equal. Set

$$A(\mathbf{x}) = \left\{ \sum_{(m/n)} [D_{(1,\cdots,m)}(g^{j_1}, \cdots, g^{j_m})(x_0) + r(g^{j_1}, \cdots, g^{j_m}; \mathbf{x})]^2 \right\}^{1/2};$$

$$B(\mathbf{x}) = \left\{ \sum_{(m/n)} [D_{(1,\cdots,m)}(g^{j_1}, \cdots, g^{j_m})(x_0)]^2 \right\}^{1/2}; \tag{9}$$

$$C(\mathbf{x}) = \left\{ \sum_{(m/n)} [r(g^{j_1}, \cdots, g^{j_m}; x)]^2 \right\}^{1/2}.$$

By the triangle inequality in (4) in Section 91 in Appendix 2,

$$A(\mathbf{x}) \leqq B(\mathbf{x}) + C(\mathbf{x}). \tag{10}$$

Next, write the terms in $B(\mathbf{x})$ as follows:

$$D_{(1,\cdots,m)}(g^{j_1}, \cdots, g^{j_m})(x_0) = \{D_{(1,\cdots,m)}(g^{j_1}, \cdots, g^{j_m})(x_0) + r(g^{j_1}, \cdots, g^{j_m}; \mathbf{x})\} - r(g^{j_1}, \cdots, g^{j_m}; \mathbf{x}),$$
$$(j_1, \cdots, j_m) \in (m/n). \tag{11}$$

Apply the same triangle inequality to the vector on the right in (11) to obtain

$$B(\mathbf{x}) \leqq A(\mathbf{x}) + C(\mathbf{x}). \tag{12}$$

Equations (10) and (12) show that

$$|A(\mathbf{x}) - B(\mathbf{x})| \leqq \left\{ \sum_{(m/n)} [r(g^{j_1}, \cdots, g^{j_m}; \mathbf{x})]^2 \right\}^{1/2}. \tag{13}$$

Theorem 11.4 shows, as stated above, that for each $\varepsilon > 0$ there exists a $k(\varepsilon)$ such that

$$|r(g^{j_1}, \cdots, g^{j_m}; \mathbf{x})| < \varepsilon \tag{14}$$

for each $(j_1, \cdots, j_m)$ in (m/n) and each $\mathbf{x}$ in $\mathscr{P}_k$ provided that $k \geqq k(\varepsilon)$. Then (13) shows that

$$|A(\mathbf{x}) - B(\mathbf{x})| < \binom{n}{m}^{1/2}\varepsilon \tag{15}$$

for each $\mathbf{x}$ in $\mathscr{P}_k$ provided $k \geqq k(\varepsilon)$. Next, by (15),

$$\begin{aligned}\left|\sum_{\mathbf{x}\in\mathscr{P}_k} A(\mathbf{x})\frac{(-1)^m}{m!}\Delta(\mathbf{x}) - \sum_{\mathbf{x}\in\mathscr{P}_k} B(\mathbf{x})\frac{(-1)^m}{m!}\Delta(\mathbf{x})\right| \\ \leqq \sum_{\mathbf{x}\in\mathscr{P}_k} |A(\mathbf{x}) - B(\mathbf{x})|\frac{(-1)^m}{m!}\Delta(\mathbf{x}) \\ < \binom{n}{m}^{1/2}\varepsilon \sum_{\mathbf{x}\in\mathscr{P}_k} \frac{(-1)^m}{m!}\Delta(\mathbf{x}), \qquad k \geqq k(\varepsilon).\end{aligned} \tag{16}$$

Now since $\mathscr{P}_k$ is a simplicial subdivision of $\mathbf{a}$, Theorem 21.3 shows that

$$\sum_{\mathbf{x}\in\mathscr{P}_k} \frac{(-1)^m}{m!}\Delta(\mathbf{x}) = M(\mathbf{a}). \tag{17}$$

Then (16) and (17) show that

$$\left|\sum_{\mathbf{x}\in\mathscr{P}_k} A(\mathbf{x})\frac{(-1)^m}{m!}\Delta(\mathbf{x}) - \sum_{\mathbf{x}\in\mathscr{P}_k} B(\mathbf{x})\frac{(-1)^m}{m!}\Delta(\mathbf{x})\right| < M(\mathbf{a})\binom{n}{m}^{1/2}\varepsilon \tag{18}$$

for $k \geqq k(\varepsilon)$. Let $I(G_m)$ denote the integral in (3). Then

$$\begin{aligned}\left|\sum_{\mathbf{x}\in\mathscr{P}_k} A(\mathbf{x})\frac{(-1)^m}{m!}\Delta(\mathbf{x}) - I(G_m)\right| \\ \leqq \left|\sum_{\mathbf{x}\in\mathscr{P}_k} A(\mathbf{x})\frac{(-1)^m}{m!}\Delta(\mathbf{x}) - \sum_{\mathbf{x}\in\mathscr{P}_k} B(\mathbf{x})\frac{(-1)^m}{m!}\Delta(\mathbf{x})\right| \\ + \left|\sum_{\mathbf{x}\in\mathscr{P}_k} B(\mathbf{x})\frac{(-1)^m}{m!}\Delta(\mathbf{x}) - I(G_m)\right|.\end{aligned} \tag{19}$$

By (18), the first term on the right approaches zero as $k \to \infty$, and by (8) the second term approaches zero as $k \to \infty$. Therefore,

$$\lim_{k\to\infty} \sum_{\mathbf{x}\in\mathscr{P}_k} A(\mathbf{x})\frac{(-1)^m}{m!}\Delta(\mathbf{x}) = I(G_m). \tag{20}$$

But the left side of this equation is $|G_m|$ by (9) and (7), and $I(G_m)$ is the integral in (3). Thus (20) is (3), and the proof of the theorem is complete. □

Exercises

45.1. Let $\mathbf{b}: [b_0, b_1, \cdots, b_m]$ be a non-degenerate simplex in $\mathbb{R}^n$, $m \leqq n$, and let $\mathbf{a}: [a_0, a_1, \cdots, a_m]$ be a simplex in $\mathbb{R}^m$ such that $(-1)^m\Delta(\mathbf{a}) > 0$. Let $g: \mathbf{a} \to \mathbf{b}$ be an affine transformation such that $g(a_i) = b_i$, $i = 0, 1. \cdots, m$. Then g defines a surface G_m which has an area $|G_m|$. Prove that $|G_m| = M(\mathbf{b})$ and that $|G_m| > 0$. [Hint. Theorem 21.4 and Section 37.]

45.2. If $\mathbf{b}$ in Exercise 45.1 is a degenerate simplex, prove that $|G_m| = 0$.

45.3. Let G_2 be the surface of a cylinder of radius a and altitude h.
(a) Show that the equations of G_2 in cylindrical coordinates (θ, z) are

$$\begin{aligned} y^1 &= g^1(\theta, z) = a\cos\theta, \\ y^2 &= g^2(\theta, z) = a\sin\theta, \qquad 0 \leqq \theta \leqq 2\pi, \qquad 0 \leqq z \leqq h, \\ y^3 &= g^3(\theta, z) = z. \end{aligned}$$

(b) Use Theorem 45.2 to show that $|G_2| = 2\pi a h$.

45.4. Let G_2 be the surface of a sphere with center at the origin and radius a.
(a) Show that the equations of G_2 in spherical coordinates (θ, φ) are

$$\begin{aligned} y^1 &= g^1(\theta, \varphi) = a\cos\theta\sin\varphi, \\ y^2 &= g^2(\theta, \varphi) = a\sin\theta\sin\varphi, \qquad 0 \leqq \theta \leqq 2\pi, \qquad 0 \leqq \varphi \leqq \pi, \\ y^3 &= g^3(\theta, \varphi) = a\cos\varphi. \end{aligned}$$

(b) Use Theorem 45.2 to show that $|G_2| = 4\pi a^2$.

45.5. Let G_3 be the surface in $\mathbb{R}^4$ whose equations are

$$\begin{aligned} y^1 &= g^1(\theta_1, \theta_2, \theta_3) = a\cos\theta_1\sin\theta_2\sin\theta_3, \\ y^2 &= g^2(\theta_1, \theta_2, \theta_3) = a\sin\theta_1\sin\theta_2\sin\theta_3, \qquad 0 \leqq \theta_1 \leqq 2\pi, \\ y^3 &= g^3(\theta_1, \theta_2, \theta_3) = a\cos\theta_2\sin\theta_3, \qquad 0 \leqq \theta_2 \leqq \pi, \\ y^4 &= g^4(\theta_1, \theta_2, \theta_3) = a\cos\theta_3. \qquad 0 \leqq \theta_3 \leqq \pi. \end{aligned}$$

(a) Show that G_3 is a 3-dimensional sphere in $\mathbb{R}^4$ whose radius is a.
(b) Use Theorem 45.2 to show that $|G_3| = 2\pi^2 a^3$.

45.6. The formulas for the area of a surface take an especially simple form when the surface G_m lies in $\mathbb{R}^m$. The situation is illustrated by familiar formulas for polar and spherical coordinates. Set $A = \{(r, \theta) : 0 \leqq r \leqq a, 0 \leqq \theta \leqq 2\pi\}$. Then the equations

$$\begin{aligned} y^1 &= g^1(r, \theta) = r\cos\theta, \\ y^2 &= g^2(r, \theta) = r\sin\theta, \end{aligned}$$

map A into the circle in $\mathbb{R}^2$ with center at the origin and radius a. The circle is a surface G_2 in $\mathbb{R}^2$.
(a) Show that the area of the circle is $\int_{G_2} d(y^1, y^2)$.
(b) Prove that $D_{(1,2)}(g^1, g^2)(r, \theta) = r$ and that

$$|G_2| = \int_{G_2} d(y^1, y^2) = \int_A r\, d(r, \theta) = \int_0^{2\pi} \left(\int_0^a r\, dr \right) d\theta = \pi a^2.$$

45.7. Set $A = \{(r, \theta, \varphi) : 0 \leqq r \leqq a, 0 \leqq \theta \leqq 2\pi, 0 \leqq \varphi \leqq \pi\}$. The equations

$$\begin{aligned} y^1 &= g^1(r, \theta, \varphi) = r\cos\theta\sin\varphi, \\ y^2 &= g^2(r, \theta, \varphi) = r\sin\theta\sin\varphi, \\ y^3 &= g^3(r, \theta, \varphi) = r\cos\varphi, \end{aligned}$$

map A into the solid sphere in $\mathbb{R}^3$ with center at the origin and radius a. The sphere is a surface G_3 in $\mathbb{R}^3$.

(a) Prove that $D_{(1,2,3)}(g^1, g^2, g^3)(r, \theta, \varphi) = -r^2 \sin \varphi$.

(b) Show that the area (actually volume in this case) of the sphere is $-\int_{G_3} d(y^1, y^2, y^3)$.

(c) Prove that

$$|G_3| = \left|\int_{G_3} d(y^1, y^2, y^3)\right| = \int_A r^2 \sin \varphi \, d(r, \theta, \varphi)$$
$$= \int_0^a r^2 \, dr \int_0^{2\pi} d\theta \int_0^{\pi} \sin \varphi \, d\varphi = \frac{4\pi a^3}{3}.$$

45.8. Let G_2 be the sphere in Exercise 45.4.

(a) If (y^1, y^2, y^3) is a point on G_2, show that $(y^1/a, y^2/a, y^3/a)$ is a unit vector which is normal to the sphere and is directed outward.

(b) Show that the vectors with the following components are tangent to the sphere:

$$D_1g^1(\theta, \varphi), \qquad D_1g^2(\theta, \varphi), \qquad D_1g^3(\theta, \varphi);$$
$$D_2g^1(\theta, \varphi), \qquad D_2g^2(\theta, \varphi), \qquad D_2g^3(\theta, \varphi).$$

(c) Show that the vector with the following components is normal to the sphere and is directed outward:

$$-D_{(1,2)}(g^2, g^3)(\theta, \varphi), \qquad D_{(1,2)}(g^1, g^3)(\theta, \varphi), \qquad -D_{(1,2)}(g^1, g^2)(\theta, \varphi).$$

(d) Show that the value of the following integral is $4\pi a^2$, which is the area $|G_2|$ of the sphere [see Exercise 45.4(b)]:

$$\int_{G_2} -\frac{y^3}{a} d(y^1, y^2) + \frac{y^2}{a} d(y^1, y^3) - \frac{y^1}{a} d(y^2, y^3).$$

(e) Let $A = \{(\theta, \varphi) : 0 \leqq \theta \leqq 2\pi, 0 \leqq \varphi \leqq \pi\}$. Show that the integral in (d) equals

$$\int_A \left\{\sum_{(2/3)} [D_{(1,2)}(g^{j_1}, g^{j_2})(\theta, \varphi)]^2\right\}^{1/2} d(\theta, \varphi),$$

and thus explain why the integral in (d) equals $|G_2|$. [Hint. Theorem 84.2 (7) in Section 84 in Appendix 1.]

46. Integrals of Uniformly Convergent Sequences of Functions

This section establishes an inequality for the absolute value of a surface integral and then uses this inequality to investigate the integrals of a uniformly convergent sequence of continuous functions.

Let $\mathbf{a} : [a_0, a_1, \cdots, a_m]$ be a non-degenerate, oriented simplex in $\mathbb{R}^m$ such that $(-1)^m\Delta(\mathbf{a}) > 0$, and let the components $(g^1, \cdots, g^n)$ of $g : \mathbf{a} \to \mathbb{R}^n$, $m \leqq n$, have continuous derivatives. The function g defines a surface G_m whose trace $T(G_m)$ is a compact set in $\mathbb{R}^n$. Let $f: T(G_m) \to \mathbb{R}^{\binom{n}{m}}$ be a function which has continuous components

$$f^{(j_1,\cdots,j_m)}: T(G_m) \to \mathbb{R}, \qquad (j_1, \cdots, j_m) \quad \text{in} \quad (m/n). \tag{1}$$

Then the norm $|f(y)|$ of $f(y)$ is defined as follows:

$$|f(y)| = \left\{ \sum_{(m/n)} [f^{(j_1,\cdots,j_m)}(y)]^2 \right\}^{1/2}. \tag{2}$$

Since the components of f in (1) are continuous on the compact set $T(G_m)$, there exists a constant M such that [see Theorems 96.12 and 96.14]

$$|f(y)| \leqq M, \qquad y \in T(G_m). \tag{3}$$

46.1 Theorem. *If G_m is defined by the function $g: \mathbf{a} \to \mathbb{R}^n$ whose components have continuous derivatives; if the components of $f: T(G_m) \to \mathbb{R}^{\binom{n}{m}}$ are continuous; and if M is a bound of f as in (3), then*

$$\left| \int_{G_m} \sum_{(m/n)} f^{(j_1,\cdots,j_m)}(y)\, d(y^{j_1}, \cdots, y^{j_m}) \right| \leqq M|G_m|. \tag{4}$$

PROOF. Since f is continuous, the integral in (4) exists by Theorem 36.4, and by this theorem and Corollary 35.10 its value is

$$\lim_{k\to\infty} \sum_{\mathbf{x}\in\mathscr{P}_k} \sum_{(m/n)} f^{(j_1,\cdots,j_m)}[g(x_0)] \frac{(-1)^m}{m!} \Delta_{(1,\cdots,m)}(g^{j_1}, \cdots, g^{j_m})(\mathbf{x}). \tag{5}$$

By Schwarz's inequality [see Corollary 86.2], equations (2) and (3) above, and the definition of $|G_m|$ in Definition 45.1,

$$\begin{aligned}
&\left| \sum_{\mathbf{x}\in\mathscr{P}_k} \sum_{(m/n)} f^{(j_1,\cdots,j_m)}[g(x_0)] \frac{(-1)^m}{m!} \Delta_{(1,\cdots,m)}(g^{j_1}, \cdots, g^{j_m})(\mathbf{x}) \right| \\
&\qquad \leqq \sum_{\mathbf{x}\in\mathscr{P}_k} \left| \sum_{(m/n)} f^{(j_1,\cdots,j_m)}[g(x_0)] \frac{(-1)^m}{m!} \Delta_{(1,\cdots,m)}(g^{j_1}, \cdots, g^{j_m})(\mathbf{x}) \right| \\
&\qquad \leqq \sum_{\mathbf{x}\in\mathscr{P}_k} \left\{ \sum_{(m/n)} [f^{(j_1,\cdots,j_m)}[g(x_0)]]^2 \right\}^{1/2} \\
&\qquad \times \left\{ \sum_{(m/n)} \left[\frac{(-1)^m}{m!} \Delta_{(1,\cdots,m)}(g^{j_1}, \cdots, g^{j_m})(\mathbf{x}) \right]^2 \right\}^{1/2} \\
&\qquad \leqq M \sum_{\mathbf{x}\in\mathscr{P}_k} \left\{ \sum_{(m/n)} \left[\frac{(-1)^m}{m!} \Delta_{(1,\cdots,m)}(g^{j_1}, \cdots, g^{j_m})(\mathbf{x}) \right]^2 \right\}^{1/2}.
\end{aligned}$$

Take the limits as $k \to \infty$ of the first and last terms in this series of inequalities; the limits exist as integrals, and the result is the inequality in (4). The proof is complete. □

46.2 Definition. The sequence $f_k: T(G_m) \to \mathbb{R}^{\binom{n}{m}}$, $k = 1, 2, \cdots$, of continuous functions *converges uniformly* to the continuous function $f: T(G_m) \to \mathbb{R}^{\binom{n}{m}}$ if and only if to each $\varepsilon > 0$ there corresponds a $k(\varepsilon)$, which depends only on ε, such that

$$|f_k(y) - f(y)| < \varepsilon, \qquad y \in T(G_m), \qquad k \geqq k(\varepsilon). \tag{6}$$

46.3 Theorem. *If the sequence $f_k: T(G_m) \to \mathbb{R}^{\binom{n}{m}}$, $k = 1, 2, \cdots$, of continuous functions converges uniformly to the continuous function $f: T(G_m) \to \mathbb{R}^{\binom{n}{m}}$, then*

$$\lim_{k\to\infty} \int_{G_m} \sum_{(m/n)} f_k^{(j_1,\cdots,j_m)}(y)\, d(y^{j_1}, \cdots, y^{j_m}) = \int_{G_m} \sum_{(m/n)} f^{(j_1,\cdots,j_m)}(y)\, d(y^{j_1}, \cdots, y^{j_m}). \tag{7}$$

PROOF. The absolute value of the difference of the two integrals in (7) is

$$\left| \int_{G_m} \sum_{(m/n)} [f_k^{(j_1,\cdots,j_m)}(y) - f^{(j_1,\cdots,j_m)}(y)]\, d(y^{j_1}, \cdots, y^{j_m}) \right|. \tag{8}$$

If $k \geqq k(\varepsilon)$, then (2), (6), and (4) show that (8) is less than $\varepsilon|G_m|$. The proof of (7) and of Theorem 46.3 follows from these facts. □

EXERCISE

46.1. Let $b_0: (-1, 1)$ and $b_1: (\pi/2, \pi/2)$ be two points in $\mathbb{R}^2$, and let $\mathbf{b}: [b_0, b_1]$ be the oriented simplex with vertices b_0 and b_1.

(a) Let $f: \mathbf{b} \to \mathbb{R}^2$ be the function which has the following components:

$$f^1(y) = \sin(y^1 + y^2), \qquad f^2(y) = \cos(y^1 + y^2).$$

Show that $|f(y)| = 1$ for y in $\mathbf{b}$. Show also that the absolute value of the integral

$$\int_{\mathbf{b}} \sin(y^1 + y^2)\,dy^1 + \cos(y^1 + y^2)\,dy^2$$

is equal to or less than $[(\pi^2/2) + 2]^{1/2}$.

(b) Show that the affine transformation

$$y^1 = [(\pi/2) + 1]x - 1,$$
$$y^2 = [(\pi/2) - 1]x + 1,$$

maps the simplex $[e_0, e_1]$ in $\mathbb{R}$ into $[b_0, b_1]$. Use this transformation to evaluate the integral in (a) and show that its value is $[1 + (2/\pi)]$.

(c) If the components (f_k^1, f_k^2) of $f_k: \mathbf{b} \to \mathbb{R}^2$ are

$$f_k^1(y) = \frac{\sin(y^1 + y^2)}{k}, \qquad f_k^2(y) = \frac{\cos(y^1 + y^2)}{k},$$

show that $|f_k(y)| \leqq 1/k$ for y in $\mathbf{b}$, and hence that the sequence f_k, $k = 1, 2, \cdots$, converges uniformly to the function whose components are the zero function.

(d) Show that

$$\int_b \frac{\sin(y^1 + y^2)}{k}\,dy^1 + \frac{\cos(y^1 + y^2)}{k}\,dy^2 = \frac{1 + (2/\pi)}{k}.$$

Verify Theorem 46.3 for the integrals of the sequence f_k, $k = 1, 2, \cdots$.

CHAPTER 7

Zero Integrals, Equal Integrals, and the Transformation of Integrals

47. Introduction

This chapter contains a further investigation of the properties of surface integrals. The most general surface integral was defined in Section 36; it is an integral of the form

$$\int_{G_m} \sum_{(m/n)} f^{(j_1, \cdots, j_m)}(y)\, d(y^{j_1}, \cdots, y^{j_m}), \tag{1}$$

which, by Theorem 36.4, is evaluated by the integral

$$\int_{\mathbf{a}} \sum_{(m/n)} f^{(j_1, \cdots, j_m)}[g(x)] D_{(1, \cdots, m)}(g^{j_1}, \cdots, g^{j_m})(x)\, d(x^{j_1}, \cdots, x^{j_m}). \tag{2}$$

These formulas emphasize that the value of the integral depends both on the integrand f and also on the surface G_m. The integral clearly equals zero if

$$f^{(j_1, \cdots, j_m)}[g(x)] = 0, \qquad x \in \mathbf{a}, \qquad (j_1, \cdots, j_m) \in (m/n), \tag{3}$$

but it may equal zero in other cases also. This chapter investigates the conditions under which a surface integral equals zero and also the conditions under which two surface integrals are equal. The change-of-variable theorem is a theorem about surface integrals, and this chapter establishes a very general form of this theorem.

If the integrand of the surface integral is a derivative, then the integrals in (1) and (2) are these:

$$\int_{G_m} \sum_{(m/n)} D_{(j_1, \cdots, j_m)}(f^1, \cdots, f^m)(y)\, d(y^{j_1}, \cdots, y^{j_m}), \tag{4}$$

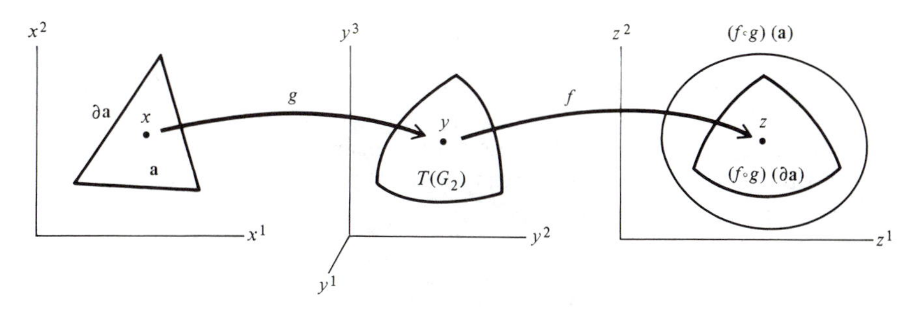

Figure 47.1. The mappings by g and f.

$$\int_{\mathbf{a}} \sum_{(m/n)} D_{(j_1,\cdots,j_m)}(f^1,\cdots,f^m)[g(x)]D_{(1,\cdots,m)}(g^{j_1},\cdots,g^{j_m})(x)\,d(x^1,\cdots,x^m). \tag{5}$$

The following example adds insight into some of the problems which arise in connection with the surface integrals of derivatives.

47.1 Example. Let $\mathbf{a}:[a_0, a_1, a_2]$ be a positively oriented Euclidean simplex in $\mathbb{R}^2$, and let $g:\mathbf{a}\to\mathbb{R}^3$ be a function whose components (g^1, g^2, g^3) have continuous derivatives on $\mathbf{a}$. The equations

$$y^j = g^j(x), \qquad j = 1, 2, 3, \qquad x \in \mathbf{a}, \qquad x = (x^1, x^2), \tag{6}$$

define a 2-dimensional surface G_2 in $\mathbb{R}^3$ whose trace is denoted by $T(G_2)$. Let E be an open set in $\mathbb{R}^3$ which contains $T(G_2)$, and let $f: E\to\mathbb{R}^2$ be a function whose components (f^1, f^2) have continuous derivatives. Then g maps $\mathbf{a}$ into $T(G_2)$ and f maps $T(G_2)$ into a set $(f\circ g)$ $(\mathbf{a})$ in the (z^1, z^2)-plane as follows:

$$z^j = f^j(y), \qquad j = 1, 2, \qquad y \in T(G_2). \tag{7}$$

Furthermore, the composite function $f\circ g$ maps $\mathbf{a}$ into $(f\circ g)(\mathbf{a})$ as shown in Figure 47.1. Also, $f\circ g$ maps the boundary $\partial\mathbf{a}$ of $\mathbf{a}$ into a curve $(f\circ g)(\partial\mathbf{a})$ in the (z^1, z^2)-plane. In reality, the function $f\circ g:\mathbf{a}\to\mathbb{R}^2$ defines a 2-dimensional surface whose trace lies in the (z^1, z^2)-plane. The boundary of this surface is the curve $(f\circ g)(\partial\mathbf{a})$, but this curve may or may not be the boundary of the set $(f\circ g)(\mathbf{a})$, which is the trace of the surface.

Let $\mathscr{P}_k$, $k = 1, 2, \cdots$, be a sequence of simplicial subdivisions of $\mathbf{a}$. Thus if $\mathbf{x}\in\mathscr{P}_k$, then $\mathbf{x}$ is a positively oriented simplex $[x_0, x_1, x_2]$. Also, $(f\circ g)(\mathbf{x})$ is the (abstract) simplex $((f\circ g)(x_0), (f\circ g)(x_1), (f\circ g)(x_2))$, and

$$\frac{(-1)^2}{2!}\Delta_{(1,2)}(f\circ g)(\mathbf{x}) = \frac{(-1)^2}{2!}\begin{vmatrix} (f^1\circ g)(x_0) & (f^2\circ g)(x_0) & 1 \\ \cdots & \cdots & \cdots \\ (f^1\circ g)(x_2) & (f^2\circ g)(x_2) & 1 \end{vmatrix}. \tag{8}$$

The expression on the right in (8) is the signed area of the simplex whose vertices are $(f\circ g)(x_0)$, $(f\circ g)(x_1)$, $(f\circ g)(x_2)$. The boundary of the chain

$$\sum\{(f\circ g)(\mathbf{x}) : \mathbf{x}\in\mathscr{P}_k\} \tag{9}$$

is a one-dimensional chain whose vertices are points on the curve $(f\circ g)(\partial\mathbf{a})$. Some of the simplexes $(f\circ g)(\mathbf{x})$ may be positively oriented in the (z^1, z^2)-plane and others negatively oriented; some may be inside the curve $(f\circ g)(\partial\mathbf{a})$ and others outside (of course, the curve $(f\circ g)(\partial\mathbf{a})$ need not be a simple closed curve). In all cases, however,

$$\sum_{\mathbf{x}\in\mathscr{P}_k}\frac{(-1)^2}{2!}\Delta_{(1,2)}(f\circ g)(\mathbf{x}) \tag{10}$$

is approximately what is called the "area of the curve $(f\circ g)(\partial\mathbf{a})$". Furthermore, the limit of the sum in (10) is by definition the "area of the curve $(f\circ g)(\partial\mathbf{a})$". Also, by Lemma 38.5,

$$\lim_{k\to\infty}\sum_{\mathbf{x}\in\mathscr{P}_k}\frac{(-1)^2}{2!}\Delta_{(1,2)}(f\circ g)(\mathbf{x}) = \int_{\mathbf{a}} D_{(1,2)}(f^1\circ g, f^2\circ g)(x)\,d(x^1, x^2), \tag{11}$$

and by Theorem 4.5 [see also Example 4.3] this integral equals

$$\begin{aligned}\int_{\mathbf{a}}\{&D_{(1,2)}(f^1,f^2)[g(x)]D_{(1,2)}(g^1, g^2)(x)\\ &+ D_{(1,3)}(f^1,f^2)[g(x)]D_{(1,2)}(g^1, g^3)(x)\\ &+ D_{(2,3)}(f^1,f^2)[g(x)]D_{(1,2)}(g^2, g^3)(x)\}\,d(x^1, x^2).\end{aligned} \tag{12}$$

Thus the limit in (11) is recognized as the surface integral in (12), and this example provides a geometric interpretation for the surface integral of a derivative: it is the area of the curve $(f\circ g)(\partial\mathbf{a})$. There is a similar geometric interpretation for each surface integral of a derivative.

Exercises

47.1. Let $\mathbf{a}$ be the oriented simplex $[a_0, a_1, a_2]$ in $\mathbb{R}^2$ whose vertices are $a_0 : (1, 2)$, $a_1 : (5, 4)$, and $a_2 : (3, 8)$. Let $g : \mathbf{a}\to\mathbb{R}^3$ be the function with components (g^1, g^2, g^3) such that

$$g^1(x) = x^1 + x^2 + 1,$$
$$g^2(x) = x^1 - x^2 + 2,$$
$$g^3(x) = 2x^1 + 3x^2 - 1.$$

(a) Show that g maps $\mathbf{a} : [a_0, a_1, a_2]$ into the simplex $\mathbf{b} : [b_0, b_1, b_2]$ in $\mathbb{R}^3$ whose vertices are $b_0 : (4, 1, 7)$, $b_1 : (10, 3, 21)$, and $b_2 : (12, -3, 29)$. Thus, $T(G_2) = [b_0, b_1, b_2]$.

(b) Let E be an open set in $\mathbb{R}^3$ which contains $\mathbf{b}$, and let $f : E\to\mathbb{R}^2$ be the function with components (f^1, f^2) such that

$$f^1(y) = y^1 + y^2 + y^3,$$
$$f^2(y) = y^1 - y^2 - y^3, \qquad y = (y^1, y^2, y^3).$$

Show that f maps the simplex $\mathbf{b}$ into the simplex $\mathbf{c}:[c_0, c_1, c_2]$ in $\mathbb{R}^2$ whose vertices are $c_0:(12, -4)$, $c_1:(34, -14)$, and $c_2:(38, -14)$. Show also that $f\circ g$ has components $(f^1\circ g, f^2\circ g)$ such that

$$(f^1\circ g)(x) = 4x^1 + 3x^2 + 2,$$
$$(f^2\circ g)(x) = -2x^1 - x^2.$$

Verify that $f\circ g$ maps $\mathbf{a}$ into $\mathbf{c}$, and show that this mapping is one-to-one.

(c) Show that the integral (5) for these functions is

$$\int_{\mathbf{a}} [(-2)(-2) + (-2)(1) + (0)(5)]\,d(x^1, x^2) = \int_{\mathbf{a}} 2d(x^1, x^2) = 20.$$

(d) Show that the area of the simplex $\mathbf{c}:[c_0, c_1, c_2]$ is 20 and thus verify the statement in Example 47.1 that the surface integral of a derivative in (4) is equal to the "area of the curve $(f\circ g)(\partial\mathbf{a})$".

47.2. Let $\mathbf{e}$ be the oriented simplex $[e_0, e_1, e_2]$ in $\mathbb{R}^2$ whose vertices are $e_0:(0, 0)$, $e_1:(1, 0)$, and $e_2:(0, 1)$. Let $g:\mathbf{e}\to\mathbb{R}^3$ be the function with components (g^1, g^2, g^3) such that

$$g^1(x) = (\pi/4)x^1,$$
$$g^2(x) = (\pi/4)x^2, \qquad x\in\mathbf{e},$$
$$g^3(x) = -(\pi/2)x^1 - (\pi/2)x^2 + (\pi/2).$$

(a) Show that g maps $\mathbf{e}:[e_0, e_1, e_2]$ into the simplex $\mathbf{b}:[b_0, b_1, b_2]$ in $\mathbb{R}^3$ whose vertices are $b_0:(0, 0, \pi/2)$, $b_1:(\pi/4, 0, 0)$, and $b_2:(0, \pi/4, 0)$. Thus $T(G_2) = [b_0, b_1, b_2]$.

(b) Let E be an open set in $\mathbb{R}^3$ which contains $\mathbf{b}$, and let $f: E\to\mathbb{R}^2$ be the function with components (f^1, f^2) such that

$$f^1(y) = \sin(y^1 + y^2 + y^3),$$
$$f^2(y) = \cos(y^1 + y^2 + y^3), \qquad y = (y^1, y^2, y^3).$$

Show that f maps each point in $\mathbf{b}$ into a point on the unit circle with center at the origin, and that $(f\circ g)(\mathbf{e})$ is the arc of this circle in the first quadrant which is bounded by the points $(1, 0)$ and $(\sqrt{2}/2, \sqrt{2}/2)$. Show also that $f\circ g$ has components $(f^1\circ g, f^2\circ g)$ such that

$$(f^1\circ g)(x) = \sin[(\pi/2) - (\pi/4)x^1 - (\pi/4)x^2],$$
$$(f^2\circ g)(x) = \cos[(\pi/2) - (\pi/4)x^1 - (\pi/4)x^2], \qquad x\in\mathbf{e}, \qquad x = (x^1, x^2).$$

(c) Show that

$$D_{(j_1, j_2)}(f^1, f^2)(y) = 0, \qquad y\in E, \qquad (j_1, j_2)\in(2/3).$$

Thus show that

$$\int_{\mathbf{e}} \sum_{(2/3)} D_{(j_1, j_2)}(f^1, f^2)[g(x)]D_{(1,2)}(g^{j_1}, g^{j_2})(x)\,d(x^1, x^2) = 0.$$

(d) The function f maps $\mathbf{b}$ into a set on the circle $(z^1)^2 + (z^2)^2 - 1 = 0$. Explain why this fact suggests that the integral in (c) has the value zero.

48. Some Integrals Which Have the Value Zero

This section explains several methods which can be used to prove that certain integrals have the value zero. The results are used to prove an important addition theorem for the integrals of continuous functions.

Let $\mathbf{a}$ be an m-dimensional Euclidean simplex $[a_0, a_1, \cdots, a_m]$ in $\mathbb{R}^m$; then $\Delta(\mathbf{a}) \neq 0$. Let $g : \mathbf{a} \to \mathbb{R}^n$, $m \leqq n$, be a function with components $(g^1, \cdots, g^m)$ which have continuous derivatives on $\mathbf{a}$; then g defines a surface G_m with trace $T(G_m)$. Also, let E be an open set in $\mathbb{R}^n$ which contains $T(G_m)$. Finally, let $f : E \to \mathbb{R}^m$ be a function whose components $(f^1, \cdots, f^m)$ have continuous derivatives.

48.1 Theorem. *If* $g : \mathbf{a} \to \mathbb{R}^n$ *and* $f : E \to \mathbb{R}^m$ *are the functions just described, and if*

$$\sum_{(m/n)} D_{(j_1, \cdots, j_m)}(f^1, \cdots, f^m)[g(x)] D_{(1, \cdots, m)}(g^{j_1}, \cdots, g^{j_m})(x) = 0 \tag{1}$$

for all x *in* $\mathbf{a}$, *then*

$$\int_{\partial G_m} \sum_{(m-1/n)} f^1(y) D_{(j_1, \cdots, j_{m-1})}(f^2, \cdots, f^m)(y)\, d(y^{j_1}, \cdots, y^{j_{m-1}}) = 0. \tag{2}$$

Proof. By the fundamental theorem of the integral calculus in Theorem 39.1,

$$\begin{aligned} &\int_{G_m} \sum_{(m/n)} D_{(j_1, \cdots, j_m)}(f^1, \cdots, f^m)(y)\, d(y^{j_1}, \cdots, y^{j_m}) \\ &= \int_{\partial G_m} \sum_{(m-1/n)} f^1(y) D_{(j_1, \cdots, j_{m-1})}(f^2, \cdots, f^m)(y)\, d(y^{j_1}, \cdots, y^{j_{m-1}}). \end{aligned} \tag{3}$$

The integral on the left in (3) is evaluated by the following integral.

$$\int_{\mathbf{a}} \sum_{(m/n)} D_{(j_1, \cdots, j_m)}(f^1, \cdots, f^m)[g(x)] D_{(1, \cdots, m)}(g^{j_1}, \cdots, g^{j_m})(x)\, d(x^1, \cdots, x^m). \tag{4}$$

The hypothesis in (1) states that the integrand of this integral is identically zero on $\mathbf{a}$; thus the value of this integral is zero. Therefore the integral on the right in (3) equals zero, and the proof of Theorem 48.1 is complete. □

48.2 Examples. If $f^1(y) \equiv 1$ in E, then

$$D_{(j_1, \cdots, j_m)}(f^1, \cdots, f^m)(y) = 0, \qquad y \in E, \qquad (j_1, \cdots, j_m) \in (m/n), \tag{5}$$

and hypothesis (1) in Theorem 48.1 is satisfied. In this case equation (2) is this:

$$\int_{\partial G_m} \sum_{(m-1/n)} D_{(j_1, \cdots, j_{m-1})}(f^2, \cdots, f^m)(y)\, d(y^{j_1}, \cdots, y^{j_{m-1}}) = 0. \tag{6}$$

This result has been established already in Theorem 39.5. Another example is contained in Theorem 39.3: if the trace $T(G_m)$ of G_m is contained in a plane in $\mathbb{R}^n$ which has dimension d, $d < m$, then

$$D_{(j_1,\cdots,j_m)}(g^{j_1}, \cdots, g^{j_m})(x) = 0, \qquad x \in \mathbf{a}, \qquad (j_1, \cdots, j_m) \in (m/n), \tag{7}$$

and the hypothesis in (1) is satisfied; therefore (2) is true. Furthermore, Theorem 39.3 was used to prove Theorem 43.1 and other addition theorems for integrals in Section 43.

We shall now show that Theorem 39.3 is only a special case of a much more general result: the statements in (7) and the hypothesis in (1) are satisfied not only if $T(G_m)$ is contained in a plane of dimension d but also if $T(G_m)$ is contained in a much more general manifold. Some explanations are needed before the theorem can be stated.

Let $g: \mathbf{a} \to \mathbb{R}^n$ and $f: E \to \mathbb{R}^m$, $\mathbf{a}$ in $\mathbb{R}^m$ and $m \leqq n$, be the functions described at the beginning of this section, and recall that $T(G_m) \subset E$. Let r be an integer such that $n \geqq r \geqq n - m + 1$, and let $K_t: E \to \mathbb{R}$, $t = 1, \cdots, r$, be functions which have continuous derivatives on E. Then the set $\{y: y \in E, K_t(y) = 0\}$, assumed not empty, is an $(n-1)$-dimensional manifold in E. Define the set $\mathscr{K}$ in E by the following equation:

$$\mathscr{K} = \{y: y \in E, K_t(y) = 0, t = 1, \cdots, r\}. \tag{8}$$

Assume that $\mathscr{K}$ is not empty. Then $\mathscr{K}$ is the intersection of r manifolds, each of dimension $(n-1)$, and $\mathscr{K}$ is an $(n-r)$-dimensional manifold. Then, since $0 \leqq n - r \leqq m - 1$, the manifold $\mathscr{K}$ has dimension less than m. Let $v_1, \cdots, v_m$ and $w_1, \cdots, w_r$ be the vectors in $\mathbb{R}^n$ defined by the following matrices:

$$\begin{bmatrix} v_1 \\ \cdots \\ v_m \end{bmatrix} = \begin{bmatrix} D_1 g^1(x) & D_1 g^2(x) & \cdots & D_1 g^n(x) \\ \cdots & \cdots & \cdots & \cdots \\ D_m g^1(x) & D_m g^2(x) & \cdots & D_m g^n(x) \end{bmatrix}, \qquad x \in \mathbf{a}. \tag{9}$$

$$\begin{bmatrix} w_1 \\ \cdots \\ w_r \end{bmatrix} = \begin{bmatrix} D_1 K_1(y) & D_2 K_1(y) & \cdots & D_n K_1(y) \\ \cdots & \cdots & \cdots & \cdots \\ D_1 K_r(y) & D_2 K_r(y) & \cdots & D_n K_r(y) \end{bmatrix}, \qquad y = g(x), \qquad x \in \mathbf{a}. \tag{10}$$

48.3 Theorem. *Let $\mathscr{K}$ be the set described above; assume that $\mathscr{K}$ is not empty and that*

$$\sum_{(r/n)} [D_{(j_1,\cdots,j_r)}(K_1, \cdots, K_r)(y)]^2 \neq 0, \qquad y \in \mathscr{K}. \tag{11}$$

Furthermore, assume that $T(G_m) \subset \mathscr{K}$; that is, assume that

$$K_t[g^1(x), \cdots, g^n(x)] = 0, \qquad x \in \mathbf{a}, \qquad t = 1, \cdots, r. \tag{12}$$

Then

$$D_{(1,\cdots,m)}(g^{j_1}, \cdots, g^{j_m})(x) = 0, \qquad x \in \mathbf{a}, \qquad (j_1, \cdots, j_m) \in (m/n), \tag{13}$$

hypothesis (1) in Theorem 48.1 is satisfied, and

$$\int_{\partial G_m} \sum_{(m-1/n)} f^1(y) D_{(j_1,\cdots,j_{m-1})}(f^2, \cdots, f^m)(y)\, d(y^{j_1}, \cdots, y^{j_{m-1}}) = 0. \tag{14}$$

PROOF. Since $m + r \geqq n + 1$, the $m + r$ vectors $v_1, \cdots, v_m, w_1, \cdots, w_r$ in $\mathbb{R}^n$ are linearly dependent [see Corollary 85.5 in Appendix 2]. Thus there are constants [actually functions of x and y] $c_1(x), \cdots, c_m(x), d_1(y), \cdots, d_r(y)$, not all zero, such that

$$c_1(x)v_1 + \cdots + c_m(x)v_m + d_1(y)w_1 + \cdots + d_r(y)w_r = 0, \quad x \in \mathbf{a}, \quad y = g(x),$$
$$\sum_{j=1}^{m} [c_j(x)]^2 + \sum_{j=1}^{r} [d_j(y)]^2 \neq 0. \tag{15}$$

Recall that $x = (x^1, \cdots, x^m)$; differentiate the two sides of the equations in (12) with respect to $x^1, \cdots, x^m$ to obtain

$$\sum_{i=1}^{n} D_i K_t(y) D_j g^i(x) = 0, \quad x \in \mathbf{a}, \quad y = g(x); \quad j = 1, \cdots, m; \quad t = 1, \cdots, r. \tag{16}$$

In the inner product notation, these equations are

$$(w_t, v_j) = 0, \qquad x \in \mathbf{a}, \qquad y = g(x); \qquad j = 1, \cdots, m; \qquad t = 1, \cdots, r. \tag{17}$$

Form the inner product of each of the vectors $w_1, \cdots, w_r$ with the vector on the left in (15). The results, by (17), simplify to this:

$$\begin{aligned} &(w_1, w_1)d_1(y) + (w_1, w_2)d_2(y) + \cdots + (w_1, w_r)d_r(y) = 0, \\ &\cdots\cdots\cdots\cdots\cdots\cdots\cdots\cdots\cdots\cdots\cdots\cdots \qquad y = g(x), \\ &(w_r, w_1)d_1(y) + (w_r, w_2)d_2(y) + \cdots + (w_r, w_r)d_r(y) = 0, \quad x \in \mathbf{a}. \end{aligned} \tag{18}$$

The Binet–Cauchy multiplication theorem [see Theorem 80.1 in Appendix 1] and Corollary 3.13 show that the determinant of this system of equations is equal to the sum on the left in (11); since, by hypothesis, this sum is not zero, the determinant of the system (18) is not zero and hence the system has only the trivial solution

$$d_1(y) = 0, \cdots, d_r(y) = 0, \qquad y = g(x), \qquad x \in \mathbf{a}. \tag{19}$$

Thus the statements in (15), by (19), are the following:

$$\sum_{j=1}^{m} c_j(x)v_j = 0, \qquad \sum_{j=1}^{m} [c_j(x)]^2 \neq 0, \qquad x \in \mathbf{a}. \tag{20}$$

Therefore the vectors $v_1, \cdots, v_m$ are linearly dependent, and the determinant of each m by m minor of the matrix in (9) equals zero as follows:

$$D_{(1,\cdots,m)}(g^{j_1}, \cdots, g^{j_m})(x) = \begin{vmatrix} D_1 g^{j_1}(x) & \cdots & D_1 g^{j_m}(x) \\ \cdots & \cdots & \cdots \\ D_m g^{j_1}(x) & \cdots & D_m g^{j_m}(x) \end{vmatrix} = 0, \qquad x \in \mathbf{a}, \tag{21}$$

for each $(j_1, \cdots, j_m)$ in (m/n). Then (13) is true, hypothesis (1) in Theorem 48.1 is satisfied, and the conclusion in (14) follows from (2) in Theorem 48.1. □

48.4 Example. Let **a** be a 3-dimensional Euclidean simplex $[a_0, a_1, \cdots, a_3]$ in $\mathbb{R}^3$, and let $\mathscr{K}$ be the plane $\{y: a_1y^1 + a_2y^2 + a_3y^3 + a_4 = 0\}$ in $\mathbb{R}^3$. Let $g: \mathbf{a} \to \mathbb{R}^3$ be a function which maps **a** into the plane $\mathscr{K}$. If $\Sigma_1^3 a_i^2 \neq 0$, then the hypothesis in (11) is satisfied in this case, and

$$\int_{\partial G_3} \sum_{(j_1, j_2)} f^1(y) D_{(j_1, j_2)}(f^2, f^3)(y)\, d(y^{j_1}, y^{j_2}) = 0, \qquad (j_1, j_2) \in (2/3). \quad (22)$$

48.5 Example. Let **a** be a 2-dimensional Euclidean simplex $[a_0, a_1, a_2]$ in $\mathbb{R}^2$, and let

$$\begin{aligned} K_1(y) &= a_{11}y^1 + a_{12}y^2 + a_{13}y^3 + a_{14}, \\ K_2(y) &= a_{21}y^1 + a_{22}y^2 + a_{23}y^3 + a_{24}. \end{aligned} \quad (23)$$

If

$$\begin{vmatrix} a_{11} & a_{12} \\ a_{21} & a_{22} \end{vmatrix}^2 + \begin{vmatrix} a_{11} & a_{13} \\ a_{21} & a_{23} \end{vmatrix}^2 + \begin{vmatrix} a_{12} & a_{13} \\ a_{22} & a_{23} \end{vmatrix}^2 \neq 0, \quad (24)$$

then the hypothesis in (11) is satisfied and the two planes $K_1(y) = 0$, $K_2(y) = 0$ intersect in a straight line $\mathscr{K}$. If $g: \mathbf{a} \to \mathbb{R}^3$ maps **a** into $\mathscr{K}$, then Theorem 48.3 shows that

$$\int_{\partial G_2} \sum_{j=1}^{3} f^1(y) D_j f^2(y)\, dy^j = 0. \quad (25)$$

48.6 Examples. Let **a** be a 3-dimensional Euclidean simplex $[a_0, a_1, \cdots, a_3]$ in $\mathbb{R}^3$, and let $K(y) = (y^1)^2 + (y^2)^2 + (y^3)^2 - 1$. Then $\mathscr{K}$ is the sphere $K(y) = 0$. At points $y: (y^1, y^2, y^3)$ on this sphere,

$$[D_1K(y)]^2 + [D_2K(y)]^2 + [D_3K(y)]^2 = 4[(y^1)^2 + (y^2)^2 + (y^3)^2] = 4, \quad (26)$$

and the hypothesis in (11) is satisfied. Let $g: \mathbf{a} \to \mathbb{R}^3$ be a function which maps **a** into $\mathscr{K}$. Then Theorem 48.3 shows that

$$\int_{\partial G_3} \sum_{(j_1, j_2)} f^1(y) D_{(j_1, j_2)}(f^2, f^3)(y)\, d(y^{j_1}, y^{j_2}) = 0, \qquad (j_1, j_2) \in (2/3). \quad (27$$

The mapping $g: \mathbf{a} \to \mathbb{R}^3$ collapses the 3-dimensional simplex **a** into the 2-dimensional surface of the sphere $K(y) = 0$, and Theorem 48.3 shows that $D_{(1,2)}(g^{j_1}, g^{j_2})(x) = 0$ for x in **a** and (j_1, j_2) in $(2/3)$. Then hypothesis (1) in Theorem 48.1 is satisfied, and the fundamental theorem of the integral calculus shows that the integral in (27) equals zero.

For another example, let **a** be a 2-dimensional Euclidean simplex $[a_0, a_1, a_2]$ in $\mathbb{R}^2$, and let

$$\begin{aligned} K_1(y) &= (y^1)^2 + (y^2 - 1)^2 + (y^3)^2 - 4, \\ K_2(y) &= (y^1)^2 + (y^2 + 1)^2 + (y^3)^2 - 4. \end{aligned} \quad (28)$$

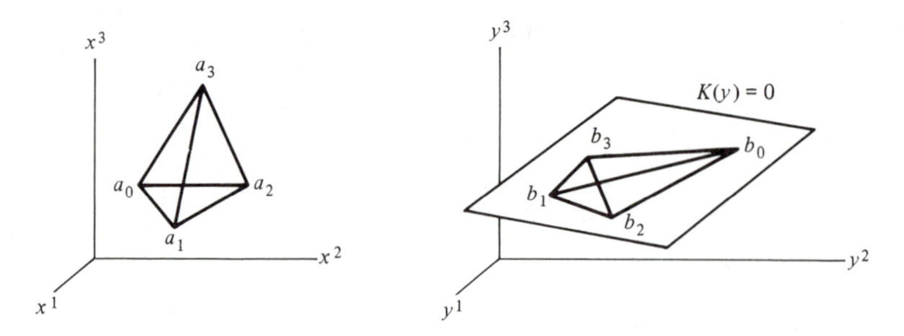

Figure 48.1. The addition theorem for integrals.

Then $K_1(y) = 0$ and $K_2(y) = 0$ are two spheres with radii equal to 2 and centers at the points (0, 1, 0) and (0, −1, 0). These spheres intersect in the circle $(y^1)^2 + (y^3)^2 - 3 = 0$ in the plane $y^2 = 0$, and $\mathscr{K}$ is the set of points on this circle. Now

$$D_{(1,2)}(K_1, K_2)(y) = \begin{vmatrix} 2y^1 & 2(y^2-1) \\ 2y^1 & 2(y^2+1) \end{vmatrix}, \qquad D_{(1,3)}(K_1, K_2)(y) = \begin{vmatrix} 2y^1 & 2y^3 \\ 2y^1 & 2y^3 \end{vmatrix},$$

$$D_{(2,3)}(K_1, K_2)(y) = \begin{vmatrix} 2(y^2-1) & 2y^3 \\ 2(y^2+1) & 2y^3 \end{vmatrix}. \tag{29}$$

If y is a point (y^1, y^2, y^3) in $\mathscr{K}$, then $y^2 = 0$ and $(y^1)^2 + (y^3)^2 = 3$, and

$$\sum_{(j_1,j_2)} [D_{(j_1,j_2)}(K_1, K_2)(y)]^2 = 64[(y^1)^2 + (y^3)^2] = 192, \quad (j_1, j_2) \in (2/3). \tag{30}$$

Thus the hypothesis in (11) is satisfied. Let $g: \mathbf{a} \to \mathbb{R}^3$ be a function which maps $\mathbf{a}$ into $\mathscr{K}$. Then Theorem 48.3 shows that

$$\int_{\partial G_2} \sum_{j=1}^{3} f^1(y) D_j f^2(y)\, dy^j = 0. \tag{31}$$

48.7 Example. This example is concerned with the addition theorem for integrals established in Section 43. It is necessary to improve, for use in later sections, the theorem established there. Let $\mathbf{a}$ be a 3-dimensional Euclidean simplex $[a_0, a_1, \cdots, a_3]$ in $\mathbb{R}^3$; then $\Delta(\mathbf{a}) \neq 0$. Let $b_0, b_1, \cdots, b_3$ be four points in a plane $K(y) = 0$ in $\mathbb{R}^3$ [see Figure 48.1]. Then $\mathbf{b}: [b_0, b_1, \cdots, b_3]$ is a degenerate simplex. Let $g: \mathbf{a} \to \mathbb{R}^3$ be an affine transformation such that $g(a_i) = b_i$, $i = 0, 1, \cdots, 3$; this affine transformation exists by Theorem 19.3 [see also equations (1) and (2) in Section 37]. Then g defines a surface G_3, and $T(G_3) = \mathbf{b}$. More precisely,

$$T(G_3) = [b_1, b_2, b_3] \cup [b_0, b_2, b_3] \cup [b_1, b_0, b_3] \cup [b_1, b_2, b_0], \tag{32}$$

and $T(G_3)$ is a compact set which lies in the plane $K(y) = 0$. Let E be an open set in $\mathbb{R}^3$ which contains $T(G_3)$, and let $f: E \to \mathbb{R}^3$ be a function whose components (f^1, f^2, f^3) have continuous derivatives. Then the hypotheses of Theorem 48.3 are satisfied, and

$$D_{(1,2,3)}(g^1, g^2, g^3)(x) = 0, \qquad x \in \mathbf{a}, \tag{33}$$

$$\int_{\partial G_3} \sum_{(j_1,j_2)} f^1(y) D_{(j_1,j_2)}(f^2, f^3)\, d(y^{j_1}, y^{j_2}) = 0. \tag{34}$$

Now $\partial G_3 = \partial[b_0, b_1, \cdots, b_3]$, and as shown in equation (5) in Section 43,

$$\begin{aligned}\partial[b_0, b_1, \cdots, b_3]& \\ &= [b_1, b_2, b_3] - [b_0, b_2, b_3] - [b_1, b_0, b_3] - [b_1, b_2, b_0].\end{aligned} \tag{35}$$

Then, as in Section 43, equation (35) can be used to convert (34) into the following statement.

$$\int_{[b_1,b_2,b_3]} I = \int_{[b_0,b_2,b_3]} I + \int_{[b_1,b_0,b_3]} I + \int_{[b_1,b_2,b_0]} I, \tag{36}$$

$$I = \sum_{(j_1,j_2)} f^1(y) D_{(j_1,j_2)}(f^2, f^3)(y)\, d(y^{j_1}, y^{j_2}), \qquad (j_1, j_2) \in (2/3). \tag{37}$$

Apply the formula in (36) and (37) to each of the functions

$$(f^{(j_1,j_2)}, y^{j_1}, y^{j_2}) : E \to \mathbb{R}^3, \qquad (j_1, j_2) \in (2/3).$$

Then as explained in the proof of Corollary 43.4, the results can be simplified to formula (36) with

$$I = f^{(j_1,j_2)}(y)\, d(y^{j_1}, y^{j_2}), \qquad (j_1, j_2) \in (2/3). \tag{38}$$

Add the three formulas in (36), (38) to obtain formula (36) with

$$I = \sum_{(j_1,j_2)} f^{(j_1,j_2)}(y)\, d(y^{j_1}, y^{j_2}). \tag{39}$$

Now the formulas in (36), $\cdots$, (39) have been derived from the fundamental theorem of the integral calculus, and for this reason it has been necessary to assume that the functions f^1, f^2, f^3 and $f^{(j_1,j_2)}$, (j_1, j_2) in (2/3), have continuous derivatives. However, the integrals in (36), (38), (39) exist if the functions $f^{(j_1,j_2)}$ are merely continuous, but it has not been shown that formula (36) holds for these continuous functions. The proof of (36), (38), (39) for continuous functions $f^{(j_1,j_2)}$ will now be given; the following theorem is needed in this proof.

48.8 Theorem (Weierstrass Polynomial Approximation Theorem). *Any real-valued continuous function on a compact subset F of $\mathbb{R}^n$ is the limit of a sequence of polynomials which converges uniformly in F.*

For this theorem and its proof, see page 133 of *Foundations of Modern Analysis*, by J. Dieudonné, published by Academic Press, New York and London, 1960.

Return to Example 48.7. Let $f^{(j_1,j_2)} : \mathbf{b} \to \mathbb{R}$, $(j_1, j_2) \in (2/3)$, be three continuous functions, and let $f_k^{(j_1,j_2)} : E \to \mathbb{R}$, $k = 1, 2, \cdots$, be sequences of polynomials such that

$$\lim_{k\to\infty} f_k^{(j_1,j_2)}(y) = f^{(j_1,j_2)}(y), \qquad y\in\mathbf{b}, \qquad (j_1,j_2)\in(2/3). \tag{40}$$

the limit being uniform on $\mathbf{b}$. These sequences of polynomials exist by Theorem 48.8. Then by Theorem 46.3,

$$\lim_{k\to\infty}\int_{[b_1,b_2,b_3]} f_k^{(j_1,j_2)}(y)\,d(y^{j_1},y^{j_2}) = \int_{[b_1,b_2,b_3]} f^{(j_1,j_2)}(y)\,d(y^{j_1},y^{j_2}) \tag{41}$$

for each (j_1,j_2) in (2/3), and similar statements hold for each of the simplexes $[b_0,b_2,b_3]$, $[b_1,b_0,b_3]$, and $[b_1,b_2,b_0]$. The integrals in (41) and the other statements exist by Theorem 36.4 since the functions $f_k^{(j_1,j_2)}$ and $f^{(j_1,j_2)}$ are continuous by hypothesis. Also, since a polynomial has continuous derivatives,

$$\begin{gathered}\int_{[b_1,b_2,b_3]} I = \int_{[b_0,b_2,b_3]} I + \int_{[b_1,b_0,b_3]} I + \int_{[b_1,b_2,b_3]} I,\\ I = f_k^{(j_1,j_2)}(y)\,d(y^{j_1},y^{j_2}), \qquad (j_1,j_2)\in(2/3).\end{gathered} \tag{42}$$

By (35) this formula can be written in more concise notation as follows:

$$\int_{\partial\mathbf{b}} f_k^{(j_1,j_2)}(y)\,d(y^{j_1},y^{j_2}) = 0. \tag{43}$$

Then for $k = 1, 2, \cdots,$

$$\begin{aligned}\int_{\partial\mathbf{b}} f^{(j_1,j_2)}(y)\,d(y^{j_1},y^{j_2}) &= \int_{\partial\mathbf{b}} f^{(j_1,j_2)}(y)\,d(y^{j_1},y^{j_2}) - \int_{\partial\mathbf{b}} f_k^{(j_1,j_2)}(y)\,d(y^{j_1},y^{j_2})\\ &= \lim_{k\to\infty}\int_{\partial\mathbf{b}} [f^{(j_1,j_2)}(y) - f_k^{(j_1,j_2)}(y)]\,d(y^{j_1},y^{j_2}).\end{aligned} \tag{44}$$

By (41), this limit equals zero. Thus

$$\int_{\partial\mathbf{b}} f^{(j_1,j_2)}(y)\,d(y^{j_1},y^{j_2}) = 0, \qquad (j_1,j_2)\in(2/3). \tag{45}$$

Add the three equations in (45) to obtain

$$\int_{\partial\mathbf{b}} \sum_{(j_1,j_2)} f^{(j_1,j_2)}(y)\,d(y^{j_1},y^{j_2}) = 0. \tag{46}$$

The proof of the following addition formulas for the integrals of continuous functions is thus complete:

$$\begin{gathered}\int_{[b_1,b_2,b_3]} I = \int_{[b_0,b_2,b_3]} I + \int_{[b_1,b_0,b_3]} I + \int_{[b_1,b_2,b_0]} I,\\ I = f^{(j_1,j_2)}(y)\,d(y^{j_1},y^{j_2}), \qquad (j_1,j_2)\in(2/3),\\ I = \sum_{(j_1,j_2)} f^{(j_1,j_2)}(y)\,d(y^{j_1},y^{j_2}).\end{gathered} \tag{47}$$

This section will be completed by proving the theorem which generalizes the formulas in (47). Let $\mathbf{a}$ be the Euclidean m-simplex $[a_0, a_1, \cdots, a_m]$ in $\mathbb{R}^m$; then $\Delta(\mathbf{a}) \neq 0$. Let n be an integer such that $m \leqq n$, and let $r = n - m + 1$. Let

$$K_t(y) = a_{t1}y^1 + a_{t2}y^2 + \cdots + a_{tn}y^n + a_{t,n+1}, \qquad t = 1, \cdots, r, \tag{48}$$

be r functions such that

$$\sum_{(r/n)} [D_{(j_1,\cdots,j_r)}(K_1, \cdots, K_r(y)]^2 \neq 0, \qquad y \in \mathbb{R}^n. \tag{49}$$

Let $\mathscr{K} = \{y : y \in \mathbb{R}^n, K_t(y) = 0, t = 1, \cdots, r\}$; then $\mathscr{K}$ is not empty and $\mathscr{K}$ is an affine subspace of dimension $m - 1$ in $\mathbb{R}^n$. Let $b_0, b_1, \cdots, b_m$ be $m + 1$ points in $\mathscr{K}$, and let $\mathbf{b}$ be the simplex $[b_0, b_1, \cdots, b_m]$. Then $\mathbf{b}$ is a degenerate simplex, but its $(m - 1)$-dimensional sides need not be degenerate [see Example 48.7]. If $r > n - m + 1$, then $\mathbf{b}$ and all of its $(m - 1)$-dimensional sides would be degenerate, and there is no interest in this case in the present context.

48.9 Theorem. *Let* $f^{(j_1,\cdots,j_m)} : \mathbf{b} \to \mathbb{R}$, $(j_1, \cdots, j_{m-1}) \in (m - 1/n)$, *be continuous functions. Then*

$$\int_{[b_1,\cdots,b_m]} I = \sum_{i=1}^{m} \int_{[b_1,\cdots,b_0/b_i,\cdots,b_m]} I, \tag{50}$$
$$I = \sum_{(m-1/n)} f^{(j_1,\cdots,j_{m-1})}(y)\, d(y^{j_1}, \cdots, y^{j_{m-1}}).$$

Proof. Example 48.7 contains a proof of this theorem in the special case $m = n = 3$, and that proof contains all of the methods and ideas required to prove Theorem 48.9. Let E be an open set which contains $\mathbf{b}$, and let $f : E \to \mathbb{R}^m$ be a function whose components $(f^1, \cdots, f^m)$ have continuous derivatives. Let $g : \mathbf{a} \to \mathbf{b}$ be an affine transformation such that $g(a_i) = b_i$, $i = 0, 1, \cdots, m$. This transformation defines a surface G_m such that $T(G_m) = \mathbf{b}$ and $\partial G_m = [b_1, \cdots, b_m] - \Sigma_1^m [b_1, \cdots, b_0/b_i, \cdots, b_m]$. Then Theorem 48.3 shows that

$$\int_{\partial \mathbf{b}} \sum_{(m-1/n)} f^1(y) D_{(j_1,\cdots,j_m)}(f^2, \cdots, f^m)(y)\, d(y^{j_1}, \cdots, y^{j_{m-1}}) = 0. \tag{51}$$

Assume next that $f^{(j_1,\cdots,j_m)} : E \to \mathbb{R}$ has continuous derivatives, and apply the formula in (51) to the function with components $(f^{(j_1,\cdots,j_{m-1})}, y^{j_1}, \cdots, y^{j_{m-1}})$. Formula (51) for this function can be simplified to

$$\int_{\partial \mathbf{b}} f^{(j_1,\cdots,j_{m-1})}(y)\, d(y^{j_1}, \cdots, y^{j_{m-1}}) = 0, \qquad (j_1, \cdots, j_{m-1}) \in (m - 1/n). \tag{52}$$

Then use the Weierstrass polynomial approximation theorem in Theorem 48.8 to show that (52) holds also if the functions $f^{(j_1,\cdots,j_{m-1})}$ are only continuous on $\mathbf{b}$; add the formulas in (52) for continuous functions to obtain (50) and complete the proof of Theorem 48.9. □

EXERCISES

48.1. The points $a_0 : (0, 0)$, $a_1 : (1, 0)$, $a_2 : (0, 1)$, and $a_3 : (1, 1)$ are the vertices of a degenerate 3-simplex in the $(x\ y)$-plane. Let $f(x, y) = x + y$. Use iterated integrals to show that

$$\int_{[a_1, a_2, a_3]} f(x, y)\, d(x, y) = -2/3,$$

$$\int_{[a_0, a_2, a_3]} f(x, y)\, d(x, y) = -1/2,$$

$$\int_{[a_1, a_0, a_3]} f(x, y)\, d(x, y) = -1/2,$$

$$\int_{[a_1, a_2, a_0]} f(x, y)\, d(x, y) = 1/3.$$

Use these values to verify the following addition formula in this special case.

$$\int_{[a_1, a_2, a_3]} I = \int_{[a_0, a_2, a_3]} I + \int_{[a_1, a_0, a_3]} I + \int_{[a_1, a_2, a_0]} I,$$

$$I = f(x, y)\, d(x, y).$$

48.2. The four points $b_0 : (0, 0, 3)$, $b_1 : (3, 0, 0)$, $b_2 : (0, 3, 0)$, and $b_3 : (1, 1, 1)$ are in the plane $y^1 + y^2 + y^3 = 3$ in (y^1, y^2, y^3)-space.

(a) Make a sketch to show the four simplexes $[b_1, b_2, b_3]$, $[b_0, b_2, b_3]$, $[b_1, b_0, b_3]$, and $[b_1, b_2, b_0]$.

(b) Verify the following addition formula by evaluating each of the four integrals.

$$\int_{[b_1, b_2, b_3]} I = \int_{[b_0, b_2, b_3]} I + \int_{[b_1, b_0, b_3]} I + \int_{[b_1, b_2, b_0]} I,$$

$$I = y^3\, d(y^1, y^2) + y^2\, d(y^1, y^3) + y^1\, d(y^2, y^3).$$

[Hint. Use Exercise 37.7 to show that the affine transformation

$$y^1 = (0 - 3)x^1 + (1 - 3)x^2 + 3,$$

$$y^2 = (3 - 0)x^1 + (1 - 0)x^2 + 0,$$

$$y^3 = (0 - 0)x^1 + (1 - 0)x^2 + 0,$$

maps $[e_0, e_1, e_2]$ into $[b_1, b_2, b_3]$. Then

$$\int_{[b_1, b_2, b_3]} I$$

$$= \int_{[e_0, e_1, e_2]} [x^2(3) + (3x^1 + x^2)(-3) + (-3x^1 - 2x^2 + 3)(3)]\, d(x^1, x^2)$$

$$= \int_{[e_0, e_1, e_2]} [-18x^1 - 6x^2 + 9]\, d(x^1, x^2) = 1/2.$$

Iterated integrals are used to evaluate the last integral. In the same way, it can be shown that

$$\int_{[b_0,b_2,b_3]} I = \int_{[e_0,e_1,e_2]} [18x^1 + 6x^2 - 9]\, d(x^1, x^2) = -1/2,$$

$$\int_{[b_1,b_0,b_3]} I = \int_{[e_0,e_1,e_2]} [6x^2 - 9]\, d(x^1, x^2) = -7/2,$$

$$\int_{[b_1,b_2,b_0]} I = \int_{[e_0,e_1,e_2]} [-54x^1 + 27]\, d(x^1, x^2) = 9/2.$$

These values verify the addition formula in (b).]

48.3. This exercise asks the reader to prove a theorem similar to Theorem 48.3, but some explanations are needed before the theorem can be stated. Let $\mathbf{a}$ be a simplex $[a_0, a_1, \cdots, a_m]$ in $\mathbb{R}^m$ such that $\Delta(\mathbf{a}) \neq 0$, and let $g: \mathbf{a} \to \mathbb{R}^n$, $m \leqq n$, be a function whose components $(g^1, \cdots, g^n)$ have continuous derivatives. Let E be an open set in $\mathbb{R}^n$ which contains $T(G_m)$, and let $f: E \to \mathbb{R}^m$ be a function whose components have continuous derivatives. Then the composite function $f \circ g$ maps $\mathbf{a}$ into $\mathbb{R}^m$; let F be an open set in $\mathbb{R}^m$ which contains the set $\{z : z = (f \circ g)(x), x \in \mathbf{a}\}$. Let r be an integer such that $m \geqq r \geqq 1$, and let $K_t: F \to \mathbb{R}$, $t = 1, \cdots, r$, be functions which have continuous derivatives in F. Define the set $\mathscr{K}$ by the following equation: $\mathscr{K} = \{z : z \in F, K_t(z) = 0, t = 1, \cdots, r\}$. Prove the following theorem.

Theorem. *If:*

(i) $\mathscr{K}$ *is not empty;*

(ii) $\sum_{(r/m)} [D_{(j_1,\cdots,j_r)}(K_1, \cdots, K_r)(z)]^2 \neq 0$, $z \in \mathscr{K}$;

(iii) $K_t[(f^1 \circ g)(x), \cdots, (f^m \circ g)(x)] = 0$, $x \in \mathbf{a}$, $t = 1, \cdots, r$;

then

(iv) $\sum_{(m/n)} D_{(j_1,\cdots,j_m)}(f^1, \cdots, f^m)[g(x)] D_{(1,\cdots,m)}(g^{j_1}, \cdots, g^{j_m})(x) = 0$, $x \in \mathbf{a}$;

(v) $\displaystyle\int_{\partial G_m} \sum_{(m-1/n)} f^1(y) D_{(j_1,\cdots,j_{m-1})}(f^2, \cdots, f^m)(y)\, d(y^{j_1}, \cdots, y^{j_{m-1}}) = 0.$

[Hint. Use the methods employed in proving Theorem 48.3 to show that $D_{(1,\cdots,m)}(f^1 \circ g, \cdots, f^m \circ g)(x) = 0$ for x in $\mathbf{a}$, which, by the chain rule, is the same as hypothesis (1) in Theorem 48.1.]

49. Integrals Over Surfaces with the Same Boundary

If a function has a continuous derivative, then the integrals of this derivative over two surfaces with the same boundary are equal. As an elementary example of this application of the fundamental theorem of the integral calculus, equation (1) in Section 44 contains the following statement:

$$\begin{aligned}\int_{C_g} \sum_{i=1}^{3} D_i f(y)\, dy^i &= f[g(b)] - f[g(a)] \\ &= f[h(b)] - f[h(a)] = \int_{C_h} \sum_{i=1}^{3} D_i f(y)\, dy^i.\end{aligned} \tag{1}$$

This section contains several theorems which state conditions under which integrals over two surfaces with the same boundary are equal.

49.1 Example. In (1) the two integrals have the same integrand, but the two integrals may be equal even when the integrands differ. Let $\mathbf{a}$ be the interval $[a_0, a_1]$ in $\mathbb{R}$, and let $g : \mathbf{a} \to \mathbb{R}^3$ and $h : \mathbf{a} \to \mathbb{R}^3$ be functions whose components (g^1, g^2, g^3) and (h^1, h^2, h^3) have continuous derivatives. Then g and h define surfaces (actually curves) G_1 and H_1 with traces $T(G_1)$ and $T(H_1)$. Let E be an open set in $\mathbb{R}^3$ which contains $T(G_1)$ and $T(H_1)$. Let $e : E \to \mathbb{R}$ and $f : E \to \mathbb{R}$ be two functions which have continuous derivatives. Assume also that

$$e[g(a_0)] = f[h(a_0)], \qquad e[g(a_1)] = f[h(a_1)]. \tag{2}$$

Then

$$\begin{aligned} \int_{G_1} \sum_{i=1}^{3} D_i e(y)\,dy^i &= e[g(a_1)] - e[g(a_0)] \\ &= f[h(a_1)] - f[h(a_0)] = \int_{H_1} \sum_{i=1}^{3} D_i f(y)\,dy^i. \end{aligned} \tag{3}$$

By specializing the functions e, f, g, h in various ways, equation (3) provides elementary examples of the theorems in this section.

49.2 Theorem. *Let* $\mathbf{a}$ *be a simplex* $[a_0, a_1, \cdots, a_m]$ *in* $\mathbb{R}^m$ *such that* $\Delta(\mathbf{a}) \neq 0$, *and let* $e : \mathbf{a} \to \mathbb{R}^m$ *and* $f : \mathbf{a} \to \mathbb{R}^m$ *be two functions whose components* $(e^1, \cdots, e^m)$ *and* $(f^1, \cdots, f^m)$ *have continuous derivatives on* $\mathbf{a}$. *Assume that*

$$e^j(x) = f^j(x), \qquad x \in \partial\mathbf{a}, \qquad j = 1, \cdots, m. \tag{4}$$

Then

$$\int_{\mathbf{a}} D_{(1,\cdots,m)}(e^1, \cdots, e^m)(x)\,d(x^1, \cdots, x^m) \tag{5}$$

$$= \int_{\partial\mathbf{a}} \sum_{(m-1/m)} e^1(x) D_{(j_1,\cdots,j_{m-1})}(e^2, \cdots, e^m)(x)\,d(x^{j_1}, \cdots, x^{j_{m-1}}) \tag{6}$$

$$= \int_{\partial\mathbf{a}} \sum_{(m-1/m)} f^1(x) D_{(j_1,\cdots,j_{m-1})}(f^2, \cdots, f^m)(x)\,d(x^{j_1}, \cdots, x^{j_{m-1}}) \tag{7}$$

$$= \int_{\mathbf{a}} D_{(1,\cdots,m)}(f^1, \cdots, f^m)(x)\,d(x^1, \cdots, x^m). \tag{8}$$

PROOF. The integrals in (5) and (6), and also those in (7) and (8), are equal by the fundamental theorem of the integral calculus in Theorem 38.4. The proof of Theorem 49.2 will be completed by proving that the integrals in (6) and (7) are equal. These integrals exist as shown in the proof of Theorem 36.4, but on a first examination it appears unlikely that they are equal because the equality of the functions in (4) does not imply the equality of the derivatives

$$\begin{gathered} D_{(j_1,\cdots,j_{m-1})}(e^2, \cdots, e^m)(x), \qquad D_{(j_1,\cdots,j_{m-1})}(f^2, \cdots, f^m)(x), \\ x \in \partial\mathbf{a}, \qquad (j_1, \cdots, j_{m-1}) \in (m-1/m). \end{gathered} \tag{9}$$

Thus the hypothesis in (4) that the functions e and f are point-wise equal in $\partial\mathbf{a}$ does not imply that the integrands of the integrals in (6) and (7) are point-wise equal in $\partial\mathbf{a}$, and it is necessary to examine the definitions of these integrals in order to prove that they are equal.

The reader will find it helpful to review Example 38.3 at this time. Let $\mathscr{P}_k$, $k = 1, 2, \cdots$, be a sequence of simplicial subdivisions of $\mathbf{a}$; these subdivisions induce simplicial subdivisions $\partial\mathscr{P}_k$ in $\partial\mathbf{a}$. Let $\mathbf{x}: (x_1, \cdots, x_m)$ denote a simplex in $\partial\mathscr{P}_k$. Then as shown in Example 38.3 and the proof of Lemma 38.6, the integrals in (6) and (7) are equal, respectively, to the limits as $k \to \infty$ of the following sums:

$$\sum_{\mathbf{x}\in\partial\mathscr{P}_k} \frac{e^1(x_1) + \cdots + e^1(x_m)}{m} \frac{(-1)^{m-1}}{(m-1)!} \begin{vmatrix} e^2(x_1) & \cdots & e^m(x_1) & 1 \\ \cdots & \cdots & \cdots & \cdots \\ e^2(x_m) & \cdots & e^m(x_m) & 1 \end{vmatrix}, \tag{10}$$

$$\sum_{\mathbf{x}\in\partial\mathscr{P}_k} \frac{f^1(x_1) + \cdots + f^1(x_m)}{m} \frac{(-1)^{m-1}}{(m-1)!} \begin{vmatrix} f^2(x_1) & \cdots & f^m(x_1) & 1 \\ \cdots & \cdots & \cdots & \cdots \\ f^2(x_m) & \cdots & f^m(x_m) & 1 \end{vmatrix}. \tag{11}$$

Now for each $\mathbf{x}$ in $\partial\mathscr{P}_k$, the points $x_1, \cdots, x_m$ are in $\partial\mathbf{a}$; then

$$e^j(x_i) = f^j(x_i), \qquad i = 1, \cdots, m, \qquad j = 1, \cdots, m, \qquad \mathbf{x}\in\partial\mathscr{P}_k, \tag{12}$$

and the sums in (10) and (11) are equal for $k = 1, 2, \cdots$. Then the limits, as $k \to \infty$, of the sums in (10) and (11) are equal, and the integrals in (6) and (7) are equal. The proof of Theorem 49.2 is complete. □

49.3 Theorem. *Let* $\mathbf{a}$ *be a simplex* $[a_0, a_1, \cdots, a_m]$ *in* $\mathbb{R}^m$ *such that* $\Delta(\mathbf{a}) \neq 0$; *let* E *be an open set in* $\mathbb{R}^n$, $m \leqq n$; *and let* $g: \mathbf{a} \to \mathbb{R}^n$ *and* $h: \mathbf{a} \to \mathbb{R}^n$ *be functions whose components* $(g^1, \cdots, g^n)$ *and* $(h^1, \cdots, h^n)$ *have continuous derivatives on* $\mathbf{a}$ *and which define surfaces* G_m *and* H_m *whose traces* $T(G_m)$ *and* $T(H_m)$ *are in* E. *Let* $f: E \to \mathbb{R}^m$ *be a function whose components* $(f^1, \cdots, f^m)$ *have continuous derivatives. Assume that*

$$g^j(x) = h^j(x), \qquad x\in\partial\mathbf{a}, \qquad j = 1, \cdots, n. \tag{13}$$

Then

$$\int_{G_m} \sum_{(m/n)} D_{(j_1,\cdots,j_m)}(f^1, \cdots, f^m)(y)\, d(y^{j_1}, \cdots, y^{j_m}) \tag{14}$$

$$= \int_{\partial G_m} \sum_{(m-1/n)} f^1(y) D_{(j_1,\cdots,j_{m-1})}(f^2, \cdots, f^m)(y)\, d(y^{j_1}, \cdots, y^{j_{m-1}}) \tag{15}$$

$$= \int_{\partial H_m} \sum_{(m-1/n)} f^1(y) D_{(j_1,\cdots,j_{m-1})}(f^2, \cdots, f^m)(y)\, d(y^{j_1}, \cdots, y^{j_{m-1}}) \tag{16}$$

$$= \int_{H_m} \sum_{(m/n)} D_{(j_1,\cdots,j_m)}(f^1, \cdots, f^m)(y)\, d(y^{j_1}, \cdots, y^{j_m}). \tag{17}$$

PROOF. The integrals in (14) and (15), and also those in (16) and (17), are equal by the fundamental theorem of the integral calculus in Theorem 39.1.

The proof will be completed by showing that the integrals in (14) and (17) are equal. Now Theorem 39.1 shows that the integrals in (14) and (17) are equal, respectively, to

$$\int_{\mathbf{a}} D_{(1,\cdots,m)}(f^1\circ g,\cdots,f^m\circ g)(x)\,d(x^1,\cdots,x^m), \tag{18}$$

$$\int_{\mathbf{a}} D_{(1,\cdots,m)}(f^1\circ h,\cdots,f^m\circ h)(x)\,d(x^1,\cdots,x^m). \tag{19}$$

As a result of the hypothesis in (13),

$$(f^j\circ g)(x)=(f^j\circ h)(x),\qquad x\in\partial\mathbf{a},\qquad j=1,\cdots,m. \tag{20}$$

Thus the hypotheses of Theorem 49.2 are satisfied, and the integrals in (18) and (19) are equal; therefore the integrals in (14) and (17) are equal and the proof of Theorem 49.3 is complete. □

The following theorem is a generalization of Theorems 44.1 and 49.2 and also of the formulas in (1).

49.4 Theorem. *Let* $\mathbf{a}$ *be a simplex* $[a_0,a_1,\cdots,a_m]$ *in* $\mathbb{R}^m$ *such that* $\Delta(\mathbf{a})\neq 0$; *let* E *be an open set in* $\mathbb{R}^n$; *and let* $g:\mathbf{a}\to\mathbb{R}^n$, $m\leqq n$, *be a function whose components* $(g^1,\cdots,g^n)$ *have continuous derivatives and which defines a surface* G_m *whose trace* $T(G_m)$ *is in* E. *Let* $e:E\to\mathbb{R}^m$ *and* $f:E\to\mathbb{R}^m$ *be two functions whose components* $(e^1,\cdots,e^m)$ *and* $(f^1,\cdots,f^m)$ *have continuous derivatives on* E. *Assume that*

$$(e^j\circ g)(x)=(f^j\circ g)(x),\qquad x\in\partial\mathbf{a},\qquad j=1,\cdots,m. \tag{21}$$

Then

$$\int_{G_m}\sum_{(m/n)} D_{(j_1,\cdots,j_m)}(e^1,\cdots,e^m)(y)\,d(y^{j_1},\cdots,y^{j_m}) \tag{22}$$

$$=\int_{\partial G_m}\sum_{(m-1/n)} e^1(y)D_{(j_1,\cdots,j_{m-1})}(e^2,\cdots,e^m)(y)\,d(y^{j_1},\cdots,y^{j_{m-1}}) \tag{23}$$

$$=\int_{\partial G_m}\sum_{(m-1/n)} f^1(y)D_{(j_1,\cdots,j_{m-1})}(f^2,\cdots,f^m)(y)\,d(y^{j_1},\cdots,y^{j_{m-1}}) \tag{24}$$

$$=\int_{G_m}\sum_{(m/n)} D_{(j_1,\cdots,j_m)}(f^1,\cdots,f^m)(y)\,d(y^{j_1},\cdots,y^{j_m}). \tag{25}$$

PROOF. By Theorem 39.1, the integrals in (22) and (25) are equal, respectively, to the following integrals:

$$\int_{\mathbf{a}} D_{(1,\cdots,m)}(e^1\circ g,\cdots,e^m\circ g)(x)\,d(x^1,\cdots,x^m), \tag{26}$$

$$\int_{\mathbf{a}} D_{(1,\cdots,m)}(f^1\circ g,\cdots,f^m\circ g)(x)\,d(x^1,\cdots,x^m). \tag{27}$$

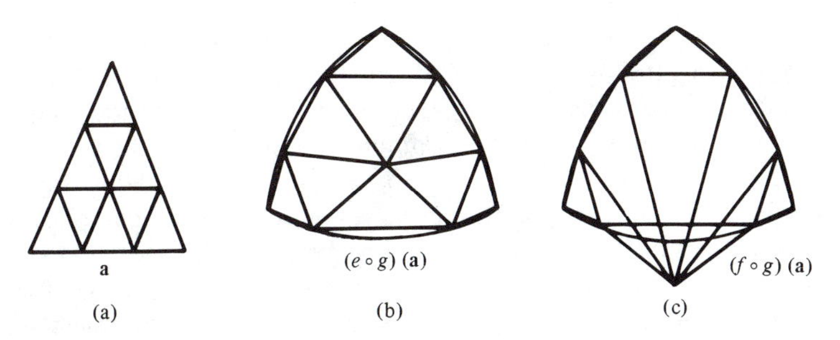

Figure 49.1. The mappings of a simplicial subdivision of **a** by $e \circ g$ and $f \circ g$.

Hypothesis (21) and Theorem 49.2 show that these integrals are equal. Theorem 39.1 shows that the integrals in (23) and (24) are equal, respectively, to those in (26) and (27). Thus the four integrals in (22), $\cdots$, (25) are equal. □

49.5 Example. This example uses the explanation which accompanies Figure 47.1 in Section 47 to provide a geometric interpretation and proof of Theorem 49.4. In that theorem there is a surface G_m in $\mathbb{R}^n$ defined by $g: \mathbf{a} \to \mathbb{R}^n$, and there are two functions $e: E \to \mathbb{R}^m$ and $f: E \to \mathbb{R}^m$ which map G_m into $\mathbb{R}^m$. Now the composite functions $e \circ g: \mathbf{a} \to \mathbb{R}^m$ and $f \circ g: \mathbf{a} \to \mathbb{R}^m$ define m-dimensional surfaces in $\mathbb{R}^m$ whose boundaries are the surfaces defined by the functions $e \circ g: \partial\mathbf{a} \to \mathbb{R}^m$ and $f \circ g: \partial\mathbf{a} \to \mathbb{R}^m$ (if $m = 2$, these boundaries are curves). An explained in Section 47 [see also Lemma 38.5],

$$\begin{aligned}\int_{G_m} \sum_{(m/n)} D_{(j_1, \cdots, j_m)}(e^1, \cdots, e^m)(y)\, d(y^{j_1}, \cdots, y^{j_m}) \\ = \lim_{k \to \infty} \sum_{\mathbf{x} \in \mathscr{P}_k} \frac{(-1)^m}{m!} \Delta_{(1, \cdots, m)}(e^1 \circ g, \cdots, e^m \circ g)(\mathbf{x}),\end{aligned} \tag{28}$$

$$\begin{aligned}\int_{G_m} \sum_{(m/n)} D_{(j_1, \cdots, j_m)}(f^1, \cdots, f^m)(y)\, d(y^{j_1}, \cdots, y^{j_m}) \\ = \lim_{k \to \infty} \sum_{\mathbf{x} \in \mathscr{P}_k} \frac{(-1)^m}{m!} \Delta_{(1, \cdots, m)}(f^1 \circ g, \cdots, f^m \circ g)(\mathbf{x}).\end{aligned} \tag{29}$$

The sum on the right in (28) is the sum of the signed volumes of the simplexes in a chain of m-simplexes in $\mathbb{R}^m$, and a similar statement holds for the sum on the right in (29). Figure 49.1 is a schematic drawing designed to indicate the nature of these sums in the special case $m = 2$. Figure 49.1(a) shows a simplicial subdivision $\mathscr{P}_k$ of **a** with nine simplexes, and Figures 49.1(b) and (c) indicate how this subdivision might be mapped by $e \circ g$ and $f \circ g$. The chains of simplexes in (b) and (c) may be quite different, but these two chains have the same boundaries because of the hypothesis that $(e \circ g)(x) = (f \circ g)(x)$ for every x in $\partial\mathbf{a}$. If two oriented 2-chains in $\mathbb{R}^2$ have the same boundary, then the sum of the signed areas of the simplexes in one of the chains is equal to the sum of the signed areas of the simplexes in the other chain [see Exercise

49.1]. Thus Figure 49.1 indicates geometrically, in the special case $m = 2$, that the sums on the right in (28) and (29) are equal, and an analytic proof of the equality can be given by using Theorem 20.1 and the methods employed in proving Theorem 43.6. For the example in Figure 49.1, the limit in (28) is the "area" of the curve $(e \circ g)(x)$, $x \in \partial \mathbf{a}$, and the limit in (29) is the "area" of the curve $(f \circ g)(x)$, $x \in \partial \mathbf{a}$ [these curves may be quite complicated in nature; they certainly are not restricted to be simple closed curves of the form shown in Figure 49.1(b) and (c)]. Since $(e \circ g)(x) = (f \circ g)(x)$ for x in $\partial \mathbf{a}$, the two limits in (28) and (29) are equal. In the general case also, the hypothesis that $(e \circ g)(x) = (f \circ g)(x)$ for x in $\partial \mathbf{a}$ [see (21)] shows that the chains $(e \circ g)(\mathscr{P}_k)$ and $(f \circ g)(\mathscr{P}_k)$ have the same boundaries $(e \circ g)(\partial \mathscr{P}_k)$ and $(f \circ g)(\partial \mathscr{P}_k)$. Then the same arguments that were used in the special case $m = 2$ [see Figure 49.1] can be employed again to show that, in the general case, the sums on the right in (28) and (29) are equal for $k = 1, 2, \cdots$. Therefore the integrals in (28) and (29), which are the same as those in (22) and (25), are equal; the fundamental theorem of the integral calculus can be used to complete the proof of Theorem 49.4.

Exercises

49.1. (a) Prove that the following two chains have the same boundary:

$$[a_0, a_1, a_2] + [a_0, a_2, a_3] + [a_0, a_3, a_1],$$

$$[b_0, a_1, a_2] + [b_0, a_2, a_3] + [b_0, a_3, a_1].$$

(b) If a_0, b_0, a_1, a_2, a_3 are points in $\mathbb{R}^2$, prove that

$$\Delta(a_0, a_1, a_2) + \Delta(a_0, a_2, a_3) + \Delta(a_0, a_3, a_1)$$
$$= \Delta(b_0, a_1, a_2) + \Delta(b_0, a_2, a_3) + \Delta(b_0, a_3, a_1).$$

[Hint. Let p be a point in $\mathbb{R}^2$. Then by Theorem 20.1,

$$\Delta(a_0, a_1, a_2) = \Delta(p, a_1, a_2) + \Delta(a_0, p, a_2) + \Delta(a_0, a_1, p).$$

Evaluate each determinant in the equation in the same way, and show that the sums on the two sides are equal.]

(c) Let $\Sigma_1^r \mathbf{a}_i$ and $\Sigma_1^s \mathbf{b}_i$ be two 2-dimensional oriented simplicial chains in $\mathbb{R}^2$ [see Definition 40.1] which have the same boundary. Prove that

$$\sum_{i=1}^{r} \Delta(\mathbf{a}_i) = \sum_{i=1}^{s} \Delta(\mathbf{b}_i).$$

[Hint. The proof is similar to the proof of Theorem 43.6.]

(d) Prove the following theorem. If $\Sigma_1^r \mathbf{a}_i$ and $\Sigma_1^s \mathbf{b}_i$ are two m-dimensional oriented simplicial chains in $\mathbb{R}^m$ which have the same boundary, then

$$\sum_{i=1}^{r} \Delta(\mathbf{a}_i) = \sum_{i=1}^{s} \Delta(\mathbf{b}_i).$$

49.2. Let $\mathbf{a}:[a_0, a_1, \cdots, a_m]$ be a simplex in $\mathbb{R}^m$ such that $\Delta(\mathbf{a}) \neq 0$. Let G_m and H_m be surfaces defined by functions $g:\mathbf{a} \to \mathbb{R}^n$ and $h:\mathbf{a} \to \mathbb{R}^n$, $m \leqq n$, whose components have continuous derivatives. Let E be an open set in $\mathbb{R}^n$ which contains $T(G_m)$ and $T(H_m)$, and let $e:E \to \mathbb{R}^m$ and $f:E \to \mathbb{R}^m$ be functions whose components have continuous derivatives.

(a) Prove that

$$\int_{\mathbf{a}} D_{(1,\cdots,m)}(e^1 \circ g, \cdots, e^m \circ g)(x)\, d(x^1, \cdots, x^m)$$
$$= \lim_{k\to\infty} \sum_{\mathbf{x}\in\mathscr{P}_k} \frac{(-1)^m}{m!} \Delta_{(1,\cdots,m)}(e^1 \circ g, \cdots, e^m \circ g)(\mathbf{x}),$$

$$\int_{\mathbf{a}} D_{(1,\cdots,m)}(f^1 \circ h, \cdots, f^m \circ h)(x)\, d(x^1, \cdots, x^m)$$
$$= \lim_{k\to\infty} \sum_{\mathbf{x}\in\mathscr{P}_k} \frac{(-1)^m}{m!} \Delta_{(1,\cdots,m)}(f^1 \circ h, \cdots, f^m \circ h)(\mathbf{x}).$$

[Hints. Lemma 38.5 and Example 47.1.]

(b) If $(e^j \circ g)(x) = (f^j \circ h)(x)$ for every x in $\partial\mathbf{a}$ and $j = 1, \cdots, m$, prove that the two integrals in (a) are equal. [Hint. Exercise 49.1(d), Example 49.5, and Exercise 49.2(a).]

(c) Prove the following statements:

$$\int_{\mathbf{a}} D_{(1,\cdots,m)}(e^1 \circ g, \cdots, e^m \circ g)(x)\, d(x^1, \cdots, x^m)$$
$$= \int_{\mathbf{a}} \sum_{(m/n)} D_{(j_1,\cdots,j_m)}(e^1, \cdots, e^m)[g(x)] D_{(1,\cdots,m)}(g^{j_1}, \cdots, g^{j_m})(x)\, d(x^1, \cdots, x^m)$$
$$= \int_{G_m} \sum_{(m/n)} D_{(j_1,\cdots,j_m)}(e^1, \cdots, e^m)(y)\, d(y^{j_1}, \cdots, y^{j_m})$$
$$= \int_{\partial G_m} \sum_{(m-1/n)} e^1(y) D_{(j_1,\cdots,j_{m-1})}(e^2, \cdots, e^m)(y)\, d(y^{j_1}, \cdots, y^{j_{m-1}}).$$

$$\int_{\mathbf{a}} D_{(1,\cdots,m)}(f^1 \circ h, \cdots, f^m \circ h)(x)\, d(x^1, \cdots, x^m)$$
$$= \int_{\mathbf{a}} \sum_{(m/n)} D_{(j_1,\cdots,j_m)}(f^1, \cdots, f^m)[h(x)] D_{(1,\cdots,m)}(h^{j_1}, \cdots, h^{j_m})(x)\, d(x^1, \cdots, x^m)$$
$$= \int_{H_m} \sum_{(m/n)} D_{(j_1,\cdots,j_m)}(f^1, \cdots, f^m)(y)\, d(y^{j_1}, \cdots, y^{j_m})$$
$$= \int_{\partial H_m} \sum_{(m-1/n)} f^1(y) D_{(j_1,\cdots,j_{m-1})}(f^2, \cdots, f^m)(y)\, d(y^{j_1}, \cdots, y^{j_{m-1}}).$$

[Hint. Theorem 4.5, equation (31) in Theorem 36.4, and Theorem 39.1]

(d) If $(e^j \circ g)(x) = (f^j \circ h)(x)$ for every x in $\partial\mathbf{a}$, use the result in (b) to show that the eight integrals in (c) are equal.

(e) Show that the theorem in this exercise includes Theorems 49.2, 49.3, and 49.4 as special cases.

50. Integrals on Affine Surfaces with the Same Boundary

The theorems in Section 49 are general in one respect but special in another. They are general in the sense that they treat integrals over general surfaces, but they are special in the sense that they apply only to integrals of derivatives. The present section establishes results, similar to those in Section 49, for more general integrals of the form

$$\int_{G_m} \sum_{(m/n)} f^{(j_1,\cdots,j_m)}(y)\,d(y^{j_1},\cdots,y^{j_m}), \qquad \int_{H_m} \sum_{(m/n)} f^{(j_1,\cdots,j_m)}(y)\,d(y^{j_1},\cdots,y^{j_m}),$$

but only for surfaces G_m and H_m whose traces $T(G_m)$ and $T(H_m)$ lie in an m-dimensional affine space in $\mathbb{R}^n$ and have the same boundary. A surface whose trace lies in an affine space is called an affine surface. The methods developed in this section are important because, in some cases at least, they can be used to establish change-of-variable theorems.

The first theorem treats an important special case; its proof, however, displays the methods and tools used to establish the general theorem.

50.1 Theorem. *Let* $\mathbf{a}:[a_0, a_1, a_2]$ *be a simplex in* $\mathbb{R}^2$ *such that* $\Delta(\mathbf{a}) \neq 0$, *and let* $g:\mathbf{a}\to\mathbb{R}^2$ *and* $h:\mathbf{a}\to\mathbb{R}^2$ *be functions whose components* (g^1, g^2) *and* (h^1, h^2) *have continuous derivatives and which define surfaces* G_2 *and* H_2. *Assume that the traces* $T(G_2)$ *and* $T(H_2)$ *of* G_2 *and* H_2 *are contained in a compact convex set* C *in* $\mathbb{R}^2$. *Assume also that* G_2 *and* H_2 *have the same boundary; thus*

$$g(x) = h(x), \qquad x\in\partial\mathbf{a}. \tag{1}$$

Finally, let $f: C\to\mathbb{R}$ *be a continuous function. Then*

$$\int_{G_2} f(y)\,d(y^1, y^2) = \int_{H_2} f(y)\,d(y^1, y^2), \tag{2}$$

$$\int_{\mathbf{a}} f[g(x)]D_{(1,2)}(g^1, g^2)(x)\,d(x^1, x^2) = \int_{\mathbf{a}} f[g(x)]D_{(1,2)}(h^1, h^2)(x)\,d(x^1, x^2). \tag{3}$$

PROOF. The integrals in (2) and (3) exist by Theorem 36.4, and only the stated equalities remain to be proved. Let $\mathscr{P}_k$, $k = 1, 2, \cdots$, be a sequence of oriented simplicial subdivisions of $\mathbf{a}$ [see Definition 21.5]. Then $\mathscr{P}_k = \{\mathbf{x}:\mathbf{x}\in\mathscr{P}_k\}$, and $\mathbf{x}$ is a Euclidean simplex $[x_0, x_1, x_2]$ which is contained in $\mathbf{a}$. Then g maps the vertices x_i, $i = 0, 1, 2$, of $\mathbf{x}$ into points $y_i = g(x_i)$, $i = 0, 1, 2$. Let the simplex with vertices y_0, y_1, y_2 be denoted by $[y_0, y_1, y_2]$. Then $\mathbf{y}:[y_0, y_1, y_2]$ is a closed convex set; g maps the vertices of $\mathbf{x}$ into the vertices of $\mathbf{y}$, but g does not necessarily map the set $\mathbf{x}$ onto the set $\mathbf{y}$. Now by Definition 36.3,

$$\int_{G_2} f(y)\,d(y^1, y^2) = \lim_{k\to\infty}\sum_{\mathbf{x}\in\mathscr{P}_k} f[g(x_0)]\frac{(-1)^2}{2!}\Delta_{(1,2)}(g^1, g^2)(\mathbf{x}), \tag{4}$$

and $\int_{H_2} f(y)\,d(y^1, y^2)$ has a similar definition. In order to prove that the two integrals in (2) are equal, it is necessary to obtain representations of them which depend only on the boundaries ∂G_2 and ∂H_2, which are the same curve by (1). The assumption that G_2 and H_2 have the same boundary does not imply that $T(G_2)$ and $T(H_2)$ are the same set; it may very well happen that the intersection $T(G_2) \cap T(H_2)$ of the traces is a proper subset of both $T(G_2)$ and $T(H_2)$.

There are two steps in the proof; the first step is the proof that

$$\lim_{k\to\infty} \sum_{\mathbf{y}\in g(\mathscr{P}_k)} \int_{\mathbf{y}} f(y)\,d(y^1, y^2) = \int_{G_2} f(y)\,d(y^1, y^2). \tag{5}$$

The statement $\mathbf{y} \in g(\mathscr{P}_k)$ means that $\mathbf{y}$ is a (non-degenerate or degenerate) Euclidean simplex $[y_0, y_1, y_2]$ whose vertices are the images $g(x_0)$, $g(x_1)$, $g(x_2)$ of the vertices x_0, x_1, x_2 of the simplex $\mathbf{x}$ in $\mathscr{P}_k$. The integral

$$\int_{\mathbf{y}} f(y)\,d(y^1, y^2) \tag{6}$$

is the integral of f on $\mathbf{y}$ in $\mathbb{R}^2$ as defined in Section 35. Observe that f is defined and continuous on C and hence on each simplex $\mathbf{y}$ in $g(\mathscr{P}_k)$. Use the mean-value theorem for integrals in Theorem 35.18 to evaluate the integral in (6) as follows: there exists a point y^* on the interior of the simplex $\mathbf{y}: [y_0, y_1, y_2]$ such that

$$\int_{\mathbf{y}} f(y)\,d(y^1, y^2) = f(y^*)\frac{(-1)^2}{2!}\Delta(\mathbf{y}). \tag{7}$$

Now

$$\Delta(\mathbf{y}) = \begin{vmatrix} y_0^1 & y_0^2 & 1 \\ y_1^1 & y_1^2 & 1 \\ y_2^1 & y_2^2 & 1 \end{vmatrix} = \begin{vmatrix} g^1(x_0) & g^2(x_0) & 1 \\ g^1(x_1) & g^2(x_1) & 1 \\ g^1(x_2) & g^2(x_2) & 1 \end{vmatrix} = \Delta_{(1,2)}(g^1, g^2)(\mathbf{x}). \tag{8}$$

Then (8) shows that (7) can be written as follows:

$$\int_{\mathbf{y}} f(y)\,d(y^1, y^2) = f(y^*)\frac{(-1)^2}{2!}\Delta_{(1,2)}(g^1, g^2)(\mathbf{x}). \tag{9}$$

Now

$$\sum_{\mathbf{y}\in g(\mathscr{P}_k)} \int_{\mathbf{y}} f(y)\,d(y^1, y^2) \tag{10}$$

$$= \sum_{\mathbf{x}\in\mathscr{P}_k} f[g(x_0)]\frac{(-1)^2}{2!}\Delta_{(1,2)}(g^1, g^2)(\mathbf{x}) \tag{11}$$

$$+ \sum_{\mathbf{x}\in\mathscr{P}_k} [f(y^*) - f(y_0)]\frac{(-1)^2}{2!}\Delta_{(1,2)}(g^1, g^2)(\mathbf{x}). \tag{12}$$

The limit, as $k \to \infty$, of the expression in (11) is $\int_{G_2} f(y)\,d(y^1, y^2)$ by (4), and we shall show that the limit of the sum in (12) equals zero. These results and equations (10), $\cdots$, (12) show that

$$\int_{G_2} f(y)\,d(y^1, y^2) = \lim_{k\to\infty} \sum_{\mathbf{y}\in g(\mathscr{P}_k)} \int_{\mathbf{y}} f(y)\,d(y^1, y^2). \tag{13}$$

This equation completes the first step in the proof, as stated in (5). To prove that the limit, as $k \to \infty$, of the sum in (12) is zero, observe first that y^* is a point in the simplex $[y_0, y_1, y_2]$, which is the simplex $[g(x_0), g(x_1), g(x_2)]$. Since the norm of $\mathscr{P}_k$ tends to zero as $k \to \infty$, and since g is continuous on $\mathbf{a}$, then the diameter of $\mathbf{y}$ is uniformly small for $\mathbf{y}$ in $g(\mathscr{P}_k)$, and it tends to zero as $k \to \infty$. Since f is uniformly continuous on C, to each $\varepsilon > 0$ there corresponds a $k(\varepsilon)$ such that

$$|f(y^*) - f(y_0)| < \varepsilon \tag{14}$$

for each $\mathbf{x}$ in $\mathscr{P}_k$ provided $k \geqq k(\varepsilon)$. Then

$$\left| \sum_{\mathbf{x}\in\mathscr{P}_k} [f(y^*) - f(y_0)] \frac{(-1)^2}{2!} \Delta_{(1,2)}(g^1, g^2)(\mathbf{x}) \right| \leqq \varepsilon \sum_{\mathbf{x}\in\mathscr{P}_k} \left| \frac{1}{2!} \Delta_{(1,2)}(g^1, g^2)(\mathbf{x}) \right| \tag{15}$$

for $k \geqq k(\varepsilon)$. By Section 45, the limit, as $k \to \infty$, of the sum on the right in (15) exists and is the area of the surface G_2, which is finite since $g : \mathbf{a} \to \mathbb{R}^2$ has continuous derivatives. Thus (15) shows that the limit, as $k \to \infty$, of the sum in (12) equals zero, and the proof of (13) is complete.

A similar analysis can be made of the integral $\int_{H_2} f(y)\,d(y^1, y^2)$ of f on the surface H_2 defined by $h : \mathbf{a} \to \mathbb{R}^2$. The function h maps the vertices x_i, $i = 0, 1, 2$, of $\mathbf{x}$ in $\mathscr{P}_k$ into points $z_i = h(x_i)$, $i = 0, 1, 2$. Let the simplex with vertices z_0, z_1, z_2 be denoted by $[z_0, z_1, z_2]$. Then $\mathbf{z} : [z_0, z_1, z_2]$ is a closed convex set; h maps the vertices of $\mathbf{x}$ into the vertices of $\mathbf{z}$, but h does not necessarily map the set $\mathbf{x}$ onto the set $\mathbf{z}$. Thus $\mathbf{z}$ is a (non-degenerate or degenerate) Euclidean simplex. The same analysis that established (13) can be used again to show that

$$\int_{H_2} f(y)\,d(y^1, y^2) = \lim_{k\to\infty} \sum_{\mathbf{z}\in h(\mathscr{P}_k)} \int_{\mathbf{z}} f(y)\,d(y^1, y^2). \tag{16}$$

The second step in the proof is to show that the limits in (13) and (16) are equal. The chains $\mathbf{c}_g = \Sigma\{\mathbf{y} : \mathbf{y} \in g(\mathscr{P}_k)\}$ and $\mathbf{c}_h = \Sigma\{\mathbf{z} : \mathbf{z} \in h(\mathscr{P}_k)\}$ are oriented simplicial chains as defined in Definition 40.1. Each of these chains is contained in the convex set C in $\mathbb{R}^2$. The vertices of the boundary chain of $\mathbf{c}_g$ are the images under g of the vertices of the boundary of the chain $\Sigma\{\mathbf{x} : \mathbf{x} \in \mathscr{P}_k\}$ in $\mathbf{a}$, and the vertices of the boundary chain of $\mathbf{c}_h$ are the images under h of the vertices of the boundary of the same chain $\Sigma\{\mathbf{x} : \mathbf{x} \in \mathscr{P}_k\}$. But $g(x) = h(x)$ for x in $\partial\mathbf{a}$ by the hypothesis in (1); therefore, the chains $\mathbf{c}_g$ and $\mathbf{c}_h$ have the same boundary and Theorem 43.6 and Example 48.7 show that

$$\sum_{\mathbf{y}\in g(\mathscr{P}_k)} \int_{\mathbf{y}} f(y)\,d(y^1, y^2) = \sum_{\mathbf{z}\in h(\mathscr{P}_k)} \int_{\mathbf{z}} f(y)\,d(y^1, y^2). \tag{17}$$

The proof of (17) depends on the addition theorem for integrals established in Corollary 43.4 and more generally in Theorem 48.9. Finally, (13), (16), and (17) show that

$$\begin{aligned}\int_{G_2} f(y)\,d(y^1, y^2) &= \lim_{k\to\infty} \sum_{\mathbf{y}\in g(\mathscr{P}_k)} \int_{\mathbf{y}} f(y)\,d(y^1, y^2)\\ &= \lim_{k\to\infty} \sum_{\mathbf{z}\in h(\mathscr{P}_k)} \int_{\mathbf{z}} f(y)\,d(y^1, y^2) \qquad (18)\\ &= \int_{H_2} f(y)\,d(y^1, y^2),\end{aligned}$$

and the proof of the conclusion in (2) is complete. The integrals in (3) are equal to the integrals in (2) by Theorem 36.4, and the proof of the entire theorem is complete. □

50.2 Theorem. *Let C be a compact convex set in a 2-dimensional affine space (a plane) in $\mathbb{R}^3$. Let $\mathbf{a}:[a_0, a_1, a_2]$ be a simplex in $\mathbb{R}^2$ such that $\Delta(\mathbf{a}) \neq 0$, and let $g:\mathbf{a}\to C$ and $h:\mathbf{a}\to C$ be functions whose components $(g^1, \cdots, g^3)$ and $(h^1, \cdots, h^3)$ have continuous derivatives and which define surfaces G_2 and H_2. Assume also that G_2 and H_2 have the same boundary; thus*

$$g(x) = h(x), \qquad x\in\partial\mathbf{a}. \qquad (19)$$

Finally, let $f: C\to\mathbb{R}^3$ be a function with continuous components $f^{(j_1,j_2)}$, $(j_1, j_2)\in(2/3)$. Then

$$\int_{G_2} \sum_{(2/3)} f^{(j_1,j_2)}(y)\,d(y^{j_1}, y^{j_2}) = \int_{H_2} \sum_{(2/3)} f^{(j_1,j_2)}(y)\,d(y^{j_1}, y^{j_2}), \qquad (20)$$

$$\begin{aligned}&\int_{\mathbf{a}} \sum_{(2/3)} f^{(j_1,j_2)}[g(x)]D_{(1,2)}(g^{j_1}, g^{j_2})(x)\,d(x^1, x^2)\\ &\quad= \int_{\mathbf{a}} \sum_{(2/3)} f^{(j_1,j_2)}[h(x)]D_{(1,2)}(h^{j_1}, h^{j_2})(x)\,d(x^1, x^2).\end{aligned} \qquad (21)$$

PROOF. The proof of Theorem 50.2 is almost identical with that of Theorem 50.1; only some of the details are different. The integrals in (20) and (21) exist by Theorem 36.4, and only the stated equalities remain to be proved. Let $\mathscr{P}_k$, $k = 1, 2, \cdots$, be a sequence of oriented simplicial subdivisions of $\mathbf{a}$ [see Definition 21.5]. Then $\mathscr{P}_k = \{\mathbf{x}:\mathbf{x}\in\mathscr{P}_k\}$, and $\mathbf{x}$ is a Euclidean simplex $[x_0, x_1, x_2]$ which is contained in $\mathbf{a}$. Then g maps the vertices x_i, $i = 0, 1, 2$, of $\mathbf{x}$ into points $y_i = g(x_i)$, $i = 0, 1, 2$. Let the simplex with vertices y_0, y_1, y_2 be denoted by $[y_0, y_1, y_2]$. Then $\mathbf{y}:[y_0, y_1, y_2]$ is a closed convex set; g maps the vertices of $\mathbf{x}$ into the vertices of $\mathbf{y}$, but g does not necessarily map the set $\mathbf{x}$ onto the set $\mathbf{y}$. Now by Definition 36.3,

$$\begin{aligned}&\int_{G_2} \sum_{(2/3)} f^{(j_1,j_2)}(y)\,d(y^{j_1}, y^{j_2})\\ &\quad= \lim_{k\to\infty} \sum_{\mathbf{x}\in\mathscr{P}_k} \sum_{(2/3)} f^{(j_1,j_2)}[g(x_0)]\frac{(-1)^2}{2!}\Delta_{(1,2)}(g^{j_1}, g^{j_2})(\mathbf{x}),\end{aligned} \qquad (22)$$

and $\int_{H_2}\Sigma_{(2/3)} f^{(j_1,j_2)}(y)\,d(y^{j_1}, y^{j_2})$ has a similar definition. In order to prove that the two integrals are equal, it is necessary to obtain representations of them which depend only on the boundaries ∂G_2 and ∂H_2, which are the same curve by (19). The assumption that G_2 and H_2 have the same boundary does not imply that $T(G_2)$ and $T(H_2)$ are the same set; it may well happen that the intersection $T(G_2)\cap T(H_2)$ of the traces is a proper subset of both $T(G_2)$ and $T(H_2)$. The hypotheses show that $T(G_2)$ and $T(H_2)$ are contained in the convex set C in a 2-dimensional affine subspace of $\mathbb{R}^3$.

There are two steps in the proof; the first step is the proof that

$$\lim_{k\to\infty}\sum_{\mathbf{y}\in g(\mathscr{P}_k)}\int_{\mathbf{y}}\sum_{(2/3)} f^{(j_1,j_2)}(y)\,d(y^{j_1}, y^{j_2}) = \int_{G_2}\sum_{(2/3)} f^{(j_1,j_2)}(y)\,d(y^{j_1}, y^{j_2}). \quad (23)$$

The statement $\mathbf{y}\in g(\mathscr{P}_k)$ means that $\mathbf{y}$ is a (non-degenerate or degenerate) Euclidean simplex $[y_0, y_1, y_2]$ whose vertices are the images $g(x_0)$, $g(x_1)$, $g(x_2)$ of the vertices x_0, x_1, x_2 of the simplex $\mathbf{x}$ in $\mathscr{P}_k$. The integral

$$\int_{\mathbf{y}}\sum_{(2/3)} f^{(j_1,j_2)}(y)\,d(y^{j_1}, y^{j_2}) \quad (24)$$

is the integral of f, with components $f^{(j_1,j_2)}$, $(j_1, j_2)\in(2/3)$, on $\mathbf{y}:[y_0, y_1, y_2]$ in $\mathbb{R}^3$ as defined in Section 37. Observe that f is defined and continuous on C and hence on each simplex $\mathbf{y}$ in $g(\mathscr{P}_k)$. Use the mean-value theorem for integrals in Exercise 37.7 to evaluate the integral in (24) as follows: there exists a point y^* on the interior of the simplex $y:[g(x_0), g(x_1), g(x_2)]$ such that

$$\int_{\mathbf{y}}\sum_{(2/3)} f^{(j_1,j_2)}(y)\,d(y^{j_1}, y^{j_2}) = \sum_{(2/3)} f^{(j_1,j_2)}(y^*)\frac{(-1)^2}{2!}\Delta_{(1,2)}(g^{j_1}, g^{j_2})(\mathbf{x}). \quad (25)$$

Here

$$\Delta_{(1,2)}(g^{j_1}, g^{j_2})(\mathbf{x}) = \begin{vmatrix} g^{j_1}(x_0) & g^{j_2}(x_0) & 1 \\ g^{j_1}(x_1) & g^{j_2}(x_1) & 1 \\ g^{j_1}(x_2) & g^{j_2}(x_2) & 1 \end{vmatrix} = \begin{vmatrix} y_0^{j_1} & y_0^{j_2} & 1 \\ y_1^{j_1} & y_1^{j_2} & 1 \\ y_2^{j_1} & y_2^{j_2} & 1 \end{vmatrix} = \Delta(\mathbf{y}^{(j_1,j_2)}). \quad (26)$$

Now

$$\sum_{\mathbf{y}\in g(\mathscr{P}_k)}\int_{\mathbf{y}}\sum_{(2/3)} f^{(j_1,j_2)}(y)\,d(y^{j_1}, y^{j_2}) \quad (27)$$

$$= \sum_{\mathbf{x}\in\mathscr{P}_k}\left\{\sum_{(2/3)} f^{(j_1,j_2)}[g(x_0)]\frac{(-1)^2}{2!}\Delta_{(1,2)}(g^{j_1}, g^{j_2})(\mathbf{x})\right\} \quad (28)$$

$$+ \sum_{\mathbf{x}\in\mathscr{P}_k}\left\{\sum_{(2/3)} [f^{(j_1,j_2)}(y^*) - f^{(j_1,j_2)}(y_0)]\frac{(-1)^2}{2!}\Delta_{(1,2)}(g^{j_1}, g^{j_2})(\mathbf{x})\right\}. \quad (29)$$

The limit, as $k\to\infty$, of the expression in (28) is $\int_{G_2}\Sigma_{(2/3)} f^{(j_1,j_2)}(y)\,d(y^{j_1}, y^{j_2})$ by (22), and we shall show that the limit of the sum in (29) equals zero. These results and equations (27), $\cdots$, (29) show that

$$\int_{G_2} \sum_{(2/3)} f^{(j_1,j_2)}(y)\,d(y^{j_1}, y^{j_2}) = \lim_{k\to\infty} \sum_{\mathbf{y}\in g(\mathscr{P}_k)} \int_{\mathbf{y}} \sum_{(2/3)} f^{(j_1,j_2)}(y)\,d(y^{j_1}, y^{j_2}). \quad (30)$$

This equation completes the first step in the proof, as stated in (23). To prove that the limit, as $k \to \infty$, of the sum in (29) equals zero, observe first that y^* is a point in the simplex $[y_0, y_1, y_2]$, which is the simplex $[g(x_0), g(x_1), g(x_2)]$. Since the norm of $\mathscr{P}_k$ tends to zero as $k \to \infty$, and since g is continuous on $\mathbf{a}$, then the diameter of $\mathbf{y}$ is uniformly small for y in $g(\mathscr{P}_k)$, and it tends to zero as $k \to \infty$. Since f is uniformly continuous on C, to each $\varepsilon > 0$ there corresponds a $k(\varepsilon)$ such that

$$\left\{ \sum_{(j_1,j_2)} [f^{(j_1,j_2)}(y^*) - f^{(j_1,j_2)}(y_0)]^2 \right\}^{1/2} < \varepsilon \quad (31)$$

for each $\mathbf{x}$ in $\mathscr{P}_k$ provided $k \geqq k(\varepsilon)$. Then by Schwarz's inequality,

$$\begin{aligned}
&\left| \sum_{\mathbf{x}\in\mathscr{P}_k} \left\{ \sum_{(2/3)} [f^{(j_1,j_2)}(y^*) - f^{(j_1,j_2)}(y_0)] \frac{(-1)^2}{2!} \Delta_{(1,2)}(g^{j_1}, g^{j_2})(\mathbf{x}) \right\} \right| \\
&\quad \leqq \sum_{\mathbf{x}\in\mathscr{P}_k} \left\{ \sum_{(2/3)} [f^{(j_1,j_2)}(y^*) - f^{(j_1,j_2)}(y_0)]^2 \right\}^{1/2} \frac{1}{2!} \left\{ \sum_{(2/3)} [\Delta_{(1,2)}(g^{j_1}, g^{j_2})(\mathbf{x})]^2 \right\}^{1/2} \\
&\quad < \varepsilon \sum_{\mathbf{x}\in\mathscr{P}_k} \frac{1}{2!} \left\{ \sum_{(2/3)} [\Delta_{(1,2)}(g^{j_1}, g^{j_2})(\mathbf{x})]^2 \right\}^{1/2}
\end{aligned} \quad (32)$$

for $k \geqq k(\varepsilon)$. By Section 45, the limit, as $k \to \infty$, of the last sum exists and is the area of the surface G_2; this area exists and is finite since g has continuous derivatives. Thus (32) shows that the limit, as $k \to \infty$, of the sum in (29) equals zero, and the proof of (30) is complete.

A similar analysis can be made of the integral $\int_{H_2} \Sigma_{(2/3)} f^{(j_1,j_2)}(y)\,d(y^{j_1}, y^{j_2})$ of f on the surface H_2 defined by $h : \mathbf{a} \to \mathbb{R}^2$. The function h maps the vertices x_i, $i = 0, 1, 2$, of $\mathbf{x}$ in $\mathscr{P}$ into points $z_i = h(x_i)$, $i = 0, 1, 2$. Let the simplex with vertices z_0, z_1, z_2 be denoted by $[z_0, z_1, z_2]$. Then $\mathbf{z} : [z_0, z_1, z_2]$ is a closed convex set; h maps the vertices of $\mathbf{x}$ into the vertices of $\mathbf{z}$, but h does not necessarily map the set $\mathbf{x}$ onto the set $\mathbf{z}$. Thus $\mathbf{z}$ is a (non-degenerate or degenerate) Euclidean simplex. The same analysis that established (30) can be used again to show that

$$\int_{H_2} \sum_{(2/3)} f^{(j_1,j_2)}(y)\,d(y^{j_1}, y^{j_2}) = \lim_{k\to\infty} \sum_{\mathbf{z}\in h(\mathscr{P}_k)} \int_{\mathbf{z}} \sum_{(2/3)} f^{(j_1,j_2)}(y)\,d(y^{j_1}, y^{j_2}). \quad (33)$$

The second step in the proof is to show that the limits in (30) and (33) are equal. The chains $\mathbf{c}_g = \Sigma\{\mathbf{y} : \mathbf{y} \in g(\mathscr{P}_k)\}$ and $\mathbf{c}_h = \Sigma\{\mathbf{z} : \mathbf{z} \in h(\mathscr{P}_k)\}$ are oriented simplicial chains as defined in Definition 40.1; these chains consist of 2-dimensional simplexes in the 2-dimensional affine space (plane) in $\mathbb{R}^3$. Each of these chains is contained in the convex set C in the affine space in $\mathbb{R}^3$. The vertices of the boundary chain of $\mathbf{c}_g$ are the images under g of the vertices of the boundary of the chain $\Sigma\{\mathbf{x} : \mathbf{x} \in \mathscr{P}_k\}$ in $\mathbf{a}$, and the vertices of the boundary chain of $\mathbf{c}_h$ are the images under h of the vertices of the boundary of the same chain $\Sigma\{\mathbf{x}:\mathbf{x}\in\mathscr{P}_k\}$. But $g(x)=h(x)$ for x in $\partial\mathbf{a}$ by the hypothesis in (19); therefore, the chains $\mathbf{c}_g$ and $\mathbf{c}_h$ have the same boundary. Now Theorem 48.9 contains the addition theorem for integrals of continuous functions defined on simplexes in an affine space, and this theorem can be used to prove that

$$\sum_{\mathbf{y}\in g(\mathscr{P}_k)}\int_{\mathbf{y}}\sum_{(2/3)} f^{(j_1,j_2)}(y)\,d(y^{j_1},y^{j_2}) = \sum_{\mathbf{z}\in h(\mathscr{P}_k)}\int_{\mathbf{z}}\sum_{(2/3)} f^{(j_1,j_2)}(y)\,d(y^{j_1},y^{j_2}) \tag{34}$$

because the two chains have the same boundary; the proof is the same as the proof of Theorem 43.6. Finally, (30), (33), and (34) show that

$$\begin{aligned}\int_{G_2}\sum_{(2/3)} f^{(j_1,j_2)}(y)\,d(y^{j_1},y^{j_2}) &= \lim_{k\to\infty}\sum_{\mathbf{y}\in g(\mathscr{P}_k)}\int_{\mathbf{y}}\sum_{(2/3)} f^{(j_1,j_2)}(y)\,d(y^{j_1},y^{j_2})\\ &= \lim_{k\to\infty}\sum_{\mathbf{z}\in h(\mathscr{P}_k)}\int_{\mathbf{z}}\sum_{(2/3)} f^{(j_1,j_2)}(y)\,d(y^{j_1},y^{j_2}) \qquad (35)\\ &= \int_{H_2}\sum_{(2/3)} f^{(j_1,j_2)}(y)\,d(y^{j_1},y^{j_2}).\end{aligned}$$

Thus the proof of conclusion (20) is complete. The integrals in (21) are equal to the integrals in (20) by Theorem 36.4, and the proof of the entire theorem is complete. □

50.3 Theorem. *Let C be a compact convex set in an m-dimensional affine space in $\mathbb{R}^n$, $m \leqq n$. Let $\mathbf{a}:[a_0,a_1,\cdots,a_m]$ be a simplex in $\mathbb{R}^m$ such that $\Delta(\mathbf{a})\neq 0$, and let $g:\mathbf{a}\to C$ and $h:\mathbf{a}\to C$ be functions whose components $(g^1,\cdots,g^n)$ and $(h^1,\cdots,h^n)$ have continuous derivatives and which define surfaces G_m and H_m. Assume also that G_m and H_m have the same boundary; thus*

$$g(x)=h(x),\qquad x\in\partial\mathbf{a}. \tag{36}$$

Finally, let $f:C\to\mathbb{R}^{\binom{n}{m}}$ be a function with continuous components $f^{(j_1,\cdots,j_m)}$, $(j_1,\cdots,j_m)\in(m/n)$. Then

$$\int_{G_m}\sum_{(m/n)} f^{(j_1,\cdots,j_m)}(y)\,d(y^{j_1},\cdots,y^{j_m}) = \int_{H_m}\sum_{(m/n)} f^{(j_1,\cdots,j_m)}(y)\,d(y^{j_1},\cdots,y^{j_m}). \tag{37}$$

PROOF. Since Theorem 50.3 differs from Theorem 50.2 only in the dimensions of the surfaces G_m, H_m and of the affine space which contains $T(G_m)$, $T(H_m)$, the proofs of the two theorems are the same in all essential respects. □

50.4 Example. The theorems in this section have been stated for surfaces defined on a simplex $\mathbf{a}$, but they can be proved in the same way for surfaces defined on other regions, for example, on the square $A=[-1,1]\times[-1,1]$. Let $u:A\to\mathbb{R}^2$ and $v:A\to\mathbb{R}^2$ be two functions whose components (u^1,u^2)

and (v^1, v^2) have continuous derivatives, and define functions $g: A \to \mathbb{R}^2$ and $h: A \to \mathbb{R}^2$ as follows:

$$\begin{aligned} g^1(x^1, x^2) &= 2x^1 + [1-(x^1)^2][1-(x^2)^2]u^1(x^1, x^2), \\ g^2(x^1, x^2) &= 2x^2 + [1-(x^1)^2][1-(x^2)^2]u^2(x^1, x^2); \end{aligned} \qquad (x^1, x^2) \in A, \qquad (38)$$

$$\begin{aligned} h^1(x^1, x^2) &= 2x^1 + [1-(x^1)^2][1-(x^2)^2]v^1(x^1, x^2), \\ h^2(x^1, x^2) &= 2x^2 + [1-(x^1)^2][1-(x^2)^2]v^2(x^1, x^2). \end{aligned} \qquad (x^1, x^2) \in A, \qquad (39)$$

If (x^1, x^2) is a point on the boundary of A, then $[1-(x^1)^2][1-(x^2)^2] = 0$ and

$$g^1(x^1, x^2) = 2x^1 = h^1(x^1, x^2), \qquad g^2(x^1, x^2) = 2x^2 = h^2(x^1, x^2). \qquad (40)$$

Thus, for every choice of u and v, the functions g and h map each point (x^1, x^2) on the boundary of A into the same point; that is, $g(x) = h(x)$ for x in ∂A, and the affine surfaces G_2 and H_2 have the same boundary. Let C be a compact convex set which contains $T(G_2)$ and $T(H_2)$, and let $f: C \to \mathbb{R}$ be a continuous function. Then Theorem 50.1, stated for A rather than a simplex $\mathbf{a}$, shows that

$$\int_{G_2} f(y)\, d(y^1, y^2) = \int_{H_2} f(y)\, d(y^1, y^2). \qquad (41)$$

Since there are many possible choices for u and v, there are many surfaces G_2 and H_2 on which the integral of f has the same value.

Exercises

50.1. In the proof of Theorem 50.1, the chains $\Sigma\{\mathbf{y}: \mathbf{y} \in g(\mathscr{P}_k)\}$ and $\Sigma\{\mathbf{z}: \mathbf{z} \in h(\mathscr{P}_k)\}$ are two chains of 2-simplexes in $\mathbb{R}^2$ which have the same boundary.

(a) Assume that $T(G_2) = T(H_2)$, and make a sketch of the two chains to indicate geometrically why the following equation (17) in the proof of Theorem 50.1 is true.

$$\sum_{\mathbf{y} \in g(\mathscr{P}_k)} \int_{\mathbf{y}} f(y)\, d(y^1, y^2) = \sum_{\mathbf{z} \in h(\mathscr{P}_k)} \int_{\mathbf{z}} f(y)\, d(y^1, y^2). \qquad (17)$$

(b) Assume that $T(G_2) \cap T(H_2)$ is a proper subset of both $T(G_2)$ and $T(H_2)$, and make a sketch of the two chains to indicate geometrically why (17) is true.

(c) Use the additive property of the integral [see Sections 43 and 48] to give an analytic proof of equation (17).

50.2. (a) Give a geometric explanation of why equation (34) is true in the proof of Theorem 50.2.

(b) Use Theorem 48.9 to give a proof of equation (34).

50.3. Consider Example 50.4 again.

(a) Let B denote the square $[-2, 2] \times [-2, 2]$ in the (y^1, y^2)-plane. Show that, for every choice of u and v in the equations (38) and (39), the functions $g: A \to \mathbb{R}^2$ and $h: A \to \mathbb{R}^2$ map the point (x^1, x^2) on the boundary of A into the point $(2x^1, 2x^2)$ on the boundary of B.

(b) If $u^1(x^1, x^2) = u^2(x^1, x^2) = 3 + (x^1)^2 + (x^2)^2$ for (x^1, x^2) in A, show that $T(G_2)$ contains points outside the square B.

(c) If $v^1(x^1, x^2) = v^2(x^1, x^2) = 0$ for (x^1, x^2) in A, show that $h: A \to \mathbb{R}^2$ is the mapping $y^1 = 2x^1$, $y^2 = 2x^2$ of A onto B. Show also that, in this case, $\int_{H_2} f(y)\,d(y^1, y^2)$ is the Riemann integral $\int_B f(y)\,d(y^1, y^2)$ of f on the square B in $\mathbb{R}^2$ as defined in Section 35.

(d) Show that for every choice of u^1, u^2 in equation (38),

$$\int_{G_2} f(y^1, y^2)\,d(y^1, y^2) = \int_B f(y^1, y^2)\,d(y^1, y^2).$$

(e) Use (d) and Theorem 36.4 to show that, if $g: A \to \mathbb{R}^2$ is a function defined by equation 38, then

$$\int_A f[g^1(x), g^2(x)] D_{(1,2)}(g^1, g^2)(x)\,d(x^1, x^2) = \int_B f(y)\,d(y^1, y^2).$$

This formula is an example of a change-of-variable theorem [see Section 51]. Observe that $D_{(1,2)}(g^1, g^2)$ may have both positive and negative values in A, and that B may be a proper subset of $g(A)$.

(f) Let $f(y^1, y^2) = y^1 + y^2$, and let g be the function in (38) for which $g^1(x^1, x^2) = 2x^1$, $g^2(x^1, x^2) = 2x^2$ for all (x^1, x^2) in A. Show that

$$\int_A f[g^1(x), g^2(x)] D_{(1,2)}(g^1, g^2)(x)\,d(x^1, x^2) = \int_A 8(x^1 + x^2)\,d(x^1, x^2) = 0,$$

$$\int_B f(y^1, y^2)\,d(y^1, y^2) = \int_B (y^1 + y^2)\,d(y^1, y^2) = 0.$$

51. The Change-of-Variable Theorem

The change-of-variable theorems are among the most fundamental theorems of calculus. Some of these theorems are a part of the theory of integrals on surfaces, and others are a part of measure theory; this section treats only those of the former type.

51.1 Example. The fundamental theorem of the integral calculus is usually used to prove the most familiar change-of-variable theorem in one-dimensional elementary calculus [see Exercise 34.4]. Let $g: [a, b] \to \mathbb{R}$ be a function which has a continuous derivative; the mapping of $[a, b]$ into $\mathbb{R}$ by g is a one-dimensional surface (curve) G_1 whose trace $T(G_1)$ is a set in $\mathbb{R}$. Let E be an open set in $\mathbb{R}$ which contains $T(G_1)$, and let $f: E \to \mathbb{R}$ be a continuous function. The function $g: [a, b] \to \mathbb{R}$ is not assumed to be one-to-one, and the interval $[g(a), g(b)]$ may be a proper subset of $T(G_1)$. The change-of-variable theorem in this case is the following formula:

$$\int_{g(a)}^{g(b)} f(y)\,dy = \int_a^b f[g(x)]g'(x)\,dx. \tag{1}$$

To begin the proof of this formula, use Theorem 36.4 to show that

$$\int_{G_1} f(y)\,dy = \int_a^b f[g(x)]g'(x)\,dx. \tag{2}$$

Since f is continuous on E, there exists a function $F: E \to \mathbb{R}$ such that $F'(y) = f(y)$ [see Exercise 34.2]. Then by the definition of F', the chain rule for derivatives [see Theorem 4.1], and the fundamental theorem of the integral calculus [see (7) in Section 34],

$$\begin{aligned}\int_{G_1} f(y)\,dy &= \int_a^b f[g(x)]g'(x)\,dx = \int_a^b F'[g(x)]g'(x)\,dx \\ &= F[g(b)] - F[g(a)] = \int_{g(a)}^{g(b)} F'(y)\,dy = \int_{g(a)}^{g(b)} f(y)\,dy.\end{aligned} \tag{3}$$

This series of equations contains the proof of (1). Although the fundamental theorem of the integral calculus provides an easy proof of the change-of-variable theorem in this case, it has little, if any, role in other cases. This proof of (1) does not suggest methods which can be used to prove other change-of-variable theorems. The next example, however, shows how (1) can be proved without using the fundamental theorem, and the methods illustrated there can be widely generalized.

51.2 Example. The purpose of this example is to give a proof of the change-of-variable formula in (1) in Example 51.1 by the methods of Section 50 and without using the fundamental theorem of the integral calculus. As in (2), Theorem 36.4 shows that

$$\int_{G_1} f(y)\,dy = \int_a^b f[g(x)]g'(x)\,dx. \tag{4}$$

Define the function $h: [a, b] \to \mathbb{R}$ as follows:

$$h(x) = \frac{g(b) - g(a)}{b - a}(x - a) + g(a). \tag{5}$$

Then $g: [a, b] \to \mathbb{R}$ defines the surface G_1 with trace $T(G_1)$, and $h: [a, b] \to \mathbb{R}$ defines a surface H_1 whose trace $T(H_1)$ is the segment $[g(a), g(b)]$. Since $h(a) = g(a)$ and $h(b) = g(b)$ by (5), the surfaces G_1 and H_1 have the same boundary. Then the methods of Section 50 show that

$$\int_{G_1} f(y)\,dy = \int_{H_1} f(y)\,dy. \tag{6}$$

Since h is an affine transformation of $[a, b]$ into $[g(a), g(b)]$, it is easy to see that

$$\int_{H_1} f(y)\,dy = \int_{g(a)}^{g(b)} f(y)\,dy. \tag{7}$$

Then (6) and (7) show that

$$\int_{G_1} f(y)\,dy = \int_{g(a)}^{g(b)} f(y)\,dy, \tag{8}$$

and (4) and (8) complete the proof that

$$\int_{g(a)}^{g(b)} f(y)\,dy = \int_a^b f[g(x)]g'(x)\,dx, \tag{9}$$

which is the desired change-of-variable formula.

Exercise 50.3 contains the following change-of-variable formula.

$$\int_A f[g^1(x), g^2(x)]D_{(1,2)}(g^1, g^2)(x)\,d(x^1, x^2) = \int_B f(y)\,d(y^1, y^2). \tag{10}$$

Here A and B are squares, and $g: A \to \mathbb{R}^2$, with components g^1, g^2, maps the boundary of A into the boundary of B so that $y^1 = 2x^1$, $y^2 = 2x^2$. The next example shows that the methods used to establish this formula can be used also to prove similar formulas for integrals on simplexes.

51.3 Example. Let $\mathbf{a}: [a_0, a_1, a_2]$ and $\mathbf{b}: [b_0, b_1, b_2]$ be 2-simplexes in (x^1, x^2)-space and (y^1, y^2)-space respectively such that $\Delta(\mathbf{a}) \neq 0$ and $\Delta(\mathbf{b}) \neq 0$. Let $L: \mathbb{R}^2 \to \mathbb{R}^2$, with components (L_1, L_2), be the affine transformation such that $L(a_i) = b_i$, $i = 0, 1, 2$. Let $T_i(x) = 0$, $i = 0, 1, 2$, be the equations of the straight lines which contain the three sides of $\mathbf{a}$. Next, let $u: \mathbf{a} \to \mathbb{R}^2$ and $v: \mathbf{a} \to \mathbb{R}^2$ be functions whose components (u^1, u^2) and (v^1, v^2) have continuous derivatives. Finally, let $g: \mathbf{a} \to \mathbb{R}^2$ and $h: \mathbf{a} \to \mathbb{R}^2$ be functions whose components are defined as follows:

$$\begin{aligned} g^1(x^1, x^2) &= L_1(x) + T_1(x)T_2(x)T_3(x)u^1(x),\\ g^2(x^1, x^2) &= L_2(x) + T_1(x)T_2(x)T_3(x)u^2(x); \end{aligned} \qquad (x^1, x^2) \in \mathbf{a}, \tag{11}$$

$$\begin{aligned} h^1(x^1, x^2) &= L_1(x) + T_1(x)T_2(x)T_3(x)v^1(x),\\ h^2(x^1, x^2) &= L_2(x) + T_1(x)T_2(x)T_3(x)v^2(x). \end{aligned} \qquad (x^1, x^2) \in \mathbf{a}, \tag{12}$$

If $x: (x^1, x^2)$ is a point on the boundary of $\mathbf{a}$, then $T_1(x)T_2(x)T_3(x) = 0$ and

$$g^1(x^1, x^2) = L_1(x) = h^1(x^1, x^2), \qquad g^2(x^1, x^2) = L_2(x) = h^2(x^1, x^2). \tag{13}$$

Thus, for every choice of u and v, the functions g and h map each point (x^1, x^2) on the boundary of $\mathbf{a}$ into the same point; that is, $g(x) = h(x)$ for x in $\partial\mathbf{a}$, and the affine surfaces G_2 and H_2 defined by g and h have the same boundary. Let C be a compact convex set which contains $T(G_2)$ and $T(H_2)$, and let $f: C \to \mathbb{R}$ be a continuous function. Then Theorem 50.1 shows that

$$\int_{G_2} f(y)\,d(y^1, y^2) = \int_{H_2} f(y)\,d(y^1, y^2). \tag{14}$$

By Theorem 36.4,

$$\int_{G_2} f(y)\,d(y^1, y^2) = \int_{\mathbf{a}} f[g^1(x), g^2(x)]D_{(1,2)}(g^1, g^2)(x)\,d(x^1, x^2). \tag{15}$$

If $v^1(x) = v^2(x) = 0$ for x in ∂A, then (12) shows that h is the affine transformation L; in this case, $\int_{H_2} f(y)\,d(y^1, y^2)$ is the Riemann integral $\int_{\mathbf{b}} f(y)\,d(y^1, y^2)$. Then (14) and (15) show that for every choice of (u^1, u^2),

$$\int_{\mathbf{a}} f[g^1(x), g^2(x)]D_{(1,2)}(g^1, g^2)(x)\,d(x^1, x^2) = \int_{\mathbf{b}} f(y)\,d(y^1, y^2). \tag{16}$$

The regions of integration differ in (10) and (16), but otherwise these two change-of-variable formulas are the same.

The formulas in (10) and (16) suggest the general theorem in Theorem 51.4 below; the setting for this theorem is the following. Let $\mathbf{a}: [a_0, a_1, a_2]$ and $\mathbf{b}: [b_0, b_1, b_2]$ be 2-simplexes in (x^1, x^2)-space and (y^1, y^2)-space, respectively, such that $\Delta(\mathbf{a}) \neq 0$ and $\Delta(\mathbf{b}) \neq 0$. Let C be a compact convex set in (y^1, y^2)-space which contains $\mathbf{b}$. Let $L: \mathbb{R}^2 \to \mathbb{R}^2$, with components (L_1, L_2), be the affine transformation such that $L(a_i) = b_i$, $i = 0, 1, 2$.

51.4 Theorem. *Let $f: C \to \mathbb{R}$ be a continuous function, and let $g: \mathbf{a} \to \mathbb{R}^2$ be a function whose components (g^1, g^2) have continuous derivatives. If g defines a surface G_2 whose trace $T(G_2)$ is in C, and if*

$$g(x) = L(x), \qquad x \in \partial\mathbf{a}, \tag{17}$$

then

$$\int_{\mathbf{a}} f[g(x)]D_{(1,2)}(g^1, g^2)(x)\,d(x^1, x^2) = \int_{\mathbf{b}} f(y)\,d(y^1, y^2). \tag{18}$$

Proof. Let $h: \mathbf{a} \to \mathbb{R}^2$ be the function such that $h^i(x) = L_i(x)$ for $i = 1, 2$. Then h defines a surface H_2 whose trace $T(H_2)$ is $\mathbf{b}$. The hypothesis in (17) states that the surfaces G_2 and H_2 have the same boundary. Then Theorem 50.1 shows that

$$\int_{G_2} f(y)\,d(y^1, y^2) = \int_{H_2} f(y)\,d(y^1, y^2). \tag{19}$$

By Theorem 36.4, the integral on the left in (19) equals the integral on the left in (18). Finally, since h is the affine transformation L, the surface integral $\int_{H_2} f(y)\,d(y^1, y^2)$ is the Riemann integral $\int_{\mathbf{b}} f(y)\,d(y^1, y^2)$. These facts and (19) complete the proof of (18) and of Theorem 51.4. □

The setting for Theorem 51.5 below is the following. Let C be a compact convex set in a 2-dimensional affine space (a plane) in (y^1, y^2, y^3)-space. Let $\mathbf{a}: [a_0, a_1, a_2]$, $\Delta(\mathbf{a}) \neq 0$, be a 2-simplex in (x^1, x^2)-space, and let $\mathbf{b}: [b_0, b_1, b_2]$ be a 2-simplex in C. Let $L: \mathbb{R}^2 \to \mathbb{R}^3$, with components (L_1, L_2, L_3) be the affine transformation such that $L(a_i) = b_i$, $i = 0, 1, 2$.

51.5 Theorem. *Let $f: C \to \mathbb{R}^3$ be a function whose components $f^{(j_1, j_2)}: C \to \mathbb{R}$, $(j_1, j_2) \in (2/3)$, are continuous, and let $g: \mathbf{a} \to \mathbb{R}$ be a function whose components (g^1, g^2, g^3) have continuous derivatives. If*

$$g(x) = L(x), \qquad x \in \partial\mathbf{a}, \tag{20}$$

then

$$\int_{\mathbf{a}} \sum_{(2/3)} f^{(j_1, j_2)}[g(x)] D_{(1,2)}(g^{j_1}, g^{j_2})(x)\, d(x^1, x^2) = \int_{\mathbf{b}} \sum_{(2/3)} f^{(j_1, j_2)}(y)\, d(y^{j_1}, y^{j_2}). \tag{21}$$

PROOF. Let $h: \mathbf{a} \to \mathbb{R}^3$ be the function such that $h^i(x) = L_i(x)$, $i = 1, 2, 3$. Then h defines a surface H_2 whose trace $T(H_2)$ is $\mathbf{b}$. The hypothesis in (20) states that the surfaces G_2 and H_2 have the same boundary. Then Theorem 50.2 shows that

$$\int_{G_2} \sum_{(2/3)} f^{(j_1, j_2)}(y)\, d(y^{j_1}, y^{j_2}) = \int_{H_2} \sum_{(2/3)} f^{(j_1, j_2)}(y)\, d(y^{j_1}, y^{j_2}). \tag{22}$$

By Theorem 36.4, the integral on the left in (22) equals the integral on the left in (21). Finally, since h is the affine transformation L, the integral on the right in (22) is the integral on the simplex $\mathbf{b}$ on the right in (21) by Theorem 37.2. These facts and (22) complete the proof of (21) and of Theorem 51.5. □

The change-of-variable formula

$$\int_{\mathbf{a}} f[g(x)] D_{(1,2)}(g^1, g^2)(x)\, d(x^1, x^2) = \int_{\mathbf{b}} f(y)\, d(y^1, y^2) \tag{23}$$

[see (18) in Theorem 51.4] is a very satisfactory theorem because in some respects the function $g: \mathbf{a} \to \mathbb{R}^2$ is a very general function. The simplex $\mathbf{b}$ may be a proper subset of $T(G_2)$, a point $g(x)$ may be the image under g of many points x in $\mathbf{a}$, and the Jacobian $D_{(1,2)}(g^1, g^2)$ may have positive, negative, and zero values in $\mathbf{a}$. The proof of formula (23) depends on the special way in which g maps the boundary of $\mathbf{a}$ into the boundary of $\mathbf{b}$, and in this respect Theorem 51.4 is very unsatisfactory because the hypothesis $g(x) = L(x)$ for x in $\partial\mathbf{a}$ in (17) is a very strong hypothesis. Theorem 51.6 below is a much better theorem because the hypothesis $g(x) = L(x)$ on the boundary of $\mathbf{a}$ is replaced by a much weaker hypothesis.

The setting for Theorem 51.6 is the following. Let $\mathbf{a}: [a_0, a_1, a_2]$ and $\mathbf{b}: [b_0, b_1, b_2]$ be 2-simplexes in (x^1, x^2)-space and (y^1, y^2)-space, respectively, such that $\Delta(\mathbf{a}) \neq 0$ and $\Delta(\mathbf{b}) \neq 0$. Let C be a compact convex set in (y^1, y^2)-space which contains $\mathbf{b}$. Let $g: \mathbf{a} \to C$ be a function with the following properties: (a) $g(a_i) = b_i$, $i = 0, 1, 2$; and (b) if x is a point on a side of $[a_0, a_1, a_2]$, then $g(x)$ is a point on the corresponding side of $[b_0, b_1, b_2]$. Then g is said to map the vertices and sides of $\mathbf{a}$ onto the corresponding vertices and sides of $\mathbf{b}$. It is not assumed that the trace $T(G_2)$ of G_2 is contained in $\mathbf{b}$, nor that the mappings of the sides of $\mathbf{a}$ onto the sides of $\mathbf{b}$ are one-to-one.

51.6 Theorem. *Let $f: C \to \mathbb{R}$ be a continuous function, and let $g: \mathbf{a} \to C$ be a function whose components (g^1, g^2) have continuous derivatives and which maps the vertices and sides of $\mathbf{a}$ onto the corresponding vertices and sides of $\mathbf{b}$. Then*

$$\int_{\mathbf{a}} f[g(x)]D_{(1,2)}(g^1, g^2)(x)\,d(x^1, x^2) = \int_{\mathbf{b}} f(y)\,d(y^1, y^2). \tag{24}$$

Proof. Much of the proof of this theorem is similar to the proof of Theorem 50.1. Let $\mathscr{P}_k$ be a sequence of simplicial subdivisions of $\mathbf{a}$. If $\mathbf{x}:[x_0, x_1, x_2]$ is a simplex in $\mathscr{P}_k$, then g maps the vertices x_i, $i = 0, 1, 2$, into points $y_i = g(x_i)$. Denote the simplex with vertices y_0, y_1, y_2 by $\mathbf{y}:[y_0, y_1, y_2]$. Now by Definition 36.3,

$$\int_{G_2} f(y)\,d(y^1, y^2) = \lim_{k\to\infty} \sum_{\mathbf{x}\in\mathscr{P}_k} f[g(x_0)]\frac{(-1)^2}{2!}\Delta_{(1,2)}(g^1, g^2)(\mathbf{x}). \tag{25}$$

There are two steps in the proof; the first step is the proof that

$$\lim_{k\to\infty} \sum_{\mathbf{y}\in g(\mathscr{P}_k)} \int_{\mathbf{y}} f(y)\,d(y^1, y^2) = \int_{G_2} f(y)\,d(y^1, y^2). \tag{26}$$

The proof of this statement uses the mean-value theorem for integrals and the definition of the surface integral in (25). The complete proof of (26) is exactly the same as the proof of equation (5) in the proof of Theorem 50.1, and the reader should review it in Section 50. The second step is the proof that

$$\sum_{\mathbf{y}\in g(\mathscr{P}_k)} \int_{\mathbf{y}} f(y)\,d(y^1, y^2) = \int_{\mathbf{b}} f(y)\,d(y^1, y^2). \tag{27}$$

Then (26) and (27) show that

$$\int_{G_2} f(y)\,d(y^1, y^2) = \int_{\mathbf{b}} f(y)\,d(y^1, y^2). \tag{28}$$

By Theorem 36.4,

$$\int_{G_2} f(y)\,d(y^1, y^2) = \int_{\mathbf{a}} f[g(x)]D_{(1,2)}(g^1, g^2)(x)\,d(x^1, x^2), \tag{29}$$

and (28) and (29) show that

$$\int_{\mathbf{a}} f[g(x)]D_{(1,2)}(g^1, g^2)(x)\,d(x^1, x^2) = \int_{\mathbf{b}} f(y)\,d(y^1, y^2). \tag{30}$$

This equation is the same as (24), which is the change-of-variable formula in the conclusion of Theorem 51.6. Thus the entire proof of the theorem can be completed by establishing equation (27).

The proof of (27) is based on the addition property of the integral in Sections 43 and 48, and Example 43.7 illustrates the way the additive property of the integral is used in the present proof. Let $\mathbf{y}:[y_0, y_1, y_2]$ be a simplex in $g(\mathscr{P}_k)$. Since f is continuous on C and since C is convex, each of the integrals

$$\int_{[b_0, y_1, y_2]} f(y)\,d(y^1, y^2), \quad \int_{[y_0, b_0, y_2]} f(y)\,d(y^1, y^2), \quad \int_{[y_0, y_1, b_0]} f(y)\,d(y^1, y^2), \tag{31}$$

is defined, and the additive property of the integral states that

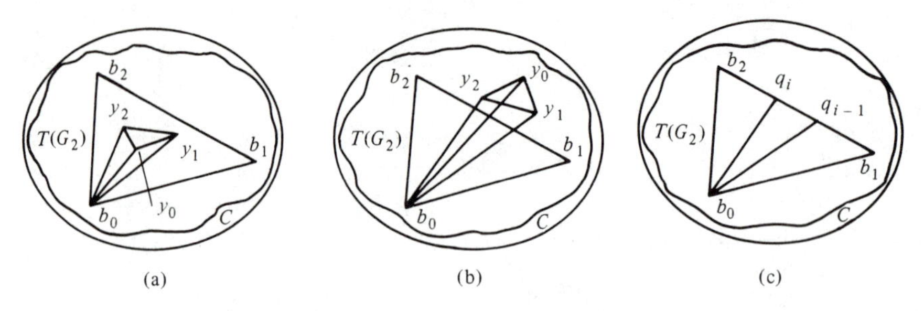

Figure 51.1. Steps in the proof of (27).

$$\int_{[y_0, y_1, y_2]} f(y)\,d(y^1, y^2) = \left\{\int_{[b_0, y_1, y_2]} + \int_{[y_0, b_0, y_2]} + \int_{[y_0, y_1, b_0]}\right\} f(y)\,d(y^1, y^2). \tag{32}$$

Figures 51.1(a) and (b) show, for two typical positions of $[y_0, y_1, y_2]$, the three simplexes $[b_0, y_1, y_2]$, $[y_0, b_0, y_2]$, and $[y_0, y_1, b_0]$ which are employed in the additive property of the integral to obtain (32). If a second simplex in $g(\mathscr{P}_k)$ contains the vertices y_1, y_2, then this simplex has the form $[y_0', y_2, y_1]$, and

$$\int_{[y_0', y_2, y_1]} f(y)\,d(y^1, y^2) = \left\{\int_{[b_0, y_2, y_1]} + \int_{[y_0', b_0, y_1]} + \int_{[y_0', y_2, b_0]}\right\} f(y)\,d(y^1, y^2). \tag{33}$$

Now $[b_0, y_2, y_1] = -[b_0, y_1, y_2]$, and

$$\int_{[b_0, y_2, y_1]} f(y)\,d(y^1, y^2) = -\int_{[b_0, y_1, y_2]} f(y)\,d(y^1, y^2). \tag{34}$$

Thus if each integral in

$$\sum_{\mathbf{y} \in g(\mathscr{P}_k)} \int_{[y_0, y_1, y_2]} f(y)\,d(y^1, y^2) \tag{35}$$

is represented as in (32), then all integrals cancel in the sum except those integrals $\int_{[b_0, y_1, y_2]} f(y)\,d(y^1, y^2)$ for which (y_1, y_2) belongs to the boundary of the chain $g(\mathscr{P}_k)$. Let $\mathbf{c}$ denote this boundary chain as follows:

$$\mathbf{c} = \sum \{(y_1, y_2) : (y_1, y_2) \in \partial g(\mathscr{P}_k)\}. \tag{36}$$

Thus equations (32), $\cdots$, (36) show that

$$\sum_{\mathbf{y} \in g(\mathscr{P}_k)} \int_{\mathbf{y}} f(y)\,d(y^1, y^2) = \sum_{(y_1, y_2) \in \mathbf{c}} \int_{[b_0, y_1, y_2]} f(y)\,d(y^1, y^2). \tag{37}$$

Now the boundary of the chain $\mathscr{P}_k$ lies in the sides of $\mathbf{a}$, and the hypothesis that g maps the sides of $\mathbf{a}$ onto the corresponding sides of $\mathbf{b}$ shows that $\mathbf{c}$ consists of one-dimensional simplexes (y_1, y_2) in the sides of $\mathbf{b}$. If (y_1, y_2) is in the side $[b_0, b_1]$ or $[b_0, b_2]$ of $\mathbf{b}$, then $[b_0, y_1, y_2]$ is a degenerate simplex,

$\Delta(b_0, y_1, y_2) = 0$, and

$$\int_{[b_0, y_1, y_2]} f(y)\, d(y^1, y^2) = 0. \tag{38}$$

Thus in the sum on the right in (37), all terms equal zero except those for which (y_1, y_2) is in $\mathbf{c}$ and also in the side $[b_1, b_2]$ of $\mathbf{b}$. Let the vertices in $\mathbf{c}$ which are on the side $[b_1, b_2]$ be denoted by q_i, $i = 0, 1, \cdots, r$. Then $q_0 = b_1$ and $q_r = b_2$, and equation (37) simplifies to the following:

$$\sum_{\mathbf{y} \in g(\mathscr{P}_k)} \int_{\mathbf{y}} f(y)\, d(y^1, y^2) = \sum_{i=1}^{r} \int_{[b_0, q_{i-1}, q_i]} f(y)\, d(y^1, y^2). \tag{39}$$

Figure 51.1(c) shows a typical simplex $[b_0, q_{i-1}, q_i]$ in the set $\{[b_0, q_{i-1}, q_i]: i = 1, \cdots, r\}$ which is used to represent the sum on the left in (39). The simplexes in this set may or may not overlap because the points $q_0, q_1, \cdots, q_r$ may or may not proceed in order from b_1 to b_2. Now make a second application of the additive property of the integral to evaluate the sum on the right in (39); represent, as follows, each integral in this sum as the sum of integrals on three simplexes, one of whose vertices is b_1:

$$\begin{aligned} &\int_{[b_0, q_{i-1}, q_i]} f(y)\, d(y^1, y^2) \\ &\qquad = \left\{ \int_{[b_1, q_{i-1}, q_i]} + \int_{[b_0, b_1, q_i]} + \int_{[b_0, q_{i-1}, b_1]} \right\} f(y)\, d(y^1, y^2). \end{aligned} \tag{40}$$

The simplex $[b_1, q_{i-1}, q_i]$ is degenerate since its three vertices are on the side $[b_1, b_2]$ of $\mathbf{b}$ [see Figure 51.1(c)]; thus $\Delta(b_1, q_{i-1}, q_i) = 0$ and the integral on $[b_1, q_{i-1}, q_i]$ equals zero. Also, $[b_0, q_{i-1}, b_1] = -[b_0, b_1, q_{i-1}]$ and

$$\int_{[b_0, q_{i-1}, b_1]} f(y)\, d(y^1, y^2) = -\int_{[b_0, b_1, q_{i-1}]} f(y)\, d(y^1, y^2). \tag{41}$$

Thus (40) simplifies to

$$\int_{[b_0, q_{i-1}, q_i]} f(y)\, d(y^1, y^2) = \left\{ \int_{[b_0, b_1, q_i]} - \int_{[b_0, b_1, q_{i-1}]} \right\} f(y)\, d(y^1, y^2). \tag{42}$$

Substitute from (42) in the sum on the right in (39); that sum becomes a telescoping sum which can be evaluated as follows:

$$\begin{aligned} &\sum_{\mathbf{y} \in g(\mathscr{P}_k)} \int_{\mathbf{y}} f(y)\, d(y^1, y^2) \\ &\qquad = \sum_{i=1}^{r} \left\{ \int_{[b_0, b_1, q_i]} f(y)\, d(y^1, y^2) - \int_{[b_0, b_1, q_{i-1}]} f(y)\, d(y^1, y^2) \right\} \\ &\qquad = \int_{[b_0, b_1, q_r]} f(y)\, d(y^1, y^2) - \int_{[b_0, b_1, q_0]} f(y)\, d(y^1, y^2). \end{aligned} \tag{43}$$

Now $q_r = b_2$ and $q_0 = b_1$; thus the first integral on the right is the integral on $[b_0, b_1, b_2]$, and the second integral equals zero since $[b_0, b_1, b_1]$ is degenerate. Thus the final form of equation (43) is this:

$$\sum_{\mathbf{y}\in g(\mathscr{P}_k)} \int_{\mathbf{y}} f(y)\,d(y^1, y^2) = \int_{[b_0, b_1, b_2]} f(y)\,d(y^1, y^2). \tag{44}$$

This equation completes the proof of equation (27) and thus that of Theorem 51.6. □

51.7 Example. Let A be a square in (x^1, x^2)-space, and let B be a square in (y^1, y^2)-space. Let $g: A \to \mathbb{R}^2$ be a function whose components (g^1, g^2) have continuous derivatives and which maps the vertices and sides of A onto the corresponding vertices and sides of B. Let C be a compact convex set which contains $T(G_2)$, and let $f: C \to \mathbb{R}$ be a continuous function. Then

$$\int_A f[g(x)]D_{(1,2)}(g^1, g^2)(x)\,d(x^1, x^2) = \int_B f(y)\,d(y^1, y^2), \tag{45}$$

and the proof of this change-of-variable formula differs only in some obvious details from the proof of Theorem 51.6. For formula (10), the hypothesis is that each side of A is mapped linearly onto the corresponding side of B; for (45), the assumption is only that g maps the vertices and sides of A onto the corresponding vertices and sides of B. If the hypotheses for (10) are satisfied, those for (45) are satisfied and (10) is a special case of (45). For a numerical example of (45), let $A = [-1, 1] \times [-1, 1]$ and $B = [-1, 1] \times [-1, 1]$, and let

$$\begin{aligned} &y^1 = g^1(x^1, x^2) = (x^1)^3, \\ &y^2 = g^2(x^1, x^2) = (x^2)^3, \qquad (x^1, x^2) \in A; \qquad f(y^1, y^2) = (y^1)^2 + (y^2)^2. \end{aligned} \tag{46}$$

Then

$$\begin{aligned} \int_A f[g(x)]D_{(1,2)}(g^1, g^2)(x)\,d(x^1, x^2) &= \int_A [(x^1)^6 + (x^2)^6]9(x^1)^2(x^2)^2\,d(x^1, x^2) \\ &= 8/3, \end{aligned} \tag{47}$$

$$\int_B f(y^1, y^2)\,d(y^1, y^2) = \int_B [(y^1)^2 + (y^2)^2]\,d(y^1, y^2) = 8/3. \tag{48}$$

An evaluation of these integrals by iterated integrals shows that each one has the value 8/3. These values verify the change-of-variable formula in (45). Given the change-of-variable formula in (45), it would clearly be easier to evaluate (47) by making the change of variables shown in (46) and then to evaluate (48).

Exercises

51.1. Let $A = [-\pi/2, \pi/2] \times [-\pi/2, \pi/2]$ in (x^1, x^2)-space, and let $b = [-1, 1] \times [-1, 1]$ in (y^1, y^2)-space.

(a) Show that the transformation

$$y^1 = g^1(x^1, x^2) = \sin x^1,$$
$$y^2 = g^2(x^1, x^2) = \sin x^2, \qquad (x^1, x^2) \in A,$$

maps A onto B and transforms the vertices and sides of A into the corresponding vertices and sides of B.

(b) Let $f(y^1, y^2) = y^1 + y^2$, and show that $\int_B f(y)\,d(y^1, y^2) = 0$.

(c) Show that

$$\int_A f[g^1(x), g^2(x)]D_{(1,2)}(g^1, g^2)(x)\,d(x^1, x^2)$$
$$= \int_A [\sin x^1 + \sin x^2]\cos x^1 \cos x^2\,d(x^1, x^2) = 0.$$

(d) Show that the values obtained in (b) and (c) verify the change-of-variable formula in (45).

51.2. Let $A = [-1, 1] \times [-1, 1]$ in (x^1, x^2)-space, and let $B = [-1, 1] \times [-1, 1]$ in (y^1, y^2)-space.

(a) Show that the transformation

$$y^1 = g^1(x^1, x^2) = 2(x^1)^3 - x^1,$$
$$y^2 = g^2(x^1, x^2) = 2(x^2)^3 - x^2, \qquad (x^1, x^2) \in A,$$

maps A onto B and transforms the vertices and sides of A into the corresponding vertices and sides of B.

(b) Let

$$f(y^1, y^2) = \sin(\pi/2)y^1 + \cos(\pi/2)y^2,$$

and show that

$$\int_B f(y^1, y^2)\,d(y^1, y^2) = 8/\pi.$$

(c) Use the change-of-variable formula in (45) to make the change of variables shown in (a) and thus to show that

$$\int_A f[g^1(x), g^2(x)]D_{(1,2)}(g^1, g^2)(x)\,d(x^1, x^2) = 8/\pi.$$

Do you think you could evaluate this integral without the help of the change-of-variable formula?

51.3. Let $A = [-1, 1] \times [-1, 1]$ in (x^1, x^2)-space, and let $B = [-2, 2] \times [-2, 2]$ in (y^1, y^2)-space. Let $u: A \to \mathbb{R}^2$ be a function whose components (u^1, u^2) have continuous derivatives. Let $g: A \to \mathbb{R}^2$ be a function whose components (g^1, g^2) are defined by the following equations:

$$y^1 = g^1(x^1, x^2) = 2x^1 + [1 - (x^1)^2][1 - (x^2)^2]u^1(x^1, x^2),$$
$$y^2 = g^2(x^1, x^2) = 2x^2 + [1 - (x^1)^2][1 - (x^2)^2]u^2(x^1, x^2), \qquad (x^1, x^2) \in A.$$

Let C be a compact convex set in (y^1, y^2)-space which contains $T(G_2)$, and let $f: C \to \mathbb{R}$ be a continuous function such that $\int_B f(y^1, y^2)\,d(y^1, y^2) = 100$.

(a) Show that g maps the vertices and sides of A onto the vertices and sides of B, and that the hypotheses for (45) stated in Example 51.7 are satisfied. Show that for some functions g, the square B is a proper subset of the trace $T(G_2)$.

(b) Show that

$$\int_A f[g^1(x), g^2(x)]D_{(1,2)}(g^1, g^2)(x)\,d(x^1, x^2) = 100$$

for every function g which satisfies the stated hypothesis. Can you evaluate this integral without using the change-of-variable theorem?

51.4. Let C be a compact convex set in a 2-dimensional affine space (a plane) in (y^1, y^2, y^3)-space. Let $\mathbf{a}: [a_0, a_1, a_2]$, $\Delta(\mathbf{a}) \neq 0$, be a 2-simplex in (x^1, x^2)-space, and let $\mathbf{b}: [b_0, b_1, b_2]$ be a 2-simplex in C.

(a) Prove the following theorem. Let $f: C \to \mathbb{R}^3$ be a function whose components $f^{(j_1, j_2)}: C \to \mathbb{R}$, $(j_1, j_2) \in (2/3)$, are continuous, and let $g: \mathbf{a} \to C$ be a function whose components (g^1, g^2, g^3) have continuous derivatives. If g maps the vertices and sides of $\mathbf{a}$ into the corresponding vertices and sides of $\mathbf{b}$, then

$$\int_{\mathbf{a}} \sum_{(2/3)} f^{(j_1, j_2)}[g(x)]D_{(1,2)}(g^{j_1}, g^{j_2})(x)\,d(x^1, x^2) = \int_{\mathbf{b}} \sum_{(2/3)} f^{(j_1, j_2)}(y)\,d(y^{j_1}, y^{j_2}).$$

(b) Determine the relation of this theorem to Theorems 51.5 and 51.6.

51.5. Let $\mathbf{a}: [a_0, a_1, \cdots, a_3]$ and $\mathbf{b}: [b_0, b_1, \cdots, b_3]$ be 3-simplexes in $(x^1, \cdots, x^3)$-space and $(y^1, \cdots, y^3)$-space, respectively, such that $\Delta(\mathbf{a}) \neq 0$ and $\Delta(\mathbf{b}) \neq 0$. Let C be a compact convex set in $(y^1, \cdots, y^3)$-space which contains $\mathbf{b}$. Prove the following theorem.

Theorem. *Let $f: C \to \mathbb{R}$ be a continuous function, and let $g: \mathbf{a} \to C$ be a function whose components $(g^1, \cdots, g^3)$ have continuous derivatives and which maps the vertices and sides of $\mathbf{a}$ onto the corresponding vertices and sides of $\mathbf{b}$. Then*

$$\int_{\mathbf{a}} f[g(x)]D_{(1,\cdots,3)}(g^1, \cdots, g^3)(x)\,d(x^1, \cdots, x^3) = \int_{\mathbf{b}} f(y)\,d(y^1, \cdots, y^3).$$

CHAPTER 8

The Evaluation of Integrals

52. Introduction

The evaluation of integrals $\int_a^b f(x)\,dx$, based on the fundamental theorem of the integral calculus, is a familiar problem in elementary calculus. If $F'(x) = f(x)$ on $[a, b]$, then $\int_a^b f(x)\,dx = \int_a^b F'(x)\,dx = F(b) - F(a)$. Here F is called a primitive of f, and $F(b) - F(a)$ is an evaluation of $\int_a^b f(x)\,dx$. If F_1 and F_2 are two primitives of f, then $F_1 - F_2$ is a constant function. Thus there is essentially one primitive of f and one evaluation of $\int_a^b f(x)\,dx$.

There are similar evaluation problems for integrals in $\mathbb{R}^m$ and on surfaces G_m. For these integrals, the fundamental theorem is again a basic tool; there are many similarities with the one-dimensional case, but the differences are striking. Some of these similarities and differences are illustrated by the following examples.

Let $A_2 = [a_1, b_1] \times [a_2, b_2]$ be a rectangular region in $\mathbb{R}^2$. The integral

$$\int_{A_2} 3[(x^1)^2 + (x^2)^2]\,d(x^1, x^2) \tag{1}$$

can be evaluated by the following statements of the fundamental theorem of the integral calculus in Theorem 38.4 and Corollary 38.7 [stated for a rectangular region rather than a simplex].

$$\int_{A_2} D_{(1,2)}(f^1, f^2)\,d(x^1, x^2) = \int_{\partial A_2} f^1(x) D_1 f^2(x)\,dx^1 + f^1(x) D_2 f^2(x)\,dx^2 \tag{2}$$

$$= -\int_{\partial A_2} f^2(x) D_1 f^1(x)\,dx^1 + f^2(x) D_2 f^1(x)\,dx^2. \tag{3}$$

Define two functions $(f^1, f^2)\colon A_2 \to \mathbb{R}^2$ as follows:

$$f^1(x) = (x^1)^3 + 3x^1(x^2)^2, \qquad f^2(x) = x^2; \tag{4}$$

$$f^1(x) = x^1, \qquad f^2(x) = 3(x^1)^2x^2 + (x^2)^3. \tag{5}$$

Then in both cases $D_{(1,2)}(f^1, f^2)(x) = 3[(x^1)^2 + (x^2)^2]$, and the integrals on the right in (2) and (3) are evaluations of (1).

If $k: A_2 \to \mathbb{R}$ is a function which has continuous derivatives, then Theorems 39.5, 40.5 and (6) in Examples 48.2 show that

$$\int_{\partial A_2} D_1k(x)\,dx^1 + D_2k(x)\,dx^2 = 0. \tag{6}$$

Thus, if the integral on the right in (2) is an evaluation of (1), then

$$\int_{\partial A_2} [f^1(x)D_1f^2(x) + D_1k(x)]\,dx^1 + [f^1(x)D_2f^2(x) + D_2k(x)]\,dx^2 \tag{7}$$

is also an evaluation of (1).

Still other evaluations of (1) can be obtained from the following Green's theorem [see equation (7) in Section 41]:

$$\int_{A_2} [D_1f^2(x) - D_2f^1(x)]\,d(x^1, x^2) = \int_{\partial A_2} f^1(x)\,dx^1 + f^2(x)\,dx^2. \tag{8}$$

If

$$f^1(x) = x^1(x^2)^2 - (x^2)^3, \qquad f^2(x) = (x^1)^3 + (x^1)^2x^2, \tag{9}$$

then the integral on the left in (8) is (1), and the integral on the right in (8) is an evaluation of it. Finally, there are two evaluations of (1) by iterated integrals [see Example 35.20].

Let $\mathbf{a}: [a_0, a_1, a_2]$, $\Delta(\mathbf{a}) \neq 0$, be a simplex in $\mathbb{R}^2$, and let $g: \mathbf{a} \to \mathbb{R}^3$ be a function whose components (g^1, g^2, g^3) have continuous derivatives and which defines a surface G_2 in $\mathbb{R}^3$. Also, let

$$f^1(y) = \sum_{j=1}^{3} (y^j)^2, \qquad f^2(y) = \sum_{j=1}^{3} y^j, \qquad y = (y^1, y^2, y^3). \tag{10}$$

Then by Theorem 39.1,

$$\int_{G_2} 2(y^1 - y^2)\,d(y^1, y^2) + 2(y^1 - y^3)\,d(y^1, y^3) + 2(y^2 - y^3)\,d(y^2, y^3) \tag{11}$$

$$= \int_{\partial G_2} \sum_{j=1}^{3} (y^j)^2\,dy^1 + \sum_{j=1}^{3} (y^j)^2\,dy^2 + \sum_{j=1}^{3} (y^j)^2\,dy^3. \tag{12}$$

As in the first example, there are many other evaluations for the integral in (11) [see the exercises].

These examples show that a function of more than one variable may have essentially different primitives, and that an integral may have many different evaluations. Some evaluations of an integral are based directly on a primitive of the integrand, but others are derived with the help of various properties of integrals.

This chapter is concerned with various problems related to the evaluation of integrals of the following types:

$$\int_{A_n} f(x)\,d(x^1, \cdots, x^n) \quad \text{or} \quad \int_{\mathbf{a}} f(x)\,d(x^1, \cdots, x^n), \tag{13}$$

$$\int_{G_m} \sum_{(m/n)} f^{(j_1, \cdots, j_m)}(y)\,d(y^{j_1}, \cdots, y^{j_m}). \tag{14}$$

This chapter treats the following topics: the existence of evaluations for a given integral; properties of the class of evaluations for a given integral; properties of the class of integrands whose integrals have evaluations; necessary and sufficient conditions for the existence of a primitive of a given integrand; properties of the class of primitives of a given integrand; and properties of the class of integrands which have primitives. As indicated already, the results in the case $n = 1$ are frequently very different from those in the cases $n > 1$.

Exercises

52.1. Evaluate the integral in (1) by means of iterated integrals.

52.2. (a) If f^1, f^2 are the functions in (4), show that $\int_{A_2} D_{(1,2)}(f^1, f^2)(x)\,d(x^1, x^2)$ is the integral in (1).

(b) Use (2) and (3) to show that the integral in (1) equals the following integrals:

$$\int_{\partial A_2} [(x^1)^3 + 3x^1(x^2)^2]\,dx^2; \quad -\int_{\partial A_2} x^2[3(x^1)^2 + 3(x^2)^2]\,dx^1 + x^2(6x^1x^2)\,dx^2.$$

(c) Evaluate each of the integrals in (b); vertify that the value of these integrals is the value of (1) obtained in Exercise 52.1.

52.3. Repeat Exercise 52.2, using the functions f^1, f^2 in (5) to evaluate (1).

52.4. (a) If

$$f^1(x^1, x^2) = x^1(x^2)^2 - (x^2)^3, \qquad f^2(x^1, x^2) = (x^1)^3 + (x^1)^2x^2,$$

show that (8) provides an evaluation of (1).

(b) Use (a) to show that

$$\int_{A_2} 3[(x^1)^2 + (x^2)^2]\,d(x^1, x^2)$$

$$= \int_{\partial A_2} [x^1(x^2)^2 - (x^2)^3]\,dx^1 + [(x^1)^3 + (x^1)^2x^2]\,dx^2.$$

(c) Let $k(x^1, x^2) = -(x^1)^2(x^2)^2/2$; use (6) and part (b) of this exercise to show that

$$\int_{A_2} 3[(x^1)^2 + (x^2)^2]\,d(x^1, x^2) = -\int_{\partial A_2} (x^2)^3\,dx^1 - (x^1)^3\,dx^2,$$

$$= (b_1 - a_1)(b_2^3 - a_2^3) + (b_2 - a_2)(b_1^3 - a_1^3).$$

(d) Compare the evaluations of (1) obtained in this exercise with those obtained in Exercises 52.2 and 52.3. Verify that all of these evaluations give the value for (1) obtained in Exercise 52.1.

52.5. Show that the integral in (11) equals

$$-2\int_{\partial G_2} y^1 \sum_{j=1}^{3} y^j\,dy^1 + y^2 \sum_{j=1}^{3} y^j\,dy^2 + y^3 \sum_{j=1}^{3} y^j\,dy^3.$$

52.6. Use the function $k(y^1, \cdots, y^3) = -[(y^1)^3 + \cdots + (y^3)^3]/3$ to show that the integral (11) equals

$$\int_{\partial G_2} [(y^2)^2 + (y^3)^2]\,dy^1 + [(y^1)^2 + (y^3)^2]\,dy^2 + [(y^1)^2 + (y^2)^2]\,dy^3.$$

52.7. Use Exercise 52.5 to show that (11) equals

$$-2\int_{\partial G_2} y^1(y^2 + y^3)\,dy^1 + y^2(y^1 + y^3)\,dy^2 + y^3(y^1 + y^2)\,dy^3.$$

52.8. By choosing special functions for k, find two evaluations, other than those in (12) and Exercises 52.6 and 52.7, for (11).

53. Definitions

This chapter is concerned with integrals on surfaces G_m in $\mathbb{R}^n$, $m \leqq n$, and on rectangular regions A_n in $\mathbb{R}^n$, and this section contains some necessary definitions. These definitions are stated for integrals on G_m; the interpretation for integrals on $\mathbf{a}$ and A_n is usually left to the reader.

Let $\mathbf{a}: [a_0, a_1, \cdots, a_m]$, $\Delta(\mathbf{a}) \neq 0$, be an m-simplex in $\mathbb{R}^m$. Let $g: \mathbf{a} \to \mathbb{R}^n$, $m \leqq n$, be a function whose components $(g^1, \cdots, g^n)$ have continuous derivatives and which defines a surface G_m with trace $T(G_m)$. Also, let E be an open set in $\mathbb{R}^n$ which contains $T(G_m)$.

53.1 Definition. Let $f: T(G_m) \to \mathbb{R}^{\binom{n}{m}}$ and $v: T(\partial G_m) \to \mathbb{R}^{\binom{n}{m-1}}$ be functions whose components $f^{(j_1, \cdots, j_m)}$ and $v^{(j_1, \cdots, j_{m-1})}$ are continuous. If

$$\int_{G_m} \sum_{(m/n)} f^{(j_1, \cdots, j_m)}(y)\,d(y^{j_1}, \cdots, y^{j_m}) \tag{1}$$

$$= \int_{\partial G_m} \sum_{(m-1/n)} v^{(j_1, \cdots, j_{m-1})}(y)\,d(y^{j_1}, \cdots, y^{j_{m-1}}), \tag{2}$$

then the integral in (2) is said to be an evaluation of the one in (1), and the function v is an evaluator of (1). The class of all evaluators of (1) is denoted by $\mathscr{V}(f, G_m)$, and the class of functions $f: T(G_m) \to \mathbb{R}^{\binom{n}{m}}$ for which (1) has an evaluation is denoted by $\mathscr{F}_e(G_m)$. Finally, the class of all evaluators for all integrals (1) with f in $\mathscr{F}_e(G_m)$ is denoted by $\mathscr{V}(G_m)$.

53.2 Example. By the fundamental theorem of the integral calculus in Theorem 39.1,

$$\int_{G_m} \sum_{(m/n)} D_{(j_1,\cdots,j_m)}(f^1, \cdots, f^m)(y)\, d(y^{j_1}, \cdots, y^{j_m}) \tag{3}$$

$$= \int_{\partial G_m} \sum_{(m-1/n)} f^1(y) D_{(j_1,\cdots,j_{m-1})}(f^2, \cdots, f^m)(y)\, d(y^{j_1}, \cdots, y^{j_{m-1}}). \tag{4}$$

Then the integral in (4) is an evaluation of the integral in (3), and the function whose components are

$$f^1 D_{(j_1,\cdots,j_{m-1})}(f^2, \cdots, f^m): T(\partial G_m) \to \mathbb{R}, \qquad (j_1, \cdots, j_{m-1}) \in (m-1/n), \tag{5}$$

is an evaluator of (3). There are other evaluations and evaluators of (3) as suggested by Corollary 38.7, Theorem 38.8, and Exercise 39.3.

53.3 Definition. Let $F: E \to \mathbb{R}^m$ be a function whose components $(F^1, \cdots, F^m)$ have continuous derivatives. If

$$D_{(j_1,\cdots,j_m)}(F^1, \cdots, F^m)(y) = f^{(j_1,\cdots,j_m)}(y), \qquad y \in E, \qquad (j_1, \cdots, j_m) \in (m/n), \tag{6}$$

then F is called a *primitive* of f on E. The class of primitives of $f: E \to \mathbb{R}^{\binom{n}{m}}$ is denoted by $\mathscr{P}(f, E)$, and the class of functions $f: E \to \mathbb{R}^{\binom{n}{m}}$ which have primitives is denoted by $\mathscr{F}_p(E)$.

The integrals in (1) and (2) may happen to be equal although the functions v and f are not related in any significant manner; nevertheless, (2) is an evaluation of (1) according to Definition 53.1. Such evaluations, although real, are trivial or accidental, and the principal interest attaches to evaluations (2) which can be derived from (1) by some mathematical process.

Exercises

53.1. (a) Let the components (F^1, F^2) of $F: A_2 \to \mathbb{R}^2$ be defined as indicated below. Show that each of these functions F is a primitive of $f(x^1, x^2) = 3[(x^1)^2 + (x^2)^2]$, and use these primitives to evaluate $\int_{A_2} f(x^1, x^2)\, d(x^1, x^2)$.

(i) $F^1(x^1, x^2) = (x^1)^3 - (x^2)^3, \qquad F^2(x^1, x^2) = x^1 + x^2.$
(ii) $F^1(x^1, x^2) = x^1 - x^2, \qquad F^2(x^1, x^2) = (x^1)^3 + (x^2)^3.$
(iii) $F^1(x^1, x^2) = 2[(x^1)^3 - (x^2)^3] + 5(x^1 + x^2),$
$F^2(x^1, x^2) = 3[(x^1)^3 - (x^2)^3] + 8(x^1 + x^2).$

(b) Let $h: [a_2, b_2] \to \mathbb{R}$ be a function of x^2 which has a continuous derivative. Show that F, with components

$$F^1(x^1, x^2) = (x^1)^3 + 3x^1(x^2)^2 + h(x^2), \qquad F^2(x^1, x^2) = x^2,$$

is a primitive of $3[(x^1)^2 + (x^2)^2]$ on A_2. Write out the two standard evaluations of $\int_{A_2} 3[(x^1)^2 + (x^2)^2]\, d(x^1, x^2)$ by means of this primitive.

(c) Find an evaluation of $\int_{A_2} 3[(x^1)^2 + (x^2)^2]\, d(x^1, x^2)$ which is not an iterated integral and which is not derived from a primitive of $3[(x^1)^2 + (x^2)^2]$. [Hint. Exercise 52.4.]

(d) Let

$$\int_{\partial A_2} v^1(x)\,dx^1 + v^2(x)\,dx^2, \qquad \int_{\partial A_2} w^1(x)\,dx^1 + w^2(x)\,dx^2,$$

be two evaluations of $\int_{A_2} 3[(x^1)^2 + (x^2)^2]\,d(x^1, x^2)$. Show that, for every t in $\mathbb{R}$,

$$\int_{\partial A_2} [tv^1(x) + (1-t)w^1(x)]\,dx^1 + [tv^2(x) + (1-t)w^2(x)]\,dx^2$$

is also an evaluation of the given integral.

(e) What is your impression of the class of all evaluations of $\int_{A_2} 3[(x^1)^2 + (x^2)^2]\,d(x^1, x^2)$?
Is this class small or large? finite or infinite?

54. Functions and Primitives

Every continuous function $f: [a, b] \to \mathbb{R}$ has a primitive $F: [a, b] \to \mathbb{R}$, and one primitive is defined as follows:

$$F(x) = \int_a^x f(t)\,dt, \qquad a \leqq x \leqq b. \tag{1}$$

The class of all primitives of f is the set $\{F + c : c \in \mathbb{R}\}$. This case is not typical of the cases $n \geqq m > 1$. This section investigates some of the properties of the class of primitives of a given function f; it shows that some continuous functions f have no primitives. However, if f has one primitive, it has a large class of primitives as the following example shows.

54.1 Example. Let E be an open connected set in $\mathbb{R}^3$, and let $f: E \to \mathbb{R}^3$ be a function which has continuous components $f^{(j_1, j_2)}$, $(j_1, j_2) \in (2/3)$. Let $F: E \to \mathbb{R}^2$ be a function whose components (F^1, F^2) have continuous derivatives such that

$$D_{(j_1, j_2)}(F^1, F^2)(y) = f^{(j_1, j_2)}(y), \qquad y \in E, \qquad (j_1, j_2) \in (2/3). \tag{2}$$

Then F is a primitive of f by Definition 53.3. Represent the components of F as a column matrix. Let $[a_{ij}]$ be a 2×2 matrix of constants such that

$$\det[a_{ij}] = 1, \qquad i, j = 1, 2, \tag{3}$$

and let $[c_i]$ be a 2×1 matrix of constants. Define a function $H: E \to \mathbb{R}^2$, with components (H^1, H^2), by the following matrix equation:

$$\begin{bmatrix} H^1 \\ H^2 \end{bmatrix} = \begin{bmatrix} a_{11} & a_{12} \\ a_{21} & a_{22} \end{bmatrix} \begin{bmatrix} F^1 \\ F^2 \end{bmatrix} + \begin{bmatrix} c_1 \\ c_2 \end{bmatrix}. \tag{4}$$

Then

$$\begin{aligned} H^1 &= a_{11}F^1 + a_{12}F^2 + c_1, \\ H^2 &= a_{21}F^1 + a_{22}F^2 + c_2, \end{aligned} \tag{5}$$

and (5) and the formula for derivatives show that

$$\begin{aligned} D_{(j_1,j_2)}(H^1, H^2)(y) &= \begin{vmatrix} D_{j_1}H^1(y) & D_{j_2}H^1(y) \\ D_{j_1}H^2(y) & D_{j_2}H^2(y) \end{vmatrix} \\ &= \begin{vmatrix} a_{11}D_{j_1}F^1(y) + a_{12}D_{j_1}F^2(y) & a_{11}D_{j_2}F^1(y) + a_{12}D_{j_2}F^2(y) \\ a_{21}D_{j_1}F^1(y) + a_{22}D_{j_1}F^2(y) & a_{21}D_{j_2}F^1(y) + a_{22}D_{j_2}F^2(y) \end{vmatrix}. \end{aligned} \tag{6}$$

Then the Binet–Cauchy multiplication theorem [see Theorem 80.1] and equations (3) and (2) show that

$$\begin{aligned} D_{(j_1,j_2)}(H^1, H^2)(y) &= \begin{vmatrix} a_{11} & a_{12} \\ a_{21} & a_{22} \end{vmatrix} \begin{vmatrix} D_{j_1}F^1(y) & D_{j_2}F^1(y) \\ D_{j_1}F^2(y) & D_{j_2}F^2(y) \end{vmatrix} \\ &= \begin{vmatrix} D_{j_1}F^1(y) & D_{j_2}F^1(y) \\ D_{j_1}F^2(y) & D_{j_2}F^2(y) \end{vmatrix} \\ &= D_{(j_1,j_2)}(F^1, F^2)(y) \\ &= f^{(j_1,j_2)}(y), \qquad y \in E, \qquad (j_1, j_2) \in (2/3). \end{aligned} \tag{7}$$

Thus, if F is a primitive of f, then for every matrix $[c_i]$ and every matrix $[a_{ij}]$ which satisfies (3), the function H defined by (4) is also a primitive of f.

54.2 Theorem. *Let $F: E \to \mathbb{R}^m$, $E \subset \mathbb{R}^n$, $m \leqq n$, with components $(F^1, \cdots, F^m)$, be a primitive of $f: E \to \mathbb{R}^{\binom{n}{m}}$; let $[a_{ij}]$ be an $m \times m$ matrix of constants whose determinant is 1; and let $[c_i]$ be an $m \times 1$ matrix of constants. If $H: E \to \mathbb{R}^m$, with components $(H^1, \cdots, H^m)$, is a function such that*

$$\begin{bmatrix} H^1 \\ \cdots \\ H^m \end{bmatrix} = \begin{bmatrix} a_{11} & a_{12} & \cdots & a_{1m} \\ \cdots & \cdots & \cdots & \cdots \\ a_{m1} & a_{m2} & \cdots & a_{mm} \end{bmatrix} \begin{bmatrix} F^1 \\ \cdots \\ F^m \end{bmatrix} + \begin{bmatrix} c_1 \\ \cdots \\ c_m \end{bmatrix}, \tag{8}$$

then H is also a primitive of f.

PROOF. Example 54.1 contains the proof of this theorem in the special case in which $m = 2$ and $n = 3$. The proof in the general case is entirely similar to the proof in this special case, and it can be written out easily by following the proof in Example 54.1 as a model. □

54.3 Example. Let $\mathscr{G}$ denote the set of $m \times m$ matrices $[a_{ij}]$ of real numbers such that

$$\det[a_{ij}] = 1, \qquad i, j = 1, \cdots, m. \tag{9}$$

Let $\otimes$ denote the operation of matrix multiplication. Then the system $(\mathscr{G}, \otimes)$ is a group. To prove this statement, observe first that the product of two $m \times m$ matrices is an $m \times m$ matrix; thus $\mathscr{G}$ is closed under matrix multiplication. Next, matrix multiplication is known to be associative. Third, the $m \times m$ identity matrix I_m belongs to $\mathscr{G}$, and I_m is the identity element of the system $(\mathscr{G}, \otimes)$. Finally, to complete the proof that $(\mathscr{G}, \otimes)$ is a group, it is necessary to show that each matrix A in $\mathscr{G}$ has an inverse A^{-1} in $\mathscr{G}$. The fact that A^{-1} exists follows from (9) and Exercise 5.1. To show that A^{-1} is in $\mathscr{G}$, observe that $AA^{-1} = A^{-1}A = I_m$ by the definition of the inverse. Then $\det[A^{-1}A] = \det I_m = 1$, and the Binet–Cauchy multiplication theorem shows that

$$\det[A^{-1}A] = \det A^{-1} \det A = 1. \tag{10}$$

Then since $\det A = 1$ by (9), equation (10) shows that $\det A^{-1} = 1$, and hence that A^{-1} is a matrix in $\mathscr{G}$. Thus the proof that $(\mathscr{G}, \otimes)$ is a group is complete. If $m = 1$, this group contains the single element $[1]$; if $m > 1$, then $(\mathscr{G}, \otimes)$ is an infinite group.

Theorem 54.2 shows that each primitive F of f can be used to generate a class of primitives of f. As in that theorem, let $F: E \to \mathbb{R}^m$, $E \subset \mathbb{R}^n$, $m \leqq n$, with components $(F^1, \cdots, F^m)$, be a primitive of $f: E \to \mathbb{R}^{\binom{n}{m}}$. Let A be an $m \times m$ matrix in the group $\mathscr{G}$ in Example 54.3, and let C denote a matrix in the class $\mathscr{C}$ of $m \times 1$ matrices of real numbers. Define the set $\{F, \mathscr{G}, \mathscr{C}\}$ by the following equation:

$$\{F, \mathscr{G}, \mathscr{C}\} = \{H : H = AF + C, A \in \mathscr{G}, C \in \mathscr{C}\} \tag{11}$$

If F is a primitive of f, then Theorem 54.2 shows that each function H in $\{F, \mathscr{G}, \mathscr{C}\}$ is a primitive of f. How many distinct classes $\{F, \mathscr{G}, \mathscr{C}\}$ are there, and what is the relation between two of these classes? The answers to these questions are different for $m = 1$ and $m > 1$.

54.4 Theorem. *Let $m = 1$, and let E be an open connected set in $\mathbb{R}^n$, $n \geqq 1$. If F is a primitive of f, then*

$$\{H : H = F + c, c \in \mathbb{R}\} \tag{12}$$

is a class of primitives of f, and this class contains all primitives of f. If $n = 1$, every function f has a primitive F.

PROOF. If $m = 1$, then (11) becomes (12) because $\mathscr{G}$ contains only the single matrix $[1]$. If H_1 and H_2 are any two primitives of f, then the derivatives of $H_1 - H_2$ are zero on E and $H_1 - H_2$ is a constant by Exercise 9.3. Each two functions H_1, H_2 in (12) are of this form; if f has a primitive F, then (12) contains all primitives of F. If $m = 1$ and $n = 1$, then E is an open interval. Let x_0 be a point in E; then the equation

$$F(x) = \int_{x_0}^{x} f(t)\,dt, \qquad x \in E, \tag{13}$$

defines a primitive of f on E. The proof is complete. □

54.5 Theorem. *Assume that $m > 1$, and let E be an open connected set in $\mathbb{R}^n$, $m \leqq n$. If the function $f: E \to \mathbb{R}^{\binom{n}{m}}$ has a primitive F, then*

$$\{H : H = AF + C, A \in \mathscr{G}, C \in \mathscr{C}\} \tag{14}$$

is a class $\{F, \mathscr{G}, \mathscr{C}\}$ of primitives of f. If F_1 and F_2 are two primitives of f, then either

$$\{F_1, \mathscr{G}, \mathscr{C}\} = \{F_2, \mathscr{G}, \mathscr{C}\} \tag{15}$$

or

$$\{F_1, \mathscr{G}, \mathscr{C}\} \cap \{F_2, \mathscr{G}, \mathscr{C}\} = \varnothing, \textit{ the empty set.} \tag{16}$$

PROOF. If f has a primitive F, then Theorem 54.2 shows that every function $H : E \to \mathbb{R}^m$ in the class $\{F, \mathscr{G}, \mathscr{C}\}$ in (14) is also a primitive of f. If F_1 and F_2 are two primitives of f, then they generate classes of primitives $\{F_1, \mathscr{G}, \mathscr{C}\}$ and $\{F_2, \mathscr{G}, \mathscr{C}\}$. If these two classes have a function in common, then there are matrices A_1, C_1 and A_2, C_2 such that

$$A_1 F_1 + C_1 = A_2 F_2 + C_2. \tag{17}$$

The matrices A in $\mathscr{G}$ form a group under matrix multiplication [see Example 54.3], and the matrices C in $\mathscr{C}$ form a group under matrix addition. Then (17) shows that

$$\begin{aligned} A_1 F_1 &= A_2 F_2 + (C_2 - C_1), \\ F_1 &= A_1^{-1} A_2 F_2 + A_1^{-1}(C_2 - C_1). \end{aligned} \tag{18}$$

Then

$$\begin{aligned} \{F_1, \mathscr{G}, \mathscr{C}\} &= \{H : H = AF_1 + C, A \in \mathscr{G}, C \in \mathscr{C}\} \\ &= \{H : H = A[A_1^{-1} A_2 F_2 + A_1^{-1}(C_2 - C_1)] + C, A \in \mathscr{G}, C \in \mathscr{C}\} \\ &= \{H : H = (AA_1^{-1} A_2) F_2 + AA_1^{-1}(C_2 - C_1) + C, A \in \mathscr{G}, C \in \mathscr{C}\}. \end{aligned} \tag{19}$$

Now $AA_1^{-1}A_2$ is an $m \times m$ matrix in $\mathscr{G}$, and $AA_1^{-1}(C_2 - C_1) + C$ is an $m \times 1$ matrix in $\mathscr{C}$. Then (19) shows that $\{F_1, \mathscr{G}, \mathscr{C}\} \subset \{F_2, \mathscr{G}, \mathscr{C}\}$. If equation (17) is solved for F_2, then a similar calculation shows that $\{F_2, \mathscr{G}, \mathscr{C}\} \subset \{F_1, \mathscr{G}, \mathscr{C}\}$. Thus if the sets $\{F_1, \mathscr{G}, \mathscr{C}\}$ and $\{F_2, \mathscr{G}, \mathscr{C}\}$ have an element (function) in common as stated in (17), they are identical as stated in (15). Therefore, either (15) or (16) is true in all cases. Example 54.7 describes a continuous function $f: E \to \mathbb{R}^6$, $E \subset \mathbb{R}^4$, which does not have a primitive. Other examples show that, in some cases, a function f may have many distinct classes of primitives $\{F, \mathscr{G}, \mathscr{C}\}$. □

54.6 Examples. Let

$$F_1^1(x^1, x^2) = (x^1)^3 - (x^2)^3, \qquad F_1^2(x^1, x^2) = x^1 + x^2. \tag{20}$$

Then (F_1^1, F_1^2) are the components of a function $F_1 : E \to \mathbb{R}^2$, $E \subset \mathbb{R}^2$, which is a primitive [see Exercise 53.1] of the function $f: E \to \mathbb{R}$ such that $f(x^1, x^2) = 3[(x^1)^2 + (x^2)^2]$. The class of primitives $\{F_1, \mathscr{G}, \mathscr{C}\}$ generated by F_1 is described by the following equation:

$$\begin{bmatrix} a_{11} & a_{12} \\ a_{21} & a_{22} \end{bmatrix}\begin{bmatrix} F_1^1 \\ F_1^2 \end{bmatrix} + \begin{bmatrix} c_1 \\ c_2 \end{bmatrix} = \begin{bmatrix} a_{11}[(x^1)^3 - (x^2)^3] + a_{12}[x^1 + x^2] + c_1 \\ a_{21}[(x^1)^3 - (x^2)^3] + a_{22}[x^1 + x^2] + c_2 \end{bmatrix}. \tag{21}$$

If

$$\det\begin{bmatrix} a_{11} & a_{12} \\ a_{21} & a_{22} \end{bmatrix} = \det\begin{bmatrix} 2 & 5 \\ 3 & 8 \end{bmatrix} = 1, \qquad \begin{bmatrix} c_1 \\ c_2 \end{bmatrix} = \begin{bmatrix} 0 \\ 0 \end{bmatrix}, \tag{22}$$

then the primitive F_2 of f defined by (21) has the following components:

$$\begin{aligned} F_2^1(x^1, x^2) &= 2[(x^1)^3 - (x^2)^3] + 5(x^1 + x^2), \\ F_2^2(x^1, x^2) &= 3[(x^1)^3 - (x^2)^3] + 8(x^1 + x^2). \end{aligned} \tag{23}$$

Now the function $F_3 : E \to \mathbb{R}^2$ with components

$$F_3^1(x^1, x^2) = (x^1)^3 + 3x^1(x^2)^2, \qquad F_3^2(x^1, x^2) = x^2, \tag{24}$$

is also a primitive of f, but (21) shows that it is not contained in $\{F_1, \mathscr{G}, \mathscr{C}\}$. Thus $\{F_2, \mathscr{G}, \mathscr{C}\} = \{F_1, \mathscr{G}, \mathscr{C}\}$ and $\{F_3, \mathscr{G}, \mathscr{C}\} \cap \{F_1, \mathscr{G}, \mathscr{C}\} = \varnothing$.

54.7 Example. Every function $f: [a, b] \to \mathbb{R}$ has a primitive as shown in Theorem 54.4, but most functions $f: E \to \mathbb{R}^k$, $E \subset \mathbb{R}^n$, $k > 1$, have no primitive. This example describes one such function. Let E be an open connected set in $\mathbb{R}^4$, and let $f: E \to \mathbb{R}^6$ be a constant function to be defined presently. If f has a primitive, then this primitive must have two components, because the only function of four variables whose derivative has six components is a function (F^1, F^2) with two components. Thus it is appropriate to describe the given function $f: E \to \mathbb{R}^6$ as a function with components $f^{(j_1, j_2)}: E \to \mathbb{R}$, $(j_1, j_2) \in (2/4)$; assume that these components have, for all y in E, the following values:

$$f^{(1,2)}(y) = f^{(3,4)}(y) = 1, \qquad f^{(1,3)}(y) = f^{(1,4)}(y) = f^{(2,3)}(y) = f^{(2,4)}(y) = 0. \tag{25}$$

This example will show that f has no primitive on E. Observe that f is a nice function—its components are continuous and even constant—which satisfies the obvious necessary conditions for the existence of a primitive. The derivative of a function $(F^1, F^2): E \to \mathbb{R}^2$, $E \subset \mathbb{R}^4$, is a function with components $D_{(j_1, j_2)}(F^1, F^2)$, $(j_1, j_2) \in (2/4)$. Therefore, the given function $f: E \to \mathbb{R}^6$ has the right number of components $f^{(j_1, j_2)}$, $(j_1, j_2) \in (2/4)$, to enable it to be the derivative of a function $F: E \to \mathbb{R}^2$, $E \subset \mathbb{R}^4$, and there is no other possibility for a primitive.

To prove that f in (25) has no primitive, assume that it has a primitive and show that a contradiction results. Then there exists a primitive $F: E \to$

$\mathbb{R}^2$, $E \subset \mathbb{R}^4$, whose components (F^1, F^2) have continuous derivatives such that

$$\begin{vmatrix} D_1F^1 & D_2F^1 \\ D_1F^2 & D_2F^2 \end{vmatrix} = 1, \quad \begin{vmatrix} D_1F^1 & D_3F^1 \\ D_1F^2 & D_3F^2 \end{vmatrix} = 0, \quad \begin{vmatrix} D_1F^1 & D_4F^1 \\ D_1F^2 & D_4F^2 \end{vmatrix} = 0, \quad (26)$$

$$\begin{vmatrix} D_2F^1 & D_3F^1 \\ D_2F^2 & D_3F^2 \end{vmatrix} = 0, \quad \begin{vmatrix} D_2F^1 & D_4F^1 \\ D_2F^2 & D_4F^2 \end{vmatrix} = 0, \quad \begin{vmatrix} D_3F^1 & D_4F^1 \\ D_3F^2 & D_4F^2 \end{vmatrix} = 1. \quad (27)$$

The second equation in (26) and the first equation in (27) are the following:

$$D_1F^1D_3F^2 - D_1F^2D_3F^1 = 0, \qquad D_2F^1D_3F^2 - D_2F^2D_3F^1 = 0. \quad (28)$$

Multiply the first of these equations by D_2F^1 and the second by D_1F^1, and then subtract the second equation from the first. Again, multiply the first equation in (28) by D_2F^2 and the second by D_1F^2, and then subtract the second of the resulting equations from the first. The two equations obtained in this manner are the following:

$$\begin{aligned} &[D_1F^1D_2F^2 - D_1F^2D_2F^1]D_3F^1 = 0, \\ &[D_1F^1D_2F^2 - D_1F^2D_2F^1]D_3F^2 = 0. \end{aligned} \quad (29)$$

If $D_3F^1(y) \neq 0$ at some point y in E, then the first equation in (29) contradicts the first equation in (26); hence, $D_3F^1(y) = 0$ for every y in E. Likewise, if $D_3F^2(y) \neq 0$ at some point y in E, then the second equation in (29) contradicts the first equation in (26); hence, $D_3F^2(y) = 0$ for every y in E. But $D_3F^1(y) = 0$ and $D_3F^2(y) = 0$ contradict the third equation in (27). Thus the assumption that f in (25) has a primitive has led to a contradiction, and the proof that f has no primitive is complete.

Example 54.7 shows that some functions $f: E \to \mathbb{R}^{\binom{n}{m}}$, $E \subset \mathbb{R}^n$, do not have primitives, and consideration of the nature of the derivatives $D_{(j_1, \cdots, j_m)}(F^1, \cdots, F^m)$, $(j_1, \cdots, j_m) \in (m/n)$, leads to the conclusion that the class of functions f which do have primitives is not easily characterized. Nevertheless, this class of functions f has one simple property which is stated in the next theorem.

54.8 Theorem. *Let $f: E \to \mathbb{R}^{\binom{n}{m}}$, $E \subset \mathbb{R}^n$, be a function which has a primitive $F: (F^1, \cdots, F^m)$ on E. If c is a constant, then the function cf, with components $cf^{(j_1, \cdots, j_m)}$, $(j_1, \cdots, j_m) \in (m/n)$, also has a primitive on E.*

Proof. The proof consists of constructing a function $H: E \to \mathbb{R}^m$ with components $(H^1, \cdots, H^m)$ such that

$$D_{(j_1, \cdots, j_m)}(H^1, \cdots, H^m)(y) = cf^{(j_1, \cdots, j_m)}(y), \qquad y \in E, \qquad (j_1, \cdots, j_m) \in (m/n). \quad (30)$$

Let $[b_{ij}]$ be an $m \times m$ matrix of constants such that $\det[b_{ij}] = c$, and let $[d_i]$ be an $m \times 1$ matrix of constants. Let H be defined by the following equation:

$$\begin{bmatrix} H^1 \\ \cdots \\ H^m \end{bmatrix} = \begin{bmatrix} b_{11} & b_{12} & \cdots & b_{1m} \\ \cdots & \cdots & \cdots & \cdots \\ b_{m1} & b_{m2} & \cdots & b_{mm} \end{bmatrix} \begin{bmatrix} F^1 \\ \cdots \\ F^m \end{bmatrix} + \begin{bmatrix} d_1 \\ \cdot\cdot \\ d_m \end{bmatrix}. \tag{31}$$

Then by calculations similar to those which established the equations in (7),

$$\begin{aligned} D_{(j_1,\cdots,j_m)}(H^1,\cdots,H^m)(y) &= \det[b_{ij}]D_{(j_1,\cdots,j_m)}(F^1,\cdots,F^m)(y) \\ &= cf^{(j_1,\cdots,j_m)}(y). \end{aligned} \tag{32}$$

Thus H is a primitive of cf, and the proof of Theorem 54.8 is complete. □

Exercises

54.1. Let $A_2 = [a_1, b_1] \times [a_2, b_2]$, and let $f: A_2 \to \mathbb{R}$ be a function such that $f(x^1, x^2) = 4x^1x^2$.

(a) Show that each of the following functions is a primitive of f.

(i) $F^1(x^1, x^2) = (x^1)^2$, $F^2(x^1, x^2) = (x^2)^2$;
(ii) $F^1(x^1, x^2) = 2(x^1)^2x^2$, $F^2(x^1, x^2) = x^2$;
(iii) $F^1(x^1, x^2) = x^1$, $F^2(x^1, x^2) = 2x^1(x^2)^2$.

(b) Show that the class of primitives of f generated from (i) by Theorem 54.2 is the following:

$$F^1(x^1, x^2) = a_{11}(x^1)^2 + a_{12}(x^2)^2 + c_1,$$
$$F^2(x^1, x^2) = a_{21}(x^1)^2 + a_{22}(x^2)^2 + c_2, \qquad \det[a_{ij}] = 1.$$

Find the classes of primitives generated from (ii) and (iii).

(c) Show that the intersection of each two of the three classes of primitives of f generated from (i), $\cdots$, (iii) is empty.

(d) Use the three classes of primitives in (a) to obtain evaluations of $\int_{A_2} 4x^1x^2\, d(x^1, x^2)$.

54.2. Let E be an open connected set in $\mathbb{R}^3$, and let $f: E \to \mathbb{R}^3$ be a function whose components $(f^1, \cdots, f^3)$ are defined as follows:

$$f^1(y) = 2(y^1 - y^2), \qquad f^2(y) = 2(y^1 - y^3), \qquad f^3(y) = 2(y^2 - y^3).$$

(a) Show that each of the following functions is a primitive of f.

(i) $F^1(y) = (y^1)^2 + (y^2)^2 + (y^3)^2$, $F^2(y) = y^1 + y^2 + y^3$;
(ii) $F^1(y) = -(y^1 + y^2 + y^3)$, $F^2(y) = (y^1)^2 + (y^2)^2 + (y^3)^2$;
(iii) $F^1(y) = 2(y^1y^2 + y^1y^3 + y^2y^3)$, $F^2(y) = -(y^1 + y^2 + y^3)$;
(iv) $F^1(y) = 2(y^1 + y^2 + y^3)$, $F^2(y) = y^1y^2 + y^1y^3 + y^2y^3$.

(b) Show that (ii) is in the class of primitives generated from (i) by Theorem 54.2, and that (i) and (ii) generate the same class of primitives. Also, show that (iv) is in the class generated by (iii), and that (iii) and (iv) generate the same class of primitives. Show that the class of primitives generated by (i) and (ii) is disjoint from the class generated by (iii) and (iv).

(c) Let G_2 be a surface whose trace $T(G_2)$ is contained in E. Use the primitives in (a) to evaluate the following integral; compare these evaluations.

$$\int_{G_2} 2(y^1 - y^2)\,d(y^1, y^2) + 2(y^1 - y^3)\,d(y^1, y^3) + 2(y^2 - y^3)\,d(y^2, y^3).$$

54.3. Let f be a function whose components are defined as follows:

$$f^1(y) = y^2 + y^3, \qquad f^2(y) = y^1 + y^3, \qquad f^3(y) = y^1 + y^2.$$

(a) Show that the following function is a primitive of f.

$$F(y) = y^1y^2 + y^1y^3 + y^2y^3.$$

(b) Show that the class of primitives of f generated from F in (a) by Theorem 54.2 consists of the following functions.

$$y^1y^2 + y^1y^3 + y^2y^3 + c, \quad c \text{ a constant.}$$

(c) Show that the class of primitives described in (b) contains all primitives of f.
(d) Let G_1 be a curve in $\mathbb{R}^3$. Use the primitives in (b) to evaluate the following integral. Compare these evaluations for $c = 0$ and $c \neq 0$.

$$\int_{G_1} (y^2 + y^3)\,dy^1 + (y^1 + y^3)\,dy^2 + (y^1 + y^2)\,dy^3.$$

54.4. Let G_m be a surface in $\mathbb{R}^n$, $m \leq n$; let E be an open set in $\mathbb{R}^n$ which contains $T(G_m)$; and let $f: E \to \mathbb{R}^{\binom{n}{m}}$ be a function whose components $f^{(j_1, \cdots, j_m)}: E \to \mathbb{R}$ are continuous. Assume that $(F^1, \cdots, F^m)$ and $(F^1 + c_1, \cdots, F^m + c_m)$ are two primitives of f such that $c_1 \neq 0$ [see Theorem 54.2]. Use the fundamental theorem of the integral calculus and the fact that each of the primitives provides an evaluation of

$$\int_{G_m} \sum_{(m/n)} f^{(j_1, \cdots, j_m)}(y)\,d(y^{j_1}, \cdots, y^{j_m})$$

to prove that

$$\int_{\partial G_m} \sum_{(m-1/n)} D_{(j_1, \cdots, j_{m-1})}(F^2, \cdots, F^m)(y)\,d(y^{j_1}, \cdots, y^{j_{m-1}}) = 0.$$

[see Theorem 39.5 and Example 48.2].

55. Integrals and Evaluations

The purpose of this section is (a) to investigate the existence of evaluators $v: T(\partial G_m) \to \mathbb{R}^{\binom{n}{m-1}}$ and evaluations

$$\int_{\partial G_m} \sum_{(m-1/n)} v^{(j_1, \cdots, j_{m-1})}(y)\,d(y^{j_1}, \cdots, y^{j_{m-1}}) \tag{1}$$

for integrals

$$\int_{G_m} \sum_{(m/n)} f^{(j_1, \cdots, j_m)}(y)\,d(y^{j_1}, \cdots, y^{j_m}); \tag{2}$$

and (b) to determine the properties of the classes $\mathscr{V}(f, G_m)$, $\mathscr{V}(G_m)$, and $\mathscr{F}_e(G_m)$ [see Section 53 for the definitions of these classes].

55.1 Theorem. *Let $f: T(G_m) \to \mathbb{R}^{\binom{n}{m}}$ be a function which has a primitive $F: E \to \mathbb{R}^m$. Then (2) has at least m evaluator functions; that is, there exist at least m functions v such that (2) equals (1).*

PROOF. Since F is a primitive of f, the integral in (2) equals the integral

$$\int_{G_m} \sum_{(m/n)} D_{(j_1, \cdots, j_m)}(F^1, \cdots, F^m)(y)\, d(y^{j_1}, \cdots, y^{j_m}). \tag{3}$$

Let i be an integer such that $1 \leqq i \leqq m$, and define the function v as follows:

$$v^{(j_1, \cdots, j_{m-1})}(y) = (-1)^{i-1} F^i(y) D_{(j_1, \cdots, j_{m-1})}(F^1, \cdots, \widehat{F^i}, \cdots, F^m)(y),$$
$$(j_1, \cdots, j_{m-1}) \in (m-1/n). \tag{4}$$

For the function v defined thus, the fundamental theorem of the integral calculus [see Theorem 39.1 and Corollary 38.7] shows that the integral in (1) equals the integral in (3) and hence the integral in (2). Since there are m possible values for i, there are at least m evaluators and evaluations of (2) if f has a primitive. □

The next theorem is concerned with functions whose integrals on G_m or ∂G_m are zero. Section 48 contains information about such functions. Let $h: T(G_m) \to \mathbb{R}^{\binom{n}{m}}$ and $k: T(\partial G_m) \to \mathbb{R}^{\binom{n}{m-1}}$ be functions such that

$$\int_{G_m} \sum_{(m/n)} h^{(j_1, \cdots, j_m)}(y)\, d(y^{j_1}, \cdots, y^{j_m}) = 0, \tag{5}$$

$$\int_{\partial G_m} \sum_{(m-1/n)} k^{(j_1, \cdots, j_{m-1})}(y)\, d(y^{j_1}, \cdots, y^{j_{m-1}}) = 0. \tag{6}$$

55.2 Theorem. *The class $\mathscr{V}(f, G_m)$ of evaluators of the integral of f on G_m in (2) has the following properties:*

(a) *If v is any function in $\mathscr{V}(f, G_m)$, then $v + k$ is an evaluator of the integral of $f + h$ over G_m; that is,*

$$\int_{G_m} \sum_{(m/n)} [f^{(j_1, \cdots, j_m)}(y) + h^{(j_1, \cdots, j_m)}(y)]\, d(y^{j_1}, \cdots, y^{j_m}) \tag{7}$$

$$= \int_{\partial G_m} \sum_{(m-1/n)} [v^{(j_1, \cdots, j_{m-1})}(y) + k^{(j_1, \cdots, j_{m-1})}(y)]\, d(y^{j_1}, \cdots, y^{j_{m-1}}). \tag{8}$$

In particular $v + k$ is a function in $\mathscr{V}(f, G_m)$, and v is an evaluator of (7) for each h which satisfies (5).

(b) *If v and w are functions in $\mathscr{V}(f, G_m)$, then $tv + (1-t)w$ is a function in $\mathscr{V}(f, G_m)$ for each t in $\mathbb{R}$.*

PROOF. Equation (7)–(8) follows from the equality of (1) and (2), the linearity of integrals, and the hypothesis in (5) and (6). If h is chosen to be the zero function, then (7)–(8) shows that $v + k$ belongs to $\mathscr{V}(f, G_m)$. If k is the zero function, then (7)–(8) shows that v is an evaluator of (7) for every h which satisfies (5). To prove (b), let v and w be any two functions in $\mathscr{V}(f, G_m)$. Then the integrals of v and w over ∂G_m each equal (2), and the integral of $tv + (1 - t)w$ over ∂G_m, for every t in $\mathbb{R}$, also equals (2) because of the linear property of integrals [see Theorem 35.12]; therefore $tv + (1 - t)w$ is in $\mathscr{V}(f, G_m)$ for each t in $\mathbb{R}$. □

55.3 Theorem. *The class $\mathscr{V}(G_m)$ of all functions $v: T(\partial G_m) \to \mathbb{R}^{\binom{n}{m-1}}$ for which the integral* (1) *is an evaluation of some integral* (2) *is a linear space of continuous functions.*

PROOF. Section 89 in Appendix 2 contains the definition of a linear space. To prove that $\mathscr{V}(G_m)$ is a linear space, observe first that the zero function belongs to $\mathscr{V}(G_m)$ since it is an evaluator for the integral of the zero function f on $T(G_m)$. Next, let v and v' be any two functions in $\mathscr{V}(G_m)$, and let a and a' be two real numbers. Since v and v' belong to $\mathscr{V}(G_m)$, then v and v' are evaluators of the integrals of some functions f and f' in $\mathscr{F}_e(G_m)$. Then, because of the linear property of the integral, $av + a'v'$ is an evaluator of the integral of $af + a'f'$. Hence, $av + a'v'$ is in $\mathscr{V}(G_m)$. These facts, together with standard properties of function spaces, show that $\mathscr{V}(G_m)$ is a linear space of continuous functions on $T(\partial G_m)$. □

55.4 Theorem. *The class $\mathscr{F}_e(G_m)$ of functions $f: T(G_m) \to \mathbb{R}^{\binom{n}{m}}$ whose integrals* (2) *have an evaluation* (1) *has the following properties:*
(a) *$\mathscr{F}_e(G_m)$ is a linear space of continuous functions on $T(G_m)$.*
(b) *If f has a primitive, then f is in $\mathscr{F}_e(G_m)$, or $\mathscr{F}_p(G_m) \subset \mathscr{F}_e(G_m)$.*

PROOF. To prove (a), observe first that $\mathscr{F}_e(G_m)$ contains the zero function f. Also, $av + a'v'$ is an evaluator of the integral of $af + a'f'$ if v, v' are evaluators of the integrals of f, f' and a, a' are constants. These facts and standard properties of function spaces prove (a). Theorem 55.1 proves (b). □

EXERCISES

55.1. Let E be an open interval in $\mathbb{R}$. Prove that $\mathscr{F}_p(E)$, the space of continuous functions on E which have primitives, is a linear function space; describe this function space.

55.2. Let $(F^1, \cdots, F^m): E \to \mathbb{R}^m$ be a primitive of $f: T(G_m) \to \mathbb{R}^{\binom{n}{m}}$. Show that the class $\{F, \mathscr{G}, \mathscr{C}\}$ of primitives generated from $(F^1, \cdots, F^m)$ by Theorem 54.2 provides a large class of evaluations of the integral (2). Give a complete description of all evaluations of (2) which can be derived from the primitive $(F^1, \cdots, F^m)$.

55.3. Exercise 54.2 and the fundamental theorem of the integral calculus show that

$$\int_{\partial G_2} \sum_{j=1}^{3} (y^j)^2 \, dy^1 + \sum_{j=1}^{3} (y^j)^2 \, dy^2 + \sum_{j=1}^{3} (y^j)^2 \, dy^3 \qquad \text{(i)}$$

is an evaluation of

$$\int_{G_2} 2(y^1 - y^2)\,d(y^1, y^2) + 2(y^1 - y^3)\,d(y^1, y^3) + 2(y^2 - y^3)\,d(y^2, y^3). \qquad \text{(ii)}$$

Show that the following integrals are also evaluations of (ii):

$$\int_{\partial G_2} [(y^2)^2 + (y^3)^2]\,dy^1 + [(y^1)^2 + (y^3)^2]\,dy^2 + [(y^1)^2 + (y^2)^2]\,dy^3;$$

$$-\int_{\partial G_2} 2y^1(y^2 + y^3)\,dy^1 + 2y^2(y^1 + y^3)\,dy^2 + 2y^3(y^1 + y^2)\,dy^3.$$

[Hints. Exercises 54.2 and 54.4, and equation (7)–(8) in Theorem 55.2.]

56. The Existence of Primitives: Derivatives of a Single Function

The solution of the simplest case is known, and it is trivial. A continuous function $f: E \to \mathbb{R}$, $E \subset \mathbb{R}$, of a single variable always has a primitive $F: E \to \mathbb{R}$ such that $D_1F(x) = f(x)$. One such primitive is defined by the equation

$$F(x) = \int_{x_0}^{x} f(t)\,dt, \qquad x_0 \in E, \qquad x \in E, \tag{1}$$

and the class of all primitives of f is $\{F + c : c \in \mathbb{R}\}$. This case is altogether special, because in almost all other cases it is necessary to require that f have certain derivatives before the existence of primitives can be established. This section establishes necessary and sufficient conditions for the representation of a given function f as the derivative of a single function $[m = 1]$ of n variables.

Let E be an open connected set in $\mathbb{R}^n$. (A connected set is one in which every two points x_1, x_2 in the set can be connected by a polygonal curve in the set.) The derivative of a function $F: E \to \mathbb{R}$ has n components $D_jF(x)$, $j = 1, \cdots, n$, since F is a function of n variables; therefore, a necessary condition that a given function f have a primitive F, that is, that f be representable as the derivative of F, is that f be a function $f: E \to \mathbb{R}^n$ with n components. Thus the first problem is the following: given a function $f: E \to \mathbb{R}^n$ with n continuous components $(f^1, \cdots, f^n)$, find a differentiable function $F: E \to \mathbb{R}$, $E \subset \mathbb{R}^n$, such that

$$D_jF(x) = f^j(x), \qquad x \in E, \qquad j = 1, \cdots, n. \tag{2}$$

No solution is known for this problem if $n > 1$.

Since we are unable to solve the first problem, we turn our attention to a second problem with stronger hypotheses as follows: given a function $f: E \to \mathbb{R}^n$ whose components have continuous derivatives, find a differentiable function $F: E \to \mathbb{R}$, $E \subset \mathbb{R}^n$, such that (2) is satisfied. For this second problem, with stronger hypotheses, there is an additional necessary condition on f as stated in the following theorem.

56.1 Theorem (Necessary Condition). *Let $f: E \to \mathbb{R}^n$ be a function whose components $(f^1, \cdots, f^n)$ have continuous derivatives on E. If there exists a differentiable function $F: E \to \mathbb{R}$ which satisfies* (2), *then*

$$D_{j_1} f^{j_2}(x) = D_{j_2} f^{j_1}(x), \qquad x \in E, \qquad (j_1, j_2) \in (2/n). \tag{3}$$

PROOF. Let F be a function which satisfies (2). Since the components of f have continuous derivatives, then [see Theorem 41.5 and Exercise 41.4]

$$D_{j_1} f^{j_2}(x) = D_{j_1}(D_{j_2} F)(x) = D_{j_2}(D_{j_1} F)(x) = D_{j_2} f^{j_1}(x), \qquad x \in E, \qquad (j_1, j_2) \in (2/n). \tag{4}$$

The proof is complete. □

A set S in $\mathbb{R}^n$ is *star-shaped* with respect to a point x^* if and only if, for every point x in S, the segment $[x^*, x]$ is in S. A convex set is star-shaped with respect to each of its points, but many sets which are star-shaped with respect to at least one point are not convex. For example, a five-pointed star, although not convex, is star-shaped with respect to its center and also with respect to each point in a small neighborhood of its center. A set S which is star-shaped with respect to x^* is connected since each two points x_1, x_2 in S are connected by the polygonal curve $[x_1, x^*] \cup [x^*, x_2]$ in S.

56.2 Theorem (Sufficient Conditions). *Let E be an open set in $\mathbb{R}^n$ which is star-shaped with respect to a point x^*, and let $f: E \to \mathbb{R}^n$ be a function whose components $(f^1, \cdots, f^n)$ have continuous derivatives which satisfy* (3). *Then f has a primitive F; that is, there exists a function $F: E \to \mathbb{R}$ which has two continuous derivatives, and $D_j F(x) = f^j(x)$ for x in E and $j = 1, \cdots, n$.*

PROOF. Let x_0 be a fixed point in E, and let $x: (x^1, \cdots, x^n)$ be an arbitrary (ultimately variable) point in E. Let $\Sigma\{[x_{k-1}, x_k]: k = 1, \cdots, p, x_p = x\}$ be a chain of one-dimensional, oriented, simplexes in E which connects x_0 to x. The proof will show that

$$\sum_{k=1}^{p} \int_{[x_{k-1}, x_k]} \sum_{j=1}^{n} f^j(y)\, dy^j \tag{5}$$

is the value $F(x^1, \cdots, x^n)$ at $x: (x^1, \cdots, x^n)$ of a function $F: E \to \mathbb{R}$ which has two continuous derivatives and which satisfies (2). The proof consists of three parts: Part I shows that there is at least one sum (5) for each x in E, and it describes the evaluation of the integrals in (5); Part II proves that (5) has the same value for every chain which connects x_0 to $x: (x^1, \cdots, x^n)$ and thus defines a function $F: E \to \mathbb{R}$; and Part III proves that this function F of $(x^1, \cdots, x^n)$ has derivatives which satisfy (2).

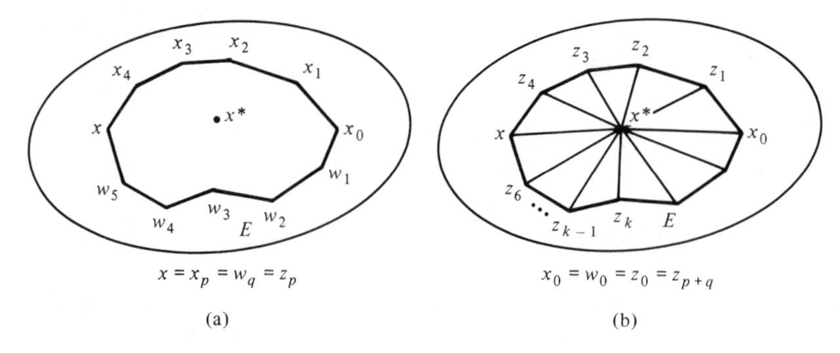

Figure 56.1. Figures for Theorem 56.2.

Part I. For every point x in E there is at least one chain $\Sigma\{[x_{k-1}, x_k]: k = 1, \cdots, p\}$ which connects x_0 to x [see Figure 56.1(a)]; for, by hypothesis, E is star shaped with respect to x^*, and $[x_0, x^*] + [x^*, x]$ is one chain which lies in E and connects x_0 to x. Furthermore, since the functions $f^1, \cdots, f^n$ are continuous on E, Section 37 shows that each of the integrals in (5) exists. Finally, Section 37 explains how to use simple integrals to evaluate these integrals. The equations

$$y^j = (x_k^j - x_{k-1}^j)t + x_{k-1}^j, \qquad j = 1, \cdots, n, \qquad 0 \leqq t \leqq 1, \tag{6}$$

define an affine transformation which maps the unit simplex $[e_0, e_1]$ in $\mathbb{R}$ onto the simplex $[x_{k-1}, x_k]$ in $\mathbb{R}^n$. Then

$$\int_{[x_{k-1}, x_k]} \sum_{j=1}^n f^j(y)\, dy^j = \int_0^1 \sum_{j=1}^n f^j[(x_k - x_{k-1})t + x_{k-1}](x_k^j - x_{k-1}^j)\, dt. \tag{7}$$

Thus the sum in (5) is well defined for every x in E, and it can be evaluated by the standard methods for evaluating integrals on simplexes in $\mathbb{R}^n$ [see Section 37].

Part II. Let $\Sigma\{[w_{k-1}, w_k]: k = 1, \cdots, q, w_0 = x_0, w_q = x\}$ be another chain in E which connects x_0 to x [see Figure 56.1(a)]. In order to show that the value of the sum in (5) depends on x alone and not on the chain which connects x_0 to x, that is, in order to show that (5) defines a function of x, it is necessary and sufficient to prove that

$$\sum_{k=1}^p \int_{[x_{k-1}, x_k]} \sum_{j=1}^n f^j(y)\, dy^j = \sum_{k=1}^q \int_{[w_{k-1}, w_k]} \sum_{j=1}^n f^j(y)\, dy^j. \tag{8}$$

Reverse the orientation in each simplex $[w_{k-1}, w_k]$ to obtain $[w_k, w_{k-1}]$. Then

$$\sum_{k=1}^p [x_{k-1}, x_k] + \sum_{k=q}^1 [w_k, w_{k-1}] \tag{9}$$

is a closed chain whose boundary is empty [see Figure 56.1(a)]. In order to simplify the notation, let the chain in (9) be denoted by $\Sigma\{[z_{k-1}, z_k]: k = 1, \cdots, p + q\}$ [see Figure 56.1(b)]. Because

$$\int_{[w_{k-1}, w_k]} \sum_{j=1}^{n} f^j(y)\,dy^j = -\int_{[w_k, w_{k-1}]} \sum_{j=1}^{n} f^j(y)\,dy^j,$$

equation (8) is equivalent to

$$\sum_{k=1}^{p+q} \int_{[z_{k-1}, z_k]} \sum_{j=1}^{n} f^j(y)\,dy^j = 0. \tag{10}$$

Thus the proof that (5) has the same value for every chain which connects x_0 to $x:(x^1, \cdots, x^n)$ will be completed by proving (10).

Let $\mathbf{b}$ be the 2-simplex $[x^*, z_{k-1}, z_k]$ in E [see Figure 56.1(b)]. Since $[z_{k-1}, z_k]$ was assumed to be in E, and since E is star-shaped with respect to x^*, the simplex $[x^*, z_{k-1}, z_k]$ is in E for $k = 1, \cdots, p+q$. Then since $f^1, \cdots, f^n$ have continuous derivatives by hypothesis, Stokes' theorem [see Theorem 41.3 and Example 41.4] states that

$$\begin{aligned}\int_{\mathbf{b}} \sum_{(j_1, j_2)} [D_{j_1} f^{j_2}(y) - D_{j_2} f^{j_1}(y)]\,d(y^{j_1}, y^{j_2}) \\ = \int_{\partial\mathbf{b}} \sum_{j=1}^{n} f^j(y)\,dy^j, \qquad (j_1, j_2) \in (2/n).\end{aligned} \tag{11}$$

Let $\mathbf{c}$ be the 2-chain which is defined as follows [see Figure 56.1(b)]:

$$\mathbf{c} = \sum_{k=1}^{p+q} [x^*, z_{k-1}, z_k]. \tag{12}$$

Then $\mathbf{c}$ is in E since each of its simplexes is in E. Moreover, (11) and (12) and Section 40 show that

$$\int_{\mathbf{c}} \sum_{(j_1, j_2)} [D_{j_1} f^{j_2}(y) - D_{j_2} f^{j_1}(y)]\,d(y^{j_1}, y^{j_2}) = \int_{\partial\mathbf{c}} \sum_{j=1}^{n} f^j(y)\,dy^j. \tag{13}$$

Since $D_{j_1} f^{j_2}(y) - D_{j_2} f^{j_1}(y) = 0$ for (j_1, j_2) in $(2/n)$ and y in E by hypothesis [see (3)], the integral on the left in (13) equals zero. The next step is to show that (13) is equivalent to (10). The definition of $\mathbf{c}$ in (12) and the definitions of the boundaries of simplexes and chains show that, since $z_{p+q} = z_0$,

$$\begin{aligned}\partial\mathbf{c} &= \sum_{k=1}^{p+q} \partial[x^*, z_{k-1}, z_k] \\ &= \sum_{k=1}^{p+q} \{(z_{k-1}, z_k] - [x^*, z_k] + [x^*, z_{k-1}]\}, \\ &= \sum_{k=1}^{p+q} [z_{k-1}, z_k] - \sum_{k=1}^{p+q} \{[x^*, z_k] - [x^*, z_{k-1}]\}, \\ &= \sum_{k=1}^{p+q} [z_{k-1}, z_k] + [x^*, z_0] - [x^*, z_{p+q}] \\ &= \sum_{k=1}^{p+q} [z_{k-1}, z_k].\end{aligned} \tag{14}$$

Thus the integral on the left in (13) equals zero, and (14) shows that the integral on the right equals

$$\sum_{k=1}^{p+q} \int_{[z_{k-1}, z_k]} \sum_{j=1}^{n} f^j(y)\,dy^j; \tag{15}$$

hence (13) is equivalent to (10), which in turn is equivalent to (8). Therefore, the proof that (5) has the same value for every chain which connects x_0 to $x:(x^1, \cdots, x^n)$ is now complete; define $F: E \to \mathbb{R}$ as follows:

$$F(x^1, \cdots, x^n) = \sum_{k=1}^{p} \int_{[x_{k-1}, x_k]} \sum_{j=1}^{n} f^j(y)\,dy^j. \tag{16}$$

Part III. The proof of Theorem 56.2 will now be completed by showing that F has derivatives such that $D_jF(x) = f^j(x)$ for x in E and $j = 1, \cdots, n$. Let $x:(x^1, \cdots, x^j, \cdots, x^n)$ be a point in E. Since E is open, x has a neighborhood $N(x, r)$ which is in E. Let w, with coordinates $(x^1, \cdots, x^j + h, \cdots, x^n)$, be a point in $N(x, r)$; a proof will show now that

$$\frac{\partial F(x)}{\partial x^j} = \lim_{h \to 0} \frac{F(w) - F(x)}{h} = f^j(x), \qquad j = 1, \cdots, n. \tag{17}$$

Since the functions $f^1, \cdots, f^n$ are continuous by hypothesis, this result and Theorem 3.19 will complete the proof that $D_jF(x)$ exists and equals $f^j(x)$ for $j = 1, \cdots, n$. Use the chain $\Sigma\{[x_{k-1}, x_k] : k = 1, \cdots, p\}$ as in (16) to calculate $F(x)$. To calculate $F(w)$, use the chain $\Sigma_1^p [x_{k-1}, x_k] + [x, w]$. Then

$$F(w) = \sum_{k=1}^{p} \int_{[x_{k-1}, x_k]} \sum_{j=1}^{n} f^j(y)\,dy^j + \int_{[x,w]} \sum_{j=1}^{n} f^j(y)\,dy^j, \tag{18}$$

and (18), (16), (7), and the mean-value theorem for integrals [see Theorem 35.18] show that

$$\begin{aligned} F(w) - F(x) &= \int_{[x,w]} \sum_{j=1}^{n} f^j(y)\,dy^j \\ &= \int_0^1 f^j(x^1, \cdots, x^j + th, \cdots, x^n)h\,dt \\ &= f^j(x^1, \cdots, x^j + t^*h, \cdots, x^n)h, \qquad 0 < t^* < 1. \end{aligned} \tag{19}$$

Since f^j is continuous at x by hypothesis, (17) follows from (19), and the proof of the entire theorem is complete. □

The function F defined in (16) depends on the point x_0, and there is a function F which corresponds to each point x_0 in E. To prove that each two of these functions differ by a constant, let F and F_1 correspond to x_0 and w_0, respectively. If $\Sigma\{[x_{k-1}, x_k] : k = 1, \cdots, p\}$ is a chain from x_0 to x, then

$$[w_0, x^*] + [x^*, x_0] + \sum\{[x_{k-1}, x_k] : k = 1, \cdots, p\} \tag{20}$$

is a chain from w_0 to x for every x in E, and the definitions of the functions F and F_1 [see (16)] show that

$$F_1(x) - F(x) = \left\{ \int_{[w_0, x^*]} + \int_{[x^*, x_0]} \right\} \sum_{j=1}^{n} f^j(x)\,dx^j \tag{21}$$

for every x in E. Thus $F_1 - F$ is a constant function.

Finally, every primitive of $f: E \to \mathbb{R}^n$ can be obtained from the primitive F defined by (16). If $H: E \to \mathbb{R}$ is a primitive of f, then $D_j H(x) = f^j(x) = D_j F(x)$ for $j = 1, \cdots, n$ and every x in E. Then $D_j(H - F)(x) = 0$ in E, and Exercise 9.3 shows that $H - F$ is a constant function. Thus $H = F + c$.

56.3 Example. Let $f: \mathbb{R}^3 \to \mathbb{R}^3$ be the function such that $f^j(x) = 2x^j$, $j = 1, 2, 3$. This function f has three variables and three components; thus f satisfies the first necessary condition for the existence of a primitive. Also, the components of f have continuous derivatives such that $D_{j_1} f^{j_2}(x) = D_{j_2} f^{j_1}(x)$ for (j_1, j_2) in $(2/3)$ and every x. Thus f satisfies the sufficient conditions in Theorem 56.2 for the existence of a primitive. The points $x_0: (0, 0, 0)$, $x_1: (x^1, 0, 0)$, $x_2: (x^1, x^2, 0)$, $x: (x^1, x^2, x^3)$ define the chain $[x_0, x_1] + [x_1, x_2] + [x_2, x]$ from x_0 to x, and (16) shows that a primitive $F: \mathbb{R}^3 \to \mathbb{R}^3$ is defined by the following equation:

$$F(x^1, x^2, x^3) = \left\{ \int_{[x_0, x_1]} + \int_{[x_1, x_2]} + \int_{[x_2, x]} \right\} \sum_{j=1}^{3} f^j(y)\,dy^j. \tag{22}$$

In this case the integrals in this equation are especially simple because each one is on a one-simplex which is parallel to one of the axes. Thus

$$\begin{aligned} F(x^1, x^2, x^3) &= \int_0^{x^1} 2y^1\,dy^1 + \int_0^{x^2} 2y^2\,dy^2 + \int_0^{x^3} 2y^3\,dy^3, \\ &= (x^1)^2 + (x^2)^2 + (x^3)^2, \end{aligned} \tag{23}$$

and it is easy to verify that F is indeed a primitive of f. Another chain which connects $x_0: (0, 0, 0)$ to $x: (x^1, x^2, x^3)$ has the single simplex $[x_0, x]$. Then

$$F(x) = \int_{[x_0, x]} \sum_{j=1}^{3} f^j(y)\,dy^j. \tag{24}$$

Equation (7) shows that the affine transformation which transforms $[e_0, e_1]$ into $[x_0, x]$ is $y^j = tx^j$, $j = 1, 2, 3$. Use this transformation to evaluate the integral in (24) as follows:

$$\begin{aligned} F(x) &= \int_0^1 [2(tx^1)x^1 + 2(tx^2)x^2 + 2(tx^3)x^3]\,dt \\ &= \int_0^1 [(x^1)^2 + (x^2)^2 + (x^3)^2] 2t\,dt \\ &= (x^1)^2 + (x^2)^2 + (x^3)^2, \end{aligned} \tag{25}$$

and this result verifies that both chains lead to the same result. Finally, a primitive of f can be found by using the points $x_1 : (x^1, x_0^2, x_0^3)$ and $x_2 : (x^1, x^2, x_0^3)$ to form the chain $[x_0, x_1] + [x_1, x_2] + [x_2, x]$ which connects $x_0 : (x_0^1, x_0^2, x_0^3)$ to $x : (x^1, x^2, x^3)$. In this case, $F(x^1, x^2, x^3)$ is found from the formula in (22) as follows:

$$\begin{aligned} F(x^1, x^2, x^3) &= \int_{x_0^1}^{x^1} 2y^1\,dy^1 + \int_{x_0^2}^{x^2} 2y^2\,dy^2 + \int_{x_0^3}^{x^3} 2y^3\,dy^3 \\ &= [(x^1)^2 + (x^2)^2 + (x^3)^2] - [(x_0^1)^2 + (x_0^2)^2 + (x_0^3)^2]. \end{aligned} \tag{26}$$

As expected, the primitives in (23) and (26) differ by a constant. A primitive can be found in this case also by using the formula in (24). Then

$$\begin{aligned} y^1 &= (x^1 - x_0^1)t + x_0^1, \\ y^2 &= (x^2 - x_0^2)t + x_0^2, \qquad 0 \leqq t \leqq 1, \\ y^3 &= (x^3 - x_0^3)t + x_0^3, \end{aligned} \tag{27}$$

and

$$\begin{aligned} F(x^1, x^2, x^3) &= \int_0^1 \sum_{j=1}^{3} 2[(x^j - x_0^j)t + x_0^j](x^j - x_0^j)\,dt \\ &= [(x^1)^2 + (x^2)^2 + (x^3)^2] - [(x_0^1)^2 + (x_0^2)^2 + (x_0^3)^2]. \end{aligned} \tag{28}$$

Exercises

56.1. Let the components (f^1, f^2, f^3) of a function $f: \mathbb{R}^3 \to \mathbb{R}^3$ be defined as follows:

$$f^1(x) = x^2 + x^3, \qquad f^2(x) = x^1 + x^3, \qquad f^3(x) = x^1 + x^2, \qquad x = (x^1, x^2, x^3).$$

(a) Show that f satisfies the sufficient conditions in Theorem 56.2 for the existence of a primitive.

(b) Take x_0 to be $(0, 0, 0)$. Find the value of the corresponding primitive F of f at $x : (x^1, x^2, x^3)$ by evaluating the sum in (16) on a chain connecting x_0 to x. For a first evaluation, let $x_1 = (x^1, 0, 0)$ and $x_2 = (x^1, x^2, 0)$, and use the chain $[x_0, x_1] + [x_1, x_2] + [x_2, x]$.

(c) In a second evaluation, let $x_1 = (0, 0, x^3)$ and $x_2 = (0, x^2, x^3)$, and again use the chain $[x_0, x_1] + [x_1, x_2] + [x_2, x]$. Verify that the sum in (16) has the same value for the two chains, and that $F(x^1, x^2, x^3) = x^1x^2 + x^1x^3 + x^2x^3$.

(d) In a third evaluation, find the value of the primitive F of f at $x : (x^1, x^2, x^3)$ by evaluating (16) on the chain with the single simplex $(x_0, x]$.

(e) Let x_0 be the point (x_0^1, x_0^2, x_0^3). Find the value of the corresponding primitive F by using an appropriate chain which connects $x_0 : (x_0^1, x_0^2, x_0^3)$ to $x : (x^1, x^2, x^3)$.

56.2. Let $f: \mathbb{R}^n \to \mathbb{R}^n$ be a function which has components $f^j(x) = m(x^j)^{m-1}, j = 1, \cdots, n$. Show that f satisfies the sufficient conditions in Theorem 56.2 for the existence of a primitive, and find a primitive.

56.3. Show that the function $f: \mathbb{R}^3 \to \mathbb{R}^3$ such that

$$f^1(x) = 2(x^1 - x^2) + 2(x^1 - x^3),$$
$$f^2(x) = -2(x^1 - x^2) + 2(x^2 - x^3), \qquad x = (x^1, x^2, x^3),$$
$$f^3(x) = -2(x^1 - x^3) - 2(x^2 - x^3),$$

satisfies the sufficient conditions in Theorem 56.2 for the existence of a primitive, and use (16) to find a primitive.

56.4. Construct an equilateral triangle on each side of a square in $\mathbb{R}^2$ to form a region with four points. Show that this region is not convex but that it is star-shaped with respect to at least one point. Describe the set of all points with respect to which the region is star-shaped.

56.5. Let $\mathbf{b}$ be a 2-simplex in $\mathbb{R}^n$, let E be an open set in $\mathbb{R}^n$ which contains $\mathbf{b}$, and let $f^j: E \to \mathbb{R}$, $j = 1, \cdots, n$, be functions which have continuous derivatives on E. Prove the statement of Stokes' theorem in (11). [Hint. Apply the fundamental theorem of the integral calculus to each of the functions $(f^j, x^j): E \to \mathbb{R}^2$, $j = 1, \cdots, n$. See Section 41.]

56.6. Let the components $(f^1, \cdots, f^n)$ of $f: \mathbb{R}^n \to \mathbb{R}^n$ be constant functions such that $f^i(x) = c_i$, $i = 1, \cdots, n$, for x in $\mathbb{R}^n$.
(a) Find a primitive of f by inspection.
(b) Show that f satisfies the sufficient conditions in Theorem 56.2 for the existence of a primitive, and use (16) to find a primitive.

56.7. Let the components (f^1, f^2) of $f: \mathbb{R}^2 \to \mathbb{R}^2$ be functions such that

$$f^1(x) = e^{x^1} \sin x^2, \qquad f^2(x) = e^{x^1} \cos x^2 \qquad \textit{for } x \text{ in } \mathbb{R}^2.$$

(a) Find a primitive of f by inspection.
(b) Show that f satisfies the sufficient conditions in Theorem 56.2 for the existence of a primitive, and use (16) to find a primitive.

56.8. Let $E_1, \cdots, E_n$ be open intervals such that $E_j = \{x^j : a_j < x^j < b_j\}$, and let $f^j: E_j \to \mathbb{R}$ be continuous functions. Let $E = E_1 \times \cdots \times E_n$, and let $f: E \to \mathbb{R}^n$ be the function whose components are $(f^1, \cdots, f^n)$. Show that f has a primitive $F: E \to \mathbb{R}$, and thus show that the conditions in Theorem 56.2 are sufficient but not necessary. [Hint. Let x_0^j be a point in E_j for $j = 1, \cdots, n$, and set

$$F(x^1, \cdots, x^n) = \int_{x_0^1}^{x^1} f^1(y^1)\,dy^1 + \cdots + \int_{x_0^n}^{x^n} f^n(y^n)\,dy^n.\Big]$$

57. The Existence of Primitives: The General Case

This section continues the investigation of the existence of primitives. More precisely, the problem is this: under what conditions can a given function $f: E \to \mathbb{R}^k$, $E \subset \mathbb{R}^n$, with components $(f^1, \cdots, f^k)$, be represented as the derivative of a function $F: E \to \mathbb{R}^m$ with components $(F^1, \cdots, F^m)$? Furthermore, if f has a primitive F, how can F be obtained (derived, constructed) from f? The solution of this problem is important because certain integrals have evaluations if their integrands have primitives [see Section 55].

Table 57.1. The number of components of the derivative of a function $F: E \to \mathbb{R}^m$, $E \subset \mathbb{R}^n$, of n variables which has m components.

n	m							
	1	2	3	4	5	...	n	...
1	1							
2	2	1						
3	3	3	1					
4	4	6	4	1				
5	5	10	10	5	1			
...	...	...	...	...	...			
n	$\binom{n}{1}$	$\binom{n}{2}$	$\binom{n}{3}$	$\binom{n}{4}$	$\binom{n}{5}$	...	$\binom{n}{n}$	
...	...	...	...	...	...	...	...	...

Section 56 has shown already that a necessary condition for the existence of a primitive concerns the number of components of the given function. Table 57.1 shows the number of components of the derivative of a function of n variables with m components.

57.1 Theorem. *If a function f which is defined on an open connected set E in $\mathbb{R}^n$ has a primitive $F: E \to \mathbb{R}^m$, $1 \leqq m \leqq n$, then f has $\binom{n}{m}$ components.*

PROOF. The derivative of a function $F: E \to \mathbb{R}^m$ is the function $D_{(j_1, \cdots, j_m)}(F^1, \cdots, F^m): E \to \mathbb{R}$, $(j_1, \cdots, j_m) \in (m/n)$. If this derivative equals the given function f on E, then f has $\binom{n}{m}$ components. □

Table 57.1 shows the number of components of the derivative of a function of n variables with m components. Thus this table shows the number of components of a function f of n variables which satisfies the necessary condition in Theorem 57.1 for the existence of a primitive F with m components. If the number of components of a function f of n variables is not one of the numbers $\binom{n}{1}, \cdots, \binom{n}{n}$ in row n of Table 57.1, then f does not have a primitive. Even if f satisfies the necessary condition in Theorem 57.1, as exhibited in Table 57.1, the function still may fail to have a primitive; Example 54.7 contains a proof of this statement. Thus the necessary condition in Theorem 57.1 is not a sufficient condition for the existence of a primitive, and it is necessary to extend the search for sufficient conditions.

Let $F: E \to \mathbb{R}^m$, $E \subset \mathbb{R}^n$, be a function whose components have continuous derivatives. Then the derivatives $D_{(j_1, \cdots, j_m)}(F^1, \cdots, F^m)(x)$ are the determinants of the $m \times m$ minors of the following matrix of derivatives:

$$\begin{bmatrix} D_1F^1(x) & D_2F^1(x) & \cdots & D_nF^1(x) \\ D_1F^2(x) & D_2F^2(x) & \cdots & D_nF^2(x) \\ \cdots & \cdots & \cdots & \cdots \\ D_1F^m(x) & D_2F^m(x) & \cdots & D_nF^m(x) \end{bmatrix}. \tag{1}$$

Thus

$$D_{(j_1,\cdots,j_m)}(F^1,\cdots,F^m)(x)$$
$$= \det\begin{bmatrix} D_{j_1}F^1(x) & \cdots & D_{j_m}F^1(x) \\ \cdots & \cdots & \cdots \\ D_{j_1}F^m(x) & \cdots & D_{j_m}F^m(x) \end{bmatrix}, \qquad (j_1,\cdots,j_m)\in(m/n). \tag{2}$$

Let $f: E \to \mathbb{R}^{\binom{n}{m}}$, $E \subset \mathbb{R}^n$, be a given function with components

$$f^{(j_1,\cdots,j_m)}: E \to \mathbb{R}, \qquad (j_1,\cdots,j_m)\in(m/n). \tag{3}$$

Then f satisfies a necessary condition for the existence of a primitive. If f has a primitive $F: E \to \mathbb{R}^m$ with components $(F^1,\cdots,F^m)$, then (1) shows that there exist functions $f_i: E \to \mathbb{R}^n$, $i = 1,\cdots,m$, with components f_i^j, $j = 1,\cdots,n$, and a matrix

$$\begin{bmatrix} f_1^1(x) & f_1^2(x) & \cdots & f_1^n(x) \\ f_2^1(x) & f_2^2(x) & \cdots & f_2^n(x) \\ \cdots & \cdots & \cdots & \cdots \\ f_m^1(x) & f_m^2(x) & \cdots & f_m^n(x) \end{bmatrix} \tag{4}$$

such that

$$f^{(j_1,\cdots,j_m)}(x) = \det\begin{bmatrix} f_1^{j_1}(x) & \cdots & f_1^{j_m}(x) \\ \cdots & \cdots & \cdots \\ f_m^{j_1}(x) & \cdots & f_m^{j_m}(x) \end{bmatrix}, \qquad (j_1,\cdots,j_m)\in(m/n). \tag{5}$$

If F is a primitive of f, then equations (1), $\cdots$, (4) show that F^1 is a primitive of $f_1, \cdots$, and that F^m is a primitive of f_m. Thus if the function f in (3) can be represented as shown in (4) and (5), then the problem of finding a primitive of f can be solved by finding primitives of $f_1,\cdots,f_m$. This problem has been treated in Section 56, and Theorem 56.2 suggests the following theorem.

57.2 Theorem (Sufficient Conditions). *Let E be an open set in $\mathbb{R}^n$ which is star-shaped with respect to at least one point x^*, and let $f: E \to \mathbb{R}^{\binom{n}{m}}$ be a function with components $f^{(j_1,\cdots,j_m)}$, $(j_1,\cdots,j_m)\in(m/n)$. If there exist functions $f_i: E \to \mathbb{R}^n$, $i = 1,\cdots,m$, whose components $(f_i^1,\cdots,f_i^n)$ satisfy (5) and have continuous derivatives such that*

$$D_{j_1}f_i^{j_2}(x) = D_{j_2}f_i^{j_1}(x), \qquad x\in E, \qquad (j_1,j_2)\in(2/n), \qquad i = 1,\cdots,m, \tag{6}$$

then f has a primitive F whose components $(F^1,\cdots,F^m)$ have two continuous derivatives such that

$$f^{(j_1,\cdots,j_m)}(x)$$
$$= \det\begin{bmatrix} D_{j_1}F^1(x) & \cdots & D_{j_m}F^1(x) \\ \cdots & \cdots & \cdots \\ D_{j_1}F^m(x) & \cdots & D_{j_m}F^m(x) \end{bmatrix}, \qquad x\in E, \qquad (j_1,\cdots,j_m)\in(m/n). \tag{7}$$

PROOF. Each function $f_i: E \to \mathbb{R}^n$, $i = 1, \cdots, m$, satisfies the hypotheses of Theorem 56.2; hence, there exists a function $F^i: E \to \mathbb{R}$ with two continuous derivatives such that

$$f_i^j = D_j F^i(x), \qquad x \in E, \qquad j = 1, \cdots, n, \qquad i = 1, \cdots, m. \tag{8}$$

Since the functions f_i^j, $i = 1, \cdots, m$ and $j = 1, \cdots, n$, satisfy (5) by hypothesis, equations (8) and (5) show that

$$f^{(j_1, \cdots, j_m)}(x) = D_{(j_1, \cdots, j_m)}(F^1, \cdots, F^m)(x), \qquad x \in E, \qquad (j_1, \cdots, j_m \in (m/n). \tag{9}$$

This equation shows that the function $F: E \to \mathbb{R}^m$, with components $(F^1, \cdots, F^m)$, is a primitive of f. The proof of Theorem 57.2 is complete. □

57.3 Example. Find a primitive of the function $f: \mathbb{R}^3 \to \mathbb{R}^3$ whose components are

$$\begin{aligned} f^{(1,2)}(x) &= 6(x^1)(x^2)^2 - 6(x^1)^2(x^2), \\ f^{(1,3)}(x) &= 6(x^1)(x^3)^2 - 6(x^1)^2(x^3), \\ f^{(2,3)}(x) &= 6(x^2)(x^3)^2 - 6(x^2)^2(x^3). \end{aligned} \tag{10}$$

Here f is a function of three independent variables, and it has three components. An examination of the row for $n = 3$ in Table 57.1 shows that f might have a primitive with one component ($m = 1$) or two components ($m = 2$). The components $f^{(1,2)}$, $f^{(1,3)}$, $f^{(2,3)}$ have continuous derivatives, but

$$\begin{gathered} D_1 f^{(1,3)}(x) \neq D_2 f^{(1,2)}(x), \qquad D_1 f^{(2,3)}(x) \neq D_3 f^{(1,2)}(x), \\ D_2 f^{(2,3)}(x) \neq D_3 f^{(1,3)}(x); \end{gathered} \tag{11}$$

therefore f does not satisfy the necessary condition in equation (3) of Theorem 56.1 for the existence of a primitive with $m = 1$. There remains the possibility that f has a primitive with two components. To check this possibility, we search for a matrix of functions such as (4) which can be used to represent f as in (5). Consider the matrix

$$\begin{bmatrix} 2x^1 & 2x^2 & 2x^3 \\ 3(x^1)^2 & 3(x^2)^2 & 3(x^3)^2 \end{bmatrix}. \tag{12}$$

Then (10) shows that

$$\begin{gathered} f^{(1,2)}(x) = \det \begin{bmatrix} 2x^1 & 2x^2 \\ 3(x^1)^2 & 3(x^2)^2 \end{bmatrix}, \qquad f^{(1,3)}(x) = \det \begin{bmatrix} 2x^1 & 2x^3 \\ 3(x^1)^2 & 3(x^3)^2 \end{bmatrix}, \\ f^{(2,3)}(x) = \det \begin{bmatrix} 2x^2 & 2x^3 \\ 3(x^2)^2 & 3(x^3)^2 \end{bmatrix}. \end{gathered} \tag{13}$$

Thus the components of f are the determinants of the 2×2 minors of the matrix in (12). Thus, Theorem 57.2 shows that there exists a primitive (F^1, F^2)

of f, whose components are given in (10), if primitives F^1 and F^2 can be found for the following functions whose components are given in the rows of (12):

$$\begin{aligned} f_1&: [2x^1 \quad 2x^2 \quad 2x^3], \\ f_2&: [3(x^1)^2 \quad 3(x^2)^2 \quad 3(x^3)^2]. \end{aligned} \tag{14}$$

Each of the functions in (14) satisfies the hypotheses of Theorems 56.2 and 57.2, which are sufficient conditions for the existence of a primitive. If

$$\begin{aligned} F^1(x) &= (x^1)^2 + (x^2)^2 + (x^3)^2, \\ F^2(x) &= (x^1)^3 + (x^2)^3 + (x^3)^3, \end{aligned} \tag{15}$$

then F^1, F^2 are primitives of f_1, f_2, respectively, and the proof of Theorem 57.2 has shown that (F^1, F^2) is a primitive of the function whose components are given in (10). Because of the simplicity of the functions in (14), the primitives F^1, F^2 in (15) can be found by inspection in this case. In all cases these functions can be defined by integrals on chains of simplexes as explained in the proof of Theorem 56.2.

The treatment of the existence of primitives is now complete in the sense that theorems have been established which contain sufficient conditions for the existence of primitives in all cases [see Theorems 56.2 and 57.2] and show how these primitives can be defined in case the sufficient conditions are satisfied [see the proofs of the two theorems]. There is one case of exceptional importance, however, which requires further treatment.

Table 57.1 shows that the derivative of n functions of n variables is a function with one component [see the entries on the diagonal in the table]. Thus there is a possibility that a single function of n variables may have a primitive.

57.4 Theorem. *Let $A_n = [a_1, b_1] \times \cdots \times [a_n, b_n]$ in $\mathbb{R}^n$, and let $f: A_n \to \mathbb{R}$ be a function which has continuous derivatives*

$$D_j f(x^1, \cdots, x^n), \qquad (x^1, \cdots, x^n) \in A_n, \qquad j = 1, \cdots, n. \tag{16}$$

Then f has n primitives $(F^1, \cdots, F^n)$, and each row in the following array contains the components of one of these primitives:

$$\begin{array}{cccc} \displaystyle\int_{a_1}^{x^1} f(t, x^2, \cdots, x^n)\,dt, & x^2, & \cdots & x^n; \\ x^1, & \displaystyle\int_{a_2}^{x^2} f(x^1, t, \cdots, x^n)\,dt, & \cdots & x^n; \\ \cdots & \cdots & \cdots & \cdots \\ x^1, & x^2, & \cdots & \displaystyle\int_{a_n}^{x^n} f(x^1, x^2, \cdots, t)\,dt. \end{array} \tag{17}$$

This theorem can be proved most easily with the help of the following lemma.

57.5 Lemma. *If $f: A_n \to \mathbb{R}$ has continuous derivatives $D_j f$, $j = 1, \cdots, n$, and if $F^j: A_n \to \mathbb{R}$ is the function such that*

$$F^j(x) = \int_{a_j}^{x^j} f(x^1, \cdots, t/x^j, \cdots, x^n)\,dt, \tag{18}$$

then F^j has continuous derivatives $D_1 F^j, \cdots, D_n F^j$, and

$$\begin{aligned} D_1 F^j(x) &= \int_{a_j}^{x^j} D_1 f(x^1, \cdots, t/x^j, \cdots, x^n)\,dt, \\ &\cdots\cdots\cdots\cdots\cdots\cdots \\ D_j F^j(x) &= f(x^1, \cdots, x^j, \cdots, x^n), \\ &\cdots\cdots\cdots\cdots\cdots\cdots \\ D_n F^j(x) &= \int_{a_n}^{x^n} D_n f(x^1, \cdots, t/x^j, \cdots, x^n)\,dt. \end{aligned} \tag{19}$$

Proof of the Lemma. The proof will be given in the special case in which $n = 2$ and

$$F^1(x) = \int_{a_1}^{x^1} f(t, x^2)\,dt. \tag{20}$$

Since f has continuous derivatives on A_2, then f is continuous on A_2. Hence, for each (x^1, x^2) in A_2, the integrand $f(t, x^2)$ of the integral in (20) is a continuous function of t on the interval $[a_1, x^1]$, and F^1 is defined on A_2. The lemma will be proved by showing that F^1 has continuous partial derivatives; then since these partial derivatives are continuous, they are derivatives by Theorem 3.19. Now

$$F^1(x^1 + h, x^2) - F^1(x^1, x^2) = \int_{x^1}^{x^1+h} f(t, x^2)\,dt, \tag{21}$$

and the mean-value theorem for integrals shows that

$$F^1(x^1 + h, x^2) - F^1(x^1, x^2) = f(x^1 + \theta h, x^2)h, \qquad 0 < \theta < 1. \tag{22}$$

Then since f is continuous,

$$\begin{aligned} \frac{\partial F^1(x^1, x^2)}{\partial x^1} &= \lim_{h\to 0} \frac{F^1(x^1 + h, x^2) - F^1(x^1, x^2)}{h}, \\ &= \lim_{h\to 0} f(x^1 + \theta h, x^2), \\ &= f(x^1, x^2). \end{aligned} \tag{23}$$

Next,

$$F^1(x^1, x^2 + k) - F^1(x^1, x^2) = \int_{a^1}^{x^1} [f(t, x^2 + k) - f(t, x^2)]\,dt. \tag{24}$$

By the mean-value theorem for functions,

$$f(t, x^2 + k) - f(t, x^2) = D_2 f(t, x^2 + \theta_t k)k, \qquad 0 < \theta_t < 1. \tag{25}$$

Then since f has continuous derivatives $D_1 f$, $D_2 f$ on A_2 by hypothesis,

$$\begin{aligned} \frac{\partial F^1(x^1, x^2)}{\partial x^2} &= \lim_{k \to 0} \frac{F^1(x^1, x^2 + k) - F^1(x^1, x^2)}{k} \\ &= \lim_{k \to 0} \int_{a_1}^{x^1} D_2 f(t, x^2 + \theta_t k)\, dt, \\ &= \int_{a_1}^{x^1} \lim_{k \to 0} D_2 f(t, x^2 + \theta_t k)\, dt, \\ &= \int_{a_1}^{x^1} D_2 f(t, x^2)\, dt. \end{aligned} \tag{26}$$

Equation (23) shows that the partial derivative of F^1 with respect to x^1 is continuous since f is continuous, and (26) and elementary properties of integrals show that the partial derivative of F^1 with respect to x^2 is continuous. Since these partial derivatives exist and are continuous, then Theorem 3.19 shows that $D_1 F^1$ and $D_2 F^1$ exist and that

$$\begin{aligned} D_1 F^1(x^1, x^2) &= \frac{\partial F^1(x^1, x^2)}{\partial x^1} = f(x^1, x^2), \\ D_2 F^1(x^1, x^2) &= \frac{\partial F^1(x^1, x^2)}{\partial x^2} = \int_{a_1}^{x^1} D_2 f(t, x^2)\, dt. \end{aligned} \tag{27}$$

Thus the proof of Lemma 57.5 is complete in the special case, and the proof of the formulas in (19) is similar to the proof of the formulas in (27) in the special case. □

Proof of Theorem 57.4. If

$$\begin{aligned} F^1(x) &= \int_{a_1}^{x^1} f(t, x^2, \cdots, x^n)\, dt, \\ F^2(x) &= x^2, \\ &\cdots\cdots\cdots \\ F^n(x) &= x^n, \end{aligned} \tag{28}$$

then Lemma 57.5 shows that

$$D_{(1, \cdots, n)}(F^1, \cdots, F^n)(x) = \begin{vmatrix} f(x) & D_2 F^1(x) & \cdots & D_n F^1(x) \\ 0 & 1 & \cdots & 0 \\ \cdots & \cdots & \cdots & \cdots \\ 0 & 0 & \cdots & 1 \end{vmatrix} = f(x); \tag{29}$$

therefore, the function $F: A_n \to \mathbb{R}^n$ whose components are defined in (28) is a primitive of $f: A_n \to \mathbb{R}$. Thus the first of the functions in (17) is a primitive of f, and a similar application of Lemma 57.5 shows that each of the n-functions in (17) is a primitive of f. The proof of Theorem 57.4 is complete. □

57.6 Example. If $f: A_3 \to \mathbb{R}$ is the function such that

$$f(x^1, x^2, x^3) = ax^1 + bx^2 + cx^3, \tag{30}$$

then it has at least three primitives by Theorem 57.4. The first of these primitives, by (17), is the function whose components are

$$\int_{a_1}^{x^1} (at + bx^2 + cx^3)\,dt, \qquad x^2, \qquad x^3. \tag{31}$$

In this case the integral in (31) can be evaluated explicitly by the fundamental theorem of the integral calculus. If $F(x^1, x^2, x^3)$ denotes this integral, then

$$F(x^1, x^2, x^3) = (1/2)a[(x^1)^2 - (a_1)^2] + b(x^1 - a_1)x^2 + c(x^1 - a_1)x^3,$$

$$D_1F(x^1, x^2, x^3) = ax^1 + bx^2 + cx^3 = f(x^1, x^2, x^3),$$

$$D_2F(x^1, x^2, x^3) = \int_{a_1}^{x^1} D_2(at + bx^2 + cx^3)\,dt = \int_{a_1}^{x^1} b\,dt = b(x^1 - a_1),$$

$$D_3F(x^1, x^2, x^3) = \int_{a_1}^{x^1} D_3(at + bx^2 + cx^3)\,dt = \int_{a_1}^{x^1} c\,dt = c(x^1 - a_1).$$

Then

$$D_{(1,2,3)}(F, x^2, x^3)(x) = \det\begin{bmatrix} f(x) & b(x^1 - a_1) & c(x^1 - a_1) \\ 0 & 1 & 0 \\ 0 & 0 & 1 \end{bmatrix} = f(x),$$

and this equation verifies that the function whose components are (F, x^2, x^3) is a primitive of f. In the same way it can be verified that each of the functions whose components are

$$x^1, \qquad \int_{a_2}^{x^2} (ax^1 + bt + cx^3)\,dt, \qquad x^3,$$

$$x^1, \qquad x^2, \qquad \int_{a_3}^{x^3} (ax^1 + bx^2 + ct)\,dt,$$

is a primitive of f [see (17) in Theorem 57.4]. Each of the three basic primitives generates a large class of primitives as explained in Theorem 54.5, but the basic primitives in (17) are especially important.

Exercises

57.1. (a) If f is defined in a set E in $\mathbb{R}^2$ and has a primitive, show that either (i) f has one component and its primitive has two components, or (ii) f has two components and its primitive has one component.

(b) If f is defined in a set E in $\mathbb{R}^3$ and has a primitive, show that either (i) f has one component and its primitive has three components, or (ii) f has three components and its primitive has one or two components.

(c) If f is defined in a set E in $\mathbb{R}^n$ and has a primitive, show that, for some integer m, $1 \leqq m \leqq n$, the function f has $\binom{n}{m}$ components and its primitive has either m or $n - m$ components. In what cases is the number of components of the primitive uniquely determined?

57.2. A function $f: \mathbb{R}^3 \to \mathbb{R}^3$ has components (f^1, f^2, f^3) which are defined as follows:

$$f^1(x) = 2(x^1 - x^2), \quad f^2(x) = 2(x^1 - x^3), \quad f^3(x) = 2(x^2 - x^3), \quad x = (x^1, x^2, x^3).$$

(*a*) Show that f does not have a primitive of the form $F: \mathbb{R}^3 \to \mathbb{R}$. [Hint. Show that f fails to satisfy the necessary conditions in Theorem 56.1 for the existence of a primitive $F: \mathbb{R}^3 \to \mathbb{R}$.]

(b) Find a primitive of f of the form $(F^1, F^2): \mathbb{R}^3 \to \mathbb{R}^2$. [Hint. Show that

$$f^1(x) = \begin{vmatrix} 2x^1 & 2x^2 \\ 1 & 1 \end{vmatrix}, \qquad f^2(x) = \begin{vmatrix} 2x^1 & 2x^3 \\ 1 & 1 \end{vmatrix}, \qquad f^3(x) = \begin{vmatrix} 2x^2 & 2x^3 \\ 1 & 1 \end{vmatrix},$$

and use Theorems 57.2 and 56.2 to find a primitive (F^1, F^2).]

(c) Find a primitive of f which is not contained in the class generated by the primitive found in (b). [Hint. Show that

$$f^1(x) = \begin{vmatrix} 2(x^2 + x^3) & 2(x^1 + x^3) \\ -1 & -1 \end{vmatrix}, \qquad f^2(x) = \begin{vmatrix} 2(x^2 + x^3) & 2(x^1 + x^2) \\ -1 & -1 \end{vmatrix},$$

$$f^3(x) = \begin{vmatrix} 2(x^1 + x^3) & 2(x^1 + x^2) \\ -1 & -1 \end{vmatrix},$$

and use Theorems 57.2 and 56.2 to find a primitive (F^1, F^2). Compare Exercise 54.2.]

57.3. Let $f: \mathbb{R}^2 \to \mathbb{R}$ be the function such that $f(x) = 3[(x^1)^2 + (x^2)^2]$. If f has a primitive, show that this primitive has two components (F^1, F^2).

(a) Use Theorems 57.2 and 56.2 and the fact that

$$3[(x^1)^2 + (x^2)^2] = \begin{vmatrix} 3(x^1)^2 & -3(x^2)^2 \\ 1 & 1 \end{vmatrix}$$

to show that f has a primitive; find this primitive.

(b) Use the fact that

$$3[(x^1)^2 + (x^2)^2] = \begin{vmatrix} 3[(x^1)^2 + (x^2)^2] & 6x^1x^2 \\ 0 & 1 \end{vmatrix} = \begin{vmatrix} 1 & 0 \\ 6x^1x^2 & 3[(x^1)^2 + (x^2)^2] \end{vmatrix}$$

to find two additional primitives (F^1, F^2) of f [see (4) and (5) in Section 52]. Show that these primitives are the two whose existence is established by Theorem 57.4, and that they are not contained in the class of primitives generated by the primitive found in (a).

57.4. Let G_3 be a surface in $\mathbb{R}^3$, let E be an open set in $\mathbb{R}^3$ which contains $T(G_3)$, and let $f^{(j_1, j_2)}: E \to \mathbb{R}$, $(j_1, j_2) \in (2/3)$, be functions which have continuous derivatives in E.

(a) Establish the following evaluation:

$$\int_{G_3} [D_3 f^{(1,2)}(y) - D_2 f^{(1,3)}(y) + D_1 f^{(2,3)}(y)]\, d(y^1, \cdots, y^3)$$

$$= \int_{\partial G_3} f^{(1,2)}(y)\, d(y^1, y^2) + f^{(1,3)}(y)\, d(y^1, y^3) + f^{(2,3)}(y)\, d(y^2, y^3).$$

[Hint. Consider $(f^{(1,2)}, y^1, y^2): E \to \mathbb{R}^3$, etc.]

(b) Derive the generalizations, for a surface G_3 in $\mathbb{R}^4$ and for a surface G_3 in $\mathbb{R}^n$, of the evaluation formula in (a).

57.5. Let $[a_{ij}]$ be an $m \times n$ matrix of constants, $m \leqq n$, and let $f: \mathbb{R}^n \to \mathbb{R}^{\binom{n}{m}}$ be a constant function such that $f^{(j_1, \cdots, j_m)}(x)$ equals the determinant of the minor of $[a_{ij}]$ in columns $(j_1, \cdots, j_m)$, for each $(j_1, \cdots, j_m)$ in (m/n). Use Theorems 57.2 and 56.2 [see also Exercise 56.6] to show that f has a primitive with m components and to find a primitive $(F^1, \cdots, F^m)$. Describe the class of primitives generated by the primitive you find.

57.6. Show that every constant function $f: \mathbb{R}^n \to \mathbb{R}^n$ has a primitive. Does every constant function $f: \mathbb{R}^n \to \mathbb{R}^{\binom{n}{m}}$, $m > 1$, have a primitive? [Hint. Exercise 56.6 and Example 54.7.]

57.7. Let $f^j: [a_j, b_j] \to \mathbb{R}$, $j = 1, \cdots, n$, be continuous functions, and let $f: A_n \to \mathbb{R}$ be the function defined on $A_n = [a_1, b_1] \times \cdots \times [a_n, b_n]$ as follows:

$$f(x^1, \cdots, x^n) = \begin{vmatrix} f^1(x^1) & 0 & \cdots & 0 \\ 0 & f^2(x^2) & \cdots & 0 \\ \cdots & \cdots & \cdots & \cdots \\ 0 & 0 & \cdots & f^n(x^n) \end{vmatrix} = f^1(x^1) f^2(x^2) \cdots f^n(x^n).$$

(a) Does f satisfy some, none, or all of the sufficient conditions in Theorem 57.2 for the existence of a primitive? Explain.

(b) Show that f has a primitive $(F^1, \cdots, F^n)$ such that

$$F^j(x^1, \cdots, x^n) = \int_{a_j}^{x^j} f^j(t)\, dt, \qquad (x^1, \cdots, x^n) \in A_n, \qquad j = 1, \cdots, n.$$

(c) Evaluate $\int_{A_n} f(x)\, d(x^1, \cdots, x^n)$ by each of the $n!$ iterated integrals and show that

$$\int_{A_n} f(x)\, d(x^1, \cdots, x^n) = \prod_{j=1}^{n} \int_{a_j}^{b_j} f^j(x^j)\, dx^j.$$

58. Iterated Integrals

The purpose of this section is to show that the fundamental theorem of the integral calculus reduces, in a special case, to the evaluation of an integral by the classical iterated integrals.

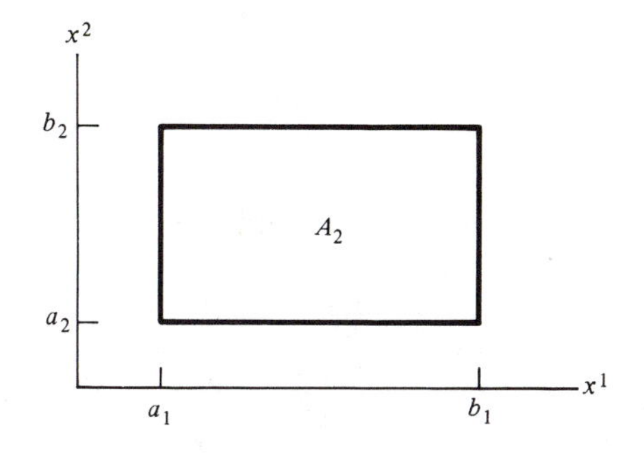

Figure 58.1. The rectangle A_2.

58.1 Theorem. *If $A_2 = [a_1, b_1] \times [a_2, b_2]$, and if $f: A_2 \to \mathbb{R}$ has continuous derivatives $D_1 f$ and $D_2 f$ on A_2, then*

$$\begin{aligned}\int_{A_2} f(x)\,d(x^1, x^2) &= \int_{a_2}^{b_2}\left(\int_{a_1}^{b_1} f(x^1, x^2)\,dx^1\right)dx^2 \\ &= \int_{a_1}^{b_1}\left(\int_{a_2}^{b_2} f(x^1, x^2)\,dx^2\right)dx^1.\end{aligned} \tag{1}$$

PROOF. Since f has continuous derivatives, Theorem 57.4 shows that f has a primitive (F^1, F^2) on A_2 such that

$$F^1(x^1, x^2) = \int_{a_1}^{x^1} f(t, x^2)\,dt, \qquad F^2(x^1, x^2) = x^2. \tag{2}$$

Then

$$\int_{A_2} f(x)\,d(x^1, x^2) = \int_{A_2} D_{(1,2)}(F^1, F^2)(x)\,d(x^1, x^2), \tag{3}$$

and the fundamental theorem of the integral calculus shows that

$$\int_{A_2} f(x)\,d(x^1, x^2) = \int_{\partial A_2} F^1(x) D_1 F^2(x)\,dx^1 + F^1(x) D_2 F^2(x)\,dx^2. \tag{4}$$

Equation (2) shows that $D_1 F^2(x) = 0$ and $D_2 F^2(x) = 1$; hence (4) simplifies to this:

$$\int_{A_2} f(x)\,d(x^1, x^2) = \int_{\partial A_2} F^1(x)\,dx^2. \tag{5}$$

An examination of Figure 58.1 shows that this equation, written out in full detail, is the following:

$$\begin{aligned}\int_{A_2} f(x)\,d(x^1, x^2) = \int_{a_1}^{b_1} F^1(x^1, a_2)\,dx^2 &+ \int_{a_2}^{b_2} F^1(b_1, x^2)\,dx^2 \\ + \int_{b_1}^{a_1} F^1(x^1, b_2)\,dx^2 &+ \int_{b_2}^{a_2} F^1(a_1, x^2)\,dx^2.\end{aligned} \tag{6}$$

Now

$$\int_{a_1}^{b_1} F^1(x^1, a_2)\,dx^2 = 0, \qquad \int_{b_1}^{a_1} F^1(x^1, b_2)\,dx^2 = 0, \tag{7}$$

because x^2 is constant on the lower and upper sides of A_2. Furthermore,

$$\int_{b_2}^{a_2} F^1(a_1, x^2)\,dx^2 = 0, \qquad \int_{a_2}^{b_2} F^1(b_1, x^2)\,dx^2 = \int_{a_2}^{b_2}\left(\int_{a_1}^{b_1} f(x^1, x^2)\,dx^1\right)dx^2, \tag{8}$$

because (2) shows that

$$F^1(a_1, x^2) = \int_{a_1}^{a_1} f(t, x^2)\,dt = 0, \tag{9}$$

$$F^1(b_1, x^2) = \int_{a_1}^{b_1} f(t, x^2)\,dt = \int_{a_1}^{b_1} f(x^1, x^2)\,dx^1. \tag{10}$$

Then (7) and (8) show that (6) simplifies to

$$\int_{A_2} f(x)\,d(x^1, x^2) = \int_{a_2}^{b_2}\left(\int_{a_1}^{b_1} f(x^1, x^2)\,dx^1\right)dx^2, \tag{11}$$

which is the first of the iterated integral formulas in (1). Theorem 57.4 shows that another primitive of $f\colon A_2 \to \mathbb{R}$ is the function (F^1, F^2) such that

$$F^1(x^1, x^2) = x^1, \qquad F^2(x^1, x^2) = \int_{a_2}^{x^2} f(x^1, t)\,dt. \tag{12}$$

This primitive and the fundamental theorem of the integral calculus can be used as in the first case to obtain an evaluation of $\int_{A_2} f(x)\,d(x^1, x^2)$; the result is the second of the iterated integral formulas in (2). The proof of Theorem 58.1 is complete. □

Theorem 58.1 shows how iterated integrals fit into the general theory of the evaluation of integrals based on the fundamental theorem of the integral calculus. Better theorems, however, can be proved by special methods. In order to prove the iterated integral formulas in (1) by using the fundamental theorem, it is necessary to assume that f has continuous derivatives; but special methods in Section 35 [see Example 35.20] have established these same formulas for functions f which are merely continuous, and even better theorems are known.

58.2 Example. This example outlines the next step in the proof by induction of the iterated integral formulas for rectangular sets A_n in $\mathbb{R}^n$, $n \geqq 3$. If $A_3 = [a_1, b_1] \times \cdots \times [a_3, b_3]$, and if $f\colon A_3 \to \mathbb{R}$ is a function which has continuous derivatives $D_1 f, \cdots, D_3 f$, then Theorem 57.4 shows that a primitive of f is the function $(F^1, \cdots, F^3)\colon A_3 \to \mathbb{R}^3$ such that

$$F^1(x) = \int_{a_1}^{x^1} f(t, x^2, x^3)\,dt, \qquad F^2(x) = x^2, \qquad F^3(x) = x^3. \tag{13}$$

Then

$$\int_{A_3} f(x)\,d(x^1, \cdots, x^3) = \int_{A_3} D_{(1,\cdots,3)}(F^1, \cdots, F^3)(x)\,d(x^1, \cdots, x^3), \tag{14}$$

and the fundamental theorem of the integral calculus states that

$$\int_{A_3} f(x)\,d(x^1, \cdots, x^3) = \int_{\partial A_3} \sum_{(2/3)} F^1(x) D_{(j_1, j_2)}(F^2, F^3)(x)\,d(x^{j_1}, x^{j_2}). \tag{15}$$

Now (13) shows that

$$\begin{aligned} &D_{(j_1, j_2)}(F^2, F^3)(x) = 0, \qquad x \in A_3, \qquad (j_1, j_2) \neq (2, 3), \qquad (j_1, j_2) \in (2/3), \\ &D_{(2,3)}(F^2, F^3)(x) = 1, \qquad x \in A_3. \end{aligned} \tag{16}$$

Thus (15) simplifies to

$$\int_{A_3} f(x)\,d(x^1, \cdots, x^3) = \int_{\partial A_3} F^1(x)\,d(x^2, x^3). \tag{17}$$

The integral on the right in (17) is extended over the six faces of A_3; the integral is zero on the faces in the planes

$$x^2 = a_2, \qquad x^2 = b_2, \qquad x^3 = a_3, \qquad x^3 = b_3, \tag{18}$$

because in each of these cases the region of integration is a line segment in (x^2, x^3)-space. The integral on the right in (17) is also zero on the face of A_3 in the plane $x^1 = a_1$ because, by (13),

$$F^1(a_1, x^2, x^3) = \int_{a_1}^{a_1} f(t, x^2, x^3)\,dt = 0. \tag{19}$$

Thus, if $A_2 = [a_2, b_2] \times [a_3, b_3]$, then (17) simplifies to the following:

$$\int_{A_3} f(x)\,d(x^1, \cdots, x^3) = \int_{A_2} F^1(b_1, x^2, x^3)\,d(x^2, x^3). \tag{20}$$

Now (13) shows that

$$F^1(b_1, x^2, x^3) = \int_{a_1}^{b_1} f(t, x^2, x^3)\,dt = \int_{a_1}^{b_1} f(x^1, x^2, x^3)\,dx^1. \tag{21}$$

Then since f has continuous derivatives by hypothesis, Lemma 57.5 shows that the function in (21) has continuous derivatives with respect to x^2 and x^3. Therefore, the integral on the right in (20) satisfies the hypotheses of Theorem 58.1, and

$$\int_{A_2} F^1(b_1, x^2, x^3)\, d(x^2, x^3) = \int_{a_3}^{b_3} \left(\int_{a_2}^{b_2} F^1(b_1, x^2, x^3)\, dx^2 \right) dx^3. \tag{22}$$

Thus (20), (21), (22) show that

$$\int_{A_3} f(x)\, d(x^1, \cdots, x^3) = \int_{a_3}^{b_3} \left(\int_{a_2}^{b_2} \left(\int_{a_1}^{b_1} f(x^1, x^2, x^3)\, dx^1 \right) dx^2 \right) dx^3. \tag{23}$$

Since Theorem 57.4 shows that there are three primitives for the first step in this evaluation and two at the second, there are $3 \cdot 2$, or $3!$, different iterated integrals which evaluate $\int_{A_3} f(x)\, d(x^1, \cdots, x^3)$. This example illustrates the essential steps in the proof by induction that $\int_{A_n} f(x)\, d(x^1, \cdots, x^n)$ can be evaluated by each of the $n!$ iterated integrals.

Exercises

58.1. Let $A_2 = [a_1, b_1] \times [a_2, b_2]$, and let $f: A_2 \to \mathbb{R}$ be the function such that $f(x^1, x^2) = 4x^1x^2$. Use each of the integrals in (1) in Theorem 58.1 to find the value of $\int_{A_2} f(x)\, d(x^1, x^2)$.

58.2. Let $A_3 = [a_1, b_1] \times \cdots \times [a_3, b_3]$, and let $f: A_3 \to \mathbb{R}$ be the function such that $f(x) = 2(x^1 + x^2 + x^3)$. Use the iterated integral in (23) to find the value of $\int_{A_3} f(x)\, d(x^1, \cdots, x^3)$. Evaluate the integral by a second iterated integral and verify that the two values are the same.

58.3. Let $f: \mathbb{R}^2 \to \mathbb{R}$ be the function such that

$$f(x^1, x^2) = \begin{vmatrix} 2x^1 & -2x^2 \\ 2x^1 & 2x^2 \end{vmatrix} = 8x^1x^2,$$

and let $A_2 = [a_1, b_1] \times [a_2, b_2]$.

(a) Use Theorem 57.2 to find a primitive (F^1, F^2) of f. Then use your primitive of f and the fundamental theorem of the integral calculus to evaluate $\int_{A_2} f(x)\, d(x^1, x^2)$.

(b) Evaluate $\int_{A_2} f(x)\, d(x^1, x^2)$ by using one of the iterated integrals in Theorem 58.1, and verify that the two values found for the integral are equal.

(c) Give reasons for believing that, in general, the iterated integral evaluations of $\int_{A_2} f(x)\, d(x^1, x^2)$ are simpler than other evaluations.

CHAPTER 9

The Kronecker Integral and the Sperner Degree

59. Preliminaries

This chapter defines the Kronecker integral and demonstrates its use in proving a form of the intermediate-value theorem. As a necessary first step in treating the Kronecker integral, Section 60 derives formulas for the volume and for the area of the surface of the sphere $S_{n-1}(a) = \{x : x \in \mathbb{R}^n, (x^1)^2 + \cdots + (x^n)^2 = a^2\}$; the formula which states the relation between the volume and the area of the surface is essentially a special case of the fundamental theorem of the integral calculus, and the formula for the area of the surface is a special case of the general Kronecker integral. As preparation for Section 60, the present section defines the gamma function and establishes its elementary properties; also, it proves a reduction formula which is used to evaluate certain definite integrals.

59.1 Definition. The *gamma function* Γ is defined as follows:

$$\Gamma(x) = \int_0^\infty t^{x-1}e^{-t}\,dt, \qquad x \in \mathbb{R}, \qquad x > 0. \tag{1}$$

Equation (1) shows that Γ is defined by an improper integral, and it is necessary to show that this improper integral converges. Now

$$\Gamma(x) = \int_0^1 t^{x-1}e^{-t}\,dt + \int_1^\infty t^{x-1}e^{-t}\,dt. \tag{2}$$

The integrand of the first integral becomes infinite at its lower limit if $0 < x < 1$, but since

$$t^{x-1}e^{-t} < t^{x-1}, \qquad 0 < x < 1, \qquad 0 < t \leqq 1,$$

$$\int_\varepsilon^1 t^{x-1}e^{-t}\,dt < \int_\varepsilon^1 t^{x-1}\,dt = \frac{1-\varepsilon^x}{x} < \frac{1}{x}, \qquad 0 < x < 1, \qquad 0 < \varepsilon < 1,$$

the first integral on the right in (2) converges for $0 < x < 1$. This integral is a proper integral for $x \geqq 1$. To show that the second integral in (2) converges, use a limit test to compare it with $\int_1^\infty t^{-2}\,dt$. L'Hospital's rule can be used to show that

$$\lim_{t\to\infty} t^{x-1}e^{-t}t^2 = \lim_{t\to\infty} \frac{t^{x+1}}{e^t} = 0$$

for every value of x. Then there is a t_0 such that

$$t^{x-1}e^{-t}t^2 < 1, \qquad t \geqq t_0,$$

$$t^{x-1}e^{-t} < \frac{1}{t^2}, \qquad t \geqq t_0, \qquad x \in \mathbb{R}.$$

Thus since $\int_1^\infty t^{-2}\,dt$ converges, $\int_1^\infty t^{x-1}e^{-t}\,dt$ converges for every x. Thus both of the integrals in (2) converge if $x > 0$, and $\Gamma(x)$ is well defined by (1).

59.2 Theorem. *The gamma function has the following properties:*

$$\Gamma(1) = 1; \tag{3}$$

$$\Gamma(x+1) = x\Gamma(x), \qquad x > 0; \tag{4}$$

$$\Gamma(n+1) = n!, \qquad n = 0, 1, \cdots; \tag{5}$$

$$\Gamma(1/2) = \sqrt{\pi}. \tag{6}$$

PROOF. Definition 59.1 shows that

$$\Gamma(1) = \lim_{b\to\infty} \int_0^b e^{-t}\,dt = \lim_{b\to\infty} -e^{-t}\big|_0^b = \lim_{b\to\infty} (1 - e^{-b}) = 1;$$

hence (3) is true. Also, use integration by parts to obtain

$$\int_0^b t^x e^{-t}\,dt = -t^x e^{-t}\big|_0^b + x\int_0^b t^{x-1}e^{-t}\,dt, \qquad x > 0. \tag{7}$$

The limit of each of the three terms exists as $b \to \infty$. Since

$$\lim_{b\to\infty} b^x e^{-b} = \lim_{b\to\infty} \frac{b^x}{e^b} = 0$$

as shown above, equation (7) in the limit becomes

$$\int_0^\infty t^x e^{-t}\,dt = x\int_0^\infty t^{x-1}e^{-t}\,dt, \qquad x > 0,$$

which, by (1), is $\Gamma(x+1) = x\Gamma(x)$. Thus the proof of (4) is complete.

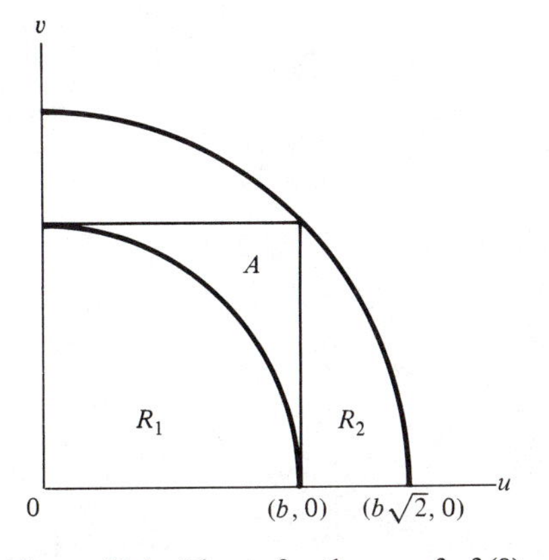

Figure 59.1. Figure for the proof of (8).

Use mathematical induction to prove (5). By definition, $0! = 1$; then (3) shows that (5) is true for $n = 0$. By (4) and (3),

$$\Gamma(1 + 1) = 1\Gamma(1) = 1!,$$

and (5) is true also for $n = 1$. Thus (5) is true for $n = 0$ and $n = 1$; assume that it is true for $n = k$. Then $\Gamma(k + 1) = k!$, and (4) shows that

$$\Gamma[(k + 1) + 1] = (k + 1)\Gamma(k + 1) = (k + 1)[k!] = (k + 1)!$$

Thus if (5) is true for $n = k$, it is also true for $n = k + 1$. Since (5) is true for $n = 0$ and $n = 1$, a complete induction shows that it is true for $n = 0, 1, \cdots$. The proof of (5) is complete.

The proof of (6) begins with the proof that

$$\int_0^\infty e^{-u^2}\,du = \frac{\sqrt{\pi}}{2}. \tag{8}$$

To prove (8), consider the double integral of $e^{-(u^2+v^2)}$ over the circular sectors R_1 and R_2 and the square A shown in Figure 59.1. Since $e^{-(u^2+v^2)} > 0$ for all (u, v),

$$\iint_{R_1} e^{-(u^2+v^2)}\,du\,dv < \iint_A e^{-(u^2+v^2)}\,du\,dv < \iint_{R_2} e^{-(u^2+v^2)}\,du\,dv. \tag{9}$$

Use iterated integrals to evaluate these integrals after first changing to polar coordinates in the integrals over R_1 and R_2. Thus

$$\begin{gathered}\int_0^b e^{-r^2} r\,dr \int_0^{\pi/2} d\theta < \int_0^b e^{-u^2}\,du \int_0^b e^{-v^2}\,dv < \int_0^{b\sqrt{2}} e^{-r^2} r\,dr \int_0^{\pi/2} d\theta,\\ \frac{\pi}{4}(1 - e^{-b^2}) < \left(\int_0^b e^{-u^2}\,du\right)^2 < \frac{\pi}{4}(1 - e^{-2b^2}).\end{gathered} \tag{10}$$

Equation (10) is true for every $b > 0$; by taking limits as $b \to \infty$ we show that

$$\left(\int_0^\infty e^{-u^2}\,du\right)^2 = \frac{\pi}{4}, \tag{11}$$

and this equation is equivalent to (8).

Now (6) can be derived from (8). By (1),

$$\Gamma(1/2) = \int_0^\infty t^{-1/2}e^{-t}\,dt.$$

Make the change of variable $t = u^2$. Then

$$\Gamma(1/2) = 2\int_0^\infty e^{-u^2}\,du = \sqrt{\pi}, \tag{12}$$

and (6) is true. The proof of Theorem 59.2 is complete. □

59.3 Theorem. *The following recurrence relation (or reduction formula) is valid for the values of n indicated:*

$$\int_0^{\pi/2} \sin^n x\,dx = \frac{n-1}{n}\int_0^{\pi/2} \sin^{n-2} x\,dx, \qquad n = 2, 3, \cdots. \tag{13}$$

PROOF. Evaluate the integral $\int_0^{\pi/2} \sin^n x\,dx$ by parts by setting

$$dv = \sin x\,dx, \qquad v = -\cos x;\ u = \sin^{n-1} x,$$
$$du = (n-1)\sin^{n-2} x\cos x\,dx.$$

This evaluation by parts is possible because $n \geqq 2$. Thus

$$\int_0^{\pi/2} \sin^n x\,dx = -\sin^{n-1} x\cos x\Big|_0^{\pi/2} + (n-1)\int_0^{\pi/2} \sin^{n-2} x\cos^2 x\,dx.$$

The first term on the right equals zero. In the integral on the right, replace $\cos^2 x$ by $1 - \sin^2 x$. Thus

$$\int_0^{\pi/2} \sin^n x\,dx = (n-1)\int_0^{\pi/2} \sin^{n-2} x\,dx - (n-1)\int_0^{\pi/2} \sin^n x\,dx. \tag{14}$$

Solve equation (14) for $\int_0^{\pi/2} \sin^n x\,dx$. The result is (13), and the proof of Theorem 59.3 is complete. □

59.4 Theorem

$$\int_0^{\pi/2} \sin^{2m} x\,dx = \frac{1\cdot 3\cdot 5\cdots 2m-1}{2\cdot 4\cdot 6\cdots 2m}\,\frac{\pi}{2}, \qquad m = 1, 2, \cdots; \tag{15}$$

$$\int_0^{\pi/2} \sin^{2m+1} x\,dx = \frac{2\cdot 4\cdot 6\cdots 2m}{1\cdot 3\cdot 5\cdots 2m+1}, \qquad m = 1, 2, \cdots. \tag{16}$$

PROOF. The proof of the theorem is obtained by repeated application of the recurrence relation in (13). Thus

$$\int_0^{\pi/2} \sin^{2m} x\,dx = \frac{3\cdot 5 \cdots 2m-1}{4\cdot 6 \cdots 2m} \int_0^{\pi/2} \sin^2 x\,dx. \tag{17}$$

But by (13),

$$\int_0^{\pi/2} \sin^2 x\,dx = \frac{1}{2}\int_0^{\pi/2} dx = \frac{1}{2}\frac{\pi}{2}. \tag{18}$$

Substitute from (18) in (17); the result is (15). In the same way, repeated application of the recurrence relation (13) to the integral $\int_0^{\pi/2} \sin^{2m+1} x\,dx$ shows that

$$\int_0^{\pi/2} \sin^{2m+1} x\,dx = \frac{2\cdot 4\cdot 6 \cdots 2m}{3\cdot 5\cdot 7 \cdots 2m+1} \int_0^{\pi/2} \sin x\,dx. \tag{19}$$

But $\int_0^{\pi/2} \sin x\,dx = 1$, and thus (19) is the same as (16). The proof of Theorem 59.4 is complete. □

Exercises

59.1. Prove the following formulas:

$$\Gamma\left(\frac{1}{2}+1\right) = \frac{1}{2}\sqrt{\pi}.$$

$$\Gamma\left(\frac{2}{2}+1\right) = 1!.$$

$$\Gamma\left(\frac{3}{2}+1\right) = \frac{3\cdot 1}{2^2}\sqrt{\pi}.$$

$$\Gamma\left(\frac{4}{2}+1\right) = 2!.$$

. . . .

$$\Gamma\left(\frac{2m}{2}+1\right) = m!, \qquad m = 0, 1, 2, \cdots.$$

$$\Gamma\left(\frac{2m+1}{2}+1\right) = \frac{(2m+1)(2m-1)\cdots 5\cdot 3\cdot 1}{2^{m+1}}\sqrt{\pi}, \qquad m = 0, 1, 2, \cdots.$$

59.2. Show that, for $n = 2$, the expression

$$\frac{\pi^{n/2} a^n}{\Gamma\left(\frac{n}{2}+1\right)}$$

equals the area in $\mathbb{R}^2$ of a circle with radius a; and that, for $n = 3$, it equals the volume in $\mathbb{R}^3$ of a sphere with radius a.

59.3. Show that, for $n = 2$, the expression

$$\frac{n\pi^{n/2} a^{n-1}}{\Gamma\left(\frac{n}{2}+1\right)}$$

equals the circumference in $\mathbb{R}^2$ of a circle with radius a; and that, for $n = 3$, it equals the area in $\mathbb{R}^3$ of the surface of a sphere with radius a.

59.4. Verify that the following formula is true for $n = 2, 3, 4$.

$$\frac{a^n}{n}\int_0^{2\pi} dx \int_0^{\pi} \sin x\, dx \int_0^{\pi} \sin^2 x\, dx \cdots \int_0^{\pi} \sin^{n-2} x\, dx = \frac{\pi^{n/2}a^n}{\Gamma\left(\frac{n}{2}+1\right)}.$$

59.5. Prove the following formula:

$$\int_0^{\pi} \sin^{n-2} x\, dx = \frac{\Gamma\left(\frac{n-1}{2}+1\right)}{(n-1)\pi^{(n-1)/2}} \frac{n\pi^{n/2}}{\Gamma\left(\frac{n}{2}+1\right)}, \qquad n = 2, 3, \cdots.$$

59.6. Use the formula in Exercise 59.5 to prove that the formula in Exercise 59.4 is true for $n = 2, 3, \cdots$.

60. The Area and the Volume of a Sphere

Let $S_{n-1}(a)$ denote the sphere in $\mathbb{R}^n$ whose center is at the origin and whose radius is a; its equation is

$$(y^1)^2 + (y^2)^2 + \cdots + (y^n)^2 = a^2. \tag{1}$$

Let $V_{n-1}(a)$ and $A_{n-1}(a)$ denote, respectively, the volume of $S_{n-1}(a)$ and the area of the surface of $S_{n-1}(a)$. The purpose of this section is to derive integral formulas for $V_{n-1}(a)$ and $A_{n-1}(a)$ and then to evaluate these integrals. The derivation of the formula for $V_{n-1}(a)$ is essentially a proof of a special case of the fundamental theorem of the integral calculus. The integrals for $V_{n-1}(a)$ and $A_{n-1}(a)$ are closely related to the Kronecker integral. The section begins with the derivation of the formulas for the circle $S_1(a)$ in $\mathbb{R}^2$ and the sphere $S_2(a)$ in $\mathbb{R}^3$.

60.1 Example. Let $B_1 = \{x : 0 \leqq x \leqq 2\pi\}$, and let

$$\begin{aligned} y^1(x) &= a \cos x, \\ y^2(x) &= a \sin x, \qquad x \in B_1. \end{aligned} \tag{2}$$

Then $[y^1(x)]^2 + [y^2(x)]^2 = a^2$, and the function $y: B_1 \to \mathbb{R}^2$ maps every point in B_1 into a point on $S_1(a)$; also, every point on $S_1(a)$ is the image of a point in B_1. If we make the usual convention that $x = 0$ and $x = 2\pi$ are the same point, then the function $y: B_1 \to \mathbb{R}^2$ is a one-to-one mapping of B_1 onto the circle $S_1(a)$ in $\mathbb{R}^2$. Define $\det[y(x), D_1 y(x)]$ by the following equation [see equation (70) in Section 38]:

$$\det[y(x), D_1 y(x)] = \begin{vmatrix} y^1(x) & y^2(x) \\ D_1 y^1(x) & D_1 y^2(x) \end{vmatrix}, \qquad x \in B_1. \tag{3}$$

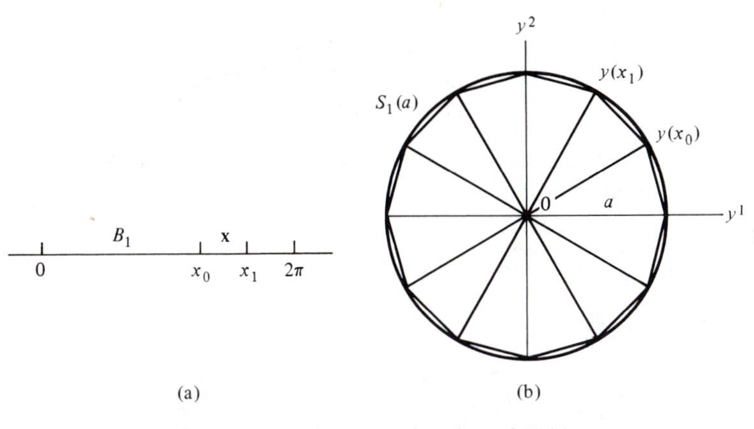

Figure 60.1. The approximation of $V_1(a)$.

The purpose of this example is to prove the following formulas and to evaluate the integral.

$$V_1(a) = \frac{1}{2}\int_{B_1} \det[y(x), D_1 y(x)]\,dx, \tag{4}$$

$$A_1(a) = \frac{2}{a} V_1(a). \tag{5}$$

Begin with the proof of (4). Let $\mathscr{P}_k$, $k = 1, 2, \cdots$, be a sequence of simplicial subdivisions of B_1, and let $\mathbf{x} : (x_0, x_1)$ denote a typical simplex in $\mathscr{P}_k$. Formula (4) will now be established by making two evaluations of the following limit:

$$\lim_{k\to\infty} \sum_{\mathbf{x}\in\mathscr{P}_k} \frac{1}{2!}\begin{vmatrix} 0 & 0 & 1 \\ y^1(x_0) & y^2(x_0) & 1 \\ y^1(x_1) & y^2(x_1) & 1 \end{vmatrix}. \tag{6}$$

Assume that $x_0 < x_1$ for each $\mathbf{x}$ in $\mathscr{P}_k$. Then $\mathscr{P}_k$ is positively oriented in $\mathbb{R}$, and the simplex $[0, y(x_0), y(x_1)]$ whose vertices are

$$0 : (0, 0), \qquad y(x_0) : (y^1(x_0), y^2(x_0)), \qquad y(x_1) : (y^1(x_1), y^2(x_1)), \tag{7}$$

is positively oriented in $\mathbb{R}^2$. Then

$$\frac{1}{2!}\begin{vmatrix} 0 & 0 & 1 \\ y^1(x_0) & y^2(x_0) & 1 \\ y^1(x_1) & y^2(x_1) & 1 \end{vmatrix} \tag{8}$$

is the area of the simplex $[0, y(x_0), y(x_1)]$ whose coordinates are shown in (7). Then the sum in (6) is the sum of the areas of the simplexes shown in Figure 60.1(b), which is approximately the area of the circle. Thus by one of the familiar definitions of the area of the circle,

$$\lim_{k\to\infty} \sum_{\mathbf{x}\in\mathscr{P}_k} \frac{1}{2!}\begin{vmatrix} 0 & 0 & 1 \\ y^1(x_0) & y^2(x_0) & 1 \\ y^1(x_1) & y^2(x_1) & 1 \end{vmatrix} = V_1(a). \tag{9}$$

We proceed now to a second evaluation of the limit in (6). By elementary properties of determinants,

$$\begin{vmatrix} 0 & 0 & 1 \\ y^1(x_0) & y^2(x_0) & 1 \\ y^1(x_1) & y^2(x_1) & 1 \end{vmatrix}$$

$$= \begin{vmatrix} y^1(x_0) & y^2(x_0) \\ y^1(x_1) & y^2(x_1) \end{vmatrix} = \begin{vmatrix} y^1(x_0) & y^2(x_0) \\ y^1(x_1) - y^1(x_0) & y^2(x_1) - y^2(x_0) \end{vmatrix} \tag{10}$$

$$= y^1(x_0)[y^2(x_1) - y^2(x_0)] - y^2(x_0)[y^1(x_1) - y^1(x_0)]$$

$$= -y^1(x_0)\begin{vmatrix} y^2(x_0) & 1 \\ y^2(x_1) & 1 \end{vmatrix} + y^2(x_0)\begin{vmatrix} y^1(x_0) & 1 \\ y^1(x_1) & 1 \end{vmatrix}.$$

Now (2) shows that the functions y^1, y^2 have continuous derivatives; therefore they satisfy the Stolz condition at x_0 as follows:

$$\begin{vmatrix} y^1(x_0) & 1 \\ y^1(x_1) & 1 \end{vmatrix} = D_1 y^1(x_0)\begin{vmatrix} x_0 & 1 \\ x_1 & 1 \end{vmatrix} + r(y^1; x_0, x_1)|x_1 - x_0|,$$

$$\begin{vmatrix} y^2(x_0) & 1 \\ y^2(x_1) & 1 \end{vmatrix} = D_1 y^2(x_0)\begin{vmatrix} x_0 & 1 \\ x_1 & 1 \end{vmatrix} + r(y^2; x_0, x_1)|x_1 - x_0|. \tag{11}$$

The formulas in (10) and (11) show that the limit in (6) equals

$$\lim_{k\to\infty} \sum_{\mathbf{x}\in\mathscr{P}_k} \frac{1}{2}\{y^1(x_0)D_1 y^2(x_0) - y^2(x_0)D_1 y^1(x_0)\}\frac{(-1)}{1!}\begin{vmatrix} x_0 & 1 \\ x_1 & 1 \end{vmatrix} \tag{12}$$

$$+ \lim_{k\to\infty} \sum_{\mathbf{x}\in\mathscr{P}_k} \{-y^1(x_0)r(y^2; x_0, x_1) + y^2(x_0)r(y^1; x_0, x_1)\}|x_1 - x_0|. \tag{13}$$

Now (2) shows that the functions y^1, y^2 are bounded; also,

$$\lim_{k\to\infty} r(y^1; x_0, x_1) = 0, \qquad \lim_{k\to\infty} r(y^2; x_0, x_1) = 0, \tag{14}$$

and these limits are uniform over the simplexes $\mathbf{x}: (x_0, x_1)$ in $\mathscr{P}_k$. Finally,

$$\sum_{\mathbf{x}\in\mathscr{P}_k} |x_1 - x_0| = 2\pi \tag{15}$$

[see Figure 60.1(a)]. These facts show that the limit in (13) is zero. By Corollary 35.10, the limit in (12) is

$$\frac{1}{2}\int_{B_1} \{y^1(x)D_1 y^2(x) - y^2(x)D_1 y^1(x)\}\,dx. \tag{16}$$

Also, by (3),

$$y^1(x)D_1y^2(x) - y^2(x)D_1y^1(x) = \begin{vmatrix} y^1(x) & y^2(x) \\ D_1y^1(x) & D_1y^2(x) \end{vmatrix} = \det[y(x), D_1y(x)], \tag{17}$$

and the integral in (16) can be written in the following form:

$$\frac{1}{2}\int_{B_1} \det[y(x), D_1y(x)]\,dx. \tag{18}$$

Thus the proof is now complete that

$$\lim_{k\to\infty} \sum_{x\in\mathscr{P}_k} \frac{1}{2!}\begin{vmatrix} 0 & 0 & 1 \\ y^1(x_0) & y^2(x_0) & 1 \\ y^1(x_1) & y^2(x_1) & 1 \end{vmatrix} = \frac{1}{2}\int_{B_1} \det[y(x), D_1y(x)]\,dx. \tag{19}$$

The proof of the formula for $V_1(a)$ in (4) follows from (9) and (19).

Next, let us establish the formula for $A_1(a)$ in (5). The integrand of the integral in (16) is the inner product of the two vectors with the following components:

$$y^1(x), y^2(x); \qquad D_1y^2(x), -D_1y^1(x). \tag{20}$$

Let θ be the angle between these two vectors; then by Theorem 84.2(7) in Appendix 1,

$$\begin{aligned} &\det[y(x), D_1y(x)] \\ &\quad = \{[y^1(x)]^2 + [y^2(x)]^2\}^{1/2}\{[D_1y^1(x)]^2 + [D_1y^2(x)]^2\}^{1/2}\cos\theta. \end{aligned} \tag{21}$$

Now $[y^1(x), y^2(x)]$ is a point on the circle whose center is at the origin and whose radius is a; hence,

$$\{[y^1(x)]^2 + [y^2(x)]^2\}^{1/2} = a. \tag{22}$$

Also,

$$\begin{aligned} y^1(x) &= a\cos x, & y^2(x) &= a\sin x, \\ D_1y^2(x) &= a\cos x = y^1(x), & -D_1y^1(x) &= a\sin x = y^2(x). \end{aligned} \tag{23}$$

Therefore the two vectors in (20) are the same vector, and the angle θ between them equals zero. Thus (21) simplifies to

$$\det[y(x), D_1y(x)] = a\{[D_1y^1(x)]^2 + [D_1y^2(x)]^2\}^{1/2}, \tag{24}$$

and the formula in (4) can be written in the following form:

$$V_1(a) = \frac{a}{2}\int_{B_1} \{[D_1y^1(x)]^2 + [D_1y^2(x)]^2\}^{1/2}\,dx. \tag{25}$$

Now by Theorem 45.2, the integral on the right is $A_1(a)$, the length of the circumference of the circle $S_1(a)$. Thus $V_1(a) = (a/2)A_1(a)$, or $A_1(a) = (2/a)V_1(a)$, and the proof of the formula in (5) is complete.

The final step in this example is to evaluate the integrals in the formulas for $V_1(a)$ and $A_1(a)$ and thus find their values. Since the integrand of the integral in (25) equals a by (23), it follows that $V_1(a) = \pi a^2$ and $A_1(a) = 2\pi a$. These same results can be obtained from the formulas in (4) and (5). For reasons which will appear later, it will be convenient to denote $\det[y(x), D_1 y(x)]$ also by I_1. Thus

$$I_1 = \det[y(x), D_1 y(x)] = \begin{vmatrix} a\cos x & a\sin x \\ -a\sin x & a\cos x \end{vmatrix} = a^2. \tag{26}$$

Now by Section 59 [see Theorem 59.2 and Exercise 59.1], $\Gamma\left(\frac{2}{2}+1\right) = 1$; then (4) and (5) show that

$$\begin{aligned} V_1(a) &= \frac{1}{2}\int_{B_1} a^2\,dx = \pi a^2 = \frac{\pi a^2}{\Gamma\left(\frac{2}{2}+1\right)}, \\ A_1(a) &= \frac{2}{a}V_1(a) = 2\pi a = \frac{2\pi a}{\Gamma\left(\frac{2}{2}+1\right)}. \end{aligned} \tag{27}$$

60.2 Example. The purpose of this example is to generalize the results in Example 60.1 by finding formulas for the volume $V_2(a)$ and area $A_2(a)$ of $S_2(a)$ in $\mathbb{R}^3$. The results will be presented so that they can be generalized later to obtain formulas for the volume $V_{n-1}(a)$ and area $A_{n-1}(a)$ of $S_{n-1}(a)$ in $\mathbb{R}^n$. The example will be developed in a series of lemmas.

It is necessary to begin the example with a treatment of spherical coordinates on $S_2(a)$ which can be generalized. Let B_2 be the set of points $x:(x^1, x^2)$ in $\mathbb{R}^2$ such that $0 \leqq x^1 \leqq 2\pi$, $0 \leqq x^2 \leqq \pi$. As usual, consider that $(0, x^2)$ and $(2\pi, x^2)$ are the same point; then topologically the set B_2 is a cylinder with radius 1 and altitude π.

60.3 Lemma. *Let $y: B_2 \to \mathbb{R}^3$ be the function such that*

$$\begin{aligned} y^1(x) &= a\cos x^1 \sin x^2, \\ y^2(x) &= a\sin x^1 \sin x^2, \qquad x = (x^1, x^2), \qquad x \in B_2. \\ y^3(x) &= a\cos x^2. \end{aligned} \tag{28}$$

Then (x^1, x^2) are the classical spherical coordinates on $S_2(a)$ in $\mathbb{R}^3$. More precisely, the function (28) maps every point in B_2 into a point on $S_2(a)$, and every point on $S_2(a)$ is the image of at least one point in B_2. All points $(x^1, 0)$ and (x^1, π), $0 \leqq x^1 \leqq 2\pi$, in B_2 are mapped into $(0, 0, a)$ and $(0, 0, -a)$ respectively, and y maps $\{(x^1, x^2): 0 \leqq x^1 \leqq 2\pi, 0 < x^2 < \pi\}$ in a one-to-one manner onto the set $\{(y^1, y^2, y^3): (y^1, y^2, y^3) \in S_2(a), -a < y^3 < a\}$.

Proof. All of the statements in the lemma are well known elementary facts about spherical coordinates, but it is desirable to give a proof of them which can be used as a starting point for an induction later. First, (28) maps B_2 into points on $S_2(a)$ because

$$\sum_{j=1}^{3} [y^j(x)]^2 = [a^2(\cos x^1)^2 + a^2(\sin x^1)^2](\sin x^2)^2 + a^2(\cos x^2)^2. \quad (29)$$

Now the expression in the square brackets on the right equals a^2 since it is the sum of the squares of the functions which define the mapping $y: B_1 \to \mathbb{R}^2$ in Example 60.1. Thus

$$\sum_{j=1}^{3} [y^j(x)]^2 = a^2(\sin x^2)^2 + a^2(\cos x^2)^2 = a^2, \quad (30)$$

and (28) maps each point in B_2 into a point on $S_2(a)$. Next, to prove that each point on $S_2(a)$ is the image of at least one point in B_2, observe first that, if (y^1, y^2, y^3) is on $S_2(a)$, then $-a \leqq y^3 \leqq a$. Since x^2 takes on all values from 0 to π in B_2, then $a \cos x^2$ takes on each value from a to $-a$ exactly once. Thus, given a point (y_0^1, y_0^2, y_0^3) on $S_2(a)$, there is a unique value x_0^2 such that $y_0^3 = a \cos x_0^2$. The points on $S_2(a)$ for which $y^3 = y_0^3$ satisfy the equation

$$(y^1)^2 + (y^2)^2 = a^2 - (y_0^3)^2 = a^2 - a^2(\cos x_0^2)^2 = (a \sin x_0^2)^2; \quad (31)$$

they are points on a circle with radius $a \sin x_0^2$. If $x^2 = x_0^2$, then the equations (28) can be written in the following form:

$$\begin{aligned} y^1(x^1, x_0^2) &= (a \sin x_0^2) \cos x^1, \\ y^2(x^1, x_0^2) &= (a \sin x_0^2) \sin x^1, \qquad 0 \leqq x^1 \leqq 2\pi, \\ y^3(x^1, x_0^2) &= a \cos x_0^2 = y_0^3. \end{aligned} \quad (32)$$

Example 60.1 shows that these equations map $\{x^1 : 0 \leqq x^1 \leqq 2\pi\}$ onto the entire circle (31) in the plane $y^3 = y_0^3$. Therefore, if (y_0^1, y_0^2, y_0^3) is on $S_2(a)$, then there is at least one point (x_0^1, x_0^2) in B_2 such that

$$\begin{aligned} y_0^1 &= (a \sin x_0^2) \cos x_0^1, \\ y_0^2 &= (a \sin x_0^2) \sin x_0^1, \qquad (x_0^1, x_0^2) \in B_2, \\ y_0^3 &= a \cos x_0^2. \end{aligned} \quad (33)$$

Thus the transformation (28) maps B_2 onto the entire surface of $S_2(a)$ as stated. All points with $x^2 = 0$ in B_2 are mapped onto $(0, 0, a)$ on $S_2(a)$, and all points with $x^2 = \pi$ are mapped onto $(0, 0, -a)$ on $S_2(a)$. If (x_1^1, x_1^2) and (x_2^1, x_2^2) are two points in B_2 such that $x_1^2 \neq x_2^2$, then y maps these points into distinct points on $S_2(a)$ since $\cos x^2$ is a strictly monotonically decreasing function on $\{x^2 : 0 \leqq x^2 \leqq \pi\}$. If (x_1^1, x_0^2) and (x_2^1, x_0^2) are two points in B_2 such that $|x_1^1 - x_2^1| \neq 0, 2\pi$ and $x_0^2 \neq 0, \pi$, then Example 60.1

shows that (28), which becomes (32) in this case, maps these points into distinct points on $S_2(a)$. Thus the mapping $y: B_2 \to S_2(a)$ is one-to-one on the set $\{(x^1, x^2): (x^1, x^2) \in B_2, 0 < x^2 < \pi\}$. The proof of the entire lemma is complete. □

Define the symbol $\det[y(x), D_1 y(x), D_2 y(x)]$ by the following equation [compare (70) in Section 38]:

$$\det[y(x), D_1 y(x), D_2 y(x)] = \begin{vmatrix} y^1(x) & y^2(x) & y^3(x) \\ D_1 y^1(x) & D_1 y^2(x) & D_1 y^3(x) \\ D_2 y^1(x) & D_2 y^2(x) & D_2 y^3(x) \end{vmatrix}, \qquad x \in B_2. \tag{34}$$

60.4 Lemma. *If* $y: B_2 \to S_2(a)$ *is the function defined in* (28), *then*

$$(-1)^3 \det[y(x), D_1 y(x), D_2 y(x)] = a^3 \sin x^2, \qquad x: (x^1, x^2) \text{ in } B_2. \tag{35}$$

PROOF. Recall from (26) that

$$I_1 = \begin{vmatrix} a \cos x^1 & a \sin x^1 \\ -a \sin x^1 & a \cos x^1 \end{vmatrix} = a^2, \qquad x^1 \in B_1. \tag{36}$$

By the definitions in (28) and (34),

$$(-1)^3 \det[y(x), D_1 y(x), D_2 y(x)]$$
$$= (-1)^3 \begin{vmatrix} a \cos x^1 \sin x^2 & a \sin x^1 \sin x^2 & a \cos x^2 \\ -a \sin x^1 \sin x^2 & a \cos x^1 \sin x^2 & 0 \\ a \cos x^1 \cos x^2 & a \sin x^1 \cos x^2 & -a \sin x^2 \end{vmatrix}. \tag{37}$$

Expand this determinant by minors of elements in the last column. In the minor of $a \cos x^2$, factor out $\sin x^2$ from the first row and $\cos x^2$ from the second row; then interchange the two rows. The resulting determinant is I_1. In the minor of $-a \sin x^2$, factor out $\sin x^2$ from each row; the resulting determinant is I_1. Thus

$$(-1)^3 \det[y(x), D_1 y(x), D_2 y(x)]$$
$$= (-1)^3 \{-a \sin x^2 (\cos x^2)^2 - a(\sin x^2)^3\} I_1 \tag{38}$$
$$= a^3 \sin x^2, \qquad x: (x^1, x^2) \text{ in } B_2.$$

The proof of Lemma 60.4 is complete. □

60.5 Lemma. *Let* $y: B_2 \to S_2(a)$ *be the function in* (28) *which defines spherical coordinates on* $S_2(a)$ *in* $\mathbb{R}^3$. *Then*

$$V_2(a) = \frac{1}{3} \int_{B_2} (-1)^3 \det[y(x), D_1 y(x), D_2 y(x)] \, d(x^1, x^2). \tag{39}$$

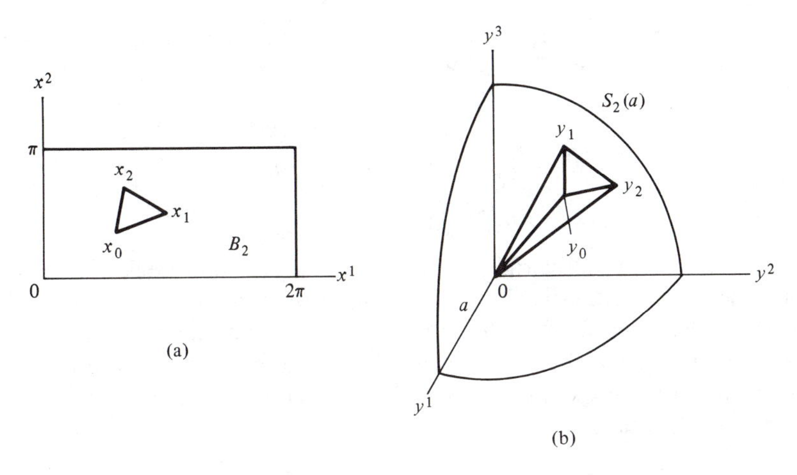

Figure 60.2. The approximation of $V_2(a)$.

PROOF. Let $\mathscr{P}_k$, $k = 1, 2, \cdots$, be a sequence of simplicial subdivisions of B_2, and let $\mathbf{x}:(x_0, x_1, x_2)$ denote a typical simplex in $\mathscr{P}_k$ [see Figure 60.2(a)]. Formula (39) will now be established by making two evaluations of the following limit:

$$\lim_{k\to\infty} \sum_{\mathbf{x}\in\mathscr{P}_k} \frac{1}{3!}\begin{vmatrix} 0 & 0 & 0 & 1 \\ y^1(x_0) & y^2(x_0) & y^3(x_0) & 1 \\ y^1(x_1) & y^2(x_1) & y^3(x_1) & 1 \\ y^1(x_2) & y^2(x_2) & y^3(x_2) & 1 \end{vmatrix}. \tag{40}$$

The term shown in the sum in (40) is either the volume or the negative of the volume of the simplex $[0, y(x_0), y(x_1), y(x_2)]$ whose vertices are

$$0:(0, 0), \qquad y(x_0):(y^1(x_0), y^2(x_0), y^3(x_0)), \cdots, y(x_2):(y^1(x_2), y^2(x_2), y^3(x_2)). \tag{41}$$

Since all points in B_2 on $x^2 = 0$, and also on $x^2 = \pi$, are mapped into a single point, some of these simplexes are degenerate and the corresponding terms are zero. We shall now show that $(1/3!)\Delta[0, y(x_0), y(x_1), y(x_2)]$ is zero or positive, at least if k is sufficiently large, and is therefore the volume of the simplex $[0, y(x_0), y(x_1), y(x_2)]$. More precisely, we shall now show that

$$\begin{aligned} \frac{1}{3!}\Delta[0, y(x_0), y(x_1), y(x_2)] \\ &= \frac{(-1)^3}{3}\det[y(x_0), D_1 y(x_0), D_2 y(x_0)]\frac{(-1)^2}{2!}\Delta(x_0, x_1, x_2) \\ &\quad + \frac{(-1)^3}{3!}r(y; \mathbf{x})\Delta(x_0, x_1, x_2), \end{aligned} \tag{42}$$

$$\lim_{x\to x_0} r(y; \mathbf{x}) = 0, \qquad \mathbf{x}\in\mathscr{P}_k.$$

Furthermore, the remainder terms $r(y; \mathbf{x})$ are uniformly small on the simplexes $\mathbf{x}$ in $\mathscr{P}_k$ as $k \to \infty$. To prove (42), observe first that

$$\frac{1}{3!}\begin{vmatrix} 0 & 0 & 0 & 1 \\ y^1(x_0) & y^2(x_0) & y^3(x_0) & 1 \\ \cdots & \cdots & \cdots & \cdots \\ y^1(x_2) & y^2(x_2) & y^3(x_2) & 1 \end{vmatrix} = \frac{(-1)^3}{3!}\begin{vmatrix} y^1(x_0) & y^2(x_0) & y^3(x_0) \\ y^1(x_1) & y^2(x_1) & y^3(x_1) \\ y^1(x_2) & y^2(x_2) & y^3(x_2) \end{vmatrix}. \quad (43)$$

In the determinant on the right, subtract the first row from each of the other rows; then expand the resulting determinant by minors of elements in the first row. Next, transform each of the resulting determinants into one which has a column of 1's; these transformations are the same as those made in (10) above. In the present case, these changes show that the determinant on the left in (43) equals

$$\frac{(-1)^3}{3!}\left\{ y^1(x_0)\begin{vmatrix} y^2(x_0) & y^3(x_0) & 1 \\ y^2(x_1) & y^3(x_1) & 1 \\ y^2(x_2) & y^3(x_2) & 1 \end{vmatrix} - y^2(x_0)\begin{vmatrix} y^1(x_0) & y^3(x_0) & 1 \\ y^1(x_1) & y^3(x_1) & 1 \\ y^1(x_2) & y^3(x_2) & 1 \end{vmatrix} \right.$$
$$\left. + y^3(x_0)\begin{vmatrix} y^1(x_0) & y^2(x_0) & 1 \\ y^1(x_1) & y^2(x_1) & 1 \\ y^1(x_2) & y^2(x_2) & 1 \end{vmatrix} \right\}. \quad (44)$$

This expression, in the notation of Theorem 11.4, is the following:

$$\frac{(-1)^3}{3!}\{y^1(x_0)\Delta(y^2, y^3)(x_0, x_1, x_2) - y^2(x_0)\Delta(y^1, y^3)(x_0, x_1, x_2)$$
$$+ y^3(x_0)\Delta(y^1, y^2)(x_0, x_1, x_2)\}. \quad (45)$$

By Theorem 11.4,

$$\Delta(y^2, y^3)(x_0, x_1, x_2) = D_{(1,2)}(y^2, y^3)(x_0)\Delta(x_0, x_1, x_2)$$
$$+ r(y^2, y^3; \mathbf{x})\Delta(x_0, x_1, x_2), \quad (46)$$

and there are similar expressions for $\Delta(y^1, y^3)(x_0, x_1, x_2)$ and $\Delta(y^1, y^2)(x_0, x_1, x_2)$. Substitute from (46), and the similar expressions, in (45). Define $r(y; \mathbf{x})$ as follows:

$$r(y; \mathbf{x}) = y^1(x_0)r(y^2, y^3; \mathbf{x}) - y^2(x_0)r(y^1, y^3; \mathbf{x}) + y^3(x_0)r(y^1, y^2; \mathbf{x}). \quad (47)$$

By Theorem 3.12 and the notation defined in (34) and (47),

$$\frac{1}{3!}\Delta[0, y(x_0), y(x_1), y(x_2)]$$
$$= \frac{(-1)^3}{3}\det[y(x_0), D_1 y(x_0), D_2 y(x_0)]\frac{(-1)^2}{2!}\Delta(x_0, x_1, x_2) \quad (48)$$
$$+ \frac{(-1)^3}{3!}r(y; \mathbf{x})\Delta(x_0, x_1, x_2).$$

Also, since $r(y^2, y^3; \mathbf{x}), \cdots, r(y^1, y^2; \mathbf{x})$ approach zero and are uniformly small on the simplexes $\mathbf{x}$ in $\mathscr{P}_k$ as $k \to \infty$, equation (47) shows that the same is true for $r(y; \mathbf{x})$. Then (48), and (35) in Lemma 60.4, show that

$$\lim_{k\to\infty} \frac{(1/3!)\Delta[0, y(x_0), y(x_1), y(x_2)]}{(1/2!)\Delta(x_0, x_1, x_2)} = \frac{(-1)^3}{3}\det[y(x_0), D_1 y(x_0), D_2 y(x_0)]$$
$$= \frac{a^3}{3}\sin x_0^2. \tag{49}$$

Now $\sin x_0^2$ is positive for $0 < x_0^2 < \pi$, and $\Delta(x_0, x_1, x_2) > 0$ for every k because $\mathscr{P}_k$ is a positively oriented subdivision of B_2. Thus $(1/3!)\Delta[0, y(x_0), y(x_1), y(x_2)] > 0$ for every simplex (x_0, x_1, x_2) in $\mathscr{P}_k$ if $0 < x_0^2 < \pi$ and k is sufficiently large. Then since (x^1, x^2) are spherical coordinates on $S_2(a)$ with the properties stated in Lemma 60.3, the sum in (40) is approximately the volume $V_2(a)$ of $S_2(a)$, and

$$\lim_{k\to\infty} \sum_{\mathbf{x}\in\mathscr{P}_k} \frac{1}{3!}\begin{vmatrix} 0 & 0 & 0 & 1 \\ y^1(x_0) & y^2(x_0) & y^3(x_0) & 1 \\ \cdots & \cdots & \cdots & \cdots \\ y^1(x_2) & y^2(x_2) & y^3(x_2) & 1 \end{vmatrix} = V_2(a). \tag{50}$$

But a second value for this limit can be obtained from (48) as follows:

$$\lim_{k\to\infty} \sum_{\mathbf{x}\in\mathscr{P}_k} \frac{1}{3!}\Delta[0, y(x_0), y(x_1), y(x_2)]$$
$$= \lim_{k\to\infty} \sum_{\mathbf{x}\in\mathscr{P}_k} \frac{(-1)^3}{3}\det[y(x_0), D_1 y(x_0), D_2 y(x_0)]\frac{(-1)^2}{2!}\Delta(x_0, x_1, x_2) \tag{51}$$
$$+ \lim_{k\to\infty} \sum_{\mathbf{x}\in\mathscr{P}_k} \frac{(-1)^3}{3!} r(y; \mathbf{x})\Delta(x_0, x_1, x_2).$$

The first limit on the right is the following integral:

$$\frac{1}{3}\int_{B_2} (-1)^3\det[y(x), D_1 y(x), D_2 y(x)]\, d(x^1, x^2). \tag{52}$$

The second limit on the right in (51) equals zero since the terms $r(y; \mathbf{x})$ are uniformly small on the simplexes $\mathbf{x}$ in $\mathscr{P}_k$ and approach zero as $k \to \infty$, and since

$$\sum_{\mathbf{x}\in\mathscr{P}_k} \frac{1}{2!}\Delta(x_0, x_1, x_2) = 2\pi^2, \tag{53}$$

the area of B_2. Now the limit in (50) equals $V_2(a)$ and also the integral in (52); then $V_2(a)$ equals the integral in (52), and the proof of (39) and of Lemma 60.5 is complete. □

60.6 Lemma. *If* $y: B_2 \to S_2(a)$ *is the function defined in* (28), *then*

$$\sum_{(j_1, j_2)} [D_{(1,2)}(y^{j_1}, y^{j_2})(x)]^2 \neq 0, \qquad x \in B_2, \qquad x^2 \neq 0, \pi, \qquad (j_1, j_2) \in (2/3). \tag{54}$$

$$\begin{aligned} -D_{(1,2)}(y^2, y^3)(x) &= (a \sin x^2) y^1(x), \\ D_{(1,2)}(y^1, y^3)(x) &= (a \sin x^2) y^2(x), \qquad x \in B_2, \\ -D_{(1,2)}(y^1, y^2)(x) &= (a \sin x^2) y^3(x). \end{aligned} \tag{55}$$

PROOF. By Theorem 3.12 and the definition of $\det[y(x), D_1 y(x), D_2 y(x)]$ in (34),

$$\begin{aligned} &(-1)^3 \det[y(x), D_1 y(x), D_2 y(x)] \\ &\quad = -y^1(x) D_{(1,2)}(y^2, y^3)(x) + y^2(x) D_{(1,2)}(y^1, y^3)(x) - y^3(x) D_{(1,2)}(y^1, y^2)(x). \end{aligned} \tag{56}$$

Since

$$(-1)^3 \det[y(x), D_1 y(x), D_2 y(x)] > 0, \qquad x \in B_2, \qquad x^2 \neq 0, \pi, \tag{57}$$

by Lemma 60.4, equation (56) shows that (54) is true. To prove (55), begin by considering the following system of equations in the unknowns u^1, u^2, u^3:

$$\begin{aligned} D_1 y^1(x) u^1 + D_1 y^2(x) u^2 + D_1 y^3(x) u^3 &= 0, \\ D_2 y^1(x) u^1 + D_2 y^2(x) u^2 + D_1 y^3(x) u^3 &= 0. \end{aligned} \tag{58}$$

Now one solution of this system of equations is

$$u^1 = y^1(x), \qquad u^2 = y^2(x), \qquad u^3 = y^3(x). \tag{59}$$

To prove this statement, differentiate the two sides of the equation $\Sigma_1^3 [y^j(x)]^2 = a^2$ as follows:

$$\begin{aligned} y^1(x) D_1 y^1(x) + y^2(x) D_1 y^2(x) + y^3(x) D_1 y^3(x) &= 0, \\ y^1(x) D_2 y^1(x) + y^2(x) D_2 y^2(x) + y^3(x) D_2 y^3(x) &= 0. \end{aligned} \tag{60}$$

These equations have a geometric meaning. The coefficients in each of the equations (58) are the components of a vector which is tangent to $S_2(a)$, and (60) states that the vector $(y^1(x), y^2(x), y^3(x))$ from the origin to the point on the sphere is normal to these tangent vectors. Now another solution of the system of equations in (58) is

$$u^1 = -D_{(1,2)}(y^2, y^3)(x), \quad u^2 = D_{(1,2)}(y^1, y^2)(x), \quad u^3 = -D_{(1,2)}(y^1, y^2)(x). \tag{61}$$

To verify this statement, substitute from (61) in (58) and observe that, in each equation, the resulting expression is the expansion of a determinant in which two rows are the same, and thus equal to zero. The system of equations in (58) is homogeneous, and (54) shows that the rank of the matrix of coefficients is 2 if $x^2 \neq 0$ or π; therefore, the solution set of (58) consists of all points (u^1, u^2, u^3) on a line which passes through the origin in $\mathbb{R}^3$. Stated geometrically, each equation in (58) represents a plane which passes through the origin, and (54) shows that, if $x^2 \neq 0$ or π, these equations represent distinct planes which intersect in a line through the origin. Since the solution in (59) is not the zero solution, every solution is a scalar multiple of this solution; in particular, the solution in (61) is a scalar multiple k of the one in (59). Thus

$$\begin{aligned} -D_{(1,2)}(y^2, y^3)(x) = ky^1(x), \qquad D_{(1,2)}(y^1, y^3)(x) = ky^2(x), \\ -D_{(1,2)}(y^1, y^2)(x) = ky^3(x). \end{aligned} \tag{62}$$

Substitute from (62) in (56) and replace $(-1)^3\det[y(x), D_1 y(x), D_2 y(x)]$ by its value $a^3 \sin x^2$ from (35) in Lemma 60.4. The result is

$$\left\{\sum_{j=1}^{3} [y^j(x)]^2\right\} k = a^3 \sin x^2. \tag{63}$$

Since $\Sigma_1^3 [y^j(x)]^2 = a^2$, equation (63) shows that $k = a \sin x^2$. Finally, (62) shows that (55) is true, at least if $x^2 \neq 0, \pi$. But the functions on the two sides of the equations are continuous, and they approach limits as x^2 tends to 0 or to π. Then these limits are equal, and the equations in (55) are true for all x in B_2. The proof of Lemma 60.6 is complete. □

60.7 Lemma

$$A_2(a) = \frac{3}{a} V_2(a).$$

PROOF. If $y: B_2 \to S_2(a)$ is the function defined in (28), then (39) in Lemma 60.5 shows that

$$V_2(a) = \frac{1}{3} \int_{B_2} (-1)^3 \det[y(x), D_1 y(x), D_2 y(x)]\, d(x^1, x^2). \tag{64}$$

Furthermore, (56) shows that the integrand of this integral is the inner product of the following two vectors:

$$\begin{aligned} y^1(x), \qquad y^2(x), \qquad y^3(x); \\ -D_{(1,2)}(y^2, y^3)(x), \qquad D_{(1,2)}(y^1, y^3)(x), \qquad -D_{(1,2)}(y^1, y^2)(x). \end{aligned} \tag{65}$$

Let θ denote the angle between these two vectors; then $\theta = 0$ since (55) in Lemma 60.6 shows that the second vector is the first vector multiplied by a positive factor. Then Theorem 84.2(7) in Appendix 1 shows that

$$\begin{aligned} &(-1)^3 \det[y(x), D_1 y(x), D_2 y(x)] \\ &= \left\{\sum_{j=1}^{3} [y^j(x)]^2\right\}^{1/2} \left\{\sum_{(j_1, j_2)} [D_{(1,2)}(y^{j_1}, y^{j_2})(x)^2\right\}^{1/2} \cos\theta, \qquad (j_1, j_2) \in (2/3), \\ &= a \left\{\sum_{(j_1, j_2)} [D_{(1,2)}(y^{j_1}, y^{j_2})(x)]^2\right\}^{1/2}. \end{aligned} \tag{66}$$

Then (64) and (66) show that

$$V_2(a) = \frac{a}{3} \int_{B_2} \left\{ \sum_{(j_1, j_2)} [D_{(1,2)}(y^{j_1}, y^{j_2})(x)]^2 \right\}^{1/2} d(x^1, x^2), \qquad (j_1, j_2) \in (2/3). \tag{67}$$

Now Theorem 45.2 shows that the integral on the right in (67) is $A_2(a)$; therefore, $V_2(a) = (a/3)A_2(a)$ and $A_2(a) = (3/a)V_2(a)$. The proof of Lemma 60.7 is complete. □

60.8 Lemma

$$V_2(a) = \frac{4\pi a^3}{3} = \frac{\pi^{3/2} a^3}{\Gamma\left(\frac{3}{2} + 1\right)}; \tag{68}$$

$$A_2(a) = 4\pi a^2 = \frac{3\pi^{3/2} a^2}{\Gamma\left(\frac{3}{2} + 1\right)}. \tag{69}$$

PROOF. Substitute from (55) in Lemma 60.6 in the integral in (67). Then

$$V_2(a) = \frac{a}{3} \int_{B_2} a^2 \sin x^2 \, d(x^1, x^2). \tag{70}$$

Use iterated integrals to evaluate this integral for $V_2(a)$. Thus

$$V_2(a) = \frac{a}{3} \int_0^{2\pi} a^2 \, dx^1 \int_0^{\pi} \sin x^2 \, dx^2 = \frac{4\pi a^3}{3}. \tag{71}$$

The values of the gamma function given in Exercise 59.1 verify that

$$\frac{4\pi a^3}{3} = \frac{\pi^{3/2} a^3}{\Gamma\left(\frac{3}{2} + 1\right)}, \tag{72}$$

and the proof of (68) is complete. Lemma 60.7 and the formulas for $V_2(a)$ in (68) show that the formulas for $A_2(a)$ in (69) are true. The proof of Lemma 60.8 is complete. □

The series of lemmas which constitute the development of Example 60.2, and which establish spherical coordinates for $S_2(a)$ and derive formulas for $V_2(a)$ and $A_2(a)$, is now complete. Examples 60.1 and 60.2 are the first steps in an induction to establish spherical coordinates for $S_{n-1}(a)$ in $\mathbb{R}^n$ and to derive formulas for $V_{n-1}(a)$ and $A_{n-1}(a)$. We now begin the second step in the induction.

60.9 Hypothesis. Let B_{n-2} be the set of points $x: (x^1, \cdots, x^{n-2})$ such that

$$0 \leqq x^1 \leqq 2\pi, \qquad 0 \leqq x^2 \leqq \pi, \cdots, 0 \leqq x^{n-2} \leqq \pi. \tag{73}$$

Define the function $z: B_{n-2} \to \mathbb{R}^{n-1}$, with components $(z^1, \cdots, z^{n-1})$, as follows:

$$\begin{aligned}
z^1 &= \cos x^1 \sin x^2 \sin x^3 \cdots \sin x^{n-3} \sin x^{n-2}, \\
z^2 &= \sin x^1 \sin x^2 \sin x^3 \cdots \sin x^{n-3} \sin x^{n-2}, \\
z^3 &= \cos x^2 \sin x^3 \cdots \sin x^{n-3} \sin x^{n-2}, \\
&\cdots\cdots\cdots\cdots\cdots\cdots \\
z^{n-2} &= \cos x^{n-3} \sin x^{n-2}, \\
z^{n-1} &= \cos x^{n-2}.
\end{aligned} \tag{74}$$

Then the transformation $y: B_{n-2} \to \mathbb{R}^{n-1}$ such that

$$\begin{aligned}
y^1 &= az^1, \\
y^2 &= az^2, \\
&\cdots\cdots \\
y^{n-1} &= az^{n-1},
\end{aligned} \tag{75}$$

maps B_{n-2} onto $S_{n-2}(a)$ and establishes $(x^1, \cdots, x^{n-2})$ in B_{n-2} as spherical coordinates on $S_{n-2}(a)$ in $\mathbb{R}^{n-1}$. More precisely, every point on $S_{n-2}(a)$ is the image of at least one point in B_{n-2}, and the mapping $y: B_{n-2} \to S_{n-2}(a)$ is one-to-one on the interior of B_{n-2}.

60.10 Theorem. *Let B_{n-1} be the set of points $x: (x^1, \cdots, x^{n-1})$ such that*

$$0 \leqq x^1 \leqq 2\pi, \qquad 0 \leqq x^2 \leqq \pi, \cdots, 0 \leqq x^{n-2} \leqq \pi, \qquad 0 \leqq x^{n-1} \leqq \pi. \tag{76}$$

Then for $n = 3, 4, \cdots$ the transformation $y: B_{n-1} \to \mathbb{R}^n$, whose components $(y^1, \cdots, y^n)$ are defined as follows, establishes $(x^1, \cdots, x^{n-1})$ as spherical coordinates on $S_{n-1}(a)$.

$$\begin{aligned}
y^1(x) &= az^1 \sin x^{n-1}, \\
y^2(x) &= az^2 \sin x^{n-1}, \\
&\cdots\cdots\cdots\cdots \qquad\qquad x \in B_{n-1}, \\
y^{n-1}(x) &= az^{n-1} \sin x^{n-1}, \\
y^n(x) &= a \cos x^{n-1}.
\end{aligned} \tag{77}$$

More precisely, every point on $S_{n-1}(a)$ is the image of at least one point in B_{n-1}, and the mapping $y: B_{n-1} \to S_{n-1}(a)$ is one-to-one on the interior of B_{n-1}.

Proof. The proof is by induction. Lemma 60.3 proves that the theorem is true for $n = 3$. The induction hypothesis is stated in Hypothesis 60.9. To complete the proof of the theorem, it is necessary and sufficient to prove that, if Hypothesis 60.9 is true for the sphere $S_{n-2}(a)$ in $\mathbb{R}^{n-1}$, then Theorem 60.10 is true for the sphere $S_{n-1}(a)$ in $\mathbb{R}^n$. A complete induction will then show that Theorem 60.10 is true for $n = 3, 4, \cdots$.

The proof is similar to that of Lemma 60.3. First,

$$\sum_{j=1}^{n} [y^j(x)]^2 = \left\{\sum_{j=1}^{n-1} [az^j]^2\right\}(\sin x^{n-1})^2 + a^2(\cos x^{n-1})^2. \tag{78}$$

Since the sum in the braces on the right side of this equation equals a^2 by the induction hypothesis in Hypothesis 60.9, the expression on the right in (78) simplifies to a^2 by the trigonometric identity $\sin^2\theta + \cos^2\theta = 1$. Thus the function y in (77) maps every point in B_{n-1} into a point on $S_{n-1}(a)$. Second, to prove that each point on $S_{n-1}(a)$ is the image of at least one point in B_{n-1}, observe that, if $(y^1, \cdots, y^n)$ is on $S_{n-1}(a)$, then $-a \leqq y^n \leqq a$. Since x^{n-1} takes on all values from 0 to π in B_{n-1}, then $a\cos x^{n-1}$ takes on each value from a to $-a$ exactly once. Thus, given a point $(y_0^1, \cdots, y_0^n)$ on $S_{n-1}(a)$, there is a unique value x_0^{n-1} such that $y_0^n = a\cos x_0^{n-1}$. The points on $S_{n-1}(a)$ for which $y^n = y_0^n$ satisfy the equation

$$\sum_{j=1}^{n-1} [y^j]^2 = a^2 - (y_0^n)^2 = a^2 - a^2(\cos x_0^{n-1})^2 = (a \sin x_0^{n-1})^2; \tag{79}$$

they are points on the sphere $S_{n-2}(a \sin x_0^{n-1})$. If $x^{n-1} = x_0^{n-1}$, then the equations (77) can be written in the following form:

$$\begin{aligned} y^1(x^1, \cdots, x^{n-2}, x_0^{n-1}) &= (a \sin x_0^{n-1})z^1, \\ &\cdots\cdots\cdots \\ y^{n-1}(x^1, \cdots, x^{n-2}, x_0^{n-1}) &= (a \sin x_0^{n-1})z^{n-1}, \qquad (x^1, \cdots, x^{n-2}) \in B_{n-2}, \\ y^n(x^1, \cdots, x^{n-2}, x_0^{n-1}) &= a \cos x_0^{n-1} = y_0^n. \end{aligned} \tag{80}$$

The induction hypothesis in Hypothesis 60.9 states that these equations map B_{n-2} onto the sphere $S_{n-2}(a \sin x_0^{n-1})$ in (79) in the plane $y^n = y_0^n$, and that each point of this sphere is the image of at least one point in B_{n-2}. Therefore, if $(y_0^1, \cdots, y_0^n)$ is on $S_{n-1}(a)$, then there is at least one point $(x_0^1, \cdots, x_0^{n-1})$ in B_{n-1} which is mapped by (77) [see also (80)] into $(y_0^1, \cdots, y_0^n)$. Thus the transformation (77) maps B_{n-1} onto the entire surface of $S_{n-1}(a)$ as stated. All points in B_{n-1} with $x^{n-1} = 0$ are mapped onto $(0, \cdots, 0, a)$ on $S_{n-1}(a)$, and all points in B_{n-1} with $x^{n-1} = \pi$ are mapped onto $(0, \cdots, 0, -a)$ on $S_{n-1}(a)$. Third, and finally, the mapping $y: B_{n-1} \to S_{n-1}(a)$ in (77) is one-to-one on the interior of B_{n-1} for the following reasons: (a) the mapping of B_{n-2} onto $S_{n-2}(a \sin x_0^{n-1})$ is one-to-one on the interior of B_{n-2} by the induction hypothesis; and (b) if $(x_1^1, \cdots, x_1^{n-1})$ and $(x_2^1, \cdots, x_2^{n-1})$ are two points in B_{n-1} such that $x_1^{n-1} \neq x_2^{n-1}$, then (77) maps these points into distinct points since $\cos x^{n-1}$ is a strictly monotonically decreasing function on $0 \leqq x^{n-1} \leqq \pi$. To summarize, the theorem is true for $n = 3$; also, if it is true for $n - 1$, it is also true for n. Thus, by a complete induction, the theorem is true for $n = 3, 4, \cdots$. The proof of Theorem 60.10 is complete. □

Let $y: B_{n-1} \to \mathbb{R}^n$ be the function defined in (77). Define $\det[y(x), D_1 y(x), \cdots, D_{n-1}y(x)]$ and I_{n-1} by the following equations:

$$\det[y(x), D_1 y(x), \cdots, D_{n-1} y(x)] = \begin{vmatrix} y^1(x) & y^2(x) & \cdots & y^n(x) \\ D_1 y^1(x) & D_1 y^2(x) & \cdots & D_1 y^n(x) \\ \cdots & \cdots & \cdots & \cdots \\ D_{n-1} y^1(x) & D_{n-1} y^2(x) & \cdots & D_{n-1} y^n(x) \end{vmatrix}. \tag{81}$$

$$I_{n-1} = \det[y(x), D_1 y(x), \cdots, D_{n-1} y(x)]. \tag{82}$$

60.11 Theorem. *If* $y: B_{n-1} \to S_{n-1}(a)$ *is the function defined in* (77), *then*

$$\begin{aligned}(-1)^n &\det[y(x), D_1 y(x), \cdots, D_{n-1} y(x)] \\ &= a^n (\sin x^2)(\sin x^3)^2 \cdots (\sin x^{n-2})^{n-3} (\sin x^{n-1})^{n-2}, \qquad x \in B_{n-1}.\end{aligned} \tag{83}$$

Proof. The proof is by induction on n. The theorem is true for $n = 3$ by Lemma 60.4 since (83) for $n = 3$ is (35). Assume that the theorem is true for $n - 1$ and prove that it is true for n. The induction hypothesis can be written in the following form:

$$I_{n-2} = (-1)^{n-1} a^{n-1} (\sin x^2)(\sin x^3)^2 \cdots (\sin x^{n-2})^{n-3}. \tag{84}$$

Now I_{n-1} is defined in terms of the function $y: B_{n-1} \to S_{n-1}(a)$ in (77); correspondingly, I_{n-2} is defined in terms of the function $y: B_{n-2} \to S_{n-2}(a)$ in (75). Thus, by the definition of I_{n-2} and I_{n-1} in (81) and (82),

$$I_{n-2} = \begin{vmatrix} az^1 & az^2 & \cdots & az^{n-1} \\ aD_1 z^1 & aD_1 z^2 & \cdots & aD_1 z^{n-1} \\ \cdots & \cdots & \cdots & \cdots \\ aD_{n-2} z^1 & aD_{n-2} z^2 & \cdots & aD_{n-2} z^{n-1} \end{vmatrix}, \tag{85}$$

$$I_{n-1} = \begin{vmatrix} az^1 \sin x^{n-1} & az^2 \sin x^{n-1} & \cdots & az^{n-1} \sin x^{n-1} & a \cos x^{n-1} \\ a(D_1 z^1) \sin x^{n-1} & a(D_1 z^2) \sin x^{n-1} & \cdots & a(D_1 z^{n-1}) \sin x^{n-1} & 0 \\ \cdots & \cdots & \cdots & \cdots & \cdots \\ a(D_{n-2} z^1) \sin x^{n-1} & a(D_{n-2} z^2) \sin x^{n-1} & \cdots & a(D_{n-2} z^{n-1}) \sin x^{n-1} & 0 \\ az^1 \cos x^{n-1} & az^2 \cos x^{n-1} & \cdots & az^{n-1} \cos x^{n-1} & -a \sin x^{n-1} \end{vmatrix}. \tag{86}$$

Expand the determinant I_{n-1} by minors of elements in the last column. In the determinant obtained from the minor of $a \cos x^{n-1}$, factor out $\sin x^{n-1}$ from each of the first $(n - 2)$ rows, and factor out $\cos x^{n-1}$ from the last row. Then move the last row across the preceding $(n - 2)$ rows so that it becomes the first row; the $(n - 2)$ interchanges of adjacent rows multiplies the determinant by $(-1)^{n-2}$. A comparison of the final determinant with the determinant in (85) shows that it is I_{n-2}. To summarize, the minor of $a \cos x^{n-1}$ is $(-1)^{n-2} I_{n-2} (\sin x^{n-1})^{n-2} \cos x^{n-1}$. Each of the $(n - 1)$ rows of the minor of $-a \sin x^{n-1}$ has the factor $\sin x^{n-1}$; the determinant which remains after these factors are removed is I_{n-2}. Thus the expansion of I_{n-1} by minors of elements in the last column shows that

$$\begin{aligned} I_{n-1} &= (-1)^{n+1}(a \cos x^{n-1})(-1)^{n-2}(\sin x^{n-1})^{n-2}(\cos x^{n-1})I_{n-2} \\ &\quad - a(\sin x^{n-1})(\sin x^{n-1})^{n-1}I_{n-2} \\ &= -a(\sin x^{n-1})^{n-2}\{(\cos x^{n-1})^2 + (\sin x^{n-1})^2\}I_{n-2} \\ &= -a(\sin x^{n-1})^{n-2}I_{n-2}. \end{aligned} \tag{87}$$

In the last equation replace I_{n-2} by its value in the induction hypothesis in (84); thus

$$I_{n-1} = (-1)^n a^n (\sin x^2)(\sin x^3)^2 \cdots (\sin x^{n-2})^{n-3}(\sin x^{n-1})^{n-2}. \tag{88}$$

But this equation is the inductive hypothesis (84) with $n - 1$ replaced by n. Since (84) is true for $n - 2 = 2$ by Lemma 60.4, a complete induction shows that (88) is true for $n = 3, 4, \cdots$. By the definition in (82), this statement is equivalent to the statement that (83) is true for $n = 3, 4, \cdots$. Thus the proof of Theorem 60.11 is complete. □

60.12 Theorem. *Let $y: B_{n-1} \to S_{n-1}(a)$ be the function in (77) which defines spherical coordinates on $S_{n-1}(a)$ in $\mathbb{R}^n$. Then*

$$V_{n-1}(a) = \frac{1}{n}\int_{B_{n-1}} (-1)^n \det[y(x), D_1 y(x), \cdots, D_{n-1} y(x)]\, d(x^1, \cdots, x^{n-1}). \tag{89}$$

PROOF. The proof is similar to the proof of Lemma 60.5. Let $\mathscr{P}_k$, $k = 1, 2, \cdots$, be a sequence of simplicial subdivisions of B_{n-1}, and let $\mathbf{x}: (x_0, x_1, \cdots, x_{n-1})$ denote a typical simplex in $\mathscr{P}_k$. Assume that the simplexes $\mathbf{x}$ in each $\mathscr{P}_k$ are positively oriented in B_{n-1} in $\mathbb{R}^{n-1}$; then

$$(-1)^{n-1}\Delta(x_0, x_1, \cdots, x_{n-1}) > 0, \qquad \mathbf{x}: (x_0, x_1, \cdots, x_{n-1}) \text{ in } \mathscr{P}_k. \tag{90}$$

Formula (89) will now be established by making two evaluations of the following limit:

$$\lim_{k\to\infty} \sum_{\mathbf{x}\in\mathscr{P}_k} \frac{1}{n!} \begin{vmatrix} 0 & 0 & \cdots & 0 & 1 \\ y^1(x_0) & y^2(x_0) & \cdots & y^n(x_0) & 1 \\ y^1(x_1) & y^2(x_1) & \cdots & y^n(x_1) & 1 \\ \cdots & \cdots & \cdots & \cdots & \cdots \\ y^1(x_{n-1}) & y^2(x_{n-1}) & \cdots & y^n(x_{n-1}) & 1 \end{vmatrix}. \tag{91}$$

The term shown in the sum in (91) is either the volume or the negative of the volume of the simplex $[0, y(x_0), y(x_1), \cdots, y(x_{n-1})]$ whose vertices are

$$0: (0, 0, \cdots, 0) \quad \text{and} \quad y(x_i): (y^1(x_i), y^2(x_i), \cdots, y^n(x_i)), \quad i = 0, 1, \cdots, n-1. \tag{92}$$

Since all points in B_{n-1} on $x^{n-1} = 0$, and also $x^{n-1} = \pi$, are mapped into a single point on $S_{n-1}(a)$, some of these simplexes are degenerate and the corresponding terms in the sum in (91) are zero. We shall now show that $(1/n!)\Delta[0, y(x_0), y(x_1), \cdots, y(x_{n-1})]$ is zero or positive, at least if k is sufficiently large, and is therefore the volume of the simplex $[0, y(x_0), y(x_1), \cdots, y(x_{n-1})]$. More precisely, we shall now show that

$$\begin{aligned}
\frac{1}{n!}&\Delta[0, y(x_0), y(x_1), \cdots, y(x_{n-1})] \\
&= \frac{(-1)^n}{n}\det[y(x_0), D_1 y(x_0), \cdots, D_{n-1} y(x_0)]\frac{(-1)^{n-1}}{(n-1)!}\Delta(x_0, x_1, \cdots, x_{n-1}) \\
&\quad + \frac{(-1)^n}{n} r(y; \mathbf{x})\frac{(-1)^{n-1}}{(n-1)!}\Delta(x_0, x_1, \cdots, x_{n-1}),
\end{aligned} \tag{93}$$

$$\lim_{\mathbf{x}\to x_0} r(y; \mathbf{x}) = 0, \qquad \mathbf{x} \in \mathscr{P}_k.$$

Furthermore, the remainder terms $r(y; \mathbf{x})$ are uniformly small on the simplexes $\mathbf{x}$ in $\mathscr{P}_k$ as $k \to \infty$. To prove (93), observe first that

$$\frac{1}{n!}\begin{vmatrix} 0 & \cdots & 0 & 1 \\ y^1(x_0) & \cdots & y^n(x_0) & 1 \\ \cdots & \cdots & \cdots & \cdots \\ y^1(x_{n-1}) & \cdots & y^n(x_{n-1}) & 1 \end{vmatrix} = \frac{(-1)^n}{n!}\begin{vmatrix} y^1(x_0) & y^2(x_0) & \cdots & y^n(x_0) \\ y^1(x_1) & y^2(x_1) & \cdots & y^n(x_1) \\ \cdots & \cdots & \cdots & \cdots \\ y^1(x_{n-1}) & y^2(x_{n-1}) & \cdots & y^n(x_{n-1}) \end{vmatrix}. \tag{94}$$

In the determinant on the right, subtract the first row from each of the other rows; then expand the resulting determinant by minors of elements in the first row. Next, transform each of the resulting determinants into one which has a column of 1's as follows:

$$\begin{vmatrix} y^2(x_1) - y^2(x_0) & \cdots & y^n(x_1) - y^n(x_0) \\ \cdots & \cdots & \cdots \\ y^2(x_{n-1}) - y^2(x_0) & \cdots & y^n(x_{n-1}) - y^n(x_0) \end{vmatrix} = (-1)^{n-1}\begin{vmatrix} y^2(x_0) & \cdots & y^n(x_0) & 1 \\ y^2(x_1) & \cdots & y^n(x_1) & 1 \\ \cdots & \cdots & \cdots & \cdots \\ y^2(x_{n-1}) & \cdots & y^n(x_{n-1}) & 1 \end{vmatrix}. \tag{95}$$

These transformations show that the expression on the left in (94) equals [compare equations (44) and (45)]

$$\frac{(-1)^n(-1)^{n-1}}{n!}\sum_{j=1}^{n}(-1)^{j+1}y^j(x_0)\Delta(y^1, \cdots, \widehat{y^j}, \cdots, y^n)(\mathbf{x}). \tag{96}$$

By Theorem 11.4,

$$\begin{aligned}\Delta(y^1, \cdots, \widehat{y^j}, \cdots, y^n)(\mathbf{x}) \\ = D_{(1,\cdots,n-1)}(y^1, \cdots, \widehat{y^j}, \cdots, y^n)(x_0)\Delta(\mathbf{x}) + r(y^1, \cdots, \widehat{y^j}, \cdots, y^n; \mathbf{x})|\Delta(\mathbf{x})|.\end{aligned} \tag{97}$$

Substitute from (97) in (96). Define $r(y; \mathbf{x})$ as follows [compare (47)]:

$$r(y; \mathbf{x}) = \sum_{j=1}^{n} (-1)^{j+1} y^j(x_0) r(y^1, \cdots, \widehat{y^j}, \cdots, y^n; \mathbf{x}). \tag{98}$$

Then (96), (97), and (98) show that the expression on the left in (94) equals

$$\begin{aligned}\frac{(-1)^n}{n} \sum_{j=1}^{n} (-1)^{j+1} y^j(x_0) D_{(1,\cdots,n-1)}(y^1, \cdots, \widehat{y^j}, \cdots, y^n)(x_0) \frac{(-1)^{n-1}}{(n-1)!}\Delta(\mathbf{x}) \\ + \frac{(-1)^n}{n} r(y; \mathbf{x}) \frac{(-1)^{n-1}}{(n-1)!}\Delta(\mathbf{x}).\end{aligned} \tag{99}$$

Now by Theorem 3.12 and the definition of the symbol $\det[y(x_0), D_1 y(x_0), \cdots, D_{n-1} y(x_0)]$ in (81),

$$\begin{aligned}\sum_{j=1}^{n} (-1)^{j+1} y^j(x_0) D_{(1,\cdots,n-1)}(y^1, \cdots, \widehat{y^j}, \cdots, y^n)(x_0) \\ = \det[y(x_0), D_1 y(x_0), \cdots, D_{n-1} y(x_0)].\end{aligned} \tag{100}$$

Thus (99) and (100) show that [compare (48) and (93)]

$$\begin{aligned}\frac{1}{n!}\Delta[0, y(x_0), y(x_1), \cdots, y(x_{n-1})] \\ = \frac{(-1)^n}{n} \det[y(x_0), D_1 y(x_0), \cdots, D_{n-1} y(x_0)] \frac{(-1)^{n-1}}{(n-1)!}\Delta(\mathbf{x}) \\ + \frac{(-1)^n}{n} r(y; x) \frac{(-1)^{n-1}}{(n-1)!}\Delta(\mathbf{x}).\end{aligned} \tag{101}$$

This equation shows that [compare (49)]

$$\begin{aligned}\lim_{\mathbf{x}\to x_0} \frac{[1/n!]\Delta[0, y(x_0), \cdots, y(x_{n-1})]}{[(-1)^{n-1}/(n-1)!]\Delta(\mathbf{x})} \\ = \frac{(-1)^n}{n} \det[y(x_0), D_1(y_0), \cdots, D_{n-1} y(x_0)].\end{aligned} \tag{102}$$

Now (83) in Theorem 60.11 shows that the right side of this equation is positive if x_0 is in the interior of B_{n-1}. Thus $(1/n!)\Delta[0, y(x_0), \cdots, y(x_{n-1})] > 0$ for every simplex $(x_0, x_1, \cdots, x_{n-1})$ in $\mathscr{P}_k$, at least if the simplex contains no point on the boundary of B_{n-1} and k is sufficiently large. All of these considerations help to justify the following statement, which for our purposes is the definition of $V_{n-1}(a)$.

$$\lim_{k\to\infty} \sum_{\mathbf{x}\in\mathscr{P}_k} \frac{1}{n!}\Delta[0, y(x_0), \cdots, y(x_{n-1})] = V_{n-1}(a). \tag{103}$$

But a second value for this limit can be obtained from (101) as follows:

$$\lim_{k\to\infty} \sum_{\mathbf{x}\in\mathscr{P}_k} \frac{1}{n!}\Delta[0, y(x_0), \cdots, y(x_{n-1})]$$
$$= \lim_{k\to\infty} \frac{1}{n} \sum_{\mathbf{x}\in\mathscr{P}_k} (-1)^n \det[y(x_0), D_1 y(x_0), \cdots, D_{n-1} y(x_0)] \frac{(-1)^{n-1}}{(n-1)!}\Delta(\mathbf{x})$$
$$+ \lim_{k\to\infty} \sum_{\mathbf{x}\in\mathscr{P}_k} \frac{(-1)^n}{n} r(y; \mathbf{x}) \frac{(-1)^{n-1}}{(n-1)!}\Delta(\mathbf{x}). \tag{104}$$

The first limit on the right is the following integral by (42) in Corollary 35.10:

$$\frac{1}{n}\int_{B_{n-1}} (-1)^n \det[y(x), D_1 y(x), \cdots, D_{n-1} y(x)]\, d(x^1, \cdots, x^{n-1}). \tag{105}$$

The second limit on the right in (104) equals zero since the terms $r(y; \mathbf{x})$ are uniformly small on the simplexes $\mathbf{x}$ in $\mathscr{P}_k$ and approach zero as $k \to \infty$, and since by (90),

$$\sum_{\mathbf{x}\in\mathscr{P}_k} \frac{(-1)^{n-1}}{(n-1)!}\Delta(\mathbf{x}) = 2\pi^{n-1}. \tag{106}$$

which is the measure of B_{n-1}. Now the limit in (103) equals $V_{n-1}(a)$ and also the integral in (105); therefore, $V_{n-1}(a)$ equals the integral in (105), and the proof of (89) and of Theorem 60.12 is complete. □

60.13 Theorem. *If $y: B_{n-1} \to S_{n-1}(a)$ is the function defined in (77), and if x is on the interior of B_{n-1}, then*

$$\sum_{(j_1,\cdots,j_{n-1})} [D_{(1,\cdots,n-1)}(y^{j_1}, \cdots, y^{j_{n-1}})(x)]^2 \neq 0, \qquad (j_1, \cdots, j_{n-1}) \in (n-1/n). \tag{107}$$

Also

$$(-1)^n(-1)^{j+1} D_{(1,\cdots,n-1)}(y^1, \cdots, \widehat{y^j}, \cdots, y^n)(x)$$
$$= \frac{(-1)^n}{a^2} \det[y(x), D_1 y(x), \cdots, D_{n-1} y(x)]\, y^j(x), \qquad j = 1, \cdots, n,\ x \in B_{n-1}. \tag{108}$$

Proof. By Theorem 3.12 and the definition of $\det[y(x), D_1 y(x), \cdots, D_{n-1} y(x)]$ in (81),

$$(-1)^n \det[y(x), D_1 y(x), \cdots, D_{n-1} y(x)]$$
$$= (-1)^n \sum_{j=1}^{n} (-1)^{j+1} y^j(x) D_{(1,\cdots,n-1)}(y^1, \cdots, \widehat{y^j}, \cdots, y^n)(x). \tag{109}$$

Since

$$(-1)^n \det[y(x), D_1 y(x), \cdots, D_{n-1} y(x)] > 0 \tag{110}$$

for every x on the interior of B_{n-1} by (83) in Theorem 60.11, equation (109) shows that (107) is true. To prove (108), begin by considering the following system of equations in the unknowns $u^1, \cdots, u^n$:

$$\sum_{j=1}^{n} D_i y^j(x) u^j = 0, \qquad i = 1, \cdots, n-1. \tag{111}$$

Now one solution of this system of equations is

$$u^1 = y^1(x), \cdots, u^n = y^n(x). \tag{112}$$

To prove this statement, differentiate the two sides of the equation $\Sigma_1^n [y^j(x)]^2 = a^2$ as follows:

$$2 \sum_{j=1}^{n} y^j(x) D_i y^j(x) = 0, \qquad i = 1, \cdots, n-1. \tag{113}$$

These equations have a geometric meaning. The coefficients in each of the equations (111) are the components of a vector which is tangent to $S_{n-1}(a)$, and (113) states that the vector $(y^1(x), \cdots, y^n(x))$ from the origin to the point on the sphere is normal to these tangent vectors. Now another solution of the system of equations in (111) is

$$u^j = (-1)^n(-1)^{j+1} D_{(1,\cdots,n-1)}(y^1, \cdots, \widehat{y^j}, \cdots, y^n)(x), \qquad j = 1, \cdots, n. \tag{114}$$

To verify this statement, substitute from (114) in (111) and observe that, in each equation, the resulting expression is $(-1)^n$ times the expansion of a determinant in which two rows are the same, and thus equal to zero. The system of equations in (111) is homogeneous, and (107) shows that the rank of the matrix of coefficients is $n-1$; that is, the matrix of coefficients of (111) has at least one $(n-1) \times (n-1)$ minor whose determinant is not zero if x is a point on the interior of B_{n-1}. Therefore, for x on the interior of B_{n-1}, the solution set of (111) consists of all points on a line which passes through the origin in $\mathbb{R}^n$. Since the solution in (112) is not the zero solution, every solution is a scalar multiple of this solution; in particular, the solution in (114) is a scalar multiple k of the one in (112). Thus

$$(-1)^n(-1)^{j+1} D_{(1,\cdots,n-1)}(y^1, \cdots, \widehat{y^j}, \cdots, y^n)(x) = k y^j(x), \qquad j = 1, \cdots, n. \tag{115}$$

Substitute from (115) in (109); the result is

$$(-1)^n \det[y(x), D_1 y(x), \cdots, D_{n-1} y(x)] = k \sum_{j=1}^{n} [y^j(x)]^2 = ka^2. \tag{116}$$

Then

$$k = \frac{(-1)^n}{a^2} \det[y(x), D_1 y(x), \cdots, D_{n-1} y(x)]. \tag{117}$$

Replace k in (115) by its value from (117); the result shows that

$$\begin{aligned}(-1)^n(-1)^{j+1}D_{(1,\cdots,n-1)}(y^1,\cdots,\widehat{y^j},\cdots,y^n)(x)\\ = \frac{(-1)^n}{a^2}\det[y(x), D_1 y(x), \cdots, D_{n-1}y(x)]\,y^j(x), \qquad j=1,\cdots,n.\end{aligned} \tag{118}$$

provided that x is a point on the interior of B_{n-1}. But the functions on the two sides of the equations in (118) are continuous, and they approach limits as x approaches a point on the boundary of B_{n-1}. Then these limits are equal, and the equations in (118), which are the same as those in (108), are true for all x in B_{n-1}. The proof of Theorem 60.13 is complete. □

60.14 Theorem. $A_{n-1}(a) = \dfrac{n}{a}V_{n-1}(a)$.

PROOF. If $y: B_{n-1} \to S_{n-1}(a)$ is the function defined in (77), then (89) in Theorem 60.12 shows that

$$V_{n-1}(a) = \frac{1}{n}\int_{B_{n-1}} (-1)^n \det[y(x), D_1 y(x), \cdots, D_{n-1}y(x)]\,d(x^1,\cdots,x^{n-1}). \tag{119}$$

Furthermore, (109) shows that the integrand of this integral is the inner product of the following two vectors:

$$\begin{aligned} &y^j(x), \qquad j=1,\cdots,n;\\ &(-1)^n(-1)^{j+1}D_{(1,\cdots,n-1)}(y^1,\cdots,\widehat{y^j},\cdots,y^n)(x), \qquad j=1,\cdots,n.\end{aligned} \tag{120}$$

Let θ denote the angle between these two vectors; then $\theta = 0$ since (108) in Theorem 60.13 shows that the second vector is the first vector multiplied by a positive scalar. Then Theorem 84.2(7) in Appendix 1 shows that

$$\begin{aligned}(-1)^n&\det[y(x), D_1 y(x), \cdots, D_{n-1}y(x)]\\ &= \left\{\sum_{j=1}^n [y^j(x)]^2\right\}^{1/2}\left\{\sum_{j=1}^n [D_{(1,\cdots,n-1)}(y^1,\cdots,\widehat{y^j},\cdots,y^n)(x)]^2\right\}^{1/2}\cos\theta\\ &= a\left\{\sum_{j=1}^n [D_{(1,\cdots,n-1)}(y^1,\cdots,\widehat{y^j},\cdots,y^n)(x)]^2\right\}^{1/2}.\end{aligned} \tag{121}$$

Then (119) and (121) show that

$$\begin{aligned}V_{n-1}(a)\\ = \frac{a}{n}\int_{B_{n-1}}\left\{\sum_{j=1}^n [D_{(1,\cdots,n-1)}(y^1,\cdots,\widehat{y^j},\cdots,y^n)(x)]^2\right\}^{1/2} d(x^1,\cdots,x^{n-1}).\end{aligned} \tag{122}$$

Now Theorem 45.2 shows that the integral on the right in (122) is $A_{n-1}(a)$; therefore, $V_{n-1}(a) = (a/n)A_{n-1}(a)$ and $A_{n-1}(a) = (n/a)V_{n-1}(a)$. The proof of Theorem 60.14 is complete. □

60.15 Theorem

$$V_{n-1}(a) = \frac{\pi^{n/2}a^n}{\Gamma\left(\frac{n}{2}+1\right)}, \qquad n = 2, 3, \cdots; \tag{123}$$

$$A_{n-1}(a) = \frac{n\pi^{n/2}a^{n-1}}{\Gamma\left(\frac{n}{2}+1\right)}, \qquad n = 2, 3, \cdots. \tag{124}$$

PROOF. If the formula in (123) is true, then the formula in (124) is true by Theorem 60.14; thus the proof can be completed by proving (123). The proof is by induction. Now (123) is true for $n = 2$ and $n = 3$ by (27) and (68). Assume that (23) is true for the sphere $S_{n-2}(a)$ in $\mathbb{R}^{n-1}$ and show that it is true for the sphere $S_{n-1}(a)$ in $\mathbb{R}^n$. Thus the induction hypothesis is

$$V_{n-2}(a) = \frac{\pi^{(n-1)/2}a^{n-1}}{\Gamma\left(\frac{n-1}{2}+1\right)}. \tag{125}$$

Now (89) in Theorem 60.12 shows that

$$V_{n-1}(a) = \frac{1}{n}\int_{B_{n-2}} (-1)^n \det[y(x), D_1 y(x), \cdots, D_{n-1} y(x)]\, d(x^1, \cdots, x^{n-1}), \tag{126}$$

and (83) in Theorem 60.11 shows that

$$\begin{aligned}(-1)^n &\det[y(x), D_1 y(x), \cdots, D_{n-1} y(x)] \\ &= a^n(\sin x^2)(\sin x^3)^2 \cdots (\sin x^{n-2})^{n-3}(\sin x^{n-1})^{n-2}, \qquad x \in B_{n-1}.\end{aligned} \tag{127}$$

Substitute from (127) in the integral in (126) and then evaluate the resulting integral by iterated integrals. Thus

$$\begin{aligned}V_{n-1}(a) = \frac{a^n}{n}\int_0^{2\pi} dx^1 \int_0^{\pi} (\sin x^2)\,dx^2 \cdots \int_0^{\pi} (\sin x^{n-2})^{n-3}\,dx^{n-2} \\ \times \int_0^{\pi} (\sin x^{n-1})^{n-2}\,dx^{n-1}.\end{aligned} \tag{128}$$

By the same formulas,

$$V_{n-2}(a) = \frac{a^{n-1}}{n-1}\int_0^{2\pi} dx^1 \int_0^{\pi} (\sin x^2)\,dx^2 \cdots \int_0^{\pi} (\sin x^{n-2})^{n-3}\,dx^{n-2}. \tag{129}$$

A comparison of (128) and (129) shows that

$$V_{n-1}(a) = \frac{a^n(n-1)}{na^{n-1}} V_{n-2}(a) \int_0^{\pi} (\sin x^{n-1})^{n-2}\,dx^{n-1}. \tag{130}$$

Now the induction hypothesis states that the value of $V_{n-2}(a)$ is given by the formula in (125). Also, by Exercise 59.5,

$$\int_0^\pi (\sin x^{n-1})^{n-2}\,dx^{n-1} = \frac{\Gamma\left(\frac{n-1}{2}+1\right)}{(n-1)\pi^{(n-1)/2}}\,\frac{n\pi^{n/2}}{\Gamma\left(\frac{n}{2}+1\right)}, \qquad n = 2, 3, \cdots. \tag{131}$$

Substitute from (125) and (131) in (130) and simplify; the result is

$$V_{n-1}(a) = \frac{\pi^{n/2}a^n}{\Gamma\left(\frac{n}{2}+1\right)}. \tag{132}$$

But this is the formula in (125) with $n-1$ replaced by n. Since the formula in (123) is true for $S_1(a)$ in $\mathbb{R}^2$ and $S_2(a)$ in $\mathbb{R}^3$, and since it is true for $S_{n-1}(a)$ in $\mathbb{R}^n$ if it is true for $S_{n-2}(a)$ in $\mathbb{R}^{n-1}$, a complete induction shows that it is true for $n = 2, 3, \cdots$. The proof of Theorem 60.15 is complete. □

60.16 Example. Some of the results in this section have important connections with the general theory of surface integrals, and it is the purpose of this example to call attention briefly to some of these connections. Let X_n be the set of points $(x^1, \cdots, x^n)$ in $\mathbb{R}^n$ such that

$$0 \leqq x^1 \leqq a, \qquad 0 \leqq x^2 \leqq 2\pi, \qquad 0 \leqq x^3 \leqq \pi, \cdots, 0 \leqq x^n \leqq \pi. \tag{133}$$

The equations [compare (77)]

$$\begin{aligned} y^1(x) &= x^1 \cos x^2 \sin x^3 \sin x^4 \cdots \sin x^{n-1} \sin x^n, \\ y^2(x) &= x^1 \sin x^2 \sin x^3 \sin x^4 \cdots \sin x^{n-1} \sin x^n, \\ &\qquad\qquad x^1 \cos x^3 \sin x^4 \cdots \sin x^{n-1} \sin x^n, \\ &\cdots\cdots\cdots\cdots\cdots\cdots \qquad x:(x^1, \cdots, x^n) \text{ in } X_n, \\ y^{n-1}(x) &= x^1 \cos x^{n-1} \sin x^n, \\ y^n(x) &= x^1 \cos x^n, \end{aligned} \tag{134}$$

define a surface G_n in $\mathbb{R}^n$ whose trace $T(G_m)$ is the solid sphere $S_{n-1}(a)$; that is,

$$T(G_n) = \{y : (y^1)^2 + \cdots + (y^n)^2 \leqq a^2\}. \tag{135}$$

To find the volume of $S_{n-1}(a)$, proceed as follows. Let $\mathscr{P}_k$, $k = 1, 2, \cdots$, be a sequence of simplicial subdivisions of X_n which are positively oriented in $\mathbb{R}^n$. If $\mathbf{x}:(x_0, x_1, \cdots, x_n)$ is a simplex in $\mathscr{P}_k$, then $(-1)^n\Delta(\mathbf{x}) > 0$. The transformation (134) maps the vertices $(x_0, x_1, \cdots, x_n)$ of $\mathbf{x}$ into the vertices $(y_0, y_1, \cdots, y_n)$ of $\mathbf{y}$ in $T(G_n)$. By Theorem 11.4,

$$\begin{aligned} &\Delta(y_0, y_1, \cdots, y_n) \\ &\quad = D_{(1,\cdots,n)}(y^1, \cdots, y^n)(x_0)\Delta(x_0, x_1, \cdots, x_n) + r(y; \mathbf{x})|\Delta(x)|. \end{aligned} \tag{136}$$

If $\Delta(y_0, y_1, \cdots, y_n)$ is positive, then

$$V_{n-1}(a) = \lim_{k\to\infty} \sum_{x\in\mathscr{P}_k} \frac{1}{n!}\Delta(y_0, y_1, \cdots, y_n). \tag{137}$$

From (136),

$$\frac{1}{n!}\Delta(y_0, y_1, \cdots, y_n) \approx (-1)^n D_{(1,\cdots,n)}(y^1, \cdots, y^n)(x_0)\frac{(-1)^n}{n!}\Delta(x_0, x_1, \cdots, x_n). \tag{138}$$

Now $(-1)^n\Delta(x_0, x_1, \cdots, x_n) > 0$ because $\mathscr{P}_k$ is positively oriented in $\mathbb{R}^n$. Equation (83) in Theorem 60.11 shows that

$$\begin{aligned}(-1)^n D_{(1,\cdots,n)}(y^1, \cdots, y^n)(x) &= (-1)^n \begin{vmatrix} D_1 y^1(x) & D_1 y^2(x) & \cdots & D_1 y^n(x) \\ D_2 y^1(x) & D_2 y^2(x) & \cdots & D_2 y^n(x) \\ \cdots & \cdots & \cdots & \cdots \\ D_n y^1(x) & D_n y^2(x) & \cdots & D_n y^n(x) \end{vmatrix} \\ &= (x^1)^{n-1}(\sin x^3)(\sin x^4)^2 \cdots (\sin x^{n-1})^{n-3}(\sin x^n)^{n-2}, \qquad x\in X_n.\end{aligned} \tag{139}$$

Therefore $(-1)^n D_{(1,\cdots,n)}(y^1, \cdots, y^n)(x) \geqq 0$ for x in X_n, and (137) and (138) show that

$$\begin{aligned}V_{n-1}(a) &= \lim_{k\to\infty} \sum_{x\in\mathscr{P}_k} (-1)^n D_{(1,\cdots,n)}(y^1, \cdots, y^n)(x_0)\frac{(-1)^n}{n!}\Delta(x_0, x_1, \cdots, x_n) \\ &= \int_{X_n} (-1)^n D_{(1,\cdots,n)}(y^1, \cdots, y^n)(x)\, d(x^1, \cdots, x^n).\end{aligned} \tag{140}$$

Thus $V_{n-1}(a)$ equals the integral of a derivative, and it can be evaluated by the fundamental theorem of the integral calculus [see Theorems 38.4 and 38.8]. As a practical matter, the easiest evaluation of the integral in (140) is obtained by using iterated integrals as already demonstrated in the proof of Theorem 60.15.

Let $f: T(G_n) \to \mathbb{R}$ be a continuous function. Then f has a surface integral on G_n, and Definition 36.3 shows that

$$\begin{aligned}&\int_{G_n} f(y)\, d(y^1, \cdots, y^n) \\ &= \lim_{k\to\infty} \sum_{x\in\mathscr{P}_k} f(y_0)\frac{(-1)^n}{n!}\Delta(y_0, y_1, \cdots, y_n) \\ &= \lim_{k\to\infty} \sum_{x\in\mathscr{P}_k} f[y(x_0)] D_{(1,\cdots,n)}(y^1, \cdots, y^n)(x_0)\frac{(-1)^n}{n!}\Delta(x_0, x_1, \cdots, x_n) \\ &= \int_{X_n} f[y(x)] D_{(1,\cdots,n)}(y^1, \cdots, y^n)(x)\, d(x^1, \cdots, x^n).\end{aligned} \tag{141}$$

This is the formula for the evaluation of an integral by spherical coordinates; in the last integral, replace $D_{(1,\cdots,n)}(y^1, \cdots, y^n)(x)$ by its value from (139) to obtain

$$\begin{aligned}\int_{G_n} f(y)\,d(y^1, \cdots, y^n) \\ = \int_X f[y(x)](-1)^n(x^1)^{n-1}(\sin x^3) \cdots (\sin x^n)^{n-2}\,d(x^1, \cdots, x^n).\end{aligned} \tag{142}$$

This formula contains, as a special case, the following familiar formula for the transformation to spherical coordinates in $\mathbb{R}^3$:

$$\int_{G_3} f(y)\,d(y^1, y^2, y^3) = \int_{X_3} f[y(x)](-1)^3(x^1)^2 \sin x^3\,d(x^1, x^2, x^3). \tag{143}$$

Exercises

60.1. Use the formulas in Theorem 60.15 to show that

$$V_4(a) = \frac{8\pi^2 a^5}{15}, \qquad A_4(a) = \frac{8\pi^2 a^4}{3}.$$

60.2. The sphere $S_0(a)$ in $\mathbb{R}$ has the equation $(y^1)^2 = a^2$, and thus it consists of the two points $y^1 = a$ and $y^1 = -a$.
- (a) Explain why it is reasonable to define the "volume" $V_0(a)$ and "area" $A_0(a)$ as follows: $V_0(a) = 2a$, $A_0(a) = 2$.
- (b) Show that the formulas for $V_{n-1}(a)$ and $A_{n-1}(a)$ in Theorem 60.15 are defined also for $n = 1$, and that for $n = 1$ these formulas give the values for $V_0(a)$ and $A_0(a)$ stated in (a).

60.3. (a) Show, by the methods of elementary calculus, that

$$V_{n-1}(a) = \int_0^a A_{n-1}(x)\,dx.$$

- (b) Assume that the formula for $A_{n-1}(a)$ in equation (124) is known. Use this formula and the formula in (a) to derive the formula for $V_{n-1}(a)$ in (123).
- (c) Assume that the formula for $V_{n-1}(a)$ in (123) is known. Use this formula and the formula in (a) to derive the formula for $A_{n-1}(a)$ in (124).

60.4. A solid sphere $S_2(a)$ in $\mathbb{R}^3$ is given. The density at the point (y^1, y^2, y^3) in the sphere is $c\{(y^1)^2 + (y^2)^2 + (y^3)^2\}^{1/2}$. Show that the mass of this sphere is $c\pi a^4$.

60.5. A solid sphere $S_{n-1}(a)$ in $\mathbb{R}^n$ is given. The density at the point $(y^1, \cdots, y^n)$ in the sphere is $c\{\Sigma_1^n (y^j)^2\}^{1/2}$. Show that the mass of the sphere is

$$\frac{ca^{n+1}}{n+1}\frac{n\pi^{n/2}}{\Gamma(\frac{n}{2}+1)}.$$

[Hint. Example 60.16 and Exercise 59.4.]

61. The Kronecker Integral

The winding number of a curve about a point in the plane is an integer—zero, positive, or negative—which indicates how, and how many times, the curve winds around the point. The Kronecker integral is an integer-valued integral which is defined for the curve and point, and whose value is the winding number of the curve about the point. In $\mathbb{R}^3$ there is a Kronecker integral which is integer-valued and which describes whether, and how, a surface surrounds a point. This section treats the Kronecker integral in $\mathbb{R}^n$; it begins with examples in $\mathbb{R}^2$ and $\mathbb{R}^3$.

61.1 Example. Let $g: [a, b] \to \mathbb{R}^2$ be a function whose components (g^1, g^2) have one continuous derivative. Then the mapping $g: [a, b] \to \mathbb{R}^2$ defines a curve G_1 in the plane; assume that this curve is closed:

$$g(a) = g(b). \tag{1}$$

Assume also that the origin $0: (0, 0)$ is not in the trace $T(G_1)$ of G_1. Define $|g(x)|$ to be the distance from the origin to the point $g(x)$ on G_1; thus

$$|g(x)| = \{[g^1(x)]^2 + [g^2(x)]^2\}^{1/2}, \qquad x \in [a, b]. \tag{2}$$

Then since the origin is not in $T(G_1)$,

$$|g(x)| > 0, \qquad x \in [a, b]. \tag{3}$$

Define the function $y: [a, b] \to \mathbb{R}^2$, with components (y^1, y^2), as follows:

$$y^j(x) = \frac{g^j(x)}{|g(x)|}, \qquad j = 1, 2, \qquad x \in [a, b]. \tag{4}$$

Then $\{[y^1(x)]^2 + [y^2(x)]^2\}^{1/2} = 1$, and the mapping $y: [a, b] \to \mathbb{R}^2$ describes a curve Y_1 whose trace $T(Y_1)$ is contained in $S_1(1)$, the unit circle in $\mathbb{R}^2$ with center at the origin. Next, compute the (signed) area subtended by the curve Y_1 at the origin. This area has a sign which is derived from the orientation in the plane, and this sign is used to establish an orientation on the unit circle $S_1(1)$. To explain this statement, let $\mathscr{P}_k$, $k = 1, 2, \cdots$, be a sequence of simplicial subdivisions of $[a, b]$, and let $\mathbf{x}: [x_0, x_1]$ be a typical simplex in $\mathscr{P}_k$. Then the approximate area subtended at the origin by the curve Y_1 is

$$\sum_{\mathbf{x} \in \mathscr{P}_k} \frac{1}{2!} \begin{vmatrix} 0 & 0 & 1 \\ y^1(x_0) & y^2(x_0) & 1 \\ y^1(x_1) & y^2(x_1) & 1 \end{vmatrix}. \tag{5}$$

Each of the determinants has a sign which it derives from the orientation of the simplex in the plane. The area subtended by the curve at the origin is the limit of the expression in (5) as $k \to \infty$, and Example 60.1 explains that this limit is

$$\frac{1}{2}\int_a^b \det[y(x), D_1 y(x)]\,dx. \tag{6}$$

Since the curve Y_1 is closed, this integral is a certain multiple w of the area of $S_1(1)$. The curve may return to its starting point without winding around the origin; then $w = 0$. If the curve winds w times around 0 in the positive (counterclockwise) direction, then the integral is positive and it equals πw since π is the area of $S_1(1)$ by Theorem 60.15. The winding number w of the curve G_1 at the origin 0 is defined as follows:

$$\frac{1}{2}\int_a^b \det[y(x), D_1 y(x)]\,dx = \pi w, \qquad w = \frac{1}{2\pi}\int_a^b \det[y(x), D_1 y(x)]\,dx. \tag{7}$$

Intuitively, at least, w is a positive, zero, or negative integer, and this statement can be verified easily in examples. Thus, if n is an integer, the curve

$$y^1(x) = \cos nx, \qquad y^2 = \sin nx, \qquad 0 \leqq x \leqq 2\pi, \tag{8}$$

is closed, $\det[y(x), D_1 y(x)] = n$, and

$$w = \frac{1}{2\pi}\int_0^{2\pi} n\,dx = n. \tag{9}$$

The integral (6) is the (signed) area subtended by the curve Y_1 at the center 0 of the circle $S_1(1)$, and the area of $S_1(1)$ is π. Thus, stated in terms of areas, w is the area in (6) divided by π, the area of $S_1(1)$. The winding number w can be described also in terms of arc lengths. Now

$$\det[y(x), D_1 y(x)] = y^1(x)D_1 y^2(x) - y^2(x)D_1 y^1(x), \tag{10}$$

and $\det[y(x), D_1 y(x)]$ is the inner product of the two vectors with the following components:

$$y^1(x), y^2(x); \qquad D_1 y^2(x), -D_1 y^1(x). \tag{11}$$

Let θ denote the angle formed by these vectors. Then since the expression on the right in (10) is an inner product,

$$\begin{aligned} &\det[y(x), D_1 y(x)] \\ &\quad = \{[y^1(x)]^2 + [y^2(x)]^2\}^{1/2}\{[D_1 y^1(x)]^2 + [D_1 y^2(x)]^2\}^{1/2}\cos\theta, \\ &\quad = \{[D_1 y^1(x)]^2 + [D_1 y^2(x)]^2\}^{1/2}\cos\theta. \end{aligned} \tag{12}$$

As in Section 60, it can be shown that the vectors in (11) lie on a line through the origin; thus $\theta = 0$ or $\theta = \pi$. Equation (12) shows that $\theta = 0$ if $\det[y(x), D_1 y(x)] > 0$, and that $\theta = \pi$ if this determinant is negative. Therefore, by (7) and (12),

$$w = \frac{1}{2\pi}\int_a^b \{[D_1 y^1(x)]^2 + [D_1 y^2(x)]^2\}^{1/2}\cos\theta\,dx. \tag{13}$$

This integral measures positive or negative arc lengths on $S_1(1)$ according as $\theta = 0$ or $\theta = \pi$. Thus the integral in (13) measures the algebraic sum of the arcs on $S_1(1)$ covered by the curve Y_1. Since Theorem 60.15 shows that $A_1(1) = 2\pi$, then (13) shows that w is the total arc length on $S_1(1)$ covered by Y_1 divided by the length of the circumference of $S_1(1)$. Since Y_1 is a closed curve, w is necessarily zero or a positive or negative integer. This discussion shows that w can be described either in terms of the number of multiples of $V_1(1)$ subtended by Y_1 at the center of $S_1(1)$, or in terms of the number of multiples of $A_1(1)$ covered by Y_1 on $S_1(1)$. The interpretations differ, but the formula is the same in the two cases.

To complete Example 61.1, find the integral for w in (7) in terms of the original functions g^1, g^2 which define the curve G_1. By (4),

$$\det[y(x), D_1 y(x)] = \begin{vmatrix} \dfrac{g^1(x)}{|g(x)|} & \dfrac{g^2(x)}{|g(x)|} \\ \dfrac{|g(x)|D_1 g^1(x) - g^1(x) D_1|g(x)|}{|g(x)|^2} & \dfrac{|g(x)|D_1 g^2(x) - g^2(x) D_1|g(x)|}{|g(x)|^2} \end{vmatrix} \tag{14}$$

Multiply the first row of this determinant by $D_1|g(x)|/|g(x)|$ and then add it to the second row. Next, factor out $1/|g(x)|$ from each row of the resulting determinant to show that

$$\det[y(x), D_1 y(x)] = \frac{1}{|g(x)|^2} \begin{vmatrix} g^1(x) & g^2(x) \\ D_1 g^1(x) & D_1 g^2(x) \end{vmatrix} = \frac{\det[g(x), D_1 g(x)]}{|g(x)|^2}. \tag{15}$$

Then (7) and (15) show that

$$w = \frac{1}{2\pi} \int_a^b \frac{\det[g(x), D_1 g(x)]}{|g(x)|^2}\, dx. \tag{16}$$

The integral in (16) is called the Kronecker integral of the curve G_1 with respect to $0:(0, 0)$; it is denoted by $K_1(g, 0)$. Thus

$$K_1(g, 0) = \frac{1}{2\pi} \int_a^b \frac{\det[g(x), D_1 g(x)]}{|g(x)|^2}\, dx, \tag{17}$$

and $w = K_1(g, 0)$.

61.2 Example. The purpose of this example is to generalize the results in Example 61.1 by describing the Kronecker integral for a surface G_2 in $\mathbb{R}^3$. First it is necessary to describe a closed surface. Let $D_3(a)$ denote the solid cube in $\mathbb{R}^3$ whose vertices are the eight points $(\pm a, \pm a, \pm a)$, and let $C_2(a)$ denote the union of the six squares which form the boundary of $D_3(a)$. In each face in $C_2(a)$, one of the coordinates is constant, and the other two are variable. One way to simplify the exposition is to assume that $C_2(a)$ is cut along enough edges so that it can be spread out in the plane $\mathbb{R}^2$. Then points in $C_2(a)$ can be described as points x with coordinates (x^1, x^2). In this case, however, it is necessary to identify the two boundary points which are the same point in the uncut boundary $C_2(a)$ of $D_3(a)$. Thus in both interpretations, $C_2(a)$ is a 2-dimensional set which has no boundary. Another way to treat the problem is to give the description in terms of the coordinates (x^1, x^2) in the face $x^3 = -a$ of $C_2(a)$, and to realize that similar results hold with suitable changes in the notation in the other faces of $C_2(a)$.

Let $g: C_2(a) \to \mathbb{R}^3$ be a function whose components (g^1, g^2, g^3) have one continuous derivative except perhaps on the edges of $C_2(a)$. Then the mapping $g: C_2(a) \to \mathbb{R}^3$ defines a surface G_2 in $\mathbb{R}^3$; this surface has no boundary—it is a closed surface—because $C_2(a)$ has no boundary. Assume that the origin $0:(0, 0, 0)$ in $\mathbb{R}^3$ is not in the trace $T(G_2)$ of G_2. Define $|g(x)|$ to be the distance from the origin to the point $g(x)$ on G_2; thus

$$|g(x)| = \left\{\sum_{j=1}^{3} [g^j(x)]^2\right\}^{1/2}, \qquad x \in C_2(a). \tag{18}$$

Then since the origin 0 is not in $T(G_2)$,

$$|g(x)| > 0, \qquad x \in C_2(a). \tag{19}$$

Define the function $y: C_2(a) \to \mathbb{R}^3$, with components (y^1, y^2, y^3), as follows:

$$y^j(x) = \frac{g^j(x)}{|g(x)|}, \qquad j = 1, 2, 3, \qquad x \in C_2(a). \tag{20}$$

Then $\{\Sigma_1^3 [y^j(x)]^2\}^{1/2} = 1$, and the mapping $y: C_2(a) \to \mathbb{R}^3$ describes a surface Y_2 whose trace $T(Y_2)$ is contained in $S_2(1)$, the unit sphere in $\mathbb{R}^3$ with center at the origin. Next, find the volume subtended by the surface Y_2 at the origin. This volume has a sign which is derived from the orientation in $\mathbb{R}^3$, and this sign is used to establish an orientation on the unit sphere $S_2(1)$. To explain this statement, let $\mathscr{P}_k$, $k = 1, 2, \cdots$, be a sequence of simplicial subdivisions of $C_2(a)$; observe that a sequence of simplicial subdivisions of $D_3(a)$ induces a sequence of simplicial subdivisions $\mathscr{P}_k$ in its boundary $C_2(a)$. Let $\mathbf{x}: [x_0, x_1, x_2]$ be a typical simplex in $\mathscr{P}_k$. Then the approximate volume subtended at the origin by the surface Y_2 is

$$\sum_{\mathbf{x} \in \mathscr{P}_k} \frac{1}{3!} \begin{vmatrix} 0 & 0 & 0 & 1 \\ y^1(x_0) & y^2(x_0) & y^3(x_0) & 1 \\ \cdots & \cdots & \cdots & \cdots \\ y^1(x_2) & y^2(x_2) & y^3(x_2) & 1 \end{vmatrix}. \tag{21}$$

Each of these determinants has a sign which it derives from the orientation of the simplex in $\mathbb{R}^3$. The volume subtended by the surface Y_2 at the origin is the limit of the sum in (21) as $k \to \infty$, and Example 60.2 explains that this limit is

$$\frac{1}{3} \int_{C_2(a)} (-1)^3 \det[y(x), D_1 y(x), D_2 y(x)] \, d(x^1, x^2). \tag{22}$$

Since the surface Y_2 is closed, this integral is a certain multiple w of the volume $V_2(1)$ of $S_2(1)$. The surface Y_2 may lie on the surface of $S_2(1)$ without enclosing it; in this case $w = 0$. Or Y_2 may encircle $S_2(1)$ a certain positive or negative number of times. In all cases there is an integer w such that (22) equals $wV_2(1)$. The winding number w of the surface G_2 at the origin 0 is defined as follows:

$$\frac{1}{3}\int_{C_2(a)} (-1)^3 \det[y(x), D_1 y(x), D_2 y(x)]\, d(x^1, x^2) = wV_2(1) = \frac{4\pi w}{3}. \quad (23)$$

$$w = \frac{1}{4\pi}\int_{C_2(a)} (-1)^3 \det[y(x), D_1 y(x), D_2 y(x)]\, d(x^1, x^2). \quad (24)$$

Intuitively, at least, w is a positive, zero, or negative integer, and this statement can be verified easily in examples. Thus if [see Example 60.2]

$$\begin{aligned} y^1(x) &= \cos nx^1 \sin x^2, \\ y^2(x) &= \sin nx^1 \sin x^2, \qquad (x^1, x^2) \in B_2 : [0, 2\pi] \times [0, \pi], \\ y^3(x) &= \cos x^2, \end{aligned} \quad (25)$$

and if n is an integer, then Y_2 is a closed surface, and

$$\det[y(x), D_1 y(x), D_2 y(x)] = -n \sin x^2, \quad (26)$$

$$w = \frac{1}{4\pi}\int_{B_2} n \sin x^2\, d(x^1, x^2) = \frac{n}{4\pi}\int_0^{2\pi} dx^1 \int_0^{\pi} \sin x^2\, dx^2 = n. \quad (27)$$

The integral (22) is the (signed) volume subtended by the surface Y_2 at the center 0 of the sphere $S_2(1)$, and the volume of $S_2(1)$ is $4\pi/3$. Thus, stated in terms of volumes, w is the volume in (22) divided by $4\pi/3$, the volume of $S_2(1)$. The number w can be described also in terms of signed areas on $S_2(1)$. Now

$$\begin{aligned} &\det[y(x), D_1 y(x), D_2 y(x)] \\ &\quad = y^1(x)D_{(1,2)}(y^2, y^3)(x) - y^2(x)D_{(1,2)}(y^1, y^3)(x) + y^3(x)D_{(1,2)}(y^1, y^2)(x), \end{aligned} \quad (28)$$

and the determinant on the left is the inner product of the two vectors with the following components:

$$y^1(x), y^2(x), y^3(x); \quad D_{(1,2)}(y^2, y^3)(x), \quad -D_{(1,2)}(y^1, y^3)(x), \quad D_{(1,2)}(y^1, y^2)(x). \quad (29)$$

Let θ denote the angle formed by these vectors. Then since the expression on the right in (28) is an inner product,

$$\begin{aligned} &\det[y(x), D_1 y(x), D_2 y(x)] \\ &\quad = \left\{\sum_{j=1}^{3} [y^j(x)]^2\right\}^{1/2} \left\{\sum_{(j_1, j_2)} [D_{(1,2)}(y^{j_1}, y^{j_2})(x)]^2\right\}^{1/2} \cos\theta \quad (30) \\ &\quad = \left\{\sum_{(j_1, j_2)} [D_{(1,2)}(y^{j_1}, y^{j_2})(x)]^2\right\}^{1/2} \cos\theta, \qquad (j_1, j_2) \in (2/3). \end{aligned}$$

As in Example 60.2, it can be shown that the two vectors in (29) lie on a line through the origin; thus $\theta = 0$ or $\theta = \pi$. Equation (30) shows that $\theta = 0$ if the determinant on the left is positive, and that $\theta = \pi$ if this determinant is negative. Therefore, by (24) and (30),

$$w = \frac{1}{4\pi} \int_{C_2(a)} (-1)^3 \left\{ \sum_{(j_1, j_2)} [D_{(1,2)}(y^{j_1}, y^{j_2})(x)]^2 \right\}^{1/2} \cos\theta \, d(x^1, x^2), \quad (j_1, j_2) \in (2/3). \tag{31}$$

The integral measures (signed) areas on the sphere $S_2(1)$; the integrand is positive on those parts of $S_2(1)$ in which $\theta = \pi$, and negative in those parts in which $\theta = 0$. Thus the expression on the right in (31) measures the algebraic sum of the positive and negative areas on $S_2(1)$ covered by the surface Y_2. Since Theorem 60.15 shows that $A_2(1) = 4\pi$, then (31) shows that w is the total area on $S_2(1)$ covered by Y_2, divided by the area of the surface of $S_2(1)$. Since Y_2 is a closed surface, w is necessarily zero or a positive or negative integer. This discussion shows that w can be described either in terms of the number of multiples of $V_2(1)$ subtended by Y_2 at the center of $S_2(1)$, or in terms of the number of multiples of $A_2(1)$ covered by Y_2 on $S_2(1)$. The interpretations differ, but the formula (24) is the same in both cases.

To complete Example 61.1, find the integral for w in (31) in terms of the original functions g^1, g^2, g^3 which define the surface G_2. A calculation entirely similar to the one in (14) and (15) shows that

$$\begin{aligned} \det[y(x), D_1 y(x), D_2 y(x)] &= \frac{1}{|g(x)|^3} \begin{vmatrix} g^1(x) & g^2(x) & g^3(x) \\ D_1 g^1(x) & D_1 g^2(x) & D_1 g^3(x) \\ D_2 g^1(x) & D_2 g^2(x) & D_2 g^3(x) \end{vmatrix} \\ &= \frac{\det[g(x), D_1 g(x), D_2 g(x)]}{|g(x)|^3}. \end{aligned} \tag{32}$$

Then (24) shows that

$$w = \frac{1}{A_2(1)} \int_{C_2(a)} \frac{(-1)^3 \det[g(x), D_1 g(x), D_2 g(x)]}{|g(x)|^3} d(x^1, x^2). \tag{33}$$

The Kronecker integral for the surface G_2, defined by the function $g: C_2(a) \to \mathbb{R}^3$, with respect to $0: (0, 0, 0)$ is denoted by $K_2(g, 0)$ and defined as follows:

$$\begin{aligned} K_2(g, 0) &= \frac{1}{A_2(1)} \int_{C_2(a)} \frac{\det[g(x), D_1 g(x), D_2 g(x)]}{|g(x)|^3} d(x^1, x^2) \\ &= \frac{1}{3V_2(1)} \int_{C_2(a)} \frac{\det[g(x), D_1 g(x), D_2 g(x)]}{|g(x)|^3} d(x^1, x^2). \end{aligned} \tag{34}$$

Then the winding number of G_2 with respect to the origin is $(-1)^3 K_2(g, 0)$. This completes Example 61.2.

61.3 Example. The general case should now be clear from the two special cases which have been treated in Examples 61.1 and 61.2. Let $D_n(a)$ be the solid cube in $\mathbb{R}^n$ whose vertices are the points $(\pm a, \pm a, \cdots, \pm a)$, and let $C_{n-1}(a)$ denote the boundary of $D_n(a)$. Let $g: C_{n-1}(a) \to \mathbb{R}^n$ be a continuous function whose components $(g^1, \cdots, g^n)$ have one continuous derivative except perhaps on the edges of $C_{n-1}(a)$. Then the function $g: C_{n-1}(a) \to \mathbb{R}^n$ defines a surface G_{n-1} in $\mathbb{R}^n$; this surface has no boundary because $C_{n-1}(a)$ has no boundary. Assume that the origin $0: (0, 0, \cdots, 0)$ in $\mathbb{R}^n$ is not a point in the trace $T(G_{n-1})$ of G_{n-1}. Define $|g(x)|$ to be the distance from the origin to the point $g(x)$ on G_{n-1}; then $|g(x)| > 0$ for all x in G_2 because the origin is not in $T(G_{n-1})$. Define the function $y: C_{n-1}(a) \to \mathbb{R}^n$, with components $(y^1, \cdots, y^n)$, as follows:

$$y^j(x) = \frac{g^j(x)}{|g(x)|}, \qquad j = 1, \cdots, n, \qquad x \in C_{n-1}(a). \tag{35}$$

Then the function $y: C_{n-1}(a) \to \mathbb{R}^n$ defines a surface Y_{n-1} whose trace $T(Y_{n-1})$ is contained in $S_{n-1}(1)$, the unit sphere in $\mathbb{R}^n$ with center at the origin. The volume subtended by Y_{n-1} at the origin is

$$\frac{1}{n}\int_{C_{n-1}(a)} (-1)^n \det[y(x), D_1 y(x), \cdots, D_{n-1}y(x)]\, d(x^1, \cdots, x^{n-1}). \tag{36}$$

Since Y_{n-1} is a closed surface, the (signed) volume subtended by Y_{n-1} at the origin is an integral multiple w of the volume $V_{n-1}(1)$ of $S_{n-1}(1)$. Thus the expression in (36) equals $wV_{n-1}(1)$, and

$$w = \frac{1}{nV_{n-1}(1)}\int_{C_{n-1}(a)} (-1)^n \det[y(x), D_1 y(x), \cdots, D_{n-1}y(x)]\, d(x^1, \cdots, x^{n-1}). \tag{37}$$

Now the same arguments that have been used before can be employed again to show that

$$\det[y(x), D_1 y(x), \cdots, D_{n-1}y(x)] = \left\{\sum_{(j_1, \cdots, j_{n-1})} [D_{(1,\cdots,n-1)}(y^{j_1}, \cdots, y^{j_{n-1}})(x)]^2\right\}^{1/2} \cos\theta. \tag{38}$$

Here $\theta = 0$ if the determinant on the left is positive, and $\theta = \pi$ if the determinant is negative. Substitute from (38) in (37); then, since $nV_{n-1}(1) = A_{n-1}(1)$ by Theorem 60.14,

$$w = \frac{1}{A_{n-1}(1)}\int_{C_{n-1}(a)} (-1)^n \left\{\sum_{(j_1, \cdots, j_{n-1})} [D_{(1,\cdots,n-1)}(y^{j_1}, \cdots, y^{j_{n-1}})(x)]^2\right\}^{1/2} \times \cos\theta\, d(x^1, \cdots, x^{n-1}). \tag{39}$$

By Theorem 45.2, this integral is the algebraic sum of the areas on $S_{n-1}(1)$ which are covered in the positive sense and in the negative sense by Y_{n-1}.

Equation (39) shows that the integral in it equals $wA_{n-1}(1)$, and that (37) can be written also as follows:

$$w = \frac{1}{A_{n-1}(1)} \int_{C_{n-1}(a)} (-1)^n \det[y(x), D_1 y(x), \cdots, D_{n-1} y(x)]\, d(x^1, \cdots, x^{n-1}). \tag{40}$$

This discussion shows that w can be described either in terms of the number of multiples of $V_{n-1}(1)$ subtended by Y_{n-1} at the center of $S_{n-1}(1)$, or in terms of the number of multiples of $A_{n-1}(1)$ covered by Y_{n-1} on $S_{n-1}(1)$. The interpretations differ, but the formulas are the same.

As in Examples 61.1 and 61.2, it is possible to express $\det[y(x), D_1 y(x), \cdots, D_{n-1} y(x)]$ in terms of the original functions $g^1, \cdots, g^n$. A calculation similar to the one in (14) and (15) shows that

$$\det[y(x), D_1 y(x), \cdots, D_{n-1} y(x)]$$

$$= \frac{1}{|g(x)|^n} \begin{vmatrix} g^1(x) & g^2(x) & \cdots & g^n(x) \\ D_1 g^1(x) & D_1 g^2(x) & \cdots & D_1 g^n(x) \\ \cdots & \cdots & \cdots & \cdots \\ D_{n-1} g^1(x) & D_{n-1} g^2(x) & & D_{n-1} g^n(x) \end{vmatrix} \tag{41}$$

$$= \frac{\det[g(x), D_1 g(x), \cdots, D_{n-1} g(x)]}{|g(x)|^n}.$$

Then (40) and (41) show that

$$w = \frac{1}{A_{n-1}(1)} \int_{C_{n-1}(a)} \frac{(-1)^n \det[g(x), D_1 g(x), \cdots, D_{n-1} g(x)]}{|g(x)|^n} d(x^1, \cdots, x^{n-1}). \tag{42}$$

The Kronecker integral for the surface G_{n-1}, defined by the function $g: C_{n-1}(a) \to \mathbb{R}^n$, with respect to the origin $0: (0, \cdots, 0)$ in $\mathbb{R}^n$ is denoted by $K_{n-1}(g, 0)$ and defined as follows:

$$K_{n-1}(g, 0) = \frac{1}{A_{n-1}(1)} \int_{C_{n-1}(a)} \frac{\det[g(x), D_1 g(x), \cdots, D_{n-1} g(x)]}{|g(x)|^n} d(x^1, \cdots, x^{n-1}). \tag{43}$$

Then the winding number w of G_{n-1} with respect to the origin in $\mathbb{R}^n$ is $(-1)^n K_{n-1}(g, 0)$.

There are also winding numbers and Kronecker integrals with respect to points other than the origin. Let G_n be the surface defined by the function $g: C_{n-1}(a) \to \mathbb{R}^n$, and let $c: (c^1, \cdots, c^n)$ be a point which is not in $T(G_n)$. Define $|g(x) - c|$ to be the distance from c to the point $g(x)$ on G_n; thus

$$|g(x) - c| = \left\{ \sum_{j=1}^{n} [g^j(x) - c^j]^2 \right\}^{1/2}, \qquad x \in C_{n-1}(a). \tag{44}$$

Then since c is not in $T(G_n)$, clearly $|g(x) - c| > 0$ for every x in $C_{n-1}(a)$. Define the function $y: C_{n-1}(a) \to \mathbb{R}^n$, with components $(y^1, \cdots, y^n)$, as follows:

$$y^j(x) = \frac{g^j(x) - c^j}{|g(x) - c|}, \qquad j = 1, \cdots, n, \qquad x \in C_{n-1}(a). \tag{45}$$

The function $y : C_{n-1}(a) \to \mathbb{R}^n$ describes a surface Y_{n-1} whose trace $T(Y_{n-1})$ is contained in $S_{n-1}(1)$, the unit sphere in $\mathbb{R}^n$ with center at the origin. The winding number w of G_{n-1} about c is defined to be the winding number of Y_{n-1} about the origin. Thus the formula in (40) holds as before. It is desirable to state this formula in terms of the function g, however, and some differences appear at this point. The definition of the function y in (45) and a calculation similar to that in (14) shows that

$$\det[y(x), D_1 y(x), \cdots, D_{n-1} y(x)]$$

$$= \frac{1}{|g(x) - c|^n} \begin{vmatrix} g^1(x) - c^1 & g^2(x) - c^2 & \cdots & g^n(x) - c^n \\ D_1 g^1(x) & D_1 g^2(x) & \cdots & D_1 g^n(x) \\ \cdots & \cdots & \cdots & \cdots \\ D_{n-1} g^1(x) & D_{n-1} g^2(x) & \cdots & D_{n-1} g^n(x) \end{vmatrix} \tag{46}$$

$$= \frac{\det[g(x) - c, D_1 g(x), \cdots, D_{n-1} g(x)]}{|g(x) - c|^n}.$$

The formula for the winding number is obtained by substituting from (46) in (40). The final formula for w, and the definition of the Kronecker integral $K_{n-1}(g, c)$, are the following:

$$w = \frac{1}{A_{n-1}(1)} \int_{C_{n-1}(a)} \frac{(-1)^n \det[g(x) - c, D_1 g(x), \cdots, D_{n-1} g(x)]}{|g(x) - c|^n} d(x^1, \cdots, x^{n-1}). \tag{47}$$

$$K_{n-1}(g, c) = \frac{1}{A_{n-1}(1)} \int_{C_{n-1}(a)} \frac{\det[g(x) - c, D_1 g(x), \cdots, D_{n-1} g(x)]}{|g(x) - c|^n} d(x^1, \cdots, x^{n-1}). \tag{48}$$

61.4 Example. This example contains a numerical example to illustrate the application of the formulas developed in this section. The boundary $C_2(a)$ of the solid cube $D_3(a)$ in $\mathbb{R}^3$ is a closed surface; define $g : C_2(a) \to \mathbb{R}^3$ to be the identity mapping such that $g(x) = x$ for each x in $C_2(a)$. Then $g : C_2(a) \to \mathbb{R}^3$ defines a surface G_2 whose trace $T(G_2)$ is $C_2(a)$. Evaluate the following formula for the surface G_2:

$$\frac{1}{A_2(1)} \int_{C_2(a)} \frac{(-1)^3 \det[g(x), D_1 g(x), D_2 g(x)]}{|g(x)|^3} d(x^1, x^2). \tag{49}$$

Now because of the symmetry, this formula has the same value on each of the six faces of $C_2(a)$; thus its value is six times its value on the face whose vertices are $(\pm a, \pm a, -a)$. Let $(x^1, x^2, -a)$ be the coordinates of points

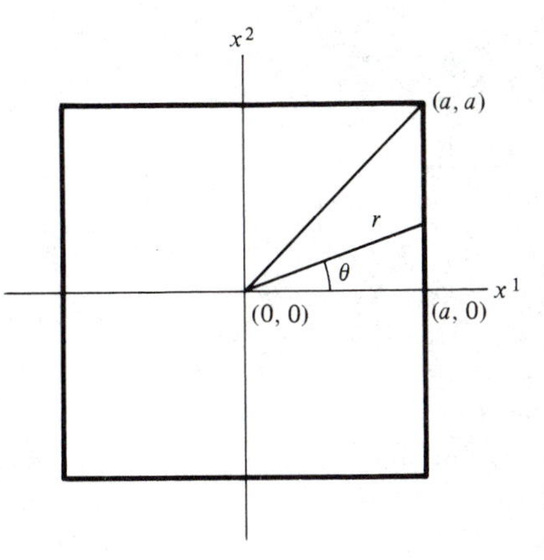

Figure 61.1. Figure for (52) and (53).

in this face. Then

$$\begin{aligned} g^1(x) &= x^1, \qquad -a \leqq x^1 \leqq a, \\ g^2(x) &= x^2, \qquad -a \leqq x^2 \leqq a, \\ g^3(x) &= -a. \end{aligned} \tag{50}$$

Also,

$$\det[g(x), D_1 g(x), D_2 g(x)] = \begin{vmatrix} x^1 & x^2 & -a \\ 1 & 0 & 0 \\ 0 & 1 & 0 \end{vmatrix} = -a, \tag{51}$$
$$|g(x)| = [a^2 + (x^1)^2 + (x^2)^2]^{1/2},$$

and the formula (49) on the face $x^3 = -a$ of $C_2(a)$, evaluated by iterated integrals, is

$$\frac{1}{4\pi}\int_{-a}^{a}\left(\int_{-a}^{a}\frac{a\,dx^1}{[a^2 + (x^1)^2 + (x^2)^2]^{3/2}}\right)dx^2. \tag{52}$$

Because of the symmetry again, the value of this integral is eight times its value over the right triangle with vertices (0, 0), $(a, 0)$, (a, a) [see Figure 61.1]. Introduce polar coordinates as indicated. Then (52) equals

$$\frac{8a}{4\pi}\int_0^{\pi/4}\left(\int_0^{a\sec\theta}\frac{r\,dr}{(a^2 + r^2)^{3/2}}\right)d\theta = \frac{1}{2} - \frac{2}{\pi}\int_0^{\pi/4}\frac{d\theta}{(1 + \sec^2\theta)^{1/2}}. \tag{53}$$

To evaluate the remaining integral on the right in (53), observe first that

$$\int_0^{\pi/4}\frac{d\theta}{(1 + \sec^2\theta)^{1/2}} = \int_0^{\pi/4}\frac{\cos\theta}{(1 + \cos^2\theta)^{1/2}}\,d\theta.$$

Next, make the following substitution:

$$\cos\theta = u, \qquad d\theta = -\frac{du}{(1-u^2)^{1/2}}.$$

Then

$$\int_0^{\pi/4} \frac{\cos\theta\, d\theta}{(1+\cos^2\theta)^{1/2}} = \int_1^{\sqrt{2}/2} \frac{-u\, du}{(1+u^2)^{1/2}(1-u^2)^{1/2}} = \int_{\sqrt{2}/2}^{1} \frac{u\, du}{(1-u^4)^{1/2}}$$
$$= \frac{1}{2} \arc\sin u^2 \Big|_{\sqrt{2}/2}^{1} = \frac{1}{2}\left(\frac{\pi}{2} - \frac{\pi}{6}\right) = \frac{\pi}{6}.$$

Substitute this value in (53); the result shows that the value of the integral in (52) is 1/6. Therefore, the value of (49) is 1. This answer is intuitively correct since the surface G_2 surrounds the origin once, and the orientations are such that the result is positive. Also, this answer shows that $K_2(g, 0) = -1$ since $K_2(g, 0)$ is the negative of the integral in (49).

EXERCISES

61.1. The equations

$$\begin{aligned} g^1(x) &= a\cos x^1 \sin x^2, \\ g^2(x) &= a \sin x^1 \sin x^2, \qquad 0 \leqq x^1 \leqq 2\pi, \qquad 0 \leqq x^2 \leqq \pi, \\ g^3(x) &= a\cos x^2, \end{aligned}$$

define a mapping $g: B_2 \to S_2(a)$ of the rectangle $B_2: [0, 2\pi] \times [0, \pi]$ onto the sphere $S_2(a)$ with center at the origin and radius a. Use the formula in (34) to show that

$$K_2(g, 0) = \frac{-1}{4\pi} \int_0^{2\pi} \left(\int_0^{\pi} \sin x^2\, dx^2 \right) dx^1 = -1.$$

61.2. The equations

$$\begin{aligned} g^1(x) &= a\cos x^1 \sin x^2, \\ g^2(x) &= a \sin x^1 \sin x^2, \qquad 0 \leqq x^1 \leqq 2\pi, \qquad 0 \leqq x^2 \leqq \pi, \\ g^3(x) &= h + a\cos x^2, \end{aligned}$$

define a mapping $g: B_2 \to \mathbb{R}^3$ of the rectangle $B_2: [0, 2\pi] \times [0, \pi]$ onto the sphere with center at $(0, 0, h)$ and radius a.

(a) Use the formula in (34) to show that

$$K_2(g, 0) = \frac{1}{2} \int_0^{\pi} \frac{-a^2 \sin x^2 [a + h\cos x^2]}{[a^2 + h^2 + 2ah \cos x^2]^{3/2}} dx^2.$$

(b) Evaluate the integral in (a) and show that

$$\begin{aligned} K_2(g, 0) &= -1, \qquad -a < h < a, \\ &= 0, \qquad |h| > a. \end{aligned}$$

(c) Explain why the results in (b) are intuitively correct.

(d) Let $c:(c^1, c^2, c^3)$ be the point $(0, 0, h)$. Use the formula in (48) to show that $K_2(g, c) = -1$ for every value of h. Why is this result intuitively correct?

61.3. Let m be an integer. The equations

$$g^1(x) = a\cos x^1 \sin mx^2,$$

$$g^2(x) = a\sin x^1 \sin mx^2, \qquad 0 \leqq x^1 \leqq 2\pi, \qquad 0 \leqq x^2 \leqq \pi,$$

$$g^3(x) = -a\cos mx^2,$$

define a mapping $g: B_2 \to \mathbb{R}^3$ of the rectangle $B_2: [0, 2\pi] \times [0, \pi]$ onto the sphere $S_2(a)$ with center at $(0, 0, 0)$ and radius a.

(a) Use the formula in (34) to show that

$$K_2(g, 0) = \frac{m}{4\pi}\int_0^{2\pi}\left(\int_0^{\pi} \sin mx^2\, dx^2\right)dx^1 = \begin{cases} 1, & \text{if } m \text{ is odd,} \\ 0, & \text{if } m \text{ is even.} \end{cases}$$

61.4. Let $z^1, z^2, \cdots, z^{n-1}$ be the functions defined in (74) in Section 60. The equations [see (77) in Section 60]

$$g^1(x) = az^1 \sin x^{n-1},$$

$$g^2(x) = az^2 \sin x^{n-1},$$

$$\cdots\cdots\cdots\cdots\cdots\cdots \qquad x \in B_{n-1},$$

$$g^{n-1}(x) = az^{n-1} \sin x^{n-1},$$

$$g^n(x) = a\cos x^{n-1},$$

define a mapping $g: B_{n-1} \to S_{n-1}(a)$ of the rectangle B_{n-1} onto the sphere $S_{n-1}(a)$ in $\mathbb{R}^n$. Use the formula in (43) and the results in Section 60 to show that

$$K_{n-1}(g, 0) = \frac{(-1)^n}{A_{n-1}(1)}\int_0^{2\pi} dx^1 \int_0^{\pi} \sin x^2\, dx^2 \cdots \int_0^{\pi} (\sin x^{n-1})^{n-2}\, dx^{n-1} = (-1)^n.$$

61.5. The equations

$$g^1(x) = \frac{ax^1}{[a^2 + (x^1)^2 + (x^2)^2]^{1/2}},$$

$$g^2(x) = \frac{ax^2}{[a^2 + (x^1)^2 + (x^2)^2]^{1/2}}, \qquad \begin{array}{l} -a \leqq x^1 \leqq a, \\ -a \leqq x^2 \leqq a, \end{array}$$

$$g^3(x) = \frac{a^2}{[a^2 + (x^1)^2 + (x^2)^2]^{1/2}},$$

define a mapping of the face of $C_2(a)$ in the plane $x^3 = -a$ onto the sphere $S_2(a)$ with center at the origin and radius a. Similar equations define the mapping g of each of the other faces of $C_2(a)$ onto $S_2(a)$. Show that

$$K_2(g, 0) = \frac{6}{4\pi}\int_{-a}^{a}\left(\int_{-a}^{a} \frac{a\, dx^1}{[a^2 + (x^1)^2 + (x^2)^2]^{3/2}}\right)dx^2 = 1.$$

Compare this problem with the one in Example 61.4 and explain the identical results.

61.6. A cylinder whose axis is on the x^3-axis in $\mathbb{R}^3$ has radius a and altitude $2a$. The cylinder has closed ends; its base and top are in the planes $x^3 = -a$ and $x^3 = a$ respectively.

(a) The equations

$$g^1(\theta, z) = a\cos\theta,$$
$$g^2(\theta, z) = a\sin\theta, \qquad 0 \leqq \theta \leqq 2\pi, \qquad -a \leqq z \leqq a,$$
$$g^3(\theta, z) = z,$$

define a mapping g of the rectangle $\{(\theta, z): 0 \leqq \theta \leqq 2\pi,\ -a \leqq z \leqq a\}$ onto the side of the cylinder. The definition of g will be completed in part (b) of this exercise. Show that the contribution of the side of the cylinder to $K_2(g, 0)$ is

$$\frac{1}{4\pi}\int_0^{2\pi} d\theta \int_{-a}^{a} \frac{a^2\,dz}{(a^2 + z^2)^{3/2}} = \frac{\sqrt{2}}{2}.$$

(b) Each end of the cylinder makes the same contribution to $K_2(g, 0)$. Introduce polar coordinates to describe the top of the cylinder. The equations

$$g^1(r, \theta) = r\cos\theta,$$
$$g^2(r, \theta) = r\sin\theta, \qquad 0 \leqq r \leqq a, \qquad 0 \leqq \theta \leqq 2\pi,$$
$$g^3(r, \theta) = a,$$

define a mapping g of the rectangle $\{(r, \theta): 0 \leqq r \leqq a, 0 \leqq \theta \leqq 2\pi\}$ onto the top end of the cylinder. Show that the contribution of the top of the cylinder to $K_2(g, 0)$ is

$$\frac{1}{4\pi}\int_0^{2\pi} d\theta \int_0^{a} \frac{ar\,dr}{(a^2 + r^2)^{3/2}} = \frac{1}{2} - \frac{\sqrt{2}}{4}.$$

(c) Show that $K_2(g, 0) = 1$.

61.7. Repeat Exercise 61.6 with the center of the cylinder at the point $c: (0, 0, h)$. Define functions g^1, g^2, g^3 which define the surface of the cylinder. Show that

$$K_2(g, 0) = \begin{cases} 1 & \text{if } |h| < a, \\ 0 & \text{if } |h| > a. \end{cases}$$

Explain why these answers are intuitively correct.

61.8. This section has defined $K_{n-1}(g, 0)$ for $n = 2, 3, \cdots$, but not for $n = 1$. This exercise suggests a definition for $K_0(g, 0)$.

(a) Since $D_1(a) = \{x: -a \leqq x \leqq a\}$, then $C_0(a) = \{a, -a\}$. Let g be a function which maps $C_0(a)$ into $\mathbb{R}$. Then in keeping with other definitions in this section, set $y(a) = g(a)/|g(a)|$ and $y(-a) = g(-a)/|g(-a)|$. Show that $y(a)$ and $y(-a)$ are points on the unit sphere $y^2 = 1$ in $\mathbb{R}$.

(b) For the definition of $K_0(g, 0)$ we need an analog of (17) in the case $n = 2$, and of (34) in the case $n = 3$. Show that the following is not a satisfactory definition:

$$K_0(g, 0) = (1/2)[y(a) + y(-a)].$$

(c) Define $K_0(g, 0)$ as follows:

$$K_0(g, 0) = (1/2)[y(a) - y(-a)] = \frac{1}{2}\left[\frac{g(a)}{|g(a)|} - \frac{g(-a)}{|g(-a)|}\right].$$

Explain the factor (1/2) [compare Exercise 60.2]. Give a reasonable explanation for the choice of $y(a) - y(-a)$ rather than $y(a) + y(-a)$. Observe that $y(a) - y(-a)$ is an integral over the boundary of an oriented one-dimensional simplex.

(d) If $K_0(g, 0)$ is defined as stated in (c), prove the following:

$$\text{if } y(a) = 1 \text{ and } y(-a) = -1, \text{ then } K_0(g, 0) = +1;$$

$$\text{if } y(a) = -1 \text{ and } y(-a) = 1, \text{ then } K_0(g, 0) = -1;$$

$$\text{in all other cases, } K_0(g, 0) = 0.$$

(e) Give as many reasons as you can why you think the definition of $K_0(g, 0)$ in (c) is the proper one.

62. The Kronecker Integral and the Sperner Degree

This section describes briefly some of the properties and applications of the Kronecker integral, and it proves that the Kronecker integral is equal to the Sperner degree at least in some cases.

The results will be established first in the plane and then generalized to $\mathbb{R}^n$. It is necessary to describe the setting for the theorems before they can be stated. Let $D_2(a)$ denote the square $[-a, a] \times [-a, a]$ in $\mathbb{R}^2$, and let $C_1(a)$ denote the boundary of $D_2(a)$ [compare Examples 61.2 and 61.3]. Then $C_1(a)$ consists of four line segments which are parallel to the axes. Correspondingly, for $0 \leqq r \leqq a$ let $D_2(r)$ denote the square $[-r, r] \times [-r, r]$, and let $C_1(r)$ denote the boundary of $D_2(r)$. Let $f: D_2(a) \to \mathbb{R}^2$ be a function whose components (f^1, f^2) each have at least one continuous derivative. Let f restricted to $C_1(r)$ be denoted by g_r. Then g_r is a function of a single variable on each of the sides of $C_1(r)$; let this variable be denoted by x. Since f has continuous derivatives on $D_2(a)$, then $g_r: C_1(r) \to \mathbb{R}^2$ has a continuous derivative on each of the sides of $C_1(r)$ for $0 < r \leqq a$. The mapping $g_r: C_1(r) \to \mathbb{R}^2$ defines a closed curve $G_1(r)$ for each r such that $0 < r \leqq a$ [see Figure 62.1]. Let $z: (z^1, z^2)$ be a point which is not in the trace of $G_1(r)$; then the Kronecker integral $K_1(g_r, z)$ of g_r with respect to z is defined [see (48) in Section 61], and

$$K_1(g_r, z) = \frac{1}{2\pi} \int_{C_1(r)} \frac{\det[g_r(x) - z, D_1 g_r(x)]}{|g_r(x) - z|^2} \, dx. \tag{1}$$

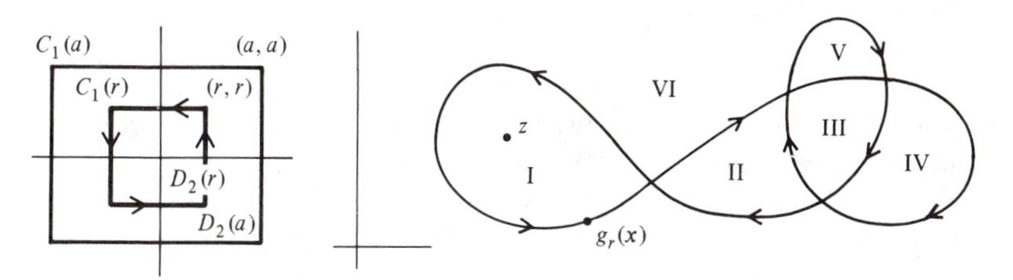

Figure 62.1. The curve $G_1(r)$.

62.1 Theorem. *For a fixed r, $0 < r \leqq a$, the Kronecker integral $K_1(g_r, z)$ is a constant function of z in each open connected set in the complement of $T[G_1(r)]$.*

PROOF. A geometric description of the proof first will aid in understanding the formal proof to be given afterward. The trace $T[G_1(r)]$ of the curve $G_1(r)$ is the curve pictured in Figure 62.1. In this figure, the curve divides the plane into six regions numbered I, $\cdots$, VI. The theorem asserts that $K_1(g_r, z)$ is a constant function of z in each of the six regions, but of course the Kronecker integral may have different values in different regions. The figure shows z as a point in region I. To find the Kronecker integral of g_r at this point z, place a unit circle with its center at z; then project the curve from z onto the circumference of the unit circle. The Kronecker integral $K_1(g_r, z)$ is a positive or negative integer which counts the number of times the projected curve winds around the unit circle in the positive or negative direction. If z moves from its original position to a nearby point but without crossing the curve from one of the six regions into another one, it is geometrically clear that the value of $K_1(g_r, z)$ does not change. For the curve shown in Figure 62.1, the values of $K_1(g_r, z)$ in the six regions are the following:

$$\text{I}, 1; \text{II}, -1; \text{III}, -2; \text{IV}, -1; \text{V}, -1; \text{VI}, 0. \tag{2}$$

For the formal proof of Theorem 62.1, let E be an open connected set in the complement of $T[G_1(r)]$, that is, one of the regions such as those shown in Figure 62.1. Then by Section 61 the Kronecker integral $K_1(g_r, z)$ is defined for each z in E, and its value is an integer. If z_0 is in E, then z_0 has a neighborhood in E. This statement is true for the following reasons. Since g_r is a continuous function of x, the set $T[G_1(r)]$ is a closed set. Then since z_0 is not in $T[G_1(r)]$, the distance from z_0 to $T[G_1(r)]$ is positive and z_0 has a neighborhood which contains no point in $T[G_1(r)]$. The integrand of $K_1(g_r, z)$ in (1) is a continuous function of z at z_0, and hence $K_1(g_r, z)$ itself is a continuous function of z at z_0. Let ε be a positive number less than 1; then there exists a $\delta > 0$ such that $N(z_0, \delta)$ is in E and

$$|K_1(g_r, z) - K_1(g_r, z_0)| < \varepsilon, \qquad |z - z_0| < \delta. \tag{3}$$

But since $K_1(g_r, z)$ is an integer, this statement implies that

$$K_1(g_r, z) = K_1(g_r, z_0), \qquad |z - z_0| < \delta. \tag{4}$$

Next, let z_1 be an arbitrary point in E; since E is connected by hypothesis, there is a polygonal curve P in E which connects z_0 and z_1. By the argument already given, each point z in P has a neighborhood in which the Kronecker integral is constant; since these neighborhoods overlap, the Kronecker integral has the same value in all of the neighborhoods. Therefore, $K_1(g_r, z_1) = K_1(g_r, z_0)$ and $K_1(g_r, z) = K_1(g_r, z_0)$ for every z in E. Thus $K_1(g_r, z)$ is constant in E, and the proof of Theorem 62.1 is complete. □

62.2 Theorem. *If z_0 is a point in $\mathbb{R}^2$, the set of points in $\{r: 0 < r < a\}$ for which $K_1(g_r, z_0)$ is defined is an open set U. If I is an open connected set (an interval) in U, then $K_1(g_r, z_0)$ is a constant function of r for r in I.*

Proof. In this theorem, the point z_0 is fixed and the curve $G_1(r)$ varies as r varies from $r = a$ to $r = 0$. Since $f: D_2(a) \to \mathbb{R}^2$ is a continuous function, then the curve $G_1(r)$ described by the function $g_r: C_1(r) \to \mathbb{R}^2$ varies continuously and uniformly as a function of r. Then geometrically, Figure 62.1 indicates that, as r varies from $r = a$ to $r = 0$, the value of $K_1(g_r, z_0)$ changes only at those values of r for which the curve $G_1(r)$ passes through z_0. These facts suggest both the truth of the theorem and its formal proof.

For the formal proof, let U be the set of points in $\{r: 0 < r < a\}$ for which $K_1(g_r, z_0)$ is defined. If U is the empty set, the theorem is true; assume henceforth that U is not empty. The first step is to prove that U is an open set. Now r is in U if and only if z_0 does not belong to $T[G_1(r)]$. For each r, the set $T[G_1(r)]$ is closed. If z_0 is not in $T[G_1(r_0)]$, then z_0 is not in $T[G_1(r)]$ for all r in some sufficiently small neighborhood of r_0. This statement is true because the function $f: D_2(a) \to \mathbb{R}^2$ is uniformly continuous. Thus every point in U has a neighborhood which is in U, and U is an open set. Let I be an interval in U; let r_0 be a point in I; and let ε be a number such that $0 < \varepsilon < 1$. Since the integrand of the integral in (1) is a continuous function of r, then the integral itself is a continuous function of r and there exists a neighborhood $N(r_0, \delta)$ so small that

$$|K_1(g_r, z_0) - K_1(g_{r_0}, z_0)| < \varepsilon, \qquad r \text{ in } I \cap N(r_0, \delta). \tag{5}$$

Since $\varepsilon < 1$ and $K_1(g_r, z_0)$ is an integer, equation (5) shows that

$$K_1(g_r, z_0) = K_1(g_{r_0}, z_0), \qquad r \text{ in } I \cap N(r_0, \delta). \tag{6}$$

Thus each point in I has a neighborhood in which $K_1(g_r, z_0)$ is constant. Then since I is an interval, $K_1(g_r, z_0)$ is a constant function of r in I, and the proof of Theorem 62.2 is complete. □

62.3 Theorem (Intermediate-Value Theorem). *Let $f: D_2(a) \to \mathbb{R}^2$ be a function whose components (f^1, f^2) have one continuous derivative. If z is a point such that $K_1(g_a, z)$ is defined, and if $K_1(g_a, z) \neq 0$, then there is a point x in the interior of $D_2(a)$ such that $f(x) = z$.*

Proof. The proof has a simple geometric motivation. Assume that the theorem is false. Then as r varies from $r = a$ to $r = 0$, the curve $G_1(r)$ never passes through z, and $K_1(g_r, z) = K_1(g_a, z) \neq 0$ for $0 < r \leqq a$. Now as r varies from $r = a$ to $r = 0$, the curve $G_1(r)$ shrinks from the curve $G_1(a)$ to a curve which consists of the single point $f(0)$. Since the theorem is false, $f(0) \neq z$. Then for all r sufficiently near 0, the Kronecker integral $K_1(g_r, z)$ is defined and its value is zero since $G_1(r)$ is a small curve near the point $f(0)$, which is distinct from z. This statement contradicts the earlier statement that $K_1(g_r, z) \neq 0$ for $0 < r \leqq a$ and proves the theorem.

The formal proof, also, begins by assuming that the theorem is false. Then for every r such that $0 \leqq r \leqq a$, the set $T[G_1(r)]$ does not contain z. Thus the Kronecker integral $K_1(g_r, z)$ is defined for $0 < r \leqq a$, and a small extension of Theorem 62.2 shows that

$$K_1(g_r, z) = K_1(g_a, z) \neq 0, \qquad 0 < r \leqq a. \tag{7}$$

Now the Kronecker integral is not defined for $r = 0$, but

$$\lim_{r \to 0} K_1(g_r, z) = \lim_{r \to 0} \frac{1}{2\pi} \int_{C_1(r)} \frac{\det[g_r(x) - z, D_1 g_r(x)]}{|g_r(x) - z|^2} dx = 0. \tag{8}$$

To see that this statement is true, observe first that the denominator in the integrand is bounded away from zero since $g_r(x)$ is near $f(0)$ and $f(0) \neq z$ for all r near $r = 0$. The numerator of the integrand is a bounded function for $0 \leqq r \leqq a$. Finally, the circumference of $C_1(r)$ tends to zero as $r \to \infty$. Then (8) follows from these facts. Then since the value of the Kronecker integral is an integer, it follows that $K_1(g_r, z) = 0$ for $r > 0$ but sufficiently near $r = 0$. But this statement contradicts (7). Since the assumption that the theorem is false has led to a contradiction, the theorem is true. □

Theorems 62.1, $\cdots$, 62.3 are the special cases $n = 2$ of general theorems in $\mathbb{R}^n$. It is necessary to describe the setting for these general theorems before they can be stated. For $0 \leqq r \leqq a$, let $D_n(r)$ denote the solid cube in $\mathbb{R}^n$ whose vertices are $(\pm r, \pm r, \cdots \pm r)$, and let $C_{n-1}(r)$ denote the boundary of $D_n(r)$. Let $f\colon D_n(a) \to \mathbb{R}^n$ be a function whose components $(f^1, \cdots, f^n)$ have one continuous derivative, and let f restricted to $C_{n-1}(r)$ be denoted by g_r. On each side of $C_{n-1}(r)$, one of the coordinates is constant and the others are variable. For simplicity let $(x^1, \cdots, x^{n-1})$ denote (symbolically) the variable coordinates in each of the sides of $C_{n-1}(r)$. Since f has continuous derivatives on $D_n(a)$, then $g_r\colon C_{n-1}(r) \to \mathbb{R}^n$ has continuous derivatives on $C_{n-1}(r)$ for $0 < r \leqq a$. The mapping $g_r\colon C_{n-1}(r) \to \mathbb{R}^n$ defines a closed surface $G_{n-1}(r)$ for each r such that $0 < r \leqq a$. Let $z\colon(z^1, \cdots, z^n)$ be a point in $\mathbb{R}^n$ which is not in $T[G_{n-1}(r)]$; then the Kronecker integral $K_{n-1}(g_r, z)$ of g_r with respect to z is defined as in (48) in Section 61, and

$$K_{n-1}(g_r, z) = \frac{1}{A_{n-1}(1)} \int_{C_{n-1}(r)} \frac{\det[g_r(x) - z, D_1 g_r(x), \cdots, D_{n-1} g_r(x)]}{|g_r(x) - z|^n} d(x^1, \cdots, x^{n-1}). \tag{9}$$

62.4 Theorem. *Let $f\colon D_n(a) \to \mathbb{R}^n$ be the function just described. For a fixed r, $0 < r \leqq a$, the Kronecker integral $K_{n-1}(g_r, z)$ is a constant function of z in each open connected set in the complement of $T[G_{n-1}(r)]$.*

PROOF. The proof of this theorem is exactly the same as the proof of Theorem 62.1, and the details are left to the reader. □

62.5 Theorem. *Let $f: D_n(a) \to \mathbb{R}^n$ be the function described above. If z_0 is a point in $\mathbb{R}^n$, the set of points in $\{r: 0 < r < a\}$ for which $K_{n-1}(g_r, z_0)$ is defined is an open set U. If I is an open connected set (an interval) in U, then $K_{n-1}(g_r, z_0)$ is a constant function of r for r in I.*

PROOF. The proof of this theorem is exactly the same as the proof of Theorem 62.2, and the details are left to the reader. □

62.6 Theorem (Intermediate-Value Theorem). *Let $f: D_n(a) \to \mathbb{R}^n$ be a function whose components $(f^1, \cdots, f^n)$ have one continuous derivative. If z is a point such that $K_{n-1}(g_a, z)$ is defined, and $K_{n-1}(g_a, z) \neq 0$, then there is a point x in the interior of $D_n(a)$ such that $f(x) = z$.*

PROOF. The proof of this theorem is exactly the same as the proof of Theorem 62.3, and the details are left to the reader. □

Consider again the function $f: D_2(a) \to \mathbb{R}^2$ in Theorem 62.3. Section 61 explains how the Kronecker integral $K_1(g_a, z)$ is determined by the mapping $g_a: C_1(a) \to \mathbb{R}^2$. The value of the Kronecker integral is an integer—positive, zero, or negative—which counts the number of times g_a winds the image of the boundary $C_1(a)$ around z. Also, Theorem 62.3 explains how the Kronecker integral can be used to prove a form of the intermediate-value theorem. Chapter 4 defines the Sperner degree $d(f, z)$ for the function f: $D_2(a) \to \mathbb{R}^2$. The value of $d(f, z)$ also is an integer—positive, zero, or negative—which counts the number of times g_a winds the image of the boundary $C_1(a)$ around z. Finally, in Section 27, the Sperner degree is used to prove an intermediate-value theorem. The following question inevitably arises: what is the relation, if any, between the Kronecker integral and the Sperner degree? The next theorem contains the answer.

62.7 Theorem. *Let $f: D_2(a) \to \mathbb{R}^2$ be a function whose components (f^1, f^2) have a continuous derivative on $[-a, a] \times [-a, a]$. If z is a point in $\mathbb{R}^2$ which is not in $T[G_1(a)]$, then both the Kronecker integral $K_1(g_a, z)$ and the Sperner degree $d(f, z)$ are defined, and $K_1(g_a, z) = d(f, z)$.*

PROOF. Section 61 has defined $K_1(g_a, z)$ and shown that it exists for the function f. The Sperner number $d(f, z)$ exists by Section 24 since f is continuous and z is not in $T[G_1(a)]$. Equations (1) in Section 25, and (22) in Section 24 show that for all sufficiently large k,

$$d(f, z) = \sum_{i \in I_k} \Delta[e_0, s_k(q_i^k), s_k(q_{i+1}^k)]. \tag{10}$$

Now the term

$$\Delta[e_0, s_k(q_i^k), s_k(q_{i+1}^k)] \tag{11}$$

is $+1$ if q_i^k is in U_1 and q_{i+1}^k is in U_2; it is -1 if q_i^k is in U_2 and q_{i+1}^k is in U_1; and it is zero in all other cases. Thus $d(f, z)$ is the algebraic sum of the number of signed crossings of the ray r_2 by the simplexes (q_i, q_{i+1}) in the boundary of the mapping of a simplicial subdivision of $D_2(a)$ by f. Theorem 25.1 shows that $d(f, z)$ equals the number of positive crossings minus the number of negative crossings of every ray from z. Now interpret these results for the Kronecker integral. The rays in Section 24 are oriented so that a simplex (q_i, q_{i+1}) which crosses r_2 in the positive direction contributes a positive arc which covers the point at which r_2 intersects the unit circle with center at z; likewise, if (q_i, q_{i+1}) crosses r_2 in the negative direction, then (q_i, q_{i+1}) contributes a negative arc which covers the intersection of r_2 and the unit circle. These considerations have established the following statement: points on the unit circle with center at z are covered by a certain number of positive and negative arcs which arise from the mapping by g_a of a simplicial subdivision of $C_1(a)$; for every point on this circle, the number of positive arcs which cover the point minus the number of negative arcs which cover the point equals the Sperner degree $d(f, z)$. Now g_a maps $C_1(a)$ onto the unit circle $S_1(1)$, and the arguments given show that

$$\int_{C_1(a)} \{[D_1 g_a^1(x)]^2 + [D_1 g_a^2(x)]^2\}^{1/2} \cos\theta\, dx = d(f, z) A_1(1). \qquad (12)$$

This equation implies that $K_1(g_a, z) = d(f, z)$, and the proof of Theorem 62.7 is complete. □

A final remark calls attention to the fact that both the Kronecker integral and the Sperner degree are near relatives of the fundamental theorem of the integral calculus.

Exercises

62.1. Assume that $g: C_1(a) \to \mathbb{R}^2$ maps $C_1(a)$ onto a single point $x_0: (x_0^1, x_0^2)$. If $z \neq x_0$, show that $K_1(g, z)$ is defined and that $K_1(g, z) = 0$. If $z = x_0$, show that $K_1(g, z)$ is not defined.

62.2. The equations

$$g^1(x) = a \cos 3x,$$

$$g^2(x) = a \sin 3x, \qquad 0 \leqq x \leqq 2\pi,$$

define a closed curve in $\mathbb{R}^2$. Find this curve.

(a) Show that $K_1(g, 0) = 3$. Next, show that $K_1(g, z) = 3$ at every point z such that $|z - 0| < a$.

(b) Use a geometric argument to show that $K_1(g, z) = 0$ at every point z such that $|z - 0| > a$.

(c) Explain why the Kronecker integral is not defined at a point z such that $|z - 0| = a$.

62.3. Let $f: D_1(a) \to \mathbb{R}$ be a continuous function, and let $g: C_0(a) \to \mathbb{R}$ be the restriction of f to the set $\{-a, a\}$. If $z \neq f(a)$ and $z \neq f(-a)$, then there is a Sperner degree $d(f, z)$ for f [compare Exercise 29.1] and a Kronecker integral $K_0(g, z)$ [compare Exercise 61.8]. Define $d(f, z)$ and $K_0(g, z)$ and describe their properties and applications; in particular, use $d(f, z)$ and $K_0(g, z)$ to give proofs of the elementary intermediate-value theorem.

62.4. Let $P_0, P_1, \cdots, P_{3n}$ [$P_{3n} = P_0$] be $3n$ equally spaced points on the unit circle with center at the origin O in $\mathbb{R}^2$, and let T_i denote the positively oriented triangle $OP_{i-1}P_i$, $i = 1, \cdots, 3n$. Similarly, let $Q_0, Q_1, \cdots, Q_n$ [$Q_n = Q_0$] be n equally spaced points on the circle with center C and radius a, and let S_i denote the positively oriented triangle $CQ_{i-1}Q_i$, $i = 1, \cdots, n$. Use an affine transformation to map each triangle T_i onto a triangle S_j so that together these affine transformations form a continuous piecewise linear transformation f which maps the union of the triangles T_i onto the union of the triangles S_i as follows: T_1, T_2, T_3 are mapped onto S_1 so that P_0P_1 and P_2P_3 go into Q_0Q_1 and P_1P_2 goes into Q_1Q_0, and so that each succeeding triple of T-triangles is mapped onto an S-triangle in the same manner. Let g denote the restriction of f to the boundary of the polygon $P_0P_1 \cdots P_{3n}$.

(a) Show that $d(f, z)$ and $K_1(g, z)$ have the value 1 if z is inside the polygon $Q_0Q_1 \cdots Q_n$, and that they have the value zero if z is outside this polygon.

(b) Let x denote a point in the plane of the polygon $P_0P_1 \cdots P_{3n}$. Find the number of solutions x of the equation $f(x) = z$ for a given point z in the plane of the polygon $Q_0Q_1 \cdots Q_n$.

(c) Work through the proof of Theorem 62.7 step by step, in the special case described in this exercise, as an aid to understanding the proof of the theorem in the general case.

62.5. Repeat Exercise 62.4 for a transformation f which maps the triangles T_k, T_{n+k}, T_{2n+k} into the triangle S_k, $k = 1, \cdots, n$, in all cases with preservation of orientation. Show that $d(f, z) = K_1(g, z) = 3$ if z is inside $Q_0Q_1 \cdots Q_n$, and that $d(f, z) = K_1(g, z) = 0$ if z is outside $Q_0Q_1 \cdots Q_n$. For each z, find the number of solutions of the equation $f(x) = z$.

CHAPTER 10

Differentiable Functions of Complex Variables

63. Introduction

This chapter contains an introduction to the theory of functions of one and of several complex variables. The effort once more is to develop the theory from a minimum number of hypotheses and by general methods which apply to complex-valued, as well as real-valued, functions.

An analytic function of one complex variable is a function which has a derivative—everything else follows from this single hypothesis. In this chapter, Goursat's proof is used to prove a special case of Cauchy's integral theorem, which is then used to prove Cauchy's integral formula. This formula is the essential tool in the proof that a differentiable (or analytic) function has an infinite number of derivatives. An analytic function of several complex variables is a function which has a derivative with respect to each of its variables. The Cauchy–Riemann equations and the properties of analytic functions of a single complex variable provide a rapid development of the theory of functions of several complex variables.

The methods used in this chapter are general methods, and several of the major methods and tools of earlier chapters recur in this chapter. The derivatives of functions of several complex variables are defined by n-vectors in the same way that derivatives of functions of real variables are defined in Chapter 1. The Stolz condition plays the same role that it does in the real case. For functions of complex variables as well as real variables, a necessary and sufficient condition that the function be differentiable is that it satisfy the Stolz condition. The fundamental theorem of the integral calculus extends without change to complex-valued functions of real variables. If the complex-valued functions are analytic, then a special case of the fundamental theorem of the integral calculus is Cauchy's integral theorem—for functions of one and of several complex variables.

This chapter is self contained in the sense that it starts at the beginning. It assumes a knowledge of the basic facts about complex numbers and an understanding of the preceding chapters of this book, but everything concerning the derivatives and integrals of functions of a complex variable is developed from the beginning in this chapter.

Exercises

63.1. The complex number $x + iy$ is denoted by z and represented by the point (x, y) in $\mathbb{R}^2$. If $r = (x^2 + y^2)^{1/2}$ and $\theta = \arctan(y/x)$, show that $x + iy = r(\cos\theta + i\sin\theta)$.

63.2. The absolute value of $z: x + iy$ is denoted by $|z|$ and defined to be $(x^2 + y^2)^{1/2}$. Prove the following statements about the absolute value.

(a) If c is a real number, then $|cz| = |c||z|$.
(b) $|z_1 z_2| = |z_1||z_2|$.
(c) $\left|\dfrac{z_1}{z_2}\right| = \dfrac{|z_1|}{|z_2|}$.
(d) The distance between the points which represent z_1 and z_2 is $|z_1 - z_2|$.
(e) The triangle inequality: $|z_1 + z_2| \leqq |z_1| + |z_2|$.
(f) If $z = x + iy$, then $|z|^2 = (x + iy)(x - iy)$.

63.3. According to Euler's definition, $e^{i\theta} = \cos\theta + i\sin\theta$.

(a) Prove that every complex number has a representation of the form $re^{i\theta}$, $r \geqq 0$.
(b) Prove that $|e^{i\theta}| = 1$ for every θ, and that $|re^{i\theta}| = r$.
(c) If $z_j = r_j(\cos\theta_j + i\sin\theta_j)$, $j = 1, 2$, then show that $z_1 z_2 = r_1 r_2[\cos(\theta_1 + \theta_2) + i\sin(\theta_1 + \theta_2)]$ and that this result is consistent with Euler's definition and the laws of exponents.
(d) Use Euler's definition to show that $e^z = e^x(\cos y + i\sin y)$.

63.4. If $f(z) = z^3$, show that there are polynomials $u(x, y)$ and $v(x, y)$ such that $f(z) = u(x, y) + iv(x, y)$, and find these polynomials.

63.5. The complex numbers are denoted by $\mathbb{C}$. Let E be an open set in $\mathbb{C}$, and let f be a complex-valued function which is defined in E.

(a) If z_0 is a point in E, explain what is meant by the statement $\lim_{z \to z_0} f(z) = a$.
(b) Explain what is meant by the statement that f is continuous at z_0.
(c) If $f(z) = z^2$, prove that f is continuous at every point in $\mathbb{C}$.

63.6. Let $[a_{ij}]$ be an $n \times n$ matrix of complex numbers. Is $\det[a_{ij}]$ defined? If it is, how? [Hint. Section 76 in Appendix 1.] Do the determinants of complex matrices have all, or some, of the properties of determinants of real matrices? In particular, are Theorems 20.1 and 20.3 true for the determinants of complex matrices? Are Sylvester's theorem of 1839 and 1851 [see Section 81] and the Bazin–Reiss–Picquet theorem [see Section 83] true for complex matrices?

63.7. If z is $x + iy$, the complex conjugate of z, denoted by $\bar{z}$, is $x - iy$. Prove the following statements:

(a) $z\bar{z} = |z|^2$.
(b) $\overline{z_1 + z_2} = \bar{z}_1 + \bar{z}_2$; $\overline{z_1 z_2} = \bar{z}_1 \bar{z}_2$.
(c) $\det[a_{ij}] \det[\bar{a}_{ij}] = |\det[a_{ij}]|^2$.

Part I. Functions of a Single Complex Variable

64. Differentiable Functions; the Cauchy–Riemann Equations

Let z denote the complex number $x + iy$; let $\mathbb{C}$ denote the set of all complex numbers z; and let $N(z_0, r) = \{z : |z - z_0| < r\}$, a neighborhood of z_0. Let $f: N(z_0, r) \to \mathbb{C}$ be a complex-valued function; then there are real-valued functions u and v of the real variables x and y such that $f(z) = u(x, y) + iv(x, y)$. The definition of the derivative $Df(z_0)$ of f at z_0 is similar to that of the derivative $f'(x_0)$ in Section 1; $Df(z_0)$ exists if and only if

$$\lim_{z \to z_0} \frac{\begin{vmatrix} f(z) & 1 \\ f(z_0) & 1 \end{vmatrix}}{\begin{vmatrix} z & 1 \\ z_0 & 1 \end{vmatrix}}, \qquad z \neq z_0, \tag{1}$$

exists; if this limit exists, its value by definition is $Df(z_0)$. If $Df(z_0)$ exists, then f is said to be differentiable at z_0. Thus f is differentiable at z_0 and its derivative equals $Df(z_0)$ if and only if to each $\varepsilon > 0$ there corresponds a $\delta(\varepsilon, z_0)$ such that

$$\left| \frac{f(z) - f(z_0)}{z - z_0} - Df(z_0) \right| < \varepsilon, \qquad 0 < |z - z_0| < \delta(\varepsilon, z_0). \tag{2}$$

64.1 Theorem. *A necessary condition that $f: N(z_0, r) \to \mathbb{C}$ be differentiable at $z_0 : x_0 + iy_0$ is that u and v have partial derivatives u_x, u_y, v_x, v_y which satisfy the following Cauchy–Riemann differential equations:*

$$u_x(x_0, y_0) = v_y(x_0, y_0), \qquad u_y(x_0, y_0) = -v_x(x_0, y_0). \tag{3}$$

PROOF. In (1) there is no restriction except $z \neq z_0$ on the way in which z approaches z_0. Find the limit twice, first with $z = x + iy_0$ and then with $z = x_0 + iy$. The two limits are the following:

$$\lim_{x \to x_0} \frac{[u(x, y_0) - u(x_0, y_0)] + i[v(x, y_0) - v(x_0, y_0)]}{(x - x_0)}$$
$$= u_x(x_0, y_0) + iv_x(x_0, y_0),$$
$$\lim_{y \to y_0} \frac{[u(x_0, y) - u(x_0, y_0)] + i[v(x_0, y) - v(x_0, y_0)]}{i(y - y_0)} \tag{4}$$
$$= -iu_y(x_0, y_0) + v_y(x_0, y_0).$$

Since f is differentiable by hypothesis, the two limits in (4) are equal. The two equations in (3) follow from equating the limits in (4). The proof is complete. □

The existence of $Df(z_0)$ does not imply that u and v are differentiable at (x_0, y_0) in the sense of Definition 2.8; Theorem 64.1 proves only that u and v have partial derivatives there. Nevertheless, this chapter will prove eventually that u and v have continuous derivatives of all orders. The next theorem states sufficient conditions for the existence of the derivative $Df(z_0)$. For convenience let p and p_0 denote the points (x, y) and (x_0, y_0) in $\mathbb{R}^2$ which correspond to $z : x + iy$ and $z_0 : x_0 + iy_0$ in $\mathbb{C}$.

64.2 Theorem. *Let u and v be real-valued functions which are defined in $N(p_0, r)$ and have derivatives $D_x u$, $D_y u$, $D_x v$, $D_y v$ [in the sense of Definition 2.8] at p_0 which satisfy the Cauchy–Riemann equations as follows:*

$$D_x u(p_0) = D_y v(p_0), \qquad D_y u(p_0) = -D_x v(p_0). \tag{5}$$

Then f is differentiable at z_0 and

$$Df(z_0) = D_x u(p_0) + iD_x v(p_0) = D_y v(p_0) - iD_y u(p_0). \tag{6}$$

PROOF. Since u and v are differentiable at p_0 by hypothesis, they satisfy the following Stolz conditions:

$$\begin{aligned} u(p) - u(p_0) &= D_x u(p_0)(x - x_0) + D_y u(p_0)(y - y_0) + r(u; p_0, p)|p - p_0|, \\ v(p) - v(p_0) &= D_x v(p_0)(x - x_0) + D_y v(p_0)(y - y_0) + r(v; p_0, p)|p - p_0|. \end{aligned} \tag{7}$$

Since the derivatives of u and v satisfy the Cauchy–Riemann equations (5) by hypothesis, a straightforward calculation shows that

$$\frac{f(z) - f(z_0)}{z - z_0} = D_x u(p_0) + iD_x v(p_0) + \frac{[r(u; p_0, p) + ir(v; p_0, p)]}{(x - x_0) + i(y - y_0)}|p - p_0|. \tag{8}$$

Since $r(u; p_0, p)$ and $r(v; p_0, p)$ approach zero as z approaches z_0,

$$Df(z_0) = \lim_{z \to z_0} \frac{f(z) - f(z_0)}{z - z_0} = D_x u(p_0) + iD_x v(p_0). \tag{9}$$

The other value given for the derivative in (6) equals the one given in (9) by the Cauchy–Riemann equations in (5). The proof of Theorem 64.2 is complete. □

64.3 Definition. The complex-valued function f is *analytic at* z_0 if and only if it is differentiable at each point in some neighborhood of z_0. Also, f is *analytic in a set* E in $\mathbb{C}$ if and only if it is differentiable at each point in some open set which contains E.

Exercises

64.1. Let functions $u: \mathbb{R}^2 \to \mathbb{R}$ and $v: \mathbb{R}^2 \to \mathbb{R}$ be defined as follows:

(i) $u(x, y) = x + y$, $v(x, y) = -x + y$;
(ii) $u(x, y) = 3x - 4y$, $v(x, y) = 4x + 3y$;
(iii) $u(x, y) = x^2 - y^2$, $v(x, y) = 2xy$.

In each case use Theorem 64.2 to show that the function $u(x, y) + iv(x, y)$ is an anlytic function of z in the entire complex plane $\mathbb{C}$. Find an algebraic expression in terms of z for each $f(z)$, and find $Df(z)$ in each case.

64.2. Let u and v be real-valued functions which are defined in a neighborhood of (x_0, y_0) and have partial derivatives there. Let f be the function such that $f(z) = u(x, y) + iv(x, y)$. If the partial derivatives of u and v are continuous at (x_0, y_0) and satisfy the Cauchy-Riemann equations there, prove that f is differentiable at $x_0 + iy_0$. [Hint. Theorem 3.19.]

64.3. Let $f: N(z_0, r) \to \mathbb{C}$ and $g: N(z_0, r) \to \mathbb{C}$ be functions which are differentiable at z_0. Prove that the functions $f + g$, fg and f/g are differentiable and have derivatives as stated in the following formulas:

$$D(f + g)(z_0) = Df(z_0) + Dg(z_0);$$

$$D(fg)(z_0) = f(z_0)Dg(z_0) + g(z_0)Df(z_0);$$

$$D(f/g)(z_0) = \frac{g(z_0)Df(z_0) - f(z_0)Dg(z_0)}{[g(z_0)]^2}, \qquad g(z_0) \neq 0.$$

64.4. Prove that the elementary functions in the following formulas are differentiable; then establish the formulas for their derivatives.

$$Dc = 0, \qquad c \text{ a constant function.}$$

$$Dz^n = nz^{n-1}, \qquad n = 1, 2, \cdots.$$

$$De^z = e^z, \qquad e^z = e^x(\cos y + i \sin y).$$

$$D \log z = \frac{1}{z}, \qquad \log z = \log(x^2 + y^2)^{1/2} + i \arcsin \frac{y}{(x^2 + y^2)^{1/2}}, \qquad z \neq 0.$$

$$D \sin z = \cos z, \qquad \sin z = \frac{(e^{iz} - e^{-iz})}{2i}.$$

$$D \cos z = -\sin z, \qquad \cos z = \frac{(e^{iz} + e^{-iz})}{2}.$$

64.5. Prove that each of the elementary functions in Exercise 64.4 has an infinite number of derivatives.

64.6. Let $f: E \to \mathbb{C}$ be a function $u + iv$ which has a derivative Df in the open set E in $\mathbb{C}$.
(a) Can you prove that Df has a derivative in E? Explain.
(b) Assume that u and v satisfy the Cauchy–Riemann equations, and that u and v have continuous second derivatives u_{xx}, u_{xy}, u_{yy} and v_{xx}, v_{xy}, v_{yy}. Prove that f has two continuous derivatives Df and $D(Df)$.
(c) Assume that u and v have continuous derivatives of all orders in E, and that f has a derivative in E. Prove that f has continuous derivatives $D^n f$, $n = 1, 2, \cdots$, of every order in E.

64.7. This section has treated the derivatives of complex-valued functions of a complex variable. There are, in addition, complex-valued functions of real variables. Give examples of such functions. Write a short essay which contains the essential facts (definitions, theorems, and formulas) about the derivatives of complex-valued functions of real variables.

65. The Stolz Condition

Let $f: N(z_0, r) \to \mathbb{C}$ be differentiable at z_0. Then

$$f(z) - f(z_0) = Df(z_0)(z - z_0) + \left[\frac{f(z) - f(z_0)}{z - z_0} - Df(z_0)\right](z - z_0). \tag{1}$$

Define $r(f; z_0, z)$ as follows:

$$\begin{aligned} r(f; z_0, z) &= \frac{f(z) - f(z_0)}{z - z_0} - Df(z_0), \qquad z \neq z_0, \\ r(f; z_0, z_0) &= 0. \end{aligned} \tag{2}$$

Then $r(f; z_0, z)$ is defined for z in $N(z_0, r)$,

$$\lim_{z \to z_0} r(f; z_0, z) = 0, \tag{3}$$

and $r(f; z_0, z)$, as a function of z, is continuous at z_0.

65.1 Definition. The function $f: N(z_0, r) \to \mathbb{C}$ satisfies the *Stolz condition at* z_0 if and only if there exists a constant A in $\mathbb{C}$ and a complex-valued function of z, denoted by $r(f; z_0, z)$ and defined in a neighborhood of z_0, such that

$$f(z) - f(z_0) = A(z - z_0) + r(f; z_0, z)(z - z_0), \tag{4}$$

$$\lim_{z \to z_0} r(f; z_0, z) = 0, \qquad r(f; z_0, z_0) = 0. \tag{5}$$

65.2 Theorem. *If $f: N(z_0, r) \to \mathbb{C}$ is differentiable at z_0, then f satisfies the Stolz condition at z_0 with $A = Df(z_0)$. If f satisfies the Stolz condition in* (4) *and* (5) *at z_0, then f is differentiable at z_0 and $Df(z_0) = A$. Thus, f is differentiable at z_0 if and only if it satisfies the Stolz condition in* (4) *and* (5) *at z_0.*

PROOF. The proof that the Stolz condition is necessary follows from (1), (2), (3) above. The proof that it is sufficient follows from (4) and (5) and the definition of the derivative. □

Let E be an open set in $\mathbb{C}$, and let f be analytic in E. If $f = u + iv$, then Df can be expressed in terms of the derivatives of u and v as stated in equation (6) in Section 64. Therefore, f has a continuous derivative Df if and only if u and v have continuous derivatives u_x, u_y, v_x, v_y.

If u and v have derivatives $D_x u$, $D_y u$, $D_x v$, $D_y v$ in E, then equation (8) in Section 64 shows that f satisfies the Stolz condition at each point z_0 in E and that

$$f(z) - f(z_0) = [D_x u(p_0) + iD_x v(p_0)](z - z_0) + r(f; z_0, z)(z - z_0), \quad (6)$$

$$r(f; z_0, z) = \frac{[r(u; p_0, p) + ir(v; p_0, p)]|z - z_0|}{z - z_0}. \quad (7)$$

Then

$$|r(f; z_0, z)| \leqq \{[r(u; p_0, p)]^2 + [r(v; p_0, p)]^2\}^{1/2}. \quad (8)$$

65.3 Theorem. *Let f be analytic in an open set E in $\mathbb{C}$, and assume that f has a continuous derivative Df in E. If F is a compact convex set in E, then to each $\varepsilon > 0$ there corresponds a $\delta(\varepsilon) > 0$ such that*

$$|r(f; z_0, z_1)| < \varepsilon \quad (9)$$

for each two points z_0, z_1 in F for which $|z_0 - z_1| < \delta(\varepsilon)$.

PROOF. Since Df is continuous in E by hypothesis, the functions u, v have continuous derivatives $D_x u$, $D_y u$, $D_x v$, $D_y v$. Thus the functions u, v satisfy the hypotheses of Theorem 9.9; hence, to each $\varepsilon > 0$ there corresponds a $\delta(\varepsilon)$ such that

$$|r(u; p_0, p_1)| < \varepsilon/\sqrt{2}, \qquad |r(v; p_0, p_1)| < \varepsilon/\sqrt{2}, \quad (10)$$

for each pair of points p_0, p_1 in F such that $|z_0 - z_1| < \delta(\varepsilon)$. Then (9) follows from (8) and (10), and the proof is complete. □

EXERCISES

65.1. Prove that f is continuous at z_0 if f has a derivative at z_0.

65.2. Let the functions $f: N(w_0, r) \to \mathbb{C}$ and $g: N(z_0, s) \to N(w_0, r)$ have derivatives $Df(w_0)$ and $Dg(z_0)$ at w_0 and z_0. If $w_0 = g(z_0)$, prove that the composite function $f \circ g$ has a derivative $D(f \circ g)(z_0)$ at z_0, and that

$$D(f \circ g)(z_0) = Df[g(z_0)]Dg(z_0).$$

[Hint. Use the Stolz condition which f satisfies at w_0 to show that

$$\frac{f[g(z)] - f[g(z_0)]}{z - z_0} = Df[g(z_0)]\frac{[g(z) - g(z_0)]}{z - z_0} + r[f; g(z_0), g(z)]\frac{g(z) - g(z_0)}{z - z_0}.\Bigg]$$

65.3. If n is a positive integer and $f(z) = (z^2 + 1)^n$, show that f is analytic in the entire complex plane; find $Df(z)$.

(a) If $f(z) = (z^2 + 1)^3$, find $Df(z)$ by two methods and show that the two answers are the same. [Hint. The binomial theorem and Exercise 65.2.]

(b) If n is a negative integer, is there a set in which f is analytic? If there is, find this set and find $Df(z)$.

65.4. If $f(z) = [e^x(\cos y + i \sin y)]^n$, prove that, for every integer n, the function f is analytic in the complex plane. Find $Df(z)$ by two methods and show that the two answers are the same.

65.5. If $u(x, y) = x^3 - 3xy^2$ and $v(x, y) = 3x^2y - y^3$, show that u and v have continuous derivatives which satisfy the Cauchy–Riemann equations. If $f(z) = u(x, y) + iv(x, y)$, show that f is analytic in the entire complex plane. Identify the function f.

65.6. Let u and v be defined and differentiable in a neighborhood of (x_0, y_0), and let $f(z) = u(x, y) + iv(x, y)$. If the derivatives of u and v satisfy the Cauchy–Riemann equations, prove that f has a derivative at $z_0 : x_0 + iy_0$, and that

$$|Df(z_0)|^2 = D_{(1,2)}(u, v)(x_0, y_0).$$

66. Integrals

This section provides an introduction to the study of integrals of complex-valued functions on curves in the complex plane $\mathbb{C}$. It treats the definitions, existence, and evaluation of these integrals, and it establishes a fundamental theorem of the integral calculus for such integrals.

Let E be an open connected set in $\mathbb{C}$, and let $f: E \to \mathbb{C}$ be a continuous function. Then $f(z) = u(x, y) + iv(x, y)$, and u and v are continuous functions of (x, y) for $x + iy$ in E. Let

$$x: [a, b] \to \mathbb{R}, \qquad y: [a, b] \to \mathbb{R}, \tag{1}$$

denote functions which have continuous derivatives x', y'; then the equation

$$z(t) = x(t) + iy(t), \qquad t \in [a, b], \tag{2}$$

defines a curve C in $\mathbb{C}$; assume that the trace $T(C)$ of C is in E. Now

$$|f[z(t)]| = \{[u(x(t), y(t))]^2 + [v(x(t), y(t))]^2\}^{1/2}, \qquad t \in [a, b]. \tag{3}$$

Thus $|f[z(t)]|$ is a continuous function of t for t in $[a, b]$, and there exists a constant M such that

$$|f[z(t)]| \leqq M, \qquad t \in [a, b]. \tag{4}$$

Since the functions x and y in (1) have continuous derivatives, the curve C has finite length L, and

$$L = \int_a^b \{[x'(t)]^2 + [y'(t)]^2\}^{1/2}\,dt. \tag{5}$$

Subdivide the interval $[a, b]$ into subintervals by points t_k such that $a = t_0 < t_1 < \cdots < t_k < \cdots < t_n = b$, and let

$$z_k = x(t_k) + iy(t_k), \qquad k = 0, 1, \cdots, n. \tag{6}$$

Then the value of the integral of f on C is approximately

$$\sum_{k=1}^{n} f(z_{k-1})(z_k - z_{k-1}). \tag{7}$$

66.1 Definition. Let $f: E \to \mathbb{C}$ be a continuous function, and let C be a curve defined by functions $x: [a, b] \to \mathbb{R}$ and $y: [a, b] \to \mathbb{R}$ which have continuous derivatives x' and y', and assume that the trace $T(C)$ of C is in E. The integral of f on C exists if and only if

$$\lim_{n\to\infty} \sum_{k=1}^{n} f(z_{k-1})(z_k - z_{k-1}) \tag{8}$$

exists, the limit being taken with respect to a sequence of subdivisions of $[a, b]$ whose norms approach zero as $n \to \infty$. If the limit in (8) exists, its value is denoted by $\int_C f(z)\,dz$ and called the *integral of f on C*. Thus

$$\int_C f(z)\,dz = \lim_{n\to\infty} \sum_{k=1}^{n} f(z_{k-1})(z_k - z_{k-1}). \tag{9}$$

Two separate proofs will be given for the existence of $\int_C f(z)\,dz$. The first of these existence theorems considers the real and the imaginary parts of the sum in (7), and it derives the existence of $\int_C f(z)\,dz$ from theorems in Section 36 which establish the existence of line and surface integrals of real-valued functions. The second existence theorem employs methods which are more characteristically complex-variable methods. Each theorem has its interest and its applications.

66.2 Theorem. *Let $f: E \to \mathbb{C}$ be a continuous function, and let C be a curve defined by functions $x: [a, b] \to \mathbb{R}$ and $y: [a, b] \to \mathbb{R}$ which have continuous derivatives x' and y'; assume that the trace $T(C)$ of C is in E. Then the integral $\int_C f(z)\,dz$ exists, and*

$$\int_C f(z)\,dz = \int_a^b \{u[x(t), y(t)] + iv[x(t), y(t)]\}\{x'(t) + iy'(t)\}\,dt \tag{10}$$

$$= \int_a^b \{u[x(t), y(t)]x'(t) - v[x(t), y(t)]y'(t)\}\,dt \tag{11}$$

$$+ i\int_a^b \{v[x(t), y(t)]x'(t) + u[x(t), y(t)]y'(t)\}\,dt. \tag{12}$$

Proof. Since $f(z) = u(x, y) + iv(x, y)$ and $z = x + iy$, the real and the imaginary parts of the sum in (7) are

$$\sum_{k=1}^{n}\left\{u[x(t_{k-1}), y(t_{k-1})]\frac{(-1)}{1!}\begin{vmatrix} x(t_{k-1}) & 1 \\ x(t_k) & 1 \end{vmatrix} - v[x(t_{k-1}), y(t_{k-1})]\frac{(-1)}{1!}\begin{vmatrix} y(t_{k-1}) & 1 \\ y(t_k) & 1 \end{vmatrix}\right\}, \tag{13}$$

$$\sum_{k=1}^{n}\left\{v[x(t_{k-1}), y(t_{k-1})]\frac{(-1)}{1!}\begin{vmatrix} x(t_{k-1}) & 1 \\ x(t_k) & 1 \end{vmatrix} + u[x(t_{k-1}), y(t_{k-1})]\frac{(-1)}{1!}\begin{vmatrix} y(t_{k-1}) & 1 \\ y(t_k) & 1 \end{vmatrix}\right\}. \tag{14}$$

As explained in Section 36, these sums lead to line integrals on the curve C. The functions u and v are continuous functions of t on C, and x and y have continuous derivatives x' and y'. Then x and y satisfy the Stolz condition, and

$$\begin{aligned} \begin{vmatrix} x(t_{k-1}) & 1 \\ x(t_k) & 1 \end{vmatrix} &= x'(t_{k-1})\begin{vmatrix} t_{k-1} & 1 \\ t_k & 1 \end{vmatrix} + r(x; t_{k-1}, t_k)|t_k - t_{k-1}|, \\ \begin{vmatrix} y(t_{k-1}) & 1 \\ y(t_k) & 1 \end{vmatrix} &= y'(t_{k-1})\begin{vmatrix} t_{k-1} & 1 \\ t_k & 1 \end{vmatrix} + r(y; t_{k-1}, t_k)|t_k - t_{k-1}|. \end{aligned} \tag{15}$$

Substitute from (15) in (13) and (14); the limits of the sums in (13) and (14) are the integrals in (11) and (12) respectively. Thus the integral $\int_C f(z)\,dz$ exists and is the complex number indicated in (11) and (12); the expression in (11) and (12) equals the integral on the right in (10). The proof of Theorem 66.2 is complete. □

Theorem 66.2 shows that the integral exists under very weak hypotheses on f. The integral exists if u and v are merely continuous functions of x and y; it is not necessary that f be analytic in E. However, it will be shown later that $\int_C f(z)\,dz$ has important special properties if f is analytic in E.

66.3 Theorem. *Let E be an open set in $\mathbb{C}$, and let $x: [a, b] \to \mathbb{R}$ and $y: [a, b] \to \mathbb{R}$ be functions which have continuous derivatives and define a curve C whose length is L and whose trace $T(C)$ is in E. Let $f: E \to \mathbb{C}$ be a continuous function such that $|f[z(t)]| \leqq M$ for $z(t) = x(t) + iy(t)$ and t in $[a, b]$. Then*

$$\left|\int_C f(z)\,dz\right| \leqq \int_a^b |f[z(t)]|\{[x'(t)]^2 + [y'(t)]^2\}^{1/2}\,dt \leqq ML. \tag{16}$$

PROOF. Since the absolute value of a sum is equal to or less than the sum of the absolute values,

$$\begin{aligned} &\left|\sum_{k=1}^{n} f(z_{k-1})(z_k - z_{k-1})\right| \\ &\qquad \leqq \sum_{k=1}^{n} |f[z(t_{k-1})]|\{[x(t_k) - x(t_{k-1})]^2 + [y(t_k) - y(t_{k-1})]^2\}^{1/2}. \end{aligned} \tag{17}$$

Since u and v are continuous functions of x and y, then (3) shows that $|f[z(t)]|$ is a continuous real-valued function of t on $[a, b]$. Also, x and y have continuous derivatives and therefore satisfy the Stolz condition as in (15). Substitute from (15) in (17). The triangle inequality shows that the absolute value of the difference of the two sums

$$\sum_{k=1}^{n} |f[z(t_{k-1})]|\{[x'(t_{k-1}) + r(x; t_{k-1}, t_k)]^2 + [y'(t_{k-1}) + r(y; t_{k-1}, t_k)]^2\}^{1/2}(t_k - t_{k-1}),$$

$$\sum_{k=1}^{n} |f[z(t_{k-1})]|\{[x'(t_{k-1})]^2 + [y'(t_{k-1})]^2\}^{1/2}(t_k - t_{k-1}),$$

is equal to or less than

$$\sum_{k=1}^{n} |f[z(t_{k-1})]|\{[r(x; t_{k-1}, t_k)]^2 + [r(y; t_{k-1}, t_k)]^2\}^{1/2}(t_k - t_{k-1}).$$

The limit of this sum is zero since

$$\lim_{n\to\infty} r(x; t_{k-1}, t_k) = 0, \qquad \lim_{n\to\infty} r(y; t_{k-1}, t_k) = 0;$$

therefore,

$$\begin{aligned}
\lim_{n\to\infty} \sum_{k=1}^{n} |f[z(t_{k-1})]|\{[x'(t_{k-1}) + r(x; t_{k-1}, t_k)]^2 + [y'(t_{k-1}) \\
+ r(y; t_{k-1}, t_k)]^2\}^{1/2}(t_k - t_{k-1}) \\
= \lim_{n\to\infty} \sum_{k=1}^{n} |f[z(t_{k-1})]|\{[x'(t_{k-1})]^2 + [y'(t_{k-1})]^2\}^{1/2}(t_k - t_{k-1}) \\
= \int_a^b |f[z(t)]|\{[x'(t)]^2 + [y'(t)]^2\}\,dt.
\end{aligned} \tag{18}$$

Then (17) shows that $|\int_C f(z)\,dz|$ is equal to or less than the limit of the sum on the right in (17), and this limit is evaluated in (18). The first inequality in (16) follows from these statements. Since $|f[z(t)]| \leqq M$ for t in $[a, b]$, and since $\{[x'(t)]^2 + [y'(t)]\}^{1/2} \geqq 0$ for t in $[a, b]$, then

$$\begin{aligned}
\int_a^b |f[z(t)]|\{[x'(t)]^2 + [y'(t)]^2\}^{1/2}\,dt &\leqq \int_a^b M\{[x'(t)]^2 + [y'(t)]^2\}^{1/2}\,dt \\
&\leqq M\int_a^b \{[x'(t)]^2 + [y'(t)]^2\}^{1/2}\,dt \\
&\leqq ML.
\end{aligned} \tag{19}$$

The proof of Theorem 66.3 is complete. □

66.4 Example. This example shows how the formulas in Theorem 66.2 can be used to evaluate the integral $\int_C z^2\,dz$ and to show that

$$\int_C z^2\,dz = \frac{[z(b)]^3}{3} - \frac{[z(a)]^3}{3}. \tag{20}$$

Since $f(z) = u(x, y) + iv(x, y)$ and $f(z) = z^2 = (x + iy)^2$, then

$$u(x, y) = x^2 - y^2, \qquad v(x, y) = 2xy. \tag{21}$$

Then by the formulas in (10), (11), (12),

$$\begin{aligned}\int_C f(z)\,dz &= \int_a^b \{[x(t)]^2 x'(t) - [y(t)]^2 x'(t) - 2x(t)y(t)y'(t)\}\,dt \\ &\quad + i\int_a^b \{2x(t)y(t)x'(t) + [x(t)]^2 y'(t) - [y(t)]^2 y'(t)\}\,dt.\end{aligned} \tag{22}$$

The first integral on the right in (22) is evaluated as follows.

$$\begin{aligned}&\int_a^b \{[x(t)]^2 x'(t) - [[y(t)]^2 x'(t) + 2x(t)y(t)y'(t)]\}\,dt \\ &\qquad = \frac{[x(t)]^3}{3} - x(t)[y(t)]^2 \Big|_a^b \\ &\qquad = \frac{[x(b)]^3 - 3x(b)[y(b)]^2}{3} - \frac{[x(a)]^3 - 3x(a)[y(a)]^2}{3}.\end{aligned} \tag{23}$$

The second integral on the right in (22) is evaluated as follows.

$$\begin{aligned}&\int_a^b \{2x(t)y(t)x'(t) + [x(t)]^2 y'(t) - [y(t)]^2 y'(t)\}\,dt \\ &\qquad = [x(t)]^2 y(t) - \frac{[y(t)]^3}{3} \Big|_a^b \\ &\qquad = \frac{3[x(b)]^2 y(b) - [y(b)]^3}{3} - \frac{3[x(a)]^2 y(a) - [y(a)]^3}{3}.\end{aligned} \tag{24}$$

Then (22), (23), (24) show that

$$\begin{aligned}\int_C z^2\,dz &= \frac{[x(b)]^3 - 3x(b)[y(b)]^2}{3} + i\frac{3[x(b)]^2 y(b) - [y(b)]^3}{3} \\ &\quad - \frac{[x(a)]^3 - 3x(a)[y(a)]^2}{3} - i\frac{3[x(a)]^2 y(a) - [y(a)]^3}{3}.\end{aligned} \tag{25}$$

It is easy to verify from these formulas that

$$\begin{aligned}\int_C z^2\,dz &= \frac{[x(b) + iy(b)]^3}{3} - \frac{[x(a) + iy(a)]^3}{3} \\ &= \frac{[z(b)]^3}{3} - \frac{[z(a)]^3}{3}.\end{aligned} \tag{26}$$

This example suggests several observations and questions. First, the value of the integral $\int_C z^2\,dz$ depends on the end points $z(a)$, $z(b)$ of C but not on the curve which connects these points. Question: under what conditions is it true that $\int_C f(z)\,dz$ depends only on the end points of C and not on the curve connecting these points? Second, if $z(a) = z(b)$, then C is a closed curve, and the formula in (26) shows that $\int_C f(z)\,dz = 0$. Question: under what conditions is $\int_C f(z)\,dz = 0$ for a closed curve C? Third, the evaluation of $\int_C z^2\,dz$ in equations (21) to (26) is not difficult, but it is long and tedious, and the fundamental theorem of the integral calculus [to be proved presently] provides a shorter and easier evaluation for $\int_C f(z)\,dz$, at least in most cases.

66.5 Example. Let C be the circle with center $z_0 : (x_0, y_0)$ and radius r. The purpose of this example is to show that

$$\int_C \frac{dz}{z - z_0} = 2\pi i. \tag{27}$$

To show that this integral exists, observe first that

$$\begin{aligned} \frac{1}{z - z_0} &= \frac{1}{(x + iy) - (x_0 + iy_0)} = \frac{1}{(x - x_0) + i(y - y_0)} \\ &= \frac{(x - x_0) - i(y - y_0)}{(x - x_0)^2 + (y - y_0)^2}. \end{aligned} \tag{28}$$

Thus, if $1/(z - z_0) = u(x, y) + iv(x, y)$, then

$$u(x, y) = \frac{(x - x_0)}{(x - x_0)^2 + (y - y_0)^2}, \qquad v(x, y) = \frac{-(y - y_0)}{(x - x_0)^2 + (y - y_0)^2}. \tag{29}$$

These expressions show that u and v are continuous functions of x and y except at the point (x_0, y_0); in particular, they are continuous in an open set E which contains C. The circle C is a curve whose equations are

$$x = x_0 + r\cos t, \qquad y = y_0 + r\sin t, \qquad 0 \leqq t \leqq 2\pi, \tag{30}$$

and the functions $x : [0, 2\pi] \to \mathbb{R}$ and $y : [0, 2\pi] \to \mathbb{R}$ which define C have continuous derivatives. Thus the integral in (27) exists by Theorem 66.2, and its value by the formula in (10) and the equations in (29) and (30) is calculated as follows:

$$\begin{aligned} \int_C \frac{dz}{z - z_0} &= \int_0^{2\pi} \frac{r\cos t - ir\sin t}{r^2}[-r\sin t + ir\cos t]\,dt \\ &= i\int_0^{2\pi} [\cos t - i\sin t][\cos t + i\sin t]\,dt \\ &= i\int_0^{2\pi} [\cos^2 t + \sin^2 t]\,dt \\ &= i\int_0^{2\pi} dt \\ &= 2\pi i. \end{aligned} \tag{31}$$

66.6 Theorem (Fundamental Theorem of the Integral Calculus). *Let $f: E \to \mathbb{C}$ be a complex-valued function which has a continuous derivative Df in the open set E in $\mathbb{C}$, and let C be a curve defined by functions $x: [a, b] \to \mathbb{R}$ and $y: [a, b] \to \mathbb{R}$ which have continuous derivatives x' and y'; assume that the trace $T(C)$ of C is in E. Then $\int_C Df(z)\,dz$ exists and*

$$\int_C Df(z)\,dz = f[x(b) + iy(b)] - f[x(a) + iy(a)]. \tag{32}$$

PROOF. Since f has a continuous derivative Df by hypothesis, Theorem 64.2 shows that $Df(z) = D_x u(x, y) + iD_x v(x, y)$. Since Df is continuous, then $D_x u$ and $D_x v$ are continuous. Also, C is defined by functions x, y which have continuous derivatives. Then the integral $\int_C Df(z)\,dz$ exists by Theorem 66.2, and

$$\begin{aligned}\int_C Df(z)\,dz = &\int_a^b \{D_x u[x(t), y(t)]x'(t) - D_x v[x(t), y(t)]y'(t)\}\,dt \\ &+ i\int_a^b \{D_x v[x(t), y(t)]x'(t) + D_x u[x(t), y(t)]y'(t)\}\,dt.\end{aligned} \tag{33}$$

Since f is analytic in E, then u and v satisfy the Cauchy–Riemann equations by Theorem 64.2 and

$$-D_x v[x(t), y(t)] = D_y u[x(t), y(t)], \quad D_x u[x(t), y(t)] = D_y v[x(t), y(t)]. \tag{34}$$

Then equation (33) can be written as follows:

$$\begin{aligned}\int_C Df(z)\,dz = &\int_a^b \{D_x u[x(t), y(t)]x'(t) + D_y u[x(t), y(t)]y'(t)\}\,dt \\ &+ i\int_a^b \{D_x v[x(t), y(t)]x'(t) + D_y v[x(t), y(t)]y'(t)\}\,dt.\end{aligned} \tag{35}$$

By the chain rule in Theorem 4.1, the integrand of each of the integrals on the right is a derivative, and the fundamental theorem of the integral calculus for real-valued functions shows that

$$\begin{aligned}\int_C Df(z)\,dz &= u[x(t), y(t)]\big|_a^b + iv[x(t), y(t)]\big|_a^b \\ &= \{u[x(b), y(b)] - u[x(a), y(a)]\} + i\{v[x(b), y(b)] \\ &\quad - v[x(a), y(a)]\} \\ &= \{u[x(b), y(b)] + iv[x(b), y(b)]\} - \{u[x(a), y(a)] \\ &\quad + iv[x(a), y(a)]\} \\ &= f[x(b) + iy(b)] - f[x(a) + iy(a)].\end{aligned} \tag{36}$$

The proof of Theorem 66.6 is complete. □

66.7 Example. Theorem 66.6 can be used to evaluate the integral $\int_C z^n\,dz$. Since

$$D\left(\frac{z^{n+1}}{n+1}\right) = z^n, \qquad n = 0, 1, \cdots,$$

by Exercise 64.4, Theorem 66.6 shows that

$$\int_C z^n\,dz = \frac{[z(b)]^{n+1}}{n+1} - \frac{[z(a)]^{n+1}}{n+1}, \qquad n = 0, 1, \cdots. \tag{37}$$

This example includes the evaluation of $\int_C z^2\,dz$ in Example 66.4 and replaces the calculation in that example by a far shorter and simpler one.

66.8 Example. Let C be a curve defined by functions $x: [a, b] \to \mathbb{R}$ and $y: [a, b] \to \mathbb{R}$ which have continuous derivatives x', y'. If C is a curve from z_a to z_b, and if c is a complex number, then

$$\int_C c\,dz = c(z_b - z_a), \tag{38}$$

$$\int_C z\,dz = (1/2)(z_b^2 - z_a^2). \tag{39}$$

The proofs of these formulas can be obtained from Theorem 66.6, but other proofs are possible. For example,

$$\sum_{k=1}^{n} c(z_k - z_{k-1}) = c\sum_{k=1}^{n} (z_k - z_{k-1}) = c(z_n - z_0) = c(z_b - z_a), \tag{40}$$

and the formula in (38) follows from (8) in Definition 66.1. For a second proof of (39), begin by proving that

$$\int_C f(z)\,dz = \lim_{n\to\infty} \sum_{k=1}^{n} f(z_k)(z_k - z_{k-1}). \tag{41}$$

By the definition in (8),

$$\int_C f(z)\,dz = \lim_{n\to\infty} \sum_{k=1}^{n} f(z_{k-1})(z_k - z_{k-1}). \tag{42}$$

Since f is continuous on E, it is uniformly continuous on C, and for each $\varepsilon > 0$ there is an $N(\varepsilon)$ such that

$$|f(z_k) - f(z_{k-1})| < \varepsilon, \qquad k = 1, \cdots, n,$$

if $n \geqq N(\varepsilon)$. Then

$$\begin{aligned}
&\left|\sum_{k=1}^{n} f(z_k)(z_k - z_{k-1}) - \sum_{k=1}^{n} f(z_{k-1})(z_k - z_{k-1})\right| \\
&\qquad \leqq \sum_{k=1}^{n} |f(z_k) - f(z_{k-1})|\,|z_k - z_{k-1}| \\
&\qquad \leqq \varepsilon \sum_{k=1}^{n} |z_k - z_{k-1}|, \qquad n \geqq N(\varepsilon), \\
&\qquad \leqq \varepsilon L, \qquad n \geqq N(\varepsilon).
\end{aligned}$$

These inequalities show that the limits in (41) and (42) are equal, and the proof of the statement in (41) is complete. Thus

$$\int_C z\,dz = \lim_{n\to\infty} \sum_{k=1}^{n} z_{k-1}(z_k - z_{k-1}) = \lim_{n\to\infty} \sum_{k=1}^{n} z_k(z_k - z_{k-1}). \tag{43}$$

From the two expressions in (43) for $\int_C z\,dz$ it follows that

$$\begin{aligned}
\int_C z\,dz &= \lim_{n\to\infty} \frac{1}{2} \sum_{k=1}^{n} (z_k + z_{k-1})(z_k - z_{k-1}) \\
&= \lim_{n\to\infty} \frac{1}{2} \sum_{k=1}^{n} (z_k^2 - z_{k-1}^2) \\
&= \lim_{n\to\infty} \frac{1}{2}(z_b^2 - z_a^2) \\
&= \frac{1}{2}(z_b^2 - z_a^2).
\end{aligned}$$

These results complete the second proof of (39). An extension of the formulas in (38) and (39) will be needed in the next section. Let C_1 and C_2 be two curves which are defined by functions x, y which have continuous derivatives x', y', and assume that the terminal point of C_1 is the initial point of C_2. Then C_1 and C_2 form a continuous curve C, but the derivatives x', y' may have a discontinuity at the point where the two curves join. If the curve C is in E, then $\int_C f(z)\,dz$ exists and

$$\int_C f(z)\,dz = \int_{C_1} f(z)\,dz + \int_{C_2} f(z)\,dz. \tag{44}$$

More generally, C may be a continuous curve which consists of curves $C_1, \cdots, C_m$, on each of which the derivatives x', y' are continuous although there may be discontinuities at the points where two curves join. Then

$$\int_C f(z)\,dz = \sum_{j=1}^{m} \int_{C_j} f(z)\,dz. \tag{45}$$

Assume that the end points of the curves $C_1, \cdots, C_m$ are the points w_0, $w_1, \cdots, w_m$ and that $w_m = w_0$. Then by the formulas in (38) and (39),

$$\int_C c\,dz = \sum_{j=1}^{m} \int_{C_j} c\,dz = \sum_{j=1}^{m} c(w_j - w_{j-1}) = c(w_m - w_0) = 0, \tag{46}$$

$$\int_C z\,dz = \sum_{j=1}^{m} \int_{C_j} z\,dz = \sum_{j=1}^{m} (1/2)(w_j^2 - w_{j-1}^2) = (1/2)(w_m^2 - w_0^2) = 0. \tag{47}$$

Theorem 66.2 establishes the existence of the integral $\int_C f(z)\,dz$ and provides a method for evaluating this integral by means of Riemann integrals of real-valued functions. The real numbers $\mathbb{R}$ form an ordered field, and the standard method of treating the Riemann integral is based on the order properties of the real numbers [see Section 35]. The complex numbers do not form an ordered field. Thus in treating the integrals $\int_C f(z)\,dz$ of complex-valued functions, there are two ways to proceed. The first is to reduce the study of these integrals to the study of integrals of real-valued functions as was done in proving Theorem 66.2, and the second is to look for entirely new methods. Other methods are available, and a second proof of Theorem 66.2, which establishes the existence of $\int_C f(z)\,dz$, will now be presented to describe one of these methods. There is interest in this new method because it can be employed to prove the existence of other types of integrals.

As before, let E be an open set in $\mathbb{C}$, and let $f: E \to \mathbb{C}$ be a continuous function. Let $x:[a, b] \to \mathbb{R}$ and $y:[a, b] \to \mathbb{R}$ be functions which have continuous derivatives and define a curve C whose trace is in E. Let $\mathscr{P}_k$, $k = 1, 2, \cdots$, be a sequence of subdivisions of $[a, b]$ whose norms approach zero. Let (t_0, t_1) denote a typical simplex (interval) in $\mathscr{P}_k$ so that $\mathscr{P}_k = \{(t_0, t_1): (t_0, t_1) \in \mathscr{P}_k\}$. Then corresponding to each (t_0, t_1) in $\mathscr{P}_k$ there are points $z(t_0)$, $z(t_1)$ on C such that

$$z(t_0) = x(t_0) + iy(t_0), \qquad z(t_1) = x(t_1) + iy(t_1), \qquad (t_0, t_1) \in \mathscr{P}_k. \tag{48}$$

Then corresponding to the subdivisions $\mathscr{P}_k$, $k = 1, 2, \cdots$, there are sums $S(f, \mathscr{P}_k)$, $k = 1, 2, \cdots$, such that

$$S(f, \mathscr{P}_k) = \sum_{(t_0, t_1) \in \mathscr{P}_k} f[z(t_0)][z(t_1) - z(t_0)], \qquad k = 1, 2, \cdots. \tag{49}$$

For each $\mathscr{P}_k$ the sum $S(f, \mathscr{P}_k)$ is a complex number, and corresponding to the sequence $\mathscr{P}_k$, $k = 1, 2, \cdots$, of subdivisions of $[a, b]$ there is a sequence

$$S(f, \mathscr{P}_k), \qquad k = 1, 2, \cdots, \tag{50}$$

of complex numbers. The proof will show that this sequence of complex numbers has a limit, and furthermore that the limit is the same for every sequence of subdivisions $\mathscr{P}_k$, $k = 1, 2, \cdots$. The proof employs the properties of Cauchy sequences; the reader will find the necessary definitions and theorems about Cauchy sequences in Section 97 in Appendix 2. There are two steps in the proof as follows:

(a) Let $\mathscr{Q}_k$, $k = 1, 2, \cdots$, be a (single specific) sequence of subdivisions of $[a, b]$ such that (i) $\mathscr{Q}_k$ is a refinement of $\mathscr{Q}_{k-1}$ for $k = 2, 3, \cdots$, and (ii) the norm of $\mathscr{Q}_k$ approaches zero as $k \to \infty$; then the sequence $S(f, \mathscr{Q}_k)$, $k = 1, 2, \cdots$, is a Cauchy sequence and therefore has a limit in $\mathbb{C}$.
(b) If $\mathscr{P}_k$, $k = 1, 2, \cdots$, is an arbitrary sequence of subdivisions of $[a, b]$ whose norms approach zero, then the sequence $S(f, \mathscr{P}_k)$, $k = 1, 2, \cdots$, has a limit and

$$\lim_{k \to \infty} S(f, \mathscr{P}_k) = \lim_{k \to \infty} S(f, \mathscr{Q}_k). \tag{51}$$

Then the integral of f on C is well defined as follows:

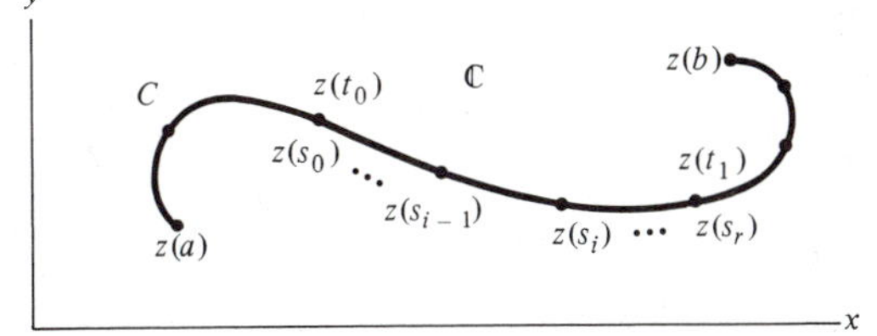

Figure 66.1. The points on C corresponding to $\mathscr{Q}_1$ and $\mathscr{Q}_2$.

$$\int_C f(z)\,dz = \lim_{k\to\infty} S(f, \mathscr{P}_k). \tag{52}$$

We are now ready to begin the proof.

The curve C has a length denoted by L. Let $\varepsilon > 0$ be given. Then there exists a number $\delta(\varepsilon)$ such that, if t_1, t_2 are any two points in $[a, b]$ for which $|t_2 - t_1| < \delta(\varepsilon)$, then

$$|f[z(t_2)] - f[z(t_1)]| < \varepsilon/L. \tag{53}$$

This statement is true because f on C is a continuous function of t on the compact set $[a, b]$; it is therefore uniformly continuous on $[a, b]$. The statement in (53) can be proved also by using the fact that the real and imaginary parts u, v of f are continuous functions of t.

66.9 Lemma. *If $\mathscr{Q}_1$ is a subdivision of $[a, b]$ whose norm is less than $\delta(\varepsilon)$, and if $\mathscr{Q}_2$ is any subdivision of $[a, b]$ which is a refinement of $\mathscr{Q}_1$, then*

$$|S(f, \mathscr{Q}_1) - S(f, \mathscr{Q}_2)| < \varepsilon. \tag{54}$$

PROOF. Let (t_0, t_1) be a typical interval in $\mathscr{Q}_1$. Then since $\mathscr{Q}_2$ is a refinement of $\mathscr{Q}_1$, there are points of subdivision $s_0, s_1, \cdots, s_r$ in $\mathscr{Q}_2$ such that

$$t_0 = s_0 < s_1 < \cdots < s_r = t_1. \tag{55}$$

Figure 66.1 shows the points on C which correspond to the points t_0, t_1 and $s_0, s_1, \cdots, s_r$ in $[a, b]$. Now compare the sums $S(f, \mathscr{Q}_1)$ and $S(f, \mathscr{Q}_2)$ on the interval (t_0, t_1) in $\mathscr{Q}_1$. The sum $S(f, \mathscr{Q}_1)$ has the single term

$$f[z(t_0)][z(t_1) - z(t_0)], \tag{56}$$

and the sum $S(f, \mathscr{Q}_2)$ has, corresponding to the points of subdivision in (55), the sum

$$\sum_{i=1}^{r} f[z(s_{i-1})][z(s_i) - z(s_{i-1})]. \tag{57}$$

Thus $|S(f, \mathscr{Q}_1) - S(f, \mathscr{Q}_2)|$ in (54) is equal to or less than the term

$$\left| f[z(t_0)][z(t_1) - z(t_0)] - \sum_{i=1}^{r} f[z(s_{i-1})][z(s_i) - z(s_{i-1})] \right| \tag{58}$$

summed over all intervals (t_0, t_1) in $\mathcal{Q}_1$. Observe that

$$\sum_{i=1}^{r} [z(s_i) - z(s_{i-1})] = z(s_r) - z(s_0) = z(t_1) - z(t_0), \tag{59}$$

$$\sum_{i=1}^{r} f[z(t_0)][z(s_i) - z(s_{i-1})] = f[z(t_0)][z(t_1) - z(t_0)]. \tag{60}$$

Replace the term $f[z(t_0)][z(t_1) - z(t_0)]$ in (58) by its value in (60); then (58) can be evaluated as follows:

$$\begin{aligned} &\left|\sum_{i=1}^{r} f[z(t_0)][z(s_i) - z(s_{i-1})] - \sum_{i=1}^{r} f[z(s_{i-1})][z(s_i) - z(s_{i-1})]\right| \\ &\qquad = \left|\sum_{i=1}^{r} \{f[z(t_0)] - f[z(s_{i-1})]\}[z(s_i) - z(s_{i-1})]\right| \\ &\qquad \leqq \sum_{i=1}^{r} |f[z(t_0)] - f[z(s_{i-1})]|\,|z(s_i) - z(s_{i-1})|. \end{aligned} \tag{61}$$

Since the norm of $\mathcal{Q}_1$ is less than $\delta(\varepsilon)$ by hypothesis, then by (53),

$$|f[z(t_0)] - f[z(s_{i-1})]| < \varepsilon/L, \qquad i = 1, \cdots, r. \tag{62}$$

Also, if $L(t_0, t_1)$ is the length of the part of the curve C from $z(t_0)$ to $z(t_1)$, then [see Figure 66.1]

$$\sum_{i=1}^{r} |z(s_i) - z(s_{i-1})| \leqq L(t_0, t_1). \tag{63}$$

Equations (61), (62), (63) show that the expression in (58) is less than $\varepsilon L(t_0, t_1)/L$; therefore,

$$|S(f, \mathcal{Q}_1) - S(f, \mathcal{Q}_2)| < \sum_{(t_0, t_1) \in \mathcal{Q}_1} \frac{\varepsilon L(t_0, t_1)}{L} = \frac{\varepsilon}{L} \sum_{(t_0, t_1) \in \mathcal{Q}_1} L(t_0, t_1) = \varepsilon. \tag{64}$$

This statement completes the proof of (54) and of Lemma 60.9. □

66.10 Lemma. *Let $\mathcal{Q}_k$, $k = 1, 2, \cdots$, be a (single specific) sequence of subdivisions of $[a, b]$ which has the following properties:* (i) *$\mathcal{Q}_k$ is a refinement of $\mathcal{Q}_{k-1}$ for $k = 2, 3, \cdots$; and* (ii) *the norms of the subdivisions $\mathcal{Q}_k$ approach zero as $k \to \infty$. Then $S(f, \mathcal{Q}_k)$, $k = 1, 2, \cdots$, is a Cauchy sequence of complex numbers, and the sequence has a limit, denoted by $S(f)$, in $\mathbb{C}$.*

PROOF. In order to prove that $S(f, \mathcal{Q}_k)$, $k = 1, 2, \cdots$, is a Cauchy sequence, it is necessary and sufficient to prove that, for each $\varepsilon > 0$ there exists a $k_0(\varepsilon)$ such that

$$|S(f, \mathcal{Q}_m) - S(f, \mathcal{Q}_n)| < \varepsilon, \qquad m > n \geqq k_0(\varepsilon). \tag{65}$$

Let $\delta(\varepsilon)$ be the number defined in the statement concerning equation (53). Choose k_0 so that the norm of $\mathcal{Q}_{k_0}$ is equal to or less than $\delta(\varepsilon)$; this choice is possible since the norms of the subdivisions $\mathcal{Q}_k$, $k = 1, 2, \cdots$, approach zero as $k \to \infty$. Let m and n be any integers such that $m > n \geqq k_0$. Then the norm of $\mathcal{Q}_n$ is equal to or less than $\delta(\varepsilon)$ since $\mathcal{Q}_n$ is either $\mathcal{Q}_{k_0}$ or a refinement of $\mathcal{Q}_{k_0}$; also $\mathcal{Q}_m$ is a refinement of $\mathcal{Q}_n$ by hypothesis since $m > n$. Then Lemma 66.9 asserts that $|S(f, \mathcal{Q}_m) - S(f, \mathcal{Q}_n)| < \varepsilon$, and the proof of (65) is complete. Thus $S(f, \mathcal{Q}_k)$, $k = 1, 2, \cdots$, is a Cauchy sequence of complex numbers. Since every Cauchy sequence of complex numbers has a limit in $\mathbb{C}$, the proof of Lemma 66.10 is complete. □

66.11 Theorem. *Let E be an open set in $\mathbb{C}$, and let $f: E \to \mathbb{C}$ be a continuous function. Let $x: [a, b] \to \mathbb{R}$ and $y: [a, b] \to \mathbb{R}$ be functions which have continuous derivatives and define a curve C whose trace is in E. If $\mathcal{P}_k$, $k = 1, 2, \cdots$, is an arbitrary sequence of subdivisions of $[a, b]$ whose norms approach zero, and if $\mathcal{Q}_k$, $k = 1, 2, \cdots$, is the specific sequence of subdivisions of $[a, b]$ described in Lemma* 66.10, *then*

$$\lim_{k\to\infty} S(f, \mathcal{P}_k) = S(f) = \lim_{k\to\infty} S(f, \mathcal{Q}_k). \tag{66}$$

Proof. Let $\varepsilon > 0$ be given, and let k_1 be an integer such that

$$|S(f, \mathcal{Q}_k) - S(f)| < \varepsilon/3, \qquad k \geqq k_1. \tag{67}$$

The integer k_1 exists since $S(f)$ is the limit of $S(f, \mathcal{Q}_k)$, $k = 1, 2, \cdots$. Let k_2 be an integer such that the norm of $\mathcal{Q}_k$ is less than $\delta(\varepsilon/3)$ for $k \geqq k_2$; the integer k_2 exists since the norms of the subdivisions $\mathcal{Q}_k$ approach zero as $k \to \infty$. Finally, let k_3 be an integer such that the norm of $\mathcal{P}_k$ is less than $\delta(\varepsilon/3)$ for $k \geqq k_3$; the integer k_3 exists since the norms of the subdivisions $\mathcal{P}_k$ approach zero as $k \to \infty$. Set $K = \max\{k_1, k_2, k_3\}$. Let $\mathcal{Q}_k\mathcal{P}_k$ denote the product subdivision of $[a, b]$; that is, $\mathcal{Q}_k\mathcal{P}_k$ is the subdivision of $[a, b]$ defined by all of the points of subdivision in both $\mathcal{Q}_k$ and $\mathcal{P}_k$. Then $\mathcal{Q}_k\mathcal{P}_k$ is a refinement of both $\mathcal{Q}_k$ and $\mathcal{P}_k$. Since $K \geqq k_2$ and $K \geqq k_3$, then

$$|S(f, \mathcal{Q}_k) - S(f, \mathcal{Q}_k\mathcal{P}_k)| < \varepsilon/3, \qquad k \geqq K, \tag{68}$$

$$|S(f, \mathcal{P}_k) - S(f, \mathcal{Q}_k\mathcal{P}_k)| < \varepsilon/3, \qquad k \geqq K, \tag{69}$$

by Lemma 66.9. Now

$$\begin{aligned} &|S(f, \mathcal{P}_k) - S(f)| \\ &\leqq |S(f, \mathcal{P}_k) - S(f, \mathcal{Q}_k\mathcal{P}_k)| + |S(f, \mathcal{Q}_k\mathcal{P}_k) - S(f, \mathcal{Q}_k)| + |S(f, \mathcal{Q}_k) - S(f)|, \end{aligned} \tag{70}$$

and, since $K \geqq k_1$, equations (67), $\cdots$, (70) show that

$$|S(f, \mathcal{P}_k) - S(f)| < \frac{\varepsilon}{3} + \frac{\varepsilon}{3} + \frac{\varepsilon}{3} = \varepsilon, \qquad k \geqq K. \tag{71}$$

Thus, for every sequence $\mathcal{P}_k$, $k = 1, 2, \cdots$, of subdivisions of $[a, b]$,

$$\lim_{k\to\infty} S(f, \mathcal{P}_k) = S(f) = \lim_{k\to\infty} S(f, \mathcal{Q}_k), \tag{72}$$

and the proof of Theorem 66.11 is complete. □

If $\mathscr{P}_k$, $k = 1, 2, \cdots$, is an arbitrary sequence of subdivisions of $[a, b]$ whose norms approach zero as $k \to \infty$, then Theorem 66.11 and Definition 66.1 show that the integral $\int_C f(z)\,dz$ exists and that

$$\int_C f(z)\,dz = \lim_{k\to\infty} S(f, \mathscr{P}_k). \tag{73}$$

Thus the integral $\int_C f(z)\,dz$ has been shown to exist by strictly complex-variable methods and without resorting to a study of the real and imaginary parts of the sums $S(f, \mathscr{P}_k)$. The methods are of interest because they can be used in the study of integrals of functions whose values are points in a vector space such as a Banach space, but the subject cannot be pursued further here.

Exercises

66.1. Let C be the curve defined by the equations $x = \cos t$, $y = \sin t$ for $0 \leqq t \leqq \pi$. Sketch this curve. Find the value of each of the following integrals:

$$\text{(i)} \int_C \sin z\,dz \qquad \text{(ii)} \int_C z\,dz \qquad \text{(iii)} \int_C e^z\,dz \qquad \text{(iv)} \int_C e^{iz}\,dz$$

66.2. Let C be the circle with center z_0 and radius r, and let f be a function of z which is analytic in an open set which contains the set $\{z : |z - z_0| \leqq r\}$.

(a) Show that the following integral exists:

$$\frac{1}{2\pi i}\int_C \frac{f(z)}{z - z_0}\,dz$$

(b) Show that $|f(z)|$ has a maximum M on C such that $|f(z)| \leqq M$ for z on C.

(c) Prove that

$$\left|\frac{1}{2\pi i}\int_C \frac{f(z)}{z - z_0}\,dz\right| \leqq M.$$

(d) Prove that the integral in the following inequality exists, and establish the inequality.

$$\left|\frac{1}{2\pi i}\int_C \frac{f(z)}{(z - z_0)^n}\,dz\right| \leqq \frac{M}{r^{n-1}}, \qquad n = 1, 2, \cdots.$$

67. A Special Case of Cauchy's Integral Theorem

The usual treatment of the theory of functions of a complex variable proceeds by the following steps: (a) prove Cauchy's integral theorem; (b) use Cauchy's integral theorem to prove Cauchy's integral formula; and (c) use Cauchy's integral formula to prove that a continuously differentiable function has derivatives of all orders and can be represented by a Taylor series. In order

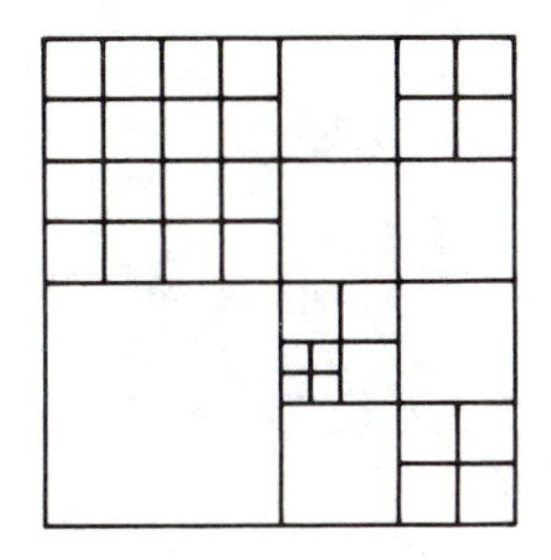

Figure 67.1. A G-subdivision of a square.

to carry out this program, the traditional treatment assumes not only that f has a derivative Df, but also that Df is continuous. The hypothesis that Df is continuous is a blemish on an otherwise beautiful theory, especially since the offending hypothesis can be avoided.

Goursat found a way to prove Cauchy's integral theorem without assuming that Df is continuous. This chapter will carry out the following program: (a) assume that f has a derivative Df in E [Df is not assumed to be continuous]; (b) use Goursat's result to prove the following special form of Cauchy's integral theorem: if f is analytic in a region bounded by two circles, then the integral of f over the boundary of the region equals zero; (c) use the result in (b) to prove Cauchy's integral formula; (d) use the result in (c) to prove that f has derivatives of all orders (the continuity of Df is thus proved rather than assumed) and that f can be represented by a Taylor series; (e) prove a general form of the fundamental theorem of the integral calculus; and (f) use the result in (e) to prove the most general form of Cauchy's integral theorem. Some definitions and preliminary results are needed before the result in (b) can be established in this section.

Let S be a square. The class of G-subdivisions of S consists of S itself, the subdivision of S into four congruent subsquares, and all of the further subdivisions of S which can be constructed as follows: subdivide S into four congruent subsquares; then subdivide one or more of these four squares into four squares, and repeat these subidvisions of subsquares a finite number of times. Then each G-subdivision of S consists of one, four, or some finite number of subsquares in S with sides parallel to the sides of S [see Figure 67.1]. The squares in a G-subdivision need not have the same size, but they are non-overlapping squares and their union is S.

Let E be an open set in $\mathbb{C}$, and let A be a compact region in E which is bounded by two circles. Let S be a square which contains A and which has its sides parallel to the axes. A G-subdivision of A is the class of non-empty subsets of A which are the intersection of A and a subsquare in a G-subdivision of S. Thus a G-subdivision of A consists of a finite number of subsets of A; they are of these two kinds: (a) a square in A bounded by four line segments; and (b) a subset of A which is contained in a subsquare of S and is bounded by arcs of one or both of the circles and by line segments. The sets in a

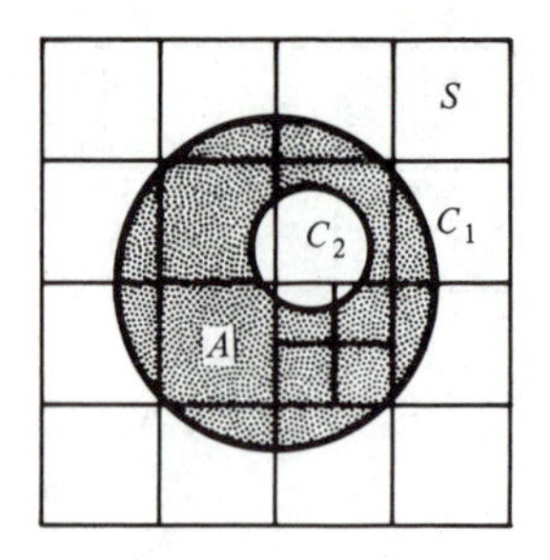

Figure 67.2. A G-subdivision of A.

G-subdivision of A are non-overlapping and their union is A [see Figure 67.2].

Let $f: E \to \mathbb{C}$ be a function which has a derivative Df on the open set E, let B be a compact set in E which is bounded by segments of straight lines and arcs of circles, and let ε be a positive number. Then, by definition, f satisfies the Goursat condition $G(\varepsilon)$ on B if and only if there exists a point z_0, in the interior of B or on the boundary ∂B of B, such that

$$|f(z) - f(z_0) - Df(z_0)(z - z_0)| \leqq \varepsilon|z - z_0|, \qquad z \in \partial B. \tag{1}$$

If B is a sufficiently small region, then f can be expected to satisfy the Goursat condition $G(\varepsilon)$ for the following reasons'. Since B is in E, then f has a derivative at each point z_0 in E and f satisfies the Stolz condition at z_0. Then by Section 65,

$$\begin{gathered} f(z) - f(z_0) - Df(z_0)(z - z_0) = r(f; z_0, z)(z - z_0), \\ r(f; z_0, z_0) = 0, \qquad \lim_{z \to z_0} r(f; z_0, z) = 0. \end{gathered} \tag{2}$$

Hence,

$$|f(z) - f(z_0) - Df(z_0)(z - z_0)| = |r(f; z_0, z)|\,|z - z_0|. \tag{3}$$

Thus if z_0 is a point in B, if $\varepsilon > 0$ is given, and if all points z on ∂B are sufficiently close to z_0, then (3) shows that f satisfies the Goursat condition $G(\varepsilon)$ on B.

67.1 Lemma (Goursat's Lemma). *Let f have a derivative Df in E, and let A be a compact region in E which is bounded by two circles. Then for each $\varepsilon > 0$ there exists a G-subdivision of A on each of whose sets f satisfies the Goursat condition $G(\varepsilon)$.*

Proof. Let S be a square which contains A and which has sides parallel to the axes [see Figure 67.2]. Assume that the lemma is false. Then there is at least one $\varepsilon > 0$ for which there is no G-subdivision of A such that f satisfies the Goursat condition $G(\varepsilon)$ on each subset of A in the G-subdivision of A; let ε_0 denote one such ε. Subdivide S into four congruent subsquares [see Figure 67.2]. Then at least one of these four subsquares fails to have a G-subdivision of its intersection with A on which f satisfies the Goursat condition $G(\varepsilon_0)$; the reason is the following: if there were a G-subdivision with the stated property on the part of A in each of the four subsquares, then the combination of these four G-subdivisions would form a G-subdivision of A on each subset of which f would satisfy the Goursat condition $G(\varepsilon_0)$. Choose one of the four subsquares of S on which there is no G-subdivision such that f satisfies the Goursat condition $G(\varepsilon_0)$ on each subset of A in the G-subdivision; let this subsquare of S be denoted by S_1. Observe that $S_1 \cap A$ is not empty. The argument can now be repeated on S_1; subdivide S_1 into four congruent subsquares. Since no subdivision of S_1 with the desired property exists, there is no subdivision of at least one of the subsquares of S_1 with the desired property; choose one of these subsquares of S_1 and call it S_2. Again, $S_2 \cap A$ is not empty. This procedure can be repeated indefinitely to form an infinite sequence S_n, $n = 1, 2, \cdots$, of closed squares such that $S \supset S_1 \supset S_2 \supset \cdots$ and $S_n \cap A \neq \varnothing$. Also, for each S_n there is no G-subdivision such that f satisfies the Goursat condition $G(\varepsilon_0)$ on each subset of A in the corresponding G-subdivision of $S_n \cap A$. For each n choose a point z_n in $S_n \cap A$. Since the diameter of S_n approaches zero as $n \to \infty$, the sequence $\{z_n : n = 1, 2, \cdots\}$ is a Cauchy sequence and it has a limit z_0 [see Section 97 in Appendix 2]. Since each z_n is a point in A and A is compact, then z_0 is a point in A. Since $S_1 \cap A \supset S_2 \cap A \supset \cdots$, all points in the sequence z_1, $z_2, \cdots$ are contained in $S_n \cap A$ beginning at least with the point z_n. Then since $S_n \cap A$ is a compact set, the point z_0 belongs to $S_n \cap A$; this statement is true for $n = 1, 2, \cdots$. Now $z_0 \in A$, $A \subset E$, and f has a derivative at each point in E by hypothesis. Then f has a derivative $Df(z_0)$ at z_0, and by (2) and (3) there exists a $\delta(\varepsilon_0)$ such that

$$|f(z) - f(z_0) - Df(z_0)(z - z_0)| < \varepsilon_0 |z - z_0|, \qquad |z - z_0| < \delta(\varepsilon_0). \tag{4}$$

Since z_0 is in the interior or on the boundary of every S_n, and since the diameter of S_n approaches zero as $n \to \infty$, there is an n_0 such that $S_{n_0} \subset \{z : |z - z_0| < \delta(\varepsilon_0)\}$. Then (4) shows that f satisfies the Goursat condition $G(\varepsilon_0)$ on $S_{n_0} \cap A$. But since $S_{n_0} \cap A$ is itself a G-subdivision of $S_{n_0} \cap A$, this statement contradicts the definition of S_{n_0}. The assumption that the lemma is false has led to a contradiction, and the proof of Lemma 67.1 is complete. □

The following theorem is a special case of Cauchy's integral theorem.

67.2 Theorem (Goursat's Theorem). *Let f have a derivative on the open set E, and let A be a compact set in E which is bounded by two circles C_1 and C_2. Then*

$$\int_{C_1} f(z)\,dz = \int_{C_2} f(z)\,dz, \tag{5}$$

the direction of integration being counterclockwise on each circle.

PROOF. Let ε be a given positive constant, and let S be a square which contains A and has its sides parallel to the axes. Then Lemma 67.1 shows that there exists a G-subdivision of A, derived from a G-subdivision of S, such that f satisfies Goursat's condition $G(\varepsilon)$ on each subset B_i in the subdivision; hence, for each B_i in the G-subdivision of A there exists a point z_i, in the interior of B_i or on its boundary ∂B_i, such that

$$|f(z) - f(z_i) - Df(z_i)(z - z_i)| \leqq \varepsilon |z - z_i|, \qquad z \in \partial B_i. \tag{6}$$

The formulas in (46) and (47) in Example 66.8 show that

$$\begin{gathered} \int_{\partial B_i} f(z_i)\,dz = 0, \qquad \int_{\partial B_i} Df(z_i) z_i\,dz = 0, \\ \int_{\partial B_i} Df(z_i) z\,dz = Df(z_i) \int_{\partial B_i} z\,dz = 0. \end{gathered} \tag{7}$$

These formulas show that

$$\begin{aligned} &\int_{\partial B_i} [f(z) - f(z_i) - Df(z_i)(z - z_i)]\,dz \\ &\qquad = \int_{\partial B_i} f(z)\,dz - \int_{\partial B_i} f(z_i)\,dz - \int_{\partial B_i} Df(z_i) z\,dz + \int_{\partial B_i} Df(z_i) z_i\,dz \qquad (8) \\ &\qquad = \int_{\partial B_i} f(z)\,dz. \end{aligned}$$

In the formulas in (7) and (8) each integral is taken in the positive (counterclockwise) direction around the boundary ∂B_i of B_i. Let C denote the complete boundary of A; that is, C denotes the curve which consists of the outer boundary circle C_1 traced in the counterclockwise direction and the inner boundary circle C_2 traced in the clockwise direction. Then

$$\int_C f(z)\,dz = \sum_{B_i} \int_{\partial B_i} f(z)\,dz. \tag{9}$$

This statement is true for the following reasons. If there is a segment of a straight line in the boundary ∂B_i of some set B_i, then in the sum on the right in (9) the integral of $f(z)$ is taken over this segment twice—once in each direction. These integrals cancel, and the only integrals which remain after all cancellations have been made are integrals over arcs of the two circles C_1 and C_2. The sum of these integrals on arcs of C_1 and C_2 is the integral on the left in (9). If the integrals on C_1 and C_2 are taken in the counterclockwise direction, then

$$\int_C f(z)\,dz = \int_{C_1} f(z)\,dz - \int_{C_2} f(z)\,dz, \tag{10}$$

and (5) can be established by proving that

$$\int_C f(z)\,dz = 0. \tag{11}$$

Thus (9) and (8) show that the proof of the theorem can be completed by proving that

$$\sum_{B_i} \int_{\partial B_i} [f(z) - f(z_i) - Df(z_i)(z - z_i)]\,dz = 0. \tag{12}$$

The sets B_i are of two types, and it will be necessary to investigate the integrals in (12) on the two types separately.

The first type of set B_i is a square bounded by four line segments; let s_i and a_i denote, respectively, the length of a side of this square B_i and its area. Now f satisfies the Goursat condition $G(\varepsilon)$ in (6) on B_i; thus the absolute value of the integrand of the integral in (12) is equal to or less than $\max\{\varepsilon|z - z_i| : z \in \partial B_i\}$, and the length of the boundary ∂B_i is $4s_i$. Thus by Theorem 66.3,

$$\left| \int_{\partial B_i} [f(z) - f(z_i) - Df(z_i)(z - z_i)]\,dz \right| \leqq 4s_i \varepsilon \max_{z \in \partial B_i} |z - z_i|. \tag{13}$$

Since z_i is a point in B_i or on its boundary, then $\max|z - z_i|$ for z in ∂B_i is equal to or less than $s_i\sqrt{2}$. Then the absolute value of the integral in (13) does not exceed $\varepsilon 4\sqrt{2}\, s_i^2$, and

$$\left| \int_{\partial B_i} [f(z) - f(z_i) - Df(z_i)(z - z_i)]\,dz \right| \leqq \varepsilon 4\sqrt{2} a_i. \tag{14}$$

The second type of set B_i is contained in a square of a G-subdivision of S and is bounded by line segments and arcs of one or both of the circles C_1 and C_2. Let c_i denote the sum of the lengths of these arcs. Then the same arguments that were used before show that, for the second type of set B_i,

$$\begin{aligned} \left| \int_{\partial B_i} [f(z) - f(z_i) - Df(z_i)(z - z_i)]\,dz \right| &\leqq \varepsilon[4s_i + c_i] \max_{z \in \partial B_i} |z - z_i| \\ &\leqq \varepsilon\sqrt{2}[4a_i + c_i s_i]. \end{aligned} \tag{15}$$

Let a denote the area of S, let c denote the sum of the lengths of the circumferences of C_1 and C_2, and let s denote the length of the side of S [the s_i need not be equal for all squares in the subdivision, but $s_i \leqq s$ for every i]. Then (14) and (15) show that

$$\left| \sum_{B_i} \int_{\partial B_i} [f(z) - f(z_i) - Df(z_i)(z - z_i)]\,dz \right| \leqq \varepsilon\sqrt{2}[4a + sc]. \tag{16}$$

Here $4a + sc$ is a constant which is independent of ε. Since (16) is true for every positive ε, equation (12) is true, and the proof is complete. □

Exercises

67.1. Let A be a region bounded by two circles C_1 and C_2 as in Figure 67.2. Assume that C_1 and C_2 have positive (counterclockwise) orientation.

(a) Show that each of the following functions is analytic in an open set E which contains A: (i) $f(z) = 2z^2 - 4z + 5$; (ii) $f(z) = \cos z - \sin z$; (iii) $f(z) = e^z$.

(b) Let C be the positively oriented boundary of A. Use Goursat's Theorem 67.2 to prove that, for each f in (a),

$$\int_{C_1} f(z)\,dz = \int_{C_2} f(z)\,dz, \qquad \int_C f(z)\,dz = 0.$$

67.2. For each of the functions in Exercise 67.1(a) find a function $F: E \to \mathbb{C}$ such that $DF(z) = f(z)$. Use Theorem 66.6 and the fact that f is analytic in E to show that

$$\int_{C_1} f(z)\,dz = 0, \qquad \int_{C_2} f(z)\,dz = 0.$$

67.3. Let C_1 be a circle in $\mathbb{C}$, and let C_2 be a circle, with center z_0, in the interior of C_1 [see Figure 67.2]. Let A be the region bounded by C_1 and C_2.

(a) If $f(z) = 1/(z - z_0)$, prove that f is analytic in an open set E which contains A.

(b) Let C_1 and C_2 have the positive orientation, and let C be the positively oriented boundary of A. Use Goursat's Theorem 67.2 to prove that

$$\int_{C_1} \frac{dz}{z - z_0} = \int_{C_2} \frac{dz}{z - z_0}, \qquad \int_C \frac{dz}{z - z_0} = 0.$$

(c) Show that $\displaystyle\int_{C_1} \frac{dz}{z - z_0} = 2\pi i$. [Hint. Prove, as in Example 66.5, that $\displaystyle\int_{C_2} \frac{dz}{z - z_0} = 2\pi i$.]

67.4. Let $f: E \to \mathbb{C}$ and $F: E \to \mathbb{C}$ be functions which have derivatives in E, and assume that $f(z) = DF(z)$ in E.

(a) Prove that $DF(z)$ is continuous in E.

(b) Let A be a compact set in E which is bounded by two circles C_1 and C_2 whose interiors are also in E. Prove that the integrals $\int_{C_1} f(z)\,dz$ and $\int_{C_2} f(z)\,dz$ exist. Then prove Goursat's Theorem 67.2 by proving that

$$\int_{C_1} f(z)\,dz = 0, \qquad \int_{C_2} f(z)\,dz = 0.$$

[Hint. Show that the integrals $\int_{C_1} DF(z)\,dz$ and $\int_{C_2} DF(z)\,dz$ exist; then use Theorem 66.6 to evaluate these integrals.]

(c) Give an example of a function for which Goursat's Theorem 67.2 cannot be proved by the method outlined in this exercise. [Hint. Exercise 67.3.]

67.5. Let A be a compact set in $\mathbb{C}$ whose boundary C consists of a curve or finite number of curves; assume that C is defined by functions which have piecewise continuous derivatives. Assume that $f: A \to \mathbb{C}$ is analytic; that is, assume that A is contained in an open set E, and that f has a derivative $Df(z)$ at each point z in E. Cauchy's integral theorem asserts that $\int_C f(z)\,dz = 0$. Exercise 67.4 suggests that one way to prove this theorem is to prove that f has a primitive $F: E \to \mathbb{C}$ such that $DF(z) = f(z)$ and then to use Theorem 66.6. Since Exercise 67.4(c) and 67.3 emphasize that this method cannot succeed in all cases, the method has only limited applicability. This exercise outlines the proof of a theorem which states sufficient conditions for the existence of a primitive F of f.

(a) Let E be an open set in $\mathbb{C}$ which is star-shaped [see Section 56] with respect to at least one point z^*. Let z_0 and z be a fixed point and an arbitrary point in E. Connect z_0 to z by a polygonal curve in E which consists of the line segments (z_0, z_1), (z_1, z_2), $\cdots$, (z_{n-1}, z_n), $z_n = z$. Show that at least one such polygonal curve exists.

(b) Show that the integrals in the sum

$$\sum_{i=1}^{n} \int_{(z_{i-1}, z_i)} f(w)\,dw$$

exist and prove that this sum has the same value on every polygonal curve which connects z_0 to z. Then this sum defines a function of z on E; call this function F. Thus

$$F(z) = \sum_{i=1}^{n} \int_{(z_{i-1}, z_i)} f(w)\,dw, \qquad z_n = z.$$

[Hints. Use Goursat's method [see Lemma 67.1 and Theorem 67.2] to prove Cauchy's integral theorem for a region A bounded by three line segments (that is, a triangle). Proceed as in Section 56, replacing Stokes' theorem by this special case of Cauchy's integral theorem.]

(c) Prove that F has a derivative, and that $DF(z) = f(z)$. [Hint. Let h be a complex number; prove that

$$DF(z) = \lim_{h \to 0} \frac{F(z+h) - F(z)}{h} = \lim_{h \to 0} \int_{(z, z+h)} f(w)\,dw = f(z).$$

Use (10) in Theorem 66.2 to represent the integral in this expression; then use the mean-value theorem to evaluate the limit.]

68. Cauchy's Integral Formula

This section uses Theorem 67.2, Goursat's form of Cauchy's integral theorem, to prove Cauchy's integral formula. As a first application of this formula, the section proves that every differentiable function has continuous derivatives of all orders and derives the classical formula for them.

68.1 Theorem (Cauchy's Integral Formula). *Let $f: E \to \mathbb{C}$ be a function which is differentiable on the open set E, and let the circle C, with center a, and its interior be contained in E. If z is any point in the interior of C, then*

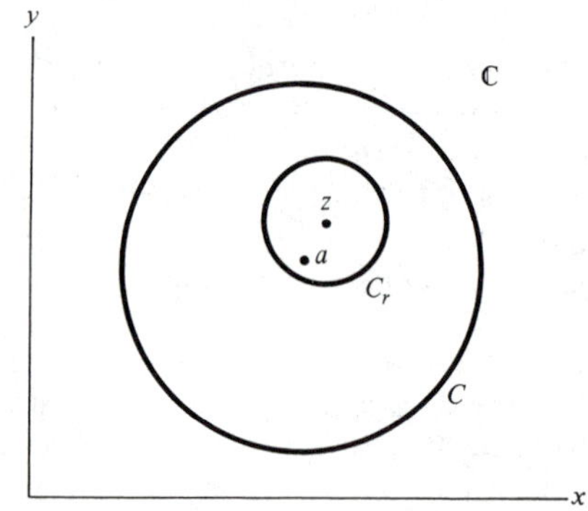

Figure 68.1. Figure for Theorem 68.1.

$$f(z) = \frac{1}{2\pi i} \int_C \frac{f(w)}{w - z}\,dw, \tag{1}$$

the direction of integration on C being positive (counterclockwise).

PROOF. Let C_r be a positively oriented circle with center z and radius r; choose r so small that C_r lies entirely inside C [see Figure 68.1]. Then for every $r > 0$ which satisfies this restriction,

$$\frac{f(w)}{w - z} \tag{2}$$

is a differentiable function of the complex variable w on an open set which contains the compact set bounded by C and C_r. Then Theorem 67.2 states that

$$\int_C \frac{f(w)}{w - z}\,dw = \int_{C_r} \frac{f(w)}{w - z}\,dw. \tag{3}$$

Since f is differentiable at z, then Section 65 shows that f satisfies Stolz's condition at z, and

$$f(w) = f(z) + Df(z)(w - z) + r(f; z, w)(w - z), \tag{4}$$

$$\int_{C_r} \frac{f(w)}{w - z}\,dw = \int_{C_r} \frac{f(z)}{w - z}\,dw + \int_{C_r} Df(z)\,dw + \int_{C_r} r(f; z, w)\,dw. \tag{5}$$

Now by Examples 66.5 and 66.8,

$$\begin{aligned} \int_{C_r} \frac{f(z)}{w - z}\,dz &= f(z) \int_{C_r} \frac{dz}{w - z} = 2\pi i f(z), \\ \int_{C_r} Df(z)\,dw &= Df(z) \int_{C_r} dw = 0. \end{aligned} \tag{6}$$

These formulas show that the equation in (5) simplifies to the following:

$$\int_{C_r} \frac{f(w)}{w - z} dw = 2\pi i f(z) + \int_{C_r} r(f; z, w)\, dw. \tag{7}$$

The value of the integral on the right in (3) does not depend on r; thus we can choose any convenient value for r in evaluating this integral in (7). Let ε be an arbitrary positive number; choose r so small that

$$|r(f; z, w)| < \varepsilon, \qquad |w - z| \leqq r. \tag{8}$$

This choice of r is possible by (3) in Section 65. Then by Theorem 66.3,

$$\left| \int_{C_r} r(f; z, w)\, dw \right| \leqq 2\pi r \varepsilon. \tag{9}$$

The value of this integral is the same for every r, and (9) shows that its value is arbitrarily small when r is small; thus

$$\int_{C_r} r(f; z, w)\, dw = 0. \tag{10}$$

Then (3), (5), (6), and (10) show that

$$\int_C \frac{f(w)}{w - z} dw = 2\pi i f(z), \tag{11}$$

and this equation is equivalent to (1). The proof of Theorem 68.1 is complete. □

In the next theorem, let $f: E \to \mathbb{C}$ be a function which is differentiable in the open set E. Let z_0 be any point in E. Then there exists a circle C with center a and radius r with these properties: (a) C and its interior are in E; and (b) z_0 is in the interior of C.

68.2 Theorem. *If f is the function just described, then f has continuous derivatives of all orders at z_0 and*

$$D^n f(z_0) = \frac{n!}{2\pi i} \int_C \frac{f(w)}{(w - z_0)^{n+1}} dw, \qquad n = 0, 1, 2, \cdots. \tag{12}$$

PROOF. Formula (12) is true for $n = 0$ since, by Theorem 68.1,

$$\begin{aligned} f(z) &= \frac{1}{2\pi i} \int_C \frac{f(w)}{w - z} dw, \qquad |z - a| < r, \\ f(z_0) &= \frac{1}{2\pi i} \int_C \frac{f(w)}{w - z_0} dw, \qquad |z_0 - a| < r. \end{aligned} \tag{13}$$

Then a simple calculation shows that

$$\begin{aligned} \frac{f(z) - f(z_0)}{z - z_0} &= \frac{1}{2\pi i} \int_C \frac{f(w)}{(w - z)(w - z_0)} dw, \\ \frac{f(z) - f(z_0)}{z - z_0} - \frac{1}{2\pi i} \int_C \frac{f(w)}{(w - z_0)^2} dw &= \frac{1}{2\pi i} \int_C \frac{z - z_0}{(w - z)(w - z_0)^2} f(w)\, dw. \end{aligned} \tag{14}$$

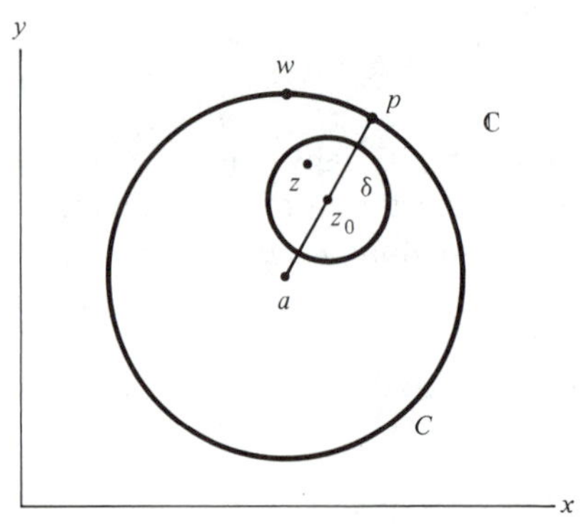

Figure 68.2. Figure for Theorem 68.2.

In Figure 68.2, the length of ap is r, the radius of the circle C. Let d denote the length of z_0p; then d is the distance from z_0 to C. Let δ be an arbitrary number such that $0 < \delta < d$, and let M be the maximum of $|f(w)|$ for w on C [see (4) in Section 66]. Then

$$|f(w)| \leqq M, \qquad |w - z_0| \geqq d, \qquad |w - z| \geqq d - \delta, \tag{15}$$

for every z and w such that $|z - z_0| < \delta$ and w is on C. Thus if $|z - z_0| < \delta$ and w is on C,

$$\left|\frac{1}{2\pi i}\frac{z - z_0}{(w - z)(w - z_0)^2}f(w)\right| \leqq \frac{1}{2\pi}\frac{M\delta}{(d - \delta)d^2}, \tag{16}$$

and (14), (16), and Theorem 66.3 show that

$$\left|\frac{f(z) - f(z_0)}{z - z_0} - \frac{1}{2\pi i}\int_C \frac{f(w)}{(w - z_0)^2}dw\right| \leqq \frac{M\delta r}{(d - \delta)d^2}, \qquad 0 < |z - z_0| < \delta. \tag{17}$$

Since this inequality is true for every δ such that $0 < \delta < d$, and since M does not depend on δ, then

$$\lim_{z \to z_0}\frac{f(z) - f(z_0)}{z - z_0} = \frac{1}{2\pi i}\int_C \frac{f(w)}{(w - z_0)^2}dw. \tag{18}$$

The derivative $Df(z_0)$ exists by hypothesis, but (18) shows that the formula in (12), in the case $n = 1$, gives the value of this derivative. The proof of (12) for $n = 2, 3, \cdots$ is completed by induction.

Assume then that

$$D^n f(z_0) = \frac{n!}{2\pi i}\int_C \frac{f(w)}{(w - z_0)^{n+1}}dw. \tag{19}$$

The proof will show that the same formula is valid if n is replaced by $n + 1$. By (19),

$$D^n f(z) = \frac{n!}{2\pi i}\int_C \frac{f(w)}{(w - z)^{n+1}}dw. \tag{20}$$

Then

$$D^n f(z) - D^n f(z_0) = \frac{n!}{2\pi i} \int_C \frac{[(w - z_0)^{n+1} - (w - z)^{n+1}]}{(w - z)^{n+1}(w - z_0)^{n+1}} f(w)\, dw. \tag{21}$$

The numerator of the integrand of the integral can be factored, and

$$\begin{aligned} [(w - z_0)^{n+1} - (w - z)^{n+1}] &= [(w - z_0) - (w - z)] \sum_{j=0}^{n} (w - z_0)^j (w - z)^{n-j} \\ &= (z - z_0) \sum_{j=0}^{n} (w - z_0)^j (w - z)^{n-j}. \end{aligned} \tag{22}$$

Thus (21) and (22) show that

$$\frac{D^n f(z) - D^n f(z_0)}{z - z_0} = \frac{n!}{2\pi i} \int_C \frac{\sum_{j=0}^{n} (w - z_0)^j (w - z)^{n-j}}{(w - z)^{n+1}(w - z_0)^{n+1}} f(w)\, dw. \tag{23}$$

We hope to show that the limit, as $z \to z_0$, of the expression on the left is the integral in (21) with n replaced by $n + 1$. Then (23) shows that

$$\begin{aligned} &\frac{D^n f(z) - D^n f(z_0)}{z - z_0} - \frac{(n + 1)!}{2\pi i} \int_C \frac{f(w)}{(w - z_0)^{n+2}} dw \\ &= \frac{(n + 1)!}{2\pi i} \int_C \left\{ \frac{\sum_{j=0}^{n} (w - z_0)^j (w - z)^{n-j}}{(n + 1)(w - z)^{n+1}(w - z_0)^{n+1}} - \frac{1}{(w - z_0)^{n+2}} \right\} f(w)\, dw. \end{aligned} \tag{24}$$

The integrand of the integral on the right in (24) is

$$\frac{\sum_{j=0}^{n} (w - z_0)^{j+1} (w - z)^{n-j} - (n + 1)(w - z)^{n+1}}{(n + 1)(w - z)^{n+1}(w - z_0)^{n+2}}. \tag{25}$$

The numerator of this fraction is a polynomial in $(w - z)$, and this polynomial vanishes if $(w - z) = (w - z_0)$. Therefore, by the factor theorem for polynomials, this numerator has the factor

$$[(w - z_0) - (w - z)] = z - z_0. \tag{26}$$

The other factor can be found by using the division algorithm to divide the numerator of (25) by the factor in (26); the quotient is

$$P(z, w) = \sum_{j=0}^{n} (n - j + 1)(w - z_0)^j (w - z)^{n-j}. \tag{27}$$

Equations (25), (26), (27) show that (24) can be written in the following form:

$$\begin{aligned} &\frac{D^n f(z) - D^n f(z_0)}{z - z_0} - \frac{(n + 1)!}{2\pi i} \int_C \frac{f(w)}{(w - z_0)^{n+2}} dw \\ &\qquad = \frac{(n + 1)!}{2\pi i} \int_C \frac{(z - z_0) P(z, w) f(w)}{(n + 1)(w - z)^{n+1}(w - z_0)^{n+2}} dw. \end{aligned} \tag{28}$$

Since $P(z, w)$ is the polynomial in (27), there is a constant N such that $|P(z, w)| \leq N$ for $|z - a| \leq r$ and w on C. Then

$$\begin{gathered} |z - z_0| < \delta, \qquad |P(z, w)| \leqq N, \qquad |f(w)| \leqq M, \\ |w - z_0| \geqq d, \qquad |w - z| \geqq d - \delta, \end{gathered} \tag{29}$$

for every z and w such that $|z - z_0| < \delta < d$ and w is on C. Thus if $|z - z_0| < \delta < d$ and w is on C, then

$$\left| \frac{(n+1)!}{2\pi i} \frac{(z - z_0)P(z, w)f(w)}{(n+1)(w - z)^{n+1}(w - z_0)^{n+2}} \right| \leqq \frac{n!}{2\pi} \frac{MN\delta}{(d - \delta)^{n+1}d^{n+2}}, \tag{30}$$

and (28), (30), and Theorem 66.3 show that

$$\left| \frac{D^n f(z) - D^n f(z_0)}{z - z_0} - \frac{(n+1)!}{2\pi i} \int_C \frac{f(w)}{(w - z_0)^{n+2}} dw \right| \leqq \frac{n!MN\delta r}{(d - \delta)^{n+1}d^{n+2}}, \tag{31}$$

for all z such that $0 < |z - z_0| < \delta < d$. Since this inequality is true for every δ such that $0 < \delta < d$, and since M and N do not depend on δ, then

$$\lim_{z \to z_0} \frac{D^n f(z) - D^n f(z_0)}{z - z_0} = \frac{(n+1)!}{2\pi i} \int_C \frac{f(w)}{(w - z_0)^{n+2}} dw. \tag{32}$$

This statement shows that $D^{n+1}f(z_0)$ exists, and that it is given by the formula in (12) with n replaced by $(n + 1)$. Thus we have shown that (12) is true for $n = 0$ and $n = 1$, and that it is true for $(n + 1)$ if it is true for n. Then by a complete induction, $D^n f(z_0)$ exists and (12) is true for $n = 0, 1, 2, \cdots$.

The proof has shown that f has an infinite number of derivatives $D^n f$ at each point z in E. Since $D^n f$ has a derivative $D^{n+1}f$, then $D^n f$ is continuous. Thus each derivative $D^n f$ is a continuous function of z at each point z in E. The proof of Theorem 68.2 is complete. □

The formula in (12) is valid for each z_0 in C; thus

$$D^n f(z) = \frac{n!}{2\pi i} \int_C \frac{f(w)\,dw}{(w - z)^{n+1}}, \qquad n = 0, 1, 2, \cdots, \qquad |z - a| < r. \tag{33}$$

This formula is easy to remember because it can be derived from Cauchy's integral formula in (1) by differentiating under the integral sign n times with respect to z.

Exercises

68.1. Recall the definition of e^z, $\sin z$, and $\cos z$ in Exercise 64.4. Verify that each of these functions has an infinite number of derivatives.

68.2. For each $z = x + iy$, the function z^n is a complex number $u_n(x, y) + iv_n(x, y)$.
- (a) Use the binomial theorem to prove that u_n and v_n are polynomials of degree n in x and y.
- (b) Prove by induction that u_n and v_n satisfy the Cauchy–Riemann equations for $n = 1, 2, \cdots$.
- (c) Prove by induction that z^n has a derivative for $n = 1, 2, \cdots$. Use this fact and Theorem 64.1 to prove that u_n and v_n satisfy the Cauchy–Riemann equations.

(d) Give three proofs that z^n has an infinite number of continuous derivatives and find these derivatives.

68.3. (a) If $f: E \to \mathbb{C}$ is differentiable in the open set E, and if $f(z) = u(x, y) + iv(x, y)$, prove that u and v have continuous derivatives of all orders in E.
(b) Prove that $u_{xx}(x, y) + u_{yy}(x, y) = 0$ and $v_{xx}(x, y) + v_{yy}(x, y) = 0$ for all $x + iy$ in E. Since u and v satisfy Laplace's equation, they are called harmonic functions.

68.4. (a) If u and v are the functions such that $z^3 = u(x, y) + iv(x, y)$, show that $u_{xx}(x, y) + u_{yy}(x, y) = 0$ and $v_{xx}(x, y) + v_{yy}(x, y) = 0$.
(b) Repeat (a) for the functions f such that $f(z) = e^z$ and $f(z) = \cos z$.

68.5. Polar coordinates can be used to represent complex numbers. If $z_1 = r_1(\cos \theta_1 + i \sin \theta_1)$ and $z_2 = r_2(\cos \theta_2 + i \sin \theta_2)$, show that $z_1 z_2 = r_1 r_2 \cos(\theta_1 + \theta_2) + i \sin(\theta_1 + \theta_2)]$.

68.6. (a) Let $f: E \to \mathbb{C}$ be a function which is differentiable in E. Show that the equation $w = f(z)$ maps E into the w-plane.
(b) If z_0 is a point in E such that $Df(z_0) \neq 0$, prove that the mapping $w = f(z)$ is one-to-one in a sufficiently small neighborhood of z_0. [Hint. Exercise 65.6 and Theorem 12.4.]
(c) If $f(z) = (z - z_0)^n$, n a positive integer greater than 1, show that $Df(z_0) = 0$. Show also that there is no neighborhood of z_0 in which the mapping $w = f(z)$ is one-to-one; describe the mapping.

68.7. (a) Let f be a function which is analytic at z_0 in E; assume that $Df(z_0) \neq 0$. Let z_1 and z_2 denote two differentiable functions of the real variable t such that $z_1(t_0) = z_0$ and $z_2(t_0) = z_0$. Show that $z = z_1(t)$ and $z = z_2(t)$ represent two curves which intersect at z_0.
(b) Show that $z_1'(t_0)$ and $z_2'(t_0)$ are the tangent vectors to the curves at z_0.
(c) Show that the mapping $w = f(z)$ maps the two curves in the z-plane into two curves $w_1(t)$ and $w_2(t)$ which intersect at $w_0 = f(z_0)$. Show that $w_i(t) = f[z_i(t)]$, $i = 1, 2$, and that the tangent vectors $w_i'(t_0)$ to the curves at w_0 satisfy the following equations:

$$w_i'(t_0) = Df[z_i(t_0)]z_i'(t_0), \qquad i = 1, 2.$$

(d) Prove that the angle between the two curves at z_0 is equal to the angle between the transformed curves at w_0. As a result of this property, the mapping is said to be *conformal* at every point z at which $Df(z) \neq 0$.

69. Taylor Series for a Differentiable Function

This section uses the results in Section 68 to establish the Taylor series representation of a differentiable function.

69.1 Theorem. *Let $f: E \to \mathbb{C}$ be a function which is differentiable in the open set E. Let C be a circle with center a and radius r such that C and its interior are contained in E. Then*

$$f(z) = \sum_{n=0}^{\infty} \frac{D^n f(a)}{n!}(z-a)^n, \qquad |z-a| < r. \tag{1}$$

PROOF. Let z be an arbitrary point, but a fixed point, on the interior of C, and let w denote a point on C. Then $|z-a| < r$ and $|w-a| = r$. Since

$$(1-t)\sum_{n=0}^{k} t^n = \sum_{n=0}^{k} (1-t)t^n = \sum_{n=0}^{k} (t^n - t^{n+1}) = 1 - t^{k+1},$$

$$1 = (1-t)\sum_{n=0}^{k} t^n + t^{k+1},$$

then

$$\frac{1}{1-t} = \sum_{n=0}^{k} t^n + \frac{t^{k+1}}{1-t}, \qquad t \neq 1. \tag{2}$$

In the identity (2), set

$$t = \frac{z-a}{w-a}. \tag{3}$$

Then $|t| < 1$, and for this value of t the identity (2) is

$$\frac{w-a}{w-z} = \sum_{n=0}^{k} \frac{(z-a)^n}{(w-a)^n} + \frac{w-a}{w-z}\frac{(z-a)^{k+1}}{(w-a)^{k+1}}, \tag{4}$$

and hence

$$\frac{1}{w-z} = \sum_{n=0}^{k} \frac{(z-a)^n}{(w-a)^{n+1}} + \frac{1}{w-z}\frac{(z-a)^{k+1}}{(w-a)^{k+1}}. \tag{5}$$

Then

$$\frac{1}{2\pi i}\int_C \frac{f(w)}{w-z}\,dw = \sum_{n=0}^{k} \frac{(z-a)^n}{2\pi i}\int_C \frac{f(w)\,dw}{(w-a)^{n+1}} + \frac{1}{2\pi i}\int_C \frac{f(w)(z-a)^{k+1}}{(w-z)(w-a)^{k+1}}\,dw. \tag{6}$$

By the results in Section 68 [see (1) and (12)], this equation is equivalent to

$$f(z) = \sum_{n=0}^{k} \frac{D^n f(a)}{n!}(z-a)^n + \frac{1}{2\pi i}\int_C \frac{f(w)}{w-z}\frac{(z-a)^{k+1}}{(w-a)^{k+1}}\,dw. \tag{7}$$

The proof of (1) will be completed by showing that

$$\lim_{k\to\infty} \frac{1}{2\pi i}\int_C \frac{f(w)}{w-z}\frac{(z-a)^{k+1}}{(w-a)^{k+1}}\,dw = 0, \tag{8}$$

for, if (8) is true, then (7) shows that

$$\lim_{k\to\infty} \sum_{n=0}^{k} \frac{D^n f(a)}{n!}(z-a)^n \tag{9}$$

exists, and that

$$f(z) = \lim_{k\to\infty} \sum_{n=0}^{k} \frac{D^n f(a)}{n!}(z-a)^n = \sum_{n=0}^{\infty} \frac{D^n f(a)}{n!}(z-a)^n. \tag{10}$$

Return to the proof of (8). Let M be the maximum of $|f(w)|$ for w on C; then since $|w - z| = |(w - a) - (z - a)| \geqq r - |z - a|$,

$$\left| \frac{1}{2\pi i} \frac{f(w)}{w - z} \frac{(z - a)^{k+1}}{(w - a)^{k+1}} \right| \leqq \frac{1}{2\pi} \frac{M}{r - |z - a|} \left(\frac{|z - a|}{r} \right)^{k+1} \tag{11}$$

for every w on C and every z such that $|z - a| < r$. Then (11) and Theorem 66.3 show that

$$\left| \frac{1}{2\pi i} \int_C \frac{f(w)}{w - z} \frac{(z - a)^{k+1}}{(w - a)^{k+1}} dw \right| \leqq \frac{Mr}{r - |z - a|} \left(\frac{|z - a|}{r} \right)^{k+1}. \tag{12}$$

Now for every z such that $|z - a| < r$,

$$\frac{|z - a|}{r} < 1, \qquad \lim_{k \to \infty} \left(\frac{|z - a|}{r} \right)^{k+1} = 0. \tag{13}$$

Then (13) and (12) show that (8) is true; therefore (9) and (10) follow from (7), and the proof of Theorem 69.1 is complete. □

The infinite series in (1) is called the *Taylor series* for f about the point $z = a$. The proof of Theorem 69.1 shows that, if f is analytic in the entire complex plane $\mathbb{C}$, then its Taylor series converges for every z. If f is analytic in the entire plane $\mathbb{C}$, then f is called an *entire function.* If f has a derivative at some points in every neighborhood of z_0 but not at z_0 itself, then z_0 is called a *singular point* of f, and f is said to have a singularity at z_0. If z_0 has a neighborhood $N(z_0, r)$ such that f has a derivative at each point of the set $\{z : z \in N(z_0, r), z \neq z_0\}$, then z_0 is called an *isolated* singular point of f. For example, if $f(z) = 1/(z - z_0)$, $z \neq z_0$, then z_0 is an isolated singular point of f and f has an isolated singularity at z_0. The Taylor series for an arbitrary function f about $z = a$ converges at least in the largest open circle which contains no singular point of f. If $f(z) = 1/(z - z_0)$, $z_0 \neq 0$, then the Taylor series for f about the point $z = 0$ converges at least for $|z| < |z_0|$.

Exercises

69.1. (a) Show that each of the following functions has an infinite number of derivatives at each point in $\mathbb{C}$: e^z, $\sin z$, and $\cos z$.

(b) Show that each of the functions in (a) can be represented by a Taylor series about each point a in $\mathbb{C}$. Show also that each of these Taylor series converges for every value of z.

(c) Represent each function in (a) by a Taylor series about the point $z = 0$; use (1) as a formula and find several terms in each series.

69.2. Find the Taylor series for $\log z$ about the point $z = 1$ [see Exercise 64.4]. Show that this series converges for $|z - 1| < 1$.

69.3. If $a_0, a_1, \cdots, a_n$ are complex numbers and $a_n \neq 0$, then $\Sigma_{i=0}^n a_i z^i$ is a polynomial of degree n.

(a) Show that a polynomial is an analytic function in $\mathbb{C}$.

(b) Let f be a polynomial in z of degree n; find the Taylor series representation for f about the point $z = a$.

(c) If f is the polynomial in (b), and if $f(a) = 0$, show that there is a polynomial g of degree $(n - 1)$ such that $f(z) = (z - a)g(z)$.

69.4. (a) Show that each of the following functions is analytic in the unit circle $|z| < 1$.

$$\text{(i) } f(z) = \frac{1}{1 - z}; \qquad \text{(ii) } f(z) = \frac{1}{z^2 - 1}; \qquad \text{(iii) } f(z) = \frac{1}{z^2 + 1}.$$

(b) Show that each of the functions in (a) has one or more isolated singular points on the unit circle $|z| = 1$; find these singular points.

(c) Find several terms of the Taylor series about $z = 0$ for each of the functions in (a), and show that each of these Taylor series converges for $|z| < 1$.

69.5. (a) Show that every polynomial $\Sigma_{i=0}^{n} a_i z^i$ is an entire function.

(b) Show that e^{z^2}, $\cos z^2$, $\sin z^2$ are entire functions of z. Find several terms of the Taylor series about $z = 0$ for each function. Verify that each of these Taylor series can be obtained by replacing z by z^2 in the corresponding series found in Exercise 69.1 (c).

(c) Let $P(z)$ and $Q(z)$ be two polynomials in z, and assume that the degree of $Q(z)$ is equal to or greater than 1. If $P(z)$ and $Q(z)$ have no zeros in common, prove that the rational function $f(z) = P(z)/Q(z)$ is not an entire function.

69.6. (a) If f is the function in Theorem 69.1, show that

$$D^n f(a) = \frac{n!}{2\pi i} \int_C \frac{f(w)}{(w - a)^{n+1}} dw.$$

(b) Let M be the maximum of $|f(z)|$ for z on C; this maximum exists since $f(z)$ is continuous on C. Prove the following inequality, known as Cauchy's inequality.

$$|D^n f(a)| \leqq \frac{n!\, M}{r^n}.$$

69.7. Prove Liouville's theorem: If f is analytic in the entire complex plane and bounded there, then f is constant. [Hint. Since r can be taken arbitrarily large, Cauchy's inequality can be used to show that $D^n f(a) = 0$ for $n = 1, 2, \cdots$. Then (1) becomes $f(z) = f(a)$ for every z in $\mathbb{C}$.]

69.8. Prove the fundamental theorem of algebra: If $P(z)$ is a polynomial in z of degree equal to or greater than 1, then the equation $P(z) = 0$ has at least one root. [Hint. Assume the theorem false. Then $1/P(z)$ is analytic in the entire complex plane. Prove that $1/P(z)$ is bounded in $\mathbb{C}$. Then, by Liouville's theorem, $1/P(z)$ is a constant.]

70. Complex-Valued Functions of Real Variables

Chapter 1 of this book treats the derivatives of real-valued functions of real variables, and Chapter 6 treats the integrals of such functions. Section 64 begins the study of analytic functions of a complex variable. From one point of view, such functions are only a special class of complex-valued functions $u + iv$ of the real variables x and y; thus the study of complex-valued functions of real variables is not really new at this point. This section, however, emphasizes some of the properties of the derivatives and integrals of complex-valued functions as functions of the real variables x and y rather than as functions of the complex variable z.

In almost all cases the real-valued functions in the former definitions and theorems can be replaced formally by complex-valued functions so that meaningful results are obtained. Some examples illustrate this statement.

70.1 Example. Let E be an open set in $\mathbb{R}^2$, and let $u: E \to \mathbb{R}$ and $v: E \to \mathbb{R}$ be functions of (x, y) in E which have derivatives as defined in Chapter 1. Then $u + iv$ is a function $h: E \to \mathbb{C}$ which is defined in E and has its values in $\mathbb{C}$. Recall from Section 76 in Appendix 1 that determinants are defined for matrices with complex elements, and that the determinants of real and of complex matrices have similar properties. Let $p_i: (x_i, y_i)$, $i = 0, 1, 2$, be three points in E such that

$$\Delta(p_0, p_1, p_2) = \begin{vmatrix} x_0 & y_0 & 1 \\ x_1 & y_1 & 1 \\ x_2 & y_2 & 1 \end{vmatrix}, \qquad |\Delta(p_0, p_1, p_2)| \geqq \rho |p_1 - p_0||p_2 - p_0|. \tag{1}$$

Then $h(p_i) = u(p_i) + iv(p_i)$, $i = 0, 1, 2$, and

$$\Delta_x h(p_0, p_1, p_2) = \begin{vmatrix} h(p_0) & y_0 & 1 \\ h(p_1) & y_1 & 1 \\ h(p_2) & y_2 & 1 \end{vmatrix} = \begin{vmatrix} u(p_0) & y_0 & 1 \\ u(p_1) & y_1 & 1 \\ u(p_2) & y_2 & 1 \end{vmatrix} + i \begin{vmatrix} v(p_0) & y_0 & 1 \\ v(p_1) & y_1 & 1 \\ v(p_2) & y_2 & 1 \end{vmatrix}; \tag{2}$$

$$\Delta_y h(p_0, p_1, p_2) = \begin{vmatrix} x_0 & h(p_0) & 1 \\ x_1 & h(p_1) & 1 \\ x_2 & h(p_2) & 1 \end{vmatrix} = \begin{vmatrix} x_0 & u(p_0) & 1 \\ x_1 & u(p_1) & 1 \\ x_2 & u(p_2) & 1 \end{vmatrix} + i \begin{vmatrix} x_0 & v(p_0) & 1 \\ x_1 & v(p_1) & 1 \\ x_2 & v(p_2) & 1 \end{vmatrix}. \tag{3}$$

Since u and v have derivatives by hypothesis, then

$$\begin{aligned} D_x h(p_0) &= \lim_{p_i \to p_0} \frac{\Delta_x h(p_0, p_1, p_2)}{\Delta(p_0, p_1, p_2)} \\ &= \lim_{p_i \to p_0} \frac{\Delta_x u(p_0, p_1, p_2)}{\Delta(p_0, p_1, p_2)} + i \lim_{p_i \to p_0} \frac{\Delta_x v(p_0, p_1, p_2)}{\Delta(p_0, p_1, p_2)} \\ &= D_x u(p_0) + i D_x v(p_0). \end{aligned} \tag{4}$$

A similar calculation shows that

$$D_y h(p_0) = D_y u(p_0) + i D_y v(p_0). \tag{5}$$

Thus the derivatives $D_x h$ and $D_y h$ of the complex-valued function h are the complex-valued functions $D_x u + iD_x v$ and $D_y u + iD_y v$. If h has a derivative at p_0, then h satisfies the following Stolz condition at p_0:

$$h(p) - h(p_0) = D_x h(p_0)(x - x_0) + D_y h(p_0)(y - y_0) + r(h; p_0, p)|p - p_0|. \quad (6)$$

This formula follows from the fact that u and v satisfy the Stolz condition, or it can be proved from first principles as in the case of real-valued functions. The properties of $r(h; p_0, p)$ in (6) can be obtained from the properties of $r(u; p_0, p)$ and $r(v; p_0, p)$.

Let $h_1 : E \to \mathbb{C}$ and $h_2 : E \to \mathbb{C}$ be two complex-valued functions which are differentiable at p_0. Set

$$\Delta_{(x,y)}(h_1, h_2)(p_0, p_1, p_2) = \begin{vmatrix} h_1(p_0) & h_2(p_0) & 1 \\ h_1(p_1) & h_2(p_1) & 1 \\ h_1(p_2) & h_2(p_2) & 1 \end{vmatrix}. \quad (7)$$

Then the Bazin–Reiss–Picquet theorem can be used [see Theorems 3.10 and 3.12] to show that

$$D_{(1,2)}(h_1, h_2)(p_0) = \lim_{p_i \to p_0} \frac{\Delta_{(x,y)}(h_1, h_2)(p_0, p_1, p_2)}{\Delta(p_0, p_1, p_2)} = \begin{vmatrix} D_x h_1(p_0) & D_y h_1(p_0) \\ D_x h_2(p_0) & D_y h_2(p_0) \end{vmatrix}. \quad (8)$$

Thus the derivative $D_{(x,y)}(h_1, h_2)(p_0)$ is the determinant of a complex matrix and is therefore a complex number; it is sometimes useful to know that the properties of determinants [see Theorems 77.3 and 77.17 in Appendix 1] can be used to write this derivative in the following form:

$$D_{(x,y)}(h_1, h_2)(p_0) = \begin{vmatrix} D_x u_1(p_0) + iD_x v_1(p_0) & D_y u_1(p_0) + iD_y v_1(p_0) \\ D_x u_2(p_0) + iD_x v_2(p_0) & D_y u_2(p_0) + iD_y v_2(p_0) \end{vmatrix} \quad (9)$$

$$= \left\{ \begin{vmatrix} D_x u_1(p_0) & D_y u_1(p_0) \\ D_x u_2(p_0) & D_y u_2(p_0) \end{vmatrix} - \begin{vmatrix} D_x v_1(p_0) & D_y v_1(p_0) \\ D_x v_2(p_0) & D_y v_2(p_0) \end{vmatrix} \right\}$$
$$+ i \left\{ \begin{vmatrix} D_x v_1(p_0) & D_y u_1(p_0) \\ D_x v_2(p_0) & D_y u_2(p_0) \end{vmatrix} + \begin{vmatrix} D_x u_1(p_0) & D_y v_1(p_0) \\ D_x u_2(p_0) & D_y v_2(p_0) \end{vmatrix} \right\}. \quad (10)$$

This brief description of the properties of the derivatives of complex-valued functions emphasizes the formal similarity of the derivatives of the two classes of functions; also, it demonstrates that, since a complex-valued function is defined by a pair of real-valued functions, the properties of derivatives of complex-valued functions can be derived from the properties of derivatives of real-valued functions.

70.2 Example. A complex-valued function $h : u + iv$ is continuous if and only if the real-valued functions u and v are continuous. Thus if $h : [a, b] \to \mathbb{C}$ is continuous, the integrals $\int_a^b u(t)\,dt$ and $\int_a^b v(t)\,dt$ exist, and $\int_a^b h(t)\,dt$ can be defined by the following equation:

$$\int_a^b h(t)\,dt = \int_a^b u(t)\,dt + i \int_a^b v(t)\,dt. \quad (11)$$

Also, the methods used in defining the integral $\int_C f(z)\,dz$ and proving Lemmas 66.9 and 66.10 and Theorem 66.11 can be used to treat the integrals of complex valued functions of real variables. The definition of $D_t h$ shows that

$$\begin{aligned}\int_a^b D_t h(t)\,dt &= \int_a^b D_t u(t)\,dt + i\int_a^b D_t v(t)\,dt \\ &= [u(b) - u(a)] + i[v(b) - v(a)] \\ &= h(b) - h(a).\end{aligned} \tag{12}$$

These equations contain a statement of the fundamental theorem of the integral calculus for the integral $\int_a^b D_t h(t)\,dt$.

The properties of determinants, oriented simplexes, and differentiable functions which were employed in Sections 38 and 39 to prove the fundamental theorem of the integral calculus for real-valued functions of several real variables are valid also for complex-valued functions of real variables, and they can be used again to prove the fundamental theorem for the new class of functions. Let $h_1 : E \to \mathbb{C}$, $h_2 : E \to \mathbb{C}$ be two functions, each of which has one continuous derivative in E. Let $\mathbf{a}$ be an oriented simplex (a_0, a_1, a_2) in E, and let $\mathscr{P}_k$, $k = 1, 2, \cdots$, be a sequence of simplicial subdivisions of $\mathbf{a}$ [see Definition 21.5]. Let $\mathbf{b}:(p_0, p_1, p_2)$ denote an oriented simplex in $\mathscr{P}_k$. Then [see Section 38]

$$\Delta_{(x,y)}(h_1, h_2)(\mathbf{b}) = \begin{vmatrix} h_1(p_0) & h_2(p_0) & 1 \\ h_1(p_1) & h_2(p_1) & 1 \\ h_1(p_2) & h_2(p_2) & 1 \end{vmatrix}, \tag{13}$$

and the proof of the fundamental theorem in this case results from two evaluations of the following limit [see the examples in Section 38]:

$$\lim_{k\to\infty} \sum_{\mathbf{b}\in\mathscr{P}_k} \frac{(-1)^2}{2!} \Delta_{(x,y)}(h_1, h_2)(\mathbf{b}). \tag{14}$$

Lemmas 38.5 and 38.6 can be repeated to show that this limit has the following two representations:

$$\int_{\mathbf{a}} D_{(x,y)}(h_1, h_2)(x, y)\,d(x, y); \tag{15}$$

$$\int_{\partial\mathbf{a}} h_1(x, y)D_x h_2(x, y)\,dx + h_1(x, y)D_y h_2(x, y)\,dy. \tag{16}$$

The fundamental theorem of the integral calculus is the statement that the integrals in (15) and (16) are equal [see Theorem 38.4].

70.3 Example. As a first numerical example of the fundamental theorem for complex-valued functions, let $h_1(x, y) = x - iy$ and $h_2(x, y) = x + iy$. Then

$$D_{(x,y)}(h_1, h_2)(x, y) = \begin{vmatrix} 1 & -i \\ 1 & i \end{vmatrix} = 2i. \tag{17}$$

Let A be the unit square with vertices (0, 0), (1, 0), (1, 1) and (0, 1). Then

$$\int_A D_{(x,y)}(h_1, h_2)(x, y)\, d(x, y) = \int_A (2i)\, d(x, y) = 2i. \tag{18}$$

$$\begin{aligned}\int_{\partial A} & h_1(x, y)D_x h_2(x, y)\, dx + h_1(x, y)D_y h_2(x, y)\, dy \\ &= \int_0^1 x\, dx + \int_0^1 (1 - iy)\, dy + \int_1^0 (x - i)\, dx + \int_1^0 (-iy)\, dy \\ &= \left(\frac{1}{2}\right) + \left(i + \frac{1}{2}\right) + \left(-\frac{1}{2} + i\right) + \left(-\frac{1}{2}\right) = 2i.\end{aligned} \tag{19}$$

Thus the integrals in (18) and (19) are equal as required by the fundamental theorem of the integral calculus. Observe that h_2 is an analytic function of z, but that h_1 is not an analytic function of z.

As a second numerical example, let

$$\begin{aligned}h_1(x, y) &= (x^2 - y^2) + i(2xy) = z^2, \\ h_2(x, y) &= x + iy = z.\end{aligned} \tag{20}$$

Then

$$\begin{aligned}D_x h_1(x, y) &= (2x) + i(2y) = D_z h_1(x, y), \\ D_y h_1(x, y) &= (-2y) + i(2x) = iD_z h_1(x, y);\end{aligned} \tag{21}$$

$$\begin{aligned}D_x h_2(x, y) &= 1 = D_z h_2(x, y), \\ D_y h_2(x, y) &= i = iD_z h_2(x, y).\end{aligned} \tag{22}$$

$$\begin{aligned}D_{(x,y)}(h_1, h_2)(x, y) &= \begin{vmatrix}(2x) + i(2y) & (-2y) + i(2x) \\ 1 & i\end{vmatrix} \\ &= \begin{vmatrix}D_z h_1(x, y) & iD_z h_1(x, y) \\ D_z h_2(x, y) & iD_z h_2(x, y)\end{vmatrix} = 0.\end{aligned} \tag{23}$$

Therefore,

$$\int_A D_{(x,y)}(h_1, h_2)(x, y)\, d(x, y) = \int_A 0\, d(x, y) = 0. \tag{24}$$

$$\begin{aligned}\int_{\partial A} & h_1(x, y)D_x h_2(x, y)\, dx + h_1(x, y)D_y h_2(x, y)\, dy \\ &= \int_0^1 x^2\, dx + \int_0^1 [(1 - y^2) + i(2y)]i\, dy \\ &\quad + \int_1^0 [(x^2 - 1) + i(2x)]\, dx + \int_1^0 (-y^2)i\, dy \\ &= \frac{1}{3} + \left(\frac{2i}{3} - 1\right) + \left(\frac{2}{3} - i\right) + \left(\frac{i}{3}\right) = 0.\end{aligned} \tag{25}$$

Thus the integrals in (24) and (25) are equal as required by the fundamental theorem of the integral calculus. Equation (23) shows that $D_{(x,y)}(h_1, h_2)(x, y) = 0$ because h_1 and h_2 are both analytic functions of z, and the fundamental theorem shows that the integral around the boundary ∂A in (25) is zero.

70.4 Example. There is also a fundamental theorem on surfaces G. Let E be an open set in $\mathbb{R}^2$, let $\mathbf{a}$ be an oriented simplex in the (t_1, t_2)-plane, and let $g: \mathbf{a} \to E$ be a function whose components (g^1, g^2) have continuous derivatives. Then g defines a plane surface G whose trace $T(G)$ lies in the (x, y)-plane, and the methods of Section 39 can be used to prove the fundamental theorem of the integral calculus in the following form:

$$\begin{aligned}\int_G & D_{(x,y)}(h_1, h_2)(x, y)\, d(x, y) \\ &= \int_{\partial G} h_1(x, y) D_x h_2(x, y)\, dx + h_1(x, y) D_y h_2(x, y)\, dy.\end{aligned} \tag{26}$$

The integral on the left in this equation is an integral on the plane surface G, and the integral on the right is a line integral on the curve ∂G which forms the boundary of G. The functions $g_1: \mathbf{a} \to \mathbb{R}$ and $g_2: \mathbf{a} \to \mathbb{R}$ are used to evaluate these integrals in the usual way.

As a first numerical example, let h_1 and h_2 be two functions which are analytic in E. Then

$$\begin{aligned} D_{(x,y)}(h_1, h_2)(x, y) &= \begin{vmatrix} D_x h_1(x, y) & D_y h_1(x, y) \\ D_x h_2(x, y) & D_y h_2(x, y) \end{vmatrix} \\ &= \begin{vmatrix} D_x h_1(x, y) & i D_x h_1(x, y) \\ D_x h_2(x, y) & i D_x h_2(x, y) \end{vmatrix} = 0, \end{aligned} \tag{27}$$

and (26) shows that

$$\int_{\partial G} h_1(x, y) D_x h_2(x, y)\, dx + h_1(x, y) D_y h_2(x, y)\, dy = 0 \tag{28}$$

for every surface G whose trace is in E.

As a second numerical example, let A be the square whose vertices are $(0, 0)$, $(1, 0)$, $(1, 1)$, $(0, 1)$ in the (t_1, t_2)-plane, and let G be the surface defined by the equations

$$x = t_1 - t_2, \qquad y = t_1 + t_2, \qquad (t_1, t_2) \in A. \tag{29}$$

Let $h_1(x, y) = x - iy$ and $h_2(x, y) = x + iy$. Then $D_{(x,y)}(h_1, h_2)(x, y) = 2i$ [see (17)]. Also, $D_{(t_1,t_2)}(x, y)(t_1, t_2) = 2$ for (t_1, t_2) in A. Then

$$\int_G D_{(x,y)}(h_1, h_2)(x, y) D_{(t_1,t_2)}(x, y)(t_1, t_2)\, d(t_1, t_2) = \int_A (2i)(2)\, d(t_1, t_2) = 4i. \tag{30}$$

$$
\begin{aligned}
\int_{\partial G} & h_1(x, y) D_x h_2(x, y)\, dx + h_1(x, y) D_y h_2(x, y)\, dy \\
&= \int_0^1 [t_1 - it_1]\, dt_1 + [t_1 - it_1] i\, dt_1 \\
&\quad + \int_0^1 [(1 - t_2) - i(1 + t_2)](-dt_2) + [(1 - t_2) - i(1 + t_2)] i\, dt_2 \\
&\quad + \int_1^0 [(t_1 - 1) - i(t_1 + 1)]\, dt_1 + [(t_1 - 1) - i(t_1 + 1)] i\, dt_1 \qquad (31) \\
&\quad + \int_1^0 [-t_2 - it_2](-dt_2) + [-t_2 - it_2] i\, dt_2 \\
&= 1 + (1 + 2i) + (-1 + 2i) + (-1) \\
&= 4i.
\end{aligned}
$$

Thus the integrals in (30) and (31) are equal as required by the fundamental theorem of the integral calculus on the surface G.

Exercises

70.1. Let A be the square in the (x, y)-plane whose vertices are $(0, 0)$, $(1, 0)$, $(1, 1)$, $(0, 1)$, and let $h_1 : \mathbb{R}^2 \to \mathbb{C}$ and $h_2 : \mathbb{R}^2 \to \mathbb{C}$ be the functions such that

$$h_1(x, y) = e^x \cos y + ie^x \sin y, \qquad h_2(x, y) = x + iy.$$

Verify the fundamental theorem of the integral calculus in this case by evaluating each of the following integrals and showing that they are equal.

(i) $\displaystyle\int_A D_{(x,y)}(h_1, h_2)(x, y)\, d(x, y)$

(ii) $\displaystyle\int_{\partial A} h_1(x, y) D_x h_2(x, y)\, dx + h_1(x, y) D_y h_2(x, y)\, dy$

70.2. Let A be the square in the (x, y)-plane whose vertices are $(1, 0)$, $(0, 1)$, $(-1, 0)$, $(0, -1)$, and let $h_1 : \mathbb{R}^2 \to \mathbb{C}$ and $h_2 : \mathbb{R}^2 \to \mathbb{C}$ be the functions such that

$$h_1(x, y) = e^x \cos y + ie^x \sin y, \qquad h_2(x, y) = (x^3 - 3xy^2) + i(3x^2 y - y^3).$$

Find the value of $\int_A D_{(x,y)}(h_1, h_2)(x, y)\, d(x, y)$, and then use this result and the fundamental theorem of the integral calculus to find the value of $\int_{\partial A} h_1(x, y) D_x h_2(x, y)\, dx + h_1(x, y) D_y h_2(x, y)\, dy$.

70.3. If $v_1(x, y) = 0$ and $v_2(x, y) = 0$ for all (x, y) in E, then $h_1(x, y) = u_1(x, y)$ and $h_2(x, y) = u_2(x, y)$. Show that, in this special case, the two forms of the fundamental theorem of the integral calculus given in this section reduce to those in Sections 38 and 39.

70.4. If $h_2(x, y) = x + iy$ for all (x, y) in E, show that the fundamental theorem of the integral calculus in (26) simplifies to

$$\int_G [iD_x h_1(x, y) - D_y h_1(x, y)]\, d(x, y) = \int_{\partial G} h_1(x, y)\, dx + ih_1(x, y)\, dy.$$

71. Cauchy's Integral Theorem

This section contains proofs of two very general forms of Cauchy's integral theorem. Each of these forms of Cauchy's theorem is essentially a corollary of a form of the fundamental theorem of the integral calculus.

Let E be a set in $\mathbb{C}$ which is open and connected but otherwise arbitrary. Let $\mathbf{c}$ be a connected chain of 2-simplexes $\mathbf{a}$ in a complex K in $\mathbb{R}^2$ which has a sequence $\mathscr{P}_k$, $k = 1, 2, \cdots$, of simplicial subdivisions with the properties stated in Definition 21.5. For example, a chain $\mathbf{c}$ with these properties can be constructed as follows. Subdivide $\mathbb{R}^2$ into unit squares by lines parallel to the axes. In each square construct the sequence of subdivisions described in Section 21 [see Figure 21.5]. From the simplexes in the first subdivision of each of the squares select a connected set of simplexes; the chain formed by these simplexes may have a very complicated structure, but it clearly has a sequence $\mathscr{P}_k$, $k = 1, 2, \cdots$, of simplicial subdivisions with the properties stated in Definition 21.5. Figure 71.1 contains an example of one such chain $\mathbf{c}$; the discussion is not restricted to this chain, and the figure is designed only to suggest the very general nature of the chains being considered. Let $g: \mathbf{c} \to E$ be a function whose components (g^1, g^2) have one continuous derivative and which defines a surface G whose trace $T(G)$ is in E. The function $g: \mathbf{c} \to E$ is not assumed to be one-to-one nor to have any special property beyond those stated; the mapping $g: \mathbf{c} \to E$ crumples $\mathbf{c}$ into E in a completely arbitrary manner. The boundary ∂G of the surface is a "curve" which is defined by functions whose derivatives are piecewise continuous.

71.1 Theorem.(Cauchy's Integral Theorem). *Let the set E in $\mathbb{C}$, the chain $\mathbf{c}$ in $\mathbb{R}^2$, and the surface G be the objects just defined. If $f_1: E \to \mathbb{C}$ and $f_2: E \to \mathbb{C}$ are analytic functions, and if $T(G)$ is in E, then*

$$\int_{\partial G} f_1(z) D_z f_2(z)\, dz = 0. \tag{1}$$

PROOF. The functions f_1 and f_2 are analytic in E. By Definition 64.3 this statement means that the functions f_1 and f_2 have derivatives $D_z f_1(z)$ and

Figure 71.1. The shaded triangles are a chain $\mathbf{c}$.

$D_z f_2(z)$ at each point in E, but these derivatives are not assumed to be continuous in E. Let A be a region in E which is bounded by two circles C_1 and C_2. Then Goursat's theorem [see Theorem 67.2] shows that

$$\int_{C_1} f_i(z)\,dz = \int_{C_2} f_i(z)\,dz, \qquad i = 1, 2.$$

This result can be used as in Section 68 to prove that each of the functions f_1 and f_2 has continuous derivatives of all orders $n = 1, 2, \cdots$ at each point z in E [see Theorem 68.2]. Then f_1 and f_2, considered as complex-valued functions of the real variables x and y, have continuous derivatives $D_x f_1$, $D_y f_1$, $D_x f_2$, and $D_y f_2$. Equation (26) in Section 70 has shown that f_1 and f_2 satisfy the fundamental theorem of the integral calculus on G as follows:

$$\begin{aligned}\int_G & D_{(x,y)}(f_1, f_2)(x, y)\,d(x, y) \\ &= \int_{\partial G} f_1(x, y) D_x f_2(x, y)\,dx + f_1(x, y) D_y f_2(x, y)\,dy.\end{aligned} \tag{2}$$

Now f_1 and f_2 are analytic in E, and (6) in Section 64 shows that

$$D_x f_i(x, y) = D_z f_i(z), \quad D_y f_i(x, y) = i D_z f_i(z), \quad (x, y) \text{ in } E, \quad i = 1, 2. \tag{3}$$

Therefore,

$$\begin{aligned} D_{(x,y)}(f_1, f_2)(x, y) &= \begin{vmatrix} D_x f_1(x, y) & D_y f_1(x, y) \\ D_x f_2(x, y) & D_y f_2(x, y) \end{vmatrix} \\ &= \begin{vmatrix} D_z f_1(z) & i D_z f_1(z) \\ D_z f_2(z) & i D_z f_2(z) \end{vmatrix} = 0, \qquad (x, y) \text{ in } E. \end{aligned} \tag{4}$$

Then $\int_G D_{(x,y)}(f_1, f_2)(x, y)\,d(x, y) = 0$ since the integrand of this integral vanishes in E, and (2), as a result of (3) and (4), can be simplified as follows:

$$\begin{gathered} \int_{\partial G} f_1(z) D_z f_2(z)\,dx + i f_1(z) D_z f_2(z)\,dy = 0, \\ \int_{\partial G} f_1(z) D_z f_2(z)[dx + i\,dy] = 0, \\ \int_{\partial G} f_1(z) D_z f_2(z)\,dz = 0. \end{gathered} \tag{5}$$

The proof of (1) and Theorem 71.1 is complete. □

71.2 Example. Let f_1 be a function which is defined as follows:

$$f_1(z) = \frac{1}{z - 3} + \frac{1}{z} + \frac{1}{z + 3}. \tag{6}$$

This function is not defined and does not have a derivative at the points $z = 3$, $z = 0$, and $z = -3$, but it is an analytic function in any region which

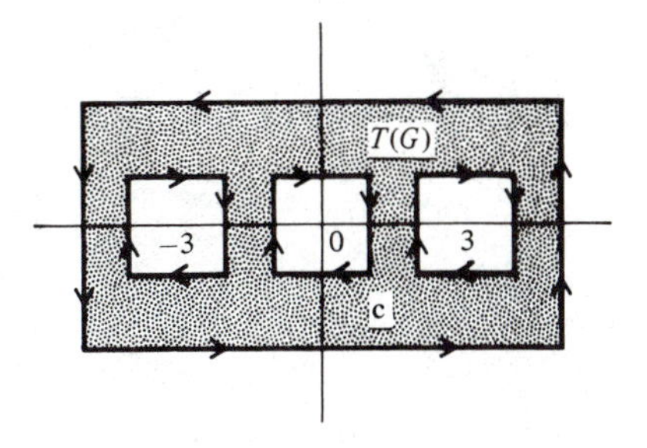

Figure 71.2. The region $T(G)$ in Example 71.2.

excludes these three points. Let $f_2(z) \equiv z$; then f_2 is analytic in the entire complex plane and $D_z f_2(z) \equiv 1$. The shaded region in Figure 71.2 can be decomposed into unit squares, each of which can be subdivided into four 2-simplexes. Thus divided, the shaded region is a chain **c** of the kind considered in Theorem 71.1. If $g: \mathbf{c} \to \mathbb{R}^2$ is the identity mapping, then the shaded region in Figure 71.2 is the trace $T(G)$ of the surface G. Since $T(G)$ does not contain the points $z = 3$, $z = 0$, $z = -3$, both of the functions f_1 and f_2 are analytic in $T(G)$. Then Theorem 71.1 asserts that $\int_{\partial G} f_1(z) D_z f_2(z)\,dz = 0$ or $\int_{\partial G} f_1(z)\,dz = 0$. The boundary ∂G consists of four separate closed curves as shown in Figure 71.2. The integral $\int_{\partial G} f_1(z)\,dz$ is taken in the positive direction around the boundary of G; the direction on each of the curves is shown by the arrows in the figure.

71.3 Lemma. *Let E be a region in $\mathbb{C}$ which is star-shaped with respect to a point z^*, and let f be a function which is analytic in E. Then there exists an analytic function $F: E \to \mathbb{C}$ such that $D_z F(z) = f(z)$ for all z in E.*

PROOF. The proof of this lemma has been outlined in Exercise 67.5 already. Let z_0 be a fixed point in E, and let z denote an arbitrary point in E. Connect z_0 to z by a polygonal curve (z_0, z_1), (z_1, z_2), $\cdots$, (z_{n-1}, z_n), $z_n = z$, in E. There is at least one such curve in E; since E is star-shaped by hypothesis, $(z_0, z^*) + (z^*, z)$ is in E and it connects z_0 to z. Since f is analytic in E, each of the integrals in the following sum exists:

$$\sum_{i=1}^{n} \int_{(z_{i-1}, z_i)} f(w)\,dw. \tag{7}$$

This sum has the same value for every polygonal curve which connects z_0 to z. To prove this statement, let (z'_0, z'_1), (z'_1, z'_2), $\cdots$, (z'_{m-1}, z'_m), $z'_0 = z_0$ and $z'_m = z$, be another such polygonal curve. The line segments (z^*, z_i), $i = 0, 1, \cdots, n$, and (z^*, z'_i), $i = 0, 1, \cdots, m$, are in E since E is star-shaped with respect to z^*. These segments and the line segments in each of the polygonal curves form triangles of the kind indicated in Figure 71.3. Each of these (solid) triangles is in E because the polygonal curves are in E and E is star-shaped with respect to z^*. Then Theorem 71.1 shows that the integral of f

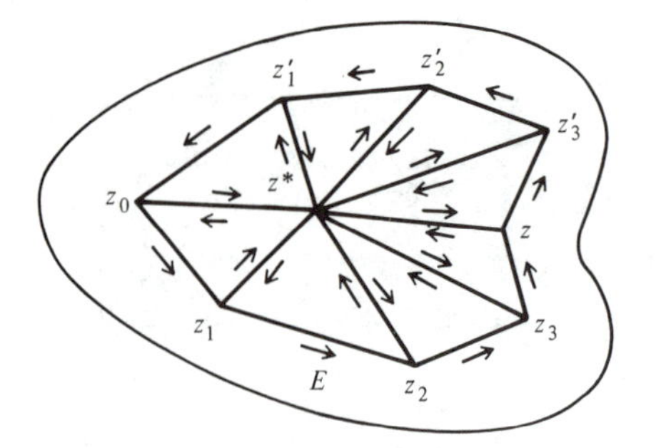

Figure 71.3. The star-shaped region E.

around each of these triangles is zero. Each of the lines which connects a point z_i or z_i' to z^* is traversed twice, once in each direction, and the sum of the two integrals is zero. Then the sum of the integrals of f around all of the triangles is zero. Thus

$$\sum_{i=1}^{n} \int_{(z_{i-1}, z_i)} f(w)\,dw + \sum_{i=1}^{m} \int_{(z_i', z_{i-1}')} f(w)\,dw = 0, \tag{8}$$

$$\sum_{i=1}^{n} \int_{(z_{i-1}, z_i)} f(w)\,dw = \sum_{i=1}^{m} \int_{(z_{i-1}', z_i')} f(w)\,dw. \tag{9}$$

These arguments show that the sum in (7) has the same value for every polygonal curve which connects z_0 to z, and this sum defines a function of z. Call this function F. Thus

$$F(z) = \sum_{i=1}^{n} \int_{(z_{i-1}, z_i)} f(w)\,dw, \qquad z_n = z. \tag{10}$$

To complete the proof of the lemma, it is necessary and sufficient to prove that $D_z F(z) = f(z)$. Let h be a small complex number. Then by definition,

$$D_z F(z) = \lim_{h \to 0} \frac{F(z+h) - F(z)}{h}. \tag{11}$$

To find the value of $F(z+h)$, use the polygonal curve with segments $(z_0, z_1), \cdots, (z_{n-1}, z), (z, z+h)$. Then

$$F(z+h) = \sum_{i=1}^{n} \int_{(z_{i-1}, z_i)} f(w)\,dw + \int_{(z, z+h)} f(w)\,dw, \tag{12}$$

and (10) and (12) show that

$$F(z+h) - F(z) = \int_{(z, z+h)} f(w)\,dw. \tag{13}$$

Since f is analytic in E, then f satisfies the Stolz condition at z as follows:

$$f(w) = f(z) + D_z f(z)(w - z) + r(f; z, w)(w - z). \tag{14}$$

Then

$$\int_{(z,z+h)} f(w)\,dw$$
$$= \int_{(z,z+h)} f(z)\,dw + \int_{(z,z+h)} D_z f(z)(w-z)\,dw + \int_{(z,z+h)} r(f;z,w)(w-z)\,dw$$
$$= f(z)h + D_z f(z)\frac{h^2}{2} + \int_{(z,z+h)} r(f;z,w)(w-z)\,dw. \tag{15}$$

Then by (11), (13), and (15),

$$D_z F(z) = f(z) + \lim_{h\to 0}\left\{D_z f(z)\frac{h}{2} + \frac{1}{h}\int_{(z,z+h)} r(f;z,w)(w-z)\,dw\right\}. \tag{16}$$

Now the limit of the first term in the braces is zero, and the proof can be completed by showing that the limit of the second term is zero. Let $\varepsilon > 0$ be given. There exists a $\delta(\varepsilon)$ such that

$$|r(f;z,w)| < \varepsilon, \qquad |w-z| < \delta(\varepsilon). \tag{17}$$

Then Theorem 66.3 shows that

$$\left|\frac{1}{h}\int_{(z,z+h)} r(f;z,w)(w-z)\,dw\right| \leqq \frac{\varepsilon|h|^2}{|h|} = \varepsilon|h|, \qquad |w-z| < \delta(\varepsilon). \tag{18}$$

Therefore,

$$\lim_{h\to 0}\frac{1}{h}\int_{(z,z+h)} r(f;z,w)(w-z)\,dw = 0, \tag{19}$$

and (16) shows that $D_z F(z) = f(z)$. The proof of Lemma 71.3 is complete. □

71.4 Theorem (Fundamental Theorem of the Integral Calculus). *Let E be an open set in $\mathbb{C}$ which is star-shaped with respect to at least one point in E, and let f be a function which is analytic in E. Let C be a curve defined by functions $x:[a,b]\to\mathbb{R}$ and $y:[a,b]\to\mathbb{R}$ which have continuous derivatives x' and y'; assume that the trace $T(C)$ of C is in E. Then f has a primitive F such that $D_z F(z) = f(z)$ for z in E, and*

$$\int_C f(z)\,dz = \int_C D_z F(z)\,dz = F[x(b)+iy(b)] - F[x(a)+iy(a)]. \tag{20}$$

If C is a closed curve, then $x(b)+iy(b) = x(a)+iy(a)$, and

$$\int_C f(z)\,dz = 0. \tag{21}$$

Proof. Since f is analytic in E and E is star-shaped with respect to at least one point, then Lemma 71.3 shows that there is an analytic function F such that

$$D_z F(z) = f(z), \qquad z \text{ in } E. \tag{22}$$

Since f and F are analytic in E, they have continuous derivatives of all orders by Theorem 68.2. Then the integrals $\int_C f(z)\,dz$ and $\int_C D_z F(z)\,dz$ exist by Section 66, and (22) shows that

$$\int_C f(z)\,dz = \int_C D_z F(z)\,dz. \tag{23}$$

Finally, Theorem 66.6 shows that

$$\int_C D_z F(z)\,dz = F[x(b) + iy(b)] - F[x(a) + iy(a)]. \tag{24}$$

Then the statements in (20) follow from (23) and (24), and (21) follows from (20) if C is a closed curve. The proof of (20) and (21) and of Theorem 71.4 is complete. □

71.5 Example. From the complex plane $\mathbb{C}$ remove the point $z = 0$ and all points on the negative x-axis; the points which remain form an open set E which is star-shaped with respect to each point on the positive x-axis. The function f such that $f(z) = 1/z$ is not defined at $z = 0$, but it is analytic in E. Then since E and f satisfy the hypotheses of Lemma 71.3, there is a function F such that $D_z F(z) = f(z)$ for all z in E. The purpose of this example is to find a simple representation for this function F.

Choose the point z_0 in Lemma 71.3 to be the point $1 + i0$ on the positive x-axis; this point is in E, and E is star-shaped with respect to this point. The choice of this point determines an additive constant in F, for if $z_0 = 1 + i0$ then $F(1 + i0) = 0$. Let $z = x + iy$. Then by Lemma 71.3,

$$F(z) = \int_{(z_0,z)} f(w)\,dw = \int_{(z_0,z)} \frac{dw}{w}. \tag{25}$$

The equations

$$\xi = 1 + t(x - 1), \qquad \eta = ty, \qquad 0 \leqq t \leqq 1, \tag{26}$$

describe the "curve" (the line segment) from z_0 to z. Equations (25) and (26) show that

$$\begin{aligned} F(z) &= \int_0^1 \frac{\xi' + i\eta'}{\xi + i\eta}\,dt = \int_0^1 \frac{[(x-1) + iy]\,dt}{[1 + t(x-1) + ity]} \\ &= \log[1 + t(x-1) + ity]\Big|_0^1 = \log(x + iy). \end{aligned} \tag{27}$$

Now (20) in Theorem 71.4 shows that the integral of f has the same value along every curve C in E from z_0 to z. Draw the circle with center at the origin which passes through z [the radius r is $(x^2 + y^2)^{1/2}$] and let z_1 denote the intersection $r + i0$ of the circle with the positive x-axis [see Figure 71.4].

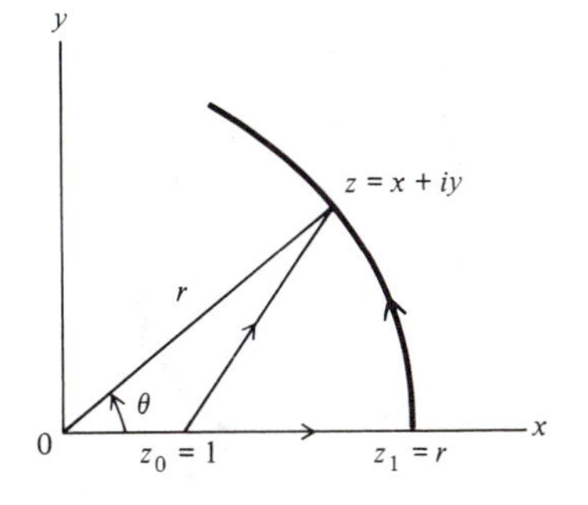

Figure 71.4. Two curves from z_0 to z.

Let arc(z_1, z) denote the curve in E from z_1 to z along the circumference of the circle, and let C be the curve which consists of the segment (z_0, z_1) from z_0 to z_1 and the arc(z_1, z) from z_1 to z. Then by Theorem 71.4,

$$F(z) = \int_{(z_0,z_1)} \frac{dw}{w} + \int_{\text{arc}(z_1,z)} \frac{dw}{w}. \tag{28}$$

Now $w = \xi$ in the first integral, and

$$\int_{(z_0,z_1)} \frac{dw}{w} = \int_1^r \frac{d\xi}{\xi} = \log \xi \Big|_1^r = \log r. \tag{29}$$

To evaluate the second integral in (28), proceed as follows [compare Example 66.5]. Let θ be the angle [see Figure 71.4] such that

$$\theta = \sin^{-1} \frac{y}{(x^2 + y^2)^{1/2}} = \cos^{-1} \frac{x}{(x^2 + y^2)^{1/2}}, \qquad -\pi < \theta < \pi. \tag{30}$$

Then the curve arc(z_1, z) is described by the equations

$$\xi = r \cos t, \qquad \eta = r \sin t, \tag{31}$$

in which t varies from $t = 0$ to $t = \theta$. Then the second integral in (28) is evaluated as follows:

$$\begin{aligned} \int_{\text{arc}(z_1,z)} \frac{dw}{w} &= \int_{\text{arc}(z_1,z)} \frac{d(\xi + i\eta)}{\xi + i\eta} = \int_{\text{arc}(z_1,z)} \frac{(\xi - i\eta)\, d(\xi + i\eta)}{\xi^2 + \eta^2} \\ &= \frac{1}{r^2} \int_0^\theta [r \cos t - ir \sin t][-r \sin t + ir \cos t]\, dt \\ &= i \int_0^\theta dt = i\theta. \end{aligned} \tag{32}$$

Substitute from (29) and (32) in (28); then

$$F(z) = \log r + i\theta, \qquad -\pi < \theta < \pi. \tag{33}$$

A comparison of (27) and (33) shows that

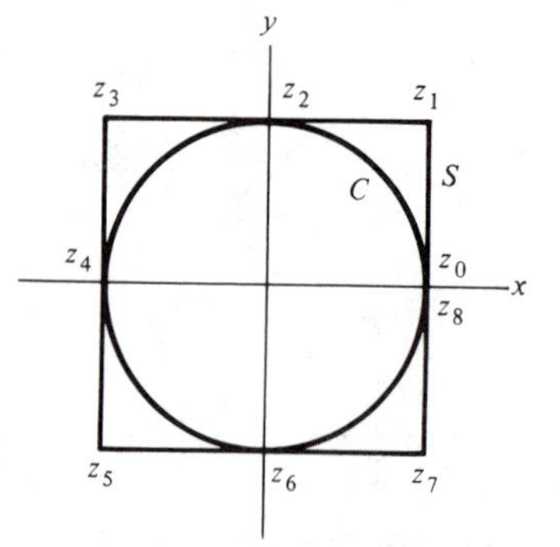

Figure 71.5. The circle C and square S.

$$\begin{aligned} \log(x+iy) &= \log(x^2+y^2)^{1/2} + i\theta, \\ \log z &= \log|z| + i\theta, \qquad -\pi < \theta < \pi, \end{aligned} \tag{34}$$

where θ is the angle in (30).

Thus one function $F: E \to \mathbb{C}$ such that $D_z F(z) = f(z) = 1/z$ in E is the function $\log z$ in (34). This function is an analytic function in the open, star-shaped region E. But there are other functions with these properties. If θ is the angle specified in (30), then for each integer n the function

$$F_n(z) = \log|z| + i(\theta + 2n\pi), \qquad n = 0, \pm 1, \pm 2, \cdots, \tag{35}$$

is analytic in E and $D_z F_n(z) = f(z) = 1/z$. Each of the functions F_n is called a branch of the function $\log z$, and the function F_0 in (35) is called the principal branch. For each of the branches of $\log z$, equation (30) shows that

$$\begin{aligned} e^{\log z} &= e^{\log|z| + i(\theta+2n\pi)} = e^{\log|z|}e^{i(\theta+2n\pi)} \\ &= |z|[\cos(\theta+2n\pi) + i\sin(\theta+2n\pi)] \\ &= r(\cos\theta + i\sin\theta) \\ &= x + iy = z. \end{aligned} \tag{36}$$

But further study of $\log z$ is beyond the scope of this book.

71.6 Example. Let C be the circle, positively oriented, with center at the origin [see Figure 71.5], and let f be the function such that $f(z) = 1/z$. Then, as shown in Example 66.5,

$$\int_C f(z)\,dz = 2\pi i. \tag{37}$$

Let S denote the curve, positively oriented, which consists of the four sides of the square which circumscribes C. Theorem 71.4 enables us to show, as follows, that

$$\int_S f(z)\,dz = \int_C f(z)\,dz = 2\pi i. \tag{38}$$

For in Figure 71.5 the segments (z_0, z_1), (z_1, z_2), and $\operatorname{arc}(z_2, z_0)$ together form a closed curve which is contained in a star-shaped region E in which $f(z)$ is analytic. Then by Theorem 71.4,

$$\int_{(z_0,z_1)} f(z)\,dz + \int_{(z_1,z_2)} f(z)\,dz + \int_{\operatorname{arc}(z_2,z_0)} f(z)\,dz = 0. \tag{39}$$

Then from (39) and three other similar equations,

$$\begin{aligned}
\left\{\int_{(z_0,z_1)} + \int_{(z_1,z_2)}\right\} f(z)\,dz &= \int_{\operatorname{arc}(z_0,z_2)} f(z)\,dz,\\
\left\{\int_{(z_2,z_3)} + \int_{(z_3,z_4)}\right\} f(z)\,dz &= \int_{\operatorname{arc}(z_2,z_4)} f(z)\,dz,\\
\left\{\int_{(z_4,z_5)} + \int_{(z_5,z_6)}\right\} f(z)\,dz &= \int_{\operatorname{arc}(z_4,z_6)} f(z)\,dz,\\
\left\{\int_{(z_6,z_7)} + \int_{(z_7,z_8)}\right\} f(z)\,dz &= \int_{\operatorname{arc}(z_6,z_8)} f(z)\,dz.
\end{aligned} \tag{40}$$

Add the four equations in (40); the sum on the left is $\int_S f(z)\,dz$, and the sum on the right is $\int_C f(z)\,dz$. Since these integrals are equal, the statement in (37) completes the proof of (38).

Let

$$f_1(z) = \frac{1}{z-3} + \frac{1}{z} + \frac{1}{z+3}, \tag{41}$$

and let S_3, S_0, S_{-3} be the curves, positively oriented, formed by the small squares around the points $z = 3$, $z = 0$, $z = -3$ in Figure 71.2. Then

$$\int_{S_3} f_1(z)\,dz = \int_{S_3} \frac{dz}{z-3} + \int_{S_3} \frac{dz}{z} + \int_{S_3} \frac{dz}{z+3}. \tag{42}$$

But each of the second and third integrals on the right equals zero by Theorem 71.4 since their integrands are analytic in an open star-shaped region which contains S_3 and its interior. This statement and Examples 66.5 and 71.6 show that

$$\int_{S_3} f_1(z)\,dz = \int_{S_3} \frac{dz}{z-3} = 2\pi i. \tag{43}$$

Similar arguments show that

$$\begin{aligned}
\int_{S_0} f_1(z)\,dz &= \int_{S_0} \frac{dz}{z} = 2\pi i,\\
\int_{S_{-3}} f_1(z)\,dz &= \int_{S_{-3}} \frac{dz}{z+3} = 2\pi i.
\end{aligned} \tag{44}$$

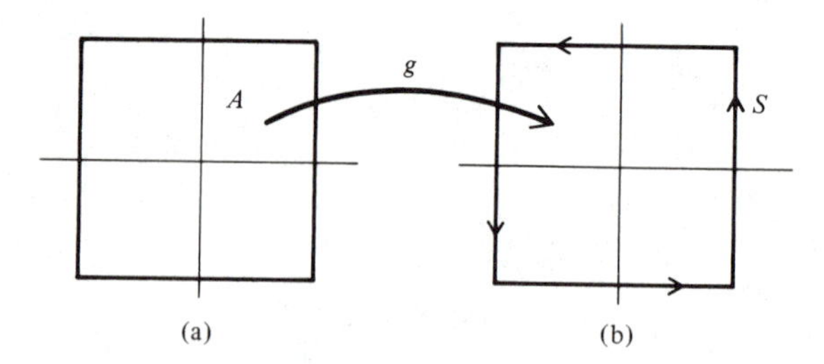

Figure 71.6. The squares A and S.

Example 71.2 shows that the integral of f_1, in the positive direction, over the total boundary of the shaded region in Figure 71.2 is zero. Thus, if R is the rectangle which forms the outer boundary of the shaded region, then (43) and (44) show that

$$\begin{aligned}\int_R f_1(z)\,dz &= \int_{S_3} \frac{dz}{z-3} + \int_{S_0} \frac{dz}{z} + \int_{S_{-3}} \frac{dz}{z+3} \\ &= 2\pi i + 2\pi i + 2\pi i = 6\pi i.\end{aligned} \tag{45}$$

71.7 Example. Let S denote the positively oriented boundary of the square in the complex plane $\mathbb{C}$ in Figure 71.6(b), and let z_0 be a point in the interior of this square. Let f be the function such that $f(z) = 1/(z - z_0)$. Methods similar to those employed in Example 71.6 can be used to show that

$$\int_S f(z)\,dz = \int_S \frac{dz}{z - z_0} = 2\pi i. \tag{46}$$

Let A be the square in $\mathbb{R}^2$ shown in Figure 71.6(a), and let g be a function whose components (g^1, g^2) have continuous derivatives, and which maps A into $\mathbb{C}$ so that each side of A is mapped into the corresponding side of S. It is not assumed that the mapping $g: A \to \mathbb{C}$ is one-to-one, nor that g maps the interior of A into the interior of S, but only that the sides of A are mapped into the corresponding sides of S. Thus g maps the positively oriented boundary of A into a curve which traces the boundary of S once in the positive direction. The mapping $g: A \to \mathbb{C}$ defines a plane surface G whose trace $T(G)$ is in $\mathbb{C}$ and whose boundary ∂G is contained in the boundary of the square in the manner described. Theorem 71.4 shows that the integral of $f(z)$ on each side of the square S is equal to the integral of $f(z)$ on the boundary of G on the same side of S. Therefore,

$$\int_{\partial G} \frac{dz}{z - z_0} = \int_S \frac{dz}{z - z_0} = 2\pi i. \tag{47}$$

This statement provides the proof for the following theorem: there exists at least one point p in A such that $g(p) = z_0$. To prove this theorem, assume that it is false. Then z_0 is not contained in $T(G)$, and since z_0 is the only point (singularity) at which f does not have a derivative, $T(G)$ is contained in an open set in which f is analytic. Therefore, Theorem 71.1 asserts that $\int_{\partial G} f(z)\,dz = 0$. But $\int_{\partial G} f(z)\,dz = 2\pi i$ by (47). Since the assumption that the theorem is false has led to a contradiction, the theorem is true as stated. Thus every point z_0 in the interior of S is contained in $T(G)$, and complex-variable theory has been used to prove an intermediate-value theorem for the mapping $g: A \to \mathbb{C}$.

Exercises

71.1. (a) Let C be the circle, positively oriented, whose center is the point $z = 1$ and whose radius is r. Find the value of each of the following integrals:

$$\int_C \frac{dz}{z-1}, \quad \int_C \frac{z^2-1}{z-1}\,dz, \quad \int_C \frac{z^2-2z+3}{z-1}\,dz, \quad \int_C \frac{e^z}{z-1}\,dz.$$

(b) Let C be the circle, positively oriented, whose center is z_0 and whose radius is r. Find the value of each of the following integrals:

$$\int_C \frac{dz}{z-z_0}, \quad \int_C \frac{z^2-1}{z-z_0}\,dz, \quad \int_C \frac{z^2-2z+3}{z-z_0}\,dz, \quad \int_C \frac{e^z}{z-z_0}\,dz.$$

(c) Let C be the circle in (b), and let f be a function which is analytic in an open set which contains C and its interior. Find the value of each of the following integrals:

$$\int_C \frac{f(z)}{z-z_0}\,dz, \quad \int_C \frac{(z-z_0)^2 f(z)}{z-z_0}\,dz, \quad \int_C \frac{f(z)\sin z}{z-z_0}\,dz, \quad \int_C \frac{f(z)e^z}{z-z_0}\,dz.$$

71.2. (a) Let C be the circle, positively oriented, whose center is the origin and whose radius is 2. Let f be the function such that

$$f(z) = \frac{2i}{z^2+1}, \qquad z^2+1 \neq 0.$$

Show that $\int_C f(z)\,dz = 0$. [Hint. Show that

$$\frac{2i}{z^2+1} = \frac{1}{z-i} - \frac{1}{z+i}.\Big]$$

(b) The sides of the square whose vertices are the points 2, $2i$, -2, *and* $-2i$ in $\mathbb{C}$ form a positively oriented curve S. If f is the function in (a), show that $\int_S f(z)\,dz = 0$.

71.3. (a) Let C be the circle, positively oriented, whose center is the point $z = 1$ and whose radius is $r = 1$. Let f be the function such that

$$f(z) = \frac{5z+1}{z^2+z-2}, \qquad z^2+z-2 \neq 0.$$

Show that $\int_C f(z)\,dz = 4\pi i$.

(b) Let f be the function in (a), and let C be the circle, positively oriented, whose center is the point $z = -1$ and whose radius is $r = 3/2$. Show that $\int_C f(z)\,dz = 6\pi i$.

(c) Let f be the function in (a), and let C be the circle, positively oriented, whose center is the point $z = 0$ and whose radius is $r = 3$. Show that $\int_C f(z)\,dz = 10\pi i$.

(d) Two circles whose centers are the origin and whose radii are 3 and 6 bound a ring region. Let C denote the positively oriented boundary of the ring region, and let f be the function in (a). Show that $\int_C f(z)\,dz = 0$.

71.4. The following examples emphasize that the conditions in Theorems 71.1 and 71.4 are sufficient but not necessary.

(a) Let E be an open connected set in $\mathbb{C}$, and let f be defined and have a derivative in E except at a finite number of points $z_1, \cdots, z_n$. Let C be a piecewise continuously differentiable closed curve whose trace does not contain any of the points $z_1, \cdots, z_n$, but is completely arbitrary otherwise. Prove that $\int_C f(z)D_z f(z)\,dz = 0$. [Hint. If C joins w_0 to w_1, show that

$$\int_C f(z)D_z f(z)\,dz = (1/2)\{[f(w_1)]^2 - [f(w_0)]^2\}.]$$

(b) Let $f(z) = 1/z$. Then

$$\int_C f(z)D_z f(z)\,dz = \int_C \frac{-1}{z^3}\,dz.$$

The functions f and $D_z f$ do not have a derivative at $z = 0$. Nevertheless, show in two ways that, if C is the unit circle with center at the origin, then $\int_C f(z)D_z f(z)\,dz = 0$.

Part II. Functions of Several Complex Variables

72. Derivatives

Part I of this chapter treats the derivatives and integrals of functions of a single complex variable, but there are, in addition, functions of several complex variables which present themselves for study. This study begins naturally with an investigation of the derivatives of these functions. Derivatives of functions of several real variables are treated in Chapter 1, and it is in order to refer to that chapter for ideas about how to proceed. A review of Section 1 shows that the treatment there appeals to geometric intuition for motivation. The graph of a function of two real variables is a surface. To find, at a point on this surface, the approximate rate of increase of the function with respect to each of its two independent variables, proceed as follows. Find the plane (graph of a linear function) through the point and two nearby points on the surface. Then the approximate rates of increase of the function are the rates of increase of the linear function, and the rates of increase (derivatives) of the function at the point are certain limits of the rates of increase of the approximating linear function. But there is little hope that this highly geometric and intuitive approach can be applied in a meaningful way to functions of several complex variables. The simplest case is a complex-valued function of two complex variables, and the graph of $w = f(z^1, z^2)$ is a set of points (z^1, z^2, w) in six-dimensional real space $\mathbb{R}^6$. Thus the intuitive geometry fails, and it is necessary to look elsewhere for motivation in developing a theory of differentiation and integration for functions of several complex variables. The solution to this problem is one which has already been employed many times in this book [compare, for example, Sections 64 and 70]: we preserve the mathematical form in a more general context than that in which it was originally developed. This principle of the *preservation of mathematical form* states, in the present case, that the definitions, the formulas, and the theorems in Chapter 1 are the pattern for developing a theory of differentiation of functions of several complex variables. The form (definitions, formulas, theorems) follows the same pattern as before, but the meaning is different (or extended) because the statements now refer to complex-valued functions of several complex variables rather than real-valued functions of several real variables.

Functions of n real variables are defined on sets in $\mathbb{R} \times \cdots \times \mathbb{R}$, denoted by $\mathbb{R}^n$, and functions of n complex variables are defined on sets in $\mathbb{C} \times \cdots \times \mathbb{C}$, denoted by $\mathbb{C}^n$. A point z in $\mathbb{C}^n$ has coordinates $(z^1, \cdots, z^n)$, and $z^1, \cdots, z^n$ are complex numbers in $\mathbb{C}$. If $z = (z^1, \cdots, z^n)$, then $z = (x^1 + iy^1, \cdots, x^n + iy^n)$, and $\mathbb{C}^n$ is embedded in $\mathbb{R}^{2n}$ in such a way that there is a one-to-one correspondence between points $(z^1, \cdots, z^n)$ in $\mathbb{C}^n$ and points $(x^1, y^1, \cdots, x^n, y^n)$ in $\mathbb{R}^{2n}$. The distance $|z_2 - z_1|$ between $z_1 : (z_1^1, \cdots, z_1^n)$ and $z_2 : (z_2^1, \cdots, z_2^n)$ is defined as follows:

$$|z_2 - z_1| = \left[\sum_{j=1}^{n} |z_2^j - z_1^j|^2 \right]^{1/2} = \left[\sum_{j=1}^{n} (x_2^j - x_1^j)^2 + (y_2^j - y_1^j)^2 \right]^{1/2}. \quad (1)$$

Thus the distance between z_1 and z_2 in $\mathbb{C}^n$ is the same as the Euclidean distance between the corresponding points in $\mathbb{R}^{2n}$, and distance in $\mathbb{C}^n$ is a norm.

72.1 Example. If $f(z)$ is a polynomial in $z^1, \cdots, z^n$ with complex coefficients, then f is a complex-valued function $f: \mathbb{C}^n \to \mathbb{C}$ of the complex variables $z^1, \cdots, z^n$. The following formulas define examples of functions $f: \mathbb{C}^n \to \mathbb{C}$:

$$f(z) = \sum_{k=0}^{m} a_k(z^1 + \cdots + z^n)^k, \qquad a_k \in \mathbb{C}, \qquad k = 0, 1, \cdots, m;$$

$$f(z) = \sin(z^1 + \cdots + z^n);$$

$$f(z) = e^{(z^1 + \cdots + z^n)}.$$

Let E be an open set in $\mathbb{C}^n$, and let $f: E \to \mathbb{C}^m$ be a function whose components $(f^1, \cdots, f^m)$ are functions $f^j: E \to \mathbb{C}$, $j = 1, \cdots, m$; then f is a function of n complex variables which maps the set E in $\mathbb{C}^n$ into the complex space $\mathbb{C}^m$.

The points $z_0, z_1, \cdots, z_n$ in $\mathbb{C}^n$ are the vertices of an oriented simplex $(z_0, z_1, \cdots, z_n)$ in $\mathbb{C}^n$; these points also determine an n-vector at z_0. An

n-vector $\mathbf{z}$ at a point z_0 in $\mathbb{C}^n$ is an ordered set $(z_1 - z_0, \cdots, z_n - z_0)$ of vectors $z_i - z_0$ whose terminal points are $z_1, \cdots, z_n$ and whose initial points are z_0. This n-vector $\mathbf{z}$ is conveniently denoted by the matrix

$$[z_i^j], \qquad i = 0, 1, \cdots, n, \qquad j = 1, \cdots, n. \tag{2}$$

The increment of z which corresponds to the n-vector $\mathbf{z}$, denoted by $\Delta(\mathbf{z})$, is the determinant of the matrix obtained by bordering (2) on the right with a column of 1's. Thus $\Delta(\mathbf{z})$ is a complex number. The following definition is analogous to Definition 2.4.

72.2 Definition. Let ρ be a number (constant) such that $0 < \rho \leqq 1$. Set

$$Z(z_0, \rho) = \left\{\mathbf{z} : |\Delta(\mathbf{z})| \geqq \rho \prod_{i=1}^{n} |z_i - z_0| > 0\right\}. \tag{3}$$

Then $Z(z_0, \rho)$ is called the *ρ-class of n-vectors at z_0*. The inequality in the definition of $Z(z_0, \rho)$ is called the *regularity condition* satisfied by $\mathbf{z}$ at z_0, and ρ is the *constant of regularity*.

Let f be a function which is defined on an open set E in $\mathbb{C}^n$ and has its values in $\mathbb{C}$. The definition of the derivative of f, in keeping with the principle of the preservation of mathematical form, is formally the same as that of real-valued functions of real-variables. Let z_0 be a point in E, and let $Z(z_0, \rho)$ be the ρ-class of n-vectors at z_0. For each $\mathbf{z}$ in $Z(z_0, \rho)$ and in E, replace the element in the i-th row and j-th column of $\Delta(\mathbf{z})$, by $f(z_i)$ for $i = 0, 1, \cdots, n$, and denote the resulting determinant by $\Delta_j f(\mathbf{z})$. Thus

$$\Delta_j f(\mathbf{z}) = \begin{vmatrix} z_0^1 & \cdots & f(z_0) & \cdots & z_0^n & 1 \\ z_1^1 & \cdots & f(z_1) & \cdots & z_1^n & 1 \\ \cdots & \cdots & \cdots & \cdots & \cdots & \cdots \\ z_n^1 & \cdots & f(z_n) & \cdots & z_n^n & 1 \end{vmatrix}, \tag{4}$$

and the elements $f(z_0), f(z_1), \cdots, f(z_n)$ are in the j-th column of the matrix.

72.3 Definition. If

$$\lim_{\mathbf{z} \to z_0} \frac{\Delta_j f(\mathbf{z})}{\Delta(\mathbf{z})}, \qquad \mathbf{z} \in Z(z_0, \rho), \qquad \mathbf{z} \text{ is in } E, \tag{5}$$

exists, then f has a j-th *derivative* at z_0; the value of this derivative is denoted by $D_j f(z_0)$ and defined to be the value of the limit in (5). Thus

$$D_j f(z_0) = \lim_{\mathbf{z} \to z_0} \frac{\Delta_j f(\mathbf{z})}{\Delta(\mathbf{z})}, \qquad \mathbf{z} \in Z(z_0, \rho), \qquad \mathbf{z} \text{ is in } E, \qquad j = 1, \cdots, n. \tag{6}$$

If the limit in (5) does not exist, then f does not have a j-th derivative. If the derivatives $D_j f(z_0)$, $j = 1, \cdots, n$, exist at z_0, then f is said to be *differentiable at z_0*; if these derivatives exist at each point in E, then f is *differentiable in E*.

72.4 Example. Let $a_1, \cdots, a_{n+1}$ be complex constants, and let

$$f(z) = a_1 z^1 + \cdots + a_n z^n + a_{n+1}, \qquad z = (z^1, \cdots, z^n). \tag{7}$$

Then elementary properties of determinants show that

$$\Delta_j f(\mathbf{z}) = a_j \Delta(\mathbf{z}), \qquad D_j f(z_0) = \lim_{\mathbf{z}\to z_0} \frac{\Delta_j f(\mathbf{z})}{\Delta(\mathbf{z})} = a_j, \qquad j = 1, \cdots, n. \tag{8}$$

Thus f has derivatives $D_1 f(z_0), \cdots, D_n f(z_0)$ at each point z_0 in $\mathbb{C}^n$, and f is differentiable in $\mathbb{C}^n$.

Just as in the case of real-valued functions, there are other kinds of derivatives. Let $f^1, \cdots, f^m$ be functions which are defined on an open set E in $\mathbb{C}^n$, $m \leqq n$, and have their values in $\mathbb{C}$. Let $(j_1, \cdots, j_m)$ be an index set in (m/n), let z_0 be a point in E, and let $\mathbf{z}$ be an n-vector in $Z(z_0, \rho)$ and in E. Replace columns $j_1, \cdots, j_m$ in $\Delta(\mathbf{z})$ in order by the columns

$$\begin{bmatrix} f^1(z_0) \\ f^1(z_1) \\ \cdots\cdots \\ f^1(z_n) \end{bmatrix}, \quad \cdots \quad \begin{bmatrix} f^m(z_0) \\ f^m(z_1) \\ \cdots\cdots \\ f^m(z_n) \end{bmatrix}, \tag{9}$$

and denote the resulting determinant by $\Delta_{(j_1, \cdots, j_m)}(f^1, \cdots, f^m)(\mathbf{z})$. If

$$\lim_{\mathbf{z}\to z_0} \frac{\Delta_{(j_1, \cdots, j_m)}(f^1, \cdots, f^m)(\mathbf{z})}{\Delta(\mathbf{z})}, \qquad \mathbf{z} \in Z(z_0, \rho), \qquad \mathbf{z} \text{ is in } E, \tag{10}$$

exists, then $(f^1, \cdots, f^m): E \to \mathbb{C}^m$ is said to have a derivative, denoted by $D_{(j_1, \cdots, j_m)}(f^1, \cdots, f^m)(z_0)$, at z_0; the value of this derivative is the value of the limit in (10). The function does not have a derivative at z_0 if the limit in (10) does not exist.

72.5 Theorem. *If the functions $f^j: E \to \mathbb{C}, j = 1, \cdots, m$, are differentiable at z_0 in E, then the derivative $D_{(j_1, \cdots, j_m)}(f^1, \cdots, f^m)(z_0)$ exists for each $(j_1, \cdots, j_m)$ in (m/n), and*

$$D_{(j_1, \cdots, j_m)}(f^1, \cdots, f^m)(z_0) = \begin{vmatrix} D_{j_1} f^1(z_0) & \cdots & D_{j_m} f^1(z_0) \\ \cdots & \cdots & \cdots \\ D_{j_1} f^m(z_0) & \cdots & D_{j_m} f^m(z_0) \end{vmatrix}. \tag{11}$$

This theorem can be proved most easily with the help of the Bazin–Reiss–Picquet theorem [see Section 83 in Appendix 1]; the proof is similar to the proof of Theorem 3.12 and Corollary 3.13.

72.6 Definition. The function $f: E \to \mathbb{C}$, $E \subset \mathbb{C}^n$, satisfies the *Stolz condition* at z_0 in E if and only if there exist (complex) constants $A_1, \cdots, A_n$ and a function of z, denoted by $r(f; z_0, z)$ and defined in a neighborhood of z_0 in E, such that

$$f(z) - f(z_0) = \sum_{j=1}^{n} A_j(z^j - z_0^j) + r(f; z_0, z)|z - z_0|, \tag{12}$$

$$\lim_{z \to z_0} r(f; z_0, z) = 0, \qquad r(f; z_0, z) = 0. \tag{13}$$

The Stolz condition is a necessary and sufficient condition for differentiability as stated in the next two theorems.

72.7 Theorem. *Let $f: E \to \mathbb{C}$, $E \subset \mathbb{C}^n$, be a function which satisfies the Stolz condition* (12), (13) *at z_0. Then the derivatives $D_j f(z_0)$ exist with respect to each class $Z(z_0, \rho)$, $0 < \rho \leqq 1$, and $D_j f(z_0) = A_j$, $j = 1, \cdots, n$.*

The proof is similar to the proof of Theorem 3.3.

72.8 Theorem. *Let $f: E \to \mathbb{C}$, $E \subset \mathbb{C}^n$, be differentiable at z_0 with respect to at least one class $Z(z_0, \rho)$, $0 < \rho \leqq 1$. Then f satisfies the Stolz condition at z_0 with $A_j = D_j f(z_0)$, $j = 1, \cdots, n$.*

The proof of this theorem is similar to the proof of Theorem 3.6. In both cases the proof employs Sylvester's interchange theorem, which is known in the literature also as Sylvester's theorem of 1839 and 1851 [see Theorem 3.5 and Section 81 in Appendix 1].

72.9 Corollary. *If $f: E \to \mathbb{C}$, $E \subset \mathbb{C}^n$, is differentiable at z_0 in E, then*

$$|f(z) - f(z_0)| \leqq \left\{ \left[\sum_{j=1}^{n} |D_j f(z_0)|^2 \right]^{1/2} + |r(f; z_0, z)| \right\} |z - z_0|, \tag{14}$$

and f is continuous at z_0.

PROOF. Since f is differentiable at z_0 by hypothesis, it satisfies the Stolz condition at z_0 by Theorem 72.8. Then (14) follows from (12), the triangle inequality for the absolute value of complex numbers, and Schwarz's inequality in Corollary 86.2. The inequality in (14) shows that f is continuous at z_0. The proof of Corollary 72.9 is similar to the proof of Corollary 3.9. □

In order to complete the theory of differentiation of functions of several complex variables, many additional results need to be established. For example, it is necessary to establish formulas for the derivatives of the sum, product, and quotient of two functions of the type $f: E \to \mathbb{C}$, $E \subset \mathbb{C}^n$. Now both the real numbers and the complex numbers form a field with respect to addition and multiplication; hence, real numbers and complex numbers have the same formal algebraic properties with respect to addition and multiplication. The derivatives of both real-valued and complex-valued functions have been defined by means of ratios of determinants. Determinants are defined for both real and complex matrices, and the properties of these determinants depend only on the properties of addition and multiplication in the fields $\mathbb{R}$ and $\mathbb{C}$. These observations suggest that any property of the derivatives of real-valued functions $f: E \to \mathbb{R}$, $E \subset \mathbb{R}^n$, which depends only on the field properties of $\mathbb{R}$ can be proved also for the derivatives of complex-valued functions $f: E \to \mathbb{C}$, $E \subset \mathbb{C}^n$. For example, the formulas for the derivatives of sums, products, and quotients are the same for real-valued and complex-valued functions; in the complex-valued case, the proof of these formulas is similar to the proof of Theorem 3.21 and is left to the reader. The real numbers are an ordered field, but the complex numbers $\mathbb{C}$ are not an ordered field. Furthermore, the topological properties of $\mathbb{R}$ and $\mathbb{C}$ are quite different. Properties of real-valued functions which depend on the order in $\mathbb{R}$ or on the topology of $\mathbb{R}$ may or may not be true for complex-valued functions $f: E \to \mathbb{C}$, $E \subset \mathbb{C}^n$; or, if true, they usually require quite different proofs. With these remarks, the remainder of the theory of differentiation of functions of several complex variables is left to the reader.

Exercises

72.1. Find the derivatives of the functions defined by each of the following formulas:

(i) $f(z^1, \cdots, z^n) = (z^1)^2 + \cdots + (z^n)^2$.
(ii) $f(z^1, z^2) = z^1 \sin z^2$.
(iii) $f(z^1, z^2) = \dfrac{1}{(z^1)^2 + (z^2)^2}, \qquad (z^1)^2 + (z^2)^2 \neq 0$.

72.2. (a) State and prove the chain rule for functions of several complex variables.
(b) Find the derivatives of the functions defined by each of the following formulas:

(i) $f(z^1, \cdots, z^n) = (z^1 + \cdots + z^n)^{10}$.
(ii) $f(z^1, \cdots, z^n) = [(z^1)^2 + \cdots + (z^n)^2]^m$.
(iii) $f(z^1, \cdots, z^n) = e^{(z^1 + \cdots + z^n)}$.
(iv) $f(z^1, z^2) = (z^1 + z^2)^2 \sin (z^1 + z^2)^2$.

72.3. Let ρ be a number such that $0 < \rho \leqq 1$, and let z_0 be a point in $\mathbb{C}^2$. Show that the class $Z(z_0, \rho)$ is not empty. [Hint. Let $\mathbb{R}^4$, with points (x^1, y^1, x^2, y^2), be the real space which corresponds to $\mathbb{C}^2$ with points $(x^1 + iy^1, x^2 + iy^2)$. Let $z_0 = (x_0^1 + iy_0^1, x_0^2 + iy_0^2)$. Show that there exist 2-vectors (z_0, z_1, z_2) in $Z(z_0, \rho)$ which are in the 2-dimensional subspace of $\mathbb{R}^4$ for which $y^1 = y_0^1$, $y^2 = y_0^2$.]

72.4. Let $\mathbf{z}: (z_0, z_1, z_2)$ be a 2-vector in $Z(z_0, \rho)$, and let the rows of the following matrix contain the coordinates of z_0, z_1, z_2:

$$\begin{bmatrix} z_0^1 & z_0^2 \\ z_1^1 & z_1^2 \\ z_2^1 & z_2^2 \end{bmatrix}$$

Since $\mathbf{z}$ is in $Z(z_0, \rho)$, then $|\Delta(\mathbf{z})| \geqq \rho |z_1 - z_0| \, |z_2 - z_0|$. Let θ_1, θ_2 be real numbers, and let (z_0, w_1, w_2) be the 2-vector described by the following matrix.

$$\begin{bmatrix} z_0^1 & z_0^2 \\ z_0^1 + e^{i\theta_1}(z_1^1 - z_0^1) & z_0^2 + e^{i\theta_1}(z_1^2 - z_0^2) \\ z_0^1 + e^{i\theta_2}(z_2^1 - z_0^1) & z_0^2 + e^{i\theta_2}(z_2^2 - z_0^2) \end{bmatrix}$$

(a) If $\theta_1 = \theta_2 = 0$, show that (z_0, w_1, w_2) is the original 2-vector $\mathbf{z}$.
(b) Show that $\Delta(z_0, w_1, w_2) = e^{i\theta_1}e^{i\theta_2}\Delta(\mathbf{z})$ and that $|\Delta(z_0, w_1, w_2)| = |\Delta(\mathbf{z})|$ for every choice of θ_1, θ_2, and hence that $|\Delta(z_0, w_1, w_2)| \geqq \rho|z_1 - z_0|\,|z_2 - z_0|$.
(c) Show that $|w_1 - z_0| = |z_1 - z_0|$ and $|w_2 - z_0| = |z_2 - z_0|$ for every choice of θ_1, θ_2 and hence that $|\Delta(z_0, w_1, w_2)| \geqq \rho|w_1 - z_0|\,|w_2 - z_0|$.
(d) Show that (z_0, w_1, w_2) is a 2-vector in $Z(z_0, \rho)$ for every choice of θ_1, θ_2.

72.5. Prove that the class $Z(z_0, \rho)$ is not empty. This statement generalizes Exercise 72.3.

72.6. Exercise 72.4 describes certain results in $\mathbb{C}^2$. State and prove the corresponding results in $\mathbb{C}^n$.

72.7. Let $h^1, \cdots, h^n$ be complex numbers, and let $(z_0, z_1, \cdots, z_n)$ be an n-vector at z_0 in $\mathbb{C}^n$ which is described by the following matrix:

$$\begin{bmatrix} z_0^1 & z_0^2 & \cdots & z_0^n \\ z_0^1 + h^1 & z_0^2 & \cdots & z_0^n \\ z_0^1 & z_0^2 + h^2 & \cdots & z_0^n \\ \cdots & \cdots & \cdots & \cdots \\ z_0^1 & z_0^2 & & z_0^n + h^n \end{bmatrix}$$

(a) Prove that $(z_0, z_1, \cdots, z_n)$ is an n-vector in $Z(z_0, \rho)$ for every ρ such that $0 < \rho \leqq 1$.
(b) Let E be an open set in $\mathbb{C}^n$, and let $f\colon E \to \mathbb{C}$ be a function which is differentiable at z_0 in E. Prove that

$$D_j f(z_0) = \lim_{h^j \to 0} \frac{f(z_j) - f(z_0)}{h^j}, \qquad j = 1, \cdots, n.$$

(c) Explain why the limit in (b) should be called the partial derivative of f with respect to the variable z^j.

72.8. Prove the following theorem. If the function $f\colon E \to \mathbb{C}$, $E \subset \mathbb{C}^n$ is differentiable at z_0 in E, and if the value of f does not depend on the variables $z^{j_1}, \cdots, z^{j_k}$, then $D_{j_1} f(z_0) = 0, \cdots, D_{j_k} f(z_0) = 0$. [Hint. Exercise 72.7(b).]

72.9. Prove the following theorem. If $f\colon E \to \mathbb{C}$, $E \subset \mathbb{C}^n$, is a constant function, then f is differentiable and $D_j f(z) = 0$ for $j = 1, \cdots, n$ and every z in E.

72.10. Let $f\colon E \to \mathbb{C}$, $E \subset \mathbb{C}^2$, be the function such that

$$f(z^1, z^2) = \frac{2z^1z^2}{(z^1)^2 + (z^2)^2}, \qquad (z^1)^2 + (z^2)^2 \neq 0,$$

$$f(z^1, z^2) = 0, \qquad (z^1)^2 + (z^2)^2 = 0.$$

(a) Show that f has partial derivatives at $(0, 0)$ in $\mathbb{C}^2$, and that these partial derivatives are both zero. [Hint. Exercise 72.7.]
(b) Show that f is not continuous at the point $(0, 0)$ in $\mathbb{C}^2$.
(c) Show that f is not differentiable at the point $(0, 0)$ in $\mathbb{C}^2$. [Hint. Corollary 72.9. Compare Example 1.1.]

73. The Cauchy–Riemann Equations and Differentiability

Let $f: E \to \mathbb{C}$, $E \subset \mathbb{C}$, be a function which has a derivative $D_z f$ at each point in E. Then f is an analytic function of z in E [see Definition 64.3], and all of the properties of analytic functions follow from the single hypothesis that $D_z f$ exists in E. The situation is similar for functions $f: E \to \mathbb{C}$, $E \subset \mathbb{C}^n$.

73.1 Definition. The complex-valued function $f: E \to \mathbb{C}$, $E \subset \mathbb{C}^n$, is *analytic at* z_0 in E if and only if the derivatives $D_{z^j} f(z)$, $j = 1, \cdots, n$, exist at each point z in some neighborhood of z_0. Also, f is *analytic in a set* E in $\mathbb{C}^n$ if and only if these derivatives exist at each point in some open set which contains E.

Let $f: E \to \mathbb{C}$, $E \subset \mathbb{C}^n$, be defined in E, and let $z^{j_1}, \cdots, z^{j_m}$ be a subset of the variables $z^1, \cdots, z^n$. If f is an analytic function of the variables $z^1, \cdots, z^n$, then f is an analytic function of the variables $z^{j_1}, \cdots, z^{j_m}$ when the remaining variables are held constant. This statement follows from the fact that a necessary and sufficient condition that f be analytic is that it satisfy the Stolz condition [see Theorems 72.7 and 72.8]. But if f satisfies the Stolz condition with respect to $z^1, \cdots, z^n$, then it also satisfies the Stolz condition with respect to $z^{j_1}, \cdots, z^{j_m}$ when the remaining variables are held constant.

73.2 Example. If $f(z) = (z^1)^2 + (z^2)^2$, then

$$\begin{aligned} f(z) &= (x^1 + iy^1)^2 + (x^2 + iy^2)^2 \\ &= [(x^1)^2 - (y^1)^2 + (x^2)^2 - (y^2)^2] + i[2x^1y^1 + 2x^2y^2] \end{aligned} \tag{1}$$

Thus in this example there are real-valued functions u and v of the four real variables x^1, y^1, x^2, y^2 such that

$$f(z) = u(x^1, y^1, x^2, y^2) + iv(x^1, y^1, x^2, y^2). \tag{2}$$

In the example in (1), the functions u and v are polynomials in their four variables; hence, they have continuous derivatives of all orders. Let $f: E \to \mathbb{C}$, $E \subset \mathbb{C}^2$, be an arbitrary analytic function; then there are real-valued functions u and v such that (2) is true. If z_0 is in E, then the derivatives $D_{z^1} f(z_0)$ and $D_{z^2} f(z_0)$ exist. Let p_0 be the point in $\mathbb{R}^4$ which corresponds to z_0 in $\mathbb{C}^2$. Then since f is analytic in each variable separately, the proof of Theorem 64.1 shows that

$$\begin{aligned} D_{z^1} f(z_0) &= u_{x^1}(p_0) + iv_{x^1}(p_0) = v_{y^1}(p_0) - iu_{y^1}(p_0), \\ D_{z^2} f(z_0) &= u_{x^2}(p_0) + iv_{x^2}(p_0) = v_{y^2}(p_0) - iu_{y^2}(p_0); \end{aligned} \tag{3}$$

$$\begin{aligned} u_{x^1}(p_0) &= v_{y^1}(p_0), \qquad u_{y^1}(p_0) = -v_{x^1}(p_0); \\ u_{x^2}(p_0) &= v_{y^2}(p_0), \qquad u_{y^2}(p_0) = -v_{x^2}(p_0). \end{aligned} \tag{4}$$

Equations (4) are the Cauchy–Riemann equations for the function f. It must be emphasized that the derivatives in (4) are classical partial derivatives and not derivatives as defined in Chapter 1 of this book. Since f has derivatives $D_{z^1}f(z_0)$ and $D_{z^2}f(z_0)$, it is possible to show directly from this fact that the equations in (4) hold for somewhat stronger derivatives than partial derivatives, but the statements in (4) are sufficient for the purposes of this section. Later results in this section will show that the partial derivatives in (3) and (4) can be replaced by derivatives, and that these derivatives are continuous.

In the general case, let $f: E \to \mathbb{C}$ be defined on an open set E in $\mathbb{C}^n$. Then there are real-valued functions u and v such that

$$f(z) = u(x^1, y^1, \cdots, x^n, y^n) + iv(x^1, y^1, \cdots, x^n, y^n),$$
$$z = (z^1, \cdots, z^n) = (x^1 + iy^1, \cdots, x^n + iy^n).$$

This section investigates the relation of the Cauchy–Riemann equations to the differentiability of f. It establishes sufficient conditions for the differentiability of f, and it proves the following: if f is differentiable on E, then u and v satisfy the Cauchy–Riemann equations in each pair of variables x^j, y^j; also, u and v have continuous derivatives $D_{x^j}u$, $D_{y^j}u$, $D_{x^j}v$, $D_{y^j}v$ for $j = 1, \cdots, n$. For convenience let p denote the point $(x^1, y^1, \cdots, x^n, y^n)$ in $\mathbb{R}^{2n}$ which corresponds to the point $z: (z^1, \cdots, z^n)$ in $\mathbb{C}^n$.

73.3 Theorem. *If $f: E \to \mathbb{C}$, $E \subset \mathbb{C}^n$, is analytic in the open set E, then u and v satisfy the following Cauchy–Riemann differential equations:*

$$u_{x^j}(p) = v_{y^j}(p), \qquad u_{y^j}(p) = -v_{x^j}(p), \qquad z \in E, \qquad j = 1, \cdots, n. \tag{5}$$

Furthermore,

$$\begin{aligned} D_j f(z) &= u_{x^j}(p) + iv_{x^j}(p), \\ D_j f(z) &= v_{y^j}(p) - iu_{y^j}(p), \end{aligned} \qquad z \in E \subset \mathbb{C}^n, \qquad j = 1, \cdots, n. \tag{6}$$

PROOF. The proof of this theorem is similar to the proof of similar results established in a special case in Example 73.2. □

73.4 Theorem. *Let $f: E \to \mathbb{C}$, $E \subset \mathbb{C}^n$, be a function $u + iv$ which is defined on E. Let u and v be differentiable at p_0 in E and satisfy there the following Cauchy–Riemann equations:*

$$D_{x^j}u(p_0) = D_{y^j}v(p_0), \qquad D_{y^j}u(p_0) = -D_{x^j}v(p_0), \qquad j = 1, \cdots, n. \tag{7}$$

Then f satisfies the Stolz condition at p_0, f is differentiable at z_0, and

$$\begin{aligned} D_j f(z_0) &= D_{x^j}u(p_0) + iD_{x^j}v(p_0), \\ iD_j f(z_0) &= D_{y^j}u(p_0) + iD_{y^j}v(p_0). \end{aligned} \tag{8}$$

PROOF. Since u and v are differentiable at p_0, they satisfy the Stolz condition at p_0 by Theorem 3.6, and

$$
\begin{aligned}
&u(p) - u(p_0) \\
&\quad = \sum_{j=1}^{n} \{D_{x^j}u(p_0)(x^j - x_0^j) + D_{y^j}u(p_0)(y^j - y_0^j)\} + r(u; p_0, p)|p - p_0|, \\
&v(p) - v(p_0) \\
&\quad = \sum_{j=1}^{n} \{D_{x^j}v(p_0)(x^j - x_0^j) + D_{y^j}v(p_0)(y^j - y_0^j)\} + r(v; p_0, p)|p - p_0|.
\end{aligned} \tag{9}
$$

Multiply the second of the equations by i and then add the two equations; since

$$
\begin{aligned}
[u(p) - u(p_0)] + i[v(p) - v(p_0)] &= [u(p) + iv(p)] - [u(p_0) + iv(p_0)] \\
&= f(z) - f(z_0),
\end{aligned}
$$

the result can be simplified, by using (7), to the following:

$$
\begin{aligned}
f(z) - f(z_0) = \sum_{j=1}^{n} &[D_{x^j}u(p_0) + iD_{x^j}v(p_0)](z^j - z_0^j) \\
&+ [r(u; p_0, p) + ir(v; p_0, p)]|p - p_0|.
\end{aligned} \tag{10}
$$

Since

$$
\lim_{p \to p_0} r(u; p_0, p) = 0, \qquad \lim_{p \to p_0} r(v; p_0, p) = 0, \tag{11}
$$

equation (10) shows that f satisfies the Stolz condition at z_0; then (10) and Theorem 72.7 show that f is differentiable at z_0 and that

$$
D_j f(z_0) = D_{x^j}u(p_0) + iD_{x^j}v(p_0), \qquad j = 1, \cdots, n. \tag{12}
$$

The second equation in (8) follows from (12) and the Cauchy–Riemann equations in (7). The proof of (8) and of Theorem 73.4 is complete. □

Let $f: E \to \mathbb{C}$, $E \subset \mathbb{C}^n$, be a function which is analytic in E, and let f be the function $u + iv$. Then Theorem 73.3 shows that u and v have partial derivatives $u_{xj}, u_{yj}, v_{xj}, v_{yj}, j = 1, \cdots, n$. Partial derivatives are weak derivatives, and in order to carry out the applications planned for the next section, it will be necessary to know that an analytic function $u + iv$ has continuous derivatives $D_{x^j}u$, $D_{y^j}u$, $D_{x^j}v$, $D_{y^j}v$, $j = 1, \cdots, n$. Cauchy's integral formula will be used to prove that u and v have continuous derivatives; the remainder of this section contains the proof. Some examples will help to explain how Cauchy's integral formula can be applied to analytic functions of several complex variables.

73.5 Example. Let $f_1 : \mathbb{C} \to \mathbb{C}$ be an analytic function of z^1, and let $f_2 : \mathbb{C} \to \mathbb{C}$ be an analytic function of z^2. Then the Stolz condition shows that f_1 and f_2 are each analytic functions of z^1 and z^2. Since the product of two analytic functions of z^1 and z^2 is an analytic function of z^1 and z^2 [see the last paragraph of Section 72], the function $f: \mathbb{C}^2 \to \mathbb{C}$ such that $f(z^1, z^2) = f_1(z^1)f_2(z^2)$ is an analytic function. Let $z_0: (z_0^1, z_0^2)$ be a point in $\mathbb{C}^2$. Let C_1 and C_2 be circles, with center z_0, in the planes $z^2 = z_0^2$ and $z^1 = z_0^1$, respectively, through z_0. Then Cauchy's integral formula [see Section 68] shows that

$$f_1(w^1)f_2(z_0^2) = \frac{1}{2\pi i}\int_{C_2} \frac{f_1(w^1)f_2(w^2)}{w^2 - z_0^2}\,dw^2,$$

or

$$f(w^1, z_0^2) = \frac{1}{2\pi i}\int_{C_2} \frac{f(w^1, w^2)}{w^2 - z_0^2}\,dw^2. \tag{13}$$

Now since f_1 is an analytic function of z^1, a second application of Cauchy's integral formula shows that

$$f_1(z_0^1)f_2(z_0^2) = \frac{1}{2\pi i}\int_{C_1} \frac{f_1(w^1)f_2(z_0^2)}{w^1 - z_0^1}\,dw^1,$$

or

$$f(z_0^1, z_0^2) = \frac{1}{2\pi i}\int_{C_1} \frac{f(w^1, z_0^2)}{w^1 - z_0^1}\,dw^1. \tag{14}$$

Equations (13) and (14) show that

$$f(z_0^1, z_0^2) = \frac{1}{2\pi i}\int_{C_1}\left(\frac{1}{2\pi i}\int_{C_2} \frac{f(w^1, w^2)}{w^2 - z_0^2}\,dw^2\right)\frac{dw^1}{w^1 - z_0^1}. \tag{15}$$

This formula has been established for a very special function f, but the next example extends its generality.

73.6 Example. Let $f: E \to \mathbb{C}$, $E \subset \mathbb{C}^2$ be an analytic function in the open set E. Let $z_0: (z_0^1, z_0^2)$ be a point in E, and let $N(z_0, r)$ be a neighborhood of z_0 in E. Let C_1 and C_2 be circles in $N(z_0, r)$, with center z_0, in the planes $z^2 = z_0^2$ and $z^1 = z_0^1$, respectively, through z_0. Then, as in Example 73.5, Cauchy's integral formula shows that $f(w^1, z_0^2)$ is given by the second formula in (13), and that $f(z_0^1, z_0^2)$ is given by the second formula in (14). Therefore the formula in (15) holds in this case also. Now (15) can be used to calculate the partial derivatives of f as follows. Let z^2 be a point near z_0^2 inside C_2. Then, as in (15),

$$f(z_0^1, z^2) = \left(\frac{1}{2\pi i}\right)^2 \int_{C_1}\left(\int_{C_2} \frac{f(w^1, w^2)}{w^2 - z^2}\,dw^2\right)\frac{dw^1}{w^1 - z_0^1}. \tag{16}$$

The formulas in (15) and (16) show that

$$\begin{aligned} f(z_0^1, z^2) &- f(z_0^1, z_0^2) \\ &= \left(\frac{1}{2\pi i}\right)^2 \int_{C_1}\left(\int_{C_2}\left[\frac{f(w^1, w^2)}{w^2 - z^2} - \frac{f(w^1, w^2)}{w^2 - z_0^2}\right]dw^2\right)\frac{dw^1}{w^1 - z_0^1} \\ &= \left(\frac{1}{2\pi i}\right)^2 \int_{C_1}\left(\int_{C_2} \frac{f(w^1, w^2)(z^2 - z_0^2)}{(w^2 - z^2)(w^2 - z_0^2)}\,dw^2\right)\frac{dw^1}{w^1 - z_0^1}. \end{aligned} \tag{17}$$

Therefore,

$$\frac{f(z_0^1, z^2) - f(z_0^1, z_0^2)}{z^2 - z_0^2} = \left(\frac{1}{2\pi i}\right)^2 \int_{C_1} \left(\int_{C_2} \frac{f(w^1, w^2)\, dw^2}{(w^2 - z^2)(w^2 - z_0^2)} \right) \frac{dw^1}{w^1 - z_0^1}. \tag{18}$$

Then arguments similar to those used in Section 68 show that

$$\begin{aligned} f_{z^2}(z_0^1, z_0^2) &= \lim_{z^2 \to z_0^2} \frac{f(z_0^1, z^2) - f(z_0^1, z_0^2)}{z^2 - z_0^2} \\ &= \left(\frac{1}{2\pi i}\right)^2 \int_{C_1} \left(\int_{C_2} \frac{f(w^1, w^2)\, dw^2}{(w^2 - z_0^2)^2} \right) \frac{dw^1}{w^1 - z_0^1}. \end{aligned} \tag{19}$$

Now the derivative $f_{z^2}(z_0^1, z_0^2)$ is the partial derivative of f with respect to z^2. It is a partial derivative and not the derivative which has been denoted by $D_2 f(z_0^1, z_0^2)$. More precisely,

$$\begin{aligned} f_{z^2}(z_0^1, z_0^2) &= u_{x^2}(x_0^1, y_0^1, x_0^2, y_0^2) + i v_{x^2}(x_0^1, y_0^1, x_0^2, y_0^2) \\ &= v_{y^2}(x_0^1, y_0^1, x_0^2, y_0^2) - i u_{y^2}(x_0^1, y_0^1, x_0^2, y_0^2). \end{aligned} \tag{20}$$

The ultimate goal of this discussion is to show that all of the partial derivatives of u and v are continuous functions of the four variables x^1, y^1, x^2, y^2, and hence, by Theorem 3.19, that u and v have continuous derivatives with respect to their four variables. The proof follows. If $z_1 : (z_1^1, z_1^2)$ is sufficiently near $z_0 : (z_0^1, z_0^2)$, then the same arguments that were used to establish (19) can be used again to show that

$$f_{z^2}(z_1^1, z_1^2) = \left(\frac{1}{2\pi i}\right)^2 \int_{C_1} \left(\int_{C_2} \frac{f(w^1, w^2)}{(w^2 - z_1^2)^2}\, dw^2 \right) \frac{dw^1}{w^1 - z_1^1}. \tag{21}$$

Let η be a small positive number which is less than one-fourth of the radii of C_1 and of C_2. Throughout the remainder of this discussion let z_1^1 be a point inside C_1 whose distance from z_0^1 is less than η, and let z_1^2 be a point in C_2 whose distance from z_0^2 is less than η. The triangle inequality and the formulas in (19) and (21) show that

$$\begin{aligned} &|f_{z^2}(z_1^1, z_1^2) - f_{z^2}(z_0^1, z_0^2)| \\ &\leq \left| \left(\frac{1}{2\pi i}\right)^2 \int_{C_1} \left(\int_{C_2} \frac{f(w^1, w^2)\, dw^2}{(w^2 - z_1^2)^2} \right) \frac{dw^1}{w^1 - z_1^1} \right. \\ &\qquad \left. - \left(\frac{1}{2\pi i}\right)^2 \int_{C_1} \left(\int_{C_2} \frac{f(w^1, w^2)\, dw^2}{(w^2 - z_1^2)^2} \right) \frac{dw^1}{w^1 - z_0^1} \right| \\ &\qquad + \left| \left(\frac{1}{2\pi i}\right)^2 \int_{C_1} \left(\int_{C_2} \frac{f(w^1, w^2)\, dw^2}{(w^2 - z_1^2)^2} \right) \frac{dw^1}{w^1 - z_0^1} \right. \\ &\qquad \left. - \left(\frac{1}{2\pi i}\right)^2 \int_{C_1} \left(\int_{C_2} \frac{f(w^1, w^2)\, dw^2}{(w^2 - z_0^2)^2} \right) \frac{dw^1}{w^1 - z_0^1} \right|. \end{aligned} \tag{22}$$

The expression inside the first set of absolute value signs on the right in (22) can be written in the following forms:

$$\begin{aligned} &\left(\frac{1}{2\pi i}\right)^2 \int_{C_1}\left(\int_{C_2} \frac{f(w^1, w^2)\,dw^2}{(w^2 - z_1^2)^2}\right)\left(\frac{1}{w^1 - z_1^1} - \frac{1}{w^1 - z_0^1}\right)dw^1 \\ &\qquad = \left(\frac{1}{2\pi i}\right)^2 \int_{C_1}\left(\int_{C_2} \frac{f(w^1, w^2)\,dw^2}{(w^2 - z_1^2)^2}\right)\frac{(z_1^1 - z_0^1)\,dw^1}{(w^1 - z_1^1)(w^1 - z_0^1)}. \end{aligned} \tag{23}$$

Now f is an analytic function; it is therefore a continuous function of z^1, z^2 by Corollary 72.9; then there exists a constant M such that

$$\left|\int_{C_2} \frac{f(w^1, w^2)}{(w^2 - z_1^2)^2}\,dw^2\right| < M \tag{24}$$

for all w^1 on C_2 and z_1^2 near z_0^2. Also, z_0^1 is the center of C_1, and z_1^1 is in C_1 and near z_0^1. Let $\varepsilon > 0$ be given. Then these facts and the second term in (23) show that there exists a $\delta_1(\varepsilon)$ such that the first term on the right in (22) is less than $\varepsilon/2$ if $|z_1^1 - z_0^1| < \delta_1(\varepsilon)$. The expression inside the second set of absolute value signs on the right in (22) can be written in the following forms:

$$\begin{aligned} &\left(\frac{1}{2\pi i}\right)^2 \int_{C_1}\left(\int_{C_2} \left[\frac{f(w^1, w^2)}{(w^2 - z_1^2)^2} - \frac{f(w^1, w^2)}{(w^2 - z_0^2)^2}\right]dw^2\right)\frac{dw^1}{w^1 - z_0^1} \\ &= \left(\frac{1}{2\pi i}\right)^2 \int_{C_1}\left(\int_{C_2} \frac{f(w^1, w^2)[(w^2 - z_0^2) + (w^2 - z_1^2)](z_1^2 - z_0^2)\,dw^2}{(w^2 - z_1^2)^2(w^2 - z_0^2)^2}\right)\frac{dw^1}{w^1 - z_0^1}. \end{aligned} \tag{25}$$

In this integral, the integral over C_2 is bounded by a constant multiplied by $|z_1^2 - z_0^2|$. Then there exists a $\delta_2(\varepsilon)$ such that the second term on the right in (22) is less than $\varepsilon/2$ if $|z_1^2 - z_0^2| < \delta_2(\varepsilon)$. Collecting terms from (22) onward, we have shown that

$$|f_{z^2}(z_1^1, z_1^2) - f_{z^2}(z_0^1, z_0^2)| < \frac{\varepsilon}{2} + \frac{\varepsilon}{2} = \varepsilon \tag{26}$$

for all (z_1^1, z_1^2) such that

$$|z_1^1 - z_0^1| < \delta_1(\varepsilon), \qquad |z_1^2 - z_0^2| < \delta_2(\varepsilon). \tag{27}$$

Therefore, the partial derivative f_{z^2} is a continuous function of (z^1, z^2), and the partial derivatives u_{x^2}, v_{x^2}, u_{y^2}, v_{y^2} in (20) are continuous functions of x^1, y^1, x^2, y^2. Now by starting over at the beginning, a similar proof can be given to show that the partial derivative f_{z^1} is a continuous function of (z^1, z^2). Since

$$\begin{aligned} f_{z^1}(z_0^1, z_0^2) &= u_{x^1}(x_0^1, y_0^1, x_0^2, y_0^2) + iv_{x^1}(x_0^1, y_0^1, x_0^2, y_0^2) \\ &= v_{y^1}(x_0^1, y_0^1, x_0^2, y_0^2) - iu_{y^1}(x_0^1, y_0^1, x_0^2, y_0^2), \end{aligned} \tag{28}$$

the partial derivatives u_{x^1}, v_{x^1}, u_{y^1}, v_{y^1} are continuous functions of x^1, y^1, x^2, y^2. Thus u and v are two functions whose partial derivatives exist and are continuous in E. Then Theorem 3.19 shows that u and v are differentiable functions. Since derivatives are equal to partial derivatives when derivatives exist, this example has shown that the derivatives

$$D_{x^j}u,\ D_{y^j}u,\ D_{x^j}v,\ D_{y^j}v, \qquad j = 1, 2, \tag{29}$$

exist and are continuous in E.

73.7 Theorem. *Let $f: E \to \mathbb{C}$, $E \subset \mathbb{C}^n$, be a function which is analytic in the open set E. If $f(z) = u(x^1, y^1, \cdots, x^n, y^n) + iv(x^1, y^1, \cdots, x^n, y^n)$, then u and v are differentiable functions of the variables $x^1, y^1, \cdots, x^n, y^n$, and the derivatives*

$$D_{x^j}u,\ D_{y^j}u,\ D_{x^j}v,\ D_{y^j}v, \qquad j = 1, \cdots, n, \tag{30}$$

are continuous in E.

PROOF. Cauchy's integral formula for an analytic function of n complex variables $z^1, z^2, \cdots, z^n$ can be written in the following form:

$$f(z^1, \cdots, z^n) = \left(\frac{1}{2\pi i}\right)^n \int_{C_1} \frac{dw^1}{w^1 - z^1} \int_{C_2} \frac{dw^2}{w^2 - z^2} \cdots \int_{C_n} \frac{f(w^1, \cdots, w^n)}{w^n - z^n}\, dw^n. \tag{31}$$

Since f is an analytic function of w^n for fixed $w^1, \cdots, w^{n-1}$, then the integral

$$\int_{C_n} \frac{f(w^1, \cdots, w^n)}{w^n - z^n}\, dw^n \tag{32}$$

exists and equals $f(w^1, \cdots, w^{n-1}, z^n)$ by Theorem 68.1. Similar statements hold for the other integrals in (31). To prove Theorem 73.7, use formula (31) to represent the partial derivatives $f_{z^1}, \cdots, f_{z^n}$ of f as in Example 73.6. Since f is analytic by hypothesis, these partial derivatives are known to exist, but their representation by formula (31) enables us to show that they are continuous. Thus we show as in Example 73.6 that u and v have continuous partial derivatives. Then Theorem 3.19 states that u and v are differentiable. Since derivatives are equal to partial derivatives when derivatives exist, it follows that the derivatives of u and v in (30) exist and are continuous in E. The proof of Theorem 73.7 is complete. □

EXERCISES

73.1. Define f thus: $f(z^1, \cdots, z^n) = (z^1 + \cdots + z^n)^2$.

(a) Prove that f is differentiable at each point z in $\mathbb{C}^n$ and find $D_j f(z)$ for $j = 1, \cdots, n$.

(b) Find $u(x^1, y^1, \cdots, x^n, y^n)$ and $v(x^1, y^1, \cdots, x^n, y^n)$ such that $f(z) = u(x^1, y^1, \cdots, x^n, y^n) + iv(x^1, y^1, \cdots, x^n, y^n)$.

(c) Prove that u and v have an infinite number of continuous derivatives in their variables $x^1, y^1, \cdots, x^n, y^n$.

(d) Verify that u and v satisfy the Cauchy–Riemann equations in each pair x^j, y^j of their variables.
(e) Find the real and complex parts of $D_j f$ and show that they satisfy the Cauchy–Riemann equations in each pair of their variables.
(f) Show that f has an infinite number of derivatives with respect to its variables and find these derivatives.

73.2. Let f be a polynomial in $z^1, \cdots, z^n$ with complex coefficients. Show that f is analytic at every point in $\mathbb{C}^n$, and explain how to find the derivatives $D_j f$, $j = 1, \cdots, n$. Prove that f has an infinite number of derivatives with respect to the variables $z^1, \cdots, z^n$.

73.3. Verify the integral representation in equation (31) for each of the following functions:
(a) $f(z^1, \cdots, z^n) = a^1 z^1 + \cdots + a^n z^n$.
(b) $f(z^1, \cdots, z^n) = az^1 \cdots z^n$.
(c) $f(z^1, \cdots, z^n) = (z^1)^2 + \cdots + (z^n)^2$.

73.4. Let $f: E \to \mathbb{C}$, $E \subset \mathbb{C}^2$, be an analytic function in the open set E. Explain how to prove that each of the derivatives $D_j f$, $j = 1, 2$, is an analytic function in E.

73.5. Let E_k be an open set in $\mathbb{C}$, and let $f_k: E_k \to \mathbb{C}$ be an analytic function for $k = 1, \cdots, n$. Let $E_1 \times \cdots \times E_n$ be the set E in $\mathbb{C}^n$, and define $f: E \to \mathbb{C}$ as follows: $f(z^1, \cdots, z^n) = f_1(z^1) + \cdots + f_n(z^n)$.
(a) Prove that f is an analytic function of $z^1, \cdots, z^n$ in E and find $D_j f(z)$, $j = 1, \cdots, n$. [Hint. Show that f satisfies the Stolz condition.]
(b) If $f = u + iv$, find u and v and show that they satisfy the Cauchy–Riemann equations in each pair of their variables.
(c) Find a Taylor series for f.

73.6. Let E be an open set in $\mathbb{C}^2$, and let $f_k: E \to \mathbb{C}$, $k = 1, 2$, be analytic functions. Then

$$f_k(z^1, z^2) = u_k(x^1, y^1, x^2, y^2) + iv_k(x^1, y^1, x^2, y^2), \qquad k = 1, 2,$$

and u_k, v_k have continuous derivatives in their four variables. For each k, $k = 1, 2$, u_k and v_k satisfy the Cauchy–Riemann equations in each pair of their variables.
(a) Prove that

$$D_{(x^1, x^2)}(f_1, f_2) = \begin{vmatrix} D_{x^1}u_1 + iD_{x^1}v_1 & D_{x^2}u_1 + iD_{x^2}v_1 \\ D_{x^1}u_2 + iD_{x^1}v_2 & D_{x^2}u_2 + iD_{x^2}v_2 \end{vmatrix}.$$

(b) Prove that

$$D_{(x^1, y^1, x^2, y^2)}(u_1, v_1, u_2, v_2) = \begin{vmatrix} D_{x^1}u_1 & -D_{x^1}v_1 & D_{x^2}u_1 & -D_{x^2}v_1 \\ D_{x^1}v_1 & D_{x^1}u_1 & D_{x^2}v_1 & D_{x^2}u_1 \\ D_{x^1}u_2 & -D_{x^1}v_2 & D_{x^2}u_2 & -D_{x^2}v_2 \\ D_{x^1}v_2 & D_{x^1}u_2 & D_{x^2}v_2 & D_{x^2}u_2 \end{vmatrix}.$$

[Hint. Start with the standard determinant for the derivative on the left; transform it into the determinant on the right by using the Cauchy–Riemann equations.]
(c) Prove that

$$D_{(x^1, y^1, x^2, y^2)}(u_1, v_1, u_2, v_2) = |D_{(x^1, x^2)}(f_1, f_2)|^2.$$

[Hint. In the determinant in (b), multiply the second and fourth rows by i and add them to the first and third rows respectively. Then multiply the first and third columns by $-i$ and add them to the second and fourth columns respectively. Expand the resulting determinant by Laplace's expansion (see Theorem 79.3 in Appendix 1). Compare Exercise 63.7(c).]

73.7. State and prove the general theorem, of which Exercises 65.6 and 73.6 are special cases.

74. Cauchy's Integral Theorem

The purpose of this section is to prove, for analytic functions of several complex variables, a generalization of Cauchy's integral theorem which is given in Part I of this chapter. The methods are the same as those used in Section 71. The present section shows that Cauchy's integral theorem for functions of several complex variables is a special case of the fundamental theorem of the integral calculus in Theorem 39.1.

74.1 Example. This example contains the simplest special case which illustrates the general theorem and the methods used in proving it. Let E be an open set in $\mathbb{C}^2$, and let $f_k : E \to \mathbb{C}$, $k = 0, 1, 2$, be analytic functions of the complex variables z^1, z^2 in E. Since

$$f_k(z^1, z^2) = u_k(x^1, y^1, x^2, y^2) + iv_k(x^1, y^1, x^2, y^2), \qquad k = 0, 1, 2, \tag{1}$$

the functions f_0, f_1, f_2 can be considered also as complex-valued functions of the real variables (x^1, y^1, x^2, y^2) in $\mathbb{R}^4$. Let $\mathbb{R}^3$ be a space with points (t^1, t^2, t^3), and let A be a cube in $\mathbb{R}^3$ with sides parallel to the coordinate planes. Let $g : A \to \mathbb{R}^4$ be a function whose components (g_1, h_1, g_2, h_2) are differentiable and have continuous derivatives; the equations

$$\begin{aligned} x^j &= g_j(t^1, \cdots, t^3), \\ y^j &= h_j(t^1, \cdots, t^3), \qquad j = 1, 2, \qquad (t^1, \cdots, t^3) \in A, \end{aligned} \tag{2}$$

define a 3-dimensional surface G_3 in $\mathbb{R}^4$. Assume that the trace $T(G_3)$ of G_3 is in E. Since f_0, f_1, f_2 are analytic in E, Section 73 shows that these functions have continuous derivatives with respect to the variables x^1, y^1, x^2, y^2. Then the following form of the fundamental theorem of the integral calculus is true by Theorem 39.1:

$$\begin{aligned} &\int_{G_3} \sum_{(j_1, \cdots, j_3)} D_{(j_1, \cdots, j_3)}(f_0, f_1, f_2)(z)\, d(w_{j_1}, \cdots, w_{j_3}), \qquad (j_1, \cdots, j_3) \in (3/4) \\ &= \int_{\partial G_3} \sum_{(j_1, j_2)} f_0(z) D_{(j_1, j_2)}(f_1, f_2)(z)\, d(w_{j_1}, w_{j_2}), \qquad (j_1, j_2) \in (2/4). \end{aligned} \tag{3}$$

Here $(j_1, \cdots, j_3)$, (j_1, j_2), $(w_{j_1}, \cdots, w_{j_3})$, and (w_{j_1}, w_{j_2}) are codes for sets of variables selected from (x^1, y^1, x^2, y^2). Let the variables x^1, y^1, x^2, y^2 be numbered in order from 1 to 4. Then $(j_1, \cdots, j_3)$ and $(w_{j_1}, \cdots, w_{j_3})$ indicate the variables which are numbered $j_1, \cdots, j_3$. For example, $D_{(1,2,3)}(f_0, f_1, f_2)$ means $D_{(x^1,y^1,x^2)}(f_0, f_1, f_2)$, and $d(w_1, w_2, w_3)$ means $d(x^1, y^1, x^2)$; also, $D_{(2,3,4)}(f_0, f_1, f_2)$ means $D_{(y^1,x^2,y^2)}(f_0, f_1, f_2)$ and $d(w_2, w_3, w_4)$ means $d(y^1, x^2, y^2)$.

The integrals in (3) call for some explanation. As explained in Section 70, they are integrals of complex-valued functions of the real variables x^1, y^1, x^2, y^2. Two methods can be used to establish the existence of such integrals. First, it is possible to consider the real and imaginary parts of all of the functions involved, and thus to reduce the study of these integrals to real-valued integrals of real-valued functions. The second method is probably better. The methods used in proving Lemma 66.10 and Theorem 66.11 can be used to establish the existence of the integrals in (3); they are surface integrals on the surface G_3 and its boundary ∂G_3.

The next step in this example is to evaluate the two integrals in (3). It will be shown that

$$\int_{G_3} \sum_{(j_1,\cdots,j_3)} D_{(j_1,\cdots,j_3)}(f_0, f_1, f_2)(z)\, d(w_{j_1}, \cdots, w_{j_3}) = 0, \qquad (j_1, \cdots, j_3) \in (3/4), \tag{4}$$

because each derivative in the integrand of this integral is zero. The first of these derivatives is

$$D_{(1,2,3)}(f_0, f_1, f_2) = \begin{vmatrix} D_{x^1} f_0 & D_{y^1} f_0 & D_{x^2} f_0 \\ D_{x^1} f_1 & D_{y^1} f_1 & D_{x^2} f_1 \\ D_{x^1} f_2 & D_{y^1} f_2 & D_{x^2} f_2 \end{vmatrix}. \tag{5}$$

From (1),

$$\begin{aligned} D_{x^j} f_k(z) &= D_{x^j} u_k(p) + i D_{x^j} v_k(p), \\ D_{y^j} f_k(z) &= D_{y^j} u_k(p) + i D_{y^j} v_k(p). \end{aligned} \qquad j = 1, 2, \qquad k = 0, 1, 2. \tag{6}$$

Equations (6) and the Cauchy–Riemann equations show that

$$D_{y^j} f_k(z) = i D_{x^j} f_k(z), \qquad j = 1, 2, \qquad k = 0, 1, 2. \tag{7}$$

Therefore, by (5) and (7),

$$D_{(1,2,3)}(f_0, f_1, f_2) = \begin{vmatrix} D_{x^1} f_0 & i D_{x^1} f_0 & D_{x^2} f_0 \\ D_{x^1} f_1 & i D_{x^1} f_1 & D_{x^2} f_1 \\ D_{x^1} f_2 & i D_{x^1} f_2 & D_{x^2} f_2 \end{vmatrix} = 0 \tag{8}$$

because two of the columns in the determinant are proportional. The other derivatives in the integrand of the integral in (4) are

$$D_{(1,2,4)}(f_0, f_1, f_2), \qquad D_{(1,3,4)}(f_0, f_1, f_2), \qquad D_{(2,3,4)}(f_0, f_1, f_2). \tag{9}$$

The first of these derivatives is zero because (1, 2, 4) contains the pair (1, 2) as in (5) and (8). The second and third derivatives in (9) are zero because (1, 3, 4) and (2, 3, 4) both contain the pair (3, 4); then (7) and arguments similar to those used in proving (8) show that the derivatives in (9) are zero. Thus the statement in (4) has been established.

Next, evaluate the integral on the right in (3). The six derivatives in the integrand of this integral are

$$\begin{array}{lll} D_{(1,2)}(f_1, f_2), & D_{(1,3)}(f_1, f_2), & D_{(1,4)}(f_1, f_2), \\ D_{(2,3)}(f_1, f_2), & D_{(2,4)}(f_1, f_2), & D_{(3,4)}(f_1, f_2). \end{array} \tag{10}$$

Now, by (7),

$$D_{(1,2)}(f_1, f_2) = \begin{vmatrix} D_{x^1}f_1 & D_{y^1}f_1 \\ D_{x^1}f_2 & D_{y^1}f_2 \end{vmatrix} = \begin{vmatrix} D_{x^1}f_1 & iD_{x^1}f_1 \\ D_{x^1}f_2 & iD_{x^1}f_2 \end{vmatrix} = 0, \tag{11}$$

and a similar evaluation shows that $D_{(3,4)}(f_1, f_2) = 0$. Thus two of the derivatives in (10) are identically zero and only four remain to be evaluated. Since

$$D_{x^j}f_k = D_{z^j}f_k, \qquad j = 1, 2, \qquad k = 0, 1, 2, \tag{12}$$

then (7) shows that

$$\begin{aligned} D_{(1,3)}(f_1, f_2) &= \begin{vmatrix} D_{x^1}f_1 & D_{x^2}f_1 \\ D_{x^1}f_2 & D_{x^2}f_2 \end{vmatrix} = \begin{vmatrix} D_{z^1}f_1 & D_{z^2}f_1 \\ D_{z^1}f_2 & D_{z^2}f_2 \end{vmatrix} = D_{(z^1,z^2)}(f_1, f_2), \\ D_{(1,4)}(f_1, f_2) &= \begin{vmatrix} D_{x^1}f_1 & D_{y^2}f_1 \\ D_{x^1}f_2 & D_{y^2}f_2 \end{vmatrix} = \begin{vmatrix} D_{x^1}f_1 & iD_{x^2}f_1 \\ D_{x^1}f_2 & iD_{x^2}f_2 \end{vmatrix} = iD_{(z^1,z^2)}(f_1, f_2). \end{aligned} \tag{13}$$

Similar calculations show that

$$D_{(2,3)}(f_1, f_2) = iD_{(z^1,z^2)}(f_1, f_2), \qquad D_{(2,4)}(f_1, f_2) = (i)^2 D_{(z^1,z^2)}(f_1, f_2). \tag{14}$$

Then equations (4), (11), (13), (14) show that equation (3) simplifies to the following:

$$\begin{aligned} &\int_{\partial G_3} f_0(z) D_{(z^1,z^2)}(f_1, f_2)(z)[d(x^1, x^2) + id(x^1, y^2) + id(y^1, x^2) + (i)^2 d(y^1, y^2)] \\ &\quad = 0. \end{aligned} \tag{15}$$

To evaluate the integral in (15), it is necessary to use the equations in (2) to express it as an integral over the boundary of A. On each side of A, one of the variables (t^1, t^2, t^3) is constant, and the other two are variable. A side on which t^3 is constant and (t^1, t^2) are the variables is typical. On this side of A, the standard procedure for evaluating the surface integral in (15) states that it should be changed into an integral in (t^1, t^2), using the equations in (2) and making the following substitutions for $d(x^1, x^2), \cdots, d(y^1, y^2)$:

$$
\begin{aligned}
d(x^1, x^2) &= D_{(1,2)}(g_1, g_2)\,d(t^1, t^2) = \begin{vmatrix} D_1g_1 & D_2g_1 \\ D_1g_2 & D_2g_2 \end{vmatrix} d(t^1, t^2). \\
d(x^1, y^2) &= D_{(1,2)}(g_1, h_2)\,d(t^1, t^2) = \begin{vmatrix} D_1g_1 & D_2g_1 \\ D_1h_2 & D_2h_2 \end{vmatrix} d(t^1, t^2). \\
d(y^1, x^2) &= D_{(1,2)}(h_1, g_2)\,d(t^1, t^2) = \begin{vmatrix} D_1h_1 & D_2h_1 \\ D_1g_2 & D_2g_2 \end{vmatrix} d(t^1, t^2). \\
d(y^1, y^2) &= D_{(1,2)}(h_1, h_2)\,d(t^1, t^2) = \begin{vmatrix} D_1h_1 & D_2h_1 \\ D_1h_2 & D_2h_2 \end{vmatrix} d(t^1, t^2).
\end{aligned} \tag{16}
$$

Then

$$
d(x^1, x^2) + i\,d(x^1, y^2) + i\,d(y^1, x^2) + (i)^2\,d(y^1, y^2) \tag{17}
$$

$$
= \left\{ \begin{vmatrix} D_1g_1 & D_2g_1 \\ D_1g_2 & D_2g_2 \end{vmatrix} + i \begin{vmatrix} D_1g_1 & D_2g_1 \\ D_1h_2 & D_2h_2 \end{vmatrix} + i \begin{vmatrix} D_1h_1 & D_2h_1 \\ D_1g_2 & D_2g_2 \end{vmatrix} \right.
$$

$$
\left. + (i)^2 \begin{vmatrix} D_1h_1 & D_2h_1 \\ D_1h_2 & D_2h_2 \end{vmatrix} \right\} d(t^1, t^2) \tag{18}
$$

$$
= \begin{vmatrix} D_1(g_1 + ih_1) & D_2(g_1 + ih_1) \\ D_1(g_2 + ih_2) & D_2(g_2 + ih_2) \end{vmatrix} d(t^1, t^2). \tag{19}
$$

Theorems 75.1 and 75.3 in Appendix 1 have been used to transform the sum in (18) into the single determinant in (19). The equations in (2) show that the equations

$$
\begin{aligned}
z^1 &= g_1(t^1, \cdots, t^3) + ih_1(t^1, \cdots, t^3), \\
z^2 &= g_2(t^1, \cdots, t^3) + ih_2(t^1, \cdots, t^3),
\end{aligned} \qquad (t^1, \cdots, t^3) \in A, \tag{20}
$$

describe the surface G_3, and that the boundary ∂G_3 of G_3 is obtained by restricting $(t^1, \cdots, t^3)$ to the boundary ∂A of A. Then on a side of A on which t^3 is constant, the expression in (19) is properly designated by $d(z^1, z^2)$. Thus (17) and (19) show that equation (15) simplifies to

$$
\int_{\partial G_3} f_0(z) D_{(z^1, z^2)}(f_1, f_2)(z)\, d(z^1, z^2) = 0. \tag{21}
$$

On the sides of A, equation (20) shows that

$$
d(z^1, z^2) = D_{(t^{j_1}, t^{j_2})}(z^1, z^2)(t)\, d(t^{j_1}, t^{j_2}), \qquad (j_1, j_2) \in (2/3). \tag{22}
$$

Thus, equations (17), (19), (20), and (22) show that (21) can be written in the following form:

$$
\int_{\partial A} f_0(z) D_{(z^1, z^2)}(f_1, f_2)(z) D_{(t^{j_1}, t^{j_2})}(z^1, z^2)\, d(t^{j_1}, t^{j_2}) = 0. \tag{23}
$$

The formula in (21) is Cauchy's integral theorem for the functions $f_k : E \to \mathbb{C}$,

$k = 0, 1, 2$, and the 2-dimensional surface ∂G_3. This formula is an exact analog of the classical Cauchy's integral theorem in Theorem 71.1, and the proof is a straightforward generalization of the proof of the earlier theorem.

Example 74.1 contains a generalization of the classical Cauchy's integral theorem, but clearly this example is only a special case of the completely general theorem. The purpose of the remainder of this section is to establish this general theorem. An understanding of the special case in the example will greatly facilitate the reading of the remainder of the section.

The analytic functions in the remainder of the section are functions of the n complex variables $z^1, \cdots, z^n$ or $x^1 + iy^1, \cdots, x^n + iy^n$. The traces of the surfaces are contained in $\mathbb{R}^{2n}$, which has coordinates $(x^1, y^1, \cdots, x^n, y^n)$. Let $\mathbb{R}^m$, $1 \leqq m \leqq 2n$, be a space with points $(t^1, \cdots, t^m)$, and let A be a cube in $\mathbb{R}^m$ with sides parallel to the coordinate planes. Let $g: A \to \mathbb{R}^{2n}$ be a function whose components $(g_1, h_1, \cdots, g_n, h_n)$ are differentiable and have continuous derivatives; the equations

$$\begin{aligned} x^j &= g_j(t^1, \cdots, t^m), \\ y^j &= h_j(t^1, \cdots, t^m), \end{aligned} \qquad j = 1, \cdots, n, \qquad (t^1, \cdots, t^m) \in A, \tag{24}$$

define an m-dimensional surface G_m in $\mathbb{R}^{2n}$.

Let E be an open set in $\mathbb{C}^n$, the complex space associated with $\mathbb{R}^{2n}$, and let $f_0, f_1, \cdots, f_{m-1}$ be analytic functions of the complex variables $z^1, \cdots, z^n$ in E. Since

$$\begin{gathered} f_k(z^1, \cdots, z^n) = u_k(x^1, y^1, \cdots, x^n, y^n) + iv_k(x^1, y^1, \cdots, x^n, y^n), \\ k = 0, 1, \cdots, m-1, \end{gathered} \tag{25}$$

the functions $f_0, f_1, \cdots, f_{m-1}$ can be considered as complex-valued functions of the real variables $(x^1, y^1, \cdots, x^n, y^n)$. Assume that the trace $T(G_m)$ of G_m is in E. Since $f_0, f_1, \cdots, f_{m-1}$ are analytic in E, Section 73 shows that they have continuous derivatives with respect to the variables $x^1, y^1, \cdots, x^n, y^n$. Thus the functions $f_0, f_1, \cdots, f_{m-1}$ in (25) and the surface G_m defined by the functions in (24) satisfy all of the hypotheses of the fundamental theorem of the integral calculus in Theorem 39.1; therefore, by that theorem,

$$\begin{aligned} &\int_{G_m} \sum_{(m/2n)} D_{(j_1, \cdots, j_m)}(f_0, f_1, \cdots, f_{m-1})(z)\, d(w_{j_1}, \cdots, w_{j_m}) \\ &= \int_{\partial G_m} \sum_{(m-1/2n)} f_0(z) D_{(j_1, \cdots, j_{m-1})}(f_1, \cdots, f_{m-1})(z)\, d(w_{j_1}, \cdots, w_{j_{m-1}}). \end{aligned} \tag{26}$$

Here $(j_1, \cdots, j_m)$, $(j_1, \cdots, j_{m-1})$, $(w_{j_1}, \cdots, w_{j_m})$, and $(w_{j_1}, \cdots, w_{j_{m-1}})$ are codes for sets of variables selected from $(x^1, y^1, \cdots, x^n, y^n)$. Let the variables $x^1, y^1, \cdots, x^n, y^n$ be numbered in order from 1 to $2n$. Then $(j_1, \cdots, j_m)$ and $(w_{j_1}, \cdots, w_{j_m})$ indicate the variables which are numbered $j_1, \cdots, j_m$. For example, if $(j_1, \cdots, j_m) = (1, \cdots, m)$, then $(j_1, \cdots, j_m)$ specifies the first m variables in the set $(x^1, y^1, \cdots, x^n, y^n)$ since $m \leqq 2n$ by hypothesis.

The program in the remainder of this section is a simple one: show that, under certain conditions, the integral on the left in (26) is zero; then the integral over the boundary ∂G_m on the right is an integral whose value is zero. Section 71 and Example 74.1 show that this program succeeds at least in some cases. Since there is a formula (26) for each m from 1 to $2n$, our hopes rise that there may be many boundaries ∂G_m for which the integral on the right in (26) is zero, but there are some surprises in store for us. The search for Cauchy integral theorems begins with an investigation of conditions under which the integral on the left in (26) is zero.

74.2 Lemma. *Let $f_0, f_1, \cdots, f_{m-1}$ be analytic functions of the complex variables $z^1, \cdots, z^n$ in the open set E in $\mathbb{C}^n$. If the set of variables in $(x^1, y^1, \cdots, x^n, y^n)$ specified by the index set $(j_1, \cdots, j_m)$ contains a pair x^j, y^j, then*

$$D_{(j_1,\cdots,j_m)}(f_0, f_1, \cdots, f_{m-1})(z) = 0, \qquad z \in E. \tag{27}$$

PROOF. Recall that

$$D_{(j_1,\cdots,j_m)}(f_0, f_1, \cdots, f_{m-1}) = \begin{vmatrix} D_{j_1}f_0 & \cdots & D_{j_m}f_0 \\ \cdots & \cdots & \cdots \\ D_{j_1}f_{m-1} & \cdots & D_{j_m}f_{m-1} \end{vmatrix}. \tag{28}$$

If $z = (x^1 + iy^1, \cdots, x^n + iy^n)$ and $p = (x^1, y^1, \cdots, x^n, y^n)$, then

$$\begin{aligned} D_{x^j}f_k(z) &= D_{x^j}u_k(p) + iD_{x^j}v_k(p), \\ D_{y^j}f_k(z) &= D_{y^j}u_k(p) + iD_{y^j}v_k(p). \end{aligned} \qquad j = 1, \cdots, n;\ k = 0, 1, \cdots, m-1. \tag{29}$$

Equations (29) and the Cauchy–Riemann equations show that

$$D_{y^j}f_k(z) = iD_{x^j}f_k(z), \qquad j = 1, \cdots, n;\ k = 0, 1, \cdots, m-1;\ z \in E. \tag{30}$$

If the set of variables in $(x^1, y^1, \cdots, x^n, y^n)$ specified by $(j_1, \cdots, j_m)$ contains x^j, y^j, then by (28),

$$D_{(j_1,\cdots,j_m)}(f_0, f_1, \cdots, f_{m-1}) = \begin{vmatrix} D_{j_1}f_0 & \cdots & D_{x^j}f_0 & D_{y^j}f_0 & \cdots & D_{j_m}f_0 \\ \cdots & \cdots & \cdots & \cdots & \cdots & \cdots \\ D_{j_1}f_{m-1} & \cdots & D_{x^j}f_{m-1} & D_{y^j}f_{m-1} & \cdots & D_{j_m}f_{m-1} \end{vmatrix}. \tag{31}$$

Equation (30) shows that the determinant in (31) has two columns which are proportional; the determinant equals zero; and the proof of (27) and of Lemma 74.2 is complete. □

74.3 Lemma. *Let $f_0, f_1, \cdots, f_{m-1}$ be analytic functions of the complex variables $z^1, \cdots, z^n$ in the open set E in $\mathbb{C}^n$. If $n + 1 \leqq m \leqq 2n$, then*

$$D_{(j_1,\cdots,j_m)}(f_0, f_1, \cdots, f_{m-1})(z) = 0, \qquad z \in E, \qquad (j_1, \cdots, j_m) \in (m/2n). \tag{32}$$

PROOF. If $m \geqq n + 1$, then every set $(j_1, \cdots, j_m)$ in $(m/2n)$ determines a set in $(x^1, y^1, \cdots, x^n, y^n)$ which contains a pair x^j, y^j, because a set which contains at most one element from each of the n pairs $(x^1, y^1), \cdots, (x^n, y^n)$ contains at most n elements and not $n + 1$ or more. Then Lemma 74.2 shows that (32) is true, and the proof of Lemma 74.3 is complete. □

74.4 Lemma. *Let $f_0, f_1, \cdots, f_{m-1}$ be analytic functions of the complex variables in the open set E in $\mathbb{C}^n$. If $n + 1 \leqq m \leqq 2n$, then the integrand of the integral on the left in* (26) *vanishes identically in E; and if $n + 1 < m \leqq 2n$, then the integrand of the integral on the right in* (26) *also vanishes identically in E.*

PROOF. If $n + 1 \leqq m \leqq 2n$, then (32) shows that each term in the integrand of the integral on the left in (26) vanishes identically in E, and the value of the integral is zero. If $n + 1 < m \leqq 2n$, then $n + 1 \leqq m - 1$ and the same arguments show that each term in the integrand of the integral on the right in (26) vanishes identically in E; hence, the integral equals zero. The proof of Lemma 74.4 is complete. □

Lemma 74.4 shows that there is only one value of m which converts the equation in (26) into a useful result of the type desired; furthermore, $n + 1$ is this value for m [in Example 74.1, $n = 2$ and $m = 3$ and hence $m = n + 1$]. The equation which results from (26) if $m = n + 1$ is

$$\int_{\partial G_{n+1}} \sum_{(n/2n)} f_0(z) D_{(j_1, \cdots, j_n)}(f_1, \cdots, f_n)(z)\, d(w_{j_1}, \cdots, w_{j_n}) = 0. \tag{33}$$

This equation is Cauchy's integral theorem for analytic functions of several complex variables. The next theorem states this result; the proof of this theorem transforms the integral in (33) into a form which is readily recognized as the analog of the classical Cauchy's integral theorem.

74.5 Theorem (Cauchy's Integral Theorem). *Let $f_0, f_1, \cdots, f_n$ be analytic functions of the complex variables $z^1, \cdots, z^n$ in the open set E in $\mathbb{C}^n$. Let A be a cube in $\mathbb{R}^{n+1}$ whose sides are parallel to the coordinate planes, and let $g: A \to \mathbb{R}^{2n}$ be a function whose components $(g_1, h_1, \cdots, g_n, h_n)$ are differentiable and have continuous derivatives, and which defines a surface G_{n+1} whose trace is in E. Then*

$$\int_{\partial G_{n+1}} f_0(z) D_{(z^1, \cdots, z^n)}(f_1, \cdots, f_n)(z)\, d(z^1, \cdots, z^n) = 0. \tag{34}$$

PROOF. Under the hypotheses stated in the theorem, the fundamental theorem of the integral calculus [see Theorem 39.1] shows that the equation in (26) is true, and Lemma 74.3 shows that this equation simplifies to the one in (33). To complete the proof of Theorem 74.5, it is necessary to show that the equation in (33) can be simplified to the one in (34).

Now many of the derivatives in the integrand of the integral in (33) are identically zero in E. The sum extends over all ordered sets of n letters which can be selected from the $2n$ letters in $(x^1, y^1, \cdots, x^n, y^n)$, but Lemma 74.2 shows that the derivative in (27) is zero except for those sets which contain one letter from each of the n pairs $(x^1, y^1), \cdots, (x^n, y^n)$. Since there are two ways to choose a letter from each pair, there are 2^n ordered sets $(j_1, \cdots, j_n)$ for which $D_{(j_1, \cdots, j_n)}(f_1, \cdots, f_n)$ is not formally identically zero in E. Let S_n denote the set of ordered index sets $(j_1, \cdots, j_n)$ for which $D_{(j_1, \cdots, j_n)}(f_1, \cdots, f_n)$ is not automatically equal to zero in E. For the set $(j_1, \cdots, j_n)$ in S_n, j_1 is either 1 or 2, j_2 is either 3 or 4, $\cdots$, and j_n is $2n - 1$ or $2n$. Thus the equation in (33) can be simplified to the following equation:

$$\int_{\partial G_{n+1}} \sum_{S_n} f_0(z) D_{(j_1, \cdots, j_n)}(f_1, \cdots, f_n)(z)\, d(w_{j_1}, \cdots, w_{j_n}) = 0. \tag{35}$$

The next step in the proof is to evaluate the derivatives in (35). Now

$$D_{(j_1, \cdots, j_n)}(f_1, \cdots, f_n) = \begin{vmatrix} D_{j_1} f_1 & \cdots & D_{j_i} f_1 & \cdots & D_{j_n} f_1 \\ \cdots & \cdots & \cdots & \cdots & \cdots \\ D_{j_1} f_n & \cdots & D_{j_i} f_n & \cdots & D_{j_n} f_n \end{vmatrix}. \tag{36}$$

The elements in the matrix of the determinant are either $D_{x^j} f_k$ or $D_{y^j} f_k$. Recall from (7) and (12) that

$$\begin{aligned} D_{x^j} f_k &= D_{z^j} f_k, \\ D_{y^j} f_k &= i D_{x^j} f_k = i D_{z^j} f_k. \end{aligned} \qquad j = 1, \cdots, n, \qquad k = 0, 1, \cdots, n. \tag{37}$$

Let $r(j_1, \cdots, j_n)$ denote the number of integers in the index set $(j_1, \cdots, j_n)$ which specify a letter y^j in the set $(x^1, y^1, \cdots, x^n, y^n)$; then the remaining integers in $(j_1, \cdots, j_n)$ specify a letter x^j in $(x^1, y^1, \cdots, x^n, y^n)$. By the definition of S_n, if $(j_1, \cdots, j_n)$ is in S_n, then j_1 specifies either x^1 or y^1, j_2 specifies either x^2 or y^2, $\cdots$, and j_n specifies either x^n or y^n. Consequently, (36) and (37) show that

$$D_{(j_1, \cdots, j_n)}(f_1, \cdots, f_n) = (i)^{r(j_1, \cdots, j_n)} D_{(z^1, \cdots, z^n)}(f_1, \cdots, f_n), \tag{38}$$

and this equation can be used to simplify (35) to the following form:

$$\int_{\partial G_{n+1}} f_0(z) D_{(z^1, \cdots, z^n)}(f_1, \cdots, f_n)(z) \sum_{S_n} (i)^{r(j_1, \cdots, j_n)} d(w_{j_1}, \cdots, w_{j_n}) = 0. \tag{39}$$

Now it is necessary to recall how the integral in (39) is evaluated. The equations in (24) show that the equations

$$\begin{gathered} z^j = g_j(t^1, \cdots, t^{n+1}) + i h_j(t^1, \cdots, t^{n+1}), \\ j = 1, \cdots, n, \qquad (t^1, \cdots, t^{n+1}) \in A, \end{gathered} \tag{40}$$

describe the surface G_{n+1}, and the boundary ∂G_{n+1} of G_{n+1} is obtained by restricting $(t^1, \cdots, t^{n+1})$ to the boundary ∂A of A. Consider a side of A on which t^{n+1} is constant and $(t^1, \cdots, t^n)$ are variables. To evaluate the integral in (39), it is necessary to replace the terms $d(w_{j_1}, \cdots, w_{j_n})$ by expressions analogous to those in (16). Thus, if $r(j_1, \cdots, j_n) = 0$, then

$$d(w_{j_1}, \cdots, w_{j_n}) = d(x^1, \cdots, x^n) = D_{(1,\cdots,n)}(g_1, \cdots, g_n)\, d(t^1, \cdots, t^n)$$

$$= \begin{vmatrix} D_1 g_1 & \cdots & D_n g_1 \\ \cdots & \cdots & \cdots \\ D_1 g_n & \cdots & D_n g_n \end{vmatrix} d(t^1, \cdots, t^n). \tag{41}$$

If $r(j_1, \cdots, j_n) = n$, then

$$d(w_{j_1}, \cdots, w_{j_n}) = d(y^1, \cdots, y^n) = D_{(1,\cdots,n)}(h_1, \cdots, h_n)\, d(t^1, \cdots, t^n),$$

$$(i)^{r(j_1,\cdots,j_n)}\, d(w_{j_1}, \cdots, w_{j_n}) = (i)^n\, d(y^1, \cdots, y^n)$$

$$= \begin{vmatrix} iD_1 h_1 & \cdots & iD_n h_1 \\ \cdots & \cdots & \cdots \\ iD_1 h_n & \cdots & iD_n h_n \end{vmatrix} d(t^1, \cdots, t^n). \tag{42}$$

In the general case, $(i)^{r(j_1,\cdots,j_n)}\, d(w_{j_1}, \cdots, w_{j_n})$ is replaced by a derivative multiplied by $d(t^1, \cdots, t^n)$. The derivative is the determinant of an n by n matrix of derivatives. Corresponding to an x^j in $(w_{j_1}, \cdots, w_{j_n})$, there is a row

$$[D_1 g_j \quad D_2 g_j \quad \cdots \quad D_n g_j] \tag{43}$$

in the matrix of the determinant, and corresponding to a y^j in $(w_{j_1}, \cdots, w_{j_n})$ there is a row

$$[iD_1 h_j \quad iD_2 h_j \quad \cdots \quad iD_n h_j] \tag{44}$$

in the matrix of the determinant. Observe that the exponent on $(i)^{r(j_1,\cdots,j_n)}$ is exactly the number of letters y^j in $(w_{j_1}, \cdots, w_{j_n})$; thus there is, in $(i)^{r(j_1,\cdots,j_n)}$, a factor i for each row of the matrix which contains an h_j. These considerations show that, in evaluating the integral in (39), the sum

$$\sum_{S_n} (i)^{r(j_1,\cdots,j_n)}\, d(w_{j_1}, \cdots, w_{j_n}) \tag{45}$$

is replaced by $d(t^1, \cdots, t^n)$ multiplied by a sum of 2^n determinants of n by n matrices. Now Theorems 75.1 and 75.3 in Appendix 1 can be used as in Example 74.1 [see equations (18) and (19)] to show that the sum in (45), when represented as described above [see (41), (42), and the following discussion], equals

$$\begin{vmatrix} D_1(g_1 + ih_1) & \cdots & D_n(g_1 + ih_1) \\ \cdots & \cdots & \cdots \\ D_1(g_n + ih_n) & \cdots & D_n(g_n + ih_n) \end{vmatrix} d(t^1, \cdots, t^n), \tag{46}$$

which is

$$D_{(1,\cdots,n)}[(g_1 + ih_1), \cdots, (g_n + ih_n)]\, d(t^1, \cdots, t^n). \tag{47}$$

Thus the integral in (39) can be evaluated as an integral over the boundary ∂A of A, and the equation in (39) can be written as the following equation:

$$\int_{\partial A} f_0(z)D_{(z^1,\cdots,z^n)}(f_1,\cdots,f_n)(z)D_{(1,\cdots,n)}[(g_1+ih_1),\cdots,$$
$$(g_n+ih_n)](t)\,d(t^{i_1},\cdots,t^n)=0. \tag{48}$$

Now because of the equations in (40), the expression in (47) is correctly denoted by $d(z^1, \cdots, z^n)$; thus

$$d(z^1,\cdots,z^n)=D_{(1,\cdots,n)}[(g_1+ih_1),\cdots,(g_n+ih_n)](t)\,d(t^{i_1},\cdots,t^{i_n}). \tag{49}$$

This equation can be used to represent the integral in (48) as a surface integral over ∂G_{n+1}; thus the equation in (48) becomes

$$\int_{\partial G_{n+1}} f_0(z)D_{(z^1,\cdots,z^n)}(f_1,\cdots,f_n)(z)\,d(z^1,\cdots,z^n)=0. \tag{50}$$

The integral in (48) provides an evaluation for the integral in (50). Since equation (50) is Cauchy's integral theorem in the present case, the proof of (34) and of Theorem 74.5 is complete. □

Cauchy's integral theorem in Theorem 74.5 has been stated and proved for a surface G_{n+1} which is the boundary of a surface G_{n+1} which is defined on a cube A in $\mathbb{R}^{n+1}$. The proof can be extended without essential change to establish Cauchy's integral theorem for boundaries ∂G_{n+1} of surfaces defined on chains of simplexes in an oriented simplicial complex [compare the corresponding generalizations of the classical theorem in Section 71].

Exercises

74.1. If $f_1(z) = z^1, \cdots, f_n(z) = z^n$ in Theorem 74.5, show that the statement of Cauchy's integral theorem in (34) can be simplified to the following equation:

$$\int_{\partial G_{n+1}} f_0(z)\,d(z^1,\cdots,z^n)=0.$$

74.2. Show that Theorem 74.5, in the special case $n = 1$, is Theorem 71.1.

74.3. Consider Theorem 74.5 again. Let $f: E \to \mathbb{C}^{n+1}$ denote the function whose components $f_0, f_1, \cdots, f_n$ are analytic in E in $\mathbb{C}^n$. Use (34) to prove the following form of Cauchy's integral theorem.

$$\int_{\partial G_{n+1}} \det[f(z), D_{z^1}f(z), \cdots, D_{z^n}f(z)]\,d(z^1,\cdots,z^n)=0,\ n=1,2,\cdots,$$

where the matrix of the determinant in the integral is

$$\begin{bmatrix} f_0(z) & f_1(z) & \cdots & f_n(z) \\ D_{z^1}f_0(z) & D_{z^1}f_1(z) & \cdots & D_{z^1}f_n(z) \\ \cdots & \cdots & \cdots & \cdots \\ D_{z^n}f_0(z) & D_{z^n}f_1(z) & \cdots & D_{z^n}f_n(z) \end{bmatrix}.$$

[Hints. Equations (74) and (75) in the proof of Theorem 38.8; Theorem 71.1.]

APPENDIX 1

Determinants

75. Introduction to Determinants

If $a_{11}a_{22} - a_{12}a_{21} \neq 0$, then the system of equations

$$\begin{aligned} a_{11}x_1 + a_{12}x_2 &= b_1, \\ a_{21}x_1 + a_{22}x_2 &= b_2, \end{aligned} \tag{1}$$

has a unique solution which can be found by eliminating in turn x_2 and x_1; the result is

$$x_1 = \frac{a_{22}b_1 - a_{12}b_2}{a_{11}a_{22} - a_{12}a_{21}}, \qquad x_2 = \frac{a_{11}b_2 - a_{21}b_1}{a_{11}a_{22} - a_{12}a_{21}}. \tag{2}$$

The denominator in these expressions can be written as the determinant of the matrix of coefficients in the system of equations; thus

$$a_{11}a_{22} - a_{12}a_{21} = \begin{vmatrix} a_{11} & a_{12} \\ a_{21} & a_{22} \end{vmatrix} = \det \begin{bmatrix} a_{11} & a_{12} \\ a_{21} & a_{22} \end{bmatrix}. \tag{3}$$

The second order determinant is a function which is defined on 2-by-2 matrices; the matrix is denoted by square brackets, the determinant is denoted by vertical lines around the array of numbers in the matrix, and the value of the determinant is shown in (3). As (3) indicates, the determinant of a matrix of real numbers is a real number, and the determinant of a matrix of complex numbers is a complex number. The following diagram is a useful memory aid in evaluating a second order determinant:

$$\begin{vmatrix} a_{11} & a_{12} \\ a_{21} & a_{22} \end{vmatrix} \begin{matrix} - \\ + \end{matrix} \tag{4}$$

Form two products as indicated by the arrows; give each product the sign shown at the point of the arrow; the value of the determinant is the sum of these signed products. Then (3) and (4) show that

$$a_{22}b_1 - a_{12}b_2 = \begin{vmatrix} b_1 & a_{12} \\ b_2 & a_{22} \end{vmatrix} = \det \begin{bmatrix} b_1 & a_{12} \\ b_2 & a_{22} \end{bmatrix},$$

$$a_{11}b_2 - a_{21}b_1 = \begin{vmatrix} a_{11} & b_1 \\ a_{21} & b_2 \end{vmatrix} = \det \begin{bmatrix} a_{11} & b_1 \\ a_{21} & b_2 \end{bmatrix}. \tag{5}$$

Equations (3) and (5) show that x_1 and x_2 in (2) are each the ratio of two determinants.

75.1 Example

$$\begin{vmatrix} 7 & 5 \\ -3 & 4 \end{vmatrix} = 28 - (-15) = 43.$$

To find x_1 in the solution of the system of equations

$$\begin{aligned} a_{11}x_1 + a_{12}x_2 + a_{13}x_3 &= b_1, \\ a_{21}x_1 + a_{22}x_2 + a_{23}x_3 &= b_2, \\ a_{31}x_1 + a_{32}x_2 + a_{33}x_3 &= b_3, \end{aligned} \tag{6}$$

multiply the three equations, respectively, by the three expressions

$$(a_{22}a_{33} - a_{32}a_{23}), \qquad -(a_{12}a_{33} - a_{32}a_{13}), \qquad (a_{12}a_{23} - a_{22}a_{13}), \tag{7}$$

and add the resulting equations. Then

$$x_1 = \frac{a_{22}a_{33}b_1 + a_{13}a_{32}b_2 + a_{12}a_{23}b_3 - a_{13}a_{22}b_3 - a_{12}a_{33}b_2 - a_{23}a_{32}b_1}{a_{11}a_{22}a_{33} + a_{13}a_{21}a_{32} + a_{12}a_{23}a_{31} - a_{13}a_{22}a_{31} - a_{12}a_{21}a_{33} - a_{11}a_{23}a_{32}} \tag{8}$$

if the denominator in this expression is not zero. This denominator is a third order determinant; it is the determinant of the matrix of coefficients in the system (6). This determinant is denoted by vertical bars around the array of numbers in the matrix; thus

$$\begin{vmatrix} a_{11} & a_{12} & a_{13} \\ a_{21} & a_{22} & a_{23} \\ a_{31} & a_{32} & a_{33} \end{vmatrix} = \det \begin{bmatrix} a_{11} & a_{12} & a_{13} \\ a_{21} & a_{22} & a_{23} \\ a_{31} & a_{32} & a_{33} \end{bmatrix}. \tag{9}$$

Corresponding to (4) for the second order determinant, the following diagram is a useful memory aid in evaluating a third order determinant:

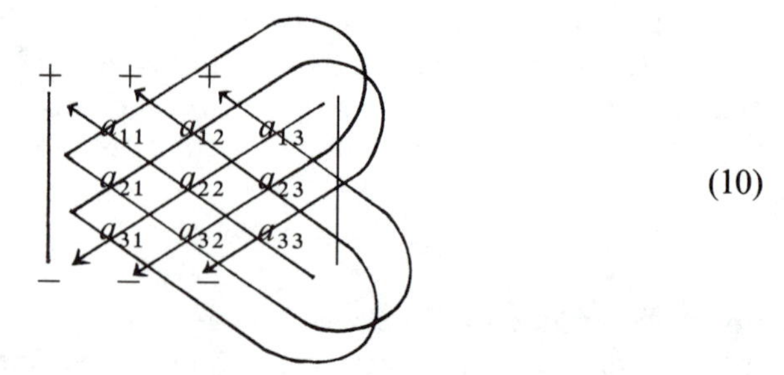

(10)

75.2 Example

$$\begin{vmatrix} 3 & 2 & 4 \\ 5 & 7 & -3 \\ 2 & 6 & 5 \end{vmatrix} = 105 + 120 + (-12) - 56 - 50 - (-54) = 161.$$

The diagrams in (4) and (10) are useful for second and third order determinants, but it must be emphasized that there are no similar diagrams for fourth and higher order determinants. The diagrams in (4) and (10) show that each product in the evaluation of the determinant contains one element from each row and each column of the matrix, but an examination of the denominator in (8) provides a more useful clue to the definition of the determinant in the general case.

In the matrix in (9), the first subscript designates the row and the second, the column. In the denominator in (8), each term contains three factors. The first subscripts of these factors are (1, 2, 3) in all cases; thus, each term contains an element from each of the three rows of the matrix. The second subscripts of the factors are 1, 2, 3 in some order; hence, each term contains an element from each column of the matrix. For example, the first subscripts of $a_{12}a_{23}a_{31}$ are (1, 2, 3) and the second subscripts are (2, 3, 1). The following are all of the ordered sets of second subscripts:

$$\begin{array}{cc} (1, 2, 3) & (3, 2, 1) \\ (3, 1, 2) & (2, 1, 3) \\ (2, 3, 1) & (1, 3, 2) \end{array} \tag{11}$$

These six arrangements are all of the permutations of the subscripts 1, 2, 3. Furthermore, these permutations determine the sign placed before each of the corresponding terms in the evaluation of the determinant. The number of inversions (from the natural order) in each permutation is the number of interchanges of adjacent elements required to return the three integers to their natural order (1, 2, 3). Each permutation in the first column in (11) contains an even number of inversions and the sign before the corresponding terms in the determinant is +; each permutation in the second column in (11) contains an odd number of inversions and the sign before the corresponding terms in the determinant is −. These observations provide a description of the third order determinant which motivates the definition of the n-th order determinant in the next section.

76. Definition of the Determinant of a Matrix

Let A be an n-by-n matrix with real or complex elements as indicated by the following statement:

$$A = \begin{bmatrix} a_{11} & a_{12} & \cdots & a_{1n} \\ a_{21} & a_{22} & \cdots & a_{2n} \\ \cdots & \cdots & \cdots & \cdots \\ a_{n1} & a_{n2} & \cdots & a_{nn} \end{bmatrix}. \tag{1}$$

As in Section 75, the determinant of A is denoted by placing vertical bars around the array of numbers in A. Another notation for A is $[a_{ij}]$, $i, j = 1, \cdots, n$; furthermore, the determinant of A is denoted also by det A or det $[a_{ij}]$.

Let $P(1, \cdots, n)$ denote the set of all permutations of the integers $1, \cdots, n$, and let $s(j_1, \cdots, j_n)$ denote the number of inversions in the permutation $(j_1, \cdots, j_n)$ in $P(1, \cdots, n)$.

76.1 Definition. The n-th order determinant is a function which is defined on the set of n-by-n matrices. The *determinant* of the n-by-n matrix A is a number which is defined by the following equation:

$$\det A = \sum_{(j_1, \cdots, j_n)} (-1)^{s(j_1, \cdots, j_n)} a_{1j_1} \cdots a_{nj_n}, \qquad (j_1, \cdots, j_n) \in P(1, \cdots, n). \tag{2}$$

76.2 Example. Show that $a_{23}a_{14}a_{42}a_{31}a_{65}a_{56}$ is a product in the expansion of det $[a_{ij}]$, $i, j = 1, \cdots, 6$, and find the sign to be placed before this product. Solution: Rearrange the factors so that the first subscripts are in the natural order; the given term is $a_{14}a_{23}a_{31}a_{42}a_{56}a_{65}$. Since the first subscripts are (1, 2, 3, 4, 5, 6) and the second subscripts are (4, 3, 1, 2, 6, 5), which is a permutation of the first six integers, the product is indeed one which occurs in the expansion of det $[a_{ij}]$. The sign to be placed before the term is found as follows:

4 precedes 3, 1, 2	3 inversions
3 precedes 1, 2	2
1	0
2	0
6 precedes 5	1
5	0
$s(4, 3, 1, 2, 6, 5) =$	6

Since $s(4, 3, 1, 2, 6, 5)$ is 6, an even number, the sign to be placed before $a_{14}a_{23}a_{31}a_{42}a_{56}a_{65}$ in the expansion of det $[a_{ij}]$ is $+$. Observe that 3 interchanges of adjacent terms changes (4, 3, 1, 2, 6, 5) into (3, 1, 2, 4, 6, 5); again, 2 interchanges of adjacent terms changes (3, 1, 2, 4, 6, 5) into (1, 2, 3, 4, 6, 5); and finally, one interchange of adjacent terms changes (1, 2, 3, 4, 6, 5) into (1, 2, 3, 4, 5, 6). These six interchanges of adjacent terms change (4, 3, 1, 2, 6, 5) into (1, 2, 3, 4, 5, 6).

The following remarks emphasize significant features or consequences of the definition of the determinant.

76.3 Remark. A matrix is a rectangular array of real or complex numbers, but a determinant is a function, defined on the set of square matrices, whose values are real or complex numbers.

76.4 Remark. The subscripts show that each product $a_{1j_1} \cdots a_{nj_n}$ contains an element from each row and from each column of A.

76.5 Remark. The expression (2) for $\det A$ contains $n!$ terms since a_{1j_1} can be chosen in n ways, a_{2j_2} can be chosen in $(n-1)$ ways after a_{1j_1} has been chosen, $\cdots$, $a_{(n-1)j_{n-1}}$ can be chosen in 2 ways after the preceding terms have been chosen, and a_{nj_n} can be chosen in one way. The total number of ways to make these choices is $n(n-1) \cdots 3 \cdot 2 \cdot 1$ or $n!$ Thus the expression for the fourth order determinant contains 24 terms and the fifth order determinant contains 120 terms. Hence, the evaluation of the formula in (2) is not a practical method for evaluating a determinant of order greater than 3, and special methods must be developed for this purpose.

76.6 Remark. A permutation in $P(1, \cdots, n)$ is called even (odd) if the number $s(j_1, \cdots, j_n)$ is even (odd). If $n \geqq 2$, then one-half of the permutations in $P(1, \cdots, n)$ are even and one-half are odd; hence, if $n \geqq 2$, then $n!/2$ terms in (2) are preceded by $+$ and $n!/2$ terms are preceded by $-$. If A is the matrix $[a]$ with one row and one column, then $\det A = a$.

76.7 Remark. If $n > 3$, there is no diagram such as those in (4) and (10) in Section 75 to simplify the evaluation of the determinant. These large determinants are evaluated either by (2) with great difficulty, or better by special methods to be described later.

77. Elementary Properties of Determinants

Let $A_1, A_2, \cdots, A_n$ denote rows $1, 2, \cdots, n$ of the n-by-n matrix A. It is frequently useful to think of the determinant as a function of the n vectors $A_1, \cdots, A_n$ which form the rows of A and to write $\det(A_1, \cdots, A_n)$ as another notation for $\det A$. This notation is used in stating some of the following properties of determinants.

77.1 Theorem. *The determinant is homogeneous in each of the rows of its matrix; or, if t is a real or complex number, then*

$$\det(A_1, \cdots, tA_k, \cdots, A_n) = t \det(A_1, \cdots, A_k, \cdots, A_n), \qquad k = 1, \cdots, n.$$

PROOF. Use equation (2) in Section 76 to find the value of $\det(A_1, \cdots, tA_k, \cdots, A_n)$. Since each term in the sum contains one factor from the

k-th row, each term in the sum for $\det(A_1, \cdots, tA_k, \cdots, A_n)$ contains the factor t. Factor out t from the sum; what remains is $\det(A_1, \cdots, A_k, \cdots, A_n)$. □

77.2 Example

$$\begin{vmatrix} 2 & 1 & 5 \\ 3t & 2t & 4t \\ 1 & 3 & 7 \end{vmatrix} = t\begin{vmatrix} 2 & 1 & 5 \\ 3 & 2 & 4 \\ 1 & 3 & 7 \end{vmatrix} = 22t.$$

77.3 Theorem *The determinant is additive in each of the rows of its matrix, or*

$$\det(A_1, \cdots, A_k + B, \cdots, A_n) = \det(A_1, \cdots, A_k, \cdots, A_n) + \det(A_1, \cdots, B, \cdots, A_n),$$

for $k = 1, 2, \cdots, n.$

PROOF. By the definition of the determinant in Section 76,

$$\begin{aligned}\det(A_1, \cdots, A_k + B, \cdots, A_n) &= \sum_{(j_1, \cdots, j_n)} (-1)^{s(j_1, \cdots, j_n)} a_{1j_1} \cdots (a_{kj_k} + b_{j_k}) \cdots a_{nj_n} \\ &= \sum_{(j_1, \cdots, j_n)} (-1)^{s(j_1, \cdots, j_n)} a_{1j_1} \cdots a_{kj_k} \cdots a_{nj_n} \\ &\quad + \sum_{(j_1, \cdots, j_n)} (-1)^{s(j_1, \cdots, j_n)} a_{1j_1} \cdots b_{j_k} \cdots a_{nj_n} \\ &= \det(A_1, \cdots, A_k, \cdots, A_n) + \det(A_1, \cdots, B, \cdots, A_n).\end{aligned}$$ □

77.4 Example

$$\begin{vmatrix} 2+3 & 4+1 & 6-5 \\ 2 & 3 & 7 \\ 1 & 2 & 1 \end{vmatrix} = \begin{vmatrix} 2 & 4 & 6 \\ 2 & 3 & 7 \\ 1 & 2 & 1 \end{vmatrix} + \begin{vmatrix} 3 & 1 & -5 \\ 2 & 3 & 7 \\ 1 & 2 & 1 \end{vmatrix} = -29.$$

The determinant on the left has the value -29; those on the right have values 4 and -33.

77.5 Theorem. *The determinant changes sign if two adjacent rows of its matrix are interchanged, or*

$$\det(A_1, \cdots, A_k, A_{k+1}, \cdots, A_n) = -\det(A_1, \cdots, A_{k+1}, A_k, \cdots, A_n),$$

for $k = 1, \cdots, n - 1.$

PROOF. The proof will be given in the case $k = 1$; the proof in the other cases is similar. By the definition of the determinant in Definition 76.1,

$$\det(A_1, A_2, \cdots, A_n) = \sum_{(j_1, \cdots, j_n)} (-1)^{s(j_1, \cdots, j_n)} a_{1j_1} a_{2j_2} \cdots a_{nj_n}, \quad (j_1, \cdots, j_n) \in P(1, \cdots, n). \tag{1}$$

The set

$$\{(j_2, j_1, \cdots, j_n) : (j_1, j_2, \cdots, j_n) \in P(1, \cdots, n)\} \tag{2}$$

contains $n!$ permutations and they are all distinct because $P(1, \cdots, n)$ contains $n!$ permutations $(j_1, j_2, \cdots, j_n)$ and they are all distinct; hence, the set (2) is $P(1, \cdots, n)$. This statement and the definition of the determinant show that

$$\det(A_2, A_1, \cdots, A_n) = \sum_{(j_1, \cdots, j_n)} (-1)^{s(j_2, j_1, \cdots, j_n)} a_{2j_2} a_{1j_1} \cdots a_{nj_n}, \quad (j_1, j_2, \cdots, j_n) \in P(1, \cdots, n). \tag{3}$$

But

$$a_{1j_1} a_{2j_2} \cdots a_{nj_n} = a_{2j_2} a_{1j_1} \cdots a_{nj_n},$$
$$(-1)^{s(j_1, j_2, \cdots, j_n)} = -(-1)^{s(j_2, j_1, \cdots, j_n)},$$

because $s(j_2, j_1, \cdots, j_n)$ differs from $s(j_1, j_2, \cdots, j_n)$ by 1; hence, each term in the sum in (1) is the negative of the corresponding term in the sum in (3). Thus

$$\det(A_1, A_2, \cdots, A_n) = -\sum_{(j_1, \cdots, j_n)} (-1)^{s(j_2, j_1, \cdots, j_n)} a_{2j_2} a_{1j_1} \cdots a_{nj_n}$$
$$= -\det(A_2, A_1, \cdots, A_n),$$

and the proof is complete. □

77.6 Example

$$\begin{vmatrix} 3 & 4 & -2 \\ 1 & 2 & 5 \\ 4 & -7 & 8 \end{vmatrix} = 48 + 14 + 80 + 16 - 32 + 105 = 231.$$

$$\begin{vmatrix} 3 & 4 & -2 \\ 4 & -7 & 8 \\ 1 & 2 & 5 \end{vmatrix} = -105 - 16 + 32 - 14 - 80 - 48 = -231.$$

The second determinant is obtained from the first one by interchanging the second and third rows of its matrix.

77.7 Theorem. *The sign of the determinant changes if any two rows of its matrix are interchanged, or*

$$\det(A_1, \cdots, A_r, \cdots, A_s, \cdots, A_n) = -\det(A_1, \cdots, A_s, \cdots, A_r, \cdots, A_n).$$

PROOF. Let k be the number of rows between A_r and A_s. Then A_r and A_s can be interchanged by $2k + 1$ interchanges of adjacent rows as follows. First, interchange A_s with adjacent rows to move it across the k rows between A_r and A_s and then across A_r itself; these interchanges result in a total of $k + 1$ changes in sign by Theorem 77.5. Next, interchange A_r with adjacent rows to move it across the k rows which stood originally between A_r and A_s; these interchanges result in k additional changes in sign. Thus the interchange of A_r and A_s results in a total of $2k + 1$ changes in sign. Since $2k + 1$ is odd, the interchange of the two rows results in changing the sign of the determinant. □

77.8 Example

$$\begin{vmatrix} 3 & 4 & -2 \\ 1 & 2 & 5 \\ 4 & -7 & 8 \end{vmatrix} = -\begin{vmatrix} 4 & -7 & 8 \\ 1 & 2 & 5 \\ 3 & 4 & -2 \end{vmatrix} = 231.$$

The determinant on the right results from interchanging the first and third rows of the matrix of the determinant on the left. Two interchanges of adjacent rows move the third row to the first row; in the new matrix interchange the second and third rows to complete the interchange of the first and third rows of the original matrix. The three interchanges of adjacent rows change the sign of the determinant by Theorem 77.5.

77.9 Theorem. *If rows A_r and A_s are equal, then* $\det(A_1, \cdots, A_r, \cdots, A_s, \cdots, A_n) = 0$. *More generally, the determinant is zero if $A_r = tA_s$.*

PROOF. By Theorem 77.7 and the hypothesis that A_r and A_s are the same,

$$\begin{aligned} \det(A_1, \cdots, A_r, \cdots, A_s, \cdots, A_n) &= -\det(A_1, \cdots, A_s, \cdots, A_r, \cdots, A_n) \\ &= -\det(A_1, \cdots, A_r, \cdots, A_s, \cdots, A_n); \end{aligned}$$

hence, $\det(A_1, \cdots, A_r, \cdots, A_s, \cdots, A_n) = 0$. If $A_r = tA_s$, then remove the factor t by Theorem 77.1 and proceed as before. □

77.10 Example

$$\begin{vmatrix} 3 & 7 & 4 \\ 1 & -2 & 3 \\ 3 & 7 & 4 \end{vmatrix} = -24 + 28 + 63 + 24 - 28 - 63 = 0.$$

77.11 Theorem. *If a row of a determinant matrix is multiplied by a constant and added to another row, the value of the determinant is unchanged. In symbols,*

$$\det(A_1, \cdots, A_r + tA_s, \cdots, A_s, \cdots, A_n) = \det(A_1, \cdots, A_r, \cdots, A_s, \cdots, A_n).$$

PROOF. By Theorem 77.3,

$$\det(A_1, \cdots, A_r + tA_s, \cdots, A_s, \cdots, A_n)$$
$$= \det(A_1, \cdots, A_r, \cdots, A_s, \cdots, A_n) + \det(A_1, \cdots, tA_s, \cdots, A_s, \cdots, A_n).$$

The second determinant on the right is zero by Theorem 77.9, and the stated result follows. □

77.12 Example

$$\begin{vmatrix} 3 & 4 & -2 \\ 1 & 2 & 5 \\ 4 & -7 & 8 \end{vmatrix} = 231. \qquad \begin{vmatrix} 11 & -10 & 14 \\ 1 & 2 & 5 \\ 4 & -7 & 8 \end{vmatrix} = 231.$$

The determinant on the right is obtained from the one on the left by multiplying the third row of its matrix by 2 and adding it to the first row.

77.13 Definition. Let $A = [a_{ij}]$, $i = 1, \cdots, m$ and $j = 1, \cdots, n$. If B is a matrix $[b_{ij}]$, $i = 1, \cdots, n$ and $j = 1, \cdots, m$, such that $b_{ij} = a_{ji}$, then B is called the *transpose of A* and denoted by A^t. Thus,

$$A = \begin{bmatrix} a_{11} & a_{12} & \cdots & a_{1n} \\ a_{21} & a_{22} & \cdots & a_{2n} \\ \cdots & \cdots & \cdots & \cdots \\ a_{m1} & a_{m2} & \cdots & a_{mn} \end{bmatrix}, \qquad A^t = \begin{bmatrix} a_{11} & a_{21} & \cdots & a_{m1} \\ a_{12} & a_{22} & \cdots & a_{m2} \\ \cdots & \cdots & \cdots & \cdots \\ a_{1n} & a_{2n} & \cdots & a_{mn} \end{bmatrix}.$$

77.14 Theorem. *Let A be an n-by-n matrix with real or complex elements, and let A^t be the transpose of A. Then* $\det A^t = \det A$.

PROOF. Let A^t be the matrix $[b_{ij}]$, $i, j = 1, \cdots, n$. Then by Definition 76.1,

$$\det A^t = \sum_{(j_1, \cdots, j_n)} (-1)^{s(j_1, \cdots, j_n)} b_{1j_1} \cdots b_{nj_n}, \qquad (j_1, \cdots, j_n) \in P(1, \cdots, n). \tag{4}$$

Since A^t is the transpose of A, then

$$b_{1j_1} \cdots b_{nj_n} = a_{j_1 1} \cdots a_{j_n n}. \tag{5}$$

Now $(j_1, \cdots, j_n)$ is a permutation of $(1, \cdots, n)$; hence, (5) shows that $b_{1j_1} \cdots b_{nj_n}$ is one of the products in the expansion of $\det A$. Thus the $n!$ products $b_{1j_1} \cdots b_{nj_n}$ which occur in $\det A^t$ are exactly the $n!$ products in $\det A$; to complete the proof it is necessary to show that $a_{j_1 1} \cdots a_{j_n n}$ has the same sign in $\det A$ that $b_{1j_1} \cdots b_{nj_n}$ has in $\det A^t$. To determine the sign of $a_{j_1 1} \cdots a_{j_n n}$ in $\det A$, rearrange the factors in this product by a succession of interchanges of adjacent letters so that the first subscripts are in the natural order $(1, \cdots, n)$. Each interchange of two adjacent factors in $a_{j_1 1} \cdots a_{j_n n}$ interchanges two of the second subscripts. Thus

$$a_{j_1 1} \cdots a_{j_n n} = a_{1k_1} \cdots a_{nk_n}, \tag{6}$$

and this equation and (5) show that

$$b_{1j_1} \cdots b_{nj_n} = a_{1k_1} \cdots a_{nk_n}. \tag{7}$$

The construction of the permutation $(k_1, \cdots, k_n)$ from $(j_1, \cdots, j_n)$ shows that $s(j_1, \cdots, j_n) = s(k_1, \cdots, k_n)$ for the following reasons. The first subscripts in $a_{j_1 1} \cdots a_{j_n n}$ can be arranged in the natural order by $s(j_1, \cdots, j_n)$ interchanges of adjacent factors, and the second subscripts in $a_{1k_1} \cdots a_{nk_n}$ can be arranged in the natural order by $s(k_1, \cdots, k_n)$ interchanges of adjacent factors. But if the interchanges used to transform $a_{j_1 1} \cdots a_{j_n n}$ into $a_{1k_1} \cdots a_{nk_n}$ are reversed and performed in the reverse order, they transform $a_{1k_1} \cdots a_{nk_n}$ into $a_{j_1 1} \cdots a_{j_n n}$; therefore,

$$s(j_1, \cdots, j_n) = s(k_1, \cdots, k_n). \tag{8}$$

Equations (7) and (8) show that

$$(-1)^{s(j_1, \cdots, j_n)} b_{1j_1} \cdots b_{nj_n} = (-1)^{s(k_1, \cdots, k_n)} a_{1k_1} \cdots a_{nk_n}. \tag{9}$$

The expression on the left in (9) is a term in $\det A^t$, and that on the right is a term in $\det A$. Thus the $n!$ terms in $\det A^t$ are term-by-term equal to the $n!$ terms in $\det A$, and $\det A^t = \det A$. □

77.15 Example. Let $\det A$ be the first determinant in Example 77.6. Then

$$\det A = \begin{vmatrix} 3 & 4 & -2 \\ 1 & 2 & 5 \\ 4 & -7 & 8 \end{vmatrix} = 48 + 14 + 80 + 16 - 32 + 105 = 231,$$

$$\det A^t = \begin{vmatrix} 3 & 1 & 4 \\ 4 & 2 & -7 \\ -2 & 5 & 8 \end{vmatrix} = 48 + 80 + 14 + 16 - 32 + 105 = 231.$$

77.16 Example. Let A be a 6-by-6 square matrix, and let $B = A^t$. Then $-b_{13}b_{24}b_{32}b_{45}b_{56}b_{61}$ is a term in $\det B$ [which is $\det A^t$] since (3, 4, 2, 5, 6, 1) is a permutation in $P(1, \cdots, 6)$, and $s(3, 4, 2, 5, 6, 1) = 7$. Now

$$\begin{aligned} -b_{13}b_{24}b_{32}b_{45}b_{56}b_{61} &= -a_{31}a_{42}a_{23}a_{54}a_{65}a_{16} \\ &= -a_{16}a_{23}a_{31}a_{42}a_{54}a_{65}. \end{aligned}$$

Here (6, 3, 1, 2, 4, 5) is a permutation in $P(1, \cdots, 6)$ and $s(6, 3, 1, 2, 4, 5) = 7$. Thus $-a_{31}a_{42}a_{23}a_{54}a_{65}a_{16}$ is a term in $\det A^t$, and the equal term $-a_{16}a_{23}a_{31}a_{42}a_{54}a_{65}$ is a term in $\det A$.

77.17 Theorem. *Let A be an n-by-n matrix of real or complex numbers. If* $\det A$ *has a certain property relative to the rows of A, then it has the corresponding property relative to the columns of A. In particular,* $\det A$ *has properties* *relative to the columns of A which correspond to the row properties described in Theorems 77.1, 77.3, 77.5, 77.7, 77.9, and 77.11.*

PROOF. Theorem 77.14 can be used to prove all of the statements in this theorem. One example is sufficient to explain the method. Let A be a matrix which has two columns the same. Then A^t has two rows the same. Since $\det A^t = 0$ by Theorem 77.9, then $\det A = 0$ since $\det A = \det A^t$ by Theorem 77.14. □

77.18 Theorem. *If I is the n-by-n identity matrix, then* $\det I = 1$. *If A is an n-by-n matrix which has a row or a column of zeros, then* $\det A = 0$.

The proofs of the statements in this theorem follow easily from the definition of the determinant [see Definition 76.1].

78. Definitions and Notation

Before proceeding further with this account of determinants, it is necessary to describe some special notation and to make several definitions.

Let M and N denote the sets of integers $1, 2, \cdots, m$ and $1, 2, \cdots, n$ respectively. Let $I^{(k)}$ denote a combination of k integers $i_1, \cdots, i_k$ selected from M, and let $I_i^{(k)}$, $i = 1, \cdots, C(m, k)$, be the complete set of combinations of k integers which can be formed from the m integers in M. Similarly, let $J_j^{(k)}, j = 1, \cdots, C(n, k)$, be the complete set of combinations of k integers $j_1, \cdots, j_k$ which can be formed from the n integers in N. The integers in both $I_i^{(k)}$ and $J_j^{(k)}$ are arranged in their natural order. Unless there is a statement to the contrary, the sets $I_i^{(k)}$ and $J_j^{(k)}$ are ordered lexicographically, that is, by the principle used in ordering the words in a dictionary, and then numbered in that order. For example, if $m = 4$ and $k = 2$, the ordered sets $I_i^{(k)}$ are the following:

$$\begin{aligned} &I_1^{(2)} = (1, 2), \quad I_2^{(2)} = (1, 3), \quad I_3^{(2)} = (1, 4), \\ &I_4^{(2)} = (2, 3), \quad I_5^{(2)} = (2, 4), \quad I_6^{(2)} = (3, 4); \end{aligned} \tag{1}$$

if $m = 4$ and $k = 3$, then

$$\begin{aligned} &I_1^{(3)} = (1, 2, 3), \quad I_2^{(3)} = (1, 2, 4), \\ &I_3^{(3)} = (1, 3, 4), \quad I_4^{(3)} = (2, 3, 4). \end{aligned} \tag{2}$$

The symbols (k/m) and (k/n) denote, respectively, the ordered sets $I_i^{(k)}$, $i = 1, \cdots, C(m, k)$, and $J_j^{(k)}$, $j = 1, \cdots, C(n, k)$. Thus, corresponding to (1) and (2),

$$\begin{aligned} (2/4) &= ((1, 2), (1, 3), (1, 4), (2, 3), (2, 4), (3, 4)), \\ (3, 4) &= ((1, 2, 3), (1, 2, 4), (1, 3, 4), (2, 3, 4)). \end{aligned} \tag{3}$$

In the general case,

$$\begin{aligned}(k/m) &= (I_i^{(k)} : i = 1, \cdots, C(m, k)),\\ (k/n) &= (J_j^{(k)} : j = 1, \cdots, C(n, k)).\end{aligned} \tag{4}$$

Finally, let $M - I_i^{(k)}$ (respectively $N - J_j^{(k)}$) denote the set of $m - k$ integers ($n - k$ integers), arranged in their natural order, in M (in N) but not in $I_i^{(k)}$ (not in $J_j^{(k)}$).

Let A be a matrix $[a_{ij}]$, $i = 1, \cdots, m$, $j = 1, \cdots, n$, with m rows and n columns. Next, let the submatrix of A formed by the k rows designated by the k integers in $I_i^{(k)}$ be denoted by $A(I_i^{(k)})$. Similarly, let the submatrix of A formed by the k columns designated by the k integers in $J_j^{(k)}$ be denoted by $A(J_j^{(k)})$. Finally, let $A(I_i^{(k)} \cdot J_j^{(k)})$ denote the submatrix of A which is contained in both $A(I_i^{(k)})$ and $A(J_j^{(k)})$. It follows that $A(I_i^{(k)} \cdot J_j^{(k)})$ is a square matrix with k rows and k columns.

78.1 Definition. The k-th *compound of* A, denoted by $A^{(k)}$, is defined for each integer k such that $1 \leqq k \leqq m$, $1 \leqq k \leqq n$, by the following statement:

$$A^{(k)} = [|A(I_i^{(k)} \cdot J_j^{(k)})|], \qquad i = 1, \cdots, C(m, k); j = 1, \cdots, C(n, k). \tag{5}$$

Thus $A^{(k)}$ is a $C(m, k)$ by $C(n, k)$ matrix; the element in the i-th row and j-th column of $A^{(k)}$ is the k-th order determinant $|A(I_i^{(k)} \cdot J_j^{(k)})|$ of a submatrix of A. For convenience define $A^{(0)}$ to be $[1]$.

78.2 Example. Let A be the 3 by 4 matrix $[a_{ij}]$, $i = 1, 2, 3$ and $j = 1, \cdots, 4$; then $A^{(k)}$ is defined for $k = 0, 1, 2, 3$. Equation (5) shows that $A^{(1)} = A$ (the first compound of A always equals A), and the second compound $A^{(2)}$ is described by the following equation:

$$A^{(2)} = \begin{bmatrix} \begin{vmatrix} a_{11} & a_{12} \\ a_{21} & a_{22} \end{vmatrix} & \cdots & \begin{vmatrix} a_{13} & a_{14} \\ a_{23} & a_{24} \end{vmatrix} \\ \cdots & \cdots & \cdots \\ \begin{vmatrix} a_{21} & a_{22} \\ a_{31} & a_{32} \end{vmatrix} & \cdots & \begin{vmatrix} a_{23} & a_{24} \\ a_{33} & a_{34} \end{vmatrix} \end{bmatrix}. \tag{6}$$

Finally, $A^{(3)}$ is the following 1-by-4 matrix whose elements are third order determinants:

$$A^{(3)} = [|A(J_1^{(3)})| \cdots |A(J_4^{(3)})|]. \tag{7}$$

Let A be an n-by-n square matrix. Then $A(N - I_i^{(k)} \cdot N - J_j^{(k)})$ denotes the $(n - k)$-by-$(n - k)$ submatrix of A which is complementary to $A(I_i^{(k)} \cdot J_j^{(k)})$; it is obtained by striking out in A the rows designated by the integers in $I_i^{(k)}$ and the columns designated by the integers in $J_j^{(k)}$. For the sake of completeness, $A(N - I_i^{(n)} \cdot N - J_1^{(n)})$ is defined to be the matrix with the single element 1.

78.3 Definition. Let A be a square matrix $[a_{ij}]$ with n rows and n columns. Let s_i and s_j denote the sums of the k integers in $I_i^{(k)}$ and $J_j^{(k)}$, respectively. The k-th *adjoint compound of* A, denoted by $\text{adj}^{(k)}A$, is defined for each k such that $1 \leqq k \leqq n$ by the following statement:

$$\begin{gathered}\text{adj}^{(k)}A = [(-1)^{s_i+s_j}|A(N - I_i^{(k)} \cdot N - J_j^{(k)})|]^t, \\ i = 1, \cdots, C(n, k), \qquad j = 1, \cdots, C(n, k).\end{gathered} \tag{8}$$

Thus $\text{adj}^{(k)}A$ is a compound matrix whose elements are signed determinants of submatrices of A, each with $n - k$ rows and columns. The superscript t indicates as usual [see Definition 77.13] that the matrix on the right in (8) is transposed. The matrix $\text{adj}^{(1)}A$ is usually denoted by $\text{adj}\, A$ and called the adjoint of A. If A is an n-by-n square matrix, then $\text{adj}^{(n)}A = [1]$.

78.4 Example. Let A be the 3-by-3 matrix $[a_{ij}]$, $i, j = 1, \cdots, 3$. Then $\text{adj}^{(k)}A$ is defined for $k = 1, \cdots, 3$, and

$$\text{adj}^{(1)}A = \text{adj}\, A = \begin{bmatrix} +\begin{vmatrix} a_{22} & a_{23} \\ a_{32} & a_{33} \end{vmatrix} & -\begin{vmatrix} a_{21} & a_{23} \\ a_{31} & a_{33} \end{vmatrix} & +\begin{vmatrix} a_{21} & a_{22} \\ a_{31} & a_{32} \end{vmatrix} \\ -\begin{vmatrix} a_{12} & a_{13} \\ a_{32} & a_{33} \end{vmatrix} & +\begin{vmatrix} a_{11} & a_{13} \\ a_{31} & a_{33} \end{vmatrix} & -\begin{vmatrix} a_{11} & a_{12} \\ a_{31} & a_{32} \end{vmatrix} \\ +\begin{vmatrix} a_{12} & a_{13} \\ a_{22} & a_{23} \end{vmatrix} & -\begin{vmatrix} a_{11} & a_{13} \\ a_{21} & a_{23} \end{vmatrix} & +\begin{vmatrix} a_{11} & a_{12} \\ a_{21} & a_{22} \end{vmatrix} \end{bmatrix}^t, \tag{9}$$

$$\text{adj}^{(2)}A = \begin{bmatrix} +a_{33} & -a_{32} & +a_{31} \\ -a_{23} & +a_{22} & -a_{21} \\ +a_{13} & -a_{12} & +a_{11} \end{bmatrix}^t. \tag{10}$$

Finally, by definition, $\text{adj}^{(3)}A = [1]$, the one by one matrix with the single element 1.

79. Expansions of Determinants

This section establishes two expansions of determinants, namely, the expansion by the elements of a row or column, and Laplace's expansion.

79.1 Theorem. *Let A be the n-by-n matrix $[a_{ij}]$, $i, j = 1, \cdots, n$. Then*

$$\det A = \sum (-1)^{i+j} a_{ij} \det[A(N - I_i^{(1)} \cdot N - J_j^{(1)})]. \tag{1}$$

The summation extends either over the set $i = 1, \cdots, n$ with j arbitrary but fixed, or over the set $j = 1, \cdots, n$ with i arbitrary but fixed.

PROOF. The proof will be given first for $i = 1$ with the summation in (1) extended over the set $j = 1, \cdots, n$. Recall that $A(N - I_i^{(1)} \cdot N - J_j^{(1)})$ is the submatrix obtained by striking out the i-th row and the j-th column of A; recall also the definition of the determinant in Definition 76.1. Then each of the following expressions contains $(n-1)!$ products in $\det A$ and they are all distinct:

$$a_{11} \det[A(N - I_1^{(1)} \cdot N - J_1^{(1)})], \cdots, a_{1n} \det[A(N - I_1^{(1)} \cdot N - J_n^{(1)})]. \tag{2}$$

Thus, these n expressions together contain all of the products in $\det A$, and there remains only the determination of the signs of the expressions. Now

$$a_{11} \det[A(N - I_1^{(1)} \cdot N - J_1^{(1)})] = \sum_{(j_2, \cdots, j_n)} (-1)^{s(1, j_2, \cdots, j_n)} a_{11} a_{2j_2} \cdots a_{nj_n}, \tag{3}$$

and the summation extends over all permutations of $(2, \cdots, n)$. Also, a_{11} is the element with $i = 1$ and $j = 1$, and (3) shows that

$$(-1)^{1+1} a_{11} \det[A(N - I_1^{(1)} \cdot N - J_1^{(1)})] \tag{4}$$

contains $(n-1)!$ terms, including sign, in $\det A$. Next, interchange the j-th column of A with the $j - 1$ columns which precede it; then (4) shows that

$$(-1)^{1+1} a_{1j} \det[A[N - I_1^{(1)} \cdot N - J_j^{(1)})] \tag{5}$$

consists of terms in $(-1)^{j-1} \det A$, or

$$(-1)^{(1+1)+(j-1)} a_{1j} \det[A(N - I_1^{(1)} \cdot N - J_j^{(1)})] \tag{6}$$

consists of terms in $\det A$. These considerations show that

$$\det A = \sum_{j=1}^{n} (-1)^{1+j} a_{1j} \det[A(N - I_1^{(1)} \cdot N - J_j^{(1)})]. \tag{7}$$

The expansion (1) for each row and column can be established in the same way. □

79.2 Example. Let $A = [a_{ij}]$, $i, j = 1, \cdots, 3$, and let A_{ij} be a shortened notation for the submatrix obtained by crossing out row i and column j in A. Then by Theorem 79.1,

$$\det A = \sum_{j=1}^{3} (-1)^{2+j} a_{2j} \det A_{2j} = \sum_{i=1}^{3} (-1)^{i+3} a_{i3} \det A_{i3}. \tag{8}$$

79.3 Theorem (Laplace's Expansion). *Let A be the n-by-n matrix $[a_{ij}]$, $i, j = 1, \cdots, n$; let k be an integer such that $1 \leqq k \leqq n$; and let s_i and s_j have the meaning given them in Definition 78.3. Then the Laplace expansion of* $\det A$ *is this:*

$$\det A = \sum (-1)^{s_i + s_j} \det[A(I_i^{(k)} \cdot J_j^{(k)})] \det[A(N - I_i^{(k)} \cdot N - J_j^{(k)})]. \tag{9}$$

The summation extends either over the set $i = 1, \cdots, C(n, k)$ with j arbitrary but fixed, or over the set $j = 1, \cdots, C(n, k)$ with i arbitrary but fixed.

PROOF. For $k = 1$, Laplace's expansion (9) is the elementary expansion (1); thus it is not surprising that the proof of Theorem 79.3 is similar to that of Theorem 79.1. Consider (9) in the case in which $i = 1$ and $j = 1, \cdots, C(n, k)$; the problem is to prove that

$$\det A = \sum_{j=1}^{C(n,k)} (-1)^{s_1+s_j}\det[A(I_1^{(k)} \cdot J_j^{(k)})]\det[A(N - I_1^{(k)} \cdot N - J_j^{(k)})]. \tag{10}$$

Since an n-th order determinant contains $n!$ products, each term in the sum in (10) contains $k!\,(n-k)!$ products. These sets of products are all distinct, and there are $C(n, k)$, or $n!/[k!(n-k)!]$, of them. Thus the sum contains the $n!$ products in $\det A$, and there remains only the verification of the signs.

The next step is to prove that the $k!(n-k)!$ terms in

$$(-1)^{s_1+s_1}\det[A(I_1^{(k)} \cdot J_1^{(k)})]\det[A(N - I_1^{(k)} \cdot N - J_1^{(k)}] \tag{11}$$

are terms (signs included) in $\det A$. First, $(-1)^{s_1+s_1} = +1$. Next, a term in the first determinant in (11) contains an element from each row and each column of $A(I_1^{(k)} \cdot J_1^{(k)})$, and a term in the second determinant in (11) contains an element from each row and column of $A(N - I_1^{(k)} \cdot N - J_1^{(k)})$; hence, each term in the product of these determinants contains an element from each row and each column of A. Thus, except possibly for sign, a term in the product in (11) is a term in $\det A$. Write all terms so that the first subscripts are in the natural order. Let

$$(-1)^{s(j_1,\cdots,j_k)}a_{1j_1}\cdots a_{kj_k}, \qquad (-1)^{s(j_{k+1},\cdots,j_n)}a_{(k+1)j_{k+1}}\cdots a_{nj_n},$$

be terms in the two determinants in (11). Here $(j_1, \cdots, j_k)$ is a permutation of $(1, \cdots, k)$ and $(j_{k+1}, \cdots, j_n)$ is a permutation of $(k+1, \cdots, n)$. Then the product of these terms is a term in $\det A$ since

$$s(j_1, \cdots, j_n) = s(j_1, \cdots, j_k) + s(j_{k+1}, \cdots, j_n).$$

Thus the $k!(n-k)!$ terms in (11) are terms (signs included) in $\det A$.

Next, consider the following general term in (10):

$$(-1)^{s_1+s_j}\det[A(I_1^{(k)} \cdot J_j^{(k)})]\det[A(N - I_1^{(k)} \cdot N - J_j^{(k)})]. \tag{12}$$

If $J_j^{(k)} = (j_1, \cdots, j_k)$, then by making $(j_1 - 1) + (j_2 - 2) + \cdots + (j_k - k)$ interchanges of adjacent columns—that is, a total of $s_j - s_1$ interchanges of adjacent columns—the columns $J_j^{(k)}$ can be brought into the $J_1^{(k)}$ positions. Then, as shown in establishing (11),

$$(-1)^{s_1+s_1}\det[A(I_1^{(k)} \cdot J_j^{(k)})]\det[A(N - I_1^{(k)} \cdot N - J_j^{(k)})] \tag{13}$$

contains terms in

$$(-1)^{s_j-s_1}\det A. \tag{14}$$

Multiply (13) and (14) each by $(-1)^{s_j-s_1}$; then (13) becomes (12) and (14) becomes $\det A$. Thus (12) contains $k!(n-k)!$ of the terms in $\det A$, and the sum in (10) contains all of the terms in $\det A$. The proof of (9) is complete in the case $i = 1$ and $j = 1, \cdots, C(n, k)$; the proof of (9) in the other cases is entirely similar. □

79.4 Example. Let A be the 4-by-4 matrix $[a_{ij}]$. Let $k = 2$, and use (9) to expand $\det A$ in terms of determinants formed from rows 2 and 3. The sets $I_i^{(2)}$, $i = 1, \cdots, 6$, are given in equation (1) in Section 78, and the sets $J_j^{(2)}$ are the same. Then $I_4^{(2)} = (2, 3)$, $N = (1, 2, 3, 4)$, and (9) is

$$\det A = \sum_{j=1}^{C(4,2)} (-1)^{5+s_j} \det[A(I_4^{(2)} \cdot J_j^{(2)})] \det[A(N - I_4^{(2)} \cdot N - J_j^{(2)})]. \tag{15}$$

Written out in full detail, this formula is the following:

$$\begin{vmatrix} a_{11} & a_{12} & \cdots & a_{14} \\ a_{21} & a_{22} & \cdots & a_{24} \\ \cdots & \cdots & \cdots & \cdots \\ a_{41} & a_{42} & \cdots & a_{44} \end{vmatrix} = \begin{vmatrix} a_{21} & a_{22} \\ a_{31} & a_{32} \end{vmatrix} \begin{vmatrix} a_{13} & a_{14} \\ a_{43} & a_{44} \end{vmatrix} - \begin{vmatrix} a_{21} & a_{23} \\ a_{31} & a_{33} \end{vmatrix} \begin{vmatrix} a_{12} & a_{14} \\ a_{42} & a_{44} \end{vmatrix}$$
$$+ \begin{vmatrix} a_{21} & a_{24} \\ a_{31} & a_{34} \end{vmatrix} \begin{vmatrix} a_{12} & a_{13} \\ a_{42} & a_{43} \end{vmatrix} + \begin{vmatrix} a_{22} & a_{23} \\ a_{32} & a_{33} \end{vmatrix} \begin{vmatrix} a_{11} & a_{14} \\ a_{41} & a_{44} \end{vmatrix}$$
$$- \begin{vmatrix} a_{22} & a_{24} \\ a_{32} & a_{34} \end{vmatrix} \begin{vmatrix} a_{11} & a_{13} \\ a_{41} & a_{43} \end{vmatrix} + \begin{vmatrix} a_{23} & a_{24} \\ a_{33} & a_{34} \end{vmatrix} \begin{vmatrix} a_{11} & a_{12} \\ a_{41} & a_{42} \end{vmatrix}.$$

79.5 Example. Let A be an m-by-m matrix; let B be an m-by-n matrix; let O be an n-by-m zero matrix; and let C be an n-by-n matrix. Then

$$\det \begin{bmatrix} A & B \\ O & C \end{bmatrix} = \det A \det C.$$

The proof is obtained by expanding the determinant on the left by Laplace's expansion using m-th order determinants of submatrices of the first m rows. There is only one non-zero term in the expansion since there is only one non-zero determinant whose matrix is formed from the last n rows.

79.6 Example. Laplace's expansion, using submatrices formed from the first m rows, shows that

$$\begin{vmatrix} a_{11} & \cdots & a_{1n} & 0 & \cdots & 0 \\ \cdots & \cdots & \cdots & \cdots & \cdots & \cdots \\ a_{m1} & \cdots & a_{mn} & 0 & \cdots & 0 \\ b_{11} & \cdots & b_{1n} & c_{11} & \cdots & c_{1m} \\ \cdots & \cdots & \cdots & \cdots & \cdots & \cdots \\ b_{n1} & \cdots & b_{nn} & c_{n1} & \cdots & c_{nm} \end{vmatrix} = \begin{vmatrix} a_{11} & \cdots & a_{1n} \\ \cdots & \cdots & \cdots \\ a_{n1} & \cdots & a_{nn} \end{vmatrix} \begin{vmatrix} c_{11} & \cdots & c_{1n} \\ \cdots & \cdots & \cdots \\ c_{n1} & \cdots & c_{nn} \end{vmatrix}$$

if $m = n$, and that the value of the determinant on the left is zero if $m > n$.

79.7 Theorem. *Let $A = [a_{ij}]$, $i, j = 1, \cdots, n$, and let k be an integer such that $1 \leqq k \leqq n$. Then*

$$A^{(k)} \operatorname{adj}^{(k)} A = \operatorname{adj}^{(k)} A A^{(k)} = |A| I, \tag{16}$$

$$|A^{(k)} \operatorname{adj}^{(k)} A| = |\operatorname{adj}^{(k)} A A^{(k)}| = |A|^{C(n,k)}. \tag{17}$$

PROOF. Recall the definitions of $A^{(k)}$ and $\operatorname{adj}^{(k)} A$ in Section 78. The i-th row of $A^{(k)}$ is

$$\det[A(I_i^{(k)} \cdot J_1^{(k)})] \ \cdots \ \det[A(I_i^{(k)} \cdot J_{C(n,k)}^{(k)})] \tag{18}$$

and the j-th column of $\operatorname{adj}^{(k)} A$ is

$$\begin{array}{c} (-1)^{s_j + s_1} \det[A(N - I_j^{(k)} \cdot N - J_1^{(k)})], \\ \vdots \\ (-1)^{s_j + s_{C(n,k)}} \det[A(N - I_j^{(k)} \cdot N - J_{C(n,k)}^{(k)})]. \end{array} \tag{19}$$

The element in the i-th row and the j-th column of $A^{(k)} \operatorname{adj}^{(k)} A$ is found by multiplying the row (18) of $A^{(k)}$ into the column (19) of $\operatorname{adj}^{(k)} A$; if $j = i$, the result by Theorem 79.3 is $\det A$. Thus each element on the main diagonal of $A^{(k)} \operatorname{adj}^{(k)} A$ is $\det A$. If $i \neq j$ in (18) and (19), the result of multiplying row (18) into column (19) is zero because Theorem 79.3 shows that the result is the expansion of a determinant whose matrix has at least two rows the same because $I_i^{(k)}$ and $N - I_j^{(k)}$ are not disjoint. Thus, each element of $A^{(k)} \operatorname{adj}^{(k)} A$ which is not on the main diagonal is zero, and $A^{(k)} \operatorname{adj}^{(k} A = |A| I$. Recall from Definitions 78.1 and 78.3 that $A^{(n)} = [|A|]$ and $\operatorname{adj}^{(n)} A = [1]$. Similar arguments show that $\operatorname{adj}^{(k)} A A^{(k)} = |A| I$. Since $|A| I$ is a diagonal matrix, $\det[|A| I] = |A|^{C(n,k)}$, and (17) follows from (16). □

79.8 Example. Let $A = [a_{ij}]$, $i, j = 1, \cdots, 3$. Equation (9) in Section 78 contains $\operatorname{adj}^{(1)} A$, written out in full. Theorem 79.1 can be used to show that $A^{(1)} \operatorname{adj}^{(1)} A = \operatorname{adj}^{(1)} A A^{(1)} = |A| I$.

80. The Multiplication Theorems

The multiplication theorems provide an evaluation of the determinant of the product of two matrices and of the determinant of the k-th compound of the product of two matrices.

80.1 Theorem (The Binet–Cauchy Multiplication Theorem). *Let A and B be two matrices with m rows and n columns, and let AB^t denote the row by column matrix product of A and B^t. Then*

$$\begin{aligned} \det AB^t &= 0 & m > n, \quad (1)\\ &= \det A \det B & m = n, \quad (2)\\ &= \sum_{j=1}^{C(n,m)} \det A(J_j^{(m)}) \det B(J_j^{(m)}) & m < n. \quad (3) \end{aligned}$$

PROOF. The most elegant proof of this theorem is based on the following identity for partitioned matrices:

$$\begin{bmatrix} I & A \\ O & I \end{bmatrix} \begin{bmatrix} A & O \\ -I & B^t \end{bmatrix} = \begin{bmatrix} O & AB^t \\ -I & B^t \end{bmatrix}. \tag{4}$$

Each of the partitioned matrices in this equation is an $(m + n)$-by-$(m + n)$ matrix. Since A and B are m-by-n matrices, the dimensions of the other matrices in (4) can be found easily; in each case I is an identity matrix and O is a zero matrix. Multiplication by the first factor on the left in (4) replaces each row of the other factor by itself plus certain multiples of other rows. Since these operations do not change the value of the determinant of a matrix,

$$\begin{vmatrix} A & O \\ -I & B^t \end{vmatrix} = \begin{vmatrix} O & AB^t \\ -I & B^t \end{vmatrix}. \tag{5}$$

Next, evaluate these two determinants by means of Laplace's expansion. In the determinant on the right, AB^t is an m-by-m matrix, and $|AB^t|$ is the only determinant whose matrix is formed from the first m rows and which is not zero. The matrix of $|AB^t|$ lies in rows $1, \cdots, m$ and columns $(n + 1), \cdots, (n + m)$; hence,

$$\begin{gathered} s_i = 1 + 2 + \cdots + m, \qquad s_j = nm + 1 + 2 + \cdots + m, \\ (-1)^{s_i + s_j} |AB^t| |-I| = (-1)^{nm+n} |AB^t|. \end{gathered} \tag{6}$$

Thus the determinant on the right in (5) equals $(-1)^{nm+n}|AB^t|$. Laplace's expansion of the determinant on the left in (5) shows that it equals $|A||B^t|$ if $m = n$, but in the other cases it must be examined more closely. Written out in full, this determinant is

$$\begin{vmatrix} a_{11} & a_{12} & \cdots & a_{1n} & 0 & 0 & \cdots & 0 \\ a_{21} & a_{22} & \cdots & a_{2n} & 0 & 0 & \cdots & 0 \\ \cdots & \cdots & \cdots & \cdots & \cdots & \cdots & \cdots & \cdots \\ a_{m1} & a_{m2} & \cdots & a_{mn} & 0 & 0 & \cdots & 0 \\ -1 & 0 & \cdots & 0 & b_{11} & b_{21} & \cdots & b_{m1} \\ 0 & -1 & \cdots & 0 & b_{12} & b_{22} & \cdots & b_{m2} \\ \cdots & \cdots & \cdots & \cdots & \cdots & \cdots & \cdots & \cdots \\ 0 & 0 & \cdots & -1 & b_{1n} & b_{2n} & \cdots & b_{mn} \end{vmatrix}. \tag{7}$$

If $m > n$, an expansion of this determinant by Laplace's method, using submatrices of m rows and columns selected from the first m rows, shows that its value is zero; (1) follows from this statement and (6). The same expansion is used when $m \leqq n$. Consider a term of the expansion in which the first factor is one of the determinants $|A(J_j^{(m)})|$. If $J_j^{(m)} = (j_1, j_2, \cdots, j_m)$, then in this expansion

$$s_i + s_j = 1 + 2 + \cdots + m + j_1 + j_2 + \cdots + j_m. \tag{8}$$

The second factor can be expanded by Laplace's method using determinants of m-by-m submatrices formed from the last m columns. There is a single term in this expansion, and its value is $|[B(J_j^{(m)})]^t||-I|$ except perhaps for sign. For this second expansion,

$$\begin{aligned} s_i + s_j &= (n - m + 1) + (n - m + 2) + \cdots + (n - m + m) \\ &\quad + j_1 + j_2 + \cdots + j_m \\ &= m(n - m) + 1 + 2 + \cdots + m + j_1 + j_2 + \cdots + j_m. \end{aligned} \tag{9}$$

Thus (8) and (9) show that the term in the expansion of (7) is $(-1)^{m(n-m)}|A(J_j^{(m)})||B(J_j^{(m)})||-I|$, and this term equals

$$(-1)^{mn+n}|A(J_j^{(m)})||B(J_j^{(m)})|, \tag{10}$$

since $|-I| = (-1)^{n-m}$. Equation (10) is valid for $j = 1, \cdots, C(n, m)$, and (7) is the sum of these terms in (10). Thus, if $m \leqq n$, then (6) and (10) show that

$$(-1)^{mn+n}|AB^t| = (-1)^{mn+n} \sum_{j=1}^{C(n,m)} |A(J_j^{(m)})||B(J_j^{(m)})|. \tag{11}$$

Since this equation simplifies to (2) and (3), the proof of Theorem 80.1 is complete. □

80.2 Example. If A and B are 2-by-3 matrices, then (7) is

$$\begin{vmatrix} a_{11} & a_{12} & a_{13} & 0 & 0 \\ a_{21} & a_{22} & a_{23} & 0 & 0 \\ -1 & 0 & 0 & b_{11} & b_{21} \\ 0 & -1 & 0 & b_{12} & b_{22} \\ 0 & 0 & -1 & b_{13} & b_{23} \end{vmatrix}. \tag{12}$$

There are three terms in Laplace's expansion of this determinant using 2-by-2 matrices formed from the first two rows. One of these terms is

$$(-1)^{1+2+2+3} \begin{vmatrix} a_{12} & a_{13} \\ a_{22} & a_{23} \end{vmatrix} \begin{vmatrix} -1 & b_{11} & b_{21} \\ 0 & b_{12} & b_{22} \\ 0 & b_{13} & b_{23} \end{vmatrix}.$$

Laplace's expansion shows that

$$\begin{vmatrix} -1 & b_{11} & b_{21} \\ 0 & b_{12} & b_{22} \\ 0 & b_{13} & b_{23} \end{vmatrix} = (-1)^{2+3+2+3} \begin{vmatrix} b_{12} & b_{22} \\ b_{13} & b_{23} \end{vmatrix} \det[-1] = -\begin{vmatrix} b_{12} & b_{13} \\ b_{22} & b_{23} \end{vmatrix}.$$

Thus one of the terms in the expansion of (12) is

$$-\begin{vmatrix} a_{12} & a_{13} \\ a_{22} & a_{23} \end{vmatrix} \begin{vmatrix} b_{12} & b_{13} \\ b_{22} & b_{23} \end{vmatrix}.$$

Since $m = 2$ and $n = 3$, equations (5) and (6) show that (12) equals $-|AB^t|$. The complete evaluation of (12) shows that

$$\begin{aligned} |AB^t| = & \begin{vmatrix} a_{11} & a_{12} \\ a_{21} & a_{22} \end{vmatrix} \begin{vmatrix} b_{11} & b_{12} \\ b_{21} & b_{22} \end{vmatrix} + \begin{vmatrix} a_{11} & a_{13} \\ a_{21} & a_{23} \end{vmatrix} \begin{vmatrix} b_{11} & b_{13} \\ b_{21} & b_{23} \end{vmatrix} \\ & + \begin{vmatrix} a_{12} & a_{13} \\ a_{22} & a_{23} \end{vmatrix} \begin{vmatrix} b_{12} & b_{13} \\ b_{22} & b_{23} \end{vmatrix}. \end{aligned} \tag{13}$$

An important extension of Theorem 80.1 concerns the k-th compound of the product of two matrices.

80.3 Theorem. *Let A be a matrix with m rows and n columns, and let B be a matrix with n rows and m columns. If k is an integer such that $1 \leqq k \leqq m$ and $1 \leqq k \leqq n$, then*

$$|(AB)^{(k)}| = |A^{(k)}B^{(k)}|, \tag{14}$$

$$|A^{(k)}B^{(k)}| = |AB|^{C(m-1,k-1)} \tag{15}$$

PROOF. The proof of (14) follows, as explained below, from these matrix identities:

$$\begin{aligned} (AB)^{(k)} &= [|A(I_i^{(k)})B(J_j^{(k)})|] \\ &= \left[\sum_{t=1}^{C(n,k)} |A(I_i^{(k)} \cdot J_t^{(k)})||B(I_t^{(k)} \cdot J_j^{(k)})|\right] \\ &= [|A(I_i^{(k)} \cdot J_j^{(k)})|][|B(I_i^{(k)} \cdot J_j^{(k)})|] \\ &= A^{(k)}B^{(k)}. \end{aligned} \tag{16}$$

The first equality here results from the observation that the element in the i-th row and j-th column of $(AB)^{(k)}$ is the determinant of the matrix obtained by multiplying the i-th combination $A(I_i^{(k)})$ of rows of A into the j-th combination $B(J_j^{(k)})$ of columns of B. The second equality follows from an application of (3) in the Binet–Cauchy multiplication theorem. The third equality results from the definition of matrix multiplication. The fourth equality results from the definition of the k-th compound. The proof of (14) now follows from equating the determinants of the first and last matrices.

The proof of (15) follows from the Sylvester–Franke theorem which states that

$$|(AB)^{(k)}| = |AB|^{C(m-1,k-1)}$$

(see Theorem 82.1 below). Although the proof will not be complete until the Sylvester–Franke theorem has been proved, the result is included here so that the statement of the multiplication theorem is complete. □

81. Sylvester's Theorem of 1839 and 1851

This theorem gives an important series of representations for the product of two determinants.

Let A and B be two square matrices. Then $A[B(J_i^{(k)})/A(J_j^{(k)})]$, or $A[B_i^{(k)}/A_j^{(k)}]$ for short, denotes the matrix obtained by replacing in order the k columns $A(J_j^{(k)})$ in A by the k columns $B(J_i^{(k)})$ in B.

81.1 Theorem (Sylvester's Theorem of 1839 and 1851). *Let A and B be two matrices with n rows and n columns, and let k be an integer such that $1 \leqq k \leqq n$. Then*

$$\begin{aligned} \sum_{t=1}^{C(n,k)} |A[B(J_t^{(k)})/A(J_i^{(k)})]||B[A(J_j^{(k)})/B(J_t^{(k)})]| &= 0, \qquad i \neq j, \\ &= |A||B|, \qquad i = j. \end{aligned} \tag{1}$$

Proof. The most elegant proof of this theorem is obtained by equating corresponding elements in two evaluations of the matrix product $\operatorname{adj}^{(k)} AB^{(k)} \operatorname{adj}^{(k)} BA^{(k)}$. First recall (16) in Theorem 79.7 and the fact that matrix multiplication is associative. Then

$$\begin{aligned} \operatorname{adj}^{(k)} AB^{(k)} \operatorname{adj}^{(k)} BA^{(k)} &= \operatorname{adj}^{(k)} A \cdot B^{(k)} \operatorname{adj}^{(k)} B \cdot A^{(k)} \\ &= \operatorname{adj}^{(k)} A \cdot |B| I \cdot A^{(k)} \\ &= \operatorname{adj}^{(k)} AA^{(k)} \cdot |B| I \\ &= |A||B| I. \end{aligned}$$

Again,

$$\begin{aligned} \operatorname{adj}^{(k)} AB^{(k)} \operatorname{adj}^{(k)} BA^{(k)} &= \operatorname{adj}^{(k)} AB^{(k)} \cdot \operatorname{adj}^{(k)} BA^{(k)} \\ &= [|A[B(J_j^{(k)})/A(J_i^{(k)})]|][|B[A(J_j^{(k)})/B(J_i^{(k)})]| \\ &= \left[\sum_{t=1}^{C(n,k)} |A[B(J_t^{(k)})/A(J_i^{(k)})]||B[A(J_j^{(k)})/B(J_t^{(k)})]| \right]. \end{aligned}$$

In the matrices on the right, the element in the i-th row and j-th column is shown in each case. Equations (1) follow by equating corresponding elements in the two evaluations of the matrix product. □

81.2 Example. Let A and B be 3-by-3 matrices, let $k = 1$, and let $i = j = 1$ in equation (1). Then (1), written out in full detail in this case, is the following:

$$\begin{vmatrix} a_{11} & a_{12} & a_{13} \\ a_{21} & a_{22} & a_{23} \\ a_{31} & a_{32} & a_{33} \end{vmatrix} \begin{vmatrix} b_{11} & b_{12} & b_{13} \\ b_{21} & b_{22} & b_{23} \\ b_{31} & b_{32} & b_{33} \end{vmatrix} = \begin{vmatrix} b_{11} & a_{12} & a_{13} \\ b_{21} & a_{22} & a_{23} \\ b_{31} & a_{32} & a_{33} \end{vmatrix} \begin{vmatrix} a_{11} & b_{12} & b_{13} \\ a_{21} & b_{22} & b_{23} \\ a_{31} & b_{32} & b_{33} \end{vmatrix}$$

$$+ \begin{vmatrix} b_{12} & a_{12} & a_{13} \\ b_{22} & a_{22} & a_{23} \\ b_{32} & a_{32} & a_{33} \end{vmatrix} \begin{vmatrix} b_{11} & a_{11} & b_{13} \\ b_{21} & a_{21} & b_{23} \\ b_{31} & a_{31} & b_{33} \end{vmatrix} + \begin{vmatrix} b_{13} & a_{12} & a_{13} \\ b_{23} & a_{22} & a_{23} \\ b_{33} & a_{32} & a_{33} \end{vmatrix} \begin{vmatrix} b_{11} & b_{12} & a_{11} \\ b_{21} & b_{22} & a_{21} \\ b_{31} & b_{32} & a_{31} \end{vmatrix}.$$

This identity can be described as follows. Replace the first column of A by the j-th column of B, and replace the j-th column of B by the first column of A; form the product of the determinants of the resulting matrices. Then $|A||B|$ equals the sum of these products formed for $j = 1, 2, 3$.

81.3 Example. Let A and B be 3-by-3 matrices, let $k = 2$, and let $i = j = 2$ in equation (1). Then (1), written out in full detail in this case, is the following:

$$\begin{vmatrix} a_{11} & a_{12} & a_{13} \\ a_{21} & a_{22} & a_{23} \\ a_{31} & a_{32} & a_{33} \end{vmatrix} \begin{vmatrix} b_{11} & b_{12} & b_{13} \\ b_{21} & b_{22} & b_{23} \\ b_{31} & b_{32} & b_{33} \end{vmatrix} = \begin{vmatrix} b_{11} & a_{12} & b_{12} \\ b_{21} & a_{22} & b_{22} \\ b_{31} & a_{32} & b_{32} \end{vmatrix} \begin{vmatrix} a_{11} & a_{13} & b_{13} \\ a_{21} & a_{23} & b_{23} \\ a_{31} & a_{33} & b_{33} \end{vmatrix}$$

$$+ \begin{vmatrix} b_{11} & a_{12} & b_{13} \\ b_{21} & a_{22} & b_{23} \\ b_{31} & a_{32} & b_{33} \end{vmatrix} \begin{vmatrix} a_{11} & b_{12} & a_{13} \\ a_{21} & b_{22} & a_{23} \\ a_{31} & b_{32} & a_{33} \end{vmatrix} + \begin{vmatrix} b_{12} & a_{12} & b_{13} \\ b_{22} & a_{22} & b_{23} \\ b_{32} & a_{32} & b_{33} \end{vmatrix} \begin{vmatrix} b_{11} & a_{11} & a_{13} \\ b_{21} & a_{21} & a_{23} \\ b_{31} & a_{31} & a_{33} \end{vmatrix}.$$

82. The Sylvester–Franke Theorem

This theorem evaluates $|A^{(k)}|$ in terms of A.

82.1 Theorem (The Sylvester–Franke Theorem). *Let A be a square matrix with n rows and n columns. Then*

$$|A^{(k)}| = |A|^{C(n-1,k-1)}, \qquad k = 1, \cdots, n. \tag{1}$$

Proof. The theorem will be proved by mathematical induction on n. The theorem is obviously true for $n = 2$. Assume that (1) is true for all determinants of order $n - 1$; a proof will now be given that (1) is true for all determinants of order n. Since (1) is true for $k = n$, the remaining proof will be given under the assumption that $1 \leqq k \leqq n - 1$. A study of Example 82.3 below [see especially the third order determinant in (10)] will help the reader to follow the steps in the proof.

Consider a_{nn} as a variable and all other elements in A as constants. Then

$|A|$ and $|A^{(k)}|$ are functions of a_{nn}, and $|A| = Da_{nn} + E$. Here D is the determinant of the submatrix of A obtained by crossing out the row and column which contain a_{nn}, and E is the determinant of the matrix obtained by replacing a_{nn} in A by 0. There is no loss of generality in assuming that the elements in $A^{(k)}$ are ordered in such a way that a_{nn} occurs only in the submatrix

$$[|A(I_i^{(k)} \cdot J_j^{(k)})|], \qquad i, j = 1, \cdots, r, \tag{2}$$

where $r = C(n-1, k-1)$; this assumption is made only for convenience in notation. Furthermore, $|A(I_i^{(k)} \cdot J_j^{(k)})| = d_{ij}a_{nn} + e_{ij}$, $i, j = 1, \cdots, r$, and d_{ij} and e_{ij} are determinants similar to D and E. It follows from the induction hypothesis, namely, the hypothesis that the theorem is true for all determinants of order $(n-1)$, that

$$|d_{ij}| = D^{C(n-2,k-2)}, \qquad i, j = 1, \cdots, r.$$

Then the determinant of the matrix in (2) is the following polynomial in a_{nn} of degree r:

$$D^{C(n-2,k-2)}a_{nn}^r + \cdots. \tag{3}$$

Next, observe that

$$[|A(I_i^{(k)} \cdot J_j^{(k)})|], \qquad i, j = r+1, \cdots, C_{(n,k)}, \tag{4}$$

is the submatrix in $A^{(k)}$ which is complementary to the one in (2), and that it is the k-th compound of the matrix of D. Now $1 \leqq k \leqq n-1$ and the matrix of D is an $(n-1)$-by-$(n-1)$ matrix; then by the induction hypothesis, the determinant of the matrix in (4) is $D^{C(n-2,k-1)}$.

Now use Laplace's method to expand $|A^{(k)}|$, using submatrices formed from the first r rows. It follows from results already established that the first term in the expansion is $D^{C(n-2,k-2)+C(n-2,k-1)}a_{nn}^r + \cdots$, a polynomial in a_{nn} of degree r. The other terms in the expansion are polynomials in a_{nn} of degree less than r. Since $C(n-2, k-2) + C(n-2, k-1) = C(n-1, k-1) = r$, it follows that

$$|A^{(k)}| = D^r a_{nn}^r + \cdots. \tag{5}$$

Furthermore, by Theorems 79.7 and 80.1,

$$|A^{(k)}||\operatorname{adj}^{(k)} A| = |A|^{C(n,k)} = (Da_{nn} + E)^{C(n,k)}. \tag{6}$$

Thus every value of a_{nn} for which $|A^{(k)}|$ vanishes is a zero of $|A|^{C(n,k)}$. But the latter, equal to $(Da_{nn} + E)^{C(n,k)}$, vanishes only for $a_{nn} = -E/D$ [there is a comment later about the assumption $D \neq 0$]. Thus all zeros of the polynomial in (5) are $a_{nn} = -E/D$, and

$$|A^{(k)}| = D^r(a_{nn} + E/D)^r = (Da_{nn} + E)^r = |A|^r = |A|^{C(n-1,k-1)}. \tag{7}$$

Thus (1) is true for determinants of order n if it is true for determinants of order $n-1$; then since (1) is true for determinants of the second order, Theorem 82.1 is true by a complete induction.

If $D = 0$, then A is the limit of a matrix B in which $D \neq 0$, and the theorem is true for the modified matrix B. Since $|B|$ and $|B^{(k)}|$ are continuous functions of the elements of B, then

$$|A^{(k)}| = \lim |B^{(k)}| = \lim |B|^{C(n-1,k-1)} = |A|^{C(n-1,k-1)}. \tag{8}$$

The proof of Theorem 82.1 is complete. □

82.2 Corollary. Let A be a square matrix with n rows and n columns. Then

$$|\mathrm{adj}^{(k)} A| = |A|^{C(n-1,k)}, \qquad k = 1, \cdots, n. \tag{9}$$

The proof follows from (6) and (1). Recall that $|\mathrm{adj}^{(n)} A| = 1$. □

82.3 Example. Equation (9) can be proved also by proving that $|\mathrm{adj}^{(k)} A| = |A^{(n-k)}|$, $k = 1, \cdots, (n-1)$, and recalling the conventions that $|\mathrm{adj}^{(n)} A| = 1$ and $|A^{(0)}| = 1$. Consider an example. Let A be the matrix $[a_{ij}]$, $i, j = 1, 2, 3$. Then equation (9) in Section 78 displays $\mathrm{adj}^{(1)} A$. Remove the factor -1 from the second row and also from the second column of $|\mathrm{adj}^{(1)} A|$; the result is the following:

$$|\mathrm{adj}^{(1)} A| = \begin{vmatrix} \begin{vmatrix} a_{22} & a_{23} \\ a_{32} & a_{33} \end{vmatrix} & \begin{vmatrix} a_{12} & a_{13} \\ a_{32} & a_{33} \end{vmatrix} & \begin{vmatrix} a_{12} & a_{13} \\ a_{22} & a_{23} \end{vmatrix} \\ \begin{vmatrix} a_{21} & a_{23} \\ a_{31} & a_{33} \end{vmatrix} & \begin{vmatrix} a_{11} & a_{13} \\ a_{31} & a_{33} \end{vmatrix} & \begin{vmatrix} a_{11} & a_{13} \\ a_{21} & a_{23} \end{vmatrix} \\ \begin{vmatrix} a_{21} & a_{22} \\ a_{31} & a_{32} \end{vmatrix} & \begin{vmatrix} a_{11} & a_{12} \\ a_{31} & a_{32} \end{vmatrix} & \begin{vmatrix} a_{11} & a_{12} \\ a_{21} & a_{22} \end{vmatrix} \end{vmatrix} \tag{10}$$

This determinant is certainly similar to $|A^{(2)}|$, but the elements of the matrix are not ordered in the standard manner specified in Section 78 for the elements of $A^{(2)}$. Nevertheless, the 3rd order determinant in (10) can be changed into $|A^{(2)}|$, with the standard ordering of elements, by the following operations: (a) interchange the first and third columns in (10) and then interchange the first and third rows; these two operations leave the value of the determinant unchanged; and (b) interchange rows and columns in the matrix; this operation also leaves the value of the determinant unchanged. Thus, the third order determinant in (10) is $|A^{(2)}|$; this example illustrates the fact that $|\mathrm{adj}^{(k)} A| = |A^{(n-k)}|$ in all cases.

The third order determinant in (10), which is $|A^{(2)}|$, is a useful example to study in connection with the proof of the Sylvester–Franke theorem. The element a_{33} occurs only in rows 1 and 2 and columns 1 and 2, and it is easy to follow the steps in the proof by referring to the example in (10). If the rows and columns of $A^{(2)}$ are ordered in the standard manner as described above, then a_{33} occurs only in rows 2 and 3 and columns 2 and 3.

Other proofs of the Sylvester–Franke theorem are known. Those in the following two papers differ from each other and from the one given above.

Leonard Tornheim, The Sylvester–Franke theorem, *American Mathematical Monthly*, **59** (1952) 389–391.

Harley Flanders, A note on the Sylvester–Franke theorem, *American Mathematical Monthly*, **60** (1953) 543–545.

83. The Bazin–Reiss–Picquet Theorem

Recall the meaning of $B[A(J_i^{(k)})/B(J_j^{(k)})]$, or $B[A_i^{(k)}/B_j^{(k)}]$ for short, from Section 81.

83.1 Theorem (The Bazin–Reiss–Picquet Theorem). *Let A and B be two square matrices with n rows and n columns each, and let k be an integer such that $1 \leqq k \leqq n$. Then*

$$|A|^{C(n-1,k-1)}|B|^{C(n-1,k)} = \|B[A_j^{(k)}/B_i^{(k)}]\|, \qquad i, j = 1, \cdots, C(n, k). \quad (1)$$

The element shown in the determinant on the right is the one in the i-th row and j-th column.

Proof. The proof once more is based on a matrix identity. An application of Laplace's expansion shows that

$$\mathrm{adj}^{(k)} BA^{(k)} = [|B[A(J_j^{(k)})/B(J_i^{(k)})]|], \qquad i, j = 1, \cdots, C(n, k). \quad (2)$$

Since the two matrices in (2) are equal, their determinants are equal. Now

$$|\mathrm{adj}^{(k)} BA^{(k)}| = |\mathrm{adj}^{(k)} B||A^{(k)}| \quad (3)$$

by the Binet–Cauchy multiplication theorem [Theorem 80.1]. Also,

$$|\mathrm{adj}^{(k)} B| = |B|^{C(n-1,k)} \quad (4)$$

by Corollary 82.2, and

$$|A^{(k)}| = |A|^{C(n-1,k-1)} \quad (5)$$

by the Sylvester–Franke theorem [Theorem 82.1]. The proof of the theorem is completed by equating the determinants of the two sides of (2) and then using (3), (4), and (5) to evaluate the determinant on the left. □

83.2 Example. Let $A = [a_{ij}]$ and $B = [b_{ij}]$, $i, j = 1, 2$. If $k = 1$, then the identity in (1) is the following:

$$|A||B| = \begin{vmatrix} \begin{vmatrix} a_{11} & b_{12} \\ a_{21} & b_{22} \end{vmatrix} & \begin{vmatrix} a_{12} & b_{12} \\ a_{22} & b_{22} \end{vmatrix} \\ \begin{vmatrix} b_{11} & a_{11} \\ b_{21} & a_{21} \end{vmatrix} & \begin{vmatrix} b_{11} & a_{12} \\ b_{21} & a_{22} \end{vmatrix} \end{vmatrix}.$$

This identity can be given various other forms; one which occurs in the applications is obtained as follows. If $|B| \neq 0$, divide the left side of the equation by $|B|^2$, and divide the right side by $|B|^2$ by dividing each row of the determinant matrix by $|B|$.

83.3 Example. Let $A = [a_{ij}]$ and $B = [b_{ij}]$, $i, j = 1, \cdots, 3$. If $k = 2$, then the identity in (1) is the following:

$$|A|^2|B| = \begin{vmatrix} |B[A_1^{(2)}/B_1^{(2)}]| & |B[A_2^{(2)}/B_1^{(2)}]| & |B[A_3^{(2)}/B_1^{(2)}]| \\ |B[A_1^{(2)}/B_2^{(2)}]| & |B[A_2^{(2)}/B_2^{(2)}]| & |B[A_3^{(2)}/B_2^{(2)}]| \\ |B[A_1^{(2)}/B_3^{(2)}]| & |B[A_2^{(2)}/B_3^{(2)}]| & |B[A_3^{(2)}/B_3^{(2)}]| \end{vmatrix}$$

83.4 Example. Let

$$A = \begin{vmatrix} a_{11} & a_{12} & 1 \\ a_{21} & a_{22} & 1 \\ a_{31} & a_{32} & 1 \end{vmatrix}, \qquad B = \begin{vmatrix} b_{11} & b_{12} & 1 \\ b_{21} & b_{22} & 1 \\ b_{31} & b_{32} & 1 \end{vmatrix}.$$

If $k = 1$, the formula in (1) is the following:

$$|A||B|^2 = \begin{vmatrix} |B[A_1^{(1)}/B_1^{(1)}]| & |B[A_2^{(1)}/B_1^{(1)}]| & |B[A_3^{(1)}/B_1^{(1)}]| \\ |B[A_1^{(1)}/B_2^{(1)}]| & |B[A_2^{(1)}/B_2^{(1)}]| & |B[A_3^{(1)}/B_2^{(1)}]| \\ |B[A_1^{(1)}/B_3^{(1)}]| & |B[A_2^{(1)}/B_3^{(1)}]| & |B[A_3^{(1)}/B_3^{(1)}]| \end{vmatrix}.$$

Now $A_3^{(1)}$ denotes the third column of A, which is a column of 1's. Then

$$|B[A_3^{(1)}/B_1^{(1)}]| = 0, \qquad |B[A_3^{(1)}/B_2^{(1)}]| = 0$$

because the matrices of these determinants have two columns which are identical. Also, $|B[A_3^{(1)}/B_3^{(1)}]| = |B|$ because the third column of A is the same as the third column of B. Use the third column to expand the third order determinant; if $|B| \neq 0$, then

$$|A||B| = \begin{vmatrix} |B[A_1^{(1)}/B_1^{(1)}]| & |B[A_2^{(1)}/B_1^{(1)}]| \\ |B[A_1^{(1)}/B_2^{(1)}]| & |B[A_2^{(1)}/B_2^{(1)}]| \end{vmatrix}.$$

If $|B| \neq 0$, another form of this formula can be obtained as follows: divide the left side of the equation by $|B|^2$, and divide the right side by $|B|^2$ by dividing each row of the determinant matrix by $|B|$. Similar formulas can be obtained in all cases in which A and B are n-by-n matrices, $|B| \neq 0$, and $k = 1$.

83.5 Example. Let A be the n-by-n identity matrix I, and let $k = 1$. Then $|A| = |I| = 1$, and the formula in (1) can be simplified to $|B|^{n-1} = |\text{adj}\, B|$. This result is equation (9) in Section 82 in the special case $k = 1$.

84. Inner Products

This section defines the inner product of two vectors in $\mathbb{R}^n$ and in $\mathbb{C}^n$, and it establishes the elementary properties of these products. As usual, $\bar{z}$ denotes the complex conjugate of the complex number z.

84.1 Definition. Let $u:(a_1, \cdots, a_n)$ and $v:(b_1, \cdots, b_n)$ be two vectors in $\mathbb{R}^n$. Then the *inner product of u and v*, denoted by (u, v), is defined as follows:

$$(u, v) = \sum_{j=1}^{n} a_j b_j. \tag{1}$$

If $u:(z_1, \cdots, z_n)$ and $v:(w_1, \cdots, w_n)$ are vectors in $\mathbb{C}^n$, then

$$(u, v) = \sum_{j=1}^{n} z_j \bar{w}_j \tag{2}$$

84.2 Theorem. *The inner product of two real vectors $u:(a_1, \cdots, a_n)$ and $v:(b_1, \cdots, b_n)$ has the following properties:*

(3) $(u, v) = (v, u)$;
(4) $(su_1 + tu_2, v) = s(u_1, v) + t(u_2, v)$ *if s and t are real numbers*;
(5) $(u, u) \geqq 0$;
(6) $(u, u) = 0$ *if and only if* $u = (0, \cdots, 0)$;
(7) *if θ is the angle between the vectors u and v, then* $(u, v) = (u, u)^{1/2}(v, v)^{1/2} \cos\theta$;
(8) $(u, v) = 0$ *if u and v are orthogonal.*

PROOF. Properties (3), $\cdots$, (6) follow easily from the definition of (u, v) in (1). To prove (7), apply the Law of Cosines to the triangle whose vertices are $(0, \cdots, 0)$, $(a_1, \cdots, a_n)$, and $(b_1, \cdots, b_n)$. The distance formula can be written in the inner product notation [see (14) below]; the lengths of the sides of the triangles are $(u, u)^{1/2}$, $(v, v)^{1/2}$, and $(u - v, u - v)^{1/2}$. Since θ is the angle between the vectors u and v, then

$$(u - v, u - v) = (u, u) + (v, v) - 2(u, u)^{1/2}(v, v)^{1/2} \cos\theta. \tag{9}$$

Since $(u - v, u - v) = (u, u) + (v, v) - 2(u, v)$ by (3) and (4), then (9) simplifies to the equation in (7). If u and v are orthogonal, then $\theta = \pi/2$, $\cos\theta = 0$, and $(u, v) = 0$. □

84.3 Theorem. *The inner product of two complex vectors $u:(z_1, \cdots, z_n)$ and $v:(w_1, \cdots, w_n)$ has the following properties:*

(10) $(u, v) = \overline{(v, u)}$;
(11) *if s, t are complex numbers, then*

$$(su_1 + tu_2, v) = s(u_1, v) + t(u_2, v),$$

$$(u, sv_1 + tv_2) = \bar{s}(u, v_1) + \bar{t}(u, v_2);$$

(12) $(u, u) \geqq 0$;
(13) $(u, u) = 0$ *if and only if* $u = (0, \cdots, 0)$.

The proofs of all these properties of (u, v) follow from the definition in (2) of the inner product of complex vectors. □

Two complex vectors u and v are defined to be orthogonal if and only if $(u, v) = 0$.

The inner product notation is useful for stating many of the familiar formulas of mathematics. For example, if $x : (x_1, \cdots, x_n)$ and $y : (y_1, \cdots, y_n)$ are two points in $\mathbb{R}^n$, then the formula for the distance $|x - y|$ between these points can be written as follows:

$$|x - y| = \left[\sum_{j=1}^{n} (x_j - y_j)^2 \right]^{1/2} = (x - y, x - y)^{1/2}. \tag{14}$$

Also, the product of two matrices can be written in the inner product notation. Let A and B be the following m-by-n matrices of real numbers:

$$A = \begin{bmatrix} a_{11} & a_{12} & \cdots & a_{1n} \\ a_{21} & a_{22} & \cdots & a_{2n} \\ \cdots & \cdots & \cdots & \cdots \\ a_{m1} & a_{m2} & \cdots & a_{mn} \end{bmatrix}, \qquad B = \begin{bmatrix} b_{11} & b_{12} & \cdots & b_{1n} \\ b_{21} & b_{22} & \cdots & b_{2n} \\ \cdots & \cdots & \cdots & \cdots \\ b_{m1} & b_{m2} & \cdots & b_{mn} \end{bmatrix}. \tag{15}$$

Let $u_1, \cdots, u_m$ and $v_1, \cdots, v_m$ denote the vectors in the rows of A and B respectively. Then

$$\det AB^t = \begin{vmatrix} (u_1, v_1) & (u_1, v_2) & \cdots & (u_1, v_m) \\ (u_2, v_1) & (u_2, v_2) & \cdots & (u_2, v_m) \\ \cdots & \cdots & \cdots & \cdots \\ (u_m, v_1) & (u_m, v_2) & \cdots & (u_m, v_m) \end{vmatrix}, \qquad A, B \text{ real}. \tag{16}$$

If A and B are matrices with complex elements, and if $\bar{B}$ denotes the matrix whose elements are the complex conjugates of those in B, then

$$\det A\bar{B}^t = \begin{vmatrix} (u_1, v_1) & (u_1, v_2) & \cdots & (u_1, v_m) \\ (u_2, v_1) & (u_2, v_2) & \cdots & (u_2, v_m) \\ \cdots & \cdots & \cdots & \cdots \\ (u_m, v_1) & (u_m, v_2) & \cdots & (u_m, v_m) \end{vmatrix}, \qquad A, B \text{ complex}. \tag{17}$$

85. Linearly Independent and Dependent Vectors; Rank of a Matrix

This section contains definitions of, and necessary and sufficient conditions for, linear dependence and linear independence of vectors. Furthermore, it defines the row rank, the column rank, and the rank of a matrix and proves that they are equal.

85.1 Definition. The vectors $u_1, \cdots, u_m$ in $\mathbb{R}^n$ (in $\mathbb{C}^n$) are said to be *linearly dependent* if and only if there exist constants $t_1, \cdots, t_m$ in $\mathbb{R}$ (in $\mathbb{C}$), not all zero, such that

$$t_1 u_1 + \cdots + t_m u_m = 0. \tag{1}$$

The vectors $u_1, \cdots, u_m$ are *linearly independent* if and only if they are not linearly dependent.

If $a_1, \cdots, a_m$ are constants and $u = a_1 u_1 + \cdots + a_m u_m$, then u is said to be a linear combination of $u_1, \cdots, u_m$. If $u_1, \cdots, u_m$ are linearly dependent, then at least one of these vectors is a linear combination of the others. If, for example, t_1 is not zero in (1), then

$$u_1 = -\left(\frac{t_2}{t_1} u_2 + \cdots + \frac{t_m}{t_1} u_m\right). \tag{2}$$

In the same way, u_k is a linear combination of the remaining vectors if t_k is not zero.

85.2 Theorem. *The vectors $u_1, \cdots, u_m$ are linearly independent if and only if $t_1 u_1 + \cdots + t_m u_m = 0$ implies that $t_1 = 0, \cdots, t_m = 0$.*

PROOF. If $u_1, \cdots, u_m$ are linearly independent and $t_1 u_1 + \cdots + t_m u_m = 0$, then $t_1 = 0, \cdots, t_m = 0$ because otherwise $u_1, \cdots, u_m$ would be linearly dependent by Definition 85.1. Also, if $t_1 u_1 + \cdots + t_m u_m = 0$ implies that $t_1 = 0, \cdots, t_m = 0$, then the vectors $u_1, \cdots, u_m$ are linearly independent since it is impossible for them to satisfy the condition for linear dependence. □

85.3 Theorem. *Let A be the square matrix $[a_{ij}]$, $i, j = 1, \cdots, n$, and let $u_1, \cdots, u_n$ and $v_1, \cdots, v_n$ be respectively the row vectors and the column vectors in A. Then*

if $\det A = 0$, *both the row and column vectors are linearly dependent*; (3)

if $\det A \neq 0$, *both the row and column vectors are linearly independent*. (4)

PROOF. The vector equation (1), with m replaced by n, is equivalent to the following system of equations:

$$\begin{aligned} a_{11} t_1 + a_{21} t_2 + \cdots + a_{n1} t_n &= 0, \\ a_{12} t_1 + a_{22} t_2 + \cdots + a_{n2} t_n &= 0, \\ \cdots\cdots\cdots\cdots\cdots\cdots\cdots\cdots\cdots \\ a_{1n} t_1 + a_{2n} t_2 + \cdots + a_{nn} t_n &= 0. \end{aligned} \tag{5}$$

The matrix of coefficients in this system of equations is A^t. If $\det A = 0$, then $\det A^t = 0$ by Theorem 77.14, and the system (5) has a non-trivial solution. Then $u_1, \cdots, u_n$ are linearly dependent by Definition 85.1. If

$\det A \neq 0$, then $\det A^t \neq 0$, and (5) has only the trivial solution $t_1 = 0, \cdots, t_n = 0$; therefore, $u_1, \cdots, u_n$ are linearly independent by Theorem 85.2. The proofs of the statements in (3) and (4) about the column vectors in A are similar to those about row vectors. □

85.4 Theorem. *Let A be an m-by-n matrix with complex elements. A necessary and sufficient condition that the row vectors $u_1, \cdots, u_m$ in A be linearly dependent is that*

$$\det A\bar{A}^t = 0. \tag{6}$$

PROOF. First, prove (6) under the assumption that $u_1, \cdots, u_m$ are linearly dependent. Then one of the vectors $u_1, \cdots, u_m$ is a linear combination of the others; consider, for example, the case in which u_1 is a linear combination of $u_2, \cdots, u_m$. Then there are constants $t_2, \cdots, t_m$ such that

$$u_1 = t_2 u_2 + \cdots + t_m u_m, \tag{7}$$

$$(u_1, u_j) = t_2(u_2, u_j) + \cdots + t_m(u_m, u_j), \qquad j = 1, \cdots, m. \tag{8}$$

Now by (17) in Section 84,

$$\det A\bar{A}^t = \begin{vmatrix} (u_1, u_1) & (u_1, u_2) & \cdots & (u_1, u_m) \\ (u_2, u_1) & (u_2, u_2) & \cdots & (u_2, u_m) \\ \cdots & \cdots & \cdots & \cdots \\ (u_m, u_1) & (u_m, u_2) & \cdots & (u_m, u_m) \end{vmatrix}, \tag{9}$$

and (8) shows that the vector in the first row of the determinant matrix is a linear combination of the vectors in the other rows. Then $\det A\bar{A}^t = 0$ by the theorems on determinants in Section 77. The proof that (6) is a necessary condition for linear dependence is complete.

Next, assume (6) and prove that $u_1, \cdots, u_m$ are linearly dependent. The determinant of the matrix of coefficients of the system of equations

$$\begin{aligned} (u_1, u_1)t_1 + (u_1, u_2)t_2 + \cdots + (u_1, u_m)t_m &= 0, \\ (u_2, u_1)t_1 + (u_2, u_2)t_2 + \cdots + (u_2, u_m)t_m &= 0, \\ \cdots\cdots\cdots\cdots\cdots\cdots\cdots\cdots\cdots\cdots \\ (u_m, u_1)t_1 + (u_m, u_2)t_2 + \cdots + (u_m, u_m)t_m &= 0, \end{aligned} \tag{10}$$

is $\det A\bar{A}^t$, which is zero by hypothesis. Then the system (10) has a non-trivial solution $t_1, \cdots, t_m$. Use the properties of the inner product in Theorem 84.3 to write equations (10) in the following form:

$$\begin{aligned} (u_1, \bar{t}_1 u_1 + \cdots + \bar{t}_m u_m) &= 0, \\ (u_2, \bar{t}_1 u_1 + \cdots + \bar{t}_m u_m) &= 0, \\ \cdots\cdots\cdots\cdots\cdots\cdots \\ (u_m, \bar{t}_1 u_1 + \cdots + \bar{t}_m u_m) &= 0. \end{aligned} \tag{11}$$

Next, multiply these equations in order by $\bar{t}_1, \cdots, \bar{t}_m$ and add them; by (11) in Theorem 84.3, the resulting equation can be simplified to this:

$$\left(\sum_{j=1}^{m} \bar{t}_j u_j, \sum_{j=1}^{m} \bar{t}_j u_j\right) = 0. \tag{12}$$

Then by (13) in Theorem 84.3,

$$\sum_{j=1}^{m} \bar{t}_j u_j = 0. \tag{13}$$

Since $\bar{t}_1, \cdots, \bar{t}_m$ are not all zero, (13) shows that the vectors $u_1, \cdots, u_m$ are linearly dependent. The proof of Theorem 85.4 is complete. □

85.5 Corollary. *Let $u_1, \cdots, u_m$ be vectors in $\mathbb{R}^n$ or $\mathbb{C}^n$. If $m > n$, then $u_1, \cdots, u_m$ are linearly dependent.*

PROOF. If $m > n$, then $\det A\bar{A}^t = 0$ by (1) in Theorem 80.1 and $u_1, \cdots, u_m$ are linearly dependent by Theorem 85.4. □

85.6 Corollary. *A necessary and sufficient condition that the row vectors $u_1, \cdots, u_m$ in Theorem 85.4 be linearly independent is that*

$$\det A\bar{A}^t \neq 0. \tag{14}$$

PROOF. First, prove (14) under the assumption that $u_1, \cdots, u_m$ are linearly independent. If $\det A\bar{A}^t$ were zero, then $u_1, \cdots, u_m$ would be linearly dependent by Theorem 85.4; thus $\det A\bar{A}^t \neq 0$. Next, assume (14) and prove that $u_1, \cdots, u_m$ are linearly independent. If $u_1, \cdots, u_m$ were linearly dependent, then $\det A\bar{A}^t$ would equal zero by Theorem 85.4; therefore, $u_1, \cdots, u_m$ are linearly independent. □

85.7 Example. If

$$A = \begin{bmatrix} 1+i & 2-i & 2+2i \\ 1+2i & 3-2i & 1+4i \\ 2+i & 1-i & 1+2i \end{bmatrix},$$

then $\det A = 4 + 21i$, the three vectors in the rows of A are linearly independent, and the three vectors in the columns of A are also linearly independent.

85.8 Remark. If A is an m-by-n matrix, then $A^{(0)}$ is defined to be $[1]$. Thus $|A(I_i^{(0)} \cdot J_j^{(0)})|$ is defined if and only if $i = j = 1$; and $|A(I_1^{(0)} \cdot J_1^{(0)})| = 1$. These conventions are employed in the statement of the next definition.

85.9 Definition. Let A be the m-by-n complex matrix $[a_{ij}]$ with row vectors $u_1, \cdots, u_m$ and column vectors $v_1, \cdots, v_n$. The *row rank* (respectively *column rank*) of A is the maximum number of linearly independent vectors in the set $u_1, \cdots, u_m$ (in the set $v_1, \cdots, v_n$). The *rank* of A is r if and only if

$$|A(I_i^{(k)} \cdot J_j^{(k)})| = 0, \qquad i = 1, \cdots, C(m, k), \qquad j = 1, \cdots, C(n, k), \quad \text{and} \quad k > r, \tag{15}$$

$$|A(I_i^{(r)} \cdot J_j^{(r)})| \neq 0 \text{ for at least one value of } i \text{ and } j. \tag{16}$$

85.10 Remark. The rank of A is r if and only if $A^{(k)}$ is the zero matrix for $k > r$ and $A^{(r)}$ is not the zero matrix.

85.11 Example. Find the row rank, the column rank, and the rank of the matrix

$$A = \begin{bmatrix} 1 & 2 & -3 & 5 \\ -4 & 7 & -18 & 4 \\ 2 & -1 & 4 & 2 \end{bmatrix}$$

Now $2u_1 - u_2 - 3u_3 = 0$; hence, the three row vectors in A are linearly dependent, and the row rank of A is less than three. If

$$B = \begin{bmatrix} 1 & 2 & -3 & 5 \\ 2 & -1 & 4 & 2 \end{bmatrix},$$

then $\det BB^t = 971$, and u_1 and u_3 are linearly independent by Corollary 85.6 (u_1, u_2 and u_2, u_3 are also linearly independent pairs). Then by Definition 85.9, the row rank of A is two. Since four vectors in $\mathbb{C}^3$ are linearly dependent by Corollary 85.5, the column rank of A is equal to or less than three. Since u_1, u_2, u_3 are linearly dependent, then $|A(J_j^{(3)})| = 0$ for $j = 1, \cdots, 4$. This fact and Theorem 85.3 show that each three of the column vectors $v_1, \cdots, v_4$ are linearly dependent. Since each two vectors in the set $v_1, \cdots, v_4$ are linearly independent (see $A^{(2)}$ below), the column rank of A is two. Finally,

$$A^{(3)} = [0 \quad 0 \quad 0 \quad 0],$$

$$A^{(2)} = \begin{bmatrix} 15 & -30 & 24 & -15 & -27 & 78 \\ -5 & 10 & -8 & 5 & 9 & -26 \\ -10 & 20 & -16 & 10 & 18 & -52 \end{bmatrix}.$$

Since $A^{(3)}$ is the zero matrix and $A^{(2)}$ is not the zero matrix, the rank of A is two. Observe that the row rank, the column rank, and the rank are equal.

85.12 Theorem. *Let A be an m-by-n complex matrix $[a_{ij}]$. Then the row rank, the column rank, and the rank of A are equal.*

PROOF. If A is the zero matrix, then the row rank, the column rank, and the rank of A are each equal to zero by Definition 85.9; hence, they are equal and the theorem is true in this case. In the remainder of the proof, assume that A is not the zero matrix.

Corollary 85.5 shows that the row and the column ranks of A are equal to or less than the minimum of m and n. Also, if $A(I_i^{(k)} \cdot J_j^{(k)})$ is a submatrix of A, then $k \leqq \min(m, n)$ and the rank of A is equal to or less than $\min(m, n)$. Let r be the rank of A. Then by Definition 85.9 there is a submatrix $A(I_i^{(r)} \cdot J_j^{(r)})$ of A such that

$$|A(I_i^{(r)} \cdot J_j^{(r)})| \neq 0, \tag{17}$$

$$|A(I_i^{(k)} \cdot J_j^{(k)})| = 0, \qquad i = 1, \cdots, C(m, k), \qquad j = 1, \cdots, C(n, k), \quad \text{and} \quad k > r. \tag{18}$$

Then the row vectors designated by $I_i^{(r)}$ are linearly independent by (17) and Corollary 85.6; also, if $k > r$, every set of k row vectors is linearly dependent by (18) and Theorem 85.4. Then the row rank of A is r by Definition 85.9. Finally, the r column vectors designated by $J_j^{(r)}$ are linearly independent by (17) and Corollary 85.6; also, if $k > r$, every set of k column vectors is linearly dependent by (18), Theorem 85.4, and Corollary 85.5. Then the column rank of A is r by Definition 85.9. Thus the row rank, the column rank, and the rank of A are each equal to r and therefore are equal, and the proof of Theorem 85.12 is complete. □

85.13 Remark. If A is an m-by-n complex matrix, then its rank is an integer r such that $0 \leqq r \leqq \min(m, n)$.

86. Schwarz's Inequality

This section proves that $\det A\bar{A}^t \geqq 0$ and, as a corollary, establishes Schwarz's inequality, a fundamental inequality for the inner product of two vectors.

86.1 Theorem. *Let A be an m-by-n matrix with complex elements. Then*

$$\det A\bar{A}^t \geqq 0. \tag{1}$$

The equality holds in (1) *if and only if the vectors $u_1, \cdots, u_m$ are linearly dependent.*

PROOF. First prove (1). If $m > n$, then $\det A\bar{A}^t = 0$ by (1) in Theorem 80.1, and (1) in Theorem 86.1 is true. If $m \leqq n$, then

$$\det A\bar{A}^t = \sum_{j=1}^{C(n,m)} \det A(J_j^{(m)}) \det \bar{A}(J_j^{(m)}) \tag{2}$$

by (2) and (3) in Theorem 80.1. Since $\overline{z_1 z_2} = \bar{z}_1 \bar{z}_2$ and $\overline{z_1 + z_2} = \bar{z}_1 + \bar{z}_2$, the definition of the determinant shows (since the determinant is a sum of products) that $\det \bar{A}(J_j^{(m)})$ is the complex conjugate of $\det A(J_j^{(m)})$. Thus (2) can be written in the following form:

$$\det A\bar{A}^t = \sum_{j=1}^{C(n,m)} \det A(J_j^{(m)}) \overline{\det A(J_j^{(m)})}. \tag{3}$$

This equation expresses $\det A\bar{A}^t$ as the inner product of a complex vector with itself; hence, $\det A\bar{A}^t \geqq 0$ by (12) in Section 84, and the proof of (1) is complete. Next, Theorem 85.4 shows that the equality holds in (1) if and only if the row vectors $u_1, \cdots, u_m$ are linearly dependent. □

86.2 Corollary (Schwarz's Inequality). *Let $u:(z_1, \cdots, z_n)$ and $v:(w_1, \cdots, w_n)$ be two vectors in $\mathbb{C}^n$. Then*

$$(u, v)(v, u) \leqq (u, u)(v, v), \tag{4}$$

and the equality holds if and only if u and v are linearly dependent.

Proof. Let A be the 2-by-n matrix whose rows are u and v. Then by Theorem 86.1,

$$\det A\bar{A}^t = \begin{vmatrix} (u, u) & (u, v) \\ (v, u) & (v, v) \end{vmatrix} \geqq 0, \tag{5}$$

and the equality holds if and only if u and v are linearly dependent. Expand the determinant in (5); then the inequality in (5) is equivalent to the one in (4). □

Since $(v, u) = \overline{(u, v)}$ by (10) in Theorem 84.3, then $(u, v)(v, u) = |(u, v)|^2$, and the inequality in (4) can be written in the following forms:

$$|(u, v)|^2 \leqq (u, u)(v, v); \tag{6}$$

$$\left| \sum_{j=1}^{n} z_j \bar{w}_j \right|^2 \leqq \sum_{j=1}^{n} |z_j|^2 \sum_{j=1}^{n} |w_j|^2. \tag{7}$$

The inequality in (4), (6), and (7) will be called Schwarz's inequality in this book. It is sometimes called Cauchy's inequality, and it is known by other names also.

86.3 Example. If $u = (2 + i, 4 - 3i, 5 + 2i)$ and $v = (3 - 2i, 5 + 3i, 7 - 4i)$, then

$$\sum_{j=1}^{3} z_j \bar{w}_j = (2 + i)(3 + 2i) + (4 - 3i)(5 - 3i) + (5 + 2i)(7 + 4i)$$

$$= 42 + 14i,$$

$$\left| \sum_{j=1}^{3} z_j \bar{w}_j \right|^2 = 1746 + 196 = 1960.$$

$$\sum_{j=1}^{3} |z_j|^2 \sum_{j=1}^{3} |w_j|^2 = (5 + 25 + 29)(13 + 34 + 65)$$

$$= (59)(112) = 6608.$$

Inequality (7) is satisfied.

87. Hadamard's Determinant Theorem

Let A be an m-by-n complex matrix. The purpose of this section is to prove a fundamental inequality for $\det A\bar{A}^t$; for square matrices this inequality simplifies to Hadamard's determinant theorem. Hadamard's theorem is an inequality for determinants which, in the simplest cases, is geometrically obvious. If A is a 3-by-3 real matrix whose rows are the vectors $u_1, \cdots, u_3$, then $|\det A|$ is the volume of the parallelepiped whose edges are $u_1, \cdots, u_3$. For fixed lengths of $u_1, \cdots, u_3$, this volume is maximum when the edges are mutually perpendicular; thus $|\det A| \leqq (u_1, u_1)^{1/2} \cdots (u_3, u_3)^{1/2}$. This example of Hadamard's determinant theorem indicates also the nature of the inequality in the general case.

87.1 Theorem. *Let A be an m-by-n complex matrix whose rows are the vectors $u_1, \cdots, u_m$ in $\mathbb{C}^n$. Then*

$$\det A\bar{A}^t \leqq (u_1, u_1) \cdots (u_m, u_m). \tag{1}$$

The equality holds in (1) *if one or more of the vectors $u_1, \cdots, u_m$ is the zero vector. If no one of the vectors is the zero vector, then the equality holds if and only if the vectors $u_1, \cdots, u_m$ are mutually orthogonal.*

Proof. If $u_1, \cdots, u_m$ are linearly dependent, then $\det A\bar{A}^t = 0$ by Theorem 85.4; since $(u_1, u_1) \cdots (u_m, u_m) \geqq 0$ by (12) in Theorem 84.3, the inequality (1) holds in this case. Furthermore, the equality holds if one or more of the vectors is the zero vector because $\det A\bar{A}^t = 0$ and one or more of the factors $(u_1, u_1), \cdots, (u_m, u_m)$ is zero in this case. Thus, in the remainder of the proof, the vectors $u_1, \cdots, u_m$ are assumed to be linearly independent (non-zero) vectors and $m \leqq n$. Then Theorem 86.1 shows that

$$\det A\bar{A}^t = \begin{vmatrix} (u_1, u_1) & (u_1, u_2) & \cdots & (u_1, u_m) \\ (u_2, u_1) & (u_2, u_2) & \cdots & (u_2, u_m) \\ \cdots & \cdots & \cdots & \cdots \\ (u_m, u_1) & (u_m, u_2) & \cdots & (u_m, u_m) \end{vmatrix} > 0. \tag{2}$$

The strategy in the proof is to modify the matrix in (2) so that the m-th row contains zeros in all positions except the m-th column. Insight into the proof which follows can be gained by studying the proof of Theorem 20.5.

Let $t_1, \cdots, t_{m-1}$ denote complex constants whose values are to be determined later; define the vector h by the following equation:

$$h = u_m - (t_1 u_1 + \cdots + t_{m-1} u_{m-1}). \tag{3}$$

Choose the constants $t_1, \cdots, t_{m-1}$ so that, if possible, h is orthogonal to $u_1, \cdots, u_{m-1}$, that is, so that

$$(h, u_1) = 0, \cdots, (h, u_{m-1}) = 0. \tag{4}$$

These equations for $t_1, \cdots, t_{m-1}$, written out in full, are the following:

$$\begin{aligned}
(u_1, u_1)t_1 &+ (u_2, u_1)t_2 + \cdots + (u_{m-1}, u_1)t_{m-1} = (u_m, u_1), \\
(u_1, u_2)t_1 &+ (u_2, u_2)t_2 + \cdots + (u_{m-1}, u_2)t_{m-1} = (u_m, u_2), \\
&\cdots\cdots\cdots\cdots\cdots\cdots\cdots\cdots\cdots\cdots\cdots\cdots \\
(u_1, u_{m-1})t_1 &+ (u_2, u_{m-1})t_2 + \cdots + (u_{m-1}, u_{m-1})t_{m-1} = (u_m, u_{m-1}).
\end{aligned} \tag{5}$$

Let B denote the matrix of coefficients of this system of equations. Since $u_1, \cdots, u_m$ are linearly independent, then $u_1, \cdots, u_{m-1}$ are linearly independent, and Theorem 86.1 shows that $\det B^t > 0$. Thus, $\det B > 0$ and the system (5) has a unique solution for $t_1, \cdots, t_{m-1}$; henceforth, let $t_1, \cdots, t_{m-1}$ denote this solution.

Multiply the first $m-1$ rows of the matrix in (2) by $t_1, \cdots, t_{m-1}$, respectively, and subtract them from the last row. Then by equations (3) and (5), Theorem 77.11, and properties of the inner product,

$$\det A\bar{A}^t = \begin{vmatrix} (u_1, u_1) & (u_1, u_2) & \cdots & (u_1, u_m) \\ \cdots & \cdots & \cdots & \cdots \\ (u_{m-1}, u_1) & (u_{m-1}, u_2) & \cdots & (u_{m-1}, u_m) \\ 0 & 0 & \cdots & (h, u_m) \end{vmatrix}. \tag{6}$$

Now by (3), $u_m = h + (t_1u_1 + \cdots + t_{m-1}u_{m-1})$; hence, by (4) and properties of the inner product,

$$\begin{aligned}
(h, u_m) &= (h, h + t_1u_1 + \cdots + t_{m-1}u_{m-1}) \\
&= (h, h) + \bar{t}_1(h, u_1) + \cdots + \bar{t}_{m-1}(h, u_{m-1}) \\
&= (h, h).
\end{aligned} \tag{7}$$

Also, similar arguments demonstrate that

$$\begin{aligned}
(u_m, u_m) &= \left(h + \sum_{j=1}^{m-1} t_ju_j, h + \sum_{j=1}^{m-1} t_ju_j\right) \\
&= (h, h) + \left(\sum_{j=1}^{m-1} t_ju_j, \sum_{j=1}^{m-1} t_ju_j\right).
\end{aligned} \tag{8}$$

Therefore

$$(h, h) = (u_m, u_m) - \left(\sum_{j=1}^{m-1} t_ju_j, \sum_{j=1}^{m-1} t_ju_j\right), \tag{9}$$

$$(h, h) \leqq (u_m, u_m). \tag{10}$$

Now (9) shows that the equality holds in (10) if and only if

$$t_1u_1 + \cdots + t_{m-1}u_{m-1} = 0. \tag{11}$$

But since $u_1, \cdots, u_{m-1}$ are linearly independent vectors, (11) is true if and only if

$$t_1 = 0, \cdots, t_{m-1} = 0. \tag{12}$$

Finally, (12) is the solution of the system (5) if and only if

$$(u_m, u_1) = 0, \cdots, (u_m, u_{m-1}) = 0. \tag{13}$$

Thus the equality holds in (10) if and only if u_m is orthogonal to each of the vectors $u_1, \cdots, u_{m-1}$. Now to summarize the results, (6) and (7) prove that

$$\det A\bar{A}^t = \begin{vmatrix} (u_1, u_1) & \cdots & (u_1, u_{m-1}) \\ \cdots & \cdots & \cdots \\ (u_{m-1}, u_1) & \cdots & (u_{m-1}, u_{m-1}) \end{vmatrix} (h, h). \tag{14}$$

Then (10), Theorem 86.1, and the linear independence of $u_1, \cdots, u_{m-1}$ show that

$$\det A\bar{A}^t \leqq \begin{vmatrix} (u_1, u_1) & \cdots & (u_1, u_{m-1}) \\ \cdots & \cdots & \cdots \\ (u_{m-1}, u_1) & \cdots & (u_{m-1}, u_{m-1}) \end{vmatrix} (u_m, u_m). \tag{15}$$

Finally, (11), (12), and (13) prove that the equality holds in (15) if and only if u_m is orthogonal to $u_1, \cdots, u_{m-1}$.

The determinant on the right in (15) satisfies the same hypotheses that were satisfied by the original determinant in (2); hence the entire argument can be repeated to prove that

$$\det A\bar{A}^t \leqq \begin{vmatrix} (u_1, u_1) & \cdots & (u_1, u_{m-2}) \\ \cdots & \cdots & \cdots \\ (u_{m-2}, u_1) & \cdots & (u_{m-2}, u_{m-2}) \end{vmatrix} (u_{m-1}, u_{m-1})(u_m, u_m). \tag{16}$$

The equality holds in (16) if and only if u_m is orthogonal to each of the vectors $u_1, \cdots, u_{m-1}$ and u_{m-1} is orthogonal to each of the vectors $u_1, \cdots, u_{m-2}$. Carried to its conclusion, this process proves that

$$\det A\bar{A}^t \leqq (u_1, u_1) \cdots (u_m, u_m), \tag{17}$$

and that the equality holds in (17) if and only if each vector in the list $u_1, \cdots, u_m$ is orthogonal to each of those which precedes it; that is, the equality holds in (17) if and only if the vectors in each pair of vectors in the set $u_1, \cdots, u_m$ are orthogonal. The proof of Theorem 87.1 is complete. □

87.2 Corollary (Hadamard's Determinant Theorem). *Let A be an n-by-n complex matrix whose rows are the vectors $u_1, \cdots, u_n$ and whose columns are the vectors $v_1, \cdots, v_n$. Then*

$$|\det A| \leqq (u_1, u_1)^{1/2} \cdots (u_n, u_n)^{1/2}, \tag{18}$$

$$|\det A| \leqq (v_1, v_1)^{1/2} \cdots (v_n, v_n)^{1/2}. \tag{19}$$

The equality holds in (18) *if one or more of the vectors $u_1, \cdots, u_n$ is the zero vector. If no one of the vectors $u_1, \cdots, u_n$ is the zero vector, then the equality holds in* (18) *if and only if the vectors $u_1, \cdots, u_n$ are mutually orthogonal. Similar statements hold for* (19) *and the vectors $v_1, \cdots, v_n$.*

Proof. Equation (2) in Theorem 80.1 and equations (2) and (3) in Section 86 show that

$$\det A\bar{A}^t = (\det A)(\overline{\det A}) = |\det A|^2.$$

Then by Theorem 87.1,

$$|\det A|^2 \leqq (u_1, u_1) \cdots (u_n, u_n), \tag{20}$$

and (18) follows. If one of the vectors $u_1, \cdots, u_n$ is the zero vector, then the equality holds in (20) and (18) since both sides of the inequality are zero. If no one of the vectors $u_1, \cdots, u_n$ is the zero vector, then Theorem 87.1 shows that the equality in (20), and hence in (18), holds if and only if the vectors $u_1, \cdots, u_n$ are mutually orthogonal. Thus the proof of (18) and the statements concerning the vectors $u_1, \cdots, u_n$ is complete. The proof of (19) and the statements concerning $v_1, \cdots, v_n$ is obtained by applying the first part of the theorem to $\det A^t$. □

87.3 Example. Let $u_1, \cdots, u_3$ and $v_1, \cdots, v_3$ denote respectively the row vectors and the column vectors in the matrix

$$A = \begin{bmatrix} 2 & 3 & 1 \\ 4 & 1 & 5 \\ 3 & -2 & 4 \end{bmatrix}.$$

No two row vectors are orthogonal, and no two column vectors are orthogonal. Now

$$\det A = 14,$$

$$[(u_1, u_1)(u_2, u_2)(u_3, u_3)]^{1/2} = [14 \cdot 42 \cdot 29]^{1/2} = [17{,}052]^{1/2} \approx 130.58,$$

$$[(v_1, v_1)(v_2, v_2)(v_3, v_3)]^{1/2} = [29 \cdot 14 \cdot 42]^{1/2} = [17{,}052]^{1/2} \approx 130.58.$$

These values verify the statements and inequalities in Corollary 87.2.

87.4 Example. Let $u_1, \cdots, u_3$ and $v_1, \cdots, v_3$ denote respectively the vectors in the rows and in the columns of the matrix

$$A = \begin{bmatrix} 3 & -2 & 5 \\ 4 & 6 & 0 \\ -15 & 10 & 13 \end{bmatrix}.$$

Since $(u_1, u_2) = (u_1, u_3) = (u_2, u_3) = 0$, the vectors $u_1, \cdots, u_3$ are mutually orthogonal. Now

$$\det A = 988,$$

$$[(u_1, u_1)(u_2, u_2)(u_3, u_3)]^{1/2} = [38 \cdot 52 \cdot 494]^{1/2} = [976{,}144]^{1/2} = 988,$$

$$[(v_1, v_1)(v_2, v_2)(v_3, v_3)]^{1/2} = [250 \cdot 140 \cdot 194]^{1/2} = [6{,}790{,}000]^{1/2}$$

$$\approx 2605.76.$$

These results verify the statements and inequalities in Corollary 87.2.

APPENDIX 2

Real Numbers, Euclidean Spaces, and Functions

88. Some Properties of the Real Numbers

The real numbers $\mathbb{R}$, although one of the most familiar of all objects, nevertheless are one of the complicated structures in mathematics. This section does not construct the real numbers nor even give a complete account of their properties; rather, it summarizes some of the most important properties of the real numbers which are needed and used in this book.

The real numbers $\mathbb{R}$ consist of the natural numbers, the integers, the rational numbers, and the irrational numbers. Two operations are defined in $\mathbb{R}$; they are called addition (denoted by $+$) and multiplication (denoted by $\times$, $\cdot$, or juxtaposition). The real numbers form a commutative group with respect to addition; the identity element is called zero and denoted by 0. The set $\mathbb{R} - \{0\}$ is a commutative group with respect to multiplication; the identity element of this group is called one and denoted by 1. Furthermore, multiplication is distributive with respect to addition; that is, if a, b, c are in $\mathbb{R}$, then

$$a(b + c) = ab + ac. \tag{1}$$

Because of these properties, the system $(\mathbb{R}, +, \times)$ is said to be a commutative (or Abelian) ring.

An order relation, denoted by $<$ and read "less than," is defined in $\mathbb{R}$. For every a, b, in $\mathbb{R}$, exactly one of the following relations holds:

$$a = b, \qquad a < b, \qquad b < a. \tag{2}$$

In addition, the order relation in $\mathbb{R}$ has the following properties for all a, b, c, d in $\mathbb{R}$:

if $a < b$ and $b < c$, then $a < c$ (that is, the relation $<$ is transitive); (3)

if $a < b$, then $a + c < b + c$; (4)

if $a < b$ and $c < d$, then $a + c < b + d$; (5)

if $a < b$ and $0 < c$, then $ac < bc$; (6)

if $a < b$, then $-b < -a$. (7)

88.1 Definition. If $E \subset \mathbb{R}$ and $b \in \mathbb{R}$, and if $x \leqq b$ for every x in E, then b is called an *upper bound of* E. If, in addition, b is an element in E, then b is the *maximum* of the set E. Similarly, if $a \in \mathbb{R}$ and $a \leqq x$ for every x in E, then a is a *lower bound of* E. If in addition, a is an element in E, then a is the *minimum* of E.

Every finite set of real numbers has a maximum and a minimum. For example, the maximum of $\{-8, -7, \cdots, 0, \cdots, 9, 10\}$ is 10, and the minimum is -8. The set $\{1, 2, \cdots\}$ has 1 for its minimum, but it has no upper bound. The bounded set $\{x : 0 < x < 1\}$ has 1 and 0 for upper and lower bounds respectively, but it has neither a maximum nor a minimum. However, 1 and 0 are the maximum and minimum respectively of the set $\{x : 0 \leqq x \leqq 1\}$. If E is a subset of $\mathbb{R}$ which has an upper bound, then the order property (3) can be used to show that E has an infinite number of upper bounds. If E has a maximum element m, then m is the least upper bound of E. If E does not have a maximum, then the situation is not intuitively clear, but the least upper bound axiom for $\mathbb{R}$ contains the answer.

88.2 Axiom (Least Upper Bound Axiom). *If E is a non-empty subset of* $\mathbb{R}$ *which has an upper bound, then E has a least upper bound.*

If E has a lower bound, then the following argument shows that E has a greatest lower bound. If $-E = \{-x : x \in E\}$, then the order properties of $\mathbb{R}$ show that $-E$ has an upper bound. Then $-E$ has a least upper bound by Axiom 88.2, and the negative of the least upper bound of $-E$ is the greatest lower bound of E.

The least upper bound and the greatest lower bound of E are denoted by l.u.b.(E) and g.l.b.(E), respectively. Many mathematicians prefer the single words *supremum* and *infimum* in place of least upper bound and greatest lower bound, respectively; in symbols,

$$\begin{aligned} \text{supremum of } E &= \sup(E) = \text{l.u.b.}(E), \\ \text{infimum of } E &= \inf(E) = \text{g.l.b.}(E). \end{aligned} \tag{8}$$

The supremum and infimum of E are generalizations of maximum and minimum of E.

88.3 Theorem. The supremum of E has the following two properties:

if x is in E, then $x \leqq \sup(E)$; (9)

for every $\varepsilon > 0$, *there exists an x in E such that* $x > \sup(E) - \varepsilon$. (10)

The infimum of E has analogous properties.

PROOF. The supremum has property (9) because $\sup(E)$ is an upper bound; it has property (10) because otherwise $\sup(E) - \varepsilon$ would be an upper bound which is smaller than the least upper bound. Also, $\inf(E) \leqq x$ for every x in E and $x < \inf(E) + \varepsilon$ for some x in E for similar reasons. □

88.4 Definition. Let $\{a_k : a_k \in \mathbb{R}, k = 1, 2, \cdots\}$ be a sequence whose values are real numbers. The sequence has a *limit* a_0 in $\mathbb{R}$ if and only if to each $\varepsilon > 0$ there corresponds a positive integer $K(\varepsilon)$ such that

$$|a_k - a_0| < \varepsilon, \qquad k \geqq K(\varepsilon). \tag{11}$$

If the sequence has a limit, we write

$$\lim_{k\to\infty} a_k = a_0 \quad \text{or} \quad \lim \{a_k : k = 1, 2, \cdots\} = a_0.$$

88.5 Theorem. *Let $\{a_k : a_k \in \mathbb{R}, k = 1, 2, \cdots\}$ be a monotonically increasing sequence which is bounded above; that is, assume that*

$$a_1 \leqq a_2 \leqq \cdots \leqq a_k \leqq \cdots, \tag{12}$$

$$a_k \leqq b, \qquad k = 1, 2, \cdots. \tag{13}$$

Let E denote the set $\{a_k : k = 1, 2, \cdots\}$. Then the sequence $\{a_k : k = 1, 2, \cdots\}$ has a limit and

$$\lim_{k\to\infty} a_k = \sup(E) \leqq b. \tag{14}$$

PROOF. The set E is bounded by (13); then $\sup(E)$ exists by Axiom 88.2 and $\sup(E) \leqq b$. Then for each $\varepsilon > 0$ there exists, by (10) in Theorem 88.3, an element a_m in E such that

$$a_m > \sup(E) - \varepsilon. \tag{15}$$

Then since the sequence is monotonically increasing by (12),

$$\sup(E) - \varepsilon < a_m \leqq a_k \leqq \sup(E), \qquad k \geqq m. \tag{16}$$

Then by Definition 88.4,

$$\lim_{k\to\infty} a_k = \sup(E), \tag{17}$$

and the proof of (14) and of the theorem is complete. □

89. Introduction to $\mathbb{R}^3$

Euclidean spaces arise in the beginning from a study of physical space. Descartes introduced coordinates (ordered triples of real numbers) for points in space and thus enabled mathematics to employ algebra in the study of geometry; he proceeded as follows. Let three mutually perpen-

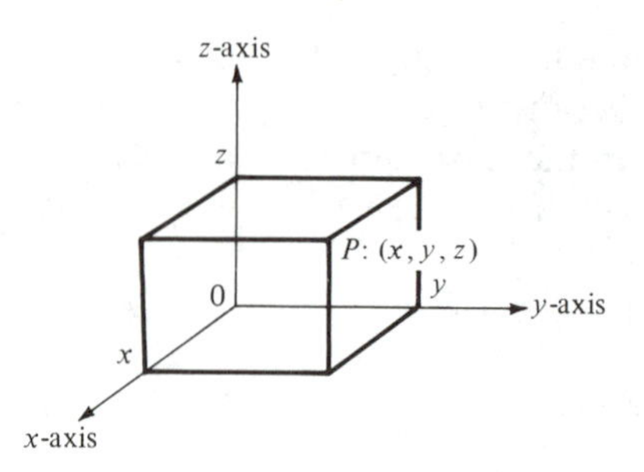

Figure 89.1. Coordinates in $\mathbb{R}^3$.

dicular planes be given; they intersect in a point called the origin, and they intersect by pairs in three straight lines which intersect at the origin. The three lines are called the axes; label them x-axis, y-axis, and z-axis. Choose a positive direction on each axis so that the three axes form a right hand system; that is, choose positive directions so that when the thumb on the right hand points in the direction of the positive z-axis, the fingers on that hand point in the direction which rotates the positive x-axis into the positive y-axis. Choose a unit distance and use it to mark off coordinates on the three axes. Use these coordinates to locate a point in space as follows. Through a point P in space construct three planes which are perpendicular, respectively, to the three axes. Let x be the coordinate, on the x-axis, of the point in which the plane through P intersects the x-axis; similarly, let y and z be the corresponding coordinates on the y-axis and the z-axis. Then (x, y, z) are the coordinates of P. Observe that to each P there corresponds a unique ordered triple (x, y, z), and that to each ordered triple (x, y, z) of real numbers there corresponds a unique point P. Space consists of the set of all points P, and there is a one-to-one correspondence between the points P in space and the set $\mathbb{R}^3$ of all triples (x, y, z) of real numbers. The construction which establishes this one-to-one correspondence is shown in Figure 89.1.

There are at least three ways of looking at (x, y, z) as follows: (a) x, y, z are the coordinates of a point P; (b) x, y, z are the components of a vector from the origin $(0, 0, 0)$ to the point $P:(x, y, z)$; and (c) x, y, z are the components of a free vector whose initial point can be placed at an arbitrary point in space. Vectors are added by adding their components, and vectors are multiplied by a scalar by multiplying their components by the scalar. Thus the following two operations are defined in the set $\mathbb{R}^3$ of all triples (x, y, z):

$$\begin{aligned}(x_1, y_1, z_1) \oplus (x_2, y_2, z_2) &= (x_1 + x_2, y_1 + y_2, z_1 + z_2). \\ a \odot (x, y, z) &= (ax, ay, az), \qquad a \in \mathbb{R}.\end{aligned} \tag{1}$$

Two triples (x_1, y_1, z_1) and (x_2, y_2, z_2) are defined to be equal if and only if $x_1 = x_2$, $y_1 = y_2$, and $z_1 = z_2$.

Let V be a set of elements v, and let $\oplus$ and $\odot$ be functions defined on $V \times V$ and $\mathbb{R} \times V$ which have the following properties for all u, v, w in V and a, b in $\mathbb{R}$:

$$u \oplus v \in V, \qquad a \odot v \in V; \qquad \text{[closure]} \tag{2}$$

$$u \oplus (v \oplus w) = (u \oplus v) \oplus w; \qquad \text{[addition is associative]} \tag{3}$$

there exists a unique element θ such that

$$v + \theta = v \text{ for every } v \text{ in } V; \qquad \text{[identity element for addition]}$$

to each element v in V there corresponds a unique element $-v$ such that (4)

$$v \oplus (-v) = \theta; \qquad \text{[additive inverse]} \tag{5}$$

$$u \oplus v = v \oplus u; \qquad \text{[addition is commutative]} \tag{6}$$

$$a \odot (b \odot v) = ab \odot v; \tag{7}$$

$$(a + b) \odot v = a \odot v \oplus b \odot v; \tag{8}$$

$$a \odot (u \oplus v) = a \odot u \oplus a \odot v; \tag{9}$$

$$1 \odot v = v. \qquad \text{[identity for scalar multiplication]} \tag{10}$$

The system $(V, \oplus, \odot)$ consisting of the set V and the operations $\oplus$ and $\odot$, called *vector addition* and *scalar multiplication* respectively, is called a *linear space* or a *vector space*.

By using the definition of equality in $\mathbb{R}^3$, the definitions of the operations $\oplus$ and $\odot$ in (1), and the properties of the real numbers, it is easy to verify that the system $(\mathbb{R}^3, \oplus, \odot)$ has properties (2), $\cdots$, (10). Because $(\mathbb{R}^3, \oplus, \odot)$ has properties (2), $\cdots$, (6), it is an example of a commutative group; because $(\mathbb{R}^3, \oplus, \odot)$ has the properties described in (2), $\cdots$, (10), it is an example of a linear space or vector space. The special symbols $\oplus$ and $\odot$ for vector addition and scalar multiplication have been used for clarity and emphasis in defining these operations in (1), but usually $(x_1, y_1, z_1) \oplus (x_2, y_2, z_2)$ and $a \odot (x, y, z)$ will be denoted by $(x_1, y_1, z_1) + (x_2, y_2, z_2)$ and $a(x, y, z)$ respectively.

Let $e_0 : (0, 0, 0)$, $e_1 : (1, 0, 0)$, $e_2 : (0, 1, 0)$, and $e_3 : (0, 0, 1)$ be respectively the origin and the unit points on the axes. Considered as vectors, e_1, e_2, e_3 are linearly independent by Theorem 85.3, and every vector in $\mathbb{R}^3$ can be represented as a linear combination of them. Thus

$$(x, y, z) = xe_1 + ye_2 + ze_3, \tag{11}$$

and e_1, e_2, e_3 form a basis for $\mathbb{R}^3$. The results in Section 85 can be used to show that $\mathbb{R}^3$ has many other bases.

The distance between the points $P_1 : (x_1, y_1, z_1)$ and $P_2 : (x_2, y_2, z_2)$ can be found by using the Pythagorean proposition. In Figure 89.2,

$$Q_1Q_2 = x_2 - x_1, \qquad P_1Q_1 = y_2 - y_1, \qquad Q_2P_2 = z_2 - z_1. \tag{12}$$

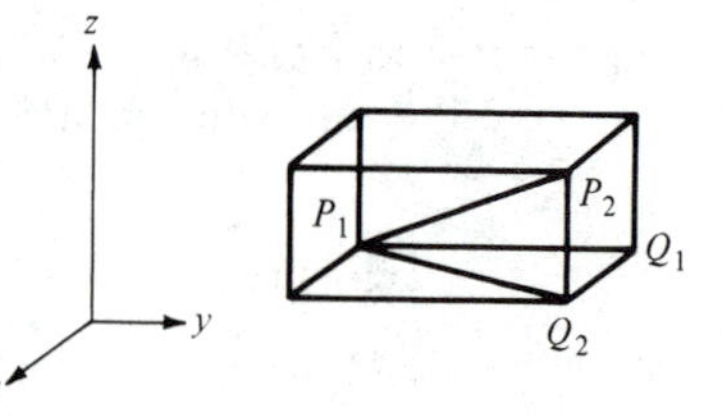

Figure 89.2. The distance between two points.

Now $P_1Q_1Q_2$ and $P_1Q_2P_2$ are right triangles and

$$\begin{aligned}(P_1P_2)^2 &= (P_1Q_2)^2 + (Q_2P_2)^2,\\ (P_1Q_2)^2 &= (Q_1Q_2)^2 + (P_1Q_1)^2,\\ (P_1P_2)^2 &= (Q_1Q_2)^2 + (P_1Q_1)^2 + (Q_2P_2)^2.\end{aligned} \tag{13}$$

Then (12) and the third equation in (13) show that

$$P_1P_2 = [(x_2 - x_1)^2 + (y_2 - y_1)^2 + (z_2 - z_1)^2]^{1/2}. \tag{14}$$

Let $(V, \oplus, \odot)$ be a vector space, and let $\| \quad \|$ denote a real-valued function which is defined on V and which has the following properties: for every u, v in V and a in $\mathbb{R}$,

$$\|u\| > 0 \quad \text{if} \quad u \neq \theta, \qquad \text{and} \quad \|\theta\| = 0; \tag{15}$$

$$\|a \odot u\| = |a|\,\|u\|; \tag{16}$$

$$\|u \oplus v\| \leqq \|u\| + \|v\| \qquad \text{[triangle inequality]}. \tag{17}$$

A function $\| \quad \|$ on V which has these properties is called a norm, and the system $(V, \oplus, \odot, \| \quad \|)$ is called a normed vector space.

Define a function $\| \quad \|$ on $\mathbb{R}^3$ as follows:

$$\|(x, y, z)\| = (x^2 + y^2 + z^2)^{1/2}. \tag{18}$$

Then $\|(x, y, z)\| > 0$ if $(x, y, z) \neq (0, 0, 0)$, and $\|(0, 0, 0)\| = 0$; thus the function defined by (18) has property (15). Also, $\|a \odot (x, y, z)\| = \|(ax, ay, az)\| = |a|\,\|(x, y, z)\|$, and the function (18) has property (16). Finally, a proof will now be given that the function in (18) has property (17). Observe that $\|(x, y, z)\|$ in (18) is the length of the vector (x, y, z) from $(0, 0, 0)$ to (x, y, z), or the distance from $(0, 0, 0)$ to (x, y, z). Let two vectors (x_1, y_1, z_1) and (x_2, y_2, z_2) be given; their sum is $(x_1 + x_2, y_1 + y_2, z_1 + z_2)$. If the three vectors are pictured with their initial point at the origin, their terminal points are P_1, P_2, and Q as shown in Figure 89.3. The vectors P_2Q and P_1Q have components (x_1, y_1, z_1) and (x_2, y_2, z_2) respectively, and the figure is a parallelogram. Because the length of each side of a triangle is equal to or less than the sum of the lengths of the other two sides, then $OQ \leqq OP_1 + P_1Q$. Since

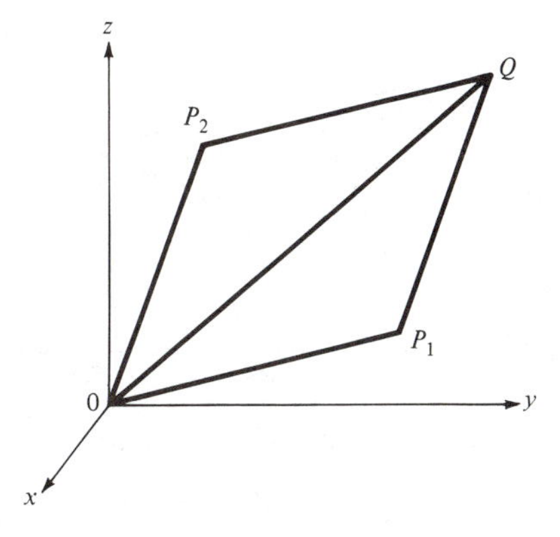

Figure 89.3. The triangle inequality.

$$OQ = \|(x_1, y_1, z_1) \oplus (x_2, y_2, z_2)\|,$$

$$OP_1 = \|(x_1, y_1, z_1)\|, \qquad P_1Q = OP_2 = \|(x_2, y_2, z_2)\|,$$

the function defined in (18) has property (17). Thus (18) defines a norm $\| \quad \|$ on $\mathbb{R}^3$, and $(\mathbb{R}^3, \oplus, \odot, \| \quad \|)$ is a normed vector space.

The inequality (17) is called the triangle inequality because of its geometric significance in Figure 89.3 for the function (norm) defined in (18). In this book, a norm will usually be denoted by $| \quad |$ rather than $\| \quad \|$. The notation $| \quad |$, which is the same as that for the absolute value of a number, usually leads to no confusion since the meaning is clear from the context.

In a normed vector space $(V, \oplus, \odot, \| \quad \|)$, the distance $d(u, v)$ from the point u to the point v is defined to be $\|u - v\|$. Thus

$$d(u, v) = \|u - v\|. \tag{19}$$

Then properties (15), $\cdots$, (17) of the norm show that the distance function has the following properties: for all u, v, w in V,

$$d(u, v) > 0 \quad \text{if} \quad u \neq v, \qquad d(u, u) = 0; \tag{20}$$

$$d(u, v) = d(v, u); \tag{21}$$

$$d(u, w) \leqq d(u, v) + d(v, w) \qquad \text{[triangle inequality]}. \tag{22}$$

To prove (22), observe that $\|u - w\| = \|(u - v) \oplus (v - w)\|$; hence,

$$d(u, w) = \|(u - v) \oplus (v - w)\| \leqq \|u - v\| + \|v - w\| = d(u, v) + d(v, w). \tag{23}$$

The definition in (18) of the norm in $\mathbb{R}^3$ shows that the distance (14) in $\mathbb{R}^3$ is defined in terms of the norm as stated in (19); hence, distance in $\mathbb{R}^3$ has properties (20), $\cdots$, (22).

Two points x_0 and x_1 on the real line are the end points of a segment, and the determinant

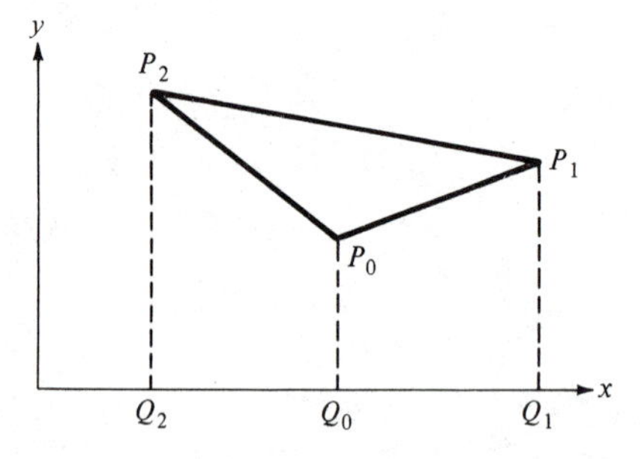

Figure 89.4. The area of the triangle $P_0P_1P_2$.

$$\frac{1}{1!}\begin{vmatrix} x_1 & 1 \\ x_0 & 1 \end{vmatrix} \tag{24}$$

is an important function associated with this segment. The absolute value of the determinant is $|x_1 - x_0|$, which is the length of the segment. If $x_1 > x_0$, then (24) is positive; if $x_1 < x_0$, then (24) is negative. It is sometimes said that (24) gives the signed length of the segment, the sign depending on the orientation of the segment in the real line.

The three points $P_0:(x_0, y_0)$, $P_1:(x_1, y_1)$, and $P_2:(x_2, y_2)$ are the vertices of the triangle $P_0P_1P_2$ in Figure 89.4. The area of this triangle equals the area of the trapezoid $Q_1P_1P_2Q_2$ minus the areas of the trapezoids $Q_1P_1P_0Q_0$ and $Q_0P_0P_2Q_2$. Thus, using (24) for distances on the x-axis, the area of the triangle is

$$\frac{y_1+y_2}{2(1!)}\begin{vmatrix} x_1 & 1 \\ x_2 & 1 \end{vmatrix} - \frac{y_1+y_0}{2(1!)}\begin{vmatrix} x_1 & 1 \\ x_0 & 1 \end{vmatrix} - \frac{y_0+y_2}{2(1!)}\begin{vmatrix} x_0 & 1 \\ x_2 & 1 \end{vmatrix}.$$

This expression simplifies to

$$-\frac{y_1}{2!}\begin{vmatrix} x_2 & 1 \\ x_0 & 1 \end{vmatrix} + \frac{y_2}{2!}\begin{vmatrix} x_1 & 1 \\ x_0 & 1 \end{vmatrix} - \frac{y_0}{2!}\begin{vmatrix} x_1 & 1 \\ x_2 & 1 \end{vmatrix},$$

which is

$$\frac{1}{2!}\begin{vmatrix} x_1 & y_1 & 1 \\ x_2 & y_2 & 1 \\ x_0 & y_0 & 1 \end{vmatrix}. \tag{25}$$

The absolute value of the determinant in (25) is the area of the triangle. It is sometimes said that (25) gives the signed area of the triangle, the sign depending on the orientation of $P_1P_2P_0$ in the plane.

The four points $P_i:(x_i, y_i, z_i)$, $i = 0, 1, \cdots, 3$, are the vertices of a tetrahedron $P_0P_1P_2P_3$ in Figure 89.5; the volume of this tetrahedron equals the sum of the volumes of the solids bounded above by the faces $P_0P_1P_3$, $P_1P_2P_3$, $P_2P_0P_3$, minus the volume of the solid bounded above by the face $P_0P_1P_2$. The volume of each of these solids equals the area of its triangular

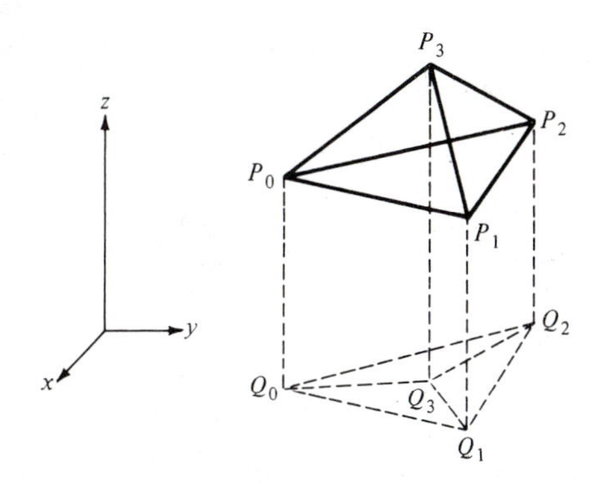

Figure 89.5. The volume of the tetrahedron $P_0P_1P_2P_3$.

base multiplied by its average altitude [this formula can be established by elementary calculus]. Each of the solids has a triangular base, and (25) can be used to find the area of the base. Figure 89.5 has been drawn so that $Q_0Q_1Q_3$, $Q_1Q_2Q_3$, and $Q_2Q_0Q_3$ are positively oriented in the plane; thus the volume of $P_0P_1P_2P_3$ is

$$\frac{z_0+z_1+z_3}{3(2!)}\begin{vmatrix} x_0 & y_0 & 1 \\ x_1 & y_1 & 1 \\ x_3 & y_3 & 1 \end{vmatrix} + \frac{z_1+z_2+z_3}{3(2!)}\begin{vmatrix} x_1 & y_1 & 1 \\ x_2 & y_2 & 1 \\ x_3 & y_3 & 1 \end{vmatrix}$$

$$+\frac{z_2+z_0+z_3}{3(2!)}\begin{vmatrix} x_2 & y_2 & 1 \\ x_0 & y_0 & 1 \\ x_3 & y_3 & 1 \end{vmatrix} - \frac{z_0+z_1+z_2}{3(2!)}\begin{vmatrix} x_0 & y_0 & 1 \\ x_1 & y_1 & 1 \\ x_2 & y_2 & 1 \end{vmatrix}.$$

This formidable expression can be simplified, with the help of the identity in Theorem 20.1, and shown to be equal to

$$\frac{z_1}{3!}\begin{vmatrix} x_2 & y_2 & 1 \\ x_3 & y_3 & 1 \\ x_0 & y_0 & 1 \end{vmatrix} - \frac{z_2}{3!}\begin{vmatrix} x_1 & y_1 & 1 \\ x_3 & y_3 & 1 \\ x_0 & y_0 & 1 \end{vmatrix} + \frac{z_3}{3!}\begin{vmatrix} x_1 & y_1 & 1 \\ x_2 & y_2 & 1 \\ x_0 & y_0 & 1 \end{vmatrix} - \frac{z_0}{3!}\begin{vmatrix} x_1 & y_1 & 1 \\ x_2 & y_2 & 1 \\ x_3 & y_3 & 1 \end{vmatrix}.$$

Theorem 79.1 now shows that this expression for the volume of the tetrahedron $P_0P_1P_2P_3$ equals

$$\frac{1}{3!}\begin{vmatrix} x_1 & y_1 & z_1 & 1 \\ x_2 & y_2 & z_2 & 1 \\ x_3 & y_3 & z_3 & 1 \\ x_0 & y_0 & z_0 & 1 \end{vmatrix}. \tag{26}$$

The pattern exhibited by the formulas in (24), (25), and (26) is clear.

Finally, consider the equations of planes and lines in $\mathbb{R}^3$. Let (a_i, b_i, c_i), $i = 0, 1, 2$, be three points in $\mathbb{R}^3$ such that the vectors

$$u_i = (a_i, b_i, c_i) - (a_0, b_0, c_0) = (a_i - a_0, b_i - b_0, c_i - c_0), \qquad i = 1, 2, \tag{27}$$

are linearly independent. The plane through the three points is the set of all points (x, y, z) in $\mathbb{R}^3$ such that u_1, u_2 and the vector

$$(x, y, z) - (a_0, b_0, c_0) = (x - a_0, y - b_0, z - c_0)$$

are linearly dependent. By Theorem 85.3, the necessary and sufficient condition that the three vectors be linearly dependent is

$$\begin{vmatrix} x - a_0 & y - b_0 & z - c_0 \\ a_1 - a_0 & b_1 - b_0 & c_1 - c_0 \\ a_2 - a_0 & b_2 - b_0 & c_2 - c_0 \end{vmatrix} = 0. \tag{28}$$

By elementary properties of determinants, this equation of the plane can be written also in the following form:

$$\begin{vmatrix} x & y & z & 1 \\ a_1 & b_1 & c_1 & 1 \\ a_2 & b_2 & c_2 & 1 \\ a_0 & b_0 & c_0 & 1 \end{vmatrix} = 0. \tag{29}$$

These equations (28) and (29) of the plane are linear equations. Since (28) and (29) are satisfied by the coordinates of the points (a_i, b_i, c_i), $i = 0, 1, 2$, the plane passes through these points. Equation (29) states that the plane consists of all points (x, y, z) such that the volume of the tetrahedron with vertices (x, y, z) and (a_i, b_i, c_i), $i = 0, 1, 2$, is zero [see (26)].

The line through (a_0, b_0, c_0) and (a_1, b_1, c_1) is the set of all points (x, y, z) such that the two vectors in the rows of

$$A = \begin{bmatrix} x - a_0 & y - b_0 & z - c_0 \\ a_1 - a_0 & b_1 - b_0 & c_1 - c_0 \end{bmatrix}$$

are linearly dependent. The necessary and sufficient condition that these vectors be linearly dependent, according to Corollary 85.6, is $\det AA^t = 0$. Thus by (3) in Theorem 80.1, the line through (a_i, b_i, c_i), $i = 0, 1$, is the set of points (x, y, z) whose coordinates satisfy the following three equations:

$$\begin{vmatrix} x - a_0 & y - b_0 \\ a_1 - a_0 & b_1 - b_0 \end{vmatrix} = 0, \qquad \begin{vmatrix} x - a_0 & z - c_0 \\ a_1 - a_0 & c_1 - c_0 \end{vmatrix} = 0,$$
$$\begin{vmatrix} y - b_0 & z - c_0 \\ b_1 - b_0 & c_1 - c_0 \end{vmatrix} = 0. \tag{30}$$

These equations of the line can be written also in the following form:

$$\begin{vmatrix} x & y & 1 \\ a_1 & b_1 & 1 \\ a_0 & b_0 & 1 \end{vmatrix} = 0, \qquad \begin{vmatrix} x & z & 1 \\ a_1 & c_1 & 1 \\ a_0 & c_0 & 1 \end{vmatrix} = 0, \qquad \begin{vmatrix} y & z & 1 \\ b_1 & c_1 & 1 \\ b_0 & c_0 & 1 \end{vmatrix} = 0. \tag{31}$$

The equations of the line in (30) and (31) are linear equations; since these equations are satisfied by the coordinates of (a_i, b_i, c_i), $i = 0, 1$, by Theorem 77.9, the line passes through these points.

90. Introduction to $\mathbb{R}^n$

Section 89 has explained how the study of three-dimensional physical space has led to the construction of the three-dimensional mathematical "space" $\mathbb{R}^3$. This section explains how $\mathbb{R}^3$ is used as a model for the construction of the n-dimensional mathematical "space" $\mathbb{R}^n$.

If A, B, C are sets with elements a, b, c, the Cartesian product $A \times B \times C$ of A, B, C is defined to be the following set of ordered triples (a, b, c):

$$A \times B \times C = \{(a, b, c) \colon a \in A, b \in B, c \in C\}.$$

Then the three-dimensional mathematical "space" $\mathbb{R}^3$ constructed in Section 89 is the Cartesian product $\mathbb{R} \times \mathbb{R} \times \mathbb{R}$, and $\mathbb{R}^3$ is used to denote this product.

But Cartesian products with more than three factors can be defined in the obvious manner. If $A_1, \cdots, A_n$ are sets with elements $a_1, \cdots, a_n$ respectively, then

$$A_1 \times A_2 \times \cdots \times A_n = \{(a_1, a_2, \cdots, a_n) : a_1 \in A_1, a_2 \in A_2, \cdots, a_n \in A_n\}. \tag{1}$$

Section 89 suggests the study of $\mathbb{R} \times \mathbb{R} \times \cdots \times \mathbb{R}$ (n factors).

90.1 Definition. The *n-dimensional Euclidean space* $\mathbb{R}^n$ is the Cartesian product $\mathbb{R} \times \mathbb{R} \times \cdots \times \mathbb{R}$ (n factors). Thus

$$\mathbb{R}^n = \{(x_1, x_2, \cdots, x_n) : x_i \in \mathbb{R}, i = 1, 2, \cdots, n\}. \tag{2}$$

There is no physical space which corresponds to $\mathbb{R}^n$ for $n > 3$, but we agree on a procedure and a program for the study of the mathematical space $\mathbb{R}^n$. Some examples will help to explain. We agree to use notation and terminology in $\mathbb{R}^n$ which is the same as—or an obvious extension of—that already employed in $\mathbb{R}^3$. The ordered triple (x, y, z) is called a point in $\mathbb{R}^3$; in a similar manner, the ordered n-tuple $(x_1, x_2, \cdots, x_n)$, or $(x^1, x^2, \cdots, x^n)$, is called a point in $\mathbb{R}^n$. The x_i-axis, $i = 1, 2, \cdots, n$, in $\mathbb{R}^n$ is the set of points $\{(0, \cdots, 0, x_i, 0, \cdots, 0) : x_i \in \mathbb{R}\}$, and there are 2-dimensional, $\cdots$, $(n-1)$-dimensional coordinate planes and hyperplanes. The origin and the unit points on the axes are the points

$$\begin{aligned} e_0 &: (0, 0, \cdots, 0), \\ e_1 &: (1, 0, \cdots, 0), \\ e_2 &: (0, 1, \cdots, 0), \\ &\cdots\cdots\cdots\cdots \\ e_n &: (0, 0, \cdots, 1). \end{aligned} \tag{3}$$

The ordered n-tuple $(x_1, x_2, \cdots, x_n)$ is considered as a point, as the vector from the origin $(0, 0, \cdots, 0)$ to the point $(x_1, x_2, \cdots, x_n)$, and as the components of a free vector whose initial point can be any point in the space $\mathbb{R}^n$. Two points or vectors $(x_1, x_2, \cdots, x_n)$ and $(y_1, y_2, \cdots, y_n)$ are equal if and only if $x_i = y_i$, $i = 1, 2, \cdots, n$. The definitions of vector addition and scalar multiplication in $\mathbb{R}^n$ are similar to the definitions of these operations in $\mathbb{R}^3$ in (1) in Section 89, and the system $(\mathbb{R}^n, \oplus, \odot)$ has the properties described in statements (2), $\cdots$, (10) in that section. The system $(\mathbb{R}^n, \oplus, \odot)$ is a linear space or a vector space. The vectors $e_1, e_2, \cdots, e_n$ are linearly independent and form a basis for $\mathbb{R}^n$; thus, for every vector $(x_1, x_2, \cdots, x_n)$ in $\mathbb{R}^n$,

$$(x_1, x_2, \cdots, x_n) = x_1 e_1 + x_2 e_2 + \cdots + x_n e_n. \tag{4}$$

Every basis for $\mathbb{R}^n$ contains n linearly independent vectors, and $\mathbb{R}^n$ is an n-dimensional vector space.

Thus the notation and the terminology are carried over from $\mathbb{R}^3$ to $\mathbb{R}^n$; likewise, the problems in $\mathbb{R}^n$ are extensions and generalizations of those which arise in $\mathbb{R}^3$. Lines and planes are studied in $\mathbb{R}^3$; in $\mathbb{R}^n$ there are not only lines but also planes of dimensions $2, \cdots, n-1$. Let $(a_i^1, a_i^2, \cdots, a_i^n)$, $i = 0, 1, \cdots, r$, be points such that the vectors

$$u_i = (a_i^1, a_i^2, \cdots, a_i^n) - (a_0^1, a_0^2, \cdots a_0^n), \qquad i = 1, \cdots, r, \tag{5}$$

are linearly independent, and let

$$v = (x^1, x^2, \cdots, x^n) - (a_0^1, a_0^2, \cdots, a_0^n). \tag{6}$$

Then

$$\begin{aligned} &\{(x^1, x^2, \cdots, x^n): \\ &\qquad (x^1, x^2, \cdots, x^n) \in \mathbb{R}^n, v, u_1, \cdots, u_r \text{ are linearly dependent}\} \end{aligned} \tag{7}$$

is called a line in $\mathbb{R}^n$ if $r = 1$ and an r-dimensional plane in $\mathbb{R}^n$ if $2 \leqq r \leqq n-1$. If $r = n$, then $u_1, u_2, \cdots, u_n$ form a basis for $\mathbb{R}^n$, and the set (7) is the entire space $\mathbb{R}^n$. The equations of the line and of the planes can be found by methods similar to those used in establishing equations (28), $\cdots$, (31) in Section 89.

In Section 89 the distance P_1P_2 between two points $P_1:(x_1, y_1, z_1)$ and $P_2:(x_2, y_2, z_2)$ in 3-dimensional space was calculated in terms of the coordinates of these points; equation (14), Section 89, shows that

$$P_1P_2 = [(x_2 - x_1)^2 + (y_2 - y_1)^2 + (z_2 - z_1)^2]^{1/2}.$$

If $P_i:(x_i^1, x_i^2, \cdots, x_i^n)$, $i = 1, 2$, are two points in $\mathbb{R}^n$, the distance P_1P_2 between these points is defined as follows:

$$P_1P_2 = \left[\sum_{j=1}^{n} (x_2^j - x_1^j)^2\right]^{1/2}. \tag{8}$$

This formula emphasizes that the proofs of the properties of distance and of $\mathbb{R}^n$ in general must be based on the properties of the real numbers $\mathbb{R}$ and of ordered n-tuples $(x^1, x^2, \cdots, x^n)$ in $\mathbb{R}^n$. There is no synthetic geometry of n-dimensional space to assist the study of $\mathbb{R}^n$; geometry has become analytic geometry. For $n > 3$ there are no figures to provide insight and to support intuition, but the study of $\mathbb{R}^3$—including figures and physical models—continues to provide strong support for the study of $\mathbb{R}^n$. The next section establishes the properties of the distance function in $\mathbb{R}^n$; it illustrates the analytic proof of results which are stated in the geometric language of three-dimensional space and whose motivation arises from the familiar properties of that space.

91. The Norm, Distance, and Triangle Inequality in $\mathbb{R}^n$

Equation (18) in Section 89 contains the definition of a norm in $\mathbb{R}^3$. Using that definition as a model, define a function $\| \quad \|$ on $\mathbb{R}^n$ as follows:

$$\|(x_1, x_2, \cdots, x_n)\| = \left[\sum_{j=1}^{n} x_j^2\right]^{1/2}, \qquad (x_1, x_2, \cdots, x_n) \in \mathbb{R}^n. \tag{1}$$

Since (1) defines a norm on $\mathbb{R}^3$ when $n = 3$, there is every reason to believe that it defines a norm for every positive integer n. Proofs must be given, however, based on the properties of the real numbers.

The function defined in (1) has the properties stated in (15) and (16) in Section 89 since

$$\begin{gathered} \|(x_1, x_2, \cdots, x_n)\| > 0 \quad \text{if} \quad (x_1, x_2, \cdots, x_n) \neq (0, 0, \cdots, 0), \\ \|(0, 0, \cdots, 0)\| = 0, \\ \|a(x_1, x_2, \cdots, x_n)\| = |a| \, \|(x_1, x_2, \cdots, x_n)\|. \end{gathered} \tag{2}$$

To prove the triangle inequality [property (17) in Section 89], let $(x_1, x_2, \cdots, x_n)$ and $(y_1, y_2, \cdots, y_n)$ be two points in $\mathbb{R}^n$. Then by Schwarz's inequality [see (7) in Section 86],

$$\sum_{j=1}^{n} x_j y_j \leqq \left[\sum_{j=1}^{n} x_j^2\right]^{1/2} \left[\sum_{j=1}^{n} y_j^2\right]^{1/2}. \tag{3}$$

Multiply this inequality by 2; then add the following expression to each side of the inequality:

$$\sum_{j=1}^{n} x_j^2 + \sum_{j=1}^{n} y_j^2.$$

The result is

$$\sum_{j=1}^{n} x_j^2 + 2\sum_{j=1}^{n} x_j y_j + \sum_{j=1}^{n} y_j^2 \leqq \sum_{j=1}^{n} x_j^2 + 2\left[\sum_{j=1}^{n} x_j^2\right]^{1/2} \left[\sum_{j=1}^{n} y_j^2\right]^{1/2} + \sum_{j=1}^{n} y_j^2.$$

This inequality can be written in the following form:

$$\sum_{j=1}^{n} (x_j + y_j)^2 \leqq \left\{\left[\sum_{j=1}^{n} x_j^2\right]^{1/2} + \left[\sum_{j=1}^{n} y_j^2\right]^{1/2}\right\}^2.$$

Take the positive square root of each side of this inequality. Then

$$\left[\sum_{j=1}^{n}(x_j+y_j)^2\right]^{1/2} \leqq \left[\sum_{j=1}^{n}x_j^2\right]^{1/2} + \left[\sum_{j=1}^{n}y_j^2\right]^{1/2}. \tag{4}$$

This inequality states that the function defined in (1) has the property described in (17) in Section 89; inequality (4) is known as the triangle inequality in $\mathbb{R}^n$. Thus the function defined in (1) has all of the properties of a norm [see (15), $\cdots$, (17) in Section 89], and the system $(\mathbb{R}^n, \oplus, \odot, \|\ \|)$ is a normed linear space or a normed vector space.

In a normed vector space the distance function is defined in terms of the norm as stated in (19) in Section 89. Thus if $x:(x_1, x_2, \cdots, x_n)$ and $y:(y_1, y_2, \cdots, y_n)$ are two points in $\mathbb{R}^n$, then

$$d(x, y) = \|x-y\| = \left[\sum_{j=1}^{n}(x_j-y_j)^2\right]^{1/2}, \tag{5}$$

and the distance function d has the properties stated in (20), $\cdots$, (22) in Section 89.

92. Open and Closed Sets and Related Matters in $\mathbb{R}^n$

This section contains a brief review of some of the basic definitions and theorems in point-set topology in $\mathbb{R}^n$.

92.1 Definition. Let ε be a positive number, and let p be a point in $\mathbb{R}^n$. Then the *ε-neighborhood of p*, denoted by $N(p, \varepsilon)$, is defined as follows:

$$N(p, \varepsilon) = \{x : x \in \mathbb{R}^n, |x-p| < \varepsilon\}. \tag{1}$$

In $\mathbb{R}^3$, $N(p, \varepsilon)$ is the interior of a sphere with center p and radius ε. For this reason, $N(p, \varepsilon)$ in $\mathbb{R}^n$ is frequently called an open sphere with center p and radius ε.

92.2 Definition. Let E be a set in $\mathbb{R}^n$, and let p be a point in $\mathbb{R}^n$. If every ε-neighborhood of p contains a point of E which is distinct from p, then p is called a *limit point* or a *point of accumulation* of E.

Examples. (a) If $E = \{x : x \text{ is a rational number and } 0 < x < 1\}$, then every point in the set $\{x : x \in \mathbb{R}, 0 \leqq x \leqq 1\}$ is a limit point of E. (b) Every point in the set $\{x : x \in \mathbb{R}^n, |x-p| = \varepsilon\}$ is a limit point of the set $N(p, \varepsilon)$. (c) No finite set has a limit point. (d) Let $\{a_k : k = 1, 2, \cdots\}$ be a monotonically increasing sequence which is bounded above, and let E denote the set $\{a_k : k = 1, 2, \cdots\}$. Then $\sup(E)$ is a limit point or point of accumulation of E if E is an infinite set.

92.3 Definition. Let E be a set in $\mathbb{R}^n$ which contains all of its points of accumulation. Then E is called a *closed* set.

92.4 Definition. If every point p in a set E in $\mathbb{R}^n$ has an ε-neighborhood $N(p, \varepsilon)$ which is contained in E, then E is called an *open* set.

Examples. (a) Every finite set is closed, but no finite set is open. (b) The entire space $\mathbb{R}^n$ is both open and closed, and the empty set is both open and closed. (c) Some sets are neither open nor closed; for example, $\{x : x \in \mathbb{R}, 0 < x \leqq 1\}$ is neither open nor closed. (d) The set $[a, b] \times [c, d]$ in $\mathbb{R}^2$ is closed. The set $\{(x, y, z) : -1 < x < 1, -1 < y < 1, -1 < z < 1\}$ is open in $\mathbb{R}^3$.

92.5 Theorem. *The complement of a closed set is an open set, and the complement of an open set is a closed set.*

PROOF. Let E be a closed set in $\mathbb{R}^n$, and assume that $p \notin E$. Then p has an ε-neighborhood which contains no point in E and is therefore in the complement of E; if every $N(p, \varepsilon)$ contained a point of E, then p would be a limit point of E and therefore in E, contrary to hypothesis. Thus the complement of E is open since each of its points has an ε-neighborhood which is contained in the complement of E.

Next, let E be an open set in $\mathbb{R}^n$, and assume that $p \in E$. Since E is open there is an ε-neighborhood of p which is contained in E. Then p is not a limit point of the complement of E. Thus all limit points of the complement of E are in the complement of E, and this set is closed. □

Examples. (a) The complement of the closed set $\{x : x \in \mathbb{R}^n, |x| = 1\}$ is an open set. (b) The complement of the open set $\{x : x \in \mathbb{R}^n, |x| > 1\}$ is a closed set (the closed ball with center at the origin and radius 1). (c) The set of points in $\mathbb{R}^n$ whose coordinates are integers is a closed set (it has no limit points), and its complement is an open set.

92.6 Theorem. *The intersection of a finite number of open sets is open, and the union of any number of open sets is open. The union of a finite number of closed sets is closed, and the intersection of any number of closed sets is closed.*

PROOF. Let $E_1, \cdots, E_m$ be open sets, and let p be a point in their intersection. Then there are ε-neighborhoods $N(p, \varepsilon_k)$ such that $N(p, \varepsilon_k) \subset E_k$, $k = 1, 2, \cdots, m$. Set $\varepsilon = \min(\varepsilon_1, \varepsilon_2, \cdots, \varepsilon_m)$. Then $\varepsilon > 0$ and $N(p, \varepsilon)$ is contained in $E_1 \cap E_2 \cap \cdots \cap E_m$, and this set is open. If p is a point in the union of any number of open sets, then p has an ε-neighborhood $N(p, \varepsilon)$ which is contained in the union; hence, the union is open.

To prove the part of the theorem about closed sets, use De Morgan's Laws, Theorem 92.5, and the part of the theorem already proved. Let $\mathscr{E}$

be a family of sets E, and let $C(E)$ denote the complement of E in $\mathbb{R}^n$. Then De Morgan's Laws are the following point-set identities:

$$\bigcup\{E : E \in \mathscr{E}\} = C(\bigcap\{C(E) : E \in \mathscr{E}\}), \tag{2}$$

$$\bigcap\{E : E \in \mathscr{E}\} = C(\bigcup\{C(E) : E \in \mathscr{E}\}). \tag{3}$$

To complete the proof of Theorem 92.6, let $\mathscr{E}$ be an arbitrary family of closed sets E. Then $C(E)$ is open by Theorem 92.5 and $\bigcup\{C(E) : E \in \mathscr{E}\}$ is open by the part of Theorem 92.6 already proved. Then (3) and Theorem 92.5 show that $\bigcap\{E : E \in \mathscr{E}\}$ is closed. In the same way, (2) can be used to show that the union of a finite number of closed sets is closed. □

92.7 Definition. Let E' denote the set of limit points of E. The *closure* of E, denoted by $\text{cl}(E)$, is the set $E \cup E'$. The *boundary* of E is the set $\text{cl}(E) \cap \text{cl}(C(E))$.

Examples. (a) Let E be the set $\{x : x \in \mathbb{R}^3, |x| < 1\}$. Then $\text{cl}(E) = \{x : x \in \mathbb{R}^3, |x| \leqq 1\}$ and $\text{cl}(C(E)) = \{x : x \in \mathbb{R}^3, |x| \geqq 1\}$. Then the boundary of E is $\{x : x \in \mathbb{R}^3, |x| = 1\}$. In this case the boundary of E has the intuitive character of a boundary. This is not always the case as the next example shows. (b) Let $E = \{x : x \in \mathbb{R}^2, |x| < 1, \text{and the coordinates of } x \text{ are rational}\}$. Then $\text{cl}(E) = \{x : x \in \mathbb{R}^2, |x| \leqq 1\}$ and $\text{cl}(C(E)) = \mathbb{R}^2$, and the boundary of E is $\{x : x \in \mathbb{R}^2, |x| \leqq 1\}$.

92.8 Definition. A set E in $\mathbb{R}^n$ is *bounded* if and only if there exists a positive number r such that $E \subset \{x : x \in \mathbb{R}^n, |x| \leqq r\}$. A set is *compact* if and only if it is closed and bounded.

Examples. (a) Every finite set in $\mathbb{R}^n$ is compact. (b) The bounded set $\{x : x \in \mathbb{R}^n, |x| < 1\}$ is not compact because it is not closed. (c) The bounded set $\{x : x \in \mathbb{R}, x = 1/n, n = 1, 2, \cdots\}$ is not compact because it is not closed. (d) The closed set $\{x : x \in \mathbb{R}, x \geqq 1\}$ is not compact because it is not bounded. (e) The set $\{x : x \in \mathbb{R}^n, |x| \leqq 1\}$ is compact because it is closed and bounded.

92.9 Definition. Let E be a set in $\mathbb{R}^n$. The *diameter* of E, denoted by $\text{diam}(E)$, is defined as follows:

$$\text{if } E \text{ is the empty set, then } \text{diam}(E) = 0; \tag{4}$$

$$\text{if } E \text{ is a non-empty bounded set, then } \text{diam}(E) = \sup\{|x - y| : x, y \text{ in } E\}; \tag{5}$$

$$\text{if } E \text{ is unbounded, then } \text{diam}(E) = \infty. \tag{6}$$

Examples. (a) If $E = \{x : x \in \mathbb{R}, 0 \leqq x \leqq 1\}$, then the diameter of E is 1 and there are two points (0 and 1) whose distance apart is equal to the diameter of E. (b) If $E = \{x : x \in \mathbb{R}, 0 < x < 1\}$, then the diameter of E is 1 but there are no two points in E whose distance apart is 1. (c) If $E = \{x : x \in \mathbb{R}^2, |x| \geqq 1\}$, then $\operatorname{diam}(E) = \infty$. (d) If E is a finite set in $\mathbb{R}^n$, then the diameter of E is the maximum distance between two points of E. (e) If E is a bounded set in $\mathbb{R}^n$, then the diameter of E is finite.

93. The Nested Interval Theorem

Let $[a, b] = \{x : x \in \mathbb{R}, a \leqq x \leqq b\}$; then $[a, b]$ is a closed interval of real numbers.

93.1 Theorem (Nested Interval Theorem in $\mathbb{R}$). *Let $[a_k, b_k]$, $k = 1, 2, \cdots$ be closed intervals in $\mathbb{R}$ such that*

$$a_k \leqq a_{k+1} \leqq b_{k+1} \leqq b_k, \qquad k = 1, 2, \cdots. \tag{1}$$

Then the intersection $\bigcap\{[a_k, b_k]: k = 1, 2, \cdots\}$ of these nested intervals is not empty. If $\lim\{b_k - a_k : k = 1, 2, \cdots\} = 0$, then there is a single point in the intersection.

PROOF. The sequence $\{a_k : a_k \in \mathbb{R}, k = 1, 2, \cdots\}$ is a monotonically increasing sequence which, by (1), is bounded above by b_1. Then $\lim\{a_k : k = 1, 2, \cdots\}$ exists by Theorem 88.5; call this limit a. Likewise, the sequence $\{b_k : b_k \in \mathbb{R}, k = 1, 2, \cdots\}$ is monotonically decreasing and bounded below by a_1; it has a limit b. Now $a \leqq b$ as a result of (1). Since

$$a_k \leqq a \leqq b \leqq b_k, \qquad k = 1, 2, \cdots, \tag{2}$$

then $[a, b] \subset [a_k, b_k]$, $k = 1, 2, \cdots$, and $\bigcap\{[a_k, b_k] : k = 1, 2, \cdots\} = [a, b]$. If $\lim\{b_k - a_k : k = 1, 2, \cdots\} = 0$, then (2) shows that $[a, b]$ cannot contain more than one point. □

93.2 Theorem (Nested Interval Theorem in $\mathbb{R}^n$). *Let*

$$I_k = [a_k^1, b_k^1] \times [a_k^2, b_k^2] \times \cdots \times [a_k^n, b_k^n], \qquad k = 1, 2, \cdots, \tag{3}$$

be closed intervals in $\mathbb{R}^n$ such that

$$a_k^j \leqq a_{k+1}^j \leqq b_{k+1}^j \leqq b_k^j, \qquad j = 1, 2, \cdots, n, \qquad k = 1, 2, \cdots. \tag{4}$$

Then the intersection $\bigcap\{I_k : k = 1, 2, \cdots\}$ of these nested intervals is not empty. If $\lim\{\operatorname{diam}(I_k) : k = 1, 2, \cdots\} = 0$, then there is a single point in the intersection.

PROOF. By Theorem 93.1,

$$\lim_{k\to\infty} a_k^j = a^j, \qquad \lim_{k\to\infty} b_k^j = b^j, \qquad \text{and} \quad a^j \leqq b^j, \qquad j = 1, 2, \cdots, n. \tag{5}$$

Then

$$[a^1, b^1] \times \cdots \times [a^n, b^n] \subset [a_k^1, b_k^1] \times \cdots \times [a_k^n, b_k^n], \qquad k = 1, 2, \cdots, \tag{6}$$

and the intersection of the intervals I_k is not empty. Now

$$\operatorname{diam}(I_k) = \left[\sum_{j=1}^{n} (b_k^j - a_k^j)^2 \right]^{1/2}, \qquad k = 1, 2, \cdots, \tag{7}$$

$$0 \leqq b_k^j - a_k^j \leqq \operatorname{diam}(I_k), \qquad j = 1, 2, \cdots, n, \qquad k = 1, 2, \cdots. \tag{8}$$

Thus, if $\lim\{\operatorname{diam}(I_k) : k = 1, 2, \cdots\} = 0$, then $a^j = b^j$, $j = 1, 2, \cdots, n$, by Theorem 93.1, and there is a single point in the intersection of the intervals I_k. □

94. The Bolzano–Weierstrass Theorem

The set of points in $\mathbb{R}^n$ whose coordinates are integers is an example of an infinite set which has no point of accumulation; observe that this set is not bounded.

94.1 Theorem (The Bolzano–Weierstrass Theorem). *Every infinite bounded set of points in $\mathbb{R}^n$ has a point of accumulation.*

PROOF. Let the infinite bounded set be denoted by E. Since E is bounded [see Definition 92.8], it is easy to show that there exists an interval

$$I_1 = [a_1^1, b_1^1] \times [a_1^2, b_1^2] \times \cdots [a_1^n, b_1^n] \tag{1}$$

which contains E. Subdivide I_1 into 2^n subintervals as follows. Subdivide $[a_1^j, b_1^j]$, $j = 1, 2, \cdots, n$, into two equal subintervals by introducing a point of division at the midpoint of the interval. Then at least one of the 2^n subintervals contains an infinite number of points in E, because otherwise E would be finite, contrary to hypothesis. From the 2^n subintervals of I_1, choose one which contains an infinite number of points of E, and call it I_2. Then

$$\begin{gathered} I_2 = [a_2^1, b_2^1] \times [a_2^2, b_2^2] \times \cdots \times [a_2^n, b_2^n], \\ a_1^j \leqq a_2^j \leqq b_2^j \leqq b_1^j, \qquad j = 1, 2, \cdots, n. \end{gathered} \tag{2}$$

This construction can be repeated to form intervals I_k, $k = 1, 2, \cdots$, such that

$$\begin{gathered} I_k = [a_k^1, b_k^1] \times [a_k^2, b_k^2] \times \cdots \times [a_k^n, b_k^n], \qquad k = 1, 2, \cdots, \\ a_k^j \leqq a_{k+1}^j \leqq b_{k+1}^j \leqq b_k^j, \qquad j = 1, 2, \cdots, n. \end{gathered} \tag{3}$$

Since $b_{k+1}^j - a_{k+1}^j = (b_k^j - a_k^j)/2$, then $\lim\{b_k^j - a_k^j : k = 1, 2, \cdots\} = 0$ for $j = 1, 2, \cdots, n$, and

$$\lim_{k\to\infty} \operatorname{diam}(I_k) = 0. \tag{4}$$

By Theorem 93.2 there is a single point in $\bigcap\{I_k : k = 1, 2, \cdots\}$; call this point p. Let $N(p, \varepsilon)$ be an ε-neighborhood of p. Choose k_0 so that diam $(I_{k_0}) < \varepsilon$; this choice is possible by (4). Then $I_{k_0} \subset N(p, \varepsilon)$. Since I_{k_0} contains an infinite number of points of E, then $N(p, \varepsilon)$ contains a point of E which is distinct from p. Since this statement is true for every $N(p, \varepsilon)$, then p is a point of accumulation of E by Definition 92.2, and the proof is complete. □

Examples. (a) If $E = \{x : x = 1/m, m = 1, 2, \cdots\}$, then E is bounded and infinite; $x = 0$ is the single point of accumulation of this set. (b) Every point in the set $\{x : x \in \mathbb{R}^n, |x| \leqq 1\}$ is a point of accumulation of $E = \{x : x \in \mathbb{R}^n, |x| < 1\}$. (c) The infinite bounded set $E = \{x : x = (-1)^m(1 - 1/m), m = 1, 2, \cdots\}$ has $+1$ and -1 as its two points of accumulation.

95. The Heine–Borel Theorem

This section contains a proof of the Heine–Borel theorem, which states an important property of compact sets.

95.1 Definition. Let E be a subset of $\mathbb{R}^n$, and let $\mathscr{E}$ be a collection of subsets of $\mathbb{R}^n$ such that E is contained in the union of the sets in $\mathscr{E}$. Then $\mathscr{E}$ is called a *covering* of E, and $\mathscr{E}$ is said to cover E.

95.2 Theorem (The Heine–Borel Theorem). *Let E be a compact set in $\mathbb{R}^n$, and let $\mathscr{E}$ be a collection of open sets which covers E. Then some finite subcollection of $\mathscr{E}$ covers E.*

PROOF. A finite set is a compact set. If $\mathscr{E}$ covers a finite set E, then each point of E is contained in an open set in $\mathscr{E}$ and the finite subcollection of $\mathscr{E}$ thus defined covers E. Thus the theorem is true in this case.

Assume then that E is an infinite compact set. Assume that the theorem is false in this case; the proof shows that this assumption leads to a contradiction. Since E is compact, then E is bounded [see Definition 92.8], and E is contained in an interval

$$I_1 = [a_1^1, b_1^1] \times [a_1^2, b_1^2] \times \cdots \times [a_1^n, b_1^n]. \tag{1}$$

Subdivide this interval into 2^n subintervals as in the proof of Theorem 94.1. If the subset of E in each of these subintervals of I_1 could be covered by a finite subcollection of $\mathscr{E}$, then the entire set E could be covered—contrary to the hypothesis that the theorem is false—by a finite subcollection of $\mathscr{E}$. Choose one of the subintervals of I_1 whose intersection with E cannot be covered by a finite subcollection of $\mathscr{E}$ and call it I_2:

$$I_2 = [a_2^1, b_2^1] \times [a_2^2, b_2^2] \times \cdots \times [a_2^n, b_2^n]. \tag{2}$$

Then $I_2 \cap E$ is an infinite subset of E because the first case in the proof shows that every finite subset of E can be covered by a finite subcollection of $\mathscr{E}$. This construction can be repeated indefinitely to form nested intervals I_k, $k = 1, 2, \cdots$, such that

$$I_1 \supset I_2 \supset \cdots \supset I_k \supset \cdots. \tag{3}$$

The intervals I_k in (3) have the following properties:

$$I_k \cap E \text{ is an infinite subset of } E; \tag{4}$$

for $k = 1, 2, \cdots$, there is no finite subcollection of $\mathscr{E}$ which covers $I_k \cap E$; (5)

$$\lim_{k \to \infty} \operatorname{diam}(I_k) = 0. \tag{6}$$

Then by Theorem 93.2, there is a single point p in all of the intervals I_k.

Now p is in E for the following reasons. Let $N(p, \varepsilon)$ be an ε-neighborhood of p. Choose k so that $I_k \subset N(p, \varepsilon)$; this choice is possible since p belongs to I_k for $k = 1, 2, \cdots$, and the diameter of I_k approaches zero by (6). Then $N(p, \varepsilon)$ contains I_k and also $I_k \cap E$; therefore $N(p, \varepsilon)$ contains an infinite number of points of E, and p is a point of accumulation of E. Since E is compact by hypothesis, it is closed; therefore, p belongs to E.

Since $p \in E$ and $\mathscr{E}$ covers E, then p belongs to an open set O in $\mathscr{E}$. Now O is open, and there is an ε-neighborhood $N(p, \varepsilon)$ of p such that $N(p, \varepsilon) \subset O$. Choose k so that $I_k \subset N(p, \varepsilon)$; this choice is possible by (6) since p is in every I_k. Thus

$$I_k \cap E \subset I_k \subset N(p, \varepsilon) \subset O. \tag{7}$$

Thus $I_k \cap E$ is covered by the single open set O in $\mathscr{E}$; since this statement contradicts (5) above, the hypothesis that the theorem is false has led to a contradiction, and the theorem is true. □

96. Functions

A function is a set of ordered pairs of a special type. Let A and B be sets. A function from A to B is a set of ordered pairs (x, y) in $A \times B$ such that, if (x, y) and (x, y') are in f, then $y = y'$. The domain D of f (range of f) is the set of elements x in A (y in B) which are first elements (second elements) of ordered pairs (x, y) in f. We say that f is a mapping from A into B and write $f: A \to B$ [read "f from A into B"]. If x in A is in the domain D of f, we write $y = f(x)$, and y is in the range of f in B. More generally, if the set E is in the domain D of f, then $f(E) = \{f(x) : x \in E\}$.

Unless there is a statement to the contrary, the functions studied in this section are functions from a set A in a Euclidean space $\mathbb{R}^n$ to a set B in the same or a different Euclidean space $\mathbb{R}^m$. If $y = f(x)$ and $x : (x_1, x_2, \cdots, x_n)$ is in A in $\mathbb{R}^n$ and $y : (y_1, y_2, \cdots, y_m)$ is in B in $\mathbb{R}^m$, then there are functions $f_j, j = 1, 2, \cdots, m$, from A to $\mathbb{R}$ such that

$$\begin{aligned} y_1 &= f_1(x_1, x_2, \cdots, x_n), \\ y_2 &= f_2(x_1, x_2, \cdots, x_n), \\ &\cdots\cdots\cdots\cdots\cdots\cdots \\ y_m &= f_m(x_1, x_2, \cdots, x_n). \end{aligned} \tag{1}$$

The functions $f_1, f_2, \cdots, f_m$ are called the components of f, and f from A in $\mathbb{R}^n$ into B in $\mathbb{R}^m$ is described as the function $f: A \to B$ or as the function $(f_1, f_2, \cdots, f_m) : A \to B$. The components of f play an essential role in the study of the properties of f.

Let $\mathscr{F}$ be the family of functions $f: A \to B$ which have the same domain D. Two functions f and g in $\mathscr{F}$ are defined to be equal if and only if

$$f(x) = g(x), \qquad x \in D. \tag{2}$$

If B is contained in a vector space in which addition and scalar multiplication are defined, then it is possible to define two operations $\oplus$ and $\odot$ as follows: if f and g are in $\mathscr{F}$ and a is in $\mathbb{R}$, then $f \oplus g$ and $a \odot f$ are functions in $\mathscr{F}$ such that

$$\begin{aligned} (f \oplus g)(x) &= f(x) + g(x), \\ (a \odot f)(x) &= af(x). \end{aligned} \tag{3}$$

The system $(\mathscr{F}, \oplus, \odot)$ is easily shown to be a vector space [see Section 89]. For simplicity, $f \oplus g$ and $a \odot f$ are usually written $f + g$ and af.

96.1 Examples. (a) The vector space $(\mathscr{F}, \oplus, \odot)$ consists of all functions $f: A \to B$ with the same domain D, but there are many subsystems of $(\mathscr{F}, \oplus, \odot)$ which are also vector spaces. For example, let $\mathscr{P}_n$ consist of all polynomials in x of degree equal to or less than n on some interval $[a, b]$ in $\mathbb{R}$. Then $\mathscr{P}_n$ is closed under addition and scalar multiplication as defined in (3), and $(\mathscr{P}_n, \oplus, \odot)$ is a vector space. In the same way, if $\mathscr{P}$ is the set of all polynomials in x on $[a, b]$, then $(\mathscr{P}, \oplus, \odot)$ is a vector space. (b) A function whose domain D is the set $\{1, 2, \cdots\}$ of positive integers is called a sequence. Let $\mathscr{S}$ denote the set of sequences whose ranges are contained in the same vector space $(V, \oplus, \odot)$, and let addition $\oplus$ and scalar multiplication $\odot$ in $\mathscr{S}$ be defined as in (3). Then $(\mathscr{S}, \oplus, \odot)$ is a vector space. (c) Let A and B be $\mathbb{R}^n$ and $\mathbb{R}^m$ respectively, and let $f: \mathbb{R}^n \to \mathbb{R}^m$ be a function with components $(f_1, f_2, \cdots, f_m)$. If there exist constants a_{ij}, $i = 1, 2, \cdots, m$, $j = 1, 2, \cdots, n$, such that

$$f_i(x) = \sum_{j=1}^{n} a_{ij} x_j, \qquad i = 1, 2, \cdots, m, \qquad x : (x_1, x_2, \cdots, x_n) \text{ in } \mathbb{R}^n,$$

then $f(ax + by) = af(x) + bf(y)$, and f is called a *linear function*. The set of linear functions f from $\mathbb{R}^n$ into $\mathbb{R}^m$ is a vector space.

96.2 Definition. The function $f: A \to B$, $A \subset \mathbb{R}^n$ and $B \subset \mathbb{R}^m$, is continuous at the point x_0 in its domain D if and only if to each $\varepsilon > 0$ there corresponds a $\delta(\varepsilon, x_0) > 0$ such that

$$|f(x) - f(x_0)| < \varepsilon \tag{4}$$

for every x in D such that $|x - x_0| < \delta(\varepsilon, x_0)$. A function is continuous on a set E if and only if it is continuous at each point in E.

96.3 Examples. (a) Let E be a set in $\mathbb{R}^n$. If $x_0 \in E$, and if x_0 has an ε-neighborhood $N(x_0, \varepsilon)$ which contains no point of E which is distinct from x_0, then x_0 is called an isolated point of E. If x_0 is an isolated point of the domain D of the function $f: A \to B$, $A \subset \mathbb{R}^n$ and $B \subset \mathbb{R}^m$, then f is continuous at x_0. (b) If $f(x) = 2x + 5$ for x in $[a, b]$ in $\mathbb{R}$, then f is continuous on $[a, b]$ since (4) can be satisfied for every x_0 in $[a, b]$ by choosing $\delta(\varepsilon, x_0) = \varepsilon/2$. (c) Let D be the set in $\mathbb{R}^n$ consisting of the closed ball $\{x : x \in \mathbb{R}^n, |x| \leqq 1\}$ and the point $p : (2, 2, \cdots, 2)$. Let $a_1, a_2, \cdots, a_n$ be constants. If $x : (x^1, x^2, \cdots, x^n)$ is in D in $\mathbb{R}^n$, set

$$f(x) = \sum_{j=1}^{n} a_j x^j.$$

The function $f: D \to \mathbb{R}$ is continuous on D. To prove this statement, let $x_0 : (x_0^1, x_0^2, \cdots, x_0^n)$ be a point in D; then by Schwarz's inequality in Corollary 86.2,

$$|f(x) - f(x_0)| \leqq \left[\sum_{j=1}^{n} a_j^2\right]^{1/2} \left[\sum_{j=1}^{n} (x^j - x_0^j)^2\right]^{1/2}.$$

For each $\varepsilon > 0$ choose $\delta(\varepsilon, x_0)$ so that

$$\delta(\varepsilon, x_0) = \varepsilon / \left[\sum_{j=1}^{n} a_j^2\right]^{1/2}.$$

Then $|f(x) - f(x_0)| < \varepsilon$ for every x_0 and x in D such that $|x - x_0| < \delta(\varepsilon, x_0)$, and f is continuous on D. Observe that $\delta(\varepsilon, x_0)$ does not depend on x_0 in this case. Observe also that f would be continuous at the isolated point $p : (2, 2, \cdots, 2)$ no matter how the function were defined at this point since p is an isolated point.

96.4 Theorem. *Let $f: A \to B$ and $g : A \to B$, $A \subset \mathbb{R}^n$ and $B \to \mathbb{R}^m$, be continuous functions at the point x_0 in their common domain D. Then $f + g$ and af are continuous at x_0.*

PROOF. The proof follows easily from these relations:

$$\begin{aligned} |(f + g)(x) - (f + g)(x_0)| &\leqq |f(x) - f(x_0)| + |g(x) - g(x_0)|, \\ |(af)(x) - (af)(x_0)| &\leqq |a|\,|f(x) - f(x_0)|. \end{aligned} \tag{5}$$

□

96.5 Example. Let $\mathscr{C}$ be the family of functions $f: A \to B$, $A \subset \mathbb{R}^n$ and $B \subset \mathbb{R}^m$, which have the same domain D and are continuous on D. Then by Theorem 96.4, equation (3), and the definition of a vector space in Section 89, the system $(\mathscr{C}, \oplus, \odot)$ is a vector space.

96.6 Theorem. Let $f: A \to B$, $A \subset \mathbb{R}^n$ and $B \subset \mathbb{R}^m$, be a function with components $(f_1, f_2, \cdots, f_m)$. Then f is continuous at x_0 in D if and only if each of the component functions $f_j: A \to \mathbb{R}$, $j = 1, 2, \cdots, m$, is continuous at x_0.

PROOF. By the definition of distance in Section 91,

$$|f(x) - f(x_0)| = \left[\sum_{j=1}^{m} |f_j(x) - f_j(x_0)|^2 \right]^{1/2}. \tag{6}$$

Then, for $j = 1, 2, \cdots, m$,

$$|f_j(x) - f_j(x_0)| \leqq |f(x) - f(x_0)| \leqq \sum_{j=1}^{m} |f_j(x) - f_j(x_0)|. \tag{7}$$

and the proof follows easily from these obvious inequalities. □

96.7 Remark. The inequalities (7) can be used to relate several other properties of f to the corresponding property of the component functions f_1, $f_2, \cdots, f_m$.

96.8 Example. Linear functions from $\mathbb{R}^n$ to $\mathbb{R}^m$ are defined in Example 96.1(c). The results in Example 96.3(c) can be used to prove that the linear functions from $\mathbb{R}^n$ to $\mathbb{R}$ are continuous. Then Theorem 96.6 shows that the linear functions from $\mathbb{R}^n$ to $\mathbb{R}^m$ are continuous.

96.9 Theorem. *Let $f: A \to \mathbb{R}$, $A \subset \mathbb{R}^n$, be continuous at x_0 in D. If $f(x_0) > 0$, then there is a δ-neighborhood $N(x_0, \delta)$ such that $f(x) > 0$ for x in $D \cap N(x_0, \delta)$. If $f(x_0) < 0$, there is a neighborhood of x_0 in which $f(x)$ is negative.*

PROOF. If $f(x_0) > 0$, choose $\varepsilon = f(x_0)/2$. Then there is a δ such that

$$|f(x) - f(x_0)| < \varepsilon \tag{8}$$

for all x in $D \cap N(x_0, \delta)$. Then

$$\begin{aligned} f(x_0) - \varepsilon < f(x) < f(x_0) + \varepsilon, \\ 0 < f(x_0)/2 < f(x) < 3f(x_0)/2, \end{aligned} \tag{9}$$

if $x \in D \cap N(x_0, \delta)$. The other statement in the theorem can be proved in a similar manner. □

96.10 Definition. Let f be a function $f: A \to B$, $A \subset \mathbb{R}^n$ and $B \subset \mathbb{R}^m$, whose domain is D. Then f is *bounded* if and only if there exists a constant M such that

$$|f(x)| < M, \qquad x \in D. \tag{10}$$

96.11 Examples. Let $x:(x_1, x_2)$ denote a point in $\mathbb{R}^2$. Define two functions $(f_1, f_2): A \to B$, $A \subset \mathbb{R}^2$ and $B \subset \mathbb{R}^2$, as follows:

(a) $f_1(x_1, x_2) = \dfrac{x_1}{1 - (x_1^2 + x_2^2)}, \qquad f_2(x_1, x_2) = \dfrac{x_2}{1 - (x_1^2 + x_2^2)}, \qquad |x| < 1.$

(b) $f_1(x_1, x_2) = \dfrac{x_1}{2 - (x_1^2 + x_2^2)}, \qquad f_2(x_1, x_2) = \dfrac{x_2}{2 - (x_1^2 + x_2^2)}, \qquad |x| \leqq 1.$

The function in (a) is unbounded because it maps its domain $D = \{x : x \in \mathbb{R}^2, |x| < 1\}$ onto the entire plane $\mathbb{R}^2$. This function is continuous, but its domain (bounded but open) is not compact. The function in (b) is bounded because it maps its domain $D = \{x : x \in \mathbb{R}^2, |x| \leqq 1\}$ onto the unit circle. This function is continuous, and its domain (bounded and closed) is compact.

96.12 Theorem. *Let f be a continuous function $f: A \to B$, $A \subset \mathbb{R}^n$ and $B \subset \mathbb{R}^m$, whose domain D is compact. Then f is bounded on D.*

PROOF. Since D is compact, D is bounded and it is contained in an interval

$$I_1 = [a_1^1, b_1^1] \times [a_1^2, b_1^2] \times \cdots \times [a_1^n, b_1^n]. \tag{11}$$

Assume that the theorem is false; that is, assume that $f(D)$ is an unbounded set in $\mathbb{R}^m$. Subdivide I_1 into 2^n intervals as in the proofs of Theorem 94.1 and 95.2. Then f is unbounded on the subset of D in at least one of the subintervals of I_1 because otherwise f would be bounded. From the 2^n subintervals of I_1, choose one which contains a subset of D on which f is unbounded, and call it I_2. Then

$$\begin{gathered} I_2 = [a_2^1, b_2^1] \times [a_2^2, b_2^2] \times \cdots \times [a_2^n, b_2^n], \\ a_1^j \leqq a_2^j \leqq b_2^j \leqq b_1^j. \end{gathered} \tag{12}$$

As in Sections 94 and 95, this construction can be repeated indefinitely to form nested intervals $I_1 \supset I_2 \supset \cdots \supset I_k \supset \cdots$ with the following properties for $k = 1, 2, \cdots$:

$$I_k \cap D \text{ is an infinite subset of } D; \tag{13}$$

$$f \text{ is unbounded on } I_k \cap D; \tag{14}$$

$$\lim_{k \to \infty} \operatorname{diam}(I_k) = 0. \tag{15}$$

Now (13) is true for the following reasons: each I_k was chosen so that (14) is true. If $I_k \cap D$ were finite, then f would be bounded on $I_k \cap D$; thus (14) implies (13). Next, by Theorem 93.2, there is a single point x_0 which is contained in I_k for $k = 1, 2, \cdots$. Now $x_0 \in D$ for the following reasons. Let $N(x_0, \delta_1)$ be a δ_1-neighborhood of x_0. Choose k_0 so that $I_{k_0} \subset N(x_0, \delta_1)$;

this choice is possible since $x_0 \in I_k$ for $k = 1, 2, \cdots$, and the diameter of I_k approaches zero by (15). Then $N(x_0, \delta_1)$ contains $I_{k_0} \cap D$, and $N(x_0, \delta_1)$ contains an infinite number of points in D by (13); hence, x_0 is a point of accumulation of D by Definition 92.2. Since D is compact, it is closed; hence, $x_0 \in D$. Then f is defined and continuous at x_0, and, given $\varepsilon > 0$, there exists a $\delta_2 > 0$ such that

$$|f(x) - f(x_0)| < \varepsilon, \qquad x \in D \cap N(x_0, \delta_2). \tag{16}$$

This statement implies that f is bounded on $D \cap N(x_0, \delta_2)$. Choose k_1 so that I_{k_1} is contained in $N(x_0, \delta_2)$; this choice is possible by the same arguments used above. Thus f is bounded on $I_{k_1} \cap D$ by (16), but it is unbounded on this set by (14). The assumption that the theorem is false has led to a contradiction, and Theorem 96.12 is true. □

96.13 Examples. The function in Example 96.11(a) is continuous but its domain is not compact; the function is unbounded. The function in Example 96.11(b) is continuous and its domain is compact; the function is bounded as required by Theorem 96.12. The following are additional examples.
(a) Define two functions as follows:

$$f(x) = 1/x, \qquad 0 < x \leqq 1.$$
$$f(x) = x^2, \qquad x \geqq 0.$$

Each of these functions is continuous but unbounded. In each case, the domain is not compact; the domain of the first function is bounded but not closed, and the domain of the second function is closed but not bounded.
(b) Define two functions as follows:

$$f(x) = 1 - x^2, \qquad -1 < x < 1.$$
$$f(x) = e^{-x^2} \sin x, \qquad x \geqq 0.$$

Each of these functions is continuous and bounded on its domain, but the domains of the functions are not compact. In the first case, the domain is bounded but not closed, and in the second it is closed but not bounded. The condition for boundedness stated in Theorem 96.12 is sufficient but not necessary.

96.14 Theorem. *Let f be a continuous function $f: A \to B$, $A \subset \mathbb{R}^n$ and $B \subset \mathbb{R}$, whose domain D is compact. Then f has a maximum and a minimum value on D; that is, there exist points x_1 and x_2 in D such that*

$$f(x_1) = \sup\{f(x) : x \in D\}, \qquad f(x_2) = \inf\{f(x) : x \in D\}. \tag{17}$$

Proof. Since D is compact and f is continuous on D, then f is bounded by Theorem 96.12, and $\sup\{f(x) : x \in D\}$ and $\inf\{f(x) : x \in D\}$ exist by Axiom 88.2 (the least upper bound axiom). Since every finite set has a maximum and a minimum, the theorem is obviously true if D is a finite set. Assume then that D is an infinite set, and prove the first statement in (17).

Since D is compact, D is bounded and it is contained in an interval I_1 [see (11)]. Subdivide I_1 into 2^n subintervals as in the proofs of Theorems 94.1 and 95.2. Then the supremum of the set of values of f on the part of D in at least one of the 2^n subintervals of I_1 is

$$\sup\{f(x): x \in D\}. \tag{18}$$

From the 2^n subintervals of I_1 choose one on which the supremum of the values of f is (18), and call it I_2. Then I_2 satisfies (12), and $I_1 \supset I_2$. This construction can be repeated indefinitely to form nested intervals $I_1 \supset I_2 \supset \cdots \supset I_k \supset \cdots$ which have the following properties for $k = 1, 2, \cdots$:

$$I_k \cap D \text{ is a non-empty subset of } D; \tag{19}$$

$$\sup\{f(x): x \in I_k \cap D\} = \sup\{f(x): x \in D\}; \tag{20}$$

$$\lim_{k \to \infty} \operatorname{diam}(I_k) = 0. \tag{21}$$

There are two possibilities in (19): (a) $I_k \cap D$ is a finite set, and (b) $I_k \cap D$ is infinite. Since every finite set in $\mathbb{R}$ has a maximum, in case (a) there is an x_1 in D such that

$$f(x_1) = \sup\{f(x): x \in I_k \cap D\} = \sup\{f(x): x \in D\}. \tag{22}$$

and the first statement in (17) is true in this case. In the remainder of the proof, assume (b): the set $I_k \cap D$ is infinite for $k = 1, 2, \cdots$.

By Theorem 93.2, there is a single point x_1 which is contained in I_k for $k = 1, 2, \cdots$. Now the same arguments that were used in the proof of Theorem 96.12 to prove that x_0 is in D can be used again to prove that x_1 is in D. Thus f is defined at x_1. Now there are exactly three possibilities [see (2) in Section 88]:

$$\begin{aligned} f(x_1) &= \sup\{f(x): x \in D\}, \\ f(x_1) &> \sup\{f(x): x \in D\}, \\ f(x_1) &< \sup\{f(x): x \in D\}. \end{aligned} \tag{23}$$

The second statement in (23) contradicts the definition of the supremum and is thus impossible. Consider the third statement in (23), and let

$$2\varepsilon = \sup\{f(x): x \in D\} - f(x_1) > 0. \tag{24}$$

Since f is continuous at x_1 by hypothesis, there exists a δ-neighborhood of x_1 such that

$$|f(x) - f(x_1)| < \varepsilon, \qquad x \in D \cap N(x_1, \delta). \tag{25}$$

Choose k_1 so that $I_{k_1} \subset N(x_1, \delta)$; this choice is possible by arguments used several times before. Then (25) and (24) show that

$$f(x_1) - \varepsilon < f(x) < f(x_1) + \varepsilon < \sup\{f(x) : x \in D\}, \qquad x \in I_{k_1} \cap D. \quad (26)$$

This statement contradicts (20), and thus the third statement in (23) is impossible. Since the second and third statements in (23) are impossible, the first statement in (23) is true, and the proof of the first statement in (17) is complete. The second statement in (17) can be established in the same way, or it can be proved by applying the first part of the theorem to the function $-f$. □

96.15 Examples. The following examples illustrate Theorem 96.14 and show that the conditions in it for the existence of a maximum and minimum are sufficient but not necessary.

(a) $f(x) = 1 - x^2$ on $|x| \leqq 1$.

The function f is continuous and its domain is compact. The maximum of f is 1 and $f(0) = 1$; the minimum of f is 0 and $f(-1) = f(1) = 0$.

(b) $f(x) = e^{-x^2} \sin x$, $x \geqq 0$.

Then f is continuous, but D is not compact; f has a positive maximum and a negative minimum (there are an infinite number of relative maxima and minima).

(c) $f(x) = e^{-x^2}$, $x \geqq 0$.

Then f is continuous, but D is not compact; the maximum of f is 1 and there is no minimum.

(d) $D = \{x : x \in \mathbb{R}^2, |x| \leqq 1\} \cup \{x : x \in \mathbb{R}^2, x = (1, 1), (2, 2)\}$,
$f(x) = |x|$, $x \in D, f(x) \in \mathbb{R}$.

Then f is continuous on its compact domain D. The maximum of f is $2\sqrt{2}$, and f has this value at the isolated point (2, 2) of D; also, f has its minimum value 0 at (0, 0).

(e) $f(-1) = -1, f(1) = 1$, and $f(x) = 0$ on $|x| < 1$, $x \in \mathbb{R}$.

Then D is a compact set in $\mathbb{R}$, but f is discontinuous at $x = 1$ and $x = -1$. Nevertheless, the maximum and minimum of f are 1 and -1 respectively.

(f) $f(1) = f(-1) = 0, f(x) = x$ on $|x| < 1$.

Then D is compact, but f is discontinuous at $x = 1$ and $x = -1$. The function has neither a maximum nor a minimum.

96.16 Definition. Let $f: A \to B$, $A \subset \mathbb{R}^n$ and $B \subset \mathbb{R}^m$, be a function which is continuous on its domain D. Then f is *uniformly continuous* on D if and only if to each $\varepsilon > 0$ there corresponds a $\delta(\varepsilon)$ such that $|f(x) - f(x_0)| < \varepsilon$ for every pair of points x_0, x in D for which $|x - x_0| < \delta(\varepsilon)$.

96.17 Examples. (a) If $f(x) = 2x + 1$, then $f(x) - f(x_0) = 2(x - x_0)$. For $\varepsilon > 0$ choose $\delta(\varepsilon) = \varepsilon/2$; then $|f(x) - f(x_0)| < \varepsilon$ for all x_0, x in $\mathbb{R}$ such that $|x - x_0| < \delta(\varepsilon)$. Thus f is uniformly continuous on the entire real line $\mathbb{R}$.
(b) Let $f(x) = 1/x$ on $D = \{x : x \in \mathbb{R}, 0 < x \leqq 1\}$. Then f is continuous on D, but it is not uniformly continuous since, for every $\delta > 0$,

$$\lim_{x \to 0} \left(\frac{1}{x} - \frac{1}{x + \delta} \right) = \lim_{x \to 0} \frac{\delta}{x(x + \delta)} = \infty.$$

Observe that D is not closed and thus not compact.
(c) Let D be an interval, and let $f: A \to B$, A and B in $\mathbb{R}$, be a function which has a bounded derivative on D. Then $f(x) - f(x_0) = f'(x^*)(x - x_0)$, $x_0 < x^* < x$, by the mean-value theorem, and f is uniformly continuous on D. For example, the sine function is uniformly continuous on every subset of $\mathbb{R}$.
(d) Let $f(x) = x^2$ on $\mathbb{R}$. If $b > 0$ and $D = \{x : 0 \leqq x \leqq b\}$, and if x and x_0 are in D, then

$$|f(x) - f(x_0)| = |(x + x_0)(x - x_0)| \leqq 2b|x - x_0|.$$

Choose $\delta(\varepsilon) = \varepsilon/2b$. Then $|f(x) - f(x_0)| < \varepsilon$ for every pair of points x, x_0 in D such that $|x - x_0| < \delta(\varepsilon)$. Thus f is uniformly continuous on the compact set D, but it is not uniformly continuous on the non-compact set $\{x : x \geqq 0\}$. The next theorem establishes the existence of a large class of uniformly continuous functions $f: A \to B$, $A \subset \mathbb{R}^n$ and $B \subset \mathbb{R}^m$.
(e) Let $f(x) = \sin(1/x)$ for $x > 0$. Then f is continuous on its domain, but it is not uniformly continuous. Observe that $f'(x) = (-1/x^2)\cos(1/x)$; hence, f' is unbounded in every neighborhood at the origin. But f' is bounded on $x \geqq \delta > 0$, and thus f is uniformly continuous on $x \geqq \delta > 0$ by part (c) of these examples.

96.18 Theorem. *Let $f: A \to B$, $A \subset \mathbb{R}^n$ and $B \subset \mathbb{R}^m$, be a function whose domain D is compact. If f is continuous on D, then f is uniformly continuous on D.*

PROOF. Since f is continuous on D, to each x_0 in D and to each $\varepsilon > 0$ there corresponds a $\delta(\varepsilon, x_0) > 0$ such that

$$|f(x) - f(x_0)| < \varepsilon/2, \qquad x \in D \cap N[x_0, \delta(\varepsilon, x_0)]. \tag{27}$$

The collection $\{N[x_0, \delta(\varepsilon, x_0)/2] : x_0 \in D\}$ of neighborhoods of points x_0 in D forms an open covering of D. Then by Theorem 95.2 (the Heine-Borel theorem), a finite subcollection of these neighborhoods covers D; let this finite subcollection be

$$N[x_i, \delta(\varepsilon, x_i)/2], \qquad i = 1, 2, \cdots, m. \tag{28}$$

Define $\delta(\varepsilon)$ as follows:

$$\delta(\varepsilon) = \min\{\delta(\varepsilon, x_i)/2 : i = 1, 2, \cdots, m\}. \tag{29}$$

Then $\delta(\varepsilon) > 0$ since $\delta(\varepsilon, x_i) > 0$ for $i = 1, 2, \cdots, m$. Let p_0 and p be two points in D such that

$$|p - p_0| < \delta(\varepsilon); \tag{30}$$

the proof will now show that

$$|f(p) - f(p_0)| < \varepsilon. \tag{31}$$

Since the sets in (28) form a covering of D, the point p_0 is contained in at least one of them; let $N[x_k, \delta(\varepsilon, x_k)/2]$ be a set which contains p_0. Then, because of (29) and (30), both p and p_0 belong to $N[x_k, \delta(\varepsilon, x_k)]$ and, by the triangle inequality,

$$|f(p) - f(p_0)| \leqq |f(p) - f(x_k)| + |f(x_k) - f(p_0)|. \tag{32}$$

But

$$|f(p) - f(x_k)| < \varepsilon/2, \qquad |f(x_k) - f(p_0)| < \varepsilon/2, \tag{33}$$

by (27). Then (32) and (33) show that $|f(p) - f(p_0)| < \varepsilon$ for every pair of points p, p_0 in D such that $|p - p_0| < \delta(\varepsilon)$. Thus f is uniformly continuous on D by Definition 96.16, and the proof is complete. □

96.19 Definition. Let $f: A \to B$ and $f_k: A \to B$, $k = 1, 2, \cdots, A \subset \mathbb{R}^n$ and $B \subset \mathbb{R}^m$, be functions all of which have the same domain D. The sequence of functions f_k *converges to* f on D if and only if to each $\varepsilon > 0$ and x in D there corresponds a positive integer $K(\varepsilon, x)$ such that

$$|f_k(x) - f(x)| < \varepsilon, \qquad k \geqq K(\varepsilon, x), \qquad x \in D. \tag{34}$$

The sequence of functions f_k *converges uniformly to* f on D if and only if to each $\varepsilon > 0$ there corresponds a positive integer $K(\varepsilon)$, which does not depend on x, such that

$$|f_k(x) - f(x)| < \varepsilon, \qquad k \geqq K(\varepsilon), \qquad \text{for every } x \text{ in } D. \tag{35}$$

96.20 Examples. (a) Let $f_k(x) = (1/k) \sin kx$, $k = 1, 2, \cdots$, on $\mathbb{R}$. Since $|f_k(x) - 0| \leqq 1/k$ for every x, this sequence of functions f_k converges uniformly to the zero function on $\mathbb{R}$.
(b) Let

$$f_k(x) = kx(1 - x^2)^k, \qquad 0 \leqq x \leqq 1, \qquad k = 1, 2, \cdots. \tag{36}$$

A little ingenuity in the evaluation of indeterminate forms shows that

$$\lim_{k\to\infty} kx(1 - x^2)^k = 0, \qquad 0 \leqq x \leqq 1. \tag{37}$$

Thus the sequence of functions f_k, $k = 1, 2, \cdots$, converges to the function which is zero on $0 \leqq x \leqq 1$, but the convergence is not uniform. The methods of elementary calculus show that

$$\max\{f_k(x) : 0 \leqq x \leqq 1\} = \frac{k}{\sqrt{1+2k}}\left[\frac{2k}{1+2k}\right]^k > \frac{1}{2}\sqrt{\frac{k}{e}}. \tag{38}$$

Then

$$\lim_{k\to\infty} \max\{f_k(x) : 0 \leqq x \leqq 1\} = +\infty, \tag{39}$$

and the convergence of f_k to the zero function cannot be uniform.

96.21 Theorem. *Let $f: A \to B$ and $f_k: A \to B$, $k = 1, 2, \cdots, A \subset \mathbb{R}^n$ and $B \subset \mathbb{R}^m$, be functions all of which have the same domain D. If f_k is continuous on D for each k, and if the sequence of functions f_k, $k = 1, 2, \cdots$, converges uniformly to f, then f is continuous on D.*

PROOF. Let $\varepsilon > 0$ be given. Choose k_0 so that

$$|f_{k_0}(x) - f(x)| < \varepsilon/3, \qquad x \in D. \tag{40}$$

This choice is possible because of the hypothesis that the sequence of functions f_k converges uniformly to f. Let x_0 be a point in D. Since f_{k_0} is continuous on D and therefore at x_0, there is a $\delta(\varepsilon, x_0)$ such that

$$|f_{k_0}(x) - f_{k_0}(x_0)| < \varepsilon/3, \qquad x \in D \cap N[x_0, \delta(\varepsilon, x_0)]. \tag{41}$$

By the triangle inequality,

$$|f(x) - f(x_0)| \leqq |f(x) - f_{k_0}(x)| + |f_{k_0}(x) - f_{k_0}(x_0)| + |f_{k_0}(x_0) - f(x_0)|. \tag{42}$$

Now by (41), the middle term on the right in (42) is less than $\varepsilon/3$ for every x in $D \cap N[x_0, \delta(\varepsilon, x_0)]$ and, by (40), the first and third terms on the right in (42) are each less than $\varepsilon/3$ for every x in D and hence for every x in $D \cap N[x_0, \delta(\varepsilon, x_0)]$. Thus (40), (41), (42) show that

$$|f(x) - f(x_0)| < \varepsilon, \qquad x \in D \cap N[x_0, \delta(\varepsilon, x_0)], \tag{43}$$

and f is continuous at x_0 by Definition 96.2. Since x_0 is an arbitrary point in D, f is continuous on D. □

96.22 Examples. (a) Theorem 96.21 shows that uniform convergence of the sequence $\{f_k : k = 1, 2, \cdots\}$ of continuous functions f_k is a sufficient condition that the limit function f be continuous, but the sequence in (36) shows that uniform convergence is not necessary.

(b) Let

$$f_k(x) = x^k, \qquad 0 \leqq x \leqq 1, \qquad k = 1, 2, \cdots. \tag{44}$$

Then f_k is continuous on $0 \leqq x \leqq 1$, and

$$\lim_{k\to\infty} f_k(x) = \begin{cases} 0, & 0 \leqq x < 1, \\ 1, & x = 1. \end{cases} \tag{45}$$

Thus the sequence (44) converges to a limit function f, but the convergence is not uniform since f is discontinuous by (45).

97. Cauchy Sequences

This section establishes a necessary and sufficient condition (called the Cauchy criterion) for the convergence of a sequence whose values are points in $\mathbb{R}^n$.

97.1 Definition. Let $\{v_i : i = 1, 2, \cdots\}$ be a sequence whose values v_i are points in a normed vector space $(V, \oplus, \odot, \|\ \ \|)$ [see Section 89]. The sequence is a *Cauchy sequence*, or it satisfies the *Cauchy criterion*, if and only if to each $\varepsilon > 0$ there corresponds a positive integer $I(\varepsilon)$ such that

$$\|v_i - v_j\| < \varepsilon, \qquad i \geqq I(\varepsilon), \qquad j \geqq I(\varepsilon). \tag{1}$$

97.2 Theorem. *Let $\{x_i : i = 1, 2, \cdots\}$ be a sequence whose values x_i are points in $\mathbb{R}^n$. A necessary and sufficient condition that the sequence have a limit in $\mathbb{R}^n$ is that the sequence be a Cauchy sequence.*

PROOF. First, the condition is necessary. Let x_0 denote the limit of the sequence. Then to each $\varepsilon > 0$ there corresponds an integer $I(\varepsilon)$ such that

$$|x_i - x_0| < \varepsilon/2, \qquad |x_j - x_0| < \varepsilon/2, \qquad i \geqq I(\varepsilon), j \geqq I(\varepsilon). \tag{2}$$

By the triangle inequality,

$$|x_i - x_j| \leqq |x_i - x_0| + |x_0 - x_j|; \tag{3}$$

hence, by (2) and (3),

$$|x_i - x_j| < \varepsilon, \qquad i \geqq I(\varepsilon), \qquad j \geqq I(\varepsilon), \tag{4}$$

and the proof that the Cauchy criterion is necessary is complete.

Next, the condition is sufficient. Assume that $\{x_i : i = 1, 2, \cdots\}$ is a Cauchy sequence and prove that it has a limit in $\mathbb{R}^n$. Let r be a positive constant; then there exists a positive integer $I(r)$ such that

$$|x_i - x_j| < r, \qquad i \geqq I(r), \qquad j \geqq I(r). \tag{5}$$

Let $j_0 = I(r)$; then

$$|x_i - x_{j_0}| < r, \qquad i \geqq I(r). \tag{6}$$

Thus the set of points $\{x_i : i = 1, 2, \cdots, j_0\}$ in $\mathbb{R}^n$ is a bounded set because it is a finite set, and the set $\{x_i : i \geqq I(r)\}$ is a bounded set by (6). Since the union of two bounded sets in $\mathbb{R}^n$ is a bounded set, the set $\{x_i : i = 1, 2, \cdots\}$ is bounded. There are two cases to consider: (a) the set $\{x_i : i = 1, 2, \cdots\}$

is a finite set of points in $\mathbb{R}^n$, and (b) the set $\{x_i : i = 1, 2, \cdots\}$ is an infinite set in $\mathbb{R}^n$. In (a), there is a point x_0 and an integer i_0 such that

$$x_i = x_0, \qquad i \geqq i_0, \tag{7}$$

because otherwise the sequence would not be a Cauchy sequence. But if (7) is true, then $\lim\{x_i : i = 1, 2, \cdots\} = x_0$, and the theorem is true in case (a). In case (b), the set $\{x_i : i = 1, 2, \cdots\}$ is an infinite bounded set in $\mathbb{R}^n$, and it has at least one point of accumulation by Theorem 94.1 [the Bolzano–Weierstrass theorem]. Let x_0 denote a point of accumulation of the set, and let $\varepsilon > 0$ be given. Then, because $\{x_i : i = 1, 2, \cdots\}$ is a Cauchy sequence, there is a positive integer $I(\varepsilon)$ such that

$$|x_i - x_j| < \varepsilon/2, \qquad i \geqq I(\varepsilon), \qquad j \geqq I(\varepsilon). \tag{8}$$

Since x_0 is a point of accumulation of the set $\{x_i : i = 1, 2, \cdots\}$, the neighborhood $N(x_0, \varepsilon/2)$ contains an infinite number of points in the set. Choose one of these points x_j for which $j \geqq I(\varepsilon)$ and call it x_{j_1}. Then

$$|x_{j_1} - x_0| < \varepsilon/2 \tag{9}$$

because $x_{j_1} \in N(x_0, \varepsilon/2)$, and

$$|x_i - x_{j_1}| < \varepsilon/2, \qquad i \geqq I(\varepsilon), \tag{10}$$

by (8) since $j_1 \geqq I(\varepsilon)$. Then, by the triangle inequality,

$$|x_i - x_0| \leqq |x_i - x_{j_1}| + |x_{j_1} - x_0| < \frac{\varepsilon}{2} + \frac{\varepsilon}{2} = \varepsilon, \qquad i \geqq I(\varepsilon), \tag{11}$$

by (9) and (10). Then $\lim\{x_i : i = 1, 2, \cdots\} = x_0$ by the definition of the limit of a sequence, and the proof of the entire theorem is complete. □

97.3 Definition. Let $(V, \oplus, \odot, \|\ \|)$ be a normed vector space. The space is *complete* if and only if every Cauchy sequence whose values are in V has a limit in V.

Theorem 97.2 proves that $\mathbb{R}^n$ with the norm defined in (1) in Section 91 is a complete vector space. The completeness of $\mathbb{R}$, and hence of $\mathbb{R}^n$ for $n = 1, 2, \cdots$, is a consequence of Axiom 88.2. There are normed vector spaces which are not complete. For example, let $(Q^n, \oplus, \odot, \|\ \|)$ be the subspace of $(\mathbb{R}^n, \oplus, \odot, \|\ \|)$ consisting of those points in $\mathbb{R}^n$ whose coordinates are rational numbers in $\mathbb{R}$. Then there are Cauchy sequences with values in Q^n which have limits in $\mathbb{R}^n$ but not in Q^n, and $(Q^n, \oplus, \odot, \|\ \|)$ is not a complete normed vector space.

As an example of the applications of the Cauchy criterion for convergence, consider the convergence of infinite series. Let $\{x_k : k = 1, 2, \cdots\}$ be a sequence whose values are points in $\mathbb{R}^n$. Set

$$s_i = \sum_{k=1}^{i} x_k. \tag{12}$$

97.4 Definition. The infinite series

$$\sum_{k=1}^{\infty} x_k \tag{13}$$

converges if

$$\lim_{i\to\infty} s_i \tag{14}$$

exists, and it *diverges* if this limit does not exist. If the series converges, then the value of the limit in (14) is called the *sum of the series* (13), and we write

$$\sum_{k=1}^{\infty} x_k = \lim_{i\to\infty} s_i. \tag{15}$$

97.5 Theorem. *If the series*

$$\sum_{k=1}^{\infty} |x_k| \tag{16}$$

converges, then the series

$$\sum_{k=1}^{\infty} x_k \tag{17}$$

converges, and

$$\left|\sum_{k=1}^{\infty} x_k\right| \leqq \sum_{k=1}^{\infty} |x_k|. \tag{18}$$

Proof. Let

$$t_i = \sum_{k=1}^{i} |x_k|, \qquad s_i = \sum_{k=1}^{i} x_k, \qquad i = 1, 2, \cdots. \tag{19}$$

Since (16) converges, the sequence $\{t_i : i = 1, 2, \cdots\}$ satisfies the Cauchy criterion by Theorem 97.2; hence, given $\varepsilon > 0$, there exists an integer $I(\varepsilon)$ such that

$$|t_i - t_j| < \varepsilon, \qquad i \geqq I(\varepsilon), \qquad j \geqq I(\varepsilon). \tag{20}$$

If $i > j$, then

$$|t_i - t_j| = \sum_{k=j+1}^{i} |x_k|, \tag{21}$$

$$|s_i - s_j| = \left|\sum_{k=j+1}^{i} x_k\right|, \tag{22}$$

and (21), (22), and the triangle inequality show that

$$|s_i - s_j| \leqq |t_i - t_j|. \tag{23}$$

Then (20) and (23) show that

$$|s_i - s_j| < \varepsilon, \qquad i \geqq I(\varepsilon), \qquad j \geqq I(\varepsilon), \tag{24}$$

and the sequence $\{s_i : i = 1, 2, \cdots\}$ satisfies the Cauchy criterion for convergence. Then $\lim\{s_i : i = 1, 2, \cdots\}$ exists by Theorem 97.2 and the infinite series (17) converges by Definition 97.4. Since

$$\sum_{k=1}^{\infty} x_k = \lim_{i\to\infty} s_i, \qquad \sum_{k=1}^{\infty} |x_k| = \lim_{i\to\infty} t_i, \tag{25}$$

$$|s_i| \leqq t_i \leqq \lim_{i\to\infty} t_i, \tag{26}$$

then

$$\left|\sum_{k=1}^{\infty} x_k\right| = \left|\lim_{i\to\infty} s_i\right| \leqq \lim_{i\to\infty} t_i = \sum_{k=1}^{\infty} |x_k|, \tag{27}$$

and the proof of (18) and of the entire theorem is complete. □

There is a definition similar to Definition 97.4 for the convergence, divergence, and sum of an infinite series in an arbitrary normed vector space; and there is a theorem similar to Theorem 97.5 for infinite series in an arbitrary complete normed vector space $(V, \oplus, \odot, \|\ \|)$.

References and Notes

1. J. Bertrand, Mémoirs sur le déterminant d'un système de fonctions, (Liouville's) *Journal de Mathématique Pures et Appliquées*, **16**(1851) 212–227.

Many years ago, in a class in elementary analytic geometry which the author was teaching, a penetrating question asked by a bright student set off a train of thought followed by serious investigations which soon established the foundations of the theory of differentiation now presented in this book. The study quickly led to some important identities in the theory of determinants, and a careful examination of the history of determinants became necessary. It showed that the identities needed in the theory of differentiation [especially Sylvester's theorem of 1839 and 1851 and the Bazin–Reiss–Picquet theorem] had all been established long ago; furthermore, the search of the literature located Bertrand's paper of 1851.

Jacobi gave, in 1834, the first significant treatment of the expressions we now call Jacobians, and in 1841 he made functional determinants the special object of an exhaustive treatment. Sylvester introduced, in 1853, the name "Jacobian" for functional determinants and the notation $J(f, g)$ for the Jacobian of f and g. Jacobi had observed the analogy of Jacobians to derivatives of functions of a single variable, and Bertrand, in the paper cited above, called attention to Jacobi's observation and then *defined* the Jacobian to be the limit in equation (25) of Section 1 of this book. He said, "Je nommerai déterminant du système des fonctions ... la limite du rapport du déterminant du système d'accroissements des fonctions au déterminant du système des accroissements correspondants des variables."

Donkin, in a paper published in 1854, pointed out the analogy of functional determinants and derivatives of functions of a single variable noted by Jacobi and Bertrand and proposed the notation

$$\frac{\partial(f, g)}{\partial(x, y)}$$

for the Jacobian to emphasize it. Donkin's notation is standard today, but Bertrand's definition of the Jacobian—which it was designed to emphasize—has been forgotten.

Bertrand's definition was widely reproduced in treatises both on the calculus and on the theory of determinants, and it was discussed and criticised by numerous important mathematicians. Nevertheless, Bertrand's *definition* was gradually changed into Bertrand's *theorem* and never became the point of departure for a significant treatment of the theory of differentiation.

2. G. Baley Price, Derivatives and Jacobians, *Annali di Matematica Pura ed Applicata* (4), **57** (1962) 311–320.

This paper contains an introduction to the theory of differentiation developed in Chapter 1 of this book. Also, it contains a brief account of the history of the Jacobian and of the role and influence of Bertrand's definition [see the note in reference 1], together with many references to the literature.

3. W. F. Donkin, On a class of differential equations, including those which occur in dynamical problems, *Philosophical Transactions of the Royal Society of London*, Part I, **144** (1854) 71–113.

For Donkin's notation for the Jacobian, see the note in reference 1.

4. Otto Stolz, *Grundzüge der Differential- und Integralrechnung*. Erster Theil: Reelle Veränderliche und Functionen, 1893; Zweiter Theil: Complexe Veränderliche und Functionen, 1896; Dritter Theil: Die Lehre von den Doppelintegralen, 1899; B. G. Teubner, Leipzig.

Stolz treated partial differentiation in the first part of this three-volume work on the calculus, and on pages 129–134 he employs the relation which is called the "Stolz condition" in this book. He offers $f(x, y) = \sqrt{|xy|}$ as an example of a continuous function which has partial derivatives at the origin but which does not satisfy the Stolz condition there. For the Stolz condition in the present book, see Sections 2, 3, 64, 65, 72, 73.

5. G. Baley Price, Some identities in the theory of determinants, *American Mathematical Monthly*, **54** (1947) 75–90.

This paper contains proofs of the unfamiliar theorems on determinants which are needed in the theory of differentiation developed in this book. In particular, it contains Sylvester's theorem of 1839 and 1851 [see Sections 3, 72, and 81], the Bazin–Reiss–Picquet theorem [see Sections 3, 72, and 83], and the Sylvester–Franke theorem [see Sections 5 and 82].

6. Paul Alexandroff und Heinz Hopf, *Topologie*, Springer-Verlag, Berlin, 1935. Chelsea reprint, 1965. xiii + 636 pages.

Alexandroff-Hopf's book is a classic on topology; for the purposes of the present book, it contains much material on simplexes, complexes, orientations, boundaries, convex sets, and other topics in n-dimensional geometry.

7. J. Dieudonné, *Foundations of Modern Analysis*, Academic Press, New York and London, 1960. xiv + 361 pages.

This book contains a proof [see page 133] of the Weierstrass polynomial approximation theorem stated in Theorem 48.8 and used in proving several formulas in Section 48 [see, for example, Theorem 48.9].

8. E. Goursat, Démonstration du théorème de Cauchy, *Acta Mathematica*, **4** (1884) 197–200.

———, Sur la définition générale des fonctions analytiques, d'après Cauchy, *Transactions of the American Mathematical Society* **1** (1900) 14–16.
These two papers by Goursat contain his proof of Cauchy's integral theorem without the hypothesis that the derivative of the function is continuous [see Section 67].

9. Tom M. Apostol, *Calculus*, Volume II, second edition, Blaisdell Publishing Company, Waltham, Massachusetts, 1969. xxi + 673 pages.

Apostol gives, on page 258, one of the modern definitions of differentiability for functions of the type $f: E \to \mathbb{R}^m$, $E \subset \mathbb{R}^n$. Theorem 6.7 proves that the class of functions which are differentiable in the restricted sense [see Section 6 of this book for the definition] is the class of functions which are differentiable according to Apostol's definition.

10. Michael Spivak, *Calculus on Manifolds*, W. A. Benjamin, New York and Amsterdam, 1965. xii + 146 pages. Paperback.

Spivak gives, on page 16, another form of the modern definition of differentiability for functions of the type $f: E \to \mathbb{R}^m$, $E \subset \mathbb{R}^n$ [see Section 6 in this book].

Index of Symbols

Index